Rheology and Processing of Polymeric Materials

Volume 1

RHEOLOGY AND PROCESSING OF POLYMERIC MATERIALS

Volume 1
Polymer Rheology

Chang Dae Han

Department of Polymer Engineering
The University of Akron

OXFORD
UNIVERSITY PRESS

2007

OXFORD
UNIVERSITY PRESS

Oxford University Press, Inc., publishes works that further
Oxford University's objective of excellence in research, scholarship,
and education.

Oxford New York
Auckland Cape Town Dar es Salaam Delhi Hong Kong Karachi
Kuala Lumpur Madrid Melbourne Mexico City Nairobi
New Delhi Shanghai Taipei Toronto

With offices in
Argentina Austria Brazil Chile Czech Republic France Greece
Guatemala Hungary Italy Japan Poland Portugal Singapore
South Korea Switzerland Thailand Turkey Ukraine Vietnam

Published by Oxford University Press, Inc.
198 Madison Avenue, New York, New York 10016
www.oup.com

Library of Congress Cataloging-in-Publication Data
Han, Chang Dae.
Rheology and processing of polymeric materials/Chang Dae Han.
v. cm.
Contents: v. 1 Polymer rheology—; v. 2 Polymer processing—
Includes bibliographical references and index.
ISBN: 978-0-19-518782-3 (vol. 1); 978-0-19-518783-0 (vol. 2)
1. Polymers–Rheology. 1. Title.

QC189.5.H36 2006
620.1′920423—dc22 2005036608

9 8 7 6 5 4 3 2 1
Printed in the United States of America
on acid-free paper

In Memory of My Parents

Preface

In the past, a number of textbooks and research monographs dealing with polymer rheology and polymer processing have been published. In the books that dealt with rheology, the authors, with a few exceptions, put emphasis on the continuum description of homogeneous polymeric fluids, while many industrially important polymeric fluids are heterogeneous, multicomponent, and/or multiphase in nature. The continuum theory, though very useful in many instances, cannot describe the effects of molecular parameters on the rheological behavior of polymeric fluids. On the other hand, the currently held molecular theory deals almost exclusively with homogeneous polymeric fluids, while there are many industrially important polymeric fluids (e.g., block copolymers, liquid-crystalline polymers, and thermoplastic polyurethanes) that are composed of more than one component exhibiting complex morphologies during flow.

In the books that dealt with polymer processing, most of the authors placed emphasis on showing how to solve the equations of momentum and heat transport during the flow of homogeneous thermoplastic polymers in a relatively simple flow geometry. In industrial polymer processing operations, more often than not, multicomponent and/or multiphase heterogeneous polymeric materials are used. Such materials include microphase-separated block copolymers, liquid-crystalline polymers having mesophase, immiscible polymer blends, highly filled polymers, organoclay nanocomposites, and thermoplastic foams. Thus an understanding of the rheology of homogeneous (neat) thermoplastic polymers is of little help to control various processing operations of heterogeneous polymeric materials. For this, one must understand the rheological behavior of each of those heterogeneous polymeric materials.

There is another very important class of polymeric materials, which are referred to thermosets. Such materials have been used for the past several decades for the fabrication of various products. Processing of thermosets requires an understanding of the rheological behavior during processing, during which low-molecular-weight oligomers (e.g., unsaturated polyester, urethanes, epoxy resins) having the molecular weight of the order of a few thousands undergo chemical reactions ultimately giving rise to cross-linked networks. Thus, a better understanding of chemorheology is vitally important to control the processing of thermosets. There are some books that dealt with the chemorheology of thermosets, or processing of some thermosets. But, very few, if any, dealt with the processing of thermosets with chemorheology in a systematic fashion.

The preceding observations have motivated me to prepare this two-volume research monograph. Volume 1 aims to present the recent developments in polymer rheology, placing emphasis on the rheological behavior of structured polymeric fluids. In so doing, I first present the fundamental principles of the rheology of polymeric fluids: (1) the kinematics and stresses of deformable bodies, (2) the continuum theory for the viscoelasticity of flexible homogeneous polymeric liquids, (3) the molecular theory for the viscoelasticity of flexible homogeneous polymeric liquids, and (4) experimental methods for measurement of the rheological properties of polymeric liquids. The materials presented are intended to set a stage for the subsequent chapters by introducing the basic concepts and principles of rheology, from both phenomenological and molecular perspectives, of structurally simple flexible and homogeneous polymeric liquids.

Next, I present the rheological behavior of various polymeric materials. Since there are so many polymeric materials, I had to make a conscious, though somewhat arbitrary, decision on the selection of the polymeric materials to be covered in this volume. Admittedly, the selection has been made on the basis of my research activities during the past three decades, since I am quite familiar with the subjects covered. Specifically, the various polymeric materials considered in Volume 1 range from rheologically simple, flexible thermoplastic homopolymers to rheologically complex polymeric materials including (1) block copolymers, (2) liquid-crystalline polymers, (3) thermoplastic polyurethanes, (4) immiscible polymer blends, (5) particulate-filled polymers, organoclay nanocomposites, and fiber-reinforced thermoplastic composites, and (6) molten polymers with solubilized gaseous component. Also, chemorheology is included in Volume 1.

Volume 2 aims to present the fundamental principles related to polymer processing operations. In presenting the materials in this volume, again, the objective was not to provide the recipes that necessarily guarantee better product quality. Rather, I put emphasis on presenting fundamental approach to effectively analyze processing problems. Polymer processing operations require combined knowledge of polymer rheology, polymer solution thermodynamics, mass transfer, heat transfer, and equipment design. Specifically, in Volume 2, I have presented the fundamental aspects of several processing operations (plasticating single-screw extrusion, wire coating extrusion, fiber spinning, tubular film blowing, injection molding, coextrusion, and foam extrusion) of thermoplastic polymers and three processing operations (reaction injection molding, pultrusion, and compression molding) of thermosets. In presenting Volume 2, I have used some materials presented in Volume 1.

In the preparation of this monograph, I have tried to present the fundamental concepts and/or principles associated with the rheology and processing of the various polymeric materials selected and I have tried to avoid presenting technological recipes. In so doing, I have pointed out an urgent need for further experimental and theoretical investigations. I sincerely hope that the materials presented in this monograph will not only encourage further experimental investigations but also stimulate future development of theory. I wish to point out that I have tried not to cite articles appearing in conference proceedings and patents unless absolutely essential, because they did not go through rigorous peer review processes.

Much of the material presented in this monograph is based on my research activities with very capable graduate students at Polytechnic University from 1967 to 1992 and at the University of Akron from 1993 to 2005. Without their participation and dedication to the various research projects that I initiated, the completion of this monograph would not have been possible. I would like to acknowledge with gratitude that Professor Jin Kon Kim at Pohang University of Science and Technology in Korea read the draft of Chapters 4, 6, 7, and 8 of Volume 1 and made very valuable comments and suggestions for improvement. Professor Ralph H. Colby at Pennsylvania State University read the draft of Chapter 7 of Volume 1 and made helpful comments and suggestions, for which I am very grateful. Professor Anthony J. McHugh at Lehigh University read the draft of Chapter 6 of Volume 2 and made many useful comments, for which I am very grateful. It is my special privilege to acknowledge the wonderful collaboration I had with Professor Takeji Hashimoto at Kyoto University in Japan for the past 18 years on phase transitions and phase behavior of block copolymers. The collaboration has enabled me to add luster to Chapter 8 of Volume 1. The collaboration was very genuine and highly professional. Such a long collaboration was made possible by mutual respect and admiration.

Chang Dae Han
The University of Akron

Akron, Ohio
June, 2005

Contents

Part II

Rheological Behavior of Polymeric Materials

Remarks on Volume 1

This volume consists of two parts. Part I describes the fundamental principles of the rheology of polymeric fluids: (1) the kinematics and stresses of deformable bodies, (2) the continuum theories for the viscoelasticity of flexible homogeneous polymeric liquids, (3) the molecular theories for the viscoelasticity of flexible homogeneous polymeric liquids, and (4) experimental methods for measurement of the rheological properties of polymeric liquids. Part I is intended to set a stage for the subsequent chapters by introducing the basic concepts and principles of rheology, from both phenomenological and molecular perspectives, of structurally simple flexible and homogeneous polymeric liquids. Part II describes the rheology of various polymeric materials, ranging from flexible ordinary thermoplastic homopolymers to thermosets, namely, (1) homopolymers, (2) miscible polymer blends, (3) block copolymers, (4) liquid-crystalline polymers, (5) thermoplastic polyurethanes, (6) immiscible polymer blends, (7) particulate-filled polymers, organoclay nanocomposites, and fiber-reinforced thermoplastic composites, (8) molten polymers with solubilized gaseous component, and (9) thermosets. In presenting the materials in Part II, I have pointed out an urgent need for further experimental and theoretical investigations. I sincerely hope that the materials presented in Part II will not only encourage further experimental investigations, but also stimulate future development of theory.

C.D.H.

Rheology and Processing of Polymeric Materials

Volume 1

1

Relationships Between Polymer Rheology and Polymer Processing

Polymer products have long been used for a variety of applications in our daily lives, as well as for some more exotic applications, such as biomedical devices, super-high-speed airplanes, and outer-space vehicles. Other applications are too numerous to mention them all here. There are many steps involved in the production of polymer products, from the synthesis of raw materials to the manufacturing of the finished products. Of the many steps involved, the fabrication (processing) step plays a pivotal role in determining the quality of the final products. Successful processing of polymeric materials requires a good understanding of their rheological behavior (Han 1976, 1981). Thus, intimate relationships exist between polymer rheology and polymer processing. In this chapter we describe briefly some of these close relationships between polymer rheology and polymer processing.

1.1 What Is Polymer Rheology?

Rheology is the science that deals with the deformation and flow of matter. Hence, polymer rheology is the science that deals with the deformation and flow of polymeric materials. Since there are a variety of polymeric materials, we can classify polymer rheology further into different categories, depending upon the nature of the polymeric materials; for instance, (1) the rheology of homogeneous polymers, (2) the rheology of miscible polymer blends, (3) the rheology of immiscible polymer blends, (4) the rheology of particulate-filled polymers, (5) the rheology of fiberglass-reinforced polymers, (6) the rheology of organoclay nanocomposites, (7) the rheology of polymeric foams, (8) the rheology of thermosets, (9) the rheology of block copolymers, and (10) the rheology of liquid-crystalline polymers. Each of these polymeric materials exhibits its own unique rheological characteristics. Thus, different theories are needed to interpret the experimental results of the rheological behavior of different polymeric materials. However, at present we do not have a comprehensive theory that can describe the rheological behavior of some polymeric materials and thus we must resort to empirical correlations to interpret the experimentally observed rheological behavior of those materials. It is then fair to state that a complete understanding of the rheological behavior of all polymeric materials remains quite a challenge indeed.

Most of the polymeric materials of practical use exhibit "viscoelastic" behavior during flow, meaning that they exhibit not only viscous behavior but also elastic (rubberlike) behavior in the liquid state. There are several different ways of describing the fluid elasticity of polymeric materials, and this subject is dealt with in Chapters 3 and 4 from a theoretical point of view and in Chapter 5 from an experimental point of view. The viscosity of a polymer is proportional to its molecular weight (M) when it is lower than a certain critical value (M_c), but the shear viscosity is proportional to the 3.4-th power of its M when $M \geq M_c$ (Berry and Fox 1968). The polymer having $M < M_c$ is referred to as "unentangled" polymer, and the polymer having $M \geq M_c$ is referred to as "entangled" polymer. It is well established that entangled polymers are highly viscoelastic, while unentangled polymers are not (Ferry 1980).

The rheological properties of polymeric materials vary with their chemical structures. Therefore, it is highly desirable to be able to relate the rheological properties of polymeric materials to their chemical structures. For instance, one may ask: Why are the rheological properties of polystyrene so different from the rheological properties of polyethylene under an identical flow condition? Unfortunately, at present there is no comprehensive molecular theory that can answer such a seemingly simple and fundamental question.

There are some molecular theories that can explain the effects of molecular weight and molecular weight distribution on the rheological properties of flexible homopolymers (Rouse 1953; Doi and Edwards 1986). This subject is discussed in Chapter 4. However, some polymers are heterogeneous when they are polymerized (e.g., block copolymers, liquid-crystalline polymers), exhibiting two phases in the liquid state. Also, one often prepares heterogeneous polymeric materials by mixing a homogenous polymer with other components (e.g., particulate fillers, chemically modified clay, glass fibers, and carbon black). The rheological behavior of such polymeric materials is quite different from that of homogenous polymers. In several chapters of Volume 1, we discuss the rheological behavior of heterogeneous polymeric materials.

1.2 How Does the Fluid Elasticity of Polymeric Liquids Manifest Itself in Flow?

There are several ways of demonstrating, experimentally, that polymeric fluids exhibit elastic characteristics. One very well known experimental observation is the behavior of liquid climb-up on a rotating rod in a polymer solution. Figure 1.1 demonstrates a dramatic difference in the behavior of liquid climb-up on a rotating rod between (a) 4 wt % aqueous solution of polyacrylamide and (b) glycerin. It is seen in Figure 1.1 that the polyacrylamide solution climbs the rod rotating within it, whereas no climb-up

Figure 1.1 Difference in liquid climb-up behavior on a rotating rod between (a) 4 wt % aqueous solution of polyacrylamide (viscoelastic fluid) and (b) glycerin (Newtonian fluid).

of glycerin is seen on the rotating rod. The phenomenon of liquid climb-up is quite contrary to what one would expect from the effect of centrifugal force (see Figure 1.1b); the faster the rod rotates, the higher the liquid climbs. The phenomenon was first observed by Garner and Nissan (1946) and later properly explained by Weissenberg (1947). The question is: What causes the liquid to climb up the rod? It is very important to notice in Figure 1.1a that the direction of liquid climb-up is perpendicular to the rotational flow direction of the liquid. That is, during the rotational flow of a liquid in the beaker, a force is generated in the direction perpendicular to the rotational direction. Apparently, such an experimental observation prompted Weissenberg (1949) to design, for the first time, a cone-and-plate rheometer, which is known today as the Weissenberg rheogoniometer, enabling one to determine first normal stress difference (N_1) in steady-state shear flow of viscoelastic polymeric fluids.

To illustrate the point, let us consider the schematic given in Figure 1.2, where a fluid is placed in the gap between the cone and the plate, and imagine the following simple experiment. Namely, place a fluid in the cone-and-plate fixture and then shear it by rotating the cone at a fixed angular speed Ω while the upper plate is held in its original position. Then, while the fluid in the cone-and-plate fixture is rotated, try to determine, via a transducer mounted at the upper plate, if a force F is generated in the direction perpendicular to the rotational direction of the cone. That is, the measurement of liquid height L in the climb-up experiment is replaced by the measurement of force F in the cone-and-plate flow experiment, the principles involved in both experiments

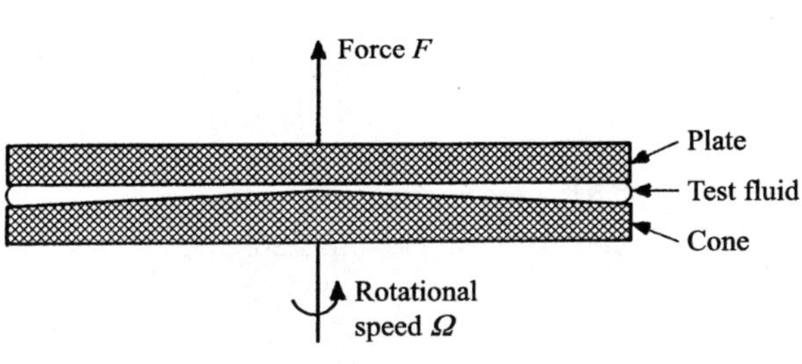

Figure 1.2 Schematic showing the flow of a test fluid placed in the cone-and-plate fixture, where force F perpendicular to the flow direction is measured as a function of rotational speed Ω.

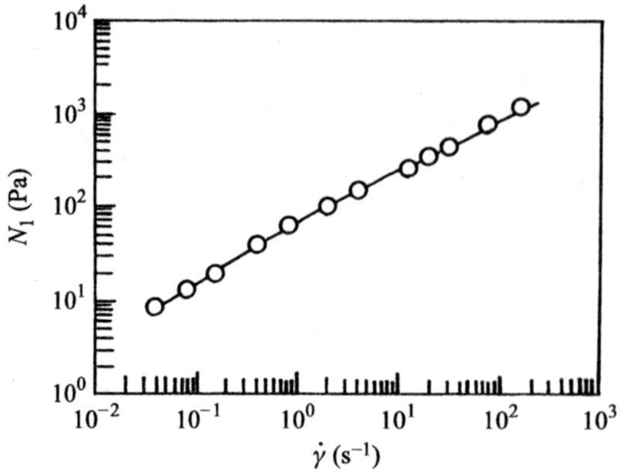

Figure 1.3 Plots of first normal stress difference (N_1) versus shear rate ($\dot{\gamma}$) for 4 wt % aqueous solution of polyacrylamide in steady-state shear flow at 25 °C.

being identical. In Chapter 5 we will elaborate quantitatively on the principle of shear flow in the cone-and-plate fixture.

Quantitative experimental observation for 4 wt % aqueous solution of polyacrylamide in the cone-and-plate fixture is presented in Figure 1.3, in which the values of N_1, which is proportional to the force F measured in the cone-and-plate fixture (see Figure 1.2), are plotted against the values of shear rate ($\dot{\gamma}$), which is proportional to the rotational speed Ω of the cone in the cone-and-plate fixture. In Chapter 5 we will present theoretical expressions that relate F to N_1, and Ω to $\dot{\gamma}$. From Figure 1.3 we can conclude that the faster the rotational speed Ω of the cone, the larger are the values of force F generated during the rotation of the cone. Conversely, no measurable force F can be detected when glycerin is placed in the cone-and-plate fixture, which is consistent with the absence of liquid climb-up of glycerin (see Figure 1.1b). In other words, the origin of the dramatic difference in the liquid climb-up behavior between the 4 wt % aqueous solution of polyacrylamide and glycerin lies in the force F generated for the 4 wt % aqueous solution of polyacrylamide in the direction perpendicular (normal) to the rotational direction of the liquid in the beaker. Such force is referred to as "normal force." Today, it is well accepted that the normal force is related to fluid elasticity. Thus, we can conclude that it is the elastic property of 4 wt % aqueous solution of polyacrylamide that gave rise to liquid climb-up on a rotating rod shown in Figure 1.1a, and that glycerin does not exhibit fluid elasticity.

Another well-known rheological experiment is illustrated in Figure 1.4, where (a) 4 wt % aqueous solution of polyacrylamide is jetting from a cylindrical tube and (b) glycerin is jetting from the same tube. It is clearly seen in Figure 1.4 that the diameter of liquid jet of 4 wt % aqueous solution of polyacrylamide swells, whereas little or no swell of the liquid jet from glycerin can be seen. The swell of 4 wt % aqueous solution of polyacrylamide, upon flowing out of a cylindrical tube, is believed to arise from the recovery of the elastic energy that was stored in the liquid while it was being sheared within the tube. Comparison of Figure 1.4 with Figure 1.1 shows very clearly that for the same liquid there is a correlation between the swell of a liquid stream and the climb-up on a rotating rod. There are other phenomena (e.g., stress relaxation, elastic recoil) observed experimentally that demonstrate the unique viscoelastic characteristics of polymeric fluids. This subject is discussed in other chapters of this book.

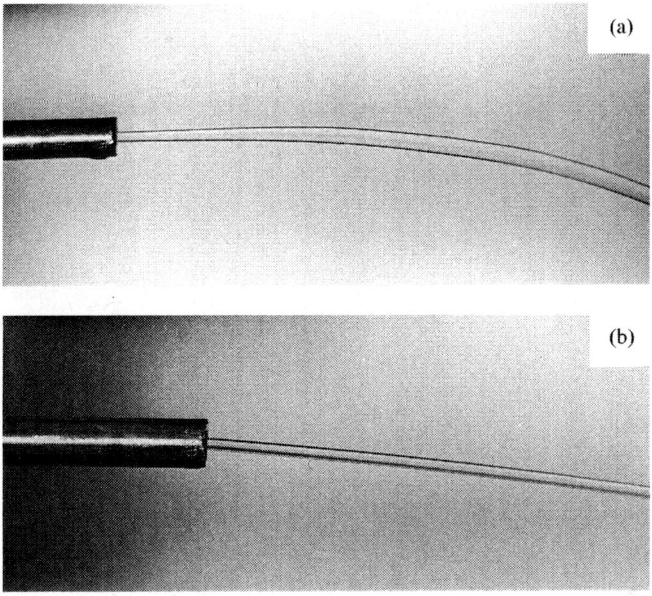

Figure 1.4 Liquid jets of (a) 4 wt % aqueous solution of polyacrylamide (viscoelastic fluid) and (b) glycerin (Newtonian fluid) upon leaving a cylindrical tube.

1.3 Shear-Thinning Behavior of Viscosity of Polymeric Liquids

Polymeric liquids, like other types of liquids, possess viscosity, which is regarded as a measure of the resistance to flow. There is a unique rheological characteristics of polymeric liquids, not seen in low-molecular-weight ordinary fluids, during flow in that the resistance to flow (viscosity) through a cylindrical tube decreases as the flow rate is increased. This is illustrated in Figure 1.5, in which the viscosity (η) of 4 wt % aqueous solution of polyacrylamide decreases with increasing shear rate ($\dot{\gamma}$),

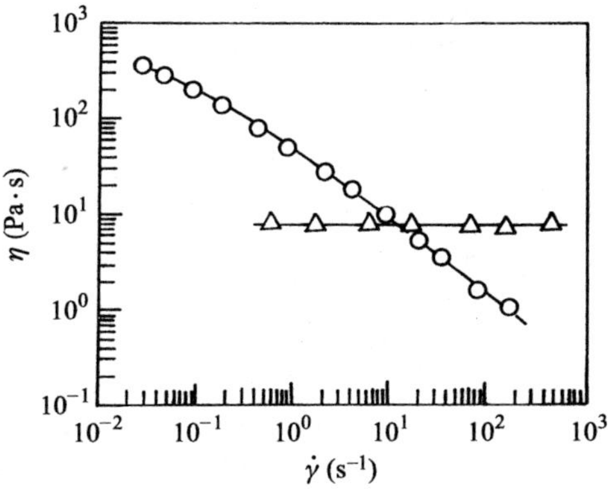

Figure 1.5 Plots of shear viscosity (η) versus shear rate ($\dot{\gamma}$) for (○) 4 wt % aqueous solution of polyacrylamide and (Δ) glycerin in steady-state shear flow at 25 °C.

which is proportional to flow rate, whereas the η of glycerin is constant and independent of $\dot{\gamma}$. The decreasing trend of η with increasing $\dot{\gamma}$, commonly referred to as "shear-thinning behavior," is believed to arise from the stretching of an "entangled" state of polymer chains to an "oriented" state when the applied shear rate is higher than a certain critical value. Conversely, because glycerin is a small molecule, it cannot possibly have an entangled state and thus no shear-thinning behavior is expected from glycerin. It should be mentioned that the shear-thinning behavior of η observed in Figure 1.5 and the shear-rate dependence of N_1 observed in Figure 1.3 are only two of the unique rheological behavior of viscoelastic polymeric liquids. Other unique rheological behavior of viscoelastic polymeric materials is discussed in several chapters of this volume.

1.4 Processing Characteristics of Polymeric Materials

Figure 1.6 gives a schematic showing the interrelationships that exist between the many steps that range from the production of a polymer to the physical/mechanical properties of the final polymer products. One must control reactor variables to produce consistent quality in a polymer, and thus one needs a "polymerization reactor simulator." The polymer produced from the reactor must be characterized in terms of its rheological properties, and thus one needs a "rheological property simulator." Since the rheological properties of polymers depend on their molecular parameters, it is highly desirable to relate the rheological properties of a polymer to its molecular parameters, and thus one must understand molecular viscoelasticity theory for polymeric materials. Since the rheological behavior of a polymer depends on temperature and pressure, and also on the geometry of a flow device, one needs a "polymer processing simulator," which is intimately related to the rheological property simulator. It should be

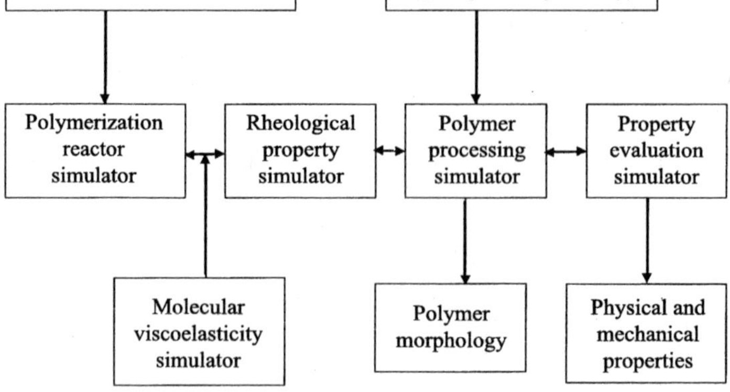

Figure 1.6 Schematic describing intimate interrelationships that exist among the reactor variables, rheological properties, processing variables, and physical/mechanical properties of polymer products.

pointed out that a polymer processing simulator must be based on the momentum and heat transfer equations at a minimum. Depending on the processes involved with the fabrication of a final product, sometimes the polymer processing simulator requires, in addition to momentum and heat transfer equations, mass transfer equations and/or reaction kinetic expression for reactive systems (including thermosets). Finally, one needs a "property evaluation simulator," which evaluates the physical, mechanical and/or optical properties of the fabricated products.

When the fabricated products do not meet the specifications of physical, mechanical and/or optical properties, there are two routes that can be pursued further; namely, either modifying the chemical structure of the polymer or modifying (or optimizing) processing conditions. A modification of the chemical structure of a polymer requires the establishment of a new or revised rheological property simulator and thus a new or revised polymer processing simulator. There are many different fabrication methods (processing techniques) for obtaining polymer products. Examples of processing techniques that are currently used in industry include extrusion, pultrusion, injection molding, compression molding, reaction injection molding, tubular film blowing, blow molding, thermoforming, fiber spinning, calendering, and foaming. Needless to say, each of these processing techniques requires a separate processing simulator. Once again, the rheological property simulator and polymer processing simulator are intimately related to each other.

The ultimate goal of the polymer fabrication industry is to manufacture products that meet the requirements for desired physical and/or mechanical properties. The end users are not interested in knowing how the polymers were synthesized or fabricated. It is the responsibility of polymer scientists and polymer engineers to provide their customers with final products that have the desired properties. It is worth mentioning that the mechanical/physical properties of a given polymer can vary in different fabricated products, depending upon the processing conditions employed. This is because different processing conditions (e.g., stretching rate in melt spinning or film blowing, or cooling rate in melt spinning or injection molding) can greatly influence the molecular orientations, the rate of solidification, and the morphological state of the solidified products, thus affecting their mechanical/physical properties. For a given polymer, understanding the relationships between processing variables and the mechanical/physical properties of fabricated products and relationships between processing variables and the morphology of fabricated products is highly desirable. However, at present the details of such relationships are rarely available in the literature. In essence, one must develop a criterion (or criteria) for processability for each processing operation; for example, fiber spinnability, tubular film blowability, injection moldability, blow moldability, and thermoformability. Processability criteria are needed to answer a fundamental question: Why is a certain polymer suitable only for producing fibers, while another polymer is suitable only for producing bottles? Establishment of such processability criteria is not a trivial task because many factors must be considered: material variables, rheological properties, processing variables, and the morphology associated with the physical/mechanical properties and the molecular orientation in the final products. For instance, in melt spinning of a given polymer, fibers of different tensile properties can be obtained by varying the rate of cooling or the rate of stretching.

After all, processing of polymeric materials requires flow through a shaping device. Thus, a rational design of a shaping device (e.g., die or mold) requires information on the rheological properties of the polymer to be processed. And, for a given process the determination of an optimum processing condition (e.g., a minimum pressure drop) requires information on the temperature dependence of shear viscosity of a polymer to be processed. It is, then, fair to state that polymer rheology is an essential part of polymer processing operations.

1.5 Application of Polymer Rheology for On-Line Control of Polymerization Reactors

In the manufacture of polymers, the control of reactor conditions is of utmost importance for the production of a polymer with consistent quality. There are two methods that can be applied to control the reactor conditions. One method is to continuously monitor the weight-average molecular weight (M_w) or number-average molecular weight (M_n) and molecular weight distribution (M_w/M_n) of the polymer leaving the reactor and then use the measured quantities to adjust, via a feedback control strategy, reactor variables (e.g., monomer/catalyst ratio, feed rate to the reactor, or reactor temperature). However, at present, on-line measurements of M_w or M_n, and M_w/M_n (using gel permeation chromatograpy for instance) are not available. Another method is to continuously monitor both the viscosity and elasticity of the polymer leaving the reactor and then use the measured quantities to adjust, via a feedback control strategy, reactor variables, as schematically shown in Figure 1.7. For this method, one must develop a rheological property simulator that relates the viscosity and elasticity of a given polymer to its molecular parameters (M_w or M_n, and M_w/M_n). A rheological property simulator can be constructed on the basis of empirical correlations when a rigorous molecular viscoelasticity theory is not available. On-line measurement of the viscoelastic properties of a polymer can be realized using a capillary or slit rheometer, the principles of which are described in Chapter 5. On-line control of the rheological properties of polymers is far more effective than on-line control of molecular weight and molecular weight distribution of polymers to control later polymer processing operations. This is because the rheological properties of a polymer dictate the optimum processing conditions for a given piece of equipment.

In summary, in Volume 1 of this book we present the fundamental aspects of polymer rheology and the rheological behavior of different types of polymeric materials. In so doing, examples will be given that show relationships between the rheological properties and the molecular parameters of specific polymeric materials. In Volume 2

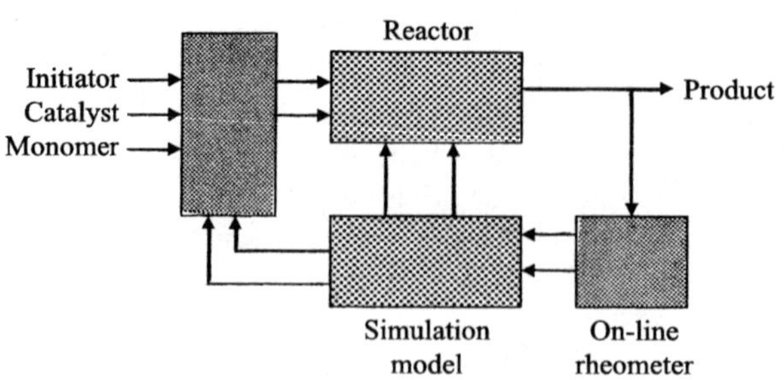

Figure 1.7 Schematic describing how on-line measurements of the rheological properties of the effluent stream from the polymerization reactor can be used to control the consistency of polymer quality.

we present the unique processing characteristics of some polymeric materials. In so doing, we choose several processing operations of thermoplastic polymers and three processing operations of thermosets or thermoset composites. No attempt is made to describe how to produce products that are better from a commercial point of view. Instead, emphasis is placed on presenting the fundamental concepts and principles, but not recipes, for each polymer processing operation chosen.

References

Berry GC, Fox RG (1968). *Adv. Polym. Sci.* 5:261.

Doi M, Edwards SF (1986). *The Theory of Polymer Dynamics*, Oxford University Press, Oxford.

Ferry JD (1980). *Viscoelastic Properties of Polymers,* 3rd ed, John Wiley & Sons, New York.

Garner FH, Nissan AH (1946). *Nature (London)*, 158:634.

Han CD (1976). *Rheology in Polymer Processing*, Academic Press, New York.

Han CD (1981). *Multiphase Flow in Polymer Processing*, Academic Press, New York.

Rouse PE (1953). *J. Chem. Phys.* 21:1272.

Weissenberg K (1947). *Nature (London)*, 159:310.

Weissenberg K (1949). In *Proceedings of the First International Congress on Rheology*, North-Holland, Amsterdam, Netherlands, p II-114.

Part I

Fundamental Principles of Polymer Rheology

2

Kinematics and Stresses of Deformable Bodies

2.1 Introduction

The form of kinematics to be used for the description of a deformation process is largely determined by the kind of mechanical response that is being described. To describe the mechanical response of purely viscous fluids it is convenient to use coordinates, which are fixed in space, since purely viscous fluids have no past memory and therefore remain in the deformed state when loads are removed. In other words, the mechanical response of purely viscous fluids is determined solely by the instantaneous values of the time rate of deformation.

However, in order to describe the deformation of a viscoelastic fluid it is necessary to follow a given material element with time as it moves to define a suitable measure of deformation that always refers to the same material element as time varies. The reason is that when a material element undergoes a finite deformation the coordinate positions of the given material element (with respect to a fixed origin) will vary. Hence, any measure of deformation defined in terms of infinitesimal deformation of fixed coordinate positions loses its physical significance since it will not always be associated with the same material element.

In this chapter, we introduce some basic concepts of the kinematics and stresses of a deformable body from the point of view of continuum mechanics, and discuss various representations of a deformation process in terms of the deformation (or strain) tensor and the rate-of-deformation (or rate-of-strain) tensor. In order to help the readers follow the material in the text, the elementary properties of second-order tensors are presented in Appendix 2A.

2.2 Description of Motion

In this section, we briefly describe the motion of a body, which consists of a set of particles (or "elements"), sometimes called "material points" (or "material elements") (Jaunzemis 1967). Let $\mathbf{X}(X^i;\ i = 1, 2, 3)$ be the particles P of the body B in some reference configuration $\boldsymbol{\kappa}$ at time $t = 0$ (i.e., undeformed state) and then we have

$$\mathbf{X} = \boldsymbol{\kappa}(\mathrm{P}) \tag{2.1}$$

in which $\boldsymbol{\kappa}$ describes the shape of the body B in the undeformed state, which in general is known to an observer. When the body B is deformed, the positions of the same particles P may be represented by (see Figure 2.1)

$$\mathbf{x}(t) = \boldsymbol{\chi}(\mathrm{P}, t) \tag{2.2}$$

in which $\mathbf{x}$ $(x^i;\ i = 1, 2, 3)$ are the positions of the particle at time t that have configurations $\boldsymbol{\chi}$. Because of the implicit assumption used that the body B is deformable, $\boldsymbol{\chi}$ describes the shape of the body at time t. If we assume that one particle can occupy only one position at a time, we can combine Eqs. (2.1) and (2.2) to give

$$\mathbf{x}(t) = \boldsymbol{\chi}_{\boldsymbol{\kappa}}(\mathbf{X}, t) \tag{2.3}$$

Equation (2.3) states that the positions of the particles P in motion at any instant may be determined from the information of the positions and configuration of the same particles in the undeformed state ($t = 0$), that is, in the reference configuration $\boldsymbol{\kappa}$. Thus $\boldsymbol{\chi}_{\boldsymbol{\kappa}}$ describes the shape of a body at time t in reference to the shape of the same body in the undeformed state (i.e., in the reference configuration). The coordinates X^i are called the "material coordinates," which describe the reference configuration

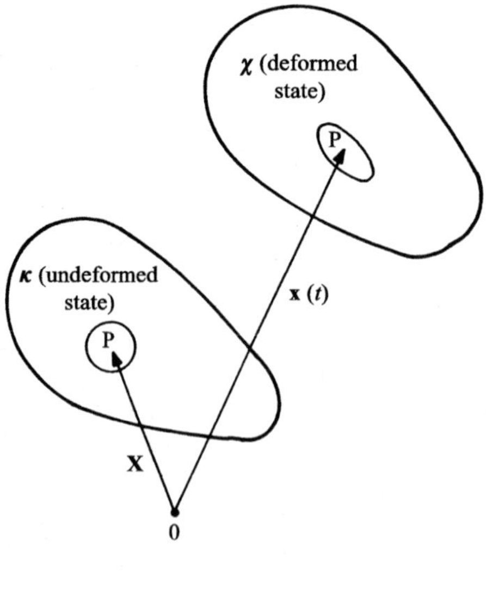

Figure 2.1 Deformation of a material element.

in Eq. (2.1), and the coordinates x^i are called the "spatial coordinates" (Jaunzemis 1967).

When dealing with motion, the present instant is usually singled out for special attention and chosen as the reference configuration. This choice is of particular interest to the description of the motion of nonperfectly elastic materials (e.g., viscoelastic fluids). The reason is that viscoelastic materials do not possess perfect memory, and therefore such materials cannot return to their original (undeformed) state when external forces are removed. It is then clear that the choice of the undeformed state as a reference configuration is not convenient for the description of the motion of viscoelastic fluids.

When the present configuration is chosen as the reference configuration, particles are identified with the positions they occupy at time t, therefore from Eq. (2.2) we have

$$\mathrm{P} = \chi^{-1}(\mathbf{x}(t), t) \tag{2.4}$$

Use of Eq. (2.4) in Eq. (2.2) gives

$$\mathbf{x}'(t') = \chi\left(\chi^{-1}(\mathbf{x}, t), t'\right) = \chi_t(\mathbf{x}, t') \tag{2.5}$$

where $\mathbf{x}'(x'^1, x'^2, x'^3)$ are the positions of the particles at time t' $(< t)$ and χ_t describes shapes of the body at time t' relative to the shape at time t; in other words, the relative configuration by means of which all other configurations are compared with the present one. Frequently, one also uses the elapsed time s, defined by $s = t - t'$, where $0 < s < \infty$ and $-\infty < t' < t$. Note further that Eq. (2.5) reduces to the trivial consequence

$$\mathbf{x}(t) = \chi_t(\mathbf{x}, t) = \mathbf{x}(t) \tag{2.6}$$

for $t' = t$.

There is another way of describing the motion of a body consisting of particles, which does not require knowledge of the paths of individual particles. In this description, called the "spatial description," the particle velocity $\mathbf{v}(t)$ at time t is considered as a dependent variable:

$$\mathbf{v}(t) = f(\mathbf{x}, t) \tag{2.7}$$

Note in Eq. (2.7) that $\mathbf{x}$ and t are independent variables; that is, in Eq. (2.7) $\mathbf{x}$ describes merely a fixed point in space.

The distinction between material and spatial descriptions is clear in that in the former $\mathbf{x}(t)$ is the dependent variable and $\mathbf{X}$ and t are the independent variables, whereas in the latter $v(t)$ is the dependent variable and $\mathbf{x}$ and t the independent variables. Frequently, the material coordinates are called "Lagrangian," and the spatial coordinates "Eulerian."

To illustrate the rules described previously, let us consider a motion described by

$$x'^1(t') = X^1(1 + t'), \quad x'^2(t') = X^1 t' + X^2, \quad x'^3(t') = X^3 \tag{2.8}$$

The spatial description of this motion may be obtained by first substituting $t = t'$ into Eq. (2.8), yielding

$$x^1(t) = X^1(1+t); \quad x^2(t) = X^1 t + X^2; \quad x^3(t) = X^3 \tag{2.9}$$

which is of the form of Eq. (2.3), and then by eliminating X^1, X^2, and X^3 with substitution of Eq. (2.9) into Eq. (2.8),

$$\begin{aligned} x'^1(t') &= \frac{(1+t')x^1(t)}{1+t} \\ x'^2(t') &= \frac{t'x^1(t)}{1+t} + x^2(t) - \frac{tx^1(t)}{1+t} \\ x'^3(t') &= x^3(t) \end{aligned} \tag{2.10}$$

which is of the form of Eq. (2.5). Equation (2.10) describes the positions of particles at time $t'(< t)$ relative to the positions of the same particles at present time t. It can be easily shown that Eq. (2.10) reduces to the identity equations for $t' = t$.

2.3 Some Representative Flow Fields

Here, we consider two important, frequently encountered flow fields: shear flow field and elongational flow field. They will be used throughout this chapter and in later chapters.

2.3.1 Steady-State Shear Flow Field

A simple flow geometry of practical interest is schematically shown in Figure 2.2. It consists of two parallel plates forming a narrow gap whose distance h is very small compared with the width w of the plates (i.e., $w \gg h$). Referring to Figure 2.2a, a fluid is placed in the gap between the two parallel plates, and then the upper plate is forced

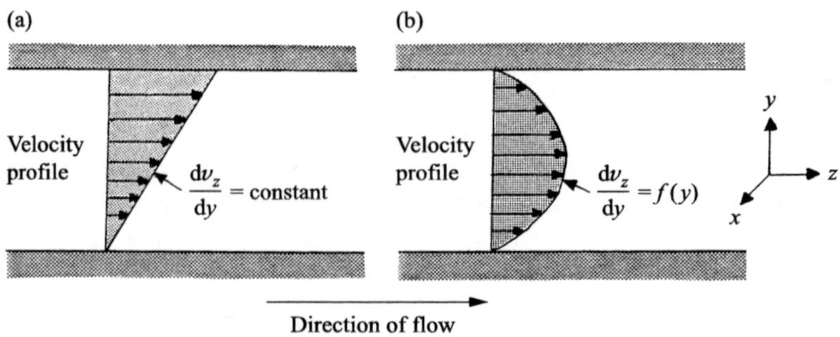

Figure 2.2 Schematic of shear flow field for (a) uniform shear flow and (b) nonuniform shear flow.

to move along the z direction while the lower plate is kept stationary. Under such situations, the velocity profile v_z is linear with respect to the y direction, giving rise to a constant velocity gradient, dv_z/dy = constant. Such a flow field is referred to as a "uniform (or simple) shear flow field." Referring to Figure 2.2b, a fluid is forced to flow through the gap between two stationary parallel plates. Under such situations, the velocity profile v_z varies with the y direction, giving rise to a parabolic velocity profile and a nonconstant velocity gradient, $dv_z/dy = f(y)$. Such a flow field is referred to as a "nonuniform shear flow field." For steady-state shear flow, the velocity field for an incompressible fluid in Cartesian coordinates (x, y, z) can be expressed as

$$v_z = \dot{\gamma} y, \quad v_y = v_x = 0 \tag{2.11}$$

where $\dot{\gamma} = dv_z/dy$ is the velocity gradient, commonly referred to as shear rate. The rate-of-strain tensor **d** for the steady-state shear flow field can be described by

$$\| \mathbf{d} \| = \left\| \begin{matrix} 0 & \dot{\gamma}/2 & 0 \\ \dot{\gamma}/2 & 0 & 0 \\ 0 & 0 & 0 \end{matrix} \right\| \tag{2.12}$$

Note that $\dot{\gamma}$ appearing in Eq. (2.12) is constant for uniform shear flow and not constant for nonuniform shear flow. In Chapter 5 we present experimental methods for the determination of the rheological properties of polymeric liquids in the uniform shear flow field using a cone-and-plate rheometer and in the nonuniform shear flow field using a capillary or slit rheometer.

2.3.2 Steady-State Elongational Flow Field

Another flow field that is also of very practical importance is the elongational (or extensional) flow field, which may be found in such polymer processing operations as fiber spinning, cast-film extrusion, film blowing, blow molding, and thermoforming. For uniaxial stretching, the velocity field $\mathbf{v}(v_x, v_y, v_z)$ of an incompressible fluid in Cartesian coordinates (x, y, z) is given by

$$dv_z/dz = \dot{\varepsilon}; \quad dv_y/dy = dv_x/dx \tag{2.13}$$

where $\dot{\varepsilon}$ is the velocity gradient in the direction of stretching z (commonly referred to as the elongation rate), y is the direction perpendicular to the stretching, and x is the neutral direction. In order to satisfy the equation of continuity, we require that

$$\frac{dv_z}{dz} + \frac{dv_y}{dy} + \frac{dv_x}{dx} = 0 \tag{2.14}$$

Using Eq. (2.13) in Eq. (2.14), we have the following expression for the rate-of-strain tensor **d** in uniaxial elongational flow

$$\|\mathbf{d}\| = \begin{Vmatrix} \dot{\varepsilon} & 0 & 0 \\ 0 & -\dot{\varepsilon}/2 & 0 \\ 0 & 0 & -\dot{\varepsilon}/2 \end{Vmatrix} \tag{2.15}$$

Note in Eq. (2.15) that $\dot{\varepsilon}$ is constant for steady-state uniform, uniaxial elongational flow and $\dot{\varepsilon}$ varies with the stretching direction z for nonuniform, uniaxial elongational flow. In Chapter 5 we present the rheological response of polymeric liquids in steady-state uniform, uniaxial elongational flow, and in Chapter 6 of Volume 2 we present the rheological response of polymeric liquids in steady-state nonuniform, uniaxial elongational flow that occurs in fiber spinning.

For equal biaxial stretching, the rate-of-strain tensor **d** can be expressed as

$$\|\mathbf{d}\| = \begin{Vmatrix} \dot{\varepsilon}_B & 0 & 0 \\ 0 & \dot{\varepsilon}_B & 0 \\ 0 & 0 & -2\dot{\varepsilon}_B \end{Vmatrix} \tag{2.16}$$

where $\dot{\varepsilon}_B$ is the elongation rate in equal biaxial stretching and is defined as

$$\dot{\varepsilon}_B = dv_z/dz = dv_y/dy \tag{2.17}$$

Note that Eqs. (2.14) and (2.17) are used to obtain Eq. (2.16).

For unequal biaxial stretching, the rate-of-strain tensor **d** is expressed as

$$\|\mathbf{d}\| = \begin{Vmatrix} \dot{\varepsilon}_a & 0 & 0 \\ 0 & \dot{\varepsilon}_b & 0 \\ 0 & 0 & -(\dot{\varepsilon}_a + \dot{\varepsilon}_b) \end{Vmatrix} \tag{2.18}$$

where $\dot{\varepsilon}_a$ and $\dot{\varepsilon}_b$ are the elongation rates in unequal biaxial stretching and are defined as

$$\dot{\varepsilon}_a = dv_z/dz; \quad \dot{\varepsilon}_b = dv_y/dy \tag{2.19}$$

In Chapter 7 of Volume 2, we present the rheological response of polymeric liquids in steady-state biaxial elongational flow.

2.4 Deformation Gradient Tensor, Strain Tensor, Velocity Gradient Tensor and Rate-of-Strain Tensor

2.4.1 Deformation Gradient Tensor

For the description of motion given by Eq. (2.3), consider two particles in the reference configuration (at $t = 0$) that are a distance $d\mathbf{X}$ apart. Then in the configuration $\boldsymbol{\chi}$

(at some other time t) these same two particles are a distance $\mathbf{dx}$ apart, given by (Jaunzemis 1967)

$$\mathrm{d}\mathbf{x}(t) = \boldsymbol{\chi}_{\boldsymbol{\kappa}}(\mathbf{X} + \mathrm{d}\mathbf{X}, t) - \boldsymbol{\chi}_{\boldsymbol{\kappa}}(\mathbf{X}, t) \tag{2.20}$$

Using Taylor's theorem we may approximate $\boldsymbol{\chi}_{\boldsymbol{\kappa}}(\mathbf{X} + \mathrm{d}\mathbf{X}, t)$ by

$$\boldsymbol{\chi}_{\boldsymbol{\kappa}}(\mathbf{X} + \mathrm{d}\mathbf{X}, t) = \boldsymbol{\chi}_{\boldsymbol{\kappa}}(\mathbf{X}, t) + (\partial \boldsymbol{\chi}_{\boldsymbol{\kappa}}/\partial \mathbf{X})\,\mathrm{d}\mathbf{X} \tag{2.21}$$

as the magnitude $|\mathrm{d}\mathbf{X}|$ of $\mathrm{d}\mathbf{X}$ approaches zero. Use of Eq. (2.21) in Eq. (2.20) gives

$$\mathrm{d}\mathbf{x}(t) = \mathbf{F}(t)\,\mathrm{d}\mathbf{X} \tag{2.22}$$

where $\mathbf{F}$ is called the deformation gradient tensor represented given by

$$\|\mathbf{F}(t)\| = \left\| \frac{\partial \mathbf{x}(\mathbf{X}, t)}{\partial \mathbf{X}} \right\| = \left\| \begin{matrix} \partial x^1/\partial X^1 & \partial x^1/\partial X^2 & \partial x^1/\partial X^3 \\ \partial x^2/\partial X^1 & \partial x^2/\partial X^2 & \partial x^2/\partial X^3 \\ \partial x^3/\partial X^1 & \partial x^3/\partial X^2 & \partial x^3/\partial X^3 \end{matrix} \right\| \tag{2.23}$$

One may also interpret $\mathbf{F}$ as a linear operator, which maps the neighborhood of the particles $\mathbf{X}$ in the reference configuration $\boldsymbol{\kappa}$ into the configuration $\boldsymbol{\chi}$.

Using the relative configuration $\boldsymbol{\chi}_t$ defined by Eq. (2.5), we can also define

$$\mathrm{d}\mathbf{x}'(t') = \mathbf{F}_t(t')\,\mathrm{d}\mathbf{x}(t) \tag{2.24}$$

where $\mathbf{F}_t(t')$ is called the relative deformation gradient tensor. It is seen in Eq. (2.24) that at $t' = t$ we have

$$\mathbf{F}_t(t) = \mathbf{I} \tag{2.25}$$

where $\mathbf{I}$ is the unit second-order tensor.

For a motion described by Eq. (2.9), for instance, we have

$$\|\mathbf{F}(t)\| = \left\| \begin{matrix} 1+t & 0 & 0 \\ t & 1 & 0 \\ 0 & 0 & 1 \end{matrix} \right\| \tag{2.26}$$

and

$$\|\mathbf{F}_t(t', t)\| = \left\| \begin{matrix} (1+t')\,/\,(1+t) & 0 & 0 \\ (t'-t)\,/\,(1+t) & 1 & 0 \\ 0 & 0 & 1 \end{matrix} \right\| \tag{2.27}$$

2.4.2 Strain Tensor

We can define other deformation tensors, also, in terms of the deformation gradient tensor **F**. According to the "polar decomposition theorem" of the second-order tensor (see Appendix 2A), the deformation gradient tensor **F**, which is an asymmetric tensor and is assumed to be nonsingular (i.e., det $\mathbf{F} \neq 0$), can be expressed as a product of a positive symmetric tensor with an orthogonal tensor (Jaunzemis 1967):

$$\mathbf{F}(t) = \mathbf{R}(t)\mathbf{U}(t) \tag{2.28}$$

where **U** is a positive symmetric tensor and **R** is an orthogonal tensor. A geometrical interpretation of Eq. (2.28) may best be illustrated in Figure 2.3. That is, the deformation $\mathbf{F} = \mathbf{RU}$ may be said to occur first by the stretches **U** in the principal direction, followed by the rotation **R**. From Eq. (2.28) one has

$$\mathbf{C}(t) = \mathbf{F}^{\mathrm{T}}(t)\mathbf{F}(t) = \mathbf{U}(t)^2 \tag{2.29}$$

where **C** is called the Cauchy–Green (deformation) tensor. Note that $\mathbf{F}^{\mathrm{T}}$ in Eq. (2.29) is the transpose of **F** and the orthogonality property of **R** (i.e., $\mathbf{RR}^{\mathrm{T}} = \mathbf{RR}^{-1} = \mathbf{I}$) has been used.

The practical significance of Eqs. (2.28) and (2.29) lies in that, because of the positive definiteness of the symmetric tensor **C**, once **F** is known one can determine the stretches **U** from

$$\mathbf{U}(t) = \left(\mathbf{C}(t)\right)^{1/2} = \left(\mathbf{F}^{\mathrm{T}}(t)\mathbf{F}(t)\right)^{1/2} \tag{2.30}$$

and the rotation **R** from

$$\mathbf{R}(t) = \mathbf{F}(t)\mathbf{U}^{-1}(t) \tag{2.31}$$

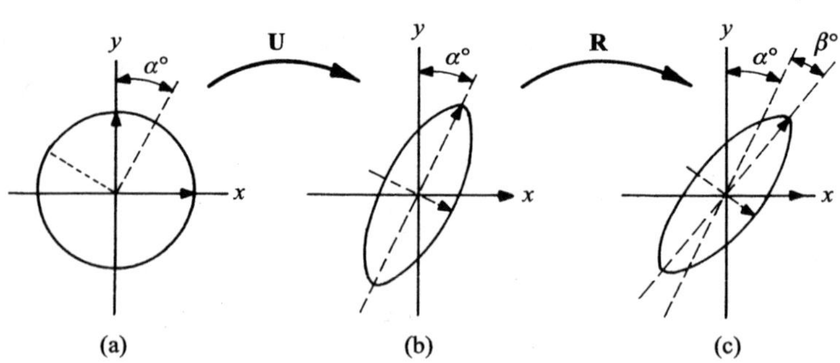

Figure 2.3 Geometrical interpretation of the polar decomposition of the deformation process, where (a) denotes undeformed state, (b) denotes the deformed state by stretches **U**, and (c) denotes the deformed state after rotation **R** following stretches **U**.

In terms of the relative deformation gradient tensor $\mathbf{F}_t(t')$, Eq. (2.29) may be written as

$$\mathbf{C}(\mathbf{x}, t, t') = \mathbf{C}_t(t') = \mathbf{F}_t^{\mathrm{T}}(t')\mathbf{F}_t(t') \tag{2.32}$$

where $\mathbf{C}_t(t')$ is called the relative Cauchy–Green (deformation) tensor, which describes the change in shape of a small material element between time t' and t. Note that at $t' = t$ we have

$$\mathbf{C}_t(t) = \mathbf{I} \tag{2.33}$$

In terms of the components of the relative deformation tensors, we can write Eq. (2.32) with the aid of Eq. (2.24) as

$$C_{ij}(\mathbf{x}, t, t') = (\partial x'^m/\partial x^i)(\partial x'^n/\partial x^j)g_{mn}(\mathbf{x}') \tag{2.34}$$

where $\mathbf{x}'(x'^1, x'^2, x'^3)$ are the spatial coordinates of the place occupied by material elements at time t' $(< t)$, and $\mathbf{x}(x^1, x^2, x^3)$ are the spatial coordinates of the place occupied by the same material elements at present time t. Note that $g_{mn}(\mathbf{x}')$ is the metric tensor (see Appendix 2B) referred to the spatial coordinates $\mathbf{x}'$ for curvilinear coordinate systems (Hawkins 1963; Jeffreys 1961). For rectangular Cartesian coordinates we have

$$g_{mn}(\mathbf{x}') = \delta_{mn}(\mathbf{x}') \tag{2.35}$$

where δ_{mn} is the Kronecker delta and is a second-order tensor. At $t = t'$ (hence $\mathbf{x} = \mathbf{x}'$), Eq. (2.34) reduces to

$$C_{ij}(\mathbf{x}) = g_{ij}(\mathbf{x}) \tag{2.36}$$

Using C_{ij} one can define the quantity E_{ij} (Eringen 1962; Jaunzemis 1967)

$$E_{ij}(\mathbf{x}, t, t') = g_{ij}(\mathbf{x}) - C_{ij}(\mathbf{x}, t, t') \tag{2.37}$$

which may be interpreted as strains that a material element located at $\mathbf{x}'$ at time t' $(< t)$ has experienced during the time period $t - t'$. $E_{ij}(\mathbf{x}, t, t')$ are the covariant components of the finite strain tensor $\mathbf{E}(x, t, t')$. Similarly, one can also define the contravariant components $E^{ij}(\mathbf{x}, t, t')$ of the finite strain tensor $\mathbf{E}(\mathbf{x}, t, t')$ by

$$E^{ij}(\mathbf{x}, t, t') = \left[C^{-1}(\mathbf{x}, t, t')\right]^{ij} - g^{ij}(\mathbf{x}) \tag{2.38}$$

where $(C^{-1})^{ij}$ are the contravariant components of the Finger deformation tensor $\mathbf{C}^{-1}(\mathbf{x}, t, t')$ defined by

$$\left[C^{-1}(\mathbf{x}, t, t')\right]^{ij} = (\partial x^i/\partial x'^m)(\partial x^j/\partial x'^n)g^{mn}(\mathbf{x}') \tag{2.39}$$

Note that the Cauchy–Green and Finger tensors are related by

$$C_{ij}(x)\left[C^{-1}(x)\right]^{jk} = \delta_i^k \tag{2.40}$$

Let us now consider steady-state simple shear flow, for which we have the velocity field of the form

$$v_1 = \dot{\gamma}x^2, \quad v_2 = v_3 = 0 \tag{2.41}$$

where $\dot{\gamma}$ is the shear rate. Now the relative deformation function $\mathbf{x}'(t')$ can be found by solving the differential equations

$$\frac{\mathrm{d}x'^1}{\mathrm{d}t'} = \dot{\gamma}x^2; \quad \frac{\mathrm{d}x'^2}{\mathrm{d}t'} = 0; \quad \frac{\mathrm{d}x'^3}{\mathrm{d}t'} = 0 \tag{2.42}$$

with the initial conditions

$$\mathbf{x}'(t')\big|_{t'=t} = \mathbf{x} \tag{2.43}$$

giving rise to

$$x'^1(t') = x^1 + (t'-t)\dot{\gamma}x^2; \quad x'^2(t') = x^2; \quad x'^3(t') = x^3 \tag{2.44}$$

The components of the relative deformation gradient tensor $\mathbf{F}_t(t')$ may be obtained by use of Eq. (2.44) in Eq. (2.24)

$$\|\mathbf{F}_t(t',t)\| = \begin{Vmatrix} 1 & (t'-t)\dot{\gamma} & 0 \\ 0 & 1 & 0 \\ 0 & 0 & 1 \end{Vmatrix} \tag{2.45}$$

The components of the relative Cauchy–Green deformation tensor $\mathbf{C}_t(t')$ may be obtained by use of Eq. (2.45) in Eq. (2.32)

$$\|\mathbf{C}_t(t',t)\| = \begin{Vmatrix} 1 & (t'-t)\dot{\gamma} & 0 \\ (t'-t)\dot{\gamma} & 1+(t'-t)^2\dot{\gamma}^2 & 0 \\ 0 & 0 & 1 \end{Vmatrix} \tag{2.46}$$

and the components of the relative Finger deformation tensor $\mathbf{C}_t^{-1}(t')$ by

$$\left\|\mathbf{C}_t^{-1}(t',t)\right\| = \begin{Vmatrix} 1+(t'-t)^2\dot{\gamma}^2 & -(t'-t)\dot{\gamma} & 0 \\ -(t'-t)\dot{\gamma} & 1 & 0 \\ 0 & 0 & 1 \end{Vmatrix} \tag{2.47}$$

Therefore, the covariant components $E_{ij}(t', t)$ of the finite strain tensor $\mathbf{E}(t', t)$ are given by

$$\left\|\mathbf{E}(t', t)\right\| = \left\| \begin{matrix} 0 & -(t'-t)\dot{\gamma} & 0 \\ -(t'-t)\dot{\gamma} & -(t'-t)^2\dot{\gamma}^2 & 0 \\ 0 & 0 & 0 \end{matrix} \right\| \tag{2.48}$$

and the contravariant components $E^{ij}(t', t)$ of the finite strain tensor $\mathbf{E}(t', t)$ are given by

$$\left\|\mathbf{E}(t', t)\right\| = \left\| \begin{matrix} (t'-t)^2\dot{\gamma}^2 & -(t'-t)\dot{\gamma} & 0 \\ -(t'-t)\dot{\gamma} & 0 & 0 \\ 0 & 0 & 0 \end{matrix} \right\| \tag{2.49}$$

We have shown here that the Cauchy–Green and Finger tensors are not equivalent measures of finite strain, which is a very important fact to remember in the formulation of constitutive equations, as is discussed in Chapter 3.

2.4.3 Velocity Gradient Tensor and Rate-of-Strain Tensor

We may take the time derivative of the deformation gradient tensor $\mathbf{F}$ (see Eq. (2.23)) as (Jaunzemis 1967)

$$\dot{\mathbf{F}}(t) = \frac{\partial \mathbf{F}(t)}{\partial t} = \frac{\partial}{\partial t}\left(\frac{\partial \mathbf{x}(\mathbf{X}, t)}{\partial \mathbf{X}}\right) \tag{2.50}$$

But since, in the material description, $\mathbf{X}$ and t are independent variables, the order of differentiation with respect to $\mathbf{X}$ and t can be interchanged:

$$\frac{\partial}{\partial t}\left(\frac{\partial x^i(\mathbf{X}, t)}{\partial X^j}\right) = \frac{\partial}{\partial X^j}\left(\frac{\partial x^i(\mathbf{X}, t)}{\partial t}\right) = \frac{\partial v^i(\mathbf{X}, t)}{\partial X^j} \tag{2.51}$$

It is seen on the right side of Eq. (2.51) that the gradient of the instantaneous velocity with respect to the coordinates X^j in the reference configuration $\boldsymbol{\kappa}$ is not a rate tensor. But the gradient of velocity with respect to the present coordinates x^j does constitute a rate tensor. We can accomplish this by using the chain rule

$$\partial v^i/\partial X^j = \left(\partial v^i/\partial x^m\right)\left(\partial x^m/\partial X^j\right) \tag{2.52}$$

or

$$\dot{\mathbf{F}}(t) = \mathbf{L}(t)\mathbf{F}(t) \tag{2.53}$$

in which $\mathbf{L}$ is the velocity gradient tensor defined as

$$L_{ij}(t) = \partial v^i(\mathbf{x}, t)/\partial x^j \tag{2.54}$$

It is seen in Eq. (2.53) that the velocity gradient tensor $\mathbf{L}(t)$ can be determined from the rate of deformation gradient tensor $\dot{\mathbf{F}}(t)$ and the inverse of the deformation gradient tensor $\mathbf{F}^{-1}(t)$, that is,

$$\mathbf{L}(t) = \dot{\mathbf{F}}(t)\mathbf{F}^{-1}(t) \tag{2.55}$$

The velocity gradient tensor $\mathbf{L}(t)$ can be determined from the relative deformation gradient tensor $\mathbf{F}_t(t')$ also, since we have

$$\mathbf{F}_t(t') = \mathbf{F}(t')\mathbf{F}^{-1}(t) \tag{2.56}$$

Therefore

$$\dot{\mathbf{F}}_t(t) = \left.\frac{\partial \mathbf{F}_t(t', t)}{\partial t'}\right|_{t'=t} = \dot{\mathbf{F}}(t)\mathbf{F}^{-1}(t) = \mathbf{L}(t) \tag{2.57}$$

and any higher-order derivative of $\mathbf{F}_t(t)$ may be defined as (Rivlin and Ericksen 1955)

$$\mathbf{F}_t^{(n)}(t) = \left.\frac{\partial^n \mathbf{F}_t(t', t)}{\partial t'^n}\right|_{t'=t} = \mathbf{L}_{(n)}(t) \tag{2.58}$$

where $\mathbf{L}_{(n)}$ is the nth acceleration gradient tensor.

We can now show further that $\mathbf{L}$ may be decomposed into symmetric and asymmetric parts

$$\mathbf{L} = \mathbf{d} + \boldsymbol{\omega} \tag{2.59}$$

in which $\mathbf{d}$ and $\boldsymbol{\omega}$ are defined as

$$\mathbf{d} = \tfrac{1}{2}(\mathbf{L} + \mathbf{L}^{\mathrm{T}}) \quad \text{or} \quad d_{ij} = \tfrac{1}{2}\left(\frac{\partial v^i}{\partial x^j} + \frac{\partial v^j}{\partial x^i}\right) \tag{2.60}$$

and

$$\boldsymbol{\omega} = \tfrac{1}{2}(\mathbf{L} - \mathbf{L}^{\mathrm{T}}) \quad \text{or} \quad \omega_{ij} = \tfrac{1}{2}\left(\frac{\partial v^i}{\partial x^j} - \frac{\partial v^j}{\partial x^i}\right) \tag{2.61}$$

respectively. Note that $d_{ij} = d_{ji}$ and $\omega_{ij} = -\omega_{ji}$. The physical interpretations of $\boldsymbol{\omega}$ and $\mathbf{d}$ are as follows. $\boldsymbol{\omega}$ is called the vorticity tensor, which is the asymmetric part of $\mathbf{L}$, and it is the material derivative of the finite rotation tensor $\mathbf{R}$ taken with respect to the present configuration (i.e., $\boldsymbol{\omega} = \dot{\mathbf{R}}_t(t)$). $\mathbf{d}$ is called the rate-of-strain tensor (or rate-of-deformation tensor), which is the symmetric part of $\mathbf{L}$, and it is the material derivative of the positive symmetric tensors $\mathbf{U}$ taken with respect to the present configuration ($\dot{\mathbf{U}}_t(t)$).

Using the relative Cauchy–Green tensor $\mathbf{C}_t(t', t)$ one can define other rate tensors, such as (Rivlin and Ericksen 1955)

$$\mathbf{A}_{(n)} = \left. \frac{\mathbf{d}^n \mathbf{C}_t(t', t)}{\mathbf{d}t'^n} \right|_{t'=t} = \sum_{k=0}^{n} \binom{n}{k} \mathbf{L}_{(k)}^T \mathbf{L}_{(n-k)} \tag{2.62}$$

in which use is made of Eqs. (2.32) and (2.58). $\mathbf{A}_{(n)}$ is called the nth-order Rivlin–Ericksen tensor. It should be noted that Rivlin–Ericksen tensors play an important role in formulating constitutive equations, which is discussed in Chapter 3.

To illustrate the usefulness of the various forms of rate tensors we have introduced, let us consider steady-state simple shear flow whose velocity field is given by Eq. (2.41) and whose motion is given by Eq. (2.44). Use of Eq. (2.45) in (2.57) gives

$$\|\mathbf{L}\| = \begin{Vmatrix} 0 & \dot{\gamma} & 0 \\ 0 & 0 & 0 \\ 0 & 0 & 0 \end{Vmatrix} \tag{2.63}$$

We now have the rate-of-strain tensor $\mathbf{d}$ given by Eq. (2.12) and the vorticity tensor $\boldsymbol{\omega}$ from Eq. (2.61):

$$\|\boldsymbol{\omega}\| = \begin{Vmatrix} 0 & \dot{\gamma}/2 & 0 \\ -\dot{\gamma}/2 & 0 & 0 \\ 0 & 0 & 0 \end{Vmatrix} \tag{2.64}$$

Further, use of Eq. (2.46) in (2.62) gives

$$\left\|\mathbf{A}_{(1)}\right\| = \begin{Vmatrix} 0 & \dot{\gamma} & 0 \\ \dot{\gamma} & 0 & 0 \\ 0 & 0 & 0 \end{Vmatrix} \tag{2.65}$$

$$\left\|\mathbf{A}_{(2)}\right\| = \begin{Vmatrix} 0 & 0 & 0 \\ 0 & 2\dot{\gamma}^2 & 0 \\ 0 & 0 & 0 \end{Vmatrix} \tag{2.66}$$

and

$$\mathbf{A}_{(n)} = 0 \quad \text{for} \quad n \geq 3 \tag{2.67}$$

Note that

$$\mathbf{A}_{(1)} = 2\mathbf{d} \tag{2.68}$$

and

$$\mathbf{C}_t(t', t) = \mathbf{I} + (t' - t)\mathbf{A}_{(1)} + \tfrac{1}{2}(t' - t)^2\mathbf{A}_{(2)} \tag{2.69}$$

It is of interest to note that the relative Cauchy–Green tensor $\mathbf{C}_t(t')$ can be expressed in terms of the Rivlin–Ericksen tensors $\mathbf{A}_{(m)}$ by (Coleman 1962; Rivlin and Ericksen 1955)

$$\mathbf{C}_t(t - s) = \mathbf{I} + \sum_{m=1}^{n-1} (-1)^m \frac{s^m}{m!} \mathbf{A}_{(m)}(t) \tag{2.70}$$

where s is the elapsed time defined as $s = t - t'$.

For the steady-state uniaxial elongational flow, the relative deformation gradient tensor $\mathbf{F}_t(t', t)$ can be written

$$\|\mathbf{F}_t(t', t)\| = \begin{Vmatrix} e^{\dot{\varepsilon}(t'-t)} & 0 & 0 \\ 0 & e^{-\dot{\varepsilon}(t'-t)/2} & 0 \\ 0 & 0 & e^{-\dot{\varepsilon}(t'-t)/2} \end{Vmatrix} \tag{2.71}$$

Use of Eq. (2.71) in (2.57) gives the rate-of-deformation tensor **d** defined by Eq. (2.15). Note that for the uniaxial elongational flow, from Eq. (2.59) we have $\mathbf{d} = \mathbf{L}$ because the vorrticity tensor vanishes. Further, we have

$$\|\mathbf{C}_t(t', t)\| = \begin{Vmatrix} e^{2\dot{\varepsilon}(t'-t)} & 0 & 0 \\ 0 & e^{-\dot{\varepsilon}(t'-t)} & 0 \\ 0 & 0 & e^{-\dot{\varepsilon}(t'-t)} \end{Vmatrix} \tag{2.72}$$

$$\left\|\mathbf{C}_t^{-1}(t', t)\right\| = \begin{Vmatrix} e^{-2\dot{\varepsilon}(t'-t)} & 0 & 0 \\ 0 & e^{\dot{\varepsilon}(t'-t)} & 0 \\ 0 & 0 & e^{\dot{\varepsilon}(t'-t)} \end{Vmatrix} \tag{2.73}$$

Since $\mathbf{E} = \mathbf{C}_t^{-1} - \mathbf{I}$, we have

$$\|\mathbf{E}(t', t)\| = \begin{Vmatrix} e^{-2\dot{\varepsilon}(t'-t)} - 1 & 0 & 0 \\ 0 & e^{\dot{\varepsilon}(t'-t)} - 1 & 0 \\ 0 & 0 & e^{\dot{\varepsilon}(t'-t)} - 1 \end{Vmatrix} \tag{2.74}$$

For steady-state equal biaxial elongational flow, the relative deformation gradient tensor $\mathbf{F}_t(t', t)$ can be written, with the aid of Eqs. (2.16) and (2.57), as

$$\|\mathbf{F}_t(t', t)\| = \begin{Vmatrix} e^{\dot{\varepsilon}_B(t'-t)} & 0 & 0 \\ 0 & e^{\dot{\varepsilon}_B(t'-t)} & 0 \\ 0 & 0 & e^{-2\dot{\varepsilon}_B(t'-t)} \end{Vmatrix} \tag{2.75}$$

Further, we have

$$\|\mathbf{C}_t(t', t)\| = \begin{Vmatrix} e^{2\dot{\varepsilon}_B(t'-t)} & 0 & 0 \\ 0 & e^{2\dot{\varepsilon}_B(t'-t)} & 0 \\ 0 & 0 & e^{-4\dot{\varepsilon}_B(t'-t)} \end{Vmatrix} \tag{2.76}$$

$$\left\|\mathbf{C}_t^{-1}(t', t)\right\| = \begin{Vmatrix} e^{-2\dot{\varepsilon}_B(t'-t)} & 0 & 0 \\ 0 & e^{-2\dot{\varepsilon}_B(t'-t)} & 0 \\ 0 & 0 & e^{4\dot{\varepsilon}_B(t'-t)} \end{Vmatrix} \tag{2.77}$$

$$\|\mathbf{E}(t', t)\| = \begin{Vmatrix} e^{-2\dot{\varepsilon}_B(t'-t)} - 1 & 0 & 0 \\ 0 & e^{-2\dot{\varepsilon}_B(t'-t)} - 1 & 0 \\ 0 & 0 & e^{4\dot{\varepsilon}_B(t'-t)} - 1 \end{Vmatrix} \tag{2.78}$$

2.5 Kinematics in Moving (Convected) Coordinates

The primary thrust of this section is to prepare ourselves in order to be able to write the kinematic quantities defined in a moving coordinate system via the transformation rules in terms of the Cartesian components in the fixed coordinate system. This is necessary, as will be discussed in the next chapter, for transforming a constitutive equation, which was first written in a moving coordinate system, into a fixed coordinates so that it can be used in conjunction with the equations of continuity, motion, and energy that are normally written in the fixed coordinates.

In describing the kinematics of a deformable body, instead of using a coordinate system fixed in space, it is convenient to use a coordinate system embedded in the moving object. This is frequently referred to as a "convected coordinate" system, and was first introduced by Oldroyd (1950). Any measure of deformation (strain) defined relative to such a coordinate system always refers to the same element of materials, and therefore should be independent of the local rate of translation or rotation. As will be shown in this section, if they are going to be useful, all kinematic variables defined in terms of the convected coordinates must be transformed to a fixed coordinate system as all physical measurements are made relative to the fixed coordinate system.

2.5.1 Convected Strain Tensor

Let a convected coordinate system be denoted by $\xi^i (i = 1, 2, 3)$ and a fixed coordinate system by $x^i (i = 1, 2, 3)$. Then, states of a material element may be described by functions

$$x^i = f^i(\boldsymbol{\xi}, t) \tag{2.79}$$

which have a unique inverse

$$\xi^i = \Phi^i(\mathbf{x}, t) \tag{2.80}$$

where t represents time.

A deformation may be said to occur when the magnitude of the distance between any two points in a material element changes. The square of the distance between the two points in a space may then be used as a quantitative measure of deformation (strain). In terms of spatial coordinates, the distance between two material points may be represented by

$$(\mathrm{d}s)^2 = \mathrm{d}\mathbf{x} \cdot \mathrm{d}\mathbf{x} = g_{mn}(\mathbf{x})\, \mathrm{d}x^m \mathrm{d}x^n \tag{2.81}$$

where g_{mn} is the spatial metric tensor (see Appendix 2B). From Eq. (2.79) we have

$$\mathrm{d}x^k = (\partial x^k / \partial \xi^i)\, \mathrm{d}\xi^i \tag{2.82}$$

Therefore, in terms of convected coordinates, the distance between two material points may be represented, by use of Eq. (2.82) in (2.81), as

$$(\mathrm{d}s)^2 = \nu_{ij}(\boldsymbol{\xi}, t)\, \mathrm{d}\xi^i\, \mathrm{d}\xi^j \tag{2.83}$$

where ν_{ij} is called the convected covariant metric, which is related to the spatial metric g_{ij} by

$$\nu_{ij}(\boldsymbol{\xi}, t) = (\partial x^m / \partial \xi^i)(\partial x^n / \partial \xi^j) g_{mn}(\mathbf{x}) \tag{2.84}$$

Similarly, the convected contravariant metric $\nu^{ij}(\boldsymbol{\xi}, t)$ is related to the spatial metric g^{ij} by

$$\nu^{ij}(\boldsymbol{\xi}, t) = (\partial \xi^i / \partial x^m)(\partial \xi^j / \partial x^n)\, g^{mn}(\mathbf{x}) \tag{2.85}$$

The change in the distance between the material points at two different times, t' and $t (> t')$, may be used as a measure of the strain, and it may be written, from Eq. (2.83), as (Eringen 1962)

$$ds^2(t) - ds^2(t') = \Gamma_{ij}\, d\xi^i d\xi^j \tag{2.86}$$

where Γ_{ij} are the components of the convected covariant strain tensor,

$$\Gamma_{ij}(\boldsymbol{\xi}, t, t') = \nu_{ij}(\boldsymbol{\xi}, t) - \nu_{ij}(\boldsymbol{\xi}, t') \tag{2.87}$$

Similarly, the components of the convected contravariant strain tensor Γ^{ij} may be defined as

$$\Gamma^{ij}(\boldsymbol{\xi}, t, t') = \nu^{ij}(\boldsymbol{\xi}, t') - \nu^{ij}(\boldsymbol{\xi}, t) \tag{2.88}$$

The definition of the convected strain tensor involves the difference between two quantities associated with a given material point at different times, and it refers to the same material point in convected coordinates. Now, we must transform the quantities $\nu_{ij}(\boldsymbol{\xi}, t)$ and $\nu_{ij}(\boldsymbol{\xi}, t')$ (also $\nu^{ij}(\boldsymbol{\xi}, t')$, and $\nu^{ij}(\boldsymbol{\xi}, t)$) in such a manner that they both refer to the same point in a coordinate system fixed in space, because physical quantities (kinematic and dynamic variables) can only be measured relative to a frame of reference fixed in space. This can be done by making use of the transformation relations between two coordinate systems.

Remembering that the coordinate systems x^i and ξ^i are arbitrary (except that x^i are fixed in space and ξ^i in the material), let us choose an arbitrary spatial coordinate system x^i and then choose a convected coordinate system ξ^i that coincides with the spatial coordinate system at present time t. Note that, in this choice, the present configuration is a reference configuration, so that all other configurations at time $t'(< t)$ are compared with the present one. From Eq. (2.84) we then have

$$\nu_{ij}(\boldsymbol{\xi}, t)\Big|_{\boldsymbol{\xi}=\mathbf{x}} = g_{ij}(\mathbf{x}) \tag{2.89}$$

and

$$\nu_{ij}(\boldsymbol{\xi}, t)\Big|_{\boldsymbol{\xi}=\mathbf{x}, t=t'} = (\partial x'^m/\partial x^i)(\partial x'^n/\partial x^j) g_{mn}(\mathbf{x}') \tag{2.90}$$

Hence use of Eqs. (2.89) and (2.90) in (2.87) gives

$$\Gamma_{ij}(\boldsymbol{\xi}, t, t')\Big|_{\boldsymbol{\xi}=\mathbf{x}} = g_{ij}(\mathbf{x}) - (\partial x'^m/\partial x^i)(\partial x'^n/\partial x^j) g_{mn}(\mathbf{x}') \tag{2.91}$$

Similarly, we can also obtain from Eq. (2.85)

$$\nu^{ij}(\boldsymbol{\xi}, t)\Big|_{\boldsymbol{\xi}=\mathbf{x}} = g^{ij}(\mathbf{x}) \tag{2.92}$$

$$\nu^{ij}(\boldsymbol{\xi}, t)\Big|_{\boldsymbol{\xi}=\mathbf{x}, t=t'} = (\partial x^i/\partial x'^m)(\partial x^j/\partial x'^n) g^{mn}(\mathbf{x}') \tag{2.93}$$

and from Eq. (2.88)

$$\Gamma^{ij}(\boldsymbol{\xi}, t, t')\Big|_{\boldsymbol{\xi}=\mathbf{x}} = (\partial x^i/\partial x'^m)(\partial x^j/\partial x'^n)g^{mn}(\mathbf{x}') - g^{ij}(\mathbf{x}) \tag{2.94}$$

We have shown how the strain tensors in the spatial coordinates may be obtained from those in the convected coordinates,

$$\nu_{ij}(\boldsymbol{\xi}, t) \xrightarrow{t'} C_{ij}(\mathbf{x}, t, t') \tag{2.95}$$

$$\nu^{ij}(\boldsymbol{\xi}, t) \xrightarrow{t'} \left[C^{-1}(\mathbf{x}, t, t')\right]^{ij} \tag{2.96}$$

$$\Gamma_{ij}(\boldsymbol{\xi}, t, t') \xrightarrow{t'} E_{ij}(\mathbf{x}, t, t') \tag{2.97}$$

$$\Gamma^{ij}(\boldsymbol{\xi}, t, t') \xrightarrow{t'} E^{ij}(\mathbf{x}, t, t') \tag{2.98}$$

2.5.2 Time Derivative of Convected Coordinates

Having defined strain tensors in convected coordinates, we now describe the rate-of-strain (or rate-of-deformation) tensor. This may be obtained by taking the derivative of a strain tensor with time, with the convected coordinates held constant. Such a derivative is commonly referred to as the "material derivative," which may be considered as the time rate of change as seen by an observer in a convected coordinate system. Using the notation D/Dt for the substantial (material) time derivative, we have from Eq. (2.86)

$$\frac{\mathrm{D}}{\mathrm{D}t}\left[\mathrm{d}s^2(t) - \mathrm{d}s^2(t')\right] = \frac{\mathrm{D}}{\mathrm{D}t}\left[\Gamma_{ij}(\boldsymbol{\xi}, t, t')\,\mathrm{d}\xi^i\,\mathrm{d}\xi^j\right] \tag{2.99}$$

Since every material point always has the same convected coordinate position at all times, regardless of the extent of deformation of the medium, the relative coordinate displacements between any two points must be constant, so that any change in the actual distance between the points must be reflected by a change in the metric ν_{ij}. That is, if the distance between two points ds changes with time, the convected metric ν_{ij} must change accordingly with time since, by definition, the convected coordinates ξ^i of a material point are independent of time. Therefore Eq. (2.99) may be rewritten with the aid of Eq. (2.86) as

$$\frac{\mathrm{D}}{\mathrm{D}t}\left[\mathrm{d}s^2(t) - \mathrm{d}s^2(t')\right] = \frac{\mathrm{D}\Gamma_{ij}}{\mathrm{D}t}\,\mathrm{d}\xi^i\,\mathrm{d}\xi^j = \frac{\mathrm{D}\nu_{ij}(\boldsymbol{\xi}, t)}{\mathrm{D}t}\,\mathrm{d}\xi^i\,\mathrm{d}\xi^j \tag{2.100}$$

Similarly, for the contravariant convected strain tensor we have

$$\frac{\mathrm{D}}{\mathrm{D}t}\left[\mathrm{d}s^2(t) - \mathrm{d}s^2(t')\right] = \frac{\mathrm{D}\Gamma^{ij}}{\mathrm{D}t}\,\mathrm{d}\xi_i\,\mathrm{d}\xi_j = -\frac{\mathrm{D}\nu^{ij}(\boldsymbol{\xi}, t)}{\mathrm{D}t}\,\mathrm{d}\xi_i\,\mathrm{d}\xi_j \tag{2.101}$$

It should be remembered that the metric tensor ν_{ij} and ν^{ij} in a convected coordinate system are related to the metric tensors g_{ij} and g^{ij} in a spatial coordinate system by Eqs. (2.84) and (2.85).

Therefore, we can write the following general rule of coordinate transformation of a second-order tensor:

$$\mathrm{A}_{mn}(\boldsymbol{\xi}, t) = (\partial x^i/\partial \xi^m)(\partial x^j/\partial \xi^n) a_{ij}(\mathbf{x}, t) \tag{2.102}$$

where $A_{mn}(\boldsymbol{\xi}, t)$ and $a_{ij}(\mathbf{x}, t)$ are covariant components of a tensor of second order in convected and fixed coordinate systems, respectively. Then the material derivative of $A_{mn}(\boldsymbol{\xi}, t)$ requires the material derivative of the right-hand side of Eq. (2.102), yielding (see Appendix 2B)

$$\frac{\mathrm{D}A_{mn}}{\mathrm{D}t} = \left(\frac{\partial x^i}{\partial \xi^m}\frac{\partial x^j}{\partial \xi^n}\right)\frac{\mathfrak{d} a_{ij}}{\mathfrak{d} t} \tag{2.103}$$

where

$$\frac{\mathfrak{d} a_{ij}}{\mathfrak{d} t} = \frac{\partial a_{ij}}{\partial t} + v^k \frac{\partial a_{ij}}{\partial x^k} + \frac{\partial v^k}{\partial x^i} a_{kj} + \frac{\partial v^k}{\partial x^j} a_{ik} \tag{2.104}$$

or in direct notation

$$\frac{\mathfrak{d}\mathbf{a}}{\mathfrak{d} t} = \frac{\mathbf{Da}}{\mathrm{D}t} + (\nabla \mathbf{v}) \cdot \mathbf{a} + \mathbf{a} \cdot (\nabla \mathbf{v})^{\mathrm{T}} \tag{2.105}$$

Here, $\mathfrak{d}/\mathfrak{d}t$ is called the "convected derivative" due to Oldroyd (1950), and it is the fixed coordinate equivalent of the material derivative of a second-order tensor referred to in convected coordinates. The physical interpretation of the right-hand side of Eq. (2.104) may be given as follows. The first two terms represent the derivative of tensor a_{ij} with time, with the fixed coordinate held constant (i.e., $\mathrm{D}a_{ij}/\mathrm{D}t$), which may be considered as the time rate of change as seen by an observer in a fixed coordinate system. The third and fourth terms represent the stretching and rotational motions of a material element referred to in a fixed coordinate system. This is because the velocity gradient $\partial v^k/\partial x^i$ (or the velocity gradient tensor $\mathbf{L}$ defined by Eq. (2.59)) may be considered as a sum of the rate of pure stretching and the material derivative of the finite rotation. For this reason, the convected derivative is sometimes referred to as the "codeformational derivative" (Bird et al. 1987).

Similarly, for contravariant components $A^{mn}(\boldsymbol{\xi}, t)$ and $a^{ij}(\mathbf{x}, t)$ of a tensor of second order in convected and fixed coordinate systems, respectively, we have

$$\frac{\mathrm{D}A^{mn}}{\mathrm{D}t} = \left(\frac{\partial \xi^m}{\partial x^i}\frac{\partial \xi^n}{\partial x^j}\right)\frac{\mathfrak{d} a^{ij}}{\mathfrak{d} t} \tag{2.106}$$

where

$$\frac{\mathfrak{d} a^{ij}}{\mathfrak{d} t} = \frac{\partial a^{ij}}{\partial t} + v^k \frac{\partial a^{ij}}{\partial x^k} - \frac{\partial v^i}{\partial x^k} a^{kj} - \frac{\partial v^j}{\partial x^k} a^{ik} \tag{2.107}$$

or in direct notation

$$\frac{\mathfrak{d} \mathbf{a}}{\mathfrak{d} t} = \frac{\mathrm{D}\mathbf{a}}{\mathrm{D}t} - (\nabla \mathbf{v})^{\mathrm{T}} \cdot \mathbf{a} - \mathbf{a} \cdot (\nabla \mathbf{v}) \tag{2.108}$$

In Chapter 3, we show that the contravariant and covariant components, respectively, of the convected derivative of the stress tensor give rise to different expressions for the material functions in steady-state simple shear flow. When compared with experimental data, it turns out that the material functions predicted from the contravariant components of the convected derivative of the stress tensor give rise to a correct trend, while the material functions predicted from the covariant components of the convected derivative of the stress tensor do not.

Now, we can apply the general rule of transformation to the material derivative of a strain tensor in the convected coordinates, given by Eqs. (2.100) and (2.101). For instance, from Eq. (2.84) we have

$$\frac{\mathrm{D}\nu_{ij}(\boldsymbol{\xi}, t)}{\mathrm{D}t} = \left(\frac{\partial x^m}{\partial \xi^i} \frac{\partial x^n}{\partial \xi^j} \right) \frac{\mathfrak{d} g_{mn}}{\mathfrak{d} t} \tag{2.109}$$

Since the spatial metric $g_{mn}(\mathbf{x})$ is independent of time, it can be easily shown that (Oldroyd 1950)

$$\frac{\mathfrak{d} g_{mn}}{\mathfrak{d} t} = 2 d_{mn} \tag{2.110}$$

where d_{mn} are the components of the rate-of-strain tensor **d** defined by Eq. (2.60).

It is important to note that there are other types of time derivatives which also transform as a tensor from convected to fixed coordinates. One particular time derivative that has received particular attention by rheologists is the so-called "Jaumann derivative," which was suggested first by Zaremba (1903) and later reformulated by other investigators (DeWitt 1955; Fromm 1947). The Jaumann derivative $\mathcal{D}/\mathcal{D}t$ of a second-order tensor a_{ij} is defined as

$$\frac{\mathcal{D} a_{ij}}{\mathcal{D} t} = \frac{\partial a_{ij}}{\partial t} + v^k \frac{\partial a_{ij}}{\partial x^k} - \omega_{ik} a_{jk} - \omega_{jk} a_{ik} \tag{2.111}$$

or in direct notation[1]

$$\frac{\mathcal{D} \mathbf{a}}{\mathcal{D} t} = \frac{\mathrm{D}\mathbf{a}}{\mathrm{D}t} - (\boldsymbol{\omega} \cdot \mathbf{a}) - (\boldsymbol{\omega} \cdot \mathbf{a})^{\mathrm{T}} \tag{2.112}$$

where $\boldsymbol{\omega}$ is the vorticity tensor defined by Eq. (2.61). The physical interpretation of the right-hand side of Eq. (2.111) may be given as follows. The first two terms represent the material derivative of a_{ij}, similar to the first two terms on the right-hand side of Eq. (2.104). However, the third and fourth terms containing only the vorticity tensor $\boldsymbol{\omega}$ represent the rotational motion of a material element referred to in a fixed coordinate system. For this reason, the Jaumann derivative is sometimes referred to as the "corotational derivative" (Bird et al. 1987). In Chapter 3 we show that the contravariant and covariant components, respectively, of the Jaumann derivative of the stress tensor give rise to identical expressions for the material functions in steady-state simple shear flow, predicting the same trend as that observed experimentally.

2.6 The Description of Stress and Material Functions

Let us consider now the stress tensor, which causes or arises from deformation. In order to give the reason why a second-order tensor is required to describe the stress, a development of Cauchy's law of motion is needed. The physical significance of the stress tensor may be illustrated best by considering the three forces acting on three faces (one force on each face) of a small cube element of fluid, as schematically shown in Figure 2.4. For instance, a force (which is the vector) acting on the face ABCD with an arbitrary direction may be resolved in three component directions: the force acting in the x_1 direction is $T_{11}\mathrm{d}x_2\mathrm{d}x_3$, the force acting in the x_2 direction is $T_{12}\mathrm{d}x_2\mathrm{d}x_3$, and the force acting in the x_3 direction is $T_{13}\mathrm{d}x_2\mathrm{d}x_3$. Similarly, the forces acting on face BCFE are $T_{21}\mathrm{d}x_1\mathrm{d}x_3$ in the x_1 direction, $T_{22}\mathrm{d}x_1\mathrm{d}x_3$ in the x_2 direction, $T_{23}\mathrm{d}x_1\mathrm{d}x_3$

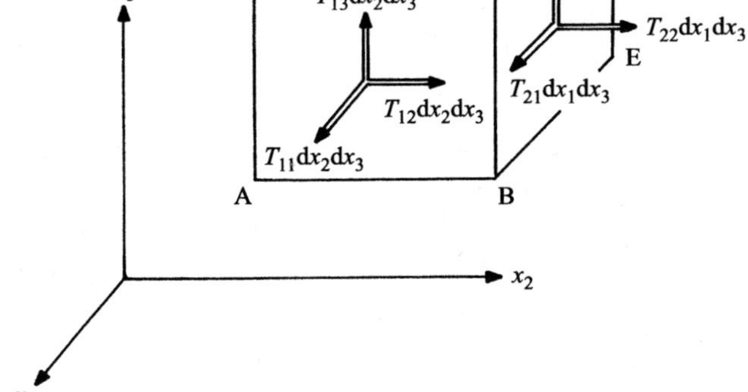

Figure 2.4 Stress components on a cube.

in the x_3 direction. Likewise, the forces acting on face DCFG are $T_{31}dx_1dx_2$ in the x_1 direction, $T_{32}dx_1dx_2$ in the x_2 direction, and $T_{33}dx_1dx_2$ in the x_3 direction.

In dealing with the state of stresses of incompressible fluids under deformation or in flow, the total stress tensor **T** is divided into two parts:

$$\left\| T_{ij} \right\| = \left\| \begin{matrix} T_{11} & T_{12} & T_{13} \\ T_{21} & T_{22} & T_{23} \\ T_{31} & T_{32} & T_{33} \end{matrix} \right\| = \left\| \begin{matrix} -p & 0 & 0 \\ 0 & -p & 0 \\ 0 & 0 & -p \end{matrix} \right\| + \left\| \begin{matrix} \sigma_{11} & \sigma_{12} & \sigma_{13} \\ \sigma_{21} & \sigma_{22} & \sigma_{23} \\ \sigma_{31} & \sigma_{32} & \sigma_{33} \end{matrix} \right\| \tag{2.113}$$

where the component T_{ij} of the stress tensor **T** is the force acting in the x_i direction on unit area of a surface normal to the x_i direction. The components T_{11}, T_{22}, and T_{33} are called normal stresses since they act normally to surfaces, the mixed components T_{12}, T_{13}, and so on, are called shear stresses. In direct notation, Eq. (2.113), using Cartesian coordinates, can be expressed by

$$\mathbf{T} = -p\boldsymbol{\delta} + \boldsymbol{\sigma} \tag{2.114}$$

where the $\boldsymbol{\delta}$ is the unit tensor, $\boldsymbol{\sigma}$ is the deviatoric stress tensor (or the extra stress tensor) that vanishes in the absence of deformation or flow, and p is the isotropic pressure. Note in Eq. (2.113) or Eq. (2.114) that p has a negative sign since it acts in the direction opposite to a normal stress (T_{11}, T_{22}, T_{33}), which by convention is chosen as pointing out of the cube (see Figure 2.4). It should be mentioned that in an incompressible liquid, the state of stress is determined by the strain or strain history only to within an additive isotropic constant, and thus p appearing in Eq. (2.113) or in Eq. (2.114) is the pressure that can be determined within the accuracy of an isotropic term. As is shown in some later chapters (e.g., Chapter 5), only pressure gradient plays a role in describing fluid motion. Thus the isotropic term, $p\boldsymbol{\delta}$ in Eq. (2.114) has no effect on fluid motion, i.e., the addition of an isotropic term of arbitrary magnitude has no consequence to the total stress tensor **T** when a fluid is in motion.

Special types of states of stress are of particular importance. In a liquid that has been at rest (i.e., there is no deformation of a fluid) for a sufficiently long time, there is no tangential component of stress on any plane of a cube and the normal component of stress is the same for all three planes, each perpendicular to the others. This is the situation where only hydrostatic pressure, $-p$, exists. In such a situation, Eq. (2.113) reduces to

$$\left\| T_{ij} \right\| = \left\| \begin{matrix} -p & 0 & 0 \\ 0 & -p & 0 \\ 0 & 0 & -p \end{matrix} \right\| \tag{2.115}$$

From Eq. (2.115) we can now define pressure as

$$-p = \tfrac{1}{3}(T_{11} + T_{22} + T_{33}) \tag{2.116}$$

Note that Eq. (2.116) can also be obtained from Eq. (2.113) with the assumption, $\sigma_{11} + \sigma_{22} + \sigma_{33} = 0$. Since such an assumption is quite arbitrary, the definition of pressure p given by Eq. (2.116) can be regarded as a somewhat arbitrary one. In fact, in general p is the thermodynamic pressure, which is related to the density ρ and the temperature through a "thermodynamic equations of state," $p = p(\rho, T)$; that is, this is taken to be the same function as that used in thermal equilibrium (Bird et al. 1987).

If we now consider the state of stress in an isotropic material, by definition the material has no preferred directions. In simple shear flow, we have

$$T_{13} = T_{31} = 0; \quad T_{23} = T_{32} = 0; \quad T_{12} = T_{21} \neq 0 \tag{2.117}$$

in which the subscript 1 denotes the direction of flow, the subscript 2 denotes the direction perpendicular to flow, and the subscript 3 denotes the remaining (neutral) direction. It follows therefore from Eq. (2.113) that the most general possible state of stress for an isotropic material in simple shear flow may be represented by

$$\begin{Vmatrix} T_{11} & T_{12} & 0 \\ T_{12} & T_{22} & 0 \\ 0 & 0 & T_{33} \end{Vmatrix} = \begin{Vmatrix} -p & 0 & 0 \\ 0 & -p & 0 \\ 0 & 0 & -p \end{Vmatrix} + \begin{Vmatrix} \sigma_{11} & \sigma_{12} & 0 \\ \sigma_{12} & \sigma_{22} & 0 \\ 0 & 0 & \sigma_{33} \end{Vmatrix} \tag{2.118}$$

Note that one cannot measure p and the components of the extra stress tensor $\boldsymbol{\sigma}$ separately during flow of a liquid. Therefore, the absolute value of any one normal component of stress is of no rheological significance. The values of the differences of normal stress components are, however, not altered by the addition of any isotropic pressure (see Eq. (2.118)), and they presumably depend on the rheological properties of the material. It follows, therefore, that there are only three independent stress quantities of rheological significance, namely, one shear component and two differences of normal components:

$$\sigma_{12}; \quad T_{11} - T_{22} = \sigma_{11} - \sigma_{22}; \quad T_{22} - T_{33} = \sigma_{22} - \sigma_{33} \tag{2.119}$$

Note that the normal stress difference $\sigma_{11} - \sigma_{33}$ becomes redundant since we have assumed $\sigma_{11} + \sigma_{22} + \sigma_{33} = 0$ in defining p by Eq. (2.116). In the rheology community, $N_1 = \sigma_{11} - \sigma_{22}$ is referred to as the first normal stress difference and $N_2 = \sigma_{11} - \sigma_{33}$ as the second normal stress difference. It now remains to be discussed how the stress quantities may be related to strain or rate of strain to describe the rheological properties of materials, in particular polymeric materials.

For steady-state shear flow, the components of the stress tensor $\mathbf{T}$ may be expressed in terms of three independent functions:

$$\sigma_{12} = \eta(\dot{\gamma})\dot{\gamma} \quad N_1 = \psi_1(\dot{\gamma})\dot{\gamma}^2 \quad N_2 = \psi_2(\dot{\gamma})\dot{\gamma}^2 \tag{2.120}$$

where $\eta(\dot{\gamma})$ is referred to as the shear-rate dependent viscosity, $\psi_1(\dot{\gamma})$ as the first normal stress difference coefficient, and $\psi_2(\dot{\gamma})$ as the second normal stress difference coefficient. Often, $\eta(\dot{\gamma})$, $\psi_1(\dot{\gamma})$, and $\psi_2(\dot{\gamma})$ are referred to as the "material functions" in steady-state shear flow. Note that N_1 and N_2, or $\psi_1(\dot{\gamma})$ and $\psi_2(\dot{\gamma})$, describe the fluid elasticity, which is elaborated on in Chapter 3.

In the past, numerous investigators have reported measurements of the rheological properties of polymeric liquids. Until now, very few polymeric fluids, if any, which exhibit a constant value of shear viscosity (i.e., $\eta(\dot{\gamma}) = \eta_0$) exhibit measurable values of N_1 and N_2. In other words, almost all polymeric fluids showing measurable values of N_1 and N_2 have also been found to exhibit shear-rate dependent (i.e., non-Newtonian) viscosity $\eta(\dot{\gamma})$. Also, polymeric liquids showing measurable values of N_1 and N_2 have been found to exhibit unusual flow behavior, such as climbing-up a rotating rod (see Figure 1.1) and extrudate swell (see Figure 1.4). It can then be understood why N_1 and N_2 are regarded as being the material functions that describes the elastic behavior of polymeric fluids. In Chapter 3 we present how the material functions vary with shear rate on the basis of continuum viscoelasticity theory, in Chapter 4 we present how the material functions vary with shear rate on the basis of molecular viscoelasticity theory, and in Chapter 5 we present experimental methods to determine the material functions.

Appendix 2A: Properties of Second-Order Tensors

Invariants

If **a** is an arbitrary vector, we can find a linear transformation **Ta**, where **T** is a linear operator, such that **Ta** has the same direction as vector **a** itself, but the two vectors **Ta** and **a** would differ in magnitude. That is,

$$\mathbf{Ta} = \lambda \mathbf{a} \tag{2A.1}$$

where λ is a real scalar to be determined. Equation (2A.1) may be rewritten as

$$(\mathbf{T} - \lambda \mathbf{I})\mathbf{a} = \mathbf{0}, \quad \text{or} \quad (T_{ij} - \lambda \delta_{ij})\mathrm{a}_j = 0 \tag{2A.2}$$

The necessary and sufficient condition for Eq. (2A.2) to have a nontrivial solution for the unknown a_j is that the determinant of the coefficients should vanish, that is,

$$\det(T_{ij} - \lambda \delta_{ij}) = 0 \tag{2A.3}$$

Expanding Eq. (2A.3), we have

$$C(\lambda) = \lambda^3 - I_1 \lambda^2 + I_2 \lambda - I_3 = 0 \tag{2A.4}$$

where

$$\begin{aligned} I_1 &= T_{11} + T_{22} + T_{33} = \mathrm{tr}\,\mathbf{T} \\ I_2 &= T_{11}T_{22} + T_{11}T_{33} + T_{22}T_{33} - T_{12}T_{21} - T_{13}T_{31} - T_{23}T_{32} \\ &= \tfrac{1}{2}\{(\mathrm{tr}\,\mathbf{T})^2 - \mathrm{tr}\,\mathbf{T}^2\} \\ &= \tfrac{1}{2}({I_1}^2 - \mathbf{T}:\mathbf{T}) \\ I_3 &= \det \mathbf{T} \end{aligned} \tag{2A.5}$$

are called the principal invariants (I_1 is the first invariant, I_2 the second invariant, and I_3 the third invariant) of **T**.

The usefulness of the invariants of a second-order tensor can be illustrated best with the rate-of-strain tensor **d**. For incompressible fluids, we have

$$I_1 = \operatorname{tr} \mathbf{d} = \frac{dv^1}{dx^1} + \frac{dv^2}{dx^2} + \frac{dv^3}{dx^3} = 0 \tag{2A.6}$$

and therefore it follows from Eq. (2A.5) that

$$I_2 = -\tfrac{1}{2}(\mathbf{d}:\mathbf{d}) = -d_{12}d_{21} - d_{13}d_{31} - d_{23}d_{32} \tag{2A.7}$$

$$I_3 = \det \mathbf{d} \tag{2A.8}$$

Note that the motion of incompressible fluids always satisfies Eq. (2A.6).

For steady-state simple shear flow, for which **d** is given by Eq. (2.12), we have

$$I_2 = -\tfrac{1}{2}(\mathbf{d}:\mathbf{d}) = -\tfrac{1}{4}\dot{\gamma}^2 \tag{2A.9}$$
$$I_3 = 0$$

where $\dot{\gamma}$ is the shear rate. It can be concluded therefore that the first invariant I_1 is a measure of fluid compressibility, the second invariant I_2 is a measure of the intensity of the rate of deformation, and that if the third invariant I_3 is zero in the motion with constant volume (i.e., isochoric motion), the motion is two-dimensional. It should be noted that, in three-dimensional flow, $I_3 \neq 0$ in general, although one frequently assumes $I_3 = 0$ for mathematical convenience.

Principal Values and Principal Directions

The roots of Eq. (2A.4), often referred to as the characteristic equation of **T**, are called the "principal values" of **T**. Two cases are possible: either (1) one root is real or (2) all three are real. If the tensor **T** is symmetric, then all principal values (λ_1, λ_2, λ_3) are real. Suppose that all principal values (λ_1, λ_2, and λ_3) are real and distinct, which implies that the tensor **T** is symmetric. Then for each of the principal values, Eq. (2A.1) must be satisfied, that is,

$$\mathbf{Ta} = \lambda_i \mathbf{a} \quad (i = 1, 2, 3) \tag{2A.10}$$

Let $\boldsymbol{e}_1$, $\boldsymbol{e}_2$, and $\boldsymbol{e}_3$ be the members of the orthogonalized set of three characteristic unit vectors corresponding to each principal value, λ_1, λ_2, and λ_3. One can construct a tensor **Q** in such a way that the elements of the unit vectors $\boldsymbol{e}_1$, $\boldsymbol{e}_2$, and $\boldsymbol{e}_3$ are the elements of the successive columns of **Q**. It can be shown (Hawkins 1963; Jeffreys 1961) that the tensor so constructed is an orthogonal tensor (i.e., $\mathbf{Q}^{\mathrm{T}} = \mathbf{Q}^{-1}$) and that the following condition is satisfied:

$$\mathbf{QTQ}^{\mathrm{T}} = \mathbf{T}' \tag{2A.11}$$

where

$$\|\mathbf{T}'\| = \begin{Vmatrix} \lambda_1 & 0 & 0 \\ 0 & \lambda_2 & 0 \\ 0 & 0 & \lambda_3 \end{Vmatrix} \tag{2A.12}$$

That is, the symmetric tensor $\mathbf{T}$ is transformed into a new symmetric tensor $\mathbf{T}'$. This tensor has three mutually perpendicular directions, called "principal directions" its off-diagonal elements are zero and its diagonal elements are the three principal values of $\mathbf{T}$.

The physical significance of Eq. (2A.11) is that the newly constructed tensor $\mathbf{Q}$ changes the axes, say x^1, x^2, x^3, of a tensor $\mathbf{T}$ into the axes x'^1, x'^2, x'^3 by rotation. The rotation of the coordinates, however, does not change the values of the three invariants, and Eq. (2A.5) may be rewritten in terms of the principal values of $\mathbf{T}$ as

$$I_1 = \lambda_1 + \lambda_2 + \lambda_3, \quad I_2 = \lambda_1\lambda_2 + \lambda_2\lambda_3 + \lambda_3\lambda_1, \quad I_3 = \lambda_1\lambda_2\lambda_3 \tag{2A.13}$$

For instance, if $\mathbf{T}$ is the stress tensor, the three principal values (λ_1, λ_2, and λ_3) are called "principal stresses," and the directions in which they are acting are called "principal directions."

The Polar Decomposition Theorem

The "polar decomposition theorem" expresses a general second order tensor as a product of a positive symmetric tensor with an orthogonal tensor. This is very useful in interpreting deformation processes in terms of a translation, a rigid rotation of the principal axes of strain, and stretching along these axes (Jaunzemis 1967).

Let $\mathbf{F}$ be a nonsingular asymmetric second-order tensor. Note that the deformation gradient tensor $\mathbf{F}$ happens to be an asymmetric tensor. Then, $\mathbf{F}$ allows the unique representations

$$\mathbf{F} = \mathbf{RU} \tag{2A.14}$$

where $\mathbf{U}$ is a positive symmetric tensor and $\mathbf{R}$ is an orthogonal tensor. Because of the orthogonality property of $\mathbf{R}$, from Eq. (2A.14) one obtains

$$\mathbf{C} = \mathbf{U}^2 = \mathbf{F}^{\mathrm{T}}\mathbf{F} \tag{2A.15}$$

where **C** is a positive definite symmetric tensor. Equation (2A.15) gives rise to the following definition:

$$\mathbf{U} = (\mathbf{C})^{1/2} = (\mathbf{F}^{\mathrm{T}}\mathbf{F})^{1/2} \tag{2A.16}$$

Equation (2A.15) shows how one could obtain **U** knowing **F**. Due to the nonsingularity of **U** (i.e., the inverse of **U** exists), Eq. (2A.14) yields

$$\mathbf{R} = \mathbf{F}\mathbf{U}^{-1} \tag{2A.17}$$

The physical significance of Eqs. (2A.16) and (2A.17) is that a deformation process can be decomposed into a pure stretching by Eq. (2A.16) and a rigid rotation by Eq. (2A.17).

Appendix 2B: Tensor Calculus

Curvilinear Coordinates and Metric Tensors

Let ξ^i (ξ^1, ξ^2, ξ^3) be the curvilinear coordinates and $x^i(x^1, x^2, x^3)$ be the Cartesian rectangular coordinates of a material point P, as depicted in Figure 2.5. The point P referred to the Cartesian coordinates x^i has position vector **P**, expressed as

$$\mathbf{P} = x^1\boldsymbol{i}_1 + x^2\boldsymbol{i}_2 + x^3\boldsymbol{i}_3 = x^k\boldsymbol{i}_k \tag{2B.1}$$

At point P referred to the curvilinear coordinates, there exists a set of base vectors $\boldsymbol{g}_i(\xi^1, \xi^2, \xi^3)$, defined as

$$\boldsymbol{g}_i = \partial\mathbf{P}/\partial\xi^i = (\partial x^m/\partial\xi^i)\boldsymbol{i}_m \tag{2B.2}$$

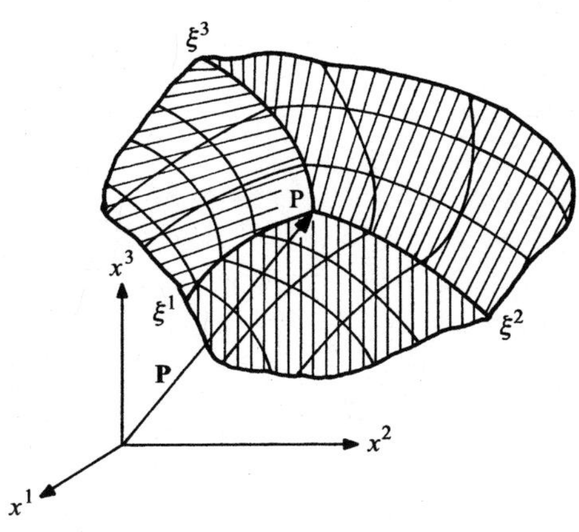

Figure 2.5 Curvilinear coordinates.

where $\boldsymbol{g}_1$, $\boldsymbol{g}_2$, and $\boldsymbol{g}_3$ are tangent to the coordinate lines ξ^1, ξ^2, and ξ^3, respectively. The metric tensor g_{ij} is defined by

$$g_{ij} = \boldsymbol{g}_i \cdot \boldsymbol{g}_j = (\partial x^m/\partial \xi^i)(\partial x^n/\partial \xi^j)\boldsymbol{i}_m \cdot \boldsymbol{i}_n = (\partial x^m/\partial \xi^i)(\partial x^n/\partial \xi^j)\delta_{mn} \tag{2B.3}$$

where δ_{mn} is the Kronecker delta.

The metric tensor g_{ij} is a covariant second-order tensor, and it is a symmetric tensor as may be seen from Eq. (2B.3),

$$g_{ij} = g_{ji} \tag{2B.4}$$

In general, curvilinear coordinates are not mutually orthogonal; that is,

$$g_{ij} = \boldsymbol{g}_i \cdot \boldsymbol{g}_j = |\boldsymbol{g}_i||\boldsymbol{g}_j| \cos(\boldsymbol{g}_i, \boldsymbol{g}_j) \neq 0; \quad (i \neq j) \tag{2B.5}$$

since the angle between the two base vectors, $\boldsymbol{g}_i$ and $\boldsymbol{g}_j$, is not necessarily 90°. A curvilinear coordinate system is said to be orthogonal if the base vectors at each point form a triplet of mutually perpendicular vectors; that is,

$$g_{ij} = 0; \quad (i \neq j) \tag{2B.6}$$

For instance, it can be easily shown that the cylindrical coordinates and spherical coordinates are orthogonal curvilinear coordinate systems.

The contravariant components g^{ij} of the metric tensor are obtained by (Hawkins 1963; Jeffreys 1961)

$$g^{ij} = \text{cofactor of } g_{ij}/\det g_{ij} \tag{2B.7}$$

Given a tensor, we can derive other tensors by raising or lowering indices; that is, by forming inner products of the given tensor with the metric tensor g_{ij} or its conjugate g^{ij} (Hawkins 1963; Jeffreys 1961). For example

$$A^{ij} = g^{im} g^{jn} A_{mn}; \quad A_{ij} = g_{im} g_{jn} A^{mn} \tag{2B.8}$$

Time Derivatives of Second-Order Tensors

Let $\xi^i(\xi^1, \xi^2, \xi^3)$ be the convected coordinates and $x^i(x^1, x^2, x^3)$ be the fixed coordinates of a material point. If a_{ij} denotes the components of a given tensor in the fixed coordinate system x^i, and A_{ij} denotes those in the convected coordinate system ξ^i, then we have the following relationship between the two for covariant components (Oldroyd 1950)

$$A_{mn}(\boldsymbol{\xi}, t) = (\partial x^i/\partial \xi^m)(\partial x^j/\partial \xi^n) a_{ij}(\mathbf{x}, t) \tag{2B.9}$$

and for contravariant components we have

$$A^{mn}(\boldsymbol{\xi}, t) = (\partial \xi^m / \partial x^i)(\partial \xi^n / \partial x^j) a^{ij}(\mathbf{x}, t) \tag{2B.10}$$

Now we seek the rule of transformation of the material derivative of tensor A_{ij}, which is a type of time derivative following the motion of the material element. It is important to realize, however, that all tensor quantities described in reference to the convected coordinate system ξ^i must be transformed into tensor quantities fixed in space (i.e., fixed coordinate system x^i) because all physical measurements are made relative to the fixed coordinate system.

The material derivative of Eq. (2B.9) is

$$\begin{aligned} \frac{\mathrm{D}A_{mn}}{\mathrm{D}t} &= \frac{\mathrm{D}}{\mathrm{D}t}\left[\left(\frac{\partial x^i}{\partial \xi^m}\right)\left(\frac{\partial x^j}{\partial \xi^n}\right) a_{ij}\right] \\ &= \left(\frac{\partial x^i}{\partial \xi^m}\right)\left(\frac{\partial x^j}{\partial \xi^n}\right)\frac{\mathrm{D}a_{ij}}{\mathrm{D}t} + \left(\frac{\partial x^j}{\partial \xi^m}\right) a_{ij} \frac{\mathrm{D}}{\mathrm{D}t}\left(\frac{\partial x^j}{\partial \xi^n}\right) + \frac{\partial x^j}{\partial \xi^n} a_{ij} \frac{\mathrm{D}}{\mathrm{D}t}\left(\frac{\partial x^i}{\partial \xi^m}\right) \end{aligned} \tag{2B.11}$$

The derivative of each term on the right-hand side of Eq. (2B.11) is given by

$$\frac{\mathrm{D}a_{ij}}{\mathrm{D}t} = \frac{\partial a_{ij}}{\partial t} + v^k \frac{\partial a_{ij}}{\partial x^k}; \quad \frac{\mathrm{D}}{\mathrm{D}t}\left(\frac{\partial x^j}{\partial \xi^n}\right) = \frac{\partial}{\partial \xi^n}\left(\frac{\mathrm{D}x^j}{\mathrm{D}t}\right) = \frac{\partial v^j}{\partial \xi^n}; \quad \frac{\mathrm{D}}{\mathrm{D}t}\left(\frac{\partial x^i}{\partial \xi^m}\right) = \frac{\partial v^i}{\partial \xi^m} \tag{2B.12}$$

Substituting Eq. (2B.12) into (2B.11) gives[2]

$$\frac{\mathrm{D}A_{mn}}{\mathrm{D}t} = \left(\frac{\partial x^i}{\partial \xi^m}\right)\left(\frac{\partial x^j}{\partial \xi^n}\right)\left[\frac{\partial a_{ij}}{\partial t} + v^k \frac{\partial a_{ij}}{\partial x^k} + \frac{\partial v^k}{\partial x^j} a_{ik} + \frac{\partial v^k}{\partial x^i} a_{kj}\right] \tag{2B.13}$$

or

$$\frac{DA_{mn}}{Dt} = \left(\frac{\partial x^i}{\partial \xi^m}\right)\left(\frac{\partial x^j}{\partial \xi^n}\right)\frac{\mathfrak{D}a_{ij}}{\mathfrak{D}t} \tag{2B.14}$$

where

$$\frac{\mathfrak{D}a_{ij}}{\mathfrak{D}t} = \frac{\partial a_{ij}}{\partial t} + v^k \frac{\partial a_{ij}}{\partial x^k} + \frac{\partial v^k}{\partial x^j} a_{ik} + \frac{\partial v^k}{\partial x^i} a_{kj} \tag{2B.15}$$

is the fixed coordinate equivalent of the material derivative of a second-order tensor referred to in convected coordinates. It is referred to as the "convected derivative."

Equation (2B.15) may be used for any fixed coordinate system (i.e., for Cartesian coordinates as well as for any curvilinear coordinates). A similar procedure can be followed for a contravariant tensor a^{ij}, yielding (Oldroyd 1950)

$$\frac{\mathfrak{d}a^{ij}}{\mathfrak{d}t} = \frac{\partial a^{ij}}{\partial t} + v^k \frac{\partial a^{ij}}{\partial x^k} - \frac{\partial v^j}{\partial x^k} a^{ik} - \frac{\partial v^i}{\partial x^k} a^{kj} \tag{2B.16}$$

The physical significance of each of the terms in Eq. (2B.15) (also in Eq. (2B.16)) is as follows. The first two terms $((\partial a_{ij}/\partial t) + v^k(\partial a_{ij}/\partial x^k))$ describe the rate of time variation while following the motion (translation) of a material element. The third and fourth terms describe the rotational motion as well as the deformation of the element, as the quantity $\partial v^k/\partial x^j$ (and $\partial v^k/\partial x^i$) is the velocity gradient that can be represented by the rate-of-deformation tensor d_{kj} and the vorticity tensor ω_{kj}; that is,

$$\partial v^k/\partial x^j = d_{kj} + \omega_{kj} \tag{2B.17}$$

The covariant and contravariant forms of a given tensor involving convected derivatives are not equivalent. That is,

$$\frac{\mathfrak{d}A_{ij}}{\mathfrak{d}t} \neq g_{im} g_{jn} \frac{\mathfrak{d}A^{mn}}{\mathfrak{d}t} \tag{2B.18}$$

(see Eq. (2B.8)). Note that the metric tensors are independent of time. In other words, the operation of convected differentiation does not commute with the operation of raising and lowering indices.

There are other time derivative operators that transform a tensor from convected to fixed coordinates, giving rise to equivalent expression of covariant and contravariant forms of a given tensor. One such time derivative operator may be formed by eliminating d_{ij} after Eq. (2B.17) is substituted into Eqs. (2B.15) and (2B.16) and adding the resulting two equations (Oldroyd 1950)

$$\frac{\mathcal{D}a_{ij}}{\mathcal{D}t} = \frac{1}{2}\left(\frac{\mathfrak{d}a_{ij}}{\mathfrak{d}t} + g_{im} g_{jn} \frac{\mathfrak{d}a^{mn}}{\mathfrak{d}t}\right) = \frac{\partial a_{ij}}{\partial t} + v_k \frac{\partial a_{ij}}{\partial x^k} - \omega_{ik} a_{jk} - \omega_{jk} a_{ik} \tag{2B.19}$$

where $\mathcal{D}/\mathcal{D}t$ is referred to as the "Jaumann derivative" (DeWitt 1955; Fromm 1947; Zaremba 1903). The physical significance of each of the terms in Eq. (2B.19) is as follows. The first two terms on the right-hand side $((\partial a_{ij}/\partial t) + v_k(\partial a_{ij}/\partial x^k))$ describe the rate of time variation, while following the translational motion of a material element, and the third and fourth terms describe the rotational motion of the material element. The Jaumann derivative gives rise to equivalent forms of a given tensor equation, which commute with the operation of raising and lowering indices, as may be seen from the definition given by Eq. (2B.19).

In the next chapter, we discuss the use of various forms of time derivatives in the formulation of constitutive equations of state. There are no clear guidelines that can be applied to determine what form of time derivative might be the best or most appropriate. Therefore, one should determine the usefulness of a given relation based on its ability to predict the experimentally observed rheological behavior of a given material.

Problems

Problem 2.1

Obtain the material and spatial descriptions of velocity and acceleration for the following motions:

(a) $x(t') = X(1+t'), y(t') = Xt' + Y, z(t') = Z$
(b) $x(t') = X + Yt', y(t') = Yt', z(t') = Z + Xt'$

Problem 2.2

Obtain the material and spatial descriptions of velocity and acceleration for the following motions:

(a) $x(\tau) = X(1+\tau^2), y(\tau) = X\tau + Y, z(\tau) = Z$
(b) $x(\tau) = Xe^{-c\tau}, y(\tau) = Ye^{-c\tau}, z(\tau) = Ze^{-c\tau}$, where c is a constant.
(c) $x(\tau) = X + Y\tau^2, y(\tau) = Y/(1+\tau), z(\tau) = Z + X\tau$, where $\tau > -1$.
(d) $x = X + K(t)Y, y = Y, z = Z$

Problem 2.3

Calculate the deformation gradient tensor $\mathbf{F}(t)$ and the relative deformation gradient tensor $\boldsymbol{F}_t(t')$ for the motion given in Problem 2.1.

Problem 2.4

Obtain velocity $v(x, t, \tau)$ and acceleration $a(x, t, \tau)$, given that the present configuration is the same as the reference one, from the descriptions of the motion given in Problem 2.2.

Problem 2.5

Calculate the deformation gradient tensor $\mathbf{F}$ for the motion given in Problem 2.2.

Problem 2.6

Calculate the relative deformation gradient tensor $\mathbf{F}_t$ for the motion given in Problem 2.2.

Problem 2.7

Calculate the following deformation and strain tensors: $\mathbf{C}(t)$, $\mathbf{C}^{-1}(t)$, $\mathbf{C}_t(t', t)$, and $\mathbf{E}_t(t', t)$ for the motion given in Problem 2.1.

Problem 2.8

Prove that $\mathbf{F}_t(t', t) = \mathbf{F}(t')\,\mathbf{F}^{-1}(t)$.

Problem 2.9

The material description of the motion is given by

$$x_1 = \sqrt{3}X_1 + X_2; \quad x_2 = 2X_2; \quad x_3 = X_3$$

(a) Obtain deformation tensor **F** and interpret the deformation geometrically.
(b) Obtain the right stretch tensor **U**.
(c) Obtain the rotation tensor **R**. [Hint: use the polar decomposition theorem for parts (b) and (c).]

Problem 2.10

A body undergoes a deformation, for which the description of motion is given by

$$z^1 = \sqrt{2X^1}; \quad z^2 = X^2; \quad z^3 = X^3$$

where (X^1, X^2, X^3) is a Cartesian coordinate system for the material description and (z^1, z^2, z^3) is a cylindrical coordinate system for the spatial description.

(a) Calculate the Cauchy–Green deformation tensor C_{ij}.
(b) Calculate the Finger deformation tensor $(C^{-1})^{ij}$.

Problem 2.11

Verify that

(a) $\mathbf{C}(\tau) = \mathbf{F}^{\mathrm{T}}(t)\,\mathbf{C}_t(\tau)\,\mathbf{F}(t)$
(b) $\det \mathbf{F} = \mathrm{d}v/\mathrm{d}V$

where $\mathrm{d}v/\mathrm{d}V$ is the ratio of the volume $\mathrm{d}v$ of elemental box in the configuration at some time t to its volume $\mathrm{d}V$ in the reference configuration at time $t = t_0$.

Problem 2.12

Given the deformation tensor

$$\|\mathbf{F}\| = \left\| \begin{matrix} 1 & K & 0 \\ 0 & 1 & 0 \\ 0 & 0 & 1 \end{matrix} \right\|$$

(a) Calculate the Cauchy–Green deformation tensor **C**.
(b) Calculate the right and left stretch tensors **U** and **V**.
(c) Calculate the rotational tensor **R**.

Problem 2.13

Obtain $\mathbf{A}_{(1)}$ and $\mathbf{A}_{(2)}$ for steady-state shear flow. [Hint: use Eqs. (2.58) and (2.62).]

Problem 2.14

For the motion given by Problem 2.2, calculate the rate-of-deformation tensor $\mathbf{d}$ and the vorticity tensor $\boldsymbol{\omega}$ from $\mathbf{L}=\dot{\mathbf{F}}\mathbf{F}^{-1}$. Then calculate $\mathbf{F}_t(\tau)$ and $\mathbf{L}=\dot{\mathbf{F}}_t(t)$.

Problem 2.15

Given the motion

$$x(\tau)=X; \quad y(\tau)=Y+(\tau-t)v(X); \quad z(\tau)=X$$

one can obtain the Cauchy–Green deformation tensor $\mathbf{C}$ in terms of Rivlin–Ericksen tensors: $\mathbf{C}_t(t-\tau)=\mathbf{I}-s\mathbf{A}_1+(1/2)s^2\mathbf{A}_2$ with $s=t-\tau$, where s is a constant, and $\mathbf{C}_t(s)$ is defined by $C_t(\tau)=F_t^{\mathrm{T}}(\tau)F_t(\tau)$. Note that $\mathbf{A}_1$ and $\mathbf{A}_2$ are the Rivlin–Ericksen tensors (Rivlin and Ericksen 1955) defined by

$$\mathbf{A}_n=\sum_{k=0}^{n}\binom{n}{k}\mathbf{L}_k^{\mathrm{T}}\mathbf{L}_{n-k}$$

where $\mathbf{L}_n(t)=\frac{\mathrm{d}^n\mathbf{F}_t(\tau)}{\mathrm{d}\tau^n}\Big|_{\tau=t}$, $n=1,2,\ldots,$ and $\mathbf{L}_0=\mathbf{I}$. Obtain $\mathbf{A}_1$ and $\mathbf{A}_2$ using the above definition.

Problem 2.16

Determine the principal values (λ_1,λ_2) and corresponding unit principal vectors $(\boldsymbol{e}_1, \boldsymbol{e}_2)$ of the matrix

$$\|\mathbf{T}\|=\begin{Vmatrix} 5 & 2 \\ 2 & 2 \end{Vmatrix}$$

and verify that $\boldsymbol{e}_1$ and $\boldsymbol{e}_2$ are orthogonal.

Problem 2.17

Find the principal values and principal directions of the tensor

$$\|\mathbf{C}\|=\begin{Vmatrix} 1 & 2 & 0 \\ 2 & 1 & 0 \\ 0 & 0 & 1 \end{Vmatrix}$$

Check your results through $\mathbf{C}'=\mathbf{Q}^{\mathrm{T}}\mathbf{C}\mathbf{Q}$, in which $\mathbf{C}'$ is the diagonal representation of $\mathbf{C}$, and the rows of the orthogonal tensor $\mathbf{Q}$ are made up of the principal vectors of $\mathbf{C}$.

Problem 2.18
Calculate $\sqrt{\mathbf{T}}$ for

$$\|\mathbf{T}\| = \begin{Vmatrix} 3 & 2 & 0 \\ 2 & 3 & 0 \\ 0 & 0 & 9 \end{Vmatrix}$$

[Hint: Use the polar decomposition theorem.]

Notes

1. In Eqs. (2.111) and (2.112) the vorticity tensor $\boldsymbol{\omega}$ is defined by Eq. (2.61), which then gives Eq. (2.64) for steady-state simple shear flow. It should be mentioned that some literature presents

$$\frac{\mathscr{D}a_{ij}}{\mathscr{D}t} = \frac{\partial a_{ij}}{\partial t} + v^k \frac{\partial a_{ij}}{\partial x^k} + \omega_{ik} a_{jk} + \omega_{jk} a_{ik} \tag{2N.1}$$

or in direct notation

$$\frac{\mathscr{D}\mathbf{a}}{\mathscr{D}t} = \frac{\mathrm{D}\mathbf{a}}{\mathrm{D}t} + (\boldsymbol{\omega}\cdot\mathbf{a}) + (\boldsymbol{\omega}\cdot\mathbf{a})^{\mathrm{T}} \tag{2N.2}$$

Note that in the definition of Eqs. (2N.1) and (2N.2) the vorticity tensor $\boldsymbol{\omega}$ is defined by

$$\omega_{ij} = \frac{1}{2}\left(\frac{\partial v^j}{\partial x^i} - \frac{\partial v^i}{\partial x^j}\right) \tag{2N.3}$$

which then gives

$$\|\boldsymbol{\omega}\| = \begin{Vmatrix} 0 & -\dot{\gamma}/2 & 0 \\ \dot{\gamma}/2 & 0 & 0 \\ 0 & 0 & 0 \end{Vmatrix} \tag{2N.4}$$

for steady-state shear flow.

2. Use was made of the following relations:

$$\frac{\partial v^j}{\partial \xi^n} = \frac{\partial x^j}{\partial \xi^n}\frac{\partial v^k}{\partial x^j} \quad \text{and} \quad \frac{\partial v^i}{\partial \xi^m} = \frac{\partial x^i}{\partial \xi^m}\frac{\partial v^k}{\partial x^i} \tag{2N.5}$$

References

Bird RB, Armstrong RC, Hassager O (1987). *Dynamics of Polymeric Liquids: Fluid Mechanics*, Vol 1, 2nd ed, John Wiley & Sons, New York.
Coleman BD (1962). *Arch. Rat. Mech. Anal.* 9:273.
DeWitt TW (1955). *J. Appl. Phys.* 26:889.

Eringen AC (1962). *Nonlinear Theory of Continuous Media*, McGraw-Hill, New York.
Fromm HZ (1947). *Z. Angew. Math. Mech.* 25/27:146.
Hawkins GA (1963). *Multilinear Analysis for Students in Engineering and Science*, John Wiley & Sons, New York.
Jaunzemis W (1967). *Continuum Mechanics*, Macmillan, New York.
Jeffreys H (1961). *Cartesian Tensors*, Cambridge University Press, Cambridge.
Oldroyd JG (1950). *Proc. Roy. Soc.* A200:523.
Rivlin RS, Ericksen JL (1955). *J. Rat. Mech. Anal.* 4:323.
Zaremba S (1903). *Bull. Intern. Acad. Sci. Cracovie,* p 594.

3 Continuum Theories for the Viscoelasticity of Flexible Homogeneous Polymeric Liquids

3.1 Introduction

There are two primary reasons for seeking a precise mathematical description of the constitutive equations for viscoelastic fluids, which relate the state of stress to the state of deformation or deformation history. The first reason is that the constitutive equations are needed to predict the rheological behavior of viscoelastic fluids for a given flow field. The second reason is that constitutive equations are needed to solve the equations of motion (momentum balance equations), energy balance equations, and/or mass balance equations in order to describe the velocity, stress, temperature, and/or concentration profiles in a given flow field that is often encountered in polymer processing operations. There are two approaches to developing constitutive equations for viscoelastic fluids: one is a continuum (phenomenological) approach and the other is a molecular approach. Depending upon the chemical structure of a polymer (e.g., flexible homopolymer, rigid rodlike polymer, microphase-separated block copolymer, segmented multicomponent polymers, highly filled polymer, miscible polymer blend, immiscible polymer blend), one may take a different approach to the formulation of the constitutive equation. In this chapter we present some representative constitutive equations for flexible, homogeneous viscoelastic liquids that have been formulated on the basis of the phenomenological approach. In the next chapter we present the molecular approach to the formulation of constitutive equations for flexible, homogeneous viscoelastic fluids.

In the formulation of the constitutive equations using a phenomenological approach, emphasis is placed on the relationship between the components of stress and the components of the rate of deformation (or strain) or deformation (or strain) history, such that the responses of a fluid to a specified flow field or stress can adequately be described. The parameters appearing in a constitutive equation are supposed to represent the characteristics of the fluid under consideration. More often than not, the parameters appearing in a phenomenological constitutive equation are determined by curve fitting to experimental results. Thus phenomenological constitutive equations shed little light on the effect of the molecular parameters of the fluid under investigation to the rheological responses of the fluid.

Basically, there are three types of continuum-based constitutive equations: differential-, integral-, and rate-type. Differential-type constitutive equations are of the form that contains a time derivative (or derivatives) of either the stress tensor, the rate-of-strain tensor, or both. Integral-type constitutive equations are of the form in which the stress tensor is represented by an integral over the strain history or rate-of-strain tensor. Rate-type constitutive equations are of the form that contains neither a time derivative nor an integral, and thus the stress tensor is expressed explicitly as a function of the rate-of-strain tensor. In this chapter we present examples of all three types of phenomenological constitutive equations for viscoelastic fluids, including the material functions from each. There are monographs (Bird et al. 1987; Larson 1988; Truesdell and Noll 1965) that are devoted entirely to the formulation of various phenomenological constitutive equations for viscoelastic fluids.

3.2 Differential-Type Constitutive Equations for Viscoelastic Fluids

3.2.1 Single-Mode Differential-Type Constitutive Equations

Let us consider the simplest mechanical model, in which one spring is attached to one dashpot, as schematically shown in Figure 3.1. When a force F is acting on the spring downward at $t = 0$ (i.e., in one-dimensional flow), the stress–strain relationship for the spring (i.e., Hookean material) may be described by

$$\sigma = G\gamma \tag{3.1}$$

where σ is the stress (the force divided by the cross-sectional area), γ is the strain defined by $(L_0 - L)/L_0$, in which L_0 is the initial length of the spring (i.e., at $t = 0$) and L is its length at time t, and G is the proportionality constant, called the "elastic modulus." Conversely, the stress response of the dashpot (i.e., purely viscous

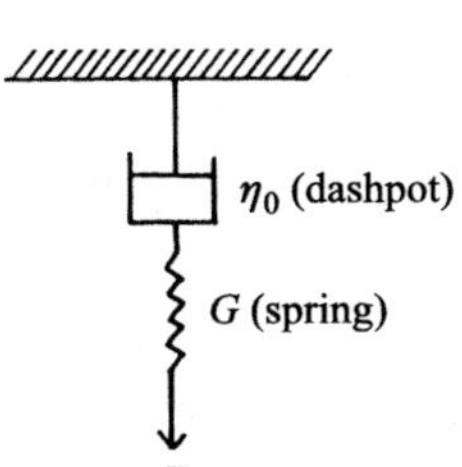

Figure 3.1 A "spring and dashpot" mechanical model for a viscoelastic fluid.

Newtonian fluid) to an applied deformation rate may be described as

$$\sigma = \eta_0 \dot{\gamma} \tag{3.2}$$

where $\dot{\gamma} = d\gamma/dt$ is the strain rate (or rate of strain) and η_0 is the proportional constant,[1] often referred to as "viscosity." In other words, the spring exhibits purely an elastic effect (i.e., as a Hookean solid) and the dashpot exhibits purely a viscous effect (i.e., as a viscous fluid).

Therefore, the total strain of the spring and dashpot at any time t is the sum of that due to the spring (reversible) and that due to the dashpot (irreversible). Combining Eqs. (3.1) and (3.2) and generalizing the resulting expression to a three-dimensional form (i.e., in tensor form), we obtain

$$\boldsymbol{\sigma} + \lambda_1 \frac{\partial \boldsymbol{\sigma}}{\partial t} = 2\eta_0 \mathbf{d} \tag{3.3}$$

where $\boldsymbol{\sigma}$ is the extra stress tensor and $\mathbf{d}$ is the rate-of-strain (or the rate-of-deformation) tensor (see Chapter 2). Equation (3.3) is referred to as the classical Maxwell mechanical model. Note that $\lambda_1 = \eta_0/G$ in Eq. (3.3) is a time constant, often referred to as the relaxation time. Equation (3.3) is capable of qualitatively explaining many well-known viscoelastic phenomena, such as stress relaxation following a sudden change in strain and elastic recovery following a sudden release of imposed stress.

However, Eq. (3.3) is valid only for extremely small strain rates because the spring and dashpot mechanical model is based on the premise that the Hookean material is subjected to an infinitesimally small displacement gradient. In order to overcome this limitation, Oldroyd (1950) proposed a generalization of Eq. (3.3) by replacing the partial derivative $\partial/\partial t$ with the convected derivative $\mathfrak{d}/\mathfrak{d}t$ (see Chapter 2), yielding

$$\boldsymbol{\sigma} + \lambda_1 \frac{\mathfrak{d} \boldsymbol{\sigma}}{\mathfrak{d} t} = 2\eta_0 \mathbf{d} \tag{3.4}$$

which is referred to as the convected Maxwell model. In Chapter 2 we discussed the physical interpretation of the convected derivative. In steady-state simple shear flow, whose velocity field is given by Eq. (2.11), Eq. (3.4) with the contravariant components of the convected derivative of $\boldsymbol{\sigma}$ (see Eq. (2.107)) yields (see Appendix 3A)

$$\left\| \begin{matrix} \sigma^{11} & \sigma^{12} & \sigma^{13} \\ \sigma^{21} & \sigma^{22} & \sigma^{23} \\ \sigma^{31} & \sigma^{32} & \sigma^{33} \end{matrix} \right\| + (-\lambda_1 \dot{\gamma}) \left\| \begin{matrix} 2\sigma^{12} & \sigma^{22} & \sigma^{23} \\ \sigma^{22} & 0 & 0 \\ \sigma^{32} & 0 & 0 \end{matrix} \right\| = 2\eta_0 \left\| \begin{matrix} 0 & \dot{\gamma}/2 & 0 \\ \dot{\gamma}/2 & 0 & 0 \\ 0 & 0 & 0 \end{matrix} \right\| \tag{3.5}$$

from which it follows

$$\eta(\dot{\gamma}) = \eta_0; \quad N_1 = 2\eta_0 \lambda_1 \dot{\gamma}^2; \quad N_2 = 0 \tag{3.6}$$

It can easily be shown that Eq. (3.4) with the covariant components of $\boldsymbol{\sigma}$ (see Eq. (2.104) for the definition of the covariant convected derivative) yields the following material functions in steady-state simple shear flow[2]

$$\eta(\dot{\gamma}) = \eta_0; \quad N_1 = 2\eta_0\lambda_1\dot{\gamma}^2; \quad N_2 = -2\eta_0\lambda_1\dot{\gamma}^2 \tag{3.7}$$

It is seen that the material functions obtained from the covariant convected derivative of $\boldsymbol{\sigma}$ are different from those obtained from the contravariant convected derivative of $\boldsymbol{\sigma}$. Experimental results reported to date indicate that the magnitude of N_2 is much smaller than that of N_1 (say $-N_2/N_1 \approx 0.2-0.3$). Therefore, the rheology community uses only the contravariant convected derivative of $\boldsymbol{\sigma}$ when using Eq. (3.4), which is referred to as the upper convected Maxwell model. However, the limitations of the upper convected Maxwell model lie in that, as shown in Eq. (3.6), (1) it predicts shear-rate independent viscosity (i.e., Newtonian viscosity, η_0), (2) N_1 is proportional to $\dot{\gamma}^2$ over the entire range of shear rate, and (3) $N_2 = 0$. There is experimental evidence (Baek et al. 1993; Christiansen and Miller 1971; Ginn and Metzner 1969; Olabisi and Williams 1972) that suggests N_2 is negative. Also, as will be shown later in this chapter, and also in Chapter 5, in steady-state shear flow for many polymeric liquids, (1) $\eta(\dot{\gamma})$ follows Newtonian behavior at low $\dot{\gamma}$ and then decreases as $\dot{\gamma}$ increases above a certain critical value, and (2) N_1 increases with $\dot{\gamma}^2$ at low $\dot{\gamma}$ and then increases with $\dot{\gamma}^n (1 < n < 2)$ as $\dot{\gamma}$ increases further above a certain critical value.

Equation (3.3) can be extended to the following form:

$$\boldsymbol{\sigma} + \lambda_1 \frac{\partial \boldsymbol{\sigma}}{\partial t} = 2\eta_0 \left(\mathbf{d} + \lambda_2 \frac{\partial \mathbf{d}}{\partial t} \right) \tag{3.8}$$

which is referred to as the classical Jeffreys model (Jeffreys 1929), which contains three constants λ_1, λ_2, and η_0. Oldroyd (1950) proposed a generalization of the Jeffreys model by replacing the partial derivative $\partial/\partial t$ in Eq. (3.8) with $\mathfrak{d}/\mathfrak{d}t$, yielding the convected Jeffreys model,

$$\boldsymbol{\sigma} + \lambda_1 \frac{\mathfrak{d} \boldsymbol{\sigma}}{\mathfrak{d} t} = 2\eta_0 \left(\mathbf{d} + \lambda_2 \frac{\mathfrak{d} \mathbf{d}}{\mathfrak{d} t} \right) \tag{3.9}$$

It can easily be shown that in steady-state simple shear flow, the use of the contravariant convected derivatives of $\boldsymbol{\sigma}$ and $\mathbf{d}$ in Eq. (3.9) yields

$$\eta(\dot{\gamma}) = \eta_0; \quad N_1 = 2\eta_0(\lambda_1 - \lambda_2)\dot{\gamma}^2; \quad N_2 = 0 \tag{3.10}$$

and the use of the covariant convected derivatives of $\boldsymbol{\sigma}$ and $\mathbf{d}$ in Eq. (3.9) yields

$$\eta(\dot{\gamma}) = \eta_0; \quad N_1 = 2\eta_0(\lambda_1 - \lambda_2)\dot{\gamma}^2; \quad N_2 = -2\eta_0(\lambda_1 - \lambda_2)\dot{\gamma}^2 \tag{3.11}$$

Again, the predicted material functions obtained using the covariant convected derivatives of $\boldsymbol{\sigma}$ and $\mathbf{d}$ in Eq. (3.9) are different from those obtained from the contravariant convected derivatives of $\boldsymbol{\sigma}$ and $\mathbf{d}$.

To overcome the deficiency of Eq. (3.4) in predicting the rheological behavior of viscoelastic fluids, a number of empirical modifications have been attempted (Santa Cruze and Deiber 1989; White and Metzner 1963). One such empirical modification of

Eq. (3.4) is to make the material constants, appearing in Eq. (3.4), become dependent upon the second invariant I_2 of the rate-of-strain tensor **d**:

$$\boldsymbol{\sigma} + \lambda(I_2)\frac{\delta \boldsymbol{\sigma}}{\delta t} = 2\eta(I_2)\mathbf{d} \tag{3.12}$$

which is referred to as the generalized upper convected Maxwell model. In describing steady-state simple shear flow, $\lambda(I_2)$ and $\eta(I_2)$ appearing in Eq. (3.12) may be expressed by

$$\eta(\dot{\gamma}) = \frac{\eta_0}{\left[1 + (\eta_1 \dot{\gamma})^2\right]^{(1-n)/2}} \tag{3.13}$$

$$\lambda(\dot{\gamma}) = \frac{\lambda_0}{\left[1 + (\lambda_1 \dot{\gamma})^2\right]^{(1-m)/2}} \tag{3.14}$$

where $\dot{\gamma}$ is shear rate, η_0 is the viscosity at $\dot{\gamma} = 0$ (which, hereafter, will be referred to as zero-shear viscosity or Newtonian viscosity), and η_1, λ_0, λ_1, m, and n are constants that must be determined by curve fitting to experimental data. Note that Eq. (3.12) predicts shear-rate dependent viscosity given by Eq. (3.13) and $N_1 = 2\eta_0\lambda(\dot{\gamma})\dot{\gamma}$, with $\lambda(\dot{\gamma})$ being given by Eq. (3.14).

If the partial derivative $\partial/\partial t$ appearing in Eq. (3.3) is replaced with the Jaumann derivative $\mathcal{D}/\mathcal{D}t$ (see Chapter 2 for the definition of $\mathcal{D}/\mathcal{D}t$), we obtain

$$\boldsymbol{\sigma} + \lambda_1 \frac{\mathcal{D}\boldsymbol{\sigma}}{\mathcal{D}t} = 2\eta_0 \mathbf{d} \tag{3.15}$$

which is referred to as the Zaremba–Fromm–DeWitt (ZFD) model (DeWitt 1955; Fromm 1947; Zaremba 1903).[3] For steady-state simple shear flow, Eq. (3.15) yields (see Appendix 3B)

$$\begin{Vmatrix} \sigma^{11} & \sigma^{12} & \sigma^{13} \\ \sigma^{21} & \sigma^{22} & \sigma^{23} \\ \sigma^{31} & \sigma^{32} & \sigma^{33} \end{Vmatrix} + \lambda_1 \dot{\gamma} \begin{Vmatrix} -\sigma^{12} & (\sigma^{11} - \sigma^{22})/2 & 0 \\ (\sigma^{11} - \sigma^{22})/2 & \sigma^{12} & 0 \\ 0 & 0 & 0 \end{Vmatrix}$$
$$= 2\eta_0 \begin{Vmatrix} 0 & \dot{\gamma}/2 & 0 \\ \dot{\gamma}/2 & 0 & 0 \\ 0 & 0 & 0 \end{Vmatrix} \tag{3.16}$$

from which we obtain

$$\eta(\dot{\gamma}) = \frac{\eta_0}{1 + \lambda_1{}^2 \dot{\gamma}^2}; \quad N_1 = \frac{2\eta_0 \lambda_1 \dot{\gamma}^2}{1 + \lambda_1{}^2 \dot{\gamma}^2}; \quad N_2 = -\frac{\eta_0 \lambda_1 \dot{\gamma}^2}{1 + \lambda_1{}^2 \dot{\gamma}^2} \tag{3.17}$$

Comparison of Eq. (3.17) with (3.6) shows that the use of the Jaumann derivative of $\boldsymbol{\sigma}$ in the classical Maxwell model gives rise to the material functions that are quite different in form as compared to when the contravariant components of the convected derivative of $\boldsymbol{\sigma}$ are used in the classical Maxwell model. It is of great interest to observe in Eq. (3.17) that the ZFD model predicts not only shear-rate dependent (non-Newtonian) viscosity but also negative value of N_2, giving rise to $-N_2/N_1 = 0.5$. Note that when using the Jaumann derivative one obtains the identical expressions for the material functions, regardless of whether the covariant or contravariant components of tensors $\boldsymbol{\sigma}$ and $\mathbf{d}$ are employed.

By replacing the partial derivative $\partial/\partial t$ in Eq. (3.8) with $\mathcal{D}/\mathcal{D}t$, one obtains the generalized Jeffreys model:

$$\boldsymbol{\sigma} + \lambda_1 \frac{\mathcal{D}\boldsymbol{\sigma}}{\mathcal{D}t} = 2\eta_0 \left(\mathbf{d} + \lambda_2 \frac{\mathcal{D}\mathbf{d}}{\mathcal{D}t} \right) \tag{3.18}$$

For steady-state simple shear flow, Eq. (3.18) yields

$$\eta(\dot{\gamma}) = \frac{\eta_0(1 + \lambda_1\lambda_2\dot{\gamma}^2)}{1 + \lambda_1{}^2\dot{\gamma}^2}; \quad N_1 = \frac{2\eta_0(\lambda_1 - \lambda_2)\,\dot{\gamma}^2}{1 + \lambda_1{}^2\dot{\gamma}^2}; \quad N_2 = -\frac{\eta_0(\lambda_1 - \lambda_2)\,\dot{\gamma}^2}{1 + \lambda_1{}^2\dot{\gamma}^2} \tag{3.19}$$

It can be seen that Eq. (3.15) is a special case of Eq. (3.18). Note that Eq. (3.18) predicts shear-rate dependent viscosity and also nonzero negative N_2, giving rise to $-N_2/N_1 = 0.5$. It should be mentioned that there is no a priori reason for preferring any single generalization of the classical Maxwell model, Eq. (3.3), or the classical Jeffreys model, Eq. (3.8). Only a careful comparison of the model predictions with experimental results will distinguish between the possible alternatives.

To address the shortcomings of the predictions associated with Eq. (3.9), Oldroyd (1958) has proposed a constitutive equation of the following form:

$$\begin{aligned} &\boldsymbol{\sigma} + \lambda_1 \frac{\mathcal{D}\boldsymbol{\sigma}}{\mathcal{D}t} - \mu_1(\boldsymbol{\sigma}\cdot\mathbf{d} + \mathbf{d}\cdot\boldsymbol{\sigma}) + \mu_0[\mathrm{tr}\,\boldsymbol{\sigma}]\,\mathbf{d} + \nu_1\big[\mathrm{tr}(\boldsymbol{\sigma}\cdot\mathbf{d})\big]\mathbf{I} \\ &\quad = 2\eta_0\left\{\mathbf{d} + \lambda_2 \frac{\mathcal{D}\mathbf{d}}{\mathcal{D}t} - 2\mu_2\mathbf{d}^2 + \nu_2\big[\mathrm{tr}\,\mathbf{d}^2\big]\mathbf{I}\right\} \end{aligned} \tag{3.20}$$

which is referred to as the Oldroyd eight-constant model, where λ_1, μ_1, μ_0, ν_1, η_0, λ_2, μ_2 and ν_2 are constants. Note that Eq. (3.20) contains nonlinear terms, $\boldsymbol{\sigma}\cdot\mathbf{d}$ and $\mathbf{d}\cdot\boldsymbol{\sigma}$, which make the predictions of the material functions more realistic, when compared with experiment, than linear differential models. While Eq. (3.20) predicts both shear-rate dependent viscosity and normal stress effects, it has the serious disadvantage in that the unique determination of the eight constants from simple experiments is not possible.

By imposing the restrictions (1) $\lambda_1 = \mu_1 = 3\nu_1/2$, (2) $\lambda_2 = \mu_2 = 3\nu_2/2$, and (3) $\mu_0 = 0$ on Eq. (3.20), Williams and Bird (1962) simplified the Oldroyd eight-constant model to

$$\begin{aligned} &\boldsymbol{\sigma} + \lambda_1\left\{\frac{\mathcal{D}\boldsymbol{\sigma}}{\mathcal{D}t} - \boldsymbol{\sigma}\cdot\mathbf{d} - \mathbf{d}\cdot\boldsymbol{\sigma} + \tfrac{2}{3}\big[\mathrm{tr}(\boldsymbol{\sigma}\cdot\mathbf{d})\big]\mathbf{I}\right\} \\ &\quad = 2\eta_0\left\{\mathbf{d} + \lambda_2\left[\frac{\mathcal{D}\mathbf{d}}{\mathcal{D}t} - 2\mathbf{d}^2 + \tfrac{2}{3}\big(\mathrm{tr}\,\mathbf{d}^2\big)\mathbf{I}\right]\right\} \end{aligned} \tag{3.21}$$

which is referred to as the Oldroyd three-constant model. For steady-state simple shear flow, Eq. (3.21) yields

$$\eta(\dot{\gamma}) = \frac{\eta_0(1 + \frac{2}{3}\lambda_1\lambda_2\dot{\gamma}^2)}{1 + \frac{2}{3}\lambda_1{}^2\dot{\gamma}^2}; \quad N_1 = \frac{2\eta_0(\lambda_1 - \lambda_2)\,\dot{\gamma}^2}{1 + \frac{2}{3}\lambda_1{}^2\dot{\gamma}^2}; \quad N_2 = 0 \tag{3.22}$$

Since experiment indicates $N_1 > 0$, from Eq. (3.22) we conclude $\lambda_1 > \lambda_2$. It is seen that Eq. (3.21) predicts the shear-rate dependent viscosity and positive N_1, but zero N_2.

Giesekus (1982) developed a class of nonlinear differential-type constitutive equations with the following form:[4]

$$\boldsymbol{\tau} + \lambda_1 \frac{\mathfrak{d}\boldsymbol{\tau}}{\mathfrak{d}t} + \frac{\alpha}{G}(\boldsymbol{\tau} \cdot \boldsymbol{\tau}) = 2\eta_0 \mathbf{d} \tag{3.23}$$

where $G = \eta_0/\lambda_1$ and $\boldsymbol{\tau}$ is defined by

$$\boldsymbol{\tau} = \boldsymbol{\sigma} - G\boldsymbol{\delta} \tag{3.24}$$

in which λ_1 is a relaxation time, G is a modulus, and α is a mobility parameter that lies in the range $0 \leq \alpha \leq 1$. Referring to Eq. (3.24), $\boldsymbol{\tau}$ differs from $\boldsymbol{\sigma}$ by the isotropic constant $G\boldsymbol{\delta}$. Since the stress tensor for incompressible fluids is only defined to within an isotropic constant (see Chapter 2), either $\boldsymbol{\tau}$ or $\boldsymbol{\sigma}$ can be used. Equation (3.23) is called the single-mode Giesekus model. Note that with $\alpha = 0$, Eq. (3.23) reduces to the upper convected Maxwell model, Eq. (3.4). It is noteworthy to mention that Eq. (3.23) contains a nonlinear term in stress, $\boldsymbol{\tau} \cdot \boldsymbol{\tau}$, while the three-constant Oldroyd model, Eq. (3.21), contains nonlinear terms, $\boldsymbol{\sigma} \cdot \mathbf{d}$, $\mathbf{d} \cdot \boldsymbol{\sigma}$, and $\mathbf{d}^2$. The presence of the nonlinear term $\boldsymbol{\tau} \cdot \boldsymbol{\tau}$ in Eq. (3.23) apparently makes the material functions of viscoelastic polymers predicted by the Giesekus model more realistic, when compared with experiment, than those from the Oldroyd three-constant model, Eq. (3.21).

In the derivation of Eq. (3.23), Giesekus (1982) used the following molecular arguments. Assuming a linear dependence of stress tensor $\boldsymbol{\sigma}$ on the configuration (strain) tensor $\mathbf{c}$,

$$\boldsymbol{\sigma} = G\mathbf{c} \tag{3.25}$$

substitution of Eq. (3.24), with the aid of Eq. (3.25), into Eq. (3.23) yields

$$\lambda_1 \frac{\mathfrak{d}\mathbf{c}}{\mathfrak{d}t} = -\,[(1 - \alpha)\boldsymbol{\delta} + \alpha\mathbf{c}] \cdot (\mathbf{c} - \boldsymbol{\delta}) \tag{3.26}$$

The configuration tensor $\mathbf{c}$ can be considered as describing the stretching and orientation of the polymer chain. It is seen that one must first solve Eq. (3.26) for the configuration tensor $\mathbf{c}$ in order to calculate stresses using Eq. (3.25). Because the stress tensor $\boldsymbol{\sigma}$ is coupled, via Eq. (3.25), with the configuration tensor $\mathbf{c}$, the Giesekus constitutive equation is different from the other constitutive equations already presented in that it can be considered as having a molecular origin.

For steady-state homogeneous flow, Eq. (3.26) reduces to

$$\alpha \mathbf{c}^2 + (1 - 2\alpha)\mathbf{c} - (1 - \alpha)\boldsymbol{\delta} = \lambda_1 \left[(\nabla \mathbf{v})^{\mathrm{T}} \cdot \mathbf{c} + \mathbf{c} \cdot (\nabla \mathbf{v}) \right] \tag{3.27}$$

since the material derivative in the convected derivative drops off (see Chapter 2 for the definition of the convected derivative). For steady-state shear flow with the rate-of-strain tensor **d** defined by Eq. (2.12), Eq. (3.27) yields[5]

$$\alpha \left\| \begin{matrix} c_{11}{}^2 + c_{12}{}^2 & c_{12}(c_{11} + c_{22}) & 0 \\ c_{12}(c_{11} + c_{22}) & c_{12}{}^2 + c_{22}{}^2 & 0 \\ 0 & 0 & c_{33}{}^2 \end{matrix} \right\| + (1 - 2\alpha) \left\| \begin{matrix} c_{11} & c_{12} & 0 \\ c_{12} & c_{22} & 0 \\ 0 & 0 & c_{33} \end{matrix} \right\|$$

$$- (1 - \alpha) \left\| \begin{matrix} 1 & 0 & 0 \\ 0 & 1 & 0 \\ 0 & 0 & 1 \end{matrix} \right\| = \lambda_1 \dot{\gamma} \left\| \begin{matrix} 2c_{12} & c_{22} & 0 \\ c_{22} & 0 & 0 \\ 0 & 0 & 0 \end{matrix} \right\| \tag{3.28}$$

After considerable algebraic manipulations, we obtain the following material functions (see Appendix 3C):

$$\eta(\dot{\gamma}) = \frac{\eta_0 (1 - g)^2}{1 + (1 - 2\alpha) g}; \quad N_1 = \frac{(2\eta_0 / \lambda_1) g (1 - \alpha g)}{\alpha (1 - g)}; \quad N_2 = -(\eta_0 / \lambda_1) g \tag{3.29}$$

where g is defined by

$$g = \frac{1 - f}{1 + (1 - 2\alpha) f} \tag{3.30}$$

with

$$f^2 = \frac{\sqrt{1 + 16\alpha(1 - \alpha)\lambda_1{}^2 \dot{\gamma}^2} - 1}{8\alpha(1 - \alpha)\lambda_1{}^2 \dot{\gamma}^2} \tag{3.31}$$

Unlike the material functions obtained from other constitutive equations already presented, the material functions given by Eq. (3.29) do not show explicitly how $\eta(\dot{\gamma})$, N_1 and N_2, respectively, vary with $\dot{\gamma}$.

For a special case, $\alpha = 1$, Eq. (3.28) reduces to

$$\left\| \begin{matrix} c_{11}{}^2 + c_{12}{}^2 & c_{12}(c_{11} + c_{22}) & 0 \\ c_{12}(c_{11} + c_{22}) & c_{12}{}^2 + c_{22}{}^2 & 0 \\ 0 & 0 & c_{33}{}^2 \end{matrix} \right\| - \left\| \begin{matrix} c_{11} & c_{12} & 0 \\ c_{12} & c_{22} & 0 \\ 0 & 0 & c_{33} \end{matrix} \right\| = \lambda_1 \dot{\gamma} \left\| \begin{matrix} 2c_{12} & c_{22} & 0 \\ c_{22} & 0 & 0 \\ 0 & 0 & 0 \end{matrix} \right\| \tag{3.32}$$

from which we obtain[6]

$$c_{12} = \frac{\lambda_1 \dot{\gamma}}{1 + \lambda_1{}^2 \dot{\gamma}^2}; \quad c_{11} = \frac{1 + 2\lambda_1{}^2 \dot{\gamma}^2}{1 + \lambda_1{}^2 \dot{\gamma}^2}; \quad c_{22} = \frac{1}{1 + \lambda_1{}^2 \dot{\gamma}^2}; \quad c_{33} = 1 \tag{3.33}$$

Use of Eq. (3.33) into Eq. (3.25) yields the material functions that are exactly the same as Eq. (3.17) obtained from the ZFD model, Eq. (3.15). However, as will be shown, the Giesekus model with $\alpha = 1$ is not the same as the ZFD model. Substituting the following relationship between the upper convected derivative and the Jaumann derivative (see Chapter 2),

$$\frac{\mathscr{D}\boldsymbol{\tau}}{\mathscr{D}t} = \frac{\mathfrak{d}\boldsymbol{\tau}}{\mathfrak{d}t} + (\boldsymbol{\tau} \cdot \mathbf{d}) + (\mathbf{d} \cdot \boldsymbol{\tau}) \tag{3.34}$$

into Eq. (3.23) with $\alpha = 1$ we obtain

$$\boldsymbol{\tau} + \lambda_1 \frac{\mathscr{D}\boldsymbol{\tau}}{\mathscr{D}t} + \frac{1}{2G}\left\{\boldsymbol{\tau} \cdot \left(\boldsymbol{\tau} - 2\eta_0 \mathbf{d}\right) + \left(\boldsymbol{\tau} - 2\eta_0 \mathbf{d}\right) \cdot \boldsymbol{\tau}\right\} = 2\eta_0 \mathbf{d} \tag{3.35}$$

It is seen that Eq. (3.35) becomes identical to Eq. (3.15) only when the symmetric part of tensor $\boldsymbol{\tau} \cdot (\boldsymbol{\tau} - 2\eta_0 \mathbf{d})$ disappears; in other words, when $\boldsymbol{\tau} - 2\eta_0 \mathbf{d} = 0$. The Giesekus model is very versatile in that it has an additional parameter, the so-called mobility parameter α, which allows for more accurate predictions of the material functions in steady-state shear flow as compared with the ZFD model. This will become clear later in this chapter when we present logarithmic plots of η versus $\dot{\gamma}$, and logarithmic plots of N_1 versus $\dot{\gamma}$ based on predictions from both the Giesekus and ZFD models. In Chapter 6 of Volume 2 we will use a modified version of the Giesekus model to simulate the high-speed melt spinning process.

3.2.2 Multimode Differential-Type Constitutive Equations

We have previously assumed that a single "spring and dashpot" mechanical model (see Figure 3.1) could describe the rheological responses of flexible homogeneous polymers. In view of the fact that flexible homogeneous polymers consist of many repeat units, a single spring and dashpot mechanical model is admittedly an oversimplification of the real situation. A more realistic approach to describing the rheological responses of long-chain macromolecules might be to regard the stress $\boldsymbol{\sigma}$ in the polymer as the superposition of individual stresses $\boldsymbol{\sigma}_i$, representing a series of springs and dashpots. This approach is the basis of multimode differential-type models.

Taking a multimode approach, Eq. (3.3) can be expressed as

$$\boldsymbol{\sigma} = \sum_{i=1}^{\infty} \boldsymbol{\sigma}_i \tag{3.36}$$

$$\boldsymbol{\sigma}_i + \lambda_i \frac{\partial \boldsymbol{\sigma}_i}{\partial t} = 2\eta_i \mathbf{d} \tag{3.37}$$

where $\boldsymbol{\sigma}_i$ is the stress tensor of the ith unit (or ith mode of action), and λ_i and η_i are material constants of the corresponding mode. Similar to the approaches already shown, the partial derivative $\partial/\partial t$ in Eq. (3.37) can be replaced by the convected derivative or the Jaumann derivative.

Spriggs (1965) proposed a multimode nonlinear differential-type constitutive equation of the form

$$\boldsymbol{\sigma}_i + \lambda_i \left\{ \frac{\mathcal{D}\boldsymbol{\sigma}_i}{\mathcal{D}t} + (1+\varepsilon)\left(\boldsymbol{\sigma}_i \cdot \mathbf{d} + \mathbf{d} \cdot \boldsymbol{\sigma}_i - \tfrac{2}{3}\left[\operatorname{tr} \boldsymbol{\sigma}_i\right]\mathbf{I}\right) \right\} = 2\eta_i \mathbf{d} \tag{3.38}$$

where ε is an adjustable parameter introduced in order to obtain empirically a nonzero second normal stress difference. Since Eq. (3.36) has an infinite set of material constants λ_i and η_i $(i = 1, \ldots, \infty)$, Spriggs introduced the following empirical forms of the constants λ_i and η_i:

$$\lambda_i = \frac{\lambda}{i^\alpha}; \quad \eta_i = \frac{\eta_0 \lambda_i}{\sum_{i=1}^{\infty} \lambda_i} = \frac{\eta_0}{i^\alpha Z(\alpha)} \tag{3.39}$$

with $Z(\alpha) = \sum_{i=1}^{\infty} 1/i^\alpha$. The Spriggs model, Eq. (3.38), has four parameters, η_0, λ, α, and ε. It is of interest to note in Eq. (3.39) that as i increases (i.e., higher modes of motion), the contributions of the corresponding material constants λ_i and η_i become less important. This is consistent with the molecular viscoelasticity theory, which will be presented in Chapter 4.

For steady-state simple shear flow, Eq. (3.38) yields

$$\eta(\dot{\gamma}) = \frac{\eta_0}{Z(\alpha)} \sum_{i=1}^{\infty} \frac{i^\alpha}{i^{2\alpha} + (c\lambda\dot{\gamma})^2} \tag{3.40a}$$

$$N_1 = \frac{2\eta_0 \lambda}{Z(\alpha)} \sum_{i=1}^{\infty} \frac{\dot{\gamma}^2}{i^{2\alpha} + (c\lambda\dot{\gamma})^2} \tag{3.40b}$$

$$N_2 = \frac{\varepsilon \eta_0 \lambda}{Z(\alpha)} \sum_{i=1}^{\infty} \frac{\dot{\gamma}^2}{i^{2\alpha} + (c\lambda\dot{\gamma})^2} \tag{3.40c}$$

with $c^2 = (2 - 2\varepsilon - \varepsilon^2)/3$. Note that the sign of ε must be negative because experiment indicates that N_2 is negative, and that the magnitude of ε is less than 1 because the available experimental data indicate $|N_2/N_1| < 0.5$.

3.3 Integral-Type Constitutive Equations for Viscoelastic Fluids

It can easily be shown that, for infinitesimally small deformations, the classical Maxwell model, Eq. (3.3), may be written in an integral form

$$\boldsymbol{\sigma}(t) = 2\int_{-\infty}^{t} m(t-t')\,\mathbf{d}(t')\,\mathrm{d}t' \tag{3.41}$$

with

$$m(t-t') = \frac{\eta_0}{\lambda_1}\,\mathrm{e}^{-(t-t')/\lambda_1} \tag{3.42}$$

which is called the memory function. The memory function for the integral representation of the classical Jeffreys model, Eq. (3.8), has the following form

$$m(t-t') = \frac{\eta_0}{\lambda_1}\left[\left(1-(\lambda_2/\lambda_1)\right)\mathrm{e}^{-(t-t')/\lambda_1} + 2\lambda_2\delta(t-t')\right] \tag{3.43}$$

Since polymeric materials consist of many segments of different submolecules, the properties of a polymeric material may be thought of being given in terms of a spectrum of these variables (e.g., λ_i and η_i for the ith submolecule). If one assumes that the components of a stress tensor are linearly related to the components of the rate-of-strain tensor, then the overall response of the N submolecules may be expressed by

$$\boldsymbol{\sigma}(t) = 2\int_{-\infty}^{t}\sum_{i=1}^{N}\frac{\eta_0}{\lambda_i}\mathrm{e}^{-(t-t')/\lambda_i}\,\mathbf{d}(t')\,\mathrm{d}t' \tag{3.44}$$

which is an integral representation of the linear multimode Maxwell model. The physical interpretation of Eq. (3.44) is as follows. If one considers the rate of strain as the cause and the stress as the resulting effect, then the observed "resulting effect" at the present time is due to the sequence of causes up to the present time t from the past. The memory function $m(t-t')$ for Eq. (3.44) is given by

$$m(t-t') = \sum_{i=1}^{N}\frac{\eta_0}{\lambda_i}\mathrm{e}^{-(t-t')/\lambda_i} \tag{3.45}$$

Integral-type constitutive equations may also be written in terms of strain tensor (or relative strain tensor), instead of rate-of-strain tensor. For infinitesimally small deformations, if the response of a system can be expressed by the linear superposition of a series of separate responses at different times to a series of step changes in the input, the stress tensor $\boldsymbol{\sigma}$ as a response can be expressed in terms of the infinitesimally small strain tensor $\mathbf{e}$ by

$$\boldsymbol{\sigma}(t) = 2\int_{-\infty}^{t}\Phi(t-t')\,\mathbf{e}(t')\,\mathrm{d}t' \tag{3.46}$$

where

$$\Phi(t) = -\mathrm{d}G(t)/\mathrm{d}t \tag{3.47}$$

in which $G(t)$ is similar to an elastic modulus, also referred to as the relaxation function. Equation (3.46) is the most general integral form of linear viscoelasticity theory.

For finite strain tensor $\mathbf{E}(t, t')$ (see Chapter 2 for the definition of $\mathbf{E}$), Eq. (3.46) can be represented by

$$\sigma(t) = \int_{-\infty}^{t} m(t - t')\mathbf{E}(t, t')\,\mathrm{d}t' \tag{3.48}$$

Lodge (1956) proposed the following form of $m(t - t')$:

$$m(t - t') = \sum_{i=1}^{N} \frac{G_i}{\lambda_i} \mathrm{e}^{-(t-t')/\lambda_i} \quad (G_i > 0,\ \lambda_i > 0) \tag{3.49}$$

in which G_i and λ_i are material constants to be determined. It can be shown that in steady-state simple shear flow, use of Eq. (3.49) in Eq. (3.48) gives[7]

$$\eta(\dot{\gamma}) = \sum_{i=1}^{N} G_i\lambda_i; \quad N_1 = 2\sum_{i=1}^{N} G_i\lambda_i{}^2\dot{\gamma}^2; \quad N_2 = 0 \tag{3.50}$$

It is seen from Eq. (3.50) that the Lodge model predicts virtually the same form for material functions as the upper convected Maxwell model does (see Eq. (3.6)).

Since the particular memory function given in Eq. (3.49) cannot predict the shear-rate dependent viscosity, which makes it not useful for describing the flow of many polymeric liquids, various semiempirical attempts (Bird and Carreau 1968; Carreau 1972; Macdonald and Bird 1966; Meister 1971; Yamamoto 1971) have been made to avoid the shortcomings of this memory function by incorporating a strain tensor $\mathbf{E}$ (see Chapter 2):

$$\sigma(t) = \int_{-\infty}^{t} m(t - t', I_2(t'))\,\mathbf{E}(t, t')\,\mathrm{d}t' \tag{3.51}$$

where I_2 is the second invariant of finite strain tensor $\mathbf{E}$. Using the following form of $m(t - t', I_2(t'))$

$$m(t - t', I_2(t')) = \sum_{i=1}^{\infty} \frac{G_i}{\lambda_i} \exp\left\{ -\int_{t'}^{t} \left[\left(1 + c\sqrt{|I_2(\xi)|\lambda_i}\right) \Big/ \lambda_i \right] \mathrm{d}\xi \right\} \tag{3.52}$$

in Eq. (3.51), Meister (1971) obtained the following material functions[8] for steady-state simple shear flow

$$\eta(\dot{\gamma}) = \sum_{i=1}^{\infty} \frac{G_i \lambda_i}{(1 + c\lambda_i \dot{\gamma})^2}; \quad N_1 = \sum_{i=1}^{\infty} \frac{2G_i \lambda_i{}^2 \dot{\gamma}^2}{(1 + c\lambda_i \dot{\gamma})^3}; \quad N_2 = 0 \tag{3.53}$$

where c is a constant. Equation (3.53) shows that the memory function, Eq. (3.52), predicts a shear-rate dependent viscosity and positive N_1, but zero N_2.

To avoid the limitation (i.e., $N_2 = 0$) described above, an empirical constitutive equation was proposed (Bird and Carreau 1968)

$$\boldsymbol{\sigma}(t) = \int_{-\infty}^{t} m(t - t', I_2(t')) \left[(1 + (\varepsilon/2))\, \mathbf{C}_t^{-1}(t', t) - (\varepsilon/2)\, \mathbf{C}_t(t', t) \right] \mathrm{d}t' \tag{3.54}$$

where $\mathbf{C}_t^{-1}(t', t)$ is the relative Finger strain tensor and $\mathbf{C}_t(t', t)$ is the relative Cauchy–Green strain tensor (see Chapter 2), and ε is an adjustable parameter that was introduced to empirically obtain a nonzero N_2. Bird and Carreau (1968) chose the following form of the memory function:

$$m\big(t - t', I_2(t')\big) = \sum_{i=1}^{\infty} \frac{\eta_i}{\lambda_{2i}{}^2} \frac{\exp\big[-(t - t')/\lambda_{2i}\big]}{1 + \frac{1}{2}\lambda_{1i}{}^2\, |I_2(t')|} \tag{3.55}$$

with

$$\eta_i = \frac{\eta_0 \lambda_{1i}}{\sum_{i=1}^{\infty} \lambda_{1i}}; \quad \lambda_{1i} = \lambda_1 \left(\frac{1 + n_1}{i + n_1} \right)^{\alpha_1}; \quad \lambda_{2i} = \lambda_2 \left(\frac{1 + n_2}{i + n_2} \right)^{\alpha_2} \tag{3.56}$$

and obtained for steady-state simple shear flow the following material functions:[9]

$$\eta(\dot{\gamma}) = \sum_{i=1}^{\infty} \frac{\eta_i}{1 + (\lambda_{1i}\dot{\gamma})^2}; \quad N_1 = \sum_{i=1}^{\infty} \frac{2\eta_i \lambda_{2i} \dot{\gamma}^2}{1 + (\lambda_{1i}\dot{\gamma})^2}; \quad N_2 = -\varepsilon \sum_{i=1}^{\infty} \frac{\eta_i \lambda_{2i} \dot{\gamma}^2}{1 + (\lambda_{1i}\dot{\gamma})^2} \tag{3.57}$$

where $\eta_0, \lambda_1, \lambda_2, \alpha_1, \alpha_2, \varepsilon, n_1$, and n_2 are constants. It should be mentioned that although nonzero N_2 is predicted from Eq. (3.54), there is no way of knowing a priori what the value of ε might be. There are other forms of empirical integral models proposed in the literature (Bogue 1966). The difficulty with using such empirical models is that there are several adjustable parameters that cannot be determined uniquely from simple experiment.

A general expression of Eq. (3.54) with $\varepsilon = 0$ may be expressed as

$$\boldsymbol{\sigma}(t) = \int_{-\infty}^{t} m(t - t', I_1, I_2)\mathbf{C}_t^{-1}(t', t)\, \mathrm{d}t' \tag{3.58}$$

where I_1 and I_2 are the first and second invariants, respectively, of $\mathbf{C}_t^{-1}(t', t)$. Wagner (1976, 1977) assumed that the nonlinear memory function $m(t - t', I_1, I_2)$ is separable into a product of two functions

$$m(t - t', I_1, I_2) = M(t - t')\, h(I_1, I_2) \tag{3.59}$$

where the memory function $M(t - t')$ can be determined from linear viscoelasticity data by assuming the form given by Eq. (3.49). Here, $h(I_1, I_2)$ is referred to as the damping function, the magnitude of which is equal to unity in the linear viscoelastic limit and is less than unity in the nonlinear region. Wagner (1979) determined $h(I_1, I_2)$ using the following expression,

$$h(I_1, I_2) = h(\gamma^2(t, t')) = h(|\gamma(t, t')|) \le 1 \tag{3.60}$$

where $\gamma(t, t')$ is the relative shear strain defined by $\gamma(t, t') = (t - t')\dot{\gamma}$, with $\dot{\gamma}$ being the rate of strain. Laun (1978) used the following approximation:

$$h(t, t') = f_1 \exp[-n_1 \,|\gamma(t, t')|] + f_2 \exp[-n_2 \,|\gamma(t, t')|] \tag{3.61}$$

and determined the numerical values of f_1, f_2, n_1, and n_2 by curve fitting to the experimental data obtained from shear flow and transient elongational flow responses of low-density polyethylene.

Kaye (1962) and Bernstein, Kersely, and Zapas (1963) independently developed the following form of a constitutive equation:

$$\sigma(t) = \int_{-\infty}^{t} \frac{\partial U(t - t', I_1, I_2)}{\partial I_1} \mathbf{C}_t^{-1}(t', t) + \frac{\partial U(t - t', I_1, I_2)}{\partial I_2} \mathbf{C}_t(t', t)\, dt' \tag{3.62}$$

where U is the strain energy function, and I_1 and I_2 are the first and second invariants, respectively, of $\mathbf{C}_t^{-1}(t', t)$. Equation (3.62) is referred to as the K–BKZ model. At present there is no rational guidance, suggesting how the scalar function U might depend on the elapsed time $t - t'$, and the two invariants I_1 and I_2. For simplicity, U may be expressed as a product of two functions,

$$U(t - t', I_1, I_2) = M(t - t') W(I_1, I_2) \tag{3.63}$$

which in form is very similar to that given by Eq. (3.59). The function W in Eq. (3.63) is called the "potential function." Substitution of Eq. (3.63) in (3.62) gives

$$\sigma(t) = \int_{-\infty}^{t} M(t - t') \left[\frac{\partial W(I_1, I_2)}{\partial I_1} \mathbf{C}_t^{-1}(t') + \frac{\partial W(I_1, I_2)}{\partial I_2} \mathbf{C}_t(t') \right] dt' \tag{3.64}$$

However, Eq. (3.64) suffers from the lack of a standard procedure for determining $W(I_1, I_2)$. Papanastasiou et al. (1983) suggested an approximate form, albeit empirical, of Eq. (3.63), and showed how to estimate the constants appearing in the expression using experimental data.

3.4 Rate-Type Constitutive Equations for Viscoelastic Fluids

Here, we present constitutive equations that do not fall within the categories of differential- and integral-type, presented in the preceding sections. For the reason of historical importance, we summarize the constitutive equations of Rivlin and Ericksen (1955) and Coleman and Noll (1959, 1960, 1961a). Because the stress tensor $\boldsymbol{\sigma}$ in the constitutive equations is expressed in terms of the Rivlin–Ericksen tensor $\mathbf{A}_{(n)} (n = 1, 2, \ldots)$ (see Chapter 2), we shall refer them to as rate-type constitutive equations.[10]

Rivlin and Ericksen (1955) proposed a nonlinear viscoelasticity theory,[11] in which the components of stress at time t in an element of material depend on the gradient of displacement, velocity, acceleration, second acceleration, ..., and $(n-1)$th acceleration in that element at time t. The fluid described by the constitutive equation based on this theory is referred to as a "memoryless fluid," the reason being that the components of stress at time t are independent of those experienced up to the time t. A general representation of this constitutive equation is given by (Rivlin and Ericksen 1955)

$$\boldsymbol{\sigma} = f(\mathbf{A}_{(1)}, \mathbf{A}_{(2)}, \ldots, \mathbf{A}_{(n)}) \tag{3.65}$$

in which components of the Rivlin–Ericksen tensors $\mathbf{A}_{(1)}$ and $\mathbf{A}_{(n)}$ are defined by

$$A_{ij}^{(1)} = 2d_{ij} = \left(\frac{\partial v_i}{\partial x_j}\right) + \left(\frac{\partial v_j}{\partial x_i}\right) \tag{3.66}$$

$$A_{ij}^{(n)} = \frac{\partial A_{ij}^{(n-1)}}{\partial t} + v_k \frac{\partial A_{ij}^{(n-1)}}{\partial x_k} + \frac{\partial v_k}{\partial x_i} A_{kj}^{(n-1)} + \frac{\partial v_k}{\partial x_j} A_{ik}^{(n-1)} \tag{3.67}$$

where $A_{ij}^{(n)}$ denotes the nth convected derivative of the Rivlin–Ericksen tensor (see Chapter 2). If the Rivlin–Ericksen tensor $\mathbf{A}_{(m)}$ is a symmetric second-order tensor and has a nonvanishing determinant, Eq. (3.65) can be simplified to the following form of constitutive equation

$$\boldsymbol{\sigma} = f(\mathbf{A}_{(1)}, \mathbf{A}_{(2)}) \tag{3.68}$$

by use of the Cayley–Hamilton theorem (see Appendix 3D). Rivlin and Ericksen (1955) have shown further that for incompressible fluids, Eq. (3.68) may be written more explicitly as

$$\begin{aligned}\boldsymbol{\sigma} = {} & \alpha_1 \mathbf{A}_{(1)} + \alpha_2 \mathbf{A}_{(1)}^2 + \alpha_3 \mathbf{A}_{(2)} + \alpha_4 \mathbf{A}_{(2)}^2 + \alpha_5 (\mathbf{A}_{(1)}\mathbf{A}_{(2)} + \mathbf{A}_{(2)}\mathbf{A}_{(1)}) \\ & + \alpha_6 (\mathbf{A}_{(1)}^2 \mathbf{A}_{(2)} + \mathbf{A}_{(2)}\mathbf{A}_{(1)}^2) + \alpha_7 (\mathbf{A}_{(2)}^2 \mathbf{A}_{(1)} + \mathbf{A}_{(1)}\mathbf{A}_{(2)}^2) \\ & + \alpha_8 (\mathbf{A}_{(1)}^2 \mathbf{A}_{(2)}^2 + \mathbf{A}_{(2)}^2 \mathbf{A}_{(1)}^2)\end{aligned} \tag{3.69}$$

It should be noted that the coefficients $\alpha_1, \ldots, \alpha_8$ appearing in Eq. (3.69) are not constant, but rather they are functions of the invariants of various quantities (e.g., $\mathbf{A}_{(1)}$ and $\mathbf{A}_{(2)}$).

For steady-state simple shear flow, it can be shown that use of Eqs. (2.65) to (2.67) (see Chapter 2) in Eq. (3.69) gives

$$\boldsymbol{\sigma} = \beta_1(\dot{\gamma})\mathbf{A}_{(1)} + \beta_2(\dot{\gamma})\mathbf{A}_{(1)}{}^2 + \beta_3(\dot{\gamma})\mathbf{A}_{(2)} \tag{3.70}$$

where β_1, β_2, and β_3 are functions of shear rate $\dot{\gamma}$, via the second invariant I_2 of the rate-of-strain tensor $\mathbf{A}_{(1)}$ (see Eq. (2.68)). The three material functions of the constitutive equation, Eq. (3.70), may be rewritten as

$$\eta(\dot{\gamma}) = \beta_1(\dot{\gamma}); \quad N_1 = -2\beta_2(\dot{\gamma})\dot{\gamma}^2; \quad N_2 = \left[\beta_1(\dot{\gamma}) + 2\beta_2(\dot{\gamma})\right]\dot{\gamma}^2 \tag{3.71}$$

As we will show, by using an entirely different approach Coleman and Markovitz (1964) obtained a constitutive equation with a formal representation very similar to Eq. (3.70) but where the coefficients are independent of the second invariant of $\mathbf{A}_{(1)}$.

Noll (1958) and Coleman and Noll (1961b) developed the "simple fluid theory." According to Coleman and Noll (1961b), the stress tensor $\boldsymbol{\sigma}$ may be expressed as an isotropic hereditary functional $\aleph$ of the strain history $\mathbf{E}(s)$:

$$\boldsymbol{\sigma} = \underset{s=0}{\overset{\infty}{\aleph}} \left\{\mathbf{E}(s)\right\} \tag{3.72}$$

where $\mathbf{E}(s)$ is considered to be the history of $\mathbf{C}_t^{-1}(s)$, defined by

$$\mathbf{E}(s) = \mathbf{C}_t^{-1}(s) - \mathbf{I} \tag{3.73}$$

with $\mathbf{I}$ being the unit second-order tensor. Here, s is the elapsed time defined by $s = t - t'$, such that a small value of s denotes the recent past and a large value of s denotes the distant past.

Using the retardation theorem (Coleman and Noll 1961b), the deformation history functional $\aleph$ may be expanded into an infinite series involving single, double, and higher integrals of the history function $\mathbf{E}(s)$. The second-order approximation that contains single and double integrals is given by

$$\begin{aligned}\boldsymbol{\sigma} = &\int_0^\infty m(s)\mathbf{E}(s)\,\mathrm{d}s + \int_0^\infty\int_0^\infty a(s_1, s_2)\mathbf{E}(s_1)\mathbf{E}(s_2)\,\mathrm{d}s_1\,\mathrm{d}s_2 \\ &+ \int_0^\infty\int_0^\infty b(s_1, s_2)\left[\mathrm{tr}(\mathbf{E}(s_1)\right]\mathbf{E}(s_2)\,\mathrm{d}s_1\,\mathrm{d}s_2\end{aligned} \tag{3.74}$$

where $m(s)$, $a(s_1, s_2)$, and $b(s_1, s_2)$ are unspecified decay functions and may be considered relaxation moduli. Equation (3.74) assumes that the motion is slow enough that the stress at the present time t might be adequately described by the deformation history weighted up to a second-degree polynomial approximation.

For a "slow flow," $\mathbf{E}(s)$ defined by Eq. (3.73) can be expressed as a series of the Rivlin–Ericksen tensors $\mathbf{A}_{(n)}$ (Coleman and Noll 1961b):

$$\mathbf{E}(s) = -s\mathbf{A}_{(1)} + (s^2/2)\mathbf{A}_{(2)} + \ldots \tag{3.75}$$

Substituting Eq. (3.75) into (3.74) and neglecting certain high-order terms resulting from the double integral terms yields (Coleman and Noll 1961b)

$$\boldsymbol{\sigma} = \eta_0\mathbf{A}_{(1)} + \beta\mathbf{A}_{(1)}^{\;2} + \nu\mathbf{A}_{(2)} \tag{3.76}$$

where

$$\eta_0 = -\int_0^\infty s\,m(s)\,ds; \quad \beta = \int_0^\infty\int_0^\infty s_1 s_2 a(s_1, s_2)\,ds_1\,ds_2; \quad \nu = \tfrac{1}{2}\int_0^\infty s^2 m(s)\,ds \tag{3.77}$$

Note that since $m(s)$ and $a(s_1, s_2)$ are functions only of time s, then η_0, β, and ν are constants. A material that can be represented by the constitutive equation given in Eq. (3.76) is called a "Coleman–Noll second-order fluid" (Coleman and Markovitz 1964; Truesdell and Noll 1965). For steady-state simple shear flow, Eq. (3.76) yields

$$\eta = \eta_0; \quad N_1 = -2\nu\,\dot{\gamma}^2; \quad N_2 = (\beta + 2\nu)\,\dot{\gamma}^2 \tag{3.78}$$

where $\dot{\gamma}$ is shear rate. It is seen from Eq. (3.78) that ν is a negative quantity since N_1 is known to be a positive quantity, and β is a positive quantity with $\beta < 2|\nu|$ since N_2 is known to be a negative quantity. The Coleman–Noll second-order fluid predicts that η is independent of $\dot{\gamma}$ and that N_1 is proportional to $\dot{\gamma}^2$. This shortcoming may be corrected by adding multiple integrals to the integral expansion of the hereditary functional $\aleph$ in Eq. (3.72). However, this will introduce more constants, which then need to be determined.

3.5 Predicted Material Functions and Experimental Observations

3.5.1 Material Functions for Steady-State Shear Flow

In the preceding sections, we have presented the material functions derived from various constitutive equations for steady-state simple shear flow. During the past three decades, numerous research groups have reported on measurements of the steady-state shear flow properties of flexible polymer solutions and melts. There are too many papers to cite them all here. The monographs by Bird et al. (1987) and Larson (1988) have presented many experimental results for steady-state shear flow of polymer solutions and melt. In this section we present some experimental results merely to show the shape of the material functions for steady-state shear flow of linear, flexible viscoelastic molten polymers and, also, the materials functions for steady-state shear flow predicted from some of the constitutive equations presented in the preceding sections.

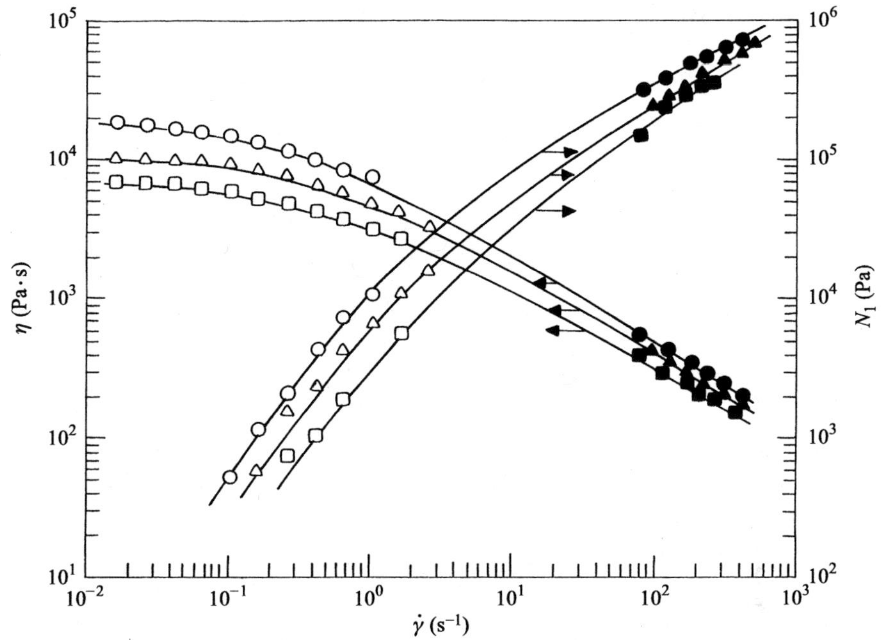

Figure 3.2 Plots of $\log \eta$ versus $\log \dot{\gamma}$ and $\log N_1$ versus $\log \dot{\gamma}$ for a low-density polyethylene at various temperatures (°C): (○, ●) 180, (△, ▲) 200, and (□, ■) 220. The data with open symbols were obtained using the cone-and-plate fixture of a rotational-type rheometer, and the data with filled symbols were obtained using a continuous-flow capillary rheometer. Refer to Chapter 5 for details of the experimental methods employed to obtain the data. (Reprinted from Han et al., *Journal of Applied Polymer Science* 28:3435. Copyright © 1983, with permission from John Wiley & Sons.)

Figure 3.2 gives logarithmic plots of η versus $\dot{\gamma}$ and logarithmic plots of N_1 versus $\dot{\gamma}$ for a low-density polyethylene at 180, 200, and 220 °C. Note that almost of all viscoelastic polymeric fluids exhibit similar shear-rate dependencies of shear viscosity and first normal stress difference. The details of the experimental methods employed to obtain the data in Figure 3.2 are presented in Chapter 5. It is seen that η stays constant at low $\dot{\gamma}$ and then decreases with increasing $\dot{\gamma}$ (i.e., non-Newtonian, shear-thinning behavior), while N_1 increases rapidly at low $\dot{\gamma}$ with a slope less than 2 and then increases gradually with further increase in $\dot{\gamma}$. Note that the many constitutive equations presented in the previous sections predict a slope of 2 in the $\log N_1$ versus $\log \dot{\gamma}$ plot as $\dot{\gamma}$ approaches zero. As is elaborated on in Chapter 6, the molecular weight distribution of a homopolymer influences the shape of the $\log N_1$ versus $\log \dot{\gamma}$ plot at low shear rates; specifically, many commercial homopolymers, which are polydisperse, would show a slope less than 2 in the $\log N_1$ versus $\log \dot{\gamma}$ plot at low shear rates (say as low as 0.01 s^{-1}). Under such circumstances, rheological measurements must be taken at shear rates much lower than 0.01 s^{-1} to observe a slope of 2 in the $\log N_1$ versus $\log \dot{\gamma}$ plot. It should be mentioned that the continuum-based constitutive equations presented in the previous sections do not (and cannot) incorporate the effect of

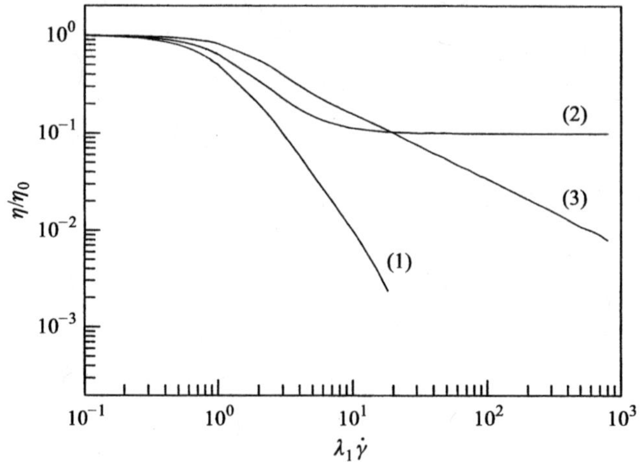

Figure 3.3 Plots of $\log \eta/\eta_0$ versus $\log \lambda_1\dot{\gamma}$ predicted from (1) the ZFD model, Eq. (3.15), (2) the Oldroyd three-constant model, Eq. (3.21), with $\lambda_1 = 1.0$ s and $\lambda_2 = 0.1$ s, and (3) the Spriggs model, Eq. (3.38), with $\lambda = 1.85$ s, $\alpha = 3.1$, and $c = 0.52$.

molecular weight distribution. The effects of molecular weight and molecular weight distribution of a polymer on its material functions for steady-state shear flow are indirectly reflected in the magnitude of the parameters (e.g., η_0 and λ_1). This subject is discussed in Chapters 4 and 6.

Figure 3.3 gives logarithmic plots of η/η_0 versus $\lambda_1\dot{\gamma}$ predicted from three models: (1) the ZFD model, (2) the Oldroyd three-constant model, and (3) the Spriggs model. It is seen in Figure 3.3 that the predicted viscosities from all three models decrease at a much faster rate than those observed experimentally (see Figure 3.2) with increasing shear rate, and that the viscosities predicted from the Oldroyd three-constant model level off as shear rate is increased, which is seldom observed experimentally.

Figure 3.4 gives logarithmic plots of $\psi_1/2\eta_0\lambda_1$ versus $\lambda_1\dot{\gamma}$ predicted from three models: (1) the ZFD model, (2) the Oldroyd three-constant model, and (3) the Spriggs model. Here ψ_1 is the first normal stress difference coefficient defined by $N_1/\dot{\gamma}^2$ (see Chapter 2) and $\psi_1/2\eta_0\lambda_1$ is a dimensionless variable. It is seen from Figure 3.4 that the predicted $\psi_1/2\eta_0\lambda_1$ from the ZFD decreases with increasing $\lambda_1\dot{\gamma}$ much faster than that from the Oldroyd three-constant and the Spriggs models, and the predicted $\psi_1/2\eta_0\lambda_1$ from the Oldroyd three-constant model still decreases faster with increasing $\lambda_1\dot{\gamma}$ than that from and the Spriggs model.

In Figure 3.5, the logarithmic plots of η versus $\dot{\gamma}$ and logarithmic plots of N_1 versus $\dot{\gamma}$ predicted from the Spriggs model are compared with the experimental results for a commercial polystyrene ($M_w = 2.8 \times 10^5$ and $M_w/M_n = 2.83$). In Figure 3.5, both η and N_1 from the Spriggs model were curve fitted, via a nonlinear least-squares method, to the experimental data. The numerically values of the four parameters (η_0, λ, α, and c) determined from curve fitting are given in the figure caption. The reason why the Spriggs model predicts steady-state flow behavior very well lies in that it is a multimode model (see Eq. (3.38)) having four parameters. Note that the more adjustable

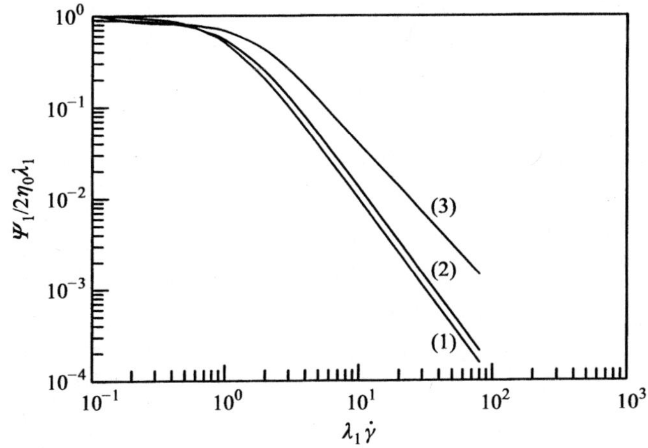

Figure 3.4 Logarithmic plots of $\psi_1/2\eta_0\lambda_1$ versus $\lambda_1\dot{\gamma}$ predicted from (1) the ZFD model, Eq. (3.15), (2) the Oldroyd three-constant model, Eq. (3.21), with $\lambda_1 = 1.0$ s and $\lambda_2 = 0.1$ s, and (3) the Spriggs model, Eq. (3.38), with $\lambda = 1.85$ s, $\alpha = 3.1$, and $c = 0.52$.

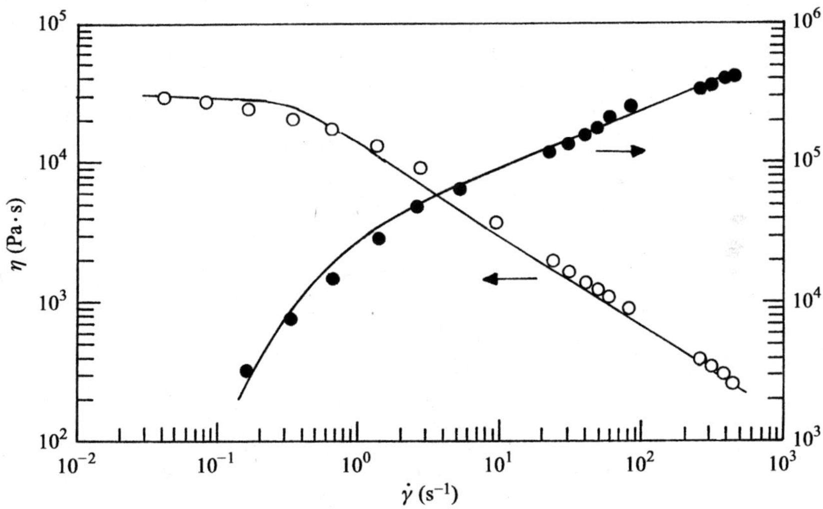

Figure 3.5 Comparison of $\log \eta$ versus $\dot{\gamma}$ plot (○) and $\log N_1$ versus $\dot{\gamma}$ plot (●) for a commercial polystyrene at 200 °C with the predictions of the Spriggs model, Eq. (3.38), with $\eta_0 = 3.1 \times 10^4$ Pa·s, $\lambda = 1.90$ s, $\alpha = 2.78$, and $c = 0.816$. The data at $\dot{\gamma} < 6$ s^{-1} were obtained using the cone-and-plate fixture of a rotational-type rheometer and the data at $\dot{\gamma} >$ 10 s^{-1} were obtained using a continuous-flow capillary rheometer. Refer to Chapter 5 for details of the experimental methods employed to obtain the data.

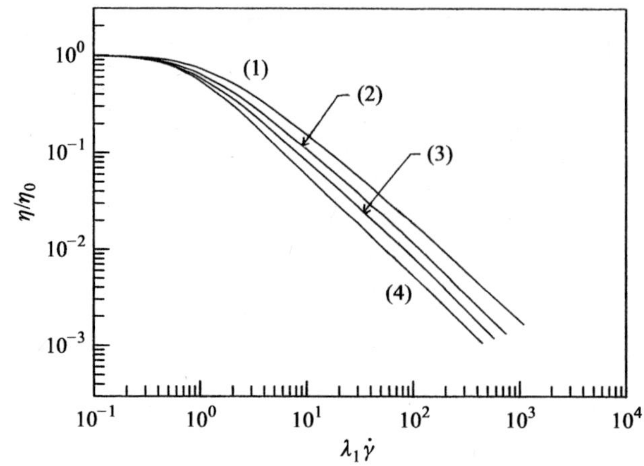

Figure 3.6 Plots of $\log \eta/\eta_0$ versus $\log \lambda_1 \dot{\gamma}$ predicted from the Giesekus model, Eqs. (3.25) and (3.26), for different values of the molecular mobility parameter α: (1) $\alpha = 0.2$, (2) $\alpha = 0.4$, (3) $\alpha = 0.6$, and (4) $\alpha = 0.8$.

parameters a model has, the better the curve fitting is expected to be. It is not difficult to surmise that a multimode model is expected to predict experimental results more accurately than a single-mode model. This observation has amply been demonstrated in the monograph by Bird et al. (1987).

Figure 3.6 gives logarithmic plots of η/η_0 versus $\log \lambda_1 \dot{\gamma}$ predicted from the Giesekus model for different values of the molecular mobility parameter α. Comparison of Figure 3.6 with Figure 3.3 indicates that the Giesekus model predicts a less steep decrease of viscosity with increasing shear rate than does the ZFD model. Figure 3.7 gives logarithmic plots of $\psi_1/2\eta_0\lambda_1$ versus $\lambda_1 \dot{\gamma}$ predicted from the Giesekus model for different values of the molecular mobility parameter α. Comparison of Figure 3.7

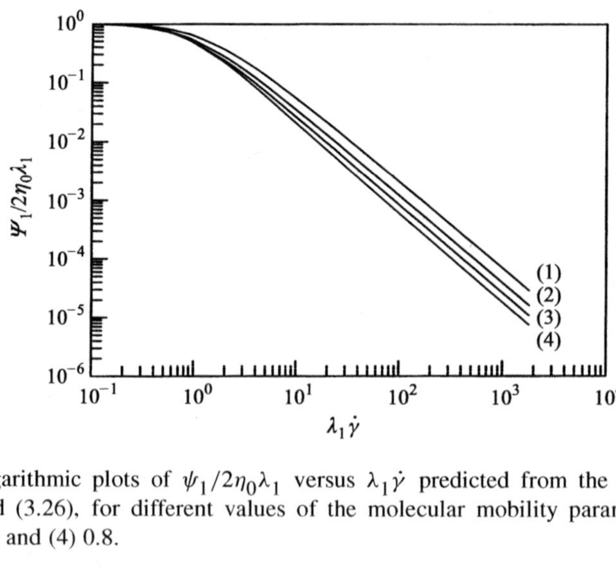

Figure 3.7 Logarithmic plots of $\psi_1/2\eta_0\lambda_1$ versus $\lambda_1 \dot{\gamma}$ predicted from the Giesekus model, Eqs. (3.25) and (3.26), for different values of the molecular mobility parameter α: (1) 0.2, (2) 0.4, (3) 0.6, and (4) 0.8.

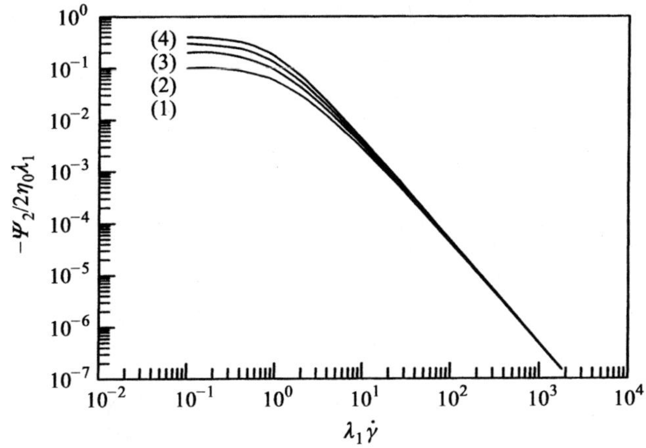

Figure 3.8 Logarithmic plots of $-\psi_2/2\eta_0\lambda_1$ versus $\lambda_1\dot{\gamma}$ predicted from the Giesekus model, Eqs. (3.25) and (3.26), for different values of the molecular mobility parameter α: (1) 0.2, (2) 0.4, (3) 0.6, and (4) 0.8.

with Figure 3.4 indicates that the Giesekus model for different values of the molecular mobility parameter α predicts a steady decrease of $\psi_1/2\eta_0\lambda_1$ with increasing $\lambda_1\dot{\gamma}$, similar to the predictions of the ZFD and Oldroyd three-constant models. Figure 3.8 gives logarithmic plots of $-\psi_2/2\eta_0\lambda_1$ versus $\lambda_1\dot{\gamma}$ predicted from the Giesekus model for different values of the molecular mobility parameter α. Here ψ_2 is the second normal stress difference coefficient defined by $N_2/\dot{\gamma}^2$ (see Chapter 2) and $\psi_2/2\eta_0\lambda_1$ is a dimensionless variable. It is very interesting to observe in Figure 3.8 that negative values of $\psi_2/2\eta_0\lambda_1$ depend conspicuously on α for small values $\lambda_1\dot{\gamma}$ (say less than about 1) but the dependence of $\psi_2/2\eta_0\lambda_1$ on α becomes progressively less as $\lambda_1\dot{\gamma}$ is increased further to about 30, and finally the dependence of $\psi_2/2\eta_0\lambda_1$ on α disappears completely at large values of $\lambda_1\dot{\gamma}$. This behavior is similar to that predicted by the ZFD model (see Eq. (3.17)) while the Oldroyd three-constant model predicts zero value of N_2 (see Eq. (3.22)). However, there is an important difference in the prediction of N_2 between the Giesekus model and the ZFD model in that the ratio $|N_2/N_1|$ is not equal to 0.5 in the Giesekus model (see Eq. (3.29)), while $|N_2/N_1| = 0.5$ at all shear rates in the ZFD model. The fact that $|N_2/N_1| \neq 0.5$ in the Giesekus model becomes very obvious when Figure 3.7 is compared with Figure 3.8. The predicted shear-rate dependencies of viscosity and normal stress differences from the Giesekus model, given in Figures 3.6 to 3.8, vary with the molecular mobility parameter α, which can be regarded as being an additional parameter that can be used for curve fitting to experimental results. As mentioned previously, the single-mode Giesekus model is a special case of the more general multimode Giesekus constitutive equation. In this regard, the multimode Giesekus model is expected to predict experimental results better than the single-mode Giesekus model, as shown in the monograph of Bird et al. (1987); namely, the eight-mode Giesekus model predicts very well steady-state shear flow of a viscoelastic polymeric liquid when compared with experimental results. Table 3.1 gives a summary of the material functions for steady-state shear flow that were derived from the constitutive equations presented in the preceding sections.

Table 3.1 Expressions for the steady-state shear viscosity and normal stress differences for some representative constitutive equations

Model	$\eta(\dot{\gamma})$	N_1	N_2
Upper convected Maxwell	η_0	$2\eta_0\lambda_1\dot{\gamma}^2$	0
Zaremba–Fromm–DeWitt (ZFD)	$\dfrac{\eta_0}{1+(\lambda_1\dot{\gamma})^2}$	$\dfrac{2\eta_0\lambda_1\dot{\gamma}^2}{1+(\lambda_1\dot{\gamma})^2}$	$-\dfrac{1}{2}N_1$
Oldroyd three-constant	$\dfrac{\eta_0\left(1+\frac{2}{3}\lambda_1\lambda_2\dot{\gamma}^2\right)}{1+\frac{2}{3}(\lambda_1\dot{\gamma})^2}$	$\dfrac{2\eta_0(\lambda_1-\lambda_2)\dot{\gamma}^2}{1+\frac{2}{3}(\lambda_1\dot{\gamma})^2}$	0
Spriggs	$\dfrac{\eta_0}{Z(\alpha)}\sum_{i=1}^{\infty}\dfrac{i^{\alpha}}{i^{2\alpha}+(c\lambda\dot{\gamma})^2}$	$\dfrac{2\eta_0\lambda}{Z(\alpha)}\sum_{i=1}^{\infty}\dfrac{\dot{\gamma}^2}{i^{2\alpha}+(c\lambda\dot{\gamma})^2}$	$\dfrac{\varepsilon}{2}N_1\ (\varepsilon<0)$
Lodge	$\sum_{i=1}^{N}G_i\lambda_i$	$2\sum_{i=1}^{N}G_i\lambda_1{}^2\dot{\gamma}^2$	0
Meister	$\sum_{i=1}^{\infty}\dfrac{G_i\lambda_i}{(1+c\lambda_i\dot{\gamma})^2}$	$\sum_{i=1}^{\infty}\dfrac{2G_i\lambda_i{}^2\dot{\gamma}^2}{(1+c\lambda_i\dot{\gamma})^3}$	0
Bird–Carreau	$\sum_{i=1}^{\infty}\dfrac{\eta_i}{1+(\lambda_{1i}\omega)^2}$	$\sum_{i=1}^{\infty}\dfrac{2\eta_i\lambda_{2i}\dot{\gamma}^2}{1+(\lambda_{1i}\omega)^2}$	$\dfrac{\varepsilon}{2}N_1\ (\varepsilon<0)$
Coleman–Noll second-order	η_0	$-2\nu\dot{\gamma}^2(\nu<0)$	$(\beta+2\nu)\dot{\gamma}^2$
Giesekus*	$\dfrac{\eta_0(1-g)^2}{1+(1-2\alpha)g}$	$\dfrac{(2\eta_0/\lambda_1)g(1-\alpha g)}{\alpha(1-g)}$	$-(\eta_0/\lambda_1)g$

*where: $g=\dfrac{1-f}{1+(1-2\alpha)f}$; $f^2=\dfrac{\sqrt{1+16\alpha(1-\alpha)\lambda_1{}^2\dot{\gamma}^2}-1}{8\alpha(1-\alpha)\lambda_1{}^2\dot{\gamma}^2}$

3.5.2 Material Functions for Oscillatory Shear Flow

Oscillatory shear flow has long been used to characterize the linear viscoelastic properties of polymer solutions and melts. In Chapter 5 we describe the basic principles of such experiments. In this section we present the material functions for small-amplitude oscillatory shear flow using the constitutive equations presented in the preceding section.

When a small-amplitude oscillatory (sinusoidal) shear strain is imposed on a linear viscoelastic fluid, we expect to observe an oscillatory response in shear stress, which can be represented by

$$\gamma_{ij}^* = Re\{\gamma_0 e^{i\omega t}\}; \quad \sigma_{ij}^* = Re\{\sigma_0 e^{i\omega t}\} \tag{3.79}$$

in which γ_{ij}^* denotes complex strain, γ_0 denotes the amplitude of the oscillatory shear strain, σ_{ij}^* denotes complex shear stress, σ_0 denotes the amplitude of the oscillatory shear stress, and ω denotes the applied angular frequency. Here, we assume that the magnitude of γ_{ij}^* is sufficiently small to only give rise to a phase lag in σ_{ij}^* with σ_0 smaller than γ_0; in other words, the system is expected to exhibit linear behavior. Under such a premise, one can derive relationships between the components of σ_{ij}^* and the components of γ_{ij}^* when small-amplitude oscillatory shear flow is imposed on a viscoelastic fluid. In this section we present some representative expressions for the material functions in oscillatory shear flow using the constitutive equations presented in the preceding sections.

Let us consider the upper convected Maxwell model given by Eq. (3.4). Since we are only interested in small-amplitude oscillations with $v_1 = v_1(t, x_2)$, all nonlinear terms appearing in the convected derivative of stress tensor $\boldsymbol{\sigma}$ (see Eq. (2.107)) can be neglected and thus Eq. (3.4) reduces to the classical Maxwell equation, Eq. (3.3). Applying Eq. (3.79) to (3.3) we obtain[12]

$$\sigma_0 \mathrm{e}^{i\omega t} + \lambda_1 i\omega\, \sigma_0 \mathrm{e}^{i\omega t} = \eta_0 i\omega\, \gamma_0 \mathrm{e}^{i\omega t} \tag{3.80}$$

or

$$\sigma_{ij}^*(1 + i\lambda_1\omega) = \eta_0 \dot{\gamma}_{ij}^* \tag{3.81}$$

with $\dot{\gamma}_{ij}^* = i\omega\gamma_0 \mathrm{e}^{i\omega t}$. Equation (3.81) enables us to define a new quantity η^*, termed complex viscosity, by

$$\eta^* = \frac{\sigma^*}{\dot{\gamma}^*} = \frac{\eta_0}{1 + i\lambda_1\omega} = \frac{\eta_0}{1 + \lambda_1{}^2\omega^2} - i\frac{\eta_0\lambda_1\omega}{1 + \lambda_1{}^2\omega^2} \tag{3.82}$$

It is seen that the complex viscosity η^* consists of two components, η' and η'':

$$\eta'(\omega) = \frac{\eta_0}{1 + \lambda_1{}^2\omega^2}; \quad \eta''(\omega) = \frac{\eta_0\lambda_1\omega}{1 + \lambda_1{}^2\omega^2} \tag{3.83}$$

$\eta'(\omega)$ is called the "dynamic viscosity."

For a Hookean material, the complex modulus G^* can be defined by

$$\sigma^* = G^*\gamma^* \tag{3.84}$$

and we obtain, with the aid of Eq. (3.80), the following expression

$$G^* = \frac{\sigma^*}{\gamma^*} = \frac{i\omega\eta_0}{1 + i\lambda_1\omega} = \frac{\eta_0\lambda_1\omega^2}{1 + \lambda_1{}^2\omega^2} + i\frac{\eta_0\omega}{1 + \lambda_1{}^2\omega^2} \tag{3.85}$$

It is seen that the G^* consists of two components, G' and G'':

$$G'(\omega) = \frac{\eta_0\lambda_1\omega^2}{1 + \lambda_1{}^2\omega^2}; \quad G''(\omega) = \frac{\eta_0\omega}{1 + \lambda_1{}^2\omega^2} \tag{3.86}$$

$G'(\omega)$ is called the "dynamic storage modulus" and $G''(\omega)$ is called the "dynamic loss modulus." Comparison of Eq. (3.83) with Eq. (3.86) establishes the following relationships:

$$G'(\omega) = \omega\eta''(\omega); \quad G''(\omega) = \omega\eta'(\omega) \tag{3.87}$$

suggesting that one can obtain expressions for $G'(\omega)$ and $G''(\omega)$ by applying small-amplitude oscillatory shear strain to a viscoelastic constitutive equation. Because of the nature of the linear analysis we have performed on the upper convected Maxwell model, $\eta'(\omega)$, $\eta''(\omega)$, $G'(\omega)$, and $G''(\omega)$ represent linear viscoelastic properties and can be regarded as being the material functions for oscillatory shear flow. Hence, Eqs. (3.83) and (3.86) describe the material functions for the upper convected Maxwell model in oscillatory shear flow.

Small-amplitude oscillatory analysis can readily be applied to any nonlinear constitutive equation. For instance, applying Eq. (3.79) to the Oldroyd three-constant model, Eq. (3.21), we obtain

$$\sigma_0 e^{i\omega t} + \lambda_1 i\omega\sigma_0 e^{i\omega t} = \eta_0(\gamma_0 e^{i\omega t} + i\omega\gamma_0 e^{i\omega t}) \tag{3.88}$$

yielding

$$\eta'(\omega) = \frac{\eta_0(1+\lambda_1\lambda_2\omega^2)}{1+\lambda_1{}^2\omega^2}; \quad \eta''(\omega) = \frac{\eta_0(\lambda_1-\lambda_2)\omega}{1+\lambda_1{}^2\omega^2} \tag{3.89}$$

thus

$$G'(\omega) = \frac{\eta_0(\lambda_1-\lambda_2)\omega^2}{1+\lambda_1{}^2\omega^2}; \quad G''(\omega) = \frac{\eta_0(1+\lambda_1\lambda_2\omega^2)\omega}{1+\lambda_1{}^2\omega^2} \tag{3.90}$$

Table 3.2 gives the expressions for the dynamic storage modulus $G'(\omega)$ and dynamic viscosity $\eta'(\omega)$ for some of the other constitutive equations presented in the preceding sections.

Table 3.2 Expressions for the dynamic storage modulus and dynamic viscosity for some representative constitutive equations

Model	$G'(\omega)$	$\eta'(\omega)$
Upper convected Maxwell	$\dfrac{\eta_0\lambda_1\omega^2}{1+(\lambda_1\omega)^2}$	$\dfrac{\eta_0}{1+(\lambda_1\omega)^2}$
Oldroyd three-constant	$\dfrac{\eta_0(\lambda_1-\lambda_2)\omega^2}{1+(\lambda_1\omega)^2}$	$\dfrac{\eta_0(1+\lambda_1\lambda_2\omega^2)}{1+(\lambda_1\omega)^2}$
Spriggs	$\dfrac{\eta_0\lambda}{Z(\alpha)}\sum_{i=1}^{\infty}\dfrac{\omega^2}{i^{2\alpha}+(\lambda\omega)^2}$	$\dfrac{\eta_0}{Z(\alpha)}\sum_{i=1}^{\infty}\dfrac{i^{\alpha}}{i^{2\alpha}+(\lambda\omega)^2}$
Bird–Carreau	$\sum_{i=1}^{\infty}\dfrac{\eta_i\lambda_{2i}}{1+(\lambda_{1i}\omega)^2}$	$\sum_{i=1}^{\infty}\dfrac{\eta_i}{1+(\lambda_{1i}\omega)^2}$
Meister	$\sum_{i=1}^{\infty}\dfrac{G_i(\lambda_i\omega)^2}{1+(\lambda_i\omega)^2}$	$\sum_{i=1}^{\infty}\dfrac{G_i\lambda_i}{1+(\lambda_i\omega)^2}$

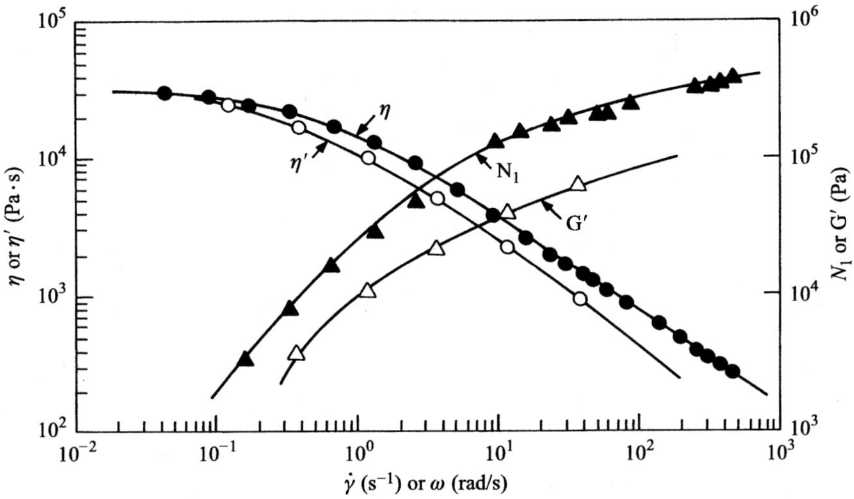

Figure 3.9 Plots of log η' versus log ω (○) and log G' versus log ω (△) in oscillatory shear flow, and plots of log η versus log $\dot{\gamma}$ (●) and log N_1 versus log $\dot{\gamma}$ (▲) in steady-state shear flow for a commercial polystyrene at 200 °C. The data for η and N_1 at low shear rates were obtained using the cone-and-plate fixture of a rotational-type rheometer, the data for η and N_1 at high shear rates were obtained using a continuous-flow capillary rheometer, and the data for η' and G' were obtained using the parallel-plate fixture of a rotational-type rheometer. Refer to Chapter 5 for details of the experimental methods employed to obtain the data. (Reprinted from Han, *Rheology in Polymer Processing*, Chapter 3. Copyright © 1976, with permission from Elsevier.)

Figure 3.9 gives logarithmic plots of η' versus ω and logarithmic plots of G' versus ω for a commercial polystyrene at 200 °C. For comparison, also given in Figure 3.9 are logarithmic plots of η versus $\dot{\gamma}$ and logarithmic plots of N_1 versus $\dot{\gamma}$ for the same polymer. The following observations are worth noting. As the value of ω decreases to a very small value, η' attains a constant value (Newtonian viscosity) agreeing with the zero-shear viscosity (η_0) obtained from steady-state shear flow experiment. Notice in Figure 3.9 that as the value of ω increases, the plot of $\log \eta'$ versus $\log \omega$ drops below the plot of $\log \eta$ versus $\log \dot{\gamma}$, and the plot of $\log G''$ versus $\log \omega$ also lies below the plots of $\log N_1$ versus $\log \dot{\gamma}$, although an increasing trend of G' increasing ω becomes less intense as compared with the increasing trend of N_1 with increasing $\dot{\gamma}$. It is interesting to observe, however, that the frequency dependence of η' is very similar to the shear-rate dependence of η, and the frequency dependence of G' is very similar to the shear-rate dependence of N_1. In the high frequency region, both η' and G' follow more or less a power-law behavior with increasing ω, again very similar to the shear-rate dependencies of η and N_1. Using a phenomenological theory, Coleman and Markovitz (1964) predicted the following relationships between small-amplitude oscillatory shear flow properties at low angular frequencies and steady-state shear flow properties at low shear rates:

$$\lim_{\omega \to 0} \eta'(\omega) = \eta_0; \quad \lim_{\omega \to 0} G'(\omega)/\omega^2 = \lim_{\dot{\gamma} \to 0} N_1(\dot{\gamma})/2\dot{\gamma}^2 \tag{3.91}$$

We present additional experimental data for G' and η' of several other homogeneous molten polymers in Chapters 4 and 6, for block copolymers in Chapter 8, and for liquid-crystalline polymers in Chapter 9.

3.5.3 Material Functions for Steady-State Elongational Flow

We now consider elongational (or extensional) flow behavior predicted from the various constitutive equations presented in the preceding sections. The subject of elongational flow of viscoelastic fluids is of fundamental interest not only to rheologists but also for polymer processing operations, such as fiber spinning and tubular film blowing (see Volume 2, Chapters 6 and 7, respectively). In Chapter 5 we present experimental methods for measuring elongational viscosity.

For uniaxial elongational flow, the elongational viscosity η_E may be defined by the ratio of tensile stress T_{11} and elongation rate $\dot{\varepsilon}$:

$$\eta_E = T_{11}/\dot{\varepsilon} \tag{3.92}$$

Note that Eq. (2.15) defines the rate-of-strain tensor **d** in steady-state uniaxial elongational flow. If the surfaces transverse to the direction of principal elongation (i.e., the direction of stretching) are unconstrained, we have

$$T_{22} = T_{33} = 0 \tag{3.93}$$

From Eq. (3.93) and (2.114) we have

$$p = \sigma_{22} = \sigma_{33} \tag{3.94}$$

Substitution of Eq. (3.94) into Eq. (2.114) yields

$$T_{11} = \sigma_{11} - \sigma_{22} \tag{3.95}$$

Thus, from Eqs. (3.92) and (3.95) we obtain

$$\eta_E = (\sigma_{11} - \sigma_{22})/\dot{\varepsilon} \tag{3.96}$$

Next, we present some representative expressions for η_E based on the constitutive equations presented in the preceding sections.

Let us consider the upper convected Maxwell model given by Eq. (3.4). For steady-state uniaxial elongation flow, for which the rate-of-strain tensor **d** is defined by Eq. (2.15), we have (see Appendix 3E)

$$\left\| \begin{matrix} \sigma^{11} & 0 & 0 \\ 0 & \sigma^{22} & 0 \\ 0 & 0 & \sigma^{33} \end{matrix} \right\| + \lambda_1 \dot{\varepsilon} \left\| \begin{matrix} -2\sigma^{11} & 0 & 0 \\ 0 & \sigma^{22} & 0 \\ 0 & 0 & \sigma^{33} \end{matrix} \right\| = 2\eta_0 \dot{\varepsilon} \left\| \begin{matrix} 1 & 0 & 0 \\ 0 & -1/2 & 0 \\ 0 & 0 & -1/2 \end{matrix} \right\| \tag{3.97}$$

From Eqs. (3.96) and (3.97) we obtain

$$\eta_E = \frac{3\eta_0}{(1 - 2\lambda_1 \dot{\varepsilon})(1 + \lambda_1 \dot{\varepsilon})} \tag{3.98}$$

For equal biaxial elongational flow, with the rate-of-strain tensor **d** defined by Eq. (2.16), we have[13]

$$\begin{Vmatrix} \sigma^{11} & 0 & 0 \\ 0 & \sigma^{22} & 0 \\ 0 & 0 & \sigma^{33} \end{Vmatrix} + \lambda_1 \dot{\varepsilon}_B \begin{Vmatrix} -2\sigma^{11} & 0 & 0 \\ 0 & -2\sigma^{22} & 0 \\ 0 & 0 & 4\sigma^{33} \end{Vmatrix} = 2\eta_0 \dot{\varepsilon}_B \begin{Vmatrix} 1 & 0 & 0 \\ 0 & 1 & 0 \\ 0 & 0 & -2 \end{Vmatrix} \tag{3.99}$$

Note that the equal biaxial elongational viscosity η_B is defined by

$$\eta_B = (\sigma^{11} - \sigma^{33})/\dot{\varepsilon}_B \tag{3.100}$$

From Eqs. (3.99) and (3.100) we obtain

$$\eta_B = \frac{6\eta_0}{(1 - 2\lambda_1 \dot{\varepsilon}_B)(1 + 4\lambda_1 \dot{\varepsilon}_B)} \tag{3.101}$$

We can also obtain the material functions for elongational flow using integral-type constitutive equations presented in the preceding sections. For this, let us derive material functions for elongational flow, η_E and η_B, for the Lodge integral model given by Eq. (3.48). For steady-state uniaxial elongational flow, substitution of the finite strain tensor **E** given by Eq. (2.74) and the memory function given by Eq. (3.49) into Eq. (3.48), we obtain (see Appendix 3F)

$$\eta_E = \sum_{i=1}^{N} \frac{3\eta_0}{(1 - 2\lambda_i \dot{\varepsilon})(1 + \lambda_i \dot{\varepsilon})} \tag{3.102}$$

For steady-state equal biaxial elongational flow we have (see Appendix 3F)

$$\eta_B = \sum_{i=1}^{N} \frac{6\eta_0}{(1 - 2\lambda_i \dot{\varepsilon}_B)(1 + 4\lambda_i \dot{\varepsilon}_B)} \tag{3.103}$$

In steady-state uniaxial elongation flow, for which the rate-of-strain tensor **d** is defined by Eq. (2.15), the Giesekus model given by Eq. (3.27) gives

$$\alpha \begin{Vmatrix} {c_{11}}^2 & 0 & 0 \\ 0 & {c_{22}}^2 & 0 \\ 0 & 0 & {c_{33}}^2 \end{Vmatrix} + (1-2\alpha) \begin{Vmatrix} c_{11} & 0 & 0 \\ 0 & c_{22} & 0 \\ 0 & 0 & c_{33} \end{Vmatrix} - (1-\alpha) \begin{Vmatrix} 1 & 0 & 0 \\ 0 & 1 & 0 \\ 0 & 0 & 1 \end{Vmatrix}$$
$$= \lambda_1 \dot{\varepsilon} \begin{Vmatrix} 2c_{11} & 0 & 0 \\ 0 & -c_{22} & 0 \\ 0 & 0 & -c_{33} \end{Vmatrix} \tag{3.104}$$

Using Eq. (3.104) in Eq. (3.25), with the aid of Eq. (3.96), we obtain

$$\eta_E = \frac{\eta_0}{2\alpha}\left\{3 + \frac{1}{\lambda_1\dot{\varepsilon}}\left[\sqrt{1-4(1-2\alpha)\lambda_1\dot{\varepsilon}+4{\lambda_1}^2\dot{\varepsilon}^2} - \sqrt{1+2(1-2\alpha)\lambda_1\dot{\varepsilon}+{\lambda_1}^2\dot{\varepsilon}^2}\right]\right\} \tag{3.105}$$

Table 3.3 gives the expressions for η_E and η_B for some of the constitutive equations presented in the preceding sections.

Figure 3.10 gives experimental observations of the strain-rate dependence of uniaxial elongational viscosity for a low-density polyethylene at 150 °C over several decades of $\dot{\varepsilon}$. It is interesting to observe in Figure 3.10 that as $\dot{\varepsilon}$ increases, η_E first increases (strain hardening), goes through a maximum, and then decreases.

Figure 3.11 gives plots of $\eta_E/3\eta_0$ versus $\lambda_1\dot{\varepsilon}$ that are predicted from two constitutive equations: (1) the upper convected Maxwell model, and (2) the Oldroyd three-constant model. It is seen in Figure 3.11 that both models predict values of η_E increasing very rapidly without bound as $\dot{\varepsilon}$ increases, in contrast to the experimental results given in Figure 3.10. As a matter of fact, all the expressions summarized in Table 3.3 predict similar elongational behavior, which is considered to be physically unrealistic.

Table 3.3 Expressions for the uniaxial and equal biaxial elongational viscosities for some representative constitutive equations

Model	η_E	η_B
Upper convected Maxwell	$\dfrac{3\eta_0}{(1-2\lambda_1\dot{\varepsilon})(1+\lambda_1\dot{\varepsilon})}$	$\dfrac{6\eta_0}{(1-2\lambda_1\dot{\varepsilon}_B)(1+4\lambda_1\dot{\varepsilon}_B)}$
Zaremba–Fromm–DeWitt (ZFD)	$3\eta_0$	$6\eta_0$
Oldroyd three–constant	$\dfrac{3\eta_0(1-\lambda_2\dot{\varepsilon})}{1-\lambda_1\dot{\varepsilon}}$	$\dfrac{6\eta_0(1+2\lambda_2\dot{\varepsilon}_B)}{1+2\lambda_1\dot{\varepsilon}_B}$
Spriggs	$\sum_{i=1}^{\infty}\dfrac{3\eta_i}{1-(1+\dot{\varepsilon})\lambda_i\dot{\varepsilon}}$	$\sum_{i=1}^{\infty}\dfrac{6\eta_i}{1+2(1+\dot{\varepsilon}_B)\lambda_i\dot{\varepsilon}_B}$
Lodge	$\sum_{i=1}^{\infty}\dfrac{3\eta_0}{(1-2\lambda_i\dot{\varepsilon})(1+\lambda_i\dot{\varepsilon})}$	$\sum_{i=1}^{\infty}\dfrac{6\eta_0}{(1-2\lambda_i\dot{\varepsilon}_B)(1+4\lambda_i\dot{\varepsilon}_B)}$
Coleman–Noll second-order	$3\eta_0[1+((\beta+\nu)/\eta_0)\dot{\varepsilon}]$	$6\eta_0[1-2((\beta+\nu)/\eta_0)\dot{\varepsilon}_B]$

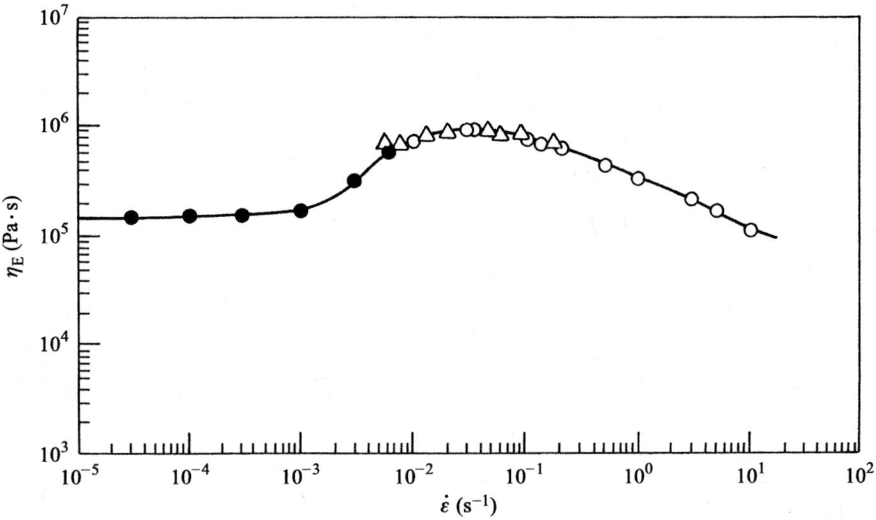

Figure 3.10 Plots of $\log \eta_E$ versus $\log \dot{\varepsilon}$ for a low-density polyethylene at 150 °C: (○) the averaged value of several measurements carried out at constant elongation rate, (●) constant elongation rate experimental data with corrections for the influence of the interfacial tension, and (△) with tensile creep measurements. Refer to Chapter 5 for details of the experimental methods to obtain steady-state elongational flow data. (Reprinted from Laun and Münstedt, *Rheologica Acta* 17:415. Copyright © 1978, with permission from Springer.)

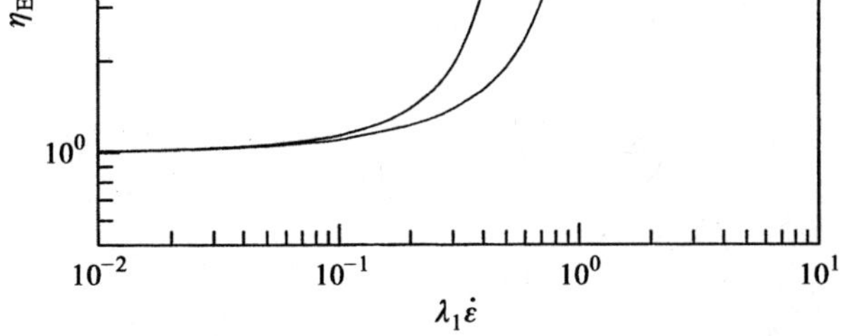

Figure 3.11 Plots of $\log \eta_E/3\eta_0$ versus $\log \lambda_1 \dot{\varepsilon}$ predicted from (1) the upper convected Maxwell model, Eq. (3.4), and (2) the Oldroyd three-constant model, Eq. (3.21), with $\lambda_1 = 1.0$ s and $\lambda_2 = 0.1$ s.

However, as can be seen in Figure 3.12, the Giesekus model predicts strain hardening for small values of the molecular mobility parameter $\alpha (< 0.6)$ and strain softening for larger values of $\alpha (> 0.6)$. Notice in Figure 3.12 that the elongational viscosity does not increase without a bound as the strain rate increases, which seems more reasonable from a physical point of view.

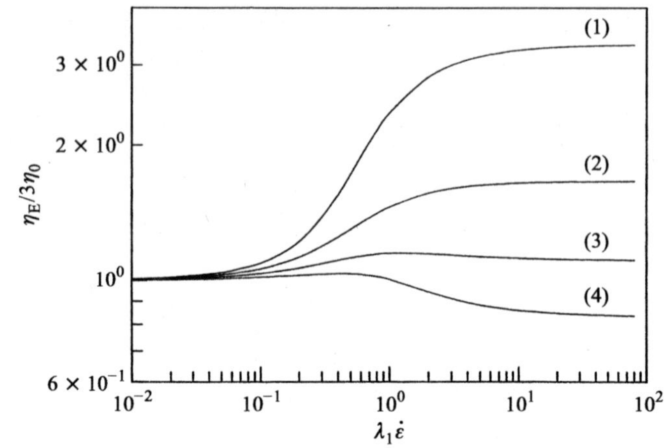

Figure 3.12 Plots of log $\eta_E/3\eta_0$ versus log $\lambda_1\dot{\varepsilon}$ predicted from the Giesekus model, Eqs. (3.25) and (3.26), for different values of the molecular mobility parameter α: (1) 0.2, (2) 0.4, (3) 0.6, and (4) 0.8.

3.6 Summary

The primary purposes of this chapter were first to introduce some representative phenomenological constitutive equations used to describe the viscoelasticity of flexible homogeneous polymeric liquids, and then to show how such constitutive equations may be used to describe relatively simple flow problems (e.g., steady-state shear flow and steady-state elongational flow). For such purposes, in this chapter we have presented only relatively simple constitutive equations. Owing to the space limitations here we have not presented other more complicated constitutive equations, which have been dealt with in the monographs by Bird et al. (1987) and Larson (1988).

There are complicated flow problems of practical interest, such as unsteady-state shear flow and unsteady-state nonviscometric flow (e.g., elongational flow). Unsteady-state shear flow problems include (1) stress buildup after start-up of shear flow, (2) decay of stress after sudden stoppage of steady-state shear flow, and (3) finite-strain relaxation (recoil) following steady-state shear flow. Again, owing to the space limitations here, we have not discussed such flows in this chapter. Since this book covers a very wide range of topics, in this chapter we are not able to discuss such complex flow problems, which have been dealt with in the monograph by Bird et al. (1987). In Volume 2 we deal with some complicated rheological problems associated with a variety of polymer processing operations.

As mentioned in the introduction to this chapter, the phenomenological constitutive equations are needed to solve the equations of motion together with, if necessary, the energy balance equation, in order to predict the stress, velocity, and temperature distributions for a given flow geometry. However, more often than not, solutions of such system equations require numerical computations, which sometimes encounter numerical instability with certain constitutive equations. Often, the numerical instability encountered is inherent with the unstable nature of the particular constitutive equation chosen. In this regard, a judicious choice of constitutive equation is vitally important for successful numerical solutions of system equations.

It is well documented in the literature that many phenomenological constitutive equations describe steady-state viscometric flow reasonably well, but nonviscometric flow rather poorly. Specifically, in this chapter we have shown that successful prediction of steady-state elongational viscosity has not been realized using several constitutive equations. In this regard, more effort is needed to develop constitutive equations that may be useful to adequately describe elongational flow.

Phenomenological constitutive equations do not contain explicit molecular and structural parameters of a flexible homogeneous polymer. For instance, they cannot predict how the viscoelastic properties of polymer may be affected by increases or decreases in molecular weight or the presence of long-chain branching. For such purposes, we need molecular viscoelasticity theory. We address this issue in the next chapter.

It should be mentioned that flexible, homogeneous polymers are a small fraction of all polymeric materials used in the polymer industry. More often than not, inhomogeneous multicomponent and/or multiphase polymer systems are processed in the polymer industry. For example, block copolymers form microdomains with an average size of 10–50 nm and therefore they must be treated as heterogeneous polymer systems. Liquid-crystalline polymers having rodlike rigid chains, forming a mesophase in the molten state, can no longer be treated as being flexible homopolymers. Under certain thermal and flow conditions, block copolymers and liquid-crystalline polymers undergo self-assembly. Thermoplastic polyurethanes are also a heterogeneous polymer system. Often two immiscible polymers, without or with a third component as a compatibilizing agent, are melt blended, forming a heterogeneous polymer system. Sometimes, particulates are added to a flexible homopolymer, giving rise to a heterogeneous polymer system. For each of the polymer systems mentioned, the continuum theory for the viscoelasticity of flexible homogeneous polymeric liquids presented in this chapter is of very limited use. Thus different types of constitutive equation are needed to describe the rheological behavior of heterogeneous, semiflexible, multicomponent, or multiphase polymer systems. In Chapters 7 through 12 we describe the rheological behavior of such complex polymer systems and summarize, where applicable, currently held constitutive equations for such polymer systems.

Appendix 3A: Derivation of Equation (3.5)

For steady-state shear flow, the contravariant convected derivative given by Eq. (2.107) reduces to

$$\frac{\mathfrak{d}\sigma^{ij}}{\mathfrak{d}t} = -\frac{\partial v^i}{\partial x^k}\sigma^{kj} - \frac{\partial v^j}{\partial x^k}\sigma^{ik} \qquad (i, j = 1, 2, 3) \tag{3A.1}$$

from which we have

$$\frac{\mathfrak{d}\sigma^{11}}{\mathfrak{d}t} = -\frac{\partial v^1}{\partial x^2}\sigma^{21} - \frac{\partial v^1}{\partial x^2}\sigma^{12} = -2\dot{\gamma}\sigma^{12} \quad \text{where } \sigma^{21} = \sigma^{12} \tag{3A.2}$$

$$\frac{\mathfrak{d}\sigma^{12}}{\mathfrak{d}t} = -\frac{\partial v^1}{\partial x^2}\sigma^{22} - \frac{\partial v^2}{\partial x^2}\sigma^{12} = -\dot{\gamma}\sigma^{22} \tag{3A.3}$$

$$\frac{\mathfrak{d}\sigma^{13}}{\mathfrak{d}t} = -\frac{\partial v^1}{\partial x^2}\sigma^{23} - \frac{\partial v^3}{\partial x^2}\sigma^{12} = -\dot{\gamma}\sigma^{23} \tag{3A.4}$$

$$\frac{\mathfrak{d}\sigma^{21}}{\mathfrak{d}t} = -\frac{\partial v^2}{\partial x^2}\sigma^{21} - \frac{\partial v^1}{\partial x^2}\sigma^{22} = -\dot{\gamma}\sigma^{22} \tag{3A.5}$$

$$\frac{\mathfrak{d}\sigma^{31}}{\mathfrak{d}t} = -\frac{\partial v^3}{\partial x^2}\sigma^{23} - \frac{\partial v^1}{\partial x^2}\sigma^{32} = -\dot{\gamma}\sigma^{32} \tag{3A.6}$$

$$\frac{\mathfrak{d}\sigma^{22}}{\mathfrak{d}t} = \frac{\mathfrak{d}\sigma^{23}}{\mathfrak{d}t} = \frac{\mathfrak{d}\sigma^{32}}{\mathfrak{d}t} = \frac{\mathfrak{d}\sigma^{33}}{\mathfrak{d}t} = 0 \tag{3A.7}$$

Thus we have

$$\frac{\mathfrak{d}\sigma^{ij}}{\mathfrak{d}t} = -\dot{\gamma}\begin{Vmatrix} 2\sigma^{12} & \sigma^{22} & \sigma^{23} \\ \sigma^{22} & 0 & 0 \\ \sigma^{32} & 0 & 0 \end{Vmatrix} \tag{3A.8}$$

Substitution of Eq. (3A.8) into (3.4) yields (3.5).

Appendix 3B: Derivation of Equation (3.16)

For steady-state shear flow, the Jaumann derivative given by Eq. (2.111) reduces to

$$\frac{\mathcal{D}\sigma_{ij}}{\mathcal{D}t} = -\omega_{ik}\sigma_{jk} - \omega_{jk}\sigma_{ik} \qquad (i, j = 1, 2) \tag{3B.1}$$

for which the vorticity tensor $\boldsymbol{\omega}$ is defined by Eq. (2.61). Note that the Jaumann derivative gives rise to the same results, regardless of whether the contravariant or covariant components of the tensor $\boldsymbol{\sigma}$ are used. From Eq. (3B.1) we have

$$\frac{\mathcal{D}\sigma_{11}}{\mathcal{D}t} = \omega_{12}\sigma_{21} - \omega_{12}\sigma_{21} = -\dot{\gamma}\sigma_{12} \quad \text{where } \sigma_{21} = \sigma_{12} \tag{3B.2}$$

$$\frac{\mathcal{D}\sigma_{12}}{\mathcal{D}t} = \omega_{21}\sigma_{11} + \omega_{12}\sigma_{22} = \frac{\dot{\gamma}}{2}(\sigma_{11} - \sigma_{22}) \tag{3B.3}$$

$$\frac{\mathcal{D}\sigma_{21}}{\mathcal{D}t} = \omega_{21}\sigma_{11} + \omega_{12}\sigma_{22} = \frac{\dot{\gamma}}{2}(\sigma_{11} - \sigma_{22}) \tag{3B.4}$$

$$\frac{\mathcal{D}\sigma_{22}}{\mathcal{D}t} = \omega_{21}\sigma_{12} + \omega_{21}\sigma_{12} = \dot{\gamma}\sigma_{12} \tag{3B.5}$$

$$\frac{\mathcal{D}\sigma_{13}}{\mathcal{D}t} = \frac{\mathcal{D}\sigma_{23}}{\mathcal{D}t} = \frac{\mathcal{D}\sigma_{31}}{\mathcal{D}t} = \frac{\mathcal{D}\sigma_{32}}{\mathcal{D}t} = \frac{\mathcal{D}\sigma_{33}}{\mathcal{D}t} = 0 \tag{3B.6}$$

Thus we have

$$\frac{\mathcal{D}\sigma_{ij}}{\mathcal{D}t} = \dot{\gamma} \begin{Vmatrix} -\sigma_{12} & (\sigma_{11}-\sigma_{22})/2 & 0 \\ (\sigma_{11}-\sigma_{22})/2 & \sigma_{12} & 0 \\ 0 & 0 & 0 \end{Vmatrix} \tag{3B.7}$$

Substitution of Eq. (3B.7) into (3.15) yields (3.16).

Appendix 3C: Derivation of Equation (3.29)

The derivation presented here follows closely that given by Giesekus (1982). From Eq. (3.28) we have

$$\alpha({c_{11}}^2 + {c_{12}}^2) + (1-2\alpha)c_{11} - (1-\alpha) - 2\lambda_1\dot{\gamma}c_{22} = 0 \tag{3C.1}$$

$$\alpha c_{12}(c_{11} + c_{22}) + (1-2\alpha)c_{12} - \lambda_1\dot{\gamma}c_{22} = 0 \tag{3C.2}$$

$$\alpha({c_{12}}^2 + {c_{22}}^2) + (1-2\alpha)c_{22} - (1-\alpha) = 0 \tag{3C.3}$$

$$\alpha {c_{33}}^2 + (1-2\alpha)c_{33} - (1-\alpha) = 0 \tag{3C.4}$$

Equation (3C.4) can be rewritten as

$$c_{33}\left[\alpha c_{33} + (1-2\alpha)\right] = 1-\alpha \tag{3C.5}$$

indicating that $c_{33} = 1$ is the only admissible solution. Note that from Eq. (3.25) we have

$$\sigma_{12} = (\eta_0/\lambda_1)c_{12} \tag{3C.6}$$

$$\sigma_{11} - \sigma_{22} = (\eta_0/\lambda_1)(c_{11} - c_{22}) \tag{3C.7}$$

$$\sigma_{22} - \sigma_{33} = (\eta_0/\lambda_1)(c_{22} - c_{33}) \tag{3C.8}$$

After long algebraic manipulations of Eqs. (3C.1)–(3C.5) we obtain

$$c_{12} = \frac{\lambda_1\dot{\gamma}(1-g)^2}{1+(1-2\alpha)g}; \quad c_{11} - c_{22} = \frac{2g(1-\alpha g)}{\alpha(1-g)}; \quad c_{22} - c_{33} = -g \tag{3C.9}$$

where g is given by Eq. (3.30). Thus, substitution of Eq. (3C.9) into Eqs. (3C.6)–(3C.8) gives Eq. (3.29).

Appendix 3D: Cayley–Hamilton Theorem

The Cayley–Hamilton theorem states that if the characteristic equation of a symmetric second-order tensor $\mathbf{A}$ is $C(\lambda) = 0$, then the tensor $\mathbf{A}$ satisfies the equation $C(\mathbf{A}) = 0$

(Hildebrand 1952). In other words, polynomial functions of $\mathbf{A}^n (n > 3)$ are defined as the linear combinations of $\mathbf{I}$, $\mathbf{A}$, and $\mathbf{A}^2$, with coefficients depending only on the invariants; that is

$$\mathbf{A}^3 - I_1\mathbf{A}^2 + I_2\mathbf{A} - I_3\mathbf{I} = 0 \tag{3D.1}$$

where I_1, I_2, and I_3 are the invariants of $\mathbf{A}$ (see Appendix 2A in Chapter 2).

The Cayley–Hamilton theorem has been widely used in the formulation of constitutive equations (Rivlin and Ericksen 1955). As an example, let the stress tensor $\boldsymbol{\sigma}$ be expressed as a function of powers of the rate-of-deformation tensor $\mathbf{d}$:

$$\boldsymbol{\sigma} = f\left(\mathbf{d}, \mathbf{d}^2, \mathbf{d}^3, \mathbf{d}^4 \ldots\right) \tag{3D.2}$$

A polynomial expansion of the right-hand side of Eq. (3D.2) may be written as

$$\boldsymbol{\sigma} = \alpha_0\mathbf{I} + \alpha_1\mathbf{d} + \alpha_2\mathbf{d}^2 + \alpha_3\mathbf{d}^3 + \cdots + \alpha_n\mathbf{d}^n \tag{3D.3}$$

where $\alpha_0, \alpha_1, \ldots, \alpha_n$ are constants. Since $\mathbf{d}$ is a symmetric tensor, use of Eq. (3D.1) yields

$$\mathbf{d}^3 = I_1\mathbf{d}^2 - I_2\mathbf{d} + I_3\mathbf{I} \tag{3D.4}$$

and

$$\mathbf{d}^4 = I_1\mathbf{d}^3 - I_2\mathbf{d}^2 + I_3\mathbf{d} = \left(I_1 - I_2\right)\mathbf{d}^2 + \left(I_3 - I_1 I_2\right)\mathbf{d} + I_1 I_3\mathbf{I} \tag{3D.5}$$

It can be concluded then that $\mathbf{d}^n (n > 3)$ can be represented in terms of $\mathbf{I}$, $\mathbf{d}$, and $\mathbf{d}^2$, with coefficients depending only on the invariants (I_1, I_2, I_3). Therefore, Eq. (3D.3) may be written as

$$\boldsymbol{\sigma} = a_0\left(I_1, I_2, I_3\right)\mathbf{I} + a_1\left(I_1, I_2, I_3\right)\mathbf{d} + a_2\left(I_1, I_2, I_3\right)\mathbf{d}^2 \tag{3D.6}$$

in which a_0, a_1, and a_2 are polynomials in the three invariants.

Appendix 3E: Derivation of Equation (3.97)

In steady-state elongational flow, the upper convected derivative defined by Eq. (2.107) reduces to

$$\frac{\mathfrak{d}\sigma^{ij}}{\mathfrak{d}t} = -\frac{\partial v^i}{\partial x^k}\sigma^{kj} - \frac{\partial v^j}{\partial x^k}\sigma^{ik} \tag{3E.1}$$

with

$$\frac{\partial v^i}{\partial x^k}\sigma^{kj} = \frac{\partial v^i}{\partial x^1}\sigma^{1j} + \frac{\partial v^i}{\partial x^2}\sigma^{2j} + \frac{\partial v^i}{\partial x^3}\sigma^{3j} \tag{3E.2}$$

$$\frac{\partial v^j}{\partial x^k}\sigma^{ik} = \frac{\partial v^j}{\partial x^1}\sigma^{i1} + \frac{\partial v^j}{\partial x^2}\sigma^{i2} + \frac{\partial v^j}{\partial x^3}\sigma^{i3} \tag{3E.3}$$

Using the rate-of-strain tensor **d** defined by Eq. (2.15) in (3E.1) with the aid of Eqs. (3E.2) and (3E.3) we obtain

$$\frac{\mathfrak{d}\sigma^{ij}}{\mathfrak{d}t} = \dot{\varepsilon}\begin{Vmatrix} -2\sigma^{11} & 0 & 0 \\ 0 & \sigma^{22} & 0 \\ 0 & 0 & \sigma^{33} \end{Vmatrix} \tag{3E.4}$$

Hence, substitution of Eq. (3E.4) into (3.4) gives (3.97).

Appendix 3F: Derivation of Equation (3.103)

Using the finite strain tensor **E** defined by Eq. (2.74) and the memory function $m(t-t')$ given by Eq. (3.49) in (3.48) we obtain

$$\sigma^{11}(t) = \int_{-\infty}^{t}\sum_{i=1}^{N}\frac{G_i}{\lambda_i}e^{-(t-t')/\lambda_i}\left[e^{-2\dot{\varepsilon}(t'-t)} - 1\right]dt' = \sum_{i=1}^{N}\frac{2G_i\lambda_i\dot{\varepsilon}}{1-2\lambda_i\dot{\varepsilon}} \tag{3F.1}$$

$$\sigma^{22}(t) = \int_{-\infty}^{t}\sum_{i=1}^{N}\frac{G_i}{\lambda_i}e^{-(t-t')/\lambda_i}\left[e^{\dot{\varepsilon}(t'-t)} - 1\right]dt' = -\sum_{i=1}^{N}\frac{G_i\lambda_i\dot{\varepsilon}}{1+\lambda_i\dot{\varepsilon}} \tag{3F.2}$$

Thus, the use of Eqs. (3F.1) and (3F.2) in (3.96) gives (3.102), in which use was made of $G_i = \eta_0/\lambda_i$. In equal biaxial elongational flow, using the finite strain tensor **E** defined by Eq. (2.75) and the memory function $m(t-t')$ given by Eq. (3.49) in (3.48), we obtain

$$\sigma^{11}(t) = \int_{-\infty}^{t}\sum_{i=1}^{N}\frac{G_i}{\lambda_i}e^{-(t-t')/\lambda_i}\left[e^{-2\dot{\varepsilon}_B(t'-t)} - 1\right]dt' = \sum_{i=1}^{N}\frac{2G_i\lambda_i\dot{\varepsilon}_B}{1-2\lambda_i\dot{\varepsilon}_B} \tag{3F.3}$$

$$\sigma^{33}(t) = \int_{-\infty}^{t}\sum_{i=1}^{N}\frac{G_i}{\lambda_i}e^{-(t-t')/\lambda_i}\left[e^{4\dot{\varepsilon}_B(t'-t)} - 1\right]dt' = -\sum_{i=1}^{N}\frac{4G_i\lambda_i\dot{\varepsilon}_B}{1+4\lambda_i\dot{\varepsilon}_B} \tag{3F.4}$$

Thus, the use of Eqs. (3F.3) and (3F.4) in (3.100) gives (3.103).

Problems

Problem 3.1
Verify Eq. (3.7).

Problem 3.2
Verify Eqs. (3.10) and (3.11).

Problem 3.3
Verify Eq. (3.19).

Problem 3.4
Verify Eq. (3.22).

Problem 3.5
Verify Eq. (3.33).

Problem 3.6
Verify Eq. (3.40).

Problem 3.7
Verify Eq. (3.50).

Problem 3.8
Verify Eq. (3.53).

Problem 3.9
Verify Eq. (3.57).

Problem 3.10
Verify Eq. (3.70).

Problem 3.11
The following form of a constitutive equation is referred to as the Reiner–Rivlin fluid:

$$\sigma_{ij} = \eta_1 d_{ij} + \eta_2 d_{ik} d_{kj} \quad (i, j = 1, 2, 3) \tag{3P.1}$$

where η_1 and η_2 are a generalized viscosity and cross-viscosity and can be arbitrary scalar functions of the three principal invariants of the rate-of-deformation tensor **d**. Assuming that a fluid described by Eq. (3P.1) is subjected to steady-state simple shear flow, for which the velocity field is given by $v_1 = f(x_2)$ and $v_2 = v_3 = 0$, where x_1 is the flow direction, x_2 is the shear direction, and x_3 is the remaining direction, obtain expressions that describe the velocity profiles of the fluid.

Problem 3.12
Consider the following form of a constitutive equation:

$$\sigma_{ij} = f\left(\mathrm{d}_{ij}, \frac{\mathcal{D}\mathrm{d}_{mn}}{\mathcal{D}t}, \frac{\mathcal{D}^2\mathrm{d}_{pq}}{\mathcal{D}t^2}, \ldots, \frac{\mathcal{D}^n\mathrm{d}_{rs}}{\mathcal{D}t^n}\right) \tag{3P.2}$$

For slow flows in which d_{ij} and its derivatives are assumed to be small, successive approximations of Eq. (3P.2) give

$$\sigma_{ij} = \alpha_1 d_{ij} + \alpha_2 \frac{\mathscr{D} d_{ij}}{\mathscr{D} t} + \alpha_3 d_{ik} d_{kj} \tag{3P.3}$$

Assuming that a fluid described by Eq. (3P.3) is subjected to steady-state simple shear flow, for which the velocity field is given by $v_1 = f(x_2)$ and $v_2 = v_3 = 0$, where x_1 is the flow direction, x_2 is the shear direction, and x_3 is the remaining direction, obtain expressions that describe the velocity profiles of such a fluid. Interpret, physically, the difference between the two constitutive equations, Eqs. (3P.1) and (3P.3).

Problem 3.13

Consider the following linear expression:

$$\left(1+\lambda_1 \frac{\partial}{\partial t}+\lambda_2 \frac{\partial^2}{\partial t^2}+\cdots+\lambda_n \frac{\partial^n}{\partial t^n}\right)\boldsymbol{\sigma}=2\eta_0\left(1+\mu_1 \frac{\partial}{\partial t}+\mu_2 \frac{\partial^2}{\partial t^2}+\cdots+\mu_m \frac{\partial^m}{\partial t^m}\right)\mathbf{d} \tag{3P.4}$$

where $n = m$ or $m + 1$, and assume that a fluid described by Eq. (3P.4) is subjected to steady-state simple shear flow, for which the velocity field is given by $v_1 = f(x_2)$ and $v_2 = v_3 = 0$, where x_1 is the flow direction, x_2 is the shear direction, and x_3 is the remaining direction.

(a) Show that if the partial derivatives $\partial/\partial t, \partial^2/\partial t^2, \ldots, \partial^n/\partial t^n$, and $\partial^m/\partial t^m$ appearing in Eq. (3P.4) are replaced by the corresponding covariant convected derivatives $\mathfrak{d}/\mathfrak{d}t, \mathfrak{d}^2/\mathfrak{d}t^2, \ldots, \mathfrak{d}^n/\mathfrak{d}t^n$, and $\mathfrak{d}^m/\mathfrak{d}t^m$, Eq. (3P.4) reduces to

$$\boldsymbol{\sigma} + \lambda_1 \frac{\mathfrak{d}\boldsymbol{\sigma}}{\mathfrak{d}t} + \lambda_2 \frac{\mathfrak{d}^2\boldsymbol{\sigma}}{\mathfrak{d}t^2} = 2\eta_0\left(\mathbf{d} + \mu_1 \frac{\mathfrak{d}\mathbf{d}}{\mathfrak{d}t}\right) \tag{3P.5}$$

and derive three material functions for steady-state simple shear flow from Eq. (3P.5).

(b) Obtain the resulting expression by replacing the partial derivatives appearing in Eq. (3P.4) with the corresponding contravariant convected derivatives when the fluid is subjected to steady-state simple shear flow and then obtain three material functions.

Problem 3.14

In modifying the second-order fluid model to be able to predict shear-rate dependent viscosity, let us assume that the double-integral term in Eq. (3.74) is negligibly small compared with the single-integral term and that the memory function is given by

$$m(s, I_2) = \sum_{i=1}^{\infty} \frac{G_i \exp(-s/\lambda_i)}{\lambda_i} \tag{3P.6}$$

with

$$1/\lambda_i = 1/\alpha_i + c\sqrt{I_2} \tag{3P.7}$$

where c, G_i, and α_i are material constants, and I_2 is the second invariant of the rate-of-strain tensor. We then have the following form of constitutive equation:

$$\sigma_{ij} = \int_0^\infty m(s, I_2) J_{ij}(s)\, ds \tag{3P.8}$$

in which $J_{ij}(s)$ is the strain history tensor, which can be written for smooth flow as a series in the first and higher derivatives of the Rivlin–Ericksen tensor:

$$J_{ij}(s) = -A_{ij}^{(1)} s + \tfrac{1}{2} A_{ij}^{(2)} s^2 + \ldots \tag{3P.9}$$

It should be remembered that $A_{ij}^{(n)} = 0$ for $n > 2$ for simple shear flow.

(a) Derive a constitutive equation in terms of $A_{ij}^{(1)}$, $A_{ij}^{(2)}$, $c, \alpha_1, \alpha_2, G_1, G_2$, and the second invariant I_2 of the rate-of-strain tensor. Consider only the first two terms ($i = 1, 2$) of the memory function given by Eq. (3P.6).

(b) Derive a constitutive equation, as in part (a), using the memory function independent of the second invariant I_2 (i.e., $1/\lambda_i \cong 1/\alpha_i$ from Eq. (3P.7)) and then obtain three material functions for steady-state simple shear flow.

Notes

1. It turns out that η_0 is the viscosity of the Newtonian fluid, often referred to as the "zero-shear viscosity," the reason for which will become clear when we present the material functions derived from various constitutive equations for viscoelastic fluids.
2. The mathematical exposition of this subject is discussed by Oldroyd (1984).
3. Zaremba (1903) first reported Eq. (3.15). Zaremba's work was unnoticed by later researchers, including Fromm (1947) and DeWitt (1955) who essentially repeated the original analysis of Zaremba.
4. Giesekus (1982) summarized nicely a series of his papers on the formulation of a new class of constitutive equations. The origin of Eq. (3.23) comes from a modification of the upper convected Maxwell model as applied to a dilute polymer solution, namely

$$\lambda_1 \frac{\mathfrak{D}\boldsymbol{\sigma}}{\mathfrak{D}t} + \mathbf{B} \cdot (\boldsymbol{\sigma} - G\boldsymbol{\delta}) = \mathbf{0} \tag{3N.1}$$

where $\mathbf{B}$ is an anisotropic tensor describing the anisotropy of the orientation of surrounding molecules and thus $\mathbf{B}$ depends on the state of stresses in the material. Giesekus assumed that

$$\mathbf{B} - \boldsymbol{\delta} = \alpha\left(\frac{\boldsymbol{\sigma}}{G} - \boldsymbol{\delta}\right) = \frac{\alpha}{G}(\boldsymbol{\sigma} - G\boldsymbol{\delta}) = \frac{\alpha}{G}\boldsymbol{\tau} \tag{3N.2}$$

where $\boldsymbol{\tau}$ is defined by $\boldsymbol{\tau} = \boldsymbol{\sigma} - G\boldsymbol{\delta}$. Substitution of Eq. (3N.2) into Eq. (3N.1) yields Eq. (3.23). In so doing, use was made of $\mathfrak{d}\boldsymbol{\delta}/\mathfrak{d}t = -2\mathbf{d} = -(\mathbf{L} + \mathbf{L}^{\mathrm{T}})$ with $\mathbf{L} = \nabla\mathbf{v}$ being the velocity gradient tensor (see Chapter 2).

5. In obtaining the components of the conformation tensor $\mathbf{c}$ from Eq. (3.27), we used the following definition of the velocity gradient tensor $\mathrm{L} = \nabla\mathbf{v}$,

$$\|\mathbf{L}\| = \begin{Vmatrix} 0 & 0 & 0 \\ \dot{\gamma} & 0 & 0 \\ 0 & 0 & 0 \end{Vmatrix} \tag{3N.3}$$

   Then, from Eq. (3.27) we obtain Eq. (3.28).

6. For $\alpha = 1$, Eq. (3.27) reduces to

$$\mathbf{c}^2 - \mathbf{c} - \lambda_1\Big[(\nabla\mathbf{v})^{\mathrm{T}} \cdot \mathbf{c} + \mathbf{c} \cdot (\nabla\mathbf{v})\Big] = \mathbf{0} \tag{3N.4}$$

   The derivation of Eq. (3.33) from (3.32) follows closely that shown above for $\alpha \neq 1$.

7. $\mathbf{E}(t', t)$, defined by Eq. (2.49), was used to integrate Eq. (3.48) by changing the variable of integration from t' to $s = t - t'$; thus $-\infty < t' \leq t$ becomes $\infty < s \leq 0$.

8. $\mathbf{E}(t', t)$, defined by Eq. (2.49), was used to integrate Eq. (3.51). In so doing, use was made of $\left| I_2 \right| = \frac{1}{2}\dot{\gamma}$ in steady-state simple shear flow.

9. $\mathbf{C}_t^{-1}(t', t)$ defined by Eq. (2.47) and $\mathbf{C}_t(t', t)$ defined by Eq. (2.46) were used to integrate Eq. (3.54). In so doing, use was made of $\left| I_2 \right| = \frac{1}{2}\dot{\gamma}$ in steady-state simple shear flow. According to Bird and Carreau (1968), time constants $\lambda_{1i} (i = 1, 2, \ldots)$ were introduced to describe the rate of creation of network junctions, and time constants $\lambda_{2i} (i = 1, 2, \ldots)$ were introduced to describe the rate of loss of network junctions.

10. The term "rate-type" is used for the reason that the Rivlin–Ericksen tensors $\mathbf{A}_{(n)}$ for $n \geq 2$ contain the time derivative term, although such classification may not be justified from a rigorous viewpoint.

11. This theory can be regarded as being an asymptote of the general viscoelasticity theory, known as simple fluid theory by Coleman and Noll (1961b) for long duration, slowly varying flow.

12. Note that from the definition of the rate-of-strain tensor $\mathbf{d}$ for shear flow, given by Eq. (2.12), we have $2d_{12} = \dot{\gamma}$. Thus the factor 2 does not appear in Eq. (3.80).

13. Using the rate-of-strain tensor $\mathbf{d}$ defined by Eq. (2.16) for equal biaxial elongational flow in Eq. (3E.1), with the aid of Eqs. (3E.2) and (3E.3), we obtain

$$\frac{\mathfrak{d}\sigma^{ij}}{\mathfrak{d}t} = \dot{\varepsilon}_{\mathrm{B}} \begin{Vmatrix} -2\sigma^{11} & 0 & 0 \\ 0 & -2\sigma^{22} & 0 \\ 0 & 0 & 4\sigma^{33} \end{Vmatrix} \tag{3N.5}$$

   Hence, substitution of Eq. (3N.5) into (3.4) gives (3.99).

References

Baek SG, Magda JJ, Larson RG (1993). *J. Rheol.* 37:1201.
Bernstein B, Kearsley EA, Zapas LJ (1963). *Trans. Soc. Rheol.* 7:391.
Bird RB, Armstrong RC, Hassager L (1987). *Dynamics of Polymeric Liquids: Fluid Mechanics*, Vol 1, 2nd ed, John Wiley & Sons, New York.
Bird RB, Carreau P (1968). *Chem. Eng. Sci.* 23:427.
Bogue DC (1966). *Ind. Eng. Chem. Fundam.* 5:253.
Carreau P (1972). *Trans. Soc. Rheol.* 16:99.
Christiansen EB, Miller MJ (1971). *Trans. Soc. Rheol.* 15:189.
Coleman BD, Markovitz H (1964). *J. Appl. Phys.* 35:1
Coleman BD, Noll W (1959). *Arch. Rat. Mech. Anal.* 3:289.
Coleman BD, Noll W (1960). *Arch. Rat. Mech. Anal.* 6:355.
Coleman BD, Noll W (1961a). *Ana. N.Y. Acad. Sci.* 89:672.
Coleman BD, Noll W (1961b). *Rev. Mod. Phys.* 33:239.
DeWitt TW (1955). *J. Appl. Phys.* 26:889.
Fromm H (1947). *Z. Angew. Math. Mech.* 25/27:146.
Giesekus H (1982). *J. Non-Newtonian Fluid Mech.* 11:69
Ginn RF, Metzner AB (1969). *Trans. Soc. Rheol.* 13:429.
Han CD (1976). *Rheology in Polymer Processing*, Academic Press, New York.
Han CD, Kim, YJ, Chuang, HK, Kwack, TH (1983). *J. Appl. Polym. Sci.* 28:3435.
Hildebrand FB (1952). *Methods of Applied Mathematics*, Prentice-Hall, Englewood Cliff, New Jersey.
Jeffreys H (1929). *The Earth*, Cambridge University Press, Cambridge.
Kaye A (1962). College of Aeronautics, Cranfield, Note No. 134.
Larson RD (1988). *Constitutive Equations for Polymer Melts and Solutions*, Butterworths, Stoneham, Massachusetts.
Laun HM (1978). *Rheol. Acta* 17:1.
Laun HM, Münstedt H (1978). *Rheol. Acta* 17:415.
Lodge AS (1956). *Trans. Faraday Soc.* 52:120.
Macdonald IF, Bird RB (1966). *Phys. Chem.* 70:2068.
Meister BJ (1971). *Trans. Soc. Rheol.* 15:63.
Noll W (1958). *Arch. Rat. Mech. Anal.* 2:197.
Olabisi O, Williams MC (1972). *Trans. Soc. Rheol.* 16:727.
Oldroyd JG (1950). *Proc. Roy. Soc.* A200:523.
Oldroyd JG (1958). *Proc. Roy. Soc.* A245:278.
Oldroyd JG (1984). *J. Non-Newtonian Fluid Mech.* 14:9.
Papanastasiou AC, Scriven LE, Macosko CW (1983). *J. Rheol.* 27:387.
Rivlin RS, Ericksen JL (1955). *J. Rat. Mech. Anal.* 4:323.
Santa Cruze ASM, Deiber JA (1989). *J. Rheol.* 33:391.
Spriggs TW (1965). *Chem. Eng. Sci.* 20:931.
Truesdell C, Noll W (1965). In *Handbuch der Physik: The Nonlinear Field Theories of Mechanics*, Springer, Berlin.
Wagner MH (1976). *Rheol. Acta* 15:136.
Wagner MH (1977). *Rheol. Acta* 16:43.
Wagner MH (1979). *Rheol. Acta* 18:681.
White JL, Metzner AB (1963). *J. Appl. Polym. Sci.* 7:1867.
Williams MC, Bird RB (1962). *Phys. Fluids* 5:1126.
Yamamoto M (1971). *Trans. Soc. Rheol.* 15:331.
Zaremba S (1903). *Bull. Intern. Acad. Sci. Cracovie*, p 594.

4

Molecular Theories for the Viscoelasticity of Flexible Homogeneous Polymeric Liquids

4.1 Introduction

The fact that a polymer consists of a number of chains of different lengths, each in turn consisting of a series of monomer units, means that the motion of one part of the polymer chain will profoundly affect the motion of other parts. Hence, for a given polymer, a description of microscopic processes occurring under a given flow field depends on hypotheses regarding the molecular structure and mechanisms of flow in the polymer.

Today, it is well-known, gained from practical experience, that the molecular weight, the molecular weight distribution, and the degree of long-chain branching influence the rheological properties of polymeric liquids. Therefore, a better understanding of the relationship between molecular parameters and rheological properties is very important from the standpoints of both polymer synthesis and polymer processing. However, the theoretical development of this aspect of the problem is far from complete, although some important progress has been made.

In the preceding chapter, we discussed the viscoelastic behavior of polymeric liquids from the phenomenological point of view, without associating the significance of theoretical predictions to molecular origin(s). Specifically, we have seen that the rheological equations of state contain parameters that vary from polymer to polymer. Since it has amply been demonstrated by experiment that the extent of a particular viscoelastic behavior is greatly influenced by the molecular parameters, such as molecular weight, molecular weight distribution, and the degree of long-chain branching, predictions of any viscoelastic behavior of polymers on the basis of phenomenological theory is of very limited use to either control the quality of polymers produced or improve the performance of polymers, unless the parameters appearing in various continuum constitutive equations are related to molecular parameters.

There is ample experimental evidence (Allen and Fox 1964; Berry and Fox 1968; Bueche 1962; Fox and Allen 1964; Graessley 1974), as illustrated in Figure 4.1, that the zero-shear viscosity (η_0) of a polymer is proportional to the molecular weight (M) below a critical value M_c, while above M_c it increases rapidly and becomes proportional to $M^{3.4}$, that is

$$\eta_0 = \begin{cases} KM & \text{for} \quad M \le M_c \\ KM^{3.4} & \text{for} \quad M > M_c \end{cases} \tag{4.1}$$

The critical molecular weight M_c is believed to correspond to a value beyond which molecular entanglements (i.e., temporary couplings between neighboring chains) begin to dominate the resistance to flow. Such characteristics, and other characteristics that will be discussed later in this chapter, are attributed to entanglement effects because they appear to derive essentially from topological restrictions on the chain motions.

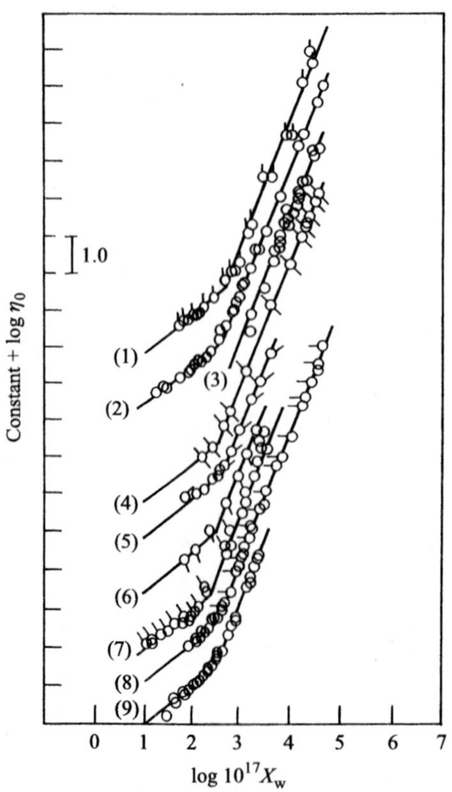

Figure 4.1 The dependence of zero-shear viscosity η_0 on molecular weight for homopolymer melts: (1) poly(dimethyl siloxane), (2) polyisobutylene, (3) polyethylene, (4) polybutadiene, (5) poly(tetramethyl *p*-silphenylene siloxane), (6) poly(methyl methacrylate), (7) poly(ethylene glycol), (8) poly(vinyl acetate), and (9) polystyrene. X_w appearing in the abscissa is a parameter which is proportional to molecular weight. (Reprinted from Berry and Fox, *Advances in Polymer Science* 5:261. Copyright © 1968, with permission from Springer.)

For concentrated solutions and molten polymers, the chain contours are extensively intermingled, so a mesh of neighboring chain contours surrounds each chain all along its length. Rearrangement of macromolecular chains on larger scales is restricted, because the chain cannot cross through its neighbors. It is well established that the molecular weight between entanglement couplings (M_e) is about one half of M_c, that is, $M_c \approx 2M_e$ (Bueche 1962; Doi and Edwards 1986; Ferry 1980; Graessley 1974). In the case of polymer solutions, both K and M_c in Eq. (4.1) change if a solvent is added to the polymer. Equation (4.1) applies to polymers with different molecular weight distributions if M is replaced by the weight-average molecular weight (M_w). From the rheological point of view, one can interpret M_c as a material constant signifying the lower limit of molecular weight for which non-Newtonian flow can be observed. It would then be expected that the onset of non-Newtonian behavior is strongly dependent on M and molecular weight distribution (or polydispersity index). It has been reported that above M_c the onset of non-Newtonian behavior occurs at lower shear rates as M and polydispersity index increase (Graessley 1974).

It then seems natural to divide the dependence of the viscoelastic behavior of polymeric liquids on M into two regimes: (1) unentangled regime, where $M < M_c$ holds, and (2) entangled regime, where $M \geq M_c$ holds. Understandably, the molecular interpretation of the viscoelastic behavior of polymeric liquids requires different concepts for the two regimes. Molecular theories for the viscoelasticity of flexible, unentangled linear polymers were developed in the early 1950s by Rouse (1953) and Zimm (1956). Two decades later, in the late 1970s, Doi and Edwards (1978a, 1978b, 1978c, 1979) developed a molecular theory for the viscoelasticity of flexible, entangled linear polymers.

In this chapter, we present currently held molecular theories for the viscoelasticity of linear, flexible macromolecular chains. We begin with a presentation of the static properties of macromolecules and the stochastic processes in the motion of macromolecular chains, as much as they will be necessary to present the molecular aspects of viscoelasticity in this and later chapters. We first present the molecular theories of Rouse (1953) and Zimm (1956), which are basically applicable to dilute polymer solutions and unentangled polymer melts, and then present the molecular theory of Doi and Edwards (1978a, 1978b, 1978c, 1979), which is applicable to concentrated polymer solutions and entangled polymer melts.

4.2 Static Properties of Macromolecules and Stochastic Processes in the Motion of Macromolecular Chains

In this section, we first introduce the most frequently used definitions of static properties, which will be used throughout this chapter and in some later chapters, and then discuss stochastic processes in the motion of macromolecular chains; that is, Brownian motion that leads to the well-known Fokker–Planck equation, which further reduces to the Smoluchowski equation and Langevin equation. These two equations play a very important role in describing the motion of macromolecular chains. Owing to the limited space available here, we do not present rigorous derivations of various expressions.

4.2.1 Static Properties of Macromolecules

4.2.1.1 *The Random Flight Model*

The general configuration of the long-chain molecules, as depicted schematically in Figure 4.2, is independent of the precise geometry of the chain, provided that the number of bonds about which free rotation can occur is sufficiently large. The particular geometry of the chain affects only the average dimensions of the chain, not its general form. An idealized model of the simplest possible kind, which does not correspond directly to any actual molecular structure, can be considered as one having N links of equal length b, in which the direction in space of any link is entirely random and bears no relation to the direction of any other link in the chain. We assume that the orientation of the ith link is random, so that the probability that it has an orientation in the range $\mathrm{d}\theta_i \mathrm{d}\phi_i$ around $\theta_i \phi_i$ in polar coordinate

$$\Psi_i \mathrm{d}\theta_i \mathrm{d}\phi_i = \frac{1}{4\pi} \sin\theta_i \, \mathrm{d}\theta_i \, \mathrm{d}\phi_i \tag{4.2}$$

is independent of the orientation of neighboring links. Such a randomly jointed chain is free from any restrictions on the freedom of motion of neighboring links and the equilibrium configuration (or distribution) function f for the entire chain is the product of the distribution functions for the individual links (Bird et al. 1987):

$$f = \prod_{i=1}^{N} \Psi_i = \left(\frac{1}{4\pi}\right)^N \prod_{i=1}^{N} \sin\theta_i \tag{4.3}$$

The function f is sometimes referred to as the "probability density" (i.e., the probability per unit volume).

4.2.1.2 *The Equilibrium Configuration (or Distribution) Function for Polymer Molecules*

In the most general form, the equilibrium (or probability) distribution function $f(\mathbf{r})$ for a chain at the position $\mathbf{r}$ can be written as (Bird et al. 1987; Huang 1987;

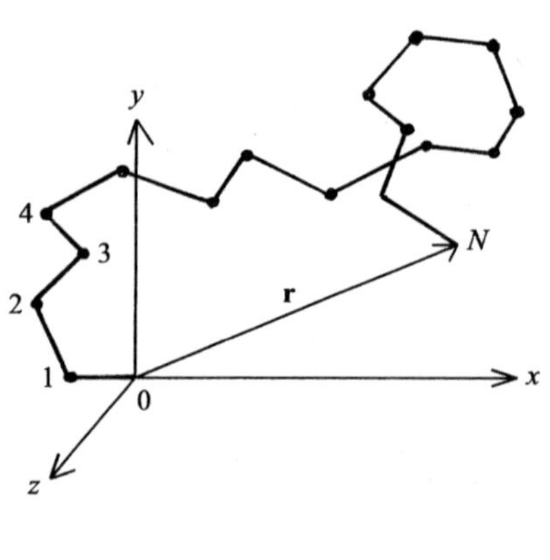

Figure 4.2 Schematic describing the random flight model, describing a random distribution of N links of equal length in the chain.

Yamakawa 1971)

$$f(\mathbf{r}) = \frac{1}{Z} e^{-U(\mathbf{r})/k_B T} \tag{4.4}$$

where $U(\mathbf{r})$ is the potential, k_B is the Boltzmann constant (1.3804×10^{-16} erg/K), T is the absolute temperature, and Z is the partition function defined by

$$Z = \int e^{-U(\mathbf{r})/k_B T} \mathrm{d}\mathbf{r} \tag{4.5}$$

For an ideal spring (Hookean behavior), the potential U is given by (see Appendix 4A)

$$U(\mathbf{r}) = \left(3k_B T/2Nb^2\right) \mathbf{r}^2 \tag{4.6}$$

Substituting Eq. (4.6) into (4.4) and determining the normalized factor such that the total probability of any value of $\mathbf{r}$ is unity (i.e., $\int f(\mathbf{r})\, \mathrm{d}\mathbf{r} = 1$) we have

$$f(\mathbf{r}) = \left(\frac{3}{2\pi N b^2}\right)^{3/2} \exp\left(-\frac{3}{2Nb^2}\mathbf{r}^2\right) \tag{4.7}$$

which is referred to as the "Gaussian distribution function." Equation (4.7) is of fundamental importance in defining the static properties of macromolecules. It is important to note, however, that Eq. (4.7) is only approximate, since its derivation involves the assumption that the distance r between the ends of the chain is much shorter than the maximum or fully extended length Nb of the chain, in other words $r \ll Nb$.

4.2.1.3 Average Value

The average value of a variable A in an environment where a large number of configurations is possible can be defined by

$$\langle A \rangle = \int A f(\mathbf{r}) \mathrm{d}\mathbf{r} = \int_0^{2\pi} \int_0^{\pi} \int_0^{\infty} A f(\mathbf{r})\, r^2 \sin\theta \, \mathrm{d}r \, \mathrm{d}\theta \, \mathrm{d}\phi \tag{4.8}$$

where the bracket $\langle\ \rangle$ denotes an average over all directions of the unit vector $\mathbf{u}$. In polar coordinates, $\mathbf{u} = [\sin\theta \ \cos\phi,\ \sin\theta \ \sin\phi, \cos\theta]$ and the volume element $\mathrm{d}\mathbf{r}$ is $r^2 \sin\theta \, \mathrm{d}r \, \mathrm{d}\theta \, \mathrm{d}\phi$.

4.2.1.4 The Mean-Square End-to-End Distance

Using Eq. (4.7), we can write the mean square of the end-to-end distance $\langle r^2 \rangle$ of a coil as[1]

$$\langle r^2 \rangle = \langle \mathbf{r} \cdot \mathbf{r} \rangle = \int r^2 f(\mathbf{r}) \, \mathrm{d}\mathbf{r} = Nb^2 \tag{4.9}$$

We now have an expression that contains two molecular parameters, b and N, which are experimentally measurable.

We introduce a very important static property, the effective step length of the chain of mean-square end-to-end distance $\langle r^2 \rangle$, often referred to as the "Kuhn length" b defined by

$$b = \langle r^2 \rangle^{1/2} / N_{\mathrm{p}}{}^{1/2} = \left(\langle r^2 \rangle^{1/2} / M^{1/2} \right) M_{\mathrm{o}}{}^{1/2} \tag{4.10}$$

in which N_{p} denotes degree of polymerization defined by $N_{\mathrm{p}} = M/M_{\mathrm{o}}$, with M being the molecular weight of polymer and M_{o} being the molecular weight of monomer. It should be emphasized that N appearing in Eq. (4.9) denotes the number of segments in a chain and it is related to N_{p} by $N = N_{\mathrm{p}}/m$ with m being the number of monomer units in a segment. Note that $\langle r^2 \rangle^{1/2}/M^{1/2}$ is independent of M and characteristic of the chemical structure of polymer. The values of $\langle r^2 \rangle^{1/2}/M^{1/2}$ for various polymers are listed in the literature (Kurata and Tsunashima 1989). For example, the value of $\langle r^2 \rangle^{1/2}/M^{1/2}$ for polystyrene is 685×10^{-4} nm. Thus, using Eq. (4.10) with $M_{\mathrm{o}} = 104$, we have $b = 0.69$ nm for polystyrene. Similarly, we have $b = 0.64$ nm for poly(methyl methacrylate), $b = 0.71$ nm for poly(α-methylstyrene), $b = 0.64$ nm for polybutadiene (98% *cis*-linkage), and $b = 0.67$ nm for polyisoprene (100% *cis*-linkage).

4.2.1.5 Properties of Gaussian Distribution Function

Equation (4.7), with the aid of Eq. (4.9), can be rewritten as

$$f(\mathbf{r}) = \left(\frac{3}{2\pi \langle r^2 \rangle} \right)^{3/2} \exp\left(-\frac{3}{2\langle r^2 \rangle} \mathbf{r}^2 \right) \tag{4.11}$$

The probability density $\varphi(\mathbf{r})$ that the components of vector $\mathbf{r}$ representing the end-to-end distance for the chain shall lie within the interval $\mathbf{r}$ to $\mathbf{r} + \mathrm{d}\mathbf{r}$ is expressed by (Yamakawa 1971)

$$\varphi(\mathbf{r}) = f(\mathbf{r}) 4\pi r^2 \tag{4.12}$$

where $f(\mathbf{r})$ is the Gaussian distribution function defined by Eq. (4.11). It should be noted that while the function $f(\mathbf{r})$ is symmetrical and has a maximum value when $r = 0$, the function $\varphi(\mathbf{r})$ is zero at $r = 0$ and reaches a maximum at a finite value of r. It can be shown that the maximum value of $\varphi(\mathbf{r})$ occurs at

$$r^* = (2/3)^{1/2} \langle r^2 \rangle^{1/2} \tag{4.13}$$

In order to observe the shapes of the two functions $f(\mathbf{r})$ and $\varphi(\mathbf{r})$ as the M of a polymer varies, let us consider two monodisperse polystyrenes (PS): (1) PS-1 with $M = 1.2 \times 10^5$, and (2) PS-2 with $M = 2.5 \times 10^5$. Assuming that the equilibrium

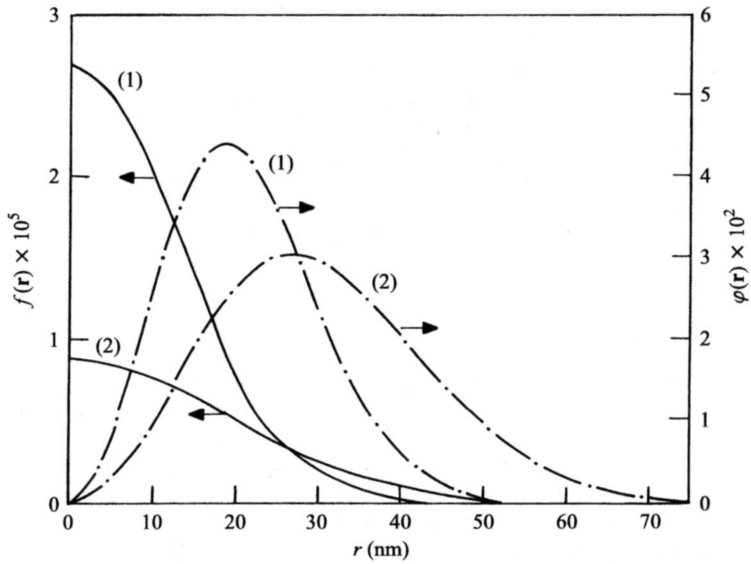

Figure 4.3 Distribution functions $f(\mathbf{r})$ and $\varphi(\mathbf{r})$ for monodisperse polystyrenes: curve 1 is for $M = 1.2 \times 10^5$ and curve 2 is for $M = 2.5 \times 10^5$. In the computations, the following parameters were used: $b = 0.68$ nm, $\langle r^2 \rangle^{1/2} = 23.1$ nm and $r^* = 18.8$ nm for curve 1, and $\langle r^2 \rangle^{1/2} = 33.3$ nm and $r^* = 27.2$ nm for curve 2, monomer molecular weight = 104.15. r^* denotes the value of r at which the maximum of $\varphi(\mathbf{r})$ occurs.

configuration of the two polystyrenes may be described by Eq. (4.11), we obtain curve 1 for PS-1 and curve 2 for PS-2 given in Figure 4.3.

4.2.1.6 The Radius of Gyration

As shown schematically in Figure 4.4, the root-mean-square distance $\langle s^2 \rangle^{1/2}$ of segments from the center of gravity of the coil, commonly referred to as the "radius of gyration," is given by (Doi and Edwards 1986; Yamakawa 1971)

$$\langle s^2 \rangle^{1/2} = \langle r^2 \rangle^{1/2} / \sqrt{6} \tag{4.14}$$

4.2.2 Stochastic Processes in the Motion of Macromolecular Chains

Every physical quantity we observe is accompanied by random fluctuations (often referred to as Brownian motion) due to thermal motion of microscopic degrees of freedom in matter. In a great many cases, such fluctuations are small in comparison with the average values of the quantity under observation and can generally be ignored. However, such fluctuations reflect the microscopic motions in the system under study, thus an analysis of such microscopic motions sheds light on the macroscopic (i.e., bulk) motions of the system. The subject of Brownian motion has been dealt with by many different scientific disciplines including polymer science (Chandrasekhar

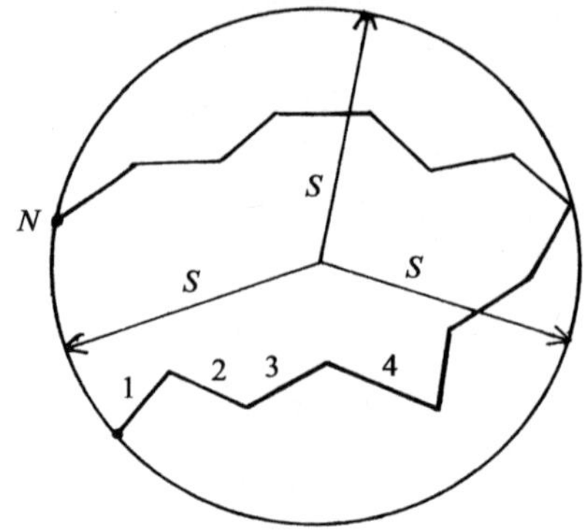

Figure 4.4 Schematic describing the radius of gyration of a random-flight chain.

1943; Kubo et al. 1985; Uhlenbeck and Ornstein 1930; van Kampen 1981; Wang and Uhlenbeck 1945). In this section, we present, without rigorous mathematical proof, two representations describing Brownian motion that are relevant to the main theme of this chapter and to some later chapters.

4.2.2.1 Smoluchowski Equation

Consider a medium that contains a large number of Brownian particles and define the particle density as $\rho(x, t)$, which, for simplicity, is assumed to depend on time t and one-dimensional position x. Brownian motion of particles makes the distribution of particles tend toward uniformity. This process is called "diffusion." Corresponding to the gradient of the density distribution, the flux (j_d) is given by

$$j_\mathrm{d} = -D\frac{\partial \rho}{\partial x} \tag{4.15}$$

where D is the diffusion coefficient. When an external field exists, a flow is produced with the terminal velocity u_o determined by the balance of the driving force F_c and the frictional force from the surrounding fluid acting on a particle. The flux (j_e) of this flow is given by

$$j_\mathrm{e} = \rho u_\mathrm{o} = \rho F_\mathrm{c}/\zeta \tag{4.16}$$

where ζ is the friction coefficient. Then the total flux (j) is

$$j = j_\mathrm{d} + j_\mathrm{e} = -D\frac{\partial \rho}{\partial x} + \frac{\rho F_\mathrm{c}}{\zeta} \tag{4.17}$$

Note that the continuity equation for ρ satisfies (Kubo et al. 1985)

$$\frac{\partial \rho(x, t)}{\partial t} = -\frac{\partial j}{\partial x} \tag{4.18}$$

Use of Eq. (4.17) in Eq. (4.18) gives

$$\frac{\partial \rho(x,t)}{\partial t} = \frac{\partial}{\partial x}\left(D\frac{\partial \rho}{\partial x}\right) - \frac{\partial}{\partial x}\left(\frac{\rho F_c}{\zeta}\right) \tag{4.19}$$

which can be rewritten as

$$\frac{\partial \rho(x,t)}{\partial t} = \frac{\partial}{\partial x}\frac{1}{\zeta}\left(k_B T\frac{\partial \rho}{\partial x} + \rho\frac{\partial U}{\partial x}\right) \tag{4.20}$$

where use was made of $F_c = -\partial U/\partial x$ with U being the external potential and the Einstein relation $D = \mu k_B T = k_B T/\zeta$ with μ being the mobility. Equation (4.20) is known as the Smoluchowski equation (Smoluchowski 1915), which is a special form of the Fokker–Planck equation (Fokker 1914; Planck 1917; Risken 1989). Note that in the absence of an external force, Eq. (4.19) reduces to the well-known diffusion equation

$$\frac{\partial \rho(x,t)}{\partial t} = D\frac{\partial^2 \rho}{\partial x^2} \tag{4.21}$$

As long as the particle density is not too high, the interaction between Brownian particles can be ignored, so that the diffusion described by Eq. (4.21) is the result of independent particle motion. Namely, the density $\rho(x, t)$ at time t and the spatial point x is (Wang and Uhlenbeck 1945)

$$\rho(x,t) = \int \rho(x_0, t_0)\, \mathrm{d}x_0\, P(x_0, t_0|x, t) \tag{4.22}$$

where $\rho(x_0, t_0)$ is the density at x_0 and t_0. The transition probability $P(x_0, t_0|x, t)$ satisfies the diffusion equation, Eq. (4.21),

$$\frac{\partial}{\partial t}P(x_0, t_0|x, t) = D\frac{\partial^2}{\partial x^2}P(x_0, t_0|x, t) \tag{4.23}$$

because Eq. (4.21) must be satisfied by $\rho(x, t)$ given by Eq. (4.22) for any arbitrary initial condition $\rho(x_0, t_0)$. Then, Eq. (4.23) simply becomes

$$\frac{\partial}{\partial t}P(x,t) = D\frac{\partial^2}{\partial x^2}P(x,t) \tag{4.24}$$

The transition probability $P(x_0, t_0|x, t)$ is a fundamental solution of Eq. (4.24) for the initial condition

$$P(x_0, t_0|x, t) = \delta(x - x_0) \tag{4.25}$$

and is given by

$$P(x_0, t_0|x, t) = \frac{1}{\left[4\pi D(t - t_0)\right]^{1/2}} \exp\left(-\frac{(x - x_0)^2}{4D(t - t_0)}\right) \tag{4.26}$$

This is the simplest possible idealization of Brownian motion.

Note that Eq. (4.20) can also be expressed in terms of the probability distribution function f (Risken 1989):

$$\frac{\partial f(x,t)}{\partial t} = \frac{\partial}{\partial x}\frac{1}{\zeta}\left(k_{\mathrm{B}}T\frac{\partial f}{\partial x} + f\frac{\partial U}{\partial x}\right) \tag{4.27}$$

4.2.2.2 Langevin Equation

If a small particle of mass m is immersed in a fluid, a friction force (F_{c}) will act on the particle with the velocity u. Stoke's law gives the simplest expression for such a friction or damping force:

$$F_{\mathrm{c}} = -\zeta u \tag{4.28}$$

where ζ is the friction coefficient. The equation of motion for the particle in the absence of additional forces is given by

$$m\frac{\mathrm{d}u(t)}{\mathrm{d}t} = -\zeta u \tag{4.29}$$

The physics behind the friction is that the molecules of the fluid collide with the particle. The momentum of the particle is transferred to the molecules of the fluid and the velocity therefore decreases to zero. Equation (4.29) is a deterministic expression in that the velocity $u(t)$ at time t is completely determined by its initial value $u(0)$ and it is valid only if the mass of the particle is so large that its velocity due to thermal fluctuations is negligible.

If we were to treat the problem exactly, we should solve the coupled equations of motion for all the molecules of the fluid, the number of which is of the order of 10^{23}. However, we cannot generally solve these many coupled equations. Furthermore, since we do not know the initial values of all the molecules of the fluid, we cannot calculate the exact motion of the small particle in the fluid. This means that we must consider an ensemble of such systems. Under such situations, the total force $F(t)$ of the molecules acting on the small particle consists of two components: a continuous damping force $F_{\mathrm{c}}(t)$, given by Eq. (4.28), and a fluctuating force $F_{\mathrm{f}}(t)$:

$$F(t) = F_{\mathrm{c}}(t) + F_{\mathrm{f}}(t) \tag{4.30}$$

where $F_{\mathrm{f}}(t)$ is a stochastic or random force, the properties of which are given only in the average (Kubo et al. 1985).

Then, the equation of motion for the particle can be written as

$$F(t) = m\frac{\mathrm{d}u(t)}{\mathrm{d}t} = -\zeta u + F_{\mathrm{f}}(t) \tag{4.31}$$

or

$$\frac{du(t)}{dt} + \tilde{\gamma} u = \Gamma(t) \tag{4.32}$$

where $\tilde{\gamma} = \zeta/m$ and $\Gamma(t) = F_{\mathrm{f}}(t)/m$. Equation (4.32) is called the Langevin equation (Kubo 1966; Kubo et al. 1985; Langevin 1908; Mori 1965; Risken 1989), and $\Gamma(t)$ is the fluctuating force per unit mass, commonly referred to as Langevin force. If we consider the force as causing the motion, then the random force $\Gamma(t)$ produces Brownian motion. Thus, our problem is to determine the stochastic processes $u(t)$ from knowing $\Gamma(t)$. Equation (4.32) is called a "stochastic differential equation," because it contains the stochastic force $\Gamma(t)$.

The statistical properties of the Langevin force $\Gamma(t)$ are usually given as follows (Kubo et al. 1985):

1. The equilibrium ensemble average of the force vanishes for all time t

$$\langle \Gamma(t) \rangle = 0 \tag{4.33}$$

because the equation of motion of the average velocity $\langle u(t) \rangle$ should be given by Eq. (4.29).

2. The force $\Gamma(t)$ at times $t > 0$ is not correlated with the initial velocity of the particle

$$\langle u(0) \cdot \Gamma(t) \rangle = 0 \tag{4.34}$$

3. For $|t - t'| > s_{\mathrm{o}}$, we have

$$\langle \Gamma(t)\Gamma(t') \rangle = 0 \tag{4.35}$$

where s_{o} is the duration of a collision between different molecules of the fluid with the small particle. Usually, s_{o} is much smaller than the relaxation time $\tau = 1/\tilde{\gamma}$ of the small particle.

4. However, in the limit of $s_{\mathrm{o}} \to 0$, we have

$$\langle \Gamma(t)\Gamma(t') \rangle = 2k_{\mathrm{B}}T\zeta\delta(t - t') \tag{4.36}$$

which is known as the "Fluctuation–dissipation theorem" (Kubo 1966; Kubo et al. 1985; Mori 1965; Risken 1989).

The random force is not a prescribed function of time, but a random function of time. Nevertheless, the solution of Eq. (4.32) can be written formally as

$$u(t) = u(0)e^{-\tilde{\gamma}t} + \int_0^t ds\, e^{-\tilde{\gamma}(t-s)}\, \Gamma(s) \tag{4.37}$$

The velocity correlation function can be determined from Eq. (4.37) by taking the dot product of $u(0)$ with each term in the equation followed by averaging over a

Maxwell distribution of initial velocities. Then

$$\langle u(0) \cdot u(t) \rangle = \langle u(0) \cdot u(0) \rangle e^{-\tilde{\gamma} t} + \int_0^t ds\, e^{-\tilde{\gamma}(t-s)} \langle u(0) \cdot \Gamma(t) \rangle \tag{4.38}$$

From Eq. (4.34), the velocity correlation function for a Brownian particle becomes

$$\phi(t) = \langle u(0) \cdot u(t) \rangle = \frac{3k_B T}{m} e^{-\tilde{\gamma} t} \tag{4.39}$$

where we have used the equipartition theorem (Kubo et al. 1985)

$$\tfrac{1}{2} m \langle u^2 \rangle = \tfrac{3}{2} k_B T \tag{4.40}$$

4.3 Molecular Theory for the Viscoelasticity of Dilute Polymer Solutions and Unentangled Polymer Melts

The linear viscoelasticity of dilute polymer solutions, first introduced by Rouse (1953), is based on the submolecular model. In his theory it is assumed that the long polymer molecule can be broken into submolecules, and that fluctuations of the end-to-end length of a polymer molecule follow a Gaussian probability function. Then, the polymer molecule under consideration is replaced by a chain of N identical segments joining $N+1$ identical beads with a completely flexible spring at each bead. Such a "bead–spring" model is shown schematically in Figure 4.5. The basic concept of this model was introduced by earlier investigators (Debye 1946; Kramers 1946). However, what was new in the Rouse theory was the introduction of the normal coordinate method, which transforms the coordinate system into diagonal form by an orthogonal transformation of the coordinates. By means of the orthogonal transformation of coordinates, the coordination of all the motions of the parts of a polymer molecule is resolved into a series of modes. Each mode has a characteristic relaxation time. Later, Zimm (1956) extended the Rouse theory to include the hydrodynamic interaction for macromolecules, using the Kirkwood and Riseman method (Kirkwood and Riseman 1948).

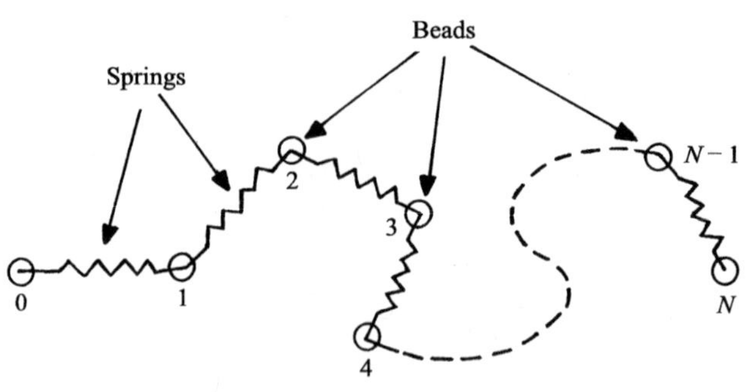

Figure 4.5 Schematic describing the bead–spring model.

The Zimm theory gives the Rouse theory as the special case of zero hydrodynamic interaction, the free-draining case. We present the Rouse–Zimm theory in the following sections. In reference to Figure 4.5, there are N submolecules or springs and $N+1$ beads. Let us assume that (1) the configuration of a submolecule is specified in terms of the vector that corresponds to its end-to-end separation, (2) the separation of the ends of the submolecules is approximated by a Gaussian probability distribution, (3) the configuration of a polymer chain that contains N submolecules is described by the corresponding set of N vectors, (4) the action of a velocity gradient disturbs the distribution of configuration of the polymer molecules away from its equilibrium form, storing free energy in the system, and (5) the coordinated thermal motion of the segment causes the configurations to drift toward their equilibrium distribution. In other words, in the bead–spring model the exact bonded structure is replaced with a series of beads connected by springs, which interact with the corresponding potential energy.

4.3.1 The Rouse Model

Let us assume that when the polymer is disturbed due to being in a shear gradient there are only two major forces, the hydrodynamic force $\mathbf{F}^{(h)}$ and the restoring (or spring) force $\mathbf{F}^{(r)}$ that the bead exerts on the liquid. The hydrodynamic force $\mathbf{F}^{(h)}$ exerted on the liquid by the bead, the components of which are assumed to be proportional to the relative velocity of the bead and solvent through the fluid, is expressed by

$$F_{xi}^{(h)} = \zeta(\mathrm{d}x_i/\mathrm{d}t - v_{xi}) \tag{4.41}$$

where $\mathrm{d}x_i/\mathrm{d}t$ is the velocity of the bead, v_{xi} is the velocity of the solvent, and ζ is the friction coefficient of the chain segment. Similar expressions can be written for y and z components. The restoring (i.e., intramolecular spring) force $\mathbf{F}^{(r)}$ on the ith bead, assuming that Hooke's law describes the spring force, can be expressed by

$$F_{0x}^{(r)} = -(3k_\mathrm{B}T/b^2)(x_0 - x_1) \tag{4.42a}$$

$$F_{ix}^{(r)} = -(3k_\mathrm{B}T/b^2)(-x_{i-1} + 2x_i - x_{i+1}) \tag{4.42b}$$

$$F_{Nx}^{(r)} = -(3k_\mathrm{B}T/b^2)(-x_{N-1} + x_N) \tag{4.42c}$$

where k_B is the Boltzmann constant, T is the absolute temperature, and b is the length of a spring. Note that $3k_\mathrm{B}T/b^2$ is the spring constant. Similar expressions can be written for y and z components.

The solvent velocity field $\mathbf{v}(v_{xi}, v_{yi}, v_{zi})$ appearing in Eq. (4.41) in the vicinity of the chain differs from the value $\mathbf{v}^\mathrm{o}$, which it would have if the chain were absent. We will assume that the hydrodynamic interaction is negligible (i.e., $\mathbf{v} = \mathbf{v}^\mathrm{o}$), and there are no external forces. Then, by combining Eqs. (4.41) and (4.42), we obtain the equation of motion for component x_i:

$$\mathrm{d}x_i/\mathrm{d}t = v_{xi}^\mathrm{o} - \alpha(-x_{i-1} + 2x_i - x_{i+1}) \quad (0 < i < N) \tag{4.43}$$

where $\alpha = 3k_B T/b^2\zeta$. The expression for $i=0$ and N are slightly modified as in Eqs. (4.42a) and (4.42c). The expressions for y_i and z_i are of the same form. The general expression for the equation of motion can be written in the form

$$\frac{d\mathbf{r}}{dt} = \mathbf{v}^o - \alpha \mathbf{A} \cdot \mathbf{r} \tag{4.44}$$

where $\mathbf{r}$ is the vector joining the neighboring beads, $d\mathbf{r}/dt$ is the velocity, $\mathbf{v}^o$ is the velocity of the solvent at the positions of the beads, and the matrix $\mathbf{A}$, often referred to as the Rouse matrix, has elements

$$A_{ij} = 2\delta_{i,j} - \delta_{i,j-1} - \delta_{i,j+1} \tag{4.45}$$

Equation (4.45) is valid for $1 \le i \le N-1$. Note that $A_{ij} = \delta_{i,j} - \delta_{i,j-1}$ for $i=0$ and $A_{ij} = \delta_{i,j} - \delta_{i,j+1}$ for $i = N$. Thus, the matrix $\mathbf{A}$ can be expressed as

$$\mathbf{A} = \left\| \begin{array}{ccccccccccc} 1 & -1 & 0 & 0 & 0 & . & . & . & 0 & 0 & 0 \\ -1 & 2 & -1 & 0 & 0 & . & . & . & 0 & 0 & 0 \\ 0 & -1 & 2 & -1 & 0 & . & . & . & 0 & 0 & \\ . & . & . & . & . & . & . & . & . & . & . \\ . & . & . & . & . & . & . & . & . & . & . \\ . & . & . & . & . & . & . & . & . & . & . \\ 0 & 0 & 0 & . & . & . & 0 & -1 & 2 & -1 & 0 \\ 0 & 0 & 0 & . & . & . & 0 & 0 & -1 & 2 & -1 \\ 0 & 0 & 0 & . & . & . & 0 & 0 & 0 & -1 & 1 \end{array} \right\| \tag{4.46}$$

For $\mathbf{v}^o = 0$ (i.e., for polymer melts), Eq. (4.44) reduces to

$$\frac{d\mathbf{r}}{dt} = -\alpha \mathbf{A} \cdot \mathbf{r} \tag{4.47}$$

Equation (4.47) can be solved using the normal coordinates.[2] Briefly stated, one ends up with solving the following differential equation for the pth mode of the normalized coordinates q_p:

$$dq_p/dt = -\alpha\lambda_p q_p \tag{4.48}$$

where λ_p are the eigenvalues of the Rouse matrix $\mathbf{A}$ and are given by

$$\lambda_p = 4\sin^2\left[p\pi/2(N+1)\right]; \quad p = 1, 2, \ldots, N \tag{4.49}$$

Integration of Eq. (4.48) gives

$$q_p = \exp(-\alpha\lambda_p t); \quad p = 1, 2, \ldots, N \tag{4.50}$$

Rouse (1953) has shown that the relaxation times τ_p (i.e., the time constants associated with the rate of stress dissipation after a given strain) can be obtained from

$$\tau_p = \frac{1}{2\alpha\lambda_p} = \frac{\zeta b^2}{6k_B T\lambda_p}; \quad p = 1, 2, \ldots, N \tag{4.51}$$

For very large values of N, Eq. (4.51) reduces to

$$\tau_p = \frac{\zeta b^2 N^2}{6\pi^2 p^2 k_B T}; \quad p = 1, 2, \ldots, N \tag{4.52}$$

Note that the largest or terminal relaxation time τ_1 for the Rouse chain (i.e., for $p = 1$ in Eq. (4.52)) is given by

$$\tau_r = \tau_1 = \frac{(N\zeta)(Nb^2)}{6\pi^2 k_B T} \tag{4.53}$$

where the quantities $N\zeta$ and Nb^2 describe the chains as a whole and are each proportional to the number of links in the chain backbone. Hereafter, τ_r will be referred to as the "Rouse relaxation time." Note that Nb^2 is related to the mean-square end-to-end distance at equilibrium $\langle r^2 \rangle$ given by Eq. (4.9).

Note further that the free-draining expression for the translational diffusion coefficient D_G (i.e., the diffusion coefficient of the center-of-mass for the Rouse chain) is given by (Doi and Edwards 1986)

$$D_G = \frac{k_B T}{N\zeta} \tag{4.54}$$

With the aid of Eqs. (4.9) and (4.54), Eq. (4.53) can be rewritten as

$$\tau_r = \frac{\langle r^2 \rangle}{6\pi^2 D_G} \tag{4.55}$$

Since N is proportional to the molecular weight M, from Eqs. (4.54) and (4.55) we have

$$D_G \propto M^{-1}; \quad \tau_r \propto M^2 \tag{4.56}$$

However, these predictions are not consistent with experimental results, which, in a theta (θ) solvent, are

$$D_G \propto M^{-1/2}; \quad \tau_r \propto M^{3/2} \tag{4.57}$$

This discrepancy comes from neglecting the hydrodynamic interactions between the solvent molecules and the polymer chains. Because of this discrepancy, the Rouse model is not suitable for describing the dynamics of polymer chains in dilute solution, but it is still very useful for describing the dynamics of undiluted polymer with $M < M_c$.

4.3.2 The Zimm Model

Debye (1946) suggested that the segments of polymer inside the polymer coil would be partially shielded from the velocity gradient of the solvent, so that the intrinsic viscosity would not increase in direct proportion to the molecular weight. Since the original work by Debye (1946), Kirkwood and Riseman (1948, 1956) have shown that hydrodynamic shielding of the inside of the polymer does reduce the dependence of viscosity on molecular weight. Zimm (1956) succeeded in combining the Kirkwood–Riseman method of taking account of the hydrodynamic interaction with the normal-coordinate formalism of Rouse.

In the Zimm theory, the force on the ith bead has, besides the terms due to the hydrodynamic and restoring forces considered in the Rouse model, additional terms due to Brownian motion and hydrodynamic shielding. The Brownian motion force $\mathbf{F}^{(b)}$ exerted on the beads is expressed by[3]

$$F_{0x}^{(b)} = -k_B T \frac{\partial \ln f}{\partial x_0} \tag{4.58a}$$

$$F_{ix}^{(b)} = -k_B T \frac{\partial \ln f}{\partial x_i} \tag{4.58b}$$

$$F_{Nx}^{(b)} = -k_B T \frac{\partial \ln f}{\partial x_N} \tag{4.58c}$$

where f is the equilibrium configuration function for the chain coordinates. Similar expressions can be written for y and z components.

Now we must take into account the velocity disturbance v'_{xi} due to the motion of other beads in the chain to the solvent velocity at the ith bead; that is, the solvent velocity v_{xi} consists of two components: $v_{xi} = v^{o}_{xi} + v'_{xi}$. The velocity disturbance v'_{xi} produced at a position i by a bead moving through the origin as it exerts a hydrodynamic force $\mathbf{F}^{(s)}$ can be expressed by $v'_{xi} = T_{ik} F_k^{(s)}$ (Kirkwood and Riseman 1948, 1956), thus

$$v_{xi} = v^{o}_{xi} + \sum_{i \neq j} T_{ij} F_j^{(s)} \tag{4.59}$$

or

$$\mathbf{v}_i = \mathbf{v}_i^{o} + \sum_{i \neq j} [\mathbf{T}_{ij} \cdot \mathbf{F}_j^{(s)}] \tag{4.60}$$

where $\mathbf{T}_{ij}$ is the Oseen tensor, which is a $((N+1) \times (N+1))$ second-order tensor defined by

$$\mathbf{T}_{ij} = \frac{1}{8\pi \eta_s r_{ij}} \left(\boldsymbol{\delta}_{ij} + \frac{\mathbf{r}_{ij}\mathbf{r}_{ij}}{r_{ij}^2} \right) \tag{4.61}$$

where $\boldsymbol{\delta}_{ij}$ are ijth components of unit tensor, $\mathbf{r}_{ij} = |\mathbf{r}_i - \mathbf{r}_j|$, and η_s is the solvent viscosity. Following Kirkwood and Riseman (1948), Zimm (1956) replaced the tensor elements of Eq. (4.61) with their equilibrium average values, yielding

$$\langle \mathbf{T}_{ij} \rangle_{\text{eq}} = \frac{1}{6\pi\eta_s} \left\langle \frac{1}{r_{ij}} \right\rangle_{\text{eq}} \boldsymbol{\delta}_{ij} \tag{4.62}$$

where the bracket $\langle\ \rangle$ is the average taken over the configuration distribution function f. Note that $\langle \mathbf{T}_{ij} \rangle_{\text{eq}}$ is no longer a function of the spatial coordinates.

$\mathbf{F}^{(s)}$ appearing in Eq. (4.60) is the sum of the restoring force $\mathbf{F}^{(r)}$ defined by Eq. (4.42) and the Brownian motion force $\mathbf{F}^{(b)}$ defined by Eq. (4.58), that is

$$\mathbf{F}^{(s)} = -k_B T \boldsymbol{\nabla} \ln f(\mathbf{r}, t) - (3k_B T/b^2)\mathbf{A} \cdot \mathbf{r} \tag{4.63}$$

where $\boldsymbol{\nabla}$ is the gradient operator and the matrix $\mathbf{A}$ is defined by Eq. (4.46). Substituting Eq. (4.63), with the aid of (4.60), into (4.41) one obtains

$$\frac{\mathrm{d}\mathbf{r}}{\mathrm{d}t} = \mathbf{v}^{\mathrm{o}} - \mathbf{M} \cdot \boldsymbol{\nabla} \ln f(\mathbf{r}, t) - (3k_B T/b^2\zeta)\mathbf{H} \cdot \mathbf{A} \cdot \mathbf{r} \tag{4.64}$$

where $\mathbf{M} = (k_B T/\zeta)\mathbf{H}$, often referred to as the mobility tensor, in which the matrix $\mathbf{H}$ is defined by

$$\mathbf{H} = \mathbf{I} + \zeta \langle \mathbf{T} \rangle_{\text{eq}} \tag{4.65}$$

where $\langle \mathbf{T} \rangle_{\text{eq}}$ is defined by Eq. (4.62). The unknown function $f(\mathbf{r}, t)$ in Eq. (4.64) is to be determined from the solution of the continuity equation

$$\frac{\partial f(\mathbf{r}, t)}{\partial t} + \boldsymbol{\nabla}^{\mathrm{T}} \cdot f(\mathbf{r}, t) \frac{\mathrm{d}\mathbf{r}}{\mathrm{d}t} = 0 \tag{4.66}$$

where $\boldsymbol{\nabla}^{\mathrm{T}}$ is the transpose of $\boldsymbol{\nabla}$. Substitution of Eq. (4.64) into Eq. (4.66) gives

$$\frac{\partial f(\mathbf{r}, t)}{\partial t} + \boldsymbol{\nabla}^{\mathrm{T}} \cdot f(\mathbf{r}, t)\mathbf{v}^{\mathrm{o}} = \boldsymbol{\nabla}^{\mathrm{T}} \cdot \mathbf{M} \cdot \left(\boldsymbol{\nabla} f(\mathbf{r}, t) + \frac{3}{b^2} f(\mathbf{r}, t)\mathbf{A} \cdot \mathbf{r} \right) \tag{4.67}$$

for nonvanishing hydrodynamic interactions in solutions. For $\mathbf{v}^{\mathrm{o}} = 0$ (i.e., for undiluted polymers), Eq. (4.67) reduces to

$$\frac{\partial f(\mathbf{r}, t)}{\partial t} = \boldsymbol{\nabla}^{\mathrm{T}} \cdot \mathbf{M} \cdot \left(\boldsymbol{\nabla} f(\mathbf{r}, t) + \frac{3}{b^2} f(\mathbf{r}, t)\, \mathbf{A} \cdot \mathbf{r} \right) \tag{4.68}$$

For constant $\mathbf{M}$ ($\equiv D$), Eq. (4.68) can be rewritten as

$$\frac{\partial f(\mathbf{r}, t)}{\partial t} = D\nabla^2 f(\mathbf{r}, t) - w(\mathbf{r}) f(\mathbf{r}, t) \tag{4.69}$$

where $\nabla^2 = \mathbf{\nabla} \cdot \mathbf{\nabla}$ and

$$w(\mathbf{r}) = -(3k_{\mathrm{B}}T/b^2\zeta)\mathbf{\nabla}^{\mathrm{T}} \cdot \mathbf{A} \cdot \mathbf{r} \tag{4.70}$$

Using normal coordinates,[2] Eq. (4.67) can be rewritten such that the values of the relaxation times τ_p may be made to depend on the eigenvalues λ_p, which are the elements of the diagonal matrix $\mathbf{\Lambda}$ that can be obtained from the following equation:

$$\mathbf{Q}^{-1}\mathbf{HAQ} = \mathbf{\Lambda} \tag{4.71}$$

where $\mathbf{Q}$ is the orthogonal matrix composed of $(N + 1)$ column vectors $\mathbf{u}$, which are eigenvectors of $\mathbf{HA}$ satisfying the following eigenvalue equation (see Appendix 4B):

$$\mathbf{HAu} = \lambda'_p\mathbf{u} \tag{4.72}$$

In other words, the solution of Eq. (4.67) now rests on the solution of Eq. (4.72).

Zimm (1956) has shown that the solution of Eq. (4.72) can be obtained by solving the following integrodifferential equation:

$$u''(r) + h\int_{-1}^{1} \frac{u''(s)}{|r - s|}\,\mathrm{d}s = -\frac{N^2\lambda}{4}u(r) \tag{4.73}$$

where the primes denote differentiation and h is the draining parameter, which is a measure of the strength of the hydrodynamic interaction and is defined by $h = \zeta N^{1/2}/(12\pi^3)^{1/2}\eta_s b$ with η_s being the viscosity of the solvent. For $h = \infty$, Zimm et al. (1956) numerically solved Eq. (4.73) to obtain the first eight eigenvalues λ'_p ($p = 0$ to 7): 0.0, 4.04, 12.79, 24.2, 37.9, 53.5, 70.7, and 89.4. They also presented the following approximate expression:

$$\lambda'_p = \frac{\pi^2}{2}p^{3/2}\left(1 - \frac{1}{2\pi p}\right) \tag{4.74}$$

for large values of p. The eigenvalues for intermediate values of h were calculated by Hearst (1962) and by Tschoegl (1963). Lodge and Wu (1971) computed eigenvalues for N up to 300.

The relaxation time τ'_p for the Zimm model is given by (Zimm 1956)

$$\tau'_p = \frac{[\eta]\eta_s M^{3/2}}{0.586\,RT\langle M^{1/2}\rangle_w \lambda'_p} \tag{4.75}$$

where $[\eta]$ is the intrinsic viscosity of the solution. In view of the fact that the ratio $[\eta]/\langle M^{1/2}\rangle_w$ is constant (Yamakawa 1971), that is

$$\frac{[\eta]}{\langle M^{1/2}\rangle_w} = \Phi\left(\frac{\langle r^2\rangle}{M}\right)^{2/3} \tag{4.76}$$

where $\langle r^2 \rangle / M$ is independent of molecular weight M and Φ is the universal constant, 2.84×10^{23}, the Zimm model correctly predicts the experimentally observed dependence of relaxation times τ'_p on M, $\tau'_p \propto M^{3/2}$.

4.3.3 Prediction of Rheological Properties

We now consider predictions of the viscoelastic properties of unentangled polymer melts based on the Rouse model. Consider the situation where a sudden strain is imposed on a polymer. Then the stress remaining in the specimen at time t can be determined from a material property referred to as the "stress relaxation modulus" $G(t)$. The $G(t)$ for the Rouse model is given by (Doi and Edwards 1986; Rouse 1953)

$$G(t) = \frac{\rho RT}{M} \sum_{p=1}^{\infty} \exp\left(-t/\tau_p\right) \tag{4.77}$$

where τ_p is given by Eq. (4.52). When $G(t)$ is known, one can obtain expressions for zero-shear viscosity η_0, steady-state compliance J_e^o, dynamic storage modulus $G'(\omega)$, and dynamic loss modulus $G''(\omega)$ from (Doi and Edwards 1986; Ferry 1980)

$$\eta_0 = \int_0^{\infty} G(t)\, \mathrm{d}t \tag{4.78}$$

$$J_e^o = \frac{1}{\eta_0^2} \int_0^{\infty} t\, G(t)\, \mathrm{d}t \tag{4.79}$$

$$G'(\omega) = \omega \int_0^{\infty} G(t)\, \sin \omega t\, \mathrm{d}t \tag{4.80}$$

$$G''(\omega) = \omega \int_0^{\infty} G(t) \cos \omega t\, \mathrm{d}t \tag{4.81}$$

Substitution of Eq. (4.77) into Eqs. (4.78)–(4.81) gives

$$\eta_0 = (\pi^2 K \rho RT/36) M \tag{4.82}$$

$$J_e^o = 2M/5\rho RT \tag{4.83}$$

$$G'(\omega) = \frac{\rho RT}{M} \sum_{p=1}^{\infty} \frac{\omega^2 \tau_p{}^2}{1 + \omega^2 \tau_p{}^2} \tag{4.84}$$

$$G''(\omega) = \frac{\rho RT}{M} \sum_{p=1}^{\infty} \frac{\omega \tau_p}{1 + \omega^2 \tau_p{}^2} \tag{4.85}$$

Note that K in Eq. (4.82) is defined by

$$K = \frac{\zeta b^2 N^2}{\pi^2 k_B T M^2} \tag{4.86}$$

Rouse (1953) also derived Eqs. (4.84) and (4.85) from the point of view of the work done on a polymer solution under oscillatory shear flow.

The Rouse model allows us to determine the stress $\boldsymbol{\sigma}$ contributed by the polymer chains from

$$\boldsymbol{\sigma} = \sum_{p=1}^{\infty} \boldsymbol{\sigma}_p \tag{4.87}$$

where $\boldsymbol{\sigma}_p$ denotes the stress contributed by the polymer chain at the pth mode ($p = 1, 2, \ldots, \infty$), which can be evaluated from:

$$\boldsymbol{\sigma}_p + \tau_p \frac{\mathfrak{d}\boldsymbol{\sigma}_p}{\mathfrak{d}t} = G_o \boldsymbol{\delta} \tag{4.88}$$

where $\mathfrak{d}/\mathfrak{d}t$ is the upper convected derivative (see Chapter 3), τ_p is the relaxation time defined by Eq. (4.52), and $G_o = \rho RT/M$.

The Rouse model gives the following integral-type constitutive equation (Doi and Edwards 1986)

$$\boldsymbol{\sigma}(t) = \int_{-\infty}^{t} \frac{\partial G(t - t')}{\partial t'} \mathbf{B}\big(\mathbf{E}(t, t')\big)\, \mathrm{d}t' \tag{4.89}$$

where $G(t)$ is the stress relaxation modulus defined by Eq. (4.77), $\mathbf{B}$ is the Finger strain tensor and $\mathbf{E}$ is the finite strain tensor (see Chapter 2), the two being related to each other by $\mathbf{B} = \mathbf{E} \cdot \mathbf{E}^T$.

In the steady-state shear flow, the Rouse model predicts

$$\sigma = \frac{\rho RT}{M} \sum_{p=1}^{\infty} \tau_p \dot{\gamma} \tag{4.90}$$

$$N_1 = \frac{2\rho RT}{M} \sum_{p=1}^{\infty} \tau_p{}^2 \dot{\gamma}^2 \tag{4.91}$$

$$N_2 = 0 \tag{4.92}$$

Use of Eq. (4.52) in (4.90) gives the viscosity[4]

$$\eta_0 = \frac{\rho \zeta b^2 N_A N^2}{36M} \tag{4.93}$$

where N_A is Avogadro's number. Since $N \propto M$, it is concluded that $\eta_0 \propto M$ and it is independent of shear rate $\dot{\gamma}$; that is, the Rouse model cannot predict shear-dependent viscosity.

Note that each of the submolecules (segments) considered in the bead–spring model has about 10 to 20 monomer units and $b^2 = mb_0{}^2$, where m is the number of monomer units in the submolecule, b_0 is the bond length of a monomer unit, and b is the effective step length (the Kuhn length) of the chain of mean-square end-to-end distance $\langle r^2 \rangle^{1/2}$. Osaki and Schrag (1971), for instance, reported that m is 11 for polystyrene and 16 for poly(α-methylstyrene). Note further that the friction coefficient ζ of a submolecule used in the development of the bead–spring model is related to the monomeric friction coefficient ζ_0 by $\zeta = m\zeta_0$. Thus, Eq. (4.93) can be rewritten as[5]

$$\eta_0 = \left(\frac{\rho b_0{}^2 \zeta_0 N_A}{36 M_o{}^2} \right) M \tag{4.94}$$

where use is made of $N = M/mM_o$ with M_o being the molecular weight of monomer units. By making use of Eq. (4.94), (4.52) can be rewritten as[6]

$$\tau_p = \frac{6\eta_0 M}{\pi^2 p^2 \rho RT} \tag{4.95}$$

and thus the terminal (Rouse) relaxation time becomes

$$\tau_r = \tau_1 = \frac{6\eta_0 M}{\pi^2 \rho RT} \tag{4.96}$$

The monomeric friction coefficient ζ_0 is one of the most important properties of macromolecules, which can be calculated, using Eq. (4.94), from the measurement of η_0.

Figure 4.6 gives plots of $\log G'_r$ versus $\log \tau_1\omega$ and $\log G''_r$ versus $\log \tau_1\omega$, which are predicted by the theories of Rouse and Zimm. Here, the reduced variables G'_r and G''_r, respectively, are defined by $G'_r = G'M/cRT$ and $G''_r = (G'' - \omega\eta_s)M/cRT$, where c is the concentration of polymer solution and τ_1 is the terminal relaxation time defined

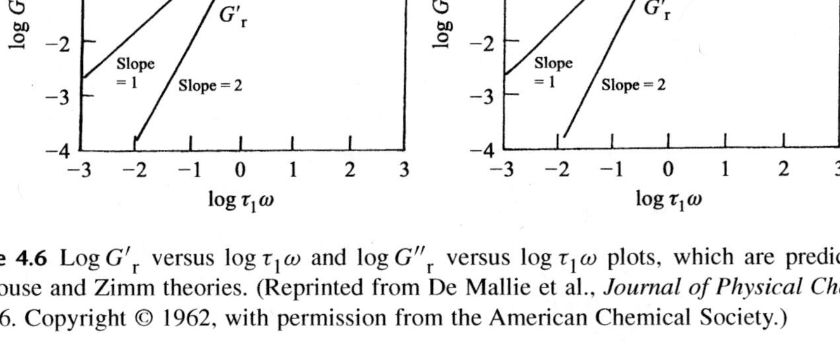

Figure 4.6 Log G'_r versus $\log \tau_1\omega$ and $\log G''_r$ versus $\log \tau_1\omega$ plots, which are predicted by the Rouse and Zimm theories. (Reprinted from De Mallie et al., *Journal of Physical Chemistry* 66:536. Copyright © 1962, with permission from the American Chemical Society.)

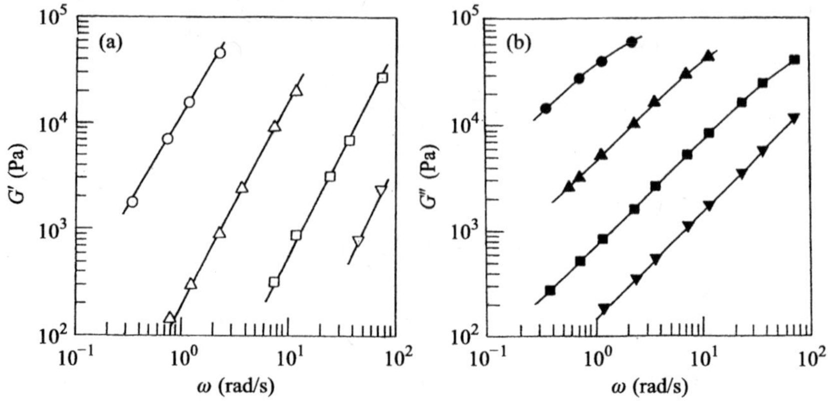

Figure 4.7 (a) Log G' versus log ω plots at various temperatures (°C): (○) 120, (△) 130, (□) 140, and (▽) 150, and (b) log G'' versus log ω plots at various temperatures (°C): (●) 120, (▲) 130, (■) 140, and (▼) 150, for a nearly monodisperse polystyrene with a molecular weight of 9×10^3, which is much lower than the M_c (3.6×10^4) of polystyrene.

by Eq. (4.53). Note that when using the Rouse model to deal with polymer melt, the concentration c is replaced by the density ρ (see Eqs. (4.84) and (4.85)). The following observations in Figure 4.6 are worth noting. At low values of $\tau_1\omega$, often referred to as the "terminal region," the slope of log G'_r versus log $\tau_1\omega$ plots is 2, while the slope of log G''_r versus log $\tau_1\omega$ plots is 1 in both the Rouse theory, where the hydrodynamic interaction is assumed to be negligible, and the Zimm theory, where the hydrodynamic interaction is included. At large values of $\tau_1\omega$, often referred to as the "transition region," the Rouse theory predicts the slope of 1/2 in both log G'_r versus log $\tau_1\omega$ and log G''_r versus log $\tau_1\omega$ plots, whereas the Zimm theory predicts the slope of 2/3 in both log G'_r versus log $\tau_1\omega$ and log G''_r versus log $\tau_1\omega$ plots.

Equations (4.84) and (4.85) have widely been used to predict the linear viscoelastic properties of polymer melts with $M < M_c$. Figure 4.7 gives log G' versus log ω plots and log G'' versus log ω plots for a nearly monodisperse polystyrene with $M = 9 \times 10^3$ at four different temperatures. Note that M_c is 3.6×10^4 for polystyrene (Ferry 1980). Thus, the polystyrene employed in Figure 4.7 is an unentangled polymer ($M < M_c$). It can be seen in Figure 4.7 that: (1) at a constant value of ω, both G' and G'' decrease with increasing temperature, (2) at a given temperature, both G' and G'' increase with increasing ω, and (3) terminal behavior is observed in both G' and G'' over the entire range of frequencies investigated.

4.4 Molecular Theory for the Viscoelasticity of Concentrated Polymer Solutions and Entangled Polymer Melts

Experiment shows that the stress relaxation modulus $G(t)$ of an entangled polymer with $M \gg M_c$ is quite different from the $G(t)$ of an unentangled polymer with $M < M_c$. This is shown schematically in Figure 4.8, where we observe two main dispersions: (1) relaxation at short times in the transition region is independent of chain length and

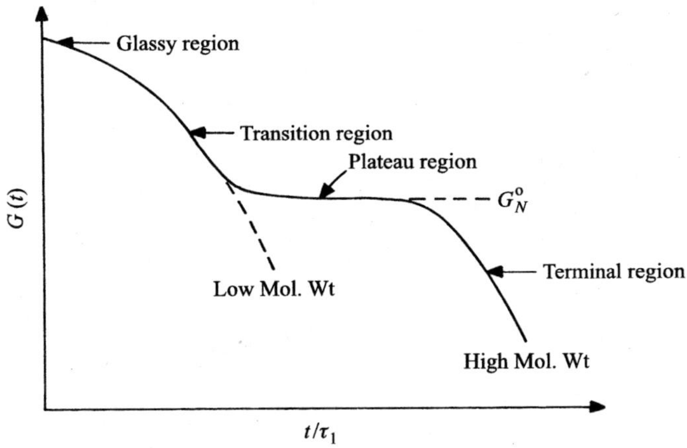

Figure 4.8 Stress relaxation modulus of unentangled and entangled polymers, displaying different regimes: transition region, plateau region, and terminal region.

appears to reflect only local rearrangements of chain conformation, and (2) relaxation at long times in the terminal region reflects the rearrangement of large-scale conformation. The location and shape of $G(t)$ depend strongly on molecular parameters: chain length and chain-length distribution. Between the terminal region and transition region in Figure 4.8 there appears a plateau region, where $G(t)$ changes only slowly with time. It has been observed that the separation in time of these two dispersions increases rapidly with chain length, but the modulus in the plateau region, G_N^o, commonly referred to as the "plateau modulus," is insensitive to molecular parameters and depends only on polymer species and concentration (Ferry 1980; Graessley 1974).

Figure 4.9 gives plots of $G(t)$ versus time at 128 °C for nearly monodisperse polystyrenes having weight-average molecular weight (M_w) ranging from 1.67×10^4 to 4.3×10^5. Note that M_c for polystyrene is 3.6×10^4 (Ferry 1980). In Figure 4.9 we observe that a plateau region begins to appear for $M_w = 4.3 \times 10^4$, which is slightly higher than the value of M_c, and a very distinct plateau region develops as M_w increases further.

There is another way of experimentally determining the plateau modulus of an entangled homopolymer. For illustration, Figure 4.10 gives reduced plots, $\log G'_r(\omega)$ versus $\log a_T\omega$ and $\log G''_r(\omega)$ versus $\log a_T\omega$, for a nearly monodisperse polybutadiene with a number-average molecular weight (M_n) of 1.3×10^5 over a very wide range of temperatures, where a_T is a shift factor that enables one to superpose experimental data at different temperatures onto a single master plot. Methods for determining a_T are discussed in Chapter 6. In Figure 4.10, G'_r and G''_r, respectively, are defined by $G'_r = (\rho_r T_r/\rho T)G'(\omega)$ and $G''_r = (\rho_r T_r/\rho T)G''(\omega)$, with ρ_r being the density at a reference temperature T_r and ρ the density at temperature T. The following observations in Figure 4.10 are worth noting. In the terminal region, the slope of $\log G'_r$ versus $\log a_T\omega$ plots is 2 and the slope of $\log G''_r$ versus $\log a_T\omega$ plots is 1. There is a very wide plateau region, from which the value of G_N^o is determined to be about 1.3×10^6 Pa.

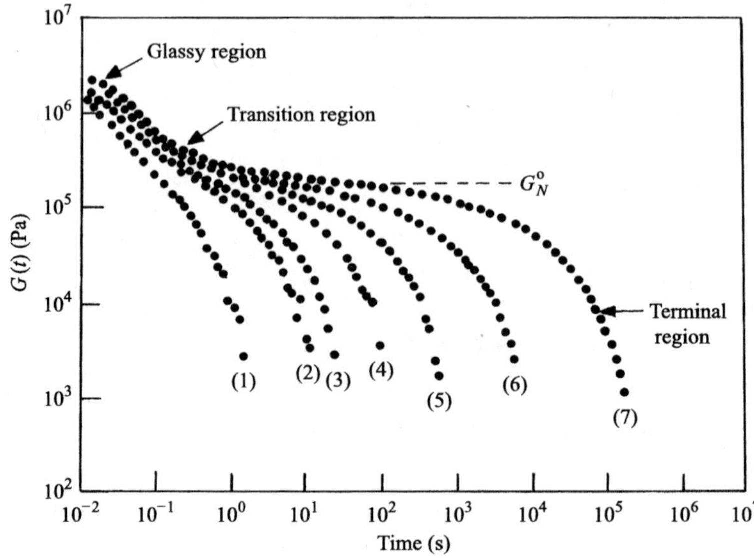

Figure 4.9 Shear stress relaxation modulus $G(t)$ versus time at 128 °C for nearly monodisperse polystyrenes having molecular weights (1) 1.67×10^4, (2) 3.51×10^4, (3) 4.39×10^4, (4) 6.80×10^4, (5) 1.02×10^5, (6) 1.79×10^5, and (7) 4.22×10^5. (Reprinted from Lin, *Macromolecules* 19:159. Copyright © 1986, with permission from the American Chemical Society.)

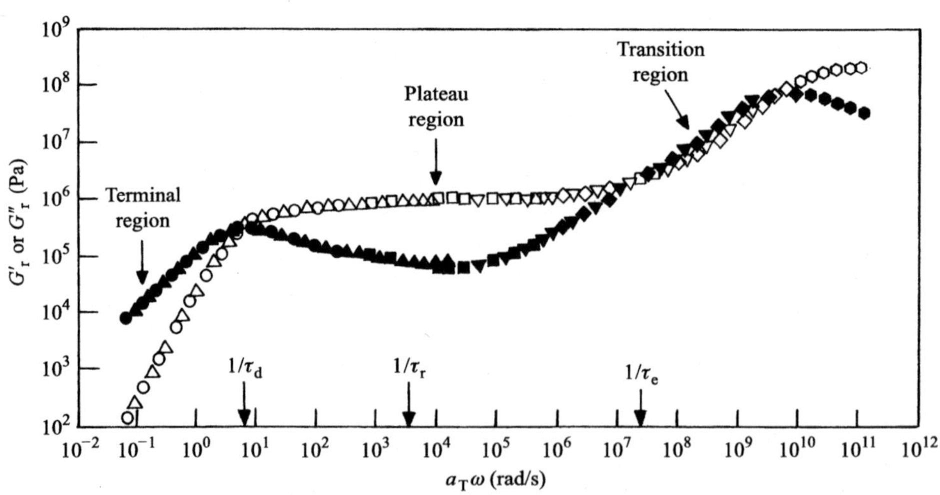

Figure 4.10 Log G'_r versus log $a_T\omega$ (open symbols) plots and log G''_r versus log $a_T\omega$ plots (filled symbols) for a nearly monodisperse polybutadiene having a molecular weight of 1.3×10^5 at various temperatures (°C): (○, ●) 25, (△, ▲) −26, (□, ■) −51, (▽, ▼) −76, (◇, ◆) −86, and (⬡, ⬢) −91. The reference temperature used is 25 °C, and the relaxation times indicated in the plot are $\tau_d = 0.124$ s, $\tau_r = 3.73 \times 10^{-4}$ s, $\tau_e = 7.16 \times 10^{-8}$ s. (Reprinted from Colby et al., *Macromolecules* 20:2226. Copyright © 1987, with permission from the American Chemical Society.)

A characteristic feature of the plateau region is that the magnitude of G'' is smaller than that of G', as illustrated in Figure 4.10. In the absence of entanglement coupling, this behavior is never seen in polymers without cross-linking and the value of G'' is always at least as large as that of G'. In the plateau region, G'' passes through a minimum. In the terminal region, the entanglements have their maximum effect in influencing the properties that reflect the longest-range molecular motions. The plateau region is one of the most important viscoelastic properties that distinguish entangled polymers from unentangled polymers. There are different ways of estimating the G_N^{o} of a polymer from a structural point of view (Aharoni 1983, 1986; Wu 1989), and values of G_N^{o} for various polymers are given in the literature (Aharoni 1983, 1986; Ferry 1980; Wu 1989).

Today, it is well-accepted that, once the value of G_N^{o} is available, one can determine the molecular weight between entanglement couplings (often referred to as "entanglement molecular weight") M_{e} from the relationship (Doi and Edwards 1986; Ferry 1980)

$$G_N^{\mathrm{o}} = \rho RT/M_{\mathrm{e}} \tag{4.97}$$

The M_{e} for polybutadiene is about 1.7×10^3 for $G_N^{\mathrm{o}} = 1.32 \times 10^6$ Pa. Note that G_N^{o} is very weakly dependent upon temperature. It is generally accepted that $M_{\mathrm{c}} \approx 2M_{\mathrm{e}}$ (Bueche 1962; Doi and Edwards 1986; Ferry 1980).

It has been suggested that the entanglement effects arise essentially from topological restrictions on the chain motions (Graessley 1965, 1967). At high concentrations of polymer solutions or in polymer melts, the chain contours are extensively intermingled, so each chain is surrounded all along its length by a mesh of neighboring chain contours. Rearrangement on a large scale is retarded because the chain cannot cross through its neighbors. To relax completely the chains must, in some sense, diffuse around one another. In this section we describe the tube model, which is based on the reptation concept for viscoelastic properties of entangled polymeric liquids.

4.4.1 Reptation Mechanism and the Tube Model

In a study of the dynamics of cross-linked chains, Edwards (1967) first conceived the idea that due to the noncrossability of chains, each chain in a highly entangled state is confined in a tube-like region. However, Edwards did not indicate how to extend this idea to the dynamics of flexible macromolecular chains. Later, de Gennes (1971, 1976) presented a key idea for the subsequent development of the "reptation" concept. He discussed the motion of a chain within the confines of fixed obstacles. Using the idea that the motion of the chain is essentially confined within a tube, de Gennes explicitly calculated the diffusion coefficient and the rotational relaxation time of the chain in the network. Later, Doi and Edwards (1978a, 1978b, 1978c, 1979) and Doi (1980a, 1980b, 1981, 1983) extended de Gennes' idea to describe the dynamics (thus viscoelastic properties) of concentrated polymer solutions and molten polymers, and they showed that many characteristic features of the nonlinear viscoelastic behavior of concentrated polymer solutions and molten polymers are naturally explained by the tube model.

In the tube model, the principal molecular motion is described by reptation; that is, motion back and forth along the chain backbone. If the characteristic length scale of motion is smaller than size a, the entanglement effect is not important and the dynamics is well described by the Rouse model (or the Zimm model if the hydrodynamic interaction is dominant). Conversely, if the length scale of the motion becomes larger than a, the dynamics of the chain is governed by reptation. Today, there is much experimental evidence indicating that reptation is the dominant motion governing the dynamics in the highly entangled state; namely, from the experimental measurement of the self-diffusion coefficient (Amis and Han 1982; Amis et al. 1984; Bartels et al. 1986; Callaghan and Pinder 1980; Green et al. 1985; Leger et al. 1981; Mills et al. 1986; Nemato et al. 1985; Smith et al. 1985, 1986; Wesson et al. 1984). The reptation model has been successful in explaining many features of the viscoelastic behavior of molten polymers in both linear and nonlinear regimes.

Doi and Edwards (1978a) introduced the concept of "primitive chain," which is the curved central axis of the tube, with the contour length L, as schematically shown in Figure 4.11. Note that the motion of the primitive chain corresponds to the overall translation of the Rouse chain along the tube. Each chain is confined in a tube and the large-scale motion of the chain will proceed through the motion along the tube. The surrounding chains also move when large-scale motion takes place along the tube. If the displacement becomes large, it distorts many surrounding chains and so is hindered.

In the tube model, the chains can stretch and contract along the tube or slide along the tube as a whole. The former motion equilibrates the fluctuation in the density of the chain segment along the tube, and the latter motion is related to the diffusion of the

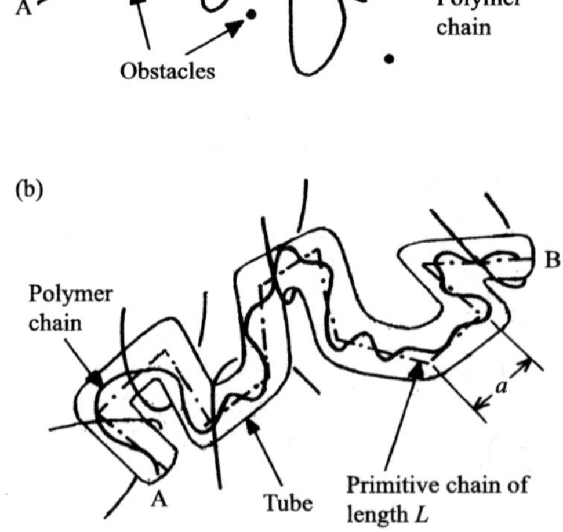

Figure 4.11 (a) Schematic representation of a flexible polymer chain with a reptative motion in an environment where other chains (denoted by dots) are assumed to lie perpendicular to the chain. (b) The tube model describing the motion of a flexible polymer chain, represented by the primitive chain, which is assumed to diffuse (i.e., undergo Brownian motion) inside a fictitious tube that consists of Z tube segments of length a.

chain as a whole. The characteristic times of these processes can be derived easily from the Rouse model consisting of N segments. In other words, the Rouse chain remains the basic model, but it is now subject to spatial constraints in the form of a tube to represent the mesh. For the chain to lose its memory of the previous deformation completely, it must slide along the tube a distance of the order of L. The characteristic time for the chain to disengage from a certain tube, τ_d, is given by (Doi and Edwards 1978a)

$$\tau_d \propto L^2/D_c \tag{4.98}$$

where D_c is the curvilinear diffusion coefficient of the primitive chain. Therefore, the primitive chain is characterized by three parameters L, a, and D_c, which must be expressed in terms of the three parameters, N, b, and ζ, appearing in the Rouse model. The parameter D_c can be identified as the diffusion coefficient of the center-of-mass of the Rouse chain (see Eq. (4.54))

$$D_c = k_B T/N\zeta \tag{4.99}$$

because the motion of the primitive chain corresponds to the overall translation of the Rouse chain along the tube. Note that the mean-square end-to-end distance of the primitive chain, which is Za^2, must be the same as that of the Rouse chain Nb^2, that is

$$\langle r^2 \rangle = Za^2 = Nb^2 \tag{4.100}$$

where Z is the number of primitive chain segments for a molecule (or the number of entanglement points per chain). Since the primitive path length L is equal to Za, we have

$$L = Nb^2/a \tag{4.101}$$

Thus, we are left with a single parameter a, which depends on the statistical nature of the network. As shown below, a is related to M_e by

$$a^2 = \frac{4}{5}\left(\frac{\langle r^2 \rangle}{M}\right) M_e \tag{4.102}$$

However, whether or not there indeed exists such a critical length a is an important question yet to be answered.

4.4.2 The Dynamics of a Primitive Chain

Let us now consider the dynamics of a primitive chain. Within the concept of reptation, in the short timescale the motion of the polymer can be regarded as wriggling around the primitive path. On a longer timescale, the conformation of the primitive path changes as the polymer moves, creating and destroying the ends of the primitive path. In the absence of an external potential, the time evolution (i.e., the dynamics) of the primitive path can be expressed by the one-dimensional diffusion equation (Doi and Edwards 1978a)

$$\frac{\partial f}{\partial t} = D_{\mathrm{c}} \frac{\partial^2 f}{\partial \xi^2} \tag{4.103}$$

where $f(\xi, t'; s)$ is the probability (or configuration distribution function) that the primitive chain moves the distance ξ, while its ends have not reached the segment s of the original tube, and D_{c} is the curvilinear diffusion coefficient. Note that Eq. (4.103) is a special case of the Smoluchowski equation, Eq. (4.20) or (4.27). The solution of Eq. (4.103), subject to the following initial and boundary conditions:

$$f(\xi, 0) = \delta(\xi); \quad f(s, t) = f(s - L, t) = 0 \tag{4.104}$$

where s is the position of the chain along some initial path ($0 < s < L$) and L is the contour length of the chain, is given by

$$f(\xi, t; s) = \sum_{p=1}^{\infty} \frac{2}{L} \sin\left(\frac{\pi p s}{L}\right) \sin\left(\frac{\pi p (s - \xi)}{L}\right) \exp(-p^2 t / \tau_{\mathrm{d}}) \tag{4.105}$$

where τ_{d} is the disengagement time (or the longest reptation time) defined by

$$\tau_{\mathrm{d}} = L^2 / D_{\mathrm{c}} \pi^2 \tag{4.106}$$

The fraction of all steps initially located at s that are still occupied after time t can be obtained by integrating Eq. (4.105), yielding

$$F(s, t) = \int_{s-L}^{s} f(\xi, t; s)\, d\xi = \sum_{\mathrm{odd}\, p}^{\infty} \frac{4}{p\pi} \sin\left(\frac{p\pi s}{L}\right) \exp(-p^2 t / \tau_{\mathrm{d}}) \tag{4.107}$$

Now the fraction of segments in the primitive chain at time t, which is still in the tube defined at time $t = 0$ (original tube), can be found by integrating Eq. (4.107), yielding

$$\Psi(t) = \frac{1}{L} \int_0^L F(s, t)\, ds = \frac{8}{\pi^2} \sum_{\mathrm{odd}\, p}^{\infty} \frac{1}{p^2} \exp(-p^2 t / \tau_{\mathrm{d}}) \tag{4.108}$$

For entangled polymers there are relaxation processes occurring in the chain segments at very short times (i.e., at $t < \tau_{\mathrm{e}}$, with τ_{e} being the equilibration time after the imposition of shear) before the stress relaxation function *G*(*t*) reaches the plateau region. In other words, the relaxation will generally involve two processes: (1) the relaxation of the contour length, and (2) the disengagement from the deformed tube, each being characterized by the time τ_{r} and τ_{d}, respectively. In the terminal region, the first relaxation process can be neglected and thus the relaxation for $t > \tau_e$ is only due to the disengagement. This can be described as follows. The timescale of the first relaxation process is essentially the Rouse relaxation times, thus from Eq. (4.53), by replacing N with N_e, we have

$$\tau_e = \frac{b^2 \zeta N_e{}^2}{6\pi^2 k_B T} \tag{4.109}$$

where N_e is the number of repeat units between two adjacent entanglement points (M_e/mM_o). Conversely, τ_d defined by Eq. (4.106) can be rewritten, with the aid of Eq. (4.99), as

$$\tau_d = \frac{Z^2 a^2 N \zeta}{\pi^2 k_B T} = \frac{Z^3 a^2 N_e \zeta}{\pi^2 k_B T} \tag{4.110}$$

where use is made of $L = Za$. From Eqs. (4.109) and (4.110) we observe $\tau_d/\tau_e = 6Z^3$,[7] indicating that $\tau_d \gg \tau_e$ for long chains, which usually is the case for entangled polymers. Note further that $\tau_d/\tau_r = 6Z$.[8]

For a very short time, say at $t < \tau_e$, the chain segment does not feel the constraints of the tube, so that the mean-square displacement of a primitive chain segment is the same as that calculated for the Rouse model in free space. At time $t \approx \tau_e$, the whole polymer is confined in a deformed tube. As time passes (i.e., at $t > \tau_e$), the Rouse behavior is stopped, because the chain feels the constraints imposed by the tube, and therefore the reptation behavior starts; that is, there exists a time τ_e at which the chain begins to feel the onset of the effect of tube constraints. For $t \gg \tau_e$, part of the polymer near the ends has disengaged from the deformed tube, while the part in the middle is still confined in the tube. Since only the segments in the deformed tube are oriented and contribute to the stress, the $G(t)$ in the terminal region is proportional to the fraction of the polymers still confined in the deformed tube $\Psi(t)$ (Doi and Edwards 1986), that is

$$G(t) = G_N^o \Psi(t) \tag{4.111}$$

Substituting Eq. (4.111), with the aid of (4.108), into Eqs. (4.78)–(4.81) we obtain expressions for η_0, J_e^o, $G'(\omega)$ and $G''(\omega)$:

$$\eta_0 = (\pi^2/12) G_N^o \tau_d \tag{4.112}$$

$$J_e^o = 6/5G_N^o \tag{4.113}$$

$$G'(\omega) = \frac{8G_N^o}{\pi^2} \sum_{\text{odd } p}^{\infty} \frac{1}{p^2} \frac{(\omega\tau_d/p^2)^2}{1 + (\omega\tau_d/p^2)^2} \tag{4.114}$$

$$G''(\omega) = \frac{8G_N^o}{\pi^2} \sum_{\text{odd } p}^{\infty} \frac{1}{p^2} \frac{\omega\tau_d/p^2}{1 + (\omega\tau_d/p^2)^2} \tag{4.115}$$

Note that the disengagement time τ_d defined by Eq. (4.106) can be rewritten, with the aid of Eqs. (4.99) and (4.101), as[9]

$$\tau_d = \frac{\zeta b^4 N^3}{\pi^2 a^2 k_B T} = (K/M_e) M^3 \tag{4.116}$$

where K is defined by Eq. (4.86). In obtaining Eq. (4.116) we made use of the relationships (1) $Z = M/M_e = N/N_e$, (2) $a^2 = N_e b^2$, (3) $M = mM_o N$, and (4) $M_e = mM_o N_e$, with m being the number of monomer units in the submolecule (i.e., in the Rouse segment). η_0 given by Eq. (4.112) can be rewritten, with the aid of Eqs. (4.86), (4.97), and (4.116), as[10]

$$\eta_0 = \left(\frac{0.10416\, \rho\, N_A \zeta_0 b_0{}^2}{M_e{}^2 M_o{}^2} \right) M^3 \tag{4.117}$$

We therefore have the following relationships:

$$\tau_d \propto M^3; \quad \eta_0 \propto M^3; \quad J_e^o \propto M^0 \tag{4.118}$$

Another important parameter in the tube model is the diffusion coefficient of the center of mass D_G of the primitive chain, which is related to the curvilinear diffusion coefficient D_c by

$$D_G = \frac{D_c}{3Z} = \frac{a^2 k_B T}{3N^2 \zeta b^2} \propto M^{-2} \tag{4.119}$$

indicating that $D_G \propto M^{-2}$ for entangled polymers. In the preceding section we showed that $D_G \propto M^{-1}$ for the Rouse model (see Eq. (4.56)).

Having shown how to compute various relaxation times, we can calculate values of τ_d, τ_r, and τ_e for a nearly monodisperse polybutadiene whose dynamic viscoelastic properties are given in Figure 4.10. Using the values $M = 1.3 \times 10^5$, $M_e = 1.8 \times 10^3$ (Ferry 1980), $\zeta_0 = 0.35 \times 10^{-6}$ dyne·s/cm at 25 °C (Ferry 1980), and $b = 6.7 \times 10^{-8}$ cm, we have determined: (1) $\tau_d = 0.124$ s using Eq. (4.116) and (4.86) with $m = 10$, (2) $\tau_r = 3.73 \times 10^{-4}$ s using Eq. (4.53) with $m = 10$, and (3) $\tau_e = 7.16 \times 10^{-8}$ s using $\tau_r/\tau_e = Z^2$, which follows from $\tau_d/\tau_e = 6Z^3$ and $\tau_d/\tau_r = 6Z$. These values are indicated in Figure 4.10.

4.4.3 Contour Length Fluctuation and Constraint Release Mechanism

As we have seen in the preceding section, the tube model as was first developed by Doi and Edwards (1978a, 1978b, 1978c, 1979) explains some important features of entangled monodisperse homopolymers, which are quite different from those of dilute polymer solutions and unentangled homopolymer melts. Table 4.1 summarizes a comparison between predictions of the bead–spring model and the tube model. In order to confirm experimentally whether the self-diffusion coefficient is indeed inversely proportional to the square of molecular weight, a number of research groups reported on

Table 4.1 Comparison of the bead–spring model with the tube model

Bead–Spring Model ($M < M_c$) (N, b, ζ)	*Tube Model* ($M \gg M_c$)(L, a, D_c)
$\eta_0 \propto M$	$\eta_0 \propto M^3$
$\tau_r \propto M^2$	$\tau_d \propto M^3$
$G_o = \rho RT/M \propto M^{-1}$	$G_N^o = \rho RT/M_e \propto M^0$
$D_G \propto M^{-1}$	$D_G \propto M^{-2}$
The memory is lost at every point of the chain.	The memory is lost only at the chain ends.

measurements of the self-diffusion coefficients for concentrated polymer solutions and polymer melts and found that the prediction of the reptation/tube model is essentially correct. However, one of the most obvious deficiencies in the tube model, which did not require experimental proof, was the dependence of η_0 on M. Specifically, while experiment shows 3.4 power dependence of η_0 on M (see Figure 4.1), the tube model predicts 3.0-power dependence. Therefore, some serious efforts were made to improve the prediction of the tube model.

4.4.3.1 Contour Length Fluctuations

While the Rouse model considers only intramolecular motions, the tube model deals with intermolecular interactions due to entanglement couplings and neglects intramolecular motions. The neglect of intramolecular motions that may occur on the timescale shorter than the timescale of reptation motions was thought to be responsible for the 3.0-power dependence of η_0 on M.

Doi (1981) incorporated fluctuations of contour length into the tube model and obtained the following expression for η_0:

$$\eta_0 = \left(\frac{K\rho RT}{3}\right)\left(\frac{M^3}{M_e^{\,2}}\right)\left\{\left[1 - (M_e/M)^{0.5}\right]^3 + 0.2(M_e/M)^{1.5}\right\} \tag{4.120}$$

where K is defined by Eq. (4.86). Doi (1983) derived another expression:

$$\eta_0 = \left(\frac{\pi^2 K\rho RT}{15}\right)\left(\frac{M^3}{M_e^{\,2}}\right)\left[1 - 1.47(M_e/M)^{0.5}\right]^3 \tag{4.121}$$

It can be shown that values of η_0 predicted from Eq. (4.121) are numerically close to the 3.4-power of M for $20M_e < M < 200M_e$.

Lin (1984) also incorporated fluctuations of contour length into the tube model and obtained the following expression for η_0:

$$\eta_0 = \left(\frac{\pi^2 K\rho RT}{15}\right)\left(\frac{M^3}{M_e^{\,2}}\right)\left\{\left[1 - (M_e/M)^{0.5}\right]^3 + (1/3)(M_e/M)^{1.5}\right\} \tag{4.122}$$

Note that for $M_e/M \ll 1$, Eqs. (4.121) and (4.122) reduce to

$$\eta_0 = \left(\frac{\pi^2 K \rho R T}{15 M_e^2}\right) M^3 \tag{4.123}$$

This observation suggests that the fluctuations of the contour length can be neglected only for $M/M_e \gg 1$. In other words, the 3.0-power dependence of η_0 on M may be valid only for very high molecular weight polymers.

Some experimental studies have attempted to verify such predictions. By measuring viscoelastic properties of a series of nearly monodisperse polybutadienes with M ranging from 1×10^3 to 1.65×10^7 ($0.5 < M/M_e < 8{,}000$), Colby et al. (1987) observed that the 3.4-power law, $\eta_0 \propto M^{3.4}$, was obeyed for M up to the M/M_e ratio of about 150 and that significant departures from the 3.4-power law were found for the M/M_e ratio of about 200 and greater. The experimental study of Colby et al. appears to support the view that there is a threshold value of M/M_e below which the 3.4-power law holds, and above which the 3.0-power-law holds. Intuitively, one would expect that such a transition will occur gradually, rather than sharply at a particular value of M/M_e.

As a matter of fact, by combining the molecular network theory and the tube model, Meister (1989) advanced an idea to predict a gradual transition from 3.4-power to 3.0-power dependence of η_0 on M as M/M_e ratio increases. Specifically, Meister obtained an expression for a 3.5-power dependence of η_0 on M:

$$\eta_0 = \left(\frac{0.001496\, \rho N_A \zeta_0 b^2}{M_e^{2.5} M_o}\right) M^{3.5} \tag{4.124}$$

It should be pointed out that Eq. (4.124) was derived without invoking fluctuations of contour length (i.e., without considering the Rouse motion in a reptating chain). The main idea behind the derivation of Eq. (4.124) is that since experimental data for η_0 is usually obtained from shear flow measurements, stress effects must be included in the reptation model; that is, when a polymer is subjected to shear flow, a relaxation of polymer chains to reptate around the entangled junctions must be taken into consideration, in addition to the reptation of the overall center-of-mass motion.

For very high molecular weight polymers, Meister (1989) modified the prediction of η_0 by the tube model to obtain the following expression:

$$\eta_0 = \left(\frac{0.02916\, \rho N_A \zeta_0 b^2}{M_e^2 M_o}\right) M^3 \tag{4.125}$$

for which two factors were taken into account: (1) the primitive path length has a distribution, and (2) only the diffusion length of the shortest segment is required to disengage the junction. Meister showed that the combination of Eqs. (4.124) and (4.125) predicts a gradual transition from a 3.4-power law to a 3.0-power law at M/M_e ratios between 200 and 2,000, which compares very favorably with the experimental data of

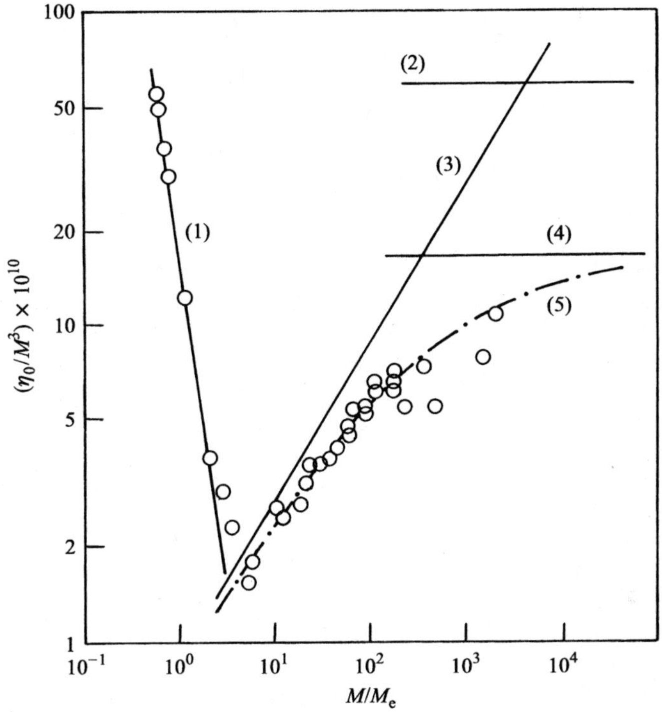

Figure 4.12 Plots of η_0/M^3 versus M/M_e for entangled polymers with various predictions: (1) from Eq. (4.94), (2) from Eq. (4.117), (3) from Eq. (4.124), (4) from Eq. (4.125), (5) from the combination of Eqs. (4.124) and (4.125). (Reprinted from Meister, *Macromolecules* 22:3611. Copyright © 1989, with permission from the American Chemical Society.)

Colby et al. (1987). This is shown in Figure 4.12, where we observe that the predictions of Eqs. (4.124) and (4.125) cross each other at $M/M_e = 479$.

4.4.3.2 Constraint Release Mechanism

In the tube model, it has been assumed that the tube is fixed in the material and that its conformational changes occur only at the end of the tube. But conformational change in the tube can occur in the middle of the tube due to both constraint release and tube deformation. Graessley (1982) proposed a constraint release mechanism, which is based on the "bond-flip model" of Orwoll and Stockmayer (1969), according to which the path of a chain is represented by a random walk of N steps with step length d. It should be recalled that in the tube model, the primitive path consists of Z steps with a step length a. According to Graessley, the surrounding mesh is represented by a regular lattice of independently reptating chains with a lattice space a. Thus a chain with Z steps occupies Z lattice cells. Each cell is bonded by neighboring chains, which represent "bars" of the cell. The bars disappear and reform as the surrounding chains move with a life time τ_w, referred to as the "waiting time," of primitive path steps located randomly along the surrounding chains. τ_w represents the average time required for the release of a constraint that allows a length preserving jump and is defined by (Graessley 1982)

$$\tau_w = \int_0^\infty [\Psi(t)]^z \, dt \tag{4.126}$$

where $\Psi(t)$ is defined by Eq. (4.108) and z is a constraint release parameter. Notice that for $z=0$, τ_w is just the average life time of the primitive path step (i.e., τ_d).

Assuming that the constraint release mechanism is independent of the pure reptation mechanism, Graessley (1982) suggested the following expression for the stress relaxation modulus:

$$G(t) = G_N^o \Psi(t) R(t) \tag{4.127}$$

where $\Psi(t)$ is defined by Eq. (4.108) and $R(t)$ is a reduced relaxation function associated with constraint release defined by

$$R(t) = \frac{1}{Z} \sum_{j=1}^{Z} \exp\left[-\lambda_j(Z)t/2\tau_w\right] \tag{4.128}$$

where $\lambda_j(Z)$ is defined by Eq. (4.49) with N replaced by Z, and τ_w is defined by Eq. (4.126). Substituting Eq. (4.127), with the aid of Eqs. (4.108) and (4.128), into (4.78)–(4.81), we obtain expressions for η_0, J_e^o, $G'(\omega)$ and $G''(\omega)$ as follows:

$$\eta_0 = \frac{8G_N^o}{\pi^2} \sum_{\text{odd } p}^{\infty} \frac{1}{p^3} \frac{\tau_{d,j}}{(p^2 + r)^{1/2}} \tag{4.129}$$

$$J_e^o = \frac{4G_N^o}{\pi^2 \eta_0{}^2} \sum_{\text{odd } p}^{\infty} \frac{1}{p^5} \frac{\tau_{d,j}{}^2 (2p^2 + r)}{(p^2 + r)^{3/2}} \tag{4.130}$$

where $r = 2\tau_d/\tau_w$ with τ_d being defined by Eq. (4.106) and

$$G'(\omega) = \frac{8G_N^o}{\pi^2} \sum_{\text{odd } p}^{\infty} \frac{1}{p^2} \frac{(\omega\tau_{d,j}/p^2)^2}{1 + (\omega\tau_{d,j}/p^2)^2} \tag{4.131}$$

$$G''(\omega) = \frac{8G_N^o}{\pi^2} \sum_{\text{odd } p}^{\infty} \frac{1}{p^2} \frac{\omega\tau_{d,j}}{1 + (\omega\tau_{d,j}/p^2)^2} \tag{4.132}$$

where $\tau_{d,j}$ is defined by

$$\frac{1}{\tau_{d,j}} = \frac{p^2}{\tau_d} + \frac{\lambda_j}{2\tau_w} \tag{4.133}$$

with λ_j being defined by (see Eq. (4.49))

$$\lambda_j = 4\sin^2\left[j\pi/2(Z+1)\right]; \quad j = 1, 2, \ldots, Z \tag{4.134}$$

For $z=3$ in Eq. (4.126) (i.e., with reptation and constraint release contributions), from Eqs. (4.129) and (4.130) we obtain $J_e^o G_N^o = 2.15$, which is consistent with experimental results, $J_e^o G_N^o = 2-3$ (Graessley 1974). Note that with $z=0$ (i.e., with reptation contribution alone) we obtain $J_e^o G_N^o = 1.2$ (see Eq. (4.113)), Eq. (4.131) reduces to (4.114), and Eq. (4.132) reduces to (4.115).

4.4.4 Constitutive Equations of State

In order to obtain a constitutive equation, one must have an expression for the stress tensor. In developing an expression for the stress tensor within the spirit of the tube model described previously, Doi and Edwards (1978b) introduced the concept of slip-links (or small loops) through which the chain may pass, as shown schematically in Figure 4.13, which assumes that (1) the primitive chain is defined by a line joining the slip-links and it passes freely through small rings, (2) the primitive chain segment is defined by a line segment between two slip-links, and (3) in the equilibrium state the slip-links are separated by a distance a and the chain must be pulled out at its ends with a constant tensile force F, which keeps the motion of the chain within the tube. Note that these slip-links represent the effect of entanglement points in a melt. Doi and Edwards then expressed the stress tensor $\boldsymbol{\sigma}$ by

$$\sigma_{\alpha\beta} = \left\langle \sum_i F_{i\alpha} r_{i\beta} \right\rangle + p\delta_{\alpha\beta} \tag{4.135}$$

where the bracket $\langle\ \rangle$ denotes the average over the equilibrium configuration function, the summation is taken over the primitive chain segments in a single chain, $\mathbf{r}_i$ is the vector joining the neighboring slip-links (or the end-to-end vector of the ith tube segment), subscripts α and $\beta (= x, y, z)$ indicate the components of the vectors or tensors, p is the hydrostatic pressure, and $\mathbf{F}_i$ is the force acting between these slip-links expressed by

$$\mathbf{F}_i = \frac{3k_B T}{n_i b^2}\mathbf{r}_i \tag{4.136}$$

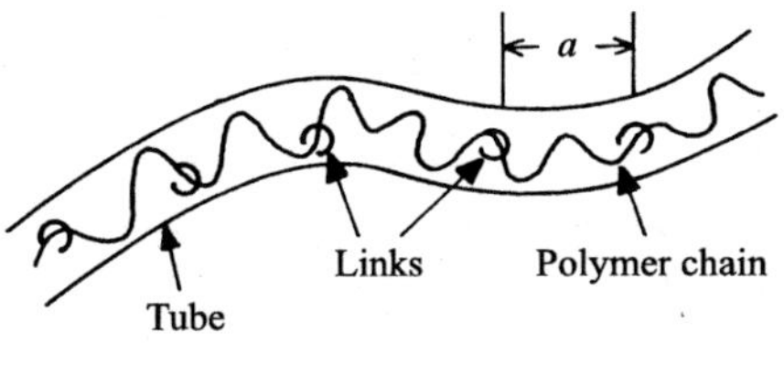

Figure 4.13 The slip-links model of Doi and Edwards, which hypothesizes that the primitive chain is defined by a line segment between slip-links and that it passes through small rings, where the slip-links are separated by a distance a.

where n_i is the number of Rouse segments in the ith tube segment and b is the bond length of a real chain. Substitution of Eq. (4.136) into (4.135) gives

$$\sigma_{\alpha\beta} = 3\nu k_{\mathrm{B}} T \left\langle \sum_i \frac{r_{i\alpha} r_{i\beta}}{n_i b^2} \right\rangle + p\delta_{\alpha\beta} \tag{4.137}$$

Consider that at time $t=0$, a flexible, linear polymer is suddenly deformed homogeneously and the deformation is kept constant thereafter. If we assume that each slip-link point changes its position affinely, Eq. (4.137) can be rewritten (Doi and Edwards 1978b) as[11]

$$\sigma_{\alpha\beta}(t) = 3\nu k_{\mathrm{B}} T \left\langle \sum_i \frac{u_{i\alpha}(t)\, u_{i\beta}(t)}{a} \right\rangle + p\delta_{\alpha\beta} \tag{4.138}$$

where use was made of $\mathbf{r}_{i\alpha} = a\mathbf{u}_\alpha(t)$, with $\mathbf{u}_\alpha(t)$ being a unit vector parallel to the primitive chain segment, and ν is the number of molecules per unit volume. Using the arc length coordinate s ($0 \le s \le \mathrm{L}$), Eq. (4.138) can be expressed by

$$\sigma_{\alpha\beta}(t) = \frac{3\nu k_{\mathrm{B}} T}{a} \int_0^L S_{\alpha\beta}(s,t)\,\mathrm{d}s + p\delta_{\alpha\beta} \tag{4.139}$$

where

$$\begin{aligned} S_{\alpha\beta}(s,t) &= \langle u_\alpha(s,t) u_\beta(s,t) - (1/3)\delta_{\alpha\beta} \rangle \\ &= \frac{1}{4\pi} \int_0^{2\pi} \int_0^{\pi} f(s,t;\mathbf{u}) \left[u_\alpha(s,t) u_\beta(s,t) - (1/3)\delta_{\alpha\beta} \right] \sin\theta\,\mathrm{d}\theta\,\mathrm{d}\phi \end{aligned} \tag{4.140}$$

in which $f(s, t; \mathbf{u})$ is the configuration distribution function.[12] By solving the Langevin equation to obtain the expression for $S_{\alpha\beta}(s, t)$, Doi and Edwards (1978b) obtained the following expression for $\boldsymbol{\sigma}(t)$:

$$\sigma_{\alpha\beta}(t) = G_{\mathrm{o}} \Psi(t) Q_{\alpha\beta}(\mathbf{E}) + p\delta_{\alpha\beta} \tag{4.141}$$

where the time-dependent function $\Psi(t)$ is defined by Eq. (4.108), G_{o}, which has the dimensions of stress, is defined by[13].

$$G_{\mathrm{o}} = 3\nu k_{\mathrm{B}} T L/a \tag{4.142}$$

and $\mathbf{Q}(\mathbf{E})$ is the orientation tensor defined by

$$Q_{\alpha\beta}(\mathbf{E}) = \left\{ \left\langle \frac{(\mathbf{E}\cdot\mathbf{u})_\alpha (\mathbf{E}\cdot\mathbf{u})_\beta}{|\mathbf{E}\cdot\mathbf{u}|} \right\rangle_{\mathrm{o}} \Big/ \langle |\mathbf{E}\cdot\mathbf{u}| \rangle_{\mathrm{o}} \right\} - (1/3)\delta_{\alpha\beta} \tag{4.143}$$

where $\mathbf{E}$ is the deformation tensor.

In order to obtain a closed form of a constitutive equation of state, Doi and Edwards (1978b) made an additional assumption, referred to as "independent-alignment approximation" (IAA), which states basically that the individual tangent vector $\mathbf{u}(s)$ of the primitive chain is assumed to undergo the same transformation, independent of other parts of the chain, as the macroscopic unit vector embedded in the material; that is, when the deformation tensor is $\mathbf{E}$, $\mathbf{u}(s)$ is changed to $\mathbf{u}'(s) = \mathbf{E}\cdot\mathbf{u}(s)/|\mathbf{E}\cdot\mathbf{u}(s)|$.[14] Note that the IAA satisfies the condition that the arc length of the primitive chain remains constant when the wriggling motion of the chain reaches an equilibrium. The IAA simplifies $\mathbf{Q}(\mathbf{E})$, defined by Eq. (4.143), into (Doi and Edwards 1978c)

$$Q_{\alpha\beta}^{\mathrm{IA}}(\mathbf{E}) = \left\langle \frac{(\mathbf{E}\cdot\mathbf{u})_\alpha\,(\mathbf{E}\cdot\mathbf{u})_\beta}{|\mathbf{E}\cdot\mathbf{u}|^2} - (1/3)\delta_{\alpha\beta} \right\rangle_{\mathrm{o}} \tag{4.144}$$

Next, we show, with an example of shear deformation, the difference in the prediction of stress relaxation between the use of $\mathbf{Q}(\mathbf{E})$ and $\mathbf{Q}^{\mathrm{IA}}(\mathbf{E})$.

Let us consider shear flow, for which the deformation tensor $\mathbf{E}$ is given by

$$\mathbf{E} = \begin{Vmatrix} 1 & \gamma & 0 \\ 0 & 1 & 0 \\ 0 & 0 & 1 \end{Vmatrix} \tag{4.145}$$

where γ is strain in shear deformation. The xy-component of $\mathbf{Q}(\mathbf{E})$ is written as

$$Q_{xy}(\mathbf{E}) = \left\{ \left\langle \frac{(\mathbf{E}\cdot\mathbf{u})_x (\mathbf{E}\cdot\mathbf{u})_y}{|\mathbf{E}\cdot\mathbf{u}|} \right\rangle_{\mathrm{o}} \Big/ \langle |\mathbf{E}\cdot\mathbf{u}| \rangle_{\mathrm{o}} \right\} - (1/3)\delta_{\alpha\beta} \tag{4.146}$$

and the xy-component of $\mathbf{Q}^{\mathrm{IA}}(\mathbf{E})$ is given by

$$Q_{xy}^{\mathrm{IA}}(\mathbf{E}) = \left\langle \frac{(\mathbf{E}\cdot\mathbf{u})_x\,(\mathbf{E}\cdot\mathbf{u})_y}{|\mathbf{E}\cdot\mathbf{u}|^2} - (1/3)\delta_{\alpha\beta} \right\rangle_{\mathrm{o}} \tag{4.147}$$

in which

$$(\mathbf{E}\cdot\mathbf{u})_x = u_x + \gamma u_y; \quad (\mathbf{E}\cdot\mathbf{u})_y = u_y; \quad |\mathbf{E}\cdot\mathbf{u}| = \left[(u_x + \gamma u_y)^2 + u_y{}^2 + u_z{}^2\right]^{1/2} \tag{4.148}$$

Thus

$$Q_{xy}(\gamma) = \frac{\left\langle \frac{(u_x + \gamma u_y)u_y}{[(u_x+\gamma u_y)^2 + u_y{}^2 + u_z{}^2]^{1/2}} \right\rangle_{\mathrm{o}}}{\left\langle [(u_x + \gamma u_y)^2 + u_y{}^2 + u_z{}^2]^{1/2} \right\rangle_{\mathrm{o}}} \tag{4.149}$$

$$Q_{xy}^{\mathrm{IA}}(\gamma) = \left\langle \frac{(u_x + \gamma u_y)u_y}{[(u_x + \gamma u_y)^2 + u_y{}^2 + u_z{}^2]} \right\rangle_{\mathrm{o}} \tag{4.150}$$

Using spherical coordinates, Eq. (4.149) can be expressed as

$$Q_{xy}(\gamma) = \frac{\frac{1}{4\pi}\int_0^{2\pi}\int_0^{\pi}\left[\frac{\sin^2\theta\cos\theta\sin\phi + \gamma\sin^2\theta\sin^2\phi}{(1+2\gamma\sin^2\theta\cos\phi\sin\phi + \gamma^2\sin^2\theta\sin^2\phi)^{1/2}}\right]\sin\theta\,d\theta\,d\phi}{\frac{1}{4\pi}\int_0^{2\pi}\int_0^{\pi}(1+2\gamma\sin^2\theta\cos\phi\sin\phi + \gamma^2\sin^2\theta\sin^2\phi)^{1/2}\sin\theta\,d\theta\,d\phi} \tag{4.151}$$

and Eq. (4.150) as

$$Q_{xy}^{\mathrm{IA}}(\gamma) = \frac{\frac{1}{4\pi}\int_0^{2\pi}\int_0^{\pi}(\sin^2\theta\cos\theta\sin\phi + \gamma\sin^2\theta\sin^2\phi)\sin\theta\,d\theta\,d\phi}{\frac{1}{4\pi}\int_0^{2\pi}\int_0^{\pi}[1+2\gamma\sin^2\theta\cos\phi\sin\phi + \gamma^2\sin^2\theta\sin^2\phi]\sin\theta\,d\theta\,d\phi} \tag{4.152}$$

Numerical integration of Eqs. (4.151) and (4.152) can be carried out, the results of which are plotted in Figure 4.14 for comparison purposes. It can be concluded from Figure 4.14 that the difference between the two, (1) $Q_{xy}(\gamma)/\gamma$ and (2) $Q_{xy}^{\mathrm{IA}}(\gamma)/\gamma$, is regarded to be negligible for all intents and purposes.

For small values of $\gamma(\ll 1)$, the denominator of Eq. (4.149) reduces to (Doi and Edwards 1986)

$$\alpha(\gamma) = \langle|\mathbf{E}\cdot\mathbf{u}|\rangle_o = \left\langle 1 + \gamma u_x u_y - 0.5\gamma^2 u_x{}^2 u_y{}^2 + 0.5\gamma^2 u_y{}^2\right\rangle_o \tag{4.153}$$

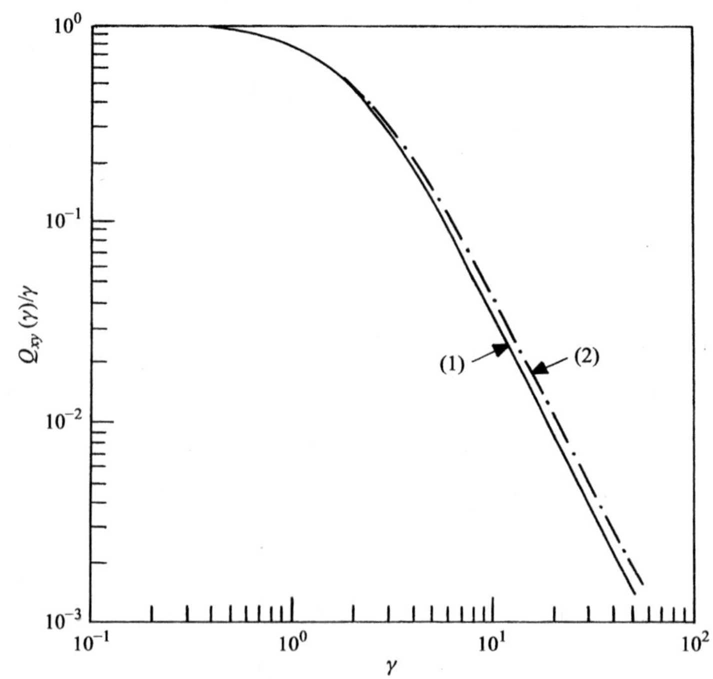

Figure 4.14 Plots of $\log Q_{xy}/\gamma$ versus $\log\gamma$ in the tube model: (1) without IAA and (2) with IAA.

Thus, Eq. (4.149) can be rewritten, with the aid of (4.153), as

$$Q_{xy}(\gamma) = \gamma \left\langle u_y{}^2 - u_x{}^2 u_y{}^2 \right\rangle_0 + 0(\gamma^2) \tag{4.154}$$

which, after appropriate integrations, gives

$$Q_{xy}(\gamma) = (4/15)\gamma \tag{4.155}$$

It can be shown from Eq. (4.150) that for small values of γ, $Q_{xy}^{\mathrm{IA}}(\gamma)$ becomes

$$Q_{xy}^{\mathrm{IA}}(\gamma) = (1/5)\gamma \tag{4.156}$$

Using IAA, Doi and Edwards (1978b) derived the following constitutive equation:

$$\boldsymbol{\sigma}(t) = G_{\mathrm{o}} \int_{-\infty}^{t} \frac{\partial \Psi(t - t')}{\partial t'} \mathbf{Q}\big(\mathbf{E}(t, t')\big)\, \mathrm{d}t' \tag{4.157}$$

where G_{o} is defined by Eq. (4.142), the time-dependent function $\Psi(t)$ is defined by Eq. (4.108), and the orientation tensor $\mathbf{Q}(\mathbf{E})$ is given by Eq. (4.144). For convenience, the superscript IA on $\mathbf{Q}$ has been omitted. Equation (4.157) can be rewritten, with the aid of (4.108), as

$$\boldsymbol{\sigma}(t) = G_{\mathrm{o}} \int_{-\infty}^{t} m(t - t')\mathbf{Q}\big(\mathbf{E}(t, t')\big)\, \mathrm{d}t' \tag{4.158}$$

where the memory function $m(t)$ is defined by

$$m(t) = \frac{8}{\pi^2} \sum_{\text{odd } p}^{\infty} \frac{1}{\tau_{\mathrm{d}}} \exp\left(-p^2 t/\tau_{\mathrm{d}}\right) \tag{4.159}$$

For shear flow defined by Eq. (4.145), Q_{xy} is given by (4.150), and $Q_{xx} - Q_{yy}$ and $Q_{yy} - Q_{zz}$, respectively, are given by

$$Q_{xx} - Q_{yy} = \left\langle \frac{(u_x + \gamma u_y)^2 - u_y{}^2}{(u_x + \gamma u_y)^2 + u_y{}^2 + u_z{}^2} \right\rangle_0 \tag{4.160}$$

$$Q_{yy} - Q_{zz} = \left\langle \frac{u_y{}^2 - u_z{}^2}{(u_x + \gamma u_y)^2 + u_y{}^2 + u_z{}^2} \right\rangle_0 \tag{4.161}$$

In spherical coordinates, by taking the reference axis θ as y-axis, Eqs. (4.150), (4.160), and (4.161), can be rewritten as

$$Q_{xy}(\gamma) = F_1(\gamma) \tag{4.162}$$

$$Q_{xx}(\gamma) - Q_{yy}(\gamma) = \gamma F_1(\gamma) \tag{4.163}$$

$$Q_{yy}(\gamma) - Q_{zz}(\gamma) = -F_2(\gamma) \tag{4.164}$$

where

$$F_1(\gamma) = \frac{1}{2\gamma}\int_0^1 \left[1 + \frac{x^2\gamma^2 - 1}{g(x,\gamma)}\right] \mathrm{d}x \tag{4.165}$$

$$F_2(\gamma) = \frac{1}{2}\int_0^1 \left[1 - \frac{(6+\gamma^2)x^2 - 1}{g(x,\gamma)}\right] \mathrm{d}x \tag{4.166}$$

in which $g(x, \gamma) = [x^4(\gamma^4 + 4\gamma^2) - 2\gamma^2 x^2 + 1]^{1/2}$.

For steady-state shear flow, three material functions, $\eta(\dot\gamma)$, $N_1(\dot\gamma)$, and $N_2(\dot\gamma)$, can be expressed by

$$\eta(\dot\gamma) = (G_\mathrm{o}/\dot\gamma)\int_0^\infty m(s)\, F_1(\dot\gamma s)\, \mathrm{d}s \tag{4.167}$$

$$N_1(\dot\gamma) = G_\mathrm{o}\dot\gamma \int_0^\infty s\, m(s)\, F_1(\dot\gamma s)\, \mathrm{d}s \tag{4.168}$$

$$N_2(\dot\gamma) = -G_\mathrm{o}\int_0^\infty m(s)\, F_2(\dot\gamma s)\, \mathrm{d}s \tag{4.169}$$

where $\dot\gamma$ denotes shear rate, $F_1(\dot\gamma s)$ and $F_2(\dot\gamma s)$ are defined by Eqs. (4.165) and (4.166), respectively, the memory function $m(s)$ by Eq. (4.159), and G_o by (4.142). One can calculate values of $\eta(\dot\gamma)$, $N_1(\dot\gamma)$, and $N_2(\dot\gamma)$ as functions of $\dot\gamma$.

Let us now consider uniaxial elongational viscosity that can be predicted by Eq. (4.158). For uniaxial elongational flow, the deformation tensor **E** is given by

$$\mathbf{E} = \left\| \begin{matrix} \varepsilon^{-1/2} & 0 & 0 \\ 0 & \varepsilon^{-1/2} & 0 \\ 0 & 0 & \varepsilon \end{matrix} \right\| \tag{4.170}$$

where ε is strain in elongational flow, and Eq. (4.158) gives

$$\sigma_{zz} - \sigma_{xx} = G_\mathrm{o}\int_{-\infty}^{t} m(t-t')F_3\left[\varepsilon(t,t')\right] \mathrm{d}t' \tag{4.171}$$

where

$$\varepsilon(t,t') = \exp\left[\int_{t'}^{t} \dot\varepsilon\,(t'')\, \mathrm{d}t''\right] \tag{4.172}$$

in which $\dot\varepsilon$ denotes steady-state elongation rate. For the orientation tensor **Q** given by Eq. (4.144), the function $F_3(\varepsilon)$ can be calculated from

$$F_3(\varepsilon) = Q_{zz} - Q_{xx} = \left\langle \frac{\varepsilon^2 {u_z}^2 - \varepsilon^{-1}{u_x}^2}{\varepsilon^2 {u_z}^2 + \varepsilon^{-1}({u_x}^2 + {u_y}^2)} \right\rangle_\mathrm{o} \tag{4.173}$$

Using spherical coordinates, Eq. (4.173) can be rewritten as

$$F_3(\varepsilon) = \frac{1}{4\pi}\int_0^{2\pi}\int_0^{\pi}\left[\frac{\varepsilon^2\cos^2\theta - \varepsilon^{-1}\sin^2\theta\cos^2\phi}{\varepsilon^{-1} + (\varepsilon^2 - \varepsilon^{-1})\cos^2\theta}\right]\sin\theta\,d\theta\,d\phi \tag{4.174}$$

Thus

$$F_3(\varepsilon) = \frac{3}{2}\left(\frac{\varepsilon^3}{\varepsilon^3 - 1}\right)\left(1 - \frac{\tan^{-1}\sqrt{\varepsilon^3 - 1}}{\sqrt{\varepsilon^3 - 1}}\right) - \frac{1}{2} \quad \text{for } \varepsilon > 1 \tag{4.175a}$$

$$F_3(\varepsilon) = \frac{3}{2}\left(\frac{\varepsilon^3}{1 - \varepsilon^3}\right)\left(\frac{\tan^{-1}\sqrt{1 - \varepsilon^3}}{\sqrt{1 - \varepsilon^3}} - 1\right) - \frac{1}{2} \quad \text{for } \varepsilon < 1 \tag{4.175b}$$

Therefore, the steady-state uniaxial elongation viscosity $\eta_E(\dot{\varepsilon})$ can be calculated from

$$\eta_E(\dot{\varepsilon}) = \left(G_o/\dot{\varepsilon}\right)\int_0^{\infty} m(s)\,F_3(e^{\dot{\varepsilon}s})\,ds \tag{4.176}$$

where use was made of $\dot{\varepsilon} = d\ln\varepsilon/dt$.

Before closing this section it should be pointed out that Eq. (4.158) can be written in the form of the K–BKZ model (see Chapter 3), by expressing the orientation tensor $\mathbf{Q}$ in terms of the Finger tensor $\mathbf{C}^{-1}$ (see Chapter 2) as (Doi and Edwards 1978c)

$$\mathbf{Q}(\mathbf{E}) = H_1(I_1, I_2, I_3)\mathbf{C}^{-1} + H_2(I_1, I_2, I_3)(\mathbf{C}^{-1})^2 + H_3(I_1, I_2, I_3)\boldsymbol{\delta} \tag{4.177}$$

where I_1, I_2, and I_3 are the first, second, and third invariants of $\mathbf{C}^{-1}$. By substituting Eq. (4.177) into (4.158), with the understanding that $I_3 = 0$ for incompressible fluids and thus the arguments of the functions H_1, H_2, and H_3 are not dependent upon I_3, we obtain (Doi and Edwards 1978c)

$$\begin{aligned}\boldsymbol{\sigma}(t) = G_o\int_{-\infty}^{t} m(t - t')\Big\{&H_1\left[I_1(t,t'), I_2(t,t')\right]\mathbf{C}^{-1}(t,t') \\ &+ H_2\left[I_1(t,t'), I_2(t,t')\right]\left(\mathbf{C}^{-1}(t,t')\right)^2\Big\}\,dt'\end{aligned} \tag{4.178}$$

It now becomes clear that Eq. (4.178) may be regarded as a special case of the K–BKZ model presented in Chapter 3.

4.4.5 Comparison of Prediction with Experiment

4.4.5.1 Steady-State Shear Flow Properties

Let us now look at steady-state shear flow properties that the constitutive equation, Eq. (4.158), predicts. Figure 4.15 gives $\log\eta/\eta_0$ versus $\log\tau_d\dot{\gamma}$ plots, which were obtained by numerical integration of Eq. (4.167) with the aid of (4.159) for the memory function $m(t)$ and Eq. (4.165) for $F_1(\dot{\gamma}t)$. Also plotted in Figure 4.15 are,

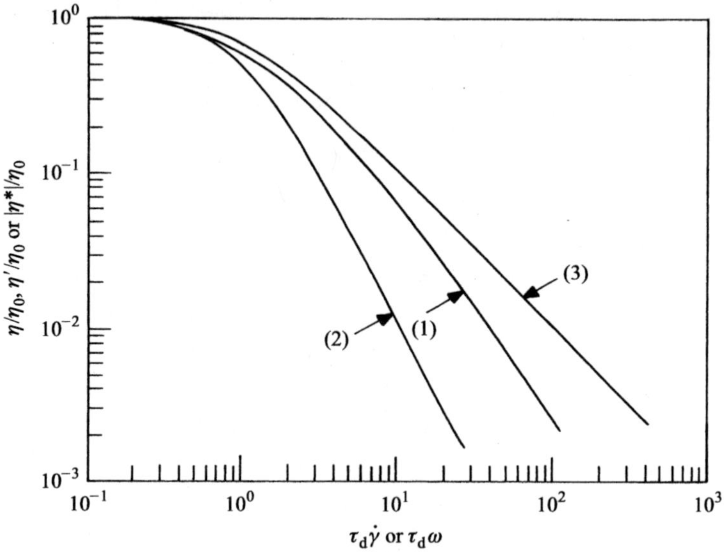

Figure 4.15 Plots of (1) $\log \eta/\eta_0$ versus $\log \tau_d\dot{\gamma}$, (2) for $\log \eta'/\eta_0$ versus $\log \tau_d\omega$, and (3) for $\log |\eta *|/\eta_0$ versus $\log \tau_d\omega$, which are predicted with the tube model.

for comparison, $\log \eta'/\eta_0$ and $\log |\eta^*|/\eta_0$ versus $\log \tau_d\omega$ plots. Note that values of $|\eta^*|$ were calculated from (Doi and Edwards 1978c):

$$|\eta^*(\omega)| = \frac{G_o}{5}\left|\sum_{\text{odd } p}^{\infty} \frac{8}{\pi^2 p^4}\left[\frac{\tau_d}{1 + i\omega\tau_d/p^2}\right]\right| \tag{4.179}$$

$\eta'(\omega)$ from

$$\eta'(\omega) = \frac{G_o}{5}\sum_{\text{odd } p}^{\infty} \frac{8}{\pi^2 p^4}\left[\frac{\tau_d}{1 + (\omega\tau_d/p^2)^2}\right] \tag{4.180}$$

and η_0 from

$$\eta_0 = \lim_{\omega\to 0} |\eta^*(\omega)| = (\pi^2/60)G_o\tau_d \tag{4.181}$$

Comparison of Eq. (4.181) with (4.112), with the aid of (4.142), gives us

$$G_N^o = (1/5)G_o = (3/5)\nu Z k_B T \tag{4.182}$$

where use is made of $Z = L/a$. It should be noted that when IAA is *not* made we have

$$G_N^o = (4/15)G_o = (4/5)\nu Z k_B T \tag{4.183}$$

We can now relate the step size a with quantities that can be measured by experiment, namely

$$a^2 = \frac{\langle r^2 \rangle}{Z} = \frac{\langle r^2 \rangle}{(5G_N^o/4\upsilon k_B T)} = \tfrac{4}{5}\left(\frac{\langle r^2 \rangle}{M}\right) M_e \tag{4.184}$$

where use is made of $\rho = \upsilon M/N_A$ and Eq. (4.97). Notice that Eq. (4.184) is identical to (4.102).

It is interesting to observe in Figure 4.15 that for very large values of $\dot{\gamma}$, η decreases with the slope of $-3/2$ in log–log plots as $\dot{\gamma}$ increases. This asymptotic behavior of η can be verified by observing that for large values of $\dot{\gamma}$, $m(t)$, defined by Eq. (4.159), can be approximated by

$$m(t) \approx \frac{4}{\pi^2 \tau_d} \int_0^\infty \exp(-p^2 t/\tau_d)\, dp = \frac{2}{\pi^{3/2}\tau_d}\left(\frac{\tau_d}{t}\right)^{1/2} \tag{4.185}$$

thus

$$\eta(\dot{\gamma}) = \frac{2G_o}{(\pi\dot{\gamma})^{3/2}\tau_d{}^{1/2}} \int_0^\infty \xi^{-1/2} F_1(\xi)\, d\xi \propto \dot{\gamma}^{-3/2} \tag{4.186}$$

Notice in Figure 4.15 that for very large values of ω, $|\eta^*|$ decreases with the slope of -1 in log–log plots as ω increases, and η' decreases with the slope of -2 in log–log plots as ω increases. The asymptotic behavior can easily be verified from Eqs. (4.179) and (4.180). It has been suggested in the literature (Cox and Merz 1958), on an empirical basis, that values of $|\eta^*|$ are almost the same as those of $\eta(\dot{\gamma})$ (see Chapter 5), but the constitutive equation, Eq. (4.158), does not support this.

Figure 4.16 gives experimentally obtained log η versus log $\dot{\gamma}$ plots at 180 and 200 °C for a nearly monodisperse polystyrene with $M_w = 1.35 \times 10^5$ and $M_w/M_n = 1.06$. Figure 4.17 shows $\log \eta/\eta_0$ versus $\log \tau_d \dot{\gamma}$ plots prepared using the data given in Figure 4.16. The values of τ_d used in Figure 4.17 are 0.093 s at 180 °C and 0.0162 s at 200 °C. Also given in Figure 4.17 is, for comparison, the theoretical prediction (solid curve) from Eq. (4.167). The experimental results given in Figure 4.17 show that the dependence of η on $\dot{\gamma}$ for the polystyrene melt is not as strong as Eq. (4.186) predicts. More experimental observations on the dependence of η on $\dot{\gamma}$ for other molten polymers are presented in Chapter 6.

Figure 4.18 gives $\log |\eta^*|/\eta_0$ and $\log \eta'/\eta_0$ plots versus $\log \tau_d \omega$ for the same polystyrene as used in figures 4.16 and 4.17. Also given in Figure 4.18 are, for comparison, theoretical predictions (solid curves) from Eq. (4.179) for $|\eta^*|$ and from (4.180) for η'. In Figure 4.18 we observe that while there is a good agreement between

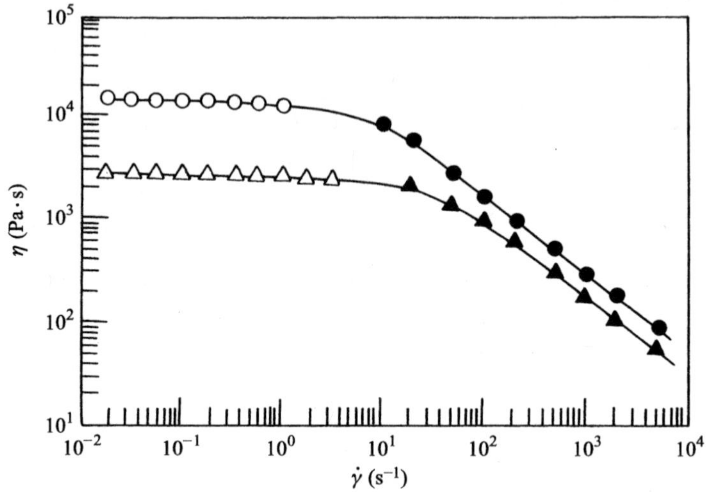

Figure 4.16 Log η versus log $\dot{\gamma}$ for a nearly monodisperse polystyrene with $M_w = 1.35 \times 10^5$ and $M_w/M_n = 1.06$, where the data at low shear rates (open symbols) were obtained using a cone-and-plate rheometer and the data at high shear rates (filled symbols) were obtained using a capillary rheometer at 180 °C (○, ●) and at 200 °C (△, ▲). η_0 at 180 °C is 1.53×10^4 Pa·s and η_0 at 200 °C is 3.06×10^3 Pa·s.

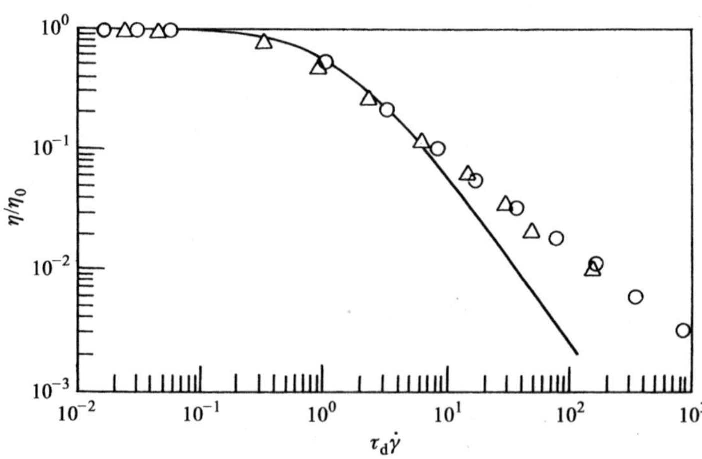

Figure 4.17 Comparison of experimentally determined plots of log η/η_0 versus log $\tau_d\dot{\gamma}$ with the prediction (solid line) made with the tube model for a nearly monodisperse polystyrene with $M_w = 1.35 \times 10^5$ and $M_w/M_n = 1.06$ at 180 °C (○) and at 200 °C (△). The values of τ_d used were 0.93×10^{-1} s at 180 °C and 1.62×10^{-2} s at 200 °C.

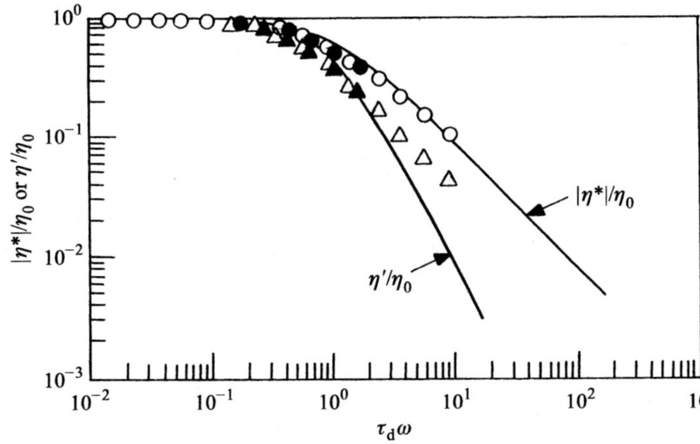

Figure 4.18 Comparison of experimentally determined $\log|\eta^*|/\eta_0$ versus $\log \tau_d\omega$ plots at 180 °C (○) and 200 °C (●), and $\log \eta'/\eta_0$ versus $\log \tau_d\omega$ plots at 180 °C (△) and 200 °C (▲), with the predictions (solid lines) made from the tube model for a nearly monodisperse polystyrene with $M_w = 1.35 \times 10^5$ and $M_w/M_n = 1.06$. The values of τ_d used are 0.93×10^{-1} s at 180 °C and 1.62×10^{-2} s at 200 °C.

experiment and theory for the $\log|\eta^*|/\eta_0$ versus $\log \tau_d\omega$ plots, the theory predicts much stronger frequency dependence of η' than the experimental results.

Equation (4.186) suggests that at very high values of $\dot{\gamma}$ the steady-state shear stress σ decreases with increasing $\dot{\gamma}$; that is, $\sigma \propto \dot{\gamma}^{-1/2}$. Since the constitutive equation, Eq. (4.158), predicts that σ increases with $\dot{\gamma}$ for small values of $\dot{\gamma}$, this observation suggests that there exists a critical value of $\dot{\gamma}$ at which σ goes through a maximum. One can find that σ will have a maximum at $\tau_d\dot{\gamma} = 3.196$, as shown in Figure 4.19. An attempt (McLeish 1988) was made to relate the existence of such a critical value of $\tau_d\dot{\gamma}$ in the $\log \sigma/G_o$ versus $\tau_d\dot{\gamma}$ plots, based on Eq. (4.158), to the onset of a flow instability, referred to as "melt fracture," which is often encountered in melt extrusion when σ exceeds a certain critical value. The experimentally obtained $\log \sigma/G_N^o$ versus $\log \tau_d\dot{\gamma}$ plot at 180 °C for the same polystyrene as used in Figures 4.16 is given in Figure 4.20, in which the following numerical values were used: $\tau_d = 0.093$ s at 180 °C and $G_N^o = 2.02 \times 10^5$ Pa. It can be seen in Figure 4.20 that σ/G_N^o increases with increasing $\tau_d\dot{\gamma}$ without going through a maximum, which is at variance with the theoretical prediction displayed in Figure 4.19.

Figure 4.21 gives logarithmic plots of Ψ_1/Ψ_{10} and $\log \Psi_2/\Psi_{20}$ versus $\tau_d\dot{\gamma}$, which were obtained by numerical integration of the right-hand side term of Eqs. (4.168) and (4.169), respectively. Here, Ψ_1 is the first normal stress difference coefficient defined by $N_1/\dot{\gamma}^2$, Ψ_2 is the second normal stress difference coefficients defined by $N_2/\dot{\gamma}^2$, Ψ_{10} is the zero-shear rate first normal stress difference coefficient defined by $\Psi_{10} = \lim_{\dot{\gamma}\to 0}\Psi_1(\dot{\gamma})$ and Ψ_{20} is the zero-shear rate second normal stress difference coefficient defined by $\Psi_{20} = \lim_{\dot{\gamma}\to 0}\Psi_2(\dot{\gamma})$. Note that (1) for very small values of $\dot{\gamma}$ we have $\Psi_{10} = \pi^4 G_o \tau_d{}^2/300$ and $\Psi_{20}/\Psi_{10} = -2/7$, and (2) for very large values of $\dot{\gamma}$ we have $\Psi_1 \propto \dot{\gamma}^{-2}$ and $\Psi_2 \propto \dot{\gamma}^{-5/2}$. On the basis of the experimental results

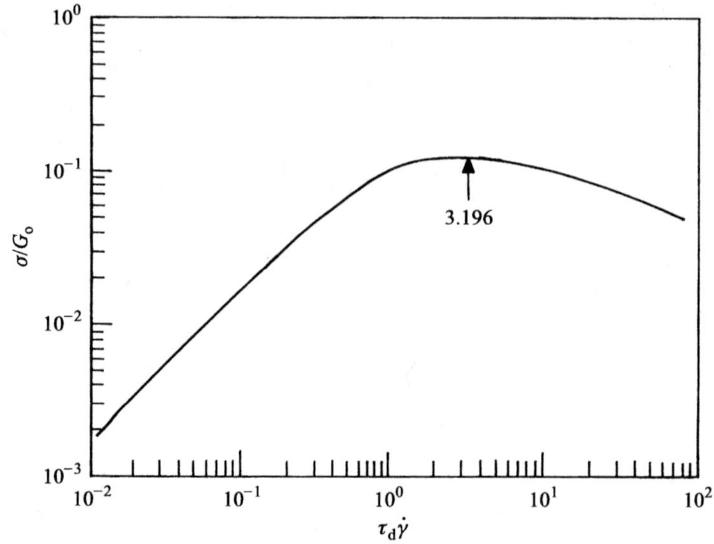

Figure 4.19 Log σ/G_o versus log $\tau_d\dot{\gamma}$ plots predicted from the tube model.

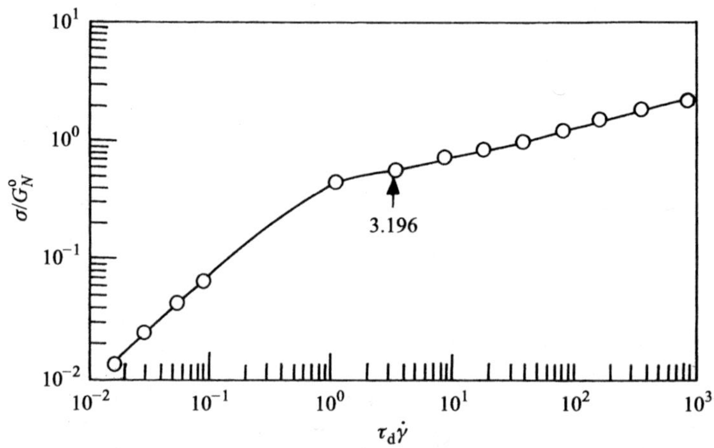

Figure 4.20 Log σ/G_N^0 versus log $\tau_d\dot{\gamma}$ plot for a nearly monodisperse polystyrene with $M_w = 1.35 \times 10^5$ and $M_w/M_n = 1.06$ at 180 °C, where the value of τ_d used was 0.93×10^{-1} s. The solid line is drawn through the data points to guide the eye.

presented in Chapter 3, we can conclude that such an asymptotic behavior of Ψ_1 and Ψ_2 is very reasonable.

4.4.5.2 *Steady-State Elongational Viscosity*

Figure 4.22 gives log $\eta_E/3\eta_0$ versus log $\tau_d\dot{\varepsilon}$ plots, which were obtained by numerical integration of Eq. (4.176). It should be remembered from Chapter 3 that as $\dot{\varepsilon}$

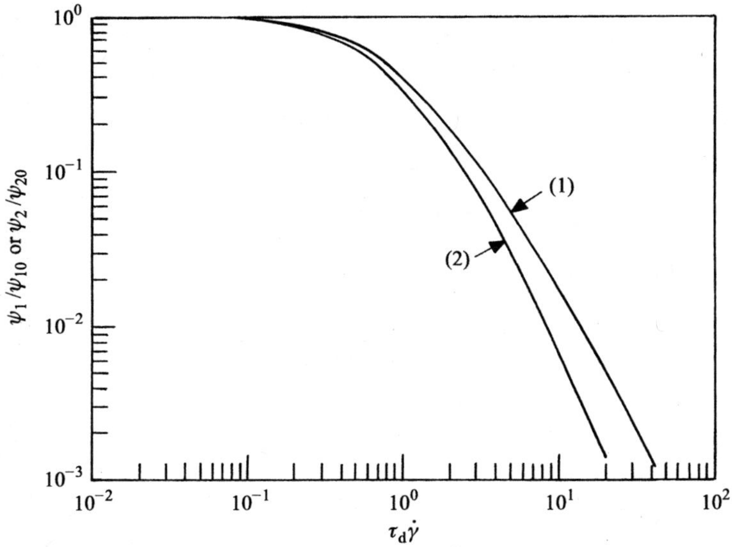

Figure 4.21 Plots for (1) log Ψ_1/Ψ_{10} versus log $\tau_d\dot{\gamma}$ and (2) log Ψ_2/Ψ_{20} versus log $\tau_d\dot{\gamma}$, which were predicted from the tube model.

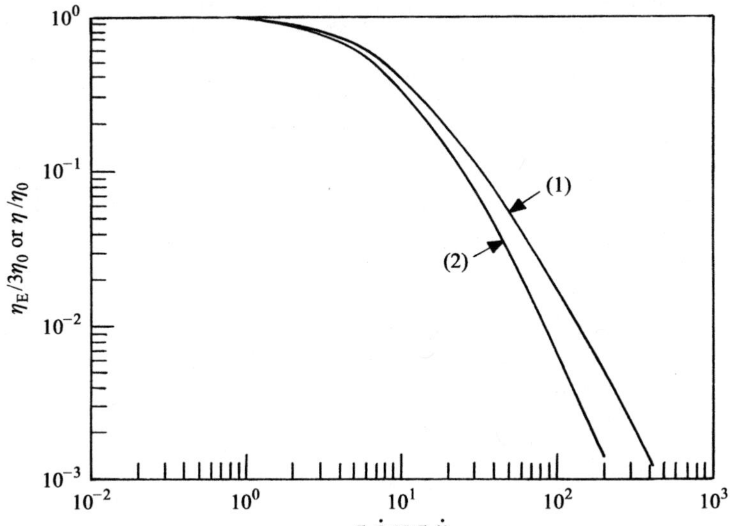

Figure 4.22 Plots for (1) log $\eta_E/3\eta_0$ versus log $\tau_d\dot{\varepsilon}$ and (2) log η/η_0 versus log $\tau_d\dot{\gamma}$ plots, which were predicted from the tube model.

approaches zero, η_E approaches $3\eta_0$. For comparison, log η/η_0 versus $\tau_d\dot{\gamma}$ plots are also given in Figure 4.22. It is interesting to observe in Figure 4.22 that the constitutive equation, Eq. (4.158), predicts a decreasing trend of η_E with increasing $\dot{\varepsilon}$, which is in consonance with some experimental results presented in Chapter 3, but it does not predict an increasing trend of η_E at low values of $\dot{\varepsilon}$. It should be pointed out that there is no adjustable parameter in Eq. (4.158).

As pointed out in Chapter 3, some phenomenological constitutive equations predict a decreasing trend of η_E with increasing $\dot{\varepsilon}$ only when a parameter is adjusted. Such phenomenological constitutive equations cannot be regarded as predictive models. Moreover, there are no molecular parameters involved in those phenomenological equations. Conversely, the constitutive equation, Eq. (4.158), contains all the important molecular parameters;[13] namely, that (1) G_o is related to M_e and (2) τ_d is related to M, b, ζ, and M_o (see Eq. (4.116)).

4.4.5.3 Time–Strain Separability

Once the expression for σ is available, the shear stress relaxation modulus $G_s(t, \gamma)$ can be calculated from

$$\sigma(t, \gamma)/\gamma = G_s(t, \gamma) \tag{4.187}$$

where t is time and γ is shear strain. A number of investigators (Osaki and Kurata 1980; Osaki et al. 1981, 1982; Takahashi et al. 1990; Vrentas and Graessley 1981, 1982) have performed stress relaxation experiments for polymer solutions or polymer melts.

Figure 4.23 gives $\log G_s(t, \gamma)$ versus $\log t$ plots at 160 °C for a nearly monodisperse polystyrene with $M_w = 2.93 \times 10^5$ and $M_w/M_n = 1.06$, at different values of γ. By shifting vertically $\log G_s(t, \gamma)$ versus $\log t$ plots, obtained at different values of γ to

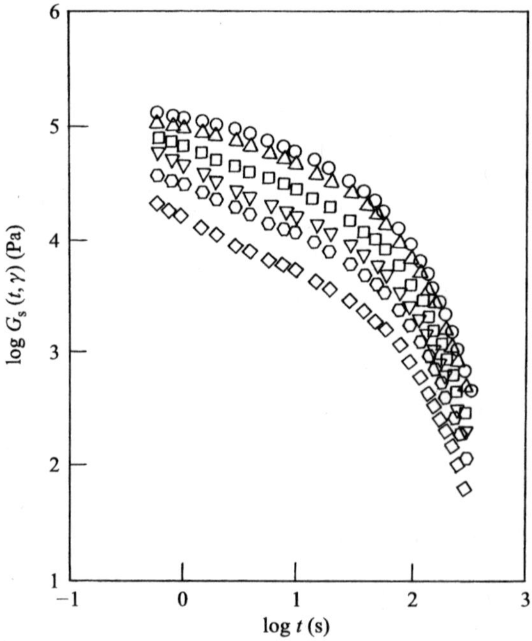

Figure 4.23 Log $G_s(t, \gamma)$ versus $\log t$ plots for polystyrene at 160 °C at various shear strains: (○) $\gamma = 0.5$, (△) $\gamma = 1$, (□) $\gamma = 2$, (▽) $\gamma = 3$, (⬡) $\gamma = 4$, and (◇) $\gamma = 6$. (Reprinted from Takahashi et al., *Nihon Reoroji Gakkaishi* 18:18. Copyright © 1990, with permission from the Society of Rheology, Japan.)

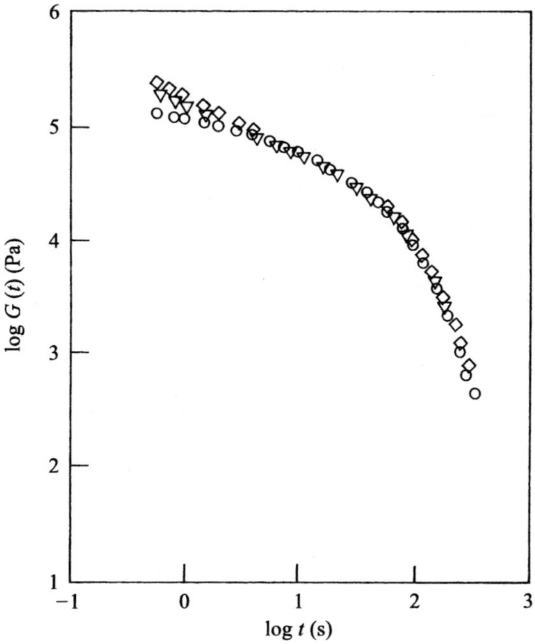

Figure 4.24 Plots of $G(t)$ versus $\log t$ for polystyrene at 160 °C at various shear strains: (○) $\gamma = 0.5$, (▽) $\gamma = 3$, and (◇) $\gamma = 6$. The reference strain used was 0.5.

a plot at $\gamma = 0.5$, we obtain Figure 4.24, which shows that at a time larger than a certain critical value, all curves are superimposed to a single curve. This experimental observation supports the idea in the constitutive equation, Eq. (4.158); namely, $G_s(\gamma, t)$ can be factored into two functions, one of time-dependent function $G(t)$ and the other of strain-dependent function $h(\gamma)$:

$$G_s(t, \gamma) = G(t)h(\gamma) \tag{4.188}$$

which is referred to as time–strain separability. For times greater than τ_e, a molecular equilibration time is much smaller than the mean relaxation time τ_m defined by

$$\tau_m = \int tG(t)\,dt \Big/ \int G(t)\,dt = \eta_0 J_e^o \tag{4.189}$$

Note that Eq. (4.188) is valid at t sufficiently large after shear deformation was imposed. Using the experimental data given in Figure 4.23 one can calculate $h(\gamma)$, commonly referred to as the "damping function." This is determined from the amount of vertical shift made in order to obtain a single master curve, as given in Figure 4.24. Figure 4.25 gives $\log h(\gamma)$ versus $\log \gamma$ plots for the polystyrene, together with the predictions made using Eqs. (4.151) and (4.152). It can be seen in Figure 4.25 that the prediction based on IAA is reasonably close to one obtained without making IAA.

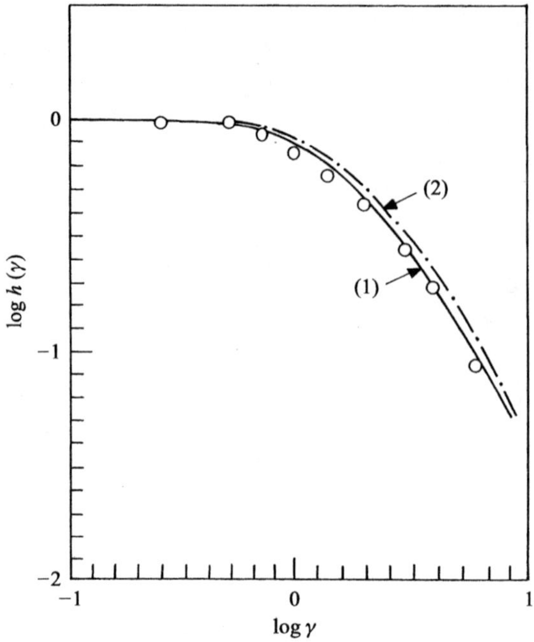

Figure 4.25 Log $h(\gamma)$ versus $\log \gamma$ plots predicted with the tube model, together with the experimental data for polystyrene at 160 °C: (1) for the prediction without IAA, (2) for the prediction with IAA, and the symbol ○ for the experimental data. (Reprinted from Takahashi et al., *Nihon Reoroji Gakkaishi* 18:18. Copyright © 1990, with permission from the Society of Rheology, Japan.)

4.4.5.4 Stress Relaxation Modulus

Basically, there are four relaxation processes involved. The first is associated with fast relaxation and describes the Rouse motion of a polymer chain between two adjacent entanglement points, and the second is associated with slow relaxation and describes the reptative motion of a polymer chain. In explaining experimental results obtained for the stress relaxation after the imposition of shear deformation, Lin (1986) proposed the following expression for the stress relaxation modulus $G(t)$:

$$G(t) = \frac{4\rho RT}{5M_e}\left[1 + \mu_A(t/\tau_A)\right]\left[1 + \tfrac{1}{4}\exp(-t/\tau_X)\right] \times \left\{\left(\frac{M_e}{M}\right)^{1/2}\mu_B(t/\tau_B) + \left[1 - \left(\frac{M_e}{M}\right)^{1/2}\right]\mu_C(t/\tau_C)\right\} \tag{4.190}$$

where $\mu_A(t/\tau_A)$ describes the fast relaxation process, which describes the Rouse chain motion between two adjacent crosslink points, and is defined by

$$\mu_A(t/\tau_A) = \sum_{p}^{N_e - 1} \exp(-t/\tau_A^p) \tag{4.191}$$

with

$$\tau_{\mathrm{A}}^{p} = \frac{\mathrm{K}\pi^2}{24\sin^2(\pi p/2N_{\mathrm{e}})}\left(\frac{M_{\mathrm{e}}}{N_{\mathrm{e}}}\right)^2 \tag{4.192}$$

It is seen from Eq. (4.191) that $\mu_{\mathrm{A}}(t/\tau_{\mathrm{A}})$ is independent of molecular weight, $\mu_{\mathrm{B}}(t/\tau_{\mathrm{B}})$ is a Rouse-type relaxation process and is defined by

$$\mu_{\mathrm{B}}(t/\tau_{\mathrm{B}}) = \frac{8}{\pi^2}\sum_{\mathrm{odd}\ p}^{\infty}\frac{1}{p^2}\exp\left(-p^2 t/\tau_{\mathrm{B}}\right) \tag{4.193}$$

with

$$\tau_{\mathrm{B}} = (1/3)KM^2 \tag{4.194}$$

It is seen from Eq. (4.193) that $\mu_{\mathrm{B}}(t/\tau_{\mathrm{B}})$ depends on molecular weight and τ_{B} is equivalent to the Rouse relaxation time τ_{r} (see Eq. (4.53)). $\mu_{\mathrm{C}}(t/\tau_{\mathrm{C}})$ describes the slow relaxation process, which is associated with the reptation motion corrected for contour length fluctuations (see Eqs. (4.120) and (4.121)), and is defined by

$$\mu_{\mathrm{C}}(t/\tau_{\mathrm{C}}) = \frac{8}{\pi^2}\sum_{\mathrm{odd}\ p}^{\infty}\frac{1}{p^2}\exp\left(-p^2 t/\tau_{\mathrm{C}}\right) \tag{4.195}$$

with

$$\tau_{\mathrm{C}} = K\left(\frac{M^3}{M_{\mathrm{e}}}\right)\left[1-\left(\frac{M_{\mathrm{e}}}{M}\right)^{1/2}\right]^2 \tag{4.196}$$

K appearing in Eqs. (4.192), (4.194), and (4.196) is given by Eq. (4.86). The term $(1/4)\left[\exp(-t/\tau_{\mathrm{X}})\right]$ in Eq. (4.190) is related to the slippage of a polymer chain through entanglement links and τ_{X} is equivalent to the Doi–Edwards equilibration time τ_{e} defined by Eq. (4.109). Thus $\tau_{\mathrm{X}} < \tau_{\mathrm{B}} < \tau_{\mathrm{C}}$, which is equivalent to $\tau_{\mathrm{e}} < \tau_{\mathrm{r}} < \tau_{\mathrm{d}}$ considered in previous sections.

Figure 4.26 gives a comparison of predictions with experimental results for the stress relaxation modulus $G(t)$ given by Eq. (4.190) at 127 °C for a nearly monodisperse polystyrene with $M_{\mathrm{w}} = 1.02 \times 10^5$ and $M_{\mathrm{w}}/M_{\mathrm{n}} = 1.02$, in which A, B, C, and X denote the relaxation processes associated with $\mu_{\mathrm{A}}(t)$, $\mu_{\mathrm{B}}(t)$, $\mu_{\mathrm{C}}(t)$, and $\mu_{\mathrm{X}}(t)$, respectively. It can be seen in Figure 4.26 that the fast relaxation process $\mu_{\mathrm{A}}(t)$ is very important to correct interpretation of experimental results for the stress relaxation modulus at very short times.

By substituting Eq. (4.190) into (4.80) and (4.81) one can calculate $G'(\omega)$ and $G''(\omega)$ as functions of angular frequency ω. Figure 4.27 gives comparisons of prediction with experiment for the frequency dependence of $G'(\omega)$ and $G''(\omega)$ for a monodisperse polystyrene. In Figure 4.27, A, B, C, and X denote the relaxation processes associated with $\mu_{\mathrm{A}}(t)$, $\mu_{\mathrm{B}}(t)$, $\mu_{\mathrm{C}}(t)$, and $\mu_{\mathrm{X}}(t)$, respectively. It can be seen in Figure 4.27 that the

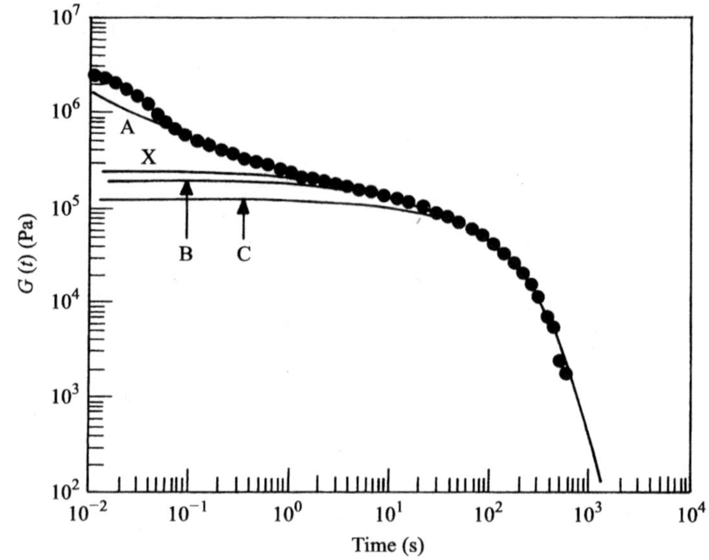

Figure 4.26 Log $G(t)$ versus log t plots for a monodisperse polystyrene with $M_w = 1.02 \times 10^5$ and $M_w/M_n = 1.02$ at 127 °C; A denotes the relaxation process associated with μ_A, B denotes the relaxation process associated with μ_B, C denotes the relaxation process associated with μ_C, and X denotes the relaxation process associated with μ_X. (Reprinted from Lin, *Macromolecules* 19:159. Copyright © 1986, with permission from the American Chemical Society.)

inclusion of the fast relaxation process $\mu_A(t)$ in $G(t)$ is essential to predict $G'(\omega)$ and $G''(\omega)$ at high frequencies (i.e., at very short times).

4.5 Summary

In this chapter, we have presented the fundamentals of molecular theory for the viscoelasticity of flexible homogeneous polymers, namely the Rouse/Zimm theory for dilute polymer solutions and unentangled polymer melts, and the Doi–Edwards theory for concentrated polymer solutions and entangled polymer melts. In doing so, we have shown how the constitutive equations from each theory have been derived and then have compared theoretical prediction with experiment. The material presented in this chapter is very important for understanding how the molecular parameters of polymers are related to the rheological properties of homopolymers.

One of the salient features of the Doi–Edwards theory is that it relates the self-diffusion coefficient of an entangled molten polymer to its viscoelastic properties. This has been made possible by the virtue of the intuitive notion that the chains of polymer molecules in the entangled state proceed by random snakelike motions, each parallel to its own contour; that is, within the confined region of a tube. This approach enables one to estimate the self-diffusion coefficient of a polymer from rheological measurements, specifically from the measurement of plateau modulus (Graessley 1980). Such a possibility enables one to investigate, on the basis of rheological measurement, polymer–polymer interdiffusion between two miscible homopolymers, which

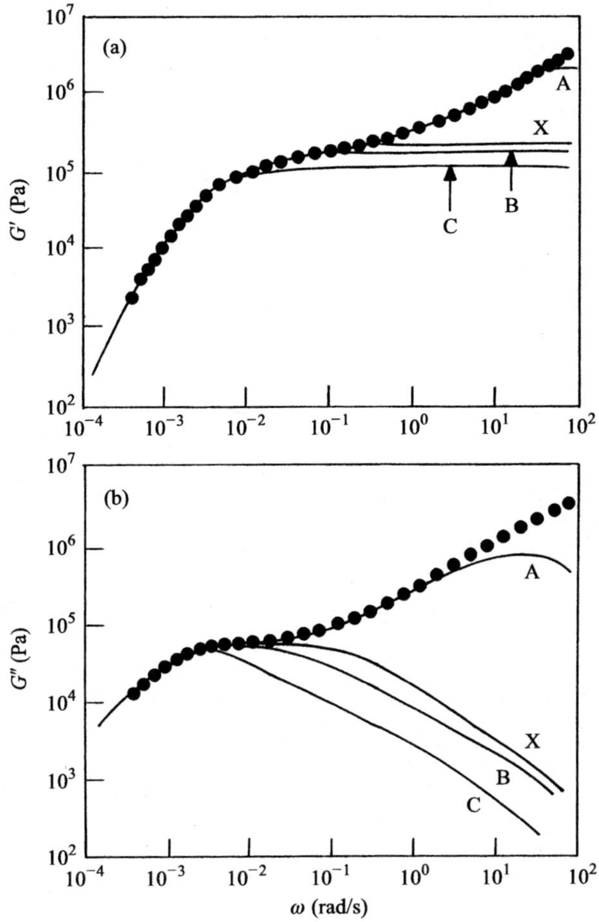

Figure 4.27 (a) $\log G'$ versus $\log \omega$ plots and (b) $\log G''$ versus $\log \omega$ plots at 127 °C for a nearly monodisperse polystyrene with $M_w = 1.02 \times 10^5$ and $M_w/M_n = 1.02$, where A denotes the relaxation process associated with μ_A, B denotes the relaxation process associated with μ_B, C denotes the relaxation process associated with μ_C, and X denotes the relaxation process associated with μ_X. (Reprinted from Lin, *Macromolecules* 19:159. Copyright © 1986, with permission from the American Chemical Society.)

is relevant for solving important practical processing problems associated with coextruding two miscible homopolymers. This subject is elaborated on in Chapter 9 of Volume 2, which deals with coextrusion.

It should be emphasized that the molecular theories presented in this chapter are valid only for flexible homopolymers and thus they cannot describe the rheological behavior of structured polymer systems, including multicomponent and/or multiphase polymers, such as block copolymers, liquid-crystalline polymers, thermoplastic polyurethanes, immiscible polymer blends, highly filled polymers, and nanocomposites. We discuss this subject in the remaining chapters of this volume.

Appendix 4A: Derivation of Equation (4.6)

The number of conformations available to the chain is proportional to the probability density f multiplied by the size of the volume element dV. According to the general principles of statistical thermodynamics, the entropy S will be proportional to the logarithm of the number of configurations available to the system; that is, to the logarithm of the number of possible configurations f corresponding to any specified state. Thus, for a chain whose ends are located at specified points separated by a distance r, the entropy S of the chain is given by

$$S = \log f(r)\,\mathrm{d}V = \log\left[B\exp(-\alpha^2 r^2)\right]\mathrm{d}V \tag{4A.1}$$

where $\alpha^2 = 3/2\langle r^2\rangle$. Thus, we have the relationship

$$S = C - \alpha^2 r^2 \tag{4A.2}$$

where B and C are constants, and the corresponding Helmholtz free energy E can be determined by $E = -k_\mathrm{B}TS$ with k_B being the Boltzmann constant and T being the absolute temperature.

The work W required to move one end of the chain from a distance r to a distance $r = r + \mathrm{d}r$ with respect to the other end is equal to the change in Helmholtz free energy, that is

$$\mathrm{d}E = \mathrm{d}W \tag{4A.3}$$

Conversely, the work done by the applied stress, corresponding, for example, to tensile force F acting on a specimen of length ℓ, is given by

$$\mathrm{d}W = F\,\mathrm{d}\ell \tag{4A.4}$$

By making use of Eqs. (4A.3) and (4A.4), the tensile force may then be expressed in the form

$$F = (\partial W/\partial \ell)_T = (\partial E/\partial \ell)_T \tag{4A.5}$$

which shows that the tensile force is equal to the change in Helmholtz energy per unit increase in length of the specimen. Therefore, we have

$$\mathrm{d}W/\mathrm{d}r = \mathrm{d}E/\mathrm{d}r = -k_\mathrm{B}T(\mathrm{d}S/\mathrm{d}r) \tag{4A.6}$$

Substitution of Eq. (4A.2) into (4A.6) gives

$$\mathrm{d}W/\mathrm{d}r = 2k_\mathrm{B}T\alpha^2 r \tag{4A.7}$$

From Eqs. (4A.5) and (4A.7) we have

$$F = \mathrm{d}W/\mathrm{d}r = 2k_\mathrm{B}T\alpha^2 r = (3k_\mathrm{B}T/Nb^2)r \tag{4A.8}$$

in which use is made of the relationship $\alpha^2 = 3/2\langle r^2\rangle = 3/2Nb^2$ (see Eq. (4.9)). Since $\mathrm{d}E = F\mathrm{d}r$ and the Helmholtz free energy E can be equated to the potential U appearing in Eq. (4.4), we obtain Eq. (4.6).

Appendix 4B: Derivation of Equation (4.71)

Defining the normal coordinates $\boldsymbol{\xi}(X, Y, Z)$ by the relation

$$\mathbf{r} = \mathbf{Q}\boldsymbol{\xi} \tag{4B.1}$$

and choosing the matrix $\mathbf{Q}$ such that the product $\mathbf{HA}$ is diagonalized by a similarity transformation

$$\mathbf{Q}^{-1}\mathbf{HAQ} = \boldsymbol{\Lambda} \tag{4B.2}$$

where the matrix $\boldsymbol{\Lambda}$ has only diagonal elements λ_j, Eq. (4.67) can be rewritten as

$$\frac{\partial f}{\partial t} + \boldsymbol{\nabla}_{\xi}{}^{\mathrm{T}} \cdot f\mathbf{v}_{\xi}{}^{\mathrm{o}} = (k_{\mathrm{B}}T/\zeta)\, \boldsymbol{\nabla}_{\xi}{}^{\mathrm{T}} \cdot \left(\mathbf{N}\boldsymbol{\nabla}_{\xi} f + \frac{3}{b^2}\boldsymbol{\Lambda}\boldsymbol{\xi} f\right) \tag{4B.3}$$

where $\mathbf{N}$ is the diagonal matrix having the elements ν_j, which can be obtained by a similarity transformation

$$\mathbf{Q}^{-1}\mathbf{H}(\mathbf{Q}^{-1})^{\mathrm{T}} = \mathbf{N} \tag{4B.4}$$

and $\boldsymbol{\nabla}_{\xi}{}^{\mathrm{T}}$ and $\boldsymbol{\nabla}_{\xi}$ are differential operators with respect to $\boldsymbol{\xi}$, which can be obtained by the usual rules for transforming partial derivatives:

$$\frac{\partial}{\partial \mathbf{r}} = \left(\mathbf{Q}^{-1}\right)^{\mathrm{T}} \cdot \frac{\partial}{\partial \boldsymbol{\xi}} \quad \text{or} \quad \boldsymbol{\nabla}_{\mathbf{r}} = \left(\mathbf{Q}^{-1}\right)^{\mathrm{T}} \boldsymbol{\nabla}_{\xi} \tag{4B.5}$$

The transpose of the above expression becomes

$$\boldsymbol{\nabla}_{\mathbf{r}}{}^{\mathrm{T}} = \boldsymbol{\nabla}_{\xi}{}^{\mathrm{T}}\mathbf{Q}^{-1} \tag{4B.6}$$

It should be mentioned that for many purposes it is not necessary to obtain a closed-form solution of Eq. (4B.3) since the average values of certain functions of the coordinates can be obtained exactly without finding the distribution function **f** itself. For such purposes, however, one must obtain eigenvalues λ'_p defined by Eq. (4B.2), which are the elements of the diagonal matrix $\boldsymbol{\Lambda}$. The diagonal matrix can be found once the matrix $\mathbf{Q}$ is found. Note that $\mathbf{Q}$ is composed of $(N+1)$ column vectors $\mathbf{u}$, which are eigenvectors of $\mathbf{HA}$ satisfying the equation

$$\mathbf{HAu} = \lambda'_p \mathbf{u} \tag{4B.7}$$

Problems

Problem 4.1

Values of η_0 of monodisperse polyisoprenes with several different molecular weights M greater than M_c were measured and they are found to follow the empirical expression

$$\eta_0 = 3.708 \times 10^{-19} \exp(4478/T)M^{3.35}$$

where η_0 is in Pa·s and T is the absolute temperature.

(a) Calculate the monomeric friction coefficient ζ_0 of isoprene at 60 °C, using the various expressions Eqs. (4.117), (4.124), and (4.125) given in this chapter and observe how the values calculated differ from each other. For the calculations, use the following information: $M_o = 68.1$, $M_e = 6 \times 10^3$, $b = 5.9 \times 10^{-8}$ cm, and specific volume v_{sp} (cm³/g) is given by

$$v_{sp} = 1.0771 + 7.22 \times 10^{-4}(T - 273) + 2.46 \times 10^{-7}(T - 273)^2$$

(b) Calculate the plateau modulus G_N^o of polyisoprene at 20, 40, and 60 °C and observe how G_N^o varies with temperature.

Problem 4.2

Values of η_0 of a polyisoprene with $M = 9 \times 10^4$ were measured at various temperatures and they are given below.

Temperature (°C)	η_0 *(Pa·s)*
40	2.31×10^4
60	1.08×10^4
80	4.70×10^3
100	2.35×10^3
120	1.25×10^3
140	7.45×10^2

(a) Calculate the viscosity shift factor a_T at 60, 80, 100, 120, and 140 °C with 40 °C as reference temperature and plot a_T versus the absolute temperature.

(b) Obtain values of C_1 and C_2 in the Williams–Landel–Ferry (WLF) equation (see Eq. (6.7)) for polyisoprene, using the values of a_T calculated above.

(c) Calculate the monomeric friction coefficient ζ_0 of isoprene at 60 °C using the WLF equation and then compare it with the value calculated in Problem 4.1.

Problem 4.3

Calculate the η_0 of a polystyrene with $M = 5 \times 10^4$ at 170, 190, and 210 °C when the η_0 of a polystyrene with $M = 2 \times 10^5$ at 210 °C is known to be 3.38×10^3 Pa · s. For the calculation, use the following WLF equation (see Chapter 6) for polystyrene

$$\log a_T = \frac{-13.7(T - T_g)}{50.5 + T - T_g} \tag{4P.1}$$

with $T_g = 100\,°\mathrm{C}$. Here T_g denote a glass transition temperature.

Problem 4.4
The value ζ_0 at T_g (100 °C) for polystyrene is reported to be 114 dyn s/cm (or 0.114 N s/m). Using the relationship $\zeta_0(T) = a_T(T)\zeta_0(T_g)$, where $a_T(T)$ is defined in Problem 4.3, calculate η_0 for a polystyrene with $M = 1.5 \times 10^4$ at 150, 180, and 200 °C. Note that the specific volume v_{sp} (cm^3/g) of polystyrene varies with temperature according to the expression

$$v_{sp} = 1/\rho = 0.9217 + 5.412 \times 10^{-4}(T - 273) + 1.687 \times 10^{-7}(T - 273)^2$$

Problem 4.5
Derive Eq. (4.48).

Problem 4.6
Derive Eqs. (4.82)–(4.85).

Problem 4.7
Derive Eqs. (4.90)–(4.92).

Problem 4.8
Derive Eqs. (4.112)–(4.115).

Problem 4.9
Derive Eqs. (4.129)–(4.132).

Problem 4.10
Derive Eqs. (4.151) and (4.152).

Problem 4.11
Derive Eqs. (4.167)–(4.169).

Notes

1. From Eq. (4.8) we have

$$\langle r^2 \rangle = \langle \mathbf{r} \cdot \mathbf{r} \rangle = \int r^2 f(r)\, d\mathbf{r}$$

$$= \int_0^{2\pi} \int_0^{\pi} \int_0^{\infty} r^2 \left(\frac{3}{2\pi N b^2}\right)^{3/2} \exp\left(-\frac{3r^2}{2Nb^2}\right) r^2 \sin\theta \, dr\, d\theta\, d\phi \tag{4N.1}$$

Note the following relationship:

$$\int_0^\infty r^4 \mathrm{e}^{-Ar^2} \mathrm{d}r = \frac{3}{8}\frac{\sqrt{\pi}}{A^{5/2}} \tag{4N.2}$$

where $A = 3/2Nb^2$. Combining Eqs. (4N.1) and (4N.2) we obtain

$$\langle r^2 \rangle = Nb^2 \tag{4N.3}$$

2. The normal coordinate transformation requires diagonalization of the matrix **A**, appearing in Eq. (4.46), in such a way that the transformed matrix will have only the diagonal components. The method of obtaining such a matrix is well documented in many standard textbooks. An excellent exposition of this method can be found in the textbook by Hildebrand (1952).
3. The equilibrium distribution function f for the chain coordinates can be expressed by (see Eq. (4.4))

$$f \propto \mathrm{e}^{-E/k_\mathrm{B}T} \tag{4N.4}$$

where E is the Helmholtz energy (or potential). Thus

$$\ln f = a - E/k_\mathrm{B}T \tag{4N.5}$$

or

$$E = -k_\mathrm{B}T \ln f + a' \tag{4N.6}$$

in which a and a' are constants. The force F_x is defined by the gradient of E, thus

$$F_x = \partial E/\partial x = -k_\mathrm{B}T \frac{\partial \ln f}{\partial x} \tag{4N.7}$$

4. Using the definition $\eta_0 = \sigma/\dot{\gamma}$, from Eq. (4.90) we have

$$\eta_0 = \frac{\rho RT}{M}\sum_{p=1}^{\infty}\tau_p = \frac{\rho RT}{M}\frac{\zeta b^2 N^2}{6\pi^2 k_\mathrm{B}T}\sum_{p=1}^{\infty}\frac{1}{p^2} = \frac{\rho RT}{M}\frac{\zeta b^2 N^2 N_\mathrm{A}}{36 k_\mathrm{B} T N_\mathrm{A}} = \frac{\rho\zeta b^2 N^2 N_\mathrm{A}}{36M} \tag{4N.8}$$

where use is made of $\sum_{p=1}^{\infty} 1/p^2 = \pi^2/6$ and $RT = k_\mathrm{B}TN_\mathrm{A}$, in which R is the universal gas constant (8.3166×10^7 erg/K mol), k_B is the Boltzmann constant (1.3804×10^{-16} erg/K), N_A is Avogadro's number (6.0247×10^{23} mol^{-1}), and T is the absolute temperature.
5. Note the following relationships: (1) $\zeta = m\zeta_0$, (2) $b^2 = mb_0{}^2$, and (3) $N_\mathrm{p} = mN$, where ζ_0 is the monomeric friction coefficient, b is the Kuhn length, b_0 is the bond length

of a monomer unit, m is the number of monomeric units in a submolecule, N is the number of submolecules, and N_p is the degree of polymerization that can also be expressed as M/M_o, with M being the total molecular weight and M_o being the molecular weight of monomer. Then, Eq. (4.93) can be rewritten as Eq. (4.94).

6. Equation (4.52) can be rewritten as

$$\tau_p = \frac{\zeta\, b^2 N^2}{6\pi^2 p^2 k_B T} = \frac{(m\zeta_0)(mb_0{}^2)(N_p/m)^2}{6\pi^2 p^2 k_B T} \tag{4N.9}$$

with $N_p = M/M_o$ and the relationship between ζ_0 and η_0 from Eq. (4.94)

$$\zeta_0 = 36\eta_0 M_o{}^2 \Big/ \rho b_o{}^2 M N_A \tag{4N.10}$$

Equation (4N.9) can be rewritten, with the aid of $k_B T N_A = RT$, as Eq. (4.95).

7. Note that the disengagement time τ_d is given by $\tau_d = L^2/D_c\pi^2 = Z^2 a^2 N\zeta/\pi^2 k_B T$ in which $L = Za$ and $D_c = k_B T/\zeta N$ is used. For $\tau_e \ll \tau$, we have $\tau_e = \zeta b^2 N_e{}^2/6\pi^2 k_B T$ for which the number of Rouse segments per chain N appearing in Eq. (4.52) is replaced by the number of Rouse segments between the entanglement points N_e. Since $M = NmM_o$ and $M_e = N_e mM_o$ in which M_o is the molecular weight of monomer units and m is the number of monomer units in a submolecule, we have $N = (M/M_e)N_e = ZN_e$, in which Z is the number of entanglement points per chain. Therefore, we have $\tau_d/\tau_e = 6Z^3$, in which use is made of $a^2/b^2 N_e = 1$.

8. Note that the disengagement time τ_d is given by $\tau_d = L^2/D_c\pi^2 = Z^2 a^2 N\zeta/\pi^2 k_B T$, in which $L = Za$ and $D_c = k_B T/\zeta N$ are used, and that the Rouse relaxation time τ_r is given by Eq. (4.53). Thus we have $\tau_d/\tau_r = 6Z^2 a^2/b^2 N$. Since $Za^2 = Nb^2$, we have $\tau_d/\tau_r = 6Z$.

9. The disengagement time τ_d can be rewritten, with the aid of Eqs. (4.99) and (4.101), as

$$\tau_d = \frac{L^2}{D_c\pi^2} = \frac{\zeta\, b^4}{\pi^2 k_B T}\frac{N^3}{a^2} = \frac{\zeta\, b^2 N^3}{\pi^2 k_B T}\frac{b^2}{a^2} = \frac{\zeta\, b^2}{\pi^2 k_B T}\frac{N^3}{N_e} \tag{4N.11}$$

in which use was made of $L = Nb^2/a$, $D_c = k_B T/\zeta N$, and $b^2/a^2 = 1/N_e$. The same expression can also be obtained from

$$\tau_d = \frac{(m\zeta_0)(mb_0{}^2)}{\pi^2 k_B T}\frac{(N_p/m)^3}{N_e} = \frac{\zeta_0 b_0{}^2}{\pi^2 k_B T}\frac{(M/M_o)^3 M_o}{mN_e M_o} = (K/M_e)M^3 \tag{4N.12}$$

where

$$K = \frac{\zeta_0 b_0{}^2}{\pi^2 k_B T M_o{}^2} = \frac{(m\zeta_0)(mb_0{}^2)}{(\pi^2 k_B T)(m^2 M_o{}^2)} = \frac{\zeta\, b^2 N^2}{\pi^2 k_B T M^2} \tag{4N.13}$$

in which use is made of $mM_o = M/N$. Note that Eq. (4N.13) is identical to Eq. (4.86).

10. Substituting the following relationship $G_N^o = \rho RT/M_e$ and the expression for τ_d given by Eq. (4.116) into (4.112), one obtains

$$\eta_0 = \frac{\rho\zeta N^3 b^4 N_A}{12 M_e a^2} \tag{4N.14}$$

where use is made of $RT = k_B T N_A$. Replacing a^2 in Eq. (4N.14) with the expression given by Eq. (4.102), with the aid of $\langle r^2 \rangle = Nb^2$, we obtain

$$\eta_0 = \frac{5}{48} \frac{\rho\zeta N_A b^2 N^2 M}{M_e{}^2} \tag{4N.15}$$

Since $\zeta = m\zeta_0$ and $mN = N_p = M/M_o$, Eq. (4N.15) can be rewritten as Eq. (4.117).

11. Here, it is assumed that the stress is due to the orientation of chain segments.

12. The average $\langle \mathbf{u}_\alpha \cdot \mathbf{u}_\beta \rangle$ over an anisotropic state, for which the distribution function is given by $f(\mathbf{u}; s, 0)$, can be calculated from

$$\langle \mathbf{u}_\alpha \cdot \mathbf{u}_\beta \rangle = \int u_\alpha u_\beta f(\mathbf{u}; s, 0)\, d\mathbf{u} \tag{4N.16}$$

Since $u_x = \sin\theta\cos\phi$, $u_y = \sin\theta\sin\phi$, and $u_z = \cos\theta$ in spherical coordinates, the average of A over the isotropic state, for which the distribution function f is given by $1/4\pi$, can be calculated from

$$\langle A \rangle = \frac{1}{4\pi} \int_0^{2\pi} \int_0^{\pi} A \sin\theta\, d\theta\, d\phi \tag{4N.17}$$

For the unit vector, $\mathbf{u} = (\sin\theta\cos\phi, \sin\theta\sin\phi, \cos\theta)$ in spherical coordinates, we have

$$\mathbf{uu} = \begin{pmatrix} \sin^2\theta\cos^2\phi & \sin^2\theta\cos\phi\sin\phi & \sin\theta\cos\theta\cos\phi \\ \sin^2\theta\cos\phi\sin\phi & \sin^2\theta\sin^2\phi & \sin\theta\cos\theta\sin\phi \\ \sin\theta\cos\theta\cos\phi & \sin\theta\cos\theta\sin\phi & \cos^2\theta \end{pmatrix} \tag{4N.18}$$

Thus, the average $\langle \mathbf{uu} \rangle$ is given by

$$\langle \mathbf{uu} \rangle = \int \mathbf{uu}\, d\mathbf{u} = \frac{1}{4\pi} \int_0^{2\pi} \int_0^{\pi} \mathbf{uu} \sin\theta\, d\theta\, d\phi = \tfrac{1}{3}\boldsymbol{\delta} \tag{4N.19}$$

13. Note that $G_o = \nu k_B TL/a = (N_A \rho/M) k_B TZ = \rho RTZ/M = \rho RT(M/M_e)/M = \rho RT/M_e$, where use is made of $\nu = (N_A\rho/M)$ and $Z = L/a = M/M_e$. Thus, G_o is related, via Eq. (4.97), to G_N^o and M_e.

14. IAA implies that if retraction is so effective that the individual coil molecule does not change its mean-square end-to-end distance, it undergoes a pure orientation. In the strict sense, however, $\mathbf{u}'(s)$ is a function of the entire conformation of the chains, and the transformation rule of $\mathbf{u}(s)$ is in general not simple.

References

Aharoni SM (1983). *Macromolecules* 16:1722.
Aharoni SM (1986). *Macromolecules* 19:426.
Allen VR, Fox TG (1964). *J. Chem. Phys.* 41:337.
Amis EJ, Han CC (1982). *Polym. Commun.* 23:1403.
Amis EJ, Han CC, Matsushita Y (1984). *Polymer* 25:650.
Bartels CR, Crist B, Fetters LJ, Graessley WW (1986). *Macromolecules* 19:785.
Berry GC, Fox TG (1968). *Adv. Polym. Sci.* 5:261.
Bird RB, Curtiss CF, Armstrong RC, Hassager O (1987). *Dynamics of Polymeric Liquids: Kinetic Theory*, Vol 2, 2nd ed, John Wiley & Sons, New York.
Bueche F (1962). *The Physical Properties of Polymers*, Interscience, New York.
Callaghan PT, Pinder DN (1980). *Macromolecules* 13:1085.
Chandrasekhar S (1943). *Rev. Mod. Phys.* 15:1.
Colby RH, Fetters LJ, Graessley WW (1987). *Macromolecules* 20:2226.
Cox WP, Merz EH (1958). *J. Polym. Sci.* 28:619.
Debye P (1946). *J. Chem. Phys.* 14:636.
De Gennes PG (1971). *J. Chem. Phys.* 55:572.
De Gennes PG (1976). *Macromolecules* 9:587.
De Mallie RB, Birnboim MH, Frederick JE, Tschoegl NW, Ferry JD (1962). *J. Phys. Chem.* 66:536.
Doi M (1980a). *J. Polym. Sci., Polym. Phys. Ed.* 18:1005.
Doi M (1980b). *J. Polym. Sci., Polym. Phys. Ed.* 18:2055.
Doi M (1981). *J. Polym. Sci., Polym. Lett. Ed.* 19:265.
Doi M (1983). *J. Polym. Sci., Polym. Phys. Ed.* 21:667.
Doi M, Edwards SF (1978a). *J. Chem. Soc., Faraday Trans. 2,* 74:1789.
Doi M, Edwards SF (1978b). *J. Chem. Soc., Faraday Trans. 2*, 74:1802.
Doi M, Edwards SF (1978c). *J. Chem. Soc., Faraday Trans. 2*, 74:1818.
Doi M, Edwards SF (1979). *J. Chem. Soc., Faraday Trans. 2*, 75:38.
Doi M, Edwards SF (1986). *The Theory of Polymer Dynamics*, Oxford University Press, Oxford.
Edwards SF (1967). *Proc. Phys. Soc.* 92:9.
Ferry JD (1980). *Viscoelastic Properties of Polymers*, 3rd ed, John Wiley & Sons, New York.
Fokker AD (1914). *Ann. Physik* 43:810.
Fox TG, Allen VR (1964). *J. Chem. Phys.* 41:344.
Graessley WW (1965). *J. Chem. Phys.* 43:2696.
Graessley WW (1967). *J. Chem. Phys.* 47:1942.
Graessley WW (1974). *Adv. Polym. Sci.* 16:1.
Graessley WW (1980). *J. Polym. Sci. Polym. Phys. Ed.* 18:27.
Graessley WW (1982). *Adv. Polym. Sci.* 47:67.
Green PF, Palmstrom CJ, Mayer JW, Kramer EJ (1985). *Macromolecules* 18:501.
Hearst JE (1962). *J. Chem. Phys.* 37:2547.
Hildebrand FB (1952). *Methods of Applied Mathematics*, Prentice Hall, Englewood Cliffs, New Jersey.
Huang K (1987). *Statistical Mechanics*, 2nd ed, John Wiley & Sons, New York.
Kirkwood JG, Riseman J (1948). *J. Chem. Phys.* 16:565.
Kirkwood JG, Riseman J (1956). In *Rheology*, Eirich FR (ed), Academic Press, Vol 1, New York, p 495.
Kramers HA (1946). *J. Chem. Phys.* 14:415.
Kubo R (1966). *Rep. Progr. Phys. London*, 29:255.

Kubo R, Toda M, Hashisume N (1985). *Statistical Physics II. Nonequilibrium Statistical Mechanics*, Springer, Berlin.
Kurata M, Tsunashima Y (1989). In *Polymer Handbook*, Vol VII, 3rd ed, Brandrup J, Immergut EH (eds), John Wiley & Sons, New York.
Langevin P (1908). *Comptes Rendus* 146:530.
Leger L, Hervet H, Rondelez F (1981). *Macromolecules* 14:1732.
Lin YH (1984). *Macromolecules* 17:2846.
Lin YH (1986). *Macromolecules* 19:159.
Lodge AS, Wu Y (1971). *Rheol. Acta* 10:539.
McLeish TC (1988). *Macromolecules* 21:1062.
Meister B (1989). *Macromolecules* 22:3611.
Mills PJ, Green PF, Palmstrom CJ, Maye JM, Kramer EJ (1986). *J. Polym. Sci., Polym. Phys. Ed.* 24:1.
Mori M (1965). *Progr. Theor. Phys.* 33:423.
Nemato N, Landry MR, Noh I, Kitano T, Wesson JA, Yu H (1985). *Macromolecules* 18:308.
Orwoll RA, Stockmayer WH (1969). *Adv. Chem. Phys.* 15:305.
Osaki K, Schrag JL (1971). *Polym. J.* 2:541.
Osaki K, Kurata M (1980). *Macromolecules* 13:671.
Osaki K, Kimura S, Kurata M (1981). *J. Polym. Sci. Polym. Phys. Ed.* 19:517.
Osaki K, Nishizawa K, Kurata M (1982). *Macromolecules* 15:1068.
Planck M (1917). *Sitzber Preuss. Akad. Wiss.* p 324.
Risken R (1989). *The Fokker–Planck Equation*, 2nd ed, Springer, Berlin.
Rouse PE (1953). *J. Chem. Phys.* 21:1272.
Smith BA, Samulski ET, Yu LP, Winnik MA (1985). *Macromolecules* 18:1901.
Smith BA, Mumby SJ, Samulski ET, Yu LP (1986). *Macromolecules* 19:470.
Smoluchowski MV (1915). *Ann. Physik* 48:1103.
Takahashi M, Taku K, Masuda T (1990). *Nihon Reoroji Gakkaishi* 18:18.
Tschoegl NW (1963). *J. Chem. Phys.* 39:49.
Uhlenbeck GE, Ornstein LS (1930). *Phys. Rev.* 36:823.
van Kampen NG (1981). *Stochastic Processes in Physics and Chemistry*, North-Holland, Amsterdam.
Vrentas CM, Graessley WW (1981). *J. Non-Newtonian Fluid Mech.* 9:339.
Vrentas CM, Graessley WW (1982). *J. Rheol.* 26:359.
Wang MC, Uhlenbeck GE (1945). *Rev. Mod. Phys.* 17:323.
Wesson JA, Noh I, Kitano T, Yu H (1984). *Macromolecules* 17:782.
Wu S (1989). *J. Polym. Sci., Polym. Phys. Ed.* 27:723.
Yamakawa H (1971). *Modern Theory of Polymer Solutions*, Harper Row, New York.
Zimm BH (1956). *J. Chem. Phys.* 24:269.
Zimm BH, Row GM, Epstein LF (1956). *J. Chem. Phys.* 24:279.

5

Experimental Methods for Measurement of the Rheological Properties of Polymeric Fluids

5.1 Introduction

There has been a continuing interest in developing experimental techniques for the measurement of the rheological properties of viscoelastic fluids. As discussed in Chapter 3, reliable experimental data are needed in order to evaluate the effectiveness of a constitutive equation in its ability to predict the rheological properties of viscoelastic fluids. Also, as is presented in later chapters, a better understanding of the rheological properties of polymers is very important for the determination of optimum processing conditions, as well as for the attainment of desired physical/mechanical properties in the finished product. Further, reliable measurement of the rheological properties of polymers can be used to control polymerization reactors in industry and also to control polymer processing operations.

In this chapter, we present experimental methods for measurement of the rheological properties of polymeric fluids. For this, we discuss experimental methods to determine (1) steady-state simple shear flow and oscillatory shear flow properties using cone-and-plate rheometry, (2) steady-state shear flow properties using capillary/slit rheometry, and (3) elongational flow properties of polymeric fluids. There are other rotational types of rheological instruments, such as those with concentric-cylinder and eccentric-parallel plates. However, such rheological instruments are not widely used today and thus in this chapter we do not present the principles and applications of such rheological instruments. In presenting the experimental methods for rheological measurements we refer to the fundamentals presented in Chapters 2 and 3. For further details of the experimental methods, there are monographs (Collyer and Clegg 1998; Dealy 1982; Ferry 1980; Walter 1975) that are devoted entirely to the discussion of rheological measurements.

The primary purpose of this chapter is to demonstrate how the fundamentals presented in Chapters 2–4 can be used in the measurement of the rheological properties of polymeric fluids. Optical rheometry is an important experimental technique for investigation of the relationship between any microphase morphology dynamics and the rheological behavior of complex polymeric fluids (e.g., liquid-crystalline polymers), which exhibit strong chain orientation during flow (Fuller 1995). A variety of optical methods (e.g., birefringence, small-angle light scattering, infrared dichroism, laser dopler velocimetry, Raman scattering) have been applied concurrently during rheological measurements. However, owing to the limited space available, optical rheometry is not presented in this chapter.

5.2 Cone-and-Plate Rheometry

5.2.1 Steady-State Shear Flow Measurement

Let us consider the flow of an incompressible fluid placed in the cone-and-plate fixture, in which a cone with a small vertical angle is placed on a horizontal flat plate, as schematically represented in Figure 5.1. The wedgelike space between the cone and the plate is filled with the fluid under test. One of the surfaces is fixed and the other rotates around the axis of the cone. It is desired to find the relationships between the torque and angular velocity, and between the net thrust acting on the cone (or plate) and angular velocity.

If we neglect inertial forces and edge-effects, for the flow geometry given in Figure 5.1 the equations of motion, in terms of spherical coordinates (φ, θ, r), may be written as

$$\frac{\partial \sigma_{\varphi\theta}}{\partial \theta} + 2 \cot\theta \sigma_{\varphi\theta} = 0 \tag{5.1}$$

$$-\frac{\rho v_\varphi{}^2 \cot\theta}{r} = -\frac{1}{r}\frac{\partial p}{\partial \theta} + \frac{1}{r \sin\theta}\frac{\partial}{\partial\theta}(\sin\theta\ \sigma_{\theta\theta}) - \frac{\cot\theta\ \sigma_{\varphi\varphi}}{r} \tag{5.2}$$

$$-\frac{\rho v_\varphi{}^2}{r} = -\frac{\partial p}{\partial r} - \frac{\sigma_{\varphi\varphi} + \sigma_{\theta\theta} - 2\sigma_{rr}}{r} \tag{5.3}$$

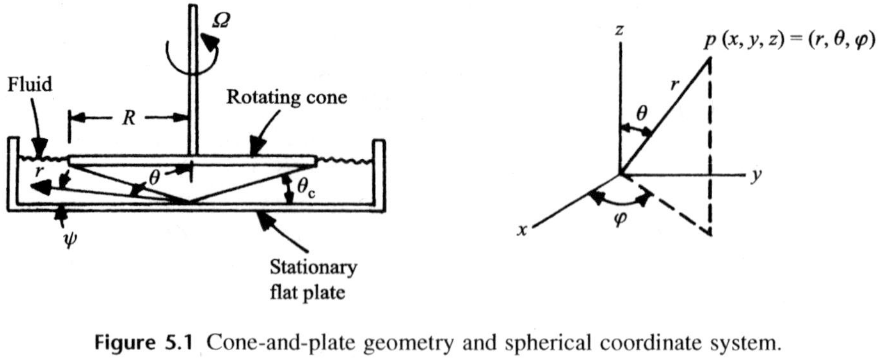

Figure 5.1 Cone-and-plate geometry and spherical coordinate system.

In Eq. (5.1) it is assumed that the stress acting on the fluid surface at the edge of the gap may be neglected. Integration of Eq. (5.1) gives

$$\sigma_{\varphi\theta} = \sigma_{12} = \frac{c_1}{\sin^2\theta} \tag{5.4}$$

where c_1 is the constant of integration. The shear stress σ_{12} can be determined from the expression for the measured torque $\Im$ on the surface of the cone:

$$\Im = \int_0^R (\tau_{\varphi\theta})_c \, 2\pi(r\cos\theta_c)^2 \, dr \tag{5.5}$$

or

$$\Im = \frac{2\pi R^3 \cos^2\theta_c (\sigma_{\varphi\theta})_c}{3} \tag{5.6}$$

in which R is the radius of the cone, θ_c is the angle between the cone and plate (see Figure 5.1), and $(\sigma_{\varphi\theta})_c$ is the shear stress at the surface of the cone. For small values of θ_c (such as $\theta_c < 2°$) $\cos\theta_c \cong 1.0$ and Eq. (5.6) may be reduced to

$$\sigma_{\varphi\theta} = \sigma_{12} = 3\Im/2\pi R^3 \tag{5.7}$$

It is of particular interest to note in Eq. (5.7) that for small values of θ_c the shear stress σ_{12} (hereafter denoted by σ) is constant across the gap between the cone and plate.

Referring to the geometry shown in Figure 5.1, the flow in the cone-and-plate instrument is assumed to have physical components of velocity:

$$v_\theta = v_r = 0; \quad v_\varphi = r\omega(\theta)\sin\theta \tag{5.8}$$

where θ is the angle between the axis of the instrument and the radius vector $\mathbf{r}$. The coordinate of the tangential direction is φ. Note that Eq. (5.8) satisfies the equation of continuity for an incompressible fluid. The rate-of-strain tensor $\mathbf{d}$ for this flow field becomes

$$\|\mathbf{d}\| = \begin{Vmatrix} 0 & \dot\gamma/2 & 0 \\ \dot\gamma/2 & 0 & 0 \\ 0 & 0 & 0 \end{Vmatrix} \tag{5.9}$$

where $\dot\gamma$ is the shear rate defined as

$$\dot\gamma = d_{\varphi\theta} = \frac{\sin\theta}{r}\frac{\partial}{\partial\theta}\left(\frac{v_\varphi}{\sin\theta}\right) \cong \frac{1}{r}\frac{\partial v_\varphi}{\partial\theta} \tag{5.10}$$

Note that use is made of $\sin\theta = \sin(\pi/2 - \Psi) \cong 1.0$ for small values of Ψ.

Integration of Eq. (5.10) gives

$$v_\varphi = r\dot{\gamma}\theta + c_2 \tag{5.11}$$

where c_2 is the constant of integration. Using the boundary condition

$$v_\varphi = 0 \quad \text{at} \quad \theta = \pi/2 \tag{5.12}$$

we have

$$v_\varphi = r\dot{\gamma}(\theta - \pi/2) \tag{5.13}$$

Use of the boundary condition (see Eq. (5.8))

$$v_\varphi = \Omega\, r \sin(\pi/2 - \theta_c) \quad \text{at} \quad \theta = \theta_c \tag{5.14}$$

in Eq. (5.13) gives for small values of θ_c

$$\dot{\gamma} = -\Omega/\theta_c \tag{5.15}$$

where Ω is the rotational speed of the cone. It is seen from Eqs. (5.7) and (5.15) that for small values of θ_c, both σ and $\dot{\gamma}$ are uniquely determined from the measurements of $\Im$ and Ω, thus permitting one to determine η (see Chapter 3).

Let us now discuss what other parameters should be measured in order to determine normal stress differences (see Chapter 2). The total stress component normal to the direction of shear (that is, the surface of the cone or plate) is

$$T_{\theta\theta} = -p + \sigma_{\theta\theta} \tag{5.16}$$

Substitution of Eq. (5.16) into Eq. (5.3) to eliminate p, with an understanding that $\sigma_{\theta\theta}$ is independent of θ, gives

$$-\rho v_\varphi{}^2 = \frac{\mathrm{d}T_{\theta\theta}}{\mathrm{d}\ln r} - N \tag{5.17}$$

where

$$N = \sigma_{\varphi\varphi} + \sigma_{\theta\theta} - 2\sigma_{rr} = \sigma_{11} + \sigma_{22} - 2\sigma_{33} \tag{5.18}$$

If all stresses are evaluated at the surface of the plate, where $v_\varphi = 0$, Eq. (5.17) reduces to

$$\frac{\mathrm{d}T_{\theta\theta}}{\mathrm{d}\ln r} = N = \text{constant} \tag{5.19}$$

It is of interest to note in Eq. (5.19) that measurement of the total normal stress $T_{\theta\theta}$ at various positions r along the stationary surface will permit one to determine N defined by Eq. (5.18).

The total force F exerted normal to the cone (or plate) may be calculated from

$$F = -\int_0^R 2\pi\, r\, T_{\theta\theta}\, \mathrm{d}r \tag{5.20}$$

Integrating by parts, Eq. (5.20) becomes

$$F = -\pi\left(R^2 T_{\theta\theta}(R) - \int_0^R r^2 \frac{\partial T_{\theta\theta}}{\partial r}\mathrm{d}r\right) \tag{5.21}$$

Introducing Eq. (5.19) into the right-hand side of Eq. (5.21), we have

$$F = -\pi R^2 \left[T_{\theta\theta}(R) - 1/2(\sigma_{\varphi\varphi} + \sigma_{\theta\theta} - 2\sigma_{rr})_R\right] \tag{5.22}$$

Noting that

$$T_{\theta\theta}(R) = -p(R) + \sigma_{\theta\theta}(R) \tag{5.23}$$

Eq. (5.22) may be written as

$$F = -\pi R^2 \left[-p(R) + \sigma_{rr}(R) - 1/2(\sigma_{\varphi\varphi} - \sigma_{\theta\theta})_R\right] \tag{5.24}$$

If the edge of the gap is in equilibrium with the atmosphere, one has

$$T_{rr}(R) = -p(R) + \sigma_{rr}(R) = -p_0 \tag{5.25}$$

Use of Eq. (5.25) in (5.24) gives

$$\sigma_{\varphi\varphi} - \sigma_{\theta\theta} = N_1 = 2F'/\pi R^2 \tag{5.26}$$

where $N_1 = \sigma_{11} - \sigma_{22}$ (see Chapter 2) and $F' = F - \pi R^2 p_0$ is the net thrust measured on the cone (or plate) in excess of that due to ambient pressure. It is seen in Eq. (5.26) that the first normal stress difference N_1 can be determined from measurement of the net thrust F' that varies with Ω. Since the measurement of Ω enables one to determine $\dot{\gamma}$ (see Eq. (5.15)), one can then establish a relationship between N_1 and $\dot{\gamma}$. When using a cone-and-plate rheometer, one must be certain that the cone angle (θ_c) is sufficiently small so that the assumption made in Eq. (5.14) is valid.

It is of particular interest to note that the definition of N given by Eq. (5.18) allows for the following relationship:

$$N - (\sigma_{\varphi\varphi} - \sigma_{\theta\theta}) = 2(\sigma_{\theta\theta} - \sigma_{rr}) \tag{5.27}$$

or

$$\sigma_{\theta\theta} - \sigma_{rr} = N_2 = (N - N_1)/2 \tag{5.28}$$

where $N_2 = \sigma_{22} - \sigma_{33}$ (see Chapter 2). That is, the second normal stress difference N_2 can be determined from Eq. (5.28) once values of N and N_1 are determined. As presented in Chapter 3, an accurate measurement of N_2 (which is negative in sign) is helpful in the evaluation of constitutive equations. To date, however, there are some studies (Baek et al. 1993; Christiansen and Miller 1971; Ginn and Metzner 1969; Olabisi and Williams 1972) reporting measurements of N_2.

Let us now look at some representative rheological measurements taken of a polymer solution and molten polymers using a cone-and-plate rheometer. Figure 5.2 gives $\log \eta$ versus $\log \dot{\gamma}$ and $\log N_1$ versus $\log \dot{\gamma}$ plots for a 4 wt % aqueous solution of polyacrylamide at 25 °C obtained from a cone-and-plate rheometer. It is seen that η decreases, while N_1 increases, with increasing $\dot{\gamma}$; typical non-Newtonian viscoelastic behavior of polymeric fluids (see Chapter 3). Notice, however, in Figure 5.2 that the polymer solution does not exhibit Newtonian behavior at the lowest $\dot{\gamma}$ employed, suggesting that much lower values of $\dot{\gamma}$ should have been applied in order to observe Newtonian behavior. Needless to say, the rheological properties of polymer solutions depend on concentration, which is discussed in Chapter 6.

Figure 5.3 gives $\log \eta$ versus $\log \dot{\gamma}$ and $\log N_1$ versus $\log \dot{\gamma}$ plots for nylon 6 at 280 °C, and for poly(ethylene terephthalate) (PET) at 300 °C obtained from a cone-and-plate rheometer. It is seen that both nylon 6 and PET melts exhibit Newtonian behavior over a very wide range of $\dot{\gamma}$ (0.07–800 s^{-1}) employed. This is not unexpected, because the molecular weights of the nylon 6 and PET are fairly low (on the order of 10^4) and thus give rise to low melt viscosities when compared with high-molecular-weight polyolefins. Note that the rheological properties of polymers depend not only

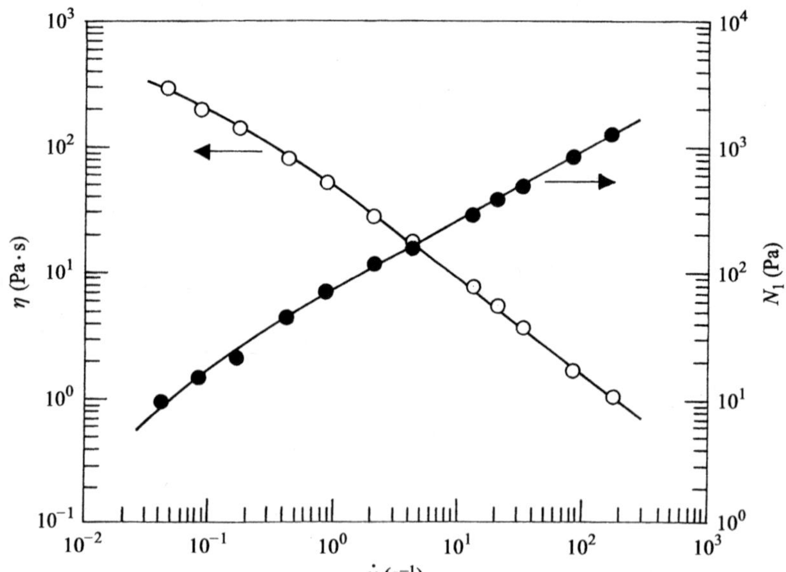

Figure 5.2 Plots of $\log \eta$ versus $\log \dot{\gamma}$ (○) and $\log N_1$ versus $\log \dot{\gamma}$ (●) for a 4 wt % aqueous solution of polyacrylamide at 25 °C.

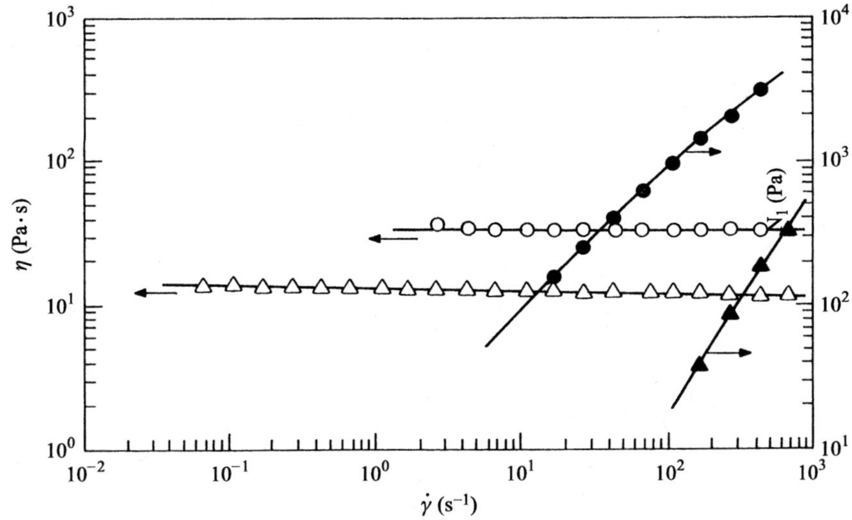

Figure 5.3 Plots of $\log \eta$ versus $\log \dot{\gamma}$ (○) and $\log N_1$ versus $\log \dot{\gamma}$ (●) for nylon 6 at 280 °C, and of $\log \eta$ versus $\log \dot{\gamma}$ (△) and $\log N_1$ versus $\log \dot{\gamma}$ (▲) for poly(ethylene terephthalate) at 300 °C.

on the chemical structure of the polymers but also on their molecular weight and molecular weight distribution. This subject is dealt with in great detail in Chapter 6. Comparison of Figure 5.2 with Figure 5.3 reveals that the viscoelasticity of the 4 wt % aqueous solution of polyacrylamide at 25 °C is much stronger than that of the nylon 6 at 280 °C and PET at 300 °C. Figure 5.4 gives $\log \eta$ versus $\log \dot{\gamma}$ and $\log N_1$ versus $\log \dot{\gamma}$ plots (open symbols) for a low-density polyethylene (LDPE) at 180, 200 and 220 °C, determined using a cone-and-plate rheometer. Filled symbols in Figure 5.4 were determined using a continuous-flow capillary rheometer, a description of which is presented in Section 5.3.2.

As will be elaborated on later in this chapter, use of the cone-and-plate rheometer in steady-state shear flow has a severe limitation in achievable $\dot{\gamma}$ for commercial thermoplastic polymer melts (e.g., polyolefins, polystyrene). Specifically, the upper limit of $\dot{\gamma}$ that can be applied to such polymer melts is about 10 s^{-1}, this is because flow instability sets in at higher $\dot{\gamma}$. Of course, such an upper limit of $\dot{\gamma}$ depends very much on the molecular weight of the polymers; for instance, the upper limit of $\dot{\gamma}$ that can be applied to very high molecular weight elastomers (e.g., natural rubber) can be as low as 0.1 s^{-1}. As can be seen in Figure 5.3, a cone-and-plate rheometer was employed to measure the rheological properties of nylon 6 and PET at $\dot{\gamma}$ up to approximately 800 s^{-1}; this was possible because their molecular weights are in the order of 10^4, which are much lower than those of LDPE and polystyrene (PS) (in the order of 10^5). Therefore, another experimental technique or different rheological instrument are needed that can be used to measure the rheological properties of high-molecular-weight polymer melts at high $\dot{\gamma}$, which are often encountered in many commercial polymer processing operations. We present such rheological techniques in the following sections.

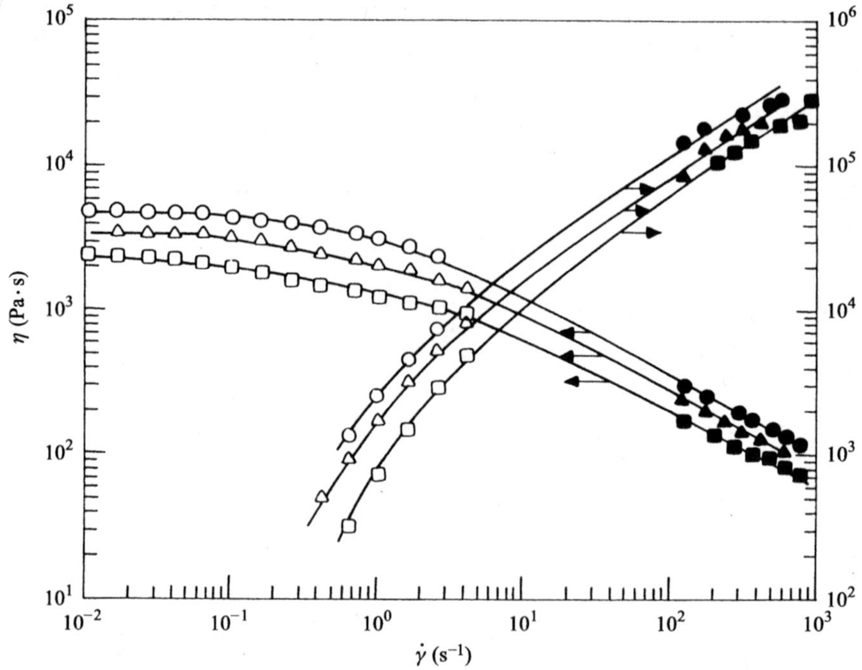

Figure 5.4 Plots of log η versus log $\dot{\gamma}$ and log N_1 versus log $\dot{\gamma}$ for an LDPE (Rexene 143) at various temperatures (°C): (○, ●) 180, (△, ▲) 200, and (□, ■) 220. The data with open symbols were obtained using a cone-and-plate rheometer, and the data with filled symbols were obtained using a continuous-flow capillary rheometer, a description of which is given in Section 5.3.2.

5.2.2 Oscillatory Shear Flow Measurement

Oscillatory shear flow properties (also referred to as dynamic viscoelastic properties) have long been used to investigate the viscoelastic properties of polymeric materials (Ferry 1980). Oscillatory shear flow measurement requires an instrument that can generate sinusoidal strain as an input to the fluid under test and record the stress resulting from the deformed fluid as an output. For such purposes, a parallel-plates fixture as well as a cone-and-plate fixture can be used; the uniform shear rate in the radial direction that is necessary when conducting steady-state shear flow experiments is no longer necessary.

In oscillatory shear flow, a sinusoidal strain is imposed on the fluid under test. If the viscoelastic behavior of the fluid is linear, the resulting stress will also vary sinusoidally, but it will be out of phase with the strain, as schematically shown in Figure 5.5. Since the sinusoidal motion can be represented in the complex domain, the following complex quantities may be defined:

$$\gamma^*(i\omega) = \gamma_o e^{i\omega t} = \gamma'(\omega) + i\gamma''(\omega) \tag{5.29}$$

$$\sigma^*(i\omega) = \sigma_o e^{i(\omega t + \varphi)} = \sigma'(\omega) + i\sigma''(\omega) \tag{5.30}$$

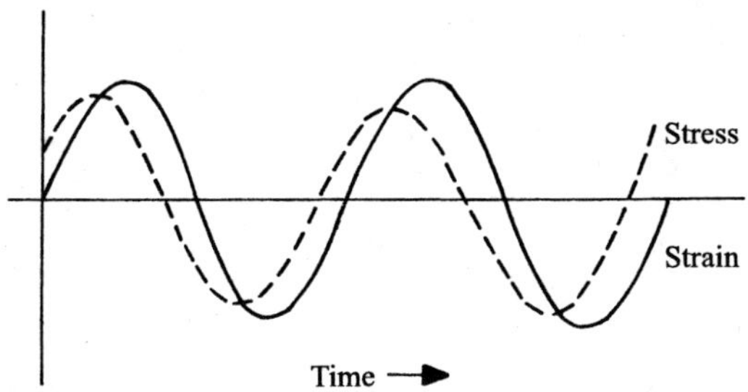

Figure 5.5 Schematic of the sinusoidally applied strain and the resulting out-of-phase sinusoidal stress response of a linear viscoelastic fluid.

where γ_0 and σ_0 are the amplitudes of the complex strain γ^* and the complex stress σ^*, respectively, and φ is the phase angle between them. The quantities with primes (γ' and σ') and double primes (γ'' and σ'') represent the real and imaginary parts of the respective complex quantities. In Eqs. (5.29) and (5.30), the response variable $\sigma^*(i\omega)$ is assumed to have the same angular frequency ω as the input variable $\gamma^*(i\omega)$. This is true only when the system (i.e., the fluid in oscillatory motion) is a linear body. It should be noted that a fluid, which would normally behave in a nonlinear fashion, can display a linear response if the experimental angular frequency ω is kept low enough.

Under an oscillatory motion, using Eqs. (5.29) and (5.30) in Eqs. (3.1) and (3.2) we obtain the following expressions for the complex modulus $G^*(i\omega)$ and the complex viscosity $\eta^*(i\omega)$:

$$G^*(i\omega) = \frac{\sigma^*(i\omega)}{\gamma^*(i\omega)} = G'(\omega) + iG''(\omega) \tag{5.31}$$

$$\eta^*(i\omega) = \frac{\sigma^*(i\omega)}{\dot{\gamma}^*(i\omega)} = \frac{\sigma^*(i\omega)}{i\omega\gamma^*(i\omega)} = \eta'(\omega) - i\eta''(\omega) \tag{5.32}$$

$\eta^*(i\omega)$ in Eq. (5.32) can be expressed in terms of $G^*(i\omega)$ as

$$\eta^*(i\omega) = \frac{G^*(i\omega)}{i\omega} = \frac{G''(\omega)}{\omega} - i\frac{G'(\omega)}{\omega} \tag{5.33}$$

From Eqs. (5.32) and (5.33) we have

$$\eta'(\omega) = G''(\omega)/\omega, \quad \eta''(\omega) = G'(\omega)/\omega \tag{5.34}$$

where $G'(\omega)$ is an in-phase elastic modulus associated with energy storage in the periodic deformation, and is called the "dynamic storage modulus." $G''(\omega)$ is an out-of-phase elastic modulus associated with the dissipation of energy as heat, and is called the "dynamic loss modulus." The real component (i.e., in-phase) of the complex viscosity is called the "dynamic viscosity."

In the actual experiment, the amplitudes of the oscillation input (γ_0) and output (σ_0), and the phase angle (φ) are measured. Therefore, each oscillatory shear flow measurement at a given ω provides two independent quantities, amplitude ratio and phase angle:

$$|G^*(i\omega)| = \sigma_o/\gamma_o = \left[(G'(\omega))^2 + (G''(\omega))^2\right]^{1/2} \tag{5.35}$$

$$\tan\varphi = G''(\omega)/G'(\omega) \tag{5.36}$$

The components of the complex modulus can be obtained by the inverse relations

$$G'(\omega) = (\sigma_o/\gamma_o)\cos\varphi, \quad G''(\omega) = (\sigma_o/\gamma_o)\sin\varphi \tag{5.37}$$

and hence from Eqs. (5.34) and (5.37) the components of $\eta^*(i\omega)$ can be written as

$$\eta'(\omega) = (\sigma_o/\gamma_o)[(\sin\varphi)/\omega], \quad \eta''(\omega) = (\sigma_o/\gamma_o)[(\cos\varphi)/\omega] \tag{5.38}$$

Representative $\log G'(\omega)$ versus $\log\omega$ and $\log G''(\omega)$ versus $\log\omega$ plots for an LDPE at three different temperatures are given in Figure 5.6. It is seen that values of G' and G'' increase with increasing ω, while they decrease with increasing temperature. Notice in Figure 5.6 that in the terminal region the slopes of $\log G''$ versus $\log\omega$ plots are less than 2 and the slopes of $\log G''$ versus $\log\omega$ plots are less than 1; behavior different from that observed for monodisperse PS in Chapter 4. This is attributable to the polydispersity of the LDPE used in Figure 5.6. The effect of polydispersity on the viscoelastic properties of polymeric fluids is presented in Chapter 6. In Chapter 3 we presented experimental data for $\log\eta'$ versus $\log\omega$, where values of η' were calculated using Eq. (5.34).

The salient feature of oscillatory shear flow measurement is that it yields information on both the viscous property $\eta'(\omega)$ and the elastic property $G'(\omega)$ of a fluid.

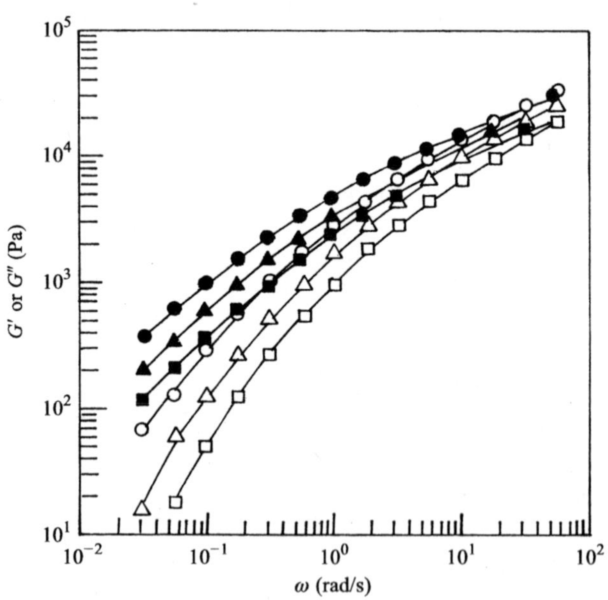

Figure 5.6 Plots of $\log G'$ versus $\log\omega$ (open symbols) and $\log G''$ versus $\log\omega$ (filled symbols) for an LDPE (Rexene 143) at various temperatures (°C): (○, ●) 180, (△, ▲) 200, and (□, ■) 220.

Also, even when the data are obtainable only over one or two decades of the logarithmic frequency scale at any one time, the viscoelastic functions can be traced out over a much larger effective range by making measurements at different temperatures, and by applying time–temperature superposition (TTS) for flexible homopolymers (see Chapter 6). In many instances, the effect of an increase in temperature is nearly equivalent to an increase in time or a decrease in frequency, as molecular viscoelastic theories suggest (see Chapter 4). When properly applied, TTS yields plots in terms of reduced variables that can be used with considerable confidence to deduce the effect of molecular parameters, and also to predict viscoelastic behavior in regions of the time or frequency scale not experimentally readily accessible (see Chapters 4 and 6).

5.3 Capillary and Slit Rheometry

In this section, we present experimental methods for measuring the rheological properties of polymeric fluids using a capillary rheometer and a slit rheometer.

5.3.1 Plunger-Type Capillary Rheometry

The plunger-type capillary rheometer has long been used because it is very simple to design and also to operate. Simply stated, polymer pellets or powders are put into the reservoir section of a capillary die, as schematically shown in Figure 5.7, which is preheated to a desired experimental temperature. The plunger placed at the top of the reservoir section of the die is then pushed down using either an inert gas or a mechanical device. In such experiments, the force applied to the plunger and the volumetric flow rate of the polymer being pushed down are measured. In this section, we present how the experimental data obtained from such a simple experiment can be used to calculate the shear viscosity of polymer. In the experiment described here, from the total force F exerted on the plunger one can calculate the stress of the fluid, which is in contact with

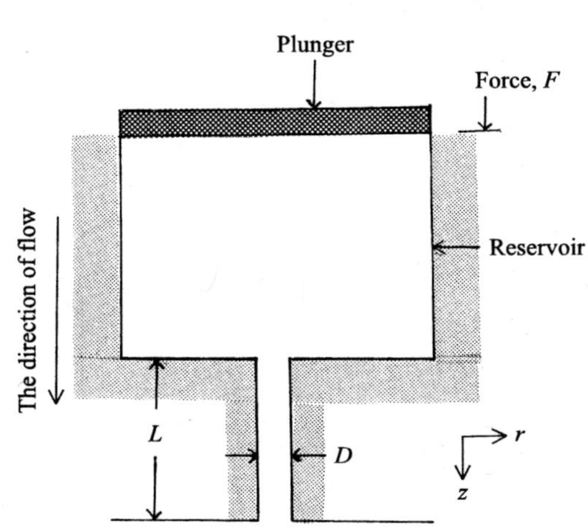

Figure 5.7 Schematic of a plunger-type capillary rheometer.

the plunger, in the flow direction from $T_{zz} = F/A$, with A being the cross-sectional area of the plunger. If we assume that the deformation of the fluid in the reservoir section is negligibly small compared with that inside the capillary, from the general expression

$$T_{zz}(r, z) = -p(r, z) + \sigma_{zz}(r, z) \tag{5.39}$$

we have

$$T_{zz}(r, z) = -p(r, z) \tag{5.40}$$

since $\sigma_{zz}(r, z) \approx 0$ inside the reservoir section. Equation (5.40) is valid in the upstream end of the reservoir section, where the extent of deformation of the fluid is negligibly small. The measurement of the total force F exerted on the plunger enables us to calculate the pressure p. Further, if we assume that the fluid exiting the die (i.e., at $z = L$) is at ambient pressure (i.e., zero gauge pressure), we can then say that the pressure calculated from Eq. (5.40) represents the total pressure drop Δp along the length L of the capillary die. Under such circumstances, we can calculate the wall shear stress σ_{w} of the fluid inside the capillary die from

$$\sigma_{\mathrm{w}} = \frac{\Delta p}{L}\frac{R}{2} \tag{5.41}$$

where R is the radius of the cylindrical tube. Hence, measurement of the force exerted on the plunger allows one to calculate σ_{w}.

In the experimental method described above, the volumetric flow rate Q of the fluid exiting the cylindrical tube is related to the velocity profile $v_z(r)$ by

$$Q = 2\pi \int_0^R r v_z(r)\,\mathrm{d}r \tag{5.42}$$

in which fully developed flow inside the capillary die is assumed. Assuming that there is no slippage at the wall (i.e., $v_z(R) = 0$), integration of Eq. (5.42) by parts gives

$$Q = -\pi \int_0^R r^2 \frac{\mathrm{d}v_z}{\mathrm{d}r}\,\mathrm{d}r \tag{5.43}$$

For shear rate $\dot{\gamma}$ defined by $\dot{\gamma} = -\mathrm{d}v_z/\mathrm{d}r$, Eq. (5.43) can be rewritten as (Han 1976)

$$\dot{\gamma}_{\mathrm{w}} = \dot{\gamma}_{\mathrm{app}} \left(\frac{3n+1}{4n} \right) \tag{5.44}$$

where $\dot{\gamma}_{\mathrm{w}}$ is the shear rate at the tube wall (i.e., at $r = R$), $\dot{\gamma}_{\mathrm{app}}$ is the "apparent" shear rate defined by

$$\dot{\gamma}_{\mathrm{app}} = 4Q/\pi R^3 \tag{5.45}$$

and n is defined by

$$n = \mathrm{d} \ln \sigma_w / \mathrm{d} \ln \dot{\gamma}_{app} \tag{5.46}$$

which can be determined from the slope of the logarithmic plot of σ_w versus $\dot{\gamma}_{app}$. The quantity inside the bracket in Eq. (5.44) is commonly referred to as the Rabinowitch–Mooney correction. Thus, values of $\dot{\gamma}_w$ can be determined from experiment. Therefore, using the experimental data for Δp and Q, one can prepare plots of shear stress σ versus shear rate $\dot{\gamma}$ (commonly referred to as "flow curve") and thus calculate the viscosity η of the fluid from the definition

$$\eta = \sigma / \dot{\gamma} \tag{5.47}$$

Here, for convenience, the subscript w in σ_w and $\dot{\gamma}_w$ has been dropped.

Figure 5.8 gives plots of Δp versus length-to-diameter (*L*/*D*) ratio for a high-density polyethylene (HDPE) at 200 °C for various values of $\dot{\gamma}$. It is seen that Δp at $z = 0$ is greater than zero and that Δp becomes zero only when an extrapolation is made inside the reservoir section. This finding was first reported by Bagley (1957) and thus Figure 5.8 is referred to as the "Bagley plot." What the Bagley plot suggests is that to calculate σ Eq. (5.41) must be modified by including an additional length of capillary die, as if a fictitious capillary die is present inside the reservoir section. Hence, Bagley (1957) suggested the following empirical expression:

$$\sigma = \frac{\Delta p}{4(L/D + n_B)} \tag{5.48}$$

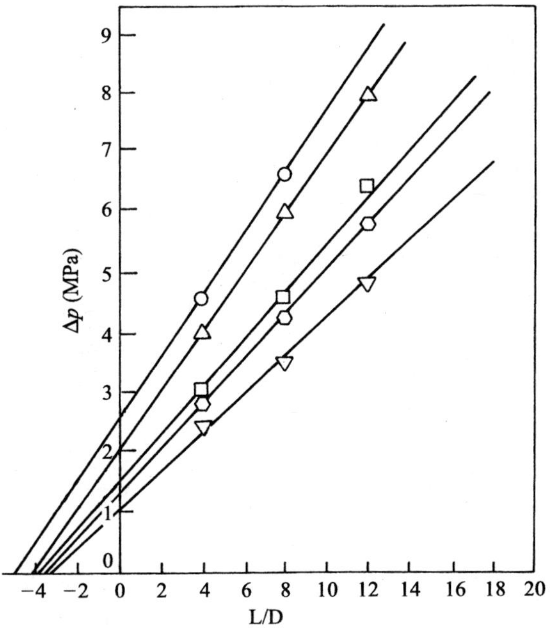

Figure 5.8 Bagley plots for an HDPE at 200 °C at various shear rates (s^{-1}): (▽) 131.2, (⬡) 234.7, (□) 289.2, (△) 568.9, and (○) 723.6. (Reprinted from Han, *Rheology in Polymer Processing*, Chapter 5. Copyright © 1976, with permission from Elsevier.)

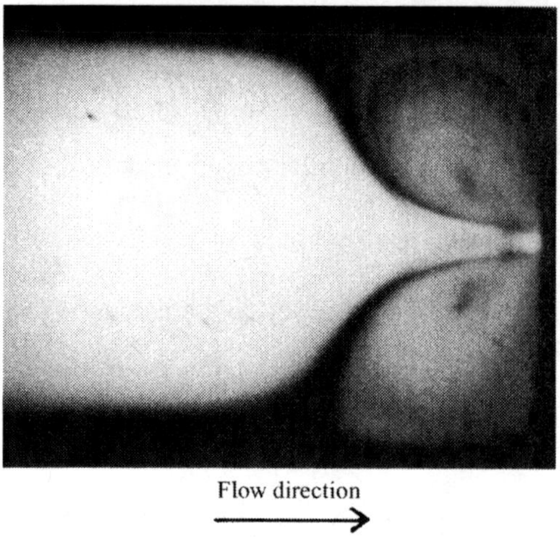

Figure 5.9 The flow patterns of an LDPE at 200 °C flowing from a large reservoir section into the slit die with a very small opening. (Reprinted from Han, *Multiphase Flow in Polymer Processing*, Chapter 2, Copyright © 1981, with permission from Elsevier.)

where n_B is referred to as the "Bagley end correction." Notice in Figure 5.8 that the value of n_B increases with shear rate. It is well documented in the literature (Han 1976) that the flow curve (thus η versus $\dot{\gamma}$ plot) will depend on L/D ratio unless the Bagley end correction is made.

The physical reason for the necessity of making the Bagley end correction in Eq. (5.48) can be understood from Figure 5.9, where a molten LDPE at 200 °C, entering into the capillary die from the large reservoir section, forms a pronounced converging melt stream. It is clear from Figure 5.9 that the formation of a converging melt streamline in the reservoir section requires a fictitious capillary length, in terms of the dimensionless quantity n_B, to be added to the true length of the capillary in order to calculate σ using Eq. (5.48).

The advantages of the experimental method described above are that (1) it is relatively easy to construct the plunger-type capillary rheometer and (2) a small amount of sample is needed. The disadvantages of the experimental method are that (1) it is very time-consuming to conduct experiments using different values of L/D ratio (three values of L/D ratio at a minimum) and (2) it allows one to determine only shear viscosity. It should be noted that the larger the value of L/D ratio, the less significant will be the contribution of n_B to the calculation of σ using Eq. (5.48). The critical value of L/D ratio, above which the contribution of n_B to the calculation of σ is negligible, depends on the elastic nature of a polymeric fluid. As a rule of thumb, $L/D \geq 20$ might be acceptable for many commercial thermoplastic polymers. However, when dealing with very high molecular weight polymers (e.g., ultra-high molecular weight HDPE, elastomers), even an $L/D = 20$ may not be sufficiently large.

5.3.2 Continuous-Flow Capillary Rheometry

In this section, we present another experimental method that also employs a capillary die. However, this method allows one to determine not only shear viscosity but also normal stress differences in steady-state shear flow by continuously supplying a polymeric

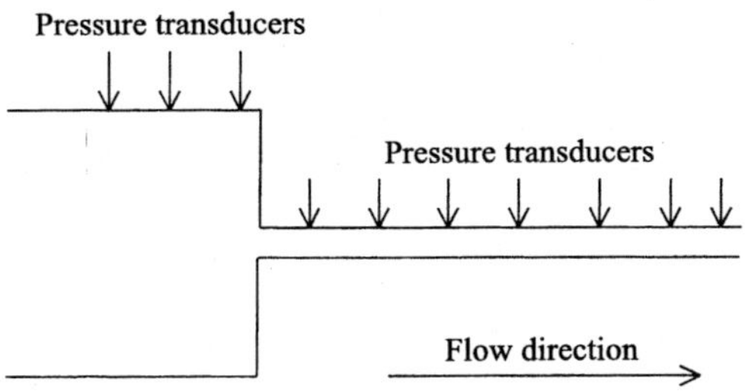

Figure 5.10 Schematic showing the side view of a continuous-flow capillary die, where a series of pressure transducers are mounted at the surfaces of the reservoir and capillary die, along the axis of the die.

fluid into a capillary die. This method is quite different, experimentally and theoretically, from the plunger-type capillary rheometry presented in the preceding section. In order to distinguish from plunger-type capillary rheometry, we refer to this method as "continuous-flow capillary rheometry."

Let us consider the situation where pressure transducers are mounted at the tube wall along the axis of a cylindrical tube, including the reservoir section, as schematically shown in Figure 5.10. Under such a situation, pressure transducers measure the outward acting total normal stress at the tube wall (i.e., at $r = R$), which will be referred to as wall normal stress, denoted by $T_{rr}(R, z)$, and which consists of two parts, as defined by

$$T_{rr}(R, z) = -p(R, z) + \sigma_{rr}(R, z) \tag{5.49}$$

in which $p(R, z)$ is the isotropic pressure and $\sigma_{rr}(R, z)$ is the deviatoric (extra) stress.

When a fluid enters a tube from a large reservoir section, the velocity profile becomes fully developed at a certain distance from the die entrance. A conventional criterion, such as the 'constant pressure gradient' in the tube, is not sufficient to determine whether a viscoelastic fluid has attained fully developed flow in a cylindrical die because the relaxation time of a viscoelastic fluid is much greater than that of a Newtonian fluid. The criteria for determining fully developed flow in viscoelastic fluids have been discussed in the literature (Han and Charles 1970).

Figure 5.11 gives the profiles of $T_{rr}(R, z)$ measured with seven pressure transducers placed at the tube wall along the axis of a long cylindrical tube ($L/D = 20$), including the reservoir section, for polybutene at 25 °C, and Figure 5.12 gives profiles of $T_{rr}(R, z)$ along the axis of a cylindrical tube including the reservoir section for an HDPE at 200 °C for two different L/D ratios (4 and 8). There is no discernible pressure drop in the entrance region of the capillary die for polybutene in Figure 5.11, whereas there are considerable pressure drops in the entrance region of the capillary die for the HDPE in Figure 5.12. The difference between the two situations is attributable to the difference in the viscoelastic properties of the two fluids; namely, as we will show, the polybutene employed in Figure 5.11 is a Newtonian fluid, while the HDPE employed in Figure 5.12 is a non-Newtonian viscoelastic fluid. There is ample experimental evidence indicating that the viscosity of the material alone cannot explain such large entrance-pressure drops (Ballenger and White 1971; Ballenger et al. 1971; Boles et al. 1970; Han 1971b; Han and Kim 1971; LaNieve and Bogue 1968).

What is most significant in the profiles of $T_{rr}(R, z)$ displayed in Figures 5.11 and 5.12 is the difference in the values of $T_{rr}(R, z)$ at the die exit; namely, in Figure 5.11

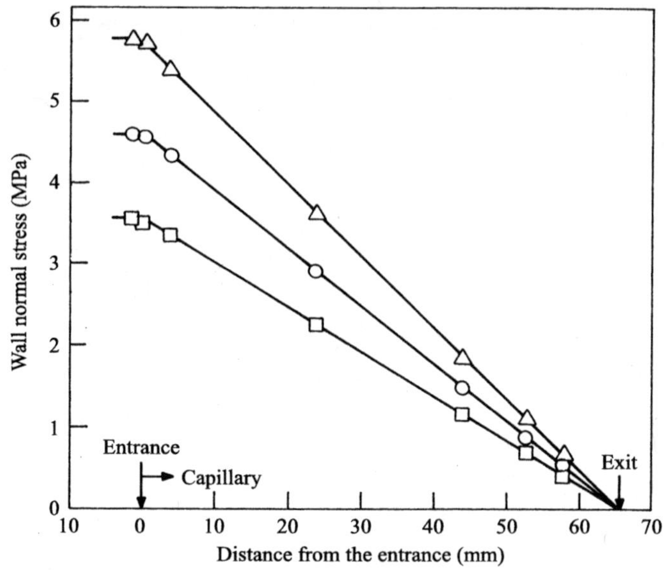

Figure 5.11 Wall normal stress profiles of a polybutene (Indopol H1900) at 25 °C in a capillary die at various shear rates (s^{-1}): (△) 50.1, (○) 40.4, and (□) 31.1. (Reprinted from Han, *Rheology in Polymer Processing*, Chapter 5. Copyright © 1976, with permission from Elsevier.)

the extrapolated values of the $T_{rr}(R, z)$ profile at the die exit (at $z = L$), $T_{rr}(R, L)$, are virtually zero (i.e., zero gauge pressure), whereas in Figure 5.12 the values of $T_{rr}(R, L)$ are nonzero (i.e., above ambient pressure). The profiles of $T_{rr}(R, z)$ displayed in Figure 5.12 may be described by

$$\Delta P = \Delta P_{\text{Ent}} + \Delta P_{\text{Cap}} + \Delta P_{\text{Exit}} \tag{5.50}$$

where ΔP_{Ent} denotes the pressure drop in the entrance region, ΔP_{Cap} denotes the pressure drop in the capillary section, and P_{Exit} denotes the value of $T_{rr}(R, L)$ and it is referred to as "exit pressure" (Arai 1970; Han 1976; Han et al. 1969, 1970, 1971; Mori and Funatsu 1968).[1] Referring to Eq. (5.50), (1) it is assumed that the ratio of reservoir diameter to capillary diameter is sufficiently large that the pressure drop due to a kinetic energy effect is negligible, (2) ΔP_{Cap} is due entirely to viscous dissipation, and (3) ΔP_{Ent} may consist of three components: (a) the pressure drop related to the viscous dissipation due to the flow converging prior to entering the capillary, (b) the pressure drop related to the viscous dissipation due to the development of the velocity profile near the entrance of the capillary, and (c) the pressure drop that may be converted into elastic energy, some of which is believed to be recoverable due to the elastic nature of the viscoelastic fluids. P_{Exit} may be related to the recoverable elastic energy, which is stored in the melt as it leaves the tube exit (Han 1976). Using Eq. (5.49), the exit pressure may be expressed as

$$P_{\text{Exit}} = -T_{rr}(R, L) \tag{5.51}$$

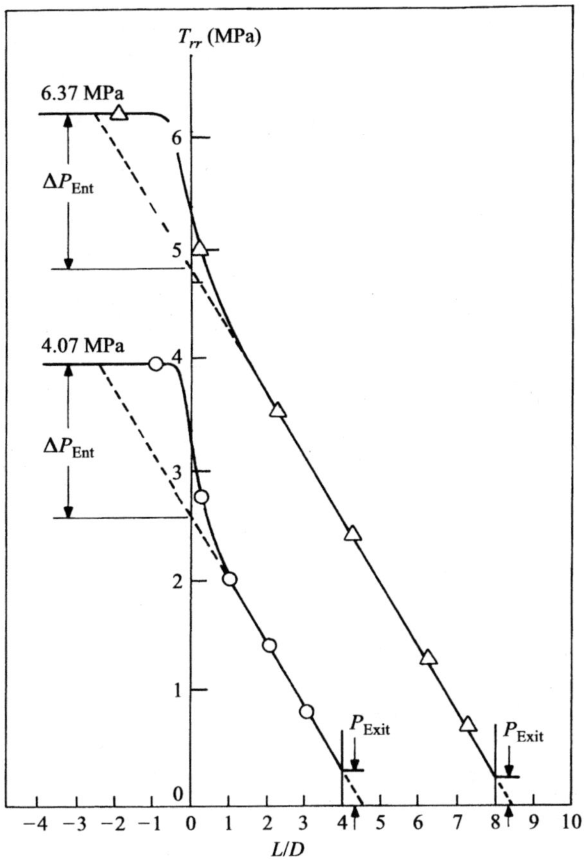

Figure 5.12 Wall normal stress profiles of an HDPE at 200 °C in capillary dies with different L/D ratios: (○) $L/D = 4$ at a shear rate 327.7 s^{-1} having $\Delta P_{Ent} = 1.414$ MPa and $P_{Exit} = 320.4$ kPa, (△) $L/D = 8$ at a shear rate 329.7 s^{-1} having $\Delta P_{Ent} = 1.414$ MPa and $P_{Exit} = 236.3$ kPa. (Reprinted from Han, *Rheology in Polymer Processing,* Chapter 5. Copyright © 1976, with permission from Elsevier.)

in which $T_{rr}(R, L)$ is the outward acting total wall normal stress at the wall ($r = R$) and at the exit of the die ($z = L$). Note that the value of $T_{rr}(R, z)$, and thus $T_{rr}(R, L)$, have negative sign (compression) because the wall normal stress itself, measured with a pressure transducer, is positive. It cannot be overemphasized that the pressure transducers mounted at the tube wall along the axis of a capillary die do *not* measure pressure $p(R, z)$ but two combined quantities, $p(R, z)$ and $\sigma_{rr}(R, z)$ (see Eq. (5.49)). There is no way of separating $p(R, z)$ and $\sigma_{rr}(R, z)$ from the measured $T_{rr}(R, z)$.

Figure 5.13 gives wall normal stress profiles in the fully developed region of a cylindrical tube with $L/D = 20$ for an LDPE (NPE 952) at 180 °C at three different values of $\dot{\gamma}$. It can be seen that P_{Exit} increases with $\dot{\gamma}$. Figure 5.14 gives log P_{Exit} versus log $\dot{\gamma}$ plots for the LDPE at three different temperatures. It is seen that P_{Exit} increases with increasing $\dot{\gamma}$ at a given temperature and decreases with increasing temperature. Similar observations have been reported for a number of different polymers (Arai 1970; Chan et al. 1990; Han et al. 1969, 1970, 1971; Han 1972, 1973a, 1973b; Han and Kim 1975; Han and Villamizar 1978; Han and Yu 1971; Han et al. 1983a, 1983b; Mori and Funatsu 1968).

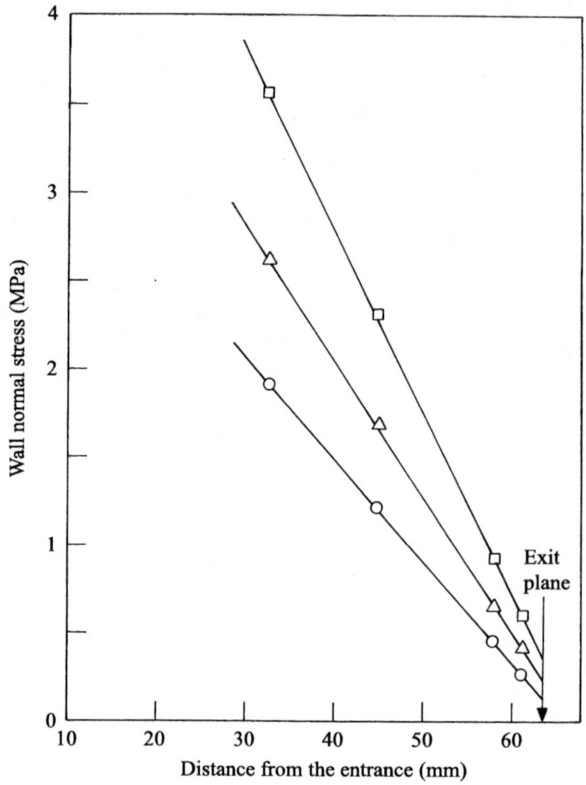

Figure 5.13 Wall normal stress profiles of an LDPE (NPE 952) at 180 °C in a capillary die ($L/D = 20$) at various shear rates (s^{-1}): (○) 82.6, (△) 185.7, and (□) 432.5.

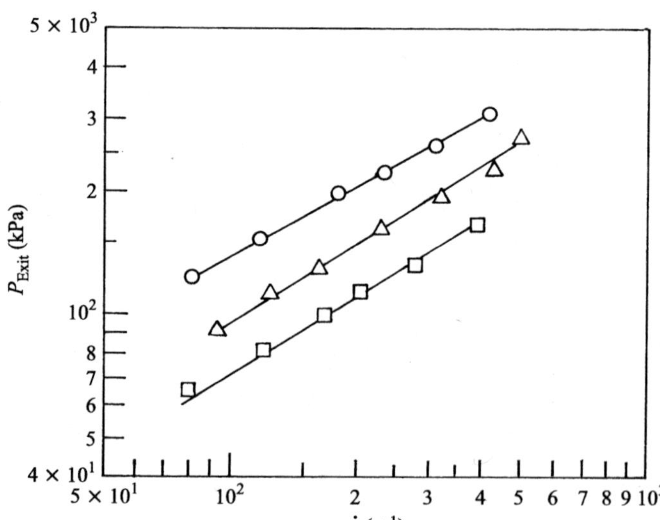

Figure 5.14 Plots of log P_{Exit} versus log $\dot{\gamma}$ for an LDPE (NPE 952) at various temperatures (°C): (○) 180, (△) 200, and (□) 220.

Of course, the measurement of $T_{rr}(R, z)$ precisely at the exit of a die is not possible mechanically. Therefore, the extrapolation of the measurement taken nearest to the exit of the die must be justified if one wishes to use the exit pressure as a means of interpreting the flow properties of the fluid tested. This is because the extrapolation tacitly assumes that the flow remains fully developed to the die exit. We will elaborate on this further later in this section.

We now present the theory (Davis et al. 1973; Han 1974) that enables one to determine both shear viscosity and normal stress differences for viscoelastic fluids from measurements of $T_{rr}(R, z)$ in a long capillary die. For steady-state fully developed flow in the cylindrical die, the equations of motion are given by

$$-\frac{\partial p}{\partial z} + \frac{1}{r}\frac{1}{\partial r}(r\sigma_{rz}) = 0 \tag{5.52}$$

$$-\frac{\partial p}{\partial r} + \frac{\partial \sigma_{rr}}{\partial r} + \frac{T_{rr} - T_{\theta\theta}}{r} = 0 \tag{5.53}$$

Integration of Eq. (5.52) from $r = 0$ to $r = R$ gives

$$\sigma = \frac{\partial p}{\partial z}\frac{R}{2} \tag{5.54}$$

where σ is the wall shear stress and $\partial p/\partial z$ is the pressure gradient in the fully developed region. For fully developed flow in which σ_{rr} no longer depends on z, differentiating both sides of Eq. (5.49) with respect to z gives

$$-\frac{\partial T_{rr}(R, z)}{\partial z} = \frac{\partial p(R, z)}{\partial z} \tag{5.55}$$

suggesting that the slope of $T_{rr}(R, z)$ profile in the fully developed region of a cylindrical tube is the same as the pressure gradient, $\partial p/\partial z$. Therefore, a direct measurement of pressure gradient in the fully developed region of a sufficiently long tube will enable one to calculate σ using Eq. (5.54) and thus η using Eq. (5.47).

Next, integration of Eq. (5.53) from $r = 0$ to $r = R$ gives

$$p(R, z) = p(0, z) + \sigma_{rr}(R) + \int_0^R (T_{rr} - T_{\theta\theta})\, \mathrm{d}\ln r \tag{5.56}$$

Making use of the definition given by Eq. (5.49), Eq. (5.56) may be rewritten as

$$T_{rr}(R, L) = -p(0, L) - \int_0^R (T_{rr} - T_{\theta\theta})\, \mathrm{d}\ln r \tag{5.57}$$

at $z = L$. Using $r = (R/\sigma_{\mathrm{w}})\sigma_{rz}$ in the integral on the right-hand side of Eq. (5.57) and differentiating both sides of the resulting equation with respect to σ gives

$$(T_{rr} - T_{\theta\theta})_R = -\sigma\frac{\mathrm{d}p(0, L)}{\mathrm{d}\sigma} + \sigma\frac{\mathrm{d}P_{\mathrm{Exit}}}{\mathrm{d}\sigma} \tag{5.58}$$

which is the second normal stress difference. Note in Eq. (5.58) that P_{Exit} is defined by Eq. (5.51). Conversely, we have the following relationship (Lodge 1964)

$$(2T_{zz} - T_{rr} - T_{\theta\theta})_R = \frac{1}{\pi R^2 \sigma} \frac{\mathrm{d}}{\mathrm{d}\sigma}[\sigma^2 (T_L + \pi R^2 P_{\text{Exit}})] \tag{5.59}$$

in which T_L is the total thrust due to elasticity, and is related to $T_{zz}(r, L)$ by

$$T_L = 2\pi \int_0^r r\, T_{zz}(r, L)\, \mathrm{d}r \tag{5.60}$$

Combining Eqs. (5.58) and (5.59) gives

$$(T_{zz} - T_{rr})_R = P_{\text{Exit}} + \frac{\sigma}{2} \frac{\mathrm{d}p(0, L)}{\mathrm{d}\sigma} + \frac{1}{2\pi R^2 \sigma} \frac{\partial}{\partial \sigma}[\sigma^2 T_L] \tag{5.61}$$

which is the first normal stress difference N_1.

In view of the fact that Reynolds numbers in polymer melt flow are very low (say, 10^{-5} to 10^{-2}) in almost all practical situations and the contribution of the thrust (T_L) terms in Eqs. (5.59) and (5.61) may be considered negligibly small compared with that of other terms for polymer melt flows, from Eq. (5.61) we have

$$(T_{zz} - T_{rr})_R = N_1 = P_{\text{Exit}} + \frac{\sigma}{2} \frac{\mathrm{d}p(0, L)}{\mathrm{d}\sigma} \tag{5.62}$$

It is seen above that in steady-state flow through a capillary die the determination of N_1 requires, in addition to P_{Exit}, information on how the center pressure $p(0, L)$ varies with σ. In practice, measurement of $p(0, L)$ is very difficult, if not impossible.

On the basis of limited experimental results, Han (1974) concluded that the exit pressures of polymer melts determined from a capillary die were the same, within experimental uncertainties, as those determined from a slit die. Such an experimental observation led Han to suggest the following relationship:

$$p(0, L) = 2P_{\text{Exit}} \tag{5.63}$$

Substitution of Eq. (5.63) into (5.58) and (5.62) gives

$$N_1 = P_{\text{Exit}} \left[1 + \frac{\mathrm{d}(\ln P_{\text{Exit}})}{\mathrm{d}(\ln \sigma)} \right] \tag{5.64}$$

and

$$N_2 = -\sigma \frac{\mathrm{d}P_{\text{Exit}}}{\mathrm{d}\sigma} \tag{5.65}$$

respectively. It can be seen that N_1 can be determined from Eq. (5.64) and N_2 from Eq. (5.65), using the measurements of P_{Exit} as a function of σ in a long capillary die.

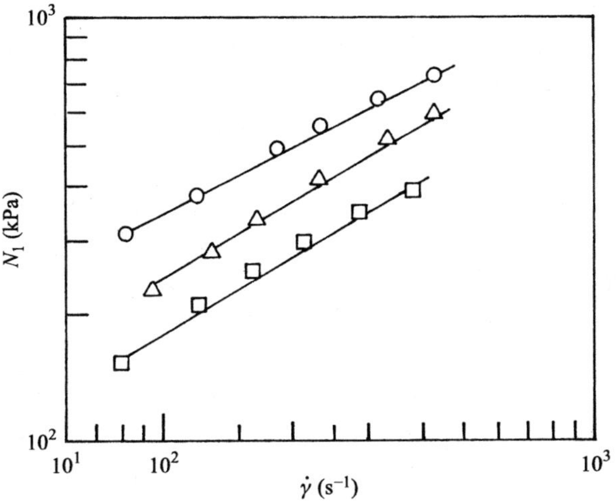

Figure 5.15 Plots of log N_1 versus log $\dot{\gamma}$ for an LDPE (NPE 952) at various temperatures (°C): (○) 180, (△) 200, and (□) 220.

Figure 5.15 gives log N_1 versus log $\dot{\gamma}$ plots for an LDPE at three different temperatures, which were obtained, via Eq. (5.64), from the P_{Exit} measurements given in Figure 5.14. Later in this chapter we compare the values of N_1 determined from continuous-flow capillary rheometry with those obtained from cone-and-plate rheometry.

5.3.3 Slit Rheometry

Another experimental method that is as important as continuous-flow capillary rheometry is slit rheometry. The basic idea of slit rheometry is the same as that of continuous-flow capillary rheometry insofar as the measurement of wall normal stress along the die axis is concerned. But, there is a significant theoretical difference between the two methods, as we will make clear, and also in the die design. The slit rheometer has some advantages over the continuous-flow capillary rheometer in the way that transducers can be mounted on the die wall, but there are also some disadvantages.

In the use of a slit die, pressure transducers can be mounted flush with the die wall, as schematically shown in Figure 5.16. The width-to-height (w/h) ratio (often referred to as "aspect ratio") of a slit die must be sufficiently large to allow the tip of the pressure transducer to be located on the die wall where isovels (contour lines connecting points having the same velocity) in a thin slit are parallel to the long sides over most of the width of the die and parallel also to the tip of the pressure transducer (see Figure 5.16). Again, pressure transducers must be mounted in a slit die where flow is fully developed. Figure 5.17 gives a photograph of isochromatic fringe patterns for PS melt in a slit die, showing that isochromatic fringe patterns are parallel to the die wall. The parallel nature of the isochromatic fringe patterns in Figure 5.17 attests that the flow is fully developed in the slit die (Han and Drexler 1973a). Notice in Figure 5.17 that the number of fringe patterns increases from the center to the die wall, indicating that the shear stress increases from the center to the die wall and

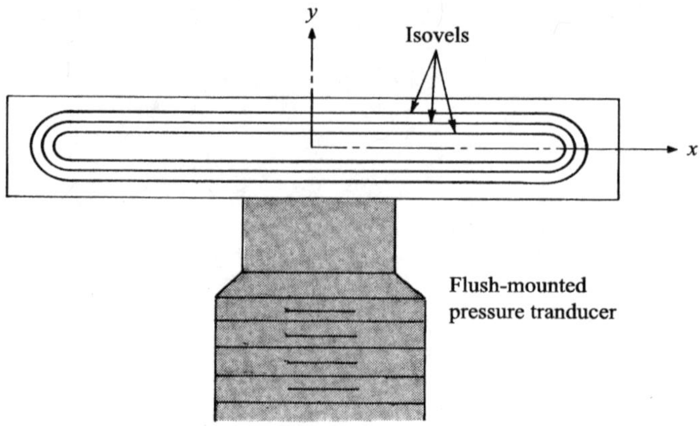

Figure 5.16 Schematic showing the isovels in the cross-section of a slit die and a pressure transducer flush mounted on the long side of the thin slit.

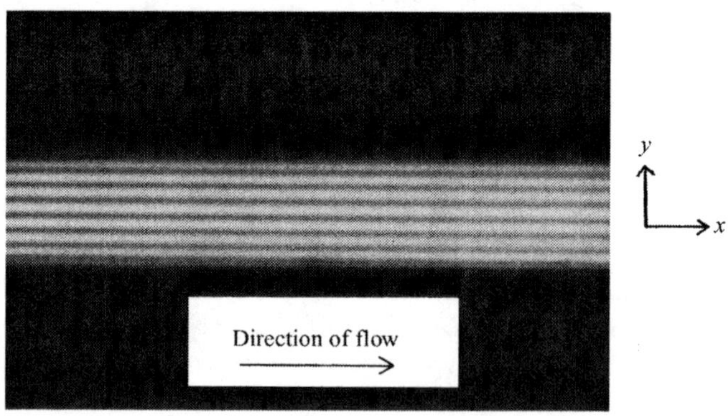

Figure 5.17 Isochromatic fringe patterns for a PS melt in the fully developed region of a slit die, where the order of fringes increases with the distance y. (Reprinted from Han and Drexler, *Journal of Applied Polymer Science* 17:2329. Copyright © 1973, with permission from John Wiley & Sons.)

shear stress is proportional to the distance y. Several research groups, notably Han and coworkers (Han 1971a, 1974; Han et al. 1973) employed slit rheometry to determine the rheological properties of polymer melts.

We now present the theory (Davis et al. 1973; Han 1974) that allows one to determine shear stress and first normal stress difference in steady-state shear flow using wall normal stress measurements along the axis of a slit die. Consider a fluid flowing through a slit die having the height h and the width w, and assume that flow has become fully developed. Then, for steady-state fully developed flow, the equations of motion in rectangular coordinates (x, y, z) are given by

$$-\frac{\partial p}{\partial z} + \frac{\partial \sigma_{yz}}{\partial y} = 0 \tag{5.66}$$

$$-\frac{\partial p}{\partial y} + \frac{\partial \sigma_{yy}}{\partial y} = 0 \tag{5.67}$$

in which z is the direction of flow and y is the direction perpendicular to z. Integrating Eq. (5.66) from $y = 0$ to $y = b$ (half of the slit height h), we obtain

$$\sigma = \frac{\partial p}{\partial z} b \tag{5.68}$$

where σ is the wall shear stress and $\partial p/\partial z$ is the pressure gradient.

Therefore, measurements of wall normal stress along the die axis, $T_{yy}(b, z)$ in the fully developed region allows one to calculate $\partial p/\partial z$ (see Eq. (5.55)), and thus σ and η. Similar to the analysis of the capillary flow presented above, we have the following expression for $\dot{\gamma}$ in slit flow (Han 1974, 1976):

$$\dot{\gamma} = \dot{\gamma}_{\text{app}} \left(\frac{2n+1}{3n} \right) \tag{5.69}$$

where $\dot{\gamma}_{\text{app}}$ is apparent shear rate in a slit die defined by

$$\dot{\gamma}_{\text{app}} = 6Q/wh^2 \tag{5.70}$$

with n being defined by Eq. (5.46).

Integrating Eq. (5.67) from $y = 0$ to $y = b$ (half of the slit height), one obtains

$$p(b, L) = p(0, L) + \sigma_{yy}(b) \tag{5.71}$$

From the definition of the deviatoric (or extra) stress, for fully developed flow one has

$$T_{yy}(b, z) = -p(b, z) + \sigma_{yy}(b) \tag{5.72}$$

Combining Eqs. (5.71) and (5.72) yields

$$p(0, L) = -T_{yy}(b, L) \tag{5.73}$$

Since $T_{yy}(b, L)$ is related to the exit pressure P_{Exit} by

$$-T_{yy}(b, L) = P_{\text{Exit}} \tag{5.74}$$

Eq. (5.73) becomes

$$p(0, L) = P_{\text{Exit}} \tag{5.75}$$

Equation (5.75) is a rather interesting result in that the pressure at the center of the slit at the exit plane is equal to the total outward acting wall normal stress at the die exit (at $z = L$).

In a similar way to Eq. (5.60), we can relate the total thrust T_L in slit flow to the axial normal stress $T_{zz}(y, L)$ by

$$T_L = 4w \int_0^b T_{zz}(y, L)\,\mathrm{d}y \tag{5.76}$$

which can be rewritten as

$$T_L = 4w \int_0^b (T_{zz} - T_{yy})\,\mathrm{d}y + 4w \int_0^b T_{yy}(y, L)\,\mathrm{d}y \tag{5.77}$$

Since it can be shown from Eq. (5.67) that

$$T_{yy}(y, L) = \text{constant} \tag{5.78}$$

Eq. (5.77) becomes

$$\int_0^b (T_{zz} - T_{yy})\,\mathrm{d}y = \frac{1}{4w}\left[T_L + 4bw\,P_{\text{Exit}}\right] \tag{5.79}$$

in which use is made of Eq. (5.74). Using $\sigma_{yz} = (y/b)\sigma$ (i.e., changing the variable y with the variable σ) in the integral on the left-hand side of Eq. (5.79) and differentiating both sides of the resulting equation with respect to σ gives

$$(T_{zz} - T_{yy})_b = P_{\text{Exit}} + \sigma\frac{\mathrm{d}P_{\text{Exit}}}{\mathrm{d}\sigma} + \frac{1}{4bw}\frac{\partial}{\partial\sigma}[\sigma T_L] \tag{5.80}$$

This is the first normal stress difference N_1 in slit flow.

For polymer melts having very low Reynolds numbers and negligible contributions of the thrust term in Eq. (5.80), we have

$$N_1 = P_{\text{Exit}}\left[1 + \frac{\mathrm{d}(\ln P_{\text{Exit}})}{\mathrm{d}(\ln\sigma)}\right] \tag{5.81}$$

It is interesting to observe in Eq. (5.81) that, in slit flow, measurement of P_{Exit} as a function of σ alone is sufficient to determine N_1.

Figure 5.18 gives the profiles of $T_{yy}(b, z)$ along the axis of a slit die ($w/h = 20$ and $h = 1.27$ mm) for a polybutene (Indopol H1900) at 35 °C at four different shear rates. It is seen that the $T_{yy}(b, z)$ profiles are linear, thus $-\partial T_{yy}(b, z)/\partial z = \partial p/\partial z$, allowing us to calculate σ of the fluid by using Eq. (5.68) and thus η using Eq. (5.47). Figure 5.19 gives a plot of $\log\eta$ versus $\log\dot{\gamma}$ for Indopol H1900 at 35 °C determined from a slit rheometer. Also given in Figure 5.19, for comparison, are values of η obtained from a cone-and-plate rheometer. It is seen that agreement between the two experimental methods is excellent.

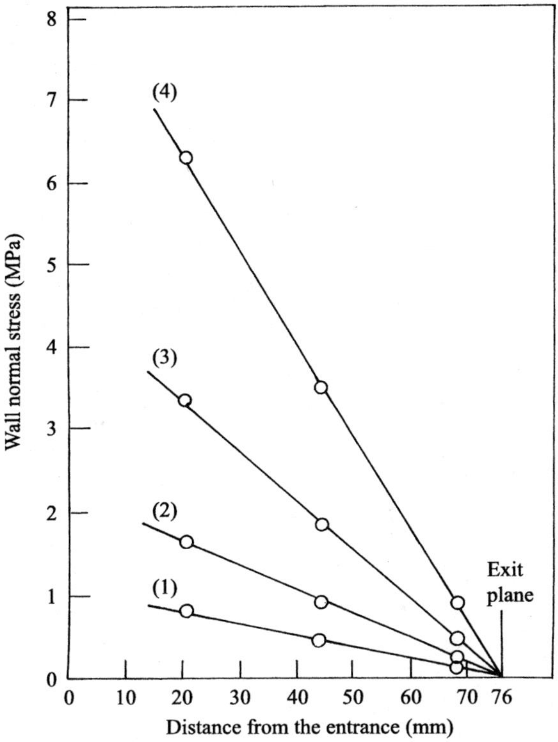

Figure 5.18 Wall normal stress profiles of a polybutene (Indopol H1900) at 35 °C in a slit die at various shear rates (s^{-1}): (1) 20.5, (2) 38.0, (3) 77.7, and (4) 157.3.

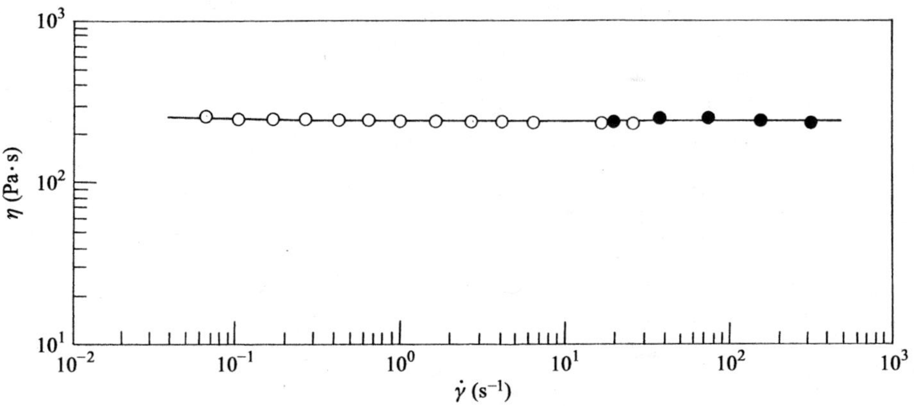

Figure 5.19 Plot of log η versus log $\dot{\gamma}$ for a polybutene (Indopol H1900) at 35 °C: (○) the data obtained from a cone-and-plate rheometer, (●) the data obtained from a slit rheometer.

Figure 5.20 gives log P_{Exit} versus log σ plots for an LDPE (NPE 952) from three different sources: from Han et al. (1983b)[2], who employed a continuous-flow capillary rheometer; from Baird et al. (1986), who employed a slit rheometer; from Lodge and de Vargas (1983), who also employed a slit rheometer. It should be pointed out that a comparison of the data displayed in Figure 5.20 from three different sources

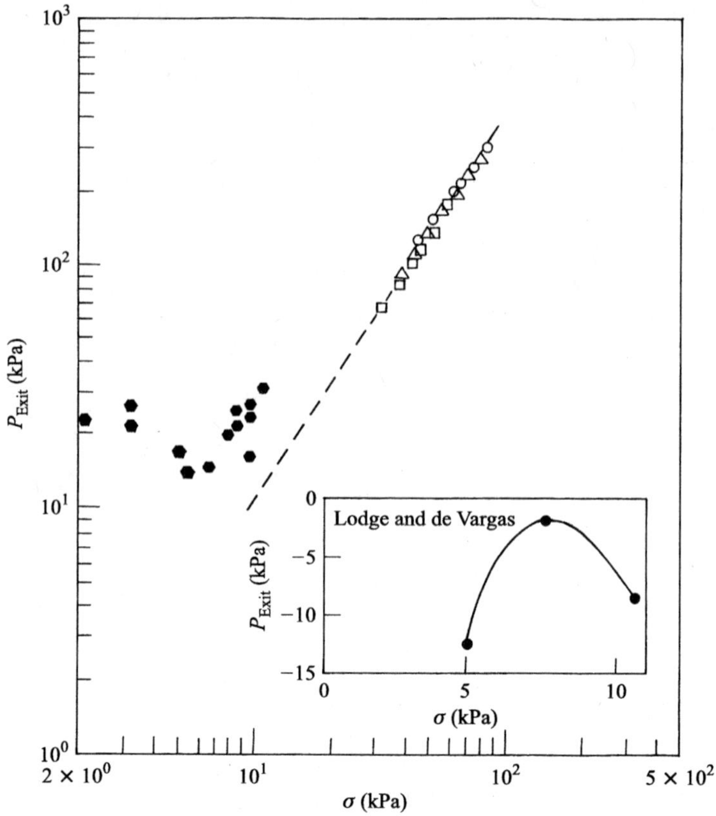

Figure 5.20 Plots of log P_{Exit} versus log σ for an LDPE (NPE 952) reported by three different research groups. The data with open symbols were reported by Han et al. (1983b), who used three pressure transducers in a continuous-flow capillary rheometer at three different temperatures (°C): (○) 180, (△) 200, and (□) 220. The data with the symbol ● were reported by Lodge and de Vargas (1983), who used only two pressure transducers in a slit rheometer at 151 °C. The data with the symbol ⬢ were reported by Baird et al. (1986), who used three pressure transducers in a slit rheometer at 151 °C.

is warranted, in spite of the fact that the experimental data from each source were obtained at different temperatures, because log P_{Exit} versus log σ plots do *not* depend on temperature (Han 1974, 1976). Referring to Figure 5.20, the P_{Exit} data of Baird et al. (1986) and Lodge and de Vargas (1983) were taken at $\sigma < 13$ kPa, while the P_{Exit} data of Han et al. (1983b) were taken at $\sigma > 30$ kPa. As we will describe, the extent of flow disturbances in the exit region may no longer be negligible when the applied σ is below a certain critical value, suggesting that the extrapolation procedure employed by Baird et al. (1986) and Lodge and de Vargas (1983) to obtain P_{Exit} might not have been warranted. The significant scatter of the P_{Exit} data of Baird et al. may be attributable to the presence of significant flow disturbances in the exit region, because the $T_{yy}(b, z)$ measurements in the slit die were taken at $\sigma < 13$ kPa. The negative values of the P_{Exit} data from Lodge and de Vargas may be attributable to two sources: (1) the presence of significant flow disturbances in the exit region because the $T_{yy}(b, z)$

measurements in the slit die were obtained at $\sigma < 13$ kPa, and (2) experimental uncertainties associated with the $T_{yy}(b, z)$ measurements because only two pressure transducers were used (i.e., the authors had no choice but to draw a straight line between two data points!). Tuna and Finlayson (1988) reported that they obtained negative values of P_{Exit} when using two pressure transducers and positive values of P_{Exit} when using three pressure transducers.

Figure 5.21 gives log N_1 versus log σ plots for another LDPE (Rexene 143) at 200 and 220 °C, which were obtained from a cone-and-plate rheometer, a slit rheometer, and a continuous-flow capillary rheometer. The values of N_1 obtained from a continuous-flow capillary rheometer were calculated using Eq. (5.64) and the values of N_1 obtained from a slit rheometer were calculated using Eq. (5.81). Notice in Figure 5.21 that log N_1 versus log σ plots obtained from capillary/slit rheometry at $\sigma \geq 25$ kPa agree very well with those obtained from cone-and-plate rheometry, suggesting that the P_{Exit} data must be taken at $\sigma \geq 25$ kPa. Table 5.1 gives a summary of P_{Exit} data obtained at $\sigma \geq 25$ kPa

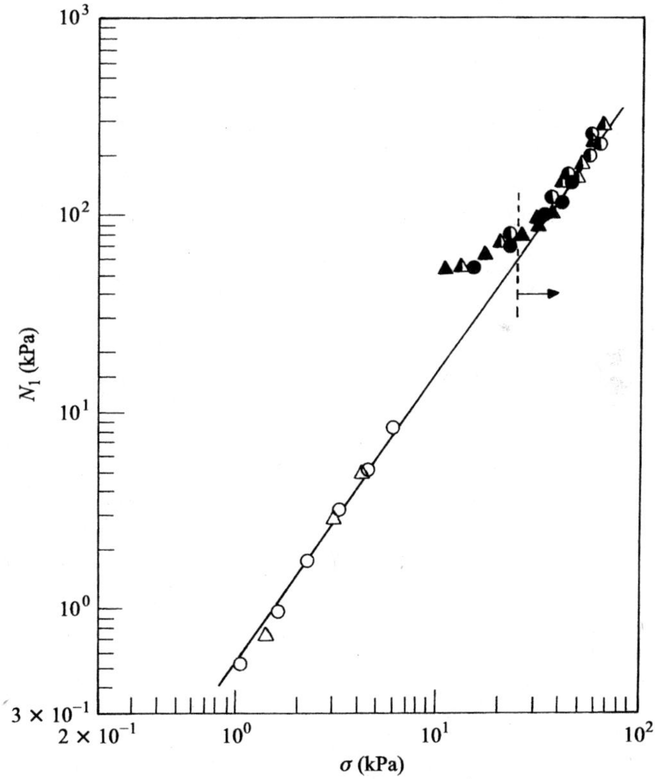

Figure 5.21 Plots of log N_1 versus log σ for an LDPE (Rexene 143) at three different temperatures. The data with open symbols were obtained using a cone-and-plate rheometer at 200 °C (○) and 220 °C (△). The data with filled symbols were obtained using a slit rheometer at 200 °C (●) and 220 °C (▲). The data with half-filled symbols were obtained using a continuous-flow capillary rheometer at 200 °C (◐) and 220 °C (◭).

Table 5.1 Summary of the materials tested, the lower limits of shear stress employed and exit pressures reported in studies by Han and coworkers

Material	*Temp.* (°*C*)	σ (*kPa*)	P_{Exit} (*psig*)	(*kPa*)	*Reference*
HDPE, PP	180	>65	>10	>69	Han et al. (1970)
HDPE, LDPE	200	>56	>20	>138	Han et al. (1971)
PS	200–240	>60	>12	>82	Han and Yu (1971)
HDPE, PS	200–240	>30	>5	>35	Han (1971a)
HDPE, PP	200	>55	>6	>41	Han (1972)
HDPE, LDPE	154	>70	>8	>55	Han (1973b)
HDPE, PS, PP, PMMA	200	>30	>7	>48	Han (1974)
HDPE, PS	200–240	>40	>15	>103	Han and Kim (1975)
HDPE, LDPE	160–200	>30	>6	>41	Han and Villamizar (1978)
HDPE, LDPE, PP, PS, PMMA	180–230	>40	>5	>35	Han et al. (1983a)
LDPE	180–220	>35	>5	>35	Han et al. (1983b)

PP = polypropylene, PS = polystyrene, PMMA = poly(methyl methacrylate).

for a number of commercial thermoplastic polymers over a wide range of temperatures. One must be aware of the fact that reproducible and reliable experimental data from a continuous-flow capillary rheometer or slit rheometer requires careful die design. We address this issue in greater detail in the following section.

5.3.4 Critical Assessment of Capillary and Slit Rheometry

There are practical limitations to each of the various experimental methods described in the preceding sections for determining the rheological properties of polymeric fluids. The use of a cone-and-plate rheometer is limited to low shear stresses (below a certain critical value) for viscoelastic polymeric fluids, because at higher shear stresses the flow becomes unstable due to the onset of secondary flow, which gives rise to erratic outputs of torque and normal force. In this regard, cone-and-plate rheometry may be regarded as being low shear-stress rheometry. In the use of a continuous-flow capillary or slit rheometer, wall normal stress must be measured at sufficiently high shear stresses so that the extrapolation of wall normal stress taken in the fully-developed region to the die exit may be warranted. In this regard, continuous-flow capillary rheometry and slit rheometry may be regarded as being high-shear-stress rheometry.

Using finite element analysis (FEA), Tuna and Finlayson (1984) determined the following relationship between P_{Exit} and N_1 for a slit die:

$$P_{Exit}/\sigma = 0.28N_1/\sigma + 0.30 \tag{5.82}$$

on the basis of the upper convected Maxwell model (see Eq. (3.4)), which predicts that η is independent of $\dot{\gamma}$, and N_1 is proportional to $\dot{\gamma}^2$. Therefore, in predicting the flow behavior in steady-state shear flow, the upper convected Maxwell model employed by Tuna and Finlayson is not different from the second-order fluid, which was also used by Reddy and Tanner (1978), who obtained the following expression:

$$P_{\text{Exit}}/\sigma = 0.26N_1/\sigma + 0.31 \tag{5.83}$$

In other words, Eqs. (5.82) and (5.83) were obtained using constitutive equations that are not capable of predicting the shear-thinning behavior of polymeric fluids. Conversely, polymer melts, which give rise to a significant degree of P_{Exit} over a certain range of σ, exhibit shear-thinning behavior (see Figure 5.4). It has been observed that when P_{Exit} is measured over a range of $\dot{\gamma}$ or σ where shear-thinning behavior is absent, the magnitude of P_{Exit} is too small to be of any rheological significance!

In spite of the fact that the upper convected Maxwell model employed by Tuna and Finlayson (1984) has some serious deficiencies, it is clear from their computational results that FEA predicts positive P_{Exit} in a slit die, which is consistent with the experimental results presented in Figure 5.14 and other experimental results (Arai 1970; Chan et al. 1990; Han 1972, 1973a, 1973b; Han and Kim 1975; Han and Villamizar 1978; Han and Yu 1971; Han et al. 1969, 1970, 1971, 1983a, 1983b; Mori and Funatsu 1968) and contradicts the experimental results of Lodge and de Vargas (1983), who, using only *two* pressure transducers, reported *negative* values of P_{Exit} (see the inset in Figure 5.20).

Using FEA, Vlachopoulos and Mitsoulis (1985) also calculated P_{Exit} and N_1 for steady-state shear flow of a viscoelastic PS in a slit die. They employed the Carreau model (see Eq. (6.11)) for shear-rate dependent viscosity and the following empirical relationship (Han et al. 1983a):

$$N_1 = A\sigma^b \tag{5.84}$$

where A and b are constants characteristic of a fluid.[3] In their numerical computations via FEA, Vlachopoulos and Mitsoulis used the following numerical values of the parameters to describe the rheological behavior of a commercial PS: $\eta_0 = 9{,}500$ Pa·s, $\lambda = 1.148$ s, $n = 0.5$, $A = 3.47 \times 10^{-3}$ Pa^{1-b}, and $b = 1.66$. They concluded that the difference between the N_1 values calculated from FEA and the N_1 values calculated from Eq. (5.81) decreased rapidly as σ increased, as summarized in Table 5.2. It is seen in Table 5.2 that at $\sigma = 28.2$ kPa, the difference between the two approaches is only 2.6%. Their conclusion supports the experimental results presented in Figure 5.21; that for $\sigma \geq 25$ kPa, the values of N_1 determined from the exit pressure method correlate very well with those determined from a cone-and-plate rheometer. Table 5.3 gives the numerical values of the parameters A and b appearing in Eq. (5.84) for some representative molten polymers. As is shown in Chapter 6, plots of $\log N_1$ versus $\log \sigma$ are independent of temperature, suggesting that the parameters A and b appearing in Eq. (5.84) are independent of temperature. Thus, Eq. (5.84) is very useful for estimating values of N_1 for specified values of σ when the numerical values of A and b are known at only one temperature.

The exit pressure method assumes that flow is fully developed at the exit plane or that the extent of flow disturbances at the exit plane is negligibly small for all intents and purposes. The subject of flow disturbances near the exit plane of a die has been

Table 5.2 Comparison of N_1 values for polystyrene predicted from Eq. (5.81) with those calculated from finite element analysis

σ (kPa)	*N_1 (kPa) From Eq. (5.81)*	*P_{Exit} (kPa)*	*N_1 (kPa) From FEA*	*Difference (%)*
5.2	5.115	3.0	8.18	37.0
15.5	31.350	13.8	34.50	9.0
20.0	47.870	20.6	51.80	7.6
28.2	84.670	34.6	86.90	2.6
35.5	124.080	49.3	126.70	2.0
43.5	173.870	68.5	173.70	−0.1
51.8	232.340	89.7	231.00	−0.6

Reprinted from Vlachopoulos and Mitsoulis, *Journal of Polymer Engineering* 5(2):173.

Table 5.3 Numerical values of the parameters A and b appearing in the empirical expression $N_1 = A\sigma^b$ for some molten polymers

Polymer	*A (Pa^{1-b})*	*b*
Polystyrene (Styron 678)	2.70×10^{-4}	1.86
Polystyrene (Styron 686)	3.12×10^{-2}	1.48
Polystyrene (Styron 685D)	2.18×10^{-4}	1.95
Low-density polyethylene (PE 529)	7.83×10^{-3}	1.57
Low-density polyethylene (NPE 952)	2.33×10^{-3}	1.74
Low-density polyethylene (Rexene 143)	8.19×10^{-3}	1.57
Polypropylene (Profax 6423)	3.44×10^{-4}	1.92
Poly(vinylidene fluoride) (Kynar 960)	1.87×10^{-4}	1.86
Nylon 6 (Capron 8207)	2.44×10^{-5}	1.94

discussed by a number of investigators (Carreau et al. 1985; Clermont et al. 1976; Schowalter and Allen 1975; Whipple and Hill 1978), who conducted experiments either with dilute polymer solutions or with polymer melts at very low $\dot{\gamma}$ or low σ. Using flow birefringence, Han and Drexler (1973b) observed that the extent of flow disturbance in the exit region of a slit die decreases with increasing $\dot{\gamma}$ (or σ). However, flow birefringence could not be used for high σ because the number of isochromatic fringe patterns increases very rapidly, making the distinction between fringe patterns virtually impossible. However there is currently no rigorous theory that predicts a critical value $\dot{\gamma}$ or σ above which the extent of flow disturbances in the exit region may be considered negligible for all intents and purposes.

It was shown, via FEA, that for a Weissenberg number (N_{We}) of 0.6, which corresponded to $\dot{\gamma} = 0.4\ s^{-1}$, a velocity rearrangement occurred as the fluid approached the die exit, but the velocity profiles remained more fully developed as the value of N_{We} increased (Tuna and Finlayson 1984). In that study, N_{We} was defined by $N_{We} = \lambda\dot{\gamma}$, with λ being a characteristic time of the fluid. Apparently, due to numerical instabilities

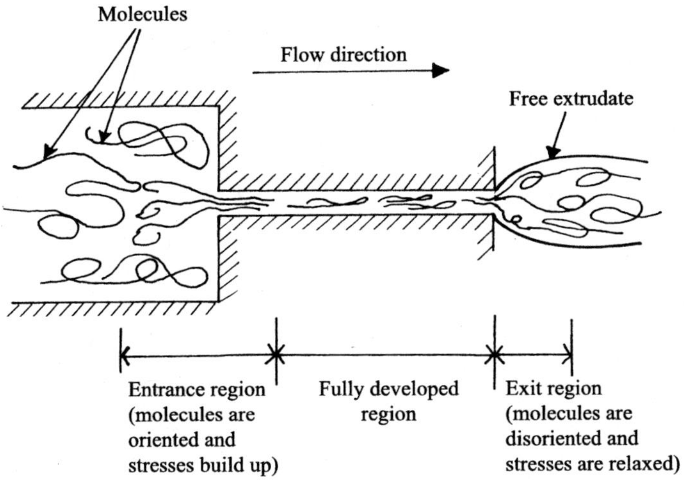

Figure 5.22 Schematic showing the deformation of polymer molecules in the entrance region, the fully developed region, and the exit region of a long capillary die. The schematic shows that polymer molecules are stretched in the entrance region, attain an equilibrium configuration in the fully developed region of the capillary die (i.e., at a distance sufficiently far away from the entrance region but before reaching the exit region), and then relax upon exiting the die, giving rise to extrudate swell.

encountered, no computations could be carried out at higher values of N_{We}, comparable to the values where significant P_{Exit} may be observed experimentally. Vlachopoulos and Mitsoulis (1985) also encountered numerical instabilities for values of $N_{\mathrm{We}} > 2.4$. Nevertheless, their computational results point out that the extent of flow disturbances in the exit region decreases with increasing N_{We}.

Let us now consider the physical origin for the existence of a critical value of σ above which the extent of flow disturbance in the exit region can be neglected for all intents and purposes. To facilitate our discussion here, a schematic is given in Figure 5.22, describing (1) the orientation of macromolecules in the entrance region of a capillary die where stresses build up, (2) the fully developed region in a capillary die where the state of macromolecular orientations remains constant, (3) the rearrangement of macromolecules as they approach the die exit, and (4) the disorientation (or stress relaxation) of macromolecules upon exiting the die. It is not difficult to imagine that the velocity profiles, owing to the rearrangement of molecules, may begin to rearrange before reaching the die exit. But, exactly where inside the capillary the rearrangement of velocity profiles might begin would depend on N_{We}, as we have discussed with reference to the numerical studies (Tuna and Finlayson 1984; Vlachopoulos and Mitsoulis 1985). Specifically, for Newtonian fluids, having $\lambda = 0$ (thus $N_{\mathrm{We}} = 0$), the rearrangement of velocity profiles is expected to begin sooner than for viscoelastic fluids.

For illustration, let us consider the schematic given in Figure 5.23 showing the fully developed velocity profile $v(y)$ inside a capillary and the exit length (L_{ex}) inside the capillary, where the rearrangement of velocity profile takes place. The extent of flow disturbances inside the capillary (i.e., the value of L_{ex}) can be described using the concept of the Deborah number (N_{De}), the ratio of the material time to the process

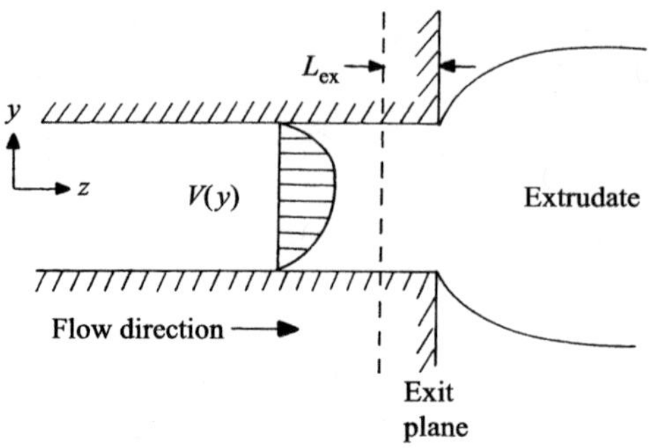

Figure 5.23 Schematic showing the flow of a polymeric liquid in the exit region of a capillary or slit die.

time (Reiner 1964). In the situation under consideration (see Figure 5.23), N_{De} can be represented by

$$N_{De} = \lambda / t_f \tag{5.85}$$

where λ is the characteristic time of a fluid and t_f is the time required for the fluid to travel through the distance L_{ex}. The t_f, in turn, can be expressed by $t_f = L_{ex}/\langle V \rangle$ with $\langle V \rangle$ being the average velocity of a fluid in the capillary. A fluid will behave more like an elastic solid when $N_{De} > 1$. In the context of the variables defined above, it is reasonable to conclude that the fluid behaves more like an elastic solid when $N_{De} > 1$ (i.e., when $\lambda > t_f$) and thus the extent of flow disturbances inside the capillary will be negligible. Conversely, when $\lambda < t_f$, the extent of flow disturbances inside the capillary will be significant, making the exit pressure method inapplicable.

To gain some physical insights into Eq. (5.85) in the situations where the extent of flow disturbances inside a slit die may be negligible, we have calculated the values of N_{De} for an LDPE (Rexene 143), corresponding to the flow conditions that appear in Figure 5.21, and the results are summarized in Table 5.4. In calculating the values of N_{De} in Table 5.4, we used the following numerical values for the various parameters involved: (1) the average velocity $\langle V \rangle$ was calculated using the expression

$$\langle V \rangle = \dot{\gamma}\left[3n/(2n+1)\right](h/6) \tag{5.86}$$

with $h = 0.177$ cm and $n = 0.5$; (2) λ was estimated using the expression $\lambda = \eta_0 M_w / \rho RT$, with $\eta_0 = 3.052 \times 10^3$ Pa·s at 200 °C and $\eta_0 = 2.025 \times 10^3$ Pa·s at 220 °C, $M_w = 1.41 \times 10^5$, $\rho = 0.716$ g/cm^3, and $R = 8.3166$ J/mol K; and (3) the value of L_{ex} was chosen to be same as the slit height h, that is $L_{ex}/h = 1$. Experimental studies (Carreau et al. 1985; Clermont et al. 1976; Schowalter and Allen 1975; Whipple and Hill 1978) indicate that the ratio $L_{ex}/h \le 1$ in a slit die for dilute polymer solutions or Newtonian liquids. It should be pointed out, however, that the ratio L_{ex}/h will depend on the type of fluid and $\dot{\gamma}$. There is evidence, both experimental and theoretical, suggesting that at the same value of $\dot{\gamma}$, the ratio L_{ex}/h is much smaller for viscoelastic polymer melts than for Newtonian fluids and dilute polymer solutions, and that the ratio L_{ex}/h decreases with increasing $\lambda\dot{\gamma}$. Therefore, $L_{ex}/h = 1$, as used in our calculations, is considered to be a very conservative value for the viscoelastic LDPE melt.

Table 5.4 Deborah number for an LDPE (Rexene 143)[a] in the exit region of a slit die

$\dot{\gamma}$ (s^{-1})	σ (kPa)	$\langle V \rangle \times 10^3$ (m/s)	N_{De}[b]
(a) 200 °C			
28.24	14.87	6.25	0.54
56.84	22.77	18.82	1.62
115.11	33.98	25.47	2.20
161.22	40.92	35.67	3.08
214.84	46.76	47.53	4.10
(b) 220 °C			
27.29	10.79	6.04	0.33
54.94	16.94	12.15	0.67
111.26	26.10	24.62	1.35
155.84	31.95	34.48	1.89
207.66	37.04	45.94	2.53

[a]Used to obtain the data reported in Figure 5.21. [b]$\lambda = 0.1528$ s at 200 °C and $\lambda = 0.0973$ s at 220 °C were used to calculate N_{De}.

Reprinted from Han 1998, *Rheological Measurement*, Collyer and Clegg (eds), Chapter 6.

It is of great interest to observe in Table 5.4 that N_{De} is much less than 1 at low σ, that N_{De} steadily increases (exceeding 1.0) with increasing σ, and that a definite relationship appears to exist between the value of N_{De} and the critical value of σ at which the log N_1 versus log σ plot begins to deviate from the correlation displayed in Figure 5.21. In other words, at $\sigma < 25$ kPa, for which $N_{De} < 1$ (see Table 5.4), the data points in Figure 5.21 deviate from the linear relationship between log N_1 and log σ. It should be emphasized that the intent here is not to determine the exact value of N_{De} at which the flow disturbances inside the die can be considered to be negligible, but rather to assess the situations where the extent of flow disturbances inside the die may be considered to be negligibly small, for all intents and purposes, using N_{De}.

Note that the seemingly low values of N_{De} in Table 5.4 stem from the fact that the LDPE melt considered has relatively small values of λ (i.e. $\lambda = 0.153$ s at 200 °C and $\lambda = 0.0973$ s at 220 °C), which were determined using $\lambda = \eta_0 M_w/\rho RT$. At very small values of λ (say, 0.01 to 0.001 s), which are characteristic of dilute polymer solutions and very weakly elastic polymer melts (e.g., nylon, polycarbonate, and poly(ethylene terephthalate)), we anticipate that $N_{De} \ll 1$, even for high values of σ. This suggests that for dilute polymer solutions and very weakly elastic polymer melts, the extent of flow disturbances inside the die may never become negligible, even at high values of σ.

It can therefore be concluded that, for all intents and purposes, the extent of flow disturbances near the die exit may be considered to be negligible for strongly viscoelastic polymer melts at high σ (say, $\sigma > 25$ kPa), justifying the extrapolation procedure used to obtain P_{Exit} and thus making the exit pressure method valid (see Table 5.1). However, this conclusion cannot be extended to dilute polymer solutions and very weakly elastic polymer melts.

Returning to Figure 5.20, using the concept of Deborah number we can now explain why the P_{Exit} data of Baird et al. (1986) and Lodge and de Vargas (1983) are of little rheological significance. We have calculated values of σ, $\langle V \rangle$, and N_{De} under the experimental conditions employed and they are summarized in Table 5.5, in which Eq. (5.86) was used to calculate $\langle V \rangle$ in a slit die, and the expression

$$\langle V \rangle = \dot{\gamma}\left[4n/(3n+1)\right](D/8) \tag{5.87}$$

was used to calculate $\langle V \rangle$ in a capillary die, where D is the capillary diameter and n is a power-law index. The values of N_{De} in Table 5.5 were calculated using $L_{\mathrm{ex}}/h = 1$ for the slit die and $L_{\mathrm{ex}}/D = 1$ for the capillary die. It can be concluded from Table 5.5 that the P_{Exit} data of Baird et al. (1986) and Lodge and de Vargas (1983) were obtained under the flow conditions that gave rise to $N_{\mathrm{De}} \ll 1$, whereas the P_{Exit} data of Han et al. (1983b) were obtained under the flow conditions that gave rise to $N_{\mathrm{De}} \gg 1$.

When using a continuous-flow capillary or slit rheometer, one must first make certain that the pressure gradients are constant (i.e., $-\partial T_{rr}/\partial z = \partial p/\partial z =$ constant in a capillary die or $-\partial T_{yy}/\partial z = \partial p/\partial z =$ constant in a slit die) in the region where wall normal stresses are measured. Nonlinear wall normal stress profiles of $T_{yy}(b, z)$ may be observed in a slit die when pressure transducers are mounted in the die area that includes the entrance region (Eswaran et al. 1963, Leblanc 1976, Rauwendaal and Fernandez 1985). Nonlinear profiles of $T_{yy}(b, z)$ in a slit die may also be observed

Table 5.5 Deborah number for an LDPE in the exit region of a slit die

Temp (°*C*)	$\dot{\gamma}$ (s^{-1})	σ (*kPa*)	$\langle V \rangle^{a}$ (*m/s*)	$N_{\mathrm{De}}{}^{b}$
Lodge and de Vargas (1983)[c]				
151	0.14	5	1.51×10^{-5}	0.07
151	1.51	11	1.63×10^{-4}	0.79
Baird et al. (1986)[d]				
150	0.05	3	1.37×10^{-5}	0.03
150	1.51	11	4.14×10^{-4}	0.79
Han et al. (1983b)[e]				
180	82.6	45.5	2.32×10^{-2}	18.4
180	432.5	83.1	1.21×10^{-1}	96.9
200	96.2	36.4	2.71×10^{-2}	11.8
200	595.7	72.9	1.67×10^{-1}	73.4
220	80.5	21.8	2.26×10^{-2}	6.7
220	391.7	39.7	1.10×10^{-1}	32.9

[a] Values of $\langle V \rangle$ for a slit die were calculated from Eq. (5.86) with $n = 0.38$, and values of $\langle V \rangle$ for a capillary die were calculated from Eq. (5.87) with $n = 0.38$. [b] Values of λ were calculated using the expression $\lambda = \eta_0 M_w/\rho RT$ with $\rho = 0.716$ g/cm^3 and $M_w = 3.7 \times 10^5$. It was found that $\lambda = 4.893$ s for $\eta_0 = 3.338 \times 10^4$ Pa·s at 150 °C, $\lambda = 2.523$ s for $\eta_0 = 1.838 \times 10^4$ Pa·s at 180 °C, $\lambda = 1.389$ s for $\eta_0 = 1.057 \times 10^4$ Pa·s at 200 °C, and $\lambda = 0.946$ s for $\eta_0 = 0.750 \times 10^4$ Pa·s at 220 °C. [c] A slit die with $h = 0.1$ cm was used. [d] A slit die with $h = 0.254$ cm was used. [e] A capillary die with $D = 0.3175$ cm was used.

when viscous shear heating becomes significant, which would occur when the applied shear rate exceeds a certain critical value. Having observed nonlinear profiles of $T_{yy}(b, z)$ in a slit die, some investigators (Laun 1983, Rauwendaal and Fernandez 1985, Tuna and Finlayson 1988) curve-fitted the data to a quadratic expression and then calculated σ by assuming that the profiles of $T_{yy}(b, z)$ were identical to the profiles of $p(b, z)$. Further, they extrapolated the nonlinear profiles of $T_{yy}(b, z)$ to the die exit and obtained P_{Exit} in order to calculate N_1 using Eq. (5.81). However, such exercises are of no rheological significance! Note that Eq. (5.81) has been derived on the premise that flow is fully developed (i.e., $\partial T_{yy}(b, z)/\partial z$ is constant) under isothermal conditions.

Laun (1983) investigated the effect of pressure on the viscosity of LDPE melt by observing nonlinear profiles of $T_{yy}(b, z)$ along the axis of a slit die. Apparently, Laun mistakenly thought that the pressure transducers mounted at the die wall measured the isotropic pressure $p(b, z)$. Using nonlinear profiles of $T_{yy}(b, z)$ along the die axis, Laun further calculated a single value of viscosity inside the slit die while claiming that the viscosity was affected by pressure. Such an approach is not justified for the following reasons. (1) As pointed out in a preceding section, there is no way that one can calculate shear viscosity of a molten polymer in a slit (or capillary) die when the profile of $T_{yy}(b, z)$ along the die axis is not constant. (2) One cannot assess the effect of pressure on the viscosity of a molten polymer flowing through a slit (or capillary) die from the measurement of $T_{yy}(b, z)$ along the die axis because one cannot obtain information on $p(b, z)$ from the measurement of $T_{yy}(b, z)$. (3) The presence of nonlinear profiles of $T_{yy}(b, z)$ along the die axis might also arise from viscous shear heating inside the die.

It should be pointed out that under any circumstance the values of $T_{yy}(b, z)$ are *not* equal to the profiles of $p(b, z)$ in a slit die, as is clear from Eq. (5.72). Hence, there is no way for one to obtain pressure profiles $p(b, z)$ from the measurements of $T_{yy}(b, z)$ in a slit die. Only when flow is fully developed in a slit die does one have, from Eq. (5.72), $-\partial T_{yy}/\partial z = \partial p/\partial z$, which then enables one to calculate σ using Eq. (5.68). When flow is not fully developed, instead of Eq. (5.72) one must use the expression

$$T_{yy}(b, z) = -p(b, z) + \sigma_{yy}(b, z) \tag{5.88}$$

to analyze the experimental data of $T_{yy}(b, z)$ profiles in a slit die. Note that Eq. (5.88) yields

$$\frac{\partial T_{yy}(b, z)}{\partial z} = -\frac{\partial p(b, z)}{\partial z} + \frac{\partial \sigma_{yy}(b, z)}{\partial z} \tag{5.89}$$

and thus $-\partial T_{yy}/\partial z \neq \partial p/\partial z$!

One must also make certain that an extrapolation of the profiles of $T_{yy}(b, z)$ to the exit plane of the die is justified for all intents and purposes. The validity of the extrapolation will depend on the type of fluid and the applied $\dot{\gamma}$ (i.e., $N_{\text{We}} = \lambda\dot{\gamma}$). Therefore, one must not generalize whether the extrapolation is valid or not based on the measurements of $T_{yy}(b, z)$ for dilute polymer solutions having small values of λ or for viscoelastic polymer melts at very low $\dot{\gamma}$.

When using a continuous-flow capillary or slit rheometer, a number of design considerations must be taken into account. (1) Pressure transducers must be mounted at positions far away from the entrance region. (2) The distance between the melt discharge from either an extruder or a pump and the entrance of the die must be sufficiently long to allow the stresses in the melt to be relaxed completely before entering the die. (3) When using a slit die, the ratio *w*/*h* must be sufficiently large, so that the isovels of the melt stream can be parallel to the long side of the die cross section. (4) The dimension of the die width *w* must be much larger than the diameter of the pressure transducer tip so that the isovels of the melt stream can be parallel to the surface of the pressure transducer tip. (5) The ratio of the reservoir diameter to the slit height *h* must be sufficiently large so that the natural streamlines of the melt in the reservoir section would not be restricted as the polymer melt approaches the die entrance. If any of these design criteria is not met, one might obtain spurious measurements of wall normal stress, yielding unreliable σ and P_{Exit}.

5.3.5 Viscous Shear Heating in a Cylindrical or Slit Die

When a very viscous molten polymer is forced to flow through a slit or capillary die, viscous shear heating can become significant above a certain critical value of $\dot{\gamma}$ or σ. Under such situations, nonlinear profiles of wall normal stress in a slit or capillary die may be observed, as described in the preceding section. Therefore, continuous-flow capillary/slit rheometry is limited to $\dot{\gamma}$ or σ, below which viscous shear heating can be neglected.

The upper limit of $\dot{\gamma}$ or σ above which the extent of viscous shear heating becomes significant can be predicted theoretically by solving the following equations of momentum and heat transfer in a slit die:

$$-\frac{\partial p}{\partial z} + \frac{\partial}{\partial y}\left(\eta \frac{\partial v_z}{\partial y}\right) = 0 \tag{5.90}$$

$$\rho c_p v_z \frac{\partial T}{\partial z} = k \frac{\partial^2 T}{\partial y^2} + \eta \left(\frac{\partial v_z}{\partial y}\right)^2 \tag{5.91}$$

and the following equations of momentum and heat transfer in a capillary die:

$$-\frac{\partial p}{\partial z} + \frac{1}{r}\frac{\partial}{\partial r}\left[r\left(\eta \frac{\partial v_z}{\partial r}\right)\right] = 0 \tag{5.92}$$

$$\rho c_p v_z \frac{\partial T}{\partial z} = k \left[\frac{1}{r}\frac{\partial}{\partial r}\left(r \frac{\partial T}{\partial r}\right)\right] + \eta \left(\frac{\partial v_z}{\partial r}\right)^2 \tag{5.93}$$

under the appropriate boundary conditions, where ρ is the density of the fluid, c_p is the specific heat, k is the thermal conductivity, and η is the viscosity. For fully developed flow, it is convenient to use the following form of the truncated power-law relationship:

$$\eta(\dot{\gamma}, T) = \begin{cases} k_o \exp(-bT) & \text{for} \quad \dot{\gamma} \le \dot{\gamma}_0 \\ k_o \exp(-bT)(\dot{\gamma}/\dot{\gamma}_0)^{n-1} & \text{for} \quad \dot{\gamma} > \dot{\gamma}_0 \end{cases} \tag{5.94}$$

where $\dot{\gamma}$ is shear rate (dv_z/dy in a slit die and dv_z/dr in a capillary die), k_o is the preexponential factor, b is a constant, T is the absolute temperature, n is a power-law index, and $\dot{\gamma}_0$ is the value of shear rate at which η deviates from Newtonian behavior. If viscous shear heating is significant, the solution of Eqs. (5.90) and (5.91) for slit flow, and Eqs. (5.92) and (5.93) for capillary flow, will yield temperature profiles that vary with both y (or r) and z.

For the LDPE (Rexene 143) used to produce Figures 5.4 and 5.21, we have estimated the temperature rise due to viscous shear heating by numerically solving Eqs. (5.90) and (5.91) with the aid of Eq. (5.94). For the numerical computation, the following values of the parameters were used: $\rho = 716$ kg/m^3, $c_p = 2{,}604$ J/kg K, $k = 0.182$ W/(mK), $k_o = 4.534 \times 10^8$ Pa·s, $b = 0.0224$ K^{-1}, $\dot{\gamma}_0 = 0.4$ s^{-1}, $n = 0.38$. The computed results indicate that (1) at $\dot{\gamma} = 400$ s^{-1}, which corresponds to $\sigma =$ 54.8 kPa, the average temperature of the melt at the exit plane of the die (with 7.62 cm die length) would have risen from 200 to 200.6 °C, and (2) at $\dot{\gamma} = 600$ s^{-1}, which corresponds to $\sigma = 67.1$ kPa, the average temperature of the melt at the die exit plane would have risen from 200 to 200.9 °C. Notice in Table 5.5 that the maximum value of σ employed to produce Figure 5.21 was 83.1 kPa. Therefore, we can conclude that for all intents and purposes, the exit pressure data in Figure 5.21 were negligibly affected by viscous shear heating. It should be noted that as $\dot{\gamma}$ or σ is increased further, one may reach a point where the extent of viscous shear heating can no longer be neglected.

5.4 Elongational Rheometry

As described in Chapter 3, the rheological response of polymeric liquids to stretching (i.e., elongational deformation) is of fundamental importance for a better understanding of materials' behavior under the influence of an external force. A number of investigators (Ballman 1965; Everage and Ballman 1976, 1977, 1978; Ide and White 1978; Laun and Münstedt 1976, 1978; Meissner 1969, 1971, 1972; Münstedt 1975; Stevenson 1972; Vinogradov et al. 1970, 1972) have reported on measurements of steady-state uniaxial elongational viscosities. Unfortunately, however, there is no single experimental technique that would be useful for all types of fluids. Some experimental techniques are better suited to solutions, whereas others are better suited to very viscous liquids, such as polymer melts. Some experimental techniques are better suited to low rates of strain (elongation rates), whereas others are better suited to high elongation rates. Here we confine our presentation to the measurement of steady-state uniaxial elongational viscosity of very viscous liquids (in particular, polymer melts), which is of practical interest to such polymer processing operations as fiber spinning and cast-film extrusion. In Chapter 3, we pointed out that measurement of steady-state uniaxial elongational viscosity is very useful for screening of constitutive equations for viscoelastic fluids.

The various experimental techniques may broadly be classified, in accordance with the degree of control of the elongational deformation, into two types: (1) controlled flow, and (2) uncontrolled flow. The former refers to an experiment where the sample is subjected to a constant elongation rate, and the latter to a non-constant elongation rate. For the purpose of screening constitutive equations for viscoelastic fluids, the controlled experiment, which generates a constant elongation rate, is preferred because

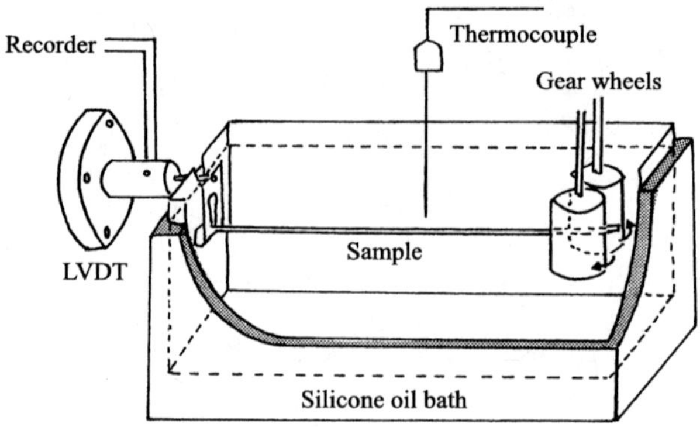

Figure 5.24 Schematic showing an elongational rheometer that enables one to conduct controlled elongational flow experiment, giving rise to constant elongation rate.

analytical expressions for elongational viscosity from various constitutive equations are available (see Chapter 3). However, as is presented in Chapter 6 of Volume 2, the uniaxial elongational flow encountered in fiber spinning does not generate a constant elongation rate.

Restricting our attention to the uniaxial elongational flow with a constant elongation rate, consider an initially unstrained, rod-shaped specimen fixed at one end and stretched in the uniaxial direction, as schematically shown in Figure 5.24, where a specimen is supported by floating or being submerged in a hot-oil bath at a predetermined temperature. A filament of uniform diameter is clamped at one end to a force transducer and at the other to two gear wheels. In such an experiment, the elongation rate $\dot{\varepsilon}$ is determined from

$$\dot{\varepsilon} = \frac{1}{L}\frac{\mathrm{d}L}{\mathrm{d}t} = \frac{V}{L} = \frac{R\Omega}{L} \tag{5.95}$$

where L is the filament length at time t, Ω is the angular velocity of the rolls, R is the radius of the wheel, and V is the wheel's linear velocity of the free end such that the following relationship is maintained:

$$\varepsilon = \ln(L/L_0) = \dot{\varepsilon}t \tag{5.96}$$

which follows from Eq. (5.95), in which ε is the total Hencky strain and L_0 is the initial length of the specimen. In this type of experiment, difficulties are often encountered in (1) avoiding the slippage of the specimen from the clamp, (2) minimizing the gravity effect, (3) minimizing the buoyancy effect (by insuring that the density of the oil is identical to that of the specimen), (4) maintaining uniformity of specimen cross section, and (5) transmitting a tensile stress across a liquid–solid interface. Precise control of the free end's velocity is also of utmost importance. Measurement of tensile force $F(t)$ enables one to determine the tensile stress $T_{zz}(t)$ from

$$T_{zz}(t) = F(t)/A(t) = \left[F(t)/A(0)\right]\exp(\dot{\varepsilon}t) \tag{5.97}$$

where $A(t)$ is the cross-sectional area of the filament at time t and $A(0)$ is the initial cross-sectional area of the filament. Hence, steady-state elongational viscosity η_E at a constant $\dot{\varepsilon}$ can be determined from

$$\eta_E = T_{zz}/\dot{\varepsilon} \tag{5.98}$$

To overcome some of the difficulties enumerated above, an alternative experimental technique was suggested (Meissner 1972), in which two sets of gripping wheels are used, instead of end loading. In this way, the specimen is stretched between the distance L fixed in space and $\dot{\varepsilon}$ is determined by Eq. (5.95).

In this type of apparatus, the tensile stress is measured through the deflection of a spring associated with one of the rotating wheels. In determining η_E with the methods described in this section, one may have to wait a long time for the stress to build up to the level where a steady-state (in both the Lagrangian and Eulerian senses) is attained because the specimen used is strain free before the test begins.

Figure 5.25 shows the transient buildup of η_E in an LDPE when it is subjected to stretching at constant $\dot{\varepsilon}$. It is seen that at low $\dot{\varepsilon}$, the η_E increases slowly to a steady-state value, and at high $\dot{\varepsilon}$, the η_E increases asymptotically. Plots of log η_E versus log $\dot{\varepsilon}$, with ε as parameter (see Eq. (5.96) for the definition of ε) are given in Figure 5.26 for an LDPE and in Figure 5.27 for a PS. It is seen that for very high $\dot{\varepsilon}$, η_E decreases with $\dot{\varepsilon}$ for all levels of ε, and that at high level of ε, η_E first increases and then decreases as $\dot{\varepsilon}$ increases.

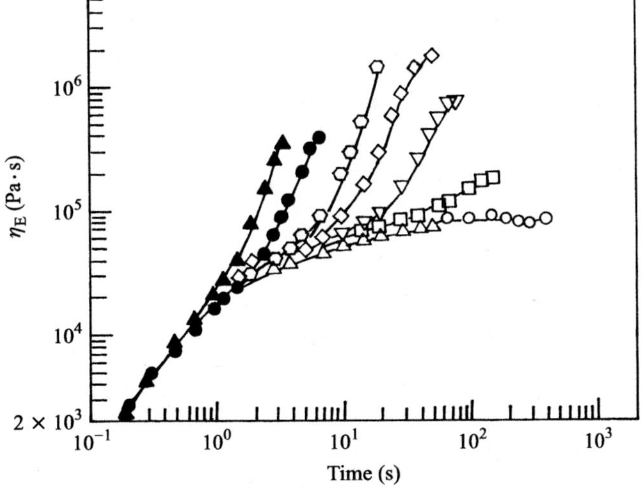

Figure 5.25 Transient growth of elongational viscosity (η_E) for an LDPE at 160 °C for different values of $\dot{\varepsilon}$ (s^{-1}): (○) 0.05, (△) 0.01, (□) 0.02, (▽) 0.05, (◇) 0.1, (⬡) 0.2, (●) 0.5, (▲) 1.0. (Reprinted from Ide and White, *Journal of Applied Polymer Science* 22:1061. Copyright © 1978, with permission from John Wiley & Sons.)

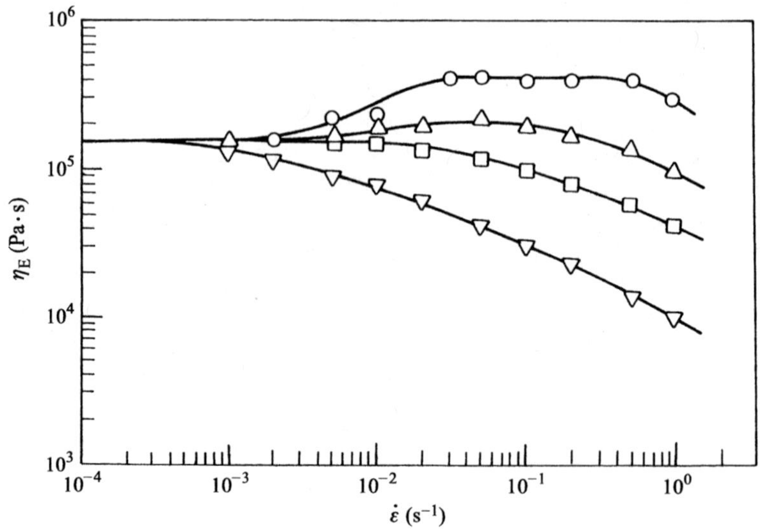

Figure 5.26 Plots of $\log \eta_E$ versus $\log \dot{\varepsilon}$ for an LDPE at 150 °C at various Hencky strains: (○) 3.0, (△) 2.0, (□) 1.0, (▽) 0.1. (Reprinted from Everage and Ballman, *Journal of Applied Polymer Science* 21:841. Copyright © 1977, with permission from John Wiley & Sons.)

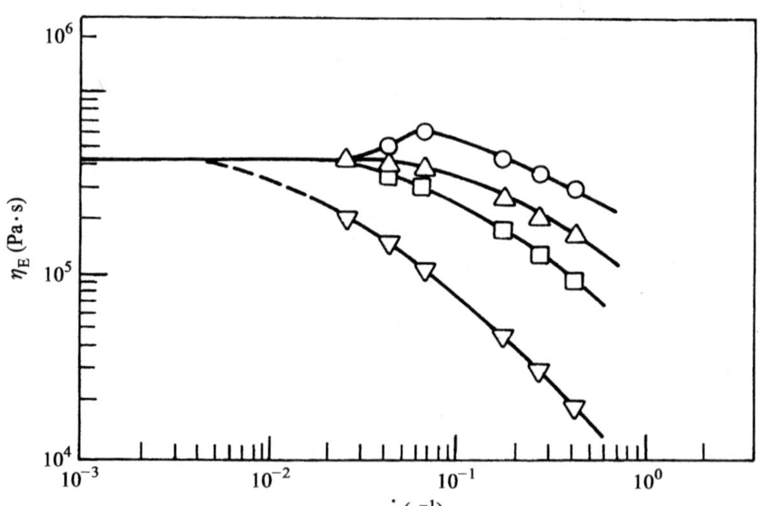

Figure 5.27 Plots of $\log \eta_E$ versus $\log \dot{\varepsilon}$ for a PS at 155 °C at various Hencky strains: (○) 1.5, (△) 0.9, (□) 0.5, (▽) 0.1. (From Everage and Ballman, *Journal of Applied Polymer Science* 21:841. Copyright © 1977, with permission from John Wiley & Sons.)

5.5 Summary

In this chapter, we have described experimental methods that are used very frequently to determine the shear flow properties of polymeric liquids. It is worth pointing out that owing to the very viscous nature of elastomers, the operating range of shear rates for elastomers is very low compared with that for thermoplastic polymers, as shown in Figure 5.28. Figure 5.29 gives $\log \eta$ versus $\log \dot{\gamma}$ plots for natural rubber, LDPE, HDPE, and polypropylene (PP), which were obtained using a slit rheometer. It is seen that the highest value of $\dot{\gamma}$ that could be applied to the natural rubber is about 30 s^{-1}, while the highest value of $\dot{\gamma}$ that could be applied to the HDPE and PP is approximately 10^3 s^{-1}. However, when the same viscosity data given in Figure 5.29 are plotted against σ, as in Figure 5.30, we observe that the highest value of σ that could be applied to the natural rubber is approximately 2×10^5 Pa, while the highest value of σ that could be applied to the HDPE and PP is approximately 8×10^4 Pa. This observation suggests that the selection of a rheometer be made on the basis of the maximum σ that could be applied and not the maximum $\dot{\gamma}$ that could be applied. Therefore, when specifying σ, one does not have to distinguish how viscous a fluid is;

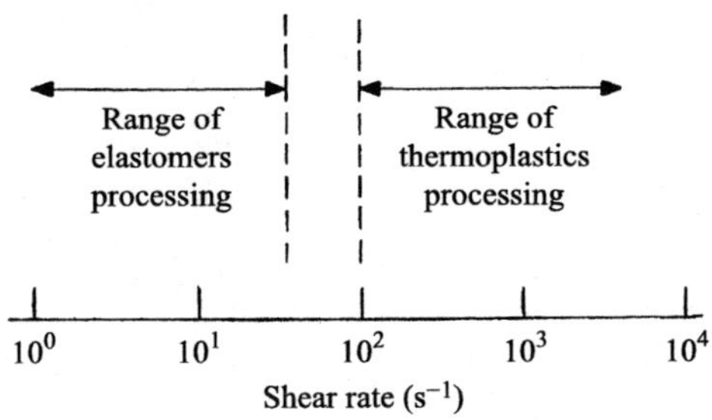

Figure 5.28 The range of shear rates encountered in the processing of elastomers and thermoplastic polymers.

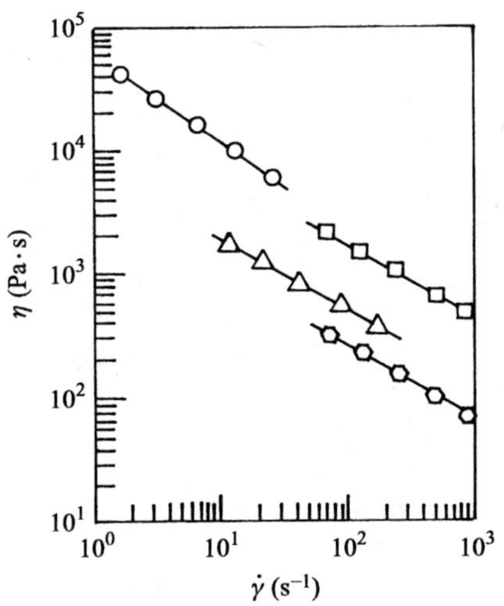

Figure 5.29 Plots of $\log \eta$ versus $\log \dot{\gamma}$ for different polymers: (○) natural rubber at 115 °C, (△) LDPE at 190 °C, (□) HDPE at 190 °C, (⬡) PP at 230 °C. The data were obtained using a slit rheometer.

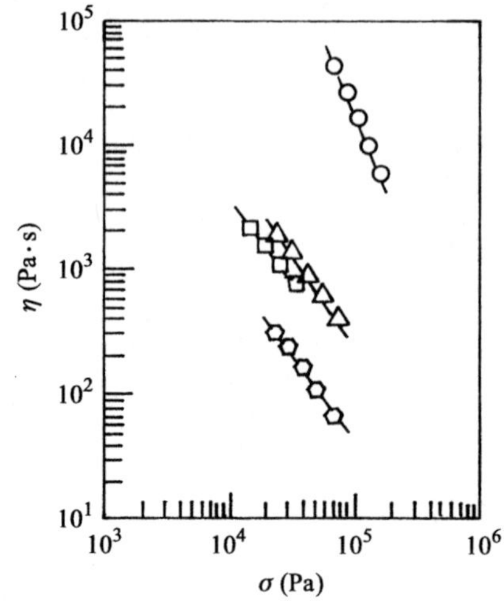

Figure 5.30 Plots of log η versus log σ for different polymers: (○) natural rubber at 115 °C, (△) LDPE at 190 °C, (□) HDPE at 190 °C, (⬡) PP at 230 °C. The data were obtained using a slit rheometer.

that is, whether the fluid under consideration is a dilute polymer solution or a molten polymer. Conversely, when specifying $\dot{\gamma}$, the upper operating limit of a rheometer varies with the type of fluid to be investigated.

On the basis of the experimental data presented in this chapter, we have prepared Figure 5.31, which describes the ranges of σ at which the cone-and-plate rheometer and capillary/slit rheometer are usually operated. Thus, a cone-and-plate rheometer may be regarded as being effective for measuring the rheological properties of polymeric fluids at low σ, while a capillary/slit rheometer may be regarded as being effective for measuring the rheological properties of polymeric fluids at high σ. Further, based on the concept of the Deborah number, we have shown that dilute polymer solutions or very weakly elastic polymer melts can give rise to significant flow disturbances near the die exit, making the exit pressure method inapplicable.

We have shown that the exit pressure method can give rise to reliable rheological information on η and N_1 over a limited range of σ for viscoelastic molten polymers, when σ is large enough to make the flow disturbances near the die exit negligible and yet low enough not to cause significant viscous shear heating. However, such a limited

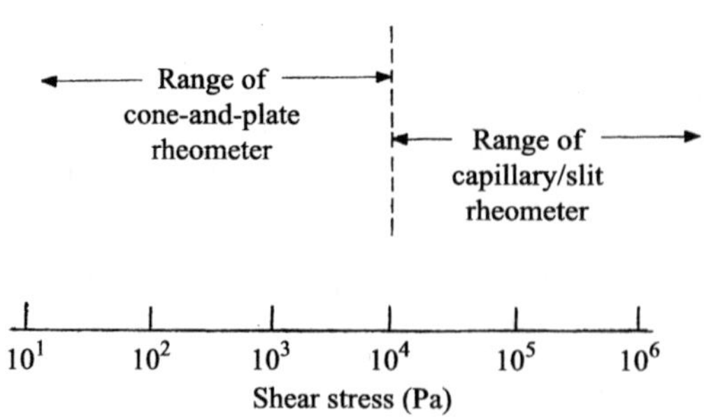

Figure 5.31 The range of shear stresses where a cone-and-plate rheometer can be used to obtain the rheological properties of polymeric liquids before flow instability sets in, and the range of shear stresses where a capillary/slit rheometer can be used to obtain the rheological properties of polymeric liquids before significant viscous shear heating is encountered.

range of σ seems to lie within the region where many polymer processing operations are practiced in industry.

Note that capillary/slit rheometry is also very useful for determining the shear flow properties of heterogeneous polymer systems, including immiscible polymer blends which are discussed in Chapter 11, highly-filled molten polymers, which are discussed in Chapter 12, and molten polymers with solubilized gaseous component, which are discussed in Chapter 13.

Problems

Problem 5.1

Table 5.6 gives the experimental data obtained for an HDPE (melt density of 0.745 gm/cm^3) at 200 °C using a continuous-flow capillary rheometer for three different *L/D* ratios.

Table 5.6 Experimental data for Problem 5.1

L/D Ratio[a]	*Total Pressure Drop* Δp *(MPa)*	*Apparent Shear Rate* (s^{-1})	*Exit Pressure (kPa)*
4	2.556	93.5	137.1
4	3.169	189.2	207.4
4	4.065	357.7	320.4
4	4.681	497.2	402.4
8	3.655	82.7	108.9
8	4.333	148.2	146.8
8	6.220	359.3	236.3
8	6.769	457.1	276.3
12	5.153	93.0	104.7
12	6.651	181.8	152.3
12	8.389	361.4	215.0
12	8.897	423.4	244.6
16	6.726	84.9	110.9
16	8.569	182.6	161.2
16	10.591	356.2	213.6

[a] $D = 0.3175$ cm.

(a) Construct the Bagley plot for the HDPE.
(b) Construct the flow curve for the HDPE after making the Bagley end corrections.
(c) Assuming that the power law holds for the HDPE, determine the flow index n and flow consistency K from the flow curve constructed in part (b).
(d) Construct the plot of true melt viscosity versus true shear rate for the HDPE.
(e) Plot the complete wall normal stress profile for the HDPE in a capillary die having $L/D = 16$, by making a correction for the exit pressure.
(f) Plot the true entrance pressure drop versus shear rate for the HDPE.

Problem 5.2

Table 5.7 gives experimental data obtained using a continuous-flow capillary rheometer for an LDPE.

Table 5.7 Experimental data for Problem 5.2

Temp (°*C*)	$\dot{\gamma}$ (s^{-1})	σ (*kPa*)	P_{Exit} (*kPa*)
180	82.8	45.5	123
180	121.3	53.1	153
180	185.7	61.7	198
180	238.0	67.1	221
180	319.0	74.0	257
180	432.5	83.1	303
200	96.2	40.3	94
200	127.8	45.3	113
200	165.2	50.2	132
200	235.2	57.3	165
200	327.3	65.0	197
200	435.4	71.4	227
200	595.7	79.3	277
220	80.5	32.1	65
220	122.5	38.4	80
220	165.2	43.4	95
220	212.6	48.1	112
220	275.9	53.2	132
220	391.7	60.6	169

(a) Prepare a log P_{Exit} versus log $\dot{\gamma}$ plot for each temperature. You will observe temperature dependence of the plots.

(b) Prepare a log P_{Exit} versus log σ plot for each temperature. You will not observe temperature dependence of the plots. Give physical reason(s) for such an observation.

(c) Prepare a log N_1 versus log $\dot{\gamma}$ plot for each temperature.

(d) Prepare a log N_1 versus log σ plot for each temperature. You will observe that the plots are virtually independent of temperature. Give physical reason(s) for such an observation.

(e) Prepare a $N_1/2\sigma$ versus log $\dot{\gamma}$ for each temperature. You will observe temperature dependence of the plot.

(f) Prepare a $N_1/2\sigma$ versus log σ plot for each temperature. You will observe that the plots are virtually independent of temperature. Give physical reason(s) for such an observation.

Problem 5.3

Table 5.8 gives the oscillatory shear flow data obtained using a cone-and-plate rheometer for an LDPE.

Table 5.8 Experimental data for Problem 5.3

Temp (°*C*)	ω (*rad/s*)	G'(*kPa*)	G''(*kPa*)
180	0.10	0.05	0.52
180	1.18	0.13	0.90
180	0.32	0.28	1.49
180	0.56	0.59	2.44
180	1.00	1.15	3.88
180	1.78	2.10	6.05
180	3.16	3.69	9.25
180	5.66	6.18	13.93
180	10.00	10.10	20.71
180	17.78	16.20	30.41
180	31.68	25.90	43.92
180	56.23	40.60	62.03
180	100.00	62.32	84.90
200	0.10	0.03	0.33
200	1.18	0.07	0.57
200	0.32	0.15	0.97
200	0.56	0.31	1.62
200	1.00	0.64	2.65
200	1.78	1.24	4.23
200	3.16	2.28	6.60
200	5.66	3.98	10.10
200	10.00	6.73	15.25
200	17.78	11.15	22.71
200	31.68	17.96	33.37
200	56.23	28.60	47.80
200	100.00	48.11	66.71
220	0.10	0.01	0.22
220	1.18	0.03	0.38
220	0.32	0.07	0.66
220	0.56	0.17	1.12
220	1.00	0.36	1.87
220	1.78	0.75	3.04
220	3.16	1.44	4.83
220	5.66	2.63	7.55
220	10.00	4.60	11.55
220	17.78	7.78	17.39
220	31.68	12.85	25.83
220	56.23	20.83	37.59
220	100.00	33.12	53.35

(a) Prepare log G' versus log ω and log G'' versus log ω plots for each temperature. You will observe temperature dependence of the plots.

(b) Prepare a log G' versus log G'' plot for each temperature. You will observe that the plots are virtually independent of temperature. Give physical reason(s) for such an observation.

Notes

1. The term "exit pressure" is a misnomer in that the transducers mounted on the tube wall do not measure the pressure $p(R, z)$ but rather the outward total normal stress at the tube wall $T_{rr}(R, z)$. Note that these two quantities are related to each other by Eq. (5.49). It is clear that $T_{rr}(R, z) = -p(R, z)$ only when the deformation of fluid is negligible (i.e., $\sigma(R, z) = 0$). Thus, in general, the profile of $T_{rr}(R, z)$ is not the same as the pressure profile, $p(R, z)$. The extrapolated value of $T_{rr}(R, z)$ at the exit plane at $z = L$, $T_{rr}(R, L)$ (referred to as the "exit pressure") represents the amount of the "elastic energy" that was stored (J/m^3) in the fluid while flowing through the die. Han (1976) explained the physical origin of $T_{rr}(R, L)$ and argued that the extrudate swell upon exiting a die is the consequence of the release of the elastic energy stored in the fluid. There would be *no* extrudate swell if a fluid has no elastic energy stored; that is, if $T_{rr}(R, L) = 0$. Han has demonstrated a correlation between extrudate swell ratio and $T_{rr}(R, L)$.
2. In a paper by Han et al. (1983b), P_{Exit} data were not presented and only a log N_1 versus log σ plot was presented. The values of N_1 in the log N_1 versus log σ plot were calculated, via Eq. (5.64), using the P_{Exit} data obtained from a continuous-flow capillary rheometer.
3. Use of Eq. (5.84) is justified by the experimental results, such as those given in Figure 5.21.

References

Arai T (1970). In *Proc. Fifth Int. Cong. Rheol.*, Onogi S (ed), University Park Press, Vol 4, Baltimore, Maryland, p 497.

Baek SG, Magda JJ, Larson RG (1993). *J. Rheol.* 37:1201.

Bagley EB (1957). *J. Appl. Phys.* 28:624.

Baird DG, Read MD, Pike RD (1986). *Polym. Eng. Sci.* 26:225.

Ballenger TF, White JL (1971). *J. Appl. Polym. Sci.* 15:1949.

Ballenger TF, Chen IJ, Crowder JW, Hagler GE, Bogue DC, White JL (1971). *Trans. Soc. Rheol.* 15:195.

Ballman RL (1965). *Rheol. Acta* 4:137.

Boles RL, Davis HL, Bogue DC (1970). *Polym. Eng. Sci.* 10:24.

Carreau PJ, Choplin L, Clermont JR (1985). *Polym. Eng. Sci.* 25:669.

Chan TW, Pan B, Yuan H (1990). *Rheol. Acta* 29:60.

Christiansen EB, Miller M (1971). *Trans. Soc. Rheol.* 15:189.

Clermont JR, Pierrard JM, Scrivener O (1976). *J. Non-Newtonian. Fluid Mech.* 1:175.
Collyer AA, Clegg DW (eds) (1998). *Rheological Measurement*, 2nd ed, Chapman and Hall, London.
Davis JM, Hutton JF, Walters K (1973). *J. Phys. D.* 6:2259.
Dealy JM (1982). *Rheometers for Molten Polymers*, Van Nostrand Reinhold, New York.
Eswaran R, Janeschitz-Kriegl H, Schiff J (1963). *Rheol. Acta* 3:83.
Everage AE, Ballman RL (1976). *J. Appl. Polym. Sci.* 20:1137.
Everage AE, Ballman RL (1977). *J. Appl. Polym. Sci.* 21:841.
Everage AE, Ballman RL (1978). *Nature* 273:213.
Ferry JD (1980). *Viscoelastic Properties of Polymers*, 3rd ed, John Wiley & Sons, New York.
Fuller GG (1995). *Optical Rheometry of Complex Fluids*, Oxford University Press, Oxford.
Ginn RF, Metzner AB. (1969). *Trans. Soc. Rheol.* 13:429.
Han CD (1971a). *J. Appl. Polym. Sci.* 15:2567.
Han CD (1971b). *AIChE J.* 17:1480.
Han CD (1972). *AIChE J.* 18:116.
Han CD (1973a). *J. Appl. Polym. Sci.* 17:1289.
Han CD (1973b). *J. Appl. Polym. Sci.* 17:1403.
Han CD (1974). *Trans. Soc. Rheol.* 18:163.
Han CD (1976). *Rheology in Polymer Processing*, Academic Press, New York, Chap 5.
Han CD (1998). In *Rheological Measurement*, 2nd ed, Collyer AA, Clegg DW (eds), Chapman and Hall, London, p 190.
Han CD, Charles M (1970). *AIChE. J.* 16:499.
Han CD, Charles M, Philippoff W (1969). *Trans. Soc. Rheol.* 13:453.
Han CD, Charles M, Philippoff W (1970). *Trans. Soc. Rheol.* 14:393.
Han CD, Drexler LH (1973a). *J. Appl. Polym. Sci.* 17:2329.
Han CD, Drexler LH (1973b). *Trans. Soc. Rheol.* 17:659.
Han CD, Kim KU (1971). *Polym. Eng. Sci.* 11:395.
Han CD, Kim KU, Siscovic N, Huang CR (1973). *J. Appl. Polym. Sci.* 17:95.
Han CD, Kim YJ, Chuang HK (1983a). *Polym. Eng. Rev.* 3:1
Han CD, Kim YJ, Chuang HK, Kwack TH (1983b). *J. Appl. Polym. Sci.* 28:3435.
Han CD, Kim YW (1975). *Trans. Soc. Rheol.* 19:245.
Han CD, Villamizar CA (1978). *J. Appl. Polym. Sci.* 22:1677.
Han CD, Yu TC (1971). *Rheol. Acta* 10:398.
Han CD, Yu TC, Kim KU (1971). *J. Appl. Polym. Sci.* 15:1149.
Ide Y, White JL (1978). *J. Appl. Polym. Sci.* 22:1061.
LaNieve HL, Bogue DC (1968). *J. Appl. Polym. Sci.* 12:353.
Laun HM (1983). *Rheol. Acta* 22:171.
Laun HM, Münstedt H (1976). *Rheol. Acta* 15:517.
Laun HM, Münstedt H (1978). *Rheol. Acta* 17:415.
Leblanc JL (1976). *Polymer* 17:235.
Lodge AS (1964). *Elastic Liquids*, Academic Press, New York.
Lodge AS, de Vargas L (1983). *Rheol. Acta* 22:151.
Meissner J (1969). *Rheol. Acta* 8:78.
Meissner J (1971). *Rheol. Acta* 10:230.
Meissner J (1972). *Trans. Soc. Rheol.* 16:405.
Mori Y, Funatsu K (1968). *Chem. High Polym. (Japan)* 25:391.
Münstedt H (1975). *Rheol. Acta* 14:1077.
Olabisi O, Williams MC (1972). *Trans. Soc. Rheol.* 16:727.
Rauwendaal C, Fernandez F (1985). *Polym. Eng. Sci.* 25:765.
Reddy KR, Tanner RI (1978). *Trans. Soc. Rheol.* 22:661.
Reiner M (1964). *Phys. Today* 17:62.

Schowalter WR, Allen RC (1975). *Trans. Soc. Rheol.* 19:129.
Stevenson JF (1972). *AIChE J.* 18:540.
Tuna NY, Finlayson BA (1984). *J. Rheol.* 28:79.
Tuna NY, Finlayson BA (1988). *J. Rheol.* 32:285.
Vinogradov GV, Radushkevich BV, Fikhman VD (1970). *J. Polym. Sci. A-2* 8:1.
Vinogradov GV, Fikhman VD, Radushkevich BV (1972). *Rheol. Acta* 11:286.
Vlachopoulos J, Mitsoulis E (1985). *J. Polym. Eng.* 5(2):173.
Walter K (1975). *Rheometry*, Chapman and Hall, London.
Whipple BA, Hill CT (1978). *AIChE J.* 24:664.

Part II

Rheological Behavior of Polymeric Materials

6 Rheology of Flexible Homopolymers

6.1 Introduction

Numerous flexible homopolymers and flexible random copolymers are commercially available. Thus, understandably, a number of research groups have reported on the rheological behavior of flexible homopolymers and flexible random copolymers[1] in the bulk and solution states. There are too many studies to cite them all here. In Chapters 3 to 5 we presented the rheological behavior, in general terms, of linear flexible homopolymers in steady-state shear flow, elongational flow, and/or oscillatory shear flow. In this chapter, we present the effects of temperature, molecular weight (although in Chapter 4 we presented theoretical predictions of the effect of molecular weight), and molecular weight distribution on the rheological behavior of linear flexible homopolymers, and also flexible homopolymers with long-chain branching. The rheological behavior of much more complex polymer systems is presented in other chapters of this volume.

From the point of view of polymer processing, temperature is one of the most important variables that greatly affect the rheological behavior of polymeric liquids. Therefore, it is very important to present the effect of temperature on rheological behavior, placing emphasis on the methods that enable one to obtain temperature-independent correlations for rheological properties. Such correlations, when available, will help one to estimate the rheological properties of the same polymer without conducting additional experiments.

With respect to polymer synthesis and polymer processing, a better understanding of the effects of molecular weight and molecular weight distribution on the rheological behavior of a polymer is of fundamental importance. In Chapter 4 we have presented molecular theory, demonstrating that the molecular weight of a linear flexible homopolymer has a profound influence on its rheological properties. Thus, information on the relationships between molecular weight and rheology, when available, will help one to choose, with little waste of time and effort, optimum processing conditions. One of the common features of all commercial homopolymers is that they are polydisperse and, therefore, it is not difficult to surmise that the molecular weight distribution of a polymer also has a profound influence on its rheological properties. Sometimes, the molecular weight distribution of a polymer greatly influences the extent of flow instability, severely limiting processability of the polymer. In general, an accurate measurement of molecular weight distribution of certain linear flexible homopolymers (e.g., polypropylene) is very difficult. Under such circumstances, one can estimate the molecular weight distribution of a polymer from rheological measurements if information on the relationships between rheology and molecular weight distribution is available. In such situations, rheological measurements can play a very important role in polymer production.

One of the as-yet completely unsolved problems in the field of molecular rheology of flexible homopolymers is the effect of long-chain branching, although active research activities have been reported in recent years. One such polymer of great commercial importance is low-density polyethylene (LDPE) having many long-chain branches. It has long been known in the polymer industry that the presence of long-chain branches in LDPE not only affects the rheological behavior (elongational viscosity in particular) but also contributes to the onset of flow instability in tubular film blowing. Therefore, it is of fundamental and practical importance to enhance our understanding of the roles that long-chain branching in LDPE plays in influencing rheological behavior.

In this chapter, we first present methods to obtain temperature-independent correlations for shear viscosity as functions of shear rate for linear flexible homopolymers. We then discuss methods for obtaining molecular-weight independent correlations for steady-state and oscillatory shear flow properties of linear flexible homopolymers. Then, we present experimental observations of rheological behavior of flexible homopolymers with long-chain branching. The main purpose of this chapter is to present methods for obtaining useful correlations, not to present exhaustive experimental data.

6.2 Rheology of Linear Flexible Homopolymers

In this section, we present the rheological behavior of linear flexible homopolymers (i.e., without side-chain branching). We first present methods for obtaining temperature-independent plots for shear viscosity (η) and first normal stress difference (N_1) in steady-state shear flow and for dynamic moduli (G'' and G'') in oscillatory shear flow. We then discuss the effect of molecular weight and molecular weight distribution on the rheological behavior of linear flexible homopolymers.

6.2.1 Temperature Dependence of Steady-State Shear Viscosity of Linear Flexible Homopolymers

To facilitate our presentation, let us look at the log η versus log $\dot{\gamma}$ and log N_1 versus log $\dot{\gamma}$ plots, given in Figure 6.1, for an LDPE at 180, 200, and 220 °C. Recall that in Chapters 3 and 4 we presented various constitutive equations describing such experimental observations. There are, however, some simple empirical expressions that have been found to be very useful to correlate viscosity η to shear rate $\dot{\gamma}$ and temperature T.

Let us examine some of those expressions that will be used in later chapters to simulate complex polymer processing operations. Referring to Figure 6.1, we observe

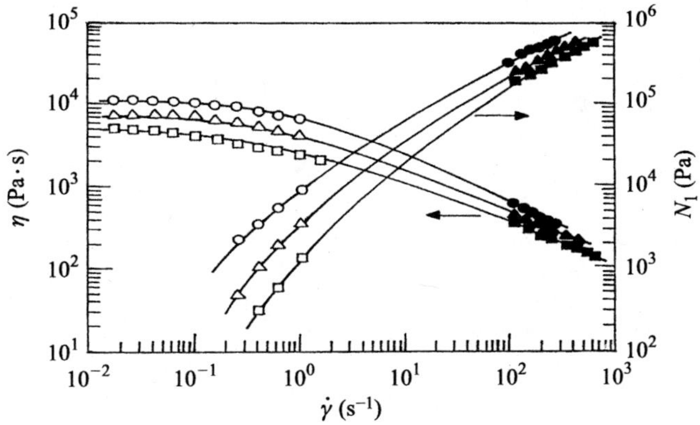

Figure 6.1 Plots of log η versus log $\dot{\gamma}$ and log N_1 versus log $\dot{\gamma}$ for an LDPE (PE 510) at various temperatures (°C): (○, ●) 180, (△, ▲) 200, and (□, ■) 220. Open symbols denote data taken with a cone-and-plate rheometer, and filled symbols denote data taken with a capillary rheometer and analyzed with the exit pressure method as described in Chapter 5.

that η remains constant (zero-shear viscosity η_0) at $\dot{\gamma}$ below a certain critical value $\dot{\gamma}_0$, and then decreases with an increasing $\dot{\gamma}$, showing that log η is proportional to log $\dot{\gamma}$ with a negative slope. Thus, the log η versus log $\dot{\gamma}$ plot at high $\dot{\gamma}$ can be approximated as

$$\eta(\dot{\gamma}, T) = K(T)\,\dot{\gamma}^{n-1} \tag{6.1}$$

where $n - 1$ represents the slope of the log η versus log $\dot{\gamma}$ plot at $\dot{\gamma} > \dot{\gamma}_0$, and K and n are empirical constants characteristic of a given polymer structure. Experimental studies suggest that K varies with temperature T, while n is virtually independent of temperature. Since the values of η given in Figure 6.1 were calculated using the definition $\eta = \sigma/\dot{\gamma}$, from Eq. (6.1) we have

$$\sigma(\dot{\gamma}, T) = K(T)\dot{\gamma}^n \tag{6.2}$$

which is referred to as the "power-law model."[2] Thus, the log η versus log $\dot{\gamma}$ plots given in Figure 6.1 can be approximated by

$$\eta(\dot{\gamma}, T) = \begin{cases} \eta_0(T) & \text{for } \dot{\gamma} \le \dot{\gamma}_0 \\ \eta_0(T)(\dot{\gamma}/\dot{\gamma}_0)^{n-1} & \text{for } \dot{\gamma} > \dot{\gamma}_0 \end{cases} \tag{6.3}$$

which is referred to as the "truncated power-law model." It can be shown that $\eta_0(T)$ in the second expression of Eq. (6.3) is related to $K(T)$, appearing in Eq. (6.1), by $K(T)[\dot{\gamma}_0(T)]^{n-1}$, in which the value of $\dot{\gamma}_0$ has been found empirically to increase with increasing T. Table 6.1 gives a summary of experimental results describing the temperature dependence of η_0, K, n, and $\dot{\gamma}_0$ for some commercial, linear flexible homopolymers. Notice in Table 6.1 that n appearing in Eq. (6.3) is virtually independent of T. Figure 6.2 describes the temperature dependence of $\dot{\gamma}_0$ for some commercial

Table 6.1 Parameters appearing in the truncated power-law model for some molten polymers

Polymer	*Temp (°C)*	η_0 *(Pa·s)*	$\dot{\gamma}_0$ (s^{-1})	*K (Pa·s^n)*	*n*
HDPE (DMDJ 5140)	200	4.00×10^4	0.12	1.25×10^4	0.45
PS (Styron 686)	200	3.10×10^4	0.44	1.77×10^4	0.31
PS (Styron 678)	200	7.36×10^3	0.92	6.93×10^3	0.33
	210	3.92×10^3	2.03	6.30×10^3	0.33
	220	2.14×10^3	4.36	5.75×10^3	0.33
	230	1.20×10^3	9.11	5.26×10^3	0.33
LDPE (NPE 952)	180	1.84×10^4	0.32	8.88×10^3	0.37
	200	1.06×10^4	0.60	7.60×10^3	0.37
	220	7.49×10^3	0.82	6.60×10^3	0.37
LDPE (PE 510)	180	1.19×10^4	0.59	8.82×10^3	0.44
	200	7.46×10^3	0.86	6.85×10^3	0.44
	220	4.90×10^3	1.20	5.43×10^3	0.44
PP (Plexar 6432)	180	1.40×10^4	0.24	5.86×10^3	0.38
	200	8.23×10^3	0.43	4.88×10^3	0.38
	220	5.04×10^3	0.72	4.12×10^3	0.38
PMMA (Plexiglas)	210	1.38×10^4	0.79	1.12×10^4	0.41
	220	6.39×10^3	1.41	7.96×10^3	0.41
	230	3.59×10^3	2.22	5.73×10^3	0.41
PVDF (Kynar)	210	6.70×10^3	0.26	4.16×10^3	0.65
	220	5.03×10^3	0.43	3.74×10^3	0.65
	230	4.04×10^3	0.60	3.37×10^3	0.65

linear flexible homopolymers. Thus, in order to use Eq. (6.3) for estimating η at any other temperatures, one must have information on $\eta_0(T)$ and an empirical correlation between $\dot{\gamma}_0$ and T.

Numerous studies have been reported on the methods that enable one to obtain temperature independent $\log \eta(\dot{\gamma}, T)$ versus $\log\ a_T\dot{\gamma}$ plots, usually referred to as reduced (or master) plots, where a_T is called a "temperature-dependent shift factor." We will show different ways of obtaining a_T. The availability of reduced plots for any given polymer will enable one to estimate the shear viscosity of the polymer at any desired shear rate and temperature. There are two different ways of obtaining reduced plots for shear viscosity, depending on whether a polymer is semicrystalline or amorphous.

For a semicrystalline, linear flexible homopolymer at temperatures above its melting point, the temperature dependence of $\eta_0(T)$ appearing in Eq. (6.3) can be described by the Arrhenius expression

$$\eta_0(T) = k_o \exp(E/RT) \tag{6.4}$$

where k_o is the preexponential factor, E is the activation energy for shear flow, R is the universal gas constant, and T is the absolute temperature. Over a narrow range of

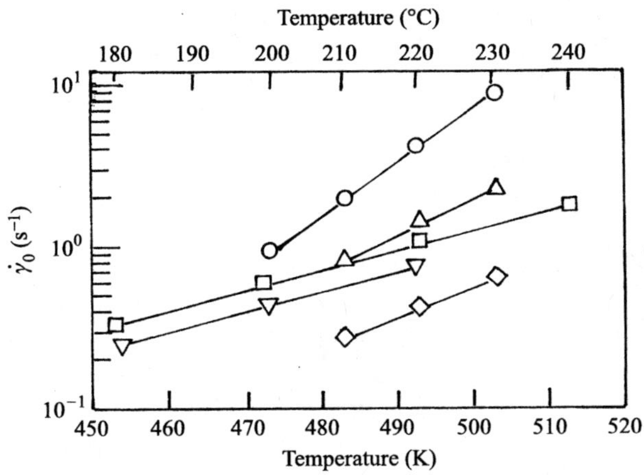

Figure 6.2 Plots of $\dot{\gamma}_0$ versus temperature for (○) polystyrene, (△) poly(methyl methacrylate), (□) LDPE, (▽) polypropylene, and (◇) poly(vinylidene fluoride).

temperatures, Eq. (6.4) is often approximated by

$$\eta_0(T) = k'_o \exp(-bT) \tag{6.5}$$

where k'_o is the preexponential constant and b is a constant. Equation (6.5) is very useful for numerically solving the differential energy balance presented in Chapter 5. Using Eq. (6.4), a_T can be defined by

$$\log a_T = \frac{E}{R}\left(\frac{1}{T_2} - \frac{1}{T_1}\right) \tag{6.6}$$

indicating that a_T represents the ratio $\eta_0(T_2)/\eta_0(T_1)$.

For an amorphous, linear flexible homopolymer at $T_g < T \leq T_g + 100\,°C$, with T_g being its glass transition temperature, the Williams–Landel–Ferry (WLF) equation (Ferry 1980; Williams et al. 1955)

$$\log a_T(T) = \frac{-C_1(T - T_r)}{C_2 + T - T_r} \tag{6.7}$$

has been found to be very useful to estimate the $\eta_0(T)$, while Eq. (6.4) may be used to estimate $\eta_0(T)$ at $T > T_g + 100\,°C$, where C_1 and C_2 are constants characteristic of a given polymer structure and T_r is an arbitrarily chosen reference temperature. The physical origin of Eq. (6.7) lies in the free volume theory (Cohen and Turnbull 1959; Doolittle and Doolittle 1957; Turnbull and Cohen 1961). Numerical values of C_1 and C_2 for various linear flexible homopolymers are available in the literature (Ferry 1980). Thus, the $\eta_0(T)$ of an amorphous polymer at $T_g < T \leq T_g + 100\,°C$ can be estimated from[3]

$$\eta_0(T) = a_T(T)\eta_0(T_r) \tag{6.8}$$

when information on $a_T(T)$ defined by Eq. (6.7) and $\eta_0(T_r)$ is available. When experimental data for $\eta_0(T)$ at various temperatures are available, values of C_1 and C_2 appearing in Eq. (6.7) can be determined by curve fitting. Once values of C_1 and C_2 appearing in Eq. (6.7) are determined for a given linear flexible homopolymer, the expression

$$\eta_0(T) = \left[\frac{a_T(T)}{a_T(T_1)}\right]\eta_0(T_1) \tag{6.9}$$

can be used to estimate $\eta_0(T)$ at any temperature T with information on $\eta_0(T_1)$, where $a_T(T)$ and $a_T(T_1)$ are defined by Eq. (6.7).

Figure 6.3 gives plots of log η_0 versus $1/T$ for some commercial flexible homopolymers, showing that indeed the experimental data are described by Eq. (6.4). Notice in Figure 6.3 that the values of η_0 for polystyrene (PS) and poly(methyl methacrylate) (PMMA) are at temperatures close to or above $T_g + 100\,°\mathrm{C}$.

The following empirical expression, the Cross equation (Cross 1965, 1969), has also been found to be very useful to describe, with reasonable accuracy, the shear-rate dependence of viscosity of molten polymers:

$$\eta(\dot{\gamma}, T) = \frac{\eta_0(T)}{1 + \left[\tau_0(T)\dot{\gamma}\right]^{1-n}} \tag{6.10}$$

where $\tau_0(T)$ is an empirical constant that depends on temperature, and n is a constant that is weakly dependent upon temperature. Plots of log η/η_0 versus log $\tau_0\dot{\gamma}$ for an LDPE at 180, 200, and 220 °C are given in Figure 6.4, where the solid line represents the calculated η/η_0, via nonlinear least-squares analysis, by adjusting the value of τ_0 appearing in Eq. (6.10) until the calculated η/η_0 matches, within a prescribed tolerance, with experimental results. Table 6.2 gives numerical values for τ_0 and n appearing in Eq. (6.10) for some commercial linear flexible homopolymers. Note that for $\tau_0\dot{\gamma} \gg 1$, Eq. (6.10) reduces to the power-law model, Eq. (6.1). Equation (6.10)

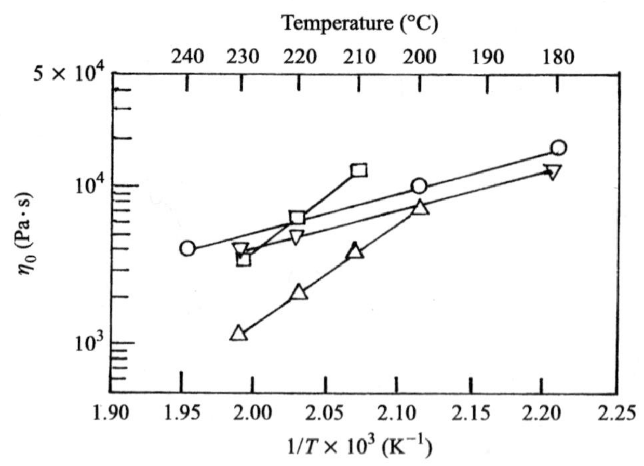

Figure 6.3 Plots of log η_0 versus $1/T$ for (○) LDPE, (△) polystyrene, (□) poly(methyl methacrylate), and (▽) polypropylene.

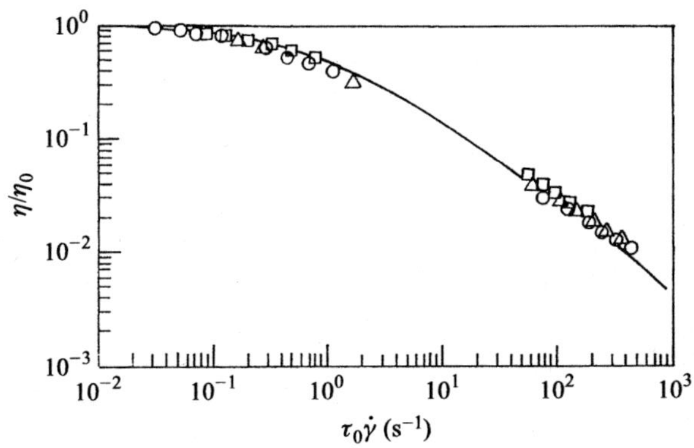

Figure 6.4 Plots of log η/η_0 versus log $\tau_0\dot{\gamma}$ for an LDPE (NPE 952) at different temperatures (°C): (○) 180, (△) 200 and (□) 220. The solid curve represents the Cross equation, for which the following numerical values were used: (1) $\eta_0 = 1.84 \times 10^4$ Pa·s, $\tau_0 = 1.03$ s, and $n = 0.220$ at 180 °C; (2) $\eta_0 = 1.06 \times 10^4$ Pa·s, $\tau_0 = 0.62$ s, and $n = 0.224$ at 200 °C; (3) $\eta_0 = 6.44 \times 10^3$ Pa·s, $\tau_0 = 0.45$ s, and $n = 0.273$ at 220 °C.

Table 6.2 Parameters appearing in the Cross equation for some molten polymers

Polymer	*Temp (°C)*	η_0 *(Pa·s)*	τ_0 *(s)*	*n*
HDPE (DMDJ 5140)	200	4.00×10^4	2.955	0.367
PS (Styron 686)	200	3.10×10^4	1.071	0.206
PS (Styron 678)	200	7.36×10^3	0.280	0.137
	220	2.14×10^3	0.223	0.345
LDPE (NPE 952)	180	1.84×10^4	1.030	0.220
	200	1.06×10^4	0.621	0.224
	220	7.49×10^3	0.456	0.273

enables one to estimate η at any desired temperatures, as long as the values of τ_0 and n are available at those temperatures. This can be realized by assuming that n is independent of temperature and by establishing an empirical correlation between τ_0 and temperature.

The following empirical expression, often referred to as the Carreau model (Carreau 1968), has also been found to describe the shear-rate dependence of viscosity of polymer melts reasonably well

$$\eta(\dot{\gamma}, T) = \frac{\eta_0(T)}{\left[1 + (\lambda(T)\dot{\gamma})^2\right]^{(1-n)/2}} \tag{6.11}$$

with $\log \eta/\eta_0$ versus $\log \lambda\dot{\gamma}$ plots giving rise to a temperature-independent correlation. Practically speaking, there is little difference between Eqs. (6.10) and (6.11). Vinogradov and Malkin (1966) reported that $\log \eta/\eta_0$ versus $\log \eta_0\dot{\gamma}$ plots yield temperature-independent correlation. The existence of such a correlation is not surprising in that, according to the tube model presented in Chapter 4 (see Eq. (4.112)), η_0 is proportional to the tube disengagement time τ_d and thus a plot of $\log \eta/\eta_0$ versus $\log \eta_0\dot{\gamma}$ can be regarded as being equivalent to a plot of $\log \eta/\eta_0$ versus $\log \tau_d\dot{\gamma}$.

Indeed, in Chapter 4 (see Figure 4.17) we have shown that $\log \eta/\eta_0$ versus $\log \tau_d\dot{\gamma}$ plots yield a temperature-independent correlation. In this regard, we can state that τ_0 appearing in Eq. (6.10) is related to the relaxation time (or the tube disengagement time, within the spirit of the tube model), which decreases with increasing temperature for a given polymer. Hence, we now understand the reason why $\log \eta/\eta_0$ versus $\log \tau_0\dot{\gamma}$ plots (see Eq. (6.10)), $\log \eta/\eta_0$ versus $\log \lambda\dot{\gamma}$ plots (see Eq. (6.11)), and $\log \eta/\eta_0$ versus $\log \eta_0\dot{\gamma}$ plots yield temperature-independent correlations.

6.2.2 Temperature Dependence of Relaxation Time and First Normal Stress Difference in Steady-State Shear Flow of Linear Flexible Homopolymers

There are several different methods for determining the elastic properties of flexible homopolymers. One simple method is the determination of relaxation time. As presented in Chapters 3 and 4, the relaxation time(s) appearing in constitutive equations represents the elastic nature of polymers. Simply put, the larger the relaxation time(s) of a polymer, the greater is its fluid elasticity. It is then not difficult to surmise that the relaxation time(s) of a polymer would depend on temperature and perhaps on the extent of applied shear rate. However, the continuum theory presented in Chapter 3 does not reveal how the relaxation time(s) might vary with temperature, whereas the molecular theory presented in Chapter 4 does.

Figure 6.5 shows the temperature dependence of the relaxation time λ, which was determined by curve fitting the experimentally obtained $\log N_1$ versus $\log \dot{\gamma}$ plot to the theoretical prediction of the modified upper convected Maxwell model, Eq. (3.12). It is seen that the value of λ decreases with increasing temperature and also with increasing $\dot{\gamma}$. Figure 6.6 shows the temperature dependence of the relaxation time λ, which was determined by curve fitting the experimentally obtained $\log N_1$ versus $\log \dot{\gamma}$ plots for five commercial linear flexible homopolymers to the theoretical prediction of the Spriggs model, Eq. (3.38). It is seen that the value of λ decreases with increasing temperature for all five homopolymers, very similar to the result seen in Figure 6.5 for

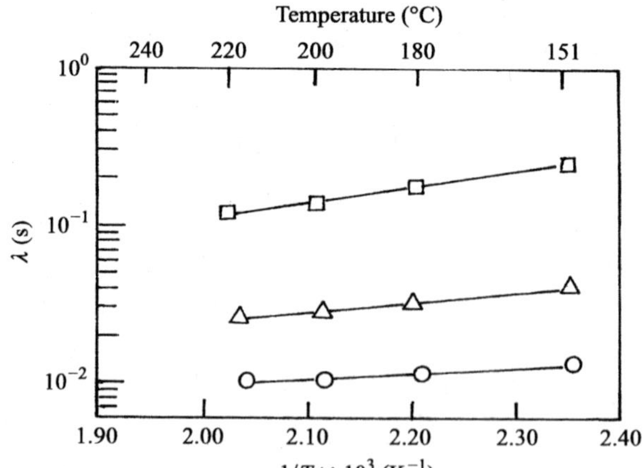

Figure 6.5 Plots of relaxation time λ versus the reciprocal of absolute temperature for an LDPE (NPE 952) at various shear rates (s^{-1}): (○) 400, (△) 100, and (□) 10. The values of λ, defined by Eq. (3.14), were determined by curve fitting the experimentally obtained $\log N_1$ versus $\log \dot{\gamma}$ plot to the theoretical prediction of the modified upper convected Maxwell model, Eq. (3.12).

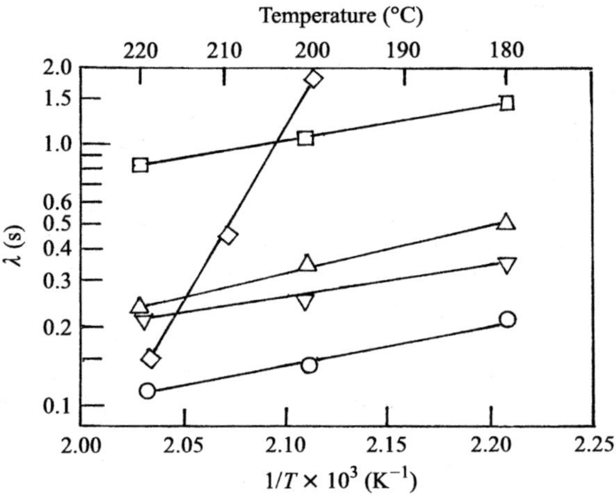

Figure 6.6 Plots of relaxation time λ versus the reciprocal of absolute temperature T for: (○) LDPE (Rexene 111), (△) LDPE (PE 510), (□) LDPE (NPE 962), (▽) polypropylene (Profax 6423), and (◇) polystyrene (Styron 678), where the values of λ, appearing in Eq. (3.39), were determined by curve fitting the experimentally obtained log N_1 versus log $\dot{\gamma}$ plot to the theoretical prediction of the Spriggs model, Eq. (3.38).

the LDPE. Note that the temperature dependence of λ and η_0 for PS is much stronger than that for LDPE and polypropylene (PP). It is worth noting that the temperature dependence of λ (i.e., the slope of $\log \lambda$ versus $1/T$ plot) observed in Figures 6.5 and 6.6 is very similar to the temperature dependence of η_0 observed in Figure 6.3. This can be explained by the molecular theory presented in Chapter 4. Specifically, according to Eq. (4.95), the Rouse relaxation time τ_p is inversely proportional to the absolute temperature and proportional to η_0. Since the temperature dependence of η_0 follows the Arrhenius expression, Eq. (6.4), we observe that the temperature dependence of relaxation time λ (or τ_p) also follows the Arrhenius expression, $\lambda \propto \exp(E'/RT)$. This empirical expression is used in Volume 2 when we discuss polymer processing operations. Also, according to Eq. (4.112), the Doi–Edwards tube disengagement time τ_d is proportional to η_0. Since τ_d in the Doi–Edwards theory can be regarded as being equivalent to τ_p in the Rouse theory, the temperature dependence of λ observed in Figures 6.5 and 6.6 can also be explained by the Doi–Edwards theory (see Chapter 4).

We present here a very effective method for obtaining a temperature-independent correlation for the N_1 of linear flexible homopolymers. For illustration purposes, we have used the data from Figure 6.1 to prepare the $\log N_1$ versus $\log \sigma$ plots given in Figure 6.7, showing a temperature-independent correlation. Such observations were reported extensively by Han and coworkers (Chuang and Han 1984; Han and Kim 1975; 1976; Han and Lem 1982; Han and Chuang 1985a, 1985b; Han and Yang 1986, 1987; Lem and Han 1982; Han et al. 1983a, 1983b), and also by others (Oda et al. 1978; Minoshima et al. 1980).

The temperature-independent correlation displayed in Figure 6.7 can be explained as follows (Han and Jhon 1986). For illustration, let us consider the Zaremba–Fromm–DeWitt (ZFD) model, Eq. (3.15) given in Chapter 3. Equation (3.17) can be rewritten as

$$\frac{\sigma}{G} = \frac{\lambda_1 \dot{\gamma}}{1 + \lambda_1{}^2 \dot{\gamma}^2}; \quad \frac{N_1}{G} = \frac{2\lambda_1{}^2 \dot{\gamma}^2}{1 + \lambda_1{}^2 \dot{\gamma}^2} \tag{6.12}$$

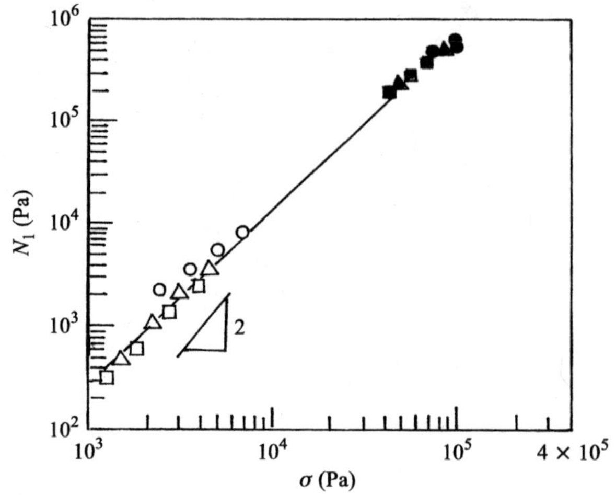

Figure 6.7 Plots of log N_1 versus log σ for an LDPE (PE 510) at various temperatures (°C): (○, ●) 180, (△, ▲) 200, and (□, ■) 220. Open symbols denote the data taken with a cone-and-plate rheometer, and filled symbols denote the data taken with a capillary rheometer and analyzed with the exit pressure method as described in Chapter 5.

with $G = \eta_0/\lambda_1$. Eliminating $\lambda_1\dot{\gamma}$ from Eq. (6.12) we obtain

$$(\sigma/G)^2 + (0.5N_1/G)^2 = 0.5N_1/G \tag{6.13}$$

which gives

$$N_1/G = 1 - \sqrt{1 - 4(\sigma/G)^2}; \quad N_1/G = 1 + \sqrt{1 - 4(\sigma/G)^2} \tag{6.14}$$

for which the restriction $\sigma/G \leq 0.5$ must be satisfied. For $\lambda_1\dot{\gamma} \ll 1$, from Eq. (6.12) we obtain[4]

$$N_1 = 2\sigma^2/G \tag{6.15}$$

Note that for entangled linear flexible homopolymers, $G = \eta_0/\lambda_1$ is related to the plateau modulus G_N^0 because, according to Eq. (4.112), η_0 is proportional to G_N^0. Using Eq. (4.97), G can be expressed by

$$G = C_3\rho T/M_e \tag{6.16}$$

where C_3 is a constant, T is the absolute temperature, ρ is the density, and M_e is the entanglement molecular weight. By combining Eqs. (6.15) and (6.16) we obtain

$$\log N_1 = 2\log\sigma + \log(2M_e/C_3\rho T) \tag{6.17}$$

which predicts that the log N_1 versus log σ plot is virtually independent of (or very weakly dependent upon) temperature. This is because T has the unit of Kelvin and thus the variations of temperature, for instance from 180 to 220 °C, would hardly be discernible in the log N_1 versus log σ plot (see Figure 6.7). Notice in Eq. (6.17) that log N_1 versus log σ plots are independent of molecular weight for entangled polymers.

Note that since ρ decreases with increasing temperature, the values of ρT may remain more or less constant over a reasonable range of temperatures. This now explains why in Figure 6.7 we observe virtually *no* temperature dependence in the log N_1 versus log σ plots.

It is worth noting that the log N_1 versus log σ plot in Figure 6.7 has the slope of less than 2, while Eq. (6.17) predicts a slope of 2. According to the experimental studies of Han and coworkers cited above, the slope of log N_1 versus log σ plots (the value of b in the expression $N_1 = A\sigma^b$ given by Eq. (5.84)) for several commercial linear flexible homopolymers lies between 1.5 and 2.0 (see Table 5.3). This is attributed to the polydisperse nature of commercial flexible homopolymers.

6.2.3 Temperature-Independent Correlations for the Linear Dynamic Viscoelastic Properties of Linear Flexible Homopolymers

Numerous research groups have applied time–temperature superposition (TTS) to the experimental data obtained from oscillatory shear measurements of linear flexible homopolymers in order to obtain reduced plots, log G' (or log G'_r) versus log $a_T\omega$ plot and/or log G'' (log G''_r) versus log $a_T\omega$ plot. Here, a_T is a temperature-dependent shift factor and G'_r are G''_r are reduced dynamic storage modulus and loss modulus, defined by $G'_r(\omega, T) = (\rho_r T_r/\rho T)G'(\omega, T)$ and $G''_r(\omega, T) = (\rho_r T_r/\rho T)G''(\omega, T)$, respectively, with ρ being the density at temperature T and ρ_r being the density at a reference temperature T_r.[5] There are different ways of obtaining a_T in order to prepare reduced plots using the data obtained from oscillatory shear measurements. One method is to shift log G' versus log ω plots or log G'' versus log ω plots along the ω axis using an arbitrarily chosen temperature as T_r. Such a procedure allows one to prepare $a_T(T)$ versus T plots. It has been found that log G'' versus log ω plots work better than log G' versus log ω plots, although ideally both plots should give the same results. Then, one can prepare log G'_r versus log $a_T\omega$ or log G''_r versus log $a_T\omega$ plots using the values of a_T already determined at each temperature. Here, we give an example of how to construct reduced plots from oscillatory shear measurements for eight nearly monodisperse polystyrenes with varying molecular weights, which were synthesized by anionic polymerization.

Figure 6.8 gives plots of log G' versus log ω and log G'' versus log ω for a polystyrene at various temperatures. Figure 6.9 shows plots of a_T versus temperature for five polystyrenes, the molecular weights of which are given in Table 6.3, where values of a_T for each polymer were determined by shifting log G'' versus log ω plots along the ω axis. Figure 6.10 gives log G'_r versus log $a_T\omega$ plots for five polystyrenes, for which the expression $1/\rho(T) = 0.9217 + 5.412 \times 10^{-4}(T - 273) + 1.687 \times 10^{-7}(T - 273)^2$ was used to calculate the density (g/cm^3) of polystyrene at each temperature T. Notice in Figure 6.10 that the log G'_r versus log $a_T\omega$ plots for each polystyrene are independent of temperature.

Often, one is interested in comparing the viscosities of flexible, amorphous homopolymers with different molecular weights and which have the same chemical structure or different chemical structures. Since the viscosities of polymers depend on both temperature and molecular weight, it is essential to suppress, if not eliminate completely, the effect of temperature on viscosity for such purposes. When the

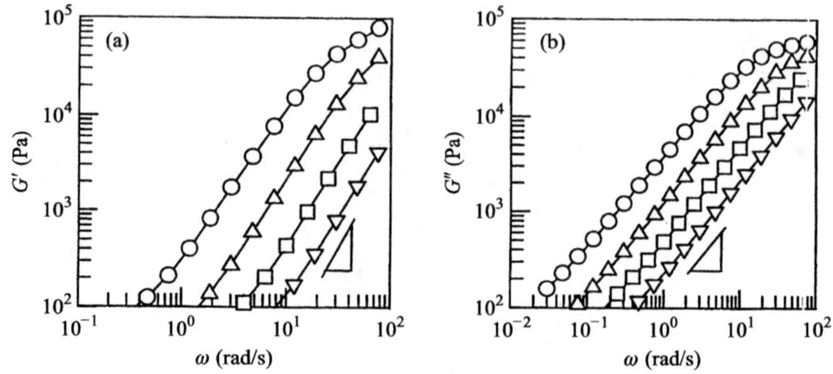

Figure 6.8 Plots of (a) log G' versus log ω and (b) log G'' versus log ω for polystyrene PS-5 at various temperatures (°C): (○) 160, (△) 170, (□) 180, and (▽) 190.

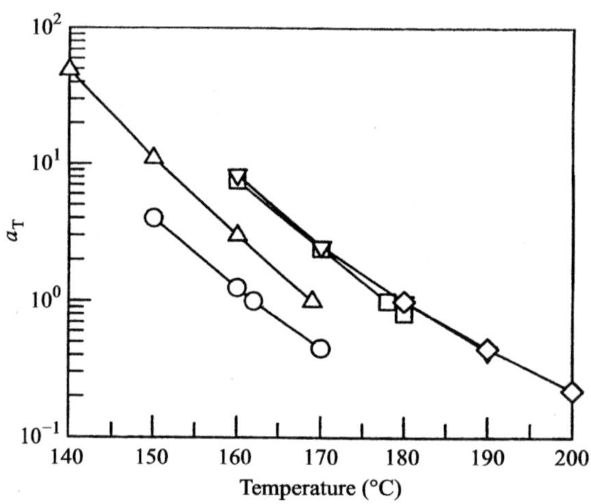

Figure 6.9 Temperature dependence of a_T for five polystyrenes with $T_r = T_g + 79\,°C$: (○) PS-1, (△) PS-2, (□) PS-4, (▽) PS-5, and (◇) PS-8.

temperature at which the viscosities are to be compared lies within the WLF region (i.e., at $T_g < T \leq T_g + 100\,°C$), one must compare the viscosities of amorphous flexible homopolymers in the iso-free volume state; that is, at the same distance from the T_g of the respective homopolymer. It should be mentioned that for a given chemical structure, the T_g of a polymer increases with molecular weight until reaching a certain critical value, as shown in Table 6.3 for eight polystyrenes. To compare the viscosities of the eight polystyrenes in an iso-free volume state, let us choose $T_r = T_g + 79\,°C$ as a reference temperature, and take the values of T_r given in Table 6.3. Figure 6.11 gives log $|\eta^*|$ versus log ω plots for the eight polystyrenes at $T_r = T_g + 79\,°C$, thus eliminating the effect of temperature on the complex viscosity $|\eta^*|$. In other words, the complex viscosities of all eight polystyrenes were determined at $T_r = T_g + 79\,°C$ and thus Figure 6.11 reflects only the effect of molecular weight.

Table 6.3 Molecular characteristics and glass transition temperatures of eight polystyrenes

Sample Code	M_w *(LS)*[a]	M_w/M_n [b]	T_g[c] *(°C)*	T_r[d] *(°C)*
PS-1	1.02×10^4	1.08	83	162
PS-2	1.83×10^4	1.09	90	169
PS-3	2.94×10^4	1.08	92	171
PS-4	3.48×10^4	1.08	99	178
PS-5	5.71×10^4	1.07	101	180
PS-6	7.97×10^4	1.03	101	180
PS-7	8.98×10^4	1.04	101	180
PS-8	10.52×10^4	1.05	101	180

[a]Determined using light scattering. [b]Determined from gel permeation chromatography. [c]Glass transition temperature determined using DSC at a heating rate of 5 °C/min. [d]Reference temperature chosen as $T_r = T_g + 79\,^\circ\mathrm{C}$.

Reprinted from Choi and Han, *Macromolecules* 37:215. Copyright © 2004, with permission from the American Chemical Society.

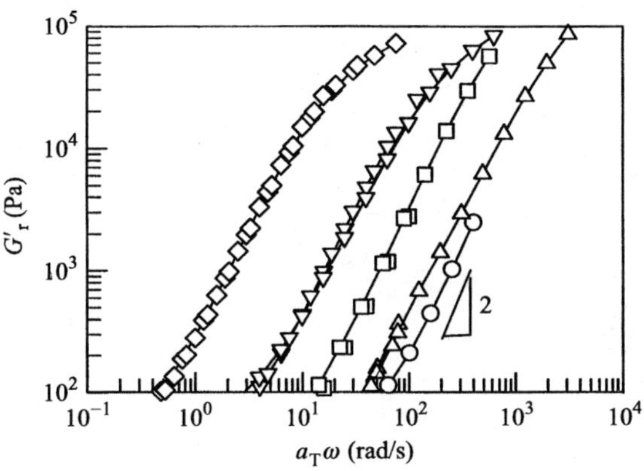

Figure 6.10 Log G'_r versus log $a_T\omega$ plots for five polystyrenes at various temperatures (°C): (○) PS-1 at 150, 160, 162, and 170, (△) PS-2 at 140, 150, 160, and 169, (□) PS-4 at 160, 170, 178, and 180, (▽) PS-5 at 160, 170, 180, and 190, and (◇) PS-8 at 180, 190, and 200.

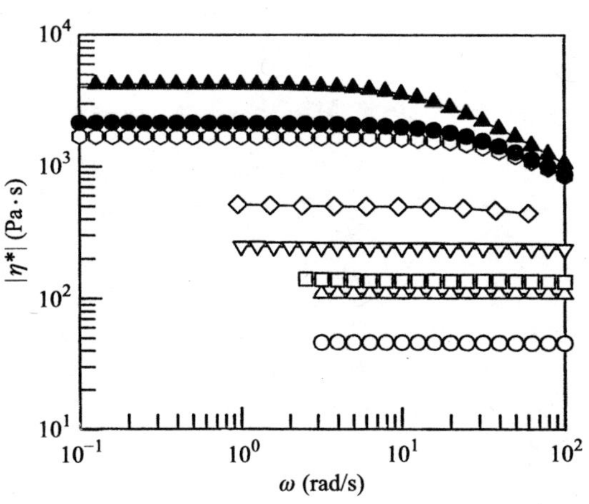

Figure 6.11 Plots of log $|\eta^*|$ versus log ω for eight polystyrenes with $T_r = T_g + 79\,^\circ\mathrm{C}$ as the reference temperature: (○) PS-1 at 162 °C, (△) PS-2 at 169 °C, (□) PS-3 at 171 °C, (▽) PS-4 at 178 °C; (◇) PS-5 at 180 °C, (⬡) PS-6 at 180 °C, (●) PS-7 at 180 °C, and (▲) PS-8 at 180 °C. (Reprinted from Choi and Han, *Macromolecules* 37:215. Copyright © 2004, with permission from the American Chemical Society.)

Another method that can be used to obtain temperature-dependent $a_T(T)$ is through the WLF equation, Eq. (6.7). When the molecular weight is sufficiently high such that the value of T_g does not depend on molecular weight, it is convenient to use T_g of the polymer as T_r in Eq. (6.7) for all molecular weights. Under such circumstances, values of C_1 and C_2 appearing in Eq. (6.7) do not vary with molecular weight and thus are only functions of molecular structure. Values of C_1 and C_2 appearing in Eq. (6.7) with $T_r = T_g$ for various polymers are given by Ferry (1980). When numerical values of C_1 and C_2 appearing in Eq. (6.7) are not available in the literature, they can be determined by curve fitting the values of empirically determined $a_T(T)$ to Eq. (6.7). For illustration, using the same values of $a_T(T)$ used to prepare Figure 6.9 for the five polystyrenes, we have determined values of C_1 and C_2, as summarized in Table 6.4. Notice in Table 6.4 that the values of C_1 and C_2 are different for the five polystyrenes because the T_g of each polymer sample is different. The values of C_1 and C_2 thus determined were used to prepare the plots of $a_T(T)$ versus $T - T_r$ given in Figure 6.12. It is seen in Figure 6.12 that values of $a_T(T)$ lie on a single line, indicating that $a_T(T)$, defined by Eq. (6.7), with the values of C_1 and C_2 given in Table 6.4 can be used to calculate, via Eq. (6.8), the zero-shear viscosity $\eta_0(T)$ at any desired temperature T. Note that values

Table 6.4 Values of C_1 and C_2 appearing in Eq. (6.7) for five polystyrenes determined by curve-fitting the empirically determined a_T

Sample Code	T_r (°C)	C_1	C_2 (K)
PS-1	162	6.37	137.2
PS-2	169	11.72	231.1
PS-4	178	15.58	337.5
PS-5	180	5.78	149.5
PS-8	180	6.33	172.6

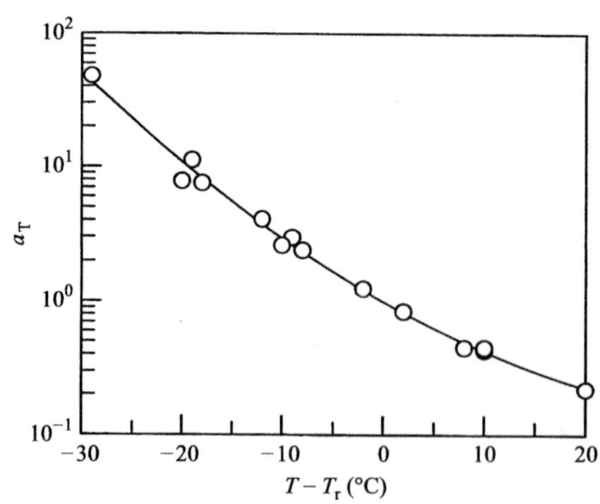

Figure 6.12 Plots of a_T versus $T - T_r$ for five polystyrenes, in which values of a_T were calculated from Eq. (6.7) with values of C_1 and C_2 given in Table 6.4.

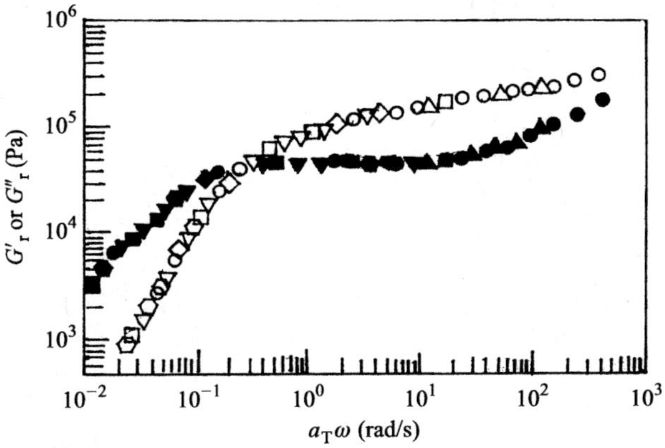

Figure 6.13 Plots of log G'_r versus log $a_T\omega$ (open symbols) and log G''_r versus log $a_T\omega$ (filled symbols) for a nearly monodisperse polystyrene ($M_w = 1.95 \times 10^5$, $M_w/M_n = 1.07$) at various temperatures (°C): (○, ●) 160, (△, ▲) 170, (□, ■) 180, (▽, ▼) 200, (◇, ◆) 210, and (⬡, ⬢) 220.

of $\eta_0(T)$ and $\eta_0(T_r)$ can be calculated from the relationship $\eta_0 = \lim_{\omega \to \infty}[G''(\omega)/\omega]$ (see Chapter 5).

Figure 6.13 gives log G'_r versus log $a_T\omega$ plots and log G''_r versus log $a_T\omega$ plots for a nearly monodisperse polystyrene ($M_w = 1.95 \times 10^5$) at various temperatures, where 160 °C is chosen as T_r. In other words, applying TTS we have obtained temperature-independent correlation. Figure 6.14 gives log G' versus log G'' plots for the same polystyrene, yielding temperature-independent correlation. Note that no manipulation of experimental data (i.e., no shift factor) is necessary to obtain Figure 6.14.

The temperature independence of log G' versus log G'' plots can be explained as follows. Recall that in Chapter 4 we presented the Doi–Edwards theory for entangled, monodisperse, linear flexible homopolymers (i.e., for $M > M_e$). In the terminal region of oscillatory shear measurements, using Eqs. (4.114) and (4.115) for $\tau_d\omega \ll 1$,

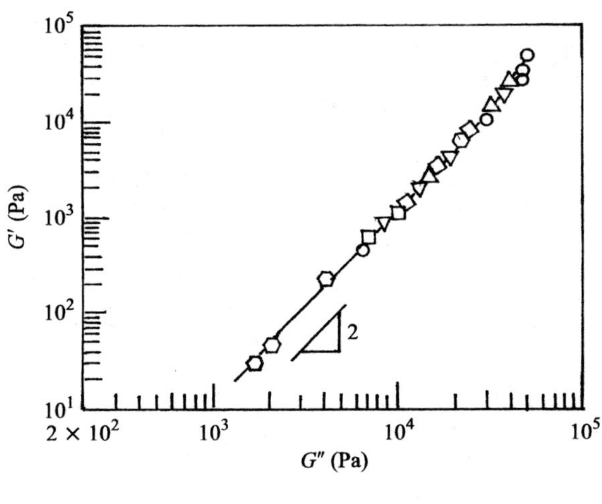

Figure 6.14 Log G' versus log G'' plots for a nearly monodisperse polystyrene ($M_w = 1.95 \times 10^5$, $M_w/M_n = 1.07$) at various temperatures (°C): (○) 160, (△) 170, (□) 180, (▽) 200, (◇) 210, and (⬡) 220.

we obtain (Han and Kim 1993)[6]

$$\log G' = 2\log G'' + \log(6/5G_N^o) \tag{6.18}$$

Equation (6.18) can be rewritten, with the aid of Eq. (4.97), as

$$\log G' = 2\log G'' + \log(6M_e/5\rho RT) \tag{6.19}$$

showing molecular weight independence. Conversely, for unentangled monodisperse homopolymers (i.e., for Rouse chains with $M < M_e$), in the terminal region of oscillatory shear measurements, using Eqs. (4.84) and (4.85) for $\tau_p\omega \ll 1$, we obtain (Han and Kim 1993)[7]

$$\log G' = 2\log G'' + \log(2M/5\rho RT) \tag{6.20}$$

showing molecular weight dependence. Equations (6.19) and (6.20) reveal that $\log G'$ versus $\log G''$ plots are very weakly dependent upon temperature, because T appears in the denominator of the logarithmic argument. For instance, a variation of temperature from 160 to 260 °C changes the value of the second term on the right-hand side of Eq. (6.19) or (6.20) so little that such contribution can hardly be discernible in the $\log G'$ versus $\log G''$ plot. This is the reason why $\log G'$ versus $\log G''$ plots given in Figure 6.14 are seen to be virtually independent of temperature! Han and coworkers have amply demonstrated experimentally (Han 1988; Han and Lem 1982; Han and Chuang 1985a, 1985b; Han and Yang, 1986, 1987; Han et al. 1983a) the temperature independence of $\log G'$ versus $\log G''$ plots for a number of linear, flexible homopolymers, and have offered a theoretical interpretation (Han and Jhon 1986; Han and Kim 1993) of the experimental observations.

We now offer a physical interpretation of the temperature independence of $\log N_1$ versus $\log \sigma$ plots (see Figure 6.7) and $\log G'$ versus $\log G''$ plots (see Figure 6.14) using the following arguments. Referring to Figure 6.15, in steady-state shear flow the shear rate ($\dot{\gamma}$) imposed on the fluid can be regarded as an input variable, while the shear stress (σ) and first normal stress difference (N_1) can be regarded as output variables (i.e., responses) of the fluid under test. Similarly, in oscillatory shear flow, the small-amplitude sinusoidal strain with an angular frequency (ω) imposed on the fluid can be regarded as an input variable, while the dynamic storage and loss moduli

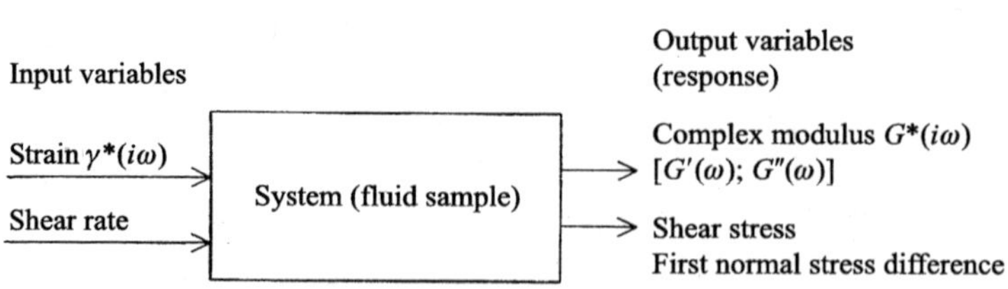

Figure 6.15 The responses $\sigma(\dot{\gamma})$ and $N_1(\dot{\gamma})$ to shear rate $\dot{\gamma}$ as an input variable in steady-state shear flow, and the responses $G'(\omega)$ and $G''(\omega)$ to small-amplitude sinusoidal strain $\gamma^*(i\omega)$ with an angular frequency ω as an input variable in oscillatory shear flow.

(G' and G'') can be regarded as output variables (i.e., responses) of the fluid under test. Both σ and G'' represent the amount of energy dissipated per unit volume of the fluid under test, while both N_1 and G' represent the amount of energy stored per unit volume of the fluid under test. Using this interpretation, log N_1 versus log σ plots and log G' versus log G'' plots can be regarded as being equivalent. Physically speaking, log N_1 versus log σ and log G' versus log G'' plots describe the amount of the energy stored against the amount of the energy dissipated per unit volume of fluid under test. Thus, such plots can be regarded as representing the extent of fluid elasticity, which depends solely on the molecular characteristics and the chemical structure of a fluid. Such an interpretation seems to suggest that the intrinsic nature of fluid elasticity is independent of temperature, as long as the molecular characteristics and the chemical structure of the fluid are not altered by increasing temperature. In this regard, log N_1 versus log σ and log G' versus log G'' plots would be very useful for investigation of the effects of molecular parameters (e.g., molecular weight, molecular weight distribution, the degree of long-chain branching) and the chemical structure of flexible homopolymers on fluid elasticity in the relative or qualitative sense. We will address those subjects in more detail below.

There are many different ways of characterizing fluid elasticity. As described in Chapters 3 and 4, and also at the beginning of this section, the relaxation time(s) of a fluid is an important variable and describes the fluid elasticity in the absolute or quantitative sense. In this regard, it is reasonable to state that the larger the relaxation time of a fluid, the greater its fluid elasticity. As shown in Figures 6.5 and 6.6, the relaxation time of a flexible homopolymer will vary with temperature and also with applied shear rate.

The theoretical interpretation, via Eq. (6.19) or (6.20), that log G' versus log G'' plots are independent of temperature for flexible homopolymers is used in Chapter 8 to determine a critical temperature at which phase transition from an anisotropic phase to the homogeneous phase takes place in block copolymers, and in Chapter 9 to determine a critical temperature at which phase transition from an anisotropic phase to the homogeneous phase takes place in thermotropic liquid-crystalline polymers. Friedrich et al. (1996) and Neumann et al. (1998) have referred to the log G' versus log G'' plot as the "Han plot."

Cole and Cole (1941), using the ordinary coordinate system, plotted the imaginary part ε'' of the complex dielectric constant against the real part ε' for a number of low-molecular-weight polar substances at various temperatures and obtained circular arcs with radii that vary with temperature. Figure 6.16 gives Cole–Cole plots, prepared with the same experimental data as that used to prepare Figure 6.14. It is seen in Figure 6.16 that the Cole–Cole plots are temperature dependent! Also, it should be noted that the Cole–Cole plot is strictly an empirical representation of data, and has *no* theoretical basis.

6.2.4 Effects of Molecular Weight and Molecular Weight Distribution on the Rheological Behavior of Linear Flexible Homopolymers

The zero-shear viscosity (η_0) of linear flexible homopolymers is perhaps the simplest, and yet a very important, fundamental rheological property. In Chapter 4, it was shown, based on molecular theory, that the η_0 of a linear flexible homopolymer is proportional

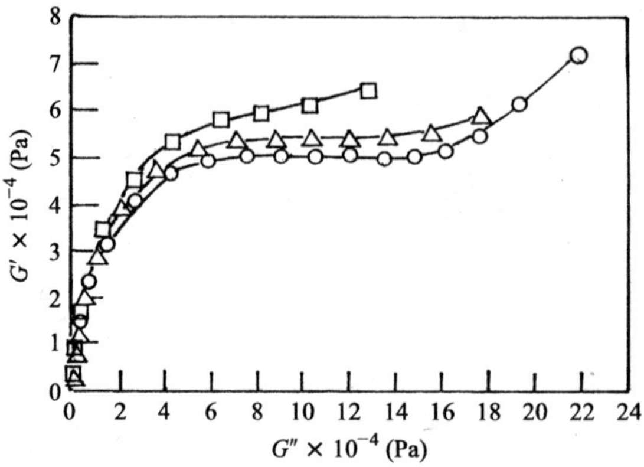

Figure 6.16 Cole–Cole plots for a nearly monodisperse polystyrene ($M_w = 1.95 \times 10^5$; $M_w/M_n = 1.07$) at various temperatures (°C): (○) 180, (△) 200, and (□) 230.

to its molecular weight (M) below a certain critical value (M_c) and proportional to the 3.4-power of M for $M \geq M_c$. Figure 6.17 shows the molecular weight dependence of η_0 for eight nearly monodisperse polystyrenes, which was obtained from the data of Figure 6.11; the molecular weights of the polystyrenes are given in Table 6.3. Applying statistical error analysis to the experimental data given in Figure 6.17, we obtain the following relationships: (1) $\eta_0 \propto M^{1.21 \pm 0.23}$ for $M < M_c$ and (2) $\eta_0 \propto M^{3.37 \pm 0.21}$ for $M \geq M_c$ for the eight polystyrenes. These are in good agreement with the theoretical prediction presented in Chapter 4 and also with the experimental findings for several flexible, linear homopolymers with different chemical structures (see Figure 4.1).

Let us now consider the molecular weight dependence of melt elasticity of linear flexible homopolymers. Figure 6.18 gives log N_1 versus log $\dot{\gamma}$ plots for a series of polyamides (nylon 6) having different molecular weights at 250 °C, and Figure 6.19 gives log $\Psi_{1,0}$ versus log M_n plots for the nylon 6 samples, where $\Psi_{1,0}$ is the zero-shear first normal stress difference coefficient, defined by $\Psi_{1,0} = \lim_{\dot{\gamma}\to 0} \Psi_1(\dot{\gamma}) = \lim_{\dot{\gamma}\to 0} N_1/\dot{\gamma}^2$, and M_n is the number-average molecular weight. The experimental

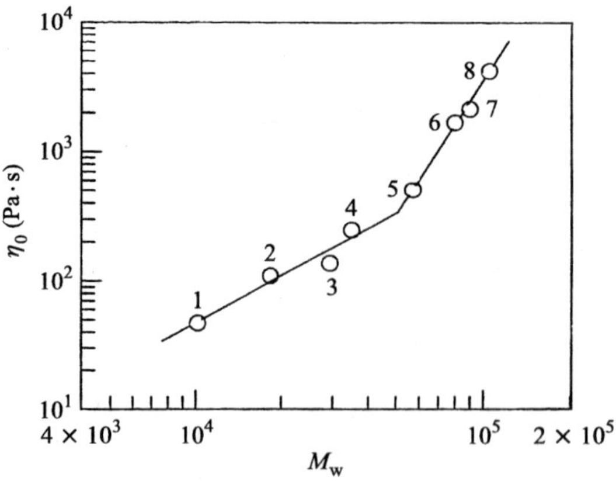

Figure 6.17 Molecular weight dependence of η_0 for polystyrene: (1) PS-1, (2) PS-2, (3) PS-3, (4) PS-4, (5) PS-5, (6) PS-6, (7) PS-7, and (8) PS-8, in which the lower slope is 1.21, the upper slope is 3.37, and $T_r = T_g + 79$ °C. The M_w of the polystyrenes are given in Table 6.3. (Reprinted from Choi and Han, *Macromolecules* 37:215. Copyright © 2004, with permission from the American Chemical Society.)

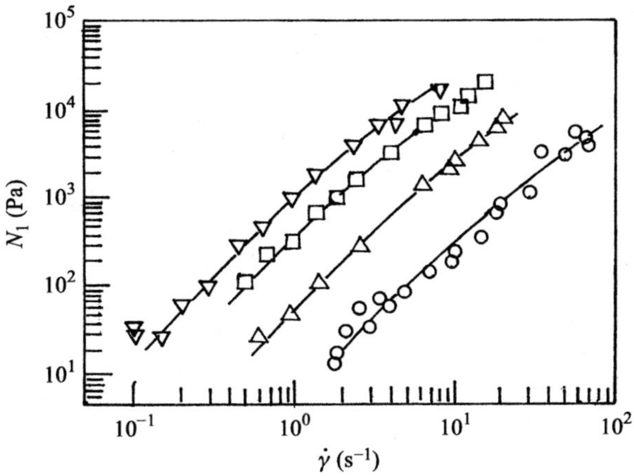

Figure 6.18 Plots of log N_1 versus log $\dot{\gamma}$ at 250 °C for polyamides (nylon 6) having different number-average molecular weights (M_n): (○) 1.8×10^4, (△) 2.6×10^4, (□) 3.3×10^4, and (▽) 4.0×10^4. (Reprinted from Laun, *Rheologica Acta* 18:478. Copyright © 1979, with permission from Springer.)

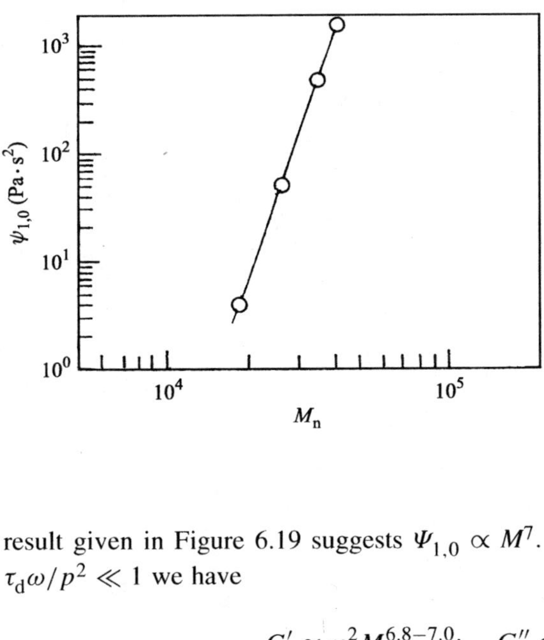

Figure 6.19 Plots of log $\Psi_{1,0}$ versus log M_n for four nylon 6 samples at 250 °C, showing $\Psi_{1,0} \propto M^7$. (Reprinted from Laun, *Rheologica Acta* 18:478. Copyright © 1979, with permission from Springer.)

result given in Figure 6.19 suggests $\Psi_{1,0} \propto M^7$. From Eqs. (4.114) and (4.115), for $\tau_d\omega/p^2 \ll 1$ we have

$$G' \propto \omega^2 M^{6.8-7.0}; \quad G'' \propto \omega M^{3.4-3.5} \tag{6.21}$$

where use was made of $\tau_d \propto M^{3.4-3.5}$. Note that $\Psi_{1,0}/2 = \lim_{\omega \to 0}[G'(\omega)/\omega^2]$ (Coleman and Markovitz 1964). This is consistent with the experimental results shown in Figure 6.19. The above observation shows that fluid elasticity has a much greater dependence on molecular weight than does shear viscosity.

Figure 6.20 gives plots of log G' versus log ω and log G'' versus log ω at 50 °C for three nearly monodisperse polybutadienes with molecular weights ranging from 4.07×10^4 to 4.35×10^5.[8] The molecular parameters for the polybutadienes are given in Table 6.5. Using $M_e = 1.7 \times 10^3$ for polybutadiene, we find that the M/M_e ratio ranges from 24 to 256 for the three polybutadienes. Figure 6.21 gives plots of log G'

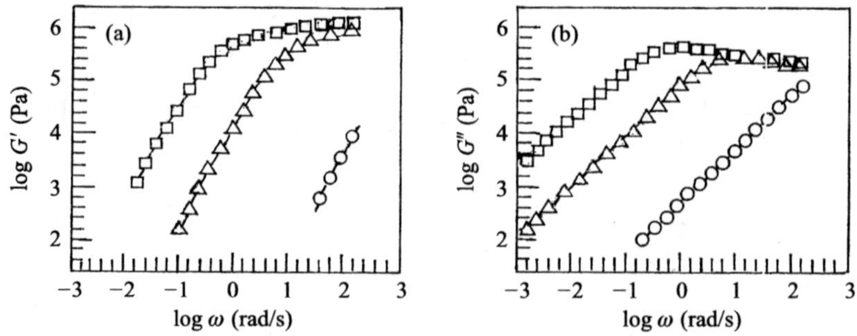

Figure 6.20 Plots of (a) log G' versus log ω and (b) log G'' versus log ω at 50 °C for three nearly monodisperse polybutadienes having different molecular weights: (○) 41L, (△) 174L, and (□) 435L. See Table 6.5 for information on the molecular weight of these polymers.

Table 6.5 Molecular weights of anionically synthesized polybutadienes

Sample Code	M[a]	M_w/M_n[b]	M[c]
41L	41×10^3	1.04	39×10^3
174L	174×10^3	1.04	181×10^3
435L	435×10^3	1.03	450×10^3

[a]Calculated from intrinsic viscosity in tetrahydrofuran. [b]Determined with gel permeation chromatography. [c]Determined with light scattering.

Reprinted from Struglinski and Graessley, *Macromolecules* 18:2630. Copyright © 1985, with permission from the American Chemical Society.

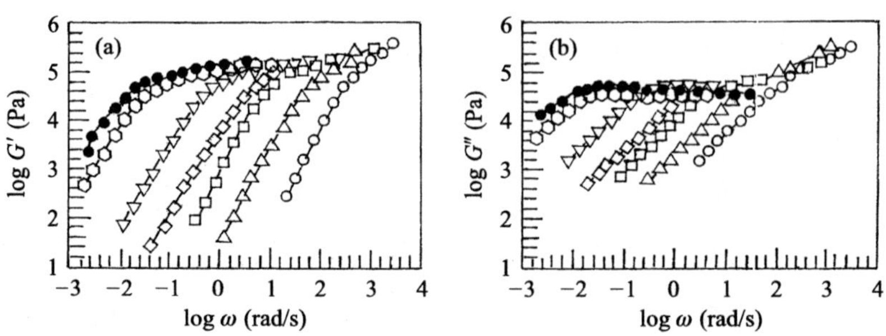

Figure 6.21 Plots of (a) log G' versus log ω and (b) log G'' versus log ω at 167 °C for seven nearly monodisperse polystyrenes having different molecular weights: (○) L23, (△) L39, (□) L72, (⬡) L89, (▽) L172, (⬡) L315, and (●) L427. See Table 6.6 for information on the molecular weight of these polymers.

versus log ω and log G'' versus log ω for seven nearly monodisperse polystyrenes with molecular weight ranging from 2.4×10^4 to 4.27×10^5.[9] The molecular parameters for the polystyrenes are given in Table 6.6. Using $M_e = 1.8 \times 10^4$ for polystyrene, we find that the M/M_e ratio ranges from 1.3 to 23.7 for the seven polystyrenes. The following observations can be made from Figures 6.20 and 6.21: (1) in the terminal

Table 6.6 Molecular weights of anionically synthesized polystyrenes

Sample Code	M_w	M_n	M_w/M_n	M_w/M_e
L23	23.4×10^3	21.8×10^3	1.08	1.3
L39	38.9×10^3	36.3×10^3	1.07	2.2
L72	72.4×10^3	68.3×10^3	1.06	4.0
L89	88.5×10^3	82.7×10^3	1.07	4.9
L172	172×10^3	161×10^3	1.07	9.6
L315	315×10^3	294×10^3	1.07	17.5
L427	427×10^3	407×10^3	1.07	23.7

region, G' shifts progressively to lower frequencies with increasing molecular weight and the plateau zone becomes broader and flatter, while all curves converge in the transition zone, and (2) G'' passes through a minimum, the extent of which become greater with increasing molecular weight. It is of interest to observe in Figure 6.21 that samples L23 and L39, the molecular weights of which are only slightly greater than M_e, exhibit neither a plateau region in G' nor a minimum in G''. One can clearly observe the effect of molecular weight on G' and G'' in Figures 6.20 and 6.21.

However, using the log G' versus log G'' plots given in Figure 6.22 we observe that the effect of molecular weight completely disappears for the polybutadienes, whereas in Figure 6.23 the effect of molecular weight persists for the polystyrenes until the M/M_e ratio becomes about 5 and larger. We can explain, with the aid of Eq. (6.19), the reason why the effect of molecular weight is not seen in Figure 6.22 for the polybutadienes. The fact that the effect of molecular weight is seen in Figure 6.23 for the polystyrenes, until the M/M_e ratio becomes approximately equal to 5 and greater, suggests that entanglement effects will not become dominant until the molecular weight

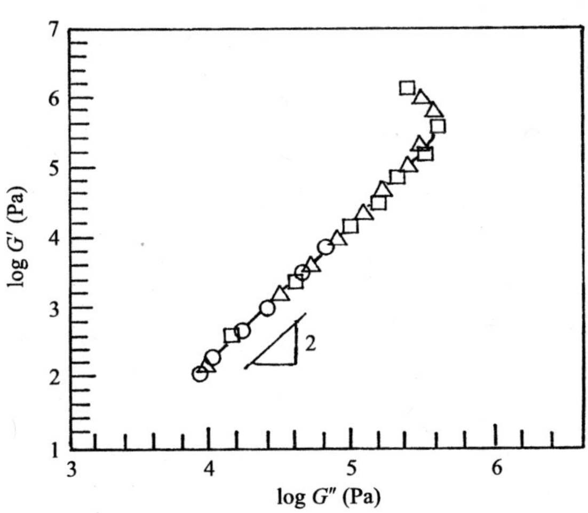

Figure 6.22 Log G' versus log G'' plots at 50 °C for three nearly monodisperse polybutadienes having different molecular weights: (○) 41L, (△) 174L, and (□) 435L. See Table 6.5 for the information on the molecular weights of these polymers.

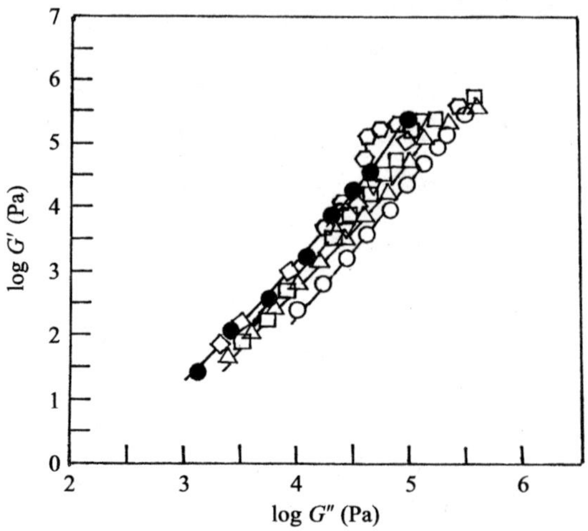

Figure 6.23 Log G' versus log G'' plots at 167 °C for seven nearly monodisperse polystyrenes having different molecular weights: (○) L23, (△) L29, (□) L72, (◇) L89, (▽) L172, (○) L315, and (●) L427. See Table 6.6 for the information on the molecular weights of these polymers.

of polystyrene becomes several times greater than the value of M_e. It should be remembered that the M/M_e ratio for the polybutadienes employed in Figure 6.22 is larger than 24, which is believed to be sufficiently large to have the entanglement effect dominant in chain dynamics. The conclusion drawn above, which is based on the log G' versus log G'' plot, with regard to a critical molecular weight for dominant entanglement effects in linear, flexible homopolymers is in agreement with the view expressed by Graessley and Struglinski (1986), who, using the measurements of steady-state recoverable compliance J_e^o, suggested that the critical molecular weight M_c' for dominant entanglement effect in linear flexible homopolymers would be 7–8 times the value of M_e.

Commercial homopolymers are invariably polydisperse. The simplest case of polydisperse, linear flexible homopolymers is a binary mixture of monodisperse homopolymers with identical chemical structures. As can be seen in Figure 6.24, the

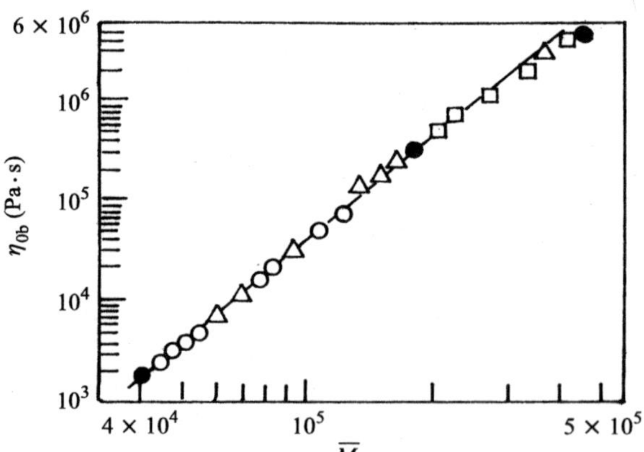

Figure 6.24 Plots of log η_{0b} versus log $\overline{M}_{w,b}$ at 25 °C for binary blends of nearly monodisperse polybutadienes: (●) pure components (41L, 174L and 435L), (○) binary blends of 41L and 174L, (△) binary blends of 41L and 435L, and (□) binary blends of 174L and 435L. See Table 6.5 for the molecular weights of the constituent components. (Reprinted from Struglinski and Graessley, *Macromolecules* 18:2630. Copyright © 1985, with permission from the American Chemical Society.)

zero-shear viscosity of such a binary blend (η_{0b}) can be calculated by the average molecular weight $\overline{M}_{w,b}$ defined by

$$\overline{M}_{w,b} = \phi_1 M_1 + \phi_2 M_2 \tag{6.22}$$

where M_1 and M_2 are the molecular weights of components 1 and 2, respectively, and ϕ_1 and ϕ_2 are the volume fractions (or weight fractions, if their densities are identical) of components 1 and 2, respectively. When dealing with binary blends consisting of dissimilar chemical structures, which will be discussed in the next chapter, the use of Eq. (6.22) is not justified. As can be seen in Figure 6.24, when the zero-shear viscosities of the constituent components follow a 3.4-power law, $\eta_{0i} \propto M_i^{3.4}$ ($i = 1, 2$), the η_{0b} for binary blends can be calculated by a 3.4-power blending law:

$$\eta_{0b} = \left[\phi_1 \eta_{01}{}^{1/3.4} + \phi_2 \eta_{02}{}^{1/3.4}\right]^{3.4} \tag{6.23}$$

There are other types of empirical blending laws suggested in the literature (Akovali 1967; Bogue et al. 1970; Friedman and Porter 1975; Graessley 1971; Graessley and Struglinski 1986; Han and Kim 1989a, 1989b; Kurata et al. 1974; Masuda et al. 1970; Mills 1975; Mills and Nevin 1971; Montfort et al. 1978, 1979, 1984, 1986a, 1986b; Ninomiya 1959, 1962; Ninomiya and Ferry 1963; Ninomiya et al. 1963; Onogi et al. 1970; Prest 1973; Prest and Porter 1973; Watanabe and Kotaka 1984; Watanabe et al. 1985).

Figure 6.25 gives plots of log η_{0b} versus blend composition for binary blends of nearly monodisperse polybutadiene, 41L/435L and 174L/435L blends, in which ϕ_L denotes the volume fraction of 435L in the respective blends and the solid lines were calculated from Eq. (6.23). It can be seen from Figure 6.25 that the

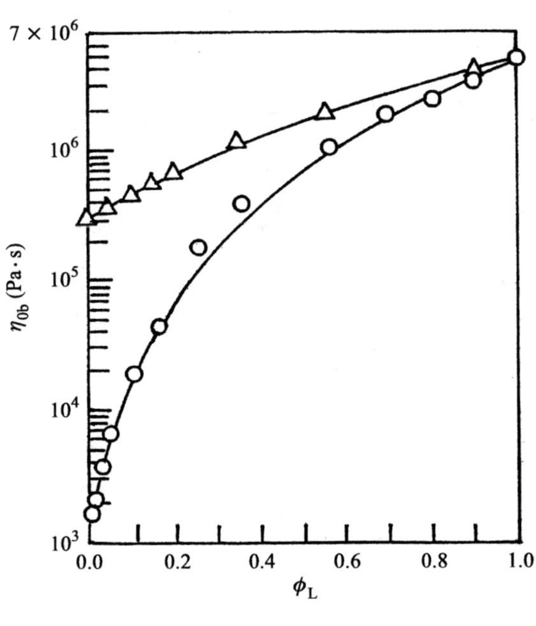

Figure 6.25 Plots of log η_{0b} versus ϕ_L for binary blends of polybutadiene at 25 °C: (○) binary blends of 41L and 435L and (△) binary blends of 174L and 435L. See Table 6.5 for the molecular weights of the constituent components. The solid lines were calculated from Eq. (6.23). (Reprinted from Struglinski and Graessley, *Macromolecules* 18:2630. Copyright © 1985, with permission from the American Chemical Society.)

empirical 3.4-power blending law, Eq. (6.23), describes the experimental results reasonably well.

Assuming that the log-normal distribution function represents the molecular weight distribution of a polydisperse flexible homopolymer, and that its stress relaxation modulus of the homopolymer is expressed by the 3.4-power law in terms of the constituent fractions, it can be shown that in the terminal region (where $\tau_d\omega \ll 1$), the following expression (Han and Kim 1989a) holds:

$$\log G_b' = 2\log G_b'' + \log(6/5G_N^o) + 3.4\log(M_z/M_w) \tag{6.24}$$

where G_b' and G_b'' are the dynamic storage and loss moduli, respectively, of the polydisperse homopolymer, and M_z and M_w are z-average and weight-average molecular weights, respectively. A comparison of Eq. (6.24) with Eq. (6.18) indicates that the values of G_b' for a polydisperse homopolymer is shifted upward by a factor of 3.4 log (M_z/M_w) above the values of G' for monodisperse flexible homopolymers; thus the shift will be greater with an increase in polydispersity index, M_z/M_w.

Graessley and Struglinski (1986) developed a theory, that predicts the linear dynamic viscoelastic properties of binary blends of monodisperse flexible homopolymers, based on the tube model, by incorporating constraint release and path fluctuations into the reptation motion (see Chapter 4). They assumed that the stress relaxation modulus of a binary blend, $G_b(t)$, may be represented by

$$G_b(t) = G_N^o\left[\phi_1\Psi_1(t)R_1(t) + \phi_2\Psi_2(t)R_2(t)\right] \tag{6.25}$$

where $\Psi_i(t)$ $(i = 1, 2)$ are defined by Eq. (4.108), and $R_i(t)$ $(i = 1, 2)$ are defined by Eq. (4.128), with the waiting time τ_w for a binary blend having the same chemical structure defined by

$$\tau_w = \int_0^\infty \left[\phi_1\Psi_1(t) + \phi_2\Psi_2(t)\right]^z \mathrm{d}t \tag{6.26}$$

where z is a constraint release parameter (see Chapter 4), and ϕ_1 and ϕ_2 are the weight fractions of components 1 and 2, respectively.

Substituting Eq. (6.25) into Eqs. (4.78)–(4.81), we obtain the following expressions for η_{0b} and J_{eb}^o (Graessley and Struglinski 1986):[10]

$$\eta_{0b} = \frac{8G_N^o}{\pi^2}\sum_{\text{odd}\,p}^{\infty}\frac{1}{p^3}\left[\frac{\phi_1\tau_{d1}}{(p^2+r_1)^{1/2}} + \frac{\phi_2\tau_{d2}}{(p^2+r_2)^{1/2}}\right] \tag{6.27}$$

$$J_{eb}^o = \frac{4G_N^o}{\pi^2\eta_{0b}^2}\sum_{\text{odd}\,p}^{\infty}\frac{1}{p^5}\left[\frac{\phi_1\tau_{d1}^2(2p^2+r_1)}{(p^2+r_1)^{3/2}} + \frac{\phi_2\tau_{d2}^2(2p^2+r_2)}{(p^2+r_2)^{3/2}}\right] \tag{6.28}$$

where $r_1 = 2\tau_{d1}/\tau_w$ and $r_2 = 2\tau_{d2}/\tau_w$ with $\tau_{d1} = L_1{}^2/D_c\pi^2$ and $\tau_{d2} = L_2{}^2/D_c\pi^2$ (see Eq. (4.106)), and for $G_b{}'(\omega)$ and $G_b{}''(\omega)$

$$G_b{}'(\omega) = \frac{8G_N^o}{\pi^2}\left[\sum_{\text{odd}\,p}^{\infty}\frac{1}{p^2}\left(\frac{1}{Z_1}\sum_{j=1}^{Z_1}\frac{(\omega\tau_{d1,j})^2}{1+(\omega\tau_{d1,j})^2}\phi_1(Z_1)+\frac{1}{Z_2}\sum_{j=1}^{Z_2}\frac{(\omega\tau_{d2,j})^2}{1+(\omega\tau_{d2,j})^2}\phi_2(Z_2)\right)\right] \tag{6.29}$$

$$G_b{}''(\omega) = \frac{8G_N^o}{\pi^2}\left[\sum_{\text{odd}\,p}^{\infty}\frac{1}{p^2}\left(\frac{1}{Z_1}\sum_{j=1}^{Z_1}\frac{\omega\tau_{d1,j}}{1+(\omega\tau_{d1,j})^2}\phi_1(Z_1)+\frac{1}{Z_2}\sum_{j=1}^{Z_2}\frac{\omega\tau_{d2,j}}{1+(\omega\tau_{d2,j})^2}\phi_2(Z_2)\right)\right] \tag{6.30}$$

We can extend this approach to polydisperse flexible homopolymers as the most general situation, for which $G_b{}'(\omega)$ and $G_b{}''(\omega)$ may be written as (Han and Kim 1989a)

$$G_b{}'(\omega) = \frac{8G_N^o}{\pi^2}\sum_{i=1}^{n}\left[\sum_{\text{odd}\,p}^{\infty}\frac{1}{p^2}\frac{1}{Z_i}\sum_{j=1}^{Z_i}\frac{(\omega\tau_{di,j})^2}{1+(\omega\tau_{di,j})^2}\right]\phi_i(Z_i) \tag{6.31}$$

and

$$G_b{}''(\omega) = \frac{8G_N^o}{\pi^2}\sum_{i=1}^{n}\left[\sum_{\text{odd}\,p}^{\infty}\frac{1}{p^2}\frac{1}{Z_i}\sum_{j=1}^{Z_i}\frac{\omega\tau_{di,j}}{1+(\omega\tau_{di,j})^2}\right]\phi_i(Z_i) \tag{6.32}$$

where n is the number of the fractions (to be chosen in the computation) in a polydisperse polymer, Z_i is the number of steps along the primitive path, $\phi_i(Z_i)$ is the fractional weight of the polymer with path steps in the range between Z_i and $Z_i + dZ_i$, and $\tau_{di,j}$ is defined by

$$\tau_{di,j} = \frac{1}{p^2/\tau_{di} + \lambda_{i,j}/2\tau_w} \tag{6.33}$$

with

$$\lambda_{i,j} = 4\sin^2\left[\frac{\pi j}{2(Z_i+1)}\right] \tag{6.34}$$

In the terminal region, where $\tau_{di,j} \ll 1$, Eqs. (6.31) and (6.32) for large values of Z_i reduce to

$$\log G_{\mathrm{b}}' = 2\log G_{\mathrm{b}}'' + \frac{\pi^2}{16}\frac{1}{G_N^{\mathrm{o}}}\left\{\frac{\sum_{i=1}^{n}\phi_i \sum_{\mathrm{odd}\,p}^{\infty}\frac{\tau_{di}{}^2}{p^5}\frac{2p^2+r_i}{(p^2+r_i)^{3/2}}}{\left[\sum_{i=1}^{n}\phi_i \sum_{\mathrm{odd}\,p}^{\infty}\frac{1}{p^3}\frac{\tau_{di}}{(p^2+r_i)^{1/2}}\right]^2}\right\} \tag{6.35}$$

where $r_i = 2\tau_{di}/\tau_{\mathrm{w}}$ $(i = 1, 2, \ldots, n)$. Note that for monodisperse linear flexible homopolymers, Eq. (6.35) reduces to

$$\log G_{\mathrm{b}}' = 2\log G_{\mathrm{b}}'' + \frac{\pi^2}{16}\frac{1}{G_N^{\mathrm{o}}}\left\{\frac{\sum_{\mathrm{odd}\,p}^{\infty}\frac{1}{p^5}\frac{2p^2+r}{(p^2+r)^{3/2}}}{\left[\sum_{\mathrm{odd}\,p}^{\infty}\frac{1}{p^3}\frac{1}{(p^2+r)^{1/2}}\right]^2}\right\} \tag{6.36}$$

where $r = 2\tau_{\mathrm{d}}/\tau_{\mathrm{w}}$. For reptation and constraint release contributions with $z = 3$, Eq. (6.36) becomes

$$\log G_{\mathrm{b}}' = 2\log G_{\mathrm{b}}'' + \log(2.15/G_N^{\mathrm{o}}) \tag{6.37}$$

and for pure reptation ($z = 0$), Eq. (6.36) reduces to Eq. (6.18).

A comparison of Eq. (6.35) with (6.36) indicates that in the terminal region, the $\log G_{\mathrm{b}}'$ versus $\log G_{\mathrm{b}}''$ plot for polydisperse homopolymers has a slope of 2, with the values of G_{b}' shifted upward above the values of G' in the $\log G'$ versus $\log G''$ plot for monodisperse homopolymers, the extent of the shift being greater with an increase in polydispersity. In the linear region, where $0 \ll \tau_{\mathrm{d}}\omega \leq 1$ holds and therefore the denominators in Eqs. (6.31) and (6.32) are no longer negligible, we have (Han and Kim 1989b)

$$\log G_{\mathrm{b}}' = x\log G_{\mathrm{b}}'' + (1-x)\log(G_N^{\mathrm{o}}/\pi^2) \tag{6.38}$$

where $1 < x < 2$, indicating that the slope of $\log G_{\mathrm{b}}'$ versus $\log G_{\mathrm{b}}''$ plot for a polydisperse linear flexible homopolymer is less than 2. Figure 6.26 summarizes schematically the effect of polydispersity on the $\log G'$ versus $\log G''$ plot for linear flexible homopolymers.

Figure 6.27 gives $\log G_{\mathrm{b}}'$ versus $\log \omega$ and $\log G_{\mathrm{b}}''$ versus $\log \omega$ plots for binary blends consisting of the nearly monodisperse polybutadienes 41L and 435L at 50 °C.[11] It is seen in Figure 6.27 that values of G_{b}' and G_{b}'' for the binary blends lie between those of the constituent components. However, in the $\log G_{\mathrm{b}}'$ versus $\log G_{\mathrm{b}}''$ plots given in Figure 6.28 for various compositions of the binary blend, values of G_{b}' for all blend samples lie above those of the constituent components. Notice in Figure 6.28 that the $\log G_{\mathrm{b}}'$ versus $\log G_{\mathrm{b}}''$ plots for the constituent components lie on a single correlation,

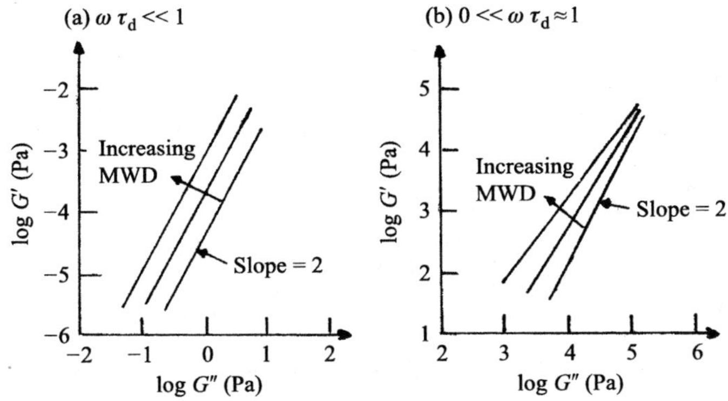

Figure 6.26 Schematic describing the dependence of log G' versus log G'' plots on molecular weight distribution (MWD) (a) in the terminal region and (b) in the linear region.

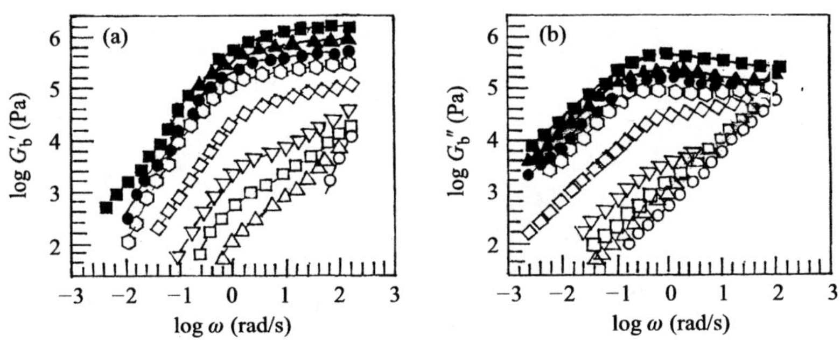

Figure 6.27 Plots of (a) log G_b' versus log ω and (b) log G_b'' versus log ω for binary blends of nearly monodisperse polybutadienes, 41L and 435L, at 50 °C for different volume fractions of 435L (ϕ): (○) 0 (neat 41L), (△) 0.025, (□) 0.05, (▽) 0.01, (◇) 0.36, (⬡) 0.56, (●) 0.7, (▲) 0.9, and (■) 1 (neat 435L). See Table 6.5 for the information on the molecular weights of the constituent components, 41L and 435L.

as expected from Eq. (6.19). This is because the constituent components are entangled homopolymers (see Table 6.5). It is interesting to observe in Figure 6.28 that at very low values of G_b'' (i.e., for $G_b'' < 10^3$ Pa) the value of G_b' for $\phi = 0.05$ is largest among the several blends and then decreases, approaching the value of G_b' of the constituent components as ϕ approaches 1. The dependence of G_b' on blend composition becomes more complicated as G_b'' increases. Figure 6.29 gives theoretical predictions, using Eqs. (6.31) and (6.32) for $z = 3$, of the dependence of log G_b' versus log G_b'' plot on blend composition. In obtaining Figure 6.29, we first calculated the value of τ_w from Eq. (6.26) to determine the value of $\tau_{di,j}$ from Eq. (6.33). A comparison of Figure 6.29 with Figure 6.28 reveals that the agreement between the two is quite good.

Figure 6.30 gives plots of log G' versus log $a_T\omega$ and log G'' versus log $a_T\omega$ for two different grades of polystyrene: L15, having $M_w = 2.29 \times 10^5$ and $M_n = 2.03 \times 10^5$,

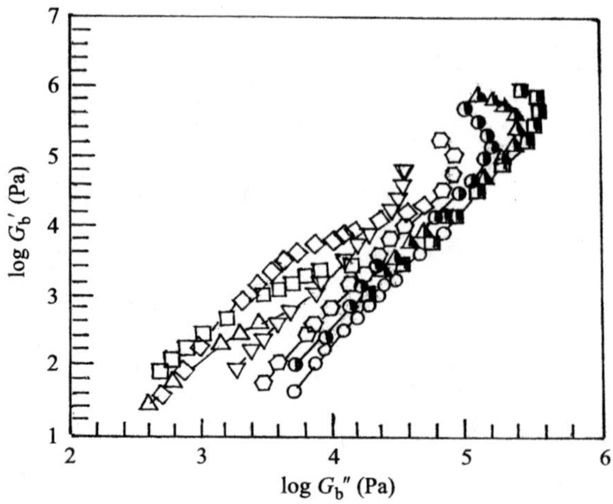

Figure 6.28 Log G_b' versus log G_b'' plots for binary blends of nearly monodisperse polybutadienes, 41L and 435L, at 50 °C for different volume fractions of 435L (ϕ): (○) 0 (neat 41L), (△) 0.025, (□) 0.05, (◇) 0.1, (▽) 0.36, (⬡) 0.56, (◑) 0.7, (▲) 0.9, and (◨) 1 (neat 435L). See Table 6.5 for the information on the molecular weights for the constituent components, 41L and 435L.

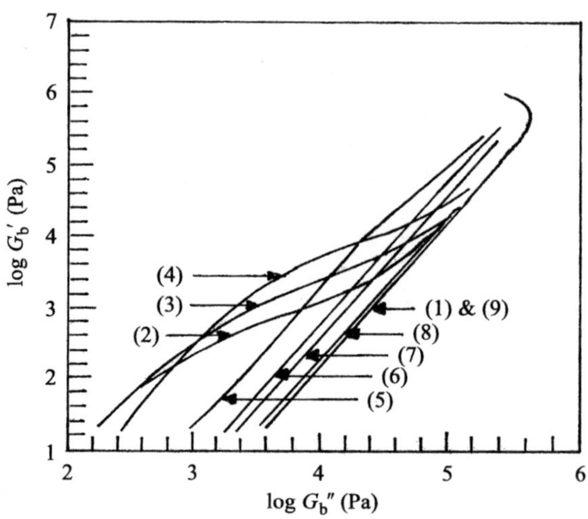

Figure 6.29 Theoretically predicted log G_b' versus log G_b'' plots, using Eqs. (6.31) and (6.32), for binary blends of nearly monodisperse polybutadienes, 41L and 435L, at 50 °C for different volume fractions of 435L (ϕ): (1) for $\phi = 0$ (neat 41L), (2) $\phi = 0.025$, (3) $\phi = 0.05$, (4) $\phi = 0.1$, (5) $\phi = 0.36$, (6) $\phi = 0.5$, (7) $\phi = 0.7$, and (8) $\phi = 0.9$, and (9) $\phi = 1$ (neat 435L).

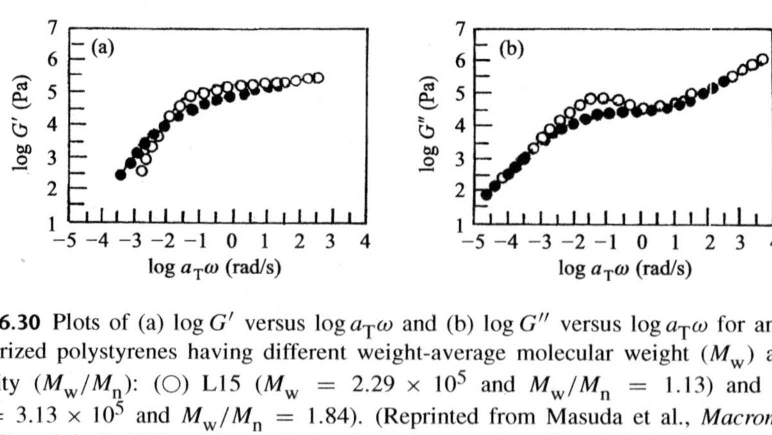

Figure 6.30 Plots of (a) log G' versus log $a_T\omega$ and (b) log G'' versus log $a_T\omega$ for anionically polymerized polystyrenes having different weight-average molecular weight (M_w) and polydispersity (M_w/M_n): (○) L15 ($M_w = 2.29 \times 10^5$ and $M_w/M_n = 1.13$) and (●) PS7 ($M_w = 3.13 \times 10^5$ and $M_w/M_n = 1.84$). (Reprinted from Masuda et al., *Macromolecules* 3:116. Copyright © 1970, with permission from the American Chemical Society.)

and PS7, having $M_w = 3.13 \times 10^5$ and $M_n = 1.7 \times 10^5$. Note that L15 has a polydispersity index (M_w/M_n) of 1.13 and PS7 has $M_w/M_n = 1.84$. From Figure 6.30 it is not possible to judge which of the two polymers is most elastic. However, the log G' versus log G'' plots given in Figure 6.31 show clearly that PS7, having a broader molecular weight distribution, is more elastic than L15, because at a fixed value of G'', the value of G' for PS7 is greater than that for L15. Notice in Figure 6.31 that the slope of the log G' versus log G'' plot for PS7 is less than that for L15, which is consistent with the schematic presented in Figure 6.26.

Let us observe further how polydispersity affects the steady-state shear flow properties of linear flexible homopolymers. Figure 6.32 gives log η versus log $\dot{\gamma}$ plots for two commercial polystyrenes (Styron 678 and Styron 685, Dow Chemical Company) at 200 and 220 °C. Note that Styron 678 has a lower molecular weight but broader

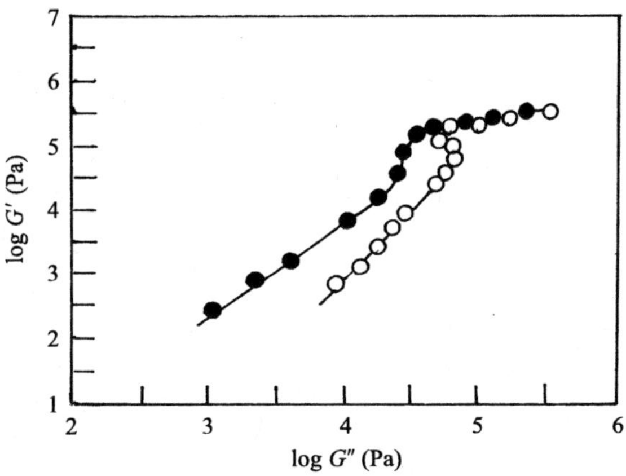

Figure 6.31 Log G' versus log G'' plots for anionically polymerized polystyrenes having different weight-average molecular weight (M_w) and polydispersity (M_w/M_n): (○) L15 ($M_w = 2.29 \times 10^5$ and $M_w/M_n = 1.13$) and (●) PS7 ($M_w = 3.13 \times 10^5$ and $M_w/M_n = 1.84$).

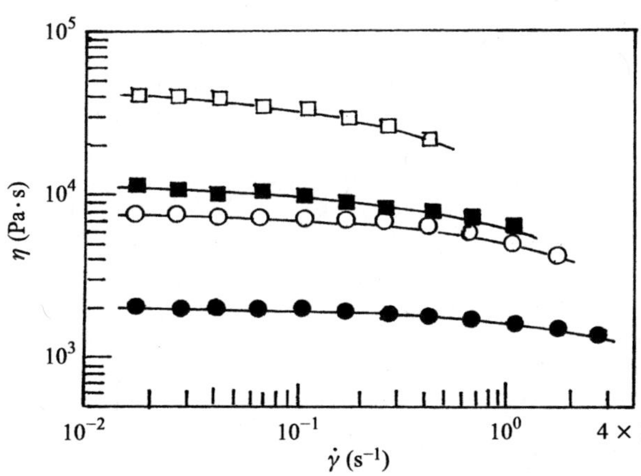

Figure 6.32 Plots of log η versus log $\dot{\gamma}$ for two commercial polystyrenes having different weight-average molecular weight (M_w) and polydispersity (M_w/M_n): (a) Styron 678 ($M_w = 2.29 \times 10^5$, and $M_w/M_n = 4.31$) at (○) 200 °C and at (●) 220 °C, and (b) Styron 685 ($M_w = 2.85 \times 10^5$; $M_w/M_n = 2.85$) at (□) 200 °C and at (■) 220 °C.

molecular weight distribution than Styron 685. It can be seen from Figure 6.32 that Styron 685 is more viscous than Styron 678, suggesting that it is molecular weight, rather than polydispersity, that plays an important role in controlling the η of the polystyrenes. Figure 6.33 gives log N_1 versus log $\dot{\gamma}$ plots for Styron 678 and Styron 685. One may be tempted to conclude from Figure 6.33 that Styron 685 is more elastic than Styron 678. However, the log N_1 versus log σ plots given in Figure 6.34 lead us to conclude that Styron 678 is more elastic than Styron 685, suggesting that the linear flexible homopolymer with the higher polydispersity index is more elastic than the homopolymer with the lower polydispersity index. Note that, according to Eq. (6.17), log N_1 versus log σ plots are independent of molecular weight for entangled, linear flexible homopolymers, and that both Styron 678 and Styron 685 are entangled linear flexible homopolymers. Therefore, the observed difference in the log N_1 versus log σ plot between Styron 678 and Styron 685, given in Figure 6.34, has little to do with the difference in molecular weight.

In investigating the effect of polydispersity on the linear viscoelastic properties of entangled homopolymers, Wasserman and Graessley (1992) used a quadratic blending

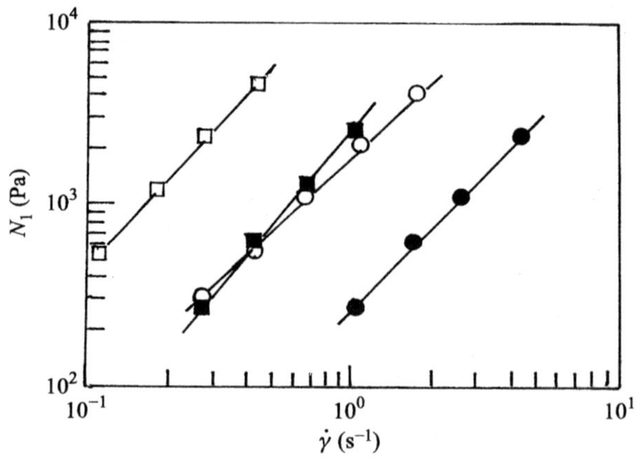

Figure 6.33 Plots of log N_1 versus log $\dot{\gamma}$ for two commercial polystyrenes having different weight-average molecular weight (M_w) and polydispersity (M_w/M_n): (a) Styron (M_w = 2.29×10^5, $M_w/M_n = 4.31$) at 200 °C (○) and at 220 °C (●), and (b) Styron 685 (M_w = 2.85×10^5, $M_w/M_n = 2.85$) at 200 °C (□) and at 220 °C (■).

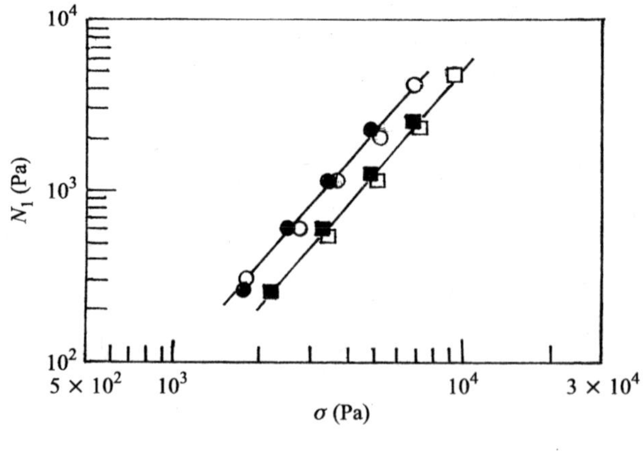

Figure 6.34 Plots of log N_1 versus log σ for two commercial polystyrenes having different weight-average molecular weight (M_w) and polydispersity (M_w/M_n): (a) Styron 678 ($M_w = 2.29 \times 10^5$, $M_w/M_n = 4.31$) at 200 °C (○) and at 220 °C (●), and (b) Styron 685 ($M_w = 2.85 \times 10^5$, M_w/M_n = 2.85) at 200 °C (□) and at 220 °C (■).

law for the stress relaxation modulus $G_b(t)$:

$$G_b(t) = G_N^o \left\{ \sum_{i=1}^{\infty} w_i \left[\Psi_i(t) \right]^{1/2} \right\}^2 \tag{6.39}$$

where w_i is the weight fraction of component i and $\Psi_i(t)$ is defined by Eq. (4.108). In doing so, they modified $\Psi_i(t)$ using the approach of des Cloizeaux (1990), which neglects the contribution of constraint release, and calculated the values of $G_b'(\omega)$ and $G_b''(\omega)$. They found that the calculated values agreed reasonably well with the experimental results for a polydisperse homopolymer. It should be mentioned that due to the nonlinear nature of Eq. (6.39), a numerical method must be used to calculate the values of η_{0b}, $G_b'(\omega)$, and $G_b''(\omega)$, while the theory presented above has analytical expressions, Eqs. (6.27)–(6.30).

6.3 Rheology of Flexible Homopolymers with Long-Chain Branching

It has long been recognized that the presence of long-chain branching greatly influences the rheological behavior of flexible homopolymers. This subject was dealt with in the 1950s through 1980s by a number of investigators (Bueche 1964; Buesse and Longworth 1962; Graessley and Prentice 1968; Guillet et al. 1965; Kraus and Gruver 1965; Long et al. 1964; Peticolas and Watkins 1957; Tung 1960; Wyman et al. 1965) and in the 1980s by others (Doi and Kuzuu 1980; Evans 1982; Graessley and Edwards 1981; Romanini et al. 1980). Bueche (1964) developed a molecular theory that relates the steady-state shear viscosity (η_{br}) of a long-chain branched flexible homopolymer to that (η_ℓ) of the linear flexible homopolymer by

$$\eta_{br}/\eta_\ell = gE(g) \tag{6.40}$$

where

$$E(g) = \begin{cases} g^{5/2} & \text{when the chains are entangled} \\ 1 & \text{when the chains are not entangled} \end{cases} \tag{6.41}$$

in which g is the ratio of the mean square radii of long-chain branched and linear homopolymers, and $E(g)$ is a factor which accounts for the change in interaction or entanglements between polymer molecules and depends on the amount of long-chain branching. Peticolas and Watkins (1957) showed that long-chain branched polyethylenes have much lower η_0 than linear polyethylenes of the same M_w. Kraus and Gruver (1965) reported that the η_{br} of narrow distribution, star-branched polybutadiene is less than that of linear polybutadiene below a certain M_w, but that the reverse is true for high-molecular-weight polybutadienes. It has been reported that the melt viscosity of flexible homopolymers having randomly distributed long-chain branching increases with the extent of long-chain branching (Chartoff and Maxwell 1970; Long et al. 1964;

Mendelson et al. 1970). The Bueche theory predicts that the viscosity of branched flexible homopolymers is lower than that of linear homopolymers of the same M_w. The reduction of melt viscosity in the presence of long-chain branching may be attributed to (1) a reduction in relaxation times and (2) a reduction in chain entanglements or other interactions between polymer molecules in the bulk. It appears that the reduction in relaxation times may be related to g given in Eq. (6.40), and the reduction in chain entanglements or other interactions between polymer molecules may be related to $E(g)$ given in Eq. (6.41).

The early research activities in the 1950s and 1960s on the rheology of flexible homopolymers with long-chain branching were apparently motivated in large part by the availability of commercial LDPEs, which were (and are still) produced from ethylene monomer using a tubular reactor or an autoclave under very high pressure (14–21 MPa). Until the mid-1970s, the LDPE produced from such a process almost monopolized the markets for very thin plastic films for use in, for instance, packaging. For illustration, let us compare LDPE with high-density polyethylene (HDPE). The unique features of LDPE include high extrusion rate and high melt strength during stretching operations as compared with HDPE; for instance, in the tubular film blowing process (see Chapter 7 of Volume 2). The high extrusion rate of LDPE arises from strong shear-thinning behavior (giving rise to low shear viscosity) compared with HDPE of comparable molecular weight. The high melt strength of LDPE arises from the presence of long-chain branching distributed randomly along the linear polyethylene chains, while the chains of HDPE have no long-chain branching. However, the production cost of LDPE is much higher than that of HDPE, essentially due to the higher energy cost of having to polymerize under such high pressures.

In the mid-1970s, a big breakthrough came to the LDPE manufacturing industry in the form of a new gas-phase fluidized-bed process using the Ziegler–Natta catalyst, referred to as the "Unipol process" (Jenkins et al. 1985, 1986), which was commercialized by the then Union Carbide Corporation. This was a big breakthrough because the Unipol process does not require high pressure and thus the production cost was reduced considerably compared with the high-pressure process. The Unipol process is known to use co-monomers (e.g., butane and hexane) and produce polyethylene under a very low pressure. Because the low-density polyethylene produced by the Unipol process was found to have virtually no long-chain branching, a new terminology emerged to represent this kind of low-density polyethylene, namely, linear low-density polyethylene (LLDPE). It has been reported that the rheological properties of LLDPE are quite different from those of LDPE (Han and Kwack 1983; Kanai and White 1984; Kwack and Han 1983).

For illustration, Figure 6.35 shows the differences in shear-rate dependence of melt viscosity between an LDPE and an LLDPE. It can be seen in Figure 6.35 that the LDPE has a stronger shear-thinning behavior than the LLDPE. This difference has very important practical implications in that the extrusion rate of LLDPE will be lower, owing to higher melt viscosity, than that of LDPE. Figure 6.36 gives plots of log N_1 versus log σ for the same LDPE and LLDPE, indicating that the LDPE is more elastic than the LLDPE. This observation suggests that the presence of long-chain branching in LDPE contributes significantly to its melt elasticity.

Another very important rheological property that is significant for the processing of LDPEs (for example, using the tubular film blowing process) is elongational viscosity

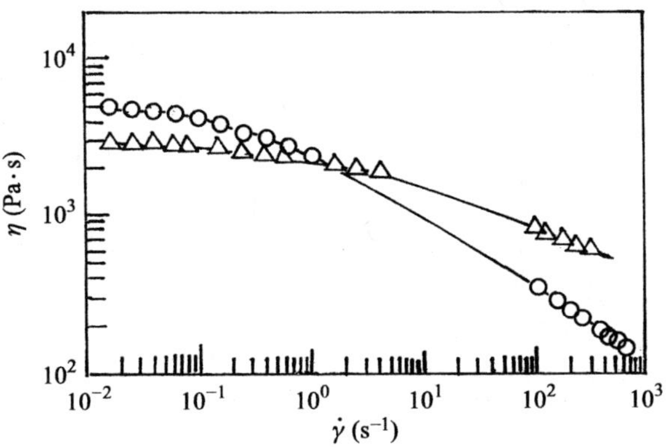

Figure 6.35 Plots of log η versus log $\dot{\gamma}$ for an LDPE (○) and an LLDPE (△) at 200 °C. (Reprinted from Kwack and Han, *Journal of Applied Polymer Science* 28:3419. Copyright © 1983, with permission from John Wiley & Sons.)

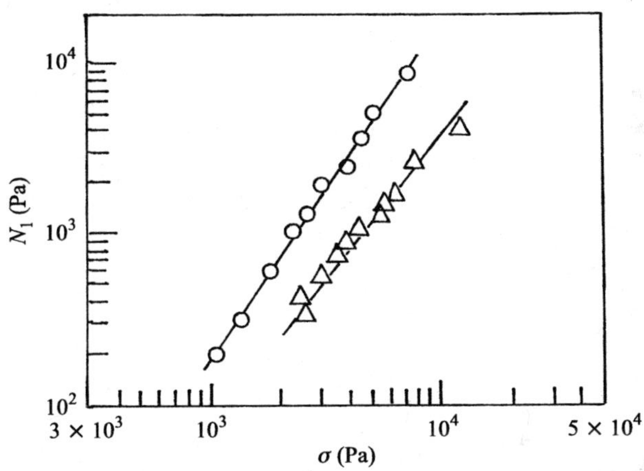

Figure 6.36 Plots of N_1 versus log σ for an LDPE (○) and an LLDPE (△) at 200 °C. (Reprinted from Kwack and Han, *Journal of Applied Polymer Science* 28:3419. Copyright © 1983, with permission from John Wiley & Sons.)

which determines the drawability (or stretchability) of a thin film during blowing. Figure 6.37 shows the transient uniaxial elongational viscosity η_E for an LDPE and an LLDPE at 200 °C, in which we observe that the LDPE exhibits strong strain-hardening compared with the LLDPE. The strong strain-hardening of LDPE limits its drawability and is related to bubble stability in tubular film blowing. In the 1970s, Han and coworkers (Han and Park 1975; Han and Shetty 1977) conducted seminal studies of bubble instability in tubular film blowing. Subsequently, other investigators (Ghaneh-Fard et al. 1996; Kanai and White 1984; Minoshima and White 1986) observed that LDPE had better bubble stability than LLDPE during tubular film blowing. This can be explained by the higher melt strength of LDPE compared with LLDPE, which is manifested by the stronger strain-hardening of LDPE in transient uniaxial elongational flow (see Figure 6.37). The above observations seem to suggest

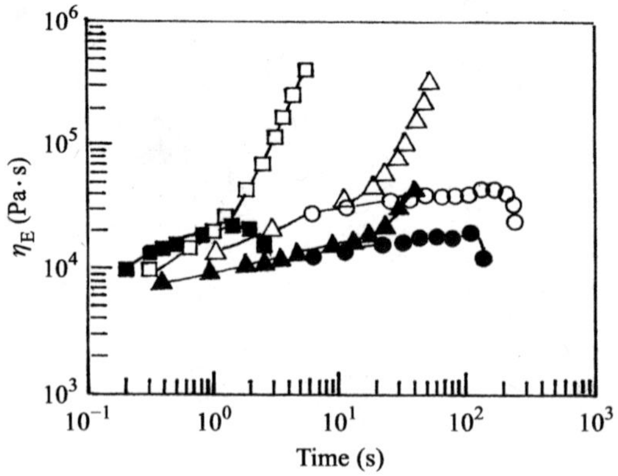

Figure 6.37 Time evolution of uniaxial elongational viscosity (η_E) at 200 °C for an LDPE (open symbols) and an LLDPE (filled symbols) at various elongation rates (s^{-1}): (○, ●) 0.01, (△, ▲) 0.1, and (□, ■) 1.0.

that the presence of a certain degree of long-chain branching in LDPE is necessary in order to have good bubble stability in tubular film blowing. In Chapter 7 of Volume 2 we elaborate on this subject in relation to tubular film blowability.

In the 1990s, another breakthrough in the manufacture of LDPE came with the discovery of metallocene catalysts (Devore et al. 1997; Welborn and Ewen 1994). It has been reported that the metallocene-catalyzed LDPEs have virtually no long-chain branching, like those produced from the Unipol process, and so they are regarded as being LLDPEs. The unique feature of metallocene-catalyzed LLDPE lies in that short chains can be distributed uniformly along the polyethylene backbone chains, and thus much improved mechanical properties of the fabricated films can be obtained. Most importantly, one can control the desired amounts of long-chain branching, as well short-chain branching, in metallocene-catalyzed LLDPEs. These brief historical remarks on the developments of LDPE and LLDPE will give the readers a better perspective on the motivation of continuing investigation of the rheology of branched flexible homopolymers.

Thus, there are many good reasons to investigate the rheological behavior of LDPEs with varying degrees of long-chain branching. For instance, the effects of long-chain branching, when entanglements dominate the terminal zone, are not well understood theoretically. Long-chain branching in a flexible homopolymer would strongly inhibit the reptation mechanism for configurational rearrangements (de Gennes 1975), and hence greatly increases the relaxation time (or the tube disengagement time, within the spirit of the Doi–Edwards tube model presented in Chapter 4) compared with that of a linear flexible homopolymer of comparable molecular weight. At present, our understanding of the subject from a molecular point of view is far from complete.

For almost two decades following the early 1960s there had been relatively limited research activities on the rheology of branched flexible homopolymers. However, in 1988 McLeish (1988) extended the concept of the Doi–Edwards tube model, which had been developed for linear flexible homopolymers (see Chapter 4), to describe the dynamics of branched flexible homopolymers. Since then, during the past several years, other investigators (Blackwell et al. 2000; Bourrigaud et al. 2003, Inkson et al. 1999; McLeish and Larson 1998; McLeish et al. 1999; Shie et al. 2003; Verbeeten et al. 2001) have actively engaged in further development of this theory. Such efforts have resulted in a new theory, referred to as the "pom-pom model" (McLeish and Larson 1998). It is too early to tell how well the pom-pom model describes experimental results, although some experimental evidence (Archer and Varshney 1998) seems to support the predictions of the pom-pom model. For the reason of very limited space available in this chapter, we will not present the pom-pom model.

We present here some experimental studies on the rheological behavior of branched flexible homopolymers. Our presentation follows the same spirit as that presented previously for linear flexible homopolymers; that is, we will use log N_1 versus log σ and log G' versus log G'' plots to discuss the effect of long-chain branching on the rheological behavior of flexible homopolymers.

Figure 6.38 gives log G' versus log G'' plots for three 4-star branched polybutadienes measured over a wide range of angular frequencies and temperatures, and Table 6.7 gives the molecular characteristics of the polymers. It is seen in Figure 6.38 that the effect of temperature is not discernible, and the values of G' become greater as the polydispersity index is increased. A comparison of Table 6.7 with Table 6.5 reveals that the polydispersity index of the star-branched polybutadienes is more sensitive to an increase in M_w than that of the linear polybutadienes. In other words, the introduction of star branching to linear polybutadienes has increased their polydispersity index. It should be remembered that log G' versus log G'' plots are independent of M_w for linear flexible entangled homopolymers (see Eq. (6.18)).

Figure 6.39 gives log G' versus log G'' plots for four 4-arm star-branched polystyrenes measured over a wide range of angular frequencies and temperatures,[12] and Table 6.8 gives a summary of the molecular characteristics of the star-branched polystyrenes.[13] In Figure 6.39 we observe that values of G' increase with increasing M_w of the star-branched polystyrene. Since the log G' versus log G'' plots are independent of M_w for entangled homopolymers, we can conclude that the observed differences in the log G' versus log G'' plots among the four star-branched polystyrenes, given in Figure 6.39, are attributable to the differences in polydispersity index.

Figure 6.40 gives molecular weight distribution curves for three different grades of commercial LDPE, and Table 6.9 gives the molecular characteristics of the LDPEs, showing that the degree of long-chain branching increases with polydispersity index.

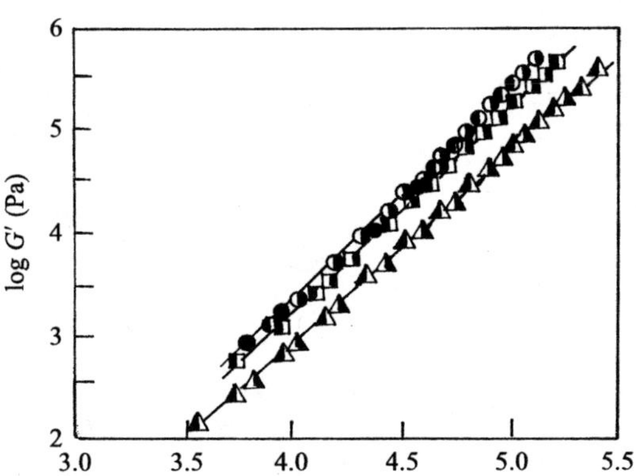

Figure 6.38 Log G' versus log G'' plots for three 4-arm star-branched polybutadienes at various temperatures: 88S4 at 24.5 °C (◭) and 50 °C (▲), 173S4 at 24.5 °C (◧) and 50 °C (◨), and 217S4 at 24.5 °C (◐), 50 °C (●), and 75 °C (◑). Data taken from Rochefort et al. (1979). See Table 6.7 for information on the molecular weights of these polymers.

Table 6.7 Molecular weights of star-branched polybutadienes

Sample Code	$(M_w)_{LC}$	$(M_n)_{OS}$	$(M_w)_{GPC}$	$(M_w/M_n)_{GPC}$
42S4	24.0×10^3	25.5×10^3	42.3×10^3	1.08
88S4	84.5×10^3	80.3×10^3	87.5×10^3	1.09
173S4	–	–	173×10^3	1.14
217S4	259×10^3	164×10^3	217×10^3	1.24

LS: light scattering weight-average molecular weight.
OS: membrane osmometer number-average molecular weight.
GPC: Determined with gel permeation chromatography.

Reprinted from Rochefort et al., *Journal of Polymer Science, Polymer Physics Edition* 17:1197. Copyright © 1979, with permission from John Wiley & Sons.

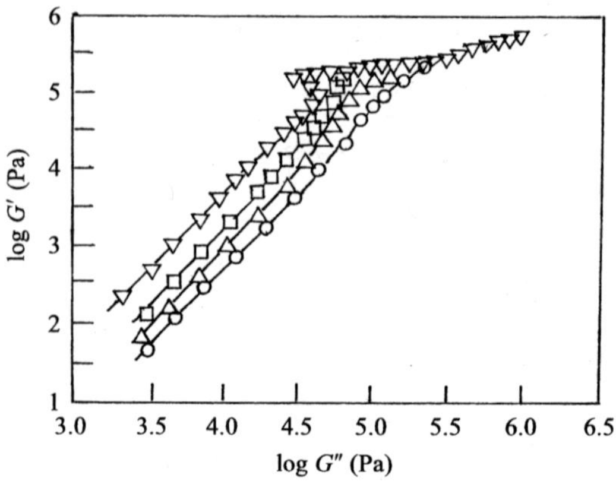

Figure 6.39 Log G' versus log G'' plots at 170 °C for four 4-arm star-branched polystyrenes: (○) S121A, (△) S111A, (□) S161A, and (▽) S181A. See Table 6.8 for information on the molecular weights of these polymers.

Table 6.8 Molecular characteristics of four-arm star-branched polystyrenes

Sample Code	M_w
S121A	0.94×10^5
S111A	1.54×10^5
S161A	3.51×10^5
S181A	10.27×10^5

Reprinted from Graessley and Roovers, *Macromolecules* 12:959. Copyright © 1979, with permission from the American Chemical Society.

Figure 6.41 gives log N_1 versus log σ and log G' versus log G'' plots for LDPE-A at 180, 200, and 220 °C. Similar plots are given in Figure 6.42 for LDPE-B and in Figure 6.43 for LDPE-C. In Figures 6.41–6.43 we observe temperature independence in both the log N_1 versus log σ and log G' versus log G'' plots. To facilitate our discussion here, composite plots are shown in Figure 6.44, which indicate that LDPE-A, having

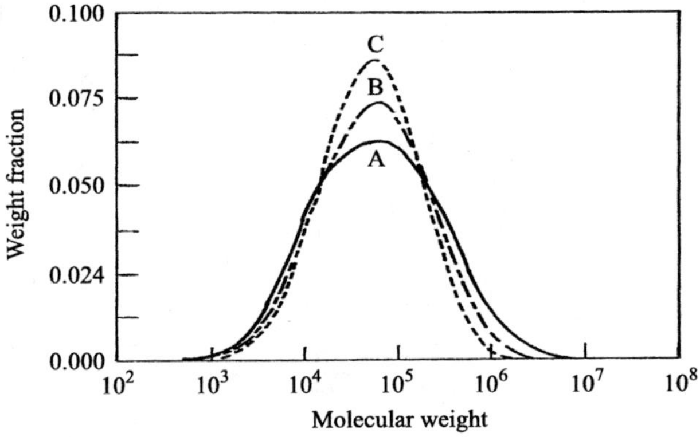

Figure 6.40 Molecular weight distribution curves for three different grades of commercial LDPEs. (Reprinted from Han and Jhon, *Journal of Applied Polymer Science* 32:3809. Copyright © 1986, with permission from John Wiley & Sons.)

Table 6.9 Molecular characteristics of commercial LDPEs

Sample Code	M_w	M_n	M_w/M_n	λ_N
A	202×10^3	21.3×10^3	9.43	3.4
B	143×10^3	22.5×10^3	6.03	2.5
C	110×10^3	26.3×10^3	4.18	1.6

λ_N represents the long-chain branching frequency, defined as the number-averaged number of branch point per 1,000 carbon atoms.

Reprinted from Han and Jhon, *Journal of Applied Polymer Science* 32:3809. Copyright © 1986, with permission from John Wiley & Sons.

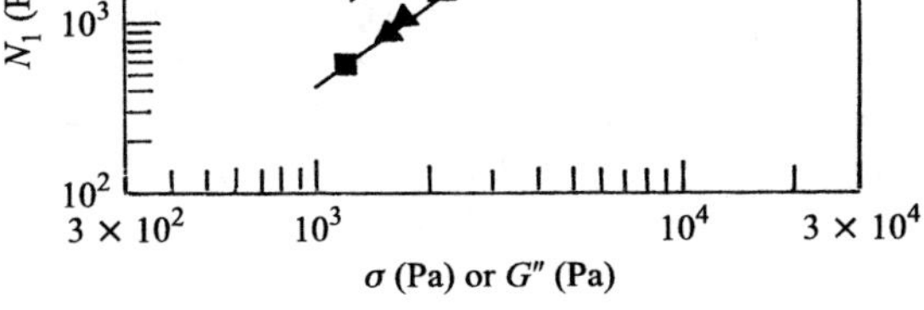

Figure 6.41 Plots of log N_1 versus log σ (open symbols) and log G' versus log G'' plots (filled symbols) for LDPE-A at various temperatures (°C): (○, ●) 180, (△, ▲) 200, and (□, ■) 220. (Reprinted from Han and Jhon, *Journal of Applied Polymer Science* 32:3809. Copyright © 1986, with permission from John Wiley & Sons.)

the highest degree of long-chain branching of the three LDPEs, is more elastic than LDPE-B and LDPE-C. This observation is consistent with the observation made on the star-branched polybutadienes (see Figure 6.38) and the star-branched polystyrenes (see Figure 6.39). Therefore, it can be concluded that the greater the degree of long-chain branching in a flexible homopolymer, the greater is the melt elasticity of the polymer.

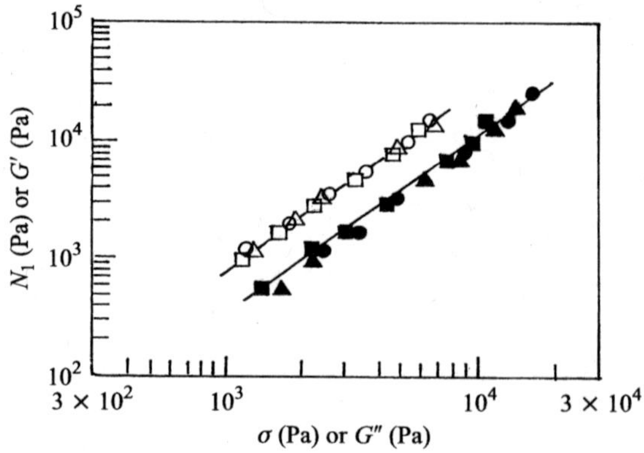

Figure 6.42 Plots of log N_1 versus log σ (open symbols) and log G' versus log G'' plots (filled symbols) for LDPE-B at various temperatures (°C): (○, ●) 180, (△, ▲) 200, and (□, ■) 220. (Reprinted from Han and Jhon, *Journal of Applied Polymer Science* 32:3809. Copyright © 1986, with permission from John Wiley & Sons.)

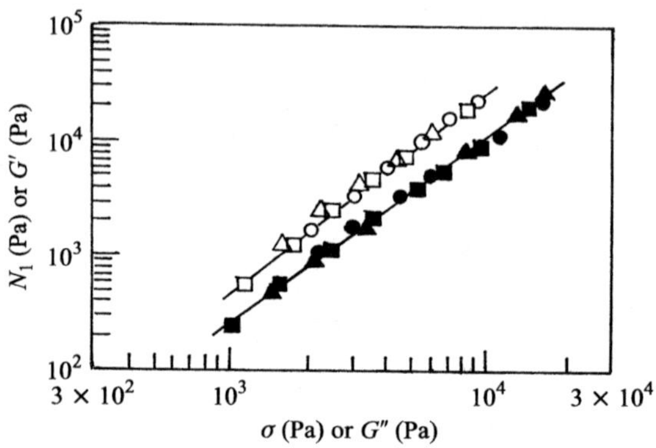

Figure 6.43 Plots of log N_1 versus log σ (open symbols) and log G' versus log G'' plots (filled symbols) for LDPE-C at various temperatures (°C): (○, ●) 180, (△, ▲) 200, and (□, ■) 220. (Reprinted from Han and Jhon, *Journal of Applied Polymer Science* 32:3809. Copyright © 1986, with permission from John Wiley & Sons.)

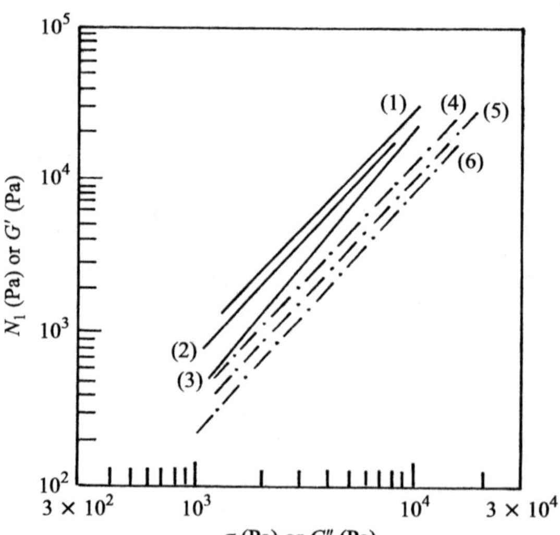

Figure 6.44 Comparison of log N_1 versus log σ plots (1) for LDPE-A, (2) for LDPE-B, and (3) for LDPE-C with log G' versus log G'' plots (4) for LDPE-A, (5) for LDPE-B, and (6) for LDPE-C for the three grades of LDPE given in Figures 6.41–6.43. (Reprinted from Han and Jhon, *Journal of Applied Polymer Science* 32:3809. Copyright © 1986, with permission from John Wiley & Sons.)

6.4 Summary

In this chapter, we have presented the rheological behavior of homopolymers, placing emphasis on the relationships between the molecular parameters and rheological behavior. We have presented a temperature-independent correlation for steady-state shear viscosity, namely, plots of log $\eta(T, \dot{\gamma})/\eta_0(T)$ versus log $\tau_0\dot{\gamma}$ or $\log a_T\dot{\gamma}$, where τ_0 is a temperature-dependent empirical constant appearing in the Cross equation and a_T is a shift factor that can be determined from the Arrhenius relation for crystalline polymers in the molten state or from the WLF relation for glassy polymers at temperatures between T_g and $T_g + 100\,°C$.

In interpreting the effect of molecular weight, molecular weight distribution, and long-chain branching on the rheological behavior of flexible homopolymers, we have advocated the use of log N_1 versus $\log \sigma$ obtained from steady-state shear flow experiments or $\log G'$ versus $\log G''$ plots obtained from oscillatory shear experiments. We have presented a theoretical basis that explains the reason why both log N_1 versus $\log \sigma$ and log G' versus log G'' plots are virtually independent of temperature and are independent of molecular weight for flexible entangled homopolymers, but are dependent upon molecular weight for flexible unentangled homopolymers.

An inverse problem of determining the molecular weight distribution of a flexible homopolymer from rheological measurements is of practical importance in the situations when measurement of molecular weight is practically very difficult, if not impossible, because a suitable solvent cannot be found for the preparation of solutions. One such situation is the determination of the molecular weight distribution of polypropylene. Many investigators (Léonardi et al. 2002; Liu et al. 1998; Maier et al. 1998; McGrory and Tuminello 1990; Mead 1994; Thimm et al. 2000; Tuminello 1986; Tuminello and Cudré-Mauroux 1991; Wasserman 1995; Wu 1985) have shown how to calculate, with different degrees of sophistication, the molecular weight distribution of flexible homopolymers from rheological measurements. For the reason of limited space, we will not elaborate on this subject here.

Problems

Problem 6.1

Table 6.10 gives oscillatory shear data taken at 170, 180, and 190 °C for a nearly monodisperse polystyrene having $M_w = 5.7 \times 10^4$. You are asked to do the following tasks.

(a) Prepare $\log G'$ versus log ω, $\log G''$ versus log ω, and log $|\eta^*|$ versus log ω plots at 170, 180, and 190 °C.

(b) Calculate shift factor a_T for the polystyrene by shifting the $\log G''$ versus log ω plots prepared in part (a) at 170 and 190 °C to the data at 180 °C as reference temperature (T_r), and then prepare a plot of a_T versus temperature T.

(c) Prepare reduced $\log G'_r$ versus $\log a_T\omega$, $\log G''_r$ versus $\log a_T\omega$, and log $|\eta^*_r|$ versus $\log a_T\omega$ plots at 170, 180, and 190 °C using the values of a_T determined in part (b) above. You will find that the reduced plots prepared will be independent of temperature. Use the values of density $\rho = 0.986$ g/cm^3

Table 6.10 Experimental data for Problem 6.1

At 170°C		
ω *(rad/s)*	G' *(Pa)*	G'' *(Pa)*
1.893	1.38×10^2	2.35×10^3
3.000	2.75×10^2	3.70×10^3
4.754	6.03×10^2	5.81×10^3
7.536	1.35×10^3	9.03×10^3
11.943	2.98×10^3	1.37×10^4
18.927	6.41×10^3	2.04×10^4
30.000	1.32×10^4	2.86×10^4
47.547	2.45×10^4	3.66×10^4
75.357	3.97×10^4	4.18×10^4
At 180°C		
ω *(rad/s)*	G' *(Pa)*	G'' *(Pa)*
2.512	7.39×10^1	1.21×10^3
3.981	1.12×10^2	2.01×10^3
6.310	2.08×10^2	3.01×10^3
10.000	4.39×10^2	4.76×10^3
15.849	9.89×10^2	7.45×10^3
25.119	2.20×10^3	1.14×10^4
39.810	4.87×10^3	1.72×10^4
63.095	1.05×10^4	2.48×10^4
At 190 °C		
ω *(rad/s)*	G' *(Pa)*	G'' *(Pa)*
7.536	9.62×10^1	1.61×10^3
11.943	1.73×10^2	2.54×10^3
18.927	3.60×10^2	3.98×10^3
30.000	8.07×10^2	6.22×10^3
47.547	1.85×10^3	9.61×10^3
75.357	4.19×10^3	1.45×10^4

at 170 °C, $\rho = 0.980$ g/cm^3 at 180 °C, and $\rho = 0.975$ g/cm^3 at 190°C to calculate G'_r, G''_r, and $|\eta^*_r|$.

(d) Use the WLF equation, Eq. (6.7), with $C_1 = 13.7$, $C_2 = 50.5$, and $T_r = 100$ °C to calculate shift factor a_T at 170, 180, and 190 °C, and then prepare a plot of a_T versus temperature.

(e) Prepare reduced $\log G'_r$ versus $\log a_T\omega$, $\log G''_r$ versus $\log a_T\omega$, and $\log |\eta^*_r|$ versus $\log a_T\omega$ plots at 170, 180, and 190 °C using the values of a_T determined in part (d) above. Assume that the density of polystyrene at T and T_r is 1.00 g/cm^3 in calculating G'_r, G''_r, and $|\eta^*_r|$. Compare the reduced plots prepared here with those prepared in part (c) above.

Notes

1. Random copolymers are not necessarily flexible. Semiflexible random liquid-crystalline polymers have been synthesized.
2. Shear rate $\dot{\gamma}$ appearing in Eq. (6.2) can be negative, for instance in flow through a capillary die depending on the side about the centerline axis of the die. Thus, the accurate representation of the power-law model must have the following form:

$$\sigma(\dot{\gamma}, T) = K(T)\,|\dot{\gamma}|^n \tag{6N.1}$$

However, for convenience, throughout the rest of this chapter and future chapters we shall use Eq. (6.2) instead of Eq. (6N.1).

3. From the molecular theory presented in Chapter 4 for linear flexible homopolymers, the shift factor $a_T(T)$ can be obtained as $a_T(T) = (\rho_r T_r/\rho T)(\eta_0(T)/\eta_0(T_r))$, where ρ_r and ρ are the densities at T_r and T, respectively, and $\eta_0(T_r)$ and $\eta_0(T)$ are the zero-shear viscosities at T_r and T, respectively. Over a narrow range of temperatures, $\rho_r T_r/\rho T \cong 1.0$, with T being the absolute temperature, and thus $a_T(T) = (\rho_r T_r/\rho T)\,(\eta_0(T)/\eta_0(T_r))$ reduces to Eq. (6.8).
4. The binomial expansion of the right-hand side of the first expression of Eq. (6.14) for $\sigma/G \ll 1$ also yields Eq. (6.15).
5. The definition $G'_r(\omega, T) = (\rho_r T_r/\rho T)G'(\omega, T)$ can be obtained from Eq. (4.114) since $G_N^o = \rho RT/M_e$ is defined by Eq. (4.97). Similarly, $G''_r(\omega, T) = (\rho_r T_r/\rho T)G''(\omega, T)$ can also be obtained from Eq. (4.115).
6. This expression is valid for all odd values of p, whereas Han and Jhon (1986) presented an expression for the special case of $p = 1$. In the derivation of Eq. (6.18), the following relationships are used:

$$\sum_{\text{odd}\,p}^{\infty} 1/p^4 = \pi^4/96; \quad \sum_{\text{odd}\,p}^{\infty} 1/p^6 = \pi^6/960 \tag{6N.2}$$

7. In the derivation of Eq. (6.20), the following relationships are used:

$$\sum_{\text{all}\,p}^{\infty} 1/p^2 = \pi^2/6; \quad \sum_{\text{all}\,p}^{\infty} 1/p^4 = \pi^4/90 \tag{6N.3}$$

Substitution of Eq. (4.95) into (4.84) and (4.85) for $\omega^2\tau_p^2 \ll 1$ yields

$$G'(\omega) = (2M/5\rho RT){\eta_0}^2\omega^2; \quad G''(\omega) = \eta_0\omega \tag{6N.4}$$

8. Figure 6.20 was prepared with dynamic frequency sweep data provided by M. J. Struglinski.
9. Figure 6.21 was prepared with dynamic frequency sweep data provided by T. Kodaka.

10. Equation (27) in a paper by Graessley and Struglinski (1986) should read Eq. (6.27), and Eqs. (6.29) and (6.30) are not given in the paper by Graessley and Struglinski (1986).
11. Figure 6.27 was prepared with dynamic frequency sweep data provided by M. J. Struglinski.
12. Plots of $\log G'$ versus $\log \omega$ and $\log G''$ versus $\log \omega$ were not reported in a paper by Graessley and Roovers (1979); Figure 6.39 was prepared using numerical data of dynamic frequency sweep experiments provided by J. Roovers.
13. The paper of Graessley and Roovers (1979) did not include polymer S111A; this polymer is included in Table 6.8. The information was provided by J. Roovers.

References

Akovali G (1967). *J. Polym. Sci. Part A-2* 5:875.
Archer LA, Varshney SK (1998). *Macromolecules* 31:6348.
Blackwell RJ, McLeish TCB, Harlen OG (2000). *J. Rheol.* 44:121.
Bogue DC, Masuda T, Einaga Y, Onogi S (1970). *Polym. J.* 1:563.
Bourrigaud S, Marin G, Poitou A (2003). *Macromolecules* 36:1388.
Bueche F (1964). *J. Chem. Phys.* 40:484.
Buesse WF, Longworth R (1962). *J. Polym. Sci.* 58:49.
Carreau PJ (1968). *Rheological Equation from Molecular Network Theory*, Doctoral Dissertation, University of Wisconsin, Madison, Wisconsin.
Chartoff RP, Maxwell B (1970). *J. Polym. Sci. A-2* 8:455.
Choi S, Han CD (2004). *Macromolecules* 37:215.
Chuang HK, Han CD (1984). *J. Appl. Polym. Sci.* 29:2205.
Cohen MH, Turnbull D (1959). *J. Chem. Phys.* 31:1164.
Cole KS, Cole RH (1941). *J. Chem. Phys.* 9:341.
Coleman BD, Markovitz H (1964). J. Appl. Phys. 35:1.
Cross MM (1965). *J. Colloid Sci.* 20:417.
Cross MM (1969). *J. Appl. Polym. Sci.* 13:765.
de Gennes P (1975). *J. Phys.* 36:1199.
des Cloizeaux J (1990). *Macromolecules* 23:4678.
Devore DD, Neithamer DR, LaPointe RE, Mussell RD (1997). *U.S. Patent* 5625087.
Doi M, Kuzuu NY (1980). *J. Polym. Sci.*, Polym. *Lett. Ed.* 18:775.
Doolittle AK, Doolittle DB (1957). *J. Appl. Phys.* 28:901.
Evans KE (1982). *J. Polym. Sci., Polym. Lett. Ed.* 20:103.
Ferry JD (1980). *Viscoelastic Properties of Polymers*, 3rd ed, John Wiley & Sons, New York.
Friedman EM, Porter RS (1975). *Trans. Soc. Rheol.* 19:493.
Friedrich Chr, Schwarzwälder C, Rieman RG (1996). *Polymer* 37:2499.
Ghaneh-Fard A, Carreau PJ, Lafleur PG (1996). *AIChE J.* 42:1388.
Graessley WW (1971). *J. Chem. Phys.* 54:5143.
Graessley WW, Edwards SF (1981). *Polymer* 22:1329.
Graessley WW, Prentice JS (1968). *J. Polym. Sci. A-2* 6:1887.
Graessley WW, Roovers J (1979). *Macromolecules* 12:959
Graessley WW, Struglinski MJ (1986). *Macromolecules* 19:1754.
Guillet JE, Combs RL, Slonaker DF, Weemes DA, Coover HW (1965). *J. Appl. Polym. Sci.* 9:757.

Han CD (1988). *J. Appl. Polym. Sci.* 35:167.
Han CD, Chuang HK (1985a). *J. Appl. Polym. Sci.* 30:2431
Han CD, Chuang HK (1985b). *J. Appl. Polym. Sci.* 30:4431.
Han CD, Jhon MS (1986). *J. Appl. Polym. Sci.* 32:3809.
Han CD, Kim YW (1975). *Trans. Soc. Rheol.* 19:245.
Han CD, Kim YW (1976). *J. Appl. Polym. Sci.* 20:2609.
Han CD, Kim JK (1989a). *Macromolecules* 22:1914.
Han CD, Kim JK (1989b), *Macromolecules* 22:4292.
Han CD, Kim JK (1993). *Polymer* 34:2533.
Han CD, Kim YJ, Chuang, HK (1983a). *Polym. Eng. Rev.* 3:1.
Han CD, Kim YJ, Chuang HK, Kwack TH (1983b). *J. Appl. Polym. Sci.* 28:3435.
Han CD, Kwack TH (1983). *J. Appl. Polym. Sci.* 28:3399.
Han CD, Lem KW (1982). *Polym. Eng. Rev.* 2:135.
Han CD, Park JY (1975). *J. Appl. Polym. Sci.* 19:3219.
Han CD, Shetty R (1977). *Ind. Eng. Chem. Fundam.* 16:49.
Han CD, Yang HH (1986). *J. Appl. Polym. Sci.* 32:1199.
Han CD, Yang HH (1987). *J. Appl. Polym. Sci.* 33:1221.
Inkson NJ, McLeish TCB, Harlen OG, Groves DJ (1999). *J. Rheol.* 43:873.
Jenkins JM, Jones RL, Jones TM (1985). *U.S. Patent* 4543399.
Jenkins JM, Jones, RL, Jones TM, Beret S (1986). *U.S. Patent* 4588790.
Kanai T, White JL (1984). *Polym. Eng. Sci.* 24:1185.
Kraus G, Gruver JT (1965). *J. Polym. Sci. A* 3:105.
Kurata M, Osaki K, Einaga Y, Sugie T (1974). *J. Polym. Sci., Polym. Phys. Ed.* 12:849.
Kwack TH, Han CD (1983). *J. Appl. Polym. Sci.* 28:3419.
Laun HM (1979). *Rheol. Acta* 18:478.
Lem KW, Han CD (1982). *J. Appl. Polym. Sci.* 27:1367.
Léonardi F, Allal A, Marin G (2002). *J. Rheol.* 46:209.
Liu Y, Shaw MT, Tuminello WH (1998). *J. Rheol.* 42:453.
Long VC, Berry GC, Hobbs LM (1964). *Polymer* 5:517.
Maier D, Eckstein A, Friedrich C, Honerkamp J (1998). *J. Rheol.* 42:1153.
Masuda T, Kitagawa K, Inoue T, Onogi S (1970). *Macromolecules* 3:116.
McGrory WJ, Tuminello WH (1990). *J. Rheol.* 34:867.
McLeish TCB (1988). *Macromolecules* 21:1062.
McLeish TCB, Larson RG (1998). *J. Rheol.* 42:81.
McLeish TCB, Allgaier J, Bick DK, Bishko G, Biswas P, Blackwell R, Blottière B, Clarke N, Gibbs B, Droves DJ, Hakiki A, Heenan RK, Johnson JM, Kant R, Read DJ, Young RN (1999). *Macromolecules* 32:6734.
Mead DW (1994). *J. Rheol.* 38:1797.
Mendelson RA, Bowles WA, Finger FL (1970). *J. Polym. Sci. A-2* 8:105.
Mills NJ (1975). *Eur. Polym. J.* 5:675.
Mills NJ, Nevin A (1971). *J. Polym. Sci. Part A-2* 9:267.
Minoshima W, White JL (1986). *J. Non-Newtonian Fluid Mech.* 19:275.
Minoshima W, White JL, Spruiell JE (1980). *Polym. Eng. Sci.* 20:1166.
Montfort JP, Marin G, Arman J, Monge Ph (1978). *Polymer* 19:277.
Montfort JP, Marin G, Arman J, Monge Ph (1979). *Rheol. Acta* 18:623.
Montfort JP, Marin G, Arman J, Monge Ph (1984). *Macromolecules* 17:1551.
Montfort JP, Marin G, Arman J, Monge Ph (1986a). *Macromolecules* 19:393.
Montfort JP, Marin G, Arman J, Monge Ph (1986b). *Macromolecules* 19:1979.
Neumann C, Loveday DR, Abetz V, Stadler R (1998). *Macromolecules* 31:2493.
Ninomiya K (1959). *J. Colloid. Sci.* 14:49.
Ninomiya K (1962). *J. Colloid. Sci.* 17:759.

Ninomiya K, Ferry JD (1963). *J. Colloid Sci.* 18:421.
Ninomiya K, Ferry JD, Oyanagi Y (1963). *J. Phys. Chem.* 67:2297.
Oda K, White JL, Clark ES (1978). *Polym. Eng. Sci.* 18:25.
Onogi S, Masuda T, Oda N, Koga K (1970). *Polym. J.* 1:542.
Peticolas WL, Watkins JM (1957). *J. Am. Chem. Soc.* 79:5083.
Prest WM (1973). *Polym. J.* 4:163.
Prest WM, Porter RS (1973). *Polym. J.* 4:154.
Romanini D, Savadori A, Gianotti G (1980). *Polymer* 21:1092.
Rochefort WE, Smith GG, Rachapudy H, Raju VR, Graessley WW (1979). *J. Polym. Sci., Polym. Phys. Ed.* 17:1197.
Shie SC, Wu CT, Hua CC (2003). *Macromolecules* 36:2141.
Struglinski MJ, Graessley WW (1985). *Macromolecules* 18:2630.
Thimm W, Friedrich C, Marth, M, Honerkamp J (2000). *J. Rheol.* 44:429.
Tuminello WH (1986). *Polym. Eng. Sci.* 26:1339.
Tuminello WH, Cudré-Mauroux X (1991). *Polym. Eng. Sci.* 31:1496.
Tung LH (1960). *J. Polym. Sci.* 46:409.
Turnbull D, Cohen MH (1961). *J. Chem. Phys.* 34:120.
Verbeeten WMH, Peters GWM, Baaijens FPT (2001). *J. Rheol.* 45:823.
Vinogradov GV, Malkin AYa (1966). *J. Polym. Sci. A-2* 4:135.
Wasserman SH (1995). *J. Rheol.* 39:601.
Wasserman SH, Graessley WW (1992). *J. Rheol.* 36:543.
Watanabe H, Kotaka T (1984). *Macromolecules* 17:2316.
Watanabe H, Sakamoto T, Kotaka T (1985). *Macromolecules* 18:1008.
Welborn HC, Ewen JA (1994). *U.S. Patent* 5324800.
Williams ML, Landel RF, Ferry JD (1955). *J. Am. Chem. Soc.* 77:3701.
Wu S (1985). *Polym. Eng. Sci.* 25:122.
Wyman DP, Elyash LJ, Frazer WJ (1965). *J. Polym. Sci. A* 3:681.

7

Rheology of Miscible Polymer Blends

7.1 Introduction

Broadly classified, there are three types of polymer blends, namely, (1) miscible polymer blends, (2) immiscible polymer blends, and (3) partially miscible polymer blends. There are many different experimental methods that can be used to investigate the miscibility of polymer blends, such as differential scanning calorimetry (DSC), dynamic mechanical thermal analysis (DMTA), dielectric measurement, cloud point measurement, microscopy, light scattering, small-angle X-ray scattering, small-angle neutron scattering, fluorescence technique, and nuclear magnetic resonance (NMR) spectroscopy. Each of these experimental methods can only probe the homogeneity (or heterogeneity) in a polymer blend at a certain scale range. Thus, the determination of the miscibility in polymer blends depends on the resolution limit of the experimental method(s) employed. For instance, DSC and DMTA have frequently been used to determine the miscibility in polymer blends by determining glass transition temperature T_g. When a single T_g value is observed in a polymer blend, the blend can be considered miscible. However, there is a general consensus among researchers that such an experimental criterion, while very useful, cannot guarantee that a polymer blend is miscible on a segmental level. Therefore, a serious question may be raised as to whether a polymer pair can be regarded as being miscible on the segmental level (say, less than approximately 5 nm). It has been reported that DMTA can resolve the size of domains (or separated phases) on the order of 5–10 nm (Molnar and Eisenberg 1992) and DSC is not as sensitive as DMTA for determining the T_g of a polymer blend (Stoelting et al. 1970). In the use of DSC to investigate the miscibility of polymer blends, one often encounters the situation where a very broad (say, 40–60 °C) single glass transition appears for certain blend compositions, such as polystyrene/poly(α-methyl styrene) (PS/PαMS) blends (Kim et al. 1998; Lin and Roe 1988; Saeki 1983) and polystyrene/poly(vinyl methyl ether) (PS/PVME) blends (Kim et al. 1998; Schneider and Wirbser 1990; Schneider et al. 1990). Under such circumstances, it is not clear how an unambiguous, single value of T_g can be read off from a DSC thermogram. NMR spectroscopy is regarded as being an effective tool for the investigation of the presence of microheterogeneity in polymer blends (Chu et al. 1990; Goldman and Shen 1966; Jack and Whittaker 1997; Schmidt-Rohr et al. 1992; Stejskal and Memory 1994; Takegoshi and Hikichi 1991). The presentation of various experimental methods for determination of the miscibility in polymer blends is outside the scope of this chapter.

Due to the two-phase or multiphase nature from the macroscopic point of view, the rheological behavior of incompatible (or heterogeneous, in the more general sense) polymer blends must be dealt with from a phenomenological (i.e., fluid mechanics) point of view. Conversely, the rheological behavior of miscible polymer blends can be dealt with (or interpreted) using a molecular approach. While there are few miscible polymer blends that have met with commercial success, numerous investigators have reported on the rheological behavior of miscible polymer blends. In this chapter, we will discuss the rheological behavior of some selected miscible polymer blends.

In the preceding chapter, we discussed the rheological behavior of binary blends consisting of compounds having identical chemical structures (i.e., polydisperse homopolymers). When dealing with the rheological behavior of a mixture consisting of two or more polymers having dissimilar chemical structure, complexity arises from the fact that one must take into account the effect of the extent (or degree) of miscibility, which naturally affects the parameter(s) related to the thermodynamic state of the mixture. Hence, a molecular interpretation of the rheological behavior of miscible polymer blends is much more complicated than the situations where two or more polymers having identical chemical structure are mixed (see Chapter 6).

In this chapter, we first present different types of phase behavior in polymer blends, followed by the experimentally observed rheological behavior of some miscible polymer blends. We then present molecular theory for the linear viscoelasticity of miscible polymer blends and compare predictions with experimental results. We present the unresolved issue of determining the plateau modulus of miscible polymer blends. The rheological behavior of immiscible polymer blends will be presented in Chapter 11.

7.2 Phase Behavior of Polymer Blend Systems

There are several criteria for determining the miscibility in polymer blends. The thermodynamic criteria date back to the seminal studies of Flory (1941, 1942) and Huggins (1941, 1942) in the 1940s. Since then, other thermodynamic theories have been developed (Flory 1965; Flory et al. 1968; Orwoll and Flory 1967; Sanchez and Lacombe 1978) for investigating the phase behavior of polymer blends. As the discussion of thermodydnamic criteria for miscibility in polymer pairs is beyond the scope of this chapter, here we illustrate briefly different phase behaviors of polymer blends in general terms, using examples that will facilitate our presentation in the rest of this chapter.

In many polymer blends, two different types of phase behavior are primarily observed: upper critical solution temperature (UCST) phase behavior and lower critical solution temperature (LCST) phase behavior, as schematically shown in Figure 7.1. As the temperature is increased, the miscibility of the polymer blend system exhibiting UCST increases (Figure 7.1a) while the miscibility of the blend system exhibiting LCST decreases (Figure 7.1b). Although not observed very often, some polymer

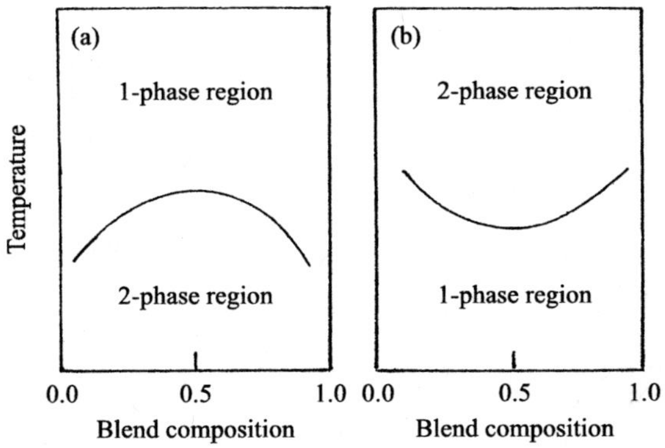

Figure 7.1 Schematic representation of the phase behavior of binary blends exhibiting (a) UCST and (b) LCST.

blend systems exhibit combined UCST and LCST phase behavior (Cong et al. 1986; Hammouda et al. 1994; Ougizawa et al. 1985; Ruzette and Mayes 2001; Ruzette et al. 2001; Ryu et al. 2002), as schematically shown in Figure 7.2a, as the temperature is increased or decreased. Note that the polymer blend residing in the region between the UCST and LCST binodal curves in Figure 7.2a is in the homogeneous state. As the molecular weight of one or both components of a polymer blend system exhibiting combined UCST and LCST phase behavior is increased, the LCST binodal curve will move downward while the UCST binodal curve will move upward. Eventually, the two binodal curves overlap, giving rise to another phase behavior, often referred to as "hourglass" type phase behavior, as schematically shown in Figure 7.2b. The thermodynamic aspects of a variety of phase diagrams for polymer blends are presented in the monograph of Koningsveld et al. (2001).

What is clear from the observations made above is that miscibility in polymer blends depends on temperature and blend composition. Thus it is essential for one to have information on the miscibility of a polymer blend system before taking rheological measurements. Needless to say, the rheological behavior of a polymer blend in the single-phase region (miscible polymer blend) will be quite different from that in the two-phase region (immiscible polymer blend).

The equilibrium phase diagrams (binodal curves) enable us to describe the extent of miscibility (or immiscibility) of a polymer blend system. A miscible polymer blend

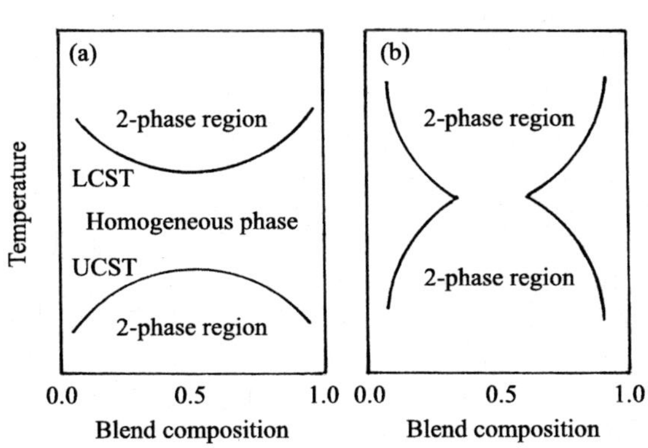

Figure 7.2 Schematic representation of the phase behavior of binary blends exhibiting (a) combined UCST and LCST and (b) "hourglass" type.

is expected to have negative values of segmental interaction parameter χ (commonly referred to as the Flory–Huggins interaction parameter). A negative value of χ implies that a polymer blend has attractive interactions, whereas a positive value of χ implies that a polymer blend has repulsive interactions. Thus, the stronger the attractive interactions that a polymer blend has, the larger will be the negative value of χ. Likewise, the stronger the repulsive interactions that a polymer blend has, the larger will be the positive value of χ, and thus it will be less miscible. Hence, it can be said that polymer blends with weak attractive interactions are expected to have very small negative values of χ.

The locations of binodal curves and thus values of χ for a blend system depend on temperature T, molecular weight M, and blend composition ϕ. Thus values of χ must be expressed in terms of T, M, and ϕ in the most general situation. However, there is scarcely any experimental data yet available for obtaining such a general expression for χ in terms of T, M, and ϕ. Certainly, the Flory–Huggins lattice theory does not take into account the effects of M and ϕ.

Extending the lattice cluster theory (Dudowicz and Freed 1991a, 1991b), Freed and Dudowicz (1992) and Dudowicz and Freed (1993) have shown that the molecular weight dependence of χ for binary blends can be expressed by

$$\chi = (a + c/N) + (b + d/N)/T \tag{7.1}$$

where a, b, c, and d are constants, T is the absolute temperature, and N denotes degree of polymerization. Note that in obtaining Eq. (7.1) it was assumed that both components have the same value of N. Notice that Eq. (7.1) reduces to

$$\chi = a + b/T \tag{7.2}$$

for very large values of N. However, it is practically very difficult to determine equilibrium phase diagram experimentally for many polymer blends having high molecular weights, making it difficult to determine the values of a, b, c, and d appearing in Eq. (7.1). Under such circumstances, one usually assumes that χ depends only on temperature and then employs Eq (7.2) to describe the effect of temperature on χ. Note that the sign of constant b appearing in Eq. (7.2) determines whether a polymer blend exhibits UCST (positive value of b) or LCST (negative value of b). Values of a and b appearing on the right-hand side of Eq. 7.2 for many different pairs of polymers are given by Balsara (1996).

Figure 7.3 gives equilibrium phase diagrams (binodal curves) exhibiting UCST of four pairs of polystyrene (PS) and polyisoprene (PI), which were obtained from cloud point measurements, in which the molecular weight of PS was increased from 2.2×10^3 to 9.5×10^3, while the molecular weight (5.1×10^3) of PI was kept constant. It is clearly seen from Figure 7.3 that the binodal curve is shifted upwards as the molecular weight of PS is increased; the critical temperature T_c is increased from 110 to 185 °C as the molecular weight of PS is increased from 2.2×10^3 to 9.5×10^3. It should be mentioned that the phase diagrams given in Figure 7.3 are limited to fairly low molecular weights of PS and PI. It is practically impossible to obtain a phase diagram

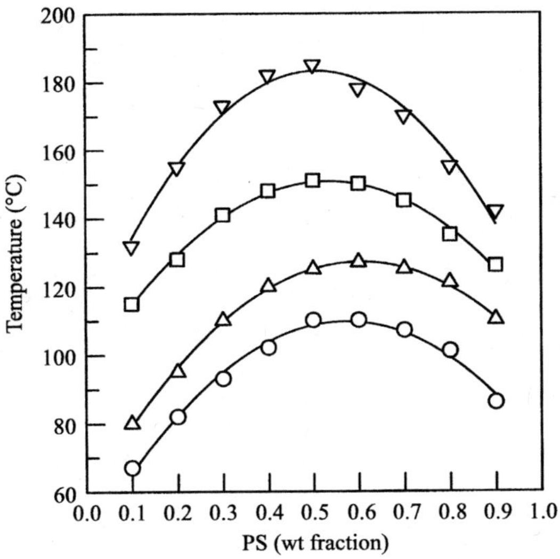

Figure 7.3 Phase diagrams for four PS/PI blends: (○) (PS-2)/(PI-5) blends with $T_c = 110\,^\circ$C, (△) (PS-3)/(PI-5) blends with $T_c = 127\,^\circ$C, (□) (PS-6)/(PI-5) blends with $T_c = 151\,^\circ$C, (▽) (PS-10)/(PI-5) blends with $T_c = 185\,^\circ$C. Note that PS-2 has $M_w = 2.2 \times 10^3$, PS-3 has $M_w = 3.1 \times 10^3$, PS-6, has $M_w = 6.0 \times 10^3$, PS-10 has $M_w = 9.5 \times 10^3$, and PI, with predominantly 1,4-addition of microstructure, has $M_w = 5.0 \times 10^3$. (Reprinted from Han et al., *Macromolecules* 35:8045. Copyright © 2002, with permission from the American Chemical Society.)

via cloud point measurement or any other methods for higher molecular weights of PS and PI. This is because PI would undergo thermal degradation/cross-linking reactions at temperatures above approximately 185 °C. By curve-fitting, via least-squares method, the experimentally determined phase diagram given in Figure 7.3 for (PS-6)/(PI-5) blends to the Flory–Huggins theory via Eq. (7.2), we obtain the following expression for α:

$$\alpha = -1.135 \times 10^{-3} + 0.7111/T \tag{7.3}$$

where α has units of mol/cm^3. Note that α is related to χ by $\chi = V_{\text{ref}}\alpha$ with V_{ref} being the molar reference volume. Note that in obtaining Eq. (7.3) the following expressions were employed for the specific volume (in units of cm^3/g):

$$v_{\text{PS}} = 0.922 + 5.412 \times 10^{-4}(T - 273) + 1.687 \times 10^{-7}(T - 273)^2 \tag{7.4}$$

for PS (Richardson and Savill 1977), and

$$v_{\text{PI}} = 1.077 + 7.221 \times 10^{-4}(T - 273) + 2.461 \times 10^{-7}(T - 273)^2 \tag{7.5}$$

for PI (Han et al. 1989).

One can also curve fit the values of α obtained at various temperatures to the empirical expression

$$\alpha = a + b/T + c\phi_B/T \tag{7.6}$$

where ϕ_B is the volume fraction of component B in A/B binary blends. The last term on the right-hand side of Eq. (7.6), $c\phi_B/T$, has a thermodynamic origin

(Roe and Zin 1980). For the (PS-6)/(PI-5) blend system whose binodal curve is given in Figure 7.3 we obtain

$$\alpha = -1.267 \times 10^{-3} + 0.6497/T + 0.0168\phi_{PS}/T \tag{7.7}$$

where ϕ_{PS} is the volume fraction of PS. It can be easily seen from Eq. (7.7) that the magnitude of the last term is much smaller than that of the second term, because $0 < \phi_{PS} < 1$ and the numerical value of 0.0168 appearing in the last term is an order of magnitude smaller than that of 0.6497 appearing in the second term. Thus, for all intents and purposes, use of Eq. (7.2) seems to be sufficient when the effect of molecular weight is neglected.

It is worth mentioning at this juncture that different values of χ may be obtained, depending upon the definition of the molar volume of a reference component V_{ref} that may be employed, since there is more than one way of defining V_{ref}. Specifically, V_{ref} may be defined by

$$V_{ref} = M_1 v_1 \tag{7.8}$$

or by

$$V_{ref} = \left[(M_1 v_1)(M_2 v_2)\right]^{1/2} \tag{7.9}$$

where M_1 and M_2 are the monomeric molecular weights of components 1 and 2, respectively, and v_1 and v_2 are the specific volumes of components 1 and 2, respectively. It is then clear that use of α is not involved with the choice of V_{ref}. Later in this chapter we will present a molecular theory for the linear viscoelasticity of miscible polymer blends, for which values of χ (or α) will be used to predict the linear dynamic viscoelastic properties of miscible polymer blends.

7.3 Experimental Observations of the Rheological Behavior of Miscible Polymer Blends

In this section, we present the experimental observations of the rheological behavior of some selected miscible polymer blends. Although there are so many pairs of miscible polymers reported in the literature, the number of studies reported on the rheological behavior of miscible polymer blend systems is rather small. Nevertheless, with the limited space available here, it is not possible to present the rheological behavior of every miscible blend system reported in the literature. Before presenting the rheological behavior of some specific miscible polymer blend systems, we first discuss the circumstances under which application of time–temperature superposition (TTS) to miscible polymer blends is warranted.

7.3.1 Time–Temperature Superposition in Miscible Polymer Blends

In the preceding chapter we discussed the application of TTS to flexible homopolymers in order to obtain generalized plots (or correlations) that may be used to estimate the

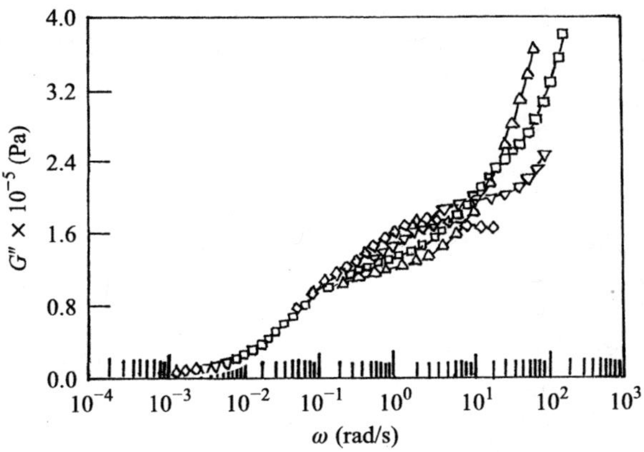

Figure 7.4 Plots of dynamic loss modulus (G'') versus logarithmic angular frequency (ω) at four temperatures (°C): (△) 120, (□) 137, (▽) 155, and (◇) 174 for 20/80 PEO/PMMA blend, which were obtained by shifting the G'' data along the ω axis to superpose with the data at 155 °C at low ω. (Reprinted from Colby, *Polymer* 30:1275. Copyright © 1989, with permission from Elsevier.)

rheological properties of the homopolymers at various temperatures without having to conduct additional experiments. However, a question may be raised as to whether application of TTS is warranted for all miscible polymer blends. We address this issue because the subject is fundamental and some confusion seems to exist in the literature.

Colby (1989) appears to be the first who reported a failure of TTS for a miscible 20/80 poly(ethylene oxide) (PEO)/poly(methyl methacrylate) (PMMA) blend, where 20/80 denotes the weight percentage of the constituent components. Figure 7.4 shows the temperature dependence of dynamic loss modulus (G'') for the 20/80 PEO/PMMA blend at 120, 137, and 174 °C, in which the data were shifted along the angular frequency axis to superpose with the data at 155 °C. Numerous investigators (Cortazar et al. 1982; Liberman et al. 1984; Ito et al. 1987; Martuscelli et al. 1984, 1986; Russell et al. 1988) reported that PEO/PMMA blends are miscible over a wide range of blend compositions. If the 20/80 PEO/PMMA blend is truly miscible on the segmental level, one should observe in Figure 7.4 a temperature-independent, single correlation. Based on Figure 7.4, Colby (1989) concluded that TTS failed in PEO/PMMA blends.

Figure 7.5 gives log G' versus log ω plots and Figure 7.6 gives log G'' versus log ω plots for the 20/80 PEO/PMMA blend at various temperatures.[1] It is very obvious from these figures that there is no way that one can obtain a temperature-independent correlation by shifting the data along the ω axis, and thus TTS would fail in the PEO/MMA blend. Notice the difference between Figure 7.4 and Figure 7.6, in that Figure 7.4 gives a linear plot of G'' over a logarithmic frequency scale while Figure 7.6 gives log-log plot. In other words, the linear plot of G'' given in Figure 7.4 shows obvious differences in G'' at three different temperatures, while the log G'' versus log ω plot given in Figure 7.6 shows a mild maximum at 174 °C.

Figure 7.7 gives log G' versus log G'' plots for the 20/80 PEO/PMMA blend, which were prepared from Figures 7.5 and 7.6. It can be seen from Figure 7.7 that for $G'' \geq$ approximately 2×10^5 Pa, log G' versus log G'' plots depend on temperature, with a slope less than 2, and for $G'' <$ approximately 2×10^5 Pa (particularly in the terminal region) log G' versus log G'' plots are more or less independent of temperature, with a slope still less than 2. Therefore, we can conclude from Figure 7.7 that TTS fails in PEO/PMMA blends. In Chapter 6 we showed that the log G' versus log G'' plot for

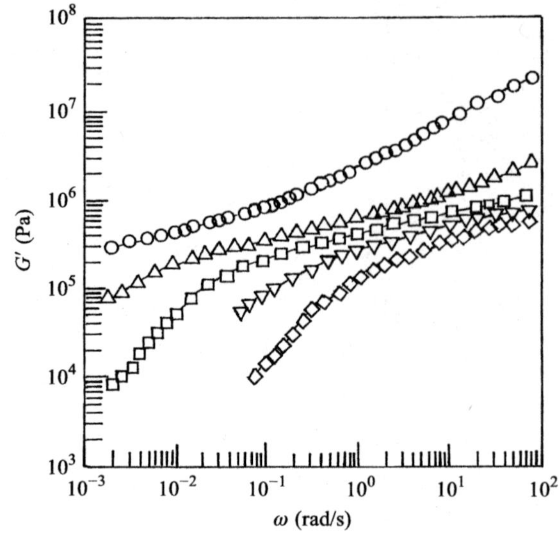

Figure 7.5 Log G' versus log ω plots for 20/80 PEO/PMMA blend at various temperatures (°C): (○) 101, (△) 120, (□) 137, (▽) 155, and (◇) 174. The plots are prepared from data provided by R. Colby.

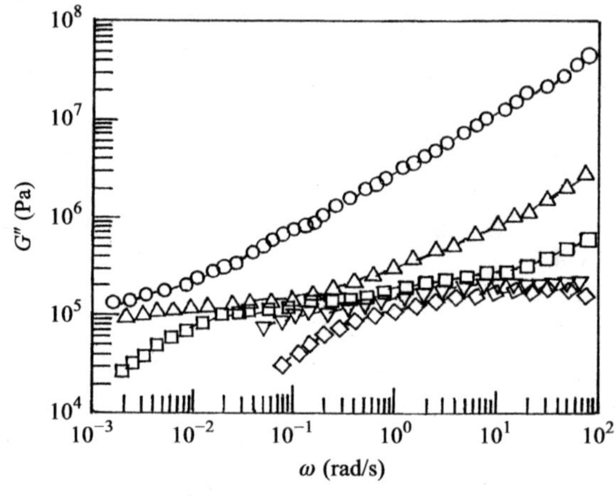

Figure 7.6 Log G'' versus log ω plots for 20/80 PEO/PMMA blend at various temperatures (°C): (○) 101, (△) 120, (□) 137, (▽) 155, and (◇) 174. The plots are prepared from data provided by R. Colby.

flexible monodisperse homopolymers is expected to be independent of temperature and its slope in the terminal region is expected to be equal to 2 at sufficiently small values of G'' (i.e., at sufficiently low values of ω in the dynamic frequency sweep experiments) if TTS works. Two things might have caused the slope of log G' versus log G'' plots in the terminal region of Figure 7.7 to be less than 2: (1) polydispersity of PEO and PMMA, and (2) the presence of microheterogeneity (since PEO/PMMA blends are miscible from a macroscopic point of view, e.g., having a single, very broad glass transition from DSC measurements). In Chapter 6 we discussed that even a polydisperse polymer is expected to have a slope of 2 in the terminal region of log G' versus log G'' plots when extremely low values of ω are applied (e.g., much lower than 10^{-2} rad/s). In view of the fact that the experimental data presented in Figures 7.4–7.6 were obtained over several decades of ω with 10^{-3} rad/s as the lowest

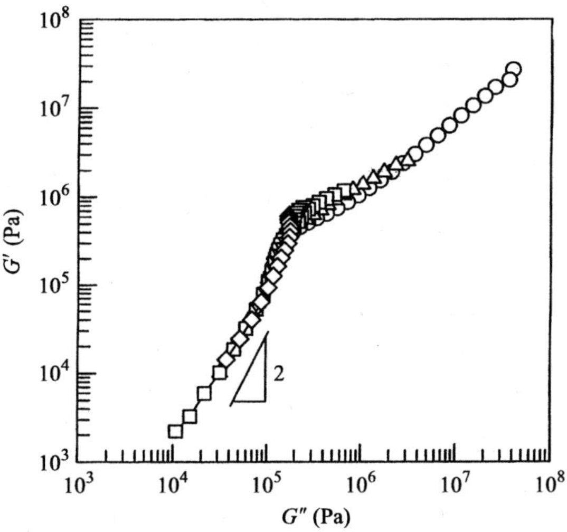

Figure 7.7 Log G' versus log G'' plots for 20/80 PEO/PMMA blend at various temperatures (°C): (○) 101, (△) 120, (□) 137C, (▽) 155, and (◇) 174. The plots are prepared from data provided by R. Colby.

value, it is highly plausible that the curvature in the terminal region of log G' versus log G'' plots for the 20/80 PEO/PMMA blend given in Figure 7.7 is most likely due to the presence of microheterogeneity. Needless to say, log G' versus log G'' plots cannot provide information on the origin of microheterogeneity.

Failure of TTS in miscible polymer blends other than PEO/PMMA blends has been reported by a number of research groups. Such miscible blends include polyisoprene/poly(vinyl ethylene) (PI/PVE) blends (Arendt et al. 1997; Chung et al. 1994a, 1994b; Roland and Ngai 1991; Roovers and Toporowski 1992; Zawada et al. 1994), PS/PVME blends (Ajji et al. 1988, 1991; Kapnistos et al. 1996; Kim et al. 1998; Pathak et al. 1999; Roland and Ngai 1992; Schneider and Wirbser 1990), and PS/PαMS blends (Kim et al. 1998). The reason for the failure of TTS in miscible polymer blends has been discussed extensively in the literature. The prevailing view is the presence of dynamic heterogeneity (Katana et al. 1995; Kumar et al. 1996; Miller et al. 1990; Zetsche and Fischer 1994), causing the failure of TTS in certain miscible polymer blends. Note that TTS does not fail in all miscible polymer blends. The next question, then, is the origin of dynamic heterogeneity in certain miscible polymer blends. Using NMR spectroscopy, Miller et al. (1990) investigated thermodynamic miscibility of PI/PVE blends exhibiting a single glass transition, as determined from DSC. They attributed a single, broad glass transition observed in the PI/PVE blends to the presence of dynamic heterogeneity. Specifically, on the basis of solid-state ^{13}C NMR measurements, Miller et al. concluded that molecular motions of the components in the PI/PVE blend had different temperature dependencies. This was because, despite the morphological homogeneity and the equivalence in free volumes, the carbons on the respective components exhibit distinct glass transitions, giving rise to a very broad glass transition observed by conventional methods. They noted that different free volume requirements for liquidlike mobility can evidently result in different glass transitions within a single-phase mixture, and thus miscible blends can exhibit dynamical heterogeneity. These experimental observations suggest that the difference in the component T_gs might play

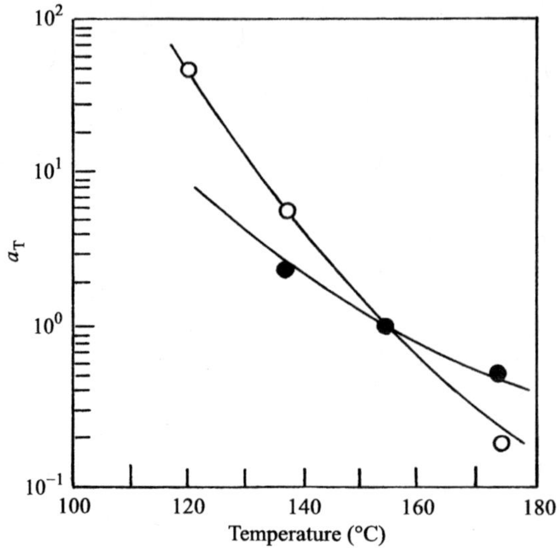

Figure 7.8 Temperature dependence of a_T for PEO (●) and PMMA (○) with the reference temperature of 155 °C, in which the solid curves were drawn using the WLF equation with $C_1 = 11.9$ and $C_2 = 69$ K for PMMA, and $C_1 = 6.9$ and $C_2 = 88$ K for PEO. (Reprinted from Colby, *Polymer* 30:1275. Copyright © 1989, with permission from Elsevier.)

an important role in determining whether a miscible polymer blend exhibits dynamic heterogeneity, and thus a failure of TTS.

Some investigators used the WLF equation, Eq. (6.7), to obtain a temperature-dependent shift factor a_T for miscible polymer blends. To illustrate the point, let us consider the temperature dependence of a_T for PEO and PMMA given in Figure 7.8. It is seen in Figure 7.8 that PEO and PMMA have quite different temperature dependences of a_T. It is not difficult to surmise that the values of a_T for various compositions in PEO/PMMA blends would be different. Note that the difference in the components T_gs (ΔT_g) between PEO and PMMA is quite large, although we recognize the fact that PEO is a semicrystalline polymer whereas PMMA is a glassy polymer having a T_g of about 130 °C. The T_g of PEO is about −52 °C, as determined via DSC from a specimen that was quenched in liquid nitrogen from 100 to −78 °C and then heated at 10 °C/min (Colby 1989). It seems, therefore, reasonable to infer that the failure of TTS in PEO/PMMA blend might be associated with the presence of dynamic heterogeneity arising from a large ΔT_g between PEO and PMMA.

Figure 7.9 shows the temperature dependence of a_T for PMMA/poly(vinylidene fluoride) (PVDF) blends and of the neat constituent components, PMMA and PVDF. Numerous research groups (Bernstein et al. 1977; Bovey et al. 1977; Coleman et al. 1977; Douglass and McBrierty 1978; Hirata and Kotaka 1981; Nishi and Wang 1975; Noland et al. 1971; Paul and Altamirano 1975; Patterson et al. 1976; Roerdink and Challa 1980; Tomura et al. 1992; Ward and Lin 1984; Wendorff 1980) have reported on the miscibility of PMMA/PVDF blends, which exhibit a single T_g over the entire range of blend compositions. It can be seen from Figure 7.9 that the temperature dependence of a_T is quite different between PMMA and PVDF, behavior very similar to that observed in Figure 7.8 for PEO and PMMA. Under such circumstances, it is not surprising to observe in Figure 7.9 that the temperature dependence of a_T for PMMA/PVDF blends varies with composition. On the basis of Figure 7.9 we can conclude that TTS would also fail in PMMA/PVDF blends. Note that PVDF is a

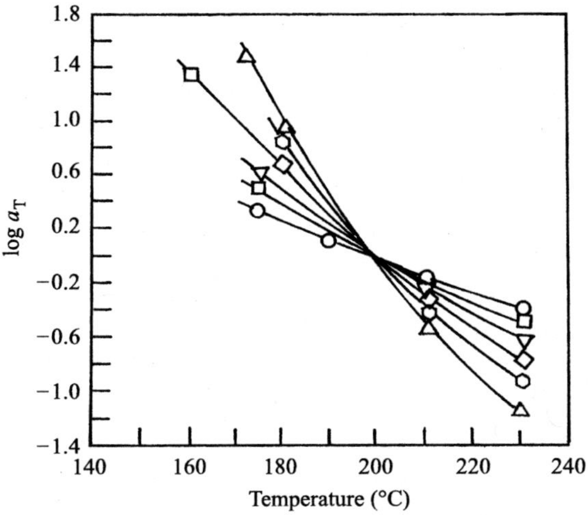

Figure 7.9 Temperature dependence of a_T for PMMA/PVDF blends: (○) PVDF, (△) PMMA, (□) 20/80 PMMA/PVDF, (▽) 40/60 PMMA/PVDF, (◇) 60/40 PMMA/PVDF, and (⬡) 80/20 PMMA/PVDF. The reference temperature T_r employed is 190 °C. (Reprinted from Yang et al., *Polymer* 35:1503. Copyright © 1994, with permission from Elsevier.)

semicrystalline polymer with a melting point of about 170 °C, while PMMA is a glassy polymer having a T_g of about 130 °C. Hence, the situation for PMMA/PVDF blends is very similar to that for PEO/PMMA blends. Later in this chapter we will discuss the rheological behavior of PMMA/PVDF blends in greater detail.

It has, however, been reported that TTS works in some miscible polymer blends with very small difference in the component T_gs (Alegria et al. 1995; Friedrich et al. 1996). Figure 7.10 shows the temperature dependence of a_T for PMMA/poly(styrene-*co*-acrylonitrile) (PSAN) blends and of the constituent components (PMMA and PSAN). It is interesting to observe in Figure 7.10 that the temperature dependence of a_T for PMMA is virtually identical to that for PSAN, and that the values of a_T for all blends lie virtually on the same curve as the constituent components. Numerous

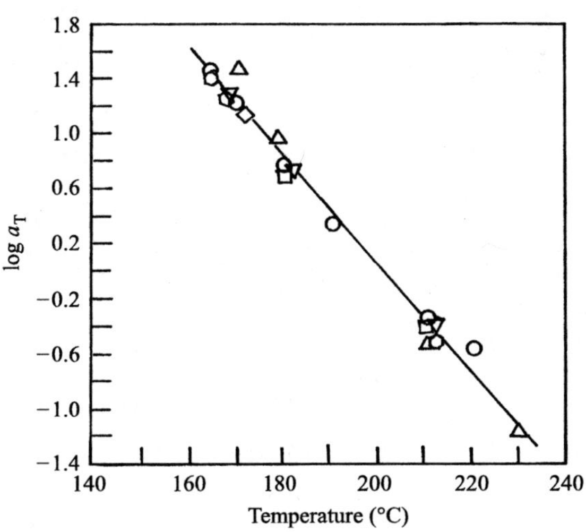

Figure 7.10 Temperature dependence of a_T for PMMA/PSAN blends: (○) PSAN, (△) PMMA, (□) 20/80 PMMA/PSAN, (▽) 40/60 PMMA/PSAN, (◇) 60/40 PMMA/PSAN, and (⬡) 80/20 PMMA/PSAN. The reference temperature T_r employed is 190 °C. (Reprinted from Yang et al., *Polymer* 35:1503. Copyright © 1994, with permission from Elsevier.)

research groups (Bernstein et al. 1977; Hahn et al. 1992; Jolene et al. 1979; McBrierty et al. 1978; McMaster 1975; Naito et al. 1978; Schmitt et al. 1980) have reported on the miscibility of PMMA/PSAN blends which exhibit a single T_g over the entire range of blend compositions. The T_g of PSAN having 25.3% acrylontrile was approximately 110 °C and the T_g of PMMA was approximately 130 °C, giving rise to $\Delta T_g \approx 20$ °C between PMMA and PSAN (Yang et al. 1994). Pathak et al. (1998) investigated the linear dynamic viscoelasticity of blends of PMMA and PSAN having 20% acrylonitrile and observed that TTS worked well. They ascribed the successful application of TTS to a small ΔT_g (about 20 °C) between the PMMA and PSAN. They concluded that the success or failure of TTS in miscible blends may be determined based on how small or how large ΔT_g of the constituent components might be, echoing the view expressed earlier by Alegria et al. (1995) and Friedrich et al. (1996). Note that the PSAN/PMMA pair has a small negative value (about −0.01) of Flory–Huggins interaction parameter χ (Hahn et al. 1992; Jolene et al. 1979; Schmitt et al. 1980); that is, the PSAN/PMMA pair has weak attractive interactions. Later in this chapter we will discuss the rheological behavior of PMMA/PSAN blends in greater detail.

Let us return to the issue of dynamic heterogeneity in some miscible polymer blends. Needless to say, direct observation of the presence of dynamic heterogeneity in miscible polymer blends would depend on the method of probing domain size used, as already mentioned when referring to the use of solid-state ^{13}C NMR measurements reported by Miller et al. (1990). In order to facilitate our discussion here, let us look at the schematic diagram given in Figure 7.11, in which the upper panels show the domain size as observed using electron microscopy (EM) and the lower panels show the mechanical responses in terms of elastic modulus (E') and loss tangent ($\tan\delta$) using dynamic mechanical spectroscopy (DMS). The arrows in the schematic represent the probe size employed for a particular experimental method chosen. In presenting Figure 7.11, Kaplan (1976) defined a compatibility number, N_c, as the ratio of the experimental probe size and the domain size. Thus, $N_c \to \infty$ for compatible polymer blends, $N_c \to 1$ for semicompatible polymer blends, and $N_c \to 0$ for incompatible polymer blends. Figure 7.11 is presented just to indicate the importance of the experimental method of probing domain size when determining the presence of dynamic heterogeneity in miscible polymer blends. After all, the widths of transitions are caused by concentration fluctuations within each phase. In this regard, the schematic given in Figure 7.11 should not cause any confusion.

For instance, the experimental probe size used in DSC or DMS to determine the T_g of a polymer blend is not small enough to detect the presence of microheterogeneity with domain sizes smaller than, say, 5–10 nm. Similarly, when the cloud points of a polymer blend are measured using light scattering, it is not possible to detect the presence of microheterogeneity with domain sizes smaller than the wavelength (approximately 450 nm) of the He–Ne laser light source. Figure 7.12 gives binodal curves for poly(styrene-*co*-maleic anhydride) (PSMA)/PVME blends exhibiting LCST behavior, which was determined using two different experimental techniques: electron spin resonance (ESR) and cloud point measurements. In Figure 7.12 we observe that the temperature determined from the ESR technique, which separates the homogeneous region from the inhomogeneous regions, is about 70 °C lower than that determined from cloud point measurements. From Figure 7.12, Müller et al. (1992) concluded that concentration inhomogeneities were present in the PSMA/PVME blend on a molecular

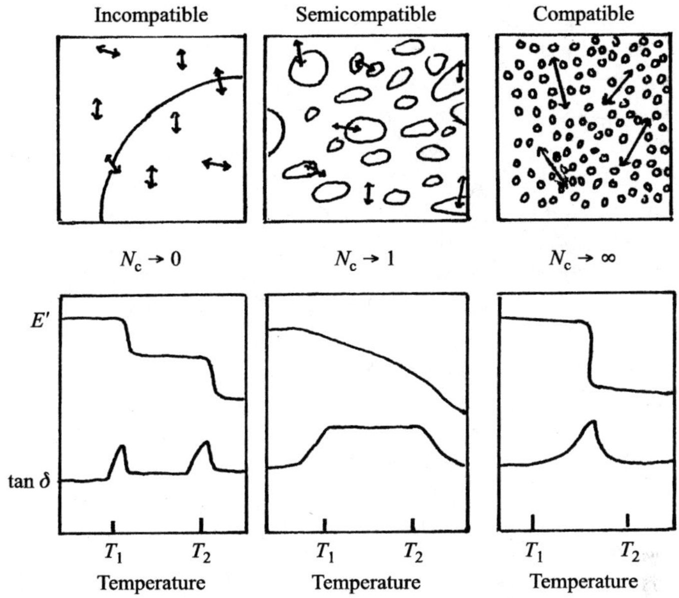

Figure 7.11 Schematic showing the experimental probe size relative to the domain size in a polymer blend when using electron microscopy and dynamic mechanical spectroscopy to determine polymer miscibility. (Reprinted from Kaplan, *Journal of Applied Polymer Science* 20:2615. Copyright © 1976, with permission from John Wiley & Sons.)

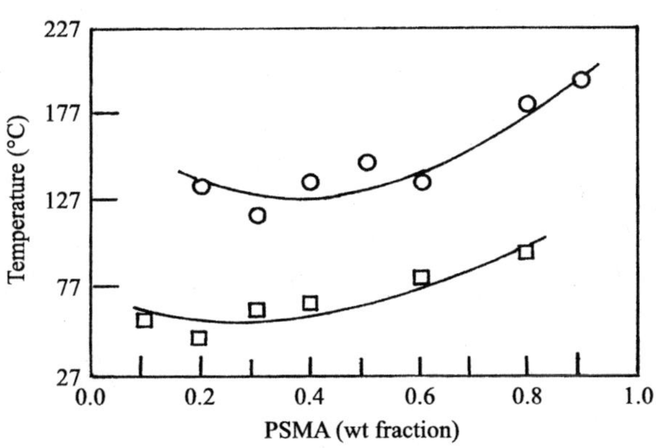

Figure 7.12 Phase diagram for PSMA/PVME blends determined from electron spin resonance technique (□) and cloud point measurements using laser light scattering (○). (Reprinted from Müller et al., *Macromolecular Rapid Communications*. Copyright © 1992, with permission from Wiley-VCH.)

scale (≤5 nm), even at temperatures far below the binodal curve determined from cloud point measurement.

Failure of TTS in some miscible polymer blends[2] has also been attributed to the presence of two species whose relaxation dynamics have different temperature dependencies (Chung et al. 1994b; Colby 1989). The slower relaxing component of the blend has a stronger temperature dependence. In other words, the relaxation time of the polymer with the higher T_g has stronger temperature dependence than the relaxation time of the blend component with the lower T_g (e.g., PEO/PMMA, PMMA/PVDF, and PS/PVME blends). In such situations, TTS has been found to fail. Again, this is the situation where the difference in component T_g (ΔT_g) plays a role in determining whether TTS might fail in miscible blends. It should be mentioned that the relaxation time τ_1 of the slower process can be determined by $\tau_1 = 1/\omega_{G''=G'}$ (i.e., the reciprocal of the cross-over frequency in the terminal region of oscillatory shear flow), and the relaxation time of the faster process can be determined by $\tau_1 = 1/\omega_{G''\max}$. As we will discuss below, miscible polymer blends invariably exhibit glass transitions that are broader than those of the constituent components, so that such breadth is not an unambiguous indication of phase segregation.

There is another view on the failure of TTS in some miscible polymer blends. Namely, TTS fails when the monomeric friction coefficients ($\zeta_{0,\mathrm{A}}$ and $\zeta_{0,\mathrm{B}}$) of the constituent components of a miscible polymer blend are different at a chosen reference temperature, $T_r = T_g +$ constant; that is, when the polymer blend is not in the iso-zeta state. In other words, when $\zeta_{0,\mathrm{A}}$ is *not* equal to $\zeta_{0,\mathrm{B}}$, the validity of a comparison of the viscosities for different blend compositions at T_r is questionable. This means that one needs information on the temperature dependence of $\zeta_{0,\mathrm{A}}$ and $\zeta_{0,\mathrm{B}}$ to determine whether a miscible polymer blend is in the iso-zeta state at T_r. For instance, Roland and Ngai (1991) concluded that TTS failed in the miscible PI/PVE blends because the friction coefficients of the constituent components were different; that is, the iso-zeta state was not attained. According to this point of view, when TTS fails in a miscible polymer blend, the use of a single average friction coefficient is not adequate to describe the rheology of miscible polymer blends.

Information on the monomeric friction coefficients of the constituent components in a miscible polymer blend is therefore vital for a better understanding of the rheological behavior (the dynamics) of miscible blends. Efforts to obtain such information have been reported by some investigators (Composto et al. 1990, 1992; Kim et al. 1994), who employed forward recoil spectrometry to measure the reptation (tracer) diffusion coefficients of the constituent components in a miscible polymer blend. Specifically, measurements of tracer diffusion coefficient (D^*) allowed them to calculate monomeric friction coefficients using the following expression, which is based on the tube model (Kim et al. 1994):[3]

$$D_i^* = D_{\mathrm{o},i}/M_i^{\,2} \tag{7.10}$$

with M_i being the molecular weight of component i and

$$D_{\mathrm{o},i} = \frac{4}{15}\frac{M_{\mathrm{o},i}\,M_{\mathrm{e,blend}}\,k_{\mathrm{B}}T}{\zeta_{0,i}} \tag{7.11}$$

where $M_{\mathrm{o},i}$ and $\zeta_{0,i}$ are the molecular weight of the monomer units and monomeric friction coefficient of component i, respectively, $M_{\mathrm{e,blend}}$ is the entanglement molecular weight of the blend as a whole, and $k_{\mathrm{B}}T$ has its usual meaning, with k_{B} being the Boltzmann constant and T being the absolute temperature. An important question arises as to how to calculate (or estimate) $M_{\mathrm{e},i}$ ($i = 1, 2$) of the constituent components in a given miscible polymer blend. The answer to this question is not trivial, and later in this chapter we will address this issue. Note in Eq. (7.10) that D_i^* depends on blend composition because $\zeta_{0,i}$ depends on blend composition. Also, Green et al. (1991)

employed elastic recoil detection to measure the reptation diffusion coefficients of the constituent components in a miscible polymer blend, which enabled them to calculate monomeric friction coefficients.

Also, the concentration fluctuations present in thermodynamically miscible polymer blends having weak interactions have been related to dynamic heterogeneity (Arendt et al. 1997; Chung et al. 1994b; Pathak et al. 1999; Roland and Ngai 1992). It can then be concluded that concentration fluctuations, dynamic heterogeneity, failure of TTS, and the broadness of glass transition in miscible polymer blends having weak interactions are interrelated.

7.3.2 Rheology of Polymer Blends Exhibiting UCST

Here, we consider the rheological behavior of PS/PαMS blends exhibiting UCST. The miscibility and phase behavior of PS/PαMS blends have been studied (Cowies and McEwen 1985; Lau et al. 1982; Lin and Roe 1987, 1988; Rameau et al. 1989; Saeki et al. 1983; Schneider and Dilger 1989; Wimpier and Maynard 1987) and the miscibility window of PS/PαMS blends has been found to be very sensitive to the molecular weights of the constituent components. We next consider the phase behavior and linear dynamic viscoelastic properties of three pairs of PS/PαMS blends, each having different molecular weights, as summarized in Table 7.1.

Figure 7.13 gives, for illustration, DSC thermograms of (PS-40)/(PαMS-18) blends at a heating rate of 20 °C/min, showing the onset points (T_{gi}), the midpoints (T_{gm}), and the final points (T_{gf}) of the glass transition. It is seen in Figure 7.13 that each blend composition has a very broad, single glass transition, indicating that (PS-40)/(PαMS-18) blends are miscible over the entire blend composition. Table 7.2 gives a summary of the values of T_{gi}, T_{gm}, and T_{gf} determined from DSC for different blend compositions of (PS-40)/(PαMS-18), (PS-38)/(PαMS-39), and (PS-40)/(PαMS-48) blend systems. What is significant in Table 7.2 is that each of the blends undergoes a very broad, single glass transition, as manifested by the width of glass transition, $\Delta T_g = T_{gf} - T_{gi}$.

Figure 7.14 gives composition-dependent T_{gi}, T_{gm}, and T_{gf} curves for (PS-40)/(PαMS-18), (PS-38)/(PαMS-39), and (PS-40)/(PαMS-48) blend systems, and shows the binodal curves calculated using the Flory–Huggins theory and the temperatures at which oscillatory shear measurements were taken for each blend. The binodal curves given in Figure 7.14 were obtained using the following expression for the interaction parameter α (Kim et al. 1998):

$$\alpha = -2.165 \times 10^{-5} + 0.029/T - 0.0021\phi_{\mathrm{P\alpha MS}}/T \tag{7.12}$$

Table 7.1 Molecular characteristics of PS/PαMS blends

Sample Codes	M_w *of PS*	M_w *of PαMS*	M_w/M_n *of PS*	M_w/M_n *of PαMS*
(PS-40)/(PαMS-18)	4.05×10^4	1.80×10^4	1.05	1.04
(PS-38)/(PαMS-39)	3.87×10^4	3.90×10^4	1.04	1.04
(PS-40)/(PαMS-48)	4.05×10^4	4.80×10^4	1.05	1.05

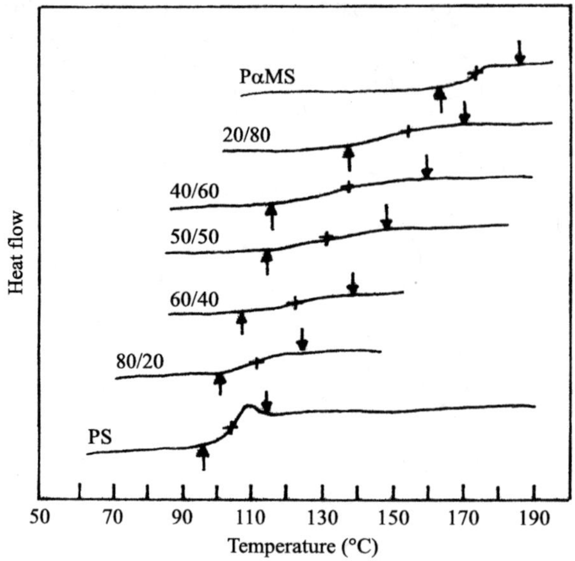

Figure 7.13 DSC thermograms for (PS-40)/(PαMS-18) blends during heating at a rate of 20 °C/min, where the arrow pointing upward represents an onset point of glass transition T_{gi}, the arrow pointing downward represents the temperature at which the glass transition is completed T_{gf}, and the symbol + represents the midpoint of glass transition T_{gm}.

Table 7.2 Summary of the glass transition temperatures for three PS/PαMS blend systems

Sample Code	T_{gi} (°C)	T_{gm} (°C)	T_{gf} (°C)	ΔT_g (°C)[a]
(a) (PS-40)/(PαMS-18) blend system				
PS-40	96	105	114	18
80/20 (PS-40)/(PαMS-18)	101	112	125	24
60/40 (PS-40)/(PαMS-18)	107	122	139	32
50/50 (PS-40)/(PαMS-18)	113	131	148	35
40/60 (PS-40)/(PαMS-18)	115	137	159	44
20/80 (PS-40)/(PαMS-18)	136	153	170	34
PαMS-18	161	173	185	24
(b) (PS-38)/(PαMS-39) blend system				
PS-38	95	103	113	18
80/20 (PS-38)/(PαMS-39)	98	108	117	19
60/40 (PS-38)/(PαMS-39)	103	117	134	31
50/50 (PS-38)/(PαMS-39)	105	127	151	46
40/60 (PS-38)/(PαMS-39)	112	137	159	45
20/80 (PS-38)/(PαMS-39)	132	153	163	31
PαMS-39	159	168	178	19
(c) (PS-40)/(PαMS-48) blend system				
PS-40	96	105	114	18
80/20 (PS-40)/(PαMS-48)	101	110	122	21
60/40 (PS-40)/(PαMS-48)	102	120	146	44
50/50 (PS-40)/(PαMS-48)	105	134	163	58
40/60 (PS-40)/(PαMS-48)	111	142	168	57
20/80 (PS-40)/(PαMS-48)	137	157	171	34
PαMS-48	169	177	189	20

[a] $\Delta T_g = T_{gf} - T_{gi}$

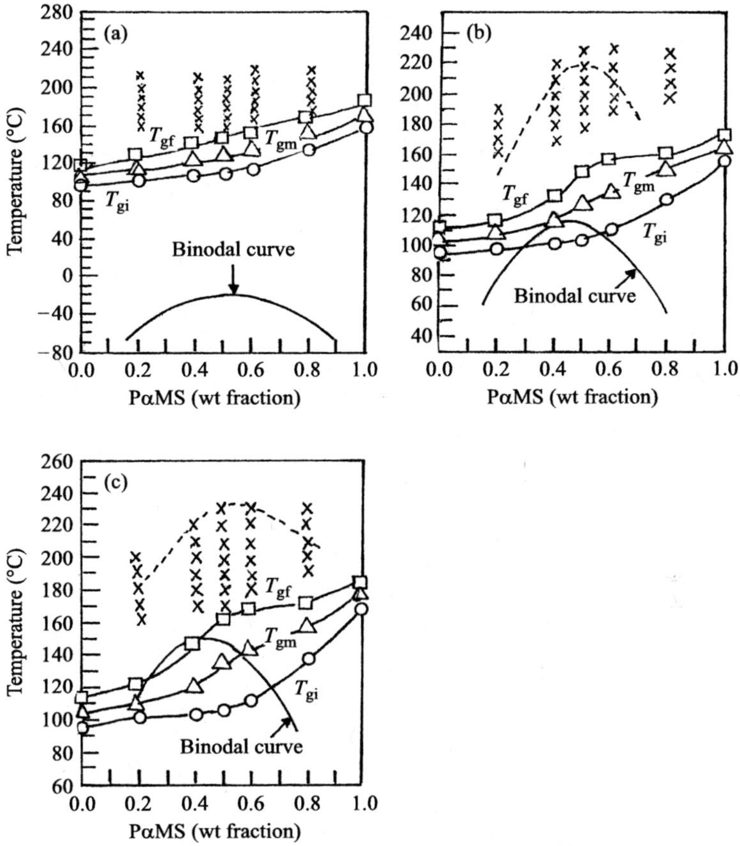

Figure 7.14 Composition-dependent glass transition temperatures T_{gi} (○), T_{gm} (△), and T_{gf} (□) for (a) (PS-40)/(PαMS-18) blend system, (b) (PS-38)/(PαMS-39) blend system, and (c) (PS-40)/(PαMS-48) blend system, in which the solid line (——) is the binodal curve calculated from the Flory–Huggins theory, the symbol × represents the temperatures at which oscillatory shear experiments were conducted, and the broken line represents the boundary separating the homogeneous phase and microheterogeneous region, as determined by log G' versus log G'' plots. (Reprinted from Kim et al., *Macromolecules* 31:8566. Copyright © 1998, with permission from the American Chemical Society.)

where $\phi_{\mathrm{P\alpha MS}}$ is the volume fraction of PαMS. Note that Eq. (7.12) was obtained by curve fitting experimental binodal curve to the Flory–Huggins theory and using the specific volume (in units of cm^3/g) for PαMS given by

$$v_{\mathrm{P\alpha MS}} = 0.87 + 5.08 \times 10^{-4}(T - 273) \tag{7.13}$$

and the specific volume for PS given by Eq. (7.4). The following observations are worth noting in Figure 7.14. The calculated binodal curve for the (PS-40)/(PαMS-18) blend system lies very far below the T_{gi} curve (Figure 7.14a), the calculated binodal curve for the (PS-38)/(PαMS-39) blend system lies just above the T_{gi} curve but below the

T_{gm} curve (Figure 7.14b), and the calculated binodal curve for the (PS-40)/(PαMS-48) blend system lies slightly above the T_{gf} curve (Figure 7.14c). That is, the calculated binodal curve moves upward as the molecular weights of the constituent components increase. The broken lines in Figures 7.14b and 7.14c represent the phase boundary separating the homogeneous and microheterogeneous regions as determined (as will be elaborated on below) by the log G' versus log G'' plots for each blend composition.

Figure 7.15 gives log G' versus log G'' plots at various temperatures for four different compositions of the (PS-40)/(PαMS-18) blend system. It can be seen in Figure 7.15 that the log G' versus log G'' plots for all four blend compositions are virtually independent of temperature and that the slope of the plots in the terminal region is very close to 2. Using the rheological criterion presented in Chapter 6 for flexible homopolymers, we can conclude from Figure 7.15 that the (PS-40)/(PαMS-18) blends are homogenous above 160 °C over the entire range of blend compositions.

Let us examine whether TTS is applicable to the (PS-40)/(PαMS-18) blend system, which gives rise to virtually temperature-independent log G' versus log G'' plots with

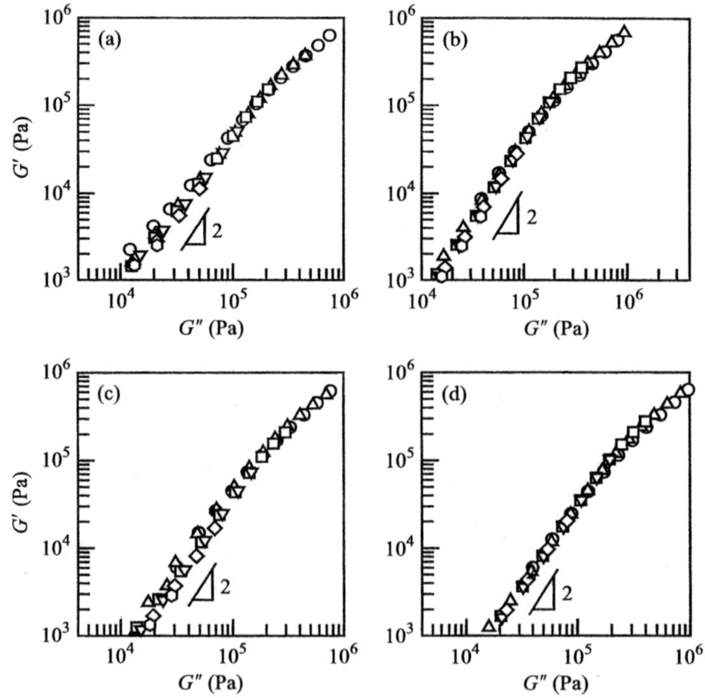

Figure 7.15 Log G′ versus log G'' plots for (a) 60/40 (PS-40)/(PαMS-18) blend at various temperatures (°C): (○) 160, (△) 170, (□) 180, (▽) 190, (◇) 200, and (⬡) 210, (b) 50/50 (PS-40)/(PαMS-18) blend at various temperatures (°C): (○) 160, (△) 170, (□) 180, (▽) 190, (◇) 200, and (⬡) 210, (c) 40/60 (PS-40)/(PαMS-18) blend at various temperatures (°C): (○) 170, (△) 180, (□) 190, (▽) 200, (◇) 210, and (⬡) 220, and (d) 20/80 (PS-40)/(PαMS-18) blend at various temperatures (°C): (○) 180, (△) 190, (□) 200, (▽) 210, and (◇) 220.

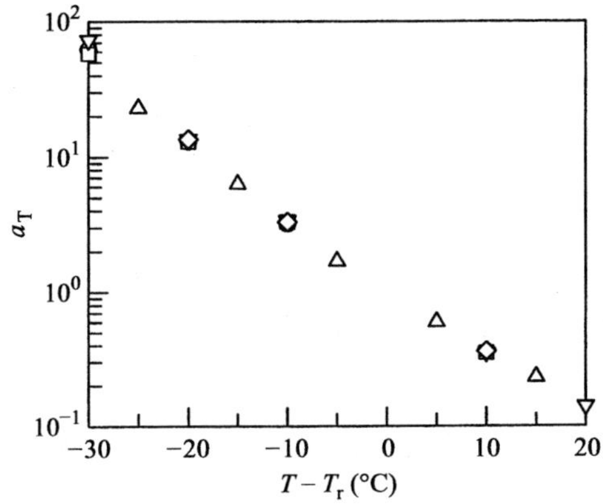

Figure 7.16 Plots of a_T versus $T - T_r$ for five blend ratios with $T_r = T_{gm} + 68\,°C$: (○) 20/80 (PS-40)/(PαMS-18) blend, (△) 40/60 (PS-40)/(PαMS-18) blend, (□) 50/50 (PS-40)/(PαMS-18) blend, (▽) 60/40 (PS-40)/(PαMS-18) blend, and (◇) 80/20 (PS-40)/(PαMS-18) blend.

a slope of 2 in the terminal region (see Figure 7.15). Figure 7.16 gives plots of a shift factor a_T, which was obtained by shifting log G'' versus log ω plots along the ω axis, versus $T - T_r$ with $T_r = T_{gm} + 68\,°C$ for the (PS-40)/(PαMS-18) blend system with varying blend ratios, showing temperature-independent correlation. Note that (PS-40)/(PαMS-18) blends undergo a single, very broad glass transition (see Figure 7.13). Thus, in obtaining Figure 7.16 we employed the midpoint of glass transition temperature (T_{gm}) of each blend to calculate reference temperature T_r (see Table 7.2 for the values of T_{gm}). It is seen in Figure 7.16 that values of a_T lie on a single line, suggesting that TTS is applicable to the (PS-40)/(PαMS-18) blend system. Using the procedures described in Chapter 6, we have determined the parameters C_1 and C_2 appearing in the WLF equation defined by Eq. (6.7), and they are summarized in Table 7.3. We have confirmed, although not presented here, that the use of WLF parameters C_1 and C_2 given in Table 7.3 has yielded a_T versus $T - T_r$ plots which are virtually identical to Figure 7.16.

What seems unusual at first glance from Figures 7.15 and 7.16 is that TTS is applicable to the (PS-40)/(PαMS-18) blends despite the fact that the difference in component T_gs between PS and PαMS is rather large, say $\Delta T_g \approx 70\,°C$. In the preceding section we mentioned that TTS may fail in miscible polymer blends when ΔT_g is larger than approximately 20 °C. The miscibility of a polymer blend depends, among other factors, on the molecular weights of the constituent components. Keeping this in mind, notice

Table 7.3 Summary of WLF parameters for (PS-40)/(PαMS-18) blends

Blend Ratio	T_r (°C)	C_1	C_2 (°C)
20/80 (PS-40)/(PαMS-18)	220	6.61	140.16
40/60 (PS-40)/(PαMS-18)	205	6.02	130.39
50/50 (PS-40)/(PαMS-18)	200	6.02	130.39
60/40 (PS-40)/(PαMS-18)	190	6.86	140.47
80/20 (PS-40)/(PαMS-18)	180	6.18	129.43

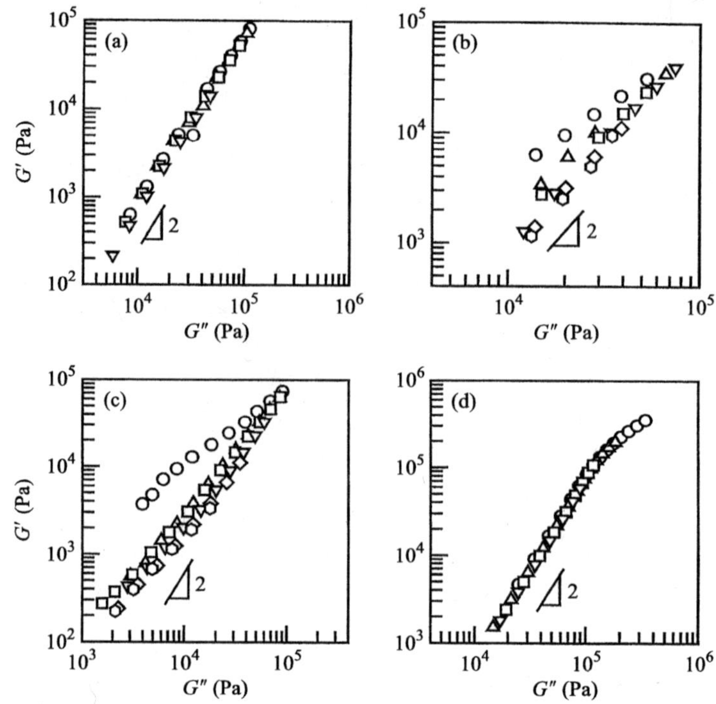

Figure 7.17 Log G' versus log G'' plots for (a) 80/20 (PS-38)/(PαMS-39) blend at various temperatures (°C): (○) 160, (△) 170, (□) 180, and (▽) 190, (b) 60/40 (PS-38)/(PαMS-39) blend at various temperatures (°C): (○) 170, (△) 180, (□) 190, (▽) 200, and (◇) 210, (c) 50/50 (PS-38)/(PαMS-39) blend at various temperatures (°C): (○) 180, (△) 190, (□) 200, (▽) 210, and (◇) 220, and (d) 20/80 (PS-38)/(PαMS-39) blend at various temperatures (°C): (○) 200 (△) 210, (□) 220, and (▽) 230.

in Figure 7.14a that the theoretical binodal curve for the (PS-40)/(PαMS-18) blends is very far below the T_gs of the blends, which is attributable to the fairly low molecular weights of PS-40 and PαMS-18.

Figure 7.17 gives log G' versus log G'' plots at various temperatures for four different compositions of the (PS-38)/(PαMS-39) blend system. It can be seen from Figures 7.17a and 7.17d that the log G' versus log G'' plots for the 20/80 (PS-38)/(PαMS-39) and 80/20 (PS-38)/(PαMS-39) blends are virtually independent of temperature and the slope of the plots in the terminal region is very close to 2. Conversely, the log G' versus log G'' plots for the 60/40 and 50/50 (PS-38)/(PαMS-39) blends show temperature dependence. Specifically, in Figure 7.17b we observe that at 170 °C the log G' versus log G'' plot in the terminal region has a slope much less than 2, but that it is steadily shifted downward with a slope closer to 2 as the temperature is increased toward 210 °C, at which point the temperature dependence of log G' versus log G'' plot begins to cease with further increase in temperature, giving rise to a slope very close to 2 in the terminal region. Thus, we can conclude that the 60/40

(PS-38)/(PαMS-39) blend forms a homogeneous phase at 220 °C and higher temperatures, but has microheterogeneity at temperatures below approximately 210 °C. Similarly, we can conclude from Figure 7.17c that the 50/50 (PS-38)/(PαMS-39) blend forms a homogeneous phase at 220 °C and higher temperatures, but has microheterogeneity at temperatures below approximately 220 °C. The conclusions drawn for the 60/40 and 50/50 (PS-38)/(PαMS-39) blends stand in contrast to the view that these two blends are homogeneous, because they undergo a single, though very broad, glass transition (see Figure 7.14b and Table 7.2). It then seems reasonable to conclude that the presence of a single, broad glass transition in the 60/40 and 50/50 (PS-38)/(PαMS-39) blends cannot discern the presence of microheterogeneity. The broken line in Figure 7.14b shows the phase boundary, which is determined from the log G' versus log G'' plots, separating the homogeneous and microheterogeneous regions.

Figure 7.18 gives reduced log G'_r versus log $a_T\omega$ plots for 60/40 (PS-38)/(PαMS-39) blend, which were obtained by first determining a temperature shift factor a_T from shifting log G'' versus log ω plots at various temperatures, with 170 °C as reference temperature T_r, along the ω axis (see Chapter 6). It is seen in Figure 7.18 that the reduced plots show temperature dependence, which is a clear indication of the failure of TTS. The temperature dependence of the log G'_r versus log $a_T\omega$ plots given in Figure 7.18 is very similar to the temperature dependence of log G' versus log G'' plots observed from Figure 7.17b.

Figure 7.19 gives log G' versus log G'' plots at various temperatures for four different compositions of the (PS-40)/(PαMS-48) blend system. It can be seen in Figure 7.19 that the log G' versus log G'' plots depend on temperature in all four blend compositions, in spite of the fact that each blend undergoes a single, broad glass transition (see Figure 7.14c). Thus, we conclude that the (PS-40)/(PαMS-48) blends have microheterogeneity over the range of temperatures investigated. Referring to Figure 7.14c, the broken line separates the homogeneous and microheterogeneous regions in the (PS-40)/(PαMS-48) blend system. Comparison of Figure 7.19 for the (PS-40)/(PαMS-48) blends with Figure 7.15 for the (PS-40)/(PαMS-18) blends and with Figure 7.17 for the (PS-38)/(PαMS-39) blends reveals that the molecular weights of the constituent

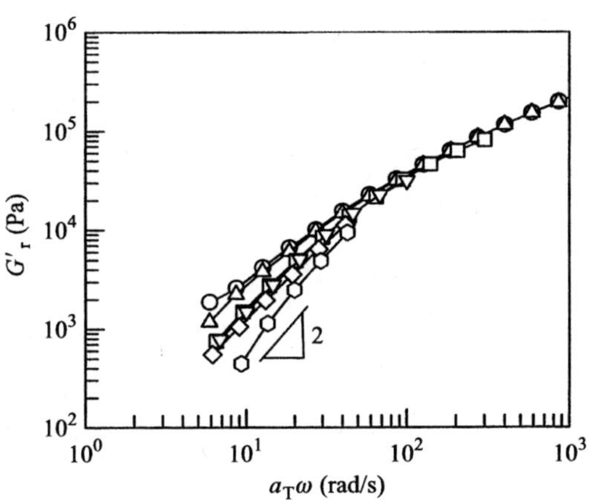

Figure 7.18 Log G'_r versus log $a_T\omega$ plots for 60/40 (PS-38)/(PαMS-39) blend with T_r = 170 °C at various temperatures (°C): (○) 170, (△) 180, (□) 190, (▽) 200, (◇) 210, and (⬡) 220.

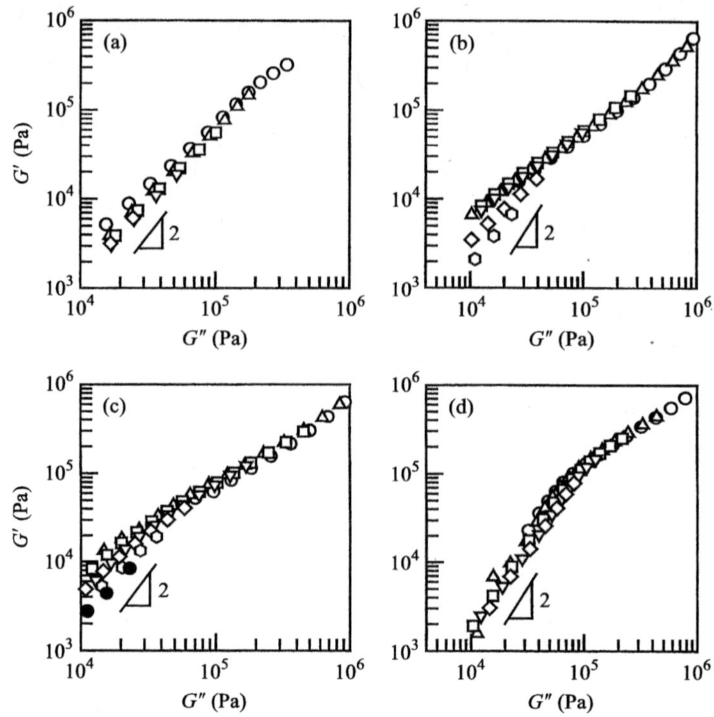

Figure 7.19 Log G' versus log G'' plots for (a) 80/20 (PS-40)/(PαMS-48) blend at various temperatures (°C): (○) 160, (△) 170, (□) 180, (▽) 190, and (◇) 200, (b) 60/40 (PS-40)/(PαMS-48) blend at various temperatures (°C): (○) 170, (△) 180, (□) 190, (▽) 200, (◇) 210, and (⬡) 220, (c) 50/50 (PS-40)/(PαMS-48) blend at various temperatures (°C): (○) 170, (△) 180, (□) 190, (▽) 200, (◇) 210, (⬡) 220, and (●) 230, and (d) 20/80 (PS-40)/(PαMS-48) blend at various temperatures (°C): (○) 190, (△) 200, (□) 210, (▽) 220, and (⬡) 230.

components have a profound influence on the linear dynamic viscoelastic properties of miscible polymer blends.[2]

Figure 7.20 shows the composition dependence of zero-shear viscosity (η_0) at $T_r = T_{gi} + 100\,°C$ as a reference temperature for (PS-40)/(PαMS-18) and (PS-40)/(PαMS-48) blends, and Figure 7.21 shows the composition dependence of η_0 at $T_r = T_{gm} + 70\,°C$ as a reference temperature for (PS-40)/(PαMS-18) and (PS-40)/(PαMS-48) blends. It is seen in Figures 7.20 and 7.21 that η_0 goes through a maximum at a certain blend composition and the dynamical behavior of chains in a PS/PαMS blend does not follow the average mobility of the blend at T_{gi} or at T_{gm}. It is clear from Figures 7.20 and 7.21 that the composition dependence of η_0 of a PS/PαMS blend varies depending on whether a reference temperature is chosen at an equal distance from T_{gi} or T_{gm}. That is, a serious question may be raised as to which of the T_g values (i.e., T_{gi}, T_{gm}, or T_{gf}) should be used to assess the composition dependence of blend viscosities when the breadth of glass transition varies significantly with blend composition (see Figure 7.14). This is a dilemma that does not seem to have an easy answer.

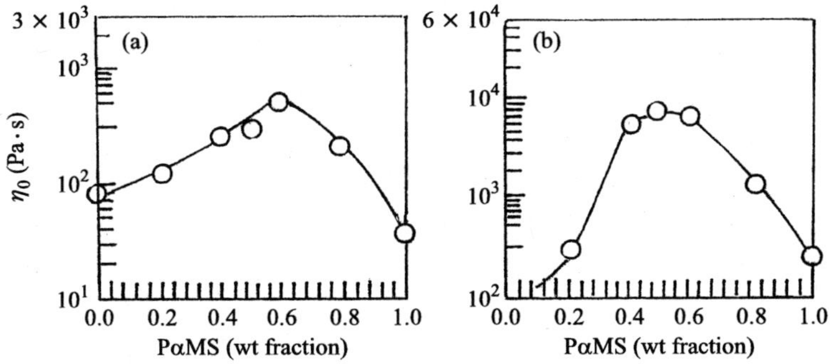

Figure 7.20 Composition dependence of η_0 for (a) the (PS-40)/(PαMS-18) blend system, and (b) the (PS-40)/(PαMS-48) blend system at $T_r = T_{gi} + 100\,°C$. (Reprinted from Kim et al., *Macromolecules* 31:8566. Copyright © 1998, with permission from the American Chemical Society.)

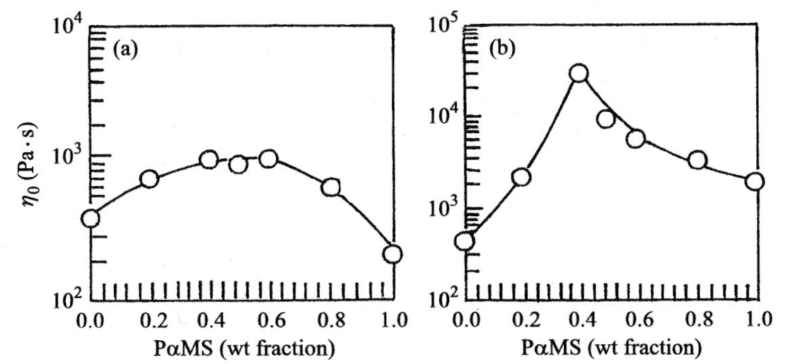

Figure 7.21 Composition dependence of η_0 for (a) the (PS-40)/(PαMS-18) blend system, and (b) the (PS-40)/(PαMS-48) blend system at $T_r = T_{gm} + 70\,°C$. (Reprinted from Kim et al., *Macromolecules* 31:8566. Copyright © 1998, with permission from the American Chemical Society.)

7.3.3 Rheology of Polymer Blends Exhibiting LCST

Here, we consider the rheological behavior of PS/PVME blends exhibiting LCST. Numerous research groups investigated phase equilibria (Nishi and Kwei 1975; Nishi et al. 1975; Yang et al. 1986), phase separation morphology (David and Kwei 1980; Nishi et al. 1975; Yang et al. 1986), the kinetics of phase separation (Halary et al. 1984; Han et al. 1986a; Okada and Han 1986; Polios et al. 1997), the temperature and composition dependences of the Flory–Huggins interaction parameter (Han et al. 1986b; Shibayama et al. 1985), and the rheological behavior (Ajji and Choplin 1991; Ajji et al. 1988, 1991; Cavaille et al. 1987; Kapnistos et al. 1996; Kim et al. 1998; Kitade et al. 1994, 1998; Pathak et al. 1999; Roland and Ngai 1992; Schneider and Wirbser 1990; Takahashi et al. 1994) of PS/PVME blends. The primary objective of this section is to present the linear dynamic viscoelasic properties of PS/PVME blends

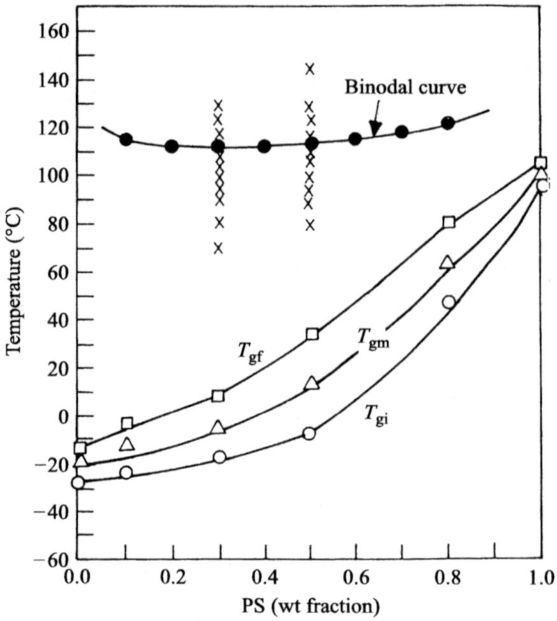

Figure 7.22 Composition-dependent cloud points (●) determined from light scattering and glass transition temperatures T_{gi} (○), T_{gm} (△), and T_{gf} (□) for (PS-110)/(PVME-95) blends, where the solid line (——) is a binodal curve calculated from curve-fitting experimental data to the Flory–Huggins theory. The symbol × denotes temperatures at which oscillatory shear experiments were conducted. (Reprinted from Kim et al., *Macromolecules* 31:8566. Copyright © 1998, with permission from the American Chemical Society.)

in relation to their phase behavior in the same spirit as presented in the previous section for PS/PαMS blends.

Figure 7.22 gives composition-dependent cloud points, composition-dependent T_{gi}, T_{gm}, and T_{gf} curves for (PS-110)/(PVME-95) blends obtained from DSC, and the temperatures at which rheological measurements were taken for the (PS-110)/(PVME-95) blends. Table 7.4 gives a summary of the values of T_{gi}, T_{gm}, and T_{gf} determined from DSC for different blend compositions of (PS-110)/(PVME-95) blends. Again, a broad glass transition in PS/PVME blends signifies the presence of dynamic heterogeneity within a miscible blend, which is attributable to a large difference in

Table 7.4 Glass transition temperatures for the PS/PVME blend system

Sample Code	T_{gi} (°C)	T_{gm} (°C)	T_{gf} (°C)	ΔT_g (°C)[a]
PS-110	97.7	102.5	107.0	9.3
80/20 (PS-110)/(PVME-95)	48.3	65.0	83.0	34.7
50/50 (PS-110)/(PVME-95)	−7.0	12.3	35.0	42.0
30/70 (PS-110)/(PVME-95)	−17.5	−6.0	8.7	26.2
10/90 (PS-110)/(PVME-95)	−23.5	−12.0	−2.5	21.0
PVME-95	−28.0	−18.0	−14.0	14.0

[a] $\Delta T_g = T_{gf} - T_{gi}$

Reprinted from Kim et al., *Macromolecules* 31:8566. Copyright ©1998, with permission from the American Chemical Society.

component T_gs ($\Delta T_g \approx 120\,°C$) between PS and PVME. The solid line in Figure 7.22 shows the binodal curve calculated from the Flory–Huggins theory using the following expression for α:

$$\alpha = 0.478 \times 10^{-3} - 0.176/T - 0.0062\phi_{PS}/T \tag{7.14}$$

where ϕ_{PS} is the volume fraction of PS, and the following expression for the specific volume (in units of cm³/g):

$$v_{PVME} = 1/\left[1.0717 - 7.67 \times 10^{-4}(T - 273) + 2.8 \times 10^{-7}(T - 273)^2\right] \tag{7.15}$$

for PVME. Notice in Figure 7.22 that the binodal curve for the (PS-110)/(PVME-95) blend system exhibits LCST at 112 ± 1 °C and the critical weight fraction of PS in the blend is 0.3.

Again, the binodal curve displayed in Figure 7.22, which was obtained from cloud point measurements, cannot describe whether the (PS-110)/(PVME-95) blends are homogeneous on the segmental level. However, we can infer from Figure 7.12 that there is dynamic heterogeneity below the binodal curve given in Figure 7.22. It is worth mentioning that based on a ^{13}C NMR study, Wagler et al. (2000) concluded that microheterogeneities from 3.5 nm to approximately 30 nm existed within the PS/PVME blends, depending on the temperature of thermal treatment applied to a specimen, and that microheterogeneities existed in the PS/PVME blends at temperatures below the binodal curve determined by cloud point measurements. Such conclusion is consistent with that made by Muller et al. (1992), who employed the electron spin resonance technique (see Figure 7.12).

Figure 7.23 gives plots of log G' versus log ω and log G'' versus log ω for a 30/70 (PS-110)/(PVME-95) blend at various temperatures. It is interesting to observe in Figure 7.23 that values of G' and G'' first decrease with increasing temperature from 70 to 100 °C, and then increase with further increase in temperature to 130 °C.

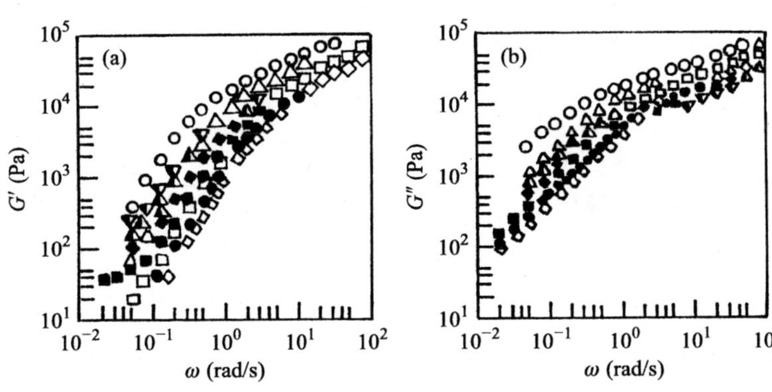

Figure 7.23 Plots of (a) log G' versus log ω and (b) log G'' versus log ω for a 30/70 (PS-110)/(PVME-95) blend at various temperatures (°C): (○) 70, (△) 80, (□) 90, (◇) 100, (●) 105, (■) 110, (◆) 113, (▲) 120, and (▼) 130.

Referring to the phase diagram given in Figure 7.22, the 30/70 (PS-110)/(PVME-95) blend at 105, 108, and 110 °C are all below the critical temperature; in other words, the blend is in the miscible region as determined from cloud point measurements. Thus, we conclude that the increasing trend of G' and G'' as the temperature approaches the critical temperature of the blend may signify the onset of composition fluctuations. Otherwise, there is no reason why both G' and G'' should increase as the temperature increases from 100 to 110 °C. Next, we elaborate on the dynamic composition fluctuations in miscible polymer blends in general.

Figure 7.24 gives $\log G'$ versus $\log G''$ plots for the 30/70 (PS-110)/(PVME-95) blend. It can be seen from Figure 7.24 that the $\log G'$ versus $\log G''$ plots in the terminal region at 105, 108, and 110 °C, which are all below the critical temperature (i.e., in the miscible region as determined from cloud point measurements), have slopes less than 2, and vary with temperature. Such characteristics of $\log G'$ versus $\log G''$ plots suggest that TTS would fail in the (PS-110)/(PVME-95) blends. As a matter of fact, a number of investigators (Ajji et al. 1988, 1991; Kapnistos et al. 1996; Pathak et al. 1999; Roland and Ngai 1992) reported failure of TTS in PS/PVME blends.

In view of the fact that a binodal curve constructed from equilibrium thermodynamics (i.e., from the Flory–Huggins theory) determines the boundary between the homogeneous and inhomogeneous regions in a mixture of two liquids, the presence of microheterogeneity discussed above in this and preceding sections for PS/PαMS and PS/PVME blend systems may be attributable to dynamic composition fluctuations near the critical point. However, the extent of dynamic composition fluctuations would vary with blend system.

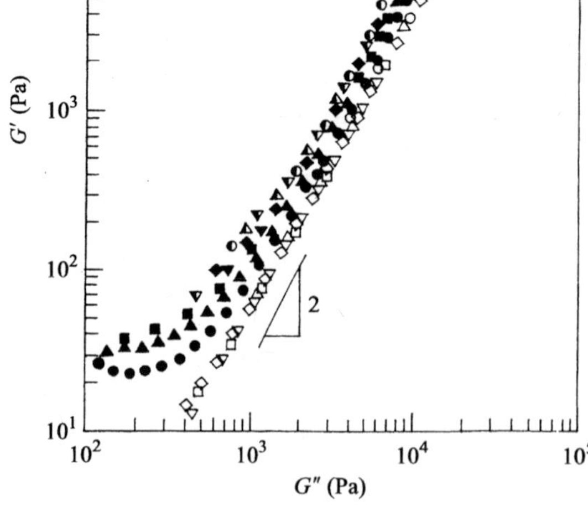

Figure 7.24 Log G' versus log G'' plots for the 30/70 (PS-110)/(PVME-95) blend at various temperatures (°C): (○) 70, (△) 80, (□) 90, (▽) 95, (◇) 100, (●) 105, (▲) 108, (■) 110, (▼) 112, (◆) 113, (◐) 117, (◭) 120, (◧) 125, and (⧨) 130. (Reprinted from Kim et al., *Macromolecules* 31:8566. Copyright © 1998, with permission from the American Chemical Society.)

The extent of concentration fluctuations, $\varepsilon = 2[1 - (\chi/\chi_c)]$, of a polymer blend near its critical temperature can be expressed by (Kim et al. 1998)

$$\varepsilon \approx \frac{4(1-\phi_{2,c})\phi_{2,c}r_1 r_2}{(1-\phi_{2,c})r_1 + \phi_{2,c}r_2}\left(\frac{d\chi}{dT}\right)_c (T_c - T) \tag{7.16}$$

where χ_c is the value of χ at the critical temperature T_c, $\phi_{2,c}$ is the critical volume fraction of reference component (PαMS for PS/PαMS blends and PS for PS/PVME blends) and r_i $(i = 1, 2)$ is the number of segments of component i. In obtaining Eq. (7.16), use was made of the following expression (de Gennes 1979)

$$2\chi_c = \frac{1}{r_1\phi_{1,c}} + \frac{1}{r_2\phi_{2,c}} \tag{7.17}$$

Use of Eq. (7.6) in Eq. (7.16) gives

$$\varepsilon \approx \frac{4\,(1-\phi_{2,c})\phi_{2,c}r_1 r_2}{(1-\phi_{2,c})r_1 + \phi_{2,c}r_2} V_{\mathrm{ref}} \left(\frac{b + c\,\phi_{2,c}}{T_c^{\,2}}\right)_c (T_c - T) \tag{7.18}$$

in which use was made of $\chi = \alpha V_{\mathrm{ref}}$, with V_{ref} being the molar volume of a reference component. A close look at Eq. (7.18) indicates that the coefficient b in Eq. (7.6) plays a predominant role in determining the extent of composition fluctuations ε near the critical point, because the magnitude of $c\phi_{2,c}$ (i.e., the contribution of the term $c\phi_B/T$ appearing on the right-hand side of Eq. (7.6)) is much smaller than that of b in Eq. (7.18).

Substituting Eq. (7.12) for PS/PαMS blends and Eq. (7.14) for PS/PVME blends into Eq. (7.16), we obtain the following relationship (Kim et al. 1998):

$$(T - T_c)_{\mathrm{PS/P\alpha MS}} \approx 15(T_c - T)_{\mathrm{PS/PVME}} \tag{7.19}$$

for the same value of ε in PS/PαMS and PS/PVME blend systems. Equation (7.19) suggests that the range of temperatures over which dynamic composition fluctuations near the critical point may persist is approximately 15 times greater in PS/PαMS blends than in PS/PVME blends.

7.4 Molecular Theory for the Linear Viscoelasticity of Miscible Polymer Blends and Comparison with Experiment

In this section, we present the molecular theory for the linear dynamic viscoelasticity of miscible polymer blends by Han and Kim (1989a, 1989b), which is based on the concept of the tube model presented in Chapter 4. Specifically, the reptation of two primitive chains of dissimilar chemical structures under an external potential will be considered, and the expressions for the linear viscoelastic properties of miscible polymer blends will be presented. We will first present the expressions for zero-shear viscosity η_{0b}, dynamic storage and loss moduli $G_b'(\omega)$ and $G_b''(\omega)$, and steady-state compliance J_{eb}^{o} for binary miscible blends of monodisperse, entangled flexible homopolymers and then consider the effect of polydispersity. There are a few other molecular theories reported on the dynamics and rheology of miscible polymer blends. Owing to the limited space available here, we do not present those theories. Interested readers are referred to the original papers (Haley and Lodge 2004; Pathak et al. 2004).

7.4.1 Linear Viscoelasticity Theory for Miscible Polymer Blends

Consider two primitive chains, 1 and 2, representing a mixture of polymers 1 and 2 with dissimilar chemical structures, which reptate in respective tubes. Let us make the following assumptions: (1) the molecular weight M of each polymer is larger than the entanglement molecular weight M_e; (2) each polymer is monodisperse; (3) for each chain, fluctuations in chain density along the tube are negligible; and (4) for each chain, the path length is constant and changes in occupation of path steps take place only by movements of the chain as a whole, and thus a step located at position x along some initial path ($0 < x < L$) will be occupied at a later time only if neither chain end has passed through it during that interval. This will be the case only if the chain has never moved farther than x in one direction along the tube or $(L - x)$ in the other. In addition, we make the following assumptions: (1) the two tubes, after mixing, retain their original diameters, a_1 and a_2, and number of segments, Z_1 and Z_2, and (2) after mixing, the motions of the two respective primitive chains are affected by the presence of other chains and the interaction between chains can be represented by an external potential U.

Then the dynamics of the primitive chain 1, representing polymer 1, can be expressed in the form of the Smoluchowski equation (Riskin 1989):[4]

$$\frac{\partial f_1}{\partial t} = D_{c1}\frac{\partial^2 f_1}{\partial {\xi_1}^2} + \frac{\partial}{\partial \xi_1}\left(\frac{D_{c1}}{k_B T}\frac{\partial U}{\partial \xi_1}\right) f_1 \tag{7.20}$$

where f_1 describes the probability that a segment of chain 1 starting at the origin at $t = 0$ will be found at a position ξ_1 at time t later, D_{c1} is the curvilinear diffusion constant of chain 1, k_B is the Boltzmann constant, T is the absolute temperature, and U is an external potential. If we assume that $\partial U/\partial \xi_1$ can be expressed in terms of the interaction parameter χ, as given by (Han and Kim 1989a)

$$\frac{\partial}{\partial \xi_1}\left(\frac{U}{k_B T}\right) = -\frac{2(-\chi)\phi_2}{a_1} \tag{7.21}$$

where ϕ_2 is the volume fraction of polymer 2 and a_1 is the tube diameter of chain 1, then Eq. (7.20) may be rewritten as

$$\frac{\partial f_1}{\partial t} = D_{c1}\frac{\partial^2 f_1}{\partial {\xi_1}^2} - \left(\frac{2(-\chi)\phi_2 D_{c1}}{a_1}\right)\frac{\partial f_1}{\partial \xi_1} \tag{7.22}$$

The boundary conditions for Eq. (7.22) are

$$f_1(\xi_1, 0) = \delta(\xi_1); \quad f_1(x_1, t) = f_1(x_1 - L_1, t) = 0 \tag{7.23}$$

where δ is the delta function, x_1 is the position of chain 1 along some initial path ($0 < x_1 < L_1$) and L_1 is the contour length of chain 1. The solution of Eq. (7.22) becomes

$$f_1(\xi_1, t;\, x_1) = \frac{2}{L_1}\sum_{p=1}^{\infty} \sin\left(\frac{p\pi x_1}{L_1}\right)\sin\left(\frac{p\pi(x_1-\xi_1)}{L_1}\right) \times \exp\left[(-\chi)\phi_2\xi_1/a_1\right]\exp(-p^2 t/\tau_{1,p}) \tag{7.24}$$

where

$$\tau_{1,p} = \frac{\tau_{d1}}{1+\left(\dfrac{(-\chi)\phi_2 Z_1}{p\pi}\right)^2} \tag{7.25}$$

in which Z_1 is the number of segments in primitive chain 1, and $\tau_{d1} = {L_1}^2/\pi^2 D_{c1}$.

The fraction of all steps initially located at x_1 which are still occupied after time t can be obtained by integrating Eq. (7.24), yielding

$$\begin{aligned} F_1(x_1, t) &= \int_{x_1-L_1}^{x_1} f_1(\xi_1, t;\, x_1)\mathrm{d}\xi_1 \\ &= \frac{2}{\pi}\sum_{p=1}^{\infty}\frac{1}{p}\sin\left(\frac{p\pi x_1}{L_1}\right)\exp\left(\frac{(-\chi)\phi_2 x_1}{a_1}\right) \\ &\quad \times \exp\left(-\frac{p^2 t}{\tau_{1,p}}\right)\left\{\frac{1-(-1)^p\exp\left[-(-\chi)\phi_2 Z_1\right]}{1+\left[(-\chi)\phi_2 Z_1/p\pi\right]^2}\right\} \end{aligned} \tag{7.26}$$

Now, the fraction of segments in chain 1 at time t which are still in tube 1 defined at time $t = 0$ (the original tube) can be obtained by integrating Eq. (7.26), yielding

$$\Psi_1(t) = \frac{1}{L_1}\int_0^{L_1} F_1(x_1, t)\mathrm{d}x_1 = \frac{4}{\pi^2}\sum_{\text{odd } p}^{\infty}\frac{H_{1,p}}{p^2}\exp(-p^2 t/\tau_{1,p}) \tag{7.27}$$

where $\tau_{1,p}$ is defined by Eq. (7.25) and $H_{1,p}$ is given by

$$H_{1,p} = \frac{1-(-1)^p\cosh\left[(-\chi)\phi_2 Z_1\right]}{\left\{1+\left[(-\chi)\phi_2 Z_1/p\pi\right]^2\right\}^2} \tag{7.28}$$

Similarly, the dynamics for chain 2, representing polymer 2, can be expressed as

$$\Psi_2(t) = \frac{4}{\pi^2}\sum_{\text{odd}\, p}^{\infty}\frac{H_{2,p}}{p^2}\exp(-p^2 t/\tau_{2,p}) \tag{7.29}$$

where $H_{2,p}$ is given by

$$H_{2,p} = \frac{1 - (-1)^p \cosh\left[(-\chi)\phi_1 Z_2\right]}{\left\{1 + \left[(-\chi)\phi_1 Z_2/p\pi\right]^2\right\}^2} \tag{7.30}$$

and $\tau_{2,p}$ is given by

$$\tau_{2,p} = \frac{\tau_{d2}}{1 + \left(\dfrac{(-\chi)\phi_1 Z_2}{p\pi}\right)^2} \tag{7.31}$$

in which Z_2 is the number of segments in chain 2 and $\tau_{d2} = L_2{}^2/\pi^2 D_{c2}$, with L_2 being the contour length of chain 2 and D_{c2} the curvilinear diffusion coefficient of chain 2.

We will now include the contribution of constraint release in the reptation motion of the miscible polymer mixture considered previously. In so doing, we will apply the constraint release mechanism introduced in Chapter 4, and will assume a linear blending law for the stress relaxation modulus $G_b(t)$ as

$$G_b(t) = G^o_{N1} w_1 \Psi_1(t) R_1(t) + G^o_{N2} w_2 \Psi_2(t) R_2(t) \tag{7.32}$$

where G^o_{Ni} $(i = 1, 2)$ are the plateau moduli, w_i $(i = 1, 2)$ are the weight (or volume) fractions, R_i $(i = 1, 2)$ are the reduced relaxation functions, $\Psi_1(t)$ is given by Eq. (7.27), and $\Psi_2(t)$ is given by Eq. (7.29). Note that R_i $(i = 1, 2)$ are associated with constraint release and defined by[5]

$$R_i(t) = \frac{1}{Z_i} \sum_{j=1}^{Z_i} \exp[-\lambda_{j,i} t/2\tau_w] \quad (i = 1, 2) \tag{7.33}$$

where the $\lambda_{j,i}$ are defined by[6]

$$\lambda_{j,i} = 4 \sin^2\left[j\pi/2(Z_i + 1)\right] \quad (i = 1, 2) \tag{7.34}$$

and τ_w is the waiting time that governs the time scale of tube renewal and is defined by

$$\tau_w = \int_0^\infty \left[w_1 \Psi_1(t) + w_2 \Psi_2(t)\right]^z dt \tag{7.35}$$

where z is a constraint release parameter which governs the strength of the constraint release contribution (see Chapter 4).

For binary blends, analytical expressions for τ_w can be obtained from Eq. (7.35) for any integer value of z. For $z = 3$, for instance, substituting Eqs. (7.27) and (7.29) into (7.35) and integrating the resulting expression, one obtains

$$\tau_w = \left(\frac{4}{\pi^2}\right)^3 \left\{ \sum_i^{\infty}\sum_j^{\infty}\sum_k^{\infty}\left[\left(\frac{H_{1,i}}{i^2}\right)\left(\frac{H_{1,j}}{j^2}\right)\left(\frac{H_{1,k}}{k^2}\right)\frac{w_1{}^3\tau_{d1}}{i^2Q_{1,i}+j^2Q_{1,j}+k^2Q_{1,k}} \right.\right.$$
$$+\left(\frac{H_{1,i}}{i^2}\right)\left(\frac{H_{1,j}}{j^2}\right)\left(\frac{H_{2,k}}{k^2}\right)\frac{3w_1{}^2w_2\tau_{d1}\tau_{d2}}{\tau_{d2}(i^2Q_{1,i}+j^2Q_{1,j})+\tau_{d1}k^2Q_{2,k}}$$
$$+\left(\frac{H_{1,i}}{i^2}\right)\left(\frac{H_{2,j}}{j^2}\right)\left(\frac{H_{2,k}}{k^2}\right)\frac{3w_1w_2{}^2\tau_{d1}\tau_{d2}}{\tau_{d2}i^2Q_{1,i}+\tau_{d1}(j^2Q_{2,j}+k^2Q_{2,k})}$$
$$\left.\left.+\left(\frac{H_{2,i}}{i^2}\right)\left(\frac{H_{2,j}}{j^2}\right)\left(\frac{H_{2,k}}{k^2}\right)\frac{w_2{}^3\tau_{d2}}{i^2Q_{2,i}+j^2Q_{2,j}+k^2Q_{2,k}}\right]\right\} \tag{7.36}$$

where

$$Q_{1,n} = 1 + \left[(-\chi)\phi_2 Z_1/\pi n\right]^2; \quad Q_{2,n} = 1 + \left[(-\chi)\phi_1 Z_2/\pi n\right]^2 \tag{7.37}$$

and $H_{m,n}$ $(m = 1, 2)$ is given by Eqs. (7.28) and (7.30).

Using the stress relaxation modulus $G_b(t)$ defined by Eq. (7.32), we can calculate the linear viscoelastic properties from the following expressions:

$$\eta_{0b} = \frac{4}{\pi^2}\sum_{i=1}^{2} w_i G_{Ni}^{o}\left\{\sum_{p=1}^{\infty}\frac{H_{i,p}}{p^2}\frac{1}{Z_i}\sum_{j=1}^{Z_i}\tau_{i,j}\right\} \tag{7.38}$$

$$J_{eb}^{o} = \frac{4}{\pi^2\eta_{0b}{}^2}\sum_{i=1}^{2} w_i G_{Ni}^{o}\left\{\sum_{p=1}^{\infty}\frac{H_{i,p}}{p^2}\frac{1}{Z_i}\sum_{j=1}^{Z_i}(\tau_{i,j})^2\right\} \tag{7.39}$$

$$G_b{}'(\omega) = \frac{4}{\pi^2}\sum_{i=1}^{2} w_i G_{Ni}^{o}\left\{\sum_{p=1}^{\infty}\frac{H_{i,p}}{p^2}\frac{1}{Z_i}\sum_{j=1}^{Z_i}\frac{(\omega\tau_{i,j})^2}{1+(\omega\tau_{i,j})^2}\right\} \tag{7.40}$$

$$G_b{}''(\omega) = \frac{4}{\pi^2}\sum_{i=1}^{2} w_i G_{Ni}^{o}\left\{\sum_{p=1}^{\infty}\frac{H_{i,p}}{p^2}\frac{1}{Z_i}\sum_{j=1}^{Z_i}\frac{\omega\tau_{i,j}}{1+(\omega\tau_{i,j})^2}\right\} \tag{7.41}$$

where

$$\frac{1}{\tau_{i,j}} = \frac{p^2}{\tau_{i,p}} + \frac{\lambda_{j,i}}{2\tau_w} \quad (i = 1, 2) \tag{7.42}$$

in which $\tau_{i,p}$ $(i = 1, 2)$ is defined by Eqs. (7.25) and (7.31), $H_{i,p}$ $(i = 1, 2)$ by Eqs. (7.28) and (7.30), $\lambda_{j,i}$ $(i = 1, 2)$ by Eq. (7.34), and τ_w by Eq. (7.36). It can be shown that Eq. (7.38) reduces to (6.27), and Eq. (7.39) reduces to (6.28), Eq. (7.40) reduces to (6.29), and Eq. (7.41) reduces to (6.30) for the following special conditions: (1) large values of Z_i (say, $Z_i > 20$), (2) $\chi = 0$ and (3) $G^o_{N1} = G^o_{N2} = G^o_N$ (i.e., for binary mixtures having components with identical chemical structure).

Since information on the tube disengagement times τ_{di} ($i = 1, 2$) is, in general, difficult to obtain, while the viscosities η_{0i} ($i = 1, 2$) of the constituent components are readily measured, it is of practical importance to express the viscosity of the blends η_{0b} in terms of η_{0i} ($i = 1, 2$). Using Eq. (7.38), we can relate τ_{di} to η_{0i} by

$$\tau_{di} = \frac{\eta_{0i}/G^o_{Ni}}{\dfrac{8}{\pi^2}\sum\limits_{\text{odd}\,p}^{\infty}\dfrac{1}{p^2}\dfrac{1}{Z_i}\sum\limits_{j=1}^{Z_i}\dfrac{1}{p^2+(\lambda_{j,i}/2\Lambda)}} \quad (i = 1, 2) \tag{7.43}$$

where Λ is defined by

$$\Lambda = \left(\frac{8}{\pi^2}\right)^z \underset{\text{all}}{\sum_i^{\infty}} \sum_j^{\infty} \underset{\text{odd}}{\sum_k^{\infty}} \cdots \sum_z^{\infty} \left(\frac{1}{i^2 j^2 k^2 \ldots z^2}\right)\left(\frac{1}{i^2 + j^2 + k^2 \ldots z^2}\right) \tag{7.44}$$

which can be approximated by

$$\Lambda = \frac{1}{z}\left(\frac{\pi^2}{12}\right)^z \tag{7.45}$$

It is clear from Eq. (7.43) that the tube disengagement time τ_{di} depends on the number of segments Z_i when constrain release is included in the tube model. By substituting Eq. (7.43) into (7.38), with the aid of Eq. (7.42), we obtain

$$\eta_{0b} = \sum_{i=1}^{2}\left[\frac{w_i\eta_{0i}\sum\limits_{p=1}^{\infty}\dfrac{H_{i,p}}{p^2}\dfrac{1}{Z_i}\sum\limits_{j=1}^{Z_i}\dfrac{1}{p^2 Q_{i,p}+\lambda_{j,i}(\tau_{di}/2\tau_w)}}{2\sum\limits_{\text{odd}\,p}^{\infty}\dfrac{1}{p^2}\dfrac{1}{Z_i}\sum\limits_{j=1}^{Z_i}\dfrac{1}{p^2+(\lambda_{j,i}/2\Lambda)}}\right] \tag{7.46}$$

For two polydisperse polymers, let us assume the following expression for the stress relaxation modulus $G_b(t)$ of the blend:

$$G_b(t) = \sum_{i=1}^{n}\left\{\sum_{j=1}^{m}\left[G^o_{N1}\frac{\phi_1}{m}w_{1i}\Psi_{1i}(t)R_{1i}(t) + G^o_{N2}\frac{\phi_2}{n}w_{2j}\Psi_{2j}(t)R_{2j}(t)\right]\right\} \tag{7.47}$$

where the upper limit n (or m) in the summation notation denotes the number of fractions, chosen for computational purposes, in the constituent component 1 (or 2), each having the molecular weight M_{1i} (or M_{2j}) and weight fraction w_{1i} (or w_{2j}), and ϕ_1 and ϕ_2 are the volume fractions of the constituent components. Note that Ψ_{pq} and R_{pq} appearing in Eq. (7.47) are given by Eqs. (7.27), (7.29), and (7.33).

Using $G_b(t)$ given by Eq. (7.47), we can calculate values of η_{0b}, $G_b{}'(\omega)$, and $G_b{}''(\omega)$ from

$$\eta_{0b} = \frac{4}{\pi^2}\sum_{i=1}^{n}\sum_{j=1}^{m}\left\{G_{N1}^{o}\frac{\phi_1}{m}w_{1i}\left[\sum_{p=1}^{\infty}\frac{H_{1i,p}}{p^2}\frac{1}{Z_{1i}}\sum_{s=1}^{Z_{1i}}\tau_{1i,s}\right] + G_{N2}^{o}\frac{\phi_2}{n}w_{2j}\left[\sum_{p=1}^{\infty}\frac{H_{2j,p}}{p^2}\frac{1}{Z_{2j}}\sum_{s=1}^{Z_{2j}}\tau_{2j,s}\right]\right\} \tag{7.48}$$

$$G_b{}'(\omega) = \frac{4}{\pi^2}\sum_{i=1}^{n}\sum_{j=1}^{m}\left\{G_{N1}^{o}\frac{\phi_1}{m}w_{1i}\left[\sum_{p=1}^{\infty}\frac{H_{1i,p}}{p^2}\frac{1}{Z_{1i}}\sum_{s=1}^{Z_{1i}}\frac{(\omega\tau_{1i,s})^2}{1+(\omega\tau_{1i,s})^2}\right] + G_{N2}^{o}\frac{\phi_2}{n}w_{2j}\left[\sum_{p=1}^{\infty}\frac{H_{2j,p}}{p^2}\frac{1}{Z_{2j}}\sum_{s=1}^{Z_{2j}}\frac{(\omega\tau_{2j,s})^2}{1+(\omega\tau_{2j,s})^2}\right]\right\} \tag{7.49}$$

$$G_b{}''(\omega) = \frac{4}{\pi^2}\sum_{i=1}^{n}\sum_{j=1}^{m}\left\{G_{N1}^{o}\frac{\phi_1}{m}w_{1i}\left[\sum_{P=1}^{\infty}\frac{H_{1i,p}}{p^2}\frac{1}{Z_{1i}}\sum_{s=1}^{Z_{1i}}\frac{\omega\tau_{1i,s}}{1+(\omega\tau_{1i,s})^2}\right] + G_{N2}^{o}\frac{\phi_2}{n}w_{2j}\left[\sum_{p=1}^{\infty}\frac{H_{2j,p}}{p^2}\frac{1}{Z_{2j}}\sum_{s=1}^{Z_{2j}}\frac{\omega\tau_{2j,s}}{1+(\omega\tau_{2j,s})^2}\right]\right\} \tag{7.50}$$

where

$$\frac{1}{\tau_{ki,s}} = \frac{p^2}{\tau_{ki,p}} + \frac{\lambda_{s,ki}}{2\tau_w} \quad (k = 1, 2) \tag{7.51}$$

7.4.2 Comparison of Theory with Experiment

7.4.2.1 Rheological Behavior of PMMA/PSAN Blends

A number of investigators (Bernstein et al. 1977; Hahn et al. 1992; Jolene et al. 1979; McBrierty et al. 1978; McMaster 1975; Naito et al. 1978; Schmitt et al. 1980) reported on the miscibility of PMMA/PSAN blends exhibiting LCST behavior. It should be mentioned that PSANs having an acrylonitrile (AN) content between approximately 9 and 32 wt % is miscible with PMMA. The critical temperature of PMMA/PSAN blends depends very much on the molecular weights of the constituent components. Note that PMMA undergoes thermal degradation at temperatures much below the LCST (e.g., 350 °C for some commercial polymers) and therefore it is very reasonable to conclude that, for all intents and purposes, PMMA/PSAN blends are miscible over a wide range of experimentally accessible temperatures. Several research groups (Han and Kim 1989a, 1989b; Kim et al. 1994; Lyngaae-Jørgensen and Sondergaard 1987; Pathak et al. 1998; Wu 1987a; Yang et al. 1994) reported on the rheological behavior of

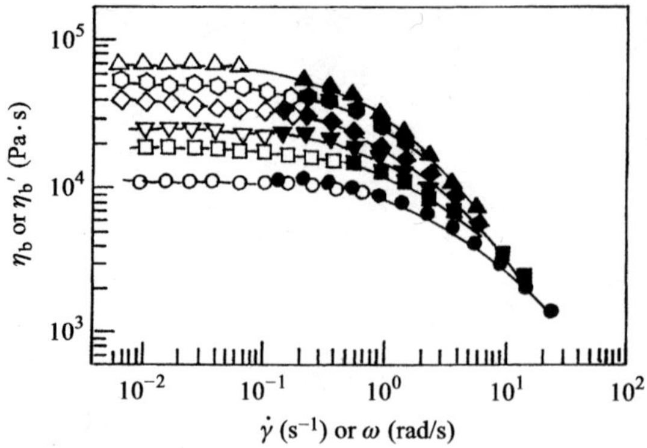

Figure 7.25 Plots of log η_b versus log $\dot{\gamma}$ (open symbols) and plots of log η_b' versus log ω (filled symbols) for PMMA/PSAN blends at 200 °C: (△, ▲) PMMA, (○, ●) PSAN, (⬡, ⬢) 80/20 PMMA/PSAN, (◇, ◆) 60/40 PMMA/PSAN, (▽, ▼) 40/60 PMMA/PSAN, and (□, ■) 20/80 PMMA/PSAN. (Reprinted from Yang et al., *Polymer* 35:1503. Copyright © 1994, with permission from Elsevier.)

PMMA/PSAN blends at temperatures below LCST. Here, we present the linear dynamic viscoelastic properties of PMMA/PSAN blends and compare them with theoretical predictions from the molecular theory presented in the preceding section.

Figure 7.25 gives plots of log η_b versus log $\dot{\gamma}$ in steady-state shear flow and plots of log η_b' versus log ω in oscillatory shear flow for PMMA/PSAN blends at 200 °C. It is interesting to observe in Figure 7.25 that values of η_b and η_b' for the blend system lie between those for the constituent components, and that the Cox–Merz rule (Cox and Merz 1958) seems to work reasonably well. This is attributable to the miscibility of PMMA/PSAN blends.

The composition dependence of Flory–Huggins interaction parameter χ for PMMA/PSAN blends at 187 °C shows that χ is negative over the entire range of blend compositions and that it becomes more negative as the weight fraction of PSAN in the blends is increased (Kim et al. 1994); for instance, $\chi = -0.011$ for the weight fraction of 0.5 at 180 °C. Schmitt et al. (1980) reported $\chi = -0.0124$ at 110 °C and $\chi = -0.0116$ at 130 °C for $M_w = 1.8 \times 10^5$, $\chi = -0.0122$ at 110 °C and $\chi = -0.0096$ at 130 °C for $M_w = 2.7 \times 10^5$, and $\chi = -0.0114$ at 110 °C and $\chi = -0.0092$ at 130 °C for $M_w = 4.4 \times 10^5$. These observations indicates that segmental interactions between PMMA and PSAN are very weak. Such information will be very useful for predicting the rheological behavior of PMMA/PSAN blends using the molecular theory presented in the preceding section.

Figure 7.26 shows the composition dependence of zero-shear viscosity (η_{0b}) of PMMA/PSAN blends at 200 and 210 °C, in which the solid lines are theoretical predictions for different values of constraint release parameter z defined in Eq. (7.35). The predictions of η_{0b} given in Figure 7.26 used the experimental values of zero-shear viscosities (η_{01} and η_{02}) of neat PMMA and PSAN. Since the molecular weights of the constituent components in the PMMA/PSAN blends are greater than the entanglement molecular weights of the respective components, an empirical relationships, $\eta_0 \propto M^{3.4}$, instead of the reptation theory that predicts $\eta_0 \propto M^3$, was used to calculate the zero-shear viscosities of neat PMMA and PSAN. The following observations are worth noting in Figure 7.26. (1) In the experimental results, the values of η_{0b} for blends show positive deviations from linearity that are very similar to the composition dependence of zero-shear viscosities of two polybutadienes having identical chemical structure (see Figure 6.25). This observation makes sense in that the PMMA/PSAN blends, having

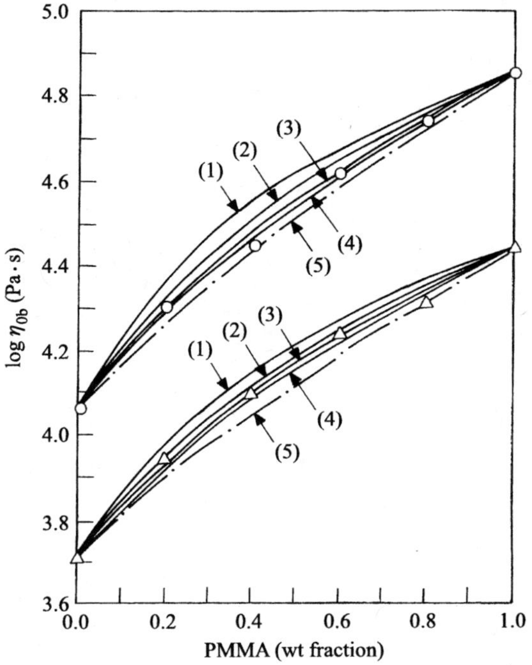

Figure 7.26 Comparison of experimental results for the composition dependence of η_{0b} for PMMA/PSAN blends with theoretical prediction for $\chi = -0.01$ at two different temperatures (°C): (○) 200 and (△) 210. The solid lines represent theoretical predictions: (1) predicted with Eq. (7.46) for $z = 0$, (2) predicted with Eq. (7.46) for $z = 1$, (3) predicted with Eq. (7.46) for $z = 3$, (4) predicted with Eq. (7.46) for $z = 6$, and (5) predicted with the 3.4-power blending law together with the reptation contribution only in the tube model given by Eq. (17) in the paper by Han and Kim (1989a). (Reprinted from Han and Kim, *Macromolecules* 22:4292. Copyright © 1989, with permission from the American Chemical Society.)

extremely small values of χ, can be regarded as if they are binary blends of two components having identical chemical structures. (2) The theoretical predictions with $z = 3$ (solid curve 3) and with $z = 6$ (solid curve 4) seem to capture the experimental results better than those with $z = 0$ (curve 1) and with 3.4-power blending law with the reptation contribution only (curve 5). This observation suggests that the inclusion of constraint release contribution in the tube model describes better the experimental results. (3) The molecular theory presented above with $\chi = -0.01$ predicts positive deviation of η_{0b} from linearity, capturing the important feature of experimental results. Next we will show how the same molecular theory predicts the composition dependence of η_{0b} for miscible polymer blends having large negative values of χ.

Figure 7.27 compares experimentally determined $\log G_b'$ versus $\log \omega$ plots, and Figure 7.28 compares experimentally determined $\log G_b''$ versus $\log \omega$ plots, with theoretical prediction made using $\chi = -0.01$ and $z = 3$ for PSMMA/PSAN blends at 210 °C. It can be seen from Figures 7.27 and 7.28 that the prediction is in good agreement with the experimental results at small values of ω (i.e., in the terminal region), but they deviate appreciably from the experimental results at large values of ω. This observation indicates that the theory presented in the preceding section is only reliable for the terminal response.

Figure 7.29 shows experimentally determined $\log G_b'$ versus $\log G_b''$ plots for PMMA/PSAN blends and also for the constituent components at 210 °C. It is interesting to observe from Figure 7.29 that the $\log G_b'$ versus $\log G_b''$ plots for some blend compositions are almost indistinguishable from those of the constituent components, and that the differences between PMMA and PSAN are very small. In Chapter 6

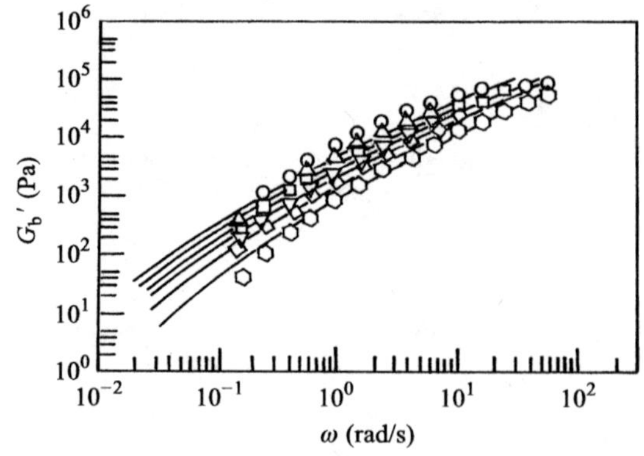

Figure 7.27 Comparison of theoretical prediction (continuous curves) with experimentally determined log $G_b{}'$ versus log ω plots for the PMMA/PSAN blend system at 210 °C: (○) PMMA, (⬡) PSAN, (△) 80/20 PMMA/PSAN blend, (□) 60/40 PMMA/PSAN blend, (▽) 40/60 PMMA/PSAN blend, and (◇) 20/80 PMMA/PSAN blend. (Reprinted from Yang et al., *Polymer* 35:1503. Copyright © 1994, with permission from Elsevier.)

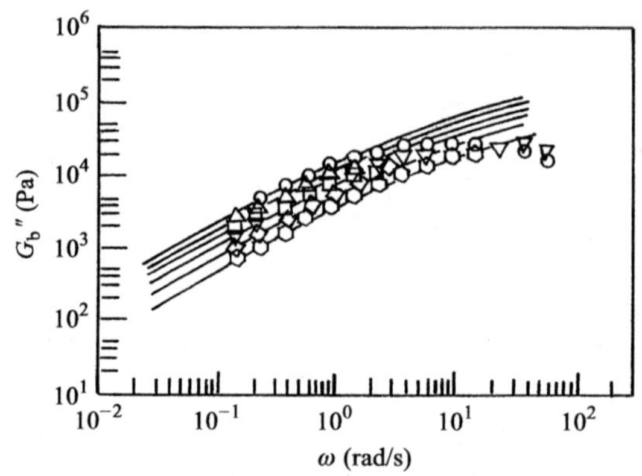

Figure 7.28 Comparison of theoretical prediction (continuous curves) with experimentally determined log $G_b{}''$ versus log ω plots for the PMMA/PSAN blend system at 210 °C: (○) PMMA, (⬡) PSAN, (△) 80/20 PMMA/PSAN blend, (□) 60/40 PMMA/PSAN blend, (▽) 40/60 PMMA/PSAN blend, and (◇) 20/80 PMMA/PSAN blend. (Reprinted from Yang et al., *Polymer* 35:1503. Copyright © 1994, with permission from Elsevier.)

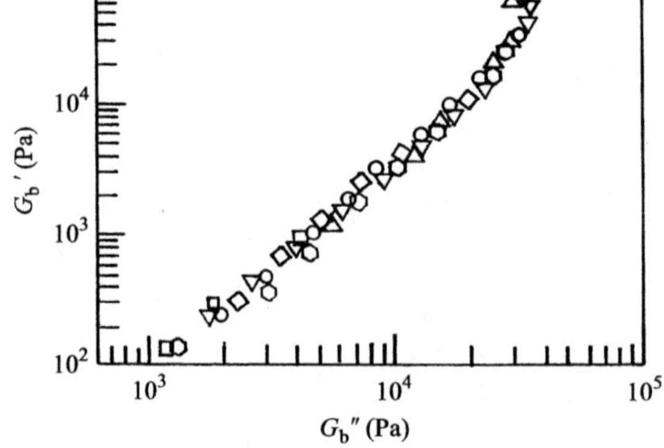

Figure 7.29 Plots of log $G_b{}'$ versus log $G_b{}''$ for the PMMA/PSAN blend system at 210 °C: (△) PMMA, (○) PSAN, (⬡) 80/20 PMMA/PSAN blend, (◇) 60/40 PMMA/PSAN blend, (▽) 40/60 PMMA/PSAN blend, and (□) 20/80 PMMA/PSAN blend. (Reprinted from Yang et al., *Polymer* 35:1503. Copyright © 1994, with permission from Elsevier.)

we discussed that for binary blends consisting of nearly monodisperse homopolymers with identical chemical structures, log G_b' versus log G_b'' plots at certain blend compositions lie above those for the constituent components. But, as the polydispersity of the constituent components, having considerable overlapping of molecular weight between the two, increases, the dependence of the log G_b' versus log G_b'' plot on blend composition is smoothed out (suppressed). In the same token, the composition dependence of log G_b' versus log G_b'' plots for miscible polymer blends would also depend on the polydispersity of the constituent components. Thus, we conclude that the experimental results given in Figure 7.29 are attributable to the polydispersity of PMMA and PSAN. When the constituent components of miscible polymer blends are monodisperse, the spread of the log G_b' versus log G_b'' plots for binary blends would be greater. However, it is possible that for certain pairs of polymers, consisting of monodisperse components of dissimilar chemical structures, log G_b' versus log G_b'' plots at certain blend compositions can lie above those of the constituent components.

7.4.2.2 Rheological Behavior of PMMA/PVDF Blends

A number of investigators (Bernstein et al. 1977; Coleman et al. 1977; Douglass and McBrierty 1978; Nishi and Wang 1975; Noland et al. 1971; Patterson et al. 1976; Paul and Altamirano 1975; Roerdink and Challa 1980; Wendorff 1980) reported on the miscibility of PMMA/PVDF blends exhibiting LCST behavior. The critical temperature of PMMA/PVDF blends depends very much on the molecular weights of the constituent components. PMMA/PVDF blends are miscible over a wide range of experimentally accessible temperatures. Several investigators (Aoki and Tanaka 1999; Han and Kim 1989a, 1989b; Wu 1987b; Yang et al. 1994) investigated the rheological behavior of PMMA/PVDF blends at temperatures below LCST.

Figure 7.30 gives plots of log η_b versus log $\dot{\gamma}$ and log η_b' versus log ω for PMMA/PVDF blends at 200 °C. It can be seen from Figure 7.30 that values of η_b and η_b' for the PMMA/PVDF blends lie between those for the constituent components, showing steady decrease of melt viscosity with increasing amounts of the less viscous PVDF. Similar to the PMMA/PSAN blends considered above, the Cox–Merz rule seems to work for the PMMA/PVDF blends.

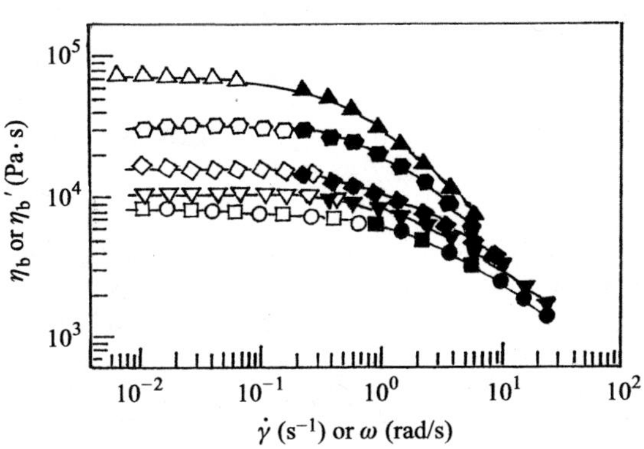

Figure 7.30 Plots of log η_b versus log $\dot{\gamma}$ (open symbols), and plots of log η_b' versus log ω (filled symbols), for the PMMA/PVDF blend system at 200 °C: (△, ▲) PMMA, (○, ●) PVDF, (⬡, ⬢) 80/20 PMMA/PVDF, (◇, ◆) 60/40 PMMA/PVDF, (▽,▼) 40/60 PMMA/PVDF, and (□, ■) 20/80 PMMA/ PVDF. (Reprinted from Yang et al., *Polymer* 35:1503. Copyright © 1994, with permission from Elsevier.)

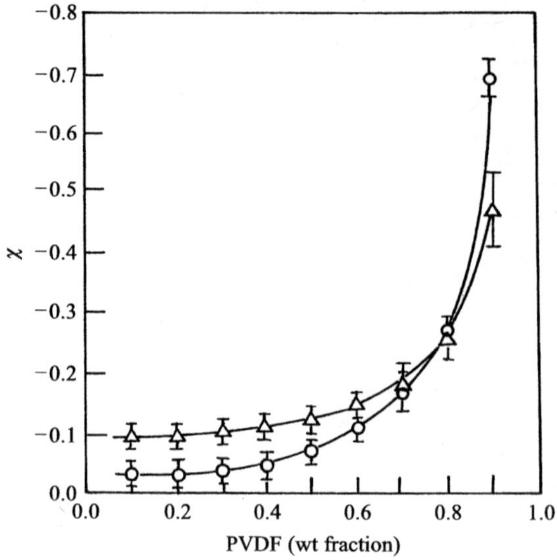

Figure 7.31 Composition dependence of interaction parameter χ for a PMMA/PVDF pair at 187 °C, which was obtained from small-angle X-ray scattering (○) and melting point depression (△). (Reprinted from Wendorff, *Journal of Polymer Science, Polymer Letter Edition* 18:439. Copyright © 1980, with permission from John Wiley & Sons.)

A few research groups (Nishi and Wang 1975; Wendorff 1980) reported on interaction parameter χ for PMMA/PVDF. Figure 7.31 gives the composition dependence of χ for PMMA/PVDF blends, showing very large negative values of χ with strong composition dependence (Wendorff 1980). It can be seen from Figure 7.31 that the PMMA/PVDF pair has very large negative values of χ, indicating that PMMA/PVDF blends have stronger segmental interactions.

Figure 7.32 shows the composition dependence of η_{0b} of PMMA/PVDF blends at three different temperatures, in which the solid lines are theoretical predictions from the molecular theory presented in the preceding section for different values of constraint release parameter z. The following observations are worth noting in Figure 7.32. (1) The experimental results show that the values of η_{0b} for blends show negative deviations from linearity. This observation is quite opposite to that made for PMMA/PSAN blends, which exhibit positive deviations from linearity (see Figure 7.26). We can conclude that the difference in the composition dependence of η_{0b} between the two blend systems may be attributable to the strong attractive segmental interactions between PMMA and PVDF (i.e., large negative values of χ) in the PMMA/PVDF blends. (2) The theoretical predictions capture the essential features of the experimental results, but the extent of agreement between theory and experiment depends on the value of constraint release parameter z.

What is significant in these observations is that the molecular theory presented in the preceding section confirms the experimental observation that the composition dependence of zero-shear viscosities of miscible polymer blends depends on interaction parameter χ, the extent of attractive segmental interactions. Without the molecular theory, we cannot explain the reason why the plots of log η_{0b} versus blend composition show positive deviations from linearity for PMMA/PSAN blends and negative deviations from linearity for PMMA/PVDF blends.

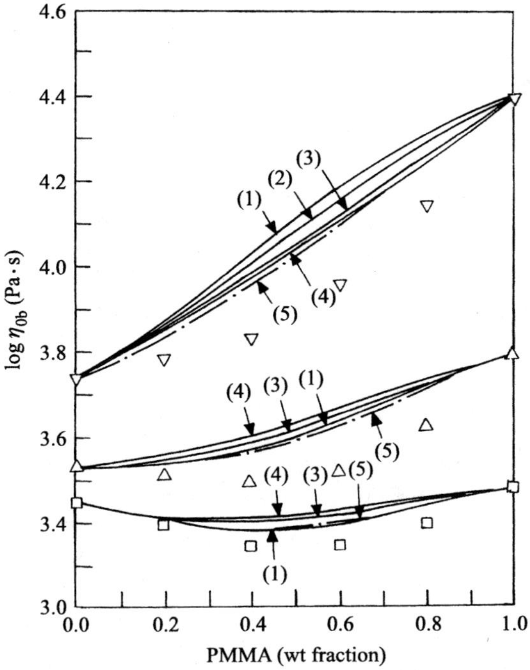

Figure 7.32 Comparison of theoretical prediction with experimental results for the dependence of η_{0b} on blend composition for PMMA/PVDF blends with $\chi = -0.3$ at three different temperatures (°C): (▽) 200, (△) 220, and (□) 230. The solid lines represent theoretical predictions: (1) predicted with Eq. (7.46) for $z = 0$, (2) predicted with Eq. (7.46) for $z = 1$, (3) predicted with Eq. (7.46) for $z = 3$, (4) predicted with Eq. (7.46) for $z = 6$, and (5) predicted with the 3.4-power blending law together with the reptation contribution only in the tube model given by Eq. (17) in the paper by Han and Kim (1989a). (Reprinted from Han and Kim, *Macromolecules* 22:4292. Copyright © 1989, with permission from the American Chemical Society.)

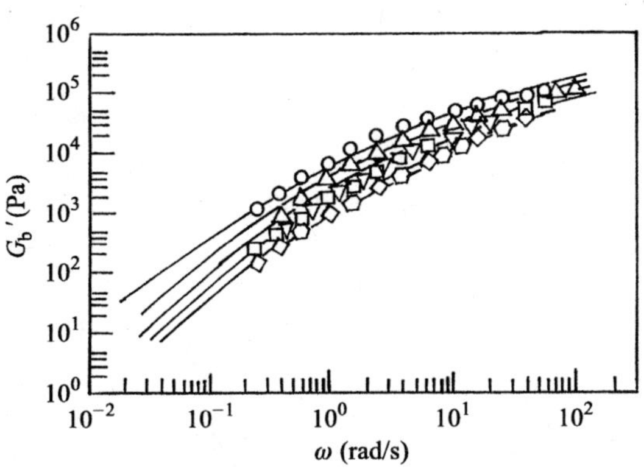

Figure 7.33 Comparison of predicted (continuous curves) with experimentally determined log $G_b{}'$ versus log ω plots for the PMMA/PVDF blend system at 210 °C: (○) PMMA, (○) PVDF, (△) 80/20 PMMA/PVDF, (□) 60/40 PMMA/PVDF, (▽) 40/60 PMMA/PVDF, and (◇) 20/80 PMMA/PVDF. In the prediction, $\chi = -0.3$ and $z = 3$ were used. (Reprinted from Yang et al., *Polymer* 35:1503. Copyright © 1994, with permission from Elsevier.)

Figure 7.33 compares experimentally determined log $G_b{}'$ versus log ω plots, and Figure 7.34 compares experimentally determined log $G_b{}''$ versus log ω plots, with theoretical prediction made with $\chi = -0.3$ and $z = 3$ for PMMA/PVDF blends at 210 °C. The predicted values of $G_b{}'$ and $G_b{}''$ at various values of ω are shown by the solids lines in Figures 7.33 and 7.34, showing that prediction is in good agreement with experiment at small values of ω (i.e., in the terminal region). But, log $G_b{}''$ versus log ω

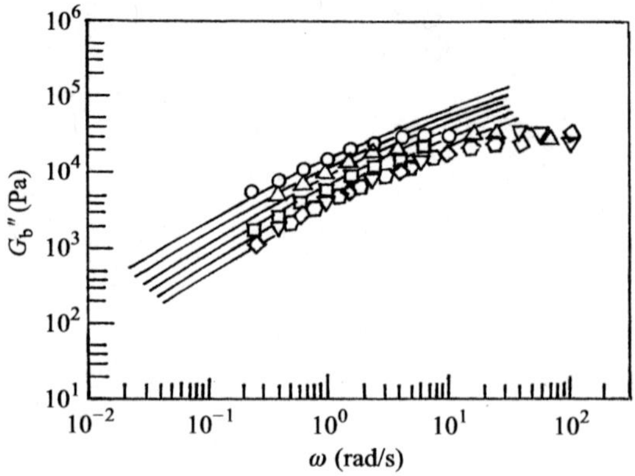

Figure 7.34 Comparison of predicted (continuous curves) with experimentally determined log G_b'' versus log ω plots for the PMMA/PVDF blend system at 210 °C: (○) PMMA, (⬡) PVDF, (△) 80/20 PMMA/PVDF, (□) 60/40 PMMA/PVDF, (▽) 40/60 PMMA/PVDF, and (◇) 20/80 PMMA/PVDF [64]. In the prediction $\chi = -0.3$ and $z = 3$ were used. (Reprinted from Yang et al., *Polymer* 35:1503. Copyright © 1994, with permission from Elsevier.)

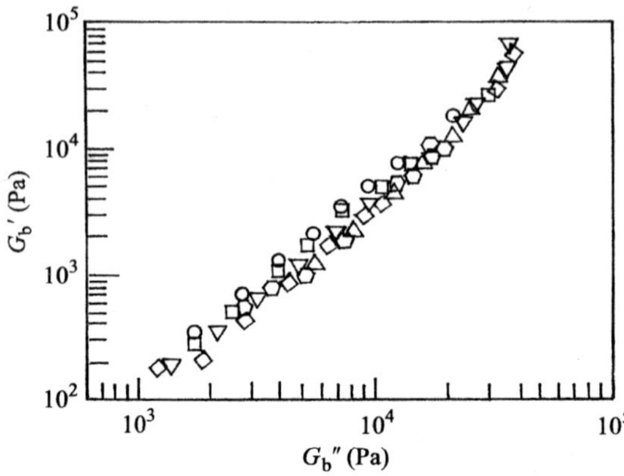

Figure 7.35 Plots of log G_b' versus log G_b'' for the PMMA/PVDF blend system at 210 °C: (△) PMMA, (○) PVDF, (⬡) 80/20 PMMA/PVDF, (◇) 60/40 PMMA/PVDF, (▽) 40/60 PMMA/PVDF, and (□) 20/80 PMMA/PVDF. (Reprinted from Yang et al., *Polymer* 35:1503. Copyright © 1994, with permission from Elsevier.)

plots deviate appreciably from the experimental results at large values of ω. Again, this observation indicates that the theory presented in the preceding section is only reliable for the terminal response.

Figure 7.35 shows experimentally determined log G_b' versus log G_b'' plots for the PMMA/PVDF blend system at 210 °C. It can be seen from Figure 7.35 that log G_b' versus log G_b'' plots for all blend compositions lie between those of the constituent components, PMMA and PVDF, suggesting that value of G_b' for PMMA/PVDF blends decreases steadily with increasing amounts of the less elastic PMMA. As we pointed out when discussing the log G_b' versus log G_b'' plots for the PMMA/PSAN blend system given in Figure 7.29, the polydispersity of PMMA and PVDF must have played a role in having the relatively narrow spread of the log G_b' versus log G_b'' plots for neat PMMA and PVDF given in Figure 7.35.

7.5 Plateau Modulus of Miscible Polymer Blends

We now derive theoretical expressions for G^o_{Nb} for miscible polymer blends using the theory presented in this chapter. For this, let us assume that G^o_{Nb} can be determined from the following expression:

$$G^{\mathrm{o}}_{N\mathrm{b}} = [G_{\mathrm{b}}(t)]_{t=\tau_{\mathrm{e}}} \tag{7.52}$$

where $G_{\mathrm{b}}(\mathrm{t})$ is the relaxation modulus of the blend and τ_{e} is the Rouse relaxation time for a polymer chain between two adjacent entanglement points. Note that τ_{e} is independent of molecular weight and that the values of τ_{e} are smaller than the values of the Rouse relaxation time for the primitive chain, τ_{r}, and also much smaller than the values of the tube disengagement time, τ_{d} (i.e., $\tau_{\mathrm{e}} < \tau_{\mathrm{r}} < \tau_{\mathrm{d}}$) (see Chapter 4). For the linear blending law defined by Eq. (7.32), we have

$$G^{\mathrm{o}}_{N\mathrm{b}} = w_1 G^{\mathrm{o}}_{N1} \Psi_1(\tau_{\mathrm{e}}) R_1(\tau_{\mathrm{e}}) + w_2 G^{\mathrm{o}}_{N2} \Psi_2(\tau_{\mathrm{e}}) R_2(\tau_{\mathrm{e}}) \tag{7.53}$$

where $\Psi_i(t)$ are given by Eqs. (7.27) and (7.29), and $R_i(t)$ by Eq. (7.33). It can be shown that

$$\Psi_i(\tau_{\mathrm{e}}) \cong 1 \tag{7.54}$$

$$R_i(\tau_{\mathrm{e}}) \approx 1 \tag{7.55}$$

Thus, from Eqs. (7.53)–(7.55) we obtain

$$G^{\mathrm{o}}_{N\mathrm{b}} = w_1 G^{\mathrm{o}}_{N1} + w_2 G^{\mathrm{o}}_{N2} \tag{7.56}$$

Figure 7.36 gives the predictions of $G^{\mathrm{o}}_{N\mathrm{b}}$ for PMMA/PVDF blends made by Eq. (7.56) together with experimental data. It can be seen from Figure 7.36 that Eq. (7.56) predicts a linear relationship between $G^{\mathrm{o}}_{N\mathrm{b}}$ and blend composition, which is far from the experimental observations.

When the 3.4-power blending law for the stress relaxation modulus $G_{\mathrm{b}}(t)$ is assumed

$$G_{\mathrm{b}}(t) = \left[w_1 [G_1(t)]^{1/3.4} + w_2 [G_2(t)]^{1/3.4} \right]^{3.4} \tag{7.57}$$

and when only the reptation contribution is included in the tube model, we obtain from Eq. (7.52), with the aid of Eqs. (7.54) and (7.55), the following expression for $G^{\mathrm{o}}_{N\mathrm{b}}$:

$$G^{\mathrm{o}}_{\mathrm{Nb}} = \left[w_1 (G^{\mathrm{o}}_{N1})^{1/3.4} + w_2 (G^{\mathrm{o}}_{N2})^{1/3.4} \right]^{3.4} \tag{7.58}$$

The prediction of $G^{\mathrm{o}}_{N\mathrm{b}}$ obtained with Eq. (7.58), for comparison, are also given in Figure 7.36 for PMMA/PVDF blends, showing that Eq. (7.58) predicts

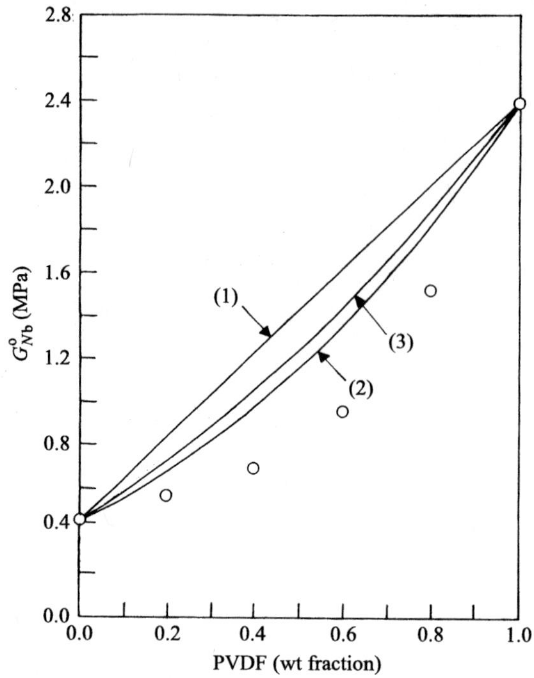

Figure 7.36 Comparison of experimental results (symbol ○) with theoretical predictions for the dependence of G^o_{Nb} on blend composition for PMMA/PVDF blends: (1) represents the prediction made with Eq. (7.56), (2) represents the prediction made with Eq. (7.58), and (3) represents the prediction made with Eq. (7.59). (Reprinted from Han and Kim, *Macromolecules* 22:4292. Copyright © 1989, with permission from the American Chemical Society.)

negative deviations of G^o_{Nb} from linearity, which is consistent with experimental observations.

Some investigators (Composto 1987; Tsenoglou 1988) derived the expression

$$G^o_{Nb} = \left[\phi_1 (G^o_{N1})^{1/2} + \phi_2 (G^o_{N2})^{1/2}\right]^2 \tag{7.59}$$

where ϕ_i is the volume fraction of component i, for the plateau modulus of PS/poly(2,6-dimethyl-1,4-phenylene ether) (PPE) blends. In the derivation of Eq. (7.59), Composto (1987) assumed that the Kuhn statistical lengths of the constituent components were the same, and obtained reasonably good agreement between predictions from Eq. (7.59) and experimental results for the PS/PPE blend system considered.

7.6 Summary

We have presented the rheological behavior of several miscible blend systems. In doing so, we discussed the phase behavior of polymer blends in general as determined primarily from DSC and/or cloud point measurement, although other experimental techniques employed are also mentioned or referred to. Since there are many miscible polymer blend systems, we had to select only representative miscible polymer blend systems for the reason of limited space available. Although we have presented the rheological behavior of certain selected miscible polymer blend systems, we have attempted to convey a view that the rheological behavior presented in this chapter may be regarded as being an archetype for all similar miscible polymer blend systems, and thus that the material presented in this chapter can help in the understanding of the attributes shared by all miscible polymer blends.

We have shown that over a wide range of temperatures investigated, $\log G'$ versus $\log G''$ plots of some miscible polymer blends for certain blend compositions show temperature dependence, while the constituent components do not. Such an experimental observation suggests that the miscible blends have microheterogeneity, which is consistent with the view of the presence of dynamic heterogeneity closely associated with the presence of concentration fluctuations. We have shown that such miscible blends fail TTS. Thus, we conclude that $\log G'$ versus $\log G''$ plot is a very useful tool that can be used to determine whether TTS is applicable to miscible polymer blends. We have described the circumstances under which application of TTS to miscible blends may be warranted and have pointed out that indiscriminate application of TTS to polymer blends, especially to partially miscible and immiscible polymer blends, should be discouraged. While the concept of the iso-free volume is useful to obtain temperature independent correlations for the rheological properties of miscible polymer blends consisting of two glassy polymers, the use of the iso-free-volume condition is not straightforward when one of the constitutive components (or both components) is crystallizable. In such a situation, the melting point of a crystalline component may play a more important role than glass transition temperature in determining the linear viscoelastic properties of miscible polymer blends in the molten state.

We have presented a molecular theory, which is based on the concept of the tube model presented in Chapter 4, for miscible polymer blends, placing emphasis on relationships between the molecular parameters and linear viscoelastic properties. We have shown that the segmental interaction parameter χ plays a central role in determining the composition dependence of linear dynamic viscoelastic properties of miscible polymer blends. The theory explains the physical origin of the linear dynamic viscoelastic properties observed experimentally in miscible polymer blends, and it predicts qualitatively the general trend of the composition dependence of linear dynamic viscoelasticity of miscible polymer blends. We compared theoretical predictions with experimental results of zero-shear viscosity (η_{0b}), dynamic storage modulus, and dynamic loss modulus of some miscible polymer blends. The theory predicts positive deviations from linearity in the plots of $\log \eta_{0b}$ versus blend composition for very small negative (or near zero) values of χ, and negative deviations from linearity in the plots of $\log \eta_{0b}$ versus blend composition for large negative values of χ. Specifically, the theory predicts reasonably well the experimental results for the miscible PMMA/PVDF blend system exhibiting negative deviations from linearity, and for the miscible PMMA/PSAN blend system exhibiting positive deviations from linearity in the plots of $\log \eta_{0b}$ versus blend composition. Note that the PMMA/PVDF pair has a fairly large negative value of χ, while the PMMA/PSAN pair has a very small negative value of χ.

In order to test the molecular theory presented in this chapter critically, it is very important to select miscible pairs of polymers that have the following features: (1) the $T_g s$ of the constituent components are not too far apart, so that an iso-free volume state is applicable, (2) the molecular weight distribution of the constituent components is as narrow as possible, making the polydispersity effect insignificant, (3) both constituent components are glassy, and (4) the dependencies of the segmental interaction parameter on blend composition and temperature are known.

The miscibility of a polymer blend depends on temperature and blend composition, making the investigation of the dynamic behavior of miscible polymer blends very challenging. Associated with the dynamics of miscible polymer blends is the mutual diffusion that, as in determining the self-diffusion coefficient in polymer melts and solutions presented in Chapter 4, can be discussed using molecular theory. Thermodynamic interactions and free-volume effects determine the mutual diffusivity in miscible polymer blends.

In this chapter, we have discussed the rheological behavior of miscible polymer blends with weak interactions. There is another class of miscible polymer blends having specific interactions (e.g., hydrogen bonding, ionic interaction). During the past two decades, several research groups (Cesteros et al. 1993; Coleman et al. 1989; Kuo et al. 2002; Lee et al. 1988; Moskala et al. 1984; Painter et al. 1988, 1989a, 1989b; Wang et al. 2000; Zhang et al. 2002a, 2002b, 2002c, 2004) reported on the miscibility of binary blends formed via hydrogen bonding. However, only a few studies (Cai et al. 2003; Hagen and Weiss 1995) have reported on the rheological behavior of miscible polymer blends formed via specific interaction. This subject is of fundamental importance from the point of view of gaining a better understanding of the role(s) that the strong interactions between the constituent components play in determining the rheological behavior of such polymer blend systems. This subject needs the attention of polymer rheologists in the future.

At present, the thermorheological complexity of chain dynamics and rheological behavior of miscible polymer blends are not well understood. It has been reported that the monomeric friction coefficient of a miscible binary blend varies with blend composition in such a very complicated manner that at present there is no general rule that enables one to predict the viscosities of binary blends, even when the monomeric friction coefficients of the constituent components are known. This is a fundamental and very important subject that should receive serious attention of polymer scientists in the future.

Problems

Problem 7.1

Using the experimental data of cloud point measurements for PS/PI pairs given in Table 7.5, obtain an expression for α defined by Eq. (7.2) for each PS/PI pair. Use Eq. (7.4) for the temperature-dependent specific volume of PS, $v_{\rm PS}$, and Eq. (7.5) for the temperature-dependent specific volume of PI, $v_{\rm PI}$. Convert α (having the units of mol/cm^3) to Flory–Huggins interaction parameter χ using the definition of molar reference volume, $V_{\rm ref} = \left[(M_{\rm S}v_{\rm PS})(M_{\rm I}v_{\rm PI})\right]^{1/2}$ with $M_{\rm S}$ being the molecular weight of styrene monomer and $M_{\rm I}$ being the molecular weight of isoprene monomer. Note that $M_{\rm S} = 104.15$ and $M_{\rm I} = 68.11$.

Table 7.5 Experimental cloud point data for Problem 7.1

(a) (PS-2)/(PI-5) pair ($T_{crit} = 110\,°C$, $w_{crit} = 0.5$, $M_{w,PS-2} = 2.2 \times 10^3$, $M_{w,PI-5} = 5.0 \times 10^3$)									
wt fraction of PS	0.1	0.2	0.3	0.4	0.5	0.6	0.7	0.8	0.9
binodal temp. (°C)	67	82	93	102	110	110	107	101	87
(b) (PS-3)/(PI-5) pair ($T_{crit} = 127\,°C$, $w_{crit} = 0.6$, $M_{w,PS-3} = 3.1 \times 10^3$, $M_{w,PI-5} = 5.0 \times 10^3$)									
wt fraction of PS	0.1	0.2	0.3	0.4	0.5	0.6	0.7	0.8	0.9
binodal temp. (°C)	80	95	110	120	125	127	125	121	110
(c) (PS-10)/(PI-5) pair ($T_{crit} = 185\,°C$, $w_{crit} = 0.5$, $M_{w,PS-10} = 9.5 \times 10^3$, $M_{w,PI-5} = 5.0 \times 10^3$)									
wt fraction of PS	0.1	0.2	0.3	0.4	0.5	0.6	0.7	0.8	0.9
binodal temp. (°C)	132	155	173	182	185	178	170	155	142

Problem 7.2
Derive Eq. (7.17).

Problem 7.3
Verify Eq. (7.19).

Problem 7.4
Derive Eqs. (7.27) and (7.29).

Problem 7.5
Derive Eqs. (7.38)–(7.41).

Problem 7.6
Derive Eq. (7.46).

Problem 7.7
Derive Eqs. (7.48)–(7.50).

Notes

1. Figures 7.5 and 7.6 were prepared using dynamic frequency sweep data provided by R. H. Colby.
2. The criterion for miscibility here is based on the experimental observation of a single, very broad glass transition from DSC.
3. See also Chapter 4.
4. In the absence of an external potential U, Eq. (7.20) reduces to Eq. (4.103).
5. See Eq. (4.128).
6. See Eq. (4.134).

References

Ajji A Choplin L (1991). *Macromolecules* 24:5221.

Ajji A, Choplin L, Prud'homme RE (1988). *J. Polym. Sci., Polym. Phys. Ed.* 26:2279.

Ajji A, Choplin L, Prud'homme RE (1991). *J. Polym. Sci., Polym. Phys. Ed.* 29:1573.

Alegria A, Elizetxea C, Cendoya I, Colmenero J (1995). *Macromolecules* 28:8819.

Aoki Y, Tanaka T (1999). *Macromolecules* 32:8560.

Arendt BH, Krishnamoorti R, Kornfield JA, Smith SD (1997). *Macromolecules* 30:1127.

Balsara NP (1996). In *Physical Properties of Polymers Handbook*, Mark JE (ed), AIP Press, Chap 19.

Bernstein RE, Cruz CA, Paul DR, Barlow JW (1977). *Macromolecules* 10:681.

Bovey FA, Schilling FC, Kwei TK, Frisch HL (1977). *Macromolecules* 10:559

Cai H, Ait-Kadi, A, Brisson J (2003). *Polymer* 44:1481.

Cavaille JY, Perez J, Jourdan C, Johari GP (1987). *J. Polym. Sci., Polym. Phys. Ed.* 25:1847.

Cesteros L, Isasi RR, Katime I (1993). *Macromolecules* 26:7256.

Chu CW, Dickinson LC, Chien JW (1990). *J. Appl. Polym. Sci.* 41:2311.

Chung GC, Kornfield JA, Smith SD (1994a). *Macromolecules* 27:964.

Chung GC, Kornfield JA, Smith SD (1994b). *Macromolecules* 27:5729.

Colby RH (1989). *Polymer* 30:1275.

Coleman MM, Zarian J, Varnell DF, Painter PC (1977). *J. Polym. Sci., Polym. Lett. Ed.* 15:745.

Coleman MM, Lichkus AM, Painter PC (1989). *Macromolecules* 22:586.

Composto RJ (1987). *Diffusion in Polymer Blends*, Doctoral Dissertation, Cornell University, Ithaca, New York.

Composto RJ, Kramer EJ, White DM (1990). *Polymer* 31:2320

Composto RJ, Kramer EJ, White DM (1992). *Macromolecules* 25:4167.

Cong G, Huang Y, MacKnight WJ, Karasz FE (1986). *Macromolecules* 19:2765.

Cortazar MM, Calahorra ME, Guzman GM (1982). *Eur. Polym. J.* 18:165.

Cowies JMG, McEwen IJ (1985). *Polymer* 26:1662.

Cox WP, Merz EM (1958). *J. Polym. Sci.* 28:619

David DD, Kwei TK (1980). *J. Polym. Sci., Polym. Phys. Ed.* 18:2337.

de Gennes PG (1979). *Scaling Concepts in Polymer Physics*, Cornell University Press, Ithaca, New York.

Douglass DC, McBrierty VJ (1978). *Macromolecules* 11:766.

Dudowicz J, Freed KF (1991a). *Macromolecules* 24:5096.

Dudowicz J, Freed KF (1991b). *Macromolecules* 24:5112.

Dudowicz J, Freed KF (1993). *Macromolecules* 26:213.

Flory PJ (1941). *J. Chem. Phys.* 9:660.

Flory PJ (1942). *J. Chem. Phys.* 10:51.

Flory PJ (1965). *J. Am. Chem. Soc.* 87:1833.

Flory PJ, Eichinger BE, Orwoll RA (1968). *Macromolecules* 1:287.

Freed KF, Dudowicz J (1992). *J. Chem. Phys.* 97:2105.

Friedrich Chr, Schwarzwälder C, Riemann RE (1996). *Polymer* 37:2499.

Goldman M, Shen L (1966). *Phys. Rev.* 144:321.

Green PF, Adolph DB, Gilliam LR (1991). *Macromolecules* 24:3377.

Hagen R, Weiss RA (1995). *Polymer* 36:4657.

Hahn K, Schmitt BJ, Kirsches M, Kirsten RG, Sallie H, Schmitt-Strike, S (1992). *Polymer* 33:5150.

Halary JL, Ubrich JM, Ninzi JM, Monnerie L, Stein RS (1984). *Polymer* 25:956.

Haley JC, Lodge TP (2004). *J. Rheol.* 48:463.

Hammouda B, Bauer, BJ, Russell TP (1994). *Macromolecules* 27:2357.

Han CD, Kim JK (1989a). *Macromolecules* 22:1914.
Han CD, Kim JK (1989b). *Macromolecules*, 22:4292.
Han CD, Kim J, Kim JK (1989). *Macromolecules* 22:383.
Han CD, Lee KM, Choi S, Hahn S (2002). *Macromolecules* 35:8045.
Han CC, Okada M, Muroga Y, McCrackin FL, Bauer BJ, Tran-Cong Q (1986a). *Polym. Eng. Sci.* 26:3.
Han CC, Okada M, Muroga Y, Bauer BJ, Tran-Cong Q (1986b). *Polym. Eng. Sci. 26*:1208.
Hirata Y, Kotaka T (1981). *Polym. J.* 13:273.
Huggins ML (1941). *J. Chem. Phys.* 9:440.
Huggins ML (1942). *J. Phys. Chem.* 46:151.
Ito H, Russell TP, Wignall GD (1987). *Macromolecules* 20:2213.
Jack KS, Whittaker AK (1997). *Macromolecules* 30:3560.
Jolene J, Kirsten RG, Schmitt BJ, Schmitt-Striker, S (1979). *Micromole. Chem.* 180:2057.
Kaplan DS (1976). *J. Appl. Polym. Sci.* 20:2615.
Kapnistos M, Hinrichs A, Vlassopoulos D, Anastasidadis SH, Stammer A, Wolf BA (1996). *Macromolecules* 29:7155.
Katana G, Fischer EW, Hack Th, Abetz V, Kremer F (1995). *Macromolecules* 28:2714.
Kim E, Kramer EJ, Wu WC, Garrett PD (1994). *Polymer* 35:5706.
Kim JK, Lee HH, Son HW, Han CD (1998). *Macromolecules* 31:8566.
Kitade S, Takahashi Y, Noda I (1994). *Macromolecules* 27:7397.
Kitade PF, de Lucca Feitas L, Krieger V, Klotz S (1998). *Polymer* 29:1643.
Koningsveld R, Stockmayer WH, Nies E (2001). *Polymer Phase Diagrams*, Oxford University Press, Oxford.
Kumar SK, Colby RH, Anastasiadis SH, Fytas G (1996). *J. Chem. Phys.* 105:3777.
Kuo SW, Lin CL, Chang FC (2002). *Polymer* 43:3943.
Lau SF, Patak, J, Underlet B (1982). *Macromolecule* 15:1278.
Lee JY, Painter PC, Coleman MM (1988). *Macromolecules* 21:954.
Liberman SA, Gomes ADS, Macchi EM (1984). *J. Polym. Sci., Polym. Chem. Ed.* 22:2809.
Lin JL, Roe RJ (1987). *Macromolecules* 20:2168.
Lin JL, Roe RJ (1988). *Polymer* 29:1227.
Lyngaae-Jørgensen J, Sondergaard K (1987). *Polym. Eng. Sci.* 27:351.
Martuscelli E, Pracella M, Yue WP (1984). *Polymer* 25:1097.
Martuscelli E, Silvestre C, Addonizio ML, Amelino L (1986). *Macromol. Chem.* 187:1557.
McBrierty VJ, Douglass DC, Kwei TK (1978). *Macromolecules* 11:1265.
McMaster LP (1975). In *Copolymers, Polyblends, and Composites*, Platzer NAJ (ed), *Adv. Chem. Series* no 142, American Chemical Society, Washington, DC, p 43.
Miller JB, McGrath KJ, Roland CM, Trask CA, Garroway AN (1990). *Macromolecules* 23:4543.
Molnar A, Eisenberg A (1992). *Macromolecules* 25:5774.
Moskala EJ, Howe SE, Painter PC, Coleman MM (1984). *Macromolecules* 17:1671.
Müller G, Staler R, Schlock S (1992). *Macromol. Chem., Rapid Common.* 13:117.
Naito K, Johnson GE, Allara DL, Kwei TK (1978). *Macromolecules* 11:1260.
Nishi T, Kwei TK (1975). *Polymer* 16:285.
Nishi T, Wang TT (1975). *Macromolecules* 8:909.
Nishi T, Wang TT, Kwei TK (1975). *Macromolecules* 8:227.
Noland S, Hsu HHC, Saxon R, Schmitt JM (1971). In *Multicomponent Polymer Systems*, Platzer NAJ (ed), *Adv. Chem. Series* no 99, American Chemical Society, Washington, DC, p 99.
Okada M, Han CC (1986). *J. Chem. Phys.* 85:5317.
Orwoll RA, Flory PJ (1967). *J. Am. Chem. Soc.* 89:6814.

Ougizawa T, Inoue T, Kammer, HW (1985). *Macromolecules* 18:2089.
Painter PC, Park T, Coleman MM (1988). *Macromolecules* 21:66.
Painter PC, Park T, Coleman MM (1989a).*Macromolecules* 22:570.
Painter PC, Park T, Coleman MM (1989b). *Macromolecules* 22:580.
Pathak JA, Colby RH, Klamath SY, Kumar S, Staler R (1998). *Macromolecules* 31:8988.
Pathak JA, Colby RH, Floudas G, Jérôme R (1999). *Macromolecules* 32:2553.
Pathak JA, Kumar SK, Colby RH (2004). *Macromolecules* 37:6994.
Patterson GD, Nishi T, Wang TT (1976). *Macromolecules* 9:603.
Paul DR, Altamirano JO (1975). In *Copolymers, Polyblends and Composites*, Platzer NAJ (ed), *Adv. Chem. Series* no 142, American Chemical. Society, Washington, DC, p 371.
Polios IS, Soliman M, Lee C, Gido SP, Schmidt-Rohr K, Winter HH (1997). *Macromolecules* 30:4470.
Rameau A, Gallot Y, Marie P, Farnoux B (1989). *Polymer* 30:386.
Richardson MJ, Savill NG (1977). *Polymer* 18:3.
Risken R (1989). *The Fokker-Planck Equation*, 2nd ed, Springer-Verlag, Berlin.
Roe RJ, Zin WC (1980). *Macromolecules* 13:1221.
Roerdink E, Challa G (1980). *Polymer* 21:509.
Roland CM, Ngai KL (1991). *Macromolecules* 24:2261.
Roland CM, Ngai KL (1992). *Macromolecules* 25:363.
Roovers J, Toporowski PM (1992). *Macromolecules* 25:3454.
Russell TP, Ito H, Wignall GD (1988). *Macromolecules* 21:1703.
Ruzette AVG, Banerjee P, Mayes AM, Russell TP (2001). *J. Chem. Phys.* 114:8205.
Ruzette AVG; Mayes AM (2001). *Macromolecules* 34:1894.
Ryu DY, Park MS, Chae SH, Jang J, Kim JK, Russell TP (2002). *Macromolecules* 35:8676.
Saeki S, Cowie JMG, McEwen IJ (1983). *Polymer* 24:60.
Sanchez IC, Lacombe RH (1978). *Macromolecules* 11:1145
Schmidt-Rohr K, Clauss J, Spiess HW (1992). *Macromolecules* 25:3273.
Schmitt BJ, Kirsten RG, Hellenic J (1980). *Micromole. Chem.* 181:1655.
Schneider HA, Dilger P (1989). *Polym. Bull.* 21:265.
Schneider HA, Cantow H, Wendland C, Bernhard L (1990). *Makromol. Chem.* 191:2377.
Schneider HA, Wirbser J (1990). *New Polymeric Mater.* 2:149.
Shibayama M, Yang H, Stein RS, Han CC (1985). *Macromolecules* 18:2179.
Stejskal EO, Memory JD (1994). *High Resolution NMR in the Solid State: Fundamentals of CP/MAS*, Oxford University Press, Oxford.
Stoelting J, Karasz FE, MacKnight W (1970). J. *Polym. Eng. Sci.* 10:133.
Takegoshi K, Hikichi K (1991). *J. Chem. Phys.* 94:3200.
Takahashi Y, Suzuki H, Nakagawa Y, Noda I (994). *Macromolecules* 27:6476.
Tomura H, Saito H, Inoue T (1992). *Macromolecules* 25:1611.
Tsenoglou C (1988). *J. Polym. Sci., Polym. Phys. Ed.*. 26:2329.
Wagler T, Rinaldi PL, Han CD, Chun H (2000). *Macromolecules* 33:1778.
Wang L, Cui S, Wang Z, Zhang X (2000). *Langmuir* 16:10490.
Ward TC, Lin TS (1984). In *Polymer Blends and Composites in Multiphase Systems*, Han CD (ed), *Adv. Chem. Series* no 206, American Chemical Society, Washington, DC, p 59.
Wendorff JH (1980). *J. Polym. Sci., Polym. Lett. Ed.* 18:439.
Wimpier JM, Maynard G (1987). *Eur. Polym. J.* 23:989.
Wu S (1987a). *Polymer* 28:1144.
Wu S (1987b). *J. Polym. Sci., Polym. Phys. Ed.*. 25:557.
Yang HH, Han CD, Kim JK (1994). *Polymer* 35:1503.
Yang H, Shibayama M, Stein RS, Shimizu N, Hashimoto T (1986). *Macromolecules* 19:1667.
Zawada JA, Fuller GG, Colby RH, Fetters LJ, Roovers J (1994). *Macromolecules* 27:6861.

Zetsche A, Fischer EW (1994). *Acta Polymer* 45:168.
Zhang S, Jin X, Painter PC, Runt J (2002a). *Macromolecules* 35:3636.
Zhang S, Painter PC, Runt J (2002b). *Macromolecules* 35:8478.
Zhang S, Painter PC, Runt J (2002c). *Macromolecules* 35:9403.
Zhang SH, Jin X, Painter PC, Runt J (2004). *Polymer* 45:3933.

8

Rheology of Block Copolymers

8.1 Introduction

Block copolymer consists of two or more long blocks with dissimilar chemical structures which are chemically connected. There are different architectures of block copolymers, namely, AB-type diblock, ABA-type triblock, ABC-type triblock, and A_mB_n radial or star-shaped block copolymers, as shown schematically in Figure 8.1. The majority of block copolymers has long been synthesized by sequential anionic polymerization, which gives rise to narrow molecular weight distribution, although other synthesis methods (e.g., cationic polymerization, atom transfer radical polymerization) have also been developed in the more recent past. Owing to immiscibility between the constituent blocks, block copolymers above a certain threshold molecular weight form microdomains (10–50 nm in size), the structure of which depends primarily on block composition (or block length ratio). The presence of microdomains confers unique mechanical properties to block copolymers. There are many papers that have dealt with the synthesis and physical/mechanical properties of block copolymers, too many to cite them all here. There are monographs describing the synthesis and physical properties of block copolymers (Aggarwal 1970; Burke and Weiss 1973; Hamley 1998; Holden et al. 1996; Hsieh and Quirk 1996; Noshay and McGrath 1977).

Figure 8.2 shows schematically four types of equilibrium microdomain structures observed in block copolymers. Referring to Figure 8.2, it is well established (Helfand and Wasserman 1982; Leibler 1980) that in microphase-separated block copolymers, spherical microdomains are observed when the volume fraction f of one of the blocks is less than approximately 0.15, hexagonally packed cylindrical microdomains are observed when the value of f is between approximately 0.15 and 0.44, and lamellar microdomains are observed when the value of f is between approximately 0.44 and 0.50. Some investigators have observed ordered bicontinuous double-diamonds (OBDD) (Thomas et al. 1986; Hasegawa et al. 1987) or bicontinuous gyroids (Hajduk et al. 1994) at a very narrow range of f (say, between approximately 0.35 and 0.40) for certain block copolymers. Figure 8.2 shows only one half of the symmetricity about $f = 0.5$.

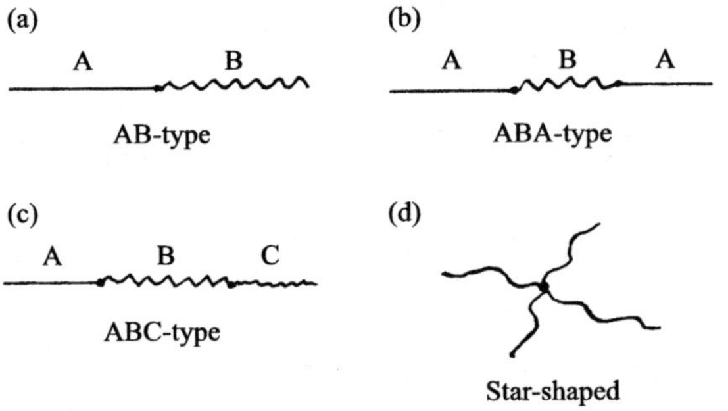

Figure 8.1 Types of block copolymers: (a) AB-type diblock copolymer, (b) ABA-type triblock copolymer, (c) ABC-type triblock copolymer, (d) A_mB_n radial or star-shaped block copolymer.

Transmission electron microscopy (TEM), small-angle X-ray scattering (SAXS), and small-angle neutron scattering (SANS) have long been used to investigate the types of microdomain structures in block copolymers.[1]

There used to be confusion about the existence of additional equilibrium microdomain structures in block copolymers. Specifically, during the period between 1993 and 1996 some investigators (Hamley et al. 1993) reported the presence of a hexagonally modulated layer (HML) microdomain structure, and others (Förster et al. 1994; Hamley et al. 1993; Khandpur et al. 1995; Schulz et al. 1996; Zhao et al. 1996) reported the presence of a hexagonally perforated layer (HPL) microdomain structure between lamellar and hexagonally packed cylindrical microdomain structures. However, in 1997 two independent research groups (Laradji et al. 1997a, 1997b; Qi and Wang 1997a, 1997b) reported theoretical studies showing that HML and HPL microdomain structures are metastable and thus they are not equilibrium microdomain structures. Subsequently, the earlier experimental claims were retracted (Hajduk et al. 1997) after it was realized that HML and HPL microdomain structures had been observed due to incomplete conversion of the initial nonequilibrium perforated layer (PL) structure to the bicontinuous gyroids. That is, the observation of PL structure made in the earlier studies was an artifact induced by the sample preparation process. The preceding discussion illustrates that extreme care is necessary to determine equilibrium microdomain structures in block copolymers.

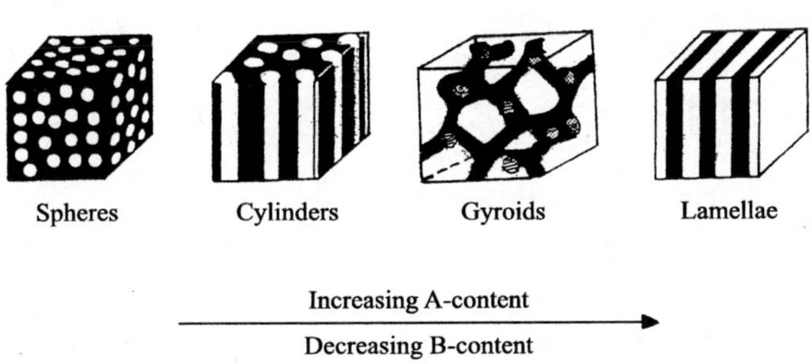

Figure 8.2 The composition dependence of the microdomain structures: spheres, cylinders, gyroids, and lamellae, in AB-type diblock copolymers.

Block copolymers have attracted much attention from polymer scientists because they possess elastomeric properties at service temperatures. Among the many block copolymers synthesized to date, diene-based block copolymers, such as polystyrene (PS)-*block*-polybutadiene (PB) (SB diblock), PS-*block*-polyisoprene (PI) (SI diblock), PS-*block*-PB-*block*-PS (SBS triblock), and PS-*block*-PI-*block*-PS (SIS triblock) copolymers, have long enjoyed commercial success. Figure 8.3 gives TEM images of three commercial SIS triblock copolymers: (1) lamella-forming Vector 4411, having a 0.42 weight fraction of PS block (w_{PS}), (2) cylinder-forming Vector 4211, having $w_{PS} = 0.33$, and (3) sphere-forming Vector 4113, having $w_{PS} = 0.13$. In Figure 8.3, the bright areas represent the PS phase and the dark areas represent the PI phase, stained with osmium tetroxide. It can easily be surmised that the formation of a microdomain structure in a block copolymer depends, among many factors, on molecular weight, block copolymer composition (block length ratio), and the extent of phase segregation between the constituent components, which in turn depends on temperature.

One of the most intriguing aspects of block copolymers is that phase transitions occur from a microphase-separated (ordered) state to the disordered state (referred to as order–disorder transition, ODT) upon heating, from the disordered state to an ordered state (referred to as disorder–order transition, DOT) upon cooling, or from one

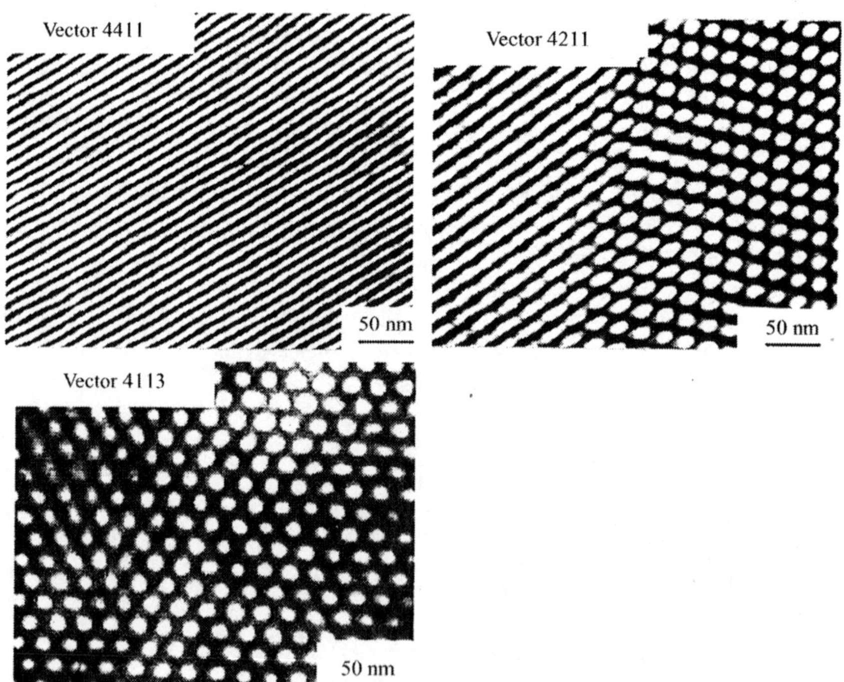

Figure 8.3 TEM images of three commercial SIS triblock copolymers (Dexco Polymers): Vector 4411 with $M_n = 8.2 \times 10^4$, $M_w/M_n = 1.05$, and $w_{PS} = 0.42$, Vector 4211 with $M_n = 1.09 \times 10^5$, $M_w/M_n = 1.05$, and $w_{PS} = 0.33$, and Vector 4113 with $M_n = 1.44 \times 10^5$, $M_w/M_n = 1.06$, and $w_{PS} = 0.13$.

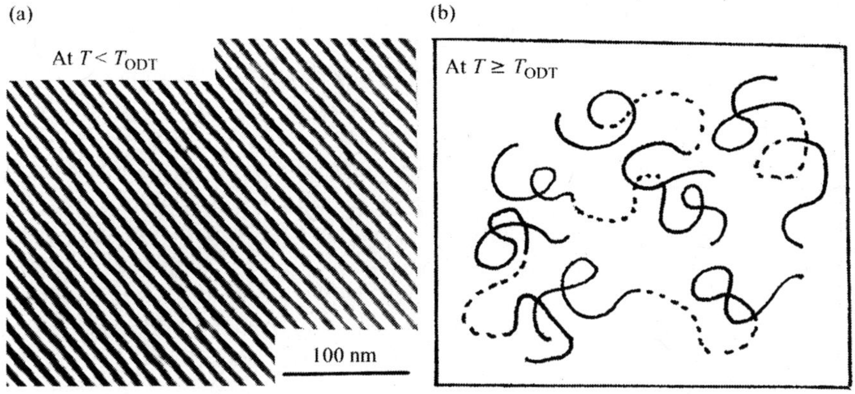

Figure 8.4 (a) TEM image of a lamella-forming SIS triblock copolymer at $T < T_{ODT}$ and (b) schematic diagram describing the flexible chains of PS and PI blocks connected in the block copolymer when it is heated to the disordered state at $T \geq T_{ODT}$.

type of microdomain structure to another type (referred to as order–order transition, OOT) upon heating or cooling. That is, such phase transitions are thermally reversible. Figure 8.4 illustrates that the lamellar microdomains in an SIS triblock copolymer at temperatures below an ODT temperature (T_{ODT}) transform into flexible chains of PS and PI that are interconnected when the temperature is increased above T_{ODT}. It can be easily surmised that the rheological behavior of a block copolymer in an ordered state (i.e., at $T < T_{ODT}$) will be quite different from that in the disordered state ($T \geq T_{ODT}$).

The connectivity between the blocks (junction effect) is expected to play an important role in determining the rheological behavior of block copolymers. For illustration, Figure 8.5 gives logarithmic plots of complex viscosity ($|\eta^*|$) versus angular frequency (ω) at various temperatures for a lamella-forming SI diblock copolymer that

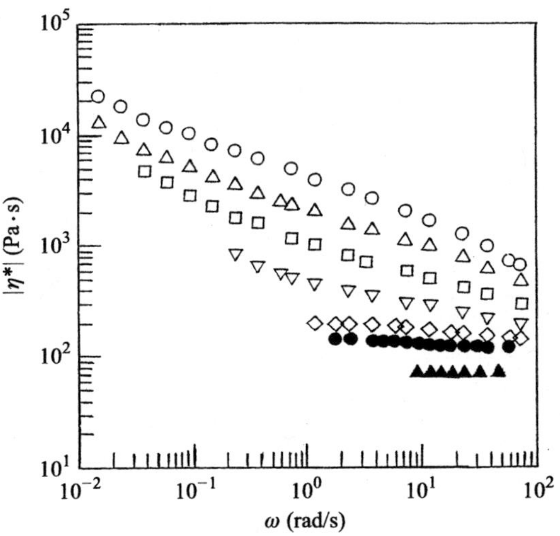

Figure 8.5 The dependence of complex viscosity ($|\eta^*|$) on angular frequency (ω) for a lamella-forming SI diblock copolymer having $M_w = 9 \times 10^3$ for PS block and $M_w = 9 \times 10^3$ for PI block at various temperatures (°C): (○) 100, (△) 105, (□) 110, (▽) 115, (◇) 117, (●) 120, and (▲) 125.

has $T_{\mathrm{ODT}} = 117\,^\circ\mathrm{C}$. It can be seen in Figure 8.5 that the SI block copolymer exhibits shear-thinning behavior over the entire range of ω at temperatures below a certain critical value, and Newtonian behavior at and above the critical temperature. Of particular note in Figure 8.5 is the dependence of $|\eta^*|$ on ω at low values of ω, exhibiting highly shear-thinning behavior. Notice in Figure 8.5 that at $\omega = 0.1$ rad/s, the values of $|\eta^*|$ at $T < 117\,^\circ\mathrm{C}$ are one or two orders of magnitude higher than those at $T \geq 117\,^\circ\mathrm{C}$. This observation is of practical significance for the processing of block copolymers. In Chapter 4 of Volume 2 we discuss the role that the T_{ODT} of a block copolymer plays in the compatibilization of two immiscible homopolymers.

Similar to the phase behavior of binary polymer blends discussed in the preceding chapter, the phase behavior of block copolymers depends on molecular weight, block composition, and the segmental interaction parameter, which in turn varies with temperature. Theories predicting phase transition temperatures of AB-type diblock copolymer (Helfand and Wasserman 1982; Leibler 1980; Matsen and Schick 1994; Vavasour and Whitmore 1992) and ABA-type triblock copolymer (Mayes and Olvera de la Cruz 1989) are well documented. Figure 8.6 gives theoretically predicted phase diagrams for AB-type diblock copolymers exhibiting upper critical order–disorder transition (UCODT),[2] in terms of χN and f, where χ denotes the Flory–Huggins interaction parameter, N denotes the number of segments in the block copolymer (also referred to as "polymerization index"), f denotes the volume fraction of one of the blocks, S denotes spherical microdomains, C denotes cylindrical microdomains, G denotes gyroids, and L denotes lamellar microdomains (see Figure 8.2). Referring to Figure 8.6, the critical value of χN (often referred to as "segregation power"), $(\chi N)_c$, for $f = 0.5$ is 10.49 for an AB-type diblock copolymer (Leibler 1980), meaning that the diblock copolymer forms lamellar microdomain for $\chi N > 10.49$ and disordered phase for $\chi N \leq 10.49$. For simplicity, one usually assumes the following temperature dependence of χ:

$$\chi = a + b/T \tag{8.1}$$

with a positive value of b for block copolymers exhibiting UCODT. Under such situations, an increase in χN implies an increase in molecular weight or a decrease in temperature (since $\chi \propto 1/T$). Thus, an increase in molecular weight or a decrease in temperature drive a block copolymer to an ordered state, while a decrease in molecular weight or an increase in temperature drives a block copolymer to the disordered state. Notice further in Figure 8.6 that $(\chi N)_c$ (thus T_{ODT}) depends on the composition (f) of block copolymer. When the sign of b in Eq. (8.1) is negative, the phase diagram of a diblock copolymer will exhibit lower critical disorder–order transition (LCDOT).[2] Later in this section we will consider the phase transition and oscillatory shear flow behavior of such a block copolymer. It should be mentioned that theory (de Gennes 1979) predicts $(\chi N)_c = 2$ for binary blends with an equal blend ratio and equal degree of polymerization of the constituent components.[3] This observation indicates that the molecular weights of constituent components required to have microphase separation in an AB-type block copolymer are much higher than those required to have macrophase separation in an AB binary blend.

In this chapter, we first present the oscillatory shear rheometry of microphase-separated block copolymers in terms of molecular weight, block composition, and

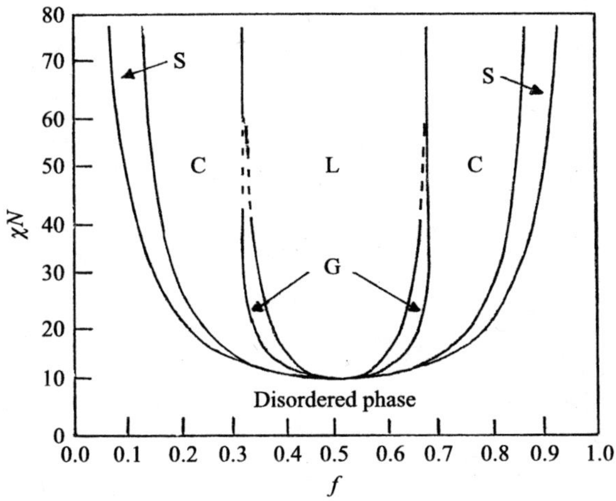

Figure 8.6 Phase diagrams of AB-type diblock copolymer in terms of χN versus f, which are based on self-consistent mean-field theory, in which S denotes spherical microdomains, C denotes hexagonally packed cylindrical microdomains, G denotes gyroids, and L denotes alternating layers of lamellae. (Reprinted from Matsen and Bates, *Macromolecules* 29:1091. Copyright © 1996, with permission from the American Chemical Society.)

the segmental interaction parameter χ of the constituent components. In so doing, we describe the rheological methods that enable one to determine phase transition temperatures of a block copolymer. We point out the importance of thermal history and sample preparation methods when determining phase transition temperatures of highly asymmetric block copolymer. We then present (1) the linear viscoelasticity of disordered block copolymer as affected by the molecular weight and block length ratio, (2) currently held molecular theory for the linear viscoelasticity of disordered block copolymer, (3) the stress relaxation modulus of microphase-separated block copolymer upon application of step strain, and (4) steady-state shear viscosity of microphase-separated block copolymers. In concluding this chapter, we will point out an urgent need for future efforts toward a better theoretical understanding of the dynamics of block copolymers. We emphasize that this chapter does not discuss the physical chemistry of block copolymers or the physics of phase transitions in block copolymers. Such subjects are outside the scope of this chapter. For reasons of space limitation here, we do not discuss the rheological behavior of the solutions of block copolymer, or the rheological behavior of mixtures of block copolymer and homopolymer. Throughout this chapter, we put emphasis on a fundamental understanding of the relationships between block composition, microdomain structure, and rheological behavior of block copolymers.

8.2 Oscillatory Shear Rheometry of Microphase-Separated Block Copolymers Exhibiting Upper Critical Order–Disorder Transition Behavior

Among several different experimental techniques reported in the literature, radiation scattering (SAXS or SANS) and oscillatory shear rheometry have been used most widely to determine phase transition temperatures of block copolymers. In this section, we present how oscillatory shear rheometry can be used to determine phase transition temperatures of a block copolymer exhibiting UCODT. Following the tradition in the literature, below we will refer to UCODT simply as ODT. Since the purpose of this chapter is to discuss the rheology of block copolymers, we will not discuss the SAXS or SANS methods for determining the phase transition temperatures of block copolymer; the subject is well documented in the literature (Bates and Fredrickson 1990; Hashimoto 1996; Mori et al. 1985; Roe et al. 1981).

8.2.1 Oscillatory Shear Rheometry of Symmetric or Nearly Symmetric Block Copolymers

Figure 8.7 gives plots of log G' versus log ω and plots of log G'' versus log ω for a nearly symmetric ($f = 0.47$) SI diblock copolymer (SI-9/9) at various temperatures ranging from 80 to 140 °C, which were obtained from dynamic frequency sweep experiments (see Chapter 5). SI-9/9 has a number-average molecular weight (M_n) of 1.8×10^4 and a polydispersity index of 1.02 and thus, for all intents and purposes, SI-9/9 can be regarded as being a nearly monodisperse diblock copolymer. In Figure 8.7a, we observe that the log G' versus log ω plot in the terminal region has a slope less than 2 at temperatures below 130 °C, changes its slope to closer to but still less than 2 at 130 °C, and then has a slope of 2 at 133 °C and higher temperatures. Conversely, in Figure 8.7b we observe that the log G'' versus log ω plot in the terminal region has a slope less than 1 at temperatures below 120 °C and changes its slope very close to 1 at 130 °C and higher temperatures. In Chapter 6, we showed that monodisperse linear flexible homopolymers have a slope of 2 in the terminal region of log G' versus log ω plots and a slope of 1 in the terminal region of log G'' versus log ω plots. Thus, from Figures 8.7a and 8.7b we can speculate that the morphological state of SI-9/9 may begin to change at a temperature somewhere between 130 and 133 °C. Since the temperature at which a change of slope in the terminal region of log G' versus log ω plots occurs is approximately 3 °C higher than the temperature at which a change of slope in the terminal region of the log G'' versus log ω plot occurs, it is reasonable to conclude that

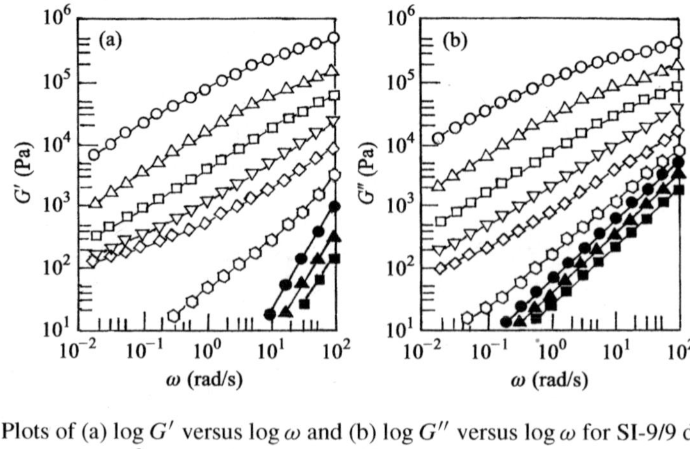

Figure 8.7 Plots of (a) log G' versus log ω and (b) log G'' versus log ω for SI-9/9 during heating at various temperatures (°C): (○) 80, (△) 90, (□) 100, (▽) 110, (◇) 120, (○) 130, (●) 133, (▲) 136, and (■) 140.

in the terminal region, the temperature dependence of the $\log G'$ versus $\log \omega$ plot is more sensitive than the temperature dependence of the $\log G''$ versus $\log \omega$ plot.

Figure 8.8 gives $\log G'$ versus $\log G''$ plots for SI-9/9 at various temperatures ranging between 80 and 140 °C, which were prepared using the data displayed in Figure 8.7. The inset on the upper left side of Figure 8.8 is a TEM image of an SI-9/9 specimen after being annealed at 110 °C for one day followed by rapid quenching in ice water, showing that SI-9/9 has lamellar microdomains. It is interesting to observe in Figure 8.8 that the $\log G'$ versus $\log G''$ plot stays more or less on a single correlation between 80 and 120 °C with a slope much less than 2 in the terminal region, makes a sudden

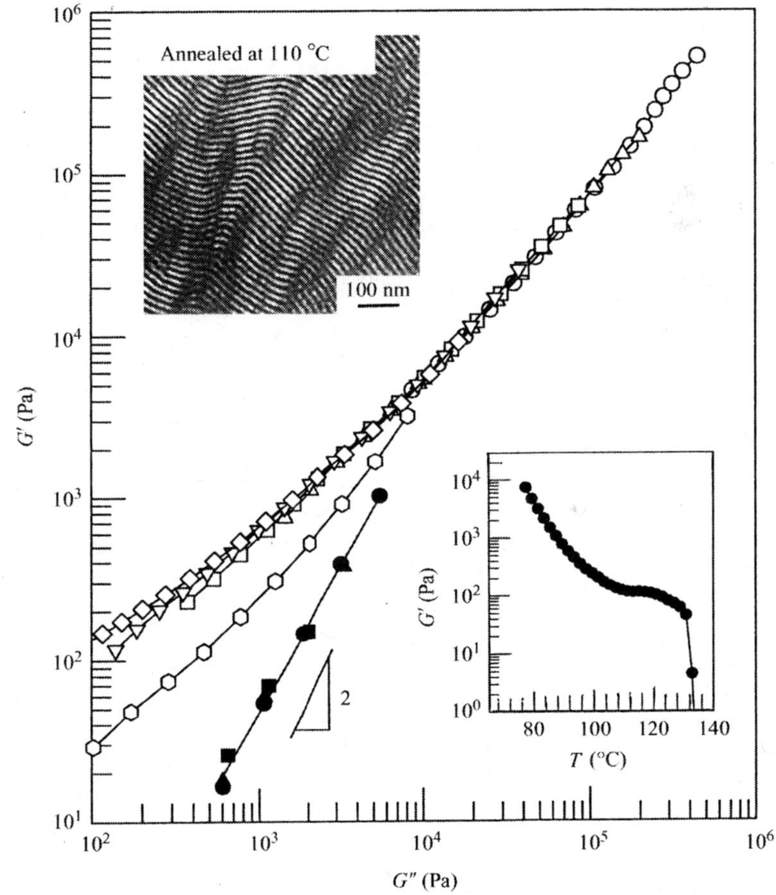

Figure 8.8 Log G' versus log G'' plots for SI-9/9 during heating at various temperatures (°C): (○) 80, (△) 90, (□) 100, (▽) 110, (◇) 120, (⬡) 130, (●) 133, (▲) 136, and (■) 140. The inset in the lower right side gives the isochronal dynamic temperature sweep experiment for SI-9/9 at $\omega = 0.01$ rad/s during heating after a specimen was annealed at 129 °C for 36 h. The inset in the upper left side gives a TEM image of SI-9/9 specimen after annealing at 128 °C for 1 day followed by rapid quenching in ice water. (Reprinted from Han et al., *Macromolecules* 33:3767. Copyright © 2000, with permission from the American Chemical Society.)

downward displacement at 130 °C with a slope still less than 2 in the terminal region, and finally makes another sudden downward displacement with a slope of 2 in the terminal region, giving rise to a temperature-independent correlation at 133, 136, and 140 °C. In Chapter 6, we amply demonstrated that the log G' versus log G'' plot for entangled flexible homopolymers is independent of temperature and also independent of molecular weight, as well as having a slope of 2 in the terminal region. Thus from Figure 8.8 we can conclude that SI-9/9 undergoes ODT at approximately 133 °C; that is, the T_{ODT} of SI-9/9 is approximately 133 °C. Such a rheological criterion has as its basis a viscoelasticity theory (see Eq. (6.18) or (6.19)). Han and coworkers (Han and Kim 1987; Han et al. 1989, 1990, 1995b, 2000) successfully applied this rheological criterion to determine the T_{ODT}s of lamella-forming and cylinder-forming block copolymers.

The physical interpretation of Figure 8.8 is as follows. (1) The log G' versus log G'' plots for SI-9/9 are not affected as the temperature is increased up to 120 °C, suggesting that the morphological state of the lamella-forming SI-9/9 is little changed because 120 °C is still very far below the T_{ODT} of SI-9/9. (2) As the temperature is increased from 120 to 130 °C, the log G' versus log G'' plot of SI-9/9 is shifted downward, suggesting that the morphological state of the lamella-forming SI-9/9 has changed somewhat because 130 °C apparently is close to the T_{ODT} of SI-9/9. (3) As the temperature is increased from 130 to 133 °C, the log G' versus log G'' plot of SI-9/9 is shifted downward further and then remains there as the temperature is increased to 136 and 140 °C, giving rise to a slope of 2 in the terminal region of log G' versus log G'' plot (liquidlike behavior). That is, the block copolymer SI-9/9 has undergone ODT at approximately 133 °C.

The inset on the lower right side of Figure 8.8 is the result of the isochronal dynamic temperature sweep experiment at $\omega = 0.01$ rad/s during heating. It can be seen that the value of G' begins to decrease rapidly at approximately 130 °C. Some investigators (Chung and Lin 1978; Gouinlock and Porter 1997; Widmaier and Meyer 1980) have advocated that the temperature at which G' begins to drop rapidly may be regarded as being the T_{ODT} of a block copolymer. Although such a rheological criterion has no theoretical basis, the rationale behind the criterion lies in the observation that microphase-separated block copolymer exhibits solidlike behavior with very high modulus until it reaches, during heating, a certain critical temperature at which the microdomains of block copolymer begin to disappear and it exhibits liquidlike behavior. Using such a rheological criterion, from the inset of Figure 8.8 we determine the T_{ODT} of SI-9/9 to be approximately 130 °C, which is in good agreement with that determined from the log G' versus log G'' plot. Therefore, we can conclude that both the isochronal dynamic temperature sweep experiments and the log G' versus log G'' plot from the dynamic frequency sweep experiments enable one to determine virtually an identical value of T_{ODT} for lamella-forming block copolymer.

One may wonder why we have not applied time–temperature superposition (TTS) to the dynamic frequency sweep data given in Figure 8.7, to determine the T_{ODT} of SI-9/9, as numerous research groups (Adams et al. 1994; Balsara et al. 1998; Bates 1984; Bates et al. 1990; Floudas et al. 1994, 1996a, 1996b; Lin et al. 1994; Modi et al. 1999; Rosedale and Bates 1990; Rosedale et al. 1995; Schulz et al. 1996; Wang et al. 2002; Winey et al. 1994) have done. As discussed in Chapter 6, application of TTS to flexible homopolymers has been practiced by two methods: (1) by empirically shifting $\log G'$ versus $\log \omega$ (and/or $\log G''$ versus $\log \omega$) plots, obtained at different temperatures, along the ω axis to the data at an arbitrarily chosen temperature as a reference temperature and then obtaining a temperature shift factor a_T, or (2) by calculating a_T using the WLF equation. In Chapter 7 we pointed out that application of TTS is not warranted, even for a miscible polymer blend, when the blend has microheterogeneity. Further, in Chapter 7 we also showed that application of TTS is not warranted for an immiscible polymer blend.

Thus, the reasons for the inapplicability of TTS to microphase-separated block copolymers in general should be very clear. Specifically, referring to Figure 8.7, there is no way one can obtain temperature-independent reduced plots by shifting the $\log G'$ versus $\log \omega$ and/or $\log G''$ versus $\log \omega$ plots of SI-9/9 in the ordered state (i.e., below 133 °C) along the ω axis by arbitrarily choosing a particular temperature as a reference temperature. Moreover, it is not warranted for one to apply the WLF equation to determine the parameters C_1 and C_2 appearing in Eq. (6.7) for microphase-separated block copolymers from viscosity measurements. This is because the WLF parameters C_1 and C_2 are commonly determined from the measurement of zero-shear viscosity (η_0) of a flexible homopolymer, while it is practically very difficult, if not impossible, to measure the η_0 of microphase-separated block copolymers (see, for example, Figure 8.5). Further, application of the WLF equation to a microphase-separated block copolymer composed of chemically dissimilar constituent components, which are invariably immiscible, is not appropriate, for the same reasons as those given for immiscible polymer blends. That is, the values of a_T determined for an ordered state of a block copolymer will not be the same as those for the disordered state, because the morphological state of a block copolymer in an ordered state is quite different from that in the disordered state. Therefore, TTS should not be applied to microphase-separated block copolymers. It cannot be overemphasized that application of TTS to any polymer system is warranted only when its morphological state remains unchanged over the entire range of temperatures investigated.

It should be mentioned that values of a_T determined for a neat homopolymer, via the WLF equation, from viscosity measurements should not be used for a block copolymer unless the monomeric friction coefficients of the constituent blocks are identical. For instance, Wang et al. (2002) determined values of a_T for neat poly(ethylenepropylene) (PEP), via the WLF equation, from viscosity measurements, and then used them to obtain $\log G'$ versus $\log a_T\omega$ and $\log G''$ versus $\log a_T\omega$ plots for PEP-*block*-poly(dimethylsiloxane) (PDMS) copolymers in both the ordered and disordered states. Such an approach is not justified for the following reasons. (1) They tacitly assumed that the values of a_T determined for a neat PEP at different temperatures were identical to those for the PEP-*block*-PDMS copolymer. Such an assumption would not be valid because the WLF parameters C_1 and C_2 for a neat PEP cannot be the same as those for microphase-separated PEP-*block*-PDMS copolymer. Note that the difference in glass transition temperature between PEP and PDMS is very large ($\Delta T_g \approx 80\,°C$)[4] (see Chapter 7). (2) The temperature dependence of the free volume, on which the WLF equation is based, of a neat PEP cannot possibly be the same as that of a PEP-*block*-PDMS copolymer that is composed of two immiscible components, PEP and PDMS. (3) They used experimentally determined η_0 for a neat PEP to calculate a_T, whereas values of η_0 of the microphase-separated PEP-*block*-PDMS copolymer were not measured. Even if values of η_0 of the microphase-separated PEP-*block*-PDMS copolymer had been measured, the temperature dependence of η_0 of a neat PEP cannot possibly be the same as that of a microphase-separated PEP-*block*-PDMS copolymer. After all, a microphase-separated block copolymer exhibiting UCODT will pass, upon heating, through a phase transition from an ordered state to a disordered state. Under such circumstances, which are very similar to the situation dealing with a binary blend exhibiting UCST discussed in Chapter 7, there is no reason why TTS should work in a microphase-separated PEP-*block*-PDMS copolymer.

For the purpose of determining the T_{ODT} of a microphase-separated block copolymer from the dynamic frequency sweep data, there is no need to obtain reduced plots of $\log G'_{\mathrm{r}}$ (or $\log G'$) versus $\log a_{\mathrm{T}}\omega$ and $\log G''_{\mathrm{r}}$ (or $\log G''$) versus $\log a_{\mathrm{T}}\omega$ plots, whether by using an empirically obtained a_{T} by shifting $\log G'$ versus $\log \omega$ or $\log G''$ versus $\log \omega$ plots along the ω axis or by using the WLF equation from viscosity measurements. Instead, as demonstrated in Figure 8.8, use of a $\log G'$ versus $\log G''$ plot, which does not require any manipulation of the dynamic frequency sweep data obtained at various temperatures, is very effective for determining the T_{ODT} of lamella-forming or cylinder-forming block copolymers.

8.2.2 Oscillatory Shear Rheometry of Highly Asymmetric Block Copolymers

Figure 8.9 gives the temperature dependence of G' for a highly asymmetric SIS triblock copolymer (Vector 4111, Dexco Polymers Company), which was annealed at 140 °C for 2 days prior to the isochronal dynamic temperature sweep experiments at an angular frequency (ω) of 0.01 rad/s in the heating process. Vector 4111 has a 0.18 weight fraction of PS block, a weight-average molecular weight of 1.4×10^5, and a polydispersity index of 1.11. Notice that the temperature dependence of G' for Vector 4111 given in Figure 8.9 is quite different from that for the lamella-forming SI-9/9 given

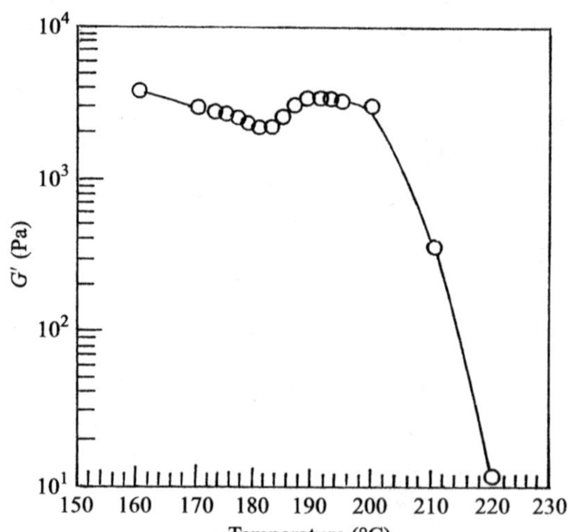

Figure 8.9 Temperature dependence of G' for a highly asymmetric SIS triblock copolymer specimen having a 0.18 weight fraction of PS block (Vector 4111, Dexco Polymers Company), which was annealed at 140 °C for 2 days prior to the isochronal dynamic temperature sweep experiments at an angular frequency of 0.01 rad/s in the heating process. (Reprinted from Sakamoto et al., *Macromolecules* 30:1621. Copyright © 1997, with permission from the American Chemical Society.)

in the inset of Figure 8.8. Referring to Figure 8.9, G' first decreases slowly with increasing temperature, going through a minimum at approximately 182 °C, and then increases, going through a maximum at approximately 190 °C, and finally decreases again, first at a moderate rate with increasing temperature to approximately 206 °C and then at a much faster rate with increasing temperature further to approximately 220 °C. That is, there are roughly four separate regions where the temperature dependence of G' differs from each other. There is no way that one can tell from Figure 8.9 what might be occurring, from the point of view of microdomain structure, in each of the four separate regions. In this regard, the oscillatory shear measurements displayed in Figure 8.9 are of very limited use in identifying the physical phenomena occurring in the Vector 4111 specimen during heating. After all, oscillatory shear measurement is not a useful experimental tool for identifying the morphology of a block copolymer. For such purposes, we need other experimental tools.

Figure 8.10 gives TEM images of Vector 4111 specimens that were annealed at 170, 185, or 200 °C and then rapidly quenched in ice water. We observe in Figure 8.10 that Vector 4111 has (1) hexagonally packed cylinders of PS phase at 170 °C, (2) the coexistence of hexagonally packed cylinders and cubic spheres of PS phase at 185 °C, and (3) cubic spheres at 200 °C. The above observations indicate that Vector 4111 undergoes an OOT, from hexagonally packed cylinders to cubic spheres, at a temperature in the vicinity of 185 °C. Thus, we can tentatively assign approximately 185 °C to be the OOT temperature (T_{OOT}) of Vector 4111. Notice in Figure 8.9 that G' goes through a minimum and then starts to increase at 182–185 °C. Hence, we can conclude from Figure 8.9 that the increasing trend of G' at 182–185 °C signifies the onset of OOT in Vector 4111. Whether the decreasing trend of G' at $T > 200$ °C, which can be seen in Figure 8.9, signifies the complete disappearance of spherical microdomains of PS in Vector 4111 remains to be seen.

Figure 8.11 gives the desmeared SAXS profiles of Vector 4111 during heating. The following observations are worth noting in Figure 8.11. (1) At $T \leq 179$ °C, the specimen has hexagonally packed cylinders (see also Figure 8.10a), with the scattering maxima at $1:\sqrt{3}:\sqrt{4}$ relative to the position of the first-order maximum. (2) An onset of OOT takes place at $T > 179$ °C and OOT is complete at $T < 185$ °C. (3) At 179 °C $< T <$ 185 °C, the specimen has mixtures of hexagonally packed cylindrical microdomains and spherical microdomains packed in a body-centered cubic (bcc) lattice, because the scattering maxima also occur at $1:\sqrt{2}:\sqrt{3}$ relative to the position of the first-order maximum; that is, the SAXS profiles consist of a composite of those from the two coexisting phases. (4) At 185 °C $\leq T \leq$ 210 °C, the specimen has spherical microdomains in a bcc lattice having considerable lattice distortion (see also Figures 8.10b and 8.10c). (5) During heating, an onset of disordering of spheres in bcc lattice takes place at $T > 210$ °C and the specimen has spherical microdomains with liquidlike spatial order (i.e., disordered spheres or disordered micelles). (6) The thick arrow on the profile at 212 °C corresponds to the broad peak from the two peaks corresponding to $\sqrt{2}$ and $\sqrt{3}$, which overlap each other. (7) At temperatures up to 210 °C we cannot find evidence that disordered micelles having only short-range order disappear completely, transforming into the disordered phase with only thermally induced fluctuations. The SAXS experiments during heating were not conducted at temperatures higher than 220 °C to avoid thermal degradation/cross-linking reactions in the SIS triblock copolymer (Sakamoto et al. 1997). Since thermal transitions are reversible during

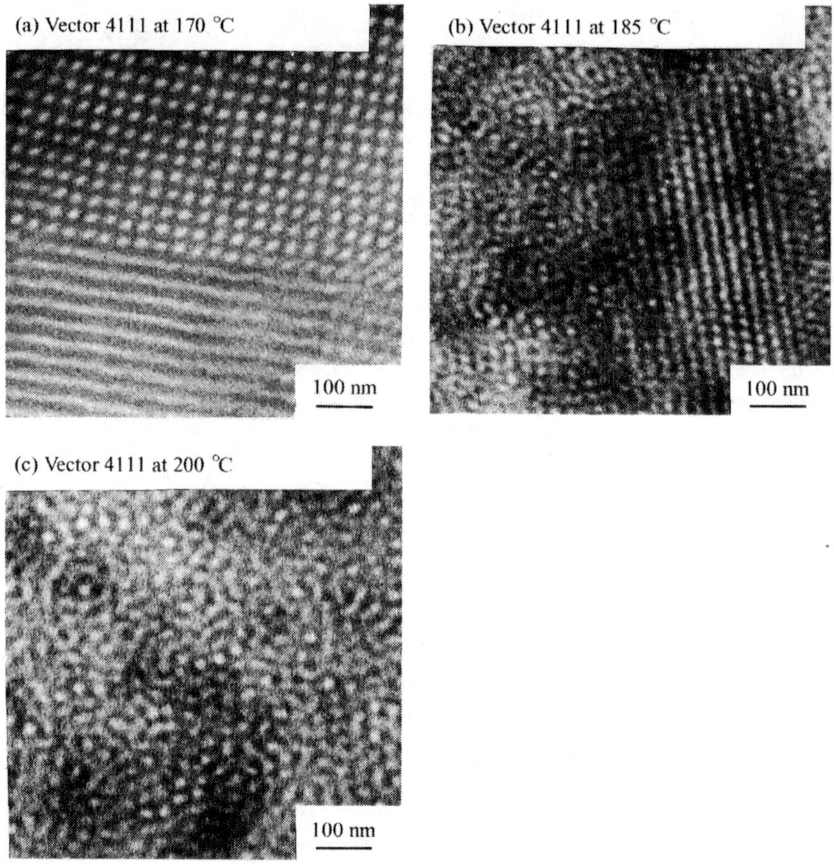

Figure 8.10 TEM images of Vector 4111 with the following thermal histories: (a) specimen was first annealed at 140 °C for 2 days, heated to 170 °C and then held there for 2 h, followed by a rapid quenching in ice water, (b) specimen was first annealed at 140 °C for 2 days, heated to a 185 °C and held there for 2 h, followed by a rapid quenching in ice water, (c) specimen was first annealed at 140 °C for 2 days, heated to 200 °C and held there for 30 min, followed by a rapid quenching in ice water. (Reprinted from Sakamoto et al., *Macromolecules* 30:1621. Copyright © 1997, with permission from the American Chemical Society.)

cooling, it is reasonable to expect that an onset of *ordering* from disordered micelles (with short-range order) to spheres in a bcc lattice (with long-range order) will take place. Such a phase transition in highly asymmetric block copolymers is referred to as "lattice disordering–ordering transition" (LDOT) (Han et al. 2000). Below we will elaborate on the reasons why a distinction must be made between LDOT in highly asymmetric block copolymers and ODT in symmetric or nearly symmetric block copolymers, although both are designated first-order transitions.

Figure 8.12 gives plots of $\log G'$ versus $\log \omega$ and plots of $\log G''$ versus $\log \omega$ during heating of a Vector 4111 specimen, which was first annealed at 140 °C for two days, for temperatures ranging from 160 to 220 °C. In obtaining the results in

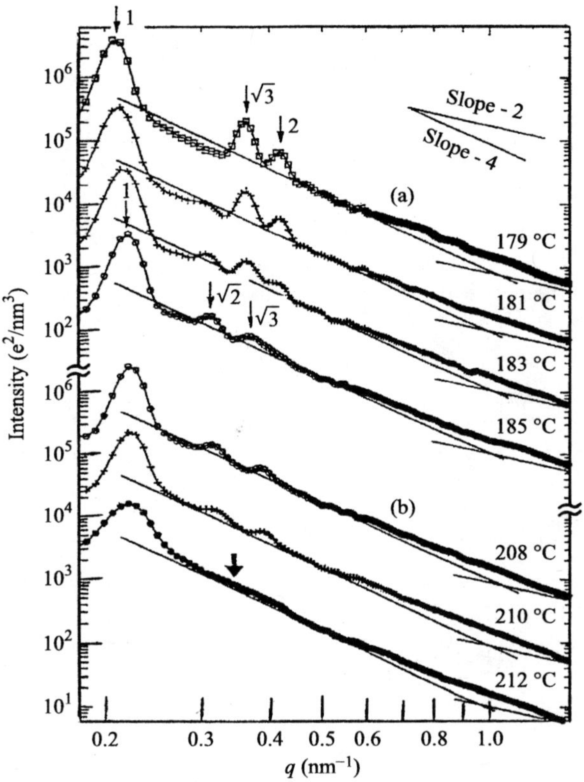

Figure 8.11 Temperature dependence of SAXS profiles for a Vector 4111 specimen during heating at various temperatures indicated on the profile. An as-cast specimen was annealed at 200 °C for 1 h prior to SAXS experiments, the temperature protocols employed during the SAXS experiments are referred to in the original paper. (Reprinted from Sakamoto et al., *Macromolecules* 30:1621. Copyright © 1997, with permission from the American Chemical Society.)

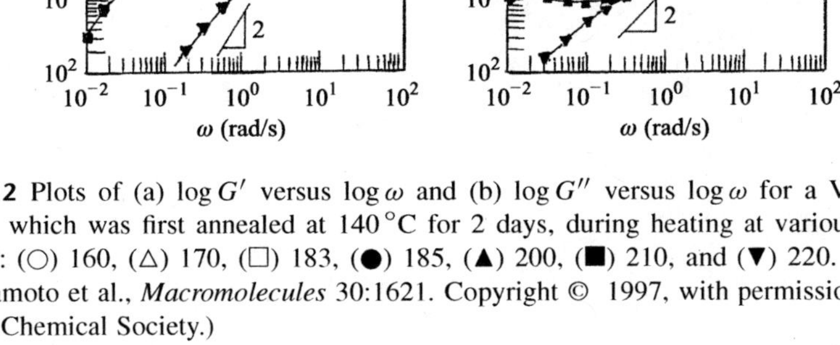

Figure 8.12 Plots of (a) log G' versus log ω and (b) log G'' versus log ω for a Vector 4111 specimen, which was first annealed at 140 °C for 2 days, during heating at various temperatures (°C): (○) 160, (△) 170, (□) 183, (●) 185, (▲) 200, (■) 210, and (▼) 220. (Reprinted from Sakamoto et al., *Macromolecules* 30:1621. Copyright © 1997, with permission from the American Chemical Society.)

Figure 8.12, the specimen was kept at a fixed temperature for about 45–60 min while the oscillatory shear flow experiment was conducted by varying ω from 0.01 to 100 rad/s (Sakamoto et al. 1997). Referring to Figure 8.12a, the following observations are worth noting: (1) the value of G' initially decreases with increasing temperature from 160 to 181 °C over the entire range of ω investigated, (2) as the temperature is increased to 185 °C, the value of G' in the terminal region becomes even greater than that at lower temperatures, 170 and 181 °C, and (3) as the temperature is increased to 210 °C, the value of G' becomes less than that at 200 °C and then exhibits a large drop as the temperature is increased further to 220 °C, having a slope in the terminal region close to 2. Referring to Figure 8.12b, the following observations are worth noting: (1) at temperatures from 185 to 210 °C, the value of G'' in the terminal region increases with increasing temperature and (2) as the temperature is increased further to 220 °C, the $\log G''$ versus $\log \omega$ plot shows the slope of approximately 1 in the terminal region. Based on the frequency dependence of G' and G'' displayed in Figure 8.12, it is very difficult to explain what might physically be occurring in Vector 4111 during heating. However, with the aid of the SAXS results presented in Figure 8.11 we can interpret the response to oscillatory shear flow given in Figure 8.12 as follows: (1) observation 1 is due to the softening of the hard hexagonally packed cylinders of PS phase with increasing temperature, (2) observation 2 is due to OOT, and increased contribution of the interface to G' in the terminal region, and (3) observation 3 is due to an onset of LDOT at approximately 210 °C. In the temperature range where only disordered micelles exist (i.e., after the completion of LDOT), the values of G' decrease with further increasing temperature.

Figure 8.13 gives $\log G'$ versus $\log G''$ plots for a Vector 4111 specimen at various temperatures during heating. In Figure 8.13 we observe a very unusual shape for $\log G'$

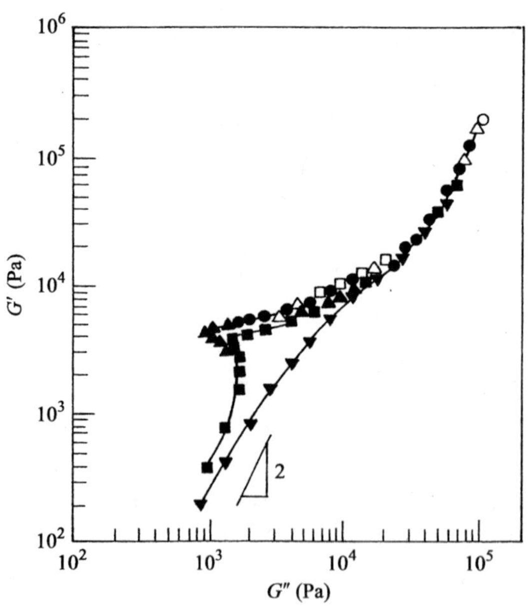

Figure 8.13 Plots of $\log G'$ versus $\log G''$ for a Vector 4111 specimen, which was first annealed at 140 °C for 2 days, during heating at various temperatures (°C): (○) 160, (△) 170, (□) 183, (●) 185, (▲) 200, (■) 210, and (▼) 220. (Reprinted from Sakamoto et al., *Macromolecules* 30:1621. Copyright © 1997, with permission from the American Chemical Society.)

versus $\log G''$ plots in that (1) at temperatures ranging from 185 to 210 °C in the terminal region we observe a discontinuous change in $\log G'$ versus $\log G''$ plots having a negative slope, and (2) the threshold temperature at which $\log G'$ versus $\log G''$ plots in the terminal region begin to exhibit a negative slope is approximately 185 °C, which corresponds roughly to the temperature at which a minimum in G' was observed in Figure 8.9. The SAXS results given in Figure 8.11 suggest that this temperature corresponds to the onset of OOT. We can thus conclude that $\log G'$ versus $\log G''$ plots can be used to determine the T_{OOT} of a block copolymer. In accordance with the SAXS results given in Figure 8.11, the $\log G'$ versus $\log G''$ plots for temperature range between 185 and 210 °C represent the region where spheres in a bcc lattice exist, and the change in the shape of the $\log G'$ versus $\log G''$ plot at 220 °C represents the onset of LDOT. Since no two curves having a slope of 2 overlap each other in the terminal region of $\log G'$ versus $\log G''$ plots given in Figure 8.13, we can conclude that Vector 4111 has not attained a disordered phase at the highest experimental temperature employed. This is attributable to the rather high molecular weight of Vector 4111.

Figure 8.14 gives plots of $\log G'$ versus $\log \omega$ and plots of $\log G''$ versus $\log \omega$ during heating of a highly asymmetric SIS triblock copolymer (SIS-110) having a 0.16 weight fraction of PS block and $M_n = 1.06 \times 10^5$ at various temperatures ranging from 140 to 214 °C. The frequency dependence of $\log G'$ versus $\log \omega$ and $\log G''$ versus $\log \omega$ plots for SIS-110 given in Figure 8.14 looks quite different from that for Vector 4111 given in Figure 8.12. Namely, in Figure 8.14 we observe that at approximately 166 °C the $\log G'$ versus $\log \omega$ plot begins to show a parallel shift with a slope of 2, while the $\log G''$ versus $\log \omega$ plot begins to show a parallel shift with a slope of 1, with increasing temperature, and both $\log G'$ versus $\log \omega$ and $\log G''$ versus $\log \omega$ plots continue to make downward parallel shifts with a further increase in temperature. Figure 8.15 gives $\log G'$ versus $\log G''$ plots for SIS-110 at various temperatures. Comparison of Figure 8.15 with Figure 8.13 reveals that the temperature dependence of the $\log G'$ versus $\log G''$ plots for SIS-110 at temperatures from 144 to 164 °C is virtually

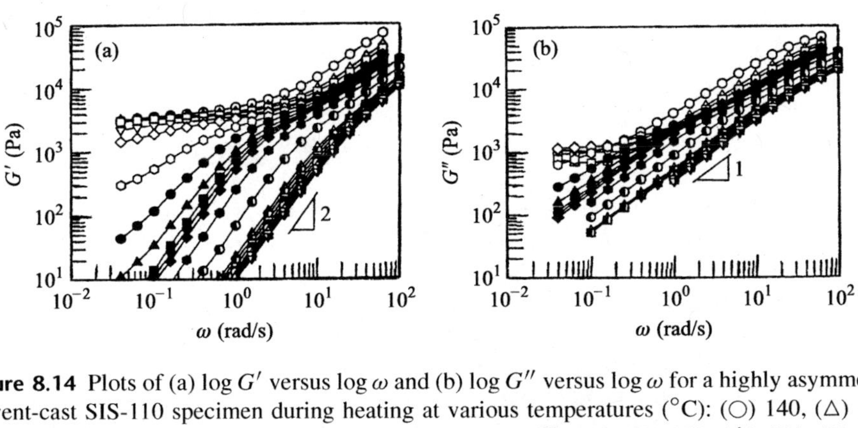

Figure 8.14 Plots of (a) $\log G'$ versus $\log \omega$ and (b) $\log G''$ versus $\log \omega$ for a highly asymmetric solvent-cast SIS-110 specimen during heating at various temperatures (°C): (○) 140, (△) 151, (□) 155, (▽) 160, (◇) 162, (⬡) 164, (●) 166, (▲) 168, (■) 170, (▼) 172, (◆) 174, (⬢) 180, (◐) 190, (◭) 200, (◧) 202, (⧨) 204, (◒) 206, (◐) 208, (◭) 210, (◨) 212, and (⧨) 214. Prior to the rheological measurements, the specimen was annealed at 110 °C for 3 days.

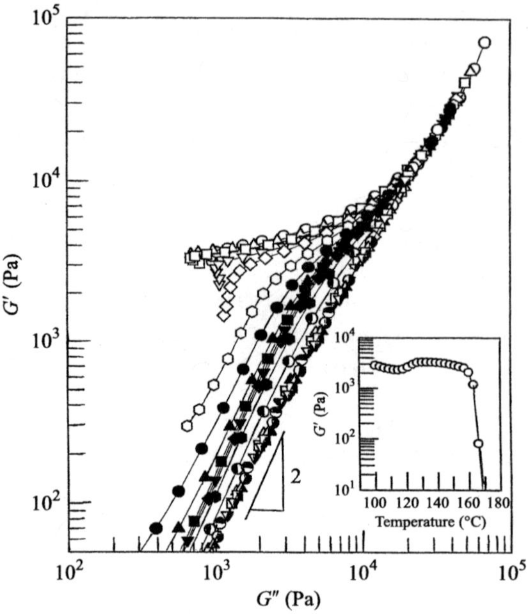

Figure 8.15 Log G' versus log G'' plots for a highly asymmetric solvent-cast SIS triblock copolymer (SIS-110) specimen during heating at various temperatures (°C): (○) 140, (△) 151, (□) 155, (▽) 160, (◇) 162, (⬡) 164, (●) 166, (▲) 168, (■) 170, (▼) 172, (◆) 174, (⬢) 180, (◐) 190, (◮) 200, (◧) 202, (⧨) 204, (◒) 206, (◑) 208, (◭) 210, (◨) 212, and (⧩) 214. Prior to the rheological measurements, the specimen was annealed at 110 °C for 3 days. The inset describes the temperature dependence of G' obtained from the isochronal dynamic temperature sweep experiment at $\omega = 0.01$ rad/s during heating. (Reprinted from Choi et al., *Macromolecules* 36:7707. Copyright © 2003, with permission from the American Chemical Society.)

identical to that for Vector 4111 at 160 to 220 °C. However, a further increase in temperature of SIS-110 caused downward parallel shifts until reaching a critical temperature, 208 °C, at which the log G' versus log G'' plot begins to be independent of temperature. That is, the lower molecular weight ($M_n = 1.06 \times 10^5$) of SIS-110, compared with the higher molecular weight ($M_n = 1.4 \times 10^5$) of Vector 4111, has enabled one to increase the measurement temperature up to 214 °C without encountering discernible thermal degradation/cross-linking reactions.

The rheological significance of Figure 8.15 is the following. (1) The log G' versus log G'' plot makes a parallel shift with a slope of 2 in the terminal region as the temperature is increased from 164 to 208 °C. The temperature at which the log G' versus log G'' plot begins to show a parallel shift signifies the LDOT temperature (T_{LDOT}), at which spherical microdomains with long-range order transform into disordered micelles only with short-range order. (2) The log G' versus log G'' plot having a slope of 2 in the terminal region becomes independent of temperature at a certain critical temperature and above. Because the data points at temperatures above 200 °C given in Figure 8.15 are very crowded on the plot, an enlarged section of the log G' versus log G'' plots given in Figure 8.15 for the values of G' ranging from 10^2 to 10^3 Pa and G'' ranging from 6×10^2 to 3×10^3 Pa is shown in Figure 8.16, in which the dynamic frequency sweep measurements were taken with a temperature interval of 2 °C from 200 to 214 °C. We observe in Figure 8.16 that SIS-110 has T_{LDOT} at approximately 166 °C, and that the log G' versus log G'' plot becomes virtually independent of temperature at $T \geq 208$ °C. The temperature 210 ± 2 °C, at which the log G' versus log G'' plot with a slope of 2 in the terminal region begins to be independent of temperature upon further heating, signifies

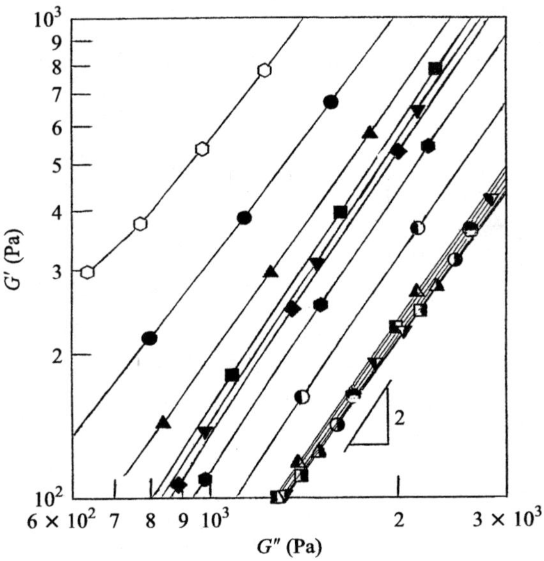

Figure 8.16 Enlarged section of log G' versus log G'' plots for SIS-110 given in Figure 8.15, for the values of G' ranging from 10^2 to 10^3 Pa and G'' ranging from 6×10^2 to 3×10^3 Pa at various temperatures (°C): (○) 164, (●) 166, (▲) 168, (■) 170, (▼) 172, (◆) 174, (⬢) 180, (◐) 190, (◭) 200, (◧) 202, (⧩) 204, (◒) 206, (◑) 208, (◬) 210, (◨) 212, and (⧨) 214. Prior to the rheological measurements, the specimen was annealed at 110 °C for 3 days. (Reprinted from Choi et al., *Macromolecules* 36:7707. Copyright © 2003, with permission from the American Chemical Society.)

the demicellization/micellization transition (DMT) temperature (T_{DMT}); that is, SIS-110 attains a micelle-free disordered phase with thermally induced fluctuations at $T \geq T_{DMT}$.

Also given in Figure 8.15 are the results of isochronal dynamic temperature sweep experiments, showing that values of G' go through a minimum at approximately 120 °C, followed by a very slight maximum, and then start to decrease precipitously at approximately 165 °C, corresponding to the T_{LDOT} determined from the log G' versus log G'' plots. It is then clear that the isochronal dynamic temperature sweep experiments enable one to determine the T_{LDOT} but not T_{DMT} for highly asymmetric block copolymers, whereas the isochronal temperature sweep experiments enable one to determine the T_{ODT} of symmetric or nearly symmetric block copolymers (compare Figure 8.15 with Figure 8.8). Needless to say, oscillatory shear rheometry cannot describe the morphological state of block copolymers. For such purposes, one needs TEM and/or SAXS measurements.

The reasons for inapplicability of TTS to symmetric or nearly symmetric copolymers, discussed in the previous section, are equally applicable to the highly asymmetric block copolymers Vector 4111 and SIS-110. Specifically, it should be very clear from Figure 8.14 that there is no way one can obtain temperature-independent reduced plots for the microphase-separated SIS-110 by shifting log G' versus log ω (and/or log G'' versus log ω) plots along the ω axis to the data at an arbitrarily chosen reference temperature below 166 °C. Thus, TTS would fail completely for the highly asymmetric SIS-110!

TEM images of SIS-110 specimens are given in Figure 8.17, from which we make the following observations. SI-110 has hexagonally packed cylinders of PS at 110 °C. As the temperature is increased to 140 °C, SIS-110 has spherical microdomains in a bcc lattice of PS. As the temperature is increased further to 175 °C, the spherical microdomains of SIS-110 have lost long-range order, giving rise to disordered micelles.

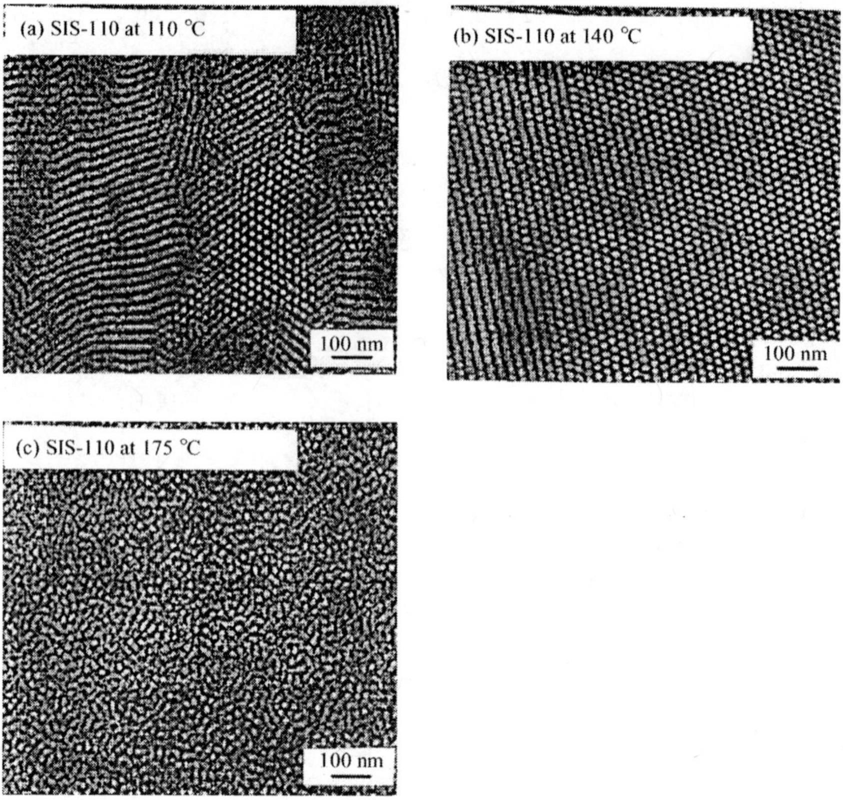

Figure 8.17 TEM images of highly asymmetric solvent-cast SIS-110 specimens, which had been annealed at 110 °C for 3 days and then annealed further at: (a) 110 °C for 1 h followed by rapid quench in ice water, (b) 140 °C for 1 h followed by rapid quench in ice water, and (c) 175 °C for 1 h followed by rapid quench in ice water. (Reprinted from Choi et al., *Macromolecules* 36:7707. Copyright © 2003, with permission from the American Chemical Society.)

Notice in Figure 8.17c that the disordered micelles have a very distinct interface at 175 °C, which lies in the region where log G' versus log G'' plots have a parallel feature (see Figures 8.15 and 8.16) exhibiting liquidlike behavior. Notice that 175 °C is about 10 °C higher than the temperature at which values of G' begin to drop precipitously in the isochronal dynamic temperature sweep experiments (see the inset of Figure 8.15). Once again, the TEM images given in Figure 8.17 confirm the assertion made above that the isochronal dynamic temperature sweep experiments enable one to determine the T_{LDOT} but not T_{DMT} of the highly asymmetric triblock copolymer SIS-110. Since the disordered micelles at 175 °C, shown in Figure 8.17c, have a distinct interface which cannot be considered as part of frozen composition fluctuations (Sakamoto and Hashimoto 1998a), the T_{LDOT} of SIS-110 must be distinguished from the T_{ODT} of nearly symmetric block copolymer SI-9/9 (see Figure 8.8).

Figure 8.18 gives the desmeared SAXS profiles for a solvent-cast SIS-110 specimen at various temperatures ranging from 110 to 220 °C in the heating process.

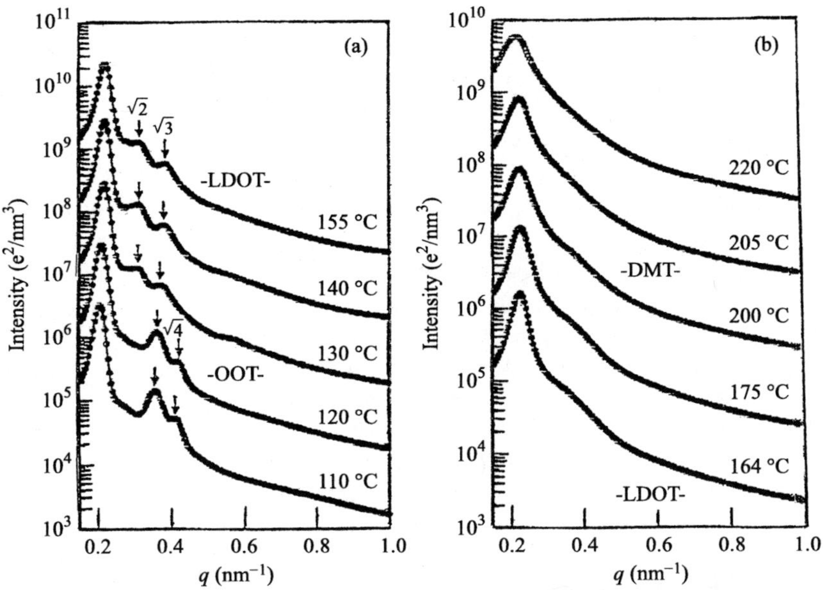

Figure 8.18 Temperature dependence of desmeared SAXS profiles for a highly asymmetric solvent-cast SIS-110 specimen in the heating process at various temperatures as indicated on the plot. Prior to the SAXS experiment, the specimen was annealed at 110 °C for 3 days. (Reprinted from Choi et al., *Macromolecules* 36:7707. Copyright © 2003, with permission from the American Chemical Society.)

The intensities of the SAXS profiles at 110 °C shown at the bottom of Figure 8.18a and at 164 °C shown at the bottom of Figure 8.18b are actually measured values, and the intensities of other SAXS profiles have been shifted up by one decade relative to the intensity profile immediately below them. The following observations are worth noting in Figure 8.18. At temperatures equal to and below 120 °C, the scattering maxima arising from the interdomain interference exist at the relative peak positions of $1{:}\sqrt{3}{:}\sqrt{4}$, indicating the existence of cylindrical microdomains on a hexagonal lattice, which supports the TEM image given in Figure 8.17a. At temperatures ranging from 130 to 155 °C, the SAXS profiles show the scattering maxima at $1{:}\sqrt{2}{:}\sqrt{3}$ relative to the first-order maximum, suggesting the existence of spherical microdomains of PS in a bcc lattice, which supports the TEM image given in Figure 8.17b. Above 164 °C, the higher-order peaks at $\sqrt{2}$ and $\sqrt{3}$ broaden and overlap into a broad shoulder. Upon increasing temperature further, the broad shoulder becomes broader and less distinct. Finally, the shoulder becomes almost indistinguishable. From the SAXS profiles given in Figure 8.18 we conclude that SIS-110 undergoes OOT from hexagonally packed cylinders to spherical microdomains of PS in a bcc lattice at temperatures between 120 and 130 °C, LDOT at temperatures between 155 and 164 °C, and DMT at temperatures between 200 to 205 °C. These observations are in good agreement with those made from the log G' versus log G'' plots given in Figures 8.15 and 8.16.

Parallel shift in log G' versus log G'' plots has also been observed in highly asymmetric SI diblock copolymers (Choi et al. 2003; Han et al. 2000) and in a highly asymmetric PS-*block*-poly(ethylene-*co*-1-butene)-*block*-PS (SEBS triblock) copolymer (Kim et al. 1999). What is interesting, however, is that Rosedale and Bates (1990) observed a parallel shift in the log G' versus log G'' plots of nearly symmetric PEP-*block*-poly(ethylethylene) (PEE) copolymers. Figure 8.19 gives log G' versus log G'' plots for a PEP-*block*-PEE copolymer, showing a parallel shift at temperatures ranging from 96 to 156 °C. Rosedale and Bates attributed the parallel shift in the log G' versus log G'' plots to the thermally induced composition fluctuations near the T_{ODT} of the particular block copolymer. However, no investigation has ever reported a parallel shift in log G' versus log G'' for symmetric or nearly symmetric SI diblock copolymers (see Figure 8.8, for example) or SIS triblock copolymers (Choi et al. 2003). This observation suggests that there is no universality of a parallel shift in log G' versus log G'' plots for symmetric or nearly symmetric block copolymers. Thus, the origin of a parallel shift observed in the log G' versus log G'' plots of highly asymmetric block copolymers must be quite different from that observed in a nearly symmetric PEP-*block*-PEE copolymer by Rosedale and Bates (1990). We offer an explanation on the origin of the parallel shift observed in the log G' versus log G'' plots of SIS-110 given in Figures 8.15 and 8.16.

A parallel shift in the log G' versus log G'' plots of SIS-110 is attributable to the presence of liquidlike disordered micelles of PS phase at $T_{\mathrm{LDOT}} \leq T < T_{\mathrm{DMT}}$ (Choi et al. 2002, 2003; Han et al. 2000). The reasoning is based on the argument that the disordered micelles of PS continue to solubilize in the matrix of PI phase with increasing temperature until reaching T_{DMT}, at which point the disordered micelles completely disappear, transforming into a micelle-free disordered phase with thermally induced composition fluctuations only. This argument is based on two independent experimental observations, namely the experimental results of TEM (see Figure 8.17) and SAXS (see Figure 8.18), which indicate that only disordered micelles with short-range order in

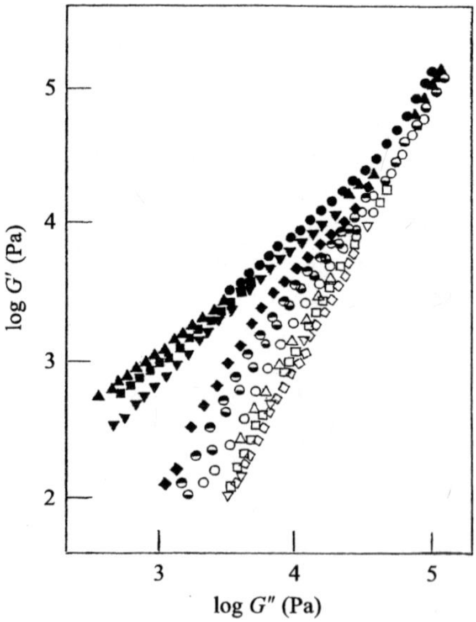

Figure 8.19 Log G' versus log G'' plots for a PEP-*block*-PEE copolymer (PEP-PEE-2) at various temperatures (°C): (●) 35, (▲) 65, (■) 85, (▼) 95, (◆) 96, (◓) 105, (◒) 112, (○) 124, (△) 135, (□) 146, (▽) 156, and (◇) 170. (Reprinted from Rosedale and Bates, *Macromolecules* 23:2329. Copyright © 1990, with permission from the American Chemical Society.)

highly asymmetric SI diblock or SIS triblock copolymers have sharp interfaces. Thus, it can be argued that disordered micelles should not be regarded as being part of the disordered phase with thermally induced composition fluctuations because independent SAXS and TEM studies have shown that the disordered phase with thermally induced fluctuations does not have a sharp interface (Sakamoto and Hashimoto 1998b). Further, the disordered micelles are thermally stable at $T_{LDOT} \leq T < T_{DMT}$, and LDOT and DMT in SIS-110 have been observed to be thermally reversible.

Schwab and Stühn (1996, 1997) also investigated, via SAXS, DOT in a highly asymmetric SI diblock copolymer; that is, phase transition and the kinetics of structure formation during the cooling process from a state of liquid order to a macrolattice in a highly asymmetric SI diblock copolymer. In between the homogeneously disordered state at high temperature and the bcc ordered array of spheres, they observed a stable state of liquidlike order, which is essentially disordered micelles, as referred to above. They noted that the size of the spheres increased continuously with decreasing temperature. Their experimental observations from SAXS support the existence of LDOT.

Figure 8.20 gives log G' versus log G'' plots for a solvent-cast SIS-110 specimen at various temperatures during cooling from 214 to 140 °C. Notice in Figure 8.20 that the specimen was cooled stepwise with 2 °C decrements initially, and then larger decrements at lower temperatures. In Figure 8.20 we observe that log G' versus log G'' plots exhibit parallel shift with decreasing temperature. Comparison of Figure 8.20 with Figure 8.15 shows clearly that demicellization during heating and micellization during cooling of the highly asymmetric triblock copolymer, SIS-110, is thermally reversible. However, a close examination of the two figures reveals that they are not exactly identical owing to a "hysteresis effect." Using SAXS, Sakamoto and Hashimoto (1998b) reported a similar hysteresis effect in another highly asymmetric SIS triblock

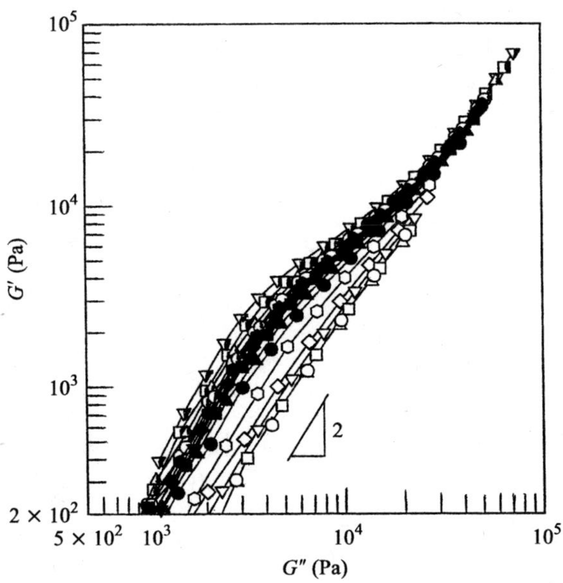

Figure 8.20 Log G' versus log G'' plots for a highly asymmetric solvent-cast SIS-110 specimen during coolingat various temperatures (°C): (○) 214, (△) 212, (□) 210, (▽) 206, (◇) 200, (⬡) 190, (●) 180, (▲) 175, (■) 170, (▼) 168, (◆) 166, (⬢) 164, (◐) 162, (◑) 160, (◭) 155, (◧) 150, and (⧨) 140. Prior to the rheological measurements, the specimen was annealed at 110 °C for 3 days. (Reprinted from Choi et al., *Macromolecules* 36:7707. Copyright © 2003, with permission from the American Chemical Society.)

copolymer, Vector 4111, and noted that the kinetics of microdomain formation in a block copolymer, upon cooling from the disordered liquid state, is very slow. The thermal reversibility in the log G' versus log G'' plots of highly asymmetric SIS-110 could not have been observed if the disordered micelles had been formed by thermally induced composition fluctuations.

The differences in phase transition mechanism between symmetric or nearly symmetric block copolymers and highly asymmetric block copolymers are summarized schematically in Figure 8.21. Namely, during heating, lamella-forming symmetric block copolymers lose long-range order at a critical temperature (T_{ODT}) and transform into a disordered phase with thermally induced composition fluctuations only, while sphere-forming highly asymmetric block copolymers lose long-range order at a critical temperature (T_{LDOT}) and transform into disordered micelles with short-range

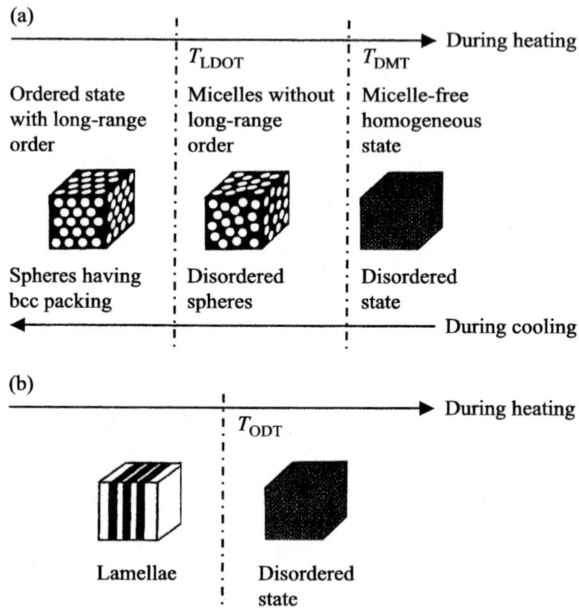

Figure 8.21 (a) Schematic showing the phase disordering and ordering processes in highly asymmetric sphere-forming block copolymers: namely, the phase transition path, during heating, from spheres with long-range order to disordered spheres (micelles) without long-range order and to the micelle-free disordered state, and the phase transition path, during cooling, from the micelle-free disordered state to disordered micelles without long-range order and to spheres with long-range order. T_{LDOT} denotes lattice disordering/ordering temperature and T_{DMT} denotes demicellization/micellization temperature. (b) Schematic showing the phase disordering process in symmetric or nearly symmetric block copolymer during heating, from lamellae to the disordered state, where T_{ODT} denotes order–disorder transition (ODT) temperature, and upon cooling from the disordered phase, when the block copolymer will undergo disorder–order transition (DOT) at T_{DOT} since the phase transition is thermally reversible. (Reprinted from Han et al., *Macromolecules* 33:3767. Copyright © 2000, with permission from the American Chemical Society.)

order which, upon further heating, transform into the micelle-free disordered phase with thermally induced composition fluctuations only at a critical temperature (T_{DMT}). LDOT is first-order transition like ODT in lamella-forming block copolymers. However, the distinction between LDOT and ODT is necessary because LDOT is followed by another phase transition, DMT, at a higher temperature (T_{DMT}), at which point disordered micelles disappear, transforming into the micelle-free disordered phase with thermally induced composition fluctuations only. Experimental studies indicate that T_{DMT} is much higher (say, as much as approximately 40 °C) than T_{LDOT} in sphere-forming SI diblock and SIS triblock copolymers (Choi et al. 2002, 2003; Han et al. 2000; Sakamoto and Hashimoto 1998a). If LDOT in a sphere-forming block copolymer is called ODT, then we are faced with a dilemma in that ODT is followed by another phase transition, DMT. By the definition of ODT, there should not be another phase transition above T_{ODT}. Under such a circumstance, the disordered micelles existing at $T_{LDOT} \leq T < T_{DMT}$ must be regarded as being part of the disordered phase associated with thermally induced composition fluctuations. Such a view has been expressed by Dormidontova and Lodge (2001) who, using a mean-field approach, predicted the existence of disordered micelles in a sphere-forming block copolymer, confirming the experimental observations (Han et al. 2000; Sakamoto et al. 1997; Sakamoto and Hashimoto 1998a), and a critical micelle temperature (T_{CMT}), lying far above T_{ODT}, which separates the disordered micelle regime from the micelle-free homogenous regime. As a matter of fact, T_{CMT} corresponds to T_{DMT}. Disordered micelles have also been observed in sphere-forming SEBS triblock copolymer by Kim et al. (1999), who employed oscillatory shear rheometry and SAXS, and in sphere-forming PS-*block*-poly(2-vinylpyridine) copolymers by Yokoyama et al. (2000) and Segalman et al. (2003), who employed dynamic secondary ion mass spectroscopy, scanning force microscopy, and SAXS. As discussed, experimental evidence indicates that disordered micelles observed in highly asymmetric SIS triblock (also SI diblock) copolymers at $T_{LDOT} \leq T < T_{DMT}$ have very sharp interfaces, and they are thermally stable and also thermally reversible. Nevertheless, the issue as to whether disordered micelles in a highly asymmetric block copolymer can be regarded as being part of a disordered phase associated with thermally induced composition fluctuations or as a separate phase remains to be resolved by future investigation.

8.2.3 Effect of Thermal History on the Oscillatory Shear Rheometry of Block Copolymers

In this section, we present the effect of thermal history on the oscillatory shear rheometry of block copolymers. We will show that the occurrence of a minimum in G' in the isochronal dynamic temperature sweep experiment does not necessarily signify OOT for highly asymmetric block copolymers; instead, it sometimes reflects imperfect bcc spheres, as determined by SAXS and TEM, due to an insufficient annealing of a specimen. Here, we will show that a minimum in G', observed for an unannealed specimen in the isochronal dynamic temperature sweep experiment, may disappear completely when the specimen is annealed for a sufficiently long time at an elevated temperature below the T_{LDOT} of a highly symmetric SI diblock copolymer.

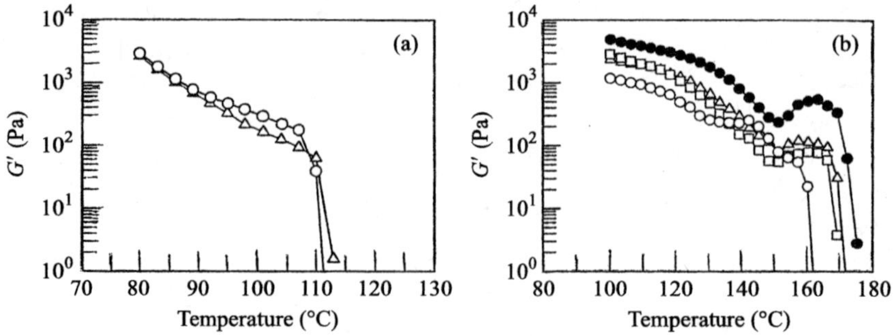

Figure 8.22 Temperature dependence of dynamic storage modulus G' of SI diblock or SIS triblock copolymers during heating in the isochronal temperature sweep experiment at $\omega = 0.01$ rad/s. (a) Symmetric SI-16 specimens prepared by compression molding at 120 °C for 20 min (△) and by solvent casting followed by isothermal annealing at 92 °C for 3 days, then rapid quenching in ice water (○). (b) Highly asymmetric SIS-110 specimens prepared by: compression molding at 150 °C for 20 min without isothermal annealing (●), compression molding at 150 °C for 20 min followed by isothermal annealing at 85 °C for 2 days (△), compression molding at 150 °C for 20 min followed by isothermal annealing at 85 °C for 3 days (□), and solvent casting followed by isothermal annealing at 85 °C for 5 days (○). (Reprinted from Choi et al., *Macromolecules* 36:7707. Copyright © 2003, with permission from the American Chemical Society.)

Figure 8.22 shows the effect of the sample preparation methods employed, compression molding or solvent casting, on the temperature dependence of G' during heating in the isochronal temperature sweep experiments at $\omega = 0.01$ rad/s for the symmetric diblock copolymer SI-16 ($M_n = 1.55 \times 10^4$, $M_w/M_n = 1.04$, $w_{PS} = 0.51$) and the highly asymmetric triblock copolymer SIS-110. Referring to Figure 8.22a, one specimen of SI-16 was prepared by compression molding at 120 °C for 20 min and another specimen of SI-16 was prepared by solvent casting followed by isothermal annealing at 92 °C for 3 days. From Figure 8.22a we determine the T_{ODT} of SI-16 to be approximately 110 °C, regardless of whether the specimen was prepared by compression molding or solvent casting. In obtaining the results given in Figure 8.22b, one specimen of SIS-110 was prepared by solvent casting followed by an isothermal annealing at 85 °C for 5 days and another specimen of SIS-110 was prepared by compression molding at 150 °C for 20 min. Subsequently, the compression-molded specimens were annealed in a vacuum oven at 85 °C for 2, 3, or 5 days and then each of the annealed specimens was subjected to isochronal dynamic temperature sweep experiment (Choi et al. 2003). The following observations are worth noting in Figure 8.22b. Over the entire range of temperatures tested, (1) the compression molded specimen without an isothermal annealing has larger values of G' compared with other specimens, and the G' goes through a large minimum at approximately 150 °C followed by a rapid increase and then begins to drop precipitously at approximately 170 °C, and has very small values at approximately 175 °C, (2) the G' of the compression molded specimen after an isothermal annealing at 85 °C for 2 days goes through a mild minimum at approximately 150 °C followed by a rapid increase and then begins to drop precipitously at approximately 168 °C, and has very small values at approximately 172 °C,

(3) the G' of the compression molded specimen after an isothermal annealing at 85 °C for 3 days also goes through a mild minimum at approximately 150 °C followed by a rapid increase and then begins to drop precipitously at approximately 168 °C, and has very small values at approximately 170 °C, and (4) the G' of the solvent-cast specimen after an isothermal annealing at 85 °C for 5 days goes through a mild minimum at approximately 135 °C and then begins to drop precipitously at approximately 161 °C, and has very small values at approximately 162 °C.

It is very clear from Figure 8.22b that the sample preparation methods employed, compression molding or solvent casting, have a profound influence on the temperature at which values of G' begin to drop precipitously for the highly asymmetric triblock copolymer SIS-110. It is particularly noteworthy that isothermal annealing of compression-molded SIS-110 specimens has brought the temperature at which values of G' become negligibly small to lower values, approaching the temperature determined from a solvent-cast specimen followed by an isothermal annealing. The above experimental results seem reasonable because a compression-molded specimen must have had some residual stresses, and the alignment imposed by the flow during compression molding would have affected the phase behavior of highly asymmetric block copolymers; in other words, the compression-molded specimen was in a nonequilibrium state. During isothermal annealing for a sufficiently long period, the residual stresses in the compression-molded specimen should have been relaxed and the alignment imposed by the flow during compression molding should have been randomized.

Figure 8.23 gives the temperature dependence of G' during the isochronal dynamic temperature sweep experiments at $\omega = 0.01$ rad/s in the heating process for highly asymmetric SI diblock copolymer SI-7/29 ($M_n = 3.6 \times 10^4$, $M_w/M_n = 1.03$, $w_{PS} = 0.19$) specimens before and after annealing at 95 °C for 15 days. It can be seen in

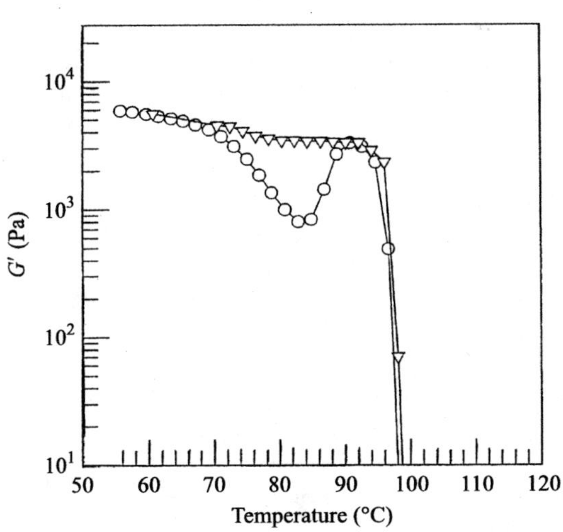

Figure 8.23 Isochronal dynamic temperature sweep experiments for a highly asymmetric SI diblock copolymer SI-7/29 at $\omega = 0.01$ rad/s during heating: (○) before annealing and (∇) after annealing at 95 °C for 15 days. (Reprinted from Yamaguchi et al., *Macromolecules* 32:7696. Copyright © 1999, with permission from the American Chemical Society.)

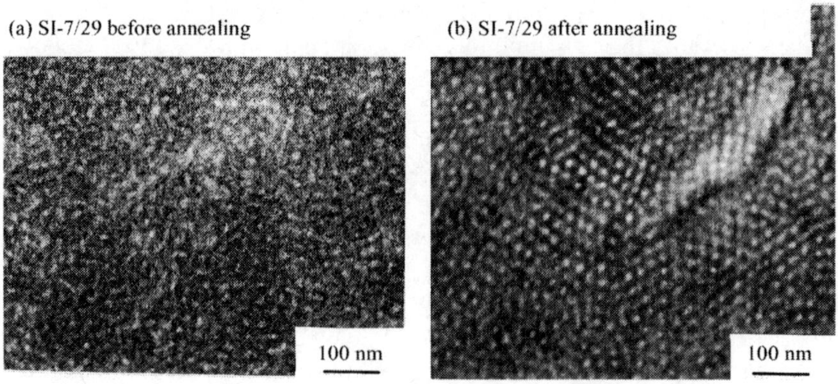

Figure 8.24 TEM images of a highly asymmetric SI diblock copolymer SI-7/29 (a) before annealing and (b) after annealing at 95 °C for 15 days. (Reprinted from Yamaguchi et al., *Macromolecules* 32:7696. Copyright © 1999, with permission from the American Chemical Society.)

Figure 8.23 that the value of G' of the unannealed specimen went through a minimum at approximately 84 °C before beginning to decrease precipitously at approximately 96 °C, whereas a minimum in G' disappeared completely when the specimen had been annealed at 95 °C for 15 days. Figure 8.24 gives TEM images of SI-7/29 specimens before annealing and after annealing at 95 °C for 15 days. It is very clear from Figure 8.24 that a prolonged annealing made the bcc lattice of the SI-7/29 very distinct with very sharp grain boundaries.

Figure 8.25 gives two-dimensional (2D) SAXS patterns of an SI diblock copolymer (SI-7/29) measured with an imaging-plate detector system before and after annealing at 95 °C for 15 days, the same annealing condition as employed in the TEM experiments described above. In Figures 8.25a and 8.25b we observe distinctly different SAXS diffraction patterns between the two specimens. Specifically, in the unannealed specimen (Figure 8.25a) we observe a diffuse ring, suggesting that the specimen consists of a large number of small, randomly oriented grains of bcc spheres composed of PS block chains in the matrix of PI block chains, such that the first-order diffraction from (110) becomes a diffraction ring. Conversely, in the annealed specimen (Figure 8.25b) we clearly observe several very bright diffraction spots, suggesting the existence of large grains composed of highly ordered bcc spheres. Only a limited number of grains are seen to give (110) and (200) diffraction spots on the 2D detector plane, and other large grains give their corresponding diffraction spots in the space other than the 2D detector plane. In Figure 8.25c we observe a considerable difference in the azimuthal-angle dependence of the first-order peak between the unannealed specimen and the annealed specimen, reflecting the difference between the diffraction pattern composed of a diffraction ring and the diffraction pattern composed of the discrete diffraction spots, respectively, shown in Figures 8.25a and 8.25b.

Figure 8.26 gives desmeared SAXS profiles with values of q that were obtained using slit collimation and a one-dimensional (1D) position sensitive detector, where $q = (4\pi/\lambda)\sin(\theta/2)$ is the magnitude of scattering vector q, with λ and θ being the

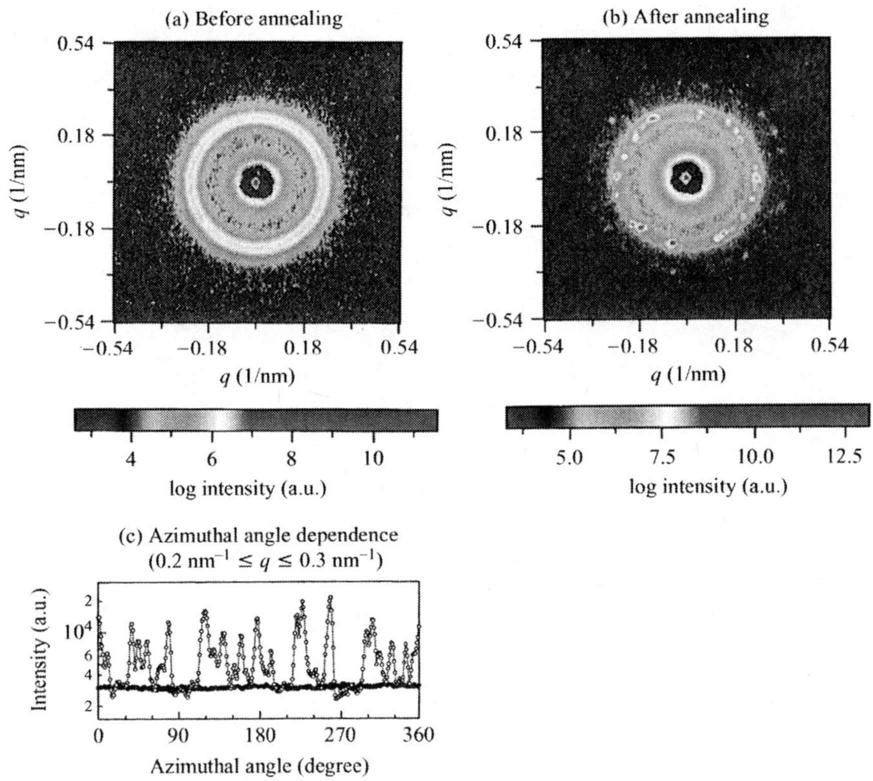

Figure 8.25 2D SAXS patterns for a highly asymmetric SI diblock copolymer SI-7/29 (a) before annealing, (b) after annealing at 95 °C for 15 days, and (c) the corresponding intensity distributions of the first-order peak with respect to azimuthal angle: (●) before annealing and (○) after annealing at 95 °C for 15 days. (Reprinted from Yamaguchi et al., *Macromolecules* 32:7696. Copyright © 1999, with permission from the American Chemical Society.) (See color plate 1.)

wavelength of X-ray and scattering angle, respectively. These results are consistent with the 2D diffraction images given in Figure 8.25. In the SAXS profiles of the annealed specimen we observe three diffraction peaks located at ratios of $1{:}\sqrt{2}{:}\sqrt{3}$ in the reciprocal space, originating from interdomain interference, and one broad peak located at approximately $q = 0.6$ nm^{-1} from the particle scattering of the isolated sphere. Moreover, we observe an extra peak (designated with an asterisk *) inside the first-order scattering maximum. This is due to an "artifact" encountered by detecting a (110) diffraction spot at a position within a window of the 1D detector. Note that once the intensity of the spot was corrected with the 1D detector, the corresponding scattering peak remained even after desmearing. The extra peak does reflect the diffraction spot image in the 2D SAXS pattern. In contrast, in the SAXS profile of the unannealed specimen we observe a broad first-order peak, a very weak second-order shoulder, and a broad maximum from the particle scattering, suggesting a less ordered bcc lattice. The SAXS study has shown no evidence of OOT taking place below T_{LDOT} from hexagonally packed cylinders of PS in the matrix of PI to the bcc spheres of PS

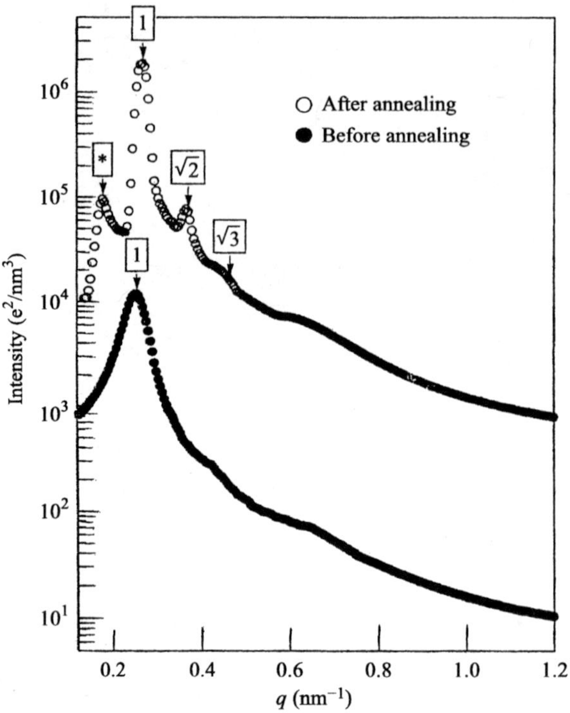

Figure 8.26 SAXS profiles for a highly asymmetric SI diblock copolymer SI-7/29 during the cooling cycle: (●) before annealing and (○) after annealing at 95 °C for 15 days. (Reprinted from Yamaguchi et al., *Macromolecules* 32:7696. Copyright © 1999, with permission from the American Chemical Society.)

in the matrix of PI at approximately 84 °C (Yamaguchi et al. 1999), which may be inferred from the isochronal dynamic temperature sweep experiments for an unannealed specimen (see Figure 8.23).

From the SAXS results presented above we can conclude that long-duration annealing of the highly asymmetric SI diblock copolymer SI-7/29 increased the grain size and hence decreased the grain-boundary area, making a more perfect bcc lattice within the grains. For such a specimen, the value of G' monotonously decreases, due to a steady softening of PS spherical microdomains and their lattices, with increasing temperature towards T_{LDOT}. Such a conclusion is supported by the TEM images (see Figure 8.24), which show that a prolonged annealing made the diffuse grain boundary very sharp, giving rise to a highly ordered bcc lattice. Conversely, the value of G' for an unannealed specimen first decreases, due to softening of PS spheres, with increasing temperature towards $T_{\mathrm{g,PS}}$. G' then increases, due to the increased perfection of the long-range order promoted by an enhanced mobility of PS chains above $T_{\mathrm{g,PS}}$, with increasing temperature to T_{LDOT}. The increase of G' with increasing temperature above approximately 84 °C is a consequence of the increase of G' due to the increased long-range order that outweighs the decrease of G' due to the softening effect.

Figure 8.27 gives $\log G'$ versus $\log G''$ plots for as-cast cylinder-forming SI diblock copolymer (SI-Z) specimens having a 0.34 weight fraction of PS block, $M_{\mathrm{w}} = 1.9 \times 10^4$, and $M_{\mathrm{w}}/M_{\mathrm{n}} = 1.09$, which were subjected to different thermal histories: No annealing, annealing at 80 °C for 1 week, and annealing at 80 °C for 2 weeks. This block copolymer has $T_{\mathrm{ODT}} = 90$ °C, as determined from $\log G'$ versus

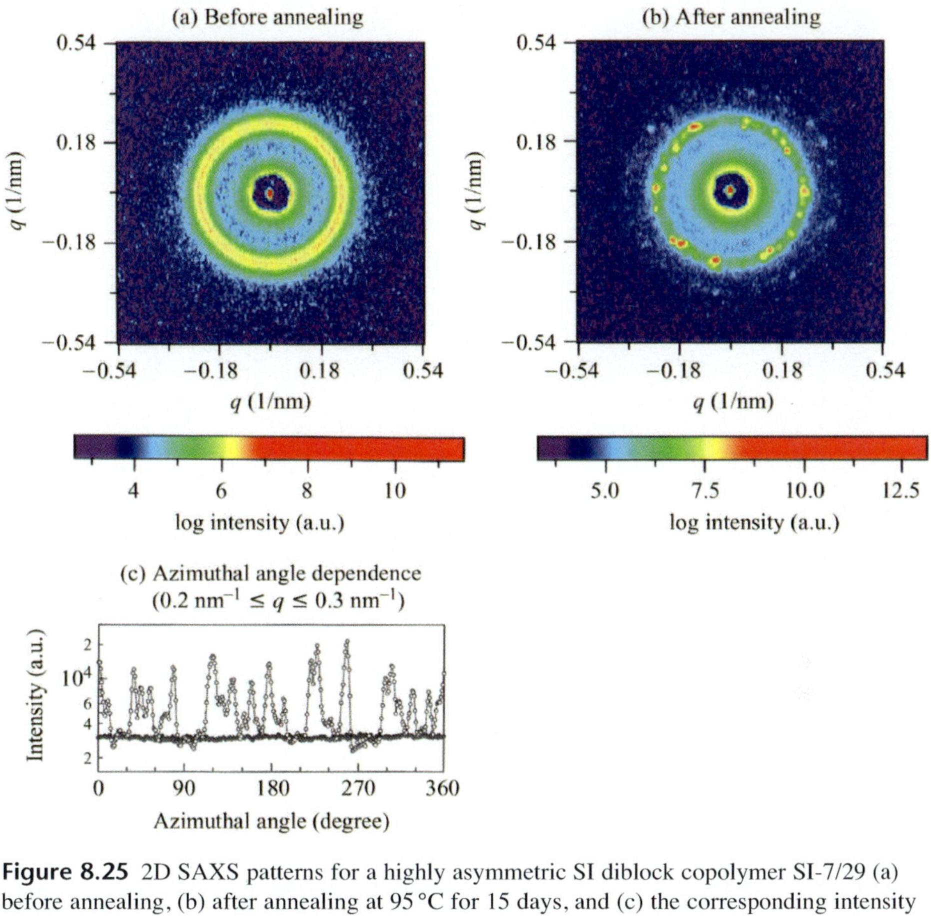

Figure 8.25 2D SAXS patterns for a highly asymmetric SI diblock copolymer SI-7/29 (a) before annealing, (b) after annealing at 95 °C for 15 days, and (c) the corresponding intensity distributions of the first-order peak with respect to azimuthal angle: (●) before annealing and (O) after annealing at 95 °C for 15 days. (Reprinted from Yamaguchi et al., *Macromolecules* 32:7696. Copyright © 1999, with permission from the American Chemical Society.)

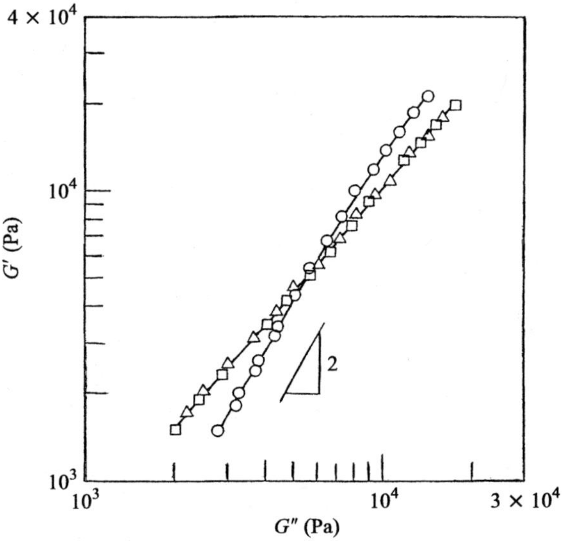

Figure 8.27 Log G' versus log G'' plots for an SI diblock copolymer SI-Z after different thermal treatments: (○) as-cast specimen without thermal treatment, (△) as-cast specimen after being annealed at 80 °C for 1 week, and (□) as-cast specimen after being annealed at 80 °C for 2 weeks. All measurements were carried out at 80 °C. (Reprinted from Hashimoto et al., *Polymer* 39:1573. Copyright © 1998, with permission from Elsevier.)

log G'' plots, and hexagonally packed cylindrical microdomains of PS, as determined from TEM and SAXS. Figure 8.27 shows a distinct difference in the rheological response between the annealed and unannealed as-cast specimens. In particular, log G' versus log G'' plots for annealed specimens have a slope much less than that of the unannealed specimens. Thus we can infer from Figure 8.27 that annealing of an as-cast specimen at 80 °C for 2 weeks helps its microdomain structure to attain an equilibrium morphology. Notice in Figure 8.27 that the slope of 2 in the log G' versus log G'' plots observed in the low-frequency region for the as-cast specimens cannot be ascribed to the thermal concentration fluctuations in the disordered state. If this were to reflect the thermal fluctuation effects in the disordered state, the slope of log G' versus log G'' plots would have been constant and independent of the duration of annealing. Thus, the liquidlike behavior, exhibiting the slope of 2 in the log G' versus log G'' plots, observed for the as-cast specimens is ascribed to the presence of defects in the ordered state. As the defects decrease with annealing, the system changes from liquidlike to soft solidlike behavior, exhibiting the slope less than 2 in the log G' versus log G'' plot. Here, we define (1) "liquidlike state" to be a state at which the cylindrical microdomain of SI-Z has not yet attained a higher degree of ordering in a grain and between grains (grains being surrounded in the disordered phase), and (2) "solidlike state" to be a state at which the cylindrical microdomains of SI-Z have already attained a higher degree of ordering in a grain and the grains fill the whole space: the domains are interconnected at grain boundaries. Thus, the change from liquidlike to soft solidlike behavior observed in as-cast specimens during annealing under quiescent conditions must be distinguished from the situation where a decrease in the amount of defects is observed under large-amplitude oscillatory shear flow, which induces shear-orientation.

Figure 8.28 gives log G' versus log G'' plots for SI-Z at 80 °C, which were obtained by conducting frequency sweep experiments at every 30 min for the period of 8 h after the specimen had been quenched from 100 to 80 °C. In Figure 8.28 we observe

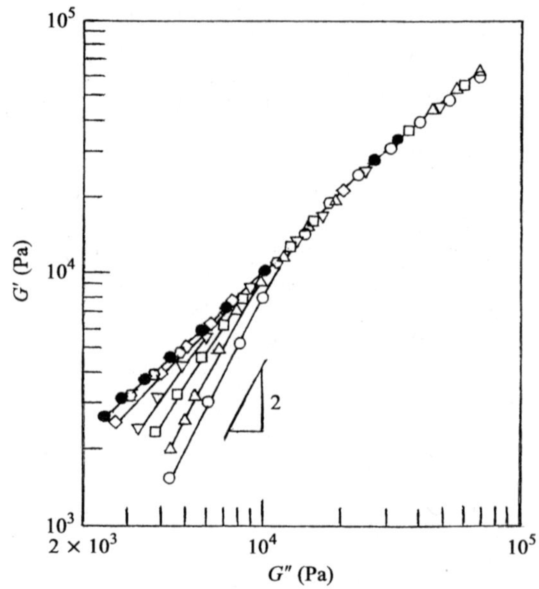

Figure 8.28 Log G' versus log G'' plots for an SI diblock copolymer SI-Z specimen, at various periods (h) after being quenched rapidly from 100 °C (>T_{ODT}) to 80 °C: (○) 0.5, (△) 2, (□) 3, (▽) 4, (◇) 6, (⬡) 7, and (●) 8. (Reprinted from Hashimoto et al., *Polymer* 39:1573. Copyright © 1998, with permission from Elsevier.)

that, upon being quenched to 80 °C, the slope of log G' versus log G'' plot in the terminal region decreases with time until approximately 7 h elapses and then remains constant. From Figure 8.28 we can conclude that it takes about 7 h after the quenching from 100 to 80 °C for the PS microdomains in SI-Z to attain a morphology close to equilibrium morphology. An independent SAXS study indicated that it took about 8 h when monitoring SAXS intensity with time (Hashimoto et al. 1998).

Figure 8.29 gives a schematic showing variation of the shape of log G' versus log G'' plots in the terminal region and variation of G' with time in the terminal region upon quenching of an SI diblock copolymer (SI-Z) from the disordered state to an ordered state. In Figure 8.29, A denotes the rheological (or morphological) state at the very early stage of ordering upon quenching from the disordered state, B and C denote the state at which an equilibrium morphology has not been yet attained, and

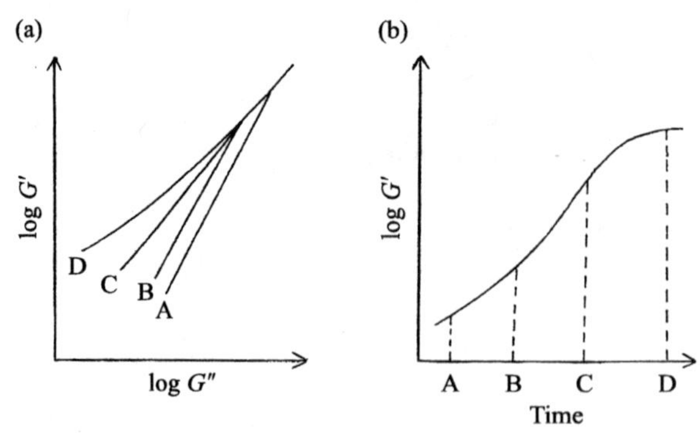

Figure 8.29 Schematic describing (a) variations of the shape of log G' versus log G'' plots in the terminal region and (b) variations of G' with time at a fixed value of G'' in the terminal region upon quenching of an SI diblock copolymer (SI-Z) from the disordered state to an ordered state. (Reprinted from Hashimoto et al., *Polymer* 39:1573. Copyright © 1998, with permission from Elsevier.)

D denotes the state at which an equilibrium morphology has been attained. Without having information on the morphology of SI-Z at the respective stages (A through D), we can only speculate on the morphology corresponding to the G' at each stage of ordering upon quenching from the disordered state. In the terminal region of the $\log G'$ versus $\log G''$ plots with reference to Figures 8.28 and 8.29, we make the following observations. At stage A (30 min after the quenching began from the disordered state at 100 °C), SI-Z has a very low value of G', implying that the morphology of the block copolymer has changed little from the disordered state, at which the slope of the $\log G'$ versus $\log G''$ plot is close to 2. At stage B (2 h after the quenching began), the value of G' has increased somewhat, indicating that microdomains have formed. Note, however, that the $\log G'$ versus $\log G''$ plot has the slope of almost 2, exhibiting a liquidlike rheological response. At stage C (3 h after the quenching began), the value of G' has increased further, indicating that ordering continues. Note that the $\log G'$ versus $\log G''$ plot still has the slope of almost 2, exhibiting a liquidlike rheological response. We speculate that from the morphological point of view, stages B and C differ from each other in the extent of ordering, and that ordered and fluctuating states might coexist. It is conceivable that grains composed of ordered cylinders in SI-Z are dispersed in the disordered matrix and that the volume fraction of these ordered regions increases with time. As the grain size and its volume fraction increase with time, the slope of the $\log G'$ versus $\log G''$ plot moves toward the left side, or upward direction, corresponding to a low frequency shift of this liquidlike rheological response with time. At stage D (7–8 h after the quenching began from the disordered state at 100 °C), the value of G' has increased considerably and no longer changes with time, indicating that an equilibrium morphology has been attained. Note that the $\log G'$ versus $\log G''$ plot has the slope of much less than 2, exhibiting a solidlike rheological response. This observation suggests that grains are volume filling at stage D, and that ordering might have attained a state close to equilibrium.

8.3 Oscillatory Shear Rheometry of Microsphase-Separated Block Copolymers Exhibiting Lower Critical Disorder–Order Transition Behavior

In this section, we present oscillatory shear rheometry of block copolymers exhibiting LCDOT and a closed-loop phase behavior with combined LCDOT and UCODT at lower and higher temperatures, respectively. Several research groups (Hasegawa et al. 1999; Karis et al. 1995; Pollard et al. 1998; Russell et al. 1994; Ruzette et al. 1998; Weidisch et al. 1999, 2000) have reported on phase transition in diblock copolymers exhibiting LCDOT. Russell et al. (1994) appear to be the first to have observed, via SAXS, LCDOT in lamella-forming poly(perdeuterated styrene)-*block*-poly(*n*-butyl methacrylate) (dPS-*block*-P*n*BMA) copolymers. Weidisch et al. (1999) also investigated phase transition in nearly symmetric dPS-*block*-P*n*BMA copolymers with varying molecular weights using SANS and isochronal dynamic temperature sweep experiments. Figure 8.30 shows the temperature dependence of G' for a nearly symmetric dPS-*block*-P*n*BMA copolymer with a 0.55 volume fraction of PS block and $M_n = 9 \times 10^4$, which was obtained from an isochronal dynamic temperature sweep experiment at $\omega = 0.01$ rad/s. It is seen from Figure 8.30 that values of G' initially

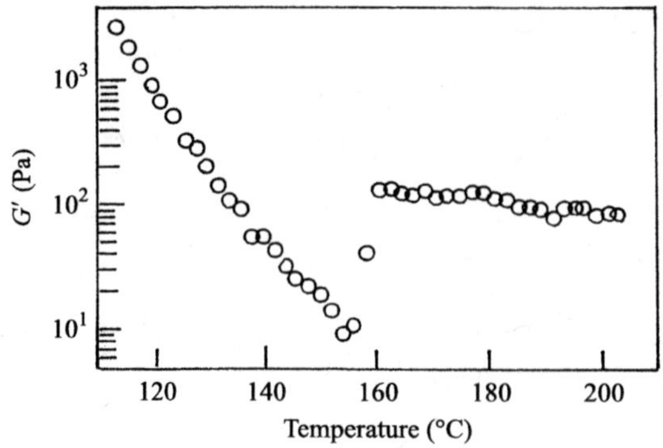

Figure 8.30 Temperature dependence of G' for a nearly symmetric dPS-*block*-P*n*BMA copolymer with a 0.55 volume fraction of PS block and $M_n = 9 \times 10^4$, which was obtained from an isochronal dynamic temperature sweep experiment at $\omega = 0.01$ rad/s. (Reprinted from Weidisch et al., *Macromolecules* 32:3405. Copyright © 1999, with permission from the American Chemical Society.)

decrease rapidly with increasing temperature, then suddenly increase very sharply at approximately 154 °C, and then remain more or less constant with a further increase in temperature. A sudden increase in G' displayed in Figure 8.30 signifies the onset of LCDOT, because a very small value of G' (from the disordered state) suddenly increases (about one order of magnitude) to a very large value (to an ordered state). The temperature dependence of G' observed in Figure 8.30 for a dPS-*block*-P*n*BMA copolymer is opposite to that observed in the inset of Figure 8.8 for an SI diblock copolymer exhibiting UCODT (or ODT).

Weidisch et al. (1999) obtained the following expression for temperature dependence of χ:

$$\chi = 0.024 - 4.56/T \tag{8.2}$$

from curve fitting the SANS profiles obtained at various temperatures to the Leibler theory (1980). Notice that the sign of the temperature coefficient appearing in the second term on the right-hand side of Eq. (8.2) is negative, which is opposite to that for polymer pairs exhibiting UCST or block copolymers exhibiting UCODT (or ODT) (e.g., SI diblock copolymer). It is worth noting that the entropic contribution to χ (the first term on the right-hand side of Eq. (8.2)) is relatively large compared with that of the enthalpic contribution (the second term on the right-hand side of Eq. (8.2)), which is quite opposite to the temperature-dependence of χ for SI diblock copolymers, for instance, exhibiting UCODT (or ODT) during heating. This observation seems to reinforce the view of Russell et al. (1994) that LCDOT is driven by entropic factors.

Figure 8.31 gives the dependence of T_{LCDOT} on the molecular weight of dPS-*block*-P*n*BMA copolymers, showing that T_{LCDOT} decreases with increasing molecular weight. This trend of molecular weight dependence of T_{LCDOT} for dPS-*block*-P*n*BMA copolymers is quite opposite to that observed for the block copolymers exhibiting UCODT (or ODT). Notice in Figure 8.31 that the T_{LCDOT} of a dPS-*block*-P*n*BMA copolymer with $M_n \approx 1.3 \times 10^5$ approaches the glass transition temperature ($T_{g,\mathrm{PS}}$) of the PS block and the dPS-*block*-P*n*BMA copolymers with higher molecular weights are microphase separated at all measured temperatures. Owing to thermal decomposition

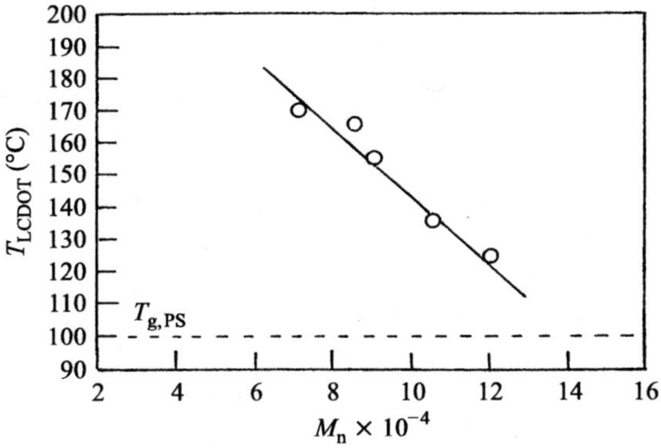

Figure 8.31 Molecular weight dependence of T_{LCDOT} for dPS-*block*-P*n*BMA copolymers. (Reprinted from Weidisch et al., *Macromolecules* 33:5495. Copyright © 2000, with permission from the American Chemical Society.)

at about 180 °C, only a narrow range of molecular weights exists where LCDOT in dPS-*block*-P*n*BMA copolymers can be observed.

Using oscillatory shear rheometry, SAXS, and SANS, Ryu et al. (2003) investigated phase transition in symmetric dPS-*block*-poly(*n*-pentyl methacrylate) (P*n*PMA) copolymers having two different molecular weights (dPS-*block*-P*n*PMA-L with $M_n = 4.5 \times 10^4$ and dPS-*block*-P*n*PMA-H with $M_n = 5.33 \times 10^4$) and three binary mixtures of the two block copolymers. Figure 8.32 shows the results of their isochronal dynamic

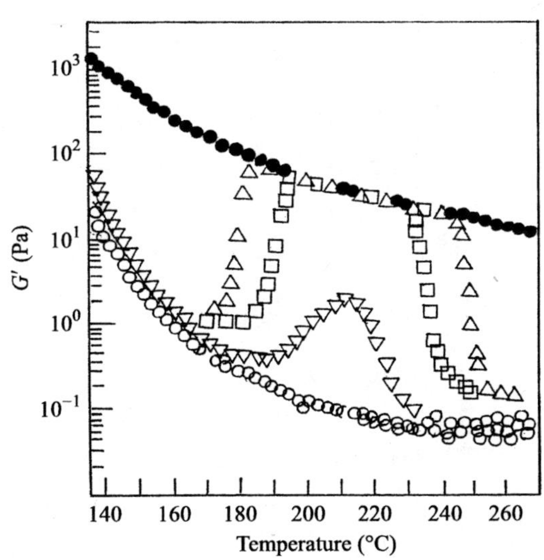

Figure 8.32 Temperature dependence of G' obtained from isochronal dynamic temperature sweep experiments at $\omega = 0.1$ rad/s for: (●) dPS-*block*-P*n*PMA-H with $M_n = 5.33 \times 10^4$, (○) dPS-*block*-P*n*PMA-L with $M_n = 4.5 \times 10^4$, (△) dPS-*block*-P*n*PMA-BH composed of 55 wt % dPS-*block*-P*n*PMA-H and 45 wt % dPS-*block*-P*n*PMA-L, (□) dPS-*block*-P*n*PMA-BM composed of 50 wt % dPS-*block*-P*n*PMA-H and 50 wt % dPS-*block*-P*n*PMA-L, and (▽) dPS-*block*-P*n*PMA-BL composed of 45 wt % dPS-*block*-P*n*PMA-H and 55 wt % dPS-*block*-P*n*PMA-L. (Reprinted from Ryu et al., *Macromolecules* 36:2894. Copyright © 2003, with permission from the American Chemical Society.)

temperature sweep experiments for the five block copolymers. The following observations are worth noting in Figure 8.32. Values of G' for dPS-*block*-P*n*PMA-L, which has the lowest molecular weight among the five polymers employed, decrease steadily with increasing temperature and level off to become extremely small. Since such a temperature dependence of G' is typically observed in ordinary flexible polymers, Ryu et al. concluded that dPS-*block*-P*n*PMA-L is in the disordered state over the entire range of temperatures investigated. Values of G' for dPS-*block*-P*n*PMA-H, which has the highest molecular weight among the five polymers employed, decrease steadily with increasing temperature, but they remain very large even as the temperature is increased to approximately 270 °C, the highest experimental temperature employed. Since the values of G' remain very large even at 270 °C, they concluded that dPS-*block*-P*n*PMA-H is in the ordered state over the entire range of temperatures investigated. Having seen the two extreme situations, they prepared binary mixtures of the two block copolymers: (1) dPS-*block*-P*n*PMA-BL composed of 45 wt % dPS-*block*-P*n*PMA-H and 55 wt % dPS-*block*-P*n*PMA-L, (2) dPS-*block*-P*n*PMA-BM composed of 50 wt % dPS-*block*-P*n*PMA-H and 50 wt % dPS-*block*-P*n*PMA-L, and (3) dPS-*block*-P*n*PMA-BH composed of 55 wt % dPS-*block*-P*n*PMA-H and 45 wt % dPS-*block*-P*n*PMA-L.

Regarding the temperature dependence of G' for dPS-*block*-P*n*PMA-BH displayed in Figure 8.32, we observe that G′ initially decreases with increasing temperature, begins to increase very rapidly at approximately 170 °C, remains there until reaching approximately 245 °C, at which point G' begins to decrease once again with a further increase in temperature, and finally values of G' become very small. The temperature dependence of G' observed above can be interpreted as follows (Ryu et al. 2003): dPS-*block*-P*n*PMA-BH first undergoes LCDOT at approximately 170 °C, because a small value of G' (from the disordered state) increases (about two orders of magnitude) to a very large value (to an ordered state), and then undergoes UCODT at approximately 245 °C, since a very large value of G' (from an ordered state) decreases (about two orders of magnitude) to a small value (to the disordered state). That is, the block copolymer dPS-*block*-P*n*PMA-BH exhibits a combined UCODT and LCDOT. Similar observations can be made from Figure 8.32 for both dPS-*block*-P*n*PMA-BM and dPS-*block*-P*n*PMA-BL. That is, Ryu et al. (2002a, 2003) observed combined UCODT and LCDOT in dPS-*block*-P*n*PMA copolymers exhibiting a closed-loop phase behavior. Note that binary polymer blend systems exhibiting a combined LCST and UCST had been observed earlier (Cong et al. 1986; Ougizawa et al. 1985; Ryu 2002b). The temperature dependence of χ for a low-molecular-weight dPS/P*n*PMA pair, which was obtained from curve fitting the SANS profiles to the Leibler theory (1980), is given by Ryu et al. (2002).

Figure 8.33 gives the molecular weight dependence of T_{LCDOT} and T_{UCODT} for dPS-*block*-P*n*BMA copolymers. It is seen in Figure 8.33 that T_{LCDOT} decreases, while T_{UCODT} increases, as molecular weight increases. It is worth noting in Figure 8.33 that a combined LCDOT and UCODT (i.e., a phase diagram of closed-loop type) can be observed over a very narrow range of molecular weights. The phase behavior of binary polymer blends exhibiting combined LCST and UCST must be distinguished from the phase behavior of dPS-*block*-P*n*PMA copolymers exhibiting LCDOT and UCODT, in that binary polymer blends are miscible at temperatures between T_{LCST} and T_{UCST} while dPS-*block*-P*n*PMA copolymers are microphase-separated at temperatures between T_{LCDOT} and T_{UCODT}.

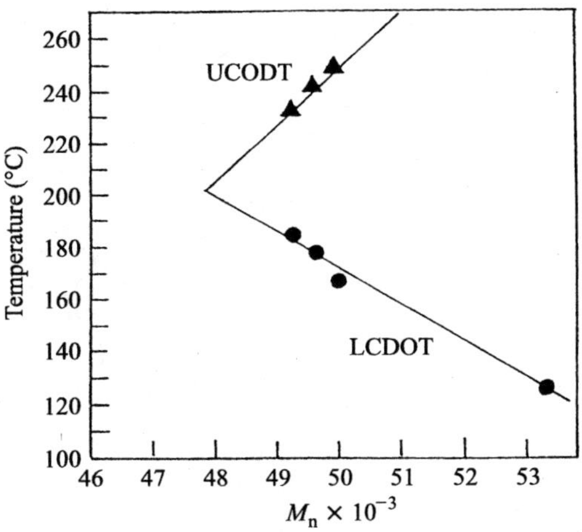

Figure 8.33 Molecular weight dependence of T_{LCDOT} (●) and T_{UCODT} (▲) for dPS-*block*-P*n*PMA copolymers, where M_n denotes number-average molecular weight. (Reprinted from Ryu et al., *Macromolecules* 36:2894. Copyright © 2003, with permission from the American Chemical Society.)

8.4 Linear Viscoelasticity of Disordered Block Copolymers

Owing to the presence of microdomain structures, the rheological behavior of block copolymers in an ordered state (at $T < T_{ODT}$) is much more complicated than that in the disordered state (at $T \geq T_{ODT}$). For instance, block copolymers with lamellar or cylindrical microdomains exhibit shear-thinning ("yield behavior") at very low shear rates in steady-state shear flow (Han et al. 1995c), very similar to thermotropic liquid-crystalline polymers with nematic mesophase (see Chapter 9), or highly filled thermoplastic composites and organoclay nanocomposites (see Chapter 12). Under such circumstances, the determination of zero-shear viscosity (η_0) is practically very difficult, if not impossible. Interestingly, Sebastian et al. (2002) reported that sphere-forming block copolymers exhibited extremely large values ($10^7 - 10^8$ Pa·s) of η_0 at extremely low shear rates (10^{-8} s^{-1}) or at very low shear stresses (<100 Pa), followed by a sudden drop in shear viscosity (almost four orders of magnitude) when shear stress was increased to 200 Pa entering a weakly shear-thinning regime.

From a theoretical point of view, there are two features that distinguish the dynamics of a block copolymer chain from the dynamics of a chain in a mixture of two homopolymers. One is the connectivity of the blocks and the other is a position-dependent external potential. When dealing with the dynamics of a microphase-separated block copolymer, one must include microdomain structures (lamellae, cylinders, gyroids, or spheres) (see Figure 8.2) in the formulation of system equations. Currently, there is no comprehensive molecular theory that enables one to predict the rheological behavior of microphase-separated block copolymers, although some attempts have been made in that direction (Fredrickson and Larson 1987; Kawasaki and Sekimoto 1987, 1988; Larson and Fredrickson 1987). Certainly, a theoretical treatment of the dynamics of disordered block copolymers would be much easier than that of microphase-separated block copolymers. Even so, there has been no theoretical study reported in the literature since the seminal studies published in 1975 by four research groups (Hall and DeWames 1975; Hansen and Shen 1975a, 1975b; Stockmayer and Kennedy 1975; Wang and DiMarzio 1975).

Some fundamental questions must be answered before considering the chain dynamics of disordered block copolymer chains of AB-type diblock or ABA-type triblock copolymers. The first question is: Should a disordered AB-type diblock copolymer, for instance, be regarded as having unentangled chains as long as the molecular weight of each block is lower than its entanglement molecular weight, even when the combined molecular weight is higher than the entanglement molecular weight of one of the two blocks? This question can be rephrased as follows. Can a disordered AB-type diblock copolymer be regarded as having entangled chains if the combined molecular weight of both blocks is higher than the entanglement molecular weight of one of the blocks, even when the molecular weight of an individual block is lower than its entanglement molecular weight? Next question is: How can one define (or determine) the entanglement molecular weight of a disordered AB-type diblock copolymer or ABA-type triblock copolymer when information on the entanglement molecular weights of the constituent blocks is available? Although the rheological behavior of disordered block copolymers is perceived to be much less complicated than that of microphase-separated block copolymers, only a small number of experimental studies (Choi and Han 2004; Han et al. 1991, 1992, 1995a; Kim and Han 1992) have reported on the rheological behavior of disordered block copolymers. In addition to the questions posed above, there are further fundamental issues that need to be addressed for a better understanding of the rheological behavior of disordered block copolymers. One such question is: Will the molecular weight (M) dependence of zero-shear viscosity (η_0) for disordered block copolymers be the same as the well-established relationship for ordinary flexible homopolymers? Namely, $\eta_0 \propto M^{1.0}$ for $M < M_c$ and $\eta_0 \propto M^{3.4}$ for $M \geq M_c$, with M_c being the viscosity critical molecular weight for flexible homopolymers (see Chapter 4). In this section, we address first the molecular weight dependence of η_0 and then linear dynamic viscoelastic properties for disordered SI diblock copolymers and PS-*block*-PαMS copolymers, placing emphasis on the effects of molecular weight and block length ratio.

8.4.1 Effect of Molecular Weight on the Zero-Shear Viscosity of Disordered Diblock Copolymers

Choi and Han (2004) carried out an experimental study to address some of the questions posed above. In so doing, they employed eight low-molecular-weight, nearly symmetric SI diblock copolymers, to investigate the molecular weight dependence of η_0 in the disordered state (at $T > T_{ODT}$). For this, they first determined the T_{ODT} of each block copolymer using isochronal dynamic temperature sweep experiments and log G' versus log G'' plots. For comparison, they also employed eight polystyrenes and then investigated the molecular weight dependence of η_0. The molecular characteristics, glass transition temperatures, and T_{ODT} of the SI diblock copolymers are given in Table 8.1. The molecular characteristics and glass transition temperatures of the polystyrenes employed are given in Table 6.3.

Figure 8.34 gives logarithmic plots of reduced zero-shear viscosity ($\eta_{0,r}$) versus weight-average molecular weight (M_w) for eight SI diblock copolymers in the disordered state, log η_0 versus log M_w plots for eight polystyrenes, both investigated by Choi and Han (2004), and log η_0 versus log M_w plots from a study of Allen and Fox (1964).

Table 8.1 Molecular characteristics, glass transition temperatures, and order–disorder transition temperature of the SI diblock copolymers

Sample Code	M_w[a]	M_w/M_n[b]	w_{PS}[c]	$T_{g,PS}$[d] (°C)	$T_{g,PI}$[e] (°C)	T_r[f] (°C)	T_{ODT}[g] (°C)
SI-1	0.79×10^4	1.10	0.55	33	−57	137	disordered[h]
SI-2	1.04×10^4	1.10	0.53	44	−60	148	disordered[h]
SI-3	1.35×10^4	1.08	0.53	55	−62	159	82
SI-4	1.50×10^4	1.09	0.54	62	−61	166	91
SI-5	1.72×10^4	1.09	0.55	67	−62	171	127
SI-6	1.91×10^4	1.05	0.55	74	−60	180	146
SI-7	2.01×10^4	1.09	0.54	76	−62	178	156
SI-8	2.10×10^4	1.07	0.51	82	−60	186	186

[a]Weight-average molecular weight determined using light scattering. [b]Polydispersity index determined from gel permeation chromatography. [c]Weight fraction of PS block determined by ^{1}H NMR spectroscopy. [d]Glass transition temperature of PS block determined using DSC at a heating rate of 20 °C/min. [e]Glass transition temperature of PI block determined using DSC at a heating rate of 20 °C/min. [f]Reference temperature chosen as $T_r = T_{g,PS} + 104$ °C, which was used to obtain a shift factor for each block copolymer. [g]Order–disorder transition temperature determined from isochronal dynamic temperature sweep experiments at $\omega = 0.1$ rad/s. [h]SI-1 and SI-2 are in the disordered state at temperatures higher than, say, 50 °C.

Reprinted from Choi and Han, *Macromolecules* 37:215. Copyright © 2004, with permission from the American Chemical Society.

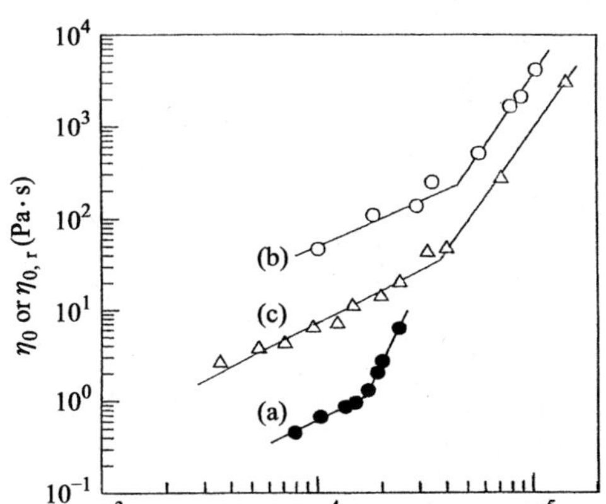

Figure 8.34 (a) The molecular weight dependence of $\eta_{0,r}$ for eight disordered SI diblock copolymers in the disordered state at $T_r = T_{g,PS} + 104$ °C as reference temperature (●), in which the lower slope is 1.15 ± 0.09 and the upper slope is 4.69 ± 0.18. The weight-average molecular weights (M_w) of the SI diblock copolymers are summarized in Table 8.1. (b) The molecular weight dependence of η_0 for the polystyrenes at $T_r = T_{g,Ps} + 79$ °C (○), in which the lower slope is 1.21 ± 0.23 and the upper slope is 3.37 ± 0.21. The M_w of the polystyrenes are summarized in Table 6.3. (c) The molecular weight dependence of η_0 for the polystyrenes at 217 °C, from a study by Allen and Fox (△), in which the lower slope is 1.21 ± 0.11 and the upper slope is 3.28 ± 0.06. (Reprinted from Choi and Han, *Macromolecules* 37:215. Copyright © 2004, with permission from the American Chemical Society.)

What is of great interest in Figure 8.34 is that at M_w greater than a certain critical value (M_c), the molecular weight dependence of $\eta_{0,r}$ for the disordered SI diblock copolymers is much stronger than the molecular weight dependence of η_0 for the polystyrenes. The following observations are worth noting in Figure 8.34: (1) $\eta_{0,r} \propto M^{1.15\pm0.09}$ for $M < M_c$ and $\eta_{0,r} \propto M^{4.69\pm0.18}$ for $M \geq M_c$ for the disordered SI diblock copolymers, (2) $\eta_0 \propto M^{1.21\pm0.23}$ for $M < M_c$ and $\eta_0 \propto M^{3.37\pm0.21}$ for $M \geq M_c$ for the polystyrenes investigated by Choi and Han (2004), and (3) $\eta_0 \propto M^{1.21\pm0.11}$ for $M < M_c$ and $\eta_0 \propto M^{3.28\pm0.06}$ for $M \geq M_c$ for the polystyrenes investigated by Allen and Fox (1964). There is excellent agreement, within experimental uncertainties, in the $\eta_0 - M$ correlations for polystyrenes between the study of Choi and Han (2004) and the study of Allen and Fox (1964). We find that the molecular weight dependence of η_0 or $\eta_{0,r}$ for $M < M_c$ is virtually identical for both polystyrene and disordered SI diblock copolymer, but the molecular weight dependence of η_0 or $\eta_{0,r}$ for $M \geq M_c$ is much greater for disordered SI diblock copolymer than for polystyrenes. The observation that $\eta_0 \propto M^{3.4}$ obtained for the polystyrenes in Figure 8.34 is in good agreement with the well-established relationship in the literature (Berry and Fox 1968).

Notice in Figure 8.34 that the value $M_c \approx 1.7 \times 10^4$ for the disordered SI diblock copolymers is close to the critical viscosity molecular weight of polyisoprene, $M_{c,PI} \approx 1.4 \times 10^4$ reported by Raju et al. (1981).[5] Here we assume $M_c \approx 2M_e$ (see Chapter 4). This observation seems to suggest that a disordered SI diblock copolymer can be regarded as having entangled chains when the combined molecular weight of both blocks exceeds the entanglement molecular weight of PI block ($M_{c,PI}$). Another interesting observation we can make from Figure 8.34 is that the viscosity critical molecular weight of PS ($M_{c,PS}$) is about 4×10^4, which is close to the value 3.6×10^4 reported by Ferry (1980). Again, here we assume $M_c \approx 2M_e$. The above observations bring us to one of the fundamental questions posed at the beginning of this section: Can a disordered AB-type diblock copolymer be regarded as having entangled chains if the combined molecular weight of both blocks is higher than the entanglement molecular weight of one of the blocks, even when the molecular weight of each block is lower than its entanglement molecular weight? According to the information on the molecular weights of PS and PI blocks for the eight SI diblock copolymers summarized in Table 8.1, the answer to the question posed above is affirmative. Notice in Table 8.1 that the molecular weight of PS block in all eight SI diblock copolymers is much lower than $M_{c,PS}(3.6 \times 10^4)$, and the molecular weight of PI block in all eight SI diblock copolymers is lower than $M_{c,PI}(1.4 \times 10^4)$. It is of great interest to observe from Figure 8.34 and Table 8.1 that four SI diblock copolymers (SI-5, SI-6, SI-7, and SI-8) which have a combined molecular weight higher than $M_{c,PI}$ give rise to the relationship $\eta_{0,r} \propto M^{4.69\pm0.18}$ (thus suggesting that they are entangled chains), and the other four SI diblock copolymers (SI-1, SI-2, SI-3, and SI-4) which have a combined molecular weight lower than $M_{c,PI}$ give rise to the relationship $\eta_{0,r} \propto M^{1.15\pm0.09}$ (thus they are Rouse chains). It should be mentioned that variations in molecular weight for SI-5, SI-6, SI-7, and SI-8 are rather small (see Table 8.1). Thus, the conclusions drawn above for the dependence of $\eta_{0,r}$ on M must be regarded with some caution.

The values of $\eta_{0,r}$ in Figure 8.34 were obtained from the logarithmic plots of reduced complex viscosity ($|\eta^*{}_r|$) versus $a_T\omega$ for the eight SI diblock copolymers in the disordered state, for which values of a_T were determined by shifting the log G'' versus log ω plots along the ω axis using $T_r = T_{g,PS} + 104\,°C$ as the reference temperature.

Note that the values of $T_{g,PS}$ are different for each SI diblock copolymer and the reference temperature T_r chosen is higher than the T_{ODT} of each SI diblock copolymer (see Table 8.1). Since in Figure 8.34 values of $\eta_{0,r}$ of the eight SI diblock copolymers in the disordered state are correlated to their M_w, one must check whether the choice of $T_r = T_{g,PS} + 104\,°C$ used to calculate a_T may be warranted for the disordered SI diblock copolymers; that is, whether the iso-zeta state has been achieved.

The following two approaches were taken for such purposes (Choi and Han 2004). First, the monomeric friction coefficients for PS and PI, $\zeta_{0,PS}$ and $\zeta_{0,PI}$, were calculated using the expressions.

$$\log \zeta_{0,PS} = 3.0 + \frac{13.7(T - T_{g,PS})}{T - T_{g,PS} + 48} \tag{8.3}$$

for $\zeta_{0,PS}$ (Allen and Fox 1964) and

$$\log \zeta_{0,PI} = \log \zeta_{00} + \frac{C_1^g C_2^g}{T - T_{g,PI} + C_2^g} \tag{8.4}$$

for $\zeta_{0,PI}$ (Klopffer et al. 1998) with $\log \zeta_{00} = -10.4$ dyn · s/cm, $C_1^g = 13.5 \pm 0.2$, and $C_2^g = 45 \pm 3K$. Note that $\zeta_{0,PS}$ and $\zeta_{0,PI}$ have the units of dyn · s/cm.[6] Figure 8.35 gives variations of $\zeta_{0,PS}$ and $\zeta_{0,PI}$ with $T_r = T_g + 104\,°C$, in which we observe that values of $\log \zeta_{0,PS}$ remain constant, while values of $\log \zeta_{0,PI}$ decrease somewhat, with increasing T_r. Note that T_r was chosen to have an equal distance (104 °C) from the $T_{g,PS}$ of each SI diblock copolymer, while the value of $T_{g,PS}$ increases from 33 °C for SI-1 to 82 °C for SI-8 (see Table 8.1). Thus, the decreasing trend of $\log \zeta_{0,PI}$ with increasing T_r observed from Figure 8.35 is due the fact that values of $T_{g,PI}$ are more or less constant in all eight SI diblock copolymers (see Table 8.1), and thus values of $T_r - T_{g,PI}$ increase from 194 °C for SI-1 to 246 °C for SI-8. The difference between $\log \zeta_{0,PS}$ and $\log \zeta_{0,PI}$ in Figure 8.35 does not remain constant. Thus, we can conclude from Figure 8.35 that, rigorously speaking, an iso-zeta state has not been achieved by simply reducing the linear dynamic viscoelastic data at $T_r = T_{g,PS} + 104\,°C$ for the

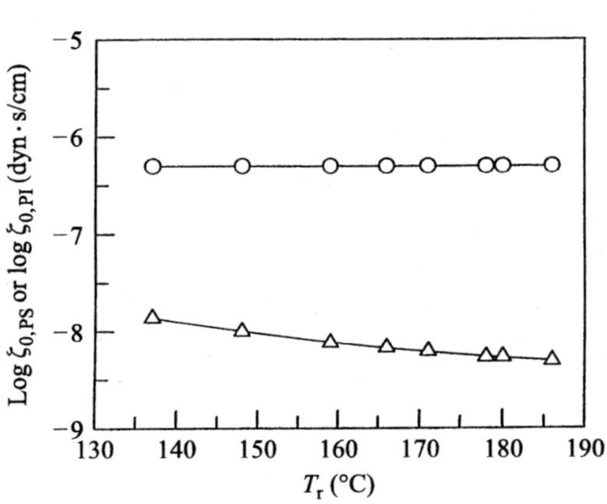

Figure 8.35 Plots of $\log \zeta_{0,PS}$ versus T_r for PS (○) and plots of $\log \zeta_{0,PI}$ versus T_r for PI (△) in eight disordered SI diblock copolymers with $T_r = T_{g,PS} + 104\,°C$ as reference temperature. (Reprinted from Choi and Han, *Macromolecules* 37:215. Copyright © 2004, with permission from the American Chemical Society.)

Table 8.2 WLF parameters for eight SI diblock copolymers

Sample Code	T_r (°C)[a]	C_1	C_2 (°C)
SI-1	137	2.46	152.96
SI-2	148	2.47	135.36
SI-3	159	3.11	132.87
SI-4	166	2.11	120.60
SI-5	171	1.59	77.86
SI-6	180	1.60	81.57
SI-7	178	1.47	71.81
SI-8	186	0.93	32.08

[a] $T_r = T_{g,PS} + 104\,°C$.

eight SI diblock copolymers in the disordered state in order to compare the molecular weight dependence of η_0.

Next, the shift factor a_T for the eight SI diblock copolymers in the disordered state (at $T > T_{ODT}$) were determined using the WLF equation, Eq. (6.7). In so doing, WLF parameters, C_1 and C_2, for all eight SI diblock copolymers were determined by fitting the experimentally determined a_T to the WLF equation, and Table 8.2 gives values of C_1 and C_2 thus determined for all eight SI diblock copolymers. Using these values of C_1 and C_2, plots of $\log a_T$ versus $T - T_r$ with $T_r = T_{g,PS} + 104\,°C$ were prepared, and they are given in Figure 8.36. It can be seen in Figure 8.36 that all eight disordered SI diblock copolymers give rise to a single correlation with some scatters at low values of $T - T_r$, suggesting that the viscoelastic coefficients at the glass transition temperature (C_1^g and C_2^g) are virtually the same for all eight disordered SI diblock copolymers and thus TTS appears to work. This was the basis upon which Choi and Han (2004) assumed that $\zeta_{0,PI}$ was approximately the same for all eight SI diblock copolymers when preparing Figure 8.34. A question may be raised as to why TTS appears to work for SI-8 in the disordered state. The answer to the question lies in

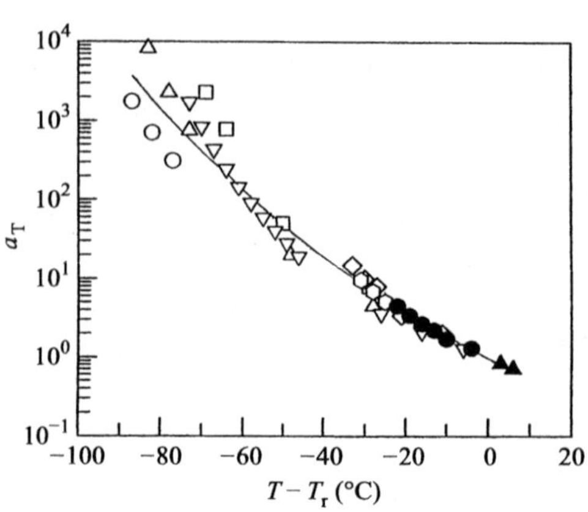

Figure 8.36 Plots of $\log a_T$ versus $T - T_r$ for eight disordered SI diblock copolymers: (○) SI-1, (△) SI-2, (□) SI-3, (▽) SI-4, (◇) SI-5, (⬡) SI-6, (●) SI-7, and (▲) SI-8, with $T_r = T_{g,PS} + 104\,°C$ as the reference temperature. The solid line is drawn as a visual guide. The weight-average molecular weights (M_w) of the SI diblock copolymers are summarized in Table 8.1. (Reprinted from Choi and Han, *Macromolecules* 37:215. Copyright © 2004, with permission from the American Chemical Society.)

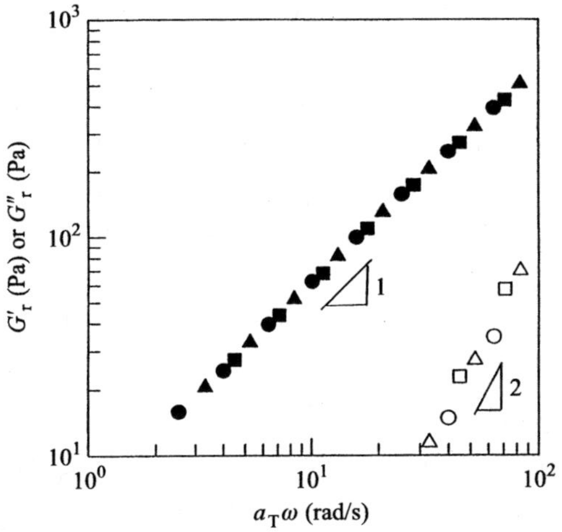

Figure 8.37 Plots of log G'_r versus log $a_T\omega$ (open symbols) and log G''_r versus log $a_T\omega$ (filled symbols) for an SI diblock copolymer, SI-8 ($M_w = 2.1 \times 10^4$), in the disordered state at various temperatures (°C): (○, ●) 186, (△, ▲) 189, and (□, ■) 192, where $T_r = T_{g,PS} + 104\,°C$ was used as reference temperature. (Reprinted from Choi and Han, *Macromolecules* 37:215. Copyright © 2004, with permission from the American Chemical Society.)

that the disordered SI diblock copolymers consist of two low-molecular-weight flexible chains, PS and PI, which are connected to each other, and apparently the disordered PS-*block*-PI chains are mixed on a segmental level. This now explains why the log G'_r versus log $a_T\omega$ plots given in Figure 8.37 for SI-8, which has the highest molecular weight among the eight block copolymers employed, at $T \geq T_{ODT}$ (in the disordered state) show temperature independence, and have a slope of 2 in the terminal region, like any flexible homopolymer, and so explains why TTS works.

It should be clear that the disordered PS-*block*-PI chain does not possess the properties of a neat PS or neat PI chain, but it has new properties influenced by the chain connectivity. For instance, the friction coefficient of a disordered PS-*block*-PI chain would not be the same as the friction coefficient of neat PS chain or the friction coefficient of neat PI chain. It is reasonable to surmise that the friction coefficient of a disordered PS-*block*-PI chain depends not only on its composition but also on chain connectivity (the junction effect) between the two chains. On the basis of the considerations presented above, we conclude that the stronger molecular weight dependence of zero-shear viscosity for disordered SI diblock copolymers ($\eta_{0,r} \propto M^{4.7}$ for $M \geq M_c$) displayed in Figure 8.34 is attributable to the presence of the styrene–isoprene (S–I) junction, which originates from the difference in the monomeric friction coefficients between PS and PI blocks. At present, however, we do not have theoretical guideline as to how the junction effect can be incorporated into the description of the dynamics of disordered block copolymers.

8.4.2 Effect of Block Length Ratio on the Linear Dynamic Viscoelasticity of Disordered Block Copolymers

In this section, we present the effect of block composition (block length ratio) on the linear dynamic viscoelasticity of disordered PS-*block*-poly(α-methylstyrene) (PαMS) copolymers and also SI diblock copolymers in the disordered state. Table 8.3 gives a summary of the molecular characteristics for seven PS-*block*-PαMS copolymers, and Figure 8.38 gives thermograms of differential scanning calorimetry (DSC) for the seven

Table 8.3 Molecular characteristics of the PS-*block*-PαMS copolymers investigated

Sample Code	M_w*(PαMS)*	M_w*(PS)*	M_w/M_n	*PαMS (wt %)*[a]
BII	5.2×10^4	7.0×10^4	1.08	47
CI	5.4×10^4	9.4×10^4	1.08	41
CII	9.9×10^4	14.5×10^4	1.09	48
CIII	13.5×10^4	12.0×10^4	1.09	57
CIV	5.0×10^4	13.0×10^4	1.08	28
CV	12.2×10^4	2.6×10^4	1.06	81
CVI	2.1×10^4	16.7×10^4	1.09	13

[a]Determined by NMR.

Reprinted from Kim and Han, *Macromolecules* 25:271. Copyright © 1992, with permission from the American Chemical Society.

PS-*block*-PαMS copolymers, showing a very broad, single glass transition, which is very similar to the DSC thermograms for the PS/PαMS binary blends presented in the preceding chapter (see Figure 7.13). One can surmise from Figure 8.38 that the seven PS-*block*-PαMS copolymers would not undergo microphase separation and thus they are disordered block copolymers. In the preceding chapter we showed that PS/PαMS binary blends undergo macrophase separation even when the M_w of the constituent components was lower than approximately 5×10^4. The above observations indicate that microphase separation in a block copolymer is less likely to occur than macrophase separation in a binary blend with comparable molecular weights. In presenting the linear dynamic viscoelasticity of the disordered PS-*block*-PαMS copolymers below, we will refer to the sample codes given in Table 8.3.

Figure 8.39 gives plots of a temperature shift factor a_T versus temperature for a disordered PS-*block*-PαMS copolymer, CII, and Figure 8.40 gives reduced plots, log G'_r

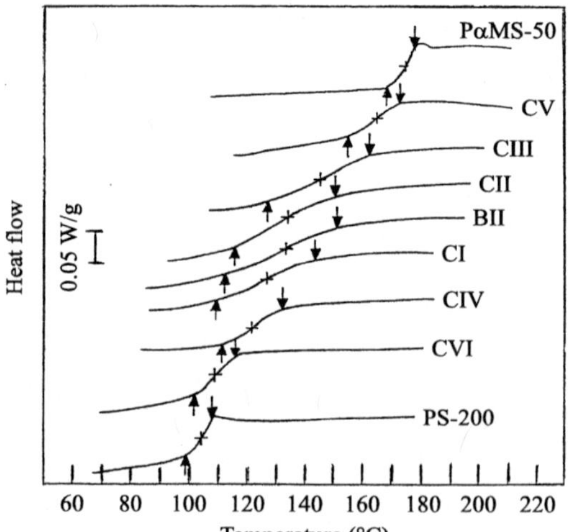

Figure 8.38 DSC thermograms obtained at a heating rate of 10 °C/min for the homopolymers, PS and PαMS, and seven PS-*block*-PαMS copolymers, the molecular characteristics of which are given in Table 8.3. On each curve, the arrow pointing upward denotes the temperature at which the transition begins and the arrow pointing downward denotes the temperature at which the transition ends. (Reprinted from Kim and Han, *Macromolecules* 25:271. Copyright © 1992, with permission from the American Chemical Society.)

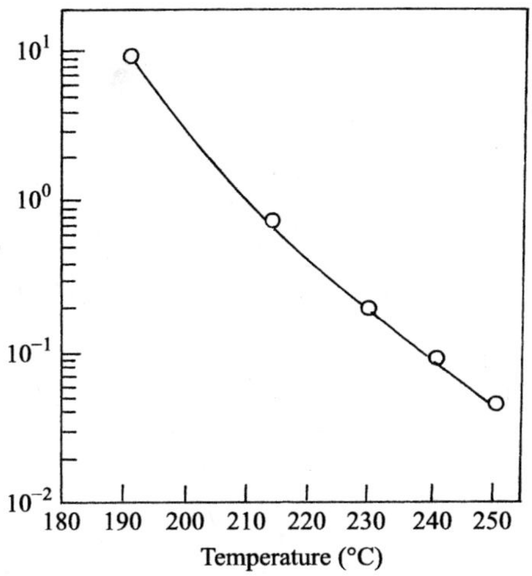

Figure 8.39 Plots of a_T versus temperature for the PS-*block*-PαMS copolymer CII (see Table 8.3 for its molecular characteristics), where the solid curve describes the prediction from the WLF equation with $C_1 = 13.7$, $C_2 = 50$, and $T_r = 210\,°C$.

versus log $a_T\omega$ and log G''_r versus log $a_T\omega$, for CII at temperatures ranging from 151 to 251 °C, for which the values of a_T given in Figure 8.39 were used. Notice in Figure 8.40 that G'_r initially increases with increasing $a_T\omega$ and then levels off, and finally increases again with increasing $a_T\omega$, while G''_r initially increases with increasing $a_T\omega$ and then goes through a minimum, and finally increases again with increasing $a_T\omega$, behavior very similar to that observed in Figure 4.10 for a monodisperse, entangled polybutadiene. Figure 8.41 gives log G'_r versus log $a_T\omega$ and log G''_r versus log $a_T\omega$ plots for a disordered PS-*block*-PαMS copolymer, BII, at temperatures ranging from

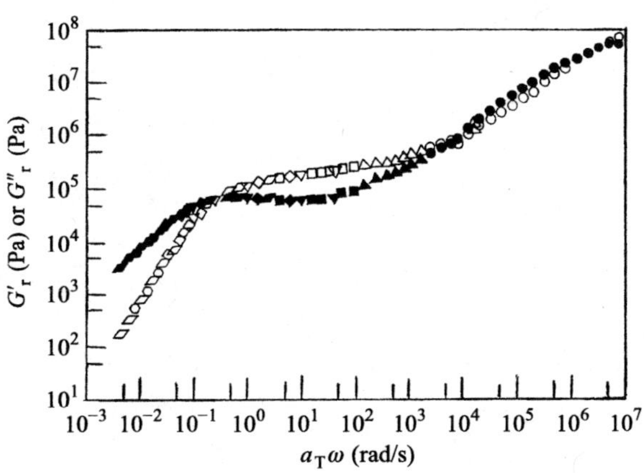

Figure 8.40 Plots of log G'_r versus log $a_T\omega$ (open symbols), and log G''_r versus log $a_T\omega$ (filled symbols) for the PS-*block*-PαMS copolymer CII (see Table 8.3 for its molecular characteristics) at various temperatures (°C): (○, ●) 151, (△, ▲) 171, (□, ■) 191, (▽, ▼) 210, (◇, ◆) 230, (⬠, ⬟) 241, and (▱, ▰) 251. The reference temperature chosen is 210 °C. (Reprinted from Kim and Han, *Macromolecules* 25:271. Copyright © 1992, with permission from the American Chemical Society.)

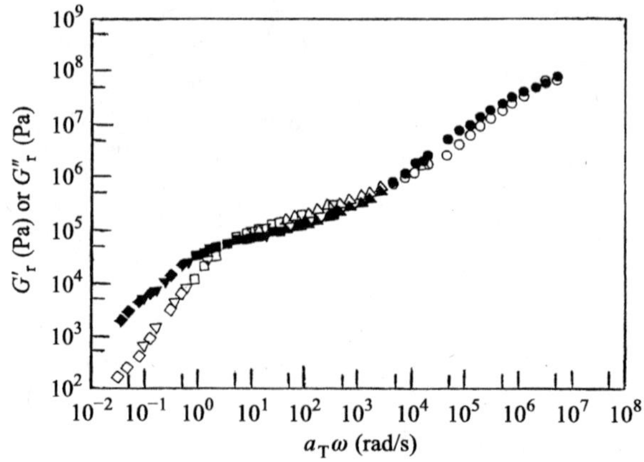

Figure 8.41 Plots of log G'_r versus log $a_T\omega$ (open symbols), and log G''_r versus log $a_T\omega$ (filled symbols) for the PS-*block*-PαMS copolymer BII (see Table 8.3 for its molecular characteristics) at various temperatures (°C): (○, ●) 151, (△, ▲) 171, (□, ■) 191, (▽, ▲) 211 and (◇, ◆) 231. The reference temperature chosen is 211 °C. (Reprinted from Kim and Han, *Macromolecules* 25:271. Copyright © 1992, with permission from the American Chemical Society.)

151 to 231 °C. It is seen in Figures 8.40 and 8.41 that TTS apparently works. Notice in Table 8.3 that these two block copolymers have almost the same block length ratios but a very large difference in total molecular weights and that the molecular weights of PαMS and PS blocks in the two block copolymers are greater than the entanglement molecular weights.

Figure 8.42 gives log G' versus log G'' plots for two disordered block copolymers, CII and BII. It is of great interest to observe in Figure 8.42 that log G' versus log G''

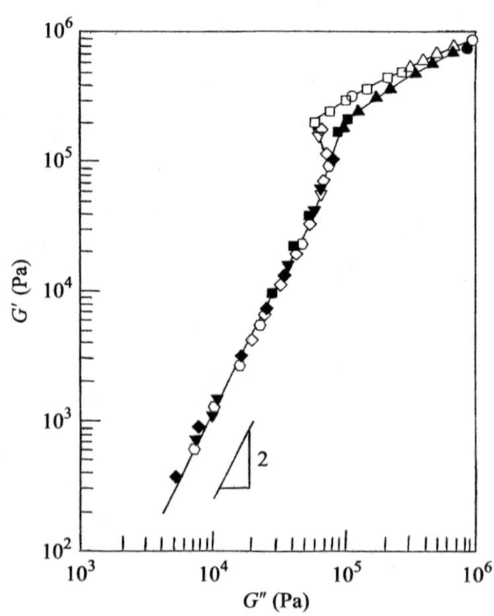

Figure 8.42 Log G' versus log G'' plots for two PS-*block*-PαMS copolymers, BII and CII, at various temperatures. BII at temperatures (°C): (●) 151, (▲) 171, (■) 191, (▼) 211, and (◆) 231. CII at temperatures (°C): (○) 151, (△) 171, (□) 191, (▽) 210, (◇) 230, and (○) 241. The molecular characteristics of BII and CII are given in Table 8.3. (Reprinted from Kim and Han, *Macromolecules* 25:271. Copyright © 1992, with permission from the American Chemical Society.)

plots for the two block copolymers give rise to a single correlation with a slope of 2 in the terminal region, and that they are not only independent of temperature but also independent of molecular weight. In Chapter 6 we made a similar observation for entangled, flexible homopolymers. We can then conclude from Figure 8.42 that both BII and CII are entangled, disordered block copolymers. If these block copolymers had been microphase separated, we would have observed temperature dependence in the log G' versus log G'' plots.

Figure 8.43 gives log G' versus log G'' plots for two disordered block copolymers, CIV and CVI, at various temperatures, showing that each lies on a single curve, independent of temperature, but that they do not lie on the same curve. Notice in Table 8.3 that CIV and CVI have almost the same total molecular weights (approximately 1.8×10^5), but different block length ratios. The observations made above from Figure 8.43 indicate that log G' versus log G'' plots are very useful for determining the effect of block length ratio on the linear dynamic viscoelasticity of disordered block copolymers. According to the interpretation of log G' versus log G'' plots presented in Chapter 6, we can conclude from Figure 8.43 that the block copolymer CVI is more elastic than the block copolymer CIV, since the log G' versus log G'' plot for CVI lies above that for CIV.

Table 8.4 gives molecular characteristics of a series of low-molecular-weight SI diblock copolymers and the T_{ODT} of each block copolymer as determined from log G' versus log G'' plots. In order to observe the effects of molecular weight (M_w) and block length ratio (ϕ) on the rheological behavior of those SI diblock copolymers given in Table 8.4, let us consider the following three cases, as schematically shown in Figure 8.44: (a) the pair of block copolymers SI-T and SI-R, (b) three block copolymers SI-M, SI-O, and SI-L, and (c) the pair of block copolymers SI-N and SI-O. Case (a) enables us to examine the effect of M_w on the rheological behavior of the two

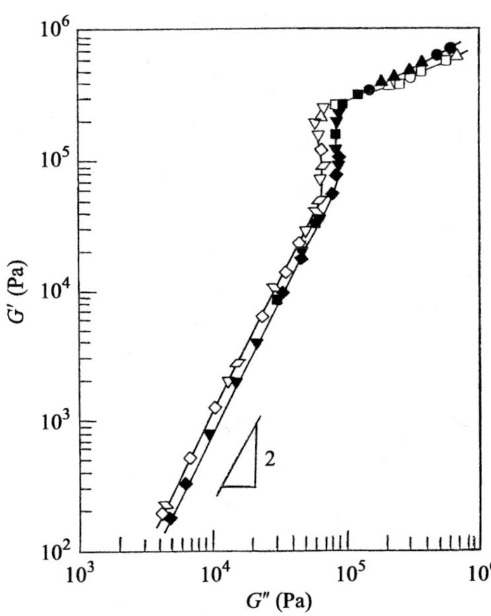

Figure 8.43 Log G' versus log G'' plots for two PS-*block*-PαMS copolymers, CIV and CVI, at various temperatures. CIV at temperatures (°C): (●) 154, (▲) 175, (■) 194, (▼) 214, and (◆) 234. CVI at temperatures (°C): (○) 131, (△) 142, (□) 162, (▽) 181, (◇) 200, (▱) 220, and (⌑) 239. The molecular characteristics of CIV and CVI are given in Table 8.3. (Reprinted from Kim and Han, *Macromolecules* 25:271. Copyright © 1992, with permission from the American Chemical Society.)

Table 8.4 Molecular characteristics and T_{ODT} of SI diblock copolymers investigated

Sample Code	$M_{w,PS}$	$M_{w,PI}$	M_w/M_n	ϕ^a	$T_{g,PS}$ (°C)	T_{ODT} (°C)[b]
SI-N	5.6×10^3	13.7×10^3	1.05	0.21	58	75
SI-T	6.2×10^3	6.2×10^3	1.07	0.40	60	≤65[c]
SI-R	8.2×10^3	6.8×10^3	1.07	0.44	65	100
SI-M	10.3×10^3	5.2×10^3	1.06	0.57	75	≤90[d]
SI-O	20.8×10^3	5.6×10^3	1.06	0.71	85	130
SI-L	34.2×10^3	6.8×10^3	1.05	0.77	94	170

[a] ϕ is block length ratio defined by $N_{PS}/(N_{PS} + N_{PI})$, where N_{PS} is the polymerization index of PS block and N_{PI} is the polymerization index of PI block in a given SI diblock copolymer. [b] The values of T_{ODT} are determined from $\log G'$ versus $\log G''$ plots. [c] The lowest possible measurement temperature is 65 °C, at and above this point $\log G'$ versus $\log G''$ plots become independent of temperature. [d] The lowest possible measurement temperature is 90 °C, at and above this point $\log G'$ versus $\log G''$ plots become independent of temperature.

Reprinted from Han et al., *Macromolecules* 28:5886. Copyright © 1995, with permission from the American Chemical Society.

block copolymers SI-T and SI-R. Note that the length of PS block in SI-R is 1.31 times that in SI-T and the length of PI block in SI-R is 1.1 times that in SI-T, but that the ϕ of the two block copolymers is about the same. Case (b) enables us to further examine the effect of M_w on the rheological properties of the three block copolymers SI-M, SI-O, and SI-L. Note that the M_w of PI block is increased moderately in going from SI-M to SI-O and to SI-L, but the M_w of PS block is doubled in going from SI-M to SI-O and tripled in going from SI-M to SI-L. In other words, the increase of M_w in these three SI diblock copolymers comes primarily from the increase in M_w of PS block. Case (c) enables us to examine the effect of ϕ on the rheological properties of the two block copolymers SI-N and SI-O. Note that while the total M_w is increased moderately, ϕ is increased considerably, in going from SI-N to SI-O; namely, $\phi = 0.21$ for SI-N and $\phi = 0.71$ for SI-O.

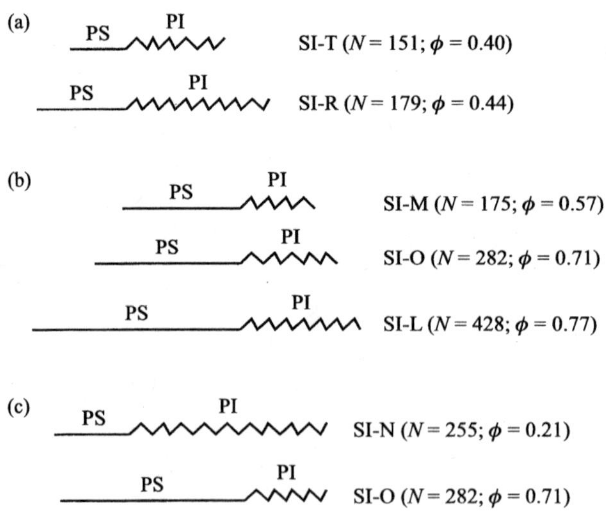

Figure 8.44 Schematic showing the block lengths of different SI diblock copolymers: (a) SI-T and SI-R, (b) SI-M, SI-O, and SI-L, and (c) SI-N and SI-O. The molecular characteristics of these SI block copolymers are given in Table 8.4. (Reprinted from Han et al., *Macromolecules* 28:5886. Copyright © 1995, with permission from the American Chemical Society.)

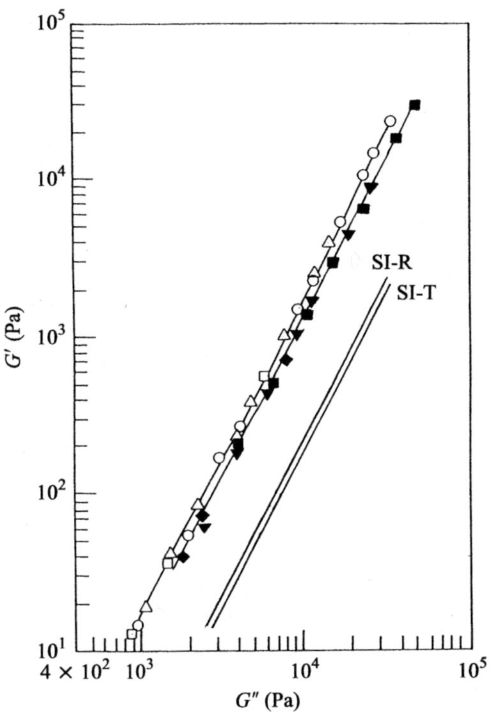

Figure 8.45 Log G' versus log G'' plots for SI-R at various temperatures (°C): (○) 100, (△) 105, and (□) 110, and for SI-T at various temperatures (°C): (■) 65, (▼) 70, and (◆) 75. The molecular characteristics of these SI block copolymers are given in Table 8.4. The solid lines are theoretical predictions, which are discussed later in this chapter. (Reprinted from Han et al., *Macromolecules* 28:5886. Copyright © 1995, with permission from the American Chemical Society.)

Figure 8.45 gives log G' versus log G'' plots for the two SI diblock copolymers SI-R and SI-T. In Figure 8.45 we observe that the temperature dependence is completely suppressed in the log G' versus log G'' plots, indicating that these two block copolymers are in the disordered state in the range of temperatures under consideration. Since the total molecular weight of SI-R is slightly higher than that of SI-T (see Table 8.4), using Eq. (6.20) we can explain why in Figure 8.45 the log G' versus log G'' plot for SI-R lies slightly above the log G' versus log G'' plot for SI-T. Here, we assume that both SI-R and SI-T are Rouse chains.

Figure 8.46 gives log G' versus log G'' plots for three SI diblock copolymers, SI-M, SI-O, and SI-L at various temperatures. If we assume that each of these block copolymers consists of Rouse chains, we can then explain, using Eq. (6.20), why the log G' versus log G'' plot for SI-L lies above the log G' versus log G'' plots for SI-O and SI-M respectively, and why the log G' versus log G'' plot for SI-O lies above the log G' versus log G'' plot for SI-M. In other words, the relative positions of the log G' versus log G'' plots for the three SI diblock copolymers given in Figure 8.47 are attributable to differences in their total molecular weights (see Table 8.4). This interpretation is based on the premise that SI-M, SI-O, and SI-L may be regarded as Rouse chains.

Figure 8.47 gives log G' versus log G'' plots for two SI diblock copolymers, SI-N and SI-O, at various temperatures. It is of interest to observe in Figure 8.47 that, in spite of the fact that the total molecular weight of SI-N is lower than that of SI-O (see Table 8.4), the log G' versus log G'' plot for SI-N lies above the log G' versus log G'' plot for SI-O. This may be attributable to the large difference in block length ratio ϕ between the two block copolymers.

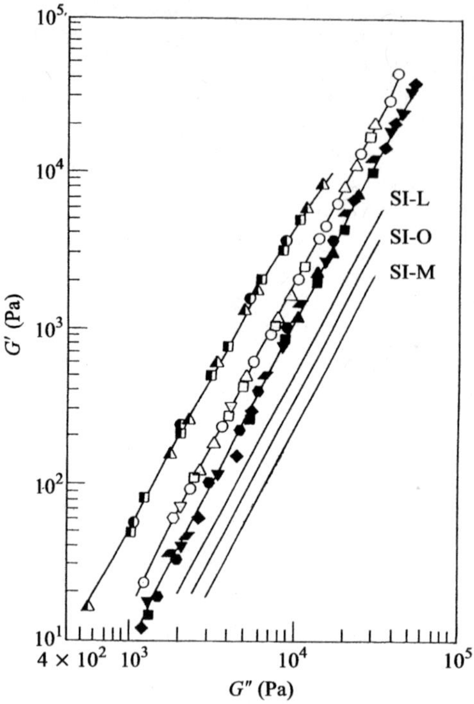

Figure 8.46 Log G' versus log G'' plots for SI-M at various temperatures (°C): (▲) 90, (■) 95, (▼) 100, (◆) 105, (●) 110, and (▰) 115, for SI-O at various temperatures (°C): (○) 130, (△) 135, (□) 140, and (▽) 145, and for SI-L at various temperatures (°C): (◐) 170, (◭) 175, and (◧) 180. The molecular characteristics of these SI block copolymers are given in Table 8.4. The solid lines are theoretical predictions, which are discussed later in this chapter. (Reprinted from Han et al., *Macromolecules* 28:5886. Copyright © 1995, with permission from the American Chemical Society.)

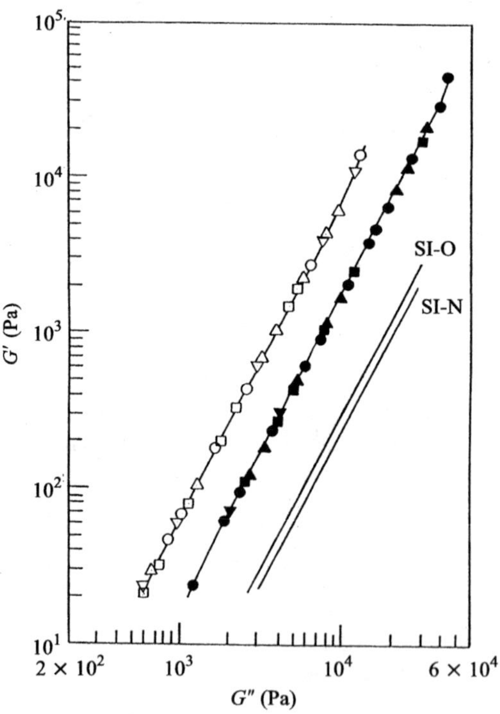

Figure 8.47 Log G' versus log G'' plots for SI-N at various temperatures (°C): (○) 75, (△) 80, (□) 85, and (▽) 90, and for SI-O at various temperatures (°C): (●) 130, (▲) 135, (■) 140, and (▼) 145. The molecular characteristics of these SI block copolymers are given in Table 8.4. The solid lines are theoretical predictions, which are discussed later in this chapter. (Reprinted from Han et al., *Macromolecules* 28:5886. Copyright © 1995, with permission from the American Chemical Society.)

8.4.3 Molecular Theory for the Linear Viscoelasticity of Disordered Block Copolymers

Since the first theoretical attempts in the 1970s (Hall and DeWames 1975; Hansen and Shen 1975a, 1975b; Stockmayer and Kennedy 1975; Wang and DiMarzio 1975) to describe the chain dynamics of disordered block copolymers, little progress has been made. The molecular theory for the linear viscoelasticity of disordered block copolymers is still in its infancy. In this section, we present first the currently held linear viscoelasticity theory for unentangled disordered block copolymer and then an extended version for disordered, entangled block copolymer.

8.4.3.1 Molecular Theory for the Linear Viscoelasticity of Disordered Unentangled Block Copolymers

Stockmayer and Kennedy (1975) conducted a seminal study on the chain dynamics of Rouse chains of AB-type diblock or ABA-type triblock copolymers by modifying the bead–spring model of Rouse for linear flexible homopolymers (see Chapter 4). They calculated the spectrum of relaxation times ($\tau_{\mathrm{p,block}}$) of the block copolymer in terms of the terminal relaxation times for the Rouse chains for the A and B blocks. Once the values of $\tau_{\mathrm{p,block}}$ are determined, one can calculate linear dynamic viscoelastic properties of disordered, unentangled block copolymers. In this section, we present the Stockmayer–Kennedy theory (1975). They considered the continuous limit of the bead–spring Rouse model (Rouse 1953):

$$\frac{\partial x}{\partial t} = \left(\frac{2}{\tau_1 \pi^2}\right)\frac{\partial^2 x}{\partial s^2} \tag{8.5}$$

where x denotes the average displacement of a bead, s is a continuous variable that lies between -1 and 1 (i.e., $-1 \leq s \leq 1$), and τ_1 is the Rouse terminal relaxation time. Figure 8.48 shows schematically the bead–spring Rouse model for an AB-type diblock copolymer. Using the boundary conditions

$$K_{\mathrm{A}}\left(\frac{\partial x_{\mathrm{A}}}{\partial s}\right)_{s=\theta} = K_{\mathrm{B}}\left(\frac{\partial x_{\mathrm{B}}}{\partial s}\right)_{s=\theta} \tag{8.6a}$$

$$x_{\mathrm{A}}(\theta, t) = x_{\mathrm{B}}(\theta, t) \tag{8.6b}$$

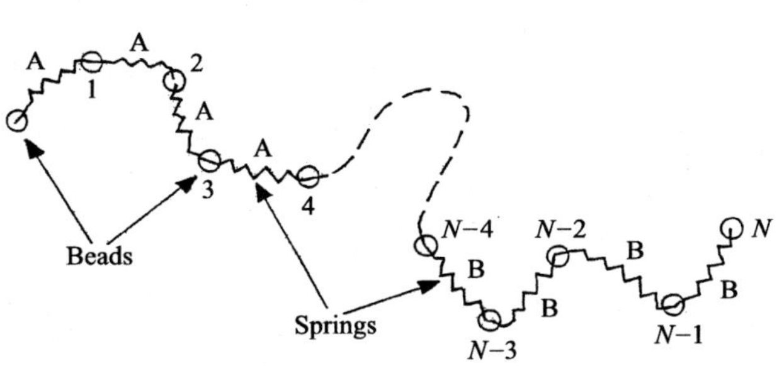

Figure 8.48 Schematic diagram describing the bead–spring Rouse model for AB-type diblock copolymer.

$$\left(\frac{\partial x(s,t)}{\partial s}\right)_{s=\pm 1} = 0 \tag{8.6c}$$

where $K_{\rm A} = 3k_{\rm B}T/b_{\rm A}{}^2$ and $K_{\rm B} = 3k_{\rm B}T/b_{\rm B}{}^2$, and where $k_{\rm B}$ is the Boltzmann constant, T is the absolute temperature, and $b_{\rm A}$ and $b_{\rm B}$ are the Kuhn lengths of block A and block B, respectively, the general solution for Eq. (8.5) can be written as

$$x_A = \exp(-t/2\tau_{p,\text{block}})(C_{\rm A} \sin \alpha s + D_{\rm A} \cos \alpha s) \tag{8.7a}$$

$$x_B = \exp(-t/2\tau_{p,\text{block}})(C_{\rm B} \sin \beta s + D_{\rm B} \cos \beta s) \tag{8.7b}$$

where $\tau_{p,\text{block}}$ is given by

1. for diblock copolymers,

$$\tau_{p,\text{block}} = \tau_{1\rm A}(\pi/\alpha_p)^2 = \tau_{1\rm B}(\pi/\beta_p)^2; \qquad p = 2, 4, 6, \ldots \tag{8.8}$$

2. for triblock copolymers,

$$\tau_{p,\text{block}} = \tau_{1\rm A}(\pi/2\alpha_p)^2 = \tau_{1\rm B}(\pi/2\beta_p)^2; \qquad p = 1, 2, 3, \ldots \tag{8.9}$$

in which $\tau_{1\rm A}$ and $\tau_{1\rm B}$ are the terminal relaxation times for block A and block B, respectively, which are defined by (see Chapter 4)

$$\tau_{1,i} = 6\eta_{0,i}M_i/\pi^2\rho RT \quad (i = \text{A, B}) \tag{8.10}$$

The values of α_p appearing in Eqs. (8.8) and (8.9) can be determined by satisfying the following characteristic equation:

$$\tan\left[\alpha_p(1-\theta)\right] = -\lambda\mu \tan \alpha_p \lambda\theta \quad \text{(even modes of } p\text{)} \tag{8.11a}$$

$$\tan\left[\alpha_p(1-\theta)\right] = \lambda\mu \cot \alpha_p \lambda\theta \quad \text{(odd modes of } p\text{)} \tag{8.11b}$$

where

$$\lambda = (\tau_{1\rm B}/\tau_{1\rm A})^{1/2}; \quad \mu = (b_{\rm A}/b_{\rm B})^2 \tag{8.12}$$

in which θ is the fractional volume of block B.

Stockmayer and Kennedy (1975) constructed generalized plots showing how the block length ratio θ and terminal relaxation times $\tau_{1\rm A}$ and $\tau_{1\rm B}$ of the respective constituent components may affect the reduced zero-shear viscosity F of a block copolymer, defined by

$$F = \frac{\sum_{p=1}^{N} \tau_{p,\text{block}}}{(1-\theta)\sum_{p=1}^{N} \tau_{p,\rm A} + \theta \sum_{p=1}^{N} \tau_{p,\rm B}} \tag{8.13}$$

where $\tau_{p,A}$ and $\tau_{p,B}$ are the relaxation times for block A and block B, respectively, and $\tau_{p,\text{block}}$ is the relaxation time for the block copolymer. Note that for an ABA-type triblock copolymer, $\sum \tau_{p,\text{block}}$ must be determined from

$$\sum \tau_{p,\text{block}} = \tau_{1,A}\left(\frac{\pi}{2}\right)^2 \sum_{\text{all } p}^{N} \frac{1}{\alpha_p{}^2} \tag{8.14}$$

and for an AB-type diblock copolymer, $\sum \tau_{p,\text{block}}$ must be determined from

$$\sum \tau_{p,\text{block}} = \tau_{1,A}\pi^2 \sum_{\text{even } p}^{N} \frac{1}{\alpha_p{}^2} \tag{8.15}$$

Note in Eqs. (8.14) and (8.15) that $\tau_{1,A}$ is the terminal relaxation time for block A, and that the summation must be taken over all integer values (for both even and odd modes of the characteristic equation, Eq. (8.11)) for an ABA-type triblock copolymer, and only over even integer values (i.e., only for even modes of the characteristic equation, Eq. (8.11a)) for an AB-type block copolymer.

Once the values of $\tau_{p,\text{block}}$ are determined, one can calculate the linear viscoelastic properties of a block copolymer from the following expression for the reduced complex dynamic modulus $G^*{}_R(\omega)$:

$$G'_R(\omega) = \sum_{\text{even}\, p}^{\infty} \frac{(\omega\tau_{p,\text{block}})^2}{1 + (\omega\tau_{p,\text{block}})^2}; \quad G''_R(\omega) = \sum_{\text{even}\, p}^{\infty} \frac{\omega\tau_{p,\text{block}}}{1 + (\omega\tau_{p,\text{block}})^2} \tag{8.16}$$

and the reduced compliance $J^o_{e,R}$ from

$$J^o_{e,R} = \lim_{\omega \to 0} \frac{G'_R(\omega)}{[G'_R(\omega)]^2 + [G''_R(\omega)]^2} \tag{8.17}$$

Once values of $G'_R(\omega)$ and $G''_R(\omega)$ are available, the storage and loss moduli, $G'(\omega)$ and $G''(\omega)$, can be calculated by

$$G'(\omega) = \frac{\rho RT}{M} G'_R(\omega); \quad G''(\omega) = \frac{\rho RT}{M} G''_R(\omega) \tag{8.18}$$

in which $M = M^b_A + M^b_B$ is the molecular weight of block copolymer, with M^b_A and M^b_B being the molecular weights of components A and B, respectively. The steady-state shear compliance J^o_e can be calculated by

$$J^o_e = \lim_{\omega \to 0} \frac{G'(\omega)}{[G'(\omega)]^2 + [G''(\omega)]^2} = \frac{M}{\rho RT} J^o_{e,R} \tag{8.19}$$

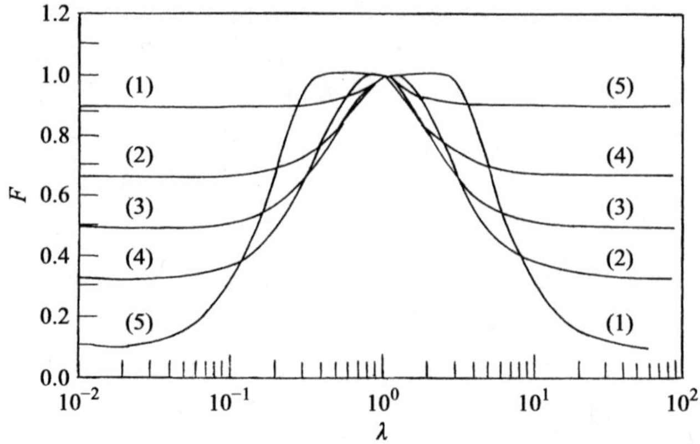

Figure 8.49 Plots of F versus λ for AB-type diblock copolymers having the same Kuhn statistical length ($\mu = 1$) for various values of block length ratio θ: (1) 1/10, (2) 1/3, (3) 1/2, (4) 2/3, and (5) 9/10. (Reprinted from Kim and Han, *Macromolecules* 25:271. Copyright © 1992, with permission from the American Chemical Society.)

Let us briefly look at the prediction of reduced viscosity F, defined by Eq. (8.13), for AB-type diblock and ABA-type triblock copolymers in terms of three important molecular parameters: the Kuhn lengths, the terminal relaxation times of the constituent blocks, and block length ratio θ. Figure 8.49 gives plots of F versus λ for an AB-type diblock copolymer at various values of θ, when the Kuhn lengths of the constituent blocks are assumed to be the same ($\mu = 1$). It is interesting to observe in Figure 8.49 that all curves pass through the same point at $\lambda = 1$ and $F = 1$. Notice that $F(\lambda, \theta) = F(1/\lambda, 1 - \theta)$, as it should be for a diblock copolymer. However, as can be seen in Figure 8.50, when $\mu \neq 1$, curves for different values of θ

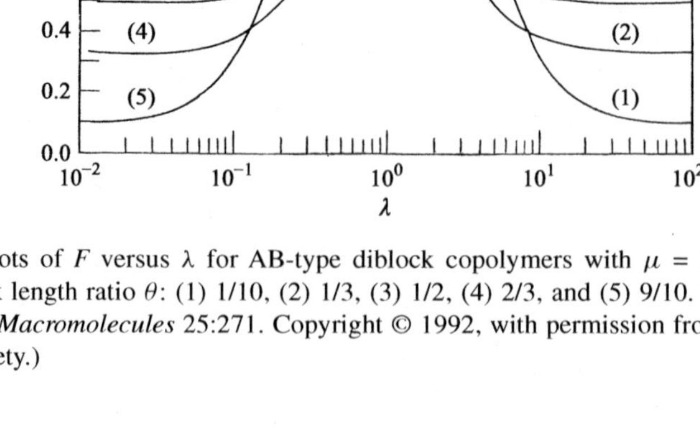

Figure 8.50 Plots of F versus λ for AB-type diblock copolymers with $\mu = 1.15$ for various values of block length ratio θ: (1) 1/10, (2) 1/3, (3) 1/2, (4) 2/3, and (5) 9/10. (Reprinted from Kim and Han, *Macromolecules* 25:271. Copyright © 1992, with permission from the American Chemical Society.)

do not pass through a common point at $\lambda = 1$. A comparison of Figure 8.50 with Figure 8.49 indicates that F is very sensitive to small differences in the statistical lengths between the constituent blocks. Also, for $\mu \neq 1$, $F(\lambda, \theta)$ is no longer the same as $F(1/\lambda, 1 - \theta)$.

Figure 8.51 gives plots of F versus λ for an ABA-type triblock copolymer with $\mu = 1.15$, corresponding to the α-methylstyrene/styrene pair, at various values of θ. It is of interest to observe in Figure 8.51 that (1) when $\lambda < 1$ (i.e., when $\tau_A > \tau_B$), the F is greater than 1 for all values of θ and the relaxation time of endblock A controls the viscosity of an ABA-type triblock copolymer, and (2) when $\lambda > 1$ (i.e., when $\tau_B > \tau_A$), the F is less than 1 for all values of θ and the relaxation time of midblock B controls the viscosity of an ABA-type triblock copolymer. It should be pointed out that for a given block copolymer or a given value of θ, a variation of λ can come from different sensitivities of relaxation time of each block to temperature. It should be noted that λ is also dependent upon θ. Such plots as those given in Figures 8.49–8.51 are very useful to the design of disordered block copolymers with desired values of zero-shear viscosity $\eta_{0,\text{block}}$.

By assuming that the effect of the presence of foreign blocks on the relaxation (or retardation) time spectrum is due exclusively to the difference in the friction coefficients, Hansen and Shen (1975a, 1975b) also extended the Rouse–Zimm theory to compute the relaxation (or retardation) time spectra of disordered block copolymers of various configurations and compositions. They obtained the equation of motion and applied normal coordinate transformation to obtain expressions for the eigenvalue problem. However, the approach taken by them requires extensive numerical computations to solve the eigenvalue problem containing a matrix, which has a very large number of

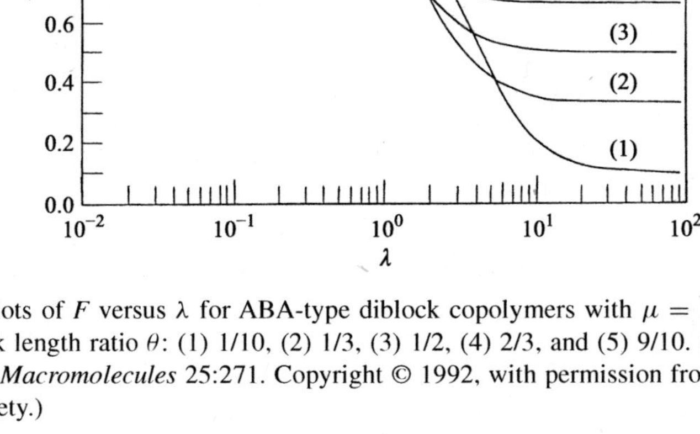

Figure 8.51 Plots of F versus λ for ABA-type diblock copolymers with $\mu = 1.15$ for various values of block length ratio θ: (1) 1/10, (2) 1/3, (3) 1/2, (4) 2/3, and (5) 9/10. (Reprinted from Kim and Han, *Macromolecules* 25:271. Copyright © 1992, with permission from the American Chemical Society.)

elements, whereas the Stockmayer–Kennedy approach described above requires the solution of a very simple transcendental equation, Eq. (8.11).

8.4.3.2 Molecular Theory for Linear Viscoelasticity of Disordered Entangled Block Copolymers

At present, there is no comprehensive theory for the chain dynamics of disordered, entangled block copolymers. Using the concept of the tube model, Kim and Han (1992) modified on an ad hoc basis the Stockmayer–Kennedy theory to interpret their experimental results for the linear viscoelastic behavior of disordered, entangled PS-*block*-PαMS copolymers. They used the expression

$$\eta_{0,\text{block}} = \frac{8}{\pi^2}\frac{\rho RT}{M_{\text{e,block}}}\sum_{p=1}^{N}\tau_{p,\text{block}} \tag{8.20}$$

to calculate the zero-shear viscosity ($\eta_{0,\text{block}}$) of an entangled block copolymer, where $M_{\text{e,block}}$ is the entangled molecular weight of the block copolymer and $\sum\tau_{p,\text{block}}$ is defined by Eq. (8.15). Note that according to the tube model (Chapter 4), the zero-shear viscosity (η_0) of a homopolymer is given by

$$\eta_0 = \frac{8}{\pi^2}\frac{\rho RT}{M_{\text{e}}}\sum_{\text{odd}\,p}^{\infty}\frac{1}{p^2}\tau_p \tag{8.21}$$

where $\tau_p = \tau_1/p^2$, with τ_1 being the terminal relaxation time. Notice that Eq. (8.20) is not quite the same as Eq. (8.21) in that the summation in Eq. (8.20) is taken over all integer values, whereas the summation in Eq. (8.21) is taken only over odd integer values. When the molecular weights of both blocks in an AB-type (or ABA-type) block copolymer are greater than the entanglement molecular weights of corresponding homopolymers, one must calculate the terminal relaxation time $\tau_{1,i}$ (i = A, B) of the constituent blocks using the tube model (see Chapter 4)

$$\tau_{1,i} = 12\eta_{0,i}M_{\text{e},i}/\pi^2\rho RT \tag{8.22}$$

where $\eta_{0,i}$ is the zero-shear viscosity of homopolymer A or homopolymer B having the same molecular weight as in the AB-type diblock copolymer, and $M_{\text{e},i}$ is the entanglement molecular weight of homopolymer A or homopolymer B. However, when the molecular weight of either block A or block B is less than the M_{e} of the corresponding homopolymer, one can calculate its terminal relaxation time using the Rouse model (see Chapter 4):

$$\tau_{1,i} = 6\eta_{0,i}M_i/\pi^2\rho RT \tag{8.23}$$

Note that the terminal relaxation times defined by Eq. (8.22) and/or Eq. (8.23) are needed in order to calculate eigenvalues α by solving Eq. (8.11).

8.4.3.3 Comparison of Theory with Experiment

Let us compare the prediction of the Stockmayer–Kennedy theory with the experimental results for SI diblock copolymers presented in a previous section. The glass transition temperature of the PS block in each SI diblock copolymer is given in Table 8.4. Using $b_{\rm PS} = 0.68$ nm, $b_{\rm PI} = 0.59$ nm, and $M = M_{\rm PS} + M_{\rm PI}$ in Eq. (8.18), we calculate $G'(\omega)$ and $G''(\omega)$ with the following procedure: we calculate $\tau_{p,\rm block}$ using Eq. (8.8), and $G'_{\rm R}(\omega)$ and $G''_{\rm R}(\omega)$ using Eq. (8.16). For this, we first calculate using Eq. (8.10) the terminal relaxation times $\tau^{\rm b}_{1,\rm PS}$ and $\tau^{\rm b}_{1,\rm PI}$ for blocks PS and PI, respectively. This requires calculations of $\eta^{\rm b}_{0,i}$, for which the following expression may be used (see Chapter 4):

$$\eta^{\rm b}_{0,i} = \zeta^{\rm b}_{0,i}\,\rho_i b_i^{\,2} N_{\rm A} M_i^{\rm b}/36 M_{{\rm o},i}^{\,2} \quad (i = \rm PS, PI) \tag{8.24}$$

where b_i is the Kuhn length for component i, $N_{\rm A}$ is Avogadro's number, and $M_{{\rm o},i}$ is the monomeric molecular weight of component i.

In Chapter 7 we pointed out that the friction coefficients for the two species in a miscible polymer blend are quite different from the friction coefficients for the corresponding homopolymers and that they vary with blend composition. It seems reasonable to assume that the friction coefficients $\zeta^{\rm b}_{0,i}$ for the two species PS and PI in the block copolymer medium might depend on the friction coefficients $\zeta^{\rm h}_{0,i}$ for the corresponding homopolymers PS and PI, and also on block copolymer composition. However, since no such information is available in the literature, let us calculate $\eta^{\rm b}_{0,\rm PS}$ using the expression

$$\eta^{\rm b}_{0,\rm PS} = \zeta^{\rm h}_{0,\rm PS}\,\rho_{\rm PS}\,b_{\rm PS}^{\,2} N_{\rm A} M^{\rm b}_{\rm PS}/36 M_{{\rm o},\rm PS}^{\,2} \tag{8.25}$$

in which $\zeta^{\rm b}_{0,\rm PS} \approx \zeta^{\rm h}_{0,\rm PS}$ is assumed.

In calculating $\eta^{\rm b}_{0,\rm PS}$ at various temperatures from Eq. (8.25), we use $\zeta^{\rm h}_{0,\rm PS} = 0.1148$ N·s/m at $T^{\rm b}_{\rm g,PS}$ and the WLF parameters $C_1 = -13.7$ and $C_2 = 50.5$, and $T_{\rm r} = T^{\rm b}_{\rm g,PS}$ for Eq. (6.7). The values of $T^{\rm b}_{\rm g,PS}$ for the six SI diblock copolymers are given in Table 8.4. In calculating $\eta^{\rm b}_{0,\rm PI}$ (in Pa·s) at various temperatures, we use the following empirical expression (Han et al. 1995a):[7]

$$\eta^{\rm b}_{0,\rm PI} = 3.70 \times 10^{-18} \exp(4478/T)(M_{\rm c,PI})^{3.4}(M^{\rm b}_{\rm PI}/M_{\rm c,PI}) \tag{8.26}$$

in which T is the absolute temperature, $M_{\rm c,PI}$ is the viscosity critical molecular weight of PI, and $M^{\rm b}_{\rm PI}$ is the molecular weight of PI block in the block copolymer. The following expressions may be used for the temperature dependence of density for PS (Richardson and Savill 1977):

$$1/\rho_{\rm PS} = 0.9199 + 5.098 \times 10^{-4}(T - 273) + 2.354 \times 10^{-7}(T - 273)^2 + \left[32.46 + 0.1017(T - 273)\right]/M^{\rm b}_{\rm PS} \tag{8.27}$$

in which M^{b}_{PS} is the molecular weight of PS block in the block copolymer, and for PI (Han et al. 1989):

$$1/\rho_{\text{PI}} = 1.0771 + 7.22 \times 10^{-4}(T - 273) + 2.46 \times 10^{-7}(T - 273)^2 \tag{8.28}$$

From Eq. (8.17) we obtain $J^{\text{o}}_{\text{e,R}} = 0.41$ for SI-R and $J^{\text{o}}_{\text{e,R}} = 0.42$ for SI-T, which are very close to $J^{\text{o}}_{\text{e},R} = 0.40$ predicted for homopolymer melts by the Rouse theory. This is not surprising because, according to Eq. (8.16), $G'_{\text{R}}(\omega)$ and $G''_{\text{R}}(\omega)$ are independent of the total molecular weight. Of course, $G'_{\text{R}}(\omega)$ and $G''_{\text{R}}(\omega)$ can vary with block length ratio ϕ. Conversely, $G'(\omega)$ and $G''(\omega)$, defined by Eq. (8.18), depend on the molecular weight of block copolymer. The theory predicts correctly the experimental trend displayed in Figure 8.45, indicating that the observed difference in $\log G'$ versus $\log G''$ plot between SI-R and SI-T is attributable to the difference in molecular weight between the two block copolymers. The theory also predicts correctly the experimental trend, displayed in Figure 8.46, of the differences in $\log G'$ versus $\log G''$ plot among the three block copolymers, SI-M, SI-O, and SI-L (Han et al. 1995a).

However, there are two questions that must be addressed. (1) How much error might have been incurred from the assumption of $\zeta^{\text{b}}_{0,\text{PS}} \approx \zeta^{\text{h}}_{0,\text{PS}}$ when calculating, using Eq. (8.18), the values of $G'(\omega)$ and $G''(\omega)$ for the SI diblock copolymers? (2) Why are the calculated values of $G'(\omega)$ and $G''(\omega)$ for the SI diblock copolymers much smaller than the experimentally measured ones? Regarding the first question, using Eq. (8.17) we calculate $J^{\text{o}}_{\text{e,R}}$ for values of λ, defined by Eq. (8.12), ranging from $0.01\lambda_{\text{o}}$ to $100\lambda_{\text{o}}$, where $\lambda_{\text{o}} = (\tau^{\text{h}}_{1,\text{PI}}/\tau^{\text{h}}_{1,\text{PS}})^{1/2}$ with $\tau^{\text{h}}_{1,\text{PI}}$ and $\tau^{\text{h}}_{1,\text{PS}}$ being the terminal relaxation times for PI and PS, respectively. Note that $\lambda/\lambda_0 = (\zeta^{\text{b}}_{0,\text{PI}}/\zeta^{\text{h}}_{0,\text{PI}})^{1/2}/(\zeta^{\text{b}}_{0,\text{PS}}/\zeta^{\text{h}}_{0,\text{PS}})^{1/2}$. We observe that the values of $J^{\text{o}}_{\text{e,R}}$ vary little (within $\pm$ 10%) as λ/λ_0 varies from 0.01 to 100. Therefore, we can conclude from Eq. (8.19) that J^{o}_{e} varies little with λ/λ_0 ranging from 0.01 to 100. From the relationship of $G'(\omega) = J^{\text{o}}_{\text{e}}[G''(\omega)]^2$ (see Eq. (8.17) for $G'_{\text{R}} \ll G''_{\text{R}}$ in the denominator) we conclude that the discrepancy observed in $\log G'$ versus $\log G''$ plots between experiment and theory cannot be ascribed solely to the errors that might have been incurred from the assumptions $\zeta^{\text{b}}_{0,\text{PS}} \approx \zeta^{\text{h}}_{0,\text{PS}}$ and $\zeta^{\text{b}}_{0,\text{PI}} \approx \zeta^{\text{h}}_{0,\text{PI}}$ made in estimating $\zeta^{\text{b}}_{0,\text{PS}}$ and $\zeta^{\text{b}}_{0,\text{PI}}$, and hence in the calculation of $\eta^{\text{b}}_{0,\text{PS}}$ from Eq. (8.25) and $\eta^{\text{b}}_{0,\text{PI}}$ from Eq. (8.26). Regarding the second question posed above, some of the SI block copolymers, if not all, under consideration may be regarded as entangled chains. This will then increase considerably the predicted values of $G'(\omega)$ and $G''(\omega)$ over those based on Rouse chains. Note that the value of $J^{\text{o}}_{\text{e,R}}$ for entangled chains is about three times that for Rouse chains (see Chapter 4).

There are several factors that are either not included or assumed to be unimportant in the Stockmayer–Kennedy theory. Hydrodynamic interactions between two chains in a block copolymer are not included in the theory. This can be a serious omission, especially when dealing with low-molecular-weight diblock copolymers, such as the SI diblock copolymers considered above. Intermolecular (thermodynamic) interactions between chemically dissimilar chains are also neglected. Inclusion of such interactions can be important, especially when the chemical affinity between two chemically dissimilar chains is rather poor, such as the case in the SI diblock copolymers considered here. In Chapter 7, which discusses the rheology of miscible polymer blends, we pointed out the importance of the segmental interaction parameter in the prediction of the linear viscoelastic properties of miscible polymer blends. In the Stockmayer–Kennedy theory, it is assumed that a diblock copolymer consisting of two Rouse chains remains as a Rouse chain. As pointed out above, this assumption may not be valid, especially when the molecular weight of at least one block is very close to or slightly lower than its M_c but the molecular weight of the combined blocks is greater than the M_c of that particular block (e.g., SI-N and SI-L). In reality, such a block copolymer may be regarded as an entangled polymer. Considering these factors, one should not be surprised to see that predicted values of G' and G'' are much smaller than experimental values.

Now, using Eq. (8.20), let us calculate the $\eta_{0,\mathrm{block}}$ of a series of disordered, entangled PS-*block*-PαMS copolymers, the linear viscoelastic properties of which are presented in Figures 8.40–8.43. Since information on $M_{\mathrm{e,block}}$ is not available, let us use the following expression

$$M_{\mathrm{e,block}} = \rho RT / G^{\mathrm{o}}_{N,\mathrm{block}} \tag{8.29}$$

where $G^{\mathrm{o}}_{N,\mathrm{block}}$ may be estimated from

$$G^{\mathrm{o}}_{N,\mathrm{block}} = \left[w_{\mathrm{A}} (G^{\mathrm{o}}_{N,\mathrm{A}})^{1/3.4} + w_{\mathrm{B}} (G^{\mathrm{o}}_{N,\mathrm{B}})^{1/3.4} \right]^{3.4} \tag{8.30}$$

in which w_{A} and w_{B} are the weight fractions of blocks A and B, respectively, and $G^{\mathrm{o}}_{N,\mathrm{A}}$ and $G^{\mathrm{o}}_{N,\mathrm{B}}$ are the plateau moduli of the constituent components, A and B, respectively. In Chapter 7, which discusses the linear viscoelasticity of miscible polymer blends, we observed that Eq. (8.30) describes experimental results reasonably well.

Let us use $\eta_{0,\mathrm{P\alpha MS}} = 1.17 \times 10^4$ Pa·s for PαMS having $M_{\mathrm{w}} = 6.7 \times 10^4$ at 230 °C to calculate, with the aid of $\eta_0 \propto M^{3.4}$, the $\eta_{0,\mathrm{P\alpha MS}}$ of the PαMS block in sample CVI (see Table 8.3), and let us use $\eta_{0,\mathrm{PS}} = 9.2 \times 10^2$ Pa·s for PS having $M_{\mathrm{w}} = 1.95 \times 10^5$ at 230 °C to calculate, with the aid of $\eta_0 \propto M^{3.4}$, the $\eta_{0,\mathrm{PS}}$ of the PS blocks in samples BII, CI, CII, CIII, CIV, and CVI, and to calculate, with the aid of $\eta_0 \propto M$, the $\eta_{0,\mathrm{PS}}$ of PS block in sample CV. A summary of the comparisons between predicted and experimental values of $\eta_{0,\mathrm{block}}$ for the seven disordered PS-*block*-PαMS copolymers listed in Table 8.3 is given in Table 8.5. The experimental values of $\eta_{0,\mathrm{block}}$ given in Table 8.5 were obtained using the relationship, $\eta_0 = \lim_{\omega \to 0} G''(\omega)/\omega$. Considering the fact that many assumptions are made in obtaining Eq. (8.20) and also that uncertainties are involved with the experimental results (e.g., molecular weight determination), the extent of agreement observed between theory and experiment is very encouraging.

To observe the effect of block copolymer composition on $\eta_{0,\mathrm{block}}$, we calculated values of $\eta_{0,\mathrm{block}}$ with the total molecular weight, reduced to a reference value, $M_{\mathrm{w}} = 2 \times 10^5$ at 230 °C. Figure 8.52 gives a comparison of prediction with experiment for seven disordered, entangled PS-*block*-PαMS copolymers, the molecular weights of which are summarized in Table 8.3. In calculating the values of $\eta_{0,\mathrm{block}}$ given in Figure 8.52, we used the relationship[8] $\eta_0 \propto M^{3.4}$ and the following numerical values for the WLF parameters: (1) $C_1 = 8.86$, $C_2 = 101.6$, and $T_{\mathrm{r}} = 204$ °C for PαMS, and (2) $C_1 = 13.7$, $C_2 = 50$, and $T_{\mathrm{r}} = 100$ °C for PS. Note in Figure 8.52 that the vertical bars on the data points indicate the range of error that would incur if there was 5% error in the determination of M_{w}. It is very encouraging to observe in Figure 8.52

Table 8.5 Comparison of prediction with experiment for the $\eta_{0,\text{block}}$ of PS-*block*-PαMS copolymers at 230 °C

Sample Code	*Measured* $\eta_{0,\text{block}}$ *(Pa·s)*	*Predicted* $\eta_{0,\text{block}}$ *(Pa·s)*
BII	1.3×10^4	2.8×10^3
CI	1.4×10^4	2.5×10^4
CII	1.5×10^5	2.3×10^4
CIII	2.0×10^5	1.1×10^5
CIV	1.2×10^4	2.1×10^3
CV	2.8×10^5	2.4×10^5
CVI	6.5×10^3	1.4×10^3

Reprinted from Kim and Han, *Macromolecules* 25:271. Copyright © 1992, with permission from the American Chemical Society.

that Eq. (8.20) gives a reasonably good idea as to how the $\eta_{0,\text{block}}$ of disordered, entangled PS-*block*-PαMS copolymer might vary with composition.

At present there is no comprehensive theory predicting the linear viscoelasticity of disordered entangled block copolymers. As shown in Chapter 7, which discusses the dynamics of miscible polymer blends, the inclusion of the intermolecular interactions (perhaps via segmental interaction parameter) between the constituent blocks in a block copolymer would be very important to the development of a comprehensive molecular theory predicting the linear viscoelasticity of disordered block copolymers.

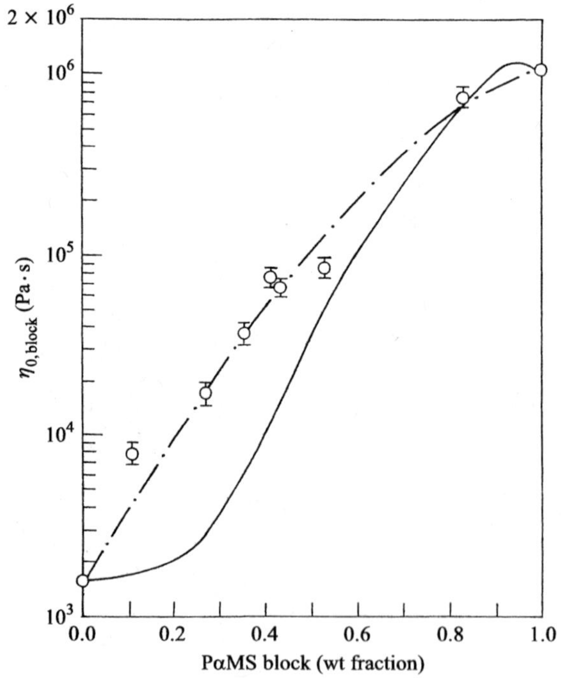

Figure 8.52 Plots of log $\eta_{0,\text{block}}$ versus weight fraction of PαMS block for the PS-*block*-PαMS copolymers, where the solid line is a prediction using Eq. (8.20) and the dotted line is drawn as a visual guide through the experimental data. The reference molecular weight chosen is 2×10^5 and the reference temperature chosen is 230 °C. (Reprinted from Kim and Han, *Macromolecules* 25:271. Copyright © 1992, with permission from the American Chemical Society.)

8.5 Stress Relaxation Modulus of Microphase-Separated Block Copolymer Upon Application of Step-Shear Strain

Stress relaxation in polymers has long been regarded as being a unique rheological response, which can be used to describe the effect of molecular architecture on the mechanical behavior of polymers. Consequently, over the years, stress relaxation, after imposition of a step-shear strain on a polymer specimen in the rest state, has extensively been studied by a number of investigators; some investigators (Einaga et al. 1972; Fukuda et al. 1975; Larson et al. 1988; Osaki et al. 1981, 1982; Vrentas and Graessley 1981, 1982; Zapas and Phillips 1971) employed homopolymer solutions, others (Laun 1978; Takahashi et al. 1990) employed homopolymer melts, and still others (Han and Kim 1994) employed thermotropic liquid-crystalline polymers. However, strange as it may seem, there are hardly any experimental studies reported on the stress relaxation modulus of microphase-separated block copolymers, although some attempts were made to describe theoretically the stress relaxation modulus $G(t, \gamma)$ of microphase-separated block copolymers (Doi et al. 1993; Ohta et al. 1993; Rubinstein and Obukhov 1993; Witten et al. 1990) upon application of step-shear strain (γ). Rubinstein and Obukhov (1993) predicted the following scaling relationships: (1) $G(t, \gamma) \sim t^{-1/2}$ for block copolymers having lamellar microdomains and (2) $G(t, \gamma) \sim t^{-1/4}$ for block copolymers having cylindrical microdomains.

Since the molecular weights of block copolymers in general are very low compared with those of commercial homopolymers, it can be easily surmised that the $G(t, \gamma)$ of block copolymers in the disordered state would be very low, making accurate measurements of $G(t, \gamma)$ very difficult. But, the $G(t, \gamma)$ of microphase-separated block copolymers might be of great interest to polymer scientists, because different microdomain structures in block copolymers might respond differently to an applied step strain. In this section, we present some experimental observations showing that the $G(t, \gamma)$ of a microphase-separated block copolymer is very sensitive to the type of microdomain structure (lamellae, cylinders, or spheres).

Figure 8.53 gives $\log G'$ versus $\log G''$ plots for a lamella-forming SIS triblock copolymer (Vector 4411, Dexco Polymers) ($M_n = 8.2 \times 10^4$, $M_w/M_n = 1.05$, $w_{PS} = 0.42$) during heating, at temperatures ranging from 140 to 220 °C, together with a TEM image. Notice in Figure 8.53 that the slope of the $\log G'$ versus $\log G''$ plots in the terminal region is much less than 2, signifying that Vector 4411 is in the ordered state over the entire range of temperatures tested. Thus, from Figure 8.53 we conclude that the T_{ODT} of Vector 4411 is much higher than 220 °C, the highest experimentally accessible temperature, before the onset of thermal degradation and/or cross-linking might take place.

Figure 8.54 shows $\log G(t, \gamma)$ versus $\log t$ plots at $\gamma = 0.04$ in the linear region for Vector 4411, at various temperatures ranging from 160 to 220 °C. The values of $G(t, \gamma)$ from step-strain experiments are determined using the expression $G(t, \gamma) = \sigma(t, \gamma)/\gamma$, where $\sigma(t, \gamma)$ is the shear stress at time t after a sudden application of shear strain γ. Note that the T_{ODT} of Vector 4411 is much higher than 220 °C (see Figure 8.53), and thus all of the experimental data given in Figure 8.54 were obtained in the ordered state. In Figure 8.54 we observe that values of $G(t, \gamma)$ decrease gradually with increasing time, behavior quite different from that reported in the literature for homopolymer melts (Laun 1978; Takahashi et al. 1990),

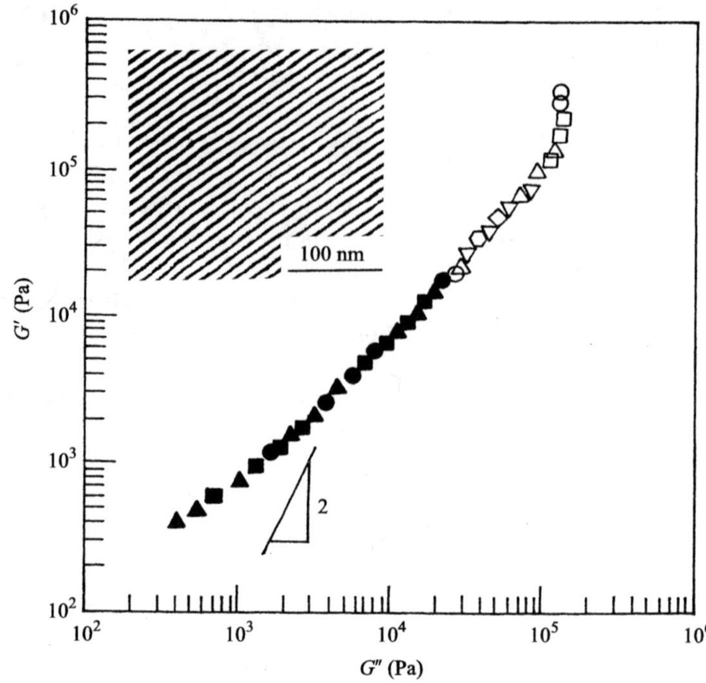

Figure 8.53 Log G' versus log G'' plots for an SIS triblock copolymer (Vector 4411) at various temperatures (°C): (○) 140, (□) 150, (△) 160, (▽) 170, (◇) 180, (⬡) 190, (●) 200, (■) 210, and (▲) 220. The inset gives a TEM image of Vector 4411, showing lamellar microdomains. The plots indicate that the T_{ODT} of Vector 4411 is much higher than 220 °C, the highest experimental temperature employed.

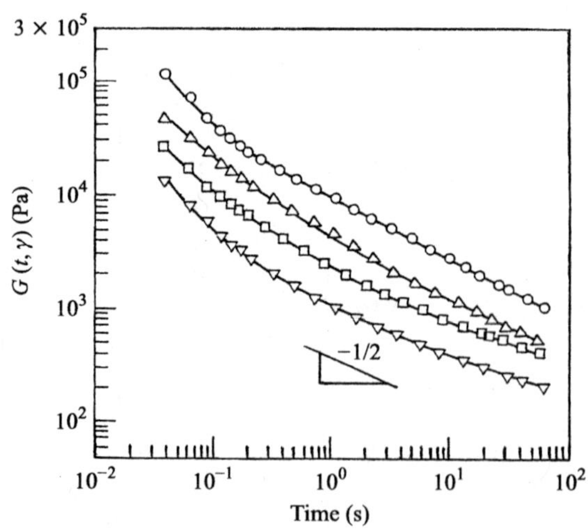

Figure 8.54 Plots of log $G(t, \gamma)$ versus log t for lamella-forming Vector 4411 at $\gamma = 0.04$ in the linear region at different temperatures (°C): (○) 160, (△) 180, (□) 200, and (▽) 220.

and the shape of $\log G(t, \gamma)$ versus $\log t$ plots remains the same over the entire range of temperatures investigated, suggesting that the lamellar microdomain structure in Vector 4411 is not altered at temperatures ranging from 160 to 220 °C. It is interesting to observe in Figure 8.54 that the scaling law predicted by Rubinstein and Obukhov (1993), $G(t, \gamma) \sim t^{-1/2}$, holds at large values of t. We have observed, although not presented here, that the $\log G(t, \gamma)$ versus $\log t$ plot in the nonlinear region ($\gamma > 1$) deviates considerably from the scaling law in the linear region, $G(t, \gamma) \sim t^{-1/2}$. Under such circumstances, it is not possible to separate the effect of time from that of γ by simply shifting the $\log G(t, \gamma)$ versus $\log t$ plot along the $G(t, \gamma)$ axis, as was often done for flexible homopolymer melts. This can be explained if we recognize the fact that the orientations of lamellar microdomains in Vector 4411 might have changed with time after large values of γ are applied.

Figure 8.55 gives $\log G'$ versus $\log G''$ plots for a cylinder-forming SIS triblock copolymer (Vector 4211, Dexco Polymers) ($M_n = 1.09 \times 10^5$, $M_w/M_n = 1.05$, $w_{PS} = 0.33$) at temperatures ranging from 160 to 220 °C, together with a TEM image showing hexagonally packed cylinders. Notice in Figure 8.55 that the slope of the $\log G'$ versus $\log G''$ plots in the terminal region is much less than 2, signifying that Vector 4211 is in the ordered state over the entire range of temperatures tested. Thus, from Figure 8.55

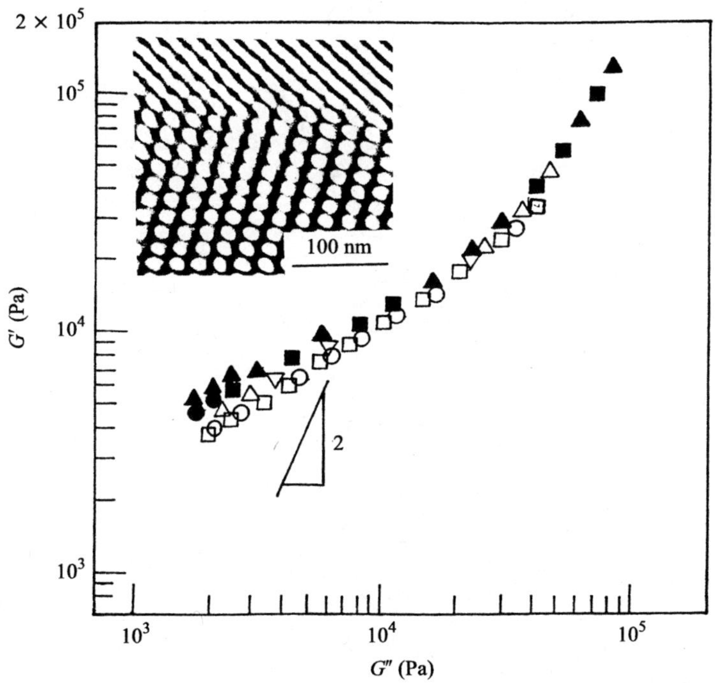

Figure 8.55 Log G' versus log G'' plots for an SIS triblock copolymer (Vector 4211) at various temperatures (°C): (▲) 160, (■) 170, (●) 180, (▽) 190, (△) 200, (□) 210, and (○) 220. The inset gives a TEM image of Vector 4211, showing hexagonally packed cylindrical microdomains. The plots indicate that the T_{ODT} of Vector 4211 is much higher than 220 °C, the highest experimental temperature employed.

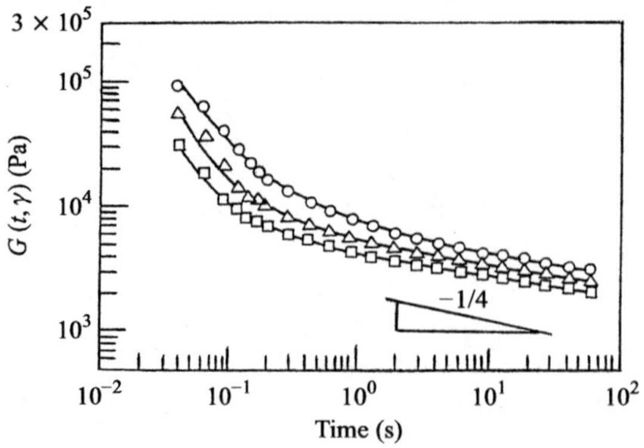

Figure 8.56 Plots of log $G(t, \gamma)$ versus log t for cylinder-forming Vector 4211 at $\gamma = 0.01$ in the linear region at different temperatures (°C): (○) 180, (△) 200, and (□) 220.

we conclude that the T_{ODT} of Vector 4211 is much higher than 220 °C, the highest experimentally accessible temperature before the onset of thermal degradation and/or cross-linking might take place.

Figure 8.56 shows log $G(t, \gamma)$ versus log t plots at $\gamma = 0.01$ in the linear region for Vector 4211 at various temperatures ranging from 180 to 220 °C. Note that Vector 4211 has a T_{ODT} much higher than 220 °C (see Figure 8.55) and thus all of the experimental data given in Figure 8.56 were obtained in the ordered state. In Figure 8.56 we observe that $G(t, \gamma)$ initially decreases rapidly, and then at a much slower rate with increasing time, as compared with the lamella-forming Vector 4411 (see Figure 8.54). The shape of log $G(t, \gamma)$ versus log t plots remains the same over the entire range of temperatures investigated, suggesting that the cylindrical microdomain structure in Vector 4211 is not altered at temperatures ranging from 180 to 220 °C. Notice in Figure 8.56 that the scaling law predicted by Rubinstein and Obukhov (1993), $G(t, \gamma) \sim t^{-1/4}$, holds at large values of t. That is, the difference in the microdomain structure of the two block copolymers, cylinder-forming Vector 4211 and lamella-forming Vector 4411, is reflected in Figures 8.54 and 8.56. We have observed, although not presented here, that the log $G(t, \gamma)$ versus log t plot in the nonlinear region ($\gamma > 1$) deviates considerably from the scaling law in the linear region, $G(t, \gamma) \sim t^{-1/4}$. Under such circumstances, it is not possible to separate the effect of time from that of γ by simply shifting the log $G(t, \gamma)$ versus log t plot along the $G(t, \gamma)$ axis, as was often done for flexible homopolymer melts.

Figure 8.57 shows log $G(t, \gamma)$ versus log t plots at $\gamma = 0.01$ in the linear region for an SIS triblock copolymer, Vector 4111 ($M_n = 1.4 \times 10^5$, $M_w/M_n = 1.11$, $w_{PS} = 0.18$), at various temperatures ranging from 170 to 220 °C. Note that Vector 4111 has hexagonally packed cylindrical microdomains at temperatures below approximately 185 °C, at which point it undergoes OOT from cylindrical to spherical microdomains, and retains spherical microdomains until reaching approximately 220 °C, at which point it undergoes LDOT at approximately 220 °C forming disordered micelles with short-range order (see Figures 8.9–8.13). It is of interest to observe in Figure 8.57 that at 170 °C the log $G(t, \gamma)$ versus log t plots follow the scaling law, $G(t, \gamma) \sim t^{-1/4}$, for cylindrical microdomains, very similar to Vector 4211 (see Figure 8.56), and at 185 and 200 °C the log $G(t, \gamma)$ versus log t plots deviate considerably from the scaling law for cylindrical microdomains. This observation makes sense because,

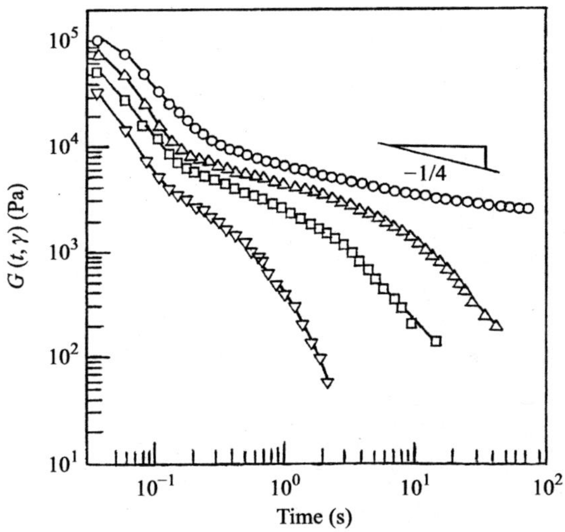

Figure 8.57 Plots of $\log G(t, \gamma)$ versus $\log t$ for an SIS triblock copolymer (Vector 4111) at $\gamma = 0.01$ in the linear region at different temperatures (°C): (○) 170, (△) 185, (□) 200, and (▽) 220.

according to Figures 8.9–8.13, Vector 4111 has cylindrical microdomains at 170 °C, and spherical microdomains at 185–200 °C. Note in Figure 8.57 that the slope of $\log G(t, \gamma)$ versus $\log t$ plot at 220 °C is larger than that at 185 and 200 °C. This observation is understandable because, according to Figures 8.9–8.13, at 220 °C Vector 4111 forms disordered micelles with short-range order; in other words, at $T \geq 220$ °C Vector 4111 exhibits liquidlike behavior in the terminal region of linear dynamic viscoelasticity.

The experimental observations presented above have hardly scratched the surface of this very complicated problem. For instance, the morphology of a lamella-forming block copolymer might be disrupted into cylinderlike fragments by large strains, and the successive stress relaxation might correspond to the relaxation of those fragments and not of the original lamellae. That is, there is a possibility for strain-induced structural changes. Thus, a concomitant in situ structural characterization during stress relaxation is highly desirable in order to fully understand the mechanisms associated with it. In other words, there can be many possibilities for strain-induced structural changes. Further, microphase-separated block copolymers have several relaxation mechanisms: orientational relaxation of individual blocks, domain relaxation, and extra slow relaxation due to defect motion and long-range hydrodynamic interactions. It is therefore necessary to specify the relaxation mechanisms that are dominant for $G(t, \gamma)$ data. In this way, the structure–rheology relationship can be understood on the basis of the structural information. This is one of many research areas in block copolymers that deserves serious attention of polymer scientists in the years ahead.

8.6 Steady-State Shear Viscosity of Microphase-Separated Block Copolymers

In the past, some investigators (Arnold and Meier 1970; Ghijsels and Raadsen 1980; Han et al. 1995c; Holden et al. 1969; Lyngaae-Jørgensen et al. 1979; Vinogradov et al.

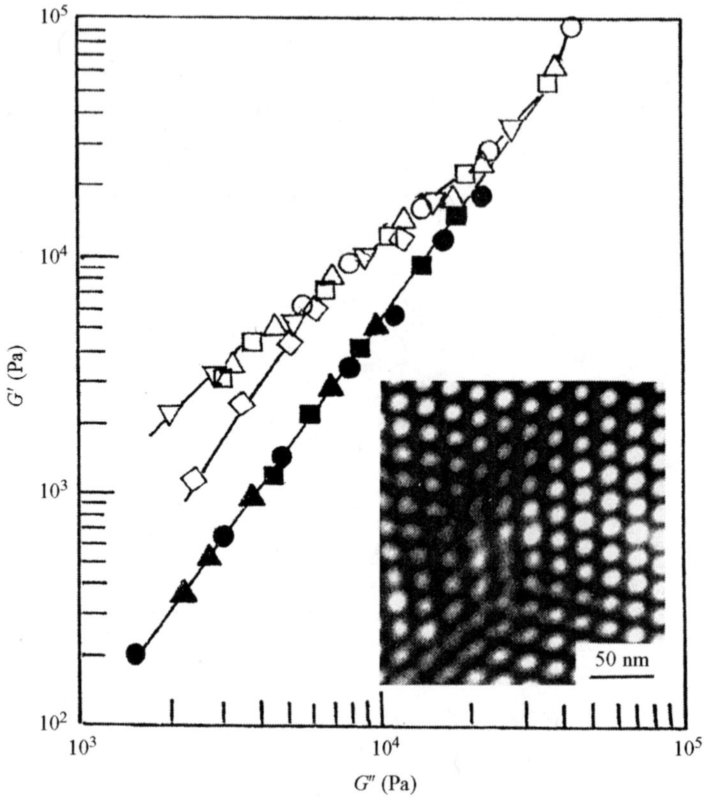

Figure 8.58 Log G' versus log G'' plots for an SBS triblock copolymer (Kraton 1102) at various temperatures (°C): (○) 160, (△) 180, (□) 190, (▽) 200, (◇) 210, (●) 220, (▲) 230, and (■) 240. The inset gives a TEM image of Kraton 1102, showing hexagonally packed cylindrical microdomains.

1978) reported on shear-rate dependent steady-state viscosity (η) of some microphase-separated block copolymers. In this section we present some representative η data of microphase-separated block copolymers, demonstrating that values of η obtained from a cone-and-plate rheometer do not always overlap those obtained from a capillary rheometer.

For this, let us consider a cylinder-forming SBS triblock copolymer (Kraton 1102,[9] Shell Development), which has a 0.29 weight fraction of PS block and a total molecular weight of 7×10^4. Figure 8.58 gives log G' versus log G'' plots for solvent-cast Kraton 1102 specimens at various temperatures ranging from 160 to 240 °C. The inset of Figure 8.58 gives a TEM image showing that Kraton 1102 has hexagonally packed cylindrical microdomains. It is seen from Figure 8.58 that the T_{ODT} of Kraton 1102 is approximately 220 °C.

Figure 8.59 gives logarithmic plots of η versus shear rate ($\dot{\gamma}$) for compression-molded Kraton 1102 specimens at 160, 170, and 180 °C, in which the temperature employed for rheological measurements coincides with that for compression molding of specimens. In Figure 8.59 we observe that values of η measured with a cone-and-plate

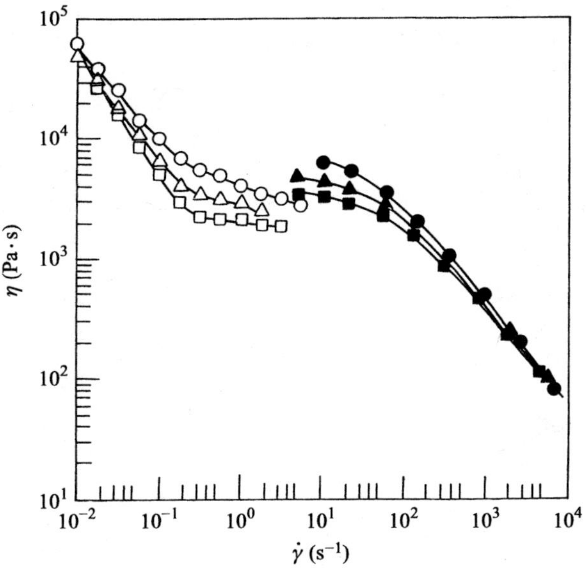

Figure 8.59 Plots of log η versus log $\dot{\gamma}$ for compression-molded Kraton 1102 specimens at three different temperatures (°C): (○, ●) 160, (△,▲) 170, and (□, ■) 180, in which open symbols are cone-and-plate data and filled symbols are capillary data. (Reprinted from Han et al., *Polymer* 36:155. Copyright © 1995, with permission from Elsevier.)

rheometer do not overlap those measured with a capillary rheometer. There is no reason why the two sets of data should overlap because the morphological state of the cylinder-forming Kraton 1102 in a cone-and-plate rheometer can be quite different from that in a capillary rheometer. For instance, the arrays of cylindrical microdomains of Kraton 1102 might have been better aligned inside a capillary die than in the gap opening of a cone-and-plate rheometer.

Figure 8.60 gives log η versus log $\dot{\gamma}$ plots for solvent-cast Kraton 1102 specimens at 180 and 200 °C, showing that the values of η measured with a cone-and-plate rheometer

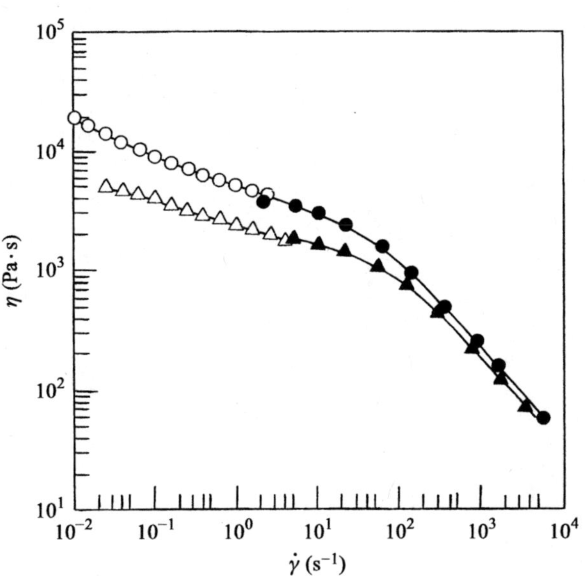

Figure 8.60 Log η versus log $\dot{\gamma}$ plots for solvent-cast Kraton 1102 specimens at two different temperatures (°C): (○, ●) 180 and (△, ▲) 200, in which open symbols are cone-and-plate data and filled symbols are capillary data. (Reprinted from Han et al., *Polymer* 36:155. Copyright © 1995, with permission from Elsevier.)

overlap reasonably well with those measured with a capillary rheometer. This stands in contrast with the observation made in Figure 8.59. Comparison of Figure 8.60 with Figure 8.59 at 180 °C reveals that at low $\dot{\gamma}$ (i.e., in the cone-and-plate rheometer) the shear-thinning behavior of compression-molded specimens is much more intense than the solvent-cast specimens, and at high $\dot{\gamma}$ (i.e., in the capillary rheometer) the values of η are higher for the compression-molded specimens than for the solvent-cast specimens. Thus, we conclude from Figures 8.59 and 8.60 that different methods of sample preparation (compression molding versus solvent casting) give rise to different rheological responses in both cone-and-plate and capillary rheometers. We have already discussed in a previous section the effects of thermal history and sample preparation method on the linear dynamic viscoelasticity and morphology of microphase-separated block copolymers.

Figure 8.61 gives a comparison of $\log \eta$ versus $\log \dot{\gamma}$ plots with $\log |\eta^*|$ versus $\log \omega$ plots for compression-molded Kraton 1102 specimens obtained from a cone-and-plate rheometer or a capillary rheometer. In Figure 8.61 we observe that the values of η obtained from steady-state shear flow are much lower than the values of $|\eta^*|$ obtained from oscillatory shear flow when both are from a cone-and-plate rheometer, and the values of η obtained from a capillary rheometer are higher than the values of $|\eta^*|$ obtained from a cone-and-plate rheometer. Even at very low $\dot{\gamma}$ (0.01–1.0 s^{-1}) in steady-state shear flow, η decreases initially very rapidly with increasing $\dot{\gamma}$ and then tends to show a plateau region as $\dot{\gamma}$ is increased further. Conversely, in oscillatory shear flow $|\eta^*|$ decreases steadily with increasing ω. Since Kraton 1102 has cylindrical microdomains (Figure 8.58), the rapid decrease in η with increasing $\dot{\gamma}$ observed in Figure 8.61 may be attributable to variations in the morphological state of the specimen taking place during shear flow. On the basis of Figure 8.61, we can conclude that the Cox–Merz rule (Cox and Merz 1958) does not hold for Kraton 1102 and hence for microphase-separated block copolymers in general.

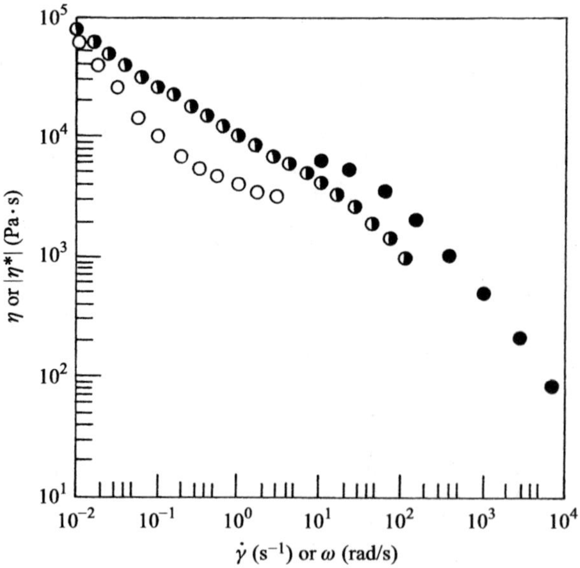

Figure 8.61 Comparison of $|\eta^*|$ with η for compression-molded Kraton 1102 specimens at 160 °C: (◑) plots of $\log |\eta^*|$ versus $\log \omega$ in the oscillatory mode, (○) plots of $\log \eta$ versus $\log \dot{\gamma}$ from a cone-and-plate rheometer, and (●) plots of $\log \eta$ versus $\log \dot{\gamma}$ from a capillary rheometer. (Reprinted from Han et al., *Polymer* 36:155. Copyright © 1995, with permission from Elsevier.)

8.7 Summary

In this chapter, we have shown that oscillatory shear rheometry is very useful for determining phase transition temperatures in microphase-separated block copolymers. Specifically, we have shown that the temperature dependence of G' from isochronal dynamic temperature sweep experiments is very useful for determining the T_{ODT} of compositionally symmetric or nearly symmetric (lamella-forming) block copolymers, and the T_{OOT} and T_{ODT} of cylinder-forming block copolymers. Although the isochronal dynamic temperature sweep experiments can detect, during heating, a temperature at which a sudden drop in G' occurs in highly asymmetric (sphere-forming) block copolymers, this rheological method cannot identify that sphere-forming block copolymers first transform into disordered micelles with short-range order at T_{LDOT} and then, with a further increase in temperature, transform into the micelle-free disordered phase, with only thermally induced composition fluctuations at a critical temperature, T_{DMT}. In this chapter, we have also shown that log G' versus log G'' plots from the dynamic frequency sweep experiment are very useful for determining the T_{ODT} of lamella-forming and cylinder-forming block copolymers, and also the T_{LDOT} and T_{DMT} of sphere-forming block copolymers. We have emphasized the importance of SAXS and TEM in combination with oscillatory shear rheometry to the investigation of phase transitions in block copolymers, because oscillatory shear rheometry is not an experimental tool for investigating the morphology of block copolymers.

We have presented experimental observations showing that the rheological behavior of a block copolymer is greatly influenced by its morphology, which in turn is influenced by sample preparation method, thermal history, and flow geometry (cone-and-plate versus capillary). The highlights of experimental observations can be summarized as follows. (1) The steady-state shear viscosities obtained using a cone-and-plate rheometer do not necessarily overlap those obtained using a capillary rheometer. (2) Over a wide range of $\dot{\gamma}$ tested, the solvent-cast specimens have lower values of η as compared with the compression-molded specimens, indicating that sample preparation methods employed may influence rheological behavior. This is attributable to the differences in morphology that existed in the respective specimens. (3) Time–temperature superposition is not applicable to microphase-separated block copolymers. (4) The complex shear viscosities do not necessarily overlap the steady-state shear viscosities at comparable shear rates and angular frequencies, indicating that the Cox–Merz rule does not hold for microphase-separated block copolymers.

We have presented experimental observations for the linear dynamic viscoelasticity of low-molecular-weight SI diblock copolymers in the disordered state. We have shown that the log G' versus log G'' plot is very useful for investigating the effects of molecular weight and block length ratio on the linear dynamic viscoelasticity of block copolymers. The experimental results were compared with currently held theory for disordered, unentangled AB-type and ABA-type block copolymers.

At present, there is no comprehensive theory that can enable one to predict the rheological behavior of a block copolymer, ordered or disordered, in terms of its molecular parameters. It can easily be surmised that prediction of the rheological properties of a block copolymer in the ordered state would be a difficult task, because the type of microdomain structure (spheres, cylinders, gyroids, or lamellae) must be incorporated into the development of a molecular theory. It should be pointed out that spherical microdomains, for instance, having the dimensions of approximately 50 nm in a block copolymer, cannot be regarded as being equal to particles suspended freely in a flexible homopolymer because the microdomains in a block copolymer are chemically attached to the chains of the other block (i.e., they are part of the entire block copolymer chains). In this regard, any attempt to predict the rheological behavior of a microphase-separated block copolymer with a continuum approach does not seem warranted. Accordingly, a molecular approach is most welcome.

The task of predicting the rheological behavior of AB-, ABA-, and ABC-type block copolymers is a real challenge. A more challenging problem would be to investigate the rheological behavior of a block copolymer consisting of a flexible chain in one block and a rigid rodlike chain in the other block. Such diblock copolymers have been synthesized in some research laboratories. Equally challenging, if not more so, would be an investigation of the rheological behavior of a block copolymer containing side-chain liquid crystals. Such block copolymers have also been synthesized in some research laboratories.

Notes

1. There are too many papers to cite here. Some specific papers will be cited later in the chapter, when we present experimental results.
2. A block copolymer that is formed from a polymer pair that exhibits upper critical solution temperature (UCST) in binary blends will show a similar phase behavior as the binary blend. Since the ordered state of such a block copolymer has microdomains (i.e., microphase separated), it is more precise to refer such phase transition occurring during heating as upper critical order–disorder transition (UCODT). The majority of the experimental studies reported in the literature dealt with block copolymers exhibiting UCODT and used the terminology ODT, instead of UCODT, since the early 1980s. A block copolymer formed from a polymer pair that exhibits lower critical solution temperature (LCST) in binary blends was first reported only in 1994. Within the spirit of distinguishing LCST and UCST in binary polymer blends, it is precise to refer to such a phase transition in block copolymers as lower critical disorder–order transition (LCDOT). Later in this chapter we will present oscillatory shear rheometry of microphase-separated block copolymers exhibiting LCDOT, or combined UCODT and LCDOT.
3. Using the notations in Eq. (7.17), we have $\phi_{1,c} = \phi_{2,c} = 0.5$ for an equal blend composition, and $r_1 = r_2 = r$ for an equal number of segments in both components. If we assume that both components have the same density, r can be equated to the degree of polymerization N, that is $N_1 = N_2 = N$. Thus, from Eq. (7.17) we obtain $(\chi N)_c = 2$.
4. The T_g of PEP having an equal mole ratio of ethylene and propylene is about $-40\,^\circ$C and the T_g of PDMS is $-118\,^\circ$C, thus yielding $\Delta T_g \approx 80\,^\circ$C.
5. The value of $M_c = 1.4 \times 10^4$ for PI was obtained using the following procedures. Raju et al. (1981) reported that the plateau modulus (G_N^o) of PI at 25 °C is 3.26×10^5 Pa. Thus, we can estimate the M_e of PI using the expression $M_e = \rho RT/G_N^o$ for entangled homopolymers (see Chapter 4), where ρ is the density (g/cm^3), R is the

universal gas constant (8.314×10^7 dyn · cm K^{-1} mol^{-1}), and T is the absolute temperature (K). Using $\rho = 0.913$ g/cm^3 at 25 °C we obtain $M_e = 7.0 \times 10^3$ for PI. Thus, we obtain $M_c \approx 1.4 \times 10^4$ for PI if we assume $M_c = 2M_e$.

6. In order not to change the axes' scales of Figure 8.35 in the paper by Klopffer et al. (1998), the units of friction coefficients have not been converted to SI units. Note that 1 dyn · s/cm = 10^{-3} N · s/m.
7. Equation 20 in the paper by Han et al. (1995a) should read Eq. (8.26).
8. The relationship $\eta_{0,\text{block}} \propto M^{3.4}$ was used by Kim and Han (1992). The predicted values of G', G'', and $\eta_{0,\text{block}}$ would have been much larger if the relationship $\eta_{0,\text{block}} \propto M^{4.7}$ (see Figure 8.34), which has been reported by Choi and Han (2004), had been used.
9. Kraton 1102 contains about 20 wt % uncoupled SB diblocks. This is because a coupling method was used to produce SBS triblock copolymer from SB diblock copolymer that was synthesized, via anionic polymerization, using a monofunctional initiator.

References

Adams JL, Graessley WW, Register RA (1994). *Macromolecules* 27:6026.
Aggarwal SL (ed) (1970). *Block Copolymers*, Plenum Press, New York.
Allen VR, Fox TG (1964). *J. Chem. Phys.* 41:337.
Arnold KR, Meier D (1970). *J. Appl. Polym. Sci.* 14:427.
Balsara NP, Garetz BA, Chang MY, Dai HJ, Newstein MC, Goveas JL, Krishnamoorti R, Rai S (1998). *Macromolecules* 31:5309.
Bates FS (1984). *Macromolecules* 17:2607.
Bates FS, Fredrickson GH (1990). *Annu. Rev. Phys. Chem.* 41:525.
Bates FS, Rosedale JH, Fredrickson GH (1990). *J. Chem. Phys.* 92:6255.
Berry GC, Fox TG (1968). *Adv. Polym. Sci.* 5:261.
Burke JJ, Weiss V (eds) (1973). *Block and Graft Copolymers*, Syracus University Press, Syracus, New York.
Choi S, Han CD (2004). *Macromolecules* 37:215.
Choi S, Lee KM, Han CD, Sota N, Hashimoto T (2002). *Macromolecules* 35:793.
Choi S, Vaidya NY, Han CD, Sota N, Hashimoto T (2003). *Macromolecules* 36:7707.
Chung CI, Lin MI (1978). *J. Polym. Sci., Polym. Phys. Ed.* 16:545.
Cong G, Huang Y, MacKnight WJ, Karasz FE (1986). *Macromolecules* 19:2765.
Cox WP, Merz EH (1958). *J. Polym. Sci.* 28:619.
de Gennes PG (1979). *Scaling Concepts in Polymer Physics*, Cornell University Press, Ithaca, New York.
Doi M, Harden JL, Ohta T (1993). *Macromolecules* 26:4935.
Dormidontova EE, Lodge TP (2001). *Macromolecules* 34:9143.
Einaga Y, Osaki K, Kurata M (1972). *Polym. J.* 2:550.
Ferry JD (1980). *Viscoelastic Properties of Polymers*, 3rd ed, Wiley, New York.
Floudas G, Hadjichristidis N, Iatrou H, Pakula T, Fischer EW (1994). *Macromolecules* 27:7735.
Floudas G, Pispas S, Hadjichristidis N, Pakula T, Erukhimovich I (1996a). *Macromolecules* 29:4142.

Floudas G, Vlassopoulos D, Pitsikalis M, Hadjichristidis N, Stamm M (1996b). *J. Chem. Phys.* 104:2083.
Förster S, Khandpur AK, Zhao J, Bates FS, Hamley IW, Ryan AJ, Bras W (1994). *Macromolecules* 27:6922.
Fredrickson GH, Larson RG (1987). *J. Chem. Phys.* 86:1553.
Fukuda M, Osaki K, Kurata M (1975). *J. Polym. Sci., Polym. Phys. Ed.* 13:1563.
Ghijsels A, Raadsen J (1980). *Pure Appl. Chem.* 52:1359.
Gouinlock EV, Porter RS (1997). *Polym. Eng. Sci.* 17:535.
Hajduk DA, Harper PE, Gruner SM, Honeker CC, Kim G, Thomas EL, Fetters LJ (1994). *Macromolecules* 27:4063.
Hajduk DA, Takenouchi H, Hillmyer MA, Bates FA, Vigild ME, Almda, K (1997). *Macromolecules* 30:3788.
Hall WF, DeWames RE (1975). *Macromolecules* 8:349.
Hamley IW, Koppi KA, Rosedale JH, Bates FS, Almdal K, Mortensen K (1993). *Macromolecules* 26:5959.
Hamley IW (1998). *The Physics of Block Copolymers*, Oxford University Press, Oxford.
Han CD, Baek DM, Kim JK (1990). *Macromolecules* 23:561.
Han CD, Baek DM, Kim JK, Hashimoto T, Okamoto S (1991). *Macromolecules* 24:5408.
Han CD, Baek DM, Kim JK, Chu SG (1992). *Polymer* 33:294.
Han CD, Baek DM, Kim JK (1995a). *Macromolecules* 28:5886.
Han CD, Baek DM, Kim JK, Ogawa T, Sakamoto N, Hashimoto T (1995b). *Macromolecules* 28:5043.
Han JH, Feng D, Choi-Feng C, Han CD (1995c). *Polymer* 36:155.
Han CD, Kim J (1987). *J. Polym. Sci., Polym. Phys. Ed.* 25:1741.
Han CD, Kim SS (1994). *J. Rheol.* 38:31.
Han CD, Kim J, Kim JK (1989). *Macromolecules* 22:383.
Han CD, Vaidya NY, Kim D, Shin G, Yamaguchi D, Hashimoto T (2000). *Macromolecules* 33:3767.
Hansen DR, Shen M (1975a). *Macromolecules* 8:343
Hansen DR, Shen M (1975b). *Macromolecules* 8:903.
Hasegawa H, Tanaka H, Yamasaki K, Hashimoto T (1987). *Macromolecules* 20:1651.
Hasegawa H, Sakamoto N, Takeno H, Jinnai H, Hashimoto T, Schwahn D, Frielinghaus H, Janssen S, Imai M, Mortensen K (1999). *Phys. Chem. Solids* 60:1307.
Hashimoto T (1996). In *Thermoplastic Elastomers*, 2nd ed, Holden G, Legge NR, Quirk R, Schroeder HE (eds), Hanser, New York, Chap 15A.
Hashimoto T, Ogawa T, Sakamoto N, Ichimiya M, Kim JK, Han CD (1998). *Polymer* 39:1573.
Helfand E, Wasserman ZR (1982). In *Developments in Block Copolymers*, Goodman I (ed), Applied Science, New York, Chap 4.
Holden G, Bishop ET, Legge NR (1969). *J. Polym. Sci. Part C* 26:37.
Holden G, Legge NR, Quirk R, Schroeder HE (eds) (1996). *Thermoplastic Elastomers*, 2nd ed, Hanser, New York.
Hsieh HL, Quirk, RP (1996). *Anionic Polymerization: Principles and Practical Applications*, Marcel Dekker, New York.
Karis TE, Russell TP, Gallot Y, Mayes AM (1995). *Macromolecules* 28:1129.
Kawasaki K, Sekimoto K (1987). *Physica* 143A:349.
Kawasaki K, Sekimoto K (1988). *Physica* 148A:361.
Khandpur AK, Förster S, Bates FS, Hamley IW, Ryan AJ, Bras W, Almdal K, Mortensen K (1995). *Macromolecules* 28:8796.
Kim JK, Han CD (1992). *Macromolecules* 25:271.
Kim JK, Lee HH, Sakurai S, Aida S, Masamoto J, Nomura S, Kitagawa Y, Suda Y (1999). *Macromolecules* 32:6707.

Klopffer MH, Bokobza L, Monnerie L (1998). *Polymer* 39:3445.
Laradji M, Shi AC, Noolandi J, Desai R (1997a). *Phys. Rev. Lett.* 78:2577.
Laradji M, Shi AC, Noolandi J, Desai R (1997b). *Macromolecules* 30:3242.
Larson RG, Fredrickson GH (1987). *Macromolecules* 20:1897.
Larson RG, Khan SA, Raju VR (1988). *J. Rheol.* 32:145.
Laun HM (1978). *Rheol. Acta* 17:1.
Leibler L (1980). *Macromolecules* 13:1602.
Lin CC, Jonnalagadda SV, Kesani PK, Dai HJ, Balsara NP (1994). *Macromolecules* 27:7769.
Lyngaae-Jørgensen J, Alle N, Martin FL (1979). In Multiphase Polymers, Cooper SL, Ester GM (eds), *Adv. Chem. Series* no 176, American Chemical Society, Washington, DC, p 541.
Matsen MW, Bates FS (1996). *Macromolecules* 29:1091.
Matsen MW, Schick M (1994). *Phys. Rev. Lett.* 72:2660.
Mayes AM, Olvera de la Cruz M (1989). *J. Chem. Phys.* 91:7228.
Modi MA, Krishnamoorti R, Ts MF, Wang HC (1999). *Macromolecules* 32:4088.
Mori K, Hasegawa H, Hashimoto T (1985). *Polym. J.* 17:799.
Noshay A, McGrath J (1977). *Block Copolymers*, Academic Press, New York.
Ohta T, Enomoto Y, Harden, JL, Doi D (1993). *Macromolecules* 26:4928.
Osaki K, Kimura S, Kurata M (1981). *J. Polym. Sci., Polym. Phys. Ed.* 19:5117.
Osaki K, Nishizawa K, Kurata M (1982). *Macromolecules* 15:1068.
Ougizawa T, Inoue T, Kammer HW (1985). *Macromolecules* 18:2089.
Pollard M, Russell TP, Ruzette AVG, Mayes AM, Gallot Y (1998). *Macromolecules* 31:6493.
Qi S, Wang ZG (1997a). *Phys. Rev. E* 55:1682.
Qi S, Wang ZG (1997b). *Macromolecules* 30:4491.
Raju BR, Menezes EV, Martin G, Graessley WW, Fetters LJ (1981). *Macromolecules* 14:1668.
Richardson MJ, Savill NG (1977). *Polymer* 18:3.
Roe RJ, Fishkis M, Chang JC (1981). *Macromolecules* 14:1091.
Rosedale JH, Bates FS (1990). *Macromolecules* 23:2329.
Rosedale JH, Bates FS, Almdal K, Mortensen K, Wignall GD (1995). *Macromolecules* 28:1429.
Rouse PE (1953). *J. Chem. Phys.* 21:1272.
Rubinstein M, Obukhov SP (1993). *Macromolecules* 26:1740.
Russell TP, Karis TE, Gallot Y, Mayes AM (1994). *Nature* 368:729.
Ruzette AVG, Benerjee P, Mayes AM, Pollard M, Russell TP, Jerome R, Slawecki T, Hjelm R, Thiyagarajan P (1998). *Macromolecules* 31:8509.
Ryu DY, Jeong U, Kim JK, Russell TP (2002a). *Nat. Mater.* 1:114.
Ryu DY, Park MS, Chae SH, Jang J, Kim JK, Russell TP (2002b). *Macromolecules* 35:8676.
Ryu DY, Jeong U, Lee DH, Kim J, Youn HS, Kim JK (2003). *Macromolecules* 36:2894.
Sakamoto N, Hashimoto T, Han CD, Kim D, Vaidya NY (1997). *Macromolecules* 30:1621.
Sakamoto N, Hashimoto T (1998a). *Macromolecules* 31:3292.
Sakamoto N, Hashimoto T (1998b). *Macromolecules* 31:8493.
Schulz MF, Khandpur AK, Bates FS, Almdal K, Mortensen K, Hajduk DA, Gruner SM (1996). *Macromolecules* 29:2857.
Schwab M, Stühn B (1996). *Phys. Rev. Lett.* 76:924.
Schwab M, Stühn B (1997). *Colloid Polym. Sci.* 275:341.
Sebastian JM, Lai C, Graessley WW, Register RA, Marchand GR (2002). *Macromolecules* 35:2700.
Segalman RA, Hexemer A, Hayward RC, Kramer EJ (2003). *Macromolecules* 36:3272.
Stockmayer WH, Kennedy JW (1975). *Macromolecules* 8:351.
Takahashi M, Taku K, Masuda T (1990). *J. Soc. Rheol. Jpn.* 18:18.
Thomas EL, Alward DB, Kinning DJ, Martin DC, Hardlin DL, Fetters LJ (1986). *Macromolecules* 19:2197.

Vavasour JD, Whitmore MD (1992). *Macromolecules* 25:5477.
Vinogrador GV, Dreval VE, Malkin AY, Yanovsky YG, Barancheeva VV, Borisenkova EK, Zabugina MP, Plotnikova EP, Sabsai OY (1978). *Rheol. Acta* 17:258.
Vrentas CM, Graessley WW (1981). *J. Non-Newtonian Fluid Mech.* 9:339.
Vrentas CM, Graessley WW (1982). *J. Rheol.* 26:359.
Wang FW, DiMarzio EA (1975). *Macromolecules* 8:356.
Wang Z, Dormidontova EE, Lodge TP (2002). *Macromolecules* 35:9687.
Weidisch R, Stamm M, Schubert DW, Arnold M, Budde H, Höring S (1999). *Macromolecules* 32:3405.
Weidisch R, Schreyeck G, Ensslen M, Michler GM, Stamm M, Schubert DW, Budde H, Höring S, Jerome, R (2000). *Macromolecules* 33:5495.
Widmaier JM, Meyer GC (1980). *J. Polym. Sci., Polym. Phys. Ed.* 18:2217.
Winey KI, Gobran DA, Xu Z, Fetters LJ, Thomas EL (1994). *Macromolecules* 27:2392.
Witten TA, Leibler L, Pincus PA (1990). *Macromolecules* 23:824.
Yamaguchi D, Hashimoto T, Vaidya NY, Han CD (1999). *Macromolecules* 32:7696.
Yokoyama H, Mates TE, Kramer EJ (2000). *Macromolecules* 33:1888.
Zapas LJ, Phillips JC (1971). *J. Res. Natl. Bur. Std.* 75A:33.
Zhao J, Majuamdar B, Schulz MK, Bates FS, Almdal K, Mortensen K, Hajduck DA, Grunner SM (1996). *Macromolecules* 29:1204.

9 Rheology of Liquid-Crystalline Polymers

9.1 Introduction

Liquid crystals (LCs) may be divided into two subgroups: (1) lyotropic LCs, formed by mixing rigid rodlike molecules with a solvent, and (2) thermotropic LCs, formed by heating. One finds in the literature such terms as mesomorphs, mesoforms, mesomorphic states, and anisotropic liquids. The molecules in LCs have an orderly arrangement, and different orders of structures (nematic, smectic, or cholesteric structure) have been observed, as schematically shown in Figure 9.1. The kinds of molecules that form LCs generally possess certain common molecular features. The structural characteristics that determine the type of mesomorphism exhibited by various molecules have been reviewed.

At present, our understanding of polymeric liquid crystals, often referred to as liquid-crystalline polymers (LCPs), is largely derived from studies of monomeric liquid crystals. However, LCPs may exhibit intrinsic differences from their monomeric counterparts because of the concatenation of monomers to form the chainlike macromolecules. The linkage of monomers inevitably means a loss of their translational and orientational independence, which in turn profoundly affects the dynamics of polymers in the liquid state. These intramolecular structural constraints are expressed in the flexibility of the polymer chain. Generally speaking, the chemical constitution of the monomer determines the flexibility and equilibrium dimensions of the polymer chain (Gray 1962). Figure 9.2 illustrates the variability of chain conformation (flexible chain, semiflexible chain, and rigid rodlike chain) forming macromolecules. Across this spectrum of chain flexibility, the persistence in the orientation of successive monomer units varies from the extreme of random orientation (flexible chains) to perfect order (the rigid rod). Hence, efforts have been made to synthesize LCPs that consist of rigid segments contributing to the formation of a mesophase and flexible segments contributing to the mobility of the entire macromolecule in the liquid state (Ober et al. 1984).

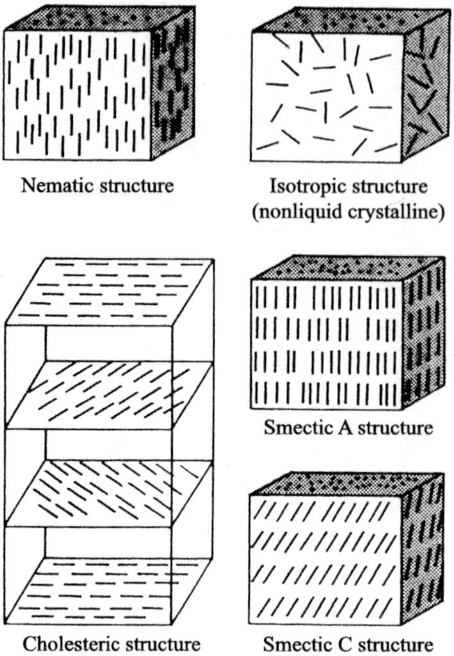

Figure 9.1 Molecular arrangements in thermotropic liquid crystals. (Reprinted from Brown and Crooker, *Chemical and Engineering News* 61(5):24. Copyright © 1983, with permission from the American Chemical Society.)

From the point of view of molecular architecture, as schematically shown in Figure 9.3, two types of LCP have been developed: (1) main-chain LCPs (MCLCPs), having the monomeric liquid crystals (i.e., mesogenic group) in the main chain of flexible links, and (2) side-chain LCPs (SCLCPs), having the monomeric liquid crystals attached, as a pendent side chain, to the main chain.

During the past three decades, LCPs have generated much interest from polymer scientists, due in large part to industrial research efforts to produce high-modulus/high-strength synthetic fibers composed of aromatic polyamides (Bair and Morgan 1972; Daniels et al. 1971; Frazer 1972; Kwolek 1971; Logullo 1971; Morgan et al. 1974; Preston 1969). For instance, Kevlar fiber (Kwolek 1971) is based on structures of poly(*p*-benzamide) (PBA) or poly(*p*-phenylene terephthalamide) (PPTA) type. Syntheses of this type of LCP, and other similar polymers, have been discussed in the literature (Black 1979; Black and Preston 1973; Kwolek et al. 1977; Morgan 1977; Preston 1975). Because of the high melting temperatures of these polymers (e.g., PBA, PPTA), which is indicative of a high degree of chain rigidity, Kevlar™ fibers are spun using solution spinning techniques. What is of particular interest here is the fact that these polymers form anisotropic (liquid-crystalline) solutions in such solvents as sulfuric acid and/or dialkylamide-salt mixtures (Kwolek et al. 1977; Morgan 1977). The use of anisotropic spinning dopes is believed to be a key factor in achieving the unusually strong mechanical properties of Kevlar fiber (Kwolek 1971; Morgan 1977); that is, Kevlar fiber is produced from a lyotropic LCP. The formation of the anisotropic phase can be detected by measuring, for instance, the occurrence of depolarization with optical microscopy. Lyotropic polymer solutions are turbid in appearance and

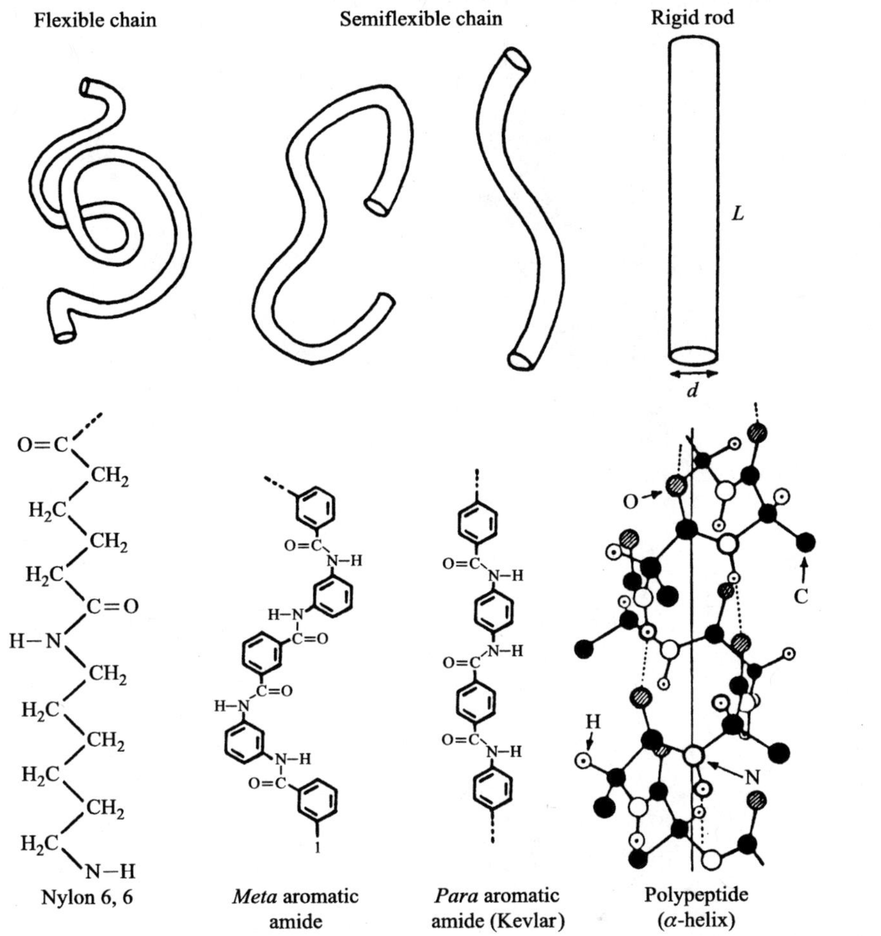

Figure 9.2 The effect of chain flexibility of macromolecules on their conformations in solution, in the molten state, or in the solid state. The flexibility of a macromolecule can be correlated to the type of crystal and physical properties of the corresponding solid polymer. (Reprinted from Samulski, *Physics Today* 35(5):40. Copyright © 1982, with permission from the American Institute of Physics.)

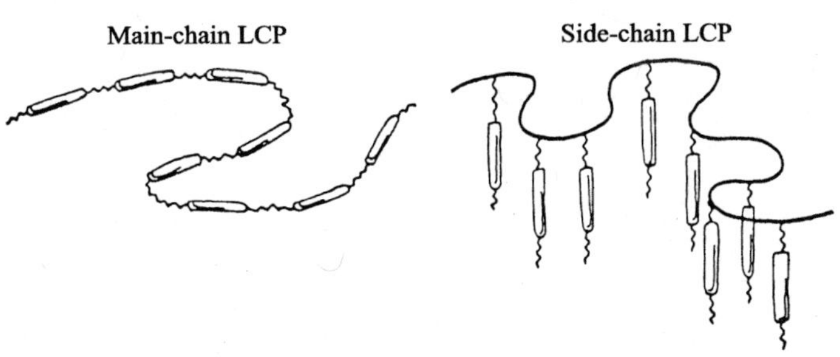

Figure 9.3 The difference in molecular architecture between main-chain LCP and side-chain LCP. (Reprinted from Samulski, *Physics Today* 35(5):40. Copyright © 1982, with permission from the American Institute of Physics.)

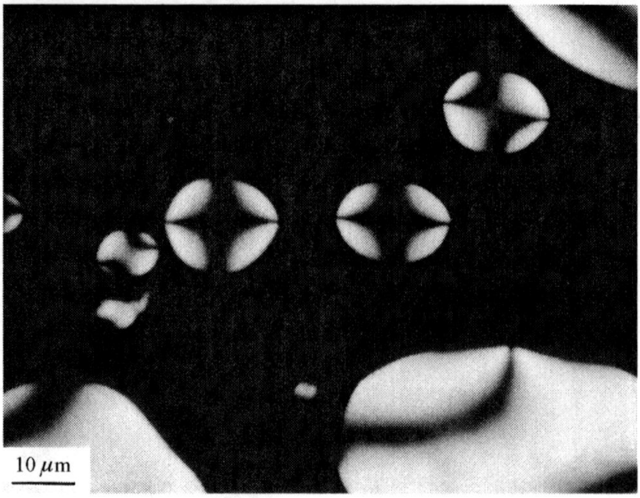

Figure 9.4 Micrograph of lyotropic solution of poly(chloro-1,4-phenylene terephthalamide) in dimethylacetamide-LiCl in a quiescent state, where droplets of liquid-crystalline phase are dispersed in a dark background of isotropic phase. (Reprinted from Morgan, *Macromolecules* 10:1381. Copyright © 1977, with permission from the American Chemical Society.)

optically birefringent in an unstrained state; that is, they depolarize plane-polarized light. Figure 9.4 shows a micrograph of droplets of a lyotropic solution of PPTA in dimethylacetamide-lithium chloride (DMAc-LiCl) in an unstrained (quiescent) state. When a lyotropic solution is subjected to an external flow field, the chains orient very easily, giving rise to unusually strong mechanical properties in the solid state. A micrograph of cast film prepared from PBA in DMAc-LiCl is shown in Figure 9.5, in which an aligned fibril structure can be seen. On a molecular basis, about 77% of the polymer chains in the crystalline state are aligned within 10° of the long axes of the fibrils (Kwolek et al. 1977). The physical reasons for such molecular alignment during flow, which then greatly influences the rheological behavior of the bulk liquid, will be presented in this chapter.

The formation of lyotropic LCPs may be construed as the occurrence of a transition from an isotropic phase to an anisotropic phase. As early as 1949, Onsager (1949) predicted, on the basis of calculation of free energy, the formation of an anisotropic phase at a critical concentration of a solution when the aspect ratio of the molecule is sufficiently large. Several years later, using the concept of lattice in mixtures of polymer and solvent, Flory (1956) also developed a theory to predict, in terms of the aspect ratio of polymer molecule, the critical concentration at which the formation of an anisotropic phase is possible. This subject has been summarized in several review articles (Flory 1984; Grosberg and Khokhlov 1981; Papkov 1984).

Also, considerable efforts have been spent on the synthesis of thermotropic liquid-crystalline polymers (TLCPs), which spontaneously form ordered structures over a certain range of temperatures. Such polymers are very attractive to industry from the processing point of view because the problem of solvent recovery would not exist.

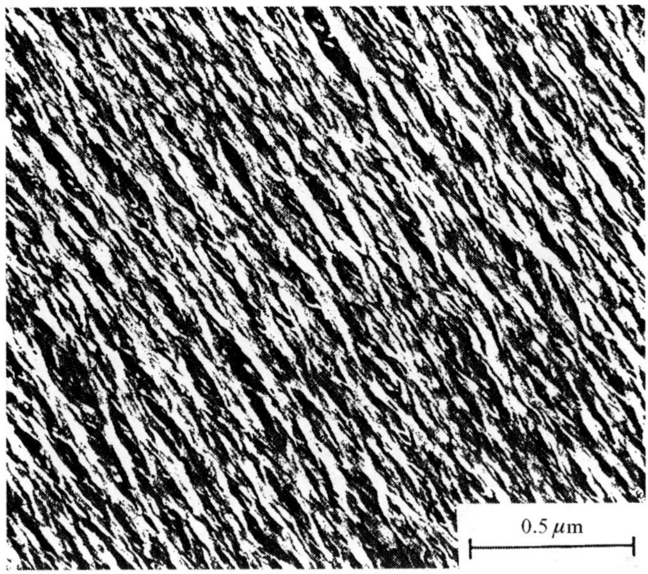

Figure 9.5 Micrograph of cast film prepared from poly(*p*-benzamide) in DMAc-LiCl, showing an aligned fibril structure. (Reprinted from Kwolek et al., *Macromolecules* 10:1390. Copyright © 1977, with permission from the American Chemical Society.)

As a means of inserting a mesogenic group in the flexible main chain, a very attractive concept has been developed; for instance, a modification of poly(ethylene terephthalate) (PET) with bisphenols, which results in a high-molecular-weight copolyester (Hamb 1972). It was Jackson and Kuhfuss (1976) who first utilized this method of synthesis successfully for the modification of PET with *p*-hydroxybenzoic acid (HBA), which yielded a TLCP. They determined the range of compositions that yielded the turbid melts which are characteristic of liquid-crystalline behavior, and found that copolyesters containing 40 mol % or higher concentrations of HBA exhibited liquid-crystalline structure. McFarlane et al. (1977) synthesized various thermotropic copolyesters, such as copolyesters by modifying PET with *p*-acetoxybenzoic acid and *p*-hydroquinone diacetate/terephthalic acid. The copolyester consisting of 60 mol % HBA and 40 mol % PET has the following chemical structure:

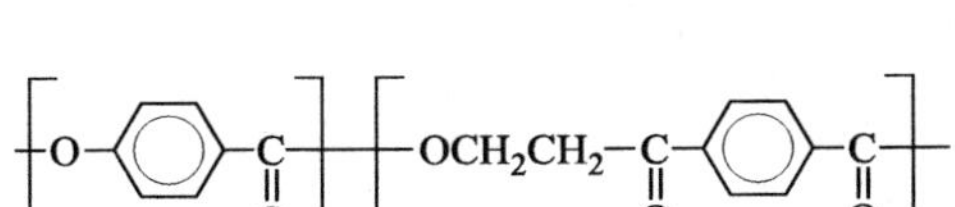

Figure 9.6 gives differential scanning calorimetry (DSC) thermograms for the 60/40 HBA/PET copolyester during heating and cooling cycles. It is very difficult to identify thermal transition temperatures from the DSC thermograms given in Figure 9.6. Although the literature reports that this polymer undergoes smectic-to-nematic (S–N) transition at approximately 320 °C, and nematic-to-isotropic (N–I) transition at approximately 420–430 °C, it is not possible for us to identify such thermal transition

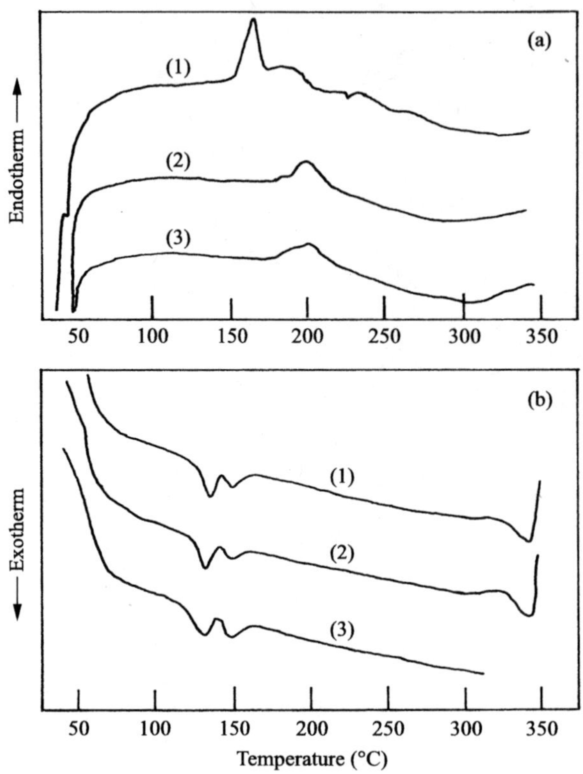

Figure 9.6 DSC thermograms for compression-molded 60/40 HBA/PET copolyester specimens during (a) the heating cycle and (b) the cooling cycle at a rate of 20 °C/min, where curve (1) is for the first heating or cooling cycle, curve (2) is for the second heating or cooling cycle, and curve (3) is for the third heating or cooling cycle.

temperatures from Figure 9.6. Further, the polymer may undergo thermal degradation at such high temperatures.

Other TLCPs, such as copolyesters of HBA and 6-hydroxy-2-naththoic acid (HNA) (Calundann 1978, 1979, 1980), and copolyesters of 4,4′-bisphenol (or hydroquinone), terephthalic acid, and HBA (Cottis et al. 1972, 1976) have also been synthesized and commercialized. For instance, copolyesters consisting of HBA and HNA have the following chemical structure:

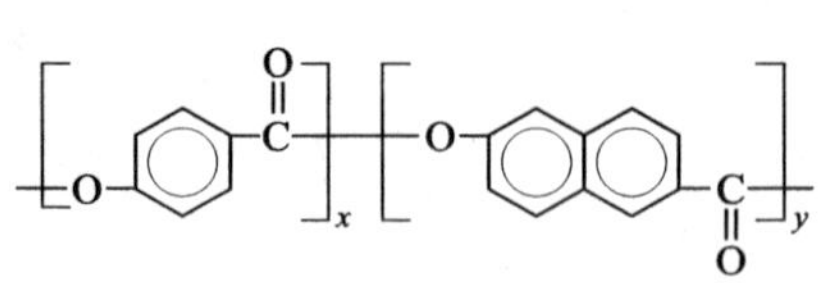

Figure 9.7 gives DSC thermograms for the 73/27 HBA/HNA copolyester during heating and cooling cycles, where 73/27 refers to the mole percent of the constituent components. The literature reports that this polymer has a melting temperature (T_m) of approximately 285 °C and undergoes an N–I transition at 370–380 °C. From Figure 9.7a we can confirm the reported T_m at approximately 285 °C, but we cannot identify an N–I transition temperature (T_{NI}) for the polymer. However, the polymer may undergo thermal

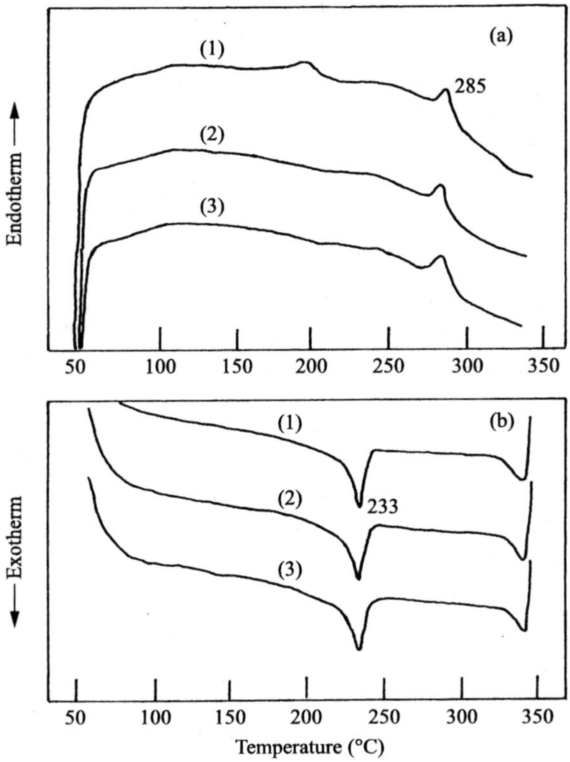

Figure 9.7 DSC thermograms for compression-molded 73/27 HBA/HNA copolyester specimens during (a) the heating cycle and (b) the cooling cycle at a rate of 20 °C/min, where curve (1) is for the first heating or cooling cycle, curve (2) is for the second heating or cooling cycle, and curve (3) is for the third heating or cooling cycle.

degradation before reaching 370–380 °C. As will be shown later in this chapter, inaccessibility to the clearing temperature of HBA/PET and HBA/HNA copolyesters poses a serious problem when attempting to obtain reproducible rheological measurements.

One can, however, synthesize thermally stable, semiflexible TLCPs that have a relatively low T_{NI}, certainly much lower than the thermal degradation temperature (Ober et al. 1984). One such example is poly[(phenylsulfonyl)-*p*-phenylene-1, 10-decamethylene)-bis(4-oxybenzoate) (PSHQ10), with the chemical structure (Furukawa and Lenz 1986; Kim and Han 1993a):

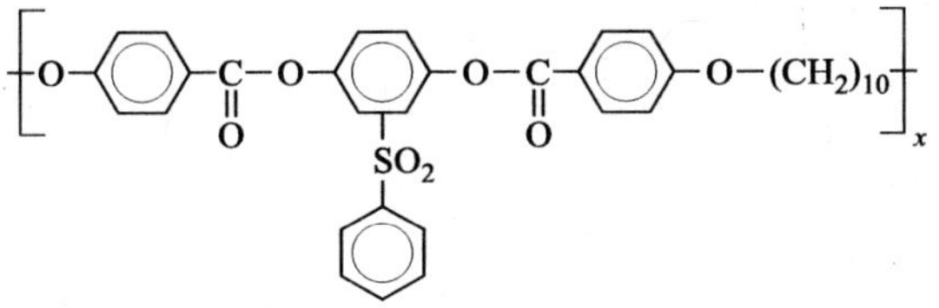

PSHQ10

Figure 9.8 shows DSC thermograms for PSHQ10 specimens that were first heated to 200 °C and then cooled down to 160 °C for isothermal annealing at various time periods, as indicated on the DSC thermograms. It can be seen in Figure 9.8 that PSHQ10 undergoes glass transition at 77–84 °C, melting transition at 107–111 °C, and N–I transition at 171–175 °C. The proof that PSHQ10 undergoes only N–I transition over a

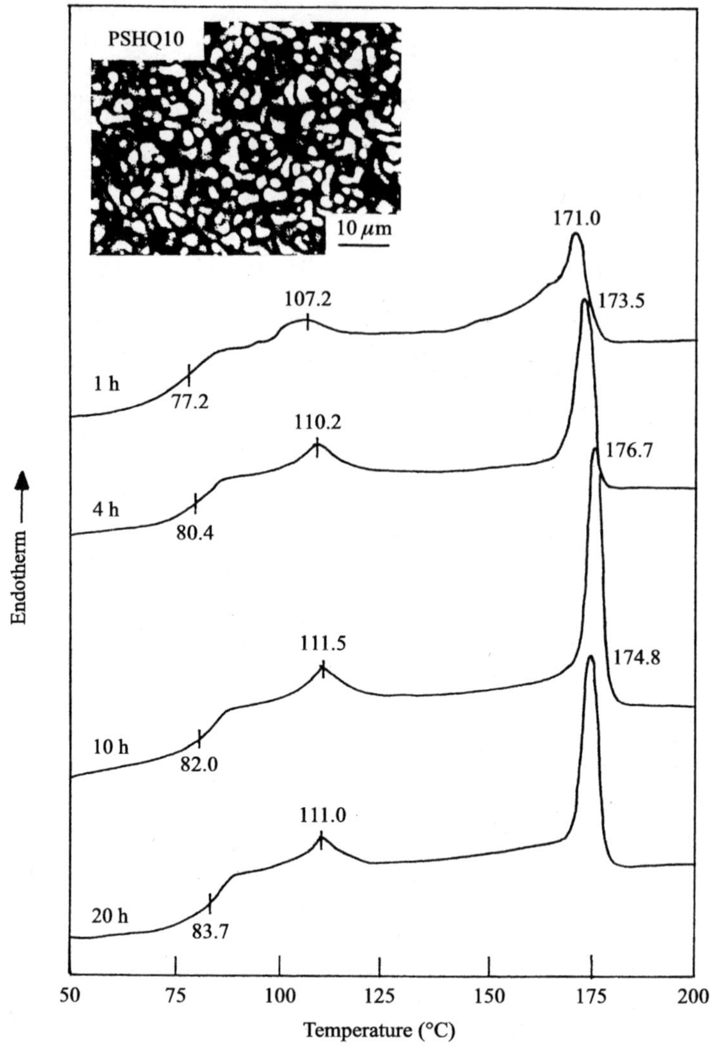

Figure 9.8 DSC thermograms for as-cast PSHQ10 specimens annealed at 160 °C for different periods as indicated on the plot, for which a fresh specimen was used for each run and the heating rate used was 20 °C/min. The inset gives a POM image of PSHQ10 taken at 160 °C in the nematic region, showing Schlieren texture. (Reprinted from Kim and Han, *Macromolecules* 26:3176. Copyright © 1993, with permission from the American Chemical Society.)

very wide range of temperatures is manifested by the image taken using polarized optical microscopy (POM) given in the inset of Figure 9.8. PSHQ10 is reported to undergo thermal degradation at approximately 350 °C (Kim and Han 1993a), suggesting that it can be used as a model compound for rheological measurements. Indeed, Han and coworkers (Han and Kim 1994a, 1994b; Han et al. 1994a; Kim and Han 1993a, 1993b, 1993c, 1994a, 1994b) conducted an extensive rheological investigation of PSHQ10 in both the nematic and isotropic regions. We present those results later in this chapter.

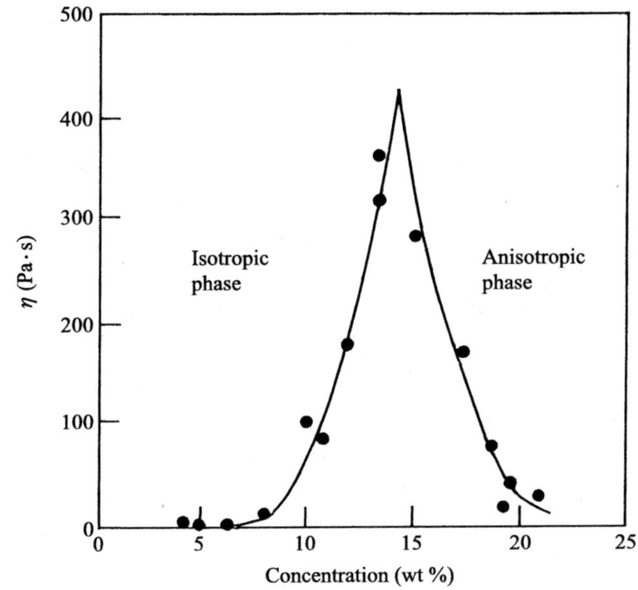

Figure 9.9 Plots of shear viscosity (η) versus concentration of poly(1,4-benzamide) in hydrofluoric acid at 0 °C. (Reprinted from Morgan, *Macromolecules* 10:1381. Copyright © 1977, with permission from the American Chemical Society.)

It should be clear that the formation and the consequent rheological properties of lyotropic LCP are dependent upon the concentration of the solution, while the formation and the consequent rheological properties of TLCP are dependent upon the temperature of the melt. To illustrate this point clearly, let us compare the concentration dependence of shear viscosity (η) of poly(1,4-benzamide) in hydrofluoric acid at 0 °C given in Figure 9.9 with the temperature dependence of η of PSHQ10 given in Figure 9.10. In Figure 9.9 we observe that η first increases with increasing concentration in the isotropic state, goes through a maximum at a certain critical concentration, and then decreases with a further increase in concentration. In Figure 9.10 we observe that η first decreases with increasing temperature, goes through a minimum followed by an increase with a further increase in temperature, goes through a maximum at a certain critical temperature, and finally decreases again with a further increase in temperature. The maximum value of η at a certain critical concentration observed in the lyotropic LCP (Figure 9.9) signifies the onset of phase transition, from the isotropic state to an anisotropic state, and the maximum value of η at a certain critical temperature in the TLCP melt (Figure 9.10) signifies the onset of phase transition, from an anisotropic state to the isotropic state. The decrease in η with increasing concentration above the critical value (Figure 9.9) and the decrease in η with increasing temperature in the nematic region of PSHQ10 (Figure 9.10) are attributable to the orientation of the mesogenic groups under the influence of applied shear flow. Referring to Figure 9.10, the increase in η, after passing a minimum, in the nematic region at approximately 162–178 °C is attributable to the existence of a broad molecular weight distribution of PSHQ10.

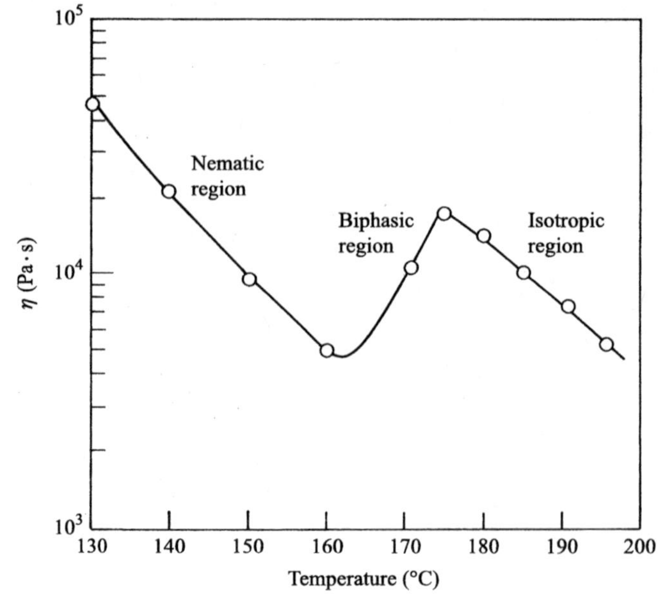

Figure 9.10 Plot of log η versus temperature for as-cast PSHQ10 specimens at $\dot{\gamma} = 0.01\ s^{-1}$. (Reprinted from Kim and Han, *Journal of Polymer Science, Polymer Physics Edition* 32:371. Copyright © 1994, with permission from John Wiley & Sons.)

This is because during heating the low-molecular-weight portion of the PSHQ10 starts to lose liquid crystallinity sooner than the high-molecular-weight portion. The PSHQ10 has a polydispersity index of about 2 (Kim and Han 1993a). It should be mentioned that the clearing temperature of TLCP depends on molecular weight below a certain critical value (Blumstein et al. 1982, 1984; Kim and Han 1993b, Laus et al. 1992; Majnusz et al. 1983; Percec et al. 1989). A TCLP with a narrow molecular weight distribution is expected to give rise to a sharp increase and decrease in η in the vicinity of the clearing temperature. There are many unusual rheological characteristics in LCPs, which are not observed in homopolymers presented in Chapter 6, nor in miscible polymer blends presented in Chapter 7. There are some similarities and also dissimilarities in rheological behaviors of LCPs and block copolymers presented in the preceding chapter.

In this chapter, we first present the currently held theories predicting the rheological behavior of monodomains and polydomains of rigid rodlike macromolecules, and we then present experimental observations of the rheological responses of lyotropic and thermotropic LCPs in transient shear flow, steady-state shear flow, oscillatory shear flow, intermittent shear flow, and reversal flow. The effects of molecular weight on the rheological behavior of semiflexible main-chain TLCPs will also be presented. The main thrusts of this chapter are to present the unique rheological characteristics of LCPs and to highlight the urgent need for further theoretical development that will enable one to predict the seemingly complicated rheological behavior of LCPs in terms of relevant molecular parameters.

9.2 Theory for the Rheology of LCPs

LCPs can occur as polydomain materials with domain sizes on the order of tens of microns. The transition from polydomain to monodomain material only involves the removal of grain boundaries or disclinations, as schematically shown in Figure 9.11, in which the arrows denote the direction of the preferred orientation by a vector **n**, called the "director." Polydomains appear to be the rule rather than the exception in LCPs. Flow processes are found to alter the structure of polydomains in more than one way. There is evidence that the flow may increase the defects or discontinuities, and that it may also decrease them to form a monodomain. As will be seen in the following sections, the director is introduced to describe the orientation of the anisotropic axis in LCPs. This vector may be regarded as an independent kinematic variable, in the sense that under certain circumstances it may vary independently of other kinematic variables, although it is intimately linked to them through the equations of motion. The alignment of directors can readily be achieved by the application of an external field: magnetic or stress. There is an intimate coupling between fluid motion and the orientation of the anisotropic axis of the material. In general, flow induces a change in orientation of the anisotropic axis, and thus affects the motion. Conversely, a change in orientation in most circumstances leads to flow, which tends to counteract or reinforce the change in alignment. In this section, we seek to elucidate the underlying fundamental principles that govern the rheological behavior of LCPs from a theoretical point of view. For this, we present the currently held theories for the dynamics of (1) rigid rodlike macromolecules with monodomains and (2) rigid rodlike macromolecules with polydomains.

9.2.1 Theory for Rigid Rodlike Macromolecules with Monodomains

Here, we present the molecular theory of Doi and coworkers (Doi 1981, 1983; Doi and Edwards 1978a, 1978b; Kuzuu and Doi 1980, 1983, 1984) for predicting the rheological behavior of concentrated solutions of monodomains consisting of rigid rodlike macromolecules.

9.2.1.1 Kinetic Equation

While considering the shear flow of a concentrated solution of rigid rods, let us restrict our attention to the nematic phase, where the orientation of a rod is determined by

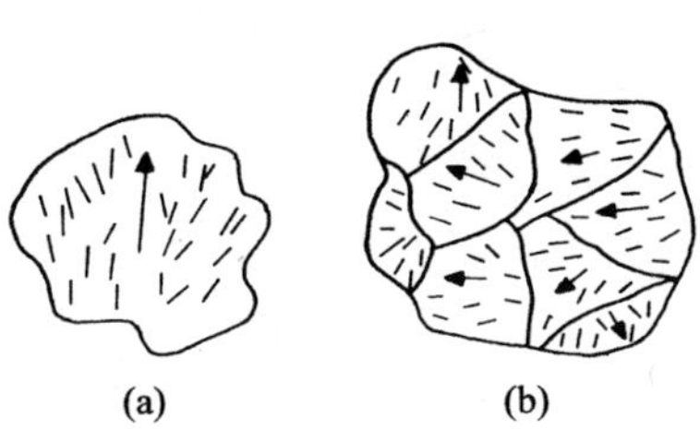

Figure 9.11 Schematic showing (a) a monodomain specimen with uniform director orientation **n** and (b) a polydomain specimen with varying director orientation **n**, where the arrows indicate the directors. The long-range order in each of the domains in (b) is similar to that in (a).

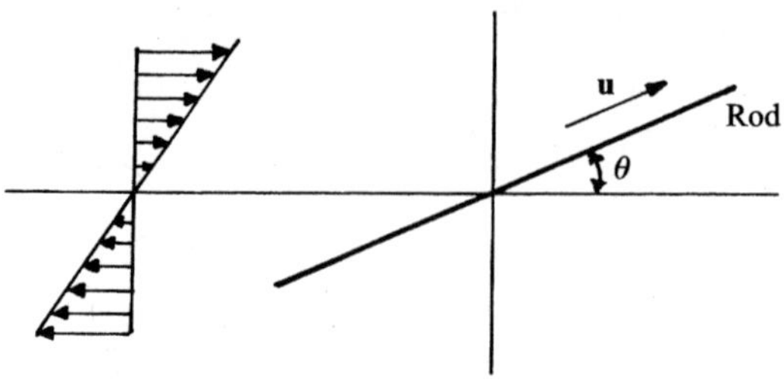

Figure 9.12 Orientation of a rod determined by the angle θ between the rod axis and the shear direction. (Reprinted from Marrucci and Maffettone, *Macromolecules* 22:4076. Copyright © 1989, with permission from the American Chemical Society.)

the angle θ between the rod axis and the shear direction, as schematically shown in Figure 9.12. Doi (1981) formulated the equation describing the dynamics of the rod population; that is, the equation describing the rate of change of the orientational probability or distribution function, $f(\mathbf{u}, t)$, due to: (1) Brownian motion (rotational diffusion), (2) the interaction among neighboring rods (intermolecular interactions represented by the mean-field potential), and (3) a velocity gradient (i.e., convective motion):

$$\frac{\partial f(\mathbf{u}, t)}{\partial t} = \nabla_{\mathbf{u}} \cdot \widetilde{D}_{\mathrm{r}} \left[\nabla_{\mathbf{u}} f(\mathbf{u}, t) + \frac{f(\mathbf{u}, t)}{k_{\mathrm{B}} T} \nabla_{\mathbf{u}} (V(\mathbf{u})) \right] - \nabla_{\mathbf{u}} \cdot \dot{\mathbf{u}} f(\mathbf{u}, t) \tag{9.1}$$

where $\mathbf{u}$ denotes the orientation of a single rod, $\widetilde{D}_{\mathrm{r}}$ is the rotary diffusion coefficient of the rod, k_{B} is the Boltzmann constant, and T is the absolute temperature. $V(\mathbf{u})$ is the total potential and consists of two parts: (1) the excluded-volume potential $V_{\mathrm{scf}}(\mathbf{u})$ of mean molecular field (i.e., intermolecular potential) and (2) the potential $V_{\mathrm{ext}}(\mathbf{u})$ of external fields (e.g., magnetic field) acting on the rod: $V(\mathbf{u}) = V_{\mathrm{scf}}(\mathbf{u}) + V_{\mathrm{ext}}(\mathbf{u})$. $\nabla_{\mathbf{u}}$ is the gradient operator on the unit sphere in the space of the unit vectors $\mathbf{u}$:

$$\nabla_{\mathbf{u}} = \frac{\partial}{\partial \mathbf{u}} = \boldsymbol{\delta}_{\theta} \frac{\partial}{\partial \theta} + \boldsymbol{\delta}_{\phi} \frac{1}{\sin\phi} \frac{\partial}{\partial \theta} \tag{9.2}$$

where $\boldsymbol{\delta}_{\theta}$ and $\boldsymbol{\delta}_{\phi}$ are the unit vectors on the sphere in the polar and azimuthal angles, respectively. Note that for a homogeneous flow, $\dot{\mathbf{u}}$ appearing in Eq. (9.1) is given by (see Appendix 9A):

$$\dot{\mathbf{u}} = \mathbf{L} \cdot \mathbf{u} - (\mathbf{u} \cdot \mathbf{L} \cdot \mathbf{u})\, \mathbf{u} = \mathbf{L} \cdot \mathbf{u} - \mathbf{L} : \mathbf{uuu} \tag{9.3}$$

where $\mathbf{L}$ is the velocity gradient tensor (see Chapter 2). If $f(\mathbf{u}, t)$ of the rodlike molecules is known, it can be used to calculate macroscopic variables, such as the stress tensor, which are averages of $\mathbf{u}$-dependent quantities.

The difference between Eq. (9.1) and the analogous equation for flexible homopolymers (Eq. (4.69)) is in the orientation-dependent rotary diffusivity $\widetilde{D}_{\mathrm{r}}$ which depends on the anisotropy of the system, as expressed by (Doi and Edwards 1978a, 1978b):

$$\frac{\widetilde{D}_{\mathrm{r}}(\mathbf{u}; [f])}{D_{\mathrm{r}}} = \frac{1}{\left[(4/\pi) \int d^2\mathbf{u}' f(\mathbf{u}', t) \sin(\mathbf{u}, \mathbf{u}') \right]^2} \tag{9.4}$$

where $(\mathbf{u}, \mathbf{u}')$ denotes the angle between the test rod and a neighboring rod with orientation vector $\mathbf{u}$, and D_r is the rotary diffusivity for a quiescent concentrated solution, which is related to D_{ro}, the diffusivity for a single isolated rod, by the relation

$$D_r = \frac{\beta D_{ro}}{\left(cL^3\right)^2} \tag{9.5}$$

where β is an undetermined numerical constant, c is the rod concentration (the number of rods per unit volume), and D_{ro} is expressed in terms of the rod geometry by

$$D_{ro} = \frac{3k_B T \ln(L/d)}{\pi \eta_s L^3} \tag{9.6}$$

in which η_s is the solvent viscosity, L is the length of the rod, and d is the diameter of the rod. Note in Eq. (9.4) that $\tilde{D}_r$ accounts for the hindered rotation of the test rod resulting from the close proximity of neighboring rods and is very small in concentrated solutions. Notice further in Eq. (9.5) that D_r decreases rapidly with increasing rod concentration ($D_r \propto c^{-2}$), and even more so with increasing rod length ($D_r \propto L^{-9}$). Once the time-dependent kinetic equation, Eq. (9.1), is solved with the appropriate initial conditions, the stress tensor $\boldsymbol{\sigma}$ can be determined.

9.2.1.2 Stress Expression and Constitutive Equation

Note that on the microscopic level, nematics are characterized by the fact that the equilibrium orientation distribution function $f(\mathbf{u})$ for the molecules in the directions $\mathbf{u}$ is not isotropic. The anisotropy is represented by the order parameter tensor $\mathbf{S}$ defined as

$$\mathbf{S} = \left\langle \mathbf{uu} - \tfrac{1}{3}\boldsymbol{\delta} \right\rangle \tag{9.7}$$

where $\boldsymbol{\delta}$ is the unit tensor and $\langle \cdots \rangle$ denotes the average over the orientation distribution function $f(\mathbf{u})$:

$$\langle \cdots \rangle = \int (\cdots)\, \mathrm{d}\mathbf{u} = \int_0^{2\pi} \int_0^{\pi} (\cdots) \sin\theta \, \mathrm{d}\theta \, \mathrm{d}\phi \tag{9.8}$$

At equilibrium, $\mathbf{S}$ depends only on the director $\mathbf{n}$, which is given by

$$\mathbf{S} = S\left(\mathbf{nn} - \tfrac{1}{3}\boldsymbol{\delta}\right) \tag{9.9}$$

where S is a scalar order parameter. From Eqs. (9.7) and (9.9), we obtain[1]

$$S = \tfrac{3}{2}\left\langle (\mathbf{u} \cdot \mathbf{n})^2 - \tfrac{1}{3} \right\rangle \tag{9.10}$$

Note that S is zero for random alignment (i.e., in the isotropic phase), and it is unity for perfect alignment. In terms of $\mathbf{S}$ defined by Eq. (9.7), the elastic stress tensor $\boldsymbol{\sigma}^{(\mathrm{E})}$ may be written as (see Appendix 9B)

$$\boldsymbol{\sigma}^{(\mathrm{E})} = 3k_{\mathrm{B}}T\mathbf{S} + \nu\langle\boldsymbol{\nabla}_{\mathbf{u}}V(\mathbf{u})\mathbf{u}\rangle \tag{9.11}$$

where ν is the number of rods in volume V and $V(\mathbf{u})$ is the total potential. The stress tensor $\boldsymbol{\sigma}$ can be divided into three parts, namely, the stress tensor due to elasticity $\boldsymbol{\sigma}^{(\mathrm{E})}$, the stress tensor due to viscosity $\boldsymbol{\sigma}^{(\mathrm{V})}$, and the stress tensor due to solvent $\boldsymbol{\sigma}^{(\mathrm{S})}$:

$$\boldsymbol{\sigma} = \boldsymbol{\sigma}^{(\mathrm{E})} + \boldsymbol{\sigma}^{(\mathrm{V})} + \boldsymbol{\sigma}^{(\mathrm{S})} \tag{9.12}$$

In order to calculate the stresses, one must first solve Eq. (9.1) for $f(\mathbf{u}, t)$ and then perform the average defined by Eq. (9.8).

In solving Eq. (9.1), $V(\mathbf{u})$ consisting of $V_{\mathrm{scf}}(\mathbf{u})$ and $V_{\mathrm{ext}}(\mathbf{u})$ must be specified. Two different expressions for $V_{\mathrm{scf}}(\mathbf{u})$ have been widely used. The Onsager potential (Onsager 1949), given by

$$V_{\mathrm{scf}}(\mathbf{u}) = Uk_{\mathrm{B}}T\int \mathrm{d}^2\mathbf{u}' f(\mathbf{u}') \sin(\mathbf{u}, \mathbf{u}') \tag{9.13}$$

is considered to be useful to describe the intermolecular interactions for dilute lyotropic solutions, where U (representing the strength of the potential) is expected to depend on the rod concentration, the integral is taken over the surface of a unit sphere in orientation space, and $(\mathbf{u}, \mathbf{u}')$ denotes the angle θ formed by the unit vectors $\mathbf{u}$ and $\mathbf{u}'$. The Maier–Saupe potential (Maier and Saupe 1958, 1959), given by

$$V_{\mathrm{scf}}(\mathbf{u}) = -\tfrac{3}{2}Uk_{\mathrm{B}}T\,\mathbf{S}:\mathbf{uu} \tag{9.14}$$

is considered to be useful to describe the intermolecular interactions for small molecules of rodlike liquid crystals. Larson (1988) has shown that the rods in the nematic phase under the Maier–Saupe potential are less aligned than under the Onsager potential, and therefore the order parameter S in the Maier–Saupe potential is less than that in the Onsager potential. Specifically, at the I–N transition, $S = 0.44$ for the Maier–Saupe potential and $S = 0.84$ for the Onsager potential. Further discussion of the order parameter S in LCPs will be deferred to the next section, where experimental results are presented.

When $V_{\mathrm{ext}}(\mathbf{u}) \ll V_{\mathrm{scf}}(\mathbf{u})$ and $V_{\mathrm{scf}}(\mathbf{u})$ is described by Eq. (9.14), $\boldsymbol{\sigma}^{(\mathrm{E})}$, $\boldsymbol{\sigma}^{(\mathrm{V})}$, and $\boldsymbol{\sigma}^{(\mathrm{S})}$ can be written as (see Appendix 9C)

$$\boldsymbol{\sigma}^{(\mathrm{E})} = 3\nu k_{\mathrm{B}}T\big[\mathbf{S} - U\big(\mathbf{S}\cdot\langle\mathbf{uu}\rangle - \mathbf{S}:\langle\mathbf{uuuu}\rangle\big)\big] \tag{9.15}$$

$$\boldsymbol{\sigma}^{(\mathrm{V})} = (1/2)\,\nu\xi_{\mathrm{r}}\mathbf{L}:\langle\mathbf{uuuu}\rangle \tag{9.16}$$

$$\boldsymbol{\sigma}^{(\mathrm{S})} = \eta_{\mathrm{s}}(\mathbf{L} + \mathbf{L}^{\mathrm{T}}) \tag{9.17}$$

where ξ_r is the rotational friction coefficient, η_s is the solvent viscosity, and $\mathbf{L}^T$ is the transpose of the velocity gradient tensor $\mathbf{L}$. According to Doi and Edwards (1986), ξ_r can be determined from

$$\xi_r = \frac{\pi \eta_s L^3}{3 \ln(L/d)} \tag{9.18}$$

where L is the length and d is the diameter of a rod.

Returning to the solution of Eq. (9.1), for simplification Doi and Edwards (1978b) replaced the orientation-dependent rotary diffusivity $\widetilde{D}_r$ (see Eq. (9.4)) with an averaged orientation-independent quantity $\overline{D}_r$:

$$\begin{aligned} \overline{D}_r &= \int d^2\mathbf{u} \widetilde{D}_r(\mathbf{u}; [f]) f(\mathbf{u}, t) \\ &= \frac{D_r}{\left[(4/\pi)\int d^2\mathbf{u} \int d^2\mathbf{u}' f(\mathbf{u}', t) f(\mathbf{u}, t) \sin(\mathbf{u}, \mathbf{u}')\right]^2} \end{aligned} \tag{9.19}$$

where D_r is given by Eq. (9.5). Using the spherical harmonic function $Y_\ell^m(\mathbf{u})$, and its complex conjugate $Y_\ell^{m*}(\mathbf{u})$, $\sin(\mathbf{u}, \mathbf{u}')$ appearing in Eq. (9.19) can be expressed by (Doi and Edwards 1978b)

$$\sin(\mathbf{u}, \mathbf{u}') = \frac{\pi}{4} - 2\pi^2 \sum_{\substack{\ell=2 \\ \text{even } \ell}}^{\infty} \left(\frac{\ell - 1}{\ell + 2}\right) \left[\frac{(\ell - 3)!!}{\ell!!}\right]^2 \sum_{m=-\ell}^{\ell} Y_\ell^{m*}(\mathbf{u}) Y_\ell^m(\mathbf{u}') \tag{9.20}$$

In simplifying the matter further, Doi (1981) approximated Eq. (9.19) by[2]

$$\frac{\overline{D}_r}{D_r} \approx \frac{1}{\left[1 - \frac{3}{2}(\mathbf{S}:\mathbf{S})\right]^2} \tag{9.21}$$

which indicates that $\overline{D}_r$ increases as the rods become aligned.

When $V_{ext}(\mathbf{u}) \ll V_{scf}(\mathbf{u})$ and $V_{scf}(\mathbf{u})$ is described by the Maier–Saupe potential, and by substituting Eq. (9.14) into (9.1) and replacing $\widetilde{D}_r$ by $\overline{D}_r$ defined by Eq. (9.21), we obtain

$$\frac{\partial f}{\partial t} = \overline{D}_r \nabla_\mathbf{u} \cdot \left[\nabla_\mathbf{u} f - \tfrac{3}{2} U f \nabla_\mathbf{u} (\mathbf{S}:\mathbf{uu})\right] - \nabla_\mathbf{u} \cdot \dot{\mathbf{u}} f \tag{9.22}$$

Now, multiplying both sides of Eq. (9.22) by $\left(\mathbf{uu} - \frac{1}{3}\boldsymbol{\delta}\right)$ and integrating over $\mathbf{u}$, we obtain (see Appendix 9D)

$$\frac{\partial \mathbf{S}}{\partial t} = \mathbf{F}(\mathbf{S}) + \mathbf{G}(\mathbf{S}) \tag{9.23}$$

where

$$\mathbf{F}(\mathbf{S}) = -6\overline{D}_\mathrm{r}\mathbf{S} + 6\overline{D}_\mathrm{r}U\big(\mathbf{S}\cdot\langle\mathbf{uu}\rangle - \mathbf{S}:\langle\mathbf{uuuu}\rangle\big) \tag{9.24}$$

$$\mathbf{G}(\mathbf{S}) = \tfrac{1}{3}(\mathbf{L} + \mathbf{L}^\mathrm{T}) + \mathbf{L}\cdot\mathbf{S} + \mathbf{S}\cdot\mathbf{L}^\mathrm{T} - 2\mathbf{L}:\langle\mathbf{uuuu}\rangle \tag{9.25}$$

At equilibrium, $\mathbf{F}(\mathbf{S}) = 0$. In this case, $\mathbf{F}(\mathbf{S})$ is a measure of the departure of the system from its equilibrium orientational distribution function. The farther the system is from equilibrium, the greater the resulting driving force (stress) to return the system to equilibrium. The tensor $\mathbf{G}(\mathbf{S})$ is a term describing the effect of the externally applied deformation on the rotational dynamics of the system.

Substitution of Eqs. (9.15)–(9.17) into (9.12) gives

$$\boldsymbol{\sigma} = 3\nu k_\mathrm{B}\mathrm{T}\Big[\mathbf{S} - U\big(\mathbf{S}\cdot\langle\mathbf{uu}\rangle - \mathbf{S}:\langle\mathbf{uuuu}\rangle\big)\Big] + (1/2)\zeta_\mathrm{r}\mathbf{L}:\langle\mathbf{uuuu}\rangle + \eta_\mathrm{s}(\mathbf{L}+\mathbf{L}^\mathrm{T}) \tag{9.26}$$

The set of Eqs. (9.23)–(9.26) provides us with an explicit form of the constitutive equation. Once the velocity gradient tensor $\mathbf{L}(t)$ is given, Eq. (9.23) enables us to calculate the order parameter tensor $\mathbf{S}(t)$ and then the stresses can be calculated from Eq. (9.26).

9.2.1.3 Prediction of the Viscoelastic Properties of LCPs

We will first use the decoupling approximation of the fourth-order tensor $\langle\mathbf{uuuu}\rangle$ by[3]

$$\langle\mathbf{uuuu}\rangle \approx \langle\mathbf{uu}\rangle\langle\mathbf{uu}\rangle = \big(\mathbf{S} + \tfrac{1}{3}\boldsymbol{\delta}\big):\big(\mathbf{S} + \tfrac{1}{3}\boldsymbol{\delta}\big) \tag{9.27}$$

and later without the decoupling approximation, in order to derive expressions for the viscoelastic properties of LCPs.

Using Eq. (9.27) in (9.23) and (9.26), we obtain (see Appendix 9E)

$$\begin{aligned}\frac{\partial\mathbf{S}}{\partial t} &= -6\overline{D}_\mathrm{r}\Big\{\big(1 - \tfrac{1}{3}U\big)\mathbf{S} - U\big(\mathbf{S}\cdot\mathbf{S} - \tfrac{1}{3}(\mathbf{S}:\mathbf{S})\boldsymbol{\delta}\big) + U(\mathbf{S}:\mathbf{S})\mathbf{S}\Big\}\\ &\quad + \tfrac{1}{3}(\mathbf{L} + \mathbf{L}^\mathrm{T}) + \mathbf{L}\cdot\mathbf{S} + \mathbf{S}\cdot\mathbf{L}^\mathrm{T} - \tfrac{2}{3}(\mathbf{L}:\mathbf{S})\boldsymbol{\delta} - 2(\mathbf{L}:\mathbf{S})\mathbf{S}\end{aligned} \tag{9.28}$$

$$\begin{aligned}\boldsymbol{\sigma} &= 3\nu k_\mathrm{B}T\Big\{\big(1 - \tfrac{1}{3}U\big)\mathbf{S} - U\big(\mathbf{S}\cdot\mathbf{S} - \tfrac{1}{3}(\mathbf{S}:\mathbf{S})\boldsymbol{\delta}\big) + U(\mathbf{S}:\mathbf{S})\mathbf{S}\Big\}\\ &\quad + (1/2)\nu\zeta_\mathrm{r}\Big\{\tfrac{1}{3}(\mathbf{L}:\mathbf{S})\boldsymbol{\delta} + (\mathbf{L}:\mathbf{S})\mathbf{S}\Big\} + \eta_\mathrm{s}(\mathbf{L} + \mathbf{L}^\mathrm{T})\end{aligned} \tag{9.29}$$

When the contributions from the viscous force and the solvent can be neglected, at steady state (i.e., $\partial \mathbf{S}/\partial t = 0$) Eq. (9.29) reduces to (see Appendix 9F)

$$\sigma_{\alpha\beta} = \frac{\nu k_B T}{2\overline{D}_r}\left\{\frac{(1-S)}{3}(L_{\alpha\beta} + L_{\beta\alpha}) + S(L_{\alpha\mu}n_\beta n_\mu + L_{\beta\mu}n_\alpha n_\mu) - 2S^2 L_{\mu\nu}n_\mu n_\nu n_\alpha n_\beta \right.$$
$$\left. + \frac{2}{3}S(S-1)\delta_{\alpha\beta}L_{\mu\nu}n_\mu n_\nu \right\} \tag{9.30}$$

in which S is the orientation order parameter defined by Eq. (9.10).

It can be shown that the steady-state shear flow properties can be predicted by (Doi 1981)[4]

$$\eta = \frac{\nu k_B T}{6\overline{D}_r}\left\{\frac{(1-S)^2(1+2S)(1+1.5S)}{(1+0.5S)^2}\right\} \tag{9.31}$$

$$N_1 = \frac{\nu k_B T}{2\overline{D}_r}\left\{\frac{S(1+2S)^{1/2}(1-S)^{3/2}(1+1.5S)}{(1+0.5S)^2}\right\}\dot{\gamma} \tag{9.32}$$

$$N_2 = \frac{\nu k_B T}{4\overline{D}_r}\left\{\frac{S^2(1+2S)^{1/2}(1-S)^{3/2}}{(1+0.5S)^2}\right\}\dot{\gamma} \tag{9.33}$$

Hence,

$$\frac{N_2}{N_1} = -\frac{S}{2(1+1.5S)} \tag{9.34}$$

in which $\overline{D}_r$ can be expressed by

$$\overline{D}_r = \frac{\beta D_{ro}}{(cL^3)^2(1-S^2)^2} \tag{9.35}$$

with β being a dimensionless coefficient, which follows from Eqs. (9.5) and (9.21). From Eqs. (9.31) and (9.32), with the aid of Eqs. (9.6) and (9.35), we observe that for $S = 0$, η and N_1 are both proportional to the 6th power of molecular weight M:

$$\eta \propto M^6; \quad N_1 \propto M^6 \tag{9.36}$$

In order to predict the nonlinear viscoelastic properties of concentrated solutions of rigid rodlike macromolecules, one must first solve Eq. (9.1) for the orientational distribution function $f(\mathbf{u}, t)$ and then Eq. (9.12) for stresses, for a given velocity field. Owing to the highly nonlinear nature of the system of equations, one must resort to numerical techniques to obtain solutions. Indeed, Kuzuu and Doi (1983, 1984) carried out numerical computations to predict nonlinear viscoelastic behavior.

Replacing $\widetilde{D}_r$ by an average quantity $\overline{D}_r$ defined by Eq. (9.21) and substituting Eq. (9.3) into (9.1) we have

$$\frac{\partial f(\mathbf{u}, t)}{\partial t} = \overline{D}_r \nabla_{\mathbf{u}} \cdot \left[\nabla_{\mathbf{u}} f(\mathbf{u}, t) + \frac{f(\mathbf{u}, t)}{k_B T} \nabla_{\mathbf{u}} \left(V_{scf}(\mathbf{u}) \right) \right] - \nabla_{\mathbf{u}} \cdot \left[(\mathbf{L} \cdot \mathbf{u} - \mathbf{L} : \mathbf{uuu}) f(\mathbf{u}, t) \right] \tag{9.37}$$

where $V(\mathbf{u}) \approx V_{scf}(\mathbf{u})$ is assumed.

The orientational distribution function $f(\mathbf{u}, t)$ can be expanded in terms of an infinite series of spherical harmonics $Y_\ell^m(\theta, \phi)$ (Doi and Edwards 1978b):

$$f(t; \theta, \phi) = \sum_{\substack{\ell=0 \\ \ell \text{ even}}}^{\infty} \sum_{m=0}^{\ell} b_{\ell m} \, |\ell m) \tag{9.38}$$

where

$$|\ell m) = Y_\ell^m(\theta, \phi) \quad \text{for } m = 0 \tag{9.39a}$$

$$|\ell m) = (1/\sqrt{2}) \left[Y_\ell^m(\theta, \phi) + (-1)^m Y_\ell^{-m}(\theta, \phi) \right] \quad \text{for } m \neq 0 \tag{9.39b}$$

Such an expansion reduces the partial differential equation, Eq. (9.37), into a set of nonlinear, ordinary first-order differential equations for the time-dependent coefficient $b_{\ell m}$ (Doi and Edwards 1978b):

$$\frac{db_{\ell m}}{dt} = \overline{D}_r \ell(\ell+1) b_{\ell m} + \frac{\overline{D}_r}{k_B T} \left\{ \ell m | \nabla_{\mathbf{u}} \cdot \left[f \nabla_{\mathbf{u}} \left(V_{scf}(\mathbf{u}) \right) \right] \right\} - \left\{ \ell m | \nabla_{\mathbf{u}} \cdot \left([\mathbf{L} \cdot \mathbf{u} - \mathbf{L} : \mathbf{uuu}] f \right) \right\} \tag{9.40}$$

where

$$\frac{D_r}{\overline{\overline{D}}_r} = \left\{ 1 - 8\pi \sum_{\substack{\ell=2 \\ \ell \text{ even}}}^{\infty} \sum_{m=0}^{\ell} \left(\frac{\ell+1}{\ell+2} \right) \left[\frac{(\ell-3)!!}{\ell!!} \right]^2 |b_{\ell m}|^2 \right\}^2 \tag{9.41}$$

in which the notation $\ell!!$ is defined by

$$\ell!! = \begin{cases} \ell(\ell-2)(\ell-4)\cdots 1, & \text{for } \ell \text{ odd} \\ \ell(\ell-2)(\ell-4)\cdots 2, & \text{for } \ell \text{ even} \\ 1 & \text{for } \ell \leq 1 \end{cases} \tag{9.42}$$

With the appropriate initial conditions $b_{\ell m}(0)$, Eq. (9.40), with the aid of Eq. (9.41), can be solved by truncating $b_{\ell m}$ at $\ell = \ell_{max}$ for a given expression for the potential $V_{scf}(\mathbf{u})$. The formula for the stress tensor defined by Eq. (9.11) can also be reduced to an expression involving the $b_{\ell m}$s (Doi and Edwards 1978b).

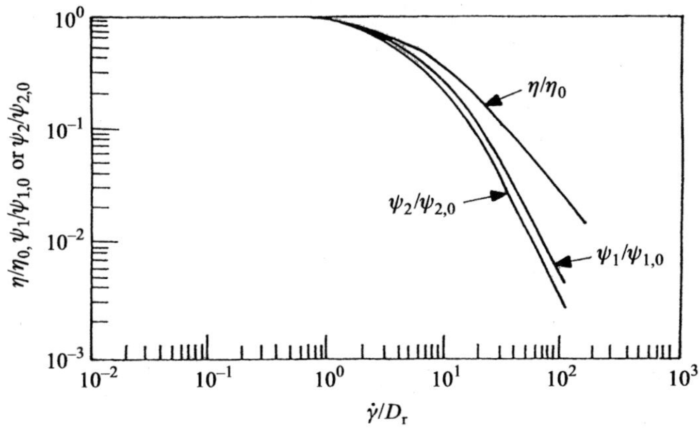

Figure 9.13 Logarithmic plots of normalized shear viscosity η/η_0 versus dimensionless shear rate $\dot{\gamma}/D_r$, normalized first normal stress difference coefficient $\Psi_1/\Psi_{1,0}$ versus $\dot{\gamma}/D_r$, and normalized second normal stress difference coefficient $\Psi_2/\Psi_{2,0}$ versus $\dot{\gamma}/D_r$ for concentrated lyotropic solutions of rodlike polymers, which are predicted by the Doi theory using the Maier–Saupe potential. Here, η_0 is zero-shear viscosity, $\Psi_{1,0}$ is first normal stress difference coefficient at $\dot{\gamma}=0$, and $\Psi_{2,0}$ is second normal stress difference coefficient at $\dot{\gamma}=0$. The plots are prepared based on Eqs. (9.43)–(9.45).

For dilute lyotropic solutions, the term containing $V_{scf}(\mathbf{u})$ in Eq. (9.37) may be neglected. Thus, neglecting the term containing $V_{scf}(\mathbf{u})$, Doi and Edwards (1978b) numerically solved Eq. (9.40), with the aid of Eq. (9.41), for steady-state shear flow. They found that the following expressions for $\sigma/3\nu k_B T$, $N_1/3\nu k_B T$, and $N_2/3\nu k_B T$ approximate with reasonable accuracy the numerical solutions:

$$\sigma/3\nu k_B T = (1/30)\Gamma(1+0.0710\Gamma^2)^{-0.525} \tag{9.43}$$

$$N_1/3\nu k_B T = (1/90)\Gamma^2(1+0.0558\Gamma^2)^{-0.874} \tag{9.44}$$

$$N_2/3\nu k_B T = -(1/315)\Gamma^2(1+0.0550\Gamma^2)^{-0.935} \tag{9.45}$$

where $\Gamma = \dot{\gamma}/D_r$. Figure 9.13 gives plots of three material functions versus $\dot{\gamma}/D_r$. Notice in Eq. (9.43) that at high shear rates the viscosity is roughly inversely proportional to $\dot{\gamma}$.

For concentrated lyotropic solutions and thermotropic melts, however, the term containing V_{scf} in Eq. (9.37) cannot be neglected. Of particular interest in this regard is the two-dimensional analysis of the Doi theory by Marrucci and Maffettone (1989). Due to the nature of the simplification made for the two-dimensional analog, they were able to obtain explicit expressions for the orientation distribution function $f(\mathbf{u})$ for steady-state shear flow without having to deal with the fourth-order tensor $\langle\mathbf{uuuu}\rangle$. They predicted that N_1 is positive at very low $\dot{\gamma}$ and then becomes negative as $\dot{\gamma}$ is increased (i.e., at an intermediate range of $\dot{\gamma}$), and it becomes positive again at very high $\dot{\gamma}$, which is in qualitative agreement with experimental observations (Baek et al. 1993b; Grizzuti et al. 1990; Kiss and Porter 1978, 1980a; Moldenaers and Mewis 1986), as will

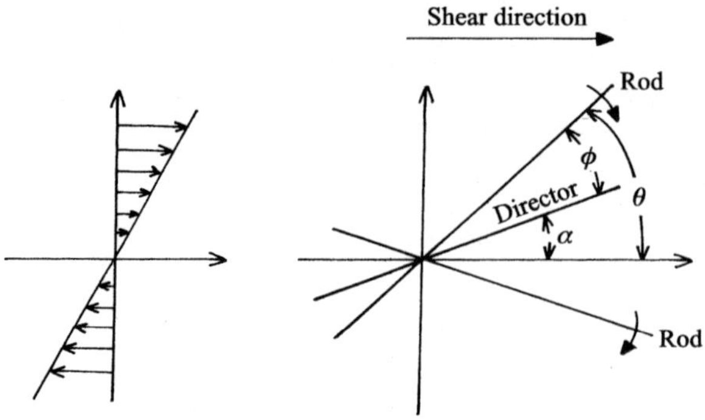

Figure 9.14 Schematic showing the angles that the rod and director, respectively, form with respect to the shear direction. (Reprinted from Marrucci and Maffettone, *Macromolecules* 22:4076. Copyright © 1989, with permission from the American Chemical Society.)

be presented later in this chapter. According to Marrucci and coworkers (Cocchini et al. 1990; Marrucci and Maffettone 1989), an oscillatory motion of the director **n** of the nematic phase is responsible for the occurrence of negative N_1 at an intermediate range of $\dot{\gamma}$, and the negative normal stress effect is attributable to an increased spread of rod orientation with respect to equilibrium. For the reason that very explicit expressions are available in the Marrucci–Maffettone analysis, we next present the expressions predicting time evolution of the director and transient behavior of σ and N_1, and then discuss some predictions made from the analysis.

Let us consider the situation, as schematically shown in Figure 9.14, in which all rods in the system are parallel to a plane and shear flow takes place in that plane. In such a situation of two-dimensional shear flow, the unit vector $\mathbf{u} = (\cos\theta, \sin\theta)$, with θ being the angle between the rod and the shear axis, can be used to specify the orientation of a rod in the plane. With reference to Figure 9.14, ϕ is the angle that a rod makes with the director, which in turn makes the angle α with the shear axis; that is,

$$\theta = \alpha + \phi \tag{9.46}$$

thus, α can be regarded as the director orientation, specified by the condition

$$\langle \sin 2\phi \rangle = 0 \tag{9.47}$$

For the shear flow under consideration, Eq. (9.37) can be rewritten as (Cocchini et al. 1990)[5]

$$\frac{\partial f}{\partial t} = \overline{D}_{\mathrm{r}} \frac{\partial}{\partial \theta}\left(\frac{\partial f}{\partial \theta} + \frac{f}{k_{\mathrm{B}}\mathrm{T}}\frac{\partial V_{\mathrm{scf}}(\theta)}{\partial \theta}\right) + \frac{\partial}{\partial \theta}(f\dot{\gamma}\sin^2\theta) \tag{9.48}$$

where $\dot{\gamma}$ is shear rate. Using the Maier–Saupe potential (given by Eq. (9.14)) in Eq. (9.48), with the aid of Eqs. (9.46) and (9.47), we obtain (see Appendix 9G)

$$\frac{\partial f}{\partial \tau} = \frac{\partial}{\partial \phi}\left(\frac{\partial f}{\partial \phi} + 2Uf\langle \cos 2\phi \rangle \langle \sin 2\phi \rangle\right) + \frac{\partial}{\partial \phi}\left(\dot{\alpha} f + \Gamma f \sin^2(\alpha + \phi)\right) \tag{9.49}$$

in which $\dot{\alpha} = d\alpha/d\tau$ represents the angular velocity of the director, $\tau = \overline{D}_r t$, and $\Gamma = \dot{\gamma}/\overline{D}_r$. By multiplying $\sin 2\phi$ on both sides of Eq. (9.49) and integrating over ϕ, we obtain (see Appendix 9H)

$$\frac{d\alpha}{d\tau} = \frac{\Gamma}{2}\left[\frac{\cos 2\alpha\langle\cos^2 2\phi\rangle}{\langle\cos 2\phi\rangle} - \frac{\sin 2\alpha\langle\sin 2\phi\cos 2\phi\rangle}{\langle\cos 2\phi\rangle} - 1\right] - 2U\langle\sin 2\phi\cos 2\phi\rangle \tag{9.50}$$

The solution of Eqs. (9.49) and (9.50) with the appropriate initial and boundary conditions gives time evolutions of the director $\alpha(t)$ and the orientational distribution function $f(t,\phi)$. The boundary conditions

$$f(0,t) = f(\pi,t); \quad (\partial f/\partial\phi)_{0,t} = (\partial f/\partial\phi)_{\pi,t} \tag{9.51}$$

and the initial condition

$$f(\phi,0) = \frac{\exp(a^*\cos 2\phi)}{\int_0^{\pi}\exp(a^*\cos 2\phi')d\phi'} \tag{9.52}$$

may be used, where the value of a^* must be determined from the following relationship:

$$\frac{a^*}{U} = \langle\cos 2\phi\rangle_0 = \frac{\int_0^{\pi}\cos 2\phi\exp(a^*\cos 2\phi)d\phi}{\int_0^{\pi}\exp(a^*\cos 2\phi')d\phi'} \tag{9.53}$$

for a given value of U; that is, a^* is an equilibrium quantity that depends on U (Cocchini et al. 1990).

Once the solution for $f(t,\phi)$ is obtained, the shear stress and the normal stress difference from the viscous and elastic contributions, respectively, can be calculated from the following expressions by taking the averages (Marrucci and Maffettone 1989):

$$\frac{\sigma^E}{ck_BT} = \langle\cos 2\phi\rangle\left[\sin 2\alpha\Big(1 - U\langle\sin^2 2\phi\rangle\Big) + \frac{U}{2}\cos 2\alpha\langle\sin 4\phi\rangle\right] \tag{9.54}$$

$$\frac{N_1^E}{ck_BT} = 2\langle\cos 2\phi\rangle\left[\cos 2\alpha\Big(1 - U\langle\sin^2 2\phi\rangle\Big) - \frac{U}{2}\sin 2\alpha\langle\sin 4\phi\rangle\right] \tag{9.55}$$

$$\frac{\sigma_1^V}{ck_BT} = B\frac{\Gamma}{2}\big[1 - \cos 4\alpha\langle\cos 4\phi\rangle + \sin 4\alpha\langle\sin 4\phi\rangle\big] \tag{9.56}$$

$$\frac{N_1^V}{ck_BT} = B\Gamma\big[\sin 4\alpha\langle\cos 4\phi\rangle + \cos 4\alpha\langle\sin 4\phi\rangle\big] \tag{9.57}$$

in which superscript E refers to elastic contribution and superscript V refers to viscous contribution. Note that B appearing in Eqs. (9.56) and (9.57) may be considered to describe the ratio of the diffusion coefficient in a concentrated solution of rods and the rotational diffusivity of rods in a dilute concentration. Thus the magnitude of B is expected to be very small because the rotation of rods in a concentrated solution

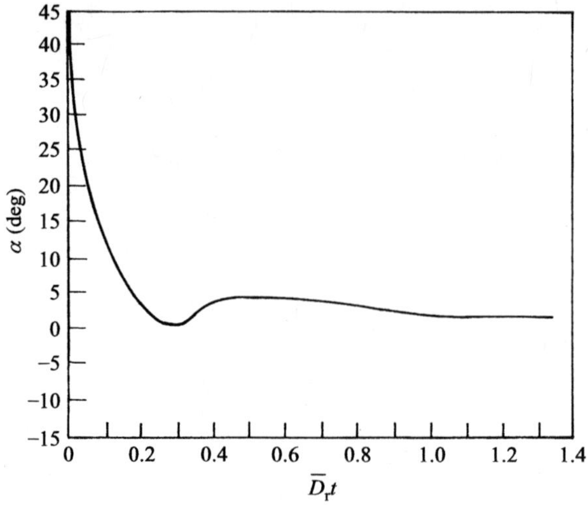

Figure 9.15 Plots of α versus $\overline{D}_r t$, describing the motion of the director in the nontumbling regime, for $U = 3.2$, $\Gamma = 30$, $\alpha_0 = 45°$, and $B = 0.01$, where $\Gamma = \dot{\gamma}/\overline{D}_r$ is dimensionless shear rate. The plots are prepared by numerical computations based on the analysis of Marrucci and Maffettone (1990).

will be hindered by the surrounding rods. This indicates that the Marrucci–Maffettone analysis is applicable to concentrated lyotropic LCPs but not to thermotropic melts. Note further that the theory is applicable to untextured lyotropic LCPs, as assumed in the Doi theory.

Marrucci and Maffettone (1990) carried out numerical integration of Eqs. (9.49) and (9.50) to obtain time evolution of the director. Figure 9.15 gives time evolution of α, indicating that the director attains steady state through damped oscillations, for $U = 3.2$, $\Gamma = 30$, $\alpha_0 = 45°$, and $B = 0.01$, where α_0 is the initial value of α at time zero (see Figure 9.14) and Γ is dimensionless shear rate $\dot{\gamma}/\overline{D}_r$. Figure 9.16 shows a

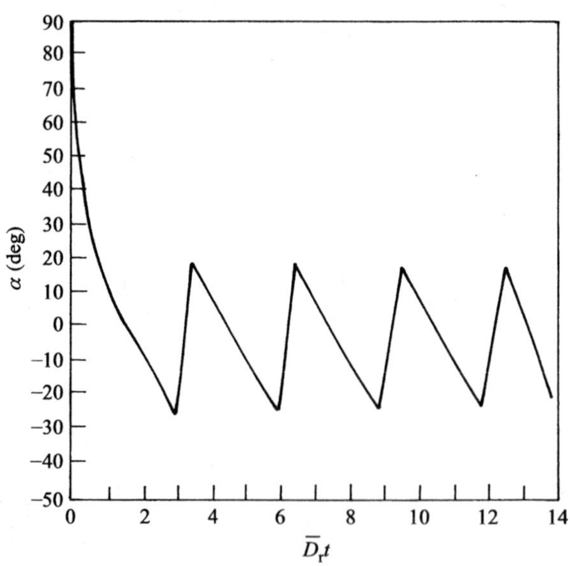

Figure 9.16 Plots of α versus $\overline{D}_r t$, describing the motion of the director in the "wagging" regime, for $U = 3.2$, $\Gamma = 4$, $\alpha_0 = 90°$, and $B = 0.01$, where $\Gamma = \dot{\gamma}/\overline{D}_r$ is dimensionless shear rate. The plots are prepared by numerical computations based on the analysis of Marrucci and Maffettone (1990).

permanent oscillation of α between two values when the value of Γ is reduced from 30 to 4. Larson (1990) referred to this type of director motion as "wagging." When the value of Γ is reduced further to 3, as can be seen in Figure 9.17, the director tumbles. What we have observed above is that the motion of the director in two-dimensional shear flow depends on the applied shear rate; namely, it tumbles at very low shear rates (where $N_1 > 0$), then exhibits permanent oscillations (or wagging) at an intermediate range of shear rates (where $N_1 < 0$), and finally attains steady state at higher shear rates (where $N_1 > 0$). The possibility of having director tumbling in steady-state shear flow of small molecules of liquid crystals has long been predicted by a number of research groups (Carlsson 1984; Carlsson and Skarp 1981; Cladis and Torza 1975; Currie and MacSithigh 1979; Manneville 1981; Pieranski and Guyon 1974), and confirmed by experiments (Burghardt and Fuller 1990, 1991).

Marrucci and Maffettone (1990) predicted the transient shear flow behavior of a polydomain by averaging the responses, which were obtained from Eqs. (9.54)–(9.57), of individual domains. In other words, a polydomain was assumed to consist of many monodomains, and thus Eqs. (9.49) and (9.50) were first solved for each monodomain using different values of initial condition α_0 (i.e., each domain was assumed to have a different director distribution at time zero) and then, in the course of time, the responses of the individual monodomains were averaged to obtain the response of a polydomain. The values of the distribution function $f(\tau, \phi)$ thus obtained were used to calculate the stresses by calculating the averages appearing in Eqs. (9.54)–(9.57).

Figure 9.18 gives plots of transient shear stress $\sigma^+(t, \dot{\gamma})$ versus shear strain ($\dot{\gamma}t$), and Figure 9.19 gives plots of transient first normal stress difference $N_1^+(t, \dot{\gamma})$ versus $\dot{\gamma}t$, for different values of Γ at a fixed value of $U = 3.2$. The following observations are worth noting in Figures 9.18 and 9.19: (1) the overshoot of $\sigma^+(t, \dot{\gamma})$ occurs at almost the same value of $\dot{\gamma}t$, regardless of the applied $\dot{\gamma}$, (2) the magnitude of $\sigma^+(t, \dot{\gamma})$ overshoot increases with increasing $\dot{\gamma}$, (3) shear stress attains steady state

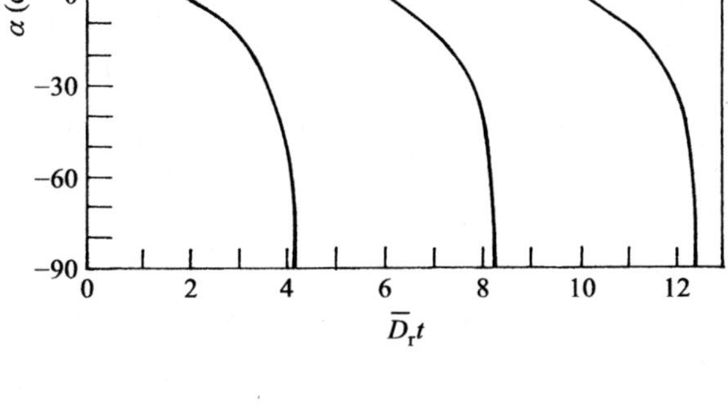

Figure 9.17 Plots of α versus $\overline{D}_r t$, describing the motion of the director in the tumbling regime, for $U = 3.2$, $\Gamma = 3$, $\alpha_0 = 90°$, and $B = 0.01$, where $\Gamma = \dot{\gamma}/\overline{D}_r$ is dimensionless shear rate. The plots are prepared by numerical computations based on the analysis of Marrucci and Maffettone (1990).

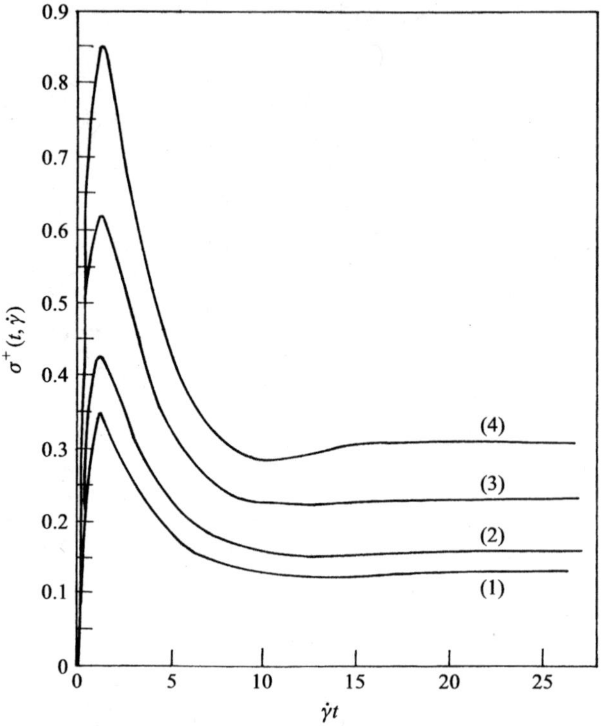

Figure 9.18 Plots of $\sigma^+(t, \dot{\gamma})$ versus $\dot{\gamma}t$ for $U = 3.2$ and various values of Γ: (1) $\Gamma = 20$, (2) $\Gamma = 30$, (3) $\Gamma = 60$, and (4) $\Gamma = 100$, with $\Gamma = \dot{\gamma}/\overline{D}_r$ being dimensionless shear rate. The plots are prepared by numerical computations based on the analysis of Marrucci and Maffettone (1990).

after one large overshoot, (4) the overshoot of $N_1^+(t, \dot{\gamma})$ occurs at different values of $\dot{\gamma}t$, depending on the applied $\dot{\gamma}$, (5) the number of overshoots of $N_1^+(t, \dot{\gamma})$ increases with increasing $\dot{\gamma}$, and (6) $N_1^+(t, \dot{\gamma})$ eventually reaches steady state with positive value as $\dot{\gamma}$ is increased. These observations are in qualitative agreement with the experimental results for lyotropic solutions (Chow and Fuller 1985; Doppert and Picken 1987; Grizzuti et al. 1990; Larson and Mead 1989; Mead and Larson 1990; Mewis and Moldenaers 1987) and semiflexible (segmented) TLCPs (Han and Kim 1994a; Han et al. 1994a; Kim and Han 1993b, 1993c). The qualitative agreement in transient shear flow behavior between the prediction (e.g., Figures 9.18 and 9.19) of the Marrucci–Maffettone analysis and the experiment results for semiflexible TLCPs (e.g., PSHQ10) in the literature must be viewed as being merely a trend for the following reasons. (1) The Marrucci–Maffettone analysis is based on the Doi theory, which assumes untextured lyotropic solutions having monodomains. Conversely, the semiflexible TLCPs employed in experiments have textures (e.g., nematic mesophase) and polydomains. Thus they cannot be regarded as being rigid rodlike molecules because they have flexible spacers. (2) The Marrucci–Maffettone analysis is based on the Maier–Saupe potential, which may be applicable to low-molecular-weight liquid crystals, whereas at present we have no theoretical expression that adequately describes the excluded-volume potential for semiflexible TLCPs. There will be further discussion of this subject later in this chapter.

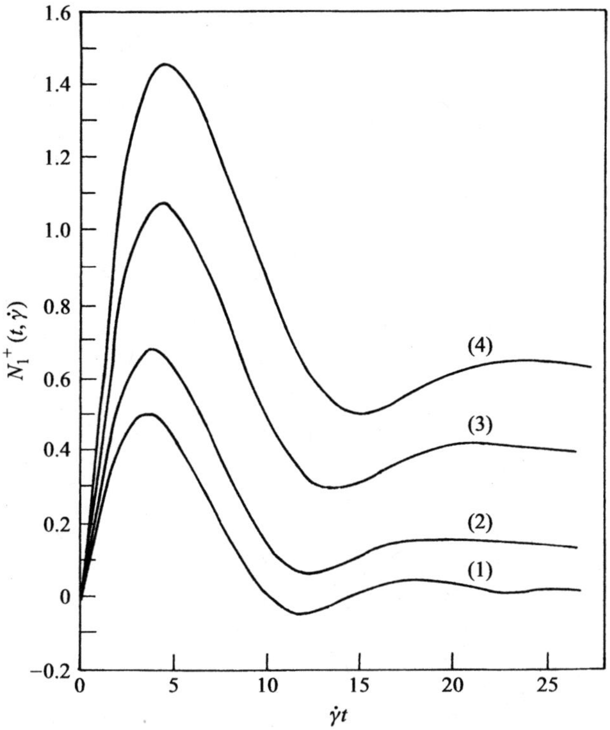

Figure 9.19 Plots of $N_1^+(t,\dot{\gamma})$ versus $\dot{\gamma}t$ for $U = 3.2$ and various values of Γ: (1) $\Gamma = 20$, (2) $\Gamma = 30$, (3) $\Gamma = 60$, and (4) $\Gamma = 100$ with $\Gamma = \dot{\gamma}/\overline{D}_r$ being dimensionless shear rate. The plots are prepared by numerical computations based on the analysis of Marrucci and Maffettone (1990).

Larson (1990) carried out a three-dimensional analysis by numerically solving Eq. (9.40) using the Onsager potential (see Eq. (9.13)) in terms of the spherical harmonic function:

$$V_{\mathrm{scf}} = \frac{\pi}{4}U - 2\pi^2 U \sum_{\substack{\ell=2 \\ \text{even } \ell}}^{\infty} \sum_{m=0}^{\ell} \left\{ \left(\frac{\ell-1}{\ell+2}\right) \left[\frac{(\ell-3)!!}{\ell!!}\right]^2 b_{\ell m}|\ell m) \right\} \tag{9.58}$$

Larson essentially confirmed the two-dimensional analysis of Marrucci and coworkers (Cocchini et al. 1990; Marrucci and Maffettone 1989, 1990). Figure 9.20 gives plots of dimensionless shear stress σ/ck_BT and dimensionless first normal stress difference $|N_1|/ck_BT$ versus $\dot{\gamma}/D_r$ for $U = 12$, where $U = 2cdL^2$ with c being the number of molecules per unit volume, d being the rod diameter, and L being the length of the rod. It can be seen in Figure 9.20 that N_1 is positive at low values of $\dot{\gamma}/D_r$, negative at intermediate values of $\dot{\gamma}/D_r$, and positive again at large values of $\dot{\gamma}/D_r$. The first change in sign of N_1 occurs in the tumbling regime, while the second change occurs at a value of $\dot{\gamma}/D_r$ for which a steady state exists. Larson (1990) noted that for $\dot{\gamma}/D_r$

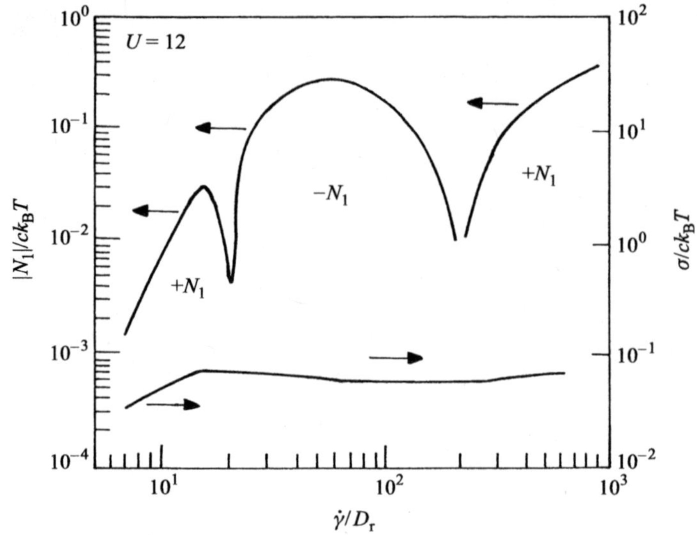

Figure 9.20 Logarithmic plots of dimensionless first normal stress difference $|N_1|/ck_BT$ versus dimensionless shear rate $\dot{\gamma}/D_r$ and dimensionless shear stress σ/ck_BT versus $\dot{\gamma}/D_r$ for lyotropic solutions of a rodlike polymer with the dimensionless concentration $U = 12$. Here, U is defined by $U = 2cdL^2$, where c is the number of molecules per unit volume, d is the rod diameter, and L is the length of the rod. (Reprinted from Larson, *Macromolecules* 23:3983. Copyright © 1990, with permission from the American Chemical Society.)

below a certain critical value, the stresses oscillate indefinitely, meaning that there is no steady state, and thus the values shown in Figure 9.20 are averages over an integral number of cycles of the oscillation in that region. It should be pointed out that the range of $\dot{\gamma}/D_r$ over which tumbling occurs would depend on the concentration U of the lyotropic solution of a rodlike polymer.

9.2.2 Theory for Rigid Rodlike Macromolecules with Polydomains

One of the severe restrictions imposed on the Doi theory presented in the preceding sections is the assumption that macromolecules are rigid rodlike and form monodomains, giving rise to uniform spatial distribution; that is, the orientation vector (director) is assumed to be spatially uniform. Thus, the Doi theory is valid for monodomains of untextured LCPs but not for polydomains, in which the director varies with position (see Figure 9.11) (i.e., spatial-dependent director). It should be emphasized that the Doi theory is an overly crude representation of LCPs, the molecules of which are often semirigid or contain flexible spacers between the rigid mesogenic groups (segmented chains). Thus, a theory describing the effects of textures on bulk rheological properties (i.e., texture-dominated bulk rheological properties) of LCPs is needed. In other words, it is important to understand the relationship between the bulk rheological behavior and the flow-induced textures. In this section, we present two currently held theories along this line. The phenomenological theory of Ericksen (1960) and Leslie

(1966, 1968, 1979) considers the spatial variation of the director, and it has been successful in describing the dynamic behavior of nematics of low-molecular-weight thermotropic liquid crystals. Although the Ericksen–Leslie theory does not predict non-linear viscoelastic behavior, which is prevalent in LCPs, it is very useful for explaining some of the experimental observations made with LCPs. It will be shown that the Doi theory, which is based on the molecular concept, connects the coefficients appearing in the Ericksen–Leslie theory to molecular parameters. A second theory, that of Larson and Doi (1991), which may also be regarded as being phenomenological, includes the presence of "polydomain" texture by using a scaling argument to relate the domain size to bulk rheological properties.

9.2.2.1 The Ericksen–Leslie Theory

When the director varies considerably throughout a sample, with the possible occurrence of discontinuities (defects or disclinations), the sample is said to have a "polydomain" structure. As a consequence, most macroscopically observable quantities, including the stresses, are in fact averages over the director spatial distribution in regions. The interplay between director and velocity gradient generates a viscous stress, which is characterized (at the phenomenological level) by five independent viscosities, known as "Leslie coefficients." The spatial distortion of the director also generates stresses, often referred to as "Frank" elastic (or distortional) stresses, named after Frank (1958), which of course bear no relationship to fluid elasticity in polymer rheology. Note that Frank elasticity can be neglected, at least in some cases for LCPs that have very high viscosities, since Frank elasticity remains essentially of the same order of magnitude as in the case of small molecules.

When limiting our attention to low-molecular-weight nematics, we may expect that, in general, flow has the following effects: (1) it alters the distribution of molecular orientations about the nematic axis (director) and (2) it affects the director itself. In other words, the velocity $\mathbf{v}(\mathbf{r})$ and the director $\mathbf{n}(\mathbf{r})$ are coupled under flow of nematic solutions. Next, we first present the expressions for stress, then discuss some important features of the Ericksen–Leslie theory, and finally show relationships existing between the six Leslie coefficients and three molecular parameters appearing in the Doi theory. The presentation of the entire Ericksen–Leslie theory (Ericksen 1960; Leslie 1966, 1968, 1979) is beyond the scope of this chapter.

Stress Expression Ericksen (1960) and Leslie (1966, 1969) derived the following expression:

$$\boldsymbol{\sigma} = \alpha_1 (\mathbf{n} \cdot \mathbf{d} \cdot \mathbf{n})\, \mathbf{nn} + \alpha_2 \mathbf{nN} + \alpha_3 \mathbf{Nn} + \alpha_4 \mathbf{d} + \alpha_5 \mathbf{nn} \cdot \mathbf{d} + \alpha_6 \mathbf{d} \cdot \mathbf{nn} \tag{9.59}$$

describing the shear stress $\boldsymbol{\sigma}$ associated with the viscous flow, in which (1) $\mathbf{n}$ is the unit vector representing the director of the nematic, (2) $\mathbf{d}$ is the rate-of-deformation tensor (see Chapter 2), (3) $\mathbf{N}$ is the vector representing the rate at which the director orientation changes with time with respect to the background fluid, expressed as:

$$\mathbf{N} = \frac{d\mathbf{n}}{dt} - \boldsymbol{\omega} \cdot \mathbf{n} \tag{9.60}$$

where $\boldsymbol{\omega}$ is the vorticity tensor (see Chapter 2), and (4) the six coefficients α_1 to α_6 (which may be positive or negative), which are referred to as the Leslie coefficients and have the units of viscosity. The following torque balance must be satisfied:

$$\mathbf{h} - (\alpha_6 - \alpha_5)(\mathbf{n}:\mathbf{d} - \mathbf{nnn}:\mathbf{d}) - (\alpha_3 - \alpha_2)\mathbf{N} = 0 \tag{9.61}$$

where $\mathbf{h}$ is the molecular field. With reference to Eq. (9.59), (1) the terms involving α_2 and α_3 are associated with rotational flows (via $\mathbf{N}$ defined by Eq. (9.60)), (2) the terms involving α_5 and α_6 with non-rotational flows (no dependence on $\boldsymbol{\omega}$), and (3) the first term describes the stretching that can be produced by a non-rotational flow.

It can be shown that only five of the six Leslie coefficients are independent because, according to Parodi (1970), the following relationship holds:

$$\alpha_2 + \alpha_3 = \alpha_6 - \alpha_5 \tag{9.62}$$

Note that for an isotropic liquid (i.e., $\alpha_1 = \alpha_2 = \alpha_3 = \alpha_5 = \alpha_6 = 0$), Eq. (9.59) reduces to the constitutive relationship for a Newtonian fluid:

$$\sigma_{ij} = \alpha_4 d_{ij} \tag{9.63}$$

with α_4 being equivalent to Newtonian viscosity η_0.

Relationships between the Six Leslie Coefficients and Three Molecular Parameters in the Doi Theory By limiting their analysis to the first-order perturbation from the equilibrium state, Kuzuu and Doi (1984) derived the Ericksen–Leslie equation from the Doi theory under weak velocity gradient and obtained the following relationships between the six Leslie coefficients and the three molecular parameters (concentration, molecular weight, and the order parameter) appearing in the Doi theory:

$$\begin{aligned} &\alpha_1 = -2S_4\bar{\eta}; \quad \alpha_2 = -S\left(1 + \tfrac{1}{\lambda}\right)\bar{\eta}; \quad \alpha_3 = -S\left(1 - \tfrac{1}{\lambda}\right)\bar{\eta}; \\ &\alpha_4 = \tfrac{2}{35}(7 - 5S - 2S_4)\bar{\eta}; \quad \alpha_5 = \tfrac{2}{7}(5S + 2S_4)\bar{\eta}; \quad \alpha_6 = -\tfrac{4}{7}(S - S_4)\bar{\eta} \end{aligned} \tag{9.64}$$

where $\bar{\eta} = \nu k_B T/2\bar{D}_r$, $S = \langle P_2(\mathbf{u}\cdot\mathbf{n})\rangle$ with P_2 being the Legendre polynomial of the second order $P_2(x) = (3x^2 - 1)/2$, and $S_4 = \langle P_4(\mathbf{u}\cdot\mathbf{n})\rangle$ with P_4 being the Legendre polynomial of the fourth order, $P_4(x) = (35x^4 - 30x^2 + 3)/8$. Note that S is the orientation order parameter defined by Eq. (9.10), and that Eq. (9.64) satisfies the equality given by Eq. (9.62). Note that λ is referred to as the tumbling parameter, which is related to S under special circumstances (Doi 1981; Marrucci 1982).

Director Tumbling A number of researchers (Carlsson 1984; Kuzuu and Doi 1984; Pieranski and Guyon 1974; Semenov 1983) have investigated, with the aid of the Ericksen–Leslie theory, shear flow of nematic liquid crystals and found that instability occurs (i.e., the director **n** of the nematic phase tumbles) when $\lambda < 1$, where λ is defined by

$$\lambda = \frac{\alpha_2 + \alpha_3}{\alpha_2 - \alpha_3} \tag{9.65}$$

in which the relationship given by Eq. (9.62) was used. Note that $|\alpha_2| > |\alpha_3|$ and α_2 is always negative. It should be noted that Eq. (9.65) is satisfied also by (9.64), in which we have the relationship $\alpha_2/\alpha_3 = (\lambda + 1)/(\lambda - 1)$. It should be pointed out that the derivation of Eq. (9.64) was made possible without using the decoupling approximation given by Eq. (9.27). However, when using the decoupling approximation, the Doi theory predicts that λ always is greater than 1, meaning that director tumbling is not possible for all values of S (Doi 1981; Marrucci 1982). However, the Doi theory without the decoupling approximation can predict director tumbling (Kuzuu and Doi 1984).

Geometrically interpreted, the parameter λ defined by Eq. (9.65) is related to the angle θ_o that the director makes with the shear direction:

$$\cos 2\theta_o = \frac{\alpha_2 - \alpha_3}{\alpha_2 + \alpha_3} = \frac{1}{\lambda} \tag{9.66}$$

It is clear from Eq. (9.65) that $\lambda < 1$ for $\alpha_3 > 0$. It is also clear from Eq. (9.66) that $\cos 2\theta_o$ will become greater than 1 for $\alpha_3 > 0$, under which the hydrodynamic torques always act to rotate the director, and the director is predicted to rotate indefinitely in shear flow, exhibiting tumbling (see Figure 9.17). Here, tumbling refers to the nonexistence of a stationary state ($d\mathbf{n}/dt \neq 0$) for the director **n**, as schematically shown in Figure 9.21, which describes what would be observed by looking perpendicular to the plane of shear. Moving along the shear direction with reference to Figure 9.21, the director maintains the angle θ_o, while staying close to the plane of shear, then jumps off the plane, makes half a turn, and returns to the plane of shear with the orientation $-\theta_o$. At the next jump, it goes back to θ_o, and so on indefinitely. In other words, tumbling occurs when the director assumes no preferred angle with respect to the flow direction, but continuously rotates around the vorticity axis until the Frank distortional stresses bring this rotation to a halt. It should be noted that $\lambda > 1$ for $\alpha_3 < 0$. Under such

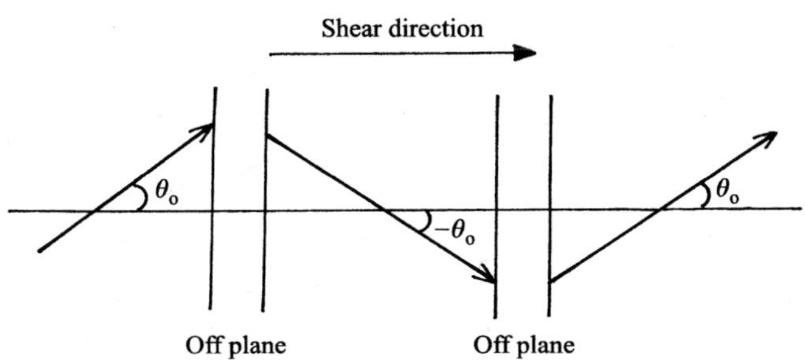

Figure 9.21 The tumbling process: what would be observed by looking perpendicular to the plane of shear. (Reprinted from Marrucci, *Pure Appl. Chem.* 57:1545. Copyright © 1985, with permission from the International Union of Pure and Applied Chemistry.)

a circumstance, a balance of hydrodynamic torque exerted on the director leads to a steady alignment of the liquid crystal at a characteristic angle with respect to the flow direction, exhibiting "flow aligning."

Negative First Normal Stress Difference N_1 Using the Ericksen–Leslie theory, Currie (1981) predicted that for $\alpha_3 < 0$, the value of N_1 for low-molecular-weight nematics can be always negative, always positive or change from negative to positive with increasing shear stress, depending on the boundary conditions. Apparently, this theoretical prediction was motivated by a need to explain the experimental observations of Kiss and Porter (1978, 1980a), who had reported negative values of N_1 in steady-state shear flow of concentrated solutions of poly(γ-benzylglutamate) (PBLG). Note that for $\alpha_3 < 0$, $\lambda > 1$, and thus the director forms a steady alignment with respect to the flow direction. Later in this chapter, we will elaborate on negative N_1 in LCP.

9.2.2.2 The Larson–Doi Theory

Although the Doi molecular theory without decoupling approximation of the fourth-order tensor $\langle \mathbf{uuuu} \rangle$ is capable, like the Ericksen–Leslie phenomenological theory, of predicting director tumbling (Kuzuu and Doi 1984), it is unable to account for spatial inhomogeneity of the director field and the associated elastic torques and stresses that will necessarily play an important role in highly textured nematics. Admittedly, the structure of a textured nematic seems to be much too complicated to allow direct modeling. Nevertheless, some attempts (Larson and Doi 1991; Marrucci 1985; Wissbrun 1985) have been made to incorporate, on an ad hoc basis, textured nematics into the prediction of the rheological behavior of LCPs by extending the low-molecular-weight nematodynamics due to Ericksen and Leslie. Marrucci and Maffettone (1989) incorporated interdomain interactions that are necessary for polydomain structure, via Frank elasticity, into a two-dimensional version of the Doi model and predicted transient rheological response of LCP. The most successful attempt made to date is one by Larson and Doi (1991), who incorporated the rate of change of the disclination in the formulation of a system of equations.

The essence of the Larson–Doi mesoscopic domain theory (Larson and Doi 1991) is summarized below. By multiplying $\mathbf{n}$ to Eqs. (9.60) and (9.61) and then taking the average over a mesoscopic length scale, they obtained the following expression for the evolution of the spatially averaged texture order parameter tensor $\overline{\mathbf{S}}$:

$$\frac{d\overline{\mathbf{S}}}{dt} = \boldsymbol{\omega}^{\mathrm{T}} \cdot \overline{\mathbf{S}} + \overline{\mathbf{S}} \cdot \boldsymbol{\omega} + \lambda \left[\tfrac{2}{3}\mathbf{d} + \mathbf{d} \cdot \overline{\mathbf{S}} + \overline{\mathbf{S}} \cdot \mathbf{d} - 2\overline{\mathbf{S}} : \mathbf{d}\left(\overline{\mathbf{S}} + \tfrac{1}{3}\boldsymbol{\delta}\right) \right] - \varepsilon \rho_{\mathrm{v}} \overline{\mathbf{S}} \qquad (9.67)$$

where λ is defined by Eq. (9.65) and $\overline{\mathbf{S}}$ is defined by

$$\overline{\mathbf{S}} = \langle [\mathbf{nn}] \rangle - \tfrac{1}{3}\boldsymbol{\delta} \qquad (9.68)$$

with $\langle [\cdots] \rangle$ denoting the average over a mesoscopic distance scale that is large compared with the texture size and yet small compared with bulk dimensions, ε is a dimensionless constant indicating the strength of distortional elasticity, and ρ_{v} is a disclination density having units of length of disclination line per unit volume. Also, by multiplying **n** to Eq. (9.59) and taking the average over a mesoscopic length scale, they obtained the following expression for the spatially averaged stress tensor $\langle[\boldsymbol{\sigma}]\rangle$:

$$\langle[\boldsymbol{\sigma}]\rangle = 2\mu\mathbf{d} + 2\mu_1\mathbf{d}:\overline{\mathbf{S}}\left(\overline{\mathbf{S}} + \tfrac{1}{3}\boldsymbol{\delta}\right) + \mu_2\left(\tfrac{2}{3}\mathbf{d} + \overline{\mathbf{S}}\cdot\mathbf{d} + \mathbf{d}\cdot\overline{\mathbf{S}}\right) - \tfrac{1}{2}\varepsilon(\alpha_2 + \alpha_3)\rho_v\overline{\mathbf{S}} \tag{9.69}$$

where

$$\mu = \alpha_4/2; \quad \mu_1 = \frac{\alpha_2}{2} - \frac{(\alpha_2 + \alpha_3)(\alpha_6 - \alpha_5)}{2(\alpha_2 - \alpha_3)}; \quad \mu_2 = \frac{\alpha_2\alpha_6 - \alpha_3\alpha_5}{\alpha_2 - \alpha_3} \tag{9.70}$$

In the derivation of Eqs. (9.69) and (9.70), the following closure approximation was used:

$$\langle[\mathbf{nnnn}]\rangle \approx \langle[\mathbf{nn}]\rangle\,\langle[\mathbf{nn}]\rangle \tag{9.71}$$

which, however, enables one to predict direct tumbling in the present formulation. Other assumptions made can be found in the original paper (Larson and Doi 1991). With reference to Eq. (9.69), distortional elasticity influences the time variation of stresses through the time variation of mesoscopic orientation $\overline{\mathbf{S}}(t)$, which will be obtained from the solution of Eq. (9.67), and also through the term containing ε on the right-hand side of Eq. (9.69).

For shear flow, the components of $\overline{\mathbf{S}}$ can be written from Eq. (9.67) (Ugaz 1999)

$$\frac{d\overline{S}_{11}}{dt} = \dot{\gamma}\overline{S}_{12}\left\{1 + \lambda\left[1 - 2\left(\overline{S}_{11} + \tfrac{1}{3}\right)\right]\right\} - \varepsilon\rho_v\overline{S}_{11} \tag{9.72}$$

$$\frac{d\overline{S}_{12}}{dt} = \dot{\gamma}\overline{S}_{12}\left\{-1 + \lambda\left[1 - 2\left(\overline{S}_{22} + \tfrac{1}{3}\right)\right]\right\} - \varepsilon\rho_v\overline{S}_{22} \tag{9.73}$$

$$\frac{d\overline{S}_{33}}{dt} = -2\dot{\gamma}\overline{S}_{12}\lambda\left(\overline{S}_{33} + \tfrac{1}{3}\right) - \varepsilon\rho_v\overline{S}_{33} \tag{9.74}$$

$$\frac{d\overline{S}_{12}}{dt} = \frac{\dot{\gamma}}{2}\left\{\left(\overline{S}_{22} - \overline{S}_{11}\right) + \lambda\left[\tfrac{2}{3} + \left(\overline{S}_{11} + \overline{S}_{22}\right) - 4\overline{S}_{12}{}^2\right]\right\} - \varepsilon\rho_v\overline{S}_{12} \tag{9.75}$$

where $\dot{\gamma}$ is shear rate. Equations (9.72)–(9.75) will become a closed set of system equations when an additional equation is provided, describing how the disclination density ρ_v varies with time. On the basis of a dimensional consideration, Larson and Doi (1991) postulated the following rate expression for ρ_v:

$$\frac{d\rho_v}{dt} = \dot{\gamma}\rho_v - \rho_v{}^2 \tag{9.76}$$

describing the dynamics of disclination lines in shear flow. And, the following expressions for shear stress σ and first normal stress difference N_1 can be written from Eq. (9.69) (Ugaz 1999):

$$\sigma = \left[\mu + 2\mu_1\overline{S}_{12}{}^2 + \frac{\mu_2}{2}\left(\tfrac{2}{3} + \overline{S}_{11} + \overline{S}_{22}\right)\right]\dot{\gamma} - \tfrac{1}{2}(\alpha_2 + \alpha_3)\varepsilon\rho_v\overline{S}_{12} \tag{9.77}$$

$$N_1 = 2\mu_1\overline{S}_{12}(\overline{S}_{11} - \overline{S}_{22})\dot{\gamma} - \tfrac{1}{2}(\alpha_2 + \alpha_3)\varepsilon\rho_v(\overline{S}_{11} - \overline{S}_{22}) \tag{9.78}$$

Thus, the time evolutions of σ and N_1 can be calculated from Eqs. (9.77) and (9.78) once the time evolutions of the mesoscopic orientation are calculated from the solution of Eqs. (9.72)–(9.76). Note that values of five (owing to the Porod relation given by Eq. (9.62)) Leslie coefficients must be provided, which is not a trivial task. In addition, the value of ε must be specified from experimental data. In this regard, ε can be regarded as an adjustable parameter.

The Larson–Doi theory predicts qualitatively the essential features of transient shear flow of some model TLCPs (Ugaz 1999). It is, however, not clear to what extent the Larson–Doi theory can describe the dynamics of TLCPs that do not exhibit tumbling. This is because the Larson–Doi theory is based on the Ericksen–Leslie theory, which determines structural responses through the tumbling parameter λ. As will be presented later in this chapter, the experimental data available to date suggest that TLCPs are flow aligning. It is fair to state that the theoretical attempts reported thus far contain, understandably, many crude approximations, and so do not warrant quantitative comparison with experimental results for textured LCPs, particularly TLCPs. Thus the development of a molecular viscoelastic theory for textured LCPs is still in its infancy.

9.3 Rheological Behavior of Lyotropic LCPs

A large number of researchers (Asada et al. 1980; Baek et al. 1993b, 1994; Berry et al. 1981; Chu et al. 1981; Einaga et al. 1985; Grizzuti et al. 1990; Helminiak and Berry 1979; Hermans 1962; Kiss and Porter 1978, 1980a; 1980b, Kiss et al. 1979; Moldenaers and Mewis 1986; Navard 1986; Navard and Haudin 1980, 1986; Onogi and Asada 1980; Papkov et al. 1974; Wong et al. 1979) have investigated the rheological behavior of lyotropic LCPs. In this section, we present some unusual rheological behavior of lyotropic LCPs, putting emphasis on the effects of molecular weight, concentration, and shear rate. Where appropriate, we will interpret experimental results using the currently held theories.

The effect of molecular weight on shear viscosity η is given in Figure 9.22 for solutions of poly(*p*-benzamide) (PBA) dissolved in a mixture of *N*,*N*-dimethyl acetamide (DMAc) and lithium chloride (LiCl), showing that for each molecular weight, η goes through a maximum at a critical concentration at which the isotropic–anisotropic transition takes place (see also Figure 9.9), and the critical concentration decreases with increasing molecular weight of PBA. However, it is of interest to observe in Figure 9.23 that a molecular weight independent correlation is obtained when reduced viscosity η/η_c is plotted against reduced concentration ϕ/ϕ^*, where η_c denotes the maximum value of η at which the concentration ϕ becomes the critical value ϕ^*.

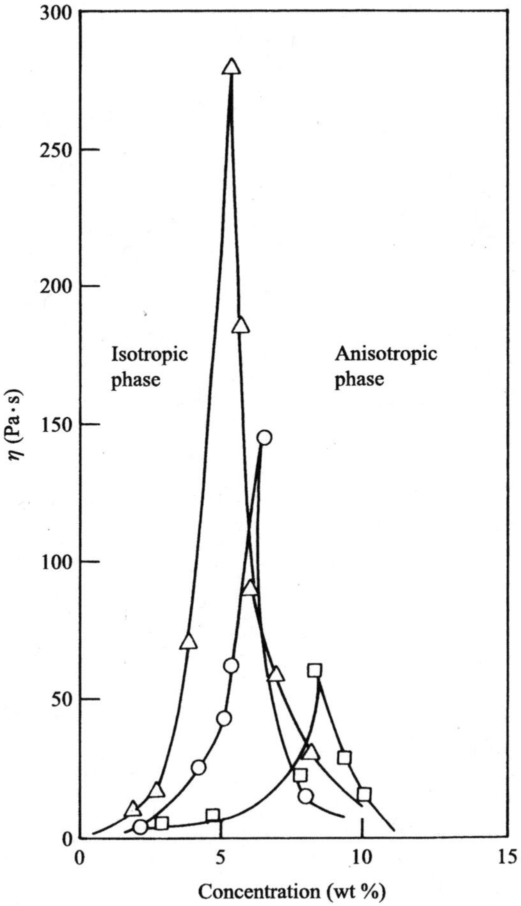

Figure 9.22 Plots of η versus concentration for solutions of PBA in DMAc-LiCl for different molecular weights of PBA: (△) 69 × 10^3, (○) 23 × 10^3, (□) 16 × 10^3. (Reprinted from Papkov et al., *Journal of Polymer Science, Polymer Physics Edition* 12:1753. Copyright © 1974, with permission from John Wiley & Sons.)

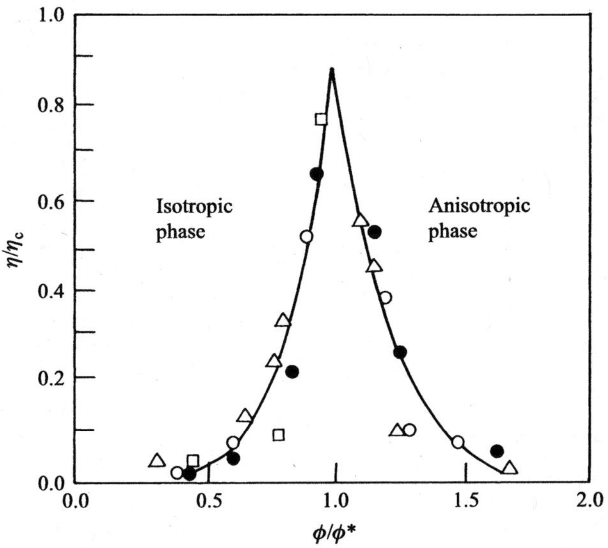

Figure 9.23 Plots of η/η_c versus ϕ/ϕ^* for PBA solutions in DMAc-LiCl for different molecular weights of PBA: (●) 11 × 10^3, (△) 17 × 10^3, (○) 22.2 × 10^3, (□) 29.2 × 10^3. (Reprinted from Papkov et al., *Journal of Polymer Science, Polymer Physics Edition* 12:1753. Copyright © 1974, with permission from John Wiley & Sons.)

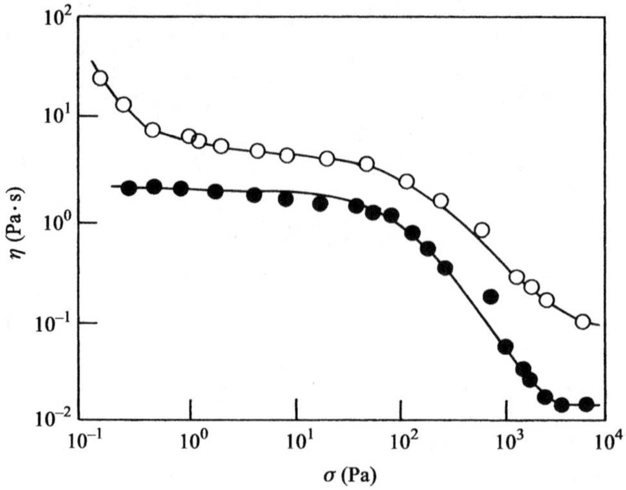

Figure 9.24 Plots of $\log \eta$ versus $\log \sigma$ for PBA ($M_w = 69 \times 10^3$) dissolved in DMAc-LiCl: (●) 3 wt % PBA (isotropic) solution and (○) 7 wt % PBA (anisotropic) solution (see Figure 9.22 for viscosity versus concentration curves). (Reprinted from Papkov et al., *Journal of Polymer Science, Polymer Physics Edition* 12:1753. Copyright © 1974, with permission from John Wiley & Sons.)

Plots of $\log \eta$ versus $\log \sigma$ for a solution of PBA dissolved in DMAc-LiCl in the isotropic state and in an anisotropic state are given in Figure 9.24, in which we observe that while the isotropic solution shows Newtonian behavior, the anisotropic solution exhibits "shear-thinning" behavior at very low values of σ, very similar to that often observed in microphase-separated block copolymers (see Chapter 8) and, also, in highly filled molten polymers, which are be presented in Chapter 12. A close examination of Figure 9.24 shows that the $\log \eta$ versus $\log \sigma$ plots have three distinct regions: (1) region I at very low values of σ exhibiting shear-thinning behavior, (2) region II at intermediate values of σ exhibiting Newtonian behavior, and (3) region III at high values of σ exhibiting again shear-thinning behavior. Such experimental observation for lyotropic LCP solutions was first noted by Onogi and Asada (1980). The physical origins of the three different regions in $\log \eta$ versus $\log \sigma$ or $\log \eta$ versus $\log \dot{\gamma}$ plots for lyotropic LCPs are believed to be associated with the variations of the morphological state of the fluid in the different regions, as schematically shown in Figure 9.25. Namely, at very low values of σ or $\dot{\gamma}$, the texture of a lyotropic solution consists of many small domains (forming a polydomain), in which the local orientations of the director vary from one small domain to another. In such a situation, each small domain may act as if it is a discrete particle suspended in a very low viscosity fluid. Thus, at extremely low values of σ or $\dot{\gamma}$ (very close to the state at rest), the polydomain hardly deforms (or flows). Such a state of polydomain would give rise to very high viscosity. As the intensity of shear flow increases, the polydomain structure begins to change by having fewer small domains due to coalescence, and at the same time the local orientations of the director begin to align in the flow direction. When the intensity of shear flow becomes sufficiently high, the local orientations of the director

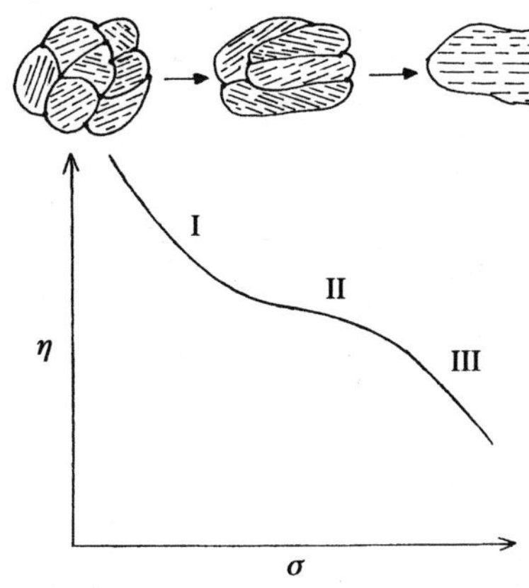

Figure 9.25 Schematic showing texture refinement in shear flow of a lyotropic LCP solution accompanied by concomitant shear-rate dependent viscosity.

in separate small domains are aligned perfectly along the flow direction, giving rise to a monodomain texture. Such variations in domain texture with increasing σ or $\dot{\gamma}$ can explain the concomitant variations of η with increasing σ or $\dot{\gamma}$, shown schematically in the plot in Figure 9.25. Direct observation of structural changes during flow by birefringence has been made, reporting texture refinement in shear flow of a textured lyotropic LCP solution (Burghardt and Hongladarom 1994). It should be mentioned that an investigation of molecular orientation of LCPs under shear flow is very important for a better understanding of the rheological behavior of LCPs. For this reason, a number of investigators have employed birefringence (Asada et al. 1980; Doppert and Picken 1987; Hongladarom et al. 1993, 1994, 1996; Kiss and Porter 1980b; Moldenaers et al. 1989) or X-ray scattering (Hongladarom et al. 1996; Picken et al. 1991; Ugaz et al. 1998), in conjunction with mechanical rheometry, to investigate rheo-optical behavior of lyotropic LCP solutions.

Plots of η versus concentration for solutions of PBLG in *m*-cresol at various $\dot{\gamma}$ are given in Figure 9.26, from which we observe that the critical concentration at which η goes through a maximum decreases with increasing $\dot{\gamma}$, and the maximum value of η increases rapidly as $\dot{\gamma}$ decreases. It is of interest to observe in Figure 9.26 that in an anisotropic region the difference in η becomes smaller as $\dot{\gamma}$ increases, which is attributable to the fact that the anisotropic phase orients in the flow direction for $\dot{\gamma}$ greater than a certain critical value.

Figure 9.27 gives plots of log N_1 versus log $\dot{\gamma}$ for a solution of 22.5 wt % PBLG in *m*-cresol, showing that the solution exhibits *negative* values of N_1 at an intermediate range of $\dot{\gamma}$. Since such unusual experimental observation was first reported by Kiss and Porter (1978, 1980a, 1980b), other investigators (Baek et al. 1993a, 1994; Grizzuti et al. 1990; Huang et al. 1999; Magda et al. 1991; Moldenaers and Mewis 1986) have also reported similar observations for varying concentrations of lyotropic solutions of PBLG or aqueous solutions of hydroxypropylcellulose (HPC). At the time when such experimental results were first reported, the physical origin of the seemingly peculiar rheological phenomenon was not understood. About a decade later, as presented in the previous section, Marrucci and coworkers (Marrucci and Maffettone 1989, Cocchini et al. 1990) and Larson (1990) independently developed a theory that enables us to understand the

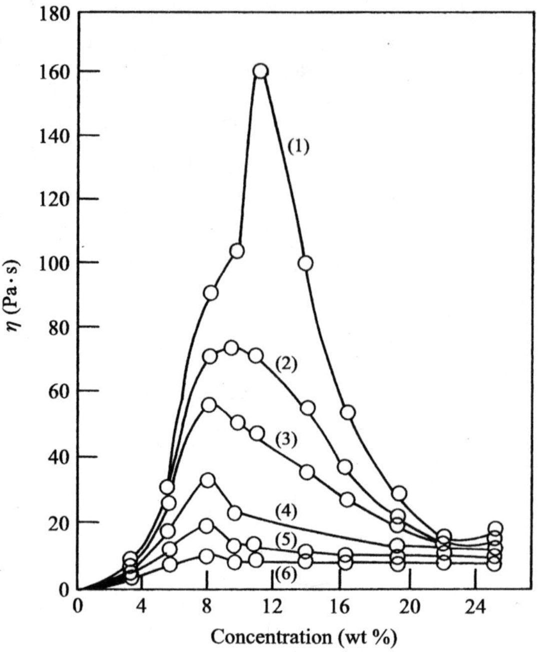

Figure 9.26 Plots of η versus concentration for a solution of PBLG in *m*-cresol at various shear rates (s^{-1}): (1) $\dot{\gamma} = 0$, (2) $\dot{\gamma} = 0.4$, (3) $\dot{\gamma} = 1$, (4) $\dot{\gamma} = 4$, (5) $\dot{\gamma} = 10$, and (6) $\dot{\gamma} = 25$. (Reprinted from Kiss and Porter, *Journal of Polymer Science, Polymer Symposia* 65:193. Copyright © 1978, with permission from John Wiley & Sons.)

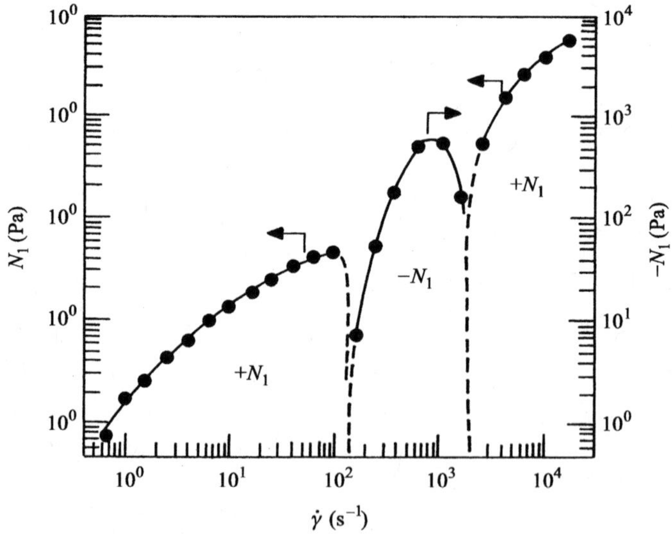

Figure 9.27 Plots of log N_1 versus log $\dot{\gamma}$ for 22.5 wt % solution of PBLG in *m*-cresol. (Reprinted from Kiss and Porter, *Journal of Polymer Science, Polymer Physics Edition* 18:361. Copyright © 1980, with permission from John Wiley & Sons.)

physical origin of negative N_1 at intermediate range of shear rates for lyotropic LCP solutions. Zhou and Han (2005) synthesized a combined main-chain/side-chain LCP (referred to as PSHQ4-7CNCOOH) that exhibits both the thermotropic and lyotropic characteristics. They used *o*-dichlorobenzene, *m*-dibromobenzene, and tetrachloroethane as solvents to prepare lyotropic solutions of PSHQ4-7CNCOOH for concentrations ranging from 22 to 70 wt % and then constructed a phase diagram. Using the lyotropic solutions of PSHQ4-7CNCOOH, Zhou and Han (2006) investigated their steady-state shear flow behaviors and made the following observations: (1) negative values of steady-state N_1 at intermediate shear rates for the lyotropic solutions of PSHQ4-7CNCOOH having concentrations of 22, 25, and 27 wt %, but only positive values of N_1 for higher concentrations over the entire range of shear rates investigated, and (2) negative values of N_1 became positive as the measurement temperature decreased below a certain critical value. Thus the sign change in N_1 depends on both the concentration and temperature of the lyotropic solutions. The details of the experimental results are referred to the original paper.

Some research groups (Grizzuti et al. 1990; Mewis and Moldenaers 1987; Moldenaers et al. 1989; Picken et al. 1991; Walker et al. 1995) investigated whether shear stress and first normal stress difference scale with strain for lyotropic solutions during transient shear flow. Specifically, Mewis and Moldenaers (1987) employed a lyotropic solution of 12 wt % PBLG in *m*-cresol, Moldenaers et al. (1989) also employed a lyotropic solution of 12 wt % PBLG in *m*-cresol, and Picken et al. (1991) employed a lyotropic solution of 19.8 wt % PPTA in sulfuric acid. These investigators observed that the ratio $\sigma^+(t, \dot{\gamma})/\sigma$ scaled with $\dot{\gamma}t$ during transient shear flow. Conversely, Grizzuti et al. (1990), who employed lyotropic solutions of HPC in water, and Walker et al. (1995), who employed a lyotropic solution of 40 wt % PBLG in *m*-cresol, failed to observe strain scaling for the ratio $\sigma^+(t, \dot{\gamma})/\sigma$ during transient shear flow. It is then fair to state that there is no general expectation for strain scaling of shear stress, even for lyotropic LCPs during transient shear flow.

Other rheological aspects of lyotropic LCPs have been reported in the literature. Specifically, reversal flow of lyotropic LCPs has been investigated by some research groups. After a reversal in flow direction, Moldenaers et al. (1989) observed that the transient shear stress $\sigma^+(t, \dot{\gamma})/\sigma$, with σ being the shear stress in steady state, scaled with strain $\dot{\gamma}t$ after the reversal in flow direction had occurred in a lyotropic solution of 12 wt % PBLG in *m*-cresol. Burghardt and Fuller (1990), who employed lyotropic solutions of PBG in *m*-cresol, made a similar observation. However, Hongladarom and Burghardt (1993) did not observe strain scaling with the ratio $\sigma^+(t, \dot{\gamma})/\sigma$ for lyotropic solutions of HPC in water after the reversal in flow direction. The situation appears to be very complicated in that Walker et al. (1995) observed strain scaling with the ratio $\sigma^+(t, \dot{\gamma})/\sigma$ for a lyotropic solution of 37 wt % PBLG in *m*-cresol but not for a lyotropic solution of 40 wt % PBLG in *m*-cresol after the reversal in flow direction. It is then fair to state that no general conclusion can be drawn on strain scaling for lyotropic LCPs after the reversal in flow direction.

Very few investigators have reported on strain scaling with first normal stress difference $N_1^+(t, \dot{\gamma})$ for lyotropic LCPs after reversal in flow direction. Chow et al. (1992) reported variations of $N_1^+(t, \dot{\gamma})$ with $\dot{\gamma}t$ for lyotropic solutions of poly(*p*-phenylenebenzobisthiazole) (PBZT) after reversal in flow direction, indicating that for this system $N_1^+(t, \dot{\gamma})$ does not follow strain scaling after flow reversal. Hongladarom and Burghardt (1993) reported that birefringence $\Delta n^{+}(t, \dot{\gamma})$ scaled with $\dot{\gamma}t$ during reversal flow of lyotropic solutions of PBG, while $\Delta n^{+}(t, \dot{\gamma})$ did not scale with $\dot{\gamma}t$ during reversal flow of lyotropic solutions of HPC in water. Further experimental and theoretical studies are needed to resolve the seemingly contradictory experimental observations.

The time evolution of G', G'', and/or $|\eta^*|$ after cessation of shear flow has been investigated for lyotropic solutions (Grizzuti et al. 1990; Mewis and Moldenaers 1987). Specifically, using aqueous solutions of HPC, Grizzuti et al. (1990) reported that $|\eta^*|$ increased with time after cessation of shear flow. Larson and Mead (1989) predicted the experimental observations by Grizzuti et al. Conversely, using solutions of PBLG dissolved in *m*-cresol, Mewis and Moldenaers (1987) reported that both G' and G'' decreased with time after cessation of shear flow. Further experimental and theoretical studies are needed to resolve the seemingly contradictory experimental observations.

9.4 Rheological Behavior of Thermotropic Main-Chain LCPs

A number of research groups (Baek et al. 1994; Blumstein et al. 1986; Chang and Han 1997a, 1997b; Cocchini et al. 1991; De'Neve et al. 1993; Driscoll et al. 1991; Gilmore et al. 1994; Gonzalez et al. 1990; Guskey and Winter 1991; Han and Kim 1994a, 1994b; Han et al. 1994a; Irwin et al. 1989; Kalika et al. 1990; Kim and Han 1993a, 1993b, 1993c, 1994a, 1994b, 2000; La Mantia and Valenza 1989; Masuda et al. 1991; Wissbrun 1980; Wissbrun and Griffin 1982; Wissbrun et al. 1987; Wunder et al. 1986; Zhou et al. 1999) have reported on the rheological behavior TLCPs. Of particular interest is the experimental observation that the rheological behavior of TLCPs is very sensitive to thermal and deformation histories (Han et al. 1994a, Kim and Han 1993a; Lin and Winter 1988, 1991), which is not the case for flexible homopolymers (discussed in Chapter 6). Therefore, unless the thermal and deformation histories of a TLCP are identical, the rheological measurements made by one research group can be different from those made by another research group. Owing to the rather complicated relationships existing between the rheological behavior and the texture in TLCPs, before taking rheological measurements one must be able to control the initial morphology (i.e., initial conditions) of the specimens; otherwise, it would not be possible for one to separate the effect of test temperature and the effect of thermal history from the overall rheological responses.

In this section, we present the rheological behaviors of TLCPs; namely, (1) transient shear flow, (2) steady-state shear flow, (3) intermittent shear flow, (4) shear flow after the change in flow direction (reversal flow), and (5) variation of dynamic moduli after cessation of shear flow. We will show that some rheological behaviors of TLCPs are different from those observed for lyotropic LCPs. Experimental results will be interpreted, where appropriate, using the currently held theories presented in this chapter.

9.4.1 Effect of Thermal History on the Rheological Behavior of Thermotropic Main-Chain LCPs

One very important aspect in the investigation of the rheological behavior of TLCPs is to control the initial conditions of specimens, which depend very much on the previous thermal and deformation histories of specimens (Han et al. 1994a; Kim and Han 1993a; Lin and Winter 1988, 1991). Han and coworkers (Han et al. 1994a; Han and Kim 1994a, 1994b; Kim and Han 1993a, 1993b, 1993c, 1994a, 1994b) have shown that the previous thermal and deformation histories of a TLCP specimen can be erased only when the specimen is heated to a temperature in the isotropic region. Such a requirement can be met only when the clearing temperature of a specimen is sufficiently below the thermal degradation temperature. Many research groups (Blumstein et al. 1986; Cocchini et al. 1991; De'Neve et al. 1993; Gonzalez et al. 1990; Guskey and Winter 1991; Irwin et al. 1989; Kalika et al. 1990; La Mantia and Valenza 1989; Masuda et al. 1991; Wissbrun 1980; Wissbrun et al. 1987; Wunder et al. 1986) used commercial TLCPs (e.g., copolyesters of HBA and PET or copolyesters of HBA and HNA) or other TLCPs, which have clearing temperatures (say, above 350 °C) close to or above the thermal degradation temperatures. In this section, we will show that reproducible rheologicial data cannot be obtained using such TLCPs. Since it is not possible to erase the previous thermal and deformation histories of such TLCPs, a serious question may be raised regarding the rheological significance of those studies. We cannot overemphasize the importance of obtaining reproducible and reliable rheological data for TLCPs.

Figure 9.28 gives DSC thermograms of 73/27 HBA/HNA copolyester, showing that it has the melting temperature of approximately 283 °C and another thermal transition temperature at 305–309 °C, the origin of which is not well understood to date, although some speculations have been made. It is interesting to observe in Figure 9.28 that the annealing conditions employed greatly affect the thermal transition in the specimen. Specifically, the annealing of a specimen at 290 °C for 50 min (curve (3) in Figure 9.28) eliminates completely the higher endothermic peak, suggesting that the structure responsible for the appearance of the endothermic peak at 305–309 °C, when the specimen was annealed at 280 or 285 °C, is not stable. It is clear from Figure 9.28 that the 73/27 HBA/HNA copolyester does not have a clearing temperature up to 340 °C. Figure 9.29 shows variations of G' and G'' with time for a 73/27 HBA/HNA copolyester specimen at 320 °C, which was placed in the cone-and-plate fixture of a

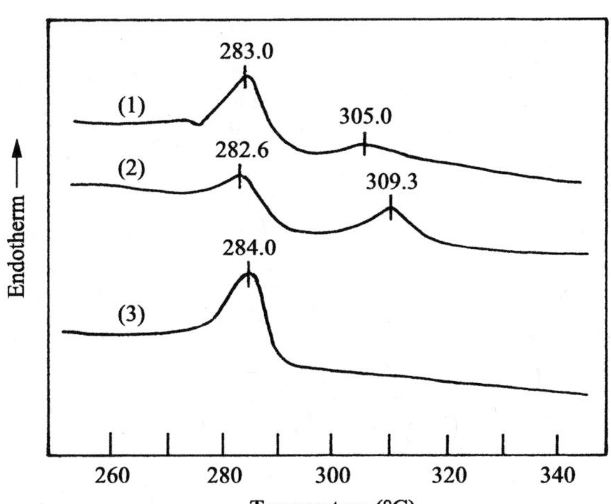

Figure 9.28 DSC thermograms for compression-molded 73/27 HBA/HNA copolyester specimens annealed at different temperatures: (1) annealed at 280 °C for 50 min, (2) annealed at 285 °C for 50 min, and (3) annealed at 290 °C for 50 min. (Reprinted from Han et al., *Molecular Crystals and Liquid Crystals* 254:335. Copyright © 1994, with permission from Taylor & Francis Group.)

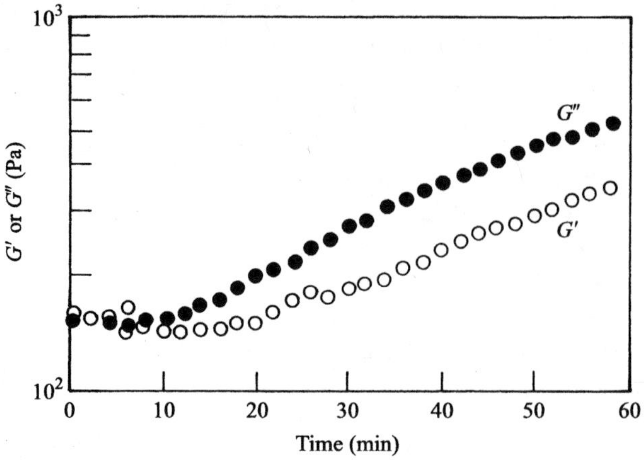

Figure 9.29 Variations of G' (○) and G'' (●) with time for a compression-molded 73/27 HBA/HNA copolyester specimen placed in the cone-and-plate fixture at 320 °C and annealed there. (Reprinted from Han et al., *Molecular Crystals and Liquid Crystals* 254:335. Copyright © 1994, with permission from Taylor & Francis Group.)

rotational rheometer that had been preheated to 320 °C. In obtaining the results given in Figure 9.29, a strain amplitude of 0.05 and an angular frequency of 1 rad/s were applied. In Figure 9.29, we observe that values of G' and G'' are more or less constant for about 10 min after the specimen is placed in the cone-and-plate fixture and then increase rapidly. The origin of such behavior was attributed to post polymerization of the specimen, although there might be other reasons unknown to us at the present time.

Figure 9.30 shows the effect of thermal history on the frequency dependence of G' and $|\eta^*|$ for 73/27 HBA/HNA copolyester. Figure 9.30a gives the temperature protocol employed, showing that a specimen was loaded in the cone-and-plate fixture at 290 °C, which took about 10 min and then a frequency sweep was conducted at 290 and 320 °C, as indicated in the temperature protocol. The following observations are worth noting in Figure 9.30. Values of G' and $|\eta^*|$ in steps 3 and 5 do not overlap each other even though the rheological measurements were taken at the same temperature, 290 °C. Values of G' and $|\eta^*|$ in steps 2 and 4 do not overlap each other even though the rheological measurements were taken at the same temperature, 320 °C. It is clear that the rheological data could not be reproduced at 290 and 320 °C, respectively. This conclusion is reinforced with the log G' versus log G'' plots given in Figure 9.31. Thus we can conclude from Figure 9.31 that the morphological state of 73/27 HBA/HNA copolyester varied during the rheological measurements, which lasted about 2 h, and thus it would not be possible to obtain reproducible rheological measurements for the 73/27 HBA/HNA copolyester.

Figure 9.32b gives, for comparison, log G' versus log G'' plots for a low-density polyethylene (LDPE) specimen subjected to the thermal history as described in the temperature protocol given in Figure 9.32a. It is clearly seen in Figure 9.32 that the log G' versus log G'' plot shows temperature independence, regardless of the thermal history to which the specimen was subjected. Such an experimental observation is expected because LDPE is a flexible homopolymer. The point we try to make here is that for a TLCP with textures, its morphology changes with temperature. In the preceding chapter we made similar observations in microphase-separated block copolymers.

Figure 9.33 shows variations of G' and G'' with time up to 6 h at a fixed angular frequency of $\omega = 0.237$ rad/s for as-cast PSHQ10 specimens at various annealing

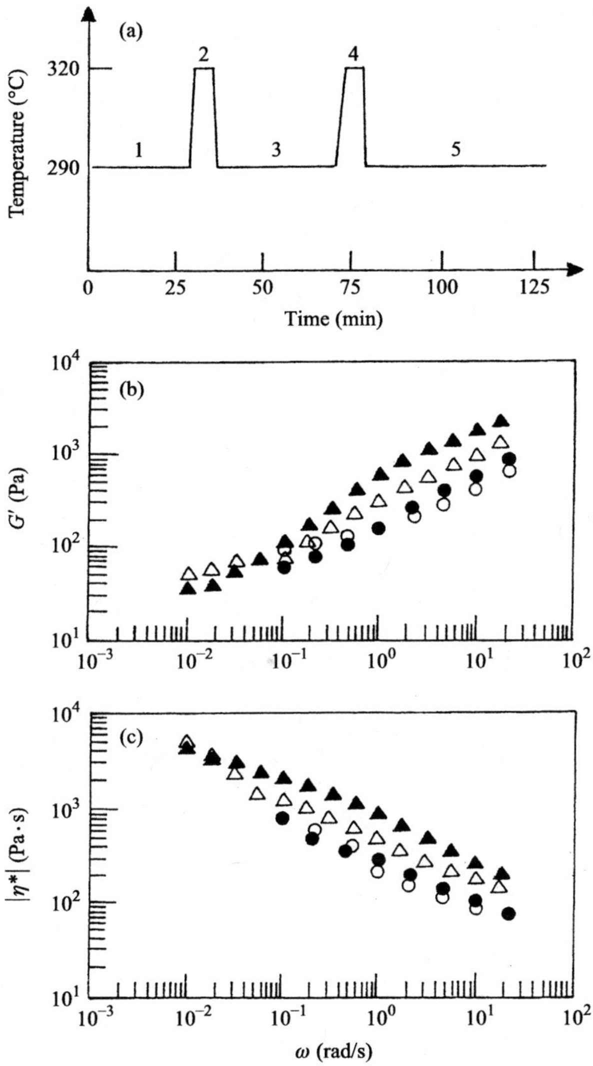

Figure 9.30 (a) Temperature protocols employed, (b) plots of log G' versus log ω, and (c) plots of log $|\eta^*|$ versus log ω for a compression-molded 73/27 HBA/HNA copolyester specimen having thermal histories as indicated in the temperature protocol: (○) step 2, (△) step 3, (●) step 4, and (▲) step 5. (Reprinted from Han et al., *Molecular Crystals and Liquid Crystals* 254:335. Copyright © 1994, with permission from Taylor & Francis Group.)

temperatures. Each measurement lasted about 50 s, and a fresh specimen was used for each test. It can be seen in Figure 9.33 that for the annealing period of up to 6 h, values of G' and G'' at 130, 135, and 140 °C increase with annealing time, the rate of increase being faster as the annealing temperature is decreased from 140 to 130 °C, whereas values of G' and G'' at 145 °C decrease initially and then remain more or less constant as annealing progresses. This observation clearly shows the effect of thermal history on the time evolution of G' and G'' during isothermal annealing, indicating that the morphological state of the specimen varied during annealing.

Figure 9.34 shows variations of $|\eta^*|$ at 130 °C with time up to 50 h for an as-cast PSHQ10 specimen, which was first placed in the cone-and-plate fixture at 190 °C in the isotropic region and then cooled slowly down to 130 °C. Also given in Figure 9.34

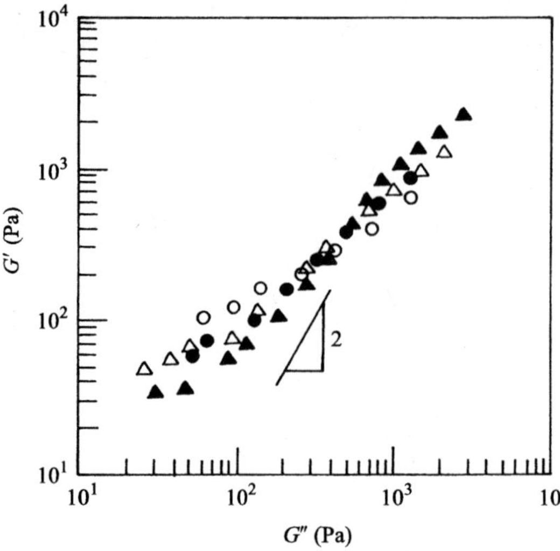

Figure 9.31 Log G' versus log G'' plots for a compression-molded 73/27 HBA/HNA copolyester specimen having the thermal histories as indicated in the temperature protocol of Figure 9.30: (○) step 2, (△) step 3, (●) step 4, and (▲) step 5. (Reprinted from Han et al., *Molecular Crystals and Liquid Crystals* 254:335. Copyright © 1994, with permission from Taylor & Francis Group.)

are, for comparison, variations of $|\eta^*|$ with time for an as-cast PSHQ10 specimen at 130 °C, which was placed in the cone-and-plate fixture preheated at 130 °C. It can be seen in Figure 9.34 that values of $|\eta^*|$ for the specimen, that had received thermal treatment at 190 °C are more or less constant for over 50 h tested.

Figure 9.35 gives log G' versus log G'' plots for a PSHQ10 specimen subjected to the thermal history as shown in the temperature protocol. The log G' versus log G'' plot shows that the rheological measurements at 140 °C in the nematic region are

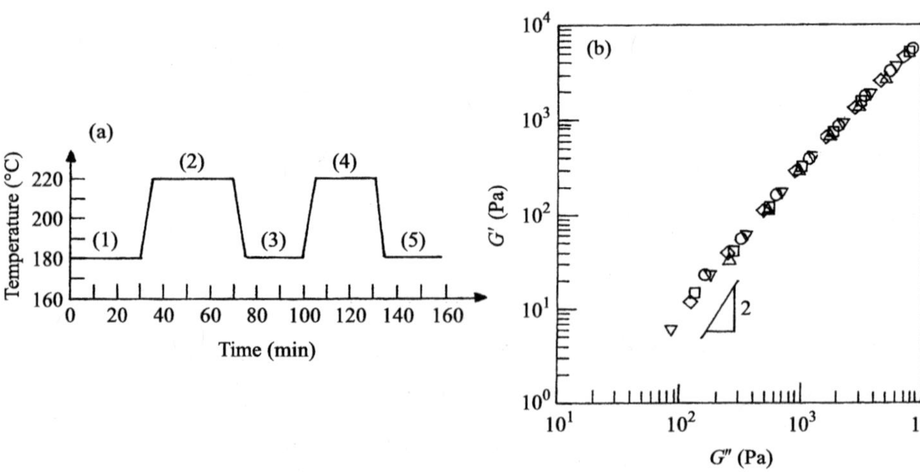

Figure 9.32 (a) Temperature protocols employed and (b) log G' versus log G'' plots for an LDPE specimen having the thermal histories as indicated in the temperature protocol: (○) step 1, (△) step 2, (□) step 3, (▽) step 4, and (◇) step 5.

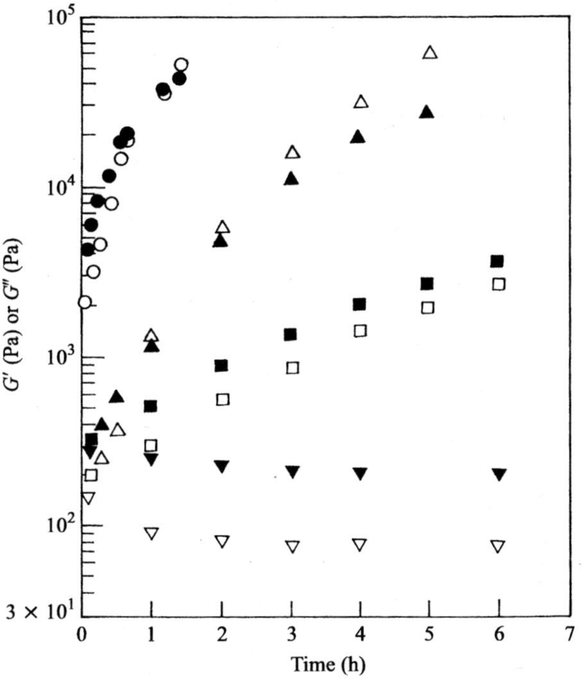

Figure 9.33 Variations of G' (open symbols) and G'' (filled symbols) with time for as-cast PSHQ10 specimens during isothermal annealing at various temperatures (°C): (○, ●) 130, (△, ▲) 135, (□, ■) 140, and (▽, ▼) 145. Small amplitude oscillatory deformations with strain amplitude of 0.05 and angular frequency of 0.237 rad/s were applied to the specimens. A fresh specimen was used for each run. (Reprinted from Han et al., *Molecular Crystals and Liquid Crystals* 254:335. Copyright © 1994, with permission from Taylor & Francis Group.)

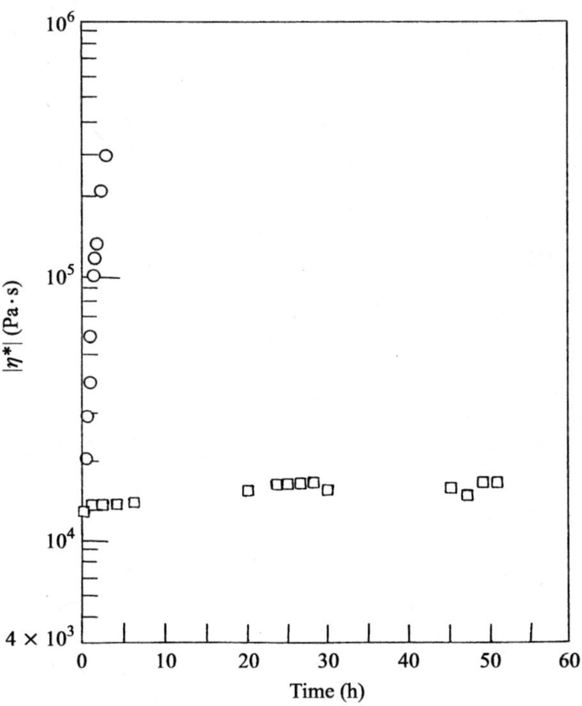

Figure 9.34 Variations of $|\eta^*|$ with time during annealing at 130 °C for two as-cast PSHQ10 specimens having the following thermal histories: (○) an as-cast specimen placed in the cone-and-plate fixture preheated at 130 °C, and (□) an as-cast specimen placed in the cone-and-plate fixture preheated at 190 °C, subjected to steady-state shear flow at $\dot{\gamma} = 0.085\ \mathrm{s}^{-1}$ for 5 min at 190 °C, and then cooled slowly down to 130 °C. Small amplitude oscillatory deformations with strain amplitude of 0.05 and angular frequency of 0.237 rad/s were applied to the specimens. (Reprinted from Han et al., *Molecular Crystals and Liquid Crystals* 254:335. Copyright © 1994, with permission from Taylor & Francis Group.)

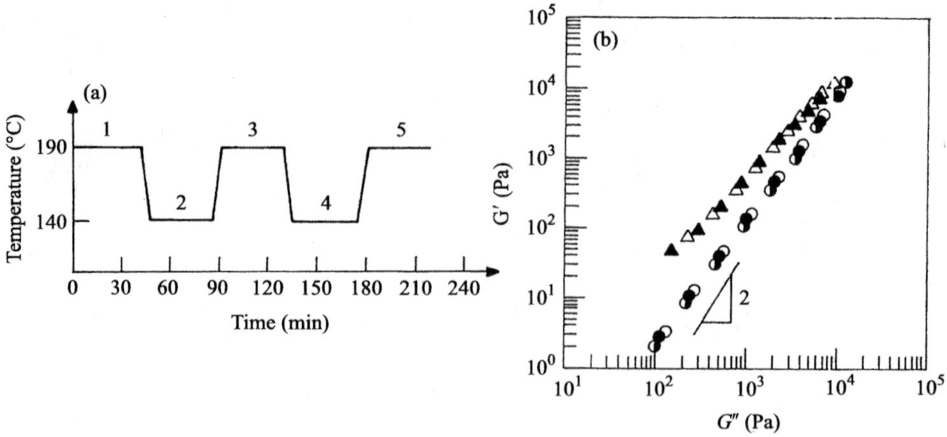

Figure 9.35 (a) Temperature protocols employed and (b) log G' versus log G'' plots for an as-cast PSHQ10 specimen having the thermal histories as indicated in the temperature protocol: (○) step 1, (△) step 2, (○) step 3, (▲) step 4, and (●) step 5. (Reprinted from Han et al., *Molecular Crystals and Liquid Crystals* 254:335. Copyright © 1994, with permission from Taylor & Francis Group.)

reproduced after each time the specimen was first cleared at 190 °C in the isotropic region. This was not the case for the 73/27 HBA/HNA copolyester, in Figures 9.30 and 9.31. It is worth noting in Figure 9.35 that the slope of log G' versus log G'' plots in the terminal region is much less than 2 when PSHQ10 is in the nematic state (at 140 °C), while the slope is 2 when PSHQ10 is in the isotropic state (at 190 °C). This observation is very similar to that made in Chapter 8 for block copolymers, namely, the slope of log G' versus log G'' plots in the terminal region is less than 2 when a block copolymer is microphase separated (at $T < T_{\mathrm{ODT}}$), while the slope is 2 when the block copolymer is in the disordered state (at $T \geq T_{\mathrm{ODT}}$).

The difference in the variation of $|\eta^*|$ with time between the two specimens, one with thermal treatment in the isotropic region and the other without thermal treatment, can be explained with the aid of DSC thermograms given in Figure 9.36. With reference to Figure 9.36a, an as-cast PSHQ10 specimen annealed at 130 °C for 20 h exhibits four thermal transitions: (1) at about 77 °C, representing the glass transition temperature (T_g), (2) at 107 °C, representing the melting temperature (T_{m2}) of crystals, (3) at 152 °C, representing the melting temperature (T_{m1}) of the crystals that were formed during isothermal annealing, referred to as high-temperature melting crystals (Kim and Han 1993a, Han et al. 1994b), and (4) at about 174 °C, representing T_{NI}. However, for an as-cast PSHQ10 specimen that first received thermal treatment at 190 °C in the isotropic region and before being annealed at 130 °C for 20 h, as shown in Figure 9.36b, we observe that the intermediate endothermic peak representing T_{m1} disappears after annealing at 130 °C for 20 h. Therefore, we can conclude that the rapid increase of G' and G'' (see Figure 9.33), and also the rapid increase of $|\eta^*|$ (see Figure 9.34), with time during isothermal annealing at various temperatures of PSHQ10 specimens without being cleared in the isotropic region is due to the formation and growth of high-temperature melting crystals. The same explanation also

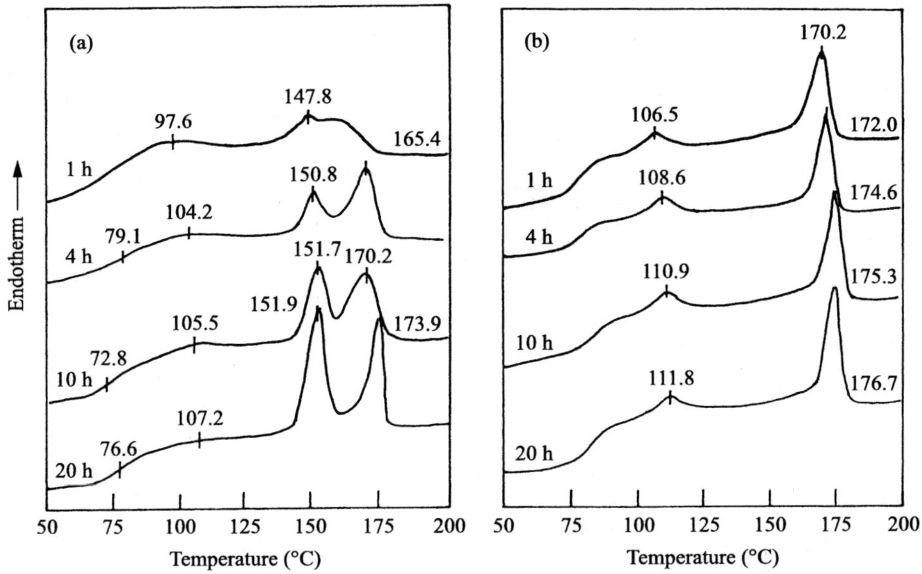

Figure 9.36 DSC thermograms for (a) as-cast PSHQ10 specimens annealed at 130 °C for different periods, as indicated on the plot, and (b) as-cast PSHQ10 specimens annealed at 130 °C for different periods, as indicated on the plot, having the following thermal histories: an as-cast specimen was first annealed at 130 °C for 30 min, then heated to 190 °C in the isotropic region and held there for 30 min, and finally cooled down slowly to 130 °C for further annealing. A fresh specimen was used for each run and the heating rate was 20 °C/min. (Reprinted from Han et al., *Molecular Crystals and Liquid Crystals* 254:335. Copyright © 1994, with permission from Taylor & Francis Group.)

applies to Figure 9.35, in which the rheological measurements at 140 °C in the nematic region are reproduced, because the specimen was first cleared in the isotropic region and thus the high-temperature melting crystals apparently were not formed within the time frame of the rheological measurements performed.

The significance of Figures 9.34–9.36 lies in that it is possible for one to take reproducible rheological measurements in the nematic region of PSHQ10 by first heating a specimen to the isotropic region (say, to 190 °C) and then cooling down slowly to a predetermined temperature below its T_{NI}. This possibility exists because PSHQ10 is thermally stable at temperatures below 350 °C in the presence of an antioxidant. However, such thermal treatment is not possible for 73/23 HBA/HNA copolyester, for instance, for the reasons presented with reference to Figure 9.28.

9.4.2 Transient Shear Flow of Thermotropic Main-Chain LCPs

Transient shear flow often provides us with very valuable information on the unique rheological characteristics of polymeric liquids, particularly multiphase polymeric liquids including LCPs. A better understanding of the transient shear flow behavior of TLCPs is very important to many processing operations. Transient shear flow of TLCPs has been investigated by a number of research groups (Chang and Han 1997a, 1997b; Cocchini et al. 1991; Guskey and Winter 1991; Han et al. 1994a; Kim and Han 1993b, 1993c, 2000; Kim and Han 2000). In this section, we present some unusual rheological responses of TLCPs during transient shear flow; namely, the growth of shear stress and first normal stress difference during transient shear flow, microstructural evolution during transient shear flow, and strain scaling in transient shear flow.

9.4.2.1 Growth of Shear Stress and First Normal Stress Difference During Transient Shear Startup of Thermotropic Main-Chain LCPs

Owing to the strong dependence of the rheological behavior of TLCP on thermal and deformation histories, the transient shear flow data for 73/27 HBA/HNA copolyester reported by two laboratories have been found to be quite different (Cocchini et al. 1991; Guskey and Winter 1991). As we have already pointed out, it is extremely important for one to control the initial morphology (i.e., initial conditions) of test specimens in order to obtain reproducible transient shear flow data for TLCPs.

Figure 9.37 gives the temperature protocol, the variations of normal stress from the time of specimen loading at 320 °C to the end of the stress relaxation after cessation

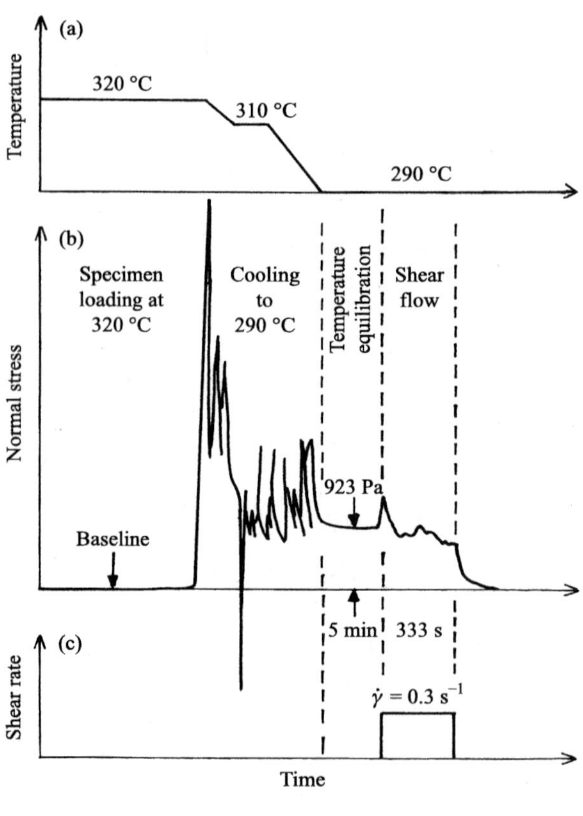

Figure 9.37 (a) Temperature protocol employed, (b) trace of normal stress during thermal equilibration, during transient and steady-state shear flow, and during stress relaxation after cessation of steady-state shear flow, and (c) the duration of the applied shear flow at a rate of 0.3 s^{-1} for a compression-molded 73/27 HBA/HNA copolyester specimen. (Reprinted from Han and Chang, *Journal of Rheology* 38:241. Copyright © 1994, with permission from the Society of Rheology.)

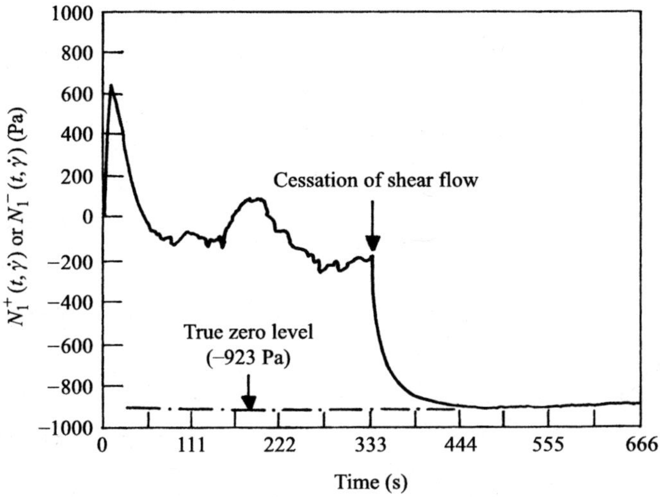

Figure 9.38 Trace of first normal stress difference of a compression-molded 73/27 HBA/HNA copolyester specimen during transient and steady-state shear flow, and during the relaxation after cessation of steady-state shear flow. The normal stress before applying a sudden shear flow to the specimen is taken to be zero. (Reprinted from Han and Chang, *Journal of Rheology* 38:241. Copyright © 1994, with permission from the Society of Rheology.)

of shear flow at 290 °C, and the duration of applied shear flow at $\dot{\gamma} = 0.3\,\text{s}^{-1}$ and 290 °C for a 73/27 HBA/HNA copolyester specimen. The following observations are worth noting in Figure 9.37: (1) after temperature equilibrium at 290 °C for 5 min, the unrelaxed normal stress of 923 Pa was present in the specimen, (2) upon startup of shear flow at $\dot{\gamma} = 0.3\,\text{s}^{-1}$, the normal stress goes through a maximum and then another mild overshoot before reaching a steady state, (3) upon cessation of shear flow, the normal stress relaxes to approaching zero, which is the baseline that we had before sample loading began. Figure 9.38 gives the trace of the growth of first normal stress difference $N_1^+(t, \dot{\gamma})$ at $\dot{\gamma} = 0.3\,\text{s}^{-1}$ upon startup of shear flow, and the decay of first normal stress difference $N_1^+(t, \dot{\gamma})$ upon cessation of shear flow, as a function of time, by regarding the normal stress before the application of a sudden shear flow to be zero. What is of great interest in Figure 9.38 is that after cessation of shear flow, $N_1^+(t, \dot{\gamma})$ decays with time, approaching the value of −923 Pa, which is actually the original baseline.

Figure 9.39 gives the trace of normal stress for a 73/27 HBA/HNA copolyester specimen, which was loaded at 320 °C and then allowed to a rest at 290 °C for 4 h, showing the unrelaxed normal stress decreased from 980 to 195 Pa at the end of 4 h waiting period. This indicates that the unrelaxed normal stress, which was introduced by the squeeze flow during specimen loading, decreases with increasing waiting time under a quiescent isothermal condition. Upon applying a sudden shear flow at $\dot{\gamma} = 0.5\,\text{s}^{-1}$, the normal stress of the specimen exhibits a very large overshoot and then relaxes to zero value after cessation of shear, which is consistent with that given in Figure 9.37. The very large peak value of $N_1^+(t, \dot{\gamma})$ observed in Figure 9.39, compared

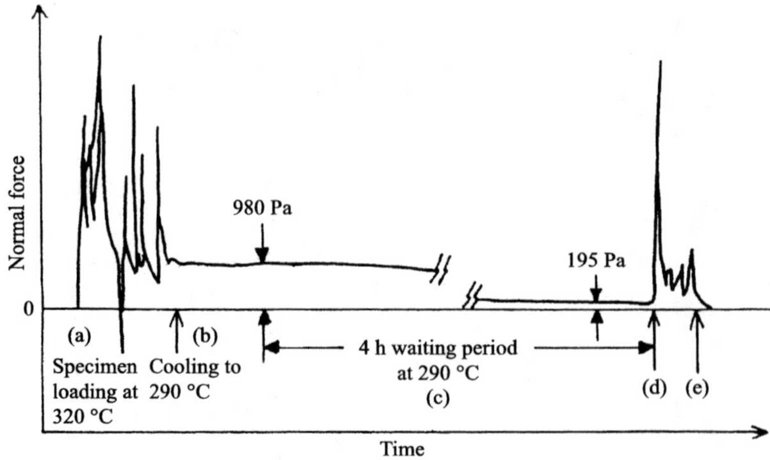

Figure 9.39 Trace of normal stress for a compression-molded 73/27 HBA/HNA copolyester specimen: (a) during sample loading at 320 °C, (b) during squeezing and cooling to 290 °C, (c) during temperature equilibration at 290 °C for 4 h, (d) transient and steady-state shear flows at $\dot{\gamma} = 0.5\ \mathrm{s}^{-1}$ for 200 s, and (e) during relaxation after cessation of shear flow. An unrelaxed normal stress of 195 Pa was present in the specimen before being subjected to shear flow. (Reprinted from Han et al., *Molecular Crystals and Liquid Crystals* 254:335. Copyright © 1994, with permission from Taylor & Francis Group.)

with the moderate size of $N_1^+(t, \dot{\gamma})$ peak observed in Figure 9.37, indicates that the morphological state of the specimen after annealing at 290 °C for 4 h (Figure 9.39) is quite different from that after annealing at 290 °C for 5 min (Figure 9.37).

Figure 9.40 gives plots of $N_1^+(t, \dot{\gamma})$ versus shear strain $\dot{\gamma}t$ at various shear rates (0.05, 0.1, and 0.3 s^{-1}) for 73/27 HBA/HNA copolyester specimens at 290 °C. Notice in Figure 9.40 that the true zero level of normal stress is regarded as the reference value, and that at $\dot{\gamma}t = 0$ the initial value of normal stress is 940 Pa when sheared at $\dot{\gamma} = 0.05\,\mathrm{s}^{-1}$, 1,060 Pa when sheared at $\dot{\gamma} = 0.1\,\mathrm{s}^{-1}$, and 923 Pa when sheared at $\dot{\gamma} = 0.3\,\mathrm{s}^{-1}$. Ideally, the initial value of normal stress should be the same in all three runs. This observation points out a real difficulty in controlling the initial condition of the specimen, because the physical squeezing of a specimen placed in the cone-and-plate fixture, however careful one might be, is hard to reproduce, thus giving rise to variations in the amount of normal stress introduced during sample loading. In Figure 9.40, we observe that values of steady-state first normal stress difference N_1 for the 73/27 HBA/HNA copolyester specimens are positive. However, information in Figure 9.40 is of little rheological significance because rheological measurement for each run was made under the condition where a substantial amount of unrelaxed normal stress was present in the specimen before being subjected to shear startup. If the presence of unrelaxed normal stress in the specimen at the time of shear startup is neglected, one can erroneously conclude from Figure 9.37 or Figure 9.39 that the value of N_1 is negative.

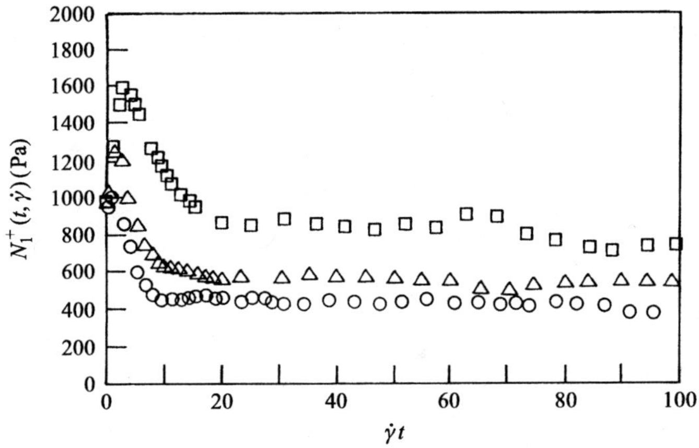

Figure 9.40 Plots of $N_1^+(t, \dot{\gamma})$ versus $\dot{\gamma}t$ for compression-molded 73/27 HBA/HNA copolyester specimens at 290 °C for various shear rates (s^{-1}): (○) 0.05, (△) 0.1, and (□) 0.3. A fresh specimen was used for each applied shear rate. Each specimen, after being cooled from 320 to 290 °C, was annealed for 5 min before being subjected to shear flow. The unrelaxed normal stress present in the specimens (i.e., the values of $N_1^+(t = 0, \dot{\gamma})$) was 940 Pa for the run at 0.05 s^{-1}, 1,060 Pa for the run at 0.1 s^{-1}, and 923 Pa for the run at 0.3 s^{-1}. (Reprinted from Han et al., *Molecular Crystals and Liquid Crystals* 254:335. Copyright © 1994, with permission from Taylor & Francis Group.)

We now describe how experimental difficulties with 73/27 HBA/HNA copolyester can be eliminated entirely in the use of PSHQ10. Figure 9.41 gives the trace of normal stress for a PSHQ10 specimen from the instant of specimen loading in the cone-and-plate fixture at 190 °C to the end of steady-state shear flow at 160 °C and 0.1 s^{-1}. In Figure 9.41 we observe that upon shear startup, the normal stress goes through a large maximum (exceeding the limit of the setting of the chart recorder) followed by a mild peak and then finally levels off to a constant value. Upon cessation of shear flow, the normal stress decreases rapidly to zero, which was the baseline as determined before sample loading. The following temperature protocol was employed by Han et al. (1994a). An as-cast specimen was placed in the cone-and-plate fixture which had been preheated to 190 °C, which is 15 °C above the T_{NI} of PSHQ10. After the temperature was equilibrated at 190 °C, the gap setting was adjusted from about 2 mm (the specimen thickness) to 160 μm by applying squeeze flow and the excess material was trimmed off. The entire process took about 20 min. Then, the temperature of the specimen was lowered from 190 to 160 °C, which took about 25 min. After temperature equilibration at 160 °C for 5 min, a sudden shear flow was applied to the specimen. It is significant to note in Figure 9.41 that just before shear startup, the normal stress was zero, which was the baseline as determined before sample loading.

As we have demonstrated, an erroneous conclusion can be drawn on the sign of first normal stress difference in TLCP if the transient or steady-state shear flow experiments were carried out in the presence of residual normal force that was generated

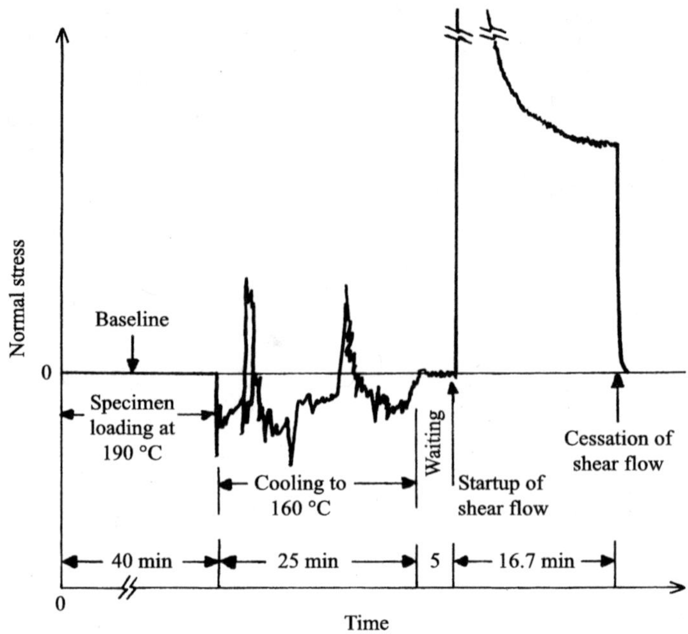

Figure 9.41 Trace of normal stress for an as-cast PSHQ10 specimen: (a) during sample loading at 190 °C, (b) during squeezing and cooling to 160 °C, (c) during temperature equilibration at 160 °C for 5 min, (d) transient and steady-state shear flows at $\dot{\gamma} = 0.1\ \mathrm{s}^{-1}$ for 2,000 s, and (e) during relaxation after cessation of shear flow. The normal stress of the specimen before being subjected to shear flow is zero, which is the baseline as determined before specimen loading. (Reprinted from Han et al., *Molecular Crystals and Liquid Crystals* 254:335. Copyright © 1994, with permission from Taylor & Francis Group.)

during the squeezing of the specimen in an anisotropic state. As a matter of fact, the experimental results reported by Guskey and Winter (1991), who employed 73/27 HBA/HNA copolyester and observed negative values of N_1, turned out to be erroneous for the very reasons we have described, as pointed out subsequently by Han and Chang (1994). The residual normal force in the specimen can be made to relax completely if one waits for a very long time after a specimen is loaded in the cone-and-plate fixture of a rheometer. However, by the time the residual normal force in the specimen relaxes completely, the polymer will no longer have the same morphological state because the morphological state of a TLCP in an anisotropic state depends very much on both thermal and deformation histories. The only way to circumvent this difficulty, as demonstrated in Figure 9.42, is to first load a specimen in the cone-and-plate fixture of a rheometer at a temperature above the clearing temperature of the specimen, wait until the normal stress becomes zero at rest, and then decrease the temperature very slowly stepwise, ascertaining that the normal stress returns to the baseline at each temperature. Notice in Figure 9.42 that upon decreasing the temperature, say 5 °C, instantly the normal stress increases and then decreases to a negative value because the gap opening between the cone and the plate must be adjusted to keep

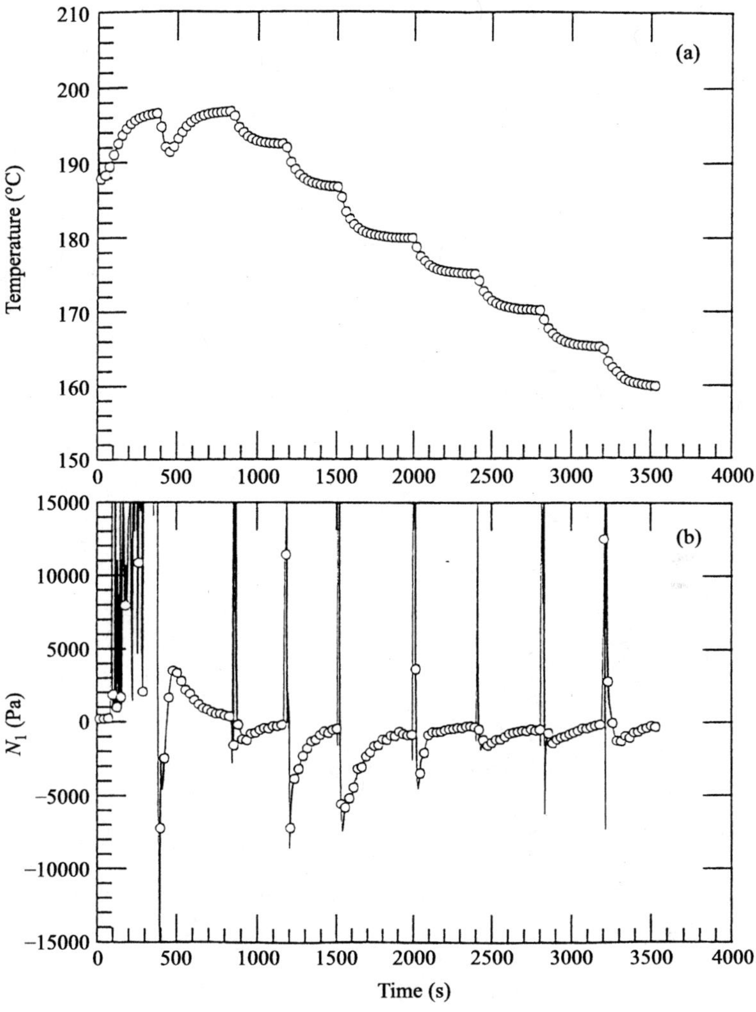

Figure 9.42 (a) Temperature protocols and (b) variations of N_1 with time during the loading of an as-cast PSHQ10 specimen at 195 °C in the isotropic region followed by subsequent stepwise cooling with a temperature interval of 5 °C down to 160 °C in the nematic region. The clearing temperature of PSHQ10 is 179 °C. Note that N_1 returns to the baseline (zero value) after a step change in temperature. (Reprinted from Kim and Han, *Macromolecules* 33:3349. Copyright © 2000, with permission from the American Chemical Society.)

the gap opening constant, but the normal stress returns slowly to the baseline after a delay. Such a procedure is essential in order to obtain reproducible initial conditions and reliable measurements of $N_1^+(t, \dot{\gamma})$ and $N_1(\dot{\gamma})$ for TLCPs in general. Needless to say, such a procedure cannot be adopted if the TLCP has a clearing temperature that is higher than the thermal degradation temperature (e.g., HBA/HNA and HBA/PET copolyesters).

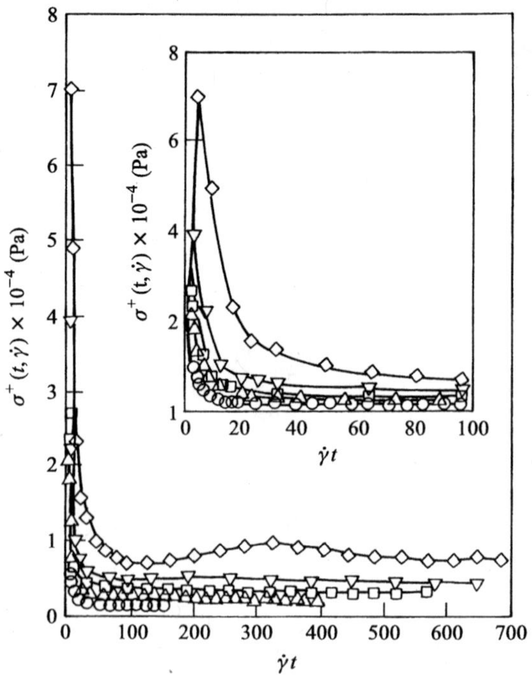

Figure 9.43 Plots of $\sigma^+(t, \dot\gamma)$ versus $\dot\gamma t$ for as-cast PSHQ10 specimens at 140 °C for various shear rates (s^{-1}): (○) 0.107, (△) 0.269, (□) 0.536, (▽) 1.07, and (◇) 2.69. A fresh specimen was used for each shear rate, and each specimen received thermal treatment in the isotropic region at 190 °C. (Reprinted from Kim and Han, *Journal of Rheology* 37:847. Copyright © 1993, with permission from the Society of Rheology.)

Figure 9.43 shows variations of $\sigma^+(t, \dot\gamma)$ with time, upon shear startup, for PSHQ10 at 140 °C for different values of $\dot\gamma$, where a fresh specimen was employed for each $\dot\gamma$. The following observation is worth noting in Figure 9.43. Upon shear startup, $\sigma^+(t, \dot\gamma)$ goes through a maximum in a very short time and multiple peaks in $\sigma^+(t, \dot\gamma)$ appear at $\dot\gamma = 2.69\,s^{-1}$, but only a single peak appears at lower $\dot\gamma$ (0.536 and 0.107 s^{-1}). As the applied $\dot\gamma$ increases, the value of $\dot\gamma t$ at which a maximum in $\sigma^+(t, \dot\gamma)$ appears to shift towards a higher value, although the extent of the shift is seen to be rather small. The ratio of the maximum to the equilibrium value of shear stress, $\sigma_{max}/\sigma_\infty$, is found to lie between 8 and 12, depending upon the applied $\dot\gamma$. Such large values of $\sigma_{max}/\sigma_\infty$ ratio are believed to be characteristic of TLCPs in general and are attributable to the existence of polydomains, which were present in the specimens in the nematic state before being subjected to a sudden shear flow. Upon shear startup, the grain boundary of polydomains disappears, tending to give rise to monodomains, in which all directors are aligned more or less in the same direction.

Figure 9.44 shows variations of $N_1^+(t, \dot\gamma)$ with time upon shear startup for PSHQ10 at 140 °C for different values of $\dot\gamma$, where a fresh specimen was employed for each $\dot\gamma$. The following observation is worth noting in Figure 9.44. Upon shear startup, $N_1^+(t, \dot\gamma)$ goes through a maximum at a critical value of $\dot\gamma t$, which becomes greater with increasing $\dot\gamma$. Also, multiple peaks in $N_1^+(t, \dot\gamma)$ appear at $\dot\gamma$ greater than a certain critical value, but only a single peak at $\dot\gamma = 0.107\,s^{-1}$. The value of $\dot\gamma t$ at which a maximum in $N_1^+(t, \dot\gamma)$ appears increases with increasing $\dot\gamma$. As shown in the inset, $N_1^+(t, \dot\gamma)$ goes through negative values at a very early stage of transient shear flow but soon turns into positive values as shearing continues, giving rise to *positive* values of N_1; that is, only

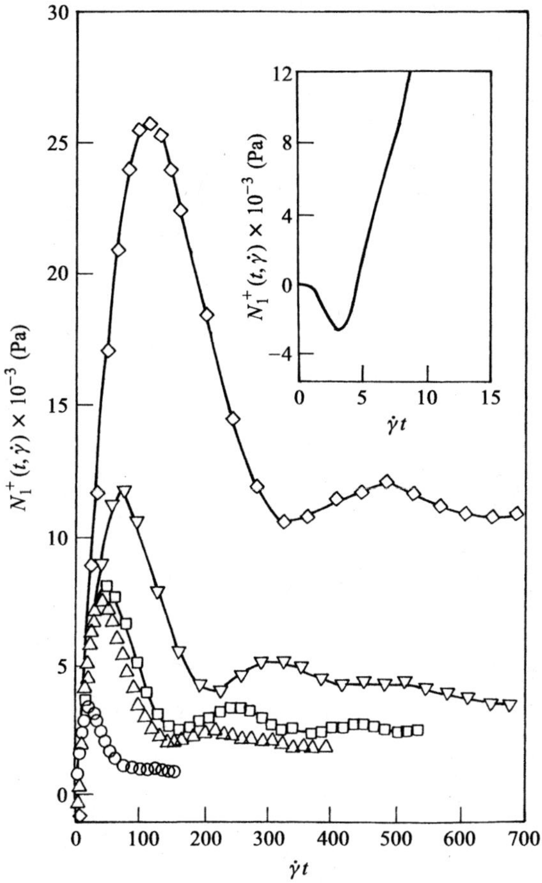

Figure 9.44 Plots of $N_1^+(t, \dot{\gamma})$ versus $\dot{\gamma}t$ for as-cast PSHQ10 specimens at 140 °C for various shear rates (s^{-1}): (○) 0.107, (△) 0.269, (□) 0.536, (▽) 1.07, and (◇) 2.69. A fresh specimen was used for each shear rate, and each specimen received thermal treatment in the isotropic region at 190 °C. (Reprinted from Kim and Han, *Journal of Rheology* 37:847. Copyright © 1993, with permission from the Society of Rheology.)

transient *negative* normal stress difference is observed. At low $\dot{\gamma}$ (say, $\dot{\gamma} < 0.536\,s^{-1}$), $N_1^+(t, \dot{\gamma})$ is seen to be positive over the entire period of shear flow tested. The ratio of the peak to the steady-state value of first normal stress difference, $N_{1,max}/N_1$, is about 2.5, depending upon the applied $\dot{\gamma}$. Kim and Han (1993c) have reported similar observation for other temperatures. Such large values of the ratio $N_{1,max}/N_1$ are believed to be the characteristics of TLCPs in general and are attributable to the existence of polydomains, which were present in the PSHQ10 specimens in the nematic state before being subjected to a sudden shear flow.

For comparison, transient rheological responses are given in Figure 9.45 for a polystyrene (PS) at 180 °C and $\dot{\gamma} = 1\,s^{-1}$, and in Figure 9.46 for a high-density polyethylene (HDPE)) at 180 °C and $\dot{\gamma} = 2\,s^{-1}$. In Figure 9.45 we observe that a very small overshoot (approximately 25% of the steady-state value) in $\sigma^+(t, \dot{\gamma})$ appears at $\dot{\gamma}t = 3$ (3 s upon shear startup) and then it decays to a steady state at $\dot{\gamma}t = 10$, whereas no overshoot in $N_1^+(t, \dot{\gamma})$ occurs. In Figure 9.46 we observe no overshoot in both $\sigma^+(t, \dot{\gamma})$ and $N_1^+(t, \dot{\gamma})$ for HDPE. It is clearly seen that the transient rheological responses for PS and HDPE (Figures 9.45 and 9.46) are quite different from those for PSHQ10 (Figures 9.43 and 9.44) in several aspects. First, a very large overshoot in both

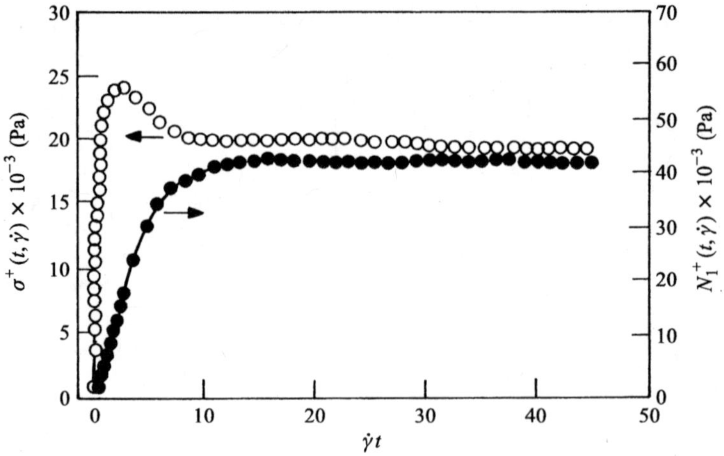

Figure 9.45 Plots of $\sigma^+(t, \dot{\gamma})$ and $N_1^+(t, \dot{\gamma})$ versus $\dot{\gamma}t$ for a PS specimen at 180 °C and $\dot{\gamma} = 1\,\text{s}^{-1}$.

$\sigma^+(t, \dot{\gamma})$ (300–400% of the steady-state value) and $N_1^+(t, \dot{\gamma})$ (approximately 200% of the steady-state value) is observed in PSHQ10, whereas very little or no overshoot in $\sigma^+(t, \dot{\gamma})$ and $N_1^+(t, \dot{\gamma})$ is observed in PS and HDPE. Second, the peak value of $N_1^+(t, \dot{\gamma})$ is 3–4 times greater than the peak value of $\sigma^+(t, \dot{\gamma})$ in PSHQ10, whereas no overshoot in $N_1^+(t, \dot{\gamma})$ occurs in both PS and HDPE. Third, the steady-state first normal stress difference (N_1) is greater than the steady-state shear stress (σ) in PSHQ10, whereas the opposite trend is seen in PS and HDPE. The origin(s) of the large overshoot in $\sigma^+(t, \dot{\gamma})$ and $N_1^+(t, \dot{\gamma})$ observed, upon shear startup, in PSHQ10 in the nematic state is attributable to the reorientation of the directors along the flow direction, which were distributed randomly in the polydomain of a solvent-cast specimen before shear startup. It has been observed that the magnitude of overshoot in $\sigma^+(t, \dot{\gamma})$

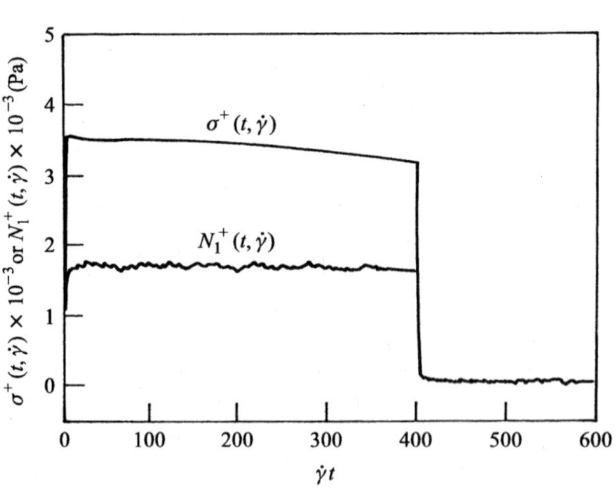

Figure 9.46 Plots of $\sigma^+(t, \dot{\gamma})$ and $N_1^+(t, \dot{\gamma})$ versus $\dot{\gamma}t$ for an HDPE specimen at 180 °C and $\dot{\gamma} = 2\,\text{s}^{-1}$.

and $N_1^+(t,\dot{\gamma})$ decreases when a TLCP specimen was presheared for a sufficiently long time (Beekmans et al. 1997). Such an experimental observation appears to support the view that reorientation of the directors is indeed responsible for the large overshoot in $\sigma^+(t,\dot{\gamma})$ and $N_1^+(t,\dot{\gamma})$ observed in a solvent-cast PSHQ10 specimen (Figures 9.43 and 9.44), because preshearing helps the directors to reorient, tending to give rise to monodomain texture from polydomain texture of a specimen (see Figure 9.25).

Let us compare the transient rheological responses in $\sigma^+(t,\dot{\gamma})$ and $N_1^+(t,\dot{\gamma})$ observed in Figures 9.43 and 9.44 with the theoretical predictions given in Figures 9.18 and 9.19. It is very interesting to observe similarities between experiment and prediction, in spite of the fact that the Marrucci–Maffettone analysis of two-dimensional shear flow is based on many simplifying assumptions, including the use of the Maier–Saupe potential, Eq. (9.14), which may not describe adequately the intermolecular interactions in PSHQ10 which has ten methylene groups as flexible spacers in the main chain and bulky phenylsulfonyl pendent side groups. Nevertheless, the qualitative feature of the Marrucci–Maffettone analysis is very encouraging. Note that the Marrucci–Maffettone analysis is based on the Doi molecular theory presented in this chapter.

What is unclear at the present time is the explanation as to why the theories based on a rheological model, which predict director tumbling and wagging, also predict the transient rheological responses in $\sigma^+(t,\dot{\gamma})$ and $N_1^+(t,\dot{\gamma})$ that resemble so much the experimental results given Figures 9.43 and 9.44. Nevertheless, it is fair to state that at present there is no theory that can adequately predict the experimentally observed transient shear flow behavior of PSHQ10 on the basis of a realistic rheological model describing the intermolecular interactions in a polymer containing flexible spacers and bulky pendent side groups. In order to predict the rheological behavior of such TLCPs, one must first develop a suitable expression for intermolecular potential, which is expected to depend not only on orientational order parameter, which in turn depends on temperature, but also on the persistence length of polymer.

9.4.2.2 Microstructural Evolution During Shear Startup of Thermotropic Main-Chain LCPs

Figure 9.47 gives the time evolution of shear stress and POM images of PSHQ10 at 160 °C and $\dot{\gamma} = 0.5\,\text{s}^{-1}$ during 15 s upon shear startup, indicating a correlation of textural yielding with the shear stress maximum. The PSHQ10 specimen was first heated to 200 °C in the isotropic phase, where it was annealed for 5 min and subsequently cooled to the test temperature of 160 °C. The specimen was then annealed for an additional 5 min to ensure thermal equilibrium (Mather et al. 2000). This thermal history resulted in the quite turbid texture shown for $t = 0$ min in Figure 9.47. In this case, little light is transmitted and the relatively featureless POM image indicates a characteristic texture scale smaller than 2 μm. Following shear startup at $\dot{\gamma} = 0.5\,\text{s}^{-1}$, the fine texture coarsens gradually during the first 10 s (5 strain units) to yield a more transparent film, indicating less scattering and significant director orientation away from the flow axis (incident polarization direction). At the peak value of $\sigma^+(\dot{\gamma},t)$, there begins a "yielding" of the tight unoriented texture to bright, stretched domains whose orientation and elongation appear to follow the decrease in $\sigma^+(\dot{\gamma},t)$ toward a steady state with time. Gradually, over the next 2–3 min (100 strain units) the overall

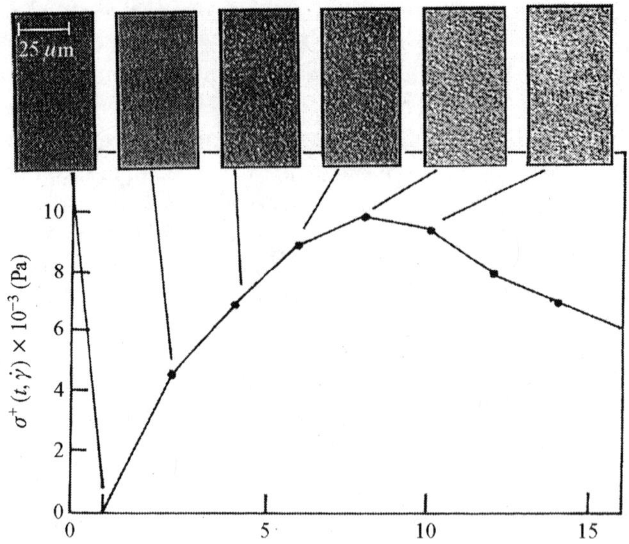

Figure 9.47 Time evolution of shear stress and POM images of an as-cast PSHQ10 specimen at 160 °C and $\dot{\gamma} = 0.5\ \mathrm{s}^{-1}$ during 15 s upon shear startup, indicating a correlation of textural yielding with the shear stress maximum. (Reprinted from Mather et al., *Macromolecules* 33:7594. Copyright © 2000, with permission from the American Chemical Society.)

brightness decreased, indicating continued enhancement of alignment along the flow axis, while the textural length scale remained fairly steady. A steady state was reached after a strain of approximately 50 units.

9.4.2.3 Strain Scaling During Transient Shear Flow of Thermotropic Main-Chain LCPs

Figure 9.48 gives plots of $\sigma^+(t,\dot{\gamma})/\sigma$ versus $\dot{\gamma}t$ and $N_1^+(t,\dot{\gamma})/N_1$ versus $\dot{\gamma}t$ for PSHQ10 at 130 and 140 °C for three values of $\dot{\gamma}$: 0.107, 0.536, and 1.07 s^{-1}. Here, σ denotes the shear stress at steady state and N_1 denotes the first normal stress difference at steady state. It should be pointed out that the transient shear flow experiments performed at 130 and 140 °C are in the nematic region. In Figure 9.48 we observe that the ratio $\sigma^+(t,\dot{\gamma})/\sigma$ scales with $\dot{\gamma}t$, while the ratio $N_1^+(t,\dot{\gamma})/N_1$ does not.

9.4.3 Flow Aligning Behavior of Thermotropic Main-Chain LCPs

One very important aspect that needs to be brought out here is the experimental evidence for flow aligning behavior of PSHQ10 or other TLCPs. In all reliable experimental results reported to date on various TLCPs (Baek et al. 1994; Chang and Han 1997a, 1997b; Han et al. 1994a; Kim and Han 1993b, 1993c, 1994a; Zhou et al. 1999), only positive values of N_1 have been found.

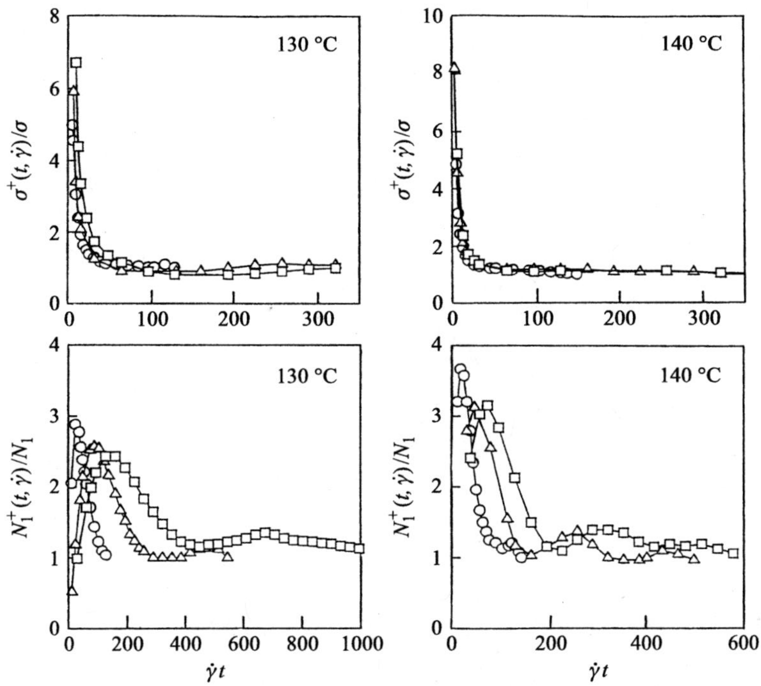

Figure 9.48 Plots of $\sigma^+(t,\dot{\gamma})/\sigma$ versus $\dot{\gamma}t$ and $N_1^+(t,\dot{\gamma})/N_1$ versus $\dot{\gamma}t$ for as-cast PSHQ10 specimens, upon shear startup at 130 °C or 140 °C and at three different shear rates (s^{-1}): (○) 0.107, (△) 0.536, and (□) 1.07. (Reprinted from Mather et al., *Macromolecules* 33:7594. Copyright © 2000, with permission from the American Chemical Society.)

Using the Ericksen–Leslie theory, Ugaz and Burghardt (1998) derived the following expression for steady-state shear flow:

$$\frac{N_1}{\sigma} = \frac{2}{\sqrt{\lambda^2 - 1}} \tag{9.79}$$

where λ is a tumbling parameter, defined by Eq. (9.65) in terms of the Leslie coefficients, describing the propensity toward tumbling ($\lambda < 1$) or flow-aligning ($\lambda > 1$) behavior. Note that as the value of λ approaches 1, the closer flow alignment is to the flow direction; that is, the smaller is the angle between the nematic director and the flow axis (see Eq. (9.66)). They also presented a relationship between λ and the local molecular order parameter S_m by

$$\lambda = \frac{2 + S_m}{3S_m} \tag{9.80}$$

in which S_m is related to the bulk orientation parameter S by

$$S = S_m \overline{S} \tag{9.81}$$

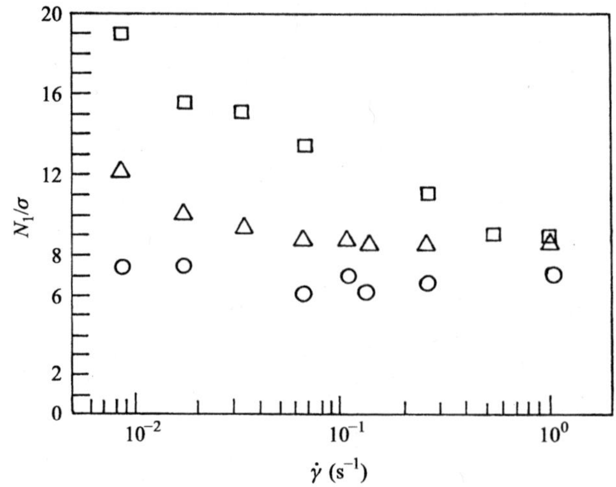

Figure 9.49 Plots of N_1/σ versus $\dot{\gamma}$ for steady-state shear flow of as-cast PSHQ10 specimens at three different temperatures (°C): (□) 140, (△) 150, and (○) 160.

where $\overline{S}$ is the mesoscopic order parameter. Note that S is a bulk orientation order parameter usually measured during shear flow, for instance by X-ray scattering.

Figure 9.49 gives plots of N_1/σ versus $\dot{\gamma}$ for as-cast PSHQ10 specimens at 140, 150, and 160 °C in the nematic region. In Figure 9.49 we observe that at 140 °C the ratio N_1/σ decreases with increasing $\dot{\gamma}$, while the shear-rate dependence of the ratio N_1/σ becomes weaker as the temperature increases to 150 and 160 °C. From Figure 9.49 we obtain, at $\dot{\gamma} = 1.07\,s^{-1}$, $N_1/\sigma = 8.3$ at 140 °C, $N_1/\sigma = 8.1$ at 150 °C, and $N_1/\sigma = 7.1$ at 160 °C. Thus, using the average value $N_1/\sigma = 7.8$, we estimate λ to be 1.03 from Eq. (9.79) and $S_m = 0.95$ from Eq. (9.80). The value of λ estimated above indicates that PSHQ10 exhibits flow-aligning behavior. The results of an X-ray scattering study indicate that for PSHQ10 at 160 °C, S increases from 0.3 to 0.6 as $\dot{\gamma}$ is increased from 0.05 to 5.0 s^{-1}. Thus the value of $S_m = 0.95$ estimated from Eq. (9.80) suggests that $\overline{S} = 0.32$–0.63 for PSHQ10 at 160 °C and $\dot{\gamma} = 0.05$–$5.0\,s^{-1}$. Using X-ray scattering, Ugaz and Burghardt (1998) have shown that a semiflexible main-chain TLCP, PSHQ6-12 (a random copolyester of PSHQ6 and PSHQ12), whose transient and steady-state shear flow behaviors had been investigated by Chang and Han (1997a), exhibits flow-aligning characteristics. Since both PSHQ10 and PSHQ6-12 are semiflexible, main-chain TLCPs having bulky phenylsulfonyl pendent side groups, Eq. (9.80) would only qualitatively describe their molecular orientations.

9.4.4 Intermittent Shear Flow of Thermotropic Main-Chain LCPs

Here, we present the transient and steady-state rheological responses after resting upon cessation of initial shear flow (referred to as intermittent shear flow).

Figure 9.50 gives variations of normal stress for PSHQ10 at 140 °C during the initial startup shear flow at $\dot{\gamma} = 0.536\,s^{-1}$, during the 2 h rest period after cessation of shear flow, and during the intermittent shear flow at $\dot{\gamma} = 0.536\,s^{-1}$. The following observations are worth noting in Figure 9.50. Upon applying a sudden shear flow to a

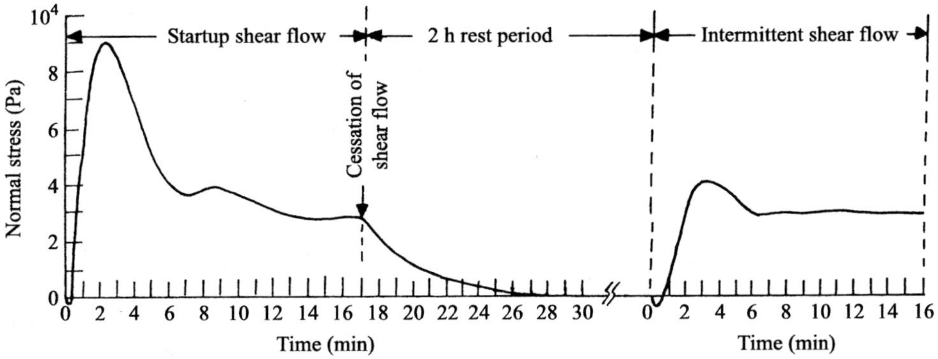

Figure 9.50 Variations of normal stress with time during the initial startup shear flow at $\dot{\gamma} = 0.536\ s^{-1}$, during relaxation after cessation of steady-state shear flow, and during the intermittent shear flow at $\dot{\gamma} = 0.536\ s^{-1}$, for an as-cast PSHQ10 specimen at 140 °C. (Reprinted from Han and Kim, *Journal of Rheology* 38:13. Copyright © 1994, with permission from the Society of Rheology.)

fresh specimen, the normal stress goes through a very large maximum, and then another mild overshoot before reaching steady state. Upon cessation of steady shear flow, it took about 11 minutes for the polymer to relax all the normal stresses. This is a rather long relaxation time, as compared with the relaxation time usually found in flexible homopolymers. Upon applying another sudden shear flow to the specimen, which had rested for 2 h, the normal stress goes through a maximum, the magnitude of which was less than one half of the peak value of the normal stress observed in the first startup shear flow. We also observe that in the intermittent shear flow, steady state is reached in about 6 min, which is much shorter than in the first startup shear flow (13 min), and that a very small transient negative normal stress region exists. The experimental results were obtained using a fresh specimen at 140 °C in the nematic state with zero value of normal stress before being subjected to a sudden shear flow. This was accomplished by first heating a fresh specimen to 190 °C in the isotropic state and then cooling very slowly to 140 °C in the nematic state.

Figure 9.51 gives variations of $N_1^+(t, \dot{\gamma})$ with time, and Figure 9.52 gives variations of $\sigma^+(t, \dot{\gamma})$ with time, at 140 °C during the intermittent (second startup) shear flow at $\dot{\gamma} = 0.536\,s^{-1}$ for PSHQ10 specimens having different rest periods (2, 17, and 67 h) after cessation of the previous (first) startup shear flow. Also given in Figures 9.51 and 9.52, for comparison, are variations of $\sigma^+(t, \dot{\gamma})$ and $N_1^+(t, \dot{\gamma})$ with time during the first shear startup (i.e., for an unsheared specimen). Again, a fresh specimen was used for each experiment. The following observations are worth noting in Figures 9.51 and 9.52: (1) the magnitude of the maximum overshoot in both $N_1^+(t, \dot{\gamma})$ and $\sigma^+(t, \dot{\gamma})$ increases with increasing rest period, (2) the maximum overshoot in $N_1^+(t, \dot{\gamma})$ appears at a strain unit of 100–150 after the first shear startup, and the time at which this maximum appears is delayed further as the rest period increases, (3) the maximum overshoot in $\sigma^+(t, \dot{\gamma})$ appears very soon (within a few strain units) after the first shear startup, (4) distinct multiple peaks appear in $N_1^+(t, \dot{\gamma})$, and (5) $N_1^+(t, \dot{\gamma})$ initially goes through small negative values and then becomes positive, increasing very rapidly with time and going through a maximum.

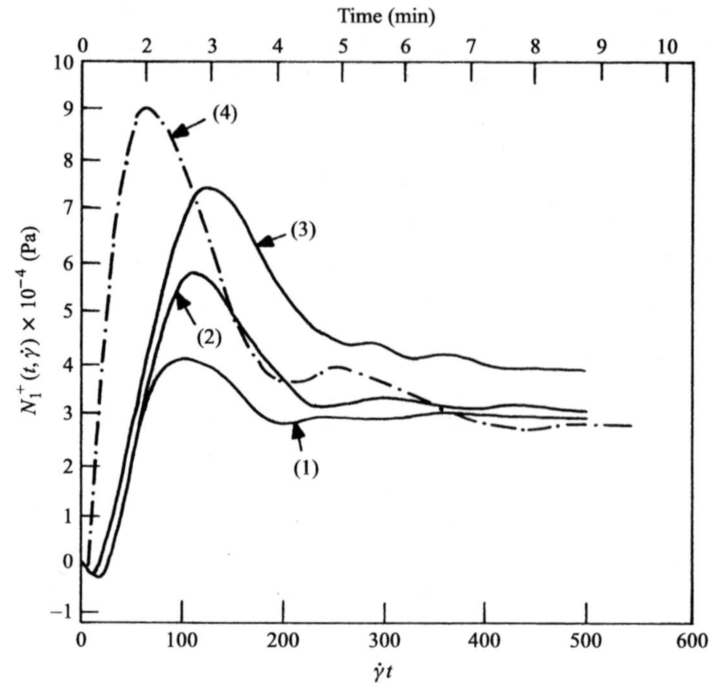

Figure 9.51 Variations of $N_1^+(t, \dot{\gamma})$ with time during the intermittent shear flow at 140 °C and $\dot{\gamma} = 0.536\,\text{s}^{-1}$ for as-cast PSHQ10 specimens: (1) with a rest period of 2 h after cessation of the initial startup shear flow, (2) with a rest period of 17 h after cessation of the initial startup shear flow, (3) with a rest period of 67 h after cessation of the initial startup shear flow. (4) Also given is the variation of $N_1^+(t, \dot{\gamma})$ with time during the *initial* startup shear flow. (Reprinted from Han and Kim, *Journal of Rheology* 38:13. Copyright © 1994, with permission from the Society of Rheology.)

9.4.5 Evolution of Dynamic Moduli of Thermotropic Main-Chain LCPs Upon Cessation of Shear Flow

Only a few studies (Guskey and Winter 1991; Han and Kim 1994a) have reported on the time variation of G', G'', and/or $|\eta^*|$ after cessation of shear flow of TLCP. Figure 9.53 shows the time evolution of G' and G'' during the rest period (t_R) up to 170 min for PSHQ10 specimens at 150 °C after cessation of steady-state shear flow, for which a fresh specimen was used for each shear rate. It can be seen in Figure 9.53 that the values of G'' steadily increase with time, while the values of G' stay more or less constant during the entire rest period investigated. The observed increase in G'' is attributable to the reorientation of the director during rest after cessation of shear flow; namely, the director oriented during shear flow begins to have random orientations, thus the texture consisting of a monodomain tends to form one consisting of polydomains. Notice in Figure 9.53 that the initial value of G'' decreases as the applied $\dot{\gamma}$ increases, but the rate of increase of G'' is greater as the applied $\dot{\gamma}$ is increased. Also, the value of G'' at the end of a rest period of 150 min is greater for the specimen sheared at $\dot{\gamma} = 1.07\,\text{s}^{-1}$,

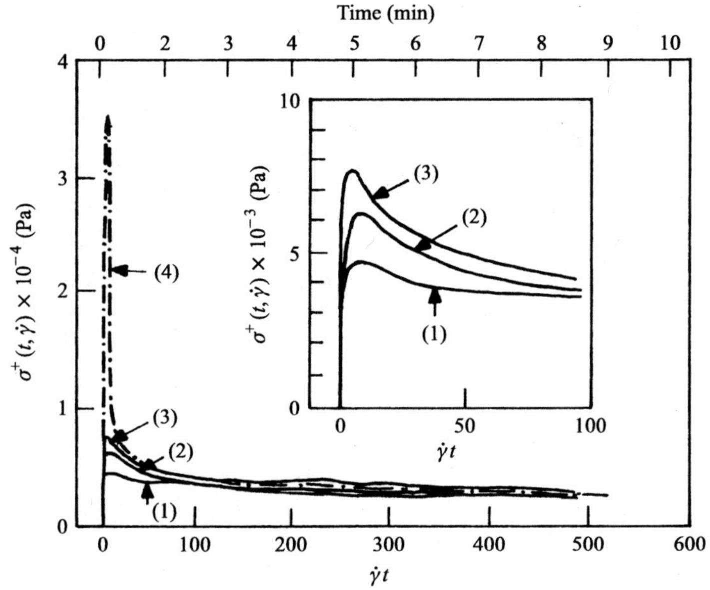

Figure 9.52 Variations of $\sigma^+(t, \dot{\gamma})$ with time during the intermittent shear flow at 140 °C and $\dot{\gamma} = 0.536\,\mathrm{s}^{-1}$ for as-cast PSHQ10 specimens: (1) with a rest period of 2 h after cessation of the initial startup shear flow, (2) with a rest period of 17 h after cessation of the initial startup shear flow, (3) with a rest period of 67 h after cessation of the initial startup shear flow. (4) Also given is the variation of $\sigma^+(t, \dot{\gamma})$ with time during the *initial* startup shear flow. (Reprinted from Han and Kim, *Journal of Rheology* 38:13. Copyright © 1994, with permission from the Society of Rheology.)

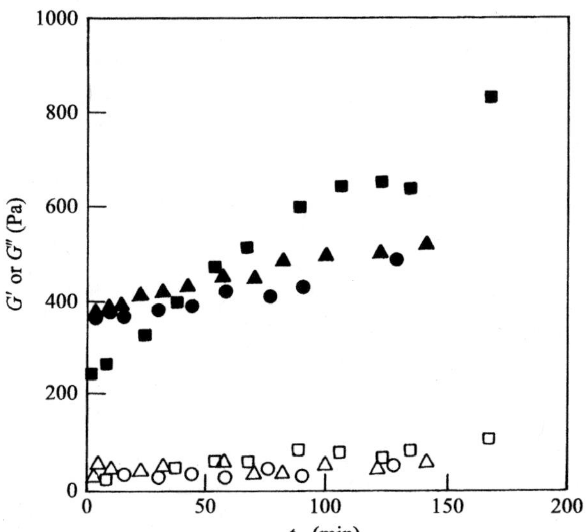

Figure 9.53 Variations of G' (open symbols) and G'' (filled symbols) versus t_R for as-cast PSHQ10 specimens at 150 °C which were subjected to steady-state shear flow at various shear rates (s^{-1}): (○, ●) 0.107, (△, ▲) 0.536, and (□, ■) 1.07. (Reprinted from Han and Kim, *Journal of Rheology* 38:13. Copyright © 1994, with permission from the Society of Rheology.)

as compared with that sheared at $\dot{\gamma} = 0.107\,s^{-1}$. This can be explained by the fact that the larger the applied shear rate, the greater the orientation of the director during shear flow.

9.4.6 Effect of Preshearing of Thermotropic Main-Chain LCPs on the Rheological Behavior

As we have shown, the rheological behavior of TLCP after the cessation of shearing is greatly affected by the extent of applied $\dot{\gamma}$, which amounts to the effect of preshearing. Such an experimental observation is unique to a polymer having mesophase structure, because preshearing affects the morphological state of such polymer. Figure 9.54 gives plots of log η versus log $\dot{\gamma}$ (open symbols) for fresh (unsheared) PSHQ10 specimens in the nematic region at 130, 140, 150, and 160 °C, showing that unsheared PSHQ10 specimens exhibit only shear-thinning behavior over the entire range of $\dot{\gamma} = 0.0085$–$0.25\,s^{-1}$. In the experiment, a fresh specimen, before being subjected to shear flow in the nematic region, first received thermal treatment in the isotropic region and then was slowly cooled down to a predetermined temperature in the nematic region. After temperature equilibration, a sudden shear flow was applied to the specimen. The results displayed (open symbols) in Figure 9.54 are at variance with the three-region log η versus log $\dot{\gamma}$ plot (see Figure 9.25); that is, there is no region II for the unsheared specimens over the entire range of $\dot{\gamma}$ investigated. Also given in Figure 9.54 are plots of log η versus log $\dot{\gamma}$ (filled symbols) for presheard PSHQ10 specimens in the nematic region at 130, 140, 150, and 160 °C, showing that presheared PSHQ10 specimens exhibit no shear-thinning behavior at very low $\dot{\gamma}$ (i.e., there is no region I with reference to Figure 9.25). Notice in Figure 9.54 that the presheared

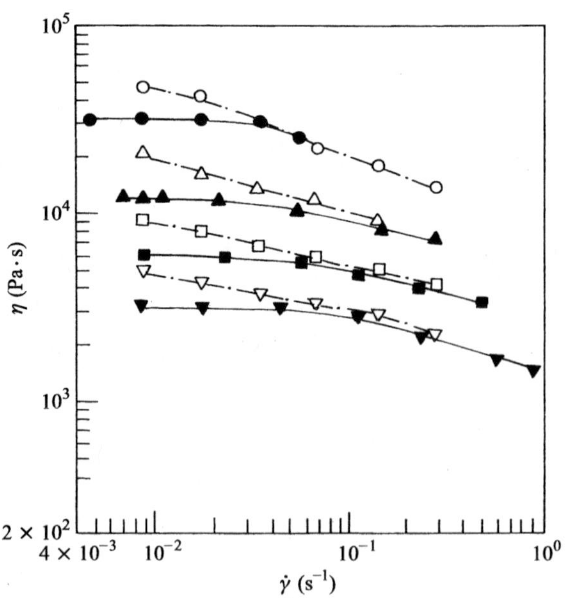

Figure 9.54 Plots of log η versus log $\dot{\gamma}$ (open symbols) for *fresh* as-cast PSHQ10 specimens at various temperatures (°C): (○) 130, (△) 140, (□) 150, and (▽) 160, and plots of log η versus log $\dot{\gamma}$ (filled symbols) for presheared PSHQ10 specimens at various temperatures (°C): (●) 130, (▲) 140, (■) 150, and (▼) 160. A single specimen was employed for each temperature, and each specimen was presheared at rates ranging from 0.0085 to 0.27 s^{-1}. (Reprinted from Kim and Han, *Journal of Polymer Science, Polymer Physics Edition* 32:371. Copyright © 1994, with permission from John Wiley & Sons.)

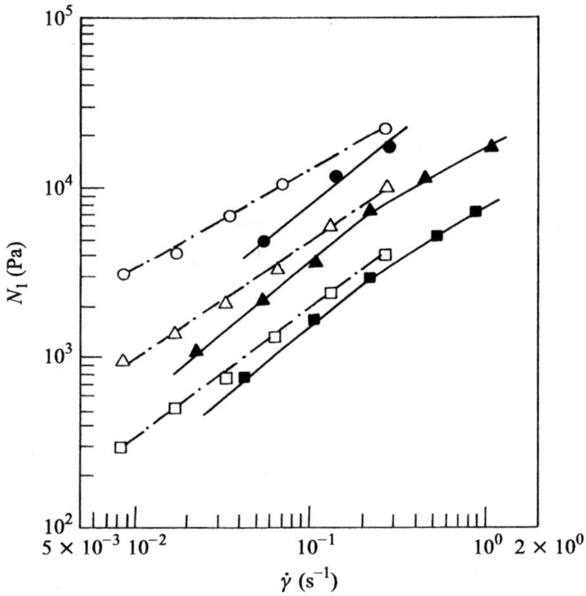

Figure 9.55 Plots of log N_1 versus log $\dot{\gamma}$ (open symbols) for fresh as-cast PSHQ10 specimens at various temperatures (°C): (○) 140, (△) 150, and (□) 160, and plots of log N_1 versus log $\dot{\gamma}$ (filled symbols) for presheared PSHQ10 specimens at various temperatures (°C): (●) 140, (▲) 150, and (■) 160. A single specimen was employed for each temperature, and each specimen was presheared at rates ranging from 0.0085 to 0.27 s^{-1}. (Reprinted from Kim and Han, *Journal of Polymer Science, Polymer Physics Edition* 32:371. Copyright © 1994, with permission from John Wiley & Sons.)

specimens have lower η at low $\dot{\gamma}$, but the effect of preshearing becomes negligibly small as $\dot{\gamma}$ increases. The absence of region I in the presheared specimens at very low $\dot{\gamma}$ is attributable to the fact that preshearing has changed the orientations of the director, transforming the texture consisting of polydomains to one consisting of monodomains.

Figure 9.55 gives plots of log N_1 versus log $\dot{\gamma}$ (open symbols) for unsheared PSHQ10 specimens in the nematic region at 140, 150, and 160 °C. Over the range of $\dot{\gamma}$ tested, using a regression analysis of the data (open symbols) given in Figure 9.55 we obtain $N_1 \propto \dot{\gamma}^n$ with $n = 0.55$ at 140 °C, $n = 0.69$ at 150 °C, and $n = 0.76$ at 160 °C. Within experimental uncertainties, the average value of $n = 0.67$ over the range of temperatures investigated. Thus, we conclude that in the nematic region, N_1 has a rather weak dependence on $\dot{\gamma}$, as compared with that of flexible homopolymers ($N_1 \propto \dot{\gamma}^2$). Also given in Figure 9.55 are plots of log N_1 versus log $\dot{\gamma}$ (filled symbols) for presheared PSHQ10 specimens in the nematic region at 140, 150, and 160 °C. Using a regression analysis of the data for the presheared specimens (filled symbols) given in Figure 9.55, we obtain $N_1 \propto \dot{\gamma}^n$ with $n = 0.80$ at 140 °C, $n = 0.74$ at 150 °C, and $n = 0.75$ at 160 °C. These observations suggest that extreme caution must be exercised when comparing the rheological data obtained from one research group with those obtained by other groups, unless the shear history of the specimens are identical.

The situation becomes quite complicated when presheared PSHQ10 specimens are subjected to shear flow. As can be seen in Figure 9.56, $N_1^+(t, \dot{\gamma})$ for the fresh specimen stays positive over the entire period of shearing, whereas $N_1^+(t, \dot{\gamma})$ for the presheared specimen first goes through negative values for a considerable period (approximately 43 min) of shearing and then becomes positive and remains there until shearing is stopped. Notice in Figure 9.56 that a presheared specimen that was allowed to rest for 88 h after being subjected to a series of preshearing has larger values of both negative and positive N_1 as compared with the specimen that was allowed to rest for only 1 h.

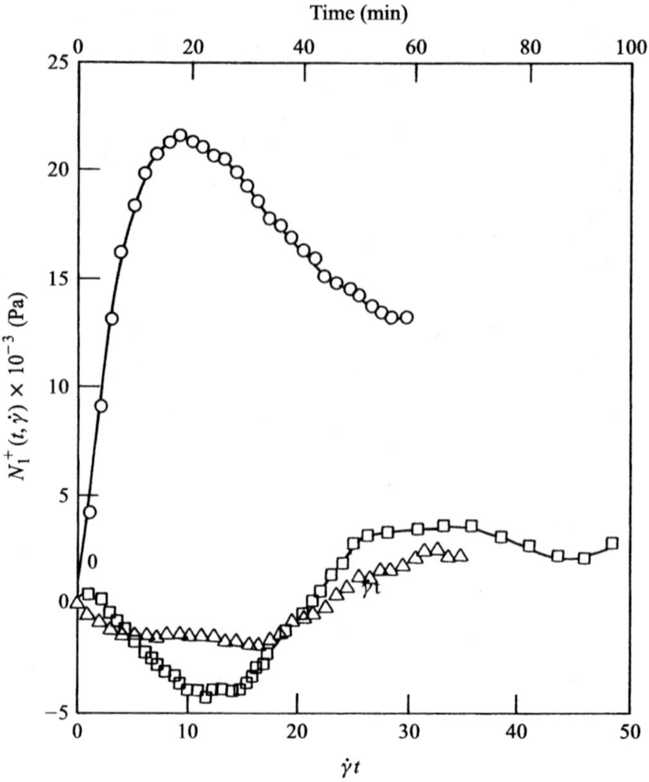

Figure 9.56 Variations of $N_1^+(t, \dot{\gamma})$ with time at 130 °C, upon startup of shear flow at $\dot{\gamma} = 0.0085\ \text{s}^{-1}$, for as-cast PSHQ10 specimens having different shear histories: (○) a fresh specimen, (△) a specimen which was subjected to shear flow at rates ranging from 0.0085 to 0.27 s^{-1} and allowed to rest for 1 h, and (□) a specimen which was subjected to shear flow at rates ranging from 0.0085 to 0.27 s^{-1} and allowed to rest for 88 h. (Reprinted from Kim and Han, *Journal of Polymer Science, Polymer Physics Edition* 32:371. Copyright © 1994, with permission from John Wiley & Sons.)

The above observations indicate that preshearing a fresh PSHQ10 specimen has a profound influence on the sign of N_1 during the subsequent transient shear flow, and that the rest period after preshearing also has a considerable influence on the magnitude of N_1 during the subsequent transient shear flow.

Figure 9.57 shows the variations of $\sigma^+(t, \dot{\gamma})$ with $\dot{\gamma}t$ upon shear startup at $\dot{\gamma} = 0.01\ \text{s}^{-1}$ for PSHQ10 specimens at 130 °C under the identical conditions employed for obtaining the variations of $N_1^+(t, \dot{\gamma})$ with $\dot{\gamma}t$ given in Figure 9.56. In Figure 9.57 we observe that a presheared specimen shows hardly any overshoot in $\sigma^+(t, \dot{\gamma})$ while a fresh specimen exhibits a very large overshoot, and that the steady-state value σ for a presheared specimen that was allowed to rest for 88 h is much greater than the σ for a presheared specimen that was allowed to rest for only 1 h. This observation indicates that the rest period of a presheared specimen before being subjected to additional shearing has a profound influence on the subsequent σ in the nematic

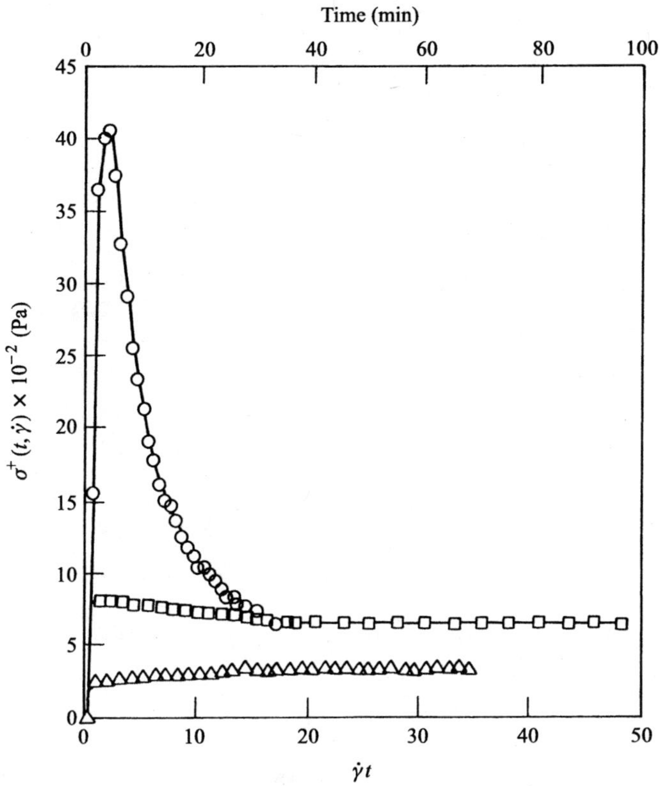

Figure 9.57 The growth of $\sigma^+(t, \dot\gamma)$ at 130 °C, upon startup of shear flow at $\dot\gamma = 0.0085\,\mathrm{s}^{-1}$, for as-cast PSHQ10 specimens having different shear histories: (○) a fresh specimen, (△) a specimen which was subjected to shear flow at rates ranging from 0.0085 to 0.27 s^{-1} and allowed to rest for 1 h, and (□) a specimen which was subjected to shear flow at rates ranging from 0.0085 to 0.27 s^{-1} and allowed to rest for 88 h. (Reprinted from Kim and Han, *Journal of Polymer Science, Polymer Physics Edition* 32:371. Copyright © 1994, with permission from John Wiley & Sons.)

region. The reason for the observed difference in σ for presheared PSHQ10 specimens having different rest periods lies in that, upon cessation of shear flow, the orientation of the directors changes, and thus the texture of the specimen changes. Again, the above observations indicate that the initial morphology of two presheared specimens of TLCP having different rest periods, upon cessation of shear flow, would not be the same.

9.4.7 Reversal Flow of Thermotropic Main-Chain LCPs

Figure 9.58 gives plots of $\sigma^+(t, \dot\gamma)$ versus $\dot\gamma t$ to shear startup for the first 200 strain units and to reversal flow for the next 200 strain units for PSHQ10 at 160 °C and at two different shear rates: 0.1 and 0.5 s^{-1}. In Figure 9.58 we observe that upon shear

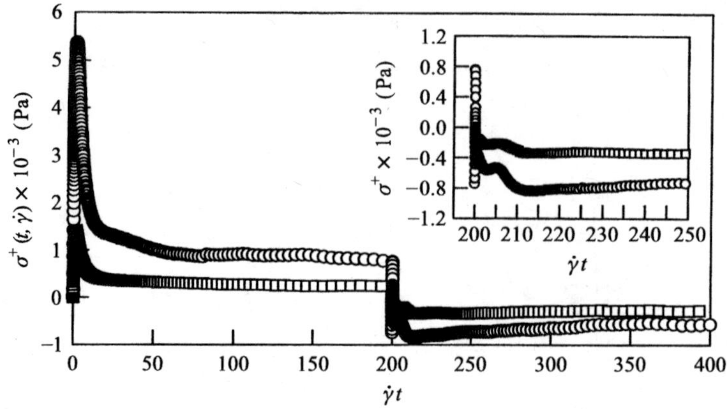

Figure 9.58 Plots of $\sigma^+(t, \dot{\gamma})$ versus $\dot{\gamma}t$ at 160 °C, upon shear startup followed by flow reversal, for as-cast PSHQ10 specimens at two different shear rates (s^{-1}): (□) 0.1 and (○) 0.5. Flow reversal began after 200 strain units upon shear startup. Note that T_{NI} is 179 °C for PSHQ10. (Reprinted from Mather et al., *Macromolecules* 33:7594. Copyright © 2000, with permission from the American Chemical Society.)

startup, $\sigma^+(t, \dot{\gamma})$ increases very rapidly, exhibiting a very large overshoot, and then decays very quickly to steady state. Upon reversal flow, in Figure 9.58 we observe a small overshoot in $\sigma^+(t, \dot{\gamma})$ (see the inset).

In the lower panel of Figure 9.59 are given the responses of $N_1^+(t, \dot{\gamma})$ to shear startup for the first 200 strain units and to reversal flow for the next 200 strain units for PSHQ10 at 160 °C at two different shear rates: 0.1 and 0.5 s^{-1}. In order to help observe the details of the overshoot and undershoot in the early stages of shear startup and reversal flow, in the upper panel of Figure 9.59 are given the responses of $N_1^+(t, \dot{\gamma})$ for the first 20 strain units upon shear startup and for the first 30 strain units upon reversal flow. The following observations are worth noting in Figure 9.59. Upon shear startup, the $N_1^+(t, \dot{\gamma})$ at $\dot{\gamma} = 0.5\,s^{-1}$ initially increases very rapidly, exhibiting a very large overshoot, followed by a very large undershoot, and then increases again, followed by a series of oscillations. However, the $N_1^+(t, \dot{\gamma})$ at $\dot{\gamma} = 0.1\,s^{-1}$ goes through a maximum and decays with oscillations. Upon reversal flow, the $N_1^+(t, \dot{\gamma})$ at both shear rates, 0.1 and 0.5 s^{-1}, exhibits both an overshoot and undershoot.

Figure 9.60 gives POM images of PSHQ10 for increasing times following reversal flow. Immediately following reversal flow, the transmitted intensity is negligible and remains so until approximately 2.5 s (1.25 strain units) have elapsed, at which point the POM image appears light gray and featureless. Over the course of the next 5–7 s, the POM pattern traverses a bright maximum in transmitted intensity, while two textural patterns appear and decay, superimposed on one another. The first texture is relatively coarse, with a characteristic length scale becoming more refined from 40 μm (at $t = 3$ s) to 10 μm (at $t = 5$ s), with no preferred orientation. The second texture consists of bright/dark alternating bands, well oriented in the direction perpendicular to the flow axis. These bands feature a characteristic spacing on the order of 10 μm, which is significantly smaller than the sample thickness of 50 μm. The contrast of the second,

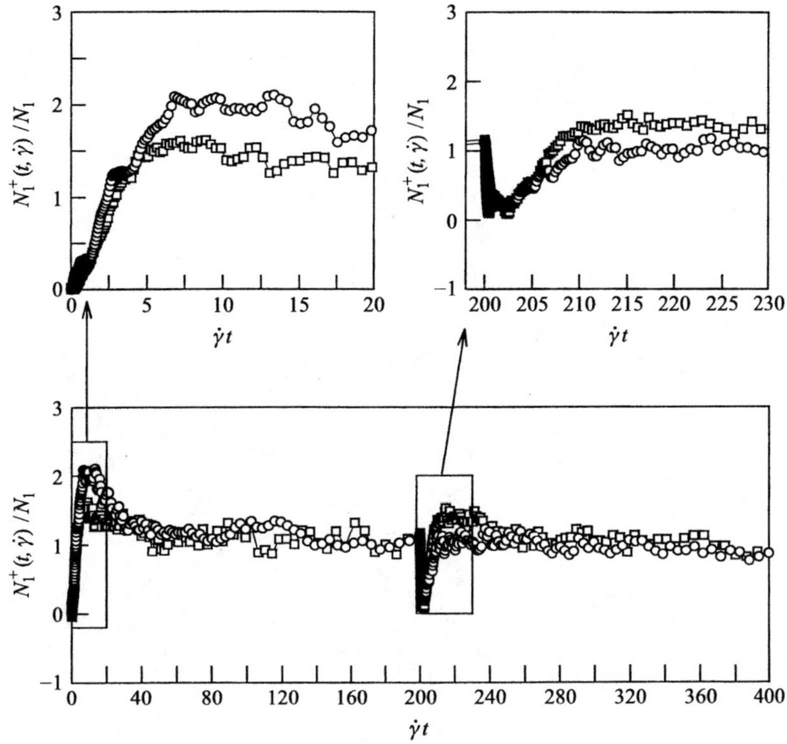

Figure 9.59 Plots of $N_1^+(t, \dot{\gamma})/N_1$ versus $\dot{\gamma}t$ at 160 °C, upon shear startup followed by flow reversal, for as-cast PSHQ10 specimens at two different shear rates (s^{-1}): (□) 0.1 and (○) 0.5. Flow reversal began after 200 strain units upon shear startup. Note that T_{NI} is 179 °C for PSHQ10. (Reprinted from Mather et al., *Macromolecules* 33:7594. Copyright © 2000, with permission from the American Chemical Society.)

banded texture is greatly enhanced by rotating the crossed polarizers by an angle of $\pi/4$. This is shown in the lower right-hand micrographs of Figure 9.60 for $t = 4.0$ and 4.5 s following reversal flow under identical conditions.

9.4.8 Effect of Molecular Weight on the Rheological Behavior of Thermotropic Main-Chain LCPs

In this section, we present experimental observations showing that the molecular weight has a much greater influence on the rheological behavior of TLCP than on the rheological behavior of homopolymers and block copolymers.

Figure 9.61 gives DSC thermograms, during the heating cycle, for six PSHQ10 samples having different molecular weights, after being annealed at 190 °C for 5 min. The molecular characteristics and the intrinsic viscosities of the polymers are summarized

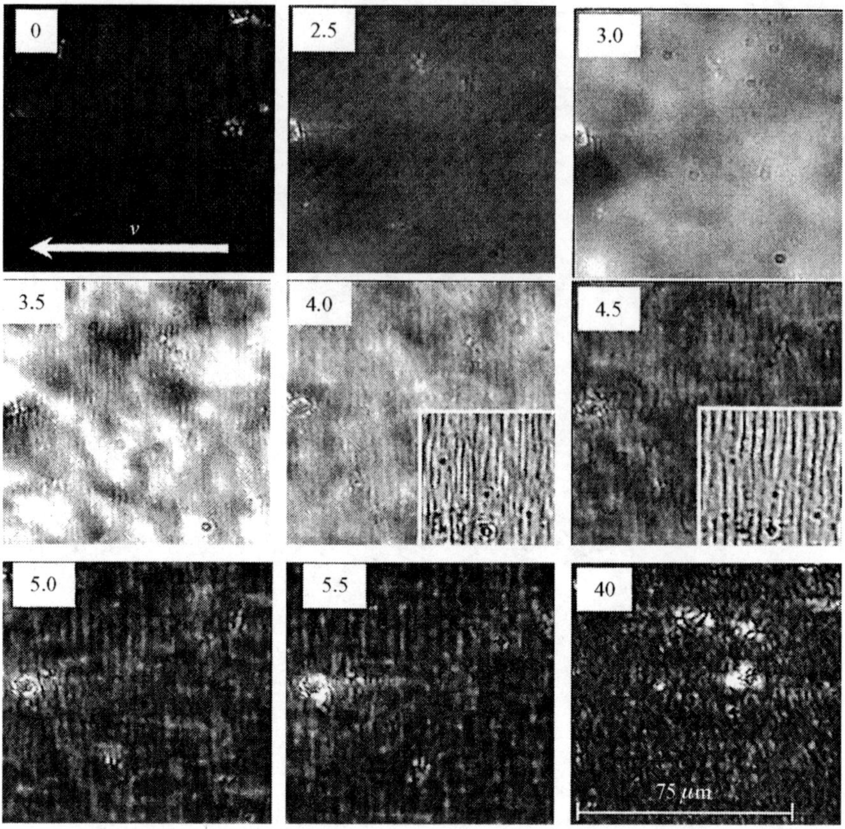

Figure 9.60 Time evolution of POM image of an as-cast PSHQ10 at 160 °C and $\dot{\gamma} = 0.5\ \text{s}^{-1}$ during a period 40 s following flow reversal, showing significant light transmission and formation of optical bands. The observation times in seconds are shown in the upper left of each POM image. The optics are arranged with the flow direction from left to right, white light polarized parallel to the flow direction, and the analyzer crossed with the polarizer. The lower right inset micrographs for $t = 4.0$ and 4.5 s used crossed polarizers, with the polarizer making an angle of π/4 with the flow direction. (Reprinted from Mather et al., *Macromolecules* 33:7594. Copyright © 2000, with permission from the American Chemical Society.)

in Table 9.1. In Figure 9.61 we observe that T_m and T_{NI} increase with increasing molecular weight (M_w). Similar behavior has also been reported for other types of TLCPs (Blumstein et al. 1982, 1984; Laus et al. 1992; Majnusz et al. 1983; Percec et al. 1989).

Figure 9.62 shows variations of $\sigma^+(t, \dot{\gamma})$ with time, and Figure 9.63 shows variations of $N_1^+(t, \dot{\gamma})$ with time for five PSHQ10 samples having different molecular weights at 140 °C and $\dot{\gamma} = 0.536\,\text{s}^{-1}$. The following observations are worth noting in Figures 9.62 and 9.63. The maximum overshoot in $\sigma^+(t, \dot{\gamma})$ increases with increasing M_w of PSHQ10, and multiple overshoots of $\sigma^+(t, \dot{\gamma})$ appear as the M_w of PSHQ10

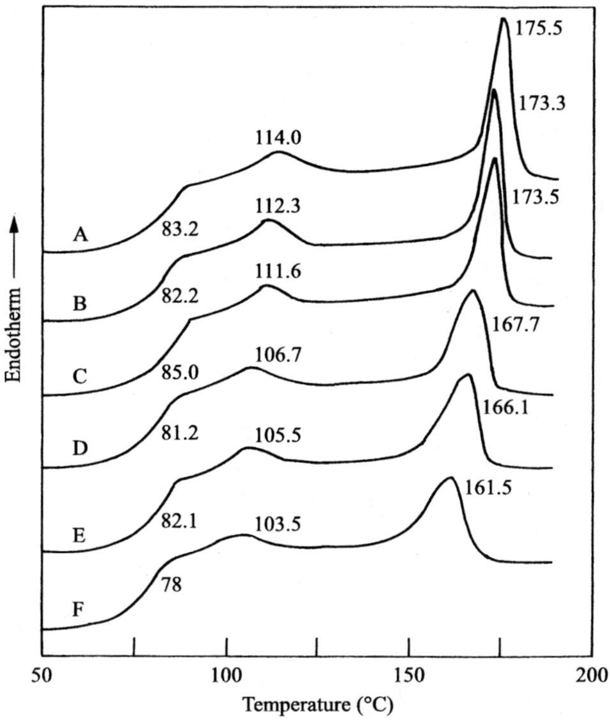

Figure 9.61 DSC thermograms for six as-cast PSHQ10 specimens having different M_w (see Table 9.1 for sample code), which were annealed for 5 min at 190 °C, showing T_g, T_m, and T_{NI}. (Reprinted from Kim and Han, *Macromolecules* 26:6633. Copyright © 1993, with permission from the American Chemical Society.)

becomes greater than a certain critical value. The appearance of multiple overshoots in $N_1^+(t, \dot{\gamma})$ is very pronounced with increasing M_w of PSHQ10, and the value of $\dot{\gamma}$ at which the first overshoot of $N_1^+(t, \dot{\gamma})$ occurs increases with increasing M_w (i.e., the higher the M_w of PSHQ10, the longer it will take for an overshoot of $N_1^+(t, \dot{\gamma})$ to appear). Notice further in Figure 9.63 that there is only a single overshoot of $N_1^+(t, \dot{\gamma})$

Table 9.1 Summary of the molecular weight and intrinsic viscosity of six PSHQ10 samples

Sample Code	M_w [a]	M_w/M_n [b]	$[\eta](dL/g)$ [c]
A	53.4×10^3	1.89	0.616
B	49.5×10^3	2.03	0.560
C	45.2×10^3	2.11	0.500
D	38.9×10^3	2.48	0.432
E	37.6×10^3	2.43	0.409
F	35.0×10^3	2.09	0.387

[a] M_w was determined against polustyrene standards. [b] M_w/M_n was determined using gel permeation chromatography. [c] $[\eta]$ was measured in dichloromethane at 23 °C using an Ubbelohde viscometer.

Reprinted from Kim and Han, *Macromolecules* 26:6633. Copyright © 1993, with permission from the American Chemical Society.

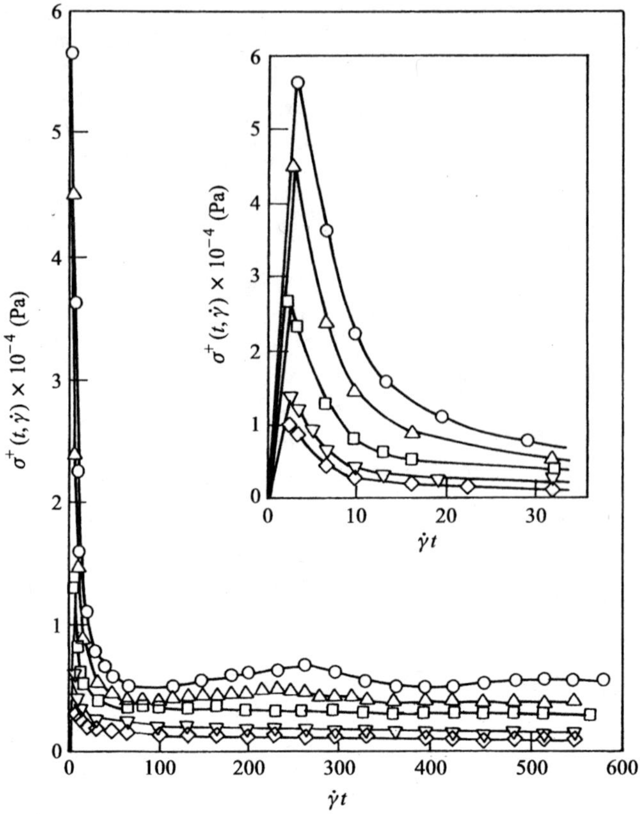

Figure 9.62 Plots of $\sigma^+(t, \dot{\gamma})$ versus $\dot{\gamma}t$ at 140 °C and $\dot{\gamma} = 0.536\,\text{s}^{-1}$ for five PSHQ10 specimens (see Table 9.1 for sample code) having different M_w: (○) sample A, (△) sample B, (□) sample C, (▽) sample D, and (◇) sample E. (Reprinted from Kim and Han, *Macromolecules* 26:6633. Copyright © 1993, with permission from the American Chemical Society.)

occurring for M_w below a certain critical value. The time required for $N_1^+(t, \dot{\gamma})$ to reach steady state is much longer than that for $\sigma^+(t, \dot{\gamma})$.

Figure 9.64 gives plots of $\log \eta$ versus $\log \dot{\gamma}$ and $\log |\eta^*|$ versus $\log \omega$ at 190 °C in the isotropic region for five PSHQ10 samples having different M_w. We observe in Figure 9.64 that the Cox–Merz rule (Cox and Merz 1958) holds for the PSHQ10 samples, which is an indication that these samples are indeed in the isotropic state. Figure 9.65 gives $\log G'$ versus $\log G''$ plots at 190 °C in the isotropic region for five PSHQ10 specimens having different M_w, showing that the dependence of M_w is not discernible in the $\log G'$ versus $\log G''$ plots. Therefore, we can conclude from Figure 9.65 that all five PSHQ10 specimens have M_w greater than the entanglement molecular weight of PSHQ10 (see Chapter 6 for molecular weight dependence of $\log G'$ versus $\log G''$ plots for linear flexible homopolymers). Further, the fact that the $\log G'$ versus $\log G''$ plots given in Figure 9.65 have the slope close to 2 in the terminal

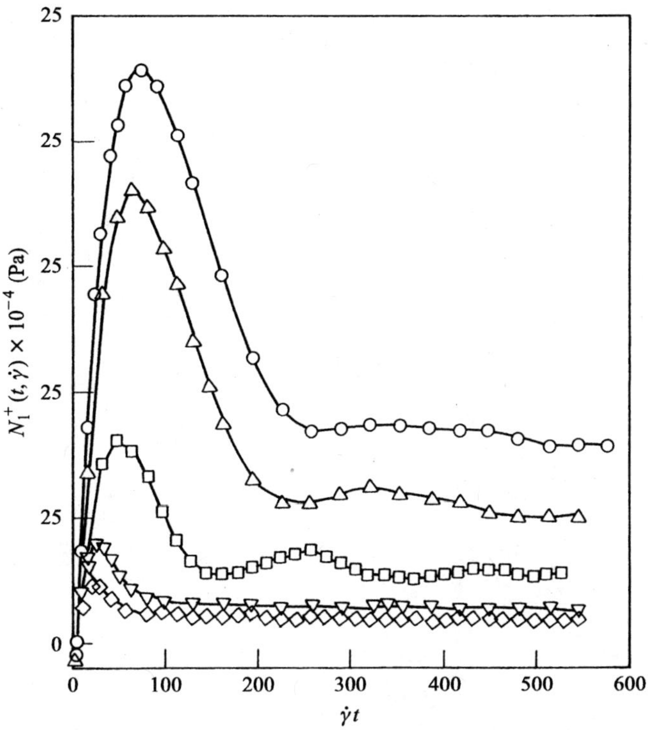

Figure 9.63 Plots of $N_1^+(t, \dot{\gamma})$ versus $\dot{\gamma}t$ at 140 °C and $\dot{\gamma} = 0.536\ s^{-1}$ for five as-cast PSHQ10 specimens (see Table 9.1 for sample code) having different M_w: (○) sample A, (△) sample B, (□) sample C, (▽) sample D, and (◇) sample E. (Reprinted from Kim and Han, *Macromolecules* 26:6633. Copyright © 1993, with permission from the American Chemical Society.)

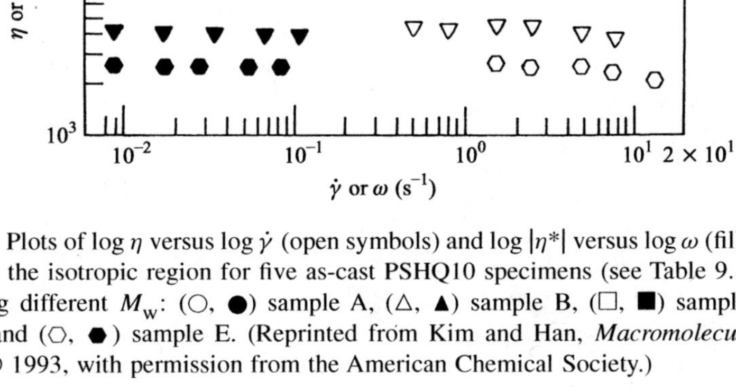

Figure 9.64 Plots of log η versus log $\dot{\gamma}$ (open symbols) and log $|\eta^*|$ versus log ω (filled symbols) at 190 °C in the isotropic region for five as-cast PSHQ10 specimens (see Table 9.1 for sample code) having different M_w: (○, ●) sample A, (△, ▲) sample B, (□, ■) sample C, (▽, ▼) sample D, and (⬡, ⬢) sample E. (Reprinted from Kim and Han, *Macromolecules* 26:6633. Copyright © 1993, with permission from the American Chemical Society.)

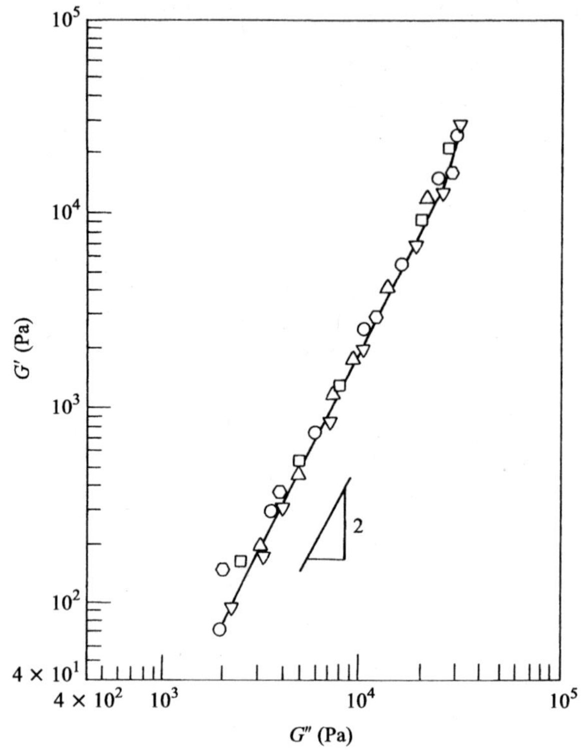

Figure 9.65 Log G' versus log G'' plots at 190 °C in the isotropic region for five as-cast PSHQ10 specimens (see Table 9.1 for sample code) having different M_w: (○) sample A, (△) sample B, (□) sample C, (▽) sample D, and (⬡) sample E. (Reprinted from Kim and Han, *Macromolecules* 26:6633. Copyright © 1993, with permission from the American Chemical Society.)

region attests to the fact that the five PSHQ10 samples are indeed in the isotropic region, because the log G' versus log G'' plot in the terminal for the nematic-forming PSHQ10 (at $T < T_{NI}$) has a slope less than 2 (see Figure 9.35b).

Figure 9.66 gives logarithmic plots of zero-shear viscosity η_0 versus M_w for five PSHQ10 samples at 140 °C in the nematic state. A regression analysis of the data in Figure 9.66 gives $\eta_0 \propto M_w^{6.1}$. It is interesting to note that the experimentally observed M_w dependence of η_0 is in good agreement with Eq. (9.36) predicted from the Doi theory. However, such an agreement may be considered to be fortuitous because Eq. (9.36) was obtained on the assumption of zero value of scalar order parameter S, $S = 0$. Since the value of S is larger than zero for the nematic-forming PSHQ10 and the value of S would depend on M_w at a given temperature, η_0 of PSHQ10 in the nematic region would not be directly proportional to $M_w^{6.1}$. At present, however, we have no information as to how S might vary with M_w. Note further that in the derivation of Eqs. (9.31)–(9.33), the decoupling approximation, Eq. (9.27), was made in treating the term $\mathbf{D} : \langle \mathbf{uuuu} \rangle$. If a higher-order approximation is made, we might obtain expressions that would have different dependence on S. Thus, we hasten to point that a seemingly good agreement between experiment and theory may be fortuitous because the Doi theory assumes that the polymer consists of rigid rodlike molecules without texture, whereas PSHQ10 has ten methylene groups as flexible spacers in the main chain and bulky pendent side groups, and it has nematic texture in an anisotropic state.

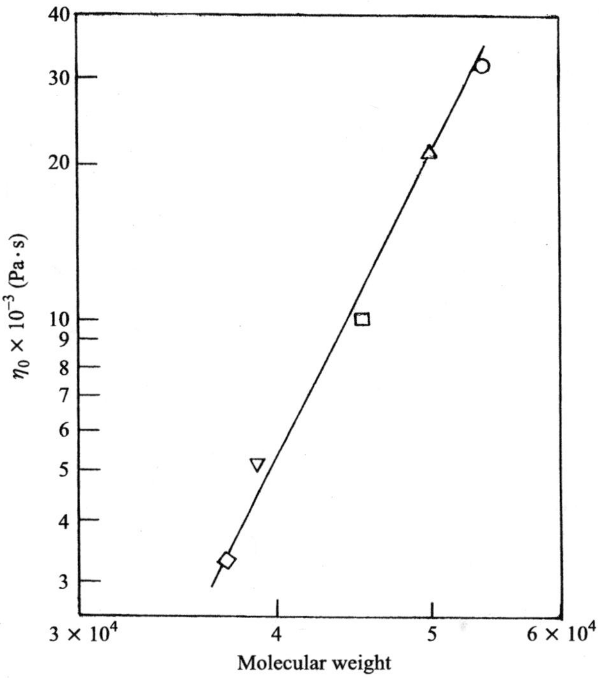

Figure 9.66 Plots of log η_0 versus M_w at 140 °C in the nematic state for five as-cast PSHQ10 specimens (see Table 9.1 for sample code): (○) sample A, (△) sample B, (□) sample C, (▽) sample D, and (◇) sample E, in which values of η_0 were obtained in the second shear-rate sweep experiment.[6]

Figure 9.67 gives plots of log N_1 versus log $\dot{\gamma}$ for five presheared PSHQ10 specimens having different M_w at 140 °C (i.e., after the first shear-rate sweep experiment). Note in Figure 9.67 that N_1 increases with M_w giving, via regression analysis, $N_1 \propto M_w^{6.7}$, which is reasonably close to the prediction, Eq. (9.36), from the Doi theory. Again, the reasonable agreement between experiment and theory may be fortuitous for the reasons we have outlined with reference to the relationship $\eta_0 \propto M_w^{6.1}$. Further, a regression analysis of the data given in Figure 9.67 gives $N_1 \propto \dot{\gamma}^{0.67}$, indicating that the shear-rate dependence of N_1 for PSHQ10 in the nematic region is weaker than the theoretical prediction $N_1 \propto \dot{\gamma}$ from Eq. (9.32) from the Doi theory.

9.4.9 Effect of Bulkiness of Pendent Side Groups on the Rheo-Optical Behavior of Thermotropic Main-Chain LCPs

It is reasonable to expect that the size and bulkiness of pendent side groups would influence the rheological behavior of TLCPs. Table 9.2 gives a summary of three TLCPs having different chemical structures of pendent side groups, in which PEHQ10 has ethoxy pendent side group, PTHQ10 has *tert*-butyl pendent side group, and PSHQ10 has phenylsulfonyl pendent side groups, each having 10 methylene units of flexible spacer. Also given in Table 9.2 are the van der Waals volumes of each pendent side group and intrinsic viscosities of the three TLCPs. It has been reported that PEHQ10 undergoes both S–N and N–I transitions, while both PTHQ10 and PSHQ10 undergo

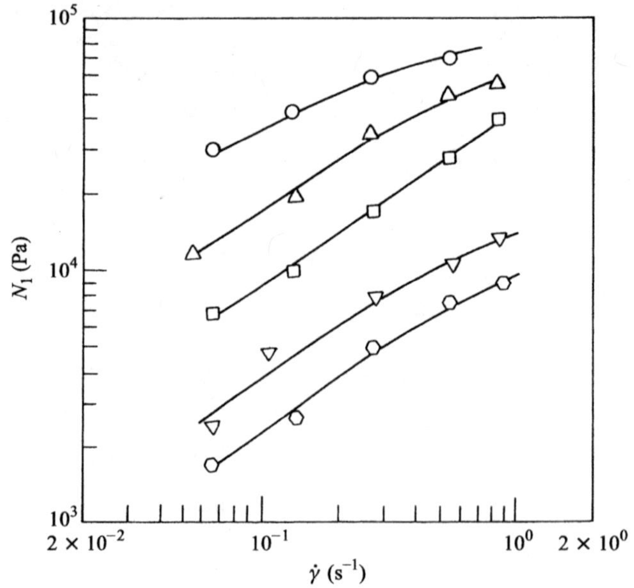

Figure 9.67 Plots of log N_1 versus log $\dot{\gamma}$ at 140 °C in the second shear-rate sweep experiment for five as-cast PSHQ10 specimens (see Table 9.1 for sample code) having different M_w: (○) sample A, (△) sample B, (□) sample C, (▽) sample D, and (⬡) sample E. (Reprinted from Kim and Han, *Macromolecules* 26:6633. Copyright © 1993, with permission from the American Chemical Society.)

only N–I transition (Kim and Han 2000). Next, we present experimental correlation between birefringence and shear rheometry in steady-state shear flow of the three TLCPs.

Figure 9.68 gives plots of birefringence Δn versus temperature and N_1/σ versus temperature for PEHQ10, PTHQ10, and PSHQ10 in steady-state shear flow at $\dot{\gamma} = 0.05\ s^{-1}$. We observe good overlap between Δn and N_1/σ over a wide range of

Table 9.2 Summary of the chemical structure and van der Waals volume of pendent side groups, intrinsic viscosity, and T_{NI} for TLCPs with varying pendent side groups

Sample Code	*Pendent Side Group*	*Van der Waals Volume of Pendent Side Group (nm^3)*	$[\eta]$ (d*L*/*g*)[a]	T_{NI} (°*C*)
PEHQ10	OC_2H_5	54.0×10^{-3}	1.393	239
PTHQ10	$C(CH_3)_3$	82.7×10^{-3}	0.771	193
PSHQ10	SO_2– 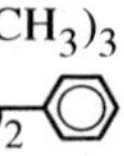	121.0×10^{-3}	0.667	179

[a] $[\eta]$ was measured in 1,1,2,2-tetrachloroethane at 23 °C.

Reprinted from Kim and Han, *Macromolecules* 33:3349. Copyright © 2000, with permission from the American Chemical Society.

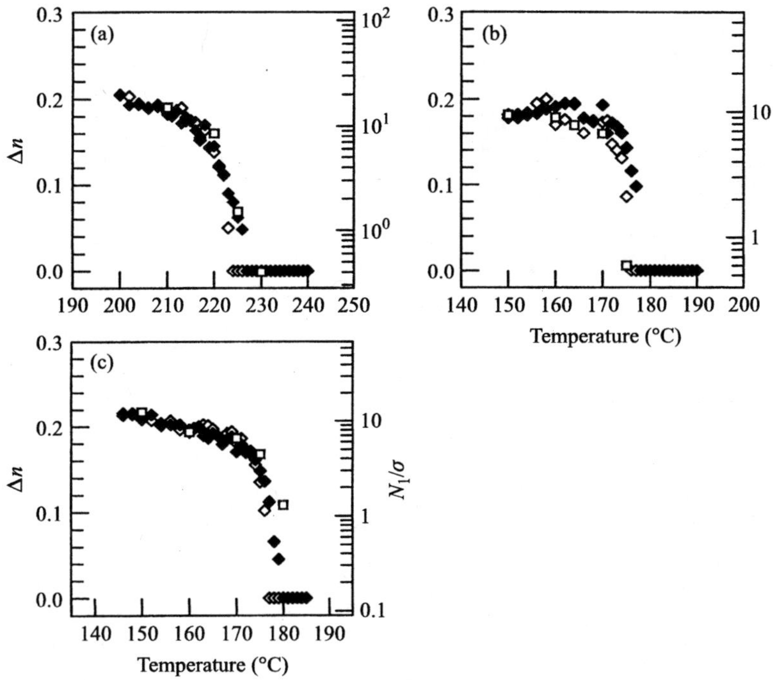

Figure 9.68 Plots of Δn versus temperature in the heating (◆) and cooling processes (◇) at $\dot{\gamma} = 0.05\,\mathrm{s}^{-1}$, and plots of N_1/σ versus temperature (□) at $\dot{\gamma} = 0.05\,\mathrm{s}^{-1}$ for as-cast specimens of (a) PEHQ10, (b) PTHQ10, and (c) PSHQ10. (Reprinted from Kim et al., *Macromolecules* 33:7922. Copyright © 2000, with permission from the American Chemical Society.)

temperatures investigated. With reference to Figure 9.68, for temperatures well below T_{NI} we observe only a weak (decreasing) temperature dependence for both Δn and the ratio N_1/σ, while for temperatures approaching T_{NI} we observe that Δn and N_1/σ begin to decrease gradually as the temperature approaches T_{NI}. At a certain critical temperature close to T_{NI}, Δn and the ratio N_1/σ drop precipitously. Of particular interest in Figure 9.68 is the observation that Δn and the ratio N_1/σ for PEHQ10 and PTHQ10 drop virtually to zero at a temperature 15–18 °C below T_{NI}, while Δn and the ratio N_1/σ for PSHQ10 become virtually zero only at temperatures much closer to T_{NI}. Since Δn describes molecular alignment of the polymer in steady-state shear flow, it is clear from Figure 9.68 that PEHQ10 and PTHQ10 begin to lose molecular alignment at temperatures 15–18 °C below T_{NI}. This suggests two possibilities. One possibility is that PEHQ10 and PTHQ10 might feature much broader molecular weight distributions than PSHQ10, so that the low molecular-weight portion of the polymers begin isotropization at temperatures 15–18 °C below T_{NI}. The second possibility is that the less bulky ethoxy pendent side groups in PEHQ10 and *tert*-butyl pendent side groups in PTHQ10, compared with the bulkier phenylsulfonyl pendent groups in PSHQ10, might destabilize molecular alignment at lower temperatures. The salient feature of Figure 9.68 is the finding of significant correlation between the optical birefringence and the mechanical stress ratio N_1/σ.

Using the values of N_1/σ in the nematic region given in Figure 9.68, we can estimate λ from Eq. (9.80) and S_m from Eq. (9.81). We obtain (1) $\lambda = 1.019$ and $S_m = 0.972$ for PEHQ10 using $N_1/\sigma = 10$, (2) $\lambda = 1.021$ and $S_m = 0.969$ for PEHQ10 using $N_1/\sigma = 9$, and (3) $\lambda = 1.031$ and $S_m = 0.956$ for PSHQ10 using $N_1/\sigma = 8$. The values of λ estimated above indicate that all three TLCPs are flow aligning and that the degree of flow alignment for PEHQ10 is higher than for PTHQ10, and the degree of flow alignment for PTHQ10 is higher than PSHQ10. Such an observation is in line with the intuitive expectation that the smaller the pendent side groups, the higher the flow alignment will be. We note from Table 9.2 that the van der Waals volumes of the three pendent side groups have the following increasing order: ethoxy group in PEHQ10 < *tert*-butyl group in PTHQ10 < phenylsulfonyl group in PSHQ10.

It is very important to investigate molecular alignment of thermotropic melts in shear flow. For this reason, many research groups have employed birefringence (Berghausen et al. 1997; Hsiao et al. 1990; Kannan and Kornfield 1994; Kim et al. 2000; Mather et al. 1997, 2000; Pujolle-Robic and Noirez 2001) and X-ray scattering (Odell et al. 1993; Ugaz and Berghardt 1998; Zhou et al. 1999), in conjunction with mechanical rheometry, to investigate the rheo-optical behavior of TLCPs.

9.5 Rheological Behavior of Thermotropic Side-Chain LCPs

Thus far, we have presented the rheological behavior of main-chain liquid-crystalline polymers (MCLCPs). In this section, we present the rheological behavior of side-chain liquid-crystalline polymers (SCLCPs). In the 1980's a number of research groups reported on the syntheses of SCLCPs and such efforts were summarized in a review article by Shibaev and Platé (1984) and in the monograph edited by McArdle (1989). Since then, many research groups have continued to synthesize SCLCPs and investigate thermal transitions and mesophase structures. The rheological investigation of SCLCPs (Berghausen et al. 1997; Colby et al. 1993, 1977; Fabre and Veyssie 1987; Grabowski and Schmidt 1994; Kannan and Kornfield 1993; Kannan et al. 1994; Lee and Han 2002, 2003; Quijada-Garrido et al. 1999, 2000; Rubin et al. 1995; Wewerka et al. 2001a, 2001b; Zentel and Wu 1986) has not been as extensive as that of MCLCPs. It can easily be surmised that the difference in the architecture between SCLCPs and MCLCPs would give rise to a profound influence on their rheological behavior. Almost all MCLCPs employed for the rheological investigations had a nematic mesophase, and only a few studies (Hudson et al. 1993; Kim and Han 2000) reported on the rheological behavior, over a very narrow range of temperatures, of MCLCPs forming smectic mesophase, because they are extremely viscous in the liquid-crystalline state. Note that smectic mesophases have two-dimensional layered structures and that there are several different forms of smectics.

Basically, an SCLCP consists of three parts: polymer backbone, side chain liquid-crystalline monomers, and flexible spacer groups that connect the polymer backbone and side-chain liquid-crystalline monomers (see Figure 9.3). It has been found that many SCLCPs have a clearing temperature far below the thermal degradation temperature, allowing one to conduct rheological investigations in the liquid-crystalline state without encountering the possibility of thermal degradation. There is experimental evidence, thought not universal, that SCLCPs having short flexible spacer groups tend to form a nematic mesophase, while SCLCPs having long flexible space groups tend to form a smectic mesophase. These observations suggest that the liquid crystallinities (nematics and smectics), hence the rheological behavior, of SCLCPs depend not only on the chemical structures of the side-chain liquid-crystalline monomers but also on the flexible spacer length.

In this section, we summarize the study of Lee and Han (2003), who investigated the rheological behavior of a homologous series of model SCLCPs having the following chemical structure (PI-*n*CNCOOH):

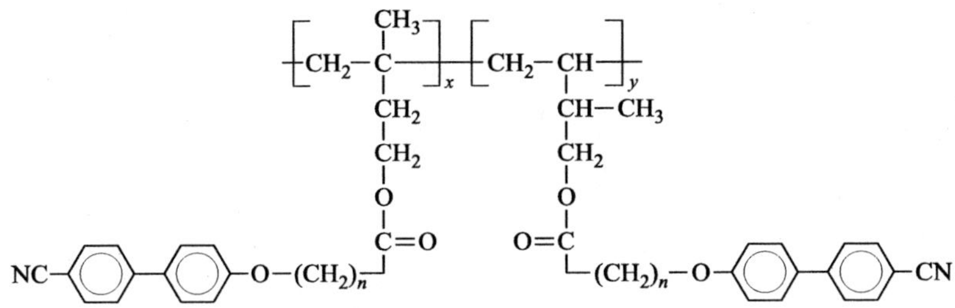

PI-*n*CNCOOH

which were obtained by grafting *n*-[[(4-cyano-4′-biphenyl)oxy]-alkyl]carboxylic acid (*n*CN-COOH) with the chemical structure

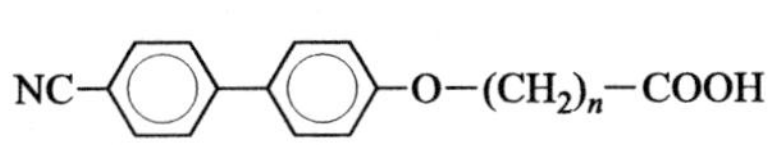

onto the chain backbone of nearly monodisperse hydroxylated polyisoprene (PIOH) with high vinyl content (34% 1,2-addition and 59% 3,4-addition). In so doing, they varied the length of methylene spacer groups from 5 to 7 and to 11. The details of the syntheses of *n*CN-COOH and PI-*n*CNCOOH can be found in the original papers of Lee and Han (2002, 2003). Table 9.3 gives a summary of the chemical structures and molecular weights of PI-*n*CNCOOH synthesized by Lee and Han (2003). According to Lee and Han, (1) PI-5CNCOOH formed nematic mesophase, while PI-7CNCOOH and PI-11CNCOOH formed smectic mesophase, (2) PI-5CNCOOH has a glass transition temperature (T_g) of 45 °C and a clearing temperature ($T_{c\ell}$) of 102 °C, (3) PI-7CNCOOH

Table 9.3 Summary of the chemical structure and molecular weight of PI-*n*CNCOOH

Sample Code	M_n	M_w/M_n [c]	*Structure*
PI	1.43×10^4 [a]	1.05	Polyisoprene backbone
PI-5CNCOOH	7.19×10^4 [b]	1.08	PI with 5CN-COOH
PI-7CNCOOH	7.78×10^4 [b]	1.07	PI with 7CN-COOH
PI-11CNCOOH	8.80×10^4 [b]	1.08	PI with 11CN-COOH

[a] Determined from membrane osmometry. [b] Calculated value based on the degree of hydroxylation. [c] Determined from GPC.

has a T_g of 40 °C and a $T_{c\ell}$ of 105 °C, and (4) PI-11CNCOOH has a T_g of 22 °C and a $T_{c\ell}$ of 113 °C. That is, $T_{c\ell}$ increases, while T_g decreases, as the number (n) of methylene groups in PI-nCNOOH increases.

Figure 9.69 gives log G' versus log ω plots for PI-5CNCOOH, PI-7CNCOOH, and PI-11CNCOOH at various temperatures. The following observations are worth noting in Figure 9.69. Over the entire range of ω tested, the slope of log G' versus log ω plots for PI-5CNCOOH is slightly less than 2 at temperatures below approximately 105 °C, and then becomes 2 when the temperature is increased to 105 °C and higher. This temperature is close to the $T_{c\ell}$ (102 °C) of PI-5CNCOOH. The frequency dependence of G' shown in Figure 9.69a makes sense because at temperatures below $T_{c\ell}$ PI-5CNCOOH has a nematic mesophase and at temperatures above $T_{c\ell}$ PI-5CNCOOH forms a homogeneous phase. Thus, the log G' versus log ω plots having a slope of 2 in the terminal region for PI-5CNCOOH at $T \geq T_{c\ell}$ exhibit liquidlike behavior because the PI backbone of PI-5CNCOOH is nearly monodisperse. As can be seen from Figure 9.69b, the log G' versus log ω plots for PI-7CNCOOH in the terminal region have a slope much less than 2 at $T < 100$ °C, but the slope of log G' versus log ω plots begins to change at 105 °C and becomes 2 at $T \geq 110$ °C in the isotropic state. Figure 9.69c shows that at $T < 105$ °C, PI-11CNCOOH exhibits solidlike behavior by having a very small slope in the terminal region of log G' versus log ω plots, and then at $T \geq 115$ °C it

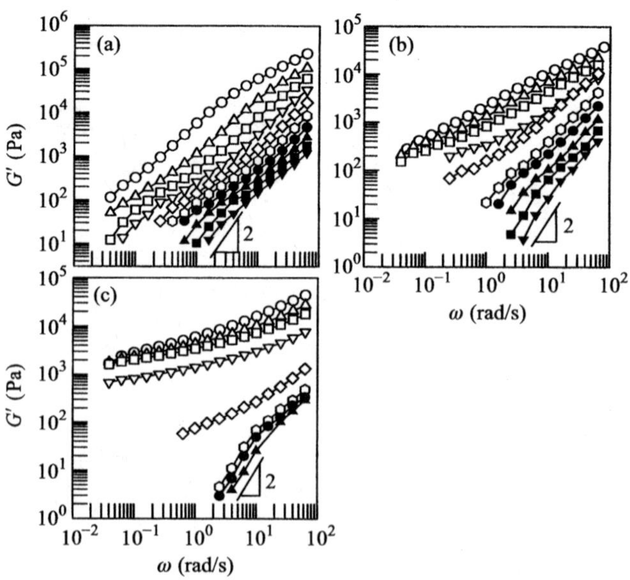

Figure 9.69 Plots of log G' versus log ω for: (a) PI-5CNCOOH at various temperatures (°C): (○) 75, (△) 80, (□) 85, (▽) 90, (◇) 95, (⬭) 100, (●) 105, (▲) 110, (■) 115, and (▼) 120, (b) PI-7CNCOOH at various temperatures (°C): (○) 80, (△) 85, (□) 90, (▽) 95, (◇) 100, (⬭) 105, (●) 110, (▲) 115, (■) 120, and (▼) 125, and (c) PI-11CNCOOH at various temperatures (°C): (○) 80, (△) 90, (□) 100, (▽) 105, (◇) 110, (⬭) 115, (●) 120, and (▲) 125. (Reprinted from Lee and Han, *Macromolecules* 36:8796. Copyright © 2003, with permission from the American Chemical Society.)

exhibits liquidlike behavior by having a slope of 2 in the terminal region of $\log G'$ versus $\log \omega$ plots. It is clear from Figure 9.69 that in the terminal region, at $T < T_{c\ell}$, the frequency dependence of $\log G'$ versus $\log \omega$ plots for PI-nCNCOOH exhibits progressively stronger solidlike behavior as the number of methylene spacer groups increases from 5 to 11. This observation is attributable to an increase in the strength of mesogenics as the number of methylene spacer groups in PI-nCNCOOH increases from 5 to 11. Notice further in Figure 9.69 that at the same distance from the $T_{c\ell}$ of each polymer, at $T_{c\ell} - T$, in the anisotropic region the magnitude of G' in the terminal region of $\log G'$ versus $\log \omega$ plots has the following order: G' (PI-11CNCOOH) > G' (PI-7CNCOOH) > G' (PI-5CNCOOG).

Figure 9.70 gives $\log |\eta^*|$ versus $\log \omega$ plots for PI-5CNCOOH, PI-7CNCOOH, and PI-11CNCOOH at various temperatures. The following observations are worth noting in Figure 9.70. The $\log |\eta^*|$ versus $\log \omega$ plots for PI-5CNCOOH show very weak frequency dependence even at $T < T_{c\ell}$(102 °C), and then they begin to show Newtonian behavior at $T \geq 110$ °C. In Figure 9.70a we observe no temperature at which an abrupt change in the frequency dependence of $\log |\eta^*|$ versus $\log \omega$ plots for PI-5CNCOOH occurs; that is, a smooth transition takes place from very weak frequency dependence in the nematic state to frequency independence in the isotropic state. In Figure 9.70b we observe that $\log |\eta^*|$ versus $\log \omega$ plots for PI-7CNCOOH

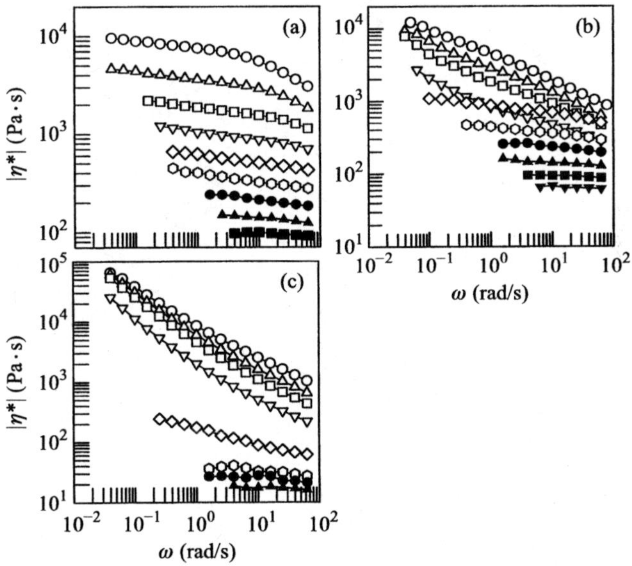

Figure 9.70 Plots of $\log |\eta^*|$ versus $\log \omega$ for (a) PI-5CNCOOH at various temperatures (°C): (○) 75, (△) 80, (□) 85, (▽) 90, (◇) 95, (⬭) 100, (●) 105, (▲) 110, (■) 115, and (▼) 120, (b) PI-7CNCOOH at various temperatures (°C): (○) 80, (△) 85, (□) 90, (▽) 95, (◇) 100, (⬭) 105, (●) 110, (▲) 115, (■) 120, and (▼) 125, and (c) PI-11CNCOOH at various temperatures (°C): (○) 80, (△) 90, (□) 100, (▽) 105, (◇) 110, (⬭) 115, (●) 120, and (▲) 125. (Reprinted from Lee and Han, *Macromolecules* 36:8796. Copyright © 2003, with permission from the American Chemical Society.)

have strong frequency dependence at $T \leq 95\,°C$, very mild frequency dependence at 105 °C, and then Newtonian behavior at $T \geq 110\,°C$. Similarly, in Figure 9.70c we observe that $\log|\eta^*|$ versus $\log\omega$ plots for PI-11CNCOOH have very strong frequency dependence at $T \leq 105\,°C$, very mild frequency dependence at 110 °C, and then Newtonian behavior at $T \geq 115\,°C$. We can conclude from Figure 9.70 that the frequency dependence of $\log|\eta^*|$ versus $\log\omega$ plots at $T < T_{c\ell}$ becomes increasingly strong as the number of methylene spacer groups in PI-*n*CNCOOH increases from 5 to 7 and to 11. Strong frequency dependence of $|\eta^*|$ at low angular frequencies in oscillatory shear flow is believed to be characteristic of multiphase polymeric liquids, including highly filled molten polymers (see Chapter 12) and microphase-separated block copolymers (see Chapter 8). What seems very interesting in Figure 9.70 is that the nematic-forming PI-5CNCOOH at $T < T_{c\ell}$ exhibits very weak frequency dependence of $|\eta^*|$ over the entire range of angular frequencies tested, as is often observed in ordinary flexible polymers, while the smectic-forming PI-7CNCOOH and PI-11CNCOOH at $T < T_{c\ell}$ exhibit very strong frequency dependence of $|\eta^*|$ at low angular frequencies.

Figure 9.71 gives $\log G'$ versus $\log G''$ plots for PI-5CNCOOH, PI-7CNCOOH, and PI-11CNCOOH at various temperatures. The following observations are worth noting in Figure 9.71. The $\log G'$ versus $\log G''$ plots for PI-5CNCOOH (Figure 9.71a) show

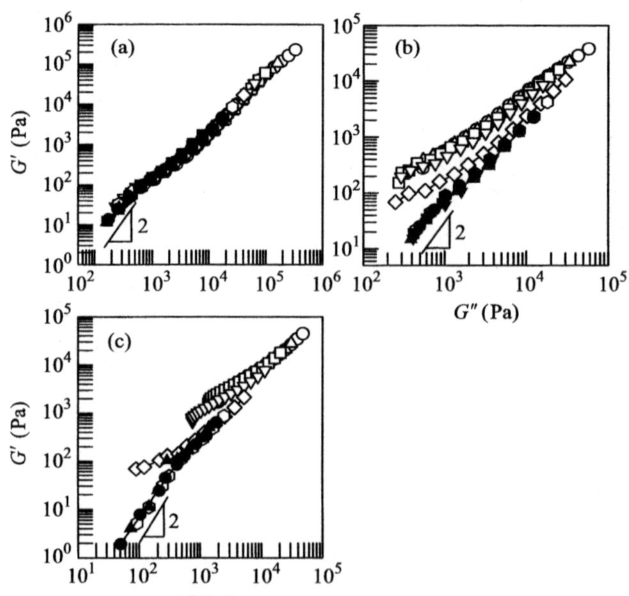

Figure 9.71 Plots of $\log G'$ versus $\log G''$ for (a) PI-5CNCOOH at various temperatures (°C): (○) 75, (△) 80, (□) 85, (▽) 90, (◇) 95, (⬡) 100, (●) 105, (▲) 110, (■) 115, and (▼) 120, (b) PI-7CNCOOH at various temperatures (°C): (○) 80, (△) 85, (□) 90, (▽) 95, (◇) 100, (⬡) 105, (●) 110, (▲) 115, (■) 120, and (▼) 125, and (c) PI-11CNCOOH at various temperatures (°C): (○) 80, (△) 90, (□) 100, (▽) 105, (◇) 110, (⬡) 115, (●) 120, and (▲) 125. (Reprinted from Lee and Han, *Macromolecules* 36:8796. Copyright © 2003, with permission from the American Chemical Society.)

very weak temperature dependence over the entire range of temperatures tested below and above $T_{c\ell}$ (102 °C). However, the log G' versus log G'' plots for PI-7CNCOOH (Figure 9.71b) show temperature dependence at $T < 110$ °C, but they become independent of temperature at $T \geq 110$ °C and have a slope of 2 in the terminal region. Similar observations can be made from Figure 9.71c for PI-11CNCOOH, namely, log G' versus log G'' plots show temperature dependence at $T \geq 110$ °C, but they become independent of temperature at $T \geq 115$ °C and have a slope of 2 in the terminal region. Notice that the temperature at which log G' versus log G'' plots for both PI-7CNCOOH and PI-11CNCOOH begin to be independent of temperature is very close to the $T_{c\ell}$. The temperature dependence of log G' versus log G'' plots in an anisotropic state of PI-7CNCOOH and PI-11CNCOOH, shown in Figure 9.71, indicates that time–temperature superposition (TTS) fails. This is expected (we have amply discussed the subject in Chapter 7 for some miscible polymer blends and in Chapter 8 for microphase-separated block copolymers). Therefore, application of TTS to TLCPs in an aniostorpic state is not justified, although some investigators (Rubin et al. 1995; Wewerka et al. 2001a, 2001b; Zhou et al. 1999) have applied TTS to TLCPs.

A question remains to be answered: Why do the log G' versus log G'' plots for PI-5CNCOOH (Figure 9.71a) show very weak temperature dependence over the entire range of temperatures tested from 75 °C (below $T_{c\ell}$) to 120 °C (above $T_{c\ell}$), while the log G' versus log G'' plots for PI-7CNCOOH and PI-11CNCOOH show distinct temperature dependence at $T < T_{c\ell}$? We can only offer the following speculative explanations on this question: (1) the nematics in PI-5CNCOOH may be very weak, while the smectics in PI-7CNCOOH and PI-11CNCOOH are very strong, and (2) five methylene spacer groups in PI-5CNCOOH may not be sufficiently long to decouple the side-chain liquid-crystalline monomers from the polymer backbone. When the number of methylene spacer groups is increased sufficiently to yield PI-7CNCOOH and PI-11CNCOOH, the side-chain liquid-crystalline monomers form layered structures and the motions of side-chain liquid-crystalline monomers may be decoupled from the motions of the polymer backbone.

Figure 9.72 gives log η versus log $\dot{\gamma}$ plots, and log N_1 versus log $\dot{\gamma}$ plots in steady-state shear flow of PI-5CNCOOH at 70 °C, PI-7CNCOOH at 75 °C, and PI-11CNCOOH at 85 °C, all in the anisotropic region, for $\dot{\gamma}$ ranging from 0.01 to 7 s^{-1}. In Figure 9.72, different temperatures were used for each polymer so that a comparison of η and N_1 among the three polymers can be made at approximately the same distance from the $T_{c\ell}$ of each polymer (i.e., at $T_{c\ell} - T$). Such an approach should minimize, if not eliminate completely, the effect of molecular weight, since $T_{c\ell}$ reflects the chemical structures of polymers and, also, depends on their molecular weights.

In Figure 9.72 there are no three separate regions of η for all three PI-*n*CNCOOHs. Specifically, PI-5CNCOOH does not exhibit shear-thinning behavior at low shear rates (i.e., no region I); instead, it exhibits a constant value of η at low $\dot{\gamma}$ (region II) and then weak shear-thinning behavior at high $\dot{\gamma}$ (region III). Conversely, PI-7CNCOOH and PI-11CNCOOH exhibit very strong shear-thinning behavior over the entire range of $\dot{\gamma}$ tested; that is, region I followed by region III, without having region II between regions I and III. Such steady-state viscosity behavior has also been observed in microphase-separated block copolymers (see Chapter 8), and highly filled molten polymers (see Chapter 12). Note that in the anisotropic region (at $T < T_{c\ell}$) of all three PI-*n*CNCOOHs, the shear-rate dependence of η in steady-state shear flow is very similar

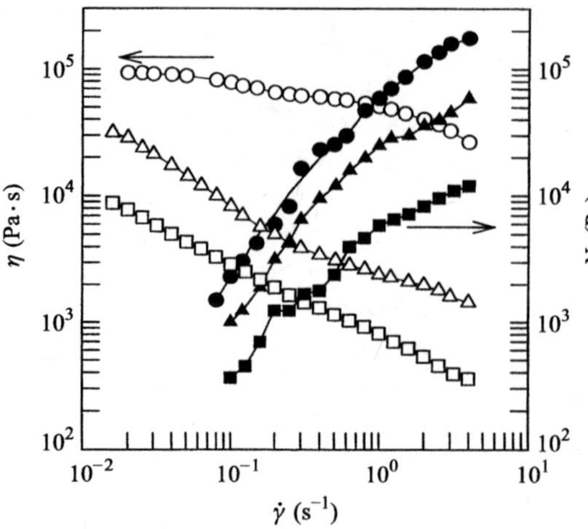

Figure 9.72 Plots of log η versus log $\dot{\gamma}$ (open symbols), and log N_1 versus log $\dot{\gamma}$ (filled symbols) for PI-5CNCOOH at 70 °C (○, ●), PI-7CNCOOH at 75 °C (△, ▲), and PI-11CNCOOH at 85 °C (□, ■). (Reprinted from Lee and Han, *Macromolecules* 36:8796. Copyright © 2003, with permission from the American Chemical Society.)

qualitatively to the angular frequency-dependence of $|\eta^*|$ in oscillatory shear flow (see Figure 9.70). Referring to Figure 9.72, the following questions may be raised: Why does PI-5CNCOOH exhibit a shear-rate dependence of η that is very similar to that of high-molecular-weight ordinary flexible polymers? What is the origin(s) of the strong shear-thinning behavior of η, especially at low $\dot{\gamma}$, in PI-7CNCOOH and PI-11CNCOOH? Notice in Figure 9.72 that values of N_1 for all three PI-*n*CNCOOHs are positive over the entire range of $\dot{\gamma}$ tested, similar to those of PSHQ10 (see Figures 9.55 and 9.67).

What is of great interest in Figure 9.72 is that at given values of $\dot{\gamma}$, both η and N_1 decrease as the number of methylene spacer groups in PI-*n*CNCOOH increases, in spite of the fact that the molecular weight of PI-*n*CNCOOH increases slightly as the number of methylene spacer groups in PI-*n*CNCOOH increases (see Table 9.3). In a preceding section we observed a very strong molecular weight dependence of η and N_1 of segmented MCLCPs ($\eta \propto M^{6.1}$ and $N_1 \propto M^{6.7}$). At present, we do not have information on the molecular weight dependence of η and N_1 of SCLCPs. Intuitively, we expect that the molecular weight dependence of η and N_1 of SCLCPs would be weaker than that of MCLCPs and would depend on the flexible spacer length of SCLCPs, because the longer the flexible spacer of SCLCP, the greater would be decoupling between the side-chain liquid-crystalline monomers and the polymer backbone. Using this argument, we can now explain why in Figure 9.72 the values of η and N_1 of PI-11CNCOOH are smaller than those of PI-7CNCOOH and PI-5CNCOOH, and why the values of η and N_1 of PI-7CNCOOH are smaller than those of PI-5CNCOOH. In other words, the contributions of side-chain liquid-crystalline monomers to the η and N_1 of PI-*n*CNCOOH would become smaller as the number of flexible spacer groups increases from 5 to 11.

Figure 9.73a shows the evolution of G' and Figure 9.73b shows the evolution of G'' as a function of rest period (t_R) upon cessation of shear flow of PI-5CNCOOH at 90 °C, PI-7CNCOOH at 95 °C, and PI-11CNCOOH at 100 °C, after each specimen had been sheared in steady-state mode at $\dot{\gamma} = 10\,s^{-1}$ for 1 h. It is seen in Figure 9.73 that values of G' and G'' for PI-5CNCOOH initially increase very rapidly and then slowly to reach a constant value at about 20 min after cessation of shear flow, while values of

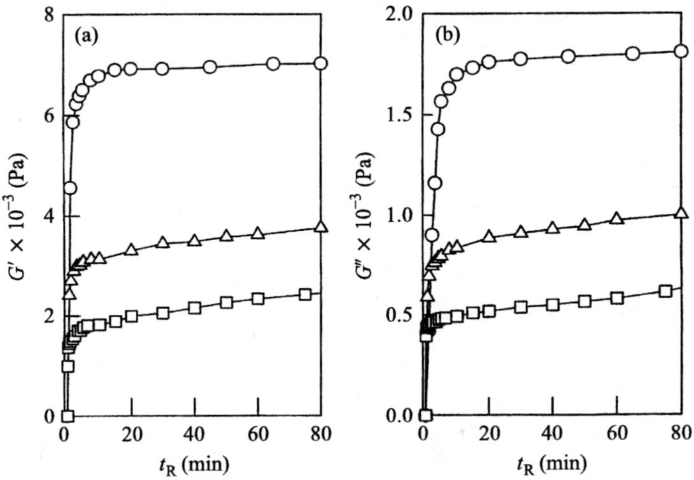

Figure 9.73 (a) Time evolution of dynamic storage modulus G' and (b) time evolution of dynamic loss modulus G'' upon cessation of steady-state shear flow at $\dot{\gamma} = 1.0\ s^{-1}$, which were monitored at an angular frequency of 1.0 rad/s for (○) PI-5CNCOOH, (△) PI-7CNCOOH, and (□) PI-11CNCOOH. (Reprinted from Lee and Han, *Macromolecules* 36:8796. Copyright © 2003, with permission from the American Chemical Society.)

G' and G'' for PI-7CNCOOH and PI-11CNCOOH do not attain a constant value, even after 80 min upon cessation of steady-state shear flow. Specifically, longer rest period upon cessation of steady-state shear flow is required as the number of methylene spacer groups in PI-*n*CNCOOH increases from 5 to 11. Note that PI-5CNCOOH has nematic mesophase, while PI-7CNCOOH and PI-11CNCOOH have smectic mesophase. Thus, we can conclude that the number of methylene spacer groups in PI-*n*CNCOOHs (thus flexible spacer length in SCLCPs, in general) has a profound influence on the rate of structural reorganization upon cessation of steady-state shear flow.

9.6 Summary

In this chapter, we have described a broad range of rheological behavior of LCPs, putting emphasis on the importance of chemical structure and thermal history of the polymers to their rheological behavior. We have shown unusual rheological characteristics of LCPs, such as (1) strong sensitivity to thermal and deformation histories, (2) a very large initial overshoot and often multiple overshoots in first normal stress difference upon shear startup, (3) very slow decay of first normal stress difference with oscillations until reaching a steady state upon shear startup, (4) very slow relaxation of first normal stress difference upon cessation of shear flow, and (5) very slow decrease in orientation as the material relaxes, as evidenced by an increase of dynamic loss modulus, upon cessation of shear flow. We have emphasized the importance of controlling the initial conditions when measuring rheological properties of LCPs. This is particularly important if rheological measurements are to be used for developing a molecular theory and/or for comparing with theoretical prediction.

One of the difficulties with taking rheological measurements of some TLCPs lies in that their rheological properties may keep changing during experiment. Under such circumstances, it is virtually impossible to obtain reproducible rheological measurements. This is a particularly serious problem when dealing with semicrystalline TLCPs, which, may undergo crystallization during rheological measurements. The most effective way of circumventing such a difficulty is to first heat a specimen to the isotropic region in order to erase all previous thermal histories, and then to cool the specimen very slowly to a preset temperature in the nematic region before commencing rheological measurements. This approach can be realized only when a specimen has a clearing temperature that is much lower than its thermal degradation temperature. It turns out that commercial TLCPs have a clearing temperature very close to or higher than their thermal degradation temperatures and thus it would not be possible to subject a specimen to thermal treatment in the isotropic region. Hence, rheological measurements of such TLCPs are of little significance for obtaining reproducible data. An erroneous conclusion can be drawn on the sign of first normal stress difference in such a TLCP if transient or steady-state shear flow experiments are carried out in the presence of residual normal force generated during the squeezing of the specimen in an anisotropic state.

Suggestion has been made that a TLCP specimen be presheared, so that the zero value of N_1 can be obtained before shear startup. But, as shown in this chapter, presshearing has a profound influence on the morphological state of a TLCP specimen, and consequently on its rheological behavior. Since the rheological behavior of a TLCP depends very much on its morphological state, one must not neglect the effect of presshearing on its rheological behavior. We have demonstrated that an attainment of zero value of N_1 before shear startup is not sufficient for one to claim that the initial condition of a TLCP specimen was controlled, because different thermal and/or deformation histories of specimens, although they may have a zero value of N_1 before shear startup, can exhibit drastically different rheological responses in transient shear flow. We have shown that when taking rheological measurements of a semicrystalline TLCP in the nematic region, one must choose experimental temperatures that lie *above* the melting point (T_{m1}) of high-temperature melting crystals in order to (at least) bring the normal force of the specimen to a baseline before shear startup.

In order to obtain rheological data that might be useful to develop a theory or to compare with a currently held theory, it is highly desirable to have model LCP compounds. The desirable characteristics of model LCP compounds for use in rheological investigations are: (1) they exhibit nematic phase and enable one to take rheological measurements over a very wide range of temperatures and (2) they have excellent thermal stability, so that specimens can be heated, without thermal degradation, to above the clearing temperature before rheological measurement begins in an anisotropic state. In this chapter, we have considered a semiflexible TLCP, PSHQ10, as a model compound. We have presented the effect of shear rate on the flow-aligning behavior of PSHQ10, and how molecular weight affects its rheological behavior.

We cannot overemphasize that the rheological behavior and the morphology of LCPs during flow are interrelated and therefore that they are inseparable, insofar as correctly interpreting rheological measurements. What is most desirable is to conduct experiments that enable one to take simultaneous measurements of the rheological properties and the morphology of an LCP specimen during flow. This subject needs to be developed further in future research efforts.

At present, there is no theory that can be used to predict the rheological behavior of semiflexible main-chain TLCPs with or without pendent side groups. In order to develop a theory for such TLCPs, one must first develop an accurate expression for an intermolecular potential. The Maier–Saupe potential in its present form is not adequate to describe the dynamics and rheological behavior of semiflexible polymer chains. This is one of the many areas that require greater attention from polymer scientists if there is going to be meaningful progress towards a better understanding of the dynamics and rheological behavior of TLCPs.

Although the Marrucci–Maffettone theory predicts the appearance of multiple overshoots in first normal stress difference in transient shear flow, the theory is based on a monodomain model that predicts a tumbling behavior, and is also based on the Maier–Saupe potential. Note that the Maier–Saupe potential may be valid for small-molecule thermotropic liquid crystals, but certainly not for semiflexible main-chain TLCPs exhibiting flow aligning behavior. The Larson–Doi mesoscopic model considers the evolution of texture on the basis of some experimental observations describing how the domain size decreases with increasing shear rate and grows upon cessation of flow. Although both the Ericksen polydomain model and the Larson–Doi mesoscopic model predict qualitatively the experimental observations for the time evolution of $\sigma^+(t, \dot{\gamma})$ and $N_1^+(t, \dot{\gamma})$ for a model compound PSHQ10 in transient shear flow, the predicted magnitude of $\sigma^+(t, \dot{\gamma})$ is lower than that of $N_1^+(t, \dot{\gamma})$, which is opposite to experimental observations. Further, the models predict a much shorter transient time (strain) of $\sigma^+(t, \dot{\gamma})$ and $N_1^+(t, \dot{\gamma})$ variations compared with experimental results. The inadequacy of the Larson–Doi model to accurately predict the time evolution of $N_1^+(t, \dot{\gamma})$ in transient shear flow of a model compound PSHQ10 may be attributable to the presence of long flexible spacers and bulky pendent side groups in the PSHQ10. Both the long flexible spacers and the bulky pendent side groups present in the model compound PSHQ10 might directly suppress molecular rotations and thus, perhaps, collective molecular rotations, or director tumbling. Although the Larson–Doi mesoscopic model is virtually the only model available in the literature that incorporates polydomain texture and distortional elastic effects, the model is based on the Leslie–Ericksen equations, which may be more appropriate for small molecule liquid crystals or lyotropic systems that have short relaxation times. Since most of the TLCPs, if not all, reported in the literature exhibit flow aligning behavior, a new theory needs to be developed that can accommodate all the features observed experimentally.

In this chapter, we have presented the rheological behavior of model SCLCP, PI-*n*CNCOOH, placing emphasis on the effect of flexible spacer length *n* on linear dynamic viscoelasticity and steady-state shear flow behavior. We have shown that the magnitude of η and N_1 in steady-state shear flow decreases as the flexible spacer length *n* in PI-*n*CNCOOH increases. The decrease in η and N_1 with increasing flexible spacer length *n* in PI-*n*CNCOOH is attributable to an enhanced decoupling between the polymer backbone and side-chain liquid-crystalline monomer as the flexible spacer length *n* in PI-*n*CNCOOH increases. We have also shown that the structural reorganization upon cessation of shear flow is much slower in the smectic-forming PI-7CNCOOH and PI-11CNCOOH than in the nematic-forming PI-5CNCOOH. The experimental observations indicate that the type of order and the length scale of the mesogenic unit have a profound influence on the rheological behavior of PI-*n*CNCOOH. As the flexible spacer length *n* in PI-*n*CNCOOH increases, decoupling of the mesogenic unit from the polymer backbone increases and thus the arrangement of the side-chain liquid-crystalline monomers becomes increasingly important to the rheological behavior of PI-*n*CNCOOH. Specifically, the degree of packing of the layered structure would play a central role in determining the rheological responses of PI-*n*CNCOOH. The experimental observations of transient shear flow behavior and intermittent shear flow behavior of the model SCLCP, PI-*n*CNCOOH, after a rest upon cessation of the initial start-up transient are referred to the original paper of Lee and Han (2003).

Appendix 9A: Derivation of Equation (9.3)

For a homogeneous flow, the velocity $\mathbf{v}(\mathbf{r}, t)$ at position $\mathbf{r}$ and at time t can be expressed by

$$\mathbf{v}(\mathbf{r}, t) = \mathbf{L}(t) \cdot \mathbf{r} \tag{9A.1}$$

where $\mathbf{L}(t)$ is the velocity gradient tensor (see Chapter 2). If a rigid rod is subjected to a homogeneous flow defined by Eq. (9A.1), the orientation vector $\mathbf{u}(t)$ will change its direction to $(\mathbf{I} + \mathbf{L}(t)\Delta t) \cdot \mathbf{u}(t)$ after a small time interval Δt, in which $\mathbf{I}$ is the identity tensor. Since $\mathbf{u}(t)$ is the unit vector at all times, we have

$$\mathbf{u}(t + \Delta t) = \frac{(\mathbf{I} + \mathbf{L}(t)\Delta t) \cdot \mathbf{u}(t)}{|(\mathbf{I} + \mathbf{L}(t)\Delta t) \cdot \mathbf{u}(t)|} \tag{9A.2}$$

Note that

$$\begin{aligned}
|(\mathbf{I} + \mathbf{L}(t)\Delta t) \cdot \mathbf{u}(t)| &= \left[\sum_{i=1}^{3}\left(u_i + \sum_{j=1}^{3} L_{ij}u_j\Delta t\right)^2\right]^{1/2} \\
&= \left[\sum_{i=1}^{3}\left(u_iu_i + 2u_i\sum_{j=1}^{3} L_{ij}u_j\Delta t\right) + \sum_{i,j=1}^{3}(L_{ij}u_j\Delta t)^2\right]^{1/2} \\
&= \left[1 + 2\mathbf{u} \cdot (\mathbf{L} \cdot \mathbf{u})\Delta t + (\mathbf{L} \cdot \mathbf{u})^2(\Delta t)^2\right]^{1/2} \\
&\approx \left[1 + 2\mathbf{u} \cdot (\mathbf{L} \cdot \mathbf{u})\Delta t\right]^{1/2}
\end{aligned} \tag{9A.3}$$

Substitution of Eq. (9A.3) into (9A.2) gives

$$\begin{aligned}
\mathbf{u}(t+\Delta t) &= \left[(\mathbf{I}+\mathbf{L}\Delta t)\cdot\mathbf{u}\right]/\left[1+2\mathbf{u}\cdot(\mathbf{L}\cdot\mathbf{u})\,\Delta t\right]^{1/2} \approx \left[(\mathbf{I}+\mathbf{L}\Delta t)\cdot\mathbf{u}\right]\left[1-\mathbf{u}\cdot(\mathbf{L}\cdot\mathbf{u})\,\Delta t\right] \\
&\approx \mathbf{u}+(\mathbf{L}\cdot\mathbf{u})\,\Delta t-\left[\mathbf{u}\cdot(\mathbf{L}\cdot\mathbf{u})\right]\mathbf{u}\Delta t
\end{aligned} \tag{9A.4}$$

which then yields

$$\lim_{\Delta t \to 0} \frac{\mathbf{u}(t+\Delta t) - \mathbf{u}(t)}{\Delta t} = \dot{\mathbf{u}} = \mathbf{L} \cdot \mathbf{u} - \mathbf{u} \cdot (\mathbf{L} \cdot \mathbf{u})\mathbf{u}$$

or

$$\dot{\mathbf{u}} = \mathbf{L} \cdot \mathbf{u} - \mathbf{u} \cdot (\mathbf{L} \cdot \mathbf{u})\mathbf{u} = \mathbf{L} \cdot \mathbf{u} - (\mathbf{L} : \mathbf{uu})\mathbf{u} = \mathbf{L} \cdot \mathbf{u} - \mathbf{L} : \mathbf{uuu} \tag{9A.5}$$

Appendix 9B: Derivation of Equation (9.11)

By introducing the rotational operator $\mathfrak{R}$, defined by

$$\mathfrak{R} = \mathbf{u} \times \nabla_{\mathbf{u}} \left(\text{or } \mathfrak{R}_i = \varepsilon_{ijk} u_j \frac{\partial}{\partial u_k} \right) \tag{9B.1}$$

where $\times$ denotes the cross product and $\varepsilon_{ijk} = \varepsilon_{jki} = \varepsilon_{kij} = \varepsilon_{ikj} = \varepsilon_{jik} = \varepsilon_{kji} = 1$ for $i \neq j \neq k$, into Eq. (9.1) we have (Doi and Edwards 1986)[7]

$$\frac{\partial f}{\partial t} = \mathfrak{R} \cdot \widetilde{D}_{\mathrm{r}} \left(\mathfrak{R} f + \frac{f}{k_{\mathrm{B}} T} \mathfrak{R}\, V(\mathrm{u}) \right) - \mathfrak{R} \cdot (\boldsymbol{\omega} f) \tag{9B.2}$$

where $\boldsymbol{\omega}$ is given by[8]

$$\boldsymbol{\omega} = \mathbf{u} \times \dot{\mathbf{u}} \tag{9B.3}$$

Now, substitution of Eqs. (9.3) and (9B.3) into (9B.2) gives

$$\frac{\partial f}{\partial t} = \mathfrak{R} \cdot \widetilde{D}_{\mathrm{r}} \left(\mathfrak{R} f + \frac{f}{k_{\mathrm{B}} T} \mathfrak{R} V(\mathbf{u}) \right) - \mathfrak{R} \cdot (\mathbf{u} \times (\mathbf{L} \cdot \mathbf{u}) f) \tag{9B.4}$$

According to Doi and Edwards (1986), the elastic stress is related to the change in the free energy per unit volume, $\Im$, for a deformation $\delta\varepsilon_{\alpha\beta}$ as

$$\delta \Im = \sigma_{\alpha\beta}^{(\mathrm{E})} \, \delta\varepsilon_{\alpha\beta} \tag{9B.5}$$

Since the free energy $\Im$ is given by (Doi and Edwards 1986)

$$\Im = \nu \int \mathrm{d}\mathbf{u}\, f\left(k_{\mathrm{B}} T \ln f + V(\mathbf{u})\right) \tag{9B.6}$$

where ν is the number of rods (or polymers in a unit volume), we have

$$\delta \Im = \nu \int \mathrm{d}\mathbf{u} \left(k_{\mathrm{B}} T\, \delta f \, \ln f + k_{\mathrm{B}} T \delta f + \delta f\, V(\mathbf{u})\right) \tag{9B.7}$$

In order to calculate δf, we can use Eq. (9B.4). Since the velocity gradient $L_{\alpha\beta} = \delta\varepsilon_{\alpha\beta}/\delta t$ becomes very large for the instantaneous deformation, the time evolution of f

in the time interval δt is dominated by $L_{\alpha\beta}$. Therefore, we can consider only the term containing $\mathbf{L}$ in Eq. (9B.4), that is

$$\delta f = -\mathfrak{R} \cdot (\mathbf{u} \times \mathbf{L} \cdot \mathbf{u}\, f)\, \delta t = -\mathfrak{R} \cdot (\mathbf{u} \times \delta\boldsymbol{\varepsilon} \cdot \mathbf{u}\, f) \tag{9B.8}$$

Substitution of Eq. (9B.8) into (9B.7) gives

$$\begin{aligned}
\delta\Im &= -\nu \int d\mathbf{u} \left\{ k_B T\, \mathfrak{R} \cdot (\mathbf{u} \times \delta\boldsymbol{\varepsilon} \cdot \mathbf{u}\, f)(\ln f + 1) + \mathfrak{R} \cdot (\mathbf{u} \times \delta\boldsymbol{\varepsilon} \cdot \mathbf{u}\, f) V(\mathbf{u}) \right\} \\
&= \nu \int d\mathbf{u} \left\{ k_B T (\mathbf{u} \times \delta\boldsymbol{\varepsilon} \cdot \mathbf{u}\, f) \cdot \mathfrak{R} (\ln f + 1) + (\mathbf{u} \times \delta\boldsymbol{\varepsilon} \cdot \mathbf{u}\, f) \cdot \mathfrak{R} V(\mathbf{u}) \right\} \\
&= \nu \int d\mathbf{u} \left\{ -k_B T\, \mathfrak{R} \cdot (\mathbf{u} \times \delta\boldsymbol{\varepsilon} \cdot \mathbf{u}) + (\mathbf{u} \times \delta\boldsymbol{\varepsilon} \cdot \mathbf{u}) \cdot \mathfrak{R} V(\mathbf{u}) \right\} f
\end{aligned} \tag{9B.9}$$

in which use is made of (1) $\mathfrak{R} (\ln f + 1) = \mathbf{u} \times \dfrac{\partial}{\partial \mathbf{u}} (\ln f + 1) = \mathbf{u} \times \dfrac{\partial f}{\partial \mathbf{u}} \dfrac{1}{f} = \mathfrak{R} f / f$ and (2) $\int (\mathfrak{R} \cdot \mathbf{A}) \mathbf{B}\, d\mathbf{u} = -\int (\mathfrak{R} \cdot \mathbf{B}) \mathbf{A}\, d\mathbf{u}$.

Equation (9B.9) can be rewritten as

$$\delta\Im_{\alpha\beta} = \nu \delta\varepsilon_{\alpha\beta} \left[3 k_B T \langle u_\alpha u_\beta - \tfrac{1}{3} \delta_{\alpha\beta} \rangle - \langle (\mathbf{u} \times \mathfrak{R} V(\mathbf{u}))_\alpha\, u_\beta \rangle \right] \tag{9B.10}$$

in which the following relationships are used:

$$\begin{aligned}
(1) \quad \mathfrak{R} \cdot (\mathbf{u} \times \delta\boldsymbol{\varepsilon} \cdot \mathbf{u}) &= \left(\mathbf{u} \times \frac{\partial}{\partial \mathbf{u}} \right) \cdot (\mathbf{u} \times \delta\boldsymbol{\varepsilon} \cdot \mathbf{u}) = \frac{\partial}{\partial \mathbf{u}} \cdot \left[(\mathbf{u} \times \delta\boldsymbol{\varepsilon} \cdot \mathbf{u}) \times \mathbf{u} \right] \\
&= \frac{\partial}{\partial \mathbf{u}} \cdot \left[(\delta\boldsymbol{\varepsilon} \cdot \mathbf{u}) - \mathbf{u} (\mathbf{u} \cdot \delta\boldsymbol{\varepsilon} \cdot \mathbf{u}) \right] = \frac{\partial}{\partial u_\alpha} \cdot \left[\delta\varepsilon_{\alpha\beta} u_\beta - u_\alpha \left(u_\beta \delta\varepsilon_{\beta q} u_q \right) \right] \\
&= \delta\varepsilon_{\alpha\beta} \delta_{\alpha\beta} - \left[u_\beta \delta\varepsilon_{\beta q} u_q + u_\alpha \delta_{\alpha\beta} \delta\varepsilon_{\beta q} u_q + u_\alpha u_\beta \delta\varepsilon_{\beta q} \delta_{\alpha q} \right] \\
&= -3 \delta\varepsilon_{\alpha\beta} \left(u_\alpha u_\beta - \tfrac{1}{3} \delta_{\alpha\beta} \right)
\end{aligned} \tag{9B.11}$$

$$\begin{aligned}
(2) \quad (\mathbf{u} \times \delta\boldsymbol{\varepsilon} \cdot \mathbf{u}) \cdot \mathfrak{R}\, V(\mathbf{u}) &= \left(\varepsilon_{jkm} u_j \delta\varepsilon_{kn} u_n \right) \left(\varepsilon_{mpq} u_p \frac{\partial V}{\partial u_q} \right) \\
&= \left(\varepsilon_{jkm} \varepsilon_{mpq} u_j u_n u_p \frac{\partial V}{\partial u_q} \right) \delta\varepsilon_{kn} = \delta\varepsilon_{kn} \left\{ -\varepsilon_{kjm} u_j \left(\varepsilon_{mpq} u_p \frac{\partial V}{\partial u_q} \right) u_n \right\} \\
&= -\delta\varepsilon_{\alpha\beta} (\mathbf{u} \times \mathfrak{R} V(\mathbf{u}))_\alpha\, u_\beta
\end{aligned} \tag{9B.12}$$

Since $\boldsymbol{\sigma}^{(E)} = \delta\Im_{\alpha\beta} / \delta\varepsilon_{\alpha\beta}$, from Eq. (9B.10) we have

$$\boldsymbol{\sigma}^{(E)} = 3 \nu k_B T \langle \mathbf{u}\mathbf{u} - \tfrac{1}{3} \boldsymbol{\delta} \rangle - \nu \left\langle \left(\mathbf{u} \times \mathfrak{R} V(\mathbf{u}) \right) \mathbf{u} \right\rangle \tag{9B.13}$$

which can be rewritten as

$$\boldsymbol{\sigma}^{(E)} = 3 \nu k_B T \langle \mathbf{u}\mathbf{u} - \tfrac{1}{3} \boldsymbol{\delta} \rangle + \nu \langle \nabla_{\mathbf{u}} V(\mathbf{u}) \mathbf{u} \rangle \tag{9B.14}$$

in which use is made of the relationship

$$\mathbf{u}\times\mathfrak{R}V(\mathbf{u})=\mathbf{u}\times\left(\mathbf{u}\times\frac{\partial}{\partial\mathbf{u}}V(\mathbf{u})\right)=\left(\mathbf{u}\cdot\frac{\partial}{\partial\mathbf{u}}V(\mathbf{u})\right)\mathbf{u}-(\mathbf{u}\cdot\mathbf{u})\frac{\partial}{\partial\mathbf{u}}V(\mathbf{u})=-\frac{\partial}{\partial\mathbf{u}}V(\mathbf{u}) \tag{9B.15}$$

where $\mathbf{u}\cdot\dfrac{\partial}{\partial\mathbf{u}}=0$. Thus we have

$$\langle\mathbf{u}\times\mathfrak{R}V(\mathbf{u})\mathbf{u}\rangle=-\int d\mathbf{u}\, f\left(\frac{\partial}{\partial\mathbf{u}}V(\mathbf{u})\right)\mathbf{u}=-\langle\nabla_{\mathbf{u}}V(\mathbf{u})\mathbf{u}\rangle \tag{9B.16}$$

Appendix 9C: Derivation of Equation (9.15)

For $V(\mathbf{u})=V_{\text{scf}}(\mathbf{u})$, with $V_{\text{scf}}(\mathbf{u})$ being defined by Eq. (9.14), $\boldsymbol{\sigma}^{(\text{E})}$ can be written from Eq. (9.11) as

$$\boldsymbol{\sigma}^{(\text{E})}=3\nu k_{\text{B}}T\,\mathbf{S}-\tfrac{3}{2}\nu Uk_{\text{B}}T\langle\nabla_{\mathbf{u}}(\mathbf{S}:\mathbf{uu})\,\mathbf{u}\rangle \tag{9C.1}$$

Using $\mathfrak{R}=\mathbf{u}\times\nabla_{\mathbf{u}}$, the second term on the right-hand side of Eq. (9C.1) can be expressed by (see Eq. (9B.16))

$$\begin{aligned}-\tfrac{3}{2}\nu Uk_{\text{B}}T\langle\nabla_{\mathbf{u}}(\mathbf{S}:\mathbf{uu})\mathbf{u}\rangle&=\tfrac{3}{2}\nu Uk_{\text{B}}T\langle\mathbf{u}\times\mathfrak{R}(\mathbf{S}:\mathbf{uu})\mathbf{u}\rangle\\&=\tfrac{3}{2}\nu Uk_{\text{B}}T\left\langle\left[\mathbf{u}\times\mathfrak{R}\left(S_{mn}u_mu_n\right)\right]_\alpha u_\beta\right\rangle\end{aligned} \tag{9C.2}$$

Now $[\mathbf{u}\times\mathfrak{R}(\mathbf{S}:\mathbf{uu})]$ can be rewritten as

$$\begin{aligned}\left[\mathbf{u}\times\mathfrak{R}\left(S_{mn}u_mu_n\right)\right]_\alpha&=\left[\mathbf{u}\times\left(\mathbf{u}\times\frac{\partial}{\partial\mathbf{u}}\left(S_{mn}u_mu_n\right)\right)\right]_\alpha\\&=\left\{\left[\mathbf{u}\cdot\frac{\partial}{\partial\mathbf{u}}\left(S_{mn}u_mu_n\right)\right]\mathbf{u}\right\}_\alpha-\left\{(\mathbf{u}\cdot\mathbf{u})\frac{\partial}{\partial\mathbf{u}}\left(S_{mn}u_mu_n\right)\right\}_\alpha\\&=u_\alpha\left\{u_i\frac{\partial}{\partial u_i}\left(S_{mn}u_mu_n\right)\right\}-\frac{\partial}{\partial u_\alpha}\left(S_{mn}u_mu_n\right)\\&=u_\alpha u_i\left(\delta_{mi}S_{mn}u_n+\delta_{ni}S_{mn}u_m\right)-\left(\delta_{m\alpha}S_{mn}u_n+\delta_{n\alpha}S_{mn}u_m\right)\\&=u_\alpha u_i\left(S_{in}u_n+S_{mi}u_m\right)-\left(S_{\alpha n}u_n+S_{\alpha m}u_m\right)\\&=2u_\alpha u_iS_{\mu i}u_\mu-2S_{\alpha\mu}u_\mu\end{aligned} \tag{9C.3}$$

Therefore we have

$$\left[\mathbf{u}\times\mathfrak{R}\left(S_{mn}u_mu_n\right)\right]_\alpha u_\beta=2u_\alpha u_\beta u_iu_\mu S_{\mu i}-2S_{\alpha\mu}u_\mu u_\beta=2u_\alpha u_\beta u_\mu u_\nu S_{\mu\nu}-2S_{\alpha\mu}u_\mu u_\beta \tag{9C.4}$$

Substitution of Eq. (9C.4) into (9C.2) gives

$$-\tfrac{3}{2}\nu U k_B T \langle \nabla_{\mathbf{u}}(\mathbf{S}:\mathbf{uu})\mathbf{u}\rangle = 3\nu k_B T \left\{ -U \left(\mathbf{S}\cdot\langle\mathbf{uu}\rangle - \mathbf{S}:\langle\mathbf{uuuu}\rangle\right)\right\} \tag{9C.5}$$

According to Doi and Edwards (1986), the viscous stress $\boldsymbol{\sigma}^{(V)}$ is related to the hydrodynamic energy dissipation W by

$$W = L_{\alpha\beta}\sigma^{(V)}_{\alpha\beta} \tag{9C.6}$$

and the work done by the frictional force can be described by

$$W = \nu\xi_{\text{str}} \left\langle (\mathbf{L}:\mathbf{uu})^2\right\rangle = \nu\xi_{\text{str}} L_{\alpha\beta} L_{\mu\nu} \left\langle u_\alpha u_\beta u_\mu u_\nu\right\rangle \tag{9C.7}$$

where ξ_{str} is a friction coefficient. From Eqs. (9C.6) and (9C.7) we can obtain Eq. (9.16). Note that Eq. (9.16) took into account the hydrodynamic interaction. Under such a situation $\xi_{\text{str}} = \xi_r/2$, with ξ_r being the rotational friction coefficient defined by Eq. (9.18).

Appendix 9D: Derivation of Equation (9.23)

Using the rotational operator $\mathfrak{R}$ defined by Eq. (9B.1) in Appendix 9B, Eq. (9.22) can be rewritten as (see the derivation of Eq. (9B.4))

$$\frac{\partial f}{\partial t} = \overline{D}_r \mathfrak{R}\cdot\left[\mathfrak{R} f - \tfrac{2}{3} U f \mathfrak{R}\,(\mathbf{S}:\mathbf{uu})\right] - \mathfrak{R}\cdot[\mathbf{u}\times(\mathbf{L}\cdot\mathbf{u})f] \tag{9D.1}$$

Multiplying $(u_\alpha u_\beta - \frac{1}{3}\delta_{\alpha\beta})$ on both sides of Eq. (9D.1) and integrating over $\mathbf{u}$, we obtain

$$(1)\quad \int d\mathbf{u}\left[\frac{\partial}{\partial t}\left(u_\alpha u_\beta - \tfrac{1}{3}\delta_{\alpha\beta}\right) f\right] = \frac{\partial S_{\alpha\beta}}{\partial t} \tag{9D.2}$$

$$(2)\quad \int d\mathbf{u}\left[\left(u_\alpha u_\beta - \tfrac{1}{3}\delta_{\alpha\beta}\right)\overline{D}_r \mathfrak{R}^2 f\right] = -6\overline{D}_r S_{\alpha\beta} \tag{9D.3}$$

which follows from

$$\mathfrak{R}^2\left(u_\alpha u_\beta - \tfrac{1}{3}\delta_{\alpha\beta}\right) = \mathfrak{R}^2 u_\alpha u_\beta = \varepsilon_{imn} u_m \frac{\partial}{\partial u_n}\left[\varepsilon_{ijk} u_j \frac{\partial}{\partial u_k} u_\alpha u_\beta\right]$$

$$= \left(\delta_{mj}\delta_{nk} - \delta_{mk}\delta_{nj}\right) u_m \frac{\partial}{\partial u_n}\left[u_j \frac{\partial}{\partial u_k}(u_\alpha u_\beta)\right]$$

$$= u_j \frac{\partial}{\partial u_k}\left[u_j \frac{\partial}{\partial u_k}(u_\alpha u_\beta)\right] - u_k \frac{\partial}{\partial u_j}\left[u_j \frac{\partial}{\partial u_k}(u_\alpha u_\beta)\right]$$

$$= u_j \frac{\partial}{\partial u_k}\left[u_j\left(u_\alpha\delta_{\beta k} + u_\beta\delta_{\alpha k}\right)\right] - u_k \frac{\partial}{\partial u_j}\left[u_j\left(u_\alpha\delta_{\beta k} + u_\beta\delta_{\alpha k}\right)\right]$$

$$= u_j\left[\frac{\partial}{\partial u_\beta}(u_j u_\alpha) + \frac{\partial}{\partial u_\alpha}(u_j u_\beta)\right] - u_\beta\left[\frac{\partial}{\partial u_j}(u_j u_\alpha)\right] - u_\alpha\left[\frac{\partial}{\partial u_j}(u_j u_\beta)\right]$$

$$= u_j u_j \delta_{\alpha\beta} + u_j u_\alpha \delta_{j\beta} + u_j u_j \delta_{\alpha\beta} + u_j u_\beta \delta_{\alpha j} - u_\beta u_\alpha \delta_{jj} - u_\beta u_j \delta_{\alpha j}$$

$$- u_\alpha u_\beta \delta_{jj} - u_\alpha u_j \delta_{j\beta}$$

$$= -6\left(u_\alpha u_\beta - \tfrac{1}{3}\delta_{\alpha\beta}\right) = -6S_{\alpha\beta}$$

(3) $$\int d\mathbf{u}\left[\left(u_\alpha u_\beta - \tfrac{1}{3}\delta_{\alpha\beta}\right)\overline{D}_r \mathfrak{R}\cdot\left\{f\,\mathfrak{R}\left(-\tfrac{3}{2}US_{\alpha\beta}u_\alpha u_\beta\right)\right\}\right]$$

$$= 6\overline{D}_r U\left(S_{\alpha\mu}\langle u_\mu u_\beta\rangle - S_{\mu\nu}\langle u_\alpha u_\beta u_\mu u_\nu\rangle\right) \quad (9D.4)$$

which follows from

$$\int d\mathbf{u}\left[\left(u_\alpha u_\beta - \tfrac{1}{3}\delta_{\alpha\beta}\right)\mathfrak{R}\cdot\left\{f\mathfrak{R}\left(-\tfrac{3}{2}US_{\alpha\beta}u_\alpha u_\beta\right)\right\}\right]$$

$$= \int d\mathbf{u}\,(-1)\left[\mathfrak{R}\left(u_\alpha u_\beta - \tfrac{1}{3}\delta_{\alpha\beta}\right)\right]\cdot f\mathfrak{R}\left(-\tfrac{3}{2}US_{\alpha\beta}u_\alpha u_\beta\right)$$

$$= \frac{3}{2}U\left\langle \mathfrak{R}\left(u_\alpha u_\beta - \tfrac{1}{3}\delta_{\alpha\beta}\right)\right\rangle\left\langle \mathfrak{R}(S_{\alpha\beta}u_\alpha u_\beta)\right\rangle$$

Now let us consider $\mathfrak{R}(u_\alpha u_\beta - \frac{1}{3}\delta_{\alpha\beta})\cdot\mathfrak{R}(S_{\alpha\beta}u_\alpha u_\beta)$, which gives[9]

$$\mathfrak{R}\left(u_\alpha u_\beta - \tfrac{1}{3}\delta_{\alpha\beta}\right)\cdot\mathfrak{R}(S_{\alpha\beta}u_\alpha u_\beta) = \varepsilon_{ijk}u_j\frac{\partial}{\partial u_k}\left(u_\alpha u_\beta - \tfrac{1}{3}\delta_{\alpha\beta}\right)\varepsilon_{i\ell m}u_\ell\frac{\partial}{\partial u_m}(S_{pq}u_p u_q)$$

$$= \left(\delta_{j\ell}\delta_{km} - \delta_{jm}\delta_{k\ell}\right)u_j\frac{\partial}{\partial u_k}\left(u_\alpha u_\beta - \tfrac{1}{3}\delta_{\alpha\beta}\right)u_\ell\frac{\partial}{\partial u_m}(S_{pq}u_p u_q)$$

$$= \frac{\partial}{\partial u_k}(u_\alpha u_\beta)\left\{u_j u_j\frac{\partial}{\partial u_k}(S_{pq}u_p u_q) - u_j u_k\frac{\partial}{\partial u_j}(S_{pq}u_p u_q)\right\}$$

$$= \frac{\partial}{\partial u_k}(u_\alpha u_\beta)\left[\delta_{kp}S_{pq}u_q + \delta_{kp}S_{pq}u_p\right] - u_j u_k\frac{\partial}{\partial u_k}(u_\alpha u_\beta)\left[\delta_{jp}S_{pq}u_q + \delta_{jq}S_{pq}u_p\right]$$

$$= (S_{kp}u_q + S_{kp}u_p)\frac{\partial}{\partial u_k}(u_\alpha u_\beta) - u_j u_k\frac{\partial}{\partial u_k}(u_\alpha u_\beta)[S_{jq}u_q + S_{pj}u_p]$$

$$= 2S_{k\mu}u_\mu\left(\delta_{\alpha k}u_\beta + \delta_{\beta k}u_\alpha\right) - 2S_{j\nu}u_\nu(u_j u_k)\left(\delta_{\alpha k}u_\beta + \delta_{\beta k}u_\alpha\right)$$

$$= 2(S_{\alpha\mu}u_\mu u_\beta + S_{\beta\mu}u_\mu u_\alpha) - 2S_{j\nu}u_\nu(u_j u_\alpha u_\beta + u_j u_\alpha u_\beta)$$

$$= 4(S_{\alpha\mu}u_\mu u_\beta - S_{\mu\nu}u_\alpha u_\beta u_\mu u_\nu)$$

(4) $\int d\mathbf{u}\left[\left(u_\alpha u_\beta - \frac{1}{3}\delta_{\alpha\beta}\right)\mathfrak{R}\cdot[\mathbf{u}\times(\mathbf{L}\cdot\mathbf{u})]f\right]$

$$= \frac{1}{3}(L_{\alpha\beta} + L_{\beta\alpha}) + L_{\alpha\mu}S_{\beta\mu} + L_{\beta\mu}S_{\alpha\mu} - 2L_{\mu\nu}\langle u_\alpha u_\beta u_\mu u_\nu\rangle \tag{9D.5}$$

which follows from

$$\mathfrak{R}\left(u_\alpha u_\beta - \frac{1}{3}\delta_{\alpha\beta}\right)\cdot[\mathbf{u}\times(\mathbf{L}\cdot\mathbf{u})] = \varepsilon_{ijk}u_j\frac{\partial}{\partial u_k}\left(u_\alpha u_\beta - \frac{1}{3}\delta_{\alpha\beta}\right)\varepsilon_{iq\nu}u_q L_{\nu\mu}u_\mu$$

$$= \left(\delta_{jq}\delta_{k\nu} - \delta_{j\nu}\delta_{kq}\right)\left[u_j\frac{\partial}{\partial u_k}(u_\alpha u_\beta)\right]u_q L_{\nu\mu}u_\mu$$

$$= u_j u_j\left[\frac{\partial}{\partial u_k}(u_\alpha u_\beta)\right]L_{k\mu}u_\mu - u_\nu u_k\frac{\partial(u_\alpha u_\beta)}{\partial u_k}L_{\nu\mu}u_\mu$$

$$= \left(u_\alpha\delta_{\beta k} + u_\beta\delta_{\alpha k}\right)L_{k\mu}u_\mu - u_\nu u_k(u_\alpha\delta_{\beta k} + u_\beta\delta_{\alpha k})L_{\nu\mu}u_\mu$$

$$= u_\alpha L_{\beta\mu}u_\mu + u_\beta L_{\alpha\beta}u_\mu - (u_\nu u_\beta u_\alpha + u_\nu u_\alpha u_\beta)L_{\nu\mu}u_\mu$$

$$= L_{\alpha\mu}u_\mu u_\beta + L_{\beta\mu}u_\mu u_\alpha - 2u_\alpha u_\beta u_\mu u_\nu L_{\mu\nu}$$

$$= \frac{1}{3}(L_{\alpha\beta} + L_{\beta\alpha}) + L_{\alpha\mu}S_{\beta\mu} + L_{\beta\mu}S_{\alpha\mu} - 2L_{\mu\nu}\langle u_\alpha u_\beta u_\mu u_\nu\rangle$$

in which use is made of $S_{ij} = \langle u_i u_j - \frac{1}{3}\delta_{ij}\rangle$. Thus, combining Eqs. (9D.2)–(9D.5) we obtain Eq. (9.23).

Appendix 9E: Derivation of Equation (9.28)

Equation (9.24), with the aid of Eqs. (9.7) and (9.27), can be rewritten as

$$F_{\alpha\beta}(\mathbf{S}) = -6\overline{D}_r\left\{S_{\alpha\beta} - U\left[S_{\alpha\mu}(S_{\beta\mu} + \frac{1}{3}\delta_{\beta\mu}) - S_{\mu\nu}(S_{\alpha\beta} + \frac{1}{3}\delta_{\alpha\beta})(S_{\mu\nu} + \frac{1}{3}\delta_{\mu\nu})\right]\right\}$$

$$= -6\overline{D}_r\Big[S_{\alpha\beta} - U(S_{\alpha\mu}S_{\beta\mu}) - \frac{1}{3}US_{\alpha\beta} + US_{\mu\nu}(S_{\alpha\beta}S_{\mu\nu} + \frac{1}{3}S_{\mu\nu}\delta_{\alpha\beta}$$

$$+ \frac{1}{3}S_{\alpha\beta}\delta_{\mu\nu} + \frac{1}{9}\delta_{\alpha\beta}\delta_{\mu\nu})\Big]$$

$$= -6\overline{D}_r\Big[(1 - \frac{1}{3}U)S_{\alpha\beta} - US_{\alpha\mu}S_{\beta\mu} + US_{\alpha\beta}S_{\mu\nu}{}^2 + \frac{1}{3}US_{\mu\nu}{}^2\delta_{\alpha\beta}$$

$$+ \frac{1}{3}US_{\alpha\beta}S_{\mu\mu} + \frac{1}{9}US_{\mu\mu}\delta_{\alpha\beta}\Big]$$

or

$$F(\mathbf{S}) = -6\overline{D}_r\left[(1 - \frac{1}{3}U)\mathbf{S} - U[\mathbf{S}\cdot\mathbf{S} - \frac{1}{3}(\mathbf{S}:\mathbf{S})\boldsymbol{\delta}] + U(\mathbf{S}:\mathbf{S})\mathbf{S}\right] \tag{9E.1}$$

in which use is made of $S_{\mu\mu} = S_{xx} + S_{yy} + S_{zz} = 0$. Using Eq. (9.27) we have

$$\begin{aligned} 2L_{\mu\nu}\langle u_\alpha u_\beta u_\mu u_\nu\rangle &= 2L_{\mu\nu}\left(S_{\alpha\beta}S_{\mu\nu} + \tfrac{1}{3}S_{\mu\nu}\delta_{\alpha\beta} + \tfrac{1}{3}S_{\alpha\beta}\delta_{\mu\nu} + \tfrac{1}{9}\delta_{\alpha\beta}\delta_{\mu\nu}\right) \\ &= 2L_{\mu\nu}S_{\alpha\beta}S_{\mu\nu} + \tfrac{2}{3}L_{\mu\mu}S_{\alpha\beta} + \tfrac{2}{3}L_{\mu\nu}S_{\mu\nu}\delta_{\alpha\beta} + \tfrac{2}{9}L_{\mu\mu}\delta_{\alpha\beta} \\ &= 2L_{\mu\nu}S_{\alpha\beta}S_{\mu\nu} + \tfrac{2}{3}L_{\mu\nu}S_{\mu\nu}\delta_{\alpha\beta} \end{aligned} \tag{9E.2}$$

in which use is made of $S_{\mu\mu} = S_{xx} + S_{yy} + S_{zz} = 0$. Thus, substitution of Eqs. (9E.1) and (9E.2) into (9.23) gives (9.28).

Appendix 9F: Derivation of Equation (9.30)

When the contributions from the viscous force and the solvent can be neglected, substitution of Eq. (9.28) into (9.29) gives

$$\boldsymbol{\sigma} = \frac{\nu k_B T}{2\overline{D}_r}\left[-\frac{\partial \mathbf{S}}{\partial t} + \tfrac{1}{3}(\mathbf{L} + \mathbf{L}^T) + \mathbf{L}\cdot\mathbf{S} + \mathbf{S}\cdot\mathbf{L}^T - \tfrac{2}{3}(\mathbf{L}\cdot\mathbf{S})\,\boldsymbol{\delta} - 2(\mathbf{L}\cdot\mathbf{S})\,\mathbf{S}\right] \tag{9F.1}$$

At steady state, $\partial\mathbf{S}/\partial t = 0$. Thus, Eq. (9F.1) can be expressed in terms of the components of the respective tensors as

$$\sigma_{\alpha\beta} = \frac{\nu k_B T}{2\overline{D}_r}\left[\tfrac{1}{3}(L_{\alpha\beta} + L_{\beta\alpha}) + L_{\alpha\mu}S_{\mu\beta} + L_{\beta\mu}S_{\mu\alpha} - \tfrac{2}{3}\delta_{\alpha\beta}L_{\mu\nu}S_{\mu\nu} - 2L_{\mu\nu}S_{\mu\nu}S_{\alpha\beta}\right] \tag{9F.2}$$

At very small values of velocity gradient $L_{\alpha\beta}$, $S_{\alpha\beta}$ appearing on the right-hand side of Eq. (9F.2) can be replaced by the equilibrium value, $S_{\alpha\beta} = S(n_\alpha n_\beta - \tfrac{1}{3}\delta_{\alpha\beta})$. Thus, we have

(1) $L_{\alpha\mu}S\left(n_\alpha n_\beta - \tfrac{1}{3}\delta_{\alpha\beta}\right) = L_{\alpha\mu}Sn_\mu n_\beta - \tfrac{1}{3}SL_{\alpha\beta}$

(2) $L_{\beta\mu}S\left(n_\mu n_\alpha - \tfrac{1}{3}\delta_{\mu\alpha}\right) = L_{\beta\mu}Sn_\mu n_\alpha - \tfrac{1}{3}SL_{\beta\alpha}$

(3) $-\tfrac{2}{3}\delta_{\alpha\beta}L_{\mu\nu}S\left(n_\mu n_\nu - \tfrac{1}{3}\delta_{\mu\nu}\right) = -\tfrac{2}{3}\delta_{\alpha\beta}L_{\mu\nu}Sn_\mu n_\nu$

(4) $-2L_{\mu\nu}S^2\left(n_\mu n_\nu - \tfrac{1}{3}\delta_{\mu\nu}\right)\left(n_\alpha n_\beta - \tfrac{1}{3}\delta_{\alpha\beta}\right)$

$= -2L_{\mu\nu}S^2 n_\mu n_\nu n_\alpha n_\beta + \tfrac{2}{3}\delta_{\alpha\beta}L_{\mu\nu}S^2 n_\mu n_\nu$

Then, substitution of the above expressions into Eq. (9F.2) gives (9.30).

Appendix 9G: Derivation of Equation (9.49)

For two-dimensional shear flow, the Maier–Saupe potential can be expressed as

$$V_{\text{scf}} = 2Uk_{\text{B}}T\langle\mathbf{uu}\rangle : \mathbf{uu} = -k_{\text{B}}T\left[\langle\cos 2\theta\rangle\cos 2\theta + \langle\sin 2\theta\rangle\sin 2\theta + 1\right] \tag{9G.1}$$

in which use is made of

$$\mathbf{uu} = \frac{1}{2}\begin{Vmatrix} 1+\cos 2\theta & \sin 2\theta \\ \sin 2\theta & 1-\cos 2\theta \end{Vmatrix} \tag{9G.2}$$

In solving Eq. (9.48), it is convenient to drop the last term appearing on the right-hand side of Eq. (9G.1) by defining a new variable V':

$$V' = V_{\text{scf}} + Uk_{\text{B}}T = -Uk_{\text{B}}T\left(\langle\cos 2\theta\rangle\cos 2\theta + \langle\sin 2\theta\rangle\sin 2\theta\right) \tag{9G.3}$$

which is equivalent to Eq. (9G.1), because only the derivative $\partial V_{\text{scf}}/\partial\theta$ is needed to solve Eq. (9.48).

Substitution of Eq. (9G.3) into (9.48) gives

$$\frac{\partial f(\tau,\theta)}{\partial\tau} = \frac{\partial}{\partial\theta}\left(\frac{\partial f}{\partial\theta} + 2U\left[\langle\cos 2\theta\rangle\sin 2\theta - \langle\sin 2\theta\rangle\cos 2\theta\right]f\right) + \Gamma\frac{\partial}{\partial\theta}\left(f\sin^2\theta\right) \tag{9G.4}$$

where $\tau = \overline{D}_{\text{r}}t$, and $\Gamma = \dot{\gamma}/\overline{D}_{\text{r}}$. Using Eqs. (9.46) and (9.47) in (9G.4), we obtain

$$\frac{\partial f(\tau,\theta)}{\partial\tau} = \frac{\partial}{\partial\theta}\left(\frac{\partial f}{\partial\theta} + 2Uf\langle\cos 2\phi\rangle\sin 2\phi\right) + \frac{\partial}{\partial\theta}\left[\Gamma f\sin^2(\alpha+\phi)\right] \tag{9G.5}$$

Using variable ϕ instead of θ, Eq. (9G.5) can be rewritten as

$$\frac{\partial f(\tau,\phi)}{\partial\tau} = \frac{\partial}{\partial\phi}\left(\frac{\partial f}{\partial\phi} + 2Uf\langle\cos 2\phi\rangle\sin 2\phi\right) + \frac{\partial}{\partial\phi}\left(\dot{\alpha}f + \Gamma f\sin^2(\alpha+\phi)\right) \tag{9G.6}$$

in which use is made of the relationship

$$\frac{\partial f(\tau,\theta)}{\partial\tau} = \frac{\partial f(\tau,\phi)}{\partial\tau} + \frac{\partial f(\tau,\phi)}{\partial\phi}\frac{\partial\phi}{\partial\tau} = \frac{\partial f(\tau,\phi)}{\partial\tau} - \dot{\alpha}\frac{\partial f(\tau,\phi)}{\partial\phi} \tag{9G.7}$$

Note that $\dot{\alpha} = \dfrac{\partial\alpha}{\partial\tau} = \dfrac{\partial\alpha}{\partial\phi}\dfrac{\partial\phi}{\partial\tau} = -\dfrac{\partial\phi}{\partial\tau}$ owing to $\dfrac{\partial\alpha}{\partial\phi} = -1$, which follows from Eq. (9.46).

Appendix 9H: Derivation of Equation (9.50)

By multiplying $\sin 2\phi$ on both sides of Eq. (9.49) and integrating over ϕ, we obtain

$$\frac{\partial \langle \sin 2\theta \rangle}{\partial \tau} = \int_0^{\pi} \left\{ \frac{\partial^2 f}{\partial \phi^2} + \frac{\partial}{\partial \phi}\left[2Uf\langle \cos 2\phi \rangle \sin 2\phi + \dot{\alpha} f + \Gamma f \sin^2(\alpha + \phi)\right] \right\} \sin 2\phi \, d\phi \qquad \text{(9H.1)}$$

which reduces to

$$0 = \int_0^{\pi} \left\{ \frac{\partial}{\partial \phi}\left[2U\langle \cos 2\phi \rangle \sin 2\phi + \dot{\alpha} + \Gamma \sin^2(\alpha + \phi)\right] \right\} f \sin 2\phi \, d\phi$$

$$= -2\int_0^{\pi} \left[2U\langle \cos 2\phi \rangle \sin 2\phi + \dot{\alpha} + \Gamma \sin^2(\alpha + \phi)\right] f \cos 2\phi \, d\phi \qquad \text{(9H.2)}$$

in which use is made, from Eq. (9.47), of the relationship

$$\langle \sin 2\phi \rangle = \int_0^{\pi} \sin 2\phi \, f(t, \phi) d\phi = -f(t, \phi) \cos 2\phi \Big|_0^{\pi} + \int_0^{\pi} \cos 2\phi \frac{\partial f}{\partial \phi} \, d\phi$$

$$= -f(t, \pi) + f(t, 0) + \int_0^{\pi} \cos 2\phi \frac{\partial f}{\partial \phi} \, d\phi = \int_0^{\pi} \sin 2\phi \frac{\partial^2 f}{\partial \phi^2} \, d\phi = 0 \qquad \text{(9H.3)}$$

Note that $f(t, \pi) = f(t, 0)$. Equation (9H.2) can be rewritten as

$$2U\langle \cos 2\phi \rangle \langle \sin 2\phi \cos 2\phi \rangle + \dot{\alpha}\langle \cos 2\phi \rangle + \Gamma \langle \sin^2(\alpha + \phi) \cos 2\phi \rangle = 0 \qquad \text{(9H.4)}$$

or

$$\dot{\alpha} = -2U\langle \sin 2\phi \cos 2\phi \rangle - \frac{\Gamma \langle \sin^2(\alpha + \phi) \cos 2\phi \rangle}{\langle \cos 2\phi \rangle} \qquad \text{(9H.5)}$$

which can be rewritten as Eq. (9.50) by using the relationship

$$\frac{\Gamma \langle \sin^2(\alpha + \phi) \cos 2\phi \rangle}{\langle \cos 2\phi \rangle} = \frac{\Gamma \left\langle \left[\frac{1}{2} - \frac{1}{2}\cos(2\alpha + 2\phi)\right] \cos 2\phi \right\rangle}{\langle \cos 2\phi \rangle}$$

$$= \frac{\Gamma}{2}\left[1 - \frac{\cos 2\alpha \langle \cos^2 2\phi \rangle}{\langle \cos 2\phi \rangle} + \frac{\sin 2\alpha \langle \sin 2\phi \cos 2\phi \rangle}{\langle \cos 2\phi \rangle}\right] \qquad \text{(9H.6)}$$

Notes

1. From Eq. (9.9), we have

$$S_{\alpha\beta}n_\alpha n_\beta = S\left(n_\alpha n_\beta n_\alpha n_\beta - \tfrac{1}{3}\delta_{\alpha\beta}n_\alpha n_\beta\right) = S\left(1 - \tfrac{1}{3}n_\alpha n_\alpha\right) = \tfrac{2}{3}S \qquad \text{(9N.1)}$$

thus, the orientational order parameter S can be expressed by

$$S = \tfrac{3}{2}n_\alpha n_\beta S_{\alpha\beta} = \tfrac{3}{2}n_\alpha n_\beta \left\langle u_\alpha u_\beta - \tfrac{1}{3}\delta_{\alpha\beta}\right\rangle = \tfrac{3}{2}\left\langle u_\alpha n_\alpha u_\beta n_\beta - \tfrac{1}{3}n_\alpha n_\beta \delta_{\alpha\beta}\right\rangle$$

$$= \tfrac{3}{2}\left\langle (\mathbf{u}\cdot\mathbf{n})^2 - \tfrac{1}{3}\right\rangle = \tfrac{1}{2}\left\langle 3\cos^2\theta - 1\right\rangle \qquad \text{(9N.2)}$$

Note that Legendre polynomial of the second order $P_2(x)$ is given by $P_2(x) = (1/2)(3x^2 - 1)$. Therefore, the orientation order parameter S is often expressed in terms of $P_2(x)$, with $x = \cos\theta$. From Eq. (9N.2), we have the following situations: (1) $S = 1$ for complete alignment ($\mathbf{u}\cdot\mathbf{n} = 1$), (2) $S = 0$ for random alignment ($\langle(\mathbf{u}\cdot\mathbf{n})^2\rangle = 1/3$), and (3) $S = -1/2$ for perpendicular alignment ($\mathbf{u}\cdot\mathbf{n} = 0$). Usually, S lies between 0 and 1 ($0 \le S \le 1$).

2. Note further that from Eq. (9.9) we have

$$\mathbf{S}:\mathbf{S} = S^2\left(\mathbf{nn} - \tfrac{1}{3}\boldsymbol{\delta}\right)_{\alpha\beta}\left(\mathbf{nn} - \tfrac{1}{3}\boldsymbol{\delta}\right)_{\beta\alpha}$$

$$= S^2\left(n_\alpha n_\beta n_\alpha n_\beta - \tfrac{2}{3}n_\alpha n_\beta \delta_{\alpha\beta} + \tfrac{1}{9}\delta_{\alpha\beta}\delta_{\alpha\beta}\right) = \tfrac{2}{3}S^2 \qquad \text{(9N.3)}$$

Thus, $S^2 = \frac{2}{3}$ ($\mathbf{S}:\mathbf{S}$). Substitution of this expression into Eq. (9.21) gives $\overline{D}_r = D_r/(1 - S^2)^2$, indicating that the value of $\overline{D}_r$ increases with increasing the orientation order parameter S.

3. There are other forms of decoupling approximations reported in the literature, some of which will be presented later in this chapter and in other chapters.

4. The factor $(1 + 1.5S)$ is missing in the numerator of Eqs. (6.20) and (6.21) in a paper of Doi (1981) and also in the numerator of Eq. (10.130) in the monograph of Doi and Edwards (1986). Further, the denominator of Eq. (10.130) in the monograph of Doi and Edwards (1986) should read $(1 + S/2)^2$.

5. By noting that for two-dimensional shear flow $\boldsymbol{\delta}_r = \mathbf{u} = (\cos\theta, \sin\theta)$, $\boldsymbol{\delta}_\theta = \partial\boldsymbol{\delta}_r/\partial\theta = \partial\mathbf{u}/\partial\theta = (-\sin\theta, \cos\theta)$, and $\boldsymbol{\delta}_\phi = (1/\sin\theta)\partial\mathbf{u}/\partial\phi = 0$, we have

$$\frac{\partial}{\partial\mathbf{u}} = \boldsymbol{\delta}_\theta\frac{\partial}{\partial\theta} + \boldsymbol{\delta}_\phi\frac{1}{\sin\theta}\frac{\partial}{\partial\phi} = \boldsymbol{\delta}_\theta\frac{\partial}{\partial\theta}. \quad \text{Thus,} \quad \frac{\partial}{\partial\mathbf{u}}\cdot\frac{\partial}{\partial\mathbf{u}} = \left(\boldsymbol{\delta}_\theta\frac{\partial}{\partial\theta}\right)\cdot\left(\boldsymbol{\delta}_\theta\frac{\partial}{\partial\theta}\right) = \frac{\partial^2}{\partial\theta^2}$$

We then have

$$\frac{\partial}{\partial\mathbf{u}}\cdot\left[(\mathbf{L}\cdot\mathbf{u} - \mathbf{L}:\mathbf{uuu})\, f\right] = \frac{\partial}{\partial\mathbf{u}}\left\{\left[\mathbf{L}\cdot\mathbf{u} - (\mathbf{u}\cdot\mathbf{L}\cdot\mathbf{u})\,\mathbf{u}\right]f\right\}$$

$$= \frac{\partial f}{\partial \theta}\Big[\boldsymbol{\delta}_\theta \cdot \mathbf{L} \cdot \mathbf{u} - (\mathbf{u} \cdot \mathbf{L} \cdot \mathbf{u})\,(\boldsymbol{\delta}_\theta \cdot \mathbf{u})\Big] + f\left\{\boldsymbol{\delta}_\theta \cdot \frac{\partial}{\partial \theta}\,(\mathbf{L} : \mathbf{u}) - \left(\boldsymbol{\delta}_\theta \cdot \frac{\partial \mathbf{u}}{\partial \theta}\right)(\mathbf{u} \cdot \mathbf{L} \cdot \mathbf{u})\right\}$$

$$= \frac{\partial f}{\partial \theta}\,(\boldsymbol{\delta}_\theta \cdot \mathbf{L} \cdot \mathbf{u}) + f\Big[\boldsymbol{\delta}_\theta \cdot \mathbf{L} \cdot \boldsymbol{\delta}_\theta - (\boldsymbol{\delta}_\theta \cdot \boldsymbol{\delta}_\theta)\,(\mathbf{L} : \mathbf{uu})\Big]$$

$$= -\dot{\gamma} \sin^2\theta \frac{\partial f}{\partial \theta} + \dot{\gamma} f\,[-\sin\theta\cos\theta - \cos\theta\sin\theta] = \frac{\partial}{\partial \theta}\left(-\dot{\gamma} f \sin^2\theta\right) \qquad (9N.4)$$

6. Figure 14 in a paper by Kim and Han (1993b) has erroneous data points, which are now corrected in this figure. The corrected data give rise to $\eta_0 \propto {M_w}^{6.1}$.
7. Note that $\boldsymbol{\omega} \times \mathbf{u} = (\mathbf{u} \times \dot{\mathbf{u}}) \times \mathbf{u} = (\mathbf{u} \cdot \mathbf{u})\dot{\mathbf{u}} - (\dot{\mathbf{u}} \cdot \mathbf{u})\mathbf{u} = \dot{\mathbf{u}}$ since $\dot{\mathbf{u}} \cdot \mathbf{u} = 0$. Thus,

$$\nabla_{\mathbf{u}} \cdot (\dot{\mathbf{u}} f) = \nabla_{\mathbf{u}} \cdot (\boldsymbol{\omega} \times \mathbf{u} f) = (\mathbf{u} \times \nabla_{\mathbf{u}}) \cdot (\boldsymbol{\omega} f) = \mathfrak{R} \cdot (\boldsymbol{\omega} f)\,. \qquad (9N.5)$$

8. $\boldsymbol{\omega} = \mathbf{u} \times \dot{\mathbf{u}} = \mathbf{u} \times [\mathbf{L} \cdot \mathbf{u} - (\mathbf{u} \cdot \mathbf{L} \cdot \mathbf{u})\mathbf{u}] = \mathbf{u} \times \mathbf{L} \cdot \mathbf{u} - \mathbf{u} \times (\mathbf{u} \cdot \mathbf{L} \cdot \mathbf{u})\mathbf{u} = \mathbf{u} \times (\mathbf{L} \cdot \mathbf{u})$. Note that $\mathbf{u} \times (\mathbf{u} \cdot \mathbf{L} \cdot \mathbf{u})\mathbf{u} = (\mathbf{u} \cdot \mathbf{L} \cdot \mathbf{u})\mathbf{u} \times \mathbf{u} = 0$.
9.

$$\mathfrak{R} \cdot \mathfrak{R} = \left(\mathbf{u} \times \frac{\partial}{\partial \mathbf{u}}\right) \cdot \left(\mathbf{u} \times \frac{\partial}{\partial \mathbf{u}}\right) = (\mathbf{u} \cdot \mathbf{u})\left(\frac{\partial}{\partial \mathbf{u}} \cdot \frac{\partial}{\partial \mathbf{u}}\right) - \left(\mathbf{u} \cdot \frac{\partial}{\partial \mathbf{u}}\right) = \nabla_{\mathbf{u}} \cdot \nabla_{\mathbf{u}}. \qquad (9N.6)$$

Note that $\left(\mathbf{u} \cdot \dfrac{\partial}{\partial \mathbf{u}}\right) = 0$.

References

Asada T, Muramatsu H, Watanabe R, Onogi S (1980). *Macromolecules* 13:867.
Baek SG, Magda JJ, Cementwala S (1993a). *J. Rheol.* 37:935.
Baek SG, Magda JJ, Larson RG (1993b). *J. Rheol.* 37:1201.
Baek SG, Magda J, Larson RG, Hudson SD (1994). *J. Rheol.* 38:1473.
Bair TI, Morgan PW (1972). U.S. Patent 3673143.
Beekmans F, Gotis, AD, Norder B (1997). *Rheol. Acta* 36:82.
Berghausen J, Fuchs J, Richtering W (1997). *Macromolecules* 30:7574.
Berry GB, Cotts PM, Chu SG (1981). *Brit. Polym. J.* 13:47.
Black WB (1979). In *Flow-Induced Crystallization in Polymer Systems*, Miller RL (ed), Gordon Breach, New York.
Black WB, Preston J (eds) (1973). *High-Modulus Wholly Aromatic Fibers*, Marcel Dekker, New York.
Blumstein A, Thomas O, Kumar S (1986). *J. Polym. Sci., Polym. Phys. Ed.* 24:27.
Blumstein RB, Stickles EM, Gautheir MM, Blumstein A, Volino F (1984). *Macromolecules* 17:177.
Blumstein A, Vilasagar S, Ponrathnam S, Clough SB, Blumstein RB, Maret G (1982) *J. Polym. Sci., Polym. Phys. Ed.* 20:877.
Brown GH, Crooker PP (1983). *Chem. Eng. News* 61(5):24.
Burghardt WR, Fuller GG (1990). *J. Rheol.* 34:959.
Burghardt WR, Fuller GG (1991). *Macromolecules* 24:2546.

Burghardt WR, Hongladarom K (1994). *Macromolecules* 27:2327.
Calundann GW (1978). U.S. Patent 4067852.
Calundann GW (1979). U.S. Patent 4161470.
Calundann GW (1980). U.S. Patent 4185995.
Carlsson T (1984). *Mol. Cryst. Liq. Cryst.* 104:307.
Carlsson T, Skarp K (1981). *Mol. Cryst. Liq. Cryst.* 78:157.
Chang S, Han CD (1997a). *Macromolecules* 30:1656.
Chang S, Han CD (1997b). *Macromolecules* 30:2021.
Chow AW, Fuller GG (1985). *Macromolecules* 18:786.
Chow AW, Hamlin RD, Ylitalo CM (1992). *Macromolecules* 25:7135.
Chu SG, Venkatraman S, Berry GC, Einaga Y (1981). *Macromolecules* 14:939.
Cladis PE, Torza S (1975). *Phys. Rev. Lett.* 35:1283.
Cocchini F, Aratari C, Marrucci G (1990). *Macromolecules* 23:4446.
Cocchini F, Nobile MR, Acierno D (1991). *J. Rheol.* 35:1171.
Colby RH, Gillmore JR, Galli G, Laus M, Ober CK, Hall E (1993). *Liq. Cryst.* 13:233.
Colby RH, Ober CK, Gilmore JR, Connelly RW, Duong T, Galli G, Laus M (1997). *Rheol. Acta* 36:498.
Cottis SG, Economy J, Nowak BE (1972). U.S. Patent 3637595.
Cottis SG, Economy J, Wohrer LC (1976). U.S. Patent 3975487.
Cox WP, Merz EH (1958). *J. Polym. Sci.* 28:619.
Currie PK (1981). *Mol. Cryst. Liq. Cryst.* 73:1.
Currie PK, MacSithigh GO (1979). *Q. J. Mech. Appl. Math.* 32:499.
Daniels BK, Peston J, Zaukelies DA (1971). U.S. Patent 3600269.
De'Neve T, Navard P, Kleman M (1993). *J. Rheol.* 37:515.
Doi M (1981). *J. Polym. Sci., Polym. Phys. Ed.* 19:229.
Doi M (1983). *Faraday Symp. Chem. Soc.* 18:49.
Doi M, Edwards SF (1978a). *J. Chem. Soc. Faraday II* 74:560.
Doi M, Edwards SF (1978b). *J. Chem. Soc. Faraday II* 74:918.
Doi M, Edwards SF (1986). *The Theory of Polymer Dynamics*, Clarendon, Oxford.
Doppert HJ, Picken SJ (1987). *Mol. Cryst. Liq. Cryst.* 153:109.
Driscoll P, Masuda T, Fujiwara K (1991). *Macromolecules* 24:1567.
Einaga GC, Berry GB, Chu SG (1985). *Polym. J.* 17:239.
Ericksen JL (1960). *Arch. Ration. Mech. Anal.* 4:231.
Fabre P, Veyssie M (1987). *Mol. Cryst. Liq. Cryst. Lett.* 4:99.
Flory PJ (1956). *Proc. R. Soc. London* A234:73.
Flory PJ (1984). *Adv. Polym. Sci.* 59:1.
Frank FC (1958). *Discuss. Faraday Soc.* 25:19.
Frazer AH (1972). U.S. Patents 3536651, 3642707.
Furukawa A, Lenz RW (1986). *Macromol. Chem., Macromol. Symp.* 2:3.
Gilmore JR, Colby RH, Hall E, Ober CK (1994). *J. Rheol.* 38:1623.
Gonzalez JM, Munoz ME, Cortazar M, Santamaria A, Pena JJ (1990). *J. Polym. Sci., Polym. Phys. Ed.* 28:1533.
Grabowski DA, Schmidt C (1994). *Macromolecules* 27:2632.
Gray GW (1962). *Molecular Structure and Properties of Liquid Crystals*, Academic Press, New York.
Grizzuti N, Cavella S, Cicarelli P (1990). *J. Rheol.* 34:1293.
Grosberg AYu, Khokhlov AR (1981). *Adv. Polym. Sci.* 41:53.
Guskey SM, Winter HH (1991). *J. Rheol.* 35:1191.
Hamb FL (1972). *J. Polym. Sci., Polym. Chem.* 10:3217.
Han CD, Chang S (1994). *J. Rheol.* 38:241.
Han CD, Kim SS (1994a). *J. Rheol.* 38:13.

Han CD, Kim SS (1994b). *J. Rheol.* 38:31.
Han CD, Chang S, Kim SS (1994a). *Mol. Cryst. Liq. Cryst.* 254:335.
Han CD, Chang S, Kim SS (1994b). *Macromolecules* 27:7699.
Helminiak TE, Berry GC (1979). *J. Polym. Sci., Polym. Symp.* 65:107.
Hermans J (1962). *J. Colloid Sci.* 17:638.
Hongladarom K, Burghardt WR (1993). *Macromolecules* 26:785.
Hongladarom K, Burghardt WR, Baek SG, Cementwala S, Magda JJ (1993). *Macromolecules* 26:772.
Hongladarom K, Secakusuma V, Burghardt WR (1994). *J. Rheol.* 38:1505.
Hongladarom K, Ugaz VM, Cinader DK, Burghardt WR, Quintana JP, Hsiao BD, Dadmun MD, Hamilton WA, Butler PD (1996). *Macromolecules* 29:5346.
Hsiao BS, Stein RS, Deutscher L, Winter HH (1990). *J. Polym. Sci. Polym. Phys. Ed.* 28:1571.
Huang CM, Magda JJ, Larson RG (1999). *J. Rheol.* 43:31.
Hudson SD, Lovinger AJ, Larson RG, Davis DD, Garay RO, Fujishiro K (1993). *Macromolecules* 26:5643.
Irwin RS, Sweeny W, Gardner KH, Gochanour CR, Weinberg M (1989). *Macromolecules* 22:1065.
Jackson WJ, Kuhfuss HF (1976). *J. Polym. Sci., Polym. Chem.* 14:2043.
Kalika DS, Giles DW, Denn MM (1990). *J. Rheol.* 34:139.
Kannan RM, Kornfield JA (1993). *Macromolecules* 26:2050.
Kannan RM, Kornfield JA (1994). *J. Rheol.* 38:1127.
Kannan RM, Rubin SF, Kornfield JA, Boeffel C (1994). *J. Rheol.* 38:1609.
Kim DO, Han CD (2000). *Macromolecules* 33:3349.
Kim DO, Han CD, Mather PT (2000). *Macromolecules* 33:7922.
Kim SS, Han CD (1993a). *Macromolecules* 26:3176.
Kim SS, Han CD (1993b). *Macromolecules* 26:6633.
Kim SS, Han CD (1993c). *J. Rheol.* 37:847.
Kim SS, Han CD (1994a). *J. Polym. Sci., Polym. Phys. Ed.* 32:371.
Kim SS, Han CD (1994b). *Polymer* 35:93.
Kiss G, Porter RS (1978). *J. Polym. Sci., Polym. Symp.* 65:193.
Kiss G, Porter RS (1980a). *J. Polym. Sci., Polym. Phys. Ed.* 18:361.
Kiss G, Porter RS (1980b). *Mol. Cryst. Liq. Cryst.* 60:267.
Kiss G, Orrell TS, Porter RS (1979). *Rheol. Acta* 18:657.
Kuzuu NY, Doi M (1980). *Polym. J.* 12:883.
Kuzuu NY, Doi M (1983). *J. Phys. Soc. Jpn.* 52:3486
Kuzuu NY, Doi M (1984). *J. Phys. Soc. Jpn.* 53:1031.
Kwolek SL (1971), U.S. Patents 3600500, 3671542.
Kwolek SL, Morgan PW, Schaefgen JR, Gurlich LW (1977). *Macromolecules* 10:1390.
La Mantia FP, Valenza A (1989). *Polym. Eng. Sci.* 29:625.
Larson RG (1988). *Constitutive Equations for Polymer Melts and Solution*, Butterworth, Stoneham, Massachusetts, p 317.
Larson RG (1990). *Macromolecules* 23:3983.
Larson RG, Doi M (1991). *J. Rheol.* 35:539.
Larson RG, Mead DW (1989). *J. Rheol.* 33:185.
Laus M, Sante Angeloni A, Galli G, Chiellini E (1992). *Macromolecules* 25:5901.
Lee KM, Han CD (2002). *Macromolecules* 35:6263.
Lee KM, Han CD (2003). *Macromolecules* 36:8796.
Leslie FM (1966). *Quart. J. Mech. Appl. Math.* 19:357.
Leslie FM (1968). *Arch. Ration. Mech. Anal.* 28:265.

Leslie FM (1979). In *Advances in Liquid Crystals,* Vol 4, Brown GH (ed), Academic Press, New York, p 1.
Lin TG, Winter HH (1988). *Macromolecules* 21:2439.
Lin TG, Winter HH (1991). *Macromolecule* 24:2877.
Logullo FM (1971). U.S. Patent 3595951.
Magda JJ, Baek SG, De Vries KL, Larson RG (1991). *Macromolecules* 24:4460.
Maier VW, Saupe A (1958). *Z. Naturforch.* A14:882.
Maier VW, Saupe A (1959). *Z. Naturforch.* A13:564.
Majnusz J, Catala JM, Lenz RW (1983). *Eur. Polym. J.* 19:1043.
Manneville P (1981). *Mol. Cryst. Liq. Cryst.* 70:223.
Marrucci G (1982). *Mol. Cryst. Liq. Cryst.* 72:153.
Marrucci G (1985). *Pure Appl. Chem.* 57:1545.
Marrucci G, Maffettone PL (1989). *Macromolecules* 22:4076.
Marrucci G, Maffettone PL (1990). *J. Rheol.* 34:1217.
Masuda T, Fujiwara K, Takahashi M (1991). *Inter. Polym. Proc.* 6:225.
Mather PT, Romo-Uribe A, Han CD, Kim SS (1997). *Macromolecules* 30:7977.
Mather PT, Jeon HG, Han CD, Chang S (2000). *Macromolecules* 33:7594.
McArdle CB (ed) (1989). *Side Chain Liquid Crystal Polymers,* Blackie and Sons, London.
McFarlane FE, Nicely VA, Davis TG (1977). In *Contemporary Topics in Polymer Science*, Pearce EM, Schaefgen JR (eds), Vol 2, Plenum, New York, p 109.
Mead DW, Larson RG (1990). *Macromolecules* 23:2524.
Mewis J, Moldenaers P (1987). *Mol. Cryst. Liq. Cryst.* 153:291.
Moldenaers P, Fuller G, Mewis J (1989). *Macromolecules* 22:960.
Moldenaers P, Mewis J (1986). *J. Rheol.* 30:567.
Morgan PW (1977). *Macromolecules* 10:1381.
Morgan HS, Preston J, Black WB (1974). U.S. Patent 3796693.
Navard P (1986). *J. Polym. Sci., Polym. Phys. Ed.* 24:435.
Navard P, Haudin JM (1980). *Brit. Polym. J.* 12:174.
Navard P, Haudin JM (1986). *J. Polym. Sci., Polym. Phys. Ed.* 24:189.
Ober CK, Jin JI, Zhou Q, Lenz RW (1984). *Adv. Polym. Sci.* 59:103.
Odell JA, Ungar G, Feijoo JL (1993). *J. Polym. Sci. Polym. Phys. Ed.* 31:141.
Onogi S, Asada T (1980). In *Proceedings of the Seventh International Congress on Rheology*, Vol 1, Astarita G, Marrucci G, Nocolias L (eds), Plenum Press, New York, p 127.
Onsager L (1949). *Ann. N.Y. Acad. Sci.* 51:627.
Papkov SP (1984). *Adv. Polym. Sci.* 59:75.
Papkov SP, Kulichikhin BG, Kalmykovam VD, Malkin AYa (1974), *J. Polym. Sci., Polym. Phys. Ed.* 12:1753.
Parodi O (1970). *J. Phys. (France)* 31:581.
Percec V, Yourd R (1989). *Macromolecules* 22:3229.
Percec V, Tomazos DC, Pugh C (1989). *Macromolecules* 22:3259.
Picken SJ, Aerts J, Doppert HL, Reuvers AJ, Northolt MG (1991). *Macromolecules* 24:1366.
Pieranski P, Guyon E (1974). *Phys. Rev. Lett.* 32:924.
Preston J (1969). U.S. Patent 3484407.
Preston J (1975). *Polym. Eng. Sci.* 15:199.
Pujolle-Robic C, Noirez L (2001). *Nature* 409:167.
Quijada-Garrido I, Siebert H, Friedrich C, Schmidt C (1999). *Rheol. Acta* 38:3495.
Quijada-Garrido I, Siebert H, Friedrich C, Schmidt C (2000). *Macromolecules* 33:3844.
Rubin SF, Kannan RM, Kornfield JA, Boeffel C (1995). *Macromolecules* 28:3521.
Samulski ET (1982). *Physics Today* 35(5):40.
Semenov AN (1983). *Sov. Phys. JETP* 58(2):321.
Shibaev VP, Platé NA (1984). *Adv. Polym. Sci.* 60/61:173.

Ugaz VM (1999). *Investigation of the Effect of Shear Flow on Molecular Orientation in Model Thermotropic Liquid Crystalline Polymers Using In Situ X-ray Scattering*, Doctoral Dissertation, Northwestern University, Evanston, Illinois.

Ugaz VM, Burghardt WR (1998). *Macromolecules* 31:8474.

Ugaz VM, Cinader DK, Burghardt WR (1998). *J. Rheol.* 42:379.

Walker LM, Wagner NJ, Larson RG, Mirau PA, Moldenaers P (1995). *J. Rheol.* 39:925.

Wewerka A, Viertler K, Vlassopoulos D, Stelzer F (2001a). *Rheol. Acta* 40:416.

Wewerka A, Floudas G, Pakula T, Stelzer F (2001b). *Macromolecules* 34:8129.

Wissbrun KF (1980). *Brit. Polym. J.* 13:163.

Wissbrun KF (1985). *Faraday Discuss. Chem. Soc.* 79:161.

Wissbrun KF, Griffin AC (1982). *J. Poly. Sci., Polym. Phys. Ed.* 20:1835.

Wissbrun KF, Kiss G, Cogswell FN (1987). *Chem. Eng. Comm.* 53:149.

Wong CP, Ohnuma H, Berry GC (1979). *J. Polym. Sci., Polym. Symp.* 65:173.

Wunder SL, Ramachandran S, Cochanour CR, Weinberg M (1986). *Macromolecules* 19:1696.

Zentel R, Wu J (1986). *Makromol. Chem.* 187:1727.

Zhou M, Han CD (2005). *Macromolecules* 38:9602.

Zhou M, Han CD (2006). *Macromolecules* 39:282.

Zhou WJ, Kornfield JA, Ugaz VM, Burghardt WR, Link DR, Clark NA (1999). *Macromolecules* 32:5581.

10

Rheology of Thermoplastic Polyurethanes

10.1 Introduction

Thermoplastic polyurethane (TPU) has received considerable attention from both the scientific and industrial communities (Hepburn 1982; Oertel 1985; Saunders and Frish 1962). Applications for TPUs include automotive exterior body panels, medical implants such as the artificial heart, membranes, ski boots, and flexible tubing. Figure 10.1 gives a schematic that shows the architecture of TPU, consisting of hard and soft segments. Hard segments, which form a crystalline phase at service temperature, are composed of diisocyanate and short-chain diols as a chain extender, while soft segments, which control low-temperature properties, are composed of difunctional long-chain polydiols with molecular weights ranging from 500 to 5000. The soft segments form a flexible matrix between the hard domains.

TPUs are synthesized by reacting difunctional long-chain diol with diisocyanate to form a prepolymer, which is then extended by a chain extender via one of two routes: (1) by a dihydric glycol chain extender or (2) by a diamine chain extender. The most commonly used diisocyanate is 4,4′-diphenylmethane diisocyanate (MDI), which reacts with a difunctional polyol forming soft segments, such as poly(tetramethylene adipate) (PTMA) or poly(oxytetramethylene) (POTM), to produce TPU, in which 1,4-butanediol (BDO) is used as a chain extender. There are two methods widely used to produce TPU: (1) one-shot reaction sequence and (2) two-stage reaction sequence. The reaction sequences for both methods are well documented in the literature (Hepburn 1982). It should be mentioned that MDI/BDO/PTMA produces ester-based TPU. One can also produce ether-based TPU when MDI reacts with POTM using BDO as a chain extender.

TPUs are often referred to as "multiblock copolymers." In order to have a better understanding of the rheological behavior of TPUs, one must first understand the relationships between the chemical structure and the morphology; thus, a complete

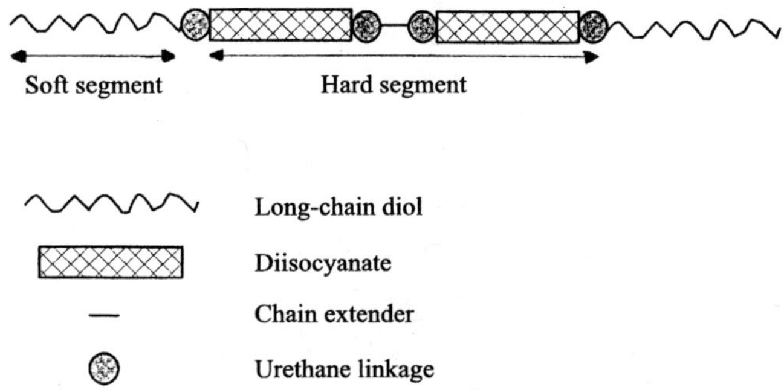

Figure 10.1 The architecture of TPU, consisting of hard and soft segments with chain extender.

characterization of the materials must be conducted. The rheological behavior of TPU depends, among many factors, on (1) the composition of the soft and hard segments, (2) the lengths of the soft and hard segments and the sequence length distribution, (3) anomalous linkages (branching, cross-linking), and (4) molecular weight.

In some TPUs, the hard segments are crystalline, and crystallinity is part of the driving force leading to phase separation. The thermodynamic incompatibility between the urethane and polyol blocks is also part of the driving force leading to phase separation. The phase separation in TPU gives rise to a relatively high modulus and a high extensibility at room temperature. The hard segments act as physical cross-links between the flexible chains and, thus, the crystalline structure has a great influence on the rheological and mechanical properties of TPU. At temperatures above the melting point of the hard segments, TPU forms mixed phases, while its original structure is recovered upon cooling below the melting point of the hard segments. Besides the chemical structures of hard and soft segments, and the volume fraction of hard segments, some of the factors affecting the morphology and consequent rheological behavior of TPUs include hydrogen bonding and thermal history.

Figure 10.2 gives a schematic showing the morphology of TPU at temperatures below the glass transition temperature (T_{gh}) of the hard segments and at temperatures above the melting point (T_{mh}) of the hard segments (Lee et al. 1987). Many research groups have investigated the morphology of melt polymerized MDI-based TPUs by small-angle X-ray scattering (SAXS) (Koberstein and Russell 1986; Leung and Koberstein 1986; Samuels and Wilkes 1973; Speckard et al. 1983; Velankar and Cooper 2000a; Wilkes and Yusek 1973), wide-angle X-ray scattering (WAXS) (Blackwell and Lee 1984), and transmission electron microscopy (TEM) (Briber and Thomas 1983, 1985; Eisenbach et al. 1984; Fridman and Thomas 1980).

Much effort has been spent on investigating thermal transitions in TPU by differential scanning calorimetry (DSC) (Chen et al. 1998; Hesketh et al. 1980; Koberstein and Russell 1986; Koberstein and Stein 1983; Kwei 1982; Leung and Koberstein 1986; Seymour et al. 1970; Seymour and Cooper 1971, 1973; Sung and Schneider 1978; Wilkes et al. 1975; Wilkes and Wildnauer 1975; Yoon and Han 2000). An abundance of experimental evidence shows that multiple thermal transitions take place during the heating and cooling of TPU, suggesting that more than one form of crystal may be

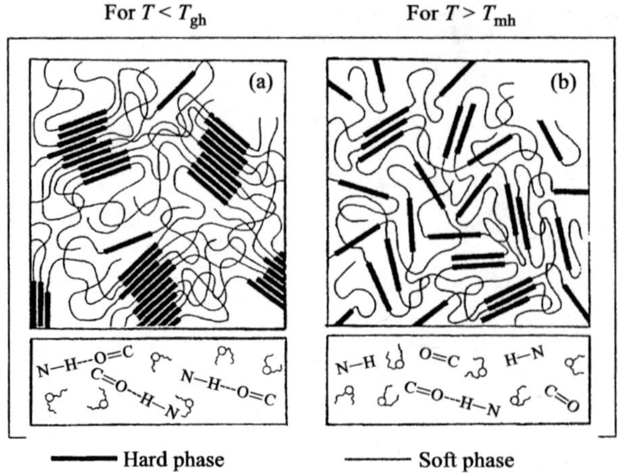

Figure 10.2 Schematic describing the morphology of TPU: (a) phase-separated morphology below the glass transition temperature (T_{gh}) of the hard segments and (b) mostly dissociated structures above the melting point (T_{mh}) of the hard segments. (Reprinted from Lee et al., *Macromolecules* 20:2089. Copyright © 1987, with permission from the American Chemical Society.)

present in TPU. The situation becomes more complicated when the thermal history of a specimen greatly influences its morphology. The thermal transitions in TPU reported in the literature may be summarized as follows: (1) a glass transition of either the hard or soft segments, (2) an endotherm of the hard segments attributable to annealing, and (3) endotherms associated with the long-range order of crystalline portions of either the soft or hard segments. In general, three endotherms associated with the hard segments are observed, as shown in Figure 10.3: (1) at 60–80 °C (endotherm I), (2) at 120–180 °C (endotherm II), and (3) at temperatures above 200 °C (endotherm III). The origin of the multiple endotherms in TPU is generally believed to be associated with different morphologies of the hard segments. These endotherms are very sensitive

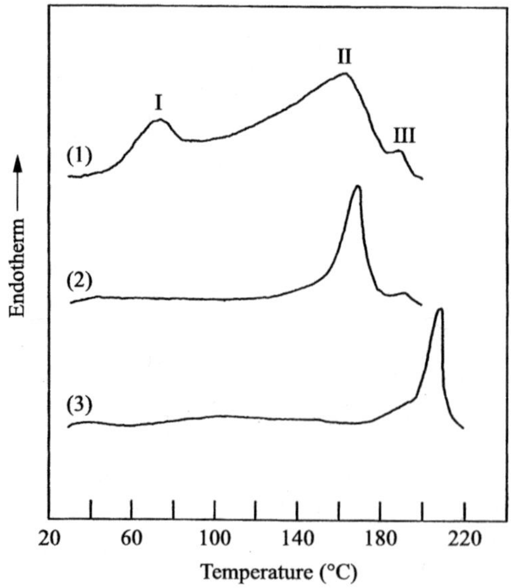

Figure 10.3 Typical DSC thermograms exhibiting multiple thermal transitions in TPU. (1) Multiple thermal transitions in an unannealed specimen, in which endotherm I represents the disordering of short-range order, endotherm II represents the disordering of long-range order, and endotherm III represents the fusion of microcrystalline phase. (2) Thermal transition in a specimen that was annealed at 130 °C. (3) Thermal transition in a specimen that was annealed at 150 °C for a long period. (Reprinted from Seymour and Cooper, *Journal of Polymer Science, Polymer Letter Edition* 9:689. Copyright © 1971, with permission from John Wiley & Sons.)

to the initial conditions (i.e., thermal and processing histories) and annealing conditions. At present, the morphological origin of multiple endotherms observed in TPU is not well understood.

TPUs are known to undergo a high degree of hydrogen bonding (Coleman et al. 1986; Goddard and Cooper 1995a; Luo et al. 1997; Senich and MacKnight 1980; Seymour et al. 1970; Srichatrapimuk and Cooper 1978; Sung et al. 1980; Sung and Schneider 1975, 1977; Tanaka et al. 1968; Yen et al. 1999; Yoon and Han 2000), as shown schematically in Figure 10.4: (1) between an ester-based soft segment and urethane, (2) between an ether-based soft segment and urethane, and (3) between urethanes. These various types of hydrogen bonding are also known to depend on temperature (Coleman et al. 1986; Goddard and Cooper 1995a). The formation of hydrogen bonds in TPU is complicated by phase intermixing and dissociation–reorganization processes. Seymour et al. (1970) investigated the hydrogen bonding in ester- and ether-based TPUs having MDI and BDO as hard segments. They observed that only 60% of NH groups were hydrogen bonded with the urethane carbonyls, and the remaining NH groups were hydrogen bonded with soft segments. The observation was explained in terms of phase intermixing and the large surface areas available of the hard domains. They also investigated the thermal mobility of TPU hydrogen bonds by following the temperature dependence of Fourier transform infrared (FTIR) absorption for the NH vibration. From the temperature dependence of the infrared (IR) spectra, they noted that hydrogen bonding between the soft and hard segments was rather weak, and that it dissociated at a lower temperature than the interurethane hydrogen bonding. Coleman et al. (1986) argued that the changes with temperature in the N–H stretching region

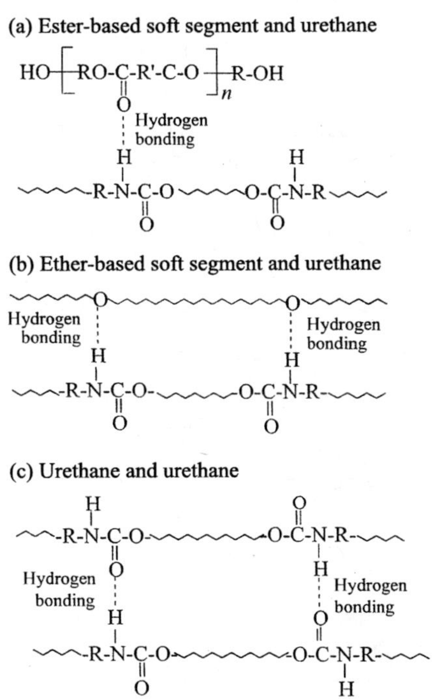

Figure 10.4 Mechanisms of hydrogen bonding in TPU.

of the FTIR spectra of polyamides and polyurethanes were misinterpreted because the strong dependence of absorptivity coefficient on the strength of the hydrogen bond was ignored. The frequency of the N–H stretching band increases with increasing temperature because the average strength of the hydrogen-bonded NH groups diminishes with increasing temperature. A decrease in hydrogen-bonded N–H band with increasing temperature is primarily due to a decrease in absorptivity coefficient with frequency or a decrease in the average strength of the hydrogen-bonded groups.

On the basis of the above observations, it is not difficult to surmise that the rheological behavior of TPUs might be much more complicated than the rheological behavior of block copolymers presented in Chapter 8 and the rheological behavior of liquid-crystalline polymers presented in Chapter 9. No wonder that so little has been reported in the literature on the rheological behavior of TPUs (Velankar and Cooper 1998, 2000b; Yoon and Han 2000). In this chapter, we present the rheological behavior of two commercial TPUs, placing emphasis on the origins of the complexity of the rheological behavior of TPU.

10.2 Effect of Thermal History on the Rheological Behavior of TPUs

In this section, we first present the effects of thermal history on the dynamic moduli of TPU during isothermal annealing and then interpret the observations using information on the formation of hydrogen bonds during isothermal annealing.

10.2.1 Time Evolution of Dynamic Moduli of TPU during Isothermal Annealing

Figure 10.5a gives a schematic showing the temperature protocol employed and Figure 10.5b gives the frequency (ω) dependence of complex viscosity ($|\eta^*|$) of an ester-based commercial MDI/BDO/PTMA TPU. The experimental data were obtained using a rotational-type rheometer with a parallel-plate fixture under the temperature protocol shown. In the experiment, an injection-molded specimen was loaded onto the parallel-plate fixture of a rheometer that had been heated to 170 °C. After temperature equilibration, the specimen was subjected to a dynamic frequency sweep experiment that lasted for 10 min (step 1). The specimen was then heated to 190 °C, which took about 5 min. After temperature equilibration at 190 °C, the specimen was subjected to a dynamic frequency sweep experiment that lasted for 10 min (step 2). The same procedure was repeated, following the temperature protocol described in Figure 10.5a. It can be seen in Figure 10.5b that the frequency dependence of $|\eta^*|$ in steps 1, 3, and 5 at 170 °C is not reproducible, while the frequency dependence of $|\eta^*|$ in steps 2 and 4 at 190 °C is almost reproducible. The above results indicate that the morphological state in step 1 was not reproduced in steps 3 and 5, suggesting that the thermal history of a TPU specimen greatly influences its rheological behavior. In the preceding chapter, we made a similar observation on thermotropic liquid-crystalline polymers (TLCPs).

Figure 10.6 shows the time evolution of dynamic storage modulus G' of an injection-molded specimen of ester-based MDI/BDO/PTMA TPU at a fixed angular frequency (ω) of 0.56 rad/s during isothermal annealing for a period of 2 h at 170,

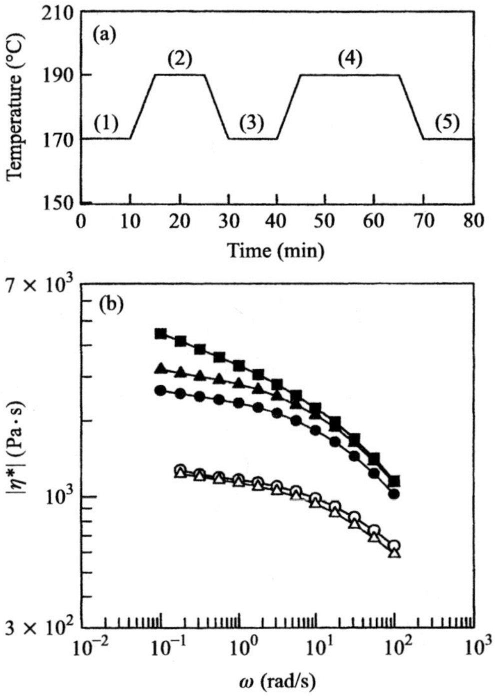

Figure 10.5 (a) Temperature protocol employed and (b) plots of log $|\eta^*|$ versus log ω for an ester-based MDI/BDO/PTMA TPU specimen at various temperatures following the temperature protocol shown: (●) 170 °C (step 1), (○) 190 °C (step 2), (▲) 170 °C (step 3), (△) 190 °C (step 4), and (■) 170 °C (step 5). A specimen prepared by compression molding at 180 °C was used for the entire experiment. (Reprinted from Yoon and Han, *Macromolecules* 33:2171. Copyright © 2000, with permission from the American Chemical Society.)

180, and 190 °C. In the experiment, each injection-molded specimen was placed on the parallel-plate fixture of a rotational-type rheometer under predetermined isothermal conditions and all three specimens were prepared by injection molding at 180 °C (Yoon and Han 2000). It can be seen in Figure 10.6 that the time evolution of G' differs at each isothermal annealing temperature. Namely, (1) at 170 °C the value of G' decreases very slowly for the first 30 min and then increases slowly with time for the rest of the experiment, which lasts for 90 min, (2) at 180 °C the value of G' increases for the

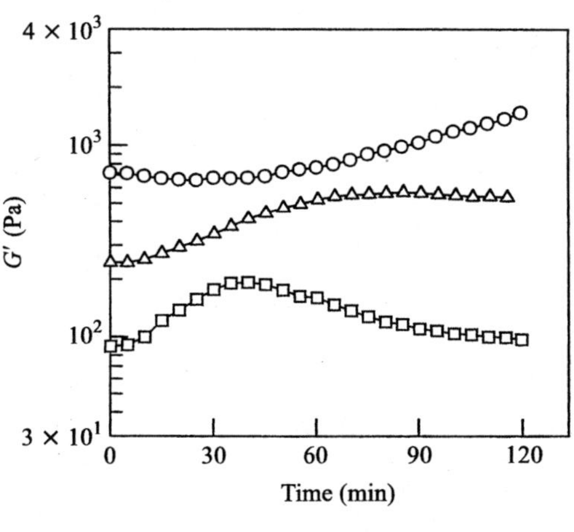

Figure 10.6 Time evolution of G' of a injection-molded specimen of ester-based MDI/BDO/PTMA TPU at $\omega = 0.56$ rad/s during isothermal annealing for a period of 2 h at three different temperatures (°C): (○) 170, (△) 180, and (□) 190. The specimens were prepared by injection molding at 180 °C. (Reprinted from Yoon and Han, *Macromolecules* 33:2171. Copyright © 2000, with permission from the American Chemical Society.)

first 70 min and then tends to level off for the remaining period of annealing, and (3) at 190 °C the value of G' initially increases rapidly, then goes through a maximum at approximately 40 min after annealing began, and then decreases slowly until it tends to level off at approximately 2 h after annealing began.

Figure 10.7 shows the time evolution of G' of five different injection-molded specimens of ester-based MDI/BDO/PTMA TPU at $\omega = 0.23$ rad/s during isothermal annealing for 2 h at 170 °C. Each specimen was injection molded at different temperatures, ranging from 180 to 220 °C. It can be seen in Figure 10.7 that the time evolution of G' during isothermal annealing at 170 °C varies strongly with the thermal history of the specimens that were prepared by injection molding. Similar observations can be made from Figure 10.8, which shows the time evolution of G' of five different specimens of ether-based MDI/BDO/POTM TPU at $\omega = 0.23$ rad/s under isothermal annealing for 2 h at 170 °C. Again, each specimen was injection molded at different temperatures, ranging from 180 to 220 °C. It is seen from Figure 10.8 the time evolution of G' of the ether-based MDI/BDO/POTM TPU is quite different from that of the ester-based MDI/BDO/PTMA TPU given in Figure 10.7. Referring to Figure 10.8, we make the following observations: (1) for the specimen prepared by injection molding at 180 °C, the value of G' initially decreases at a moderate rate for approximately 30 min after annealing began and then levels off and remains there for the subsequent 1.5 h, (2) for the specimens prepared by injection molding at 190 and 200 °C, the value of G' initially decreases very slowly and then levels off, followed by a slow increase at a moderate rate for the remaining 1.5 h, (3) for the specimen prepared by injection molding at 210 °C, the value of G' starts to increase at a moderate rate at the beginning of annealing and continues to increase for the entire annealing period, and (4) for the specimen prepared by injection molding at 220 °C, the value of G' increases very rapidly from the beginning of annealing and continues to increase very rapidly. The above observations indicate that the chemical structure (ester-based TPU versus ether-based TPU) plays an important role in variations of rheological behavior during isothermal annealing. In the following section we offer an explanation on the origin(s) of time evolution of G' observed during isothermal annealing of injection-molded

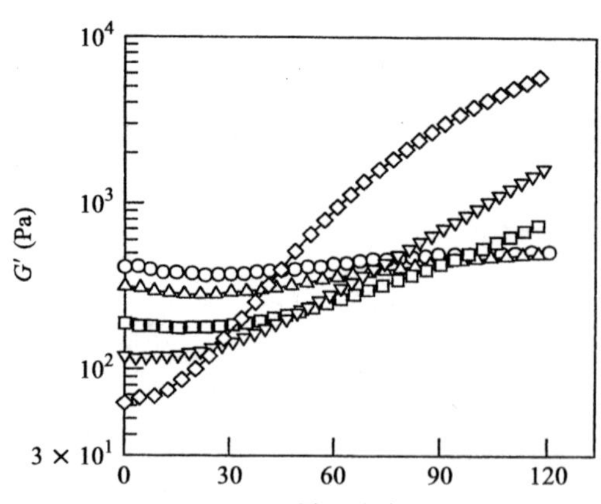

Figure 10.7 Time evolution of G' of five different injection-molded specimens of ester-based MDI/BDO/PTMA TPU at $\omega = 0.23$ rad/s during isothermal annealing for 2 h at 170 °C. The specimens were prepared via injection molding at five different temperatures (°C): (○) 180, (△) 190, (□) 200, (▽) 210, and (◇) 220. (Reprinted from Yoon and Han, *Macromolecules* 33:2171. Copyright © 2000, with permission from the American Chemical Society.)

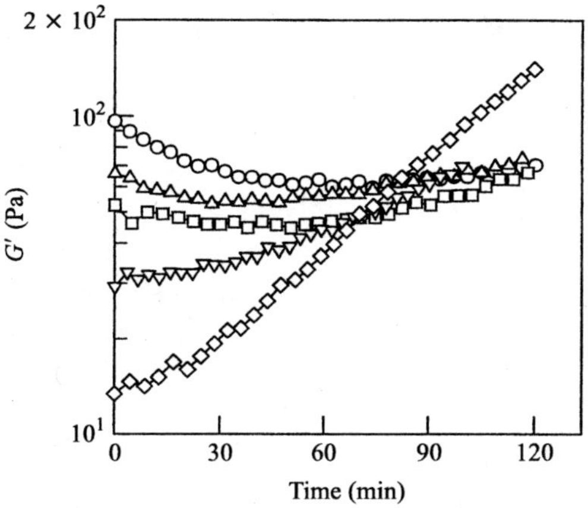

Figure 10.8 Time evolution of G' of five different injection-molded specimens of ether-based MDI/BDO/POTM TPU at $\omega = 0.23$ rad/s during isothermal annealing for 2 h at 170 °C. The specimens were prepared via injection molding at five different temperatures (°C): (○) 180, (△) 190, (□) 200, (▽) 210, and (◇) 220. (Reprinted from Yoon and Han, *Macromolecules* 33:2171. Copyright © 2000, with permission from the American Chemical Society.)

specimens of the ester-based MDI/BDO/PTMA and ether-based MDI/BDO/POTM TPUs presented in Figures 10.6–10.8.

10.2.2 Thermal Transitions in TPU during Isothermal Annealing

Figure 10.9 gives DSC thermograms of specimens of ester-based MDI/BDO/PTMA TPU that were annealed for 1 h in an isothermal chamber at the various temperatures indicated on the DSC thermograms. A fresh specimen was used for annealing at each temperature. The following observations are worth noting in Figure 10.9. The specimen

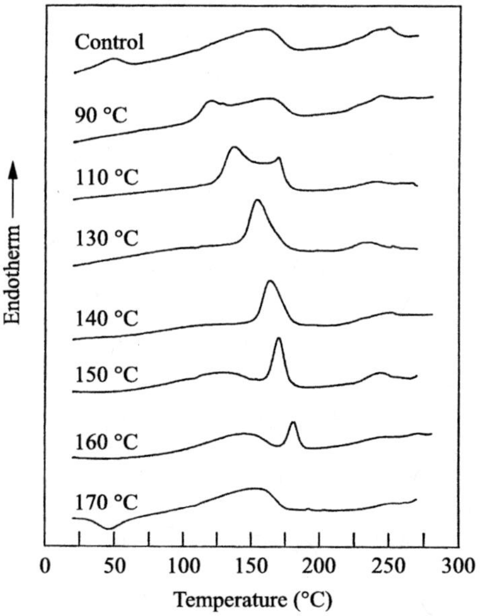

Figure 10.9 Effect of annealing temperature on thermal transitions, as determined by DSC, in ester-based MDI/BDO/PTMA TPU specimens that have been annealed for 1 h at the predetermined temperatures indicated on the DSC thermograms. DSC runs were made at a heating rate of 20 °C/min, and an as-received specimen was used for each DSC run. (Reprinted from Yoon and Han, *Macromolecules* 33:2171. Copyright © 2000, with permission from the American Chemical Society.)

without annealing (referred to as control) shows two endotherms: endotherm I at 40–60 °C and endotherm II at 120–175 °C. As the annealing temperature is increased from 90 to 130 °C, endotherm I apparently merges into endotherm II, giving rise to a single endotherm with the peak position at approximately 140 °C. As the annealing temperature is increased further to 150 °C, a new broad endothermic peak starts to appear at 120–160 °C, while the intermediate peak is shifted to a higher temperature (approximately 165 °C). When the annealing temperature is increased to 160 °C, both the lower and intermediate endothermic peaks are shifted to higher temperatures, with the area under the intermediate peak becoming much smaller. However, at an annealing temperature of 170 °C, only a single endotherm peak is observed, which is drastically different from the DSC thermograms for specimens annealed at temperatures below 170 °C.

Figure 10.10a gives DSC thermograms of specimens of ester-based MDI/BDO/PTMA TPU that were subjected to prolonged annealing (4–48 h) at various temperatures, as indicated on the DSC thermograms. From Figure 10.10a we observe that the prolonged annealing of as-received TPU specimens at 90–150 °C did not produce a significant change in the endothermic peak compared with the short-period annealing. However, it is of interest to observe in Figure 10.10a that isothermal annealing at 170 °C for 4 h produced a sharp new endothermic peak at approximately 200 °C. Figure 10.10b gives DSC thermograms of specimens of ester-based MDI/BDO/PTMA TPU, which were prepared by injection molding at 220 °C followed by isothermal annealing at 170 °C, for various periods, as indicated on the DSC thermograms. Also given, for comparison, in Figure 10.10b is the DSC thermogram of an injection-molded specimen without annealing. Note that a fresh specimen was employed for each DSC run. From Figure 10.10b we observe that isothermal annealing of the TPU

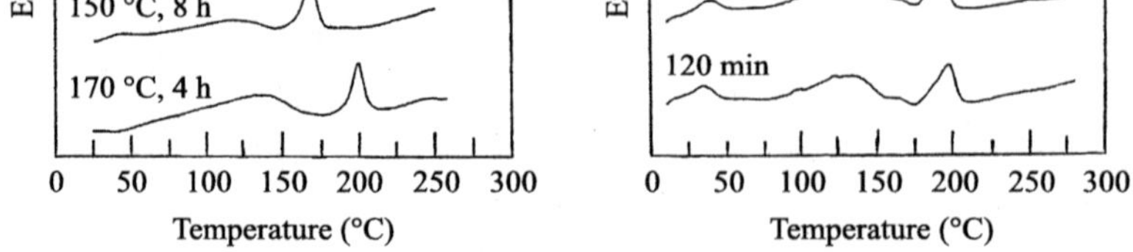

Figure 10.10 Effect of thermal history on thermal transitions, as determined by DSC, in ester-based MDI/BDO/PTMA TPU specimens: (a) DSC thermograms of as-received specimens that were annealed at various temperatures and periods, as indicated on the DSC thermograms, and (b) DSC thermograms of the specimen that was injection molded at 220 °C followed by an isothermal annealing at 170 °C for various time periods, as indicated on the DSC thermograms. A fresh specimen was used for each DSC run at a heating rate of 20 °C/min. (Reprinted from Yoon and Han, *Macromolecules* 33:2171. Copyright © 2000, with permission from the American Chemical Society.)

specimens at 170 °C for 30–120 min, after having been subjected to 220 °C, produced a new endothermic peak appearing at approximately 200 °C (endotherm III), which demonstrates the formation of a crystal-like structure. Not only does the position of endotherm III change with annealing time, but there is also an increase in the heat of fusion.

We now offer an explanation on the origin of the continuous increase in G' during isothermal annealing at 170 °C that occurred when a specimen of ester-based MID/BDO/PTMA TPU that had been prepared by injection molding at 220 °C was employed (the symbol ◊ in Figure 10.7) and also when a specimen of ether-based MDI/BDO/POTM TPU that had been prepared by injection molding at 220 °C was employed (the symbol ◊ in Figure 10.8). It is important to note that that the specimen that had been prepared by injection molding at 220 °C and then subjected to isothermal annealing at 170 °C for 2 h (see Figure 10.10b) did not dissolve in tetrahydrofuran (THF), but dissolved in *N,N'*-dimethyl formamide (Yoon and Han 2000). This observation suggests that the rapid increase in G' observed in Figure 10.7 (the symbol ◊) for the ester-based MID/BDO/PTMA TPU and also observed in Figure 10.8 (the symbol ◊) for the ether-based MDI/BDO/POTM TPU cannot be attributable to the possibility of the presence of cross-linked material in the specimen. Note in Figure 10.10b that the appearance of the endothermic peak at approximately 200 °C signifies the melting of high-temperature melting crystals that were formed during isothermal annealing. If a cross-linked material were formed during isothermal annealing, it would not melt away at 200 °C.

10.2.3 Hydrogen Bonding in TPU during Isothermal Annealing

The IR absorption bands, measured at room temperature in the range of 500–3700 cm^{-1}, are given in Figure 10.11a for the ester-based MDI/BDO/PTMA TPU and in Figure 10.11b for the ether-based MDI/BDO/POTM TPU. The inset of Figure 10.11a shows N–H stretching of the ester-based TPU from 3150 to 3500 cm^{-1}, and the inset of Figure 10.11b shows carbonyl stretching of the ether-based TPU from 1650 to 1800 cm^{-1}. The absorption bands at 2857 and 2960 cm^{-1} in Figure 10.11a are associated with symmetric and asymmetric CH_2 stretching vibrations, respectively, of the aliphatic CH_2 groups in ester-based TPU. For the ether-based TPU given in Figure 10.11b, the absorption bands of the CH_2 stretching vibrations appear at 2857 cm^{-1} (symmetric stretching) and 2940 cm^{-1} (asymmetric stretching). The area of these absorption bands was used to correct the variations in film thickness (Yoon and Han 2000). The N–H absorption band of the ester-based TPU specimen in the inset of Figure 10.11a is composed of at least four contributions. (1) The IR bands at 3440 and 3337 cm^{-1} are assigned to the N–H stretching modes of the free and hydrogen-bonded NH groups, respectively. (2) The IR band at approximately 3124 cm^{-1} is attributed to an overtone of the C–N–H stretching-bending band at 1532 cm^{-1}. (3) The weak shoulder at 3190 cm^{-1} is assigned to *cis–trans* isomerism of the hydrogen-bonded NH groups in the O=C–N–H structure. In the inset of Figure 10.11b, the carbonyl absorption band of the ether-based TPU specimen is split distinctly into two peaks, one at 1732 cm^{-1} and the other at 1702 cm^{-1}, which are attributed to free and hydrogen-bonded carbonyl

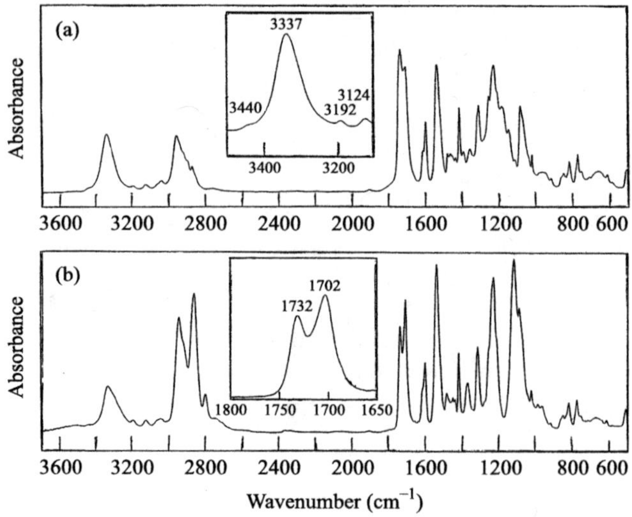

Figure 10.11 FTIR spectra of (a) as-received ester-based MDI/BDO/PTMA TPU and (b) as-received ether-based MDI/BDO/POTM TPU at room temperature. (Reprinted from Yoon and Han, *Macromolecules* 33:2171. Copyright © 2000, with permission from the American Chemical Society.)

groups, respectively. In an ether-based TPU, the carbonyl groups exist only in the hard segments. Therefore, the relative absorbances of the two carbonyl peaks should serve as an index of the degree to which this group participates in the hydrogen bonding. The details of the procedure for determining the extent of hydrogen bonding in TPU specimens are given in the literature (Coleman et al. 1986; Goddard and Cooper 1995a).

In order to investigate the effect, if any, of hydrogen bonding on the variations of G' with time during the isothermal annealing presented in Figures 10.6–10.8, IR measurements were taken during isothermal annealing at various temperatures by use of a spectrometer equipped with a hot plate (Yoon and Han 2000). Figure 10.12 shows variations of the N–H stretching bands of an ester-based MDI/BDO/PTMA TPU with increasing temperature from 30 to 250 °C in the spectrometer. In Figure 10.12, we observe that at 30 °C, most of the NH groups are hydrogen-bonded, as indicated by the large peak at about 3337 cm^{-1} and very small shoulder at 3440 cm^{-1}. With increasing temperature, the intensity of the free N–H band increases at the expense of the hydrogen-bonded N–H band, and the peak of the N–H band is shifted toward larger wavenumbers. Figure 10.13 displays the carbonyl absorption band in the IR spectra, showing distinct peaks that correspond with the free and the hydrogen-bonded groups in the ester-based MDI/BDO/PTMA TPU at 30 °C. As the temperature increases, the absorption intensity of the carbonyl peak increases. The fact that the positions of both the hydrogen-bonded NH and the carbonyl absorption are shifted to higher frequencies with increasing temperature (as can be seen in Figures 10.12 and 10.13) indicates that the strength of hydrogen bonding is weakened with increasing temperature. However, without knowledge of the temperature dependence of absorptivity

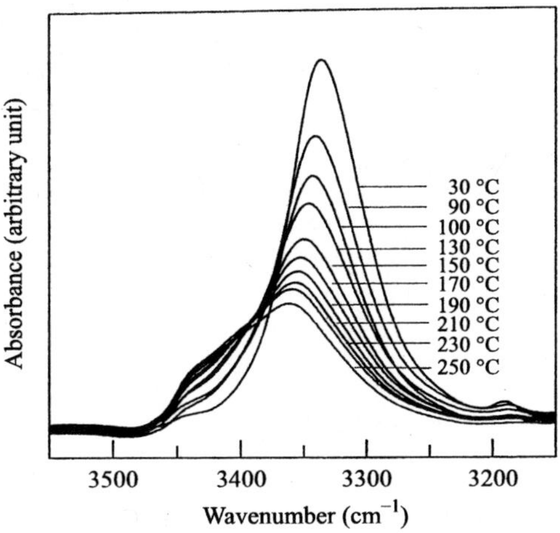

Figure 10.12 FTIR spectra in the N–H stretching region for as-received ester-based MDI/BDO/PTMA TPU at various temperatures, as indicated on the IR spectra. The resolution was within 2 cm^{-1}. (Reprinted from Yoon and Han, *Macromolecules* 33:2171. Copyright © 2000, with permission from the American Chemical Society.)

coefficient, we cannot conclude whether or not the concentrations of free NH and carbonyl groups increase with increasing temperature. Note that absorption coefficient represents an average value of the strengths of the functional groups (free NH groups, hydrogen-bonded NH groups, etc.).

Table 10.1 gives the effect of injection molding temperature on the weight-average molecular weight (M_w) of MDI/BDO/PTMA and MDI/BDO/POTM TPU specimens, as determined from gel permeation chromatography. The results show that an insignificant change in M_w (thus insignificant thermal degradation) occurs at temperatures up to 180 °C, and that M_w decreases (thus measurable thermal degradation occurs) with increasing injection molding temperature. As shown in Figures 10.7 and 10.8, during isothermal annealing, values of G' initially decreased as the molding temperature employed to prepare the specimens was increased. Table 10.2 shows the effect of annealing temperature on the M_w of an ester-based TPU. A fresh specimen was

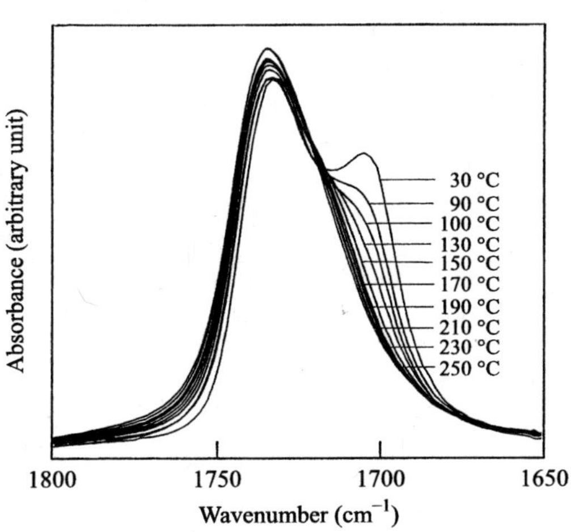

Figure 10.13 FTIR spectra in the C=O stretching region for as-received ester-based MDI/BDO/PTMA TPU at various temperatures, as indicated on the IR spectra. The resolution was within 2 cm^{-1}. (Reprinted from Yoon and Han, *Macromolecules* 33:2171. Copyright © 2000, with permission from the American Chemical Society.)

Table 10.1 Effect of processing temperature on the molecular weight of MDI/BDO/PTMA and MDI/BDO/POTM TPUs

Sample Code	*Processing Temp (°C)*[a]	M_w	M_w/M_n
Ester-based MDI/BDO/PTMA TPU			
	As-received	124×10^3	1.68
	180	124×10^3	1.98
	200	80×10^3	1.62
	220	50×10^3	1.55
Ether-based MDI/BDO/POTM TPU			
	As-received	127×10^3	2.0
	180	114×10^3	1.8
	200	78×10^3	1.7
	220	66×10^3	1.6

[a]Samples were prepared by injection molding at various temperatures.

Reprinted from Yoon and Han, *Macromolecules* 33:2171. Copyright © 2000, with permission from the American Chemical Society.

Table 10.2 Effect of annealing temperature on the molecular weight of an ester-based MDI/BDO/PTMA TPU

Annealing Temp (°C)	*Annealing Period (min)*	M_w	M_w/M_n
170			
	0	112×10^3	1.46
	30	111×10^3	1.42
	60	109×10^3	1.47
	90	114×10^3	1.50
	120	115×10^3	1.50
180			
	0	109×10^3	1.43
	30	106×10^3	1.44
	60	106×10^3	1.44
	90	112×10^3	1.45
	120	113×10^3	1.46
190			
	0	102×10^3	1.43
	30	107×10^3	1.48
	60	a	
	90	a	
	120	a	

[a]Gels partially insoluble in THF were detected.

Reprinted from Yoon and Han, *Macromolecules* 33:2171. Copyright © 2000, with permission from the American Chemical Society.

placed in the DSC cell, where it received thermal treatment for a preset period (Yoon and Han 2000). In Table 10.2, we observe that an insignificant change in M_w (thus insignificant thermal degradation) of the ester-based TPU occurs at 170 and 180 °C. However, there is a trend showing a slight decrease in M_w for the first 1 h, followed by a slight increase in M_w thereafter. Moreover, insoluble gels in THF were formed in the solution after annealing at 190 °C for 60–120 min (Yoon and Han 2000). Based on a significant increase in isocyanate groups for the ester-based TPU at temperatures above approximately 190 °C (see Figure 10.12) and evidence of the formation of insoluble gels in the solution after annealing at 190 °C for 60–120 min (see Table 10.2), we conclude that the rapid increase in G' observed during isothermal annealing at 180 and 190 °C (Figure 10.6) may be attributable to the formation of biuret or allophanate via chemical reactions between isocyanate and active hydrogen in the urethane groups.

It is highly desirable to calculate the fraction of hydrogen-bonded NH groups during isothermal annealing of TPU specimens in the FTIR spectrometer at various temperatures. This can be done when information on the molar absorptivity coefficient is available. In the absence of such information on the TPUs investigated, the A_b/A_f ratio was calculated at various temperatures from the IR spectra (Yoon and Han 2000), where A_b is the area under the hydrogen-bonded absorbance peak and A_f is the area under the free-hydrogen absorbance peak. Values of A_b and A_f were determined by curve-fitting the IR spectra recorded at various temperatures for each of the two TPUs employed. Figure 10.14a shows variations of the A_b/A_f ratio with annealing time, obtained during the in situ FTIR measurements, for specimens of ester-based MDI/BDO/PTMA TPU that were prepared by injection molding at 180 and 220 °C. Similar results are given in Figure 10.14b for the ether-based MDI/BDO/POTM TPU. It is of interest to observe in Figure 10.14 that the time evolution of A_b/A_f ratio during isothermal annealing very closely resembles the time evolution of G' during isothermal annealing given

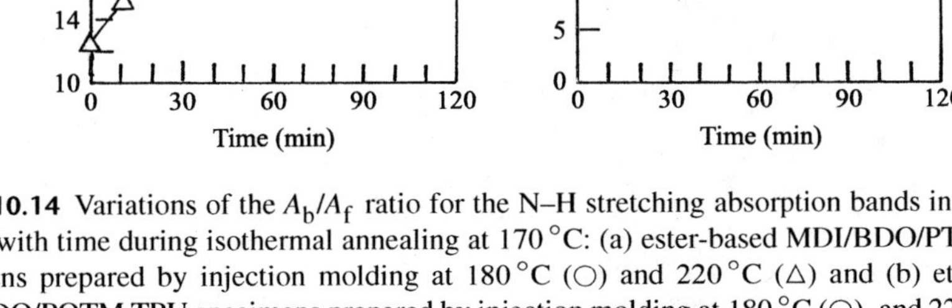

Figure 10.14 Variations of the A_b/A_f ratio for the N–H stretching absorption bands in the FTIR spectra with time during isothermal annealing at 170 °C: (a) ester-based MDI/BDO/PTMA TPU specimens prepared by injection molding at 180 °C (○) and 220 °C (△) and (b) ether-based MDI/BDO/POTM TPU specimens prepared by injection molding at 180 °C (○), and 220 °C (△). (Reprinted from Yoon and Han, *Macromolecules* 33:2171. Copyright © 2000, with permission from the American Chemical Society.)

in Figures 10.7 and 10.8. On the basis of the above observations, we can conclude that the time evolution of G' observed during isothermal annealing is related to the time evolution of the A_b/A_f ratio observed by FTIR spectroscopy during isothermal annealing. Thus, both are the consequence of concomitant variations in morphology during isothermal annealing of the TPUs investigated.

10.3 Linear Dynamic Viscoelasticity of TPUs

10.3.1 Frequency Dependence of Dynamic Moduli of TPU under Isothermal Conditions

Figure 10.15 gives plots of $\log G'$ versus $\log \omega$, and $\log G''$ versus $\log \omega$, for an ester-based MDI/BDO/PTMA TPU at various temperatures ranging from 110 to 190 °C, in which a single specimen was used for measurement at all temperatures. In Figure 10.15 we make the following observations: (1) values of G' are very large and the slope of the $\log G'$ versus $\log \omega$ plot is very small at 110 °C, which is characteristic of solidlike materials, and (2) the value of G' decreases, and at the same time the slope of the $\log G'$ versus $\log \omega$ plot increases, with increasing temperature until reaching 190 °C. The same observation can be made from the $\log G''$ versus $\log \omega$ plots.

Figure 10.16 gives $\log G'$ versus $\log G''$ plots for an ester-based MDI/BDO/PTMA TPU at various temperatures during the heating process. In order to maintain clarity, in Figure 10.16 we have divided the experimental data into two groups: part (a) gives $\log G'$ versus $\log G''$ plots that show a continuous downward shift with increasing temperature from 110 to 160 °C, and part (b) gives $\log G'$ versus $\log G''$ plots that show a continuous upward shift in the terminal region, as the temperature is increased from 170 to 190 °C. It is of great interest to observe in Figure 10.16 that the $\log G'$ versus $\log G''$ plot has a slope much less than 2 at 110 °C, moves downward with an increase

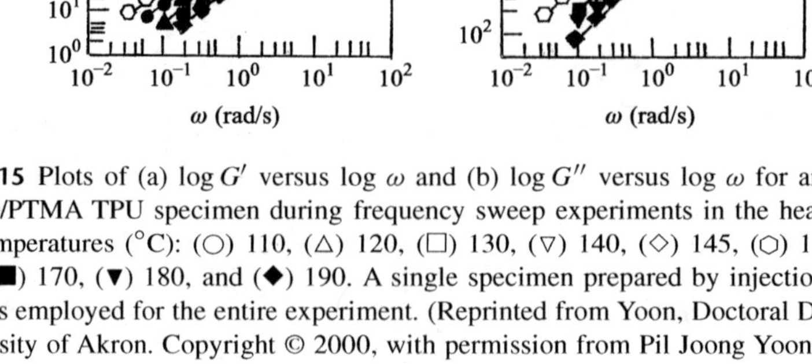

Figure 10.15 Plots of (a) $\log G'$ versus $\log \omega$ and (b) $\log G''$ versus $\log \omega$ for an ester-based MDI/BDO/PTMA TPU specimen during frequency sweep experiments in the heating cycle at various temperatures (°C): (○) 110, (△) 120, (□) 130, (▽) 140, (◇) 145, (⬡) 150, (●) 155, (▲) 160, (■) 170, (▼) 180, and (◆) 190. A single specimen prepared by injection molding at 180 °C was employed for the entire experiment. (Reprinted from Yoon, Doctoral Dissertation at the University of Akron. Copyright © 2000, with permission from Pil Joong Yoon.)

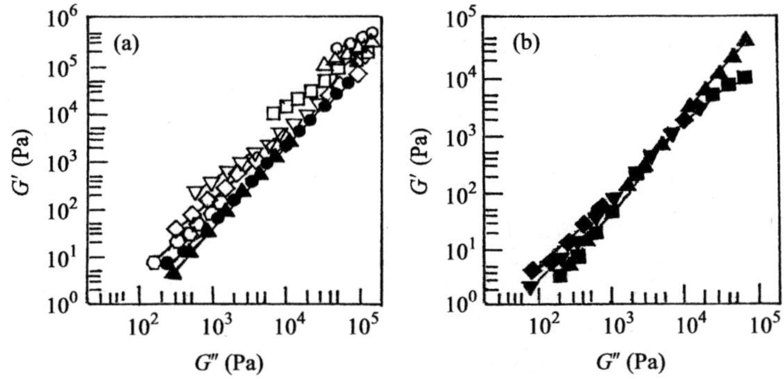

Figure 10.16 Log G' versus log G'' plots for an ester-based MDI/BDO/PTMA TPU specimen in the heating process at various temperatures (°C): (○) 110, (△) 120, (□) 130, (▽) 140, (◇) 145, (⬡) 150, (●) 155, (▲) 160, (■) 170, (▼) 180, and (◆) 190. For the sake of clarity, the plots are divided into two parts: (a) temperatures from 110 to 160 °C, and (b) temperatures from 160 to 190 °C. (Reprinted from Yoon and Han, *Macromolecules* 33:2171. Copyright © 2000, with permission from the American Chemical Society.)

in slope towards 2 as the temperature is increased from 110 to 170 °C, and then starts to move upwards as the temperature is increased further to 180 and 190 °C. Figure 10.17a gives log G' versus log G'' plots for an ether-based MDI/BDO/POTM TPU during the heating process from 140 to 190 °C, and Figure 10.17b gives log G' versus log G'' plots during the cooling process from 190 to 145 °C. Note in Figure 10.17 that during the heating process the log G' versus log G'' plot moves downward and merges into a single

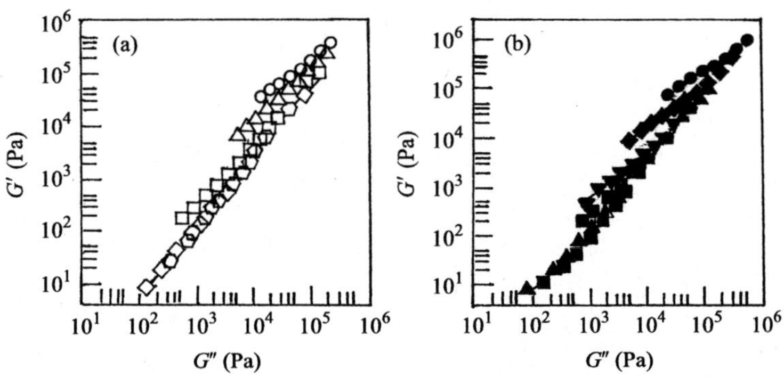

Figure 10.17 Log G' versus log G'' plots for an ether-based MDI/BDO/POTM TPU specimen: (a) the results in the heating process at various temperatures (°C): (○) 140, (△) 150, (□) 160, (▽) 170, (◇) 180, and (⬡) 190, and (b) the results in the cooling process at various temperatures (°C): (●) 190, (▲) 180, (■) 165, (▼) 155, and (◆) 145. A single specimen prepared by injection molding at 180 °C was employed for the entire experiment. (Reprinted from Yoon and Han, *Macromolecules* 33:2171. Copyright © 2000, with permission from the American Chemical Society.)

curve, which has a slope in the terminal region still less than 2 at the highest experimental temperature employed (190 °C). Also note that during the cooling process the $\log G'$ versus $\log G''$ plot moves upward with decreasing temperature, giving rise to a slope progressively smaller with decreasing temperature. Referring to Figure 10.16, we observe an increase in G' with increasing temperature above a certain critical temperature. It is quite possible that insoluble gels might have been formed during the dynamic frequency sweep experiments of the ester-based TPU at 180 and 190 °C, giving rise to an increase in G'.

The temperature dependence of the $\log G'$ versus $\log G''$ plot observed in Figures 10.16 and 10.17 suggests that the morphology of the TPUs varies with temperature over the range of the temperatures (110–190 °C) tested because the homogeneous polymers are expected to exhibit temperature independence in the $\log G'$ versus $\log G''$ plots, as demonstrated in Chapter 6. It should be mentioned that the $\log G'$ versus $\log G''$ plot cannot determine how the morphology changes and what the new morphology might be during rheological measurements. Velankar and Cooper (1998) applied time–temperature superposition (TTS) to certain TPUs and observed a failure of TTS. We conclude from Figures 10.16 and 10.17 that TTS fails for the ester-based MDI/BDO/PTMA and ether-based MDI/BDO/POTM TPUs investigated.

10.3.2 Temperature Dependence of Dynamic Moduli of TPU during Isochronal Dynamic Temperature Sweep Experiment

Let us look at the temperature dependence of dynamic moduli during the heating and cooling processes, as affected by a sample preparation of TPU. Figure 10.18 gives the temperature dependence of G' for an ester-based MDI/BDO/PTMA TPU during an isochronal dynamic temperature sweep experiment at $\omega = 0.56$ rad/s in the heating and cooling cycles. It can be seen in Figure 10.18 that a hysteresis effect is observed

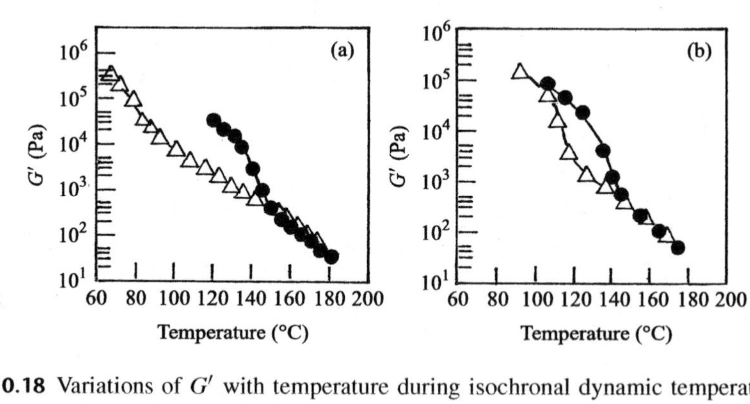

Figure 10.18 Variations of G' with temperature during isochronal dynamic temperature sweep experiments at $\omega = 0.56$ rad/s in the heating (○) and cooling (●) processes for ester-based MDI/BDO/PTMA TPU specimens: (a) a single specimen prepared by injection molding at 180 °C was used for both heating and cooling at a rate of 0.5 °C/min, and (b) a fresh specimen prepared by injection molding at 180 °C was used for heating or cooling at a rate of 0.5 °C/min. (Reprinted from Yoon and Han, *Macromolecules* 33:2171. Copyright © 2000, with permission from the American Chemical Society.)

at temperatures between 110 and 170 °C, but that there is little hysteresis effect at temperatures above 170 °C. The above observations indicate that the morphological states of the TPUs at temperatures ranging from 110 to 170 °C are different from those at temperatures above 170 °C. We observed similar hysteresis effect in block copolymers in Chapter 8 and in TLCPs in Chapter 9. The hysteresis effect observed in Figure 10.18 is not surprising because within the time scale of experiments the TPUs did not attain an equilibrium morphology.

Figure 10.19 shows variations of G' with temperature for an ester-based MDI/BDO/PTMA TPU during isochronal dynamic temperature sweep experiments at various angular frequencies. It can be seen in Figure 10.19 that G' decreases gradually with an increase in temperature and tends to level off at the lowest angular frequency employed, $\omega = 0.03$ rad/s. It has been amply demonstrated in Chapters 6–9 that the decreasing trend of G' with an increase in temperature depends strongly on the angular frequencies applied. In view of the fact that there is no sharp drop of G' at any particular temperature in Figure 10.19, we conclude that the isochronal dynamic temperature sweep experiment cannot be used to determine the order–disorder transition (ODT) temperature (T_{ODT}) or microphase separation transition (MST) temperature (T_{MST}) of the TPUs, although some investigators (Goddard and Cooper 1995b; Ryan et al. 1992) have made attempts to determine T_{MST} of TPUs from isochronal dynamic temperature sweep experiments. From both SAXS and rheological experiments, Goddard and Cooper (1995b) concluded that the determination of MST was inconclusive because considerable microphase mixing occurred near the onset of viscous flow. From temperature-resolved SAXS measurements, Velankar and Cooper (1998) observed that microphase mixing in some TPUs occurred gradually as the temperature increased. Such an observation supports the conclusion drawn from Figures 10.18 and 10.19 that the determination of T_{MST} or T_{ODT} of TPU does not seem possible from isochronal dynamic temperature sweep experiments and also from SAXS experiments for a number of reasons.

Phase mixing between hard and soft segments continues above the melting point of hard segments. Such experimental evidence can be found from variations in the peak wavenumber of N–H stretching bands in the FTIR spectra with temperature in the heating and cooling processes, as given in Figure 10.20 for an ester-based

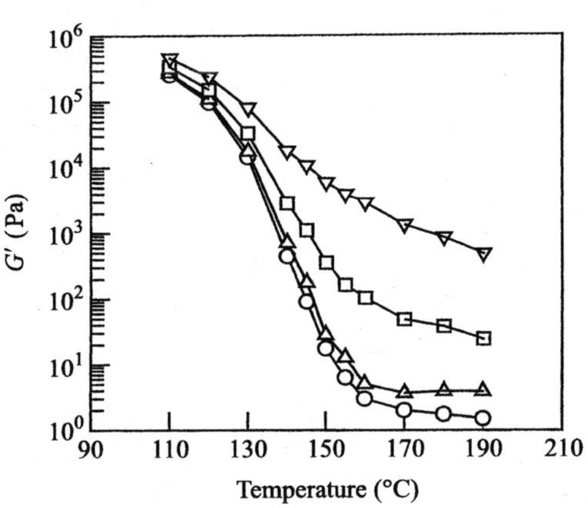

Figure 10.19 Variations of G' with temperature for an ester-based MDI/BDO/PTMA TPU specimen during isochronal dynamic temperature sweep experiments at various angular frequencies (rad/s): (○) 0.03, (△) 0.1, (□) 0.56, and (▽) 5.62, in the heating process. A single specimen prepared by injection molding at 180 °C was used for the entire experiment. (Reprinted from Yoon, Doctoral Dissertation at the University of Akron. Copyright © 2000, with permission from Pil Joong Yoon.)

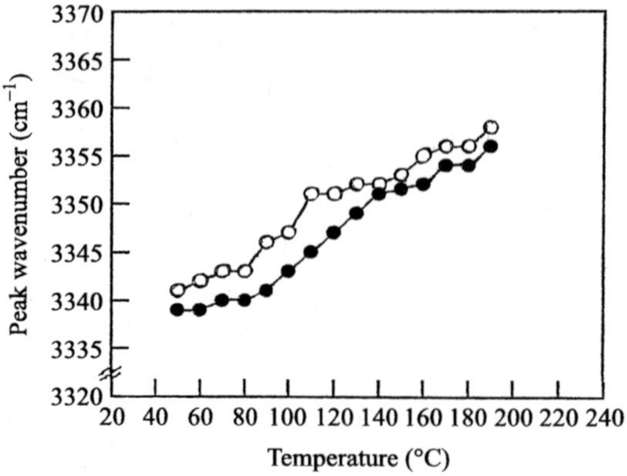

Figure 10.20 Temperature dependence of peak wavenumber of FTIR spectra in the N–H stretching region for an ester-based MDI/BDO/PTMA TPU specimen in the heating (●) and cooling (○) processes at a rate of 0.5 °C/min. A fresh specimen prepared by injection molding at 180 °C was used for each process. (Reprinted from Yoon and Han, *Macromolecules* 33:2171. Copyright © 2000, with permission from the American Chemical Society.)

MDI/BDO/PTMA TPU. In Figure 10.20 we observe a pronounced shift in the peak wavenumber over the entire range of temperatures investigated. A broad shift in peak wavenumber of N–H stretching bands with increasing temperature implies strong interactions between the soft and hard segments in the TPU specimen. This observation attests to the fact that phase mixing between the soft and hard segments continues in the temperature range tested up to 200 °C. As shown above, hydrogen bonding takes place continuously in TPU, blurring the phase boundary between hard and soft segments, although the extent of hydrogen bonding becomes weaker as the temperature is increased. Also, at elevated temperatures chemical reactions between isocyanate and active hydrogen in the urethane groups may occur, forming biuret or allophanate. The gradual decrease of G' with increasing temperature, instead of an abrupt decrease in G' at a particular temperature, observed in Figure 10.19, may be attributable in part to the hydrogen bonding taking place during rheological measurement. Thus the gradual decrease of G' with increasing temperature observed in Figure 10.19 cannot be used to determine the T_{MST} or T_{ODT} of TPU.

TPU can be regarded as being a segmental multiblock copolymer with segment lengths much shorter than those of AB- and ABA-type linear block copolymers. By taking into account all the factors listed above, we can conclude that TPU may easily lose long-range order at elevated temperatures, barring chemical reactions or thermal degradation, thus transforming into short-range liquidlike order. If such a conclusion is justified, the phase transition mechanism for TPU may be regarded as being very similar to that for highly asymmetric sphere-forming block copolymers, as presented in Chapter 8.

10.4 Steady-State Shear Viscosity of TPU

Figure 10.21 gives plots of steady-state shear viscosity η versus $\dot{\gamma}$ for an injection-molded specimen of ether-based MDI/BDO/POTM TPU specimen at 170, 180, and 190 °C, which was obtained from a cone-and-plate rheometer at low shear rates and

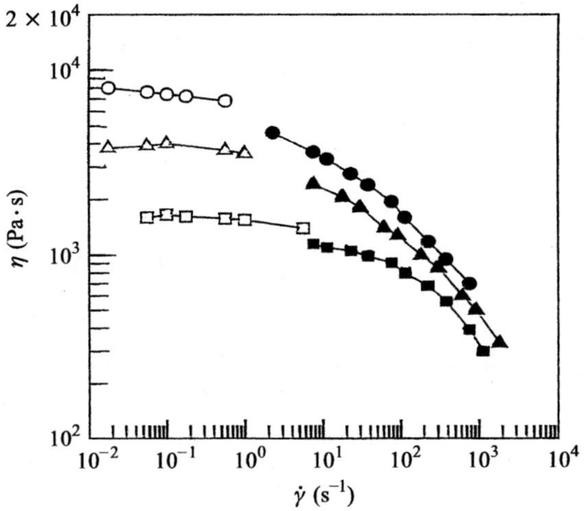

Figure 10.21 Plots of log η versus log $\dot{\gamma}$ for an ether-based MDI/BDO/POTM TPU at various temperatures (°C): (○, ●) 170, (△, ▲) 180, and (□, ■) 190. Open symbols are cone-and-plate data and filled symbols are capillary data. A fresh specimen prepared by injection molding at 180 °C was employed for each temperature. (Reprinted from Yoon, Doctoral Dissertation at the University of Akron. Copyright © 2000, with permission from Pil Joong Yoon.)

from a capillary rheometer at high shear rates. In Figure 10.21 we observe no overlap between the two sets of data, attesting to the fact that the morphological state of the TPU at low shear rates is not the same as that at high shear rates. This is very similar to the observations made in Chapter 8 for block copolymers. The lack of overlap between the capillary and cone-and-plate data can be understood because the TPUs are structured fluids with very complex morphology. There is no reason for one to expect that the morphological state of a TPU would be the same whether a TPU specimen is subjected to uniform shear flow in a cone-and-plate rheometer or to nonuniform shear flow in a capillary die.

Figure 10.22 compares steady-state shear viscosity (η) and complex viscosity ($|\eta^*|$) of ether-based MDI/BDO/POTM TPU specimens at 170, 180, and 190 °C. In Figure 10.22 we observe no overlap between the two sets of data (i.e., the

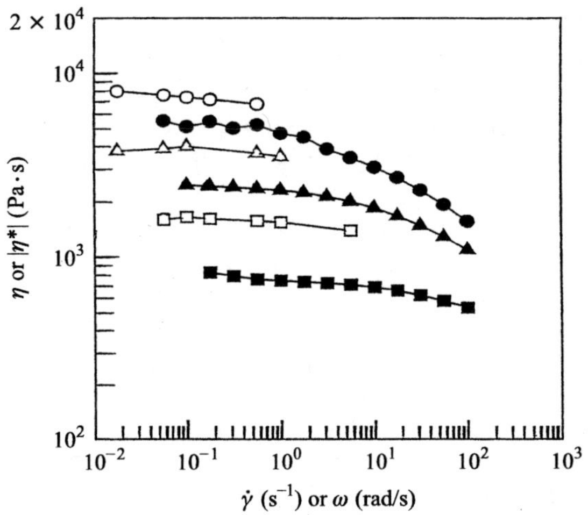

Figure 10.22 Comparison of plots of log η versus log $\dot{\gamma}$ (open symbols) with log $|\eta^*|$ versus log ω (filled symbols) for ether-based MDI/BDO/POTM TPU at various temperatures (°C): (○, ●) 170, (△, ▲) 180, and (□, ■) 190. A fresh specimen prepared by injection molding at 180 °C was employed for each temperature. (Reprinted from Yoon, Doctoral Dissertation at the University of Akron. Copyright © 2000, with permission from Pil Joong Yoon.)

Cox–Merz rule (Cox and Merz 1958) is not applicable to TPU), suggesting that the morphological state of a TPU specimen during steady-state shear flow is not the same as that during oscillatory shear flow. Similar observations were made in Chapter 8 for microphase-separated block copolymers and in Chapter 9 for TLCPs in an anisotropic state.

10.5 Summary

In this chapter, we have presented the rheological behavior of two TPUs in the molten state, showing that (1) the previous thermal history of a specimen has a profound influence on the rheological behavior of TPUs, (2) the dynamic viscoelastic properties of a TPU specimen, placed in the parallel-plate fixture of a rotational-type rheometer, vary with time during isothermal annealing, and (3) TPUs exhibit hysteresis effect during the heating and cooling processes. Such experimental observations can be understood from the point of view of the morphology of TPUs. Namely, TPUs have microdomains that can be made to flow under an external force. The complexities of the rheological behavior of TPUs stem from the inherently complex morphology of TPUs. Once we realize the fact that TPUs are structured fluids, like block copolymers and TLCPs, the complex rheological behavior observed in this chapter should not surprise us.

In this chapter, we have shown that (1) at elevated temperatures (above 190 °C) for an extended annealing period, TPUs undergo thermal degradation or chemical reactions (forming gels), as evidenced by the measurement of molecular weight via gel permeation chromatography, (2) in situ FTIR spectroscopy at elevated temperatures can be used to monitor the formation of hydrogen bonds during rheological measurement, and (3) the extent of exchange reactions in TPUs, if any, would be very low. We have concluded that hydrogen bonding may be a predominant factor responsible for the time evolution of rheological properties during experiments.

The similarities in rheological behavior between TPUs and microphase-separated block copolymers on the one hand and between TPUs and TLCPs in an anisotropic state on the other hand are not a coincidence. This is because TPUs, which consist of segmented blocks (characteristic of block copolymer), form microheterogeneity. The polar nature of the urethane segments results in strong mutual attraction, aggregation and ordering into crystalline and paracrystalline domains in the mobile soft-segment matrix. The abundance of urethane hydrogen atoms, as well as carbonyl and ether oxygen partners, permits extensive hydrogen bonding between polymer chains, which apparently restricts the mobility of the urethane chain segments in the domains. Thus, one should expect that the rheological behavior of TPU would depend strongly on its morphological state, which in turn is greatly influenced by thermal history.

However, there are important differences between TPUs and block copolymers, and between TPUs and TLCPs. As shown in this chapter, TPUs may undergo hydrogen bonding during rheological measurement, whereas block copolymers and TLCPs, unless functionalized, do not. Since hydrogen bonding rearranges chain configurations, and consequently the morphological state of TPU, it is not surprising to observe variations of rheological properties during experiments. The occurrence of hydrogen bonding is a unique characteristic of TPUs, which does not exist in block copolymers and TLCPs unless functionalized. Hydrogen bonding in TPUs is believed to pose an inherent difficulty in obtaining time-invariant, reproducible rheological measurements. Another feature of TPUs, which distinguishes them from block copolymers and TLCPs, is the strong possibility for thermal degradation before reaching a temperature at which the mesophase can be made to disappear by heating. Under such circumstances, it would not be possible to erase previous thermal history by heating a specimen to above a certain critical temperature, whereas it is possible to do so when dealing with TLCPs with clearing temperatures lower than the thermal degradation temperatures (see Chapter 9). In other words, it is very difficult to find clearing temperature in TPUs before encountering thermal degradation. It is then essential for one to synthesize a thermally stable, glassy TPU that can be heated to a sufficiently high temperature, at which the TPUs can be made to form a homogeneous phase before undergoing thermal degradation. An effort was made to synthesize a glassy low-molecular-weight TPU by Velankar and Cooper (1998, 2000b). It should be pointed out, however, that the occurrence of hydrogen bonding in such a TPU may still persist, making time-invariant rheological measurements very difficult, if not impossible.

Furthermore, in this chapter we have shown that isochronal dynamic temperature sweep experiments cannot be used to investigate ODT or MST in TPUs because intermixing between soft and hard segments takes place, and because hydrogen bonding occurs in the heating and cooling processes. Also, we have pointed out that TTS is not applicable to TPUs because the morphological state of the TPU specimens varies with temperature.

Owing to the very complicated nature of the temperature-dependent morphology of TPUs, at present it is not possible to explain completely the origins of the seemingly very complex rheological behavior presented in this chapter. What are urgently needed are simultaneous investigations of the morphology and the chain conformations of TPUs by utilizing several characterization techniques, such as SAXS, IR spectroscopy, TEM, dielectric spectroscopy, and solid-state nuclear magnetic resonance spectroscopy.

References

Blackwell J, Lee CD (1984). *J. Polym. Sci., Polym. Phys. Ed.* 22:759.
Briber RM, Thomas EL (1983). *J. Macromol. Sci. Phys.* B22:509.
Briber RM, Thomas EL (1985). *J. Polym. Sci., Polym. Phys. Ed.* 23:1985.
Chen TK, Shieh TS, Chui JY (1998). *Macromolecules* 31:1312.
Coleman MM, Lee KH, Skrovanek DJ, Painter PC (1986). *Macromolecules* 19:2149.
Cox WP, Merz EH. (1958). *J. Polym. Sci.* 28:619.
Eisenbach CD, Ribbe A, Gunter C (1984). *Macromol. Rapid Commun.* 15:395.
Fridman ID, Thomas EL (1980). *Polymer* 21:388.
Goddard RJ, Cooper SL (1995a). *Macromolecules* 28:1390.
Goddard RJ, Cooper SL (1995b). *Macromolecules* 28:1401.
Hepburn C (1982). *Polyurethane Elastomers*, Applied Science, New York.
Hesketh TR, Van Borgart JWC, Cooper SL (1980). *Polym. Eng. Sci.* 20:190.
Koberstein JT, Stein RS (1983). *J. Polym. Sci., Polym. Phys. Ed.* 21:1439.
Koberstein JT, Russell TP (1986). *Macromolecules* 19:714.
Kwei TK (1982). *J. Appl. Polym. Sci.* 27:2891.
Lee HS, Wang YK, Hsu SL (1987). *Macromolecules* 20:2089.

Leung LM, Koberstein JT (1986). *Macromolecules* 19:706.
Luo N, Wang DN, Ying SK (1997). *Macromolecules* 30:4405.
Oertel G. (ed) (1985). *Polyurethane Handbook*, Hanser, Munich.
Ryan AJ, Macosko CW, Bras W (1992). *Macromolecules* 25:6277.
Samuels SL, Wilkes GL (1973). *J. Polym. Sci. Symp.* 43:149.
Saunders JH, Frish KC (1962). *Polyurethane Chemistry and Technology, Part I*, Interscience, New York.
Senich GA, MacKnight WJ (1980). *Macromolecules* 13:106.
Seymour RW, Cooper SL (1971). *J. Polym. Sci. Polym. Lett. Ed.* 9:689.
Seymour RW, Cooper SL (1973). *Macromolecules* 6:48.
Seymour RW, Estes GM, Cooper SL (1970). *Macromolecules* 3:579.
Speckard TA, Ver Strate G, Gibson PE, Cooper SL (1983). *Polym. Eng. Sci.* 23:337.
Srichatrapimuk VW, Cooper SL (1978). *J. Macromol. Sci. B* 15:267.
Sung CSP, Schneider NS (1975). *Macromolecules* 8:68.
Sung CSP, Schneider NS (1977). *Macromolecules* 10:452.
Sung CSP, Schneider NS (1978). *J. Mat. Sci.* 13:1689.
Sung CSP, Smith TW, Sung NH (1980). *Macromolecules* 13:117.
Tanaka T, Yokoyama T, Yamaguchi Y (1968). *J. Polym. Sci. A-1* 6:2137.
Velankar S, Cooper SL (1998). *Macromolecules* 31:9181.
Velankar S, Cooper SL (2000a). *Macromolecules* 33:382.
Velankar S, Cooper SL (2000b). *Macromolecules* 33:395.
Wilkes GL, Yusek C (1973). *J. Macromol. Sci. Phys.* B7:157.
Wilkes GL, Wildnauer R (1975). *J. Appl. Phys.* 46:4148.
Wilkes GL, Bagrodia S, Humphries W, Wildnauer R (1975). *J. Polym. Sci. Polym. Lett. Ed.* 13:321.
Yen FS, Lin LL, Hong JL (1999). *Macromolecules* 32:3068.
Yoon PJ (2000). *Effect of Thermal History on the Rheological Properties of Thermoplastic Polyurethanes*, Doctoral Dissertation at the University of Akron, Akron, Ohio.
Yoon PJ, Han CD (2000). *Macromolecules* 33:2171.

11

Rheology of Immiscible Polymer Blends

11.1 Introduction

The polymer industry has been challenged to produce new polymeric materials by blending two or more homopolymers or random copolymers or by synthesizing graft copolymers. To meet the challenge, various methods have been explored, namely, (1) by synthesizing a new monomer, polymerizing it, and then blending it with an existing homopolymer or random copolymer, (2) by copolymerizing existing monomers and then blending it with an existing homopolymer or random copolymer, (3) by chemically modifying an existing homopolymer or random copolymer and then blending it with other homopolymers or copolymers already available, or (4) by synthesizing new compatibilizer(s) to improve the mechanical properties of two immiscible homopolymers or random copolymers that otherwise have unacceptable mechanical properties. There are numerous monographs (Cooper and Estes 1979; Han 1984; Paul and Newman 1978; Platzer 1971, 1975; Sperling 1974; Utracki 1990) describing various aspects of polymer blends.

In the 1970s, Han and coworkers (Han 1971, 1974; Han and Kim 1975; Han and Yu 1971a, 1971b, 1972; Han et al. 1973, 1975; Kim and Han 1976) conducted seminal experimental studies on the rheology of immiscible polymer blends and related the observed rheological behavior to blend morphology. Independently, in the same period, Vinogradov and coworkers (Ablazova et al. 1975; Brizitsky et al. 1978; Tsebrenko et al. 1974, 1976; Vinogradov et al. 1975) conducted a series of experimental studies relating the blend rheology to blend morphology. Van Oene (1972, 1978) also pursued, independently, experimental studies for a better understanding of rheology–morphology relationships in immiscible polymer blends. Since then, using different polymer pairs, numerous researchers have conducted experimental studies, which were essentially the same as, or very similar to, the previous experimental studies of Han and coworkers,

Vinogradov and coworkers, and van Oene in the 1970s. It is fair to state that those studies in the 1980s and 1990s have not revealed any significant new findings. Today, it is well recognized that the rheological behavior of an immiscible polymer blend, consisting of two homopolymers, two random copolymers, or a homopolymer and a random copolymer, depends on its morphological state, which in turn depends on, among many factors, (1) blend composition, (2) the mixing condition (temperature, the intensity of mixing, and the duration of mixing) employed, and (3) the type of mixing device employed.

When blending two immiscible homopolymers (or random copolymers) using either a batch-type mixing device (e.g., Banbury mixer) or a continuous mixing device (e.g., twin-screw compounding machine), we encounter, roughly speaking, two types of phase morphology: dispersed two-phase morphology and co-continuous morphology (see Chapter 3 of Volume 2). Co-continuous morphology is a transitory morphology and thus it is not an equilibrium morphology (Lee and Han 1999, 2000). In the situation when a dispersed two-phase morphology is obtained, a fundamental question arises as to which of the two polymers forms the discrete phase and is dispersed in the other polymer. Numerous research groups have reported on the evolution of blend morphology during compounding in an internal mixer or in a twin-screw extruder; there are too many papers to cite them all here. To date, however, there are very few theoretical studies reported that enable one to predict rheology–morphology relationships in immiscible polymer blends as affected by processing conditions. Admittedly, prediction of blend morphology during compounding is not a trivial matter.

There are three major factors that control blend morphology, which in turn control the rheological behavior of immiscible polymer blends; they are (1) thermodynamic factor, (2) interfacial tension, and (3) hydrodynamic factor. One can produce polymer blends via rapid precipitation from a solution without the influence of hydrodynamic variables, while hydrodynamic variables play a very important role in melt blending. The interfacial energy between the phases plays an important role in determining the sizes of the discrete phase (i.e., drops), while the interfacial tension between two immiscible polymers does not vary much from one pair of polymers to others, as compared with, for instance, the viscosity ratio of the constituent components. It should be mentioned that the interfacial tension can be decreased considerably, and thus decreasing the sizes of the discrete phase, by using an effective compatibilizing agent. The subject of compatibilization of two immiscible polymers will be discussed in Chapter 4 of Volume 2.

The primary goal of this chapter is to present the fundamental rheological behavior of immiscible polymer blends. In so doing, instead of presenting exhaustive experimental data from the literature, we limit our presentation to a small number of well-controlled experimental studies reported in the literature and focus on establishing fundamental rheology–morphology–processing relationships in immiscible polymer blends. We present finite element analysis of the deformation of a single drop in the entrance region of a cylindrical tube. Then we describe very briefly a theoretical approach to investigate rheology–morphology–processing relationships in immiscible polymer blends. Throughout this chapter, emphasis is placed on describing the fundamental concepts associated with the rheological behavior of immiscible polymer blend.

11.2 Experimental Observations of Rheology–Morphology Relationships in Immiscible Polymer Blends

11.2.1 Effect of Flow Geometry on the Steady-State Shear Viscosity and Morphology of Immiscible Polymer Blends

In this section, we present how the flow geometry (i.e., the type of rheometer) might influence the rheological properties of immiscible polymer blends, and methods for analyzing the experimental data obtained from different types of rheological instruments for immiscible polymer blends. This subject is of great importance for correct interpretation of rheological data for immiscible polymer blends. Specifically, we show that (1) plots of steady-state shear viscosity (η) versus shear rate ($\dot{\gamma}$) or shear stress (σ) obtained from a cone-and-plate rheometer do not necessarily overlap those obtained from a capillary (or slit) rheometer, (2) the assumptions made in the Bagley plot, which has been proven to be useful for homogeneous polymer melts (see Chapter 5), are not valid for immiscible polymer blends, and (3) the calculation of σ (thus η) using the measurements of the total pressure drop from the upstream end of the reservoir to the die exit gives rise to inaccurate information. The seemingly complicated rheological behavior of an immiscible polymer blend depends on its phase morphology, which in turn depends on the σ or $\dot{\gamma}$ in a given flow geometry at a fixed temperature.

Figure 11.1 gives the phase morphology of as-blended 70/30 poly(methyl methacrylate) (PMMA)/polystyrene (PS) specimens that were subjected to shear flow, where 70/30 refers to weight percent of the component polymers. In Figure 11.1, the dark areas represents the minor component PS and the gray/white areas represent the major component PMMA. As can be seen from Figure 11.1a, before being subjected to shear flow the PS phase in the as-blended PMMA/PS specimen forms drops dispersed in the PMMA phase forming the continuous phase. Notice in Figure 11.1a that both large and small drops are present, in which the small-size drops must have resulted from the breakup of large-size drops during the compounding in a twin extruder. When the as-blended 70/30 PMMA/PS specimen was subjected to shear flow at $\dot{\gamma} = 0.1\ s^{-1}$ in a cone-and-plate rheometer, the PS drops were elongated along the flow direction, showing evidence that breakup occurred inside the gap between the cone and the plate (see Figure 11.1b). When the as-blended 70/30 PMMA/PS specimen was subjected to shear flow at $\dot{\gamma} = 10\ s^{-1}$ in a capillary die, very long streaks of the dispersed PS phase are seen along the flow direction (see Figure 11.1c). It is of interest to observe that the areas of some dispersed PS phase in Figure 11.1c are much larger than those in Figure 11.1b. This observation suggests that, during the flow inside the capillary die, coalescence between the elongated PS drops, flowing side by side, might have taken place. When shear rate was increased further to $100\ s^{-1}$ in the capillary die, long and thin fibril-like dispersed PS phase is observed along the flow direction (Figure 11.1d).

Figure 11.2 gives the phase morphology of as-blended 50/50 PMMA/PS specimens that were subjected to shear flow. Notice that the morphology of as-blended 50/50 PMMA/PS specimen (Figure 11.2a) looks quite different from that of as-blended 70/30 PMMA/PS blend (Figure 11.1a) in that the as-blended 50/50 PMMA/PS specimen appears to have a co-continuous morphology, in which some small-size PS drops (the dark areas) are occluded inside the PMMA phase (the gray/white areas), while some small-size PMMA drops are occluded in the continuous PS phase. When an as-blended

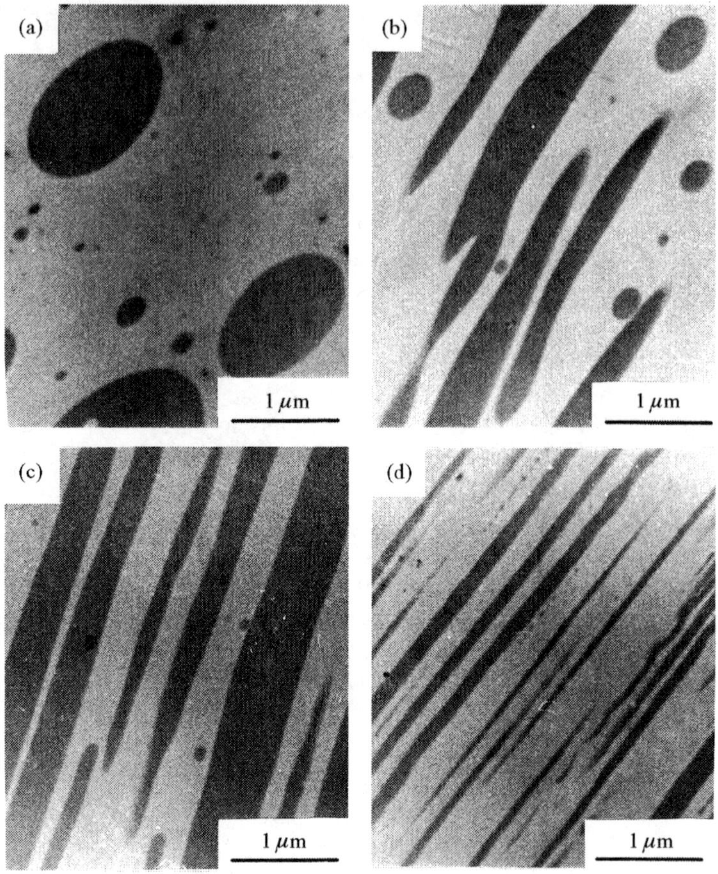

Figure 11.1 Micrographs for the 70/30 PMMA/PS blend, in which the dark areas represent the PS phase and the gray/white areas represent the PMMA phase (a) before being subjected to shear flow, (b) along the flow direction after being subjected to shear flow at $\dot{\gamma} = 0.1\ \text{s}^{-1}$ in a cone-and-plate rheometer, (c) in the flow direction after being subjected to shear flow at $\dot{\gamma} = 10\ \text{s}^{-1}$ in a capillary die, and (d) in the flow direction after being subjected to shear flow at $\dot{\gamma} = 100\ \text{s}^{-1}$ in a capillary die. The shear flow experiments were carried out at 210 °C. (Reprinted from Han et al., *Polymer* 36:2451. Copyright © 1995, with permission from Elsevier.)

50/50 PMMA/PS specimen was sheared at $\dot{\gamma} = 1\ \text{s}^{-1}$ in a cone-and-plate rheometer, the cross section of the specimen shows a very complex morphology (Figure 11.2b) in that many small-size PS drops are occluded in the PMMA phase of much larger domains. When examining the phase morphology of the specimen that was subjected to shear flow at $\dot{\gamma} = 1\ \text{s}^{-1}$ in a capillary die (Figure 11.2c), we observe clear evidence that coalescence took place during the flow inside the capillary die. When an as-blended 50/50 PS/PMMA specimen was subjected to shear flow at $\dot{\gamma} = 10\ \text{s}^{-1}$ in a capillary die (Figure 11.2d), we observe a morphology having almost co-continuous two phases, strong evidence that very extensive coalescence took place during the flow inside the capillary die.

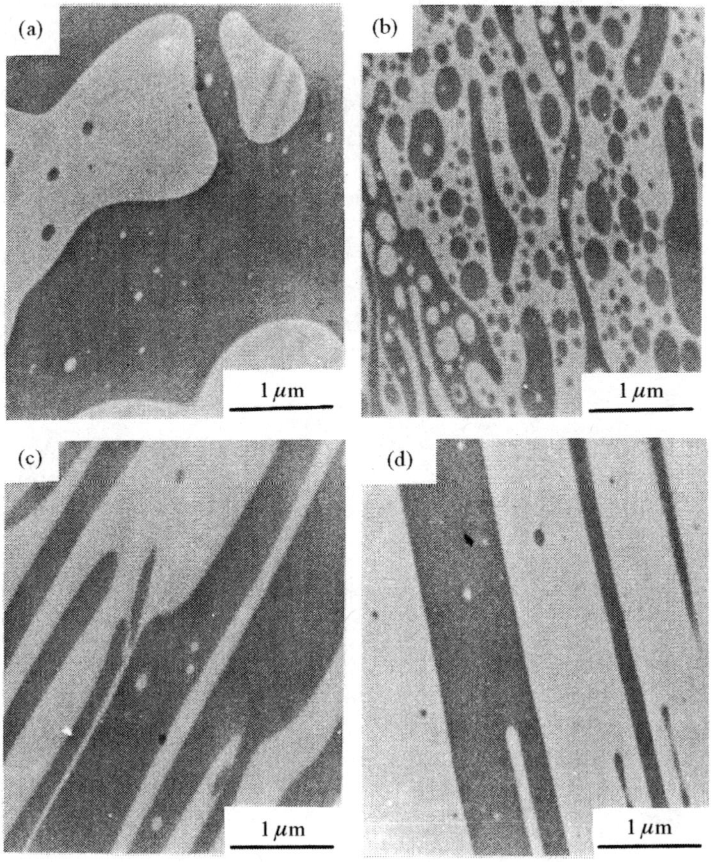

Figure 11.2 Micrographs for the 50/50 PMMA/PS blend, in which the dark areas represent the PS phase and the gray/white areas represent the PMMA phase (a) before being subjected to shear flow, (b) in the cross section of the specimen after being subjected to shear flow at $\dot{\gamma} = 1\ \mathrm{s}^{-1}$ in a cone-and-plate rheometer, (c) in the flow direction after being subjected to shear flow at $\dot{\gamma} = 1\ \mathrm{s}^{-1}$ in a capillary die, and (d) in the flow direction after being subjected to shear flow at $\dot{\gamma} = 10\ \mathrm{s}^{-1}$ in a capillary die. The shear flow experiments were carried out at 210 °C. (Reprinted from Han et al., *Polymer* 36:2451. Copyright © 1995, with permission from Elsevier.)

Figure 11.3 gives the phase morphology of as-blended 30/70 PMMA/PS specimens that were subjected to shear flow. In the as-blended 30/70 PMMA/PS specimens, the minor component PMMA (the gray/white areas) forms the discrete phase and the major component PS (the dark areas) forms the continuous phase, where both large and small drops are present (see Figure 11.3a). The small-size PMMA drops must have resulted from the breakup of the large dispersed PMMA phase during the compounding in a twin-screw extruder. Note that the state of dispersion in as-blended 30/70 PMMA/PS specimen is reversed from that in as-blended 70/30 PMMA/PS specimen (compare Figure 11.3a with Figure 11.1a). When an as-blended 30/70 PMMA/PS specimen was

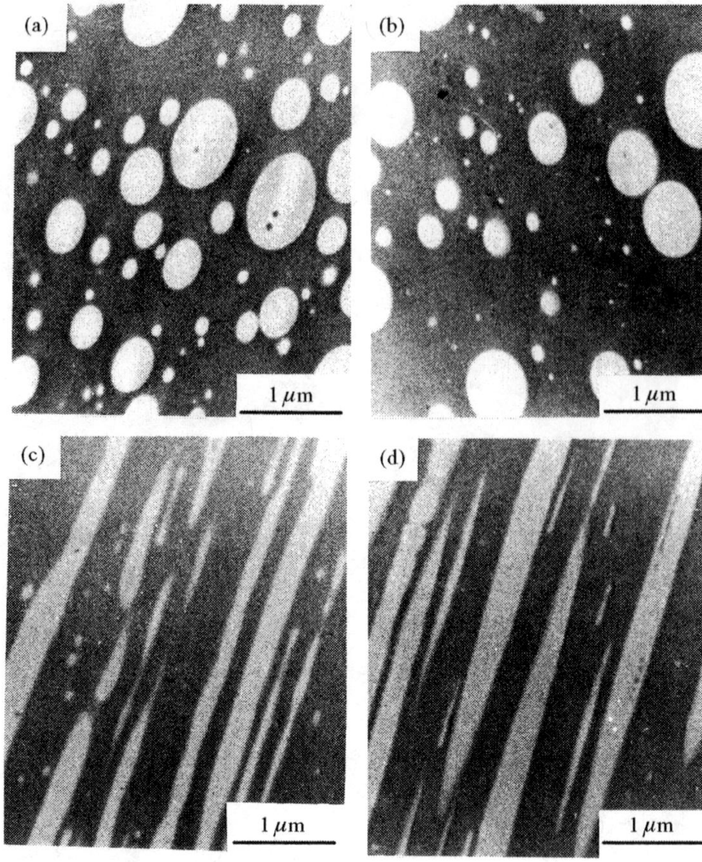

Figure 11.3 Micrographs for the 30/70 PMMA/PS blend, in which the dark areas represent the PS phase and the gray/white areas represent the PMMA phase (a) before being subjected to shear flow, (b) in the flow direction after being subjected to shear flow at $\dot{\gamma} = 0.1\ \mathrm{s}^{-1}$ in a cone-and-plate rheometer, (c) in the flow direction after being subjected to shear flow at $\dot{\gamma} = 10\ \mathrm{s}^{-1}$ in a capillary die, and (d) in the flow direction after being subjected to shear flow at $\dot{\gamma} = 100\ \mathrm{s}^{-1}$ in a capillary die. The shear flow experiments were carried out at 210 °C. (Reprinted from Han et al., *Polymer* 36:2451. Copyright © 1995, with permission from Elsevier.)

subjected to shear flow at $\dot{\gamma} = 0.1\ \mathrm{s}^{-1}$ in a cone-and-plate rheometer, little change in morphology is observed (compare Figure 11.3b with Figure 11.3a). However, when an as-blended 70/30 PMMA/PS specimen was subjected to shear flow at $\dot{\gamma} = 10$ and $100\ \mathrm{s}^{-1}$ in a capillary die, in Figures 11.3c and 11.3d, respectively, we observe that the dispersed PMMA drops were elongated considerably along the flow direction, very similar to the situations observed for the 70/30 PMMA/PS blend (see Figure 11.1c).

Figure 11.4 gives $\log \eta$ versus $\log \sigma$ plots for a 70/30 PMMA/PS blend at 210, 220, and 230 °C, using data obtained from a cone-and-plate rheometer and from a plunger-type capillary rheometer (Instron capillary rheometer). Similar plots were obtained for 50/50 PMMA/PS blend and also for 30/70 PMMA/PS blend

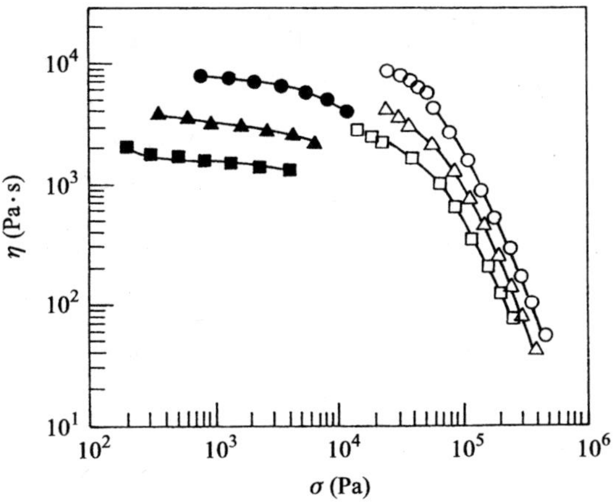

Figure 11.4 Plots of log η versus log σ for the 70/30 PMMA/PS blend at various temperatures (°C): (●, ○) 210, (▲, △) 220, and (■, □) 230. Filled symbols denote data obtained using a cone-and-plate rheometer and open symbols denote data obtained using a plunger-type capillary rheometer. (Reprinted from Han et al., *Polymer* 36:2451. Copyright © 1995, with permission from Elsevier.)

(Han et al. 1995). In Figure 11.4 we observe that there is no overlap of η between the cone-and-plate data and the capillary data; that is, the capillary data show much higher values of η than the cone-and-plate data at the same value of σ.

The lack of overlap in the plots of log η versus log σ given in Figure 11.4 is attributable to the difference in the state of dispersion (e.g., the size and shape of the dispersed PS phase) of the 70/30 PMMA/PS blend when it was subjected to shear flow in a cone-and-plate rheometer or in a plunger-type capillary rheometer (see Figure 11.1). In order to put this assertion in the right perspective, let us look at the schematic given in Figure 11.5, where part (a) shows schematically the shape of drops in a uniform shear flow field and part (b) shows schematically the shape of drops in a nonuniform shear flow field. For simplicity, in Figure 11.5 it is assumed that all drops have the same size. It is clear from Figure 11.5 that in a uniform shear flow field (i.e., in a cone-and-plate rheometer) the shape of drops would be the same at various positions across the gap opening (i.e., along the y-axis), whereas in a nonuniform shear flow field (i.e., in a capillary die) the shape of drops would be different at various positions across the capillary cross section (i.e., in the r-axis). Thus, we can surmise that, owing to the different shapes of drops that might be present in the two different types of shear flow (uniform shear flow versus nonuniform shear flow), the rheological properties of a dispersed two-phase polymer blend determined with a capillary rheometer would be different from those determined with a cone-and-plate rheometer.

There are basically two difficulties associated with the determination of η for immiscible polymer blends using a plunger-type capillary rheometer. One difficulty lies in the determination of shear rate of dispersed two-phase polymer blends (see Figures 11.1–11.3 for PMMA/PS blends). This is because one cannot define the shear rate (velocity gradient) of a dispersed two-phase polymer blend inside a capillary die. Thus, at best what one can do is to calculate an "apparent" shear rate from $\dot{\gamma}_{app} = 32Q/\pi D^3$ with Q being the volumetric flow rate and D being the capillary diameter. One can then only calculate an apparent shear viscosity from the definition $\eta_{app} = \sigma/\dot{\gamma}_{app}$. Thus, to be precise, in Figure 11.4 we should have used the notation η_{app}, instead of the notation η, to describe the melt viscosities of the 70/30 PMMA/PS blend determined from a plunger-type capillary rheometer (open symbols in Figure 11.4).

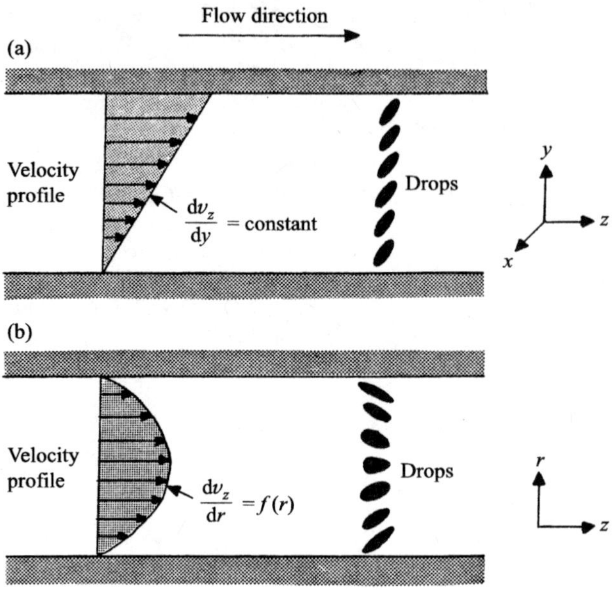

Figure 11.5 Velocity profiles and drop shapes (a) in a uniform shear flow field and (b) in a nonuniform shear flow field. (Reprinted from Han et al., *Polymer* 36:2451. Copyright © 1995, with permission from Elsevier.)

However, for convenience, throughout this chapter we use the notations $\dot{\gamma}$ and η interchangeably with the notations $\dot{\gamma}_{\text{app}}$ and η_{app} for both homopolymers and immiscible polymer blends.

The second difficulty associated with the use of a plunger-type capillary rheometer to calculate the σ of dispersed two-phase polymer blend lies in the use of the following expression:

$$\sigma = \frac{\Delta P}{4(L/D + n_{\text{B}})} \tag{11.1}$$

which is valid for homogeneous polymeric fluids (see Chapter 5), in which ΔP is the total pressure drop across the capillary die, *L*/*D* is the length-to-diameter ratio of the capillary die, and n_{B} is the Bagley end correction. In the use of Eq. (11.1), values of ΔP are obtained from the pressure measurements at the upstream end of the reservoir section of the capillary die. However, use of Eq. (11.1) is not warranted for calculating the σ of dispersed two-phase polymer blends. To explain this, let us look at the schematic given in Figure 11.6, where a dispersed two-phase polymer blend is forced to flow from the reservoir section into a long capillary die. Referring to Figure 11.6, spherical drops enter the upstream end of the reservoir section (region A), then the drops are elongated as they approach the die entrance (region B) and undergo recoil after the elongated drops pass through the entrance region (region C). The long threadlike liquid cylinders would, if they were stable, then move down the straight flow channel (region D). Such a speculation is supported by the micrograph given in Figure 11.7a, in which we observe evidence of drop recoil and also drop breakup in region C in reference to the schematic given in Figure 11.6. A close look at Figure 11.7b

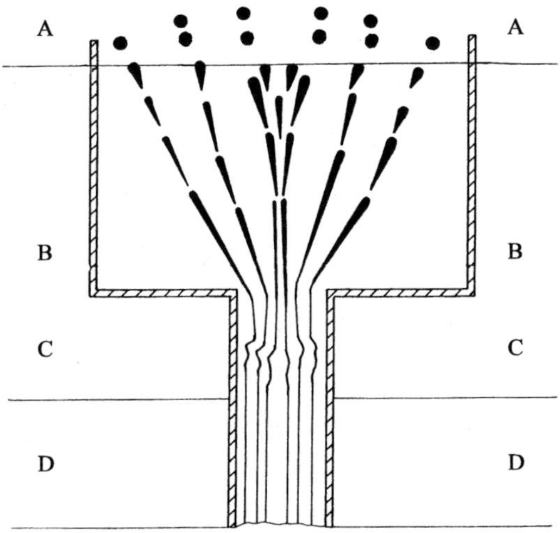

Figure 11.6 The fibrillation process occurring during flow of an immiscible polymer blend in the entrance region and in the fully developed region of a duct. (Reprinted from Tsebrenko et al., *Polymer* 17:831. Copyright © 1976, with permission from Elsevier.)

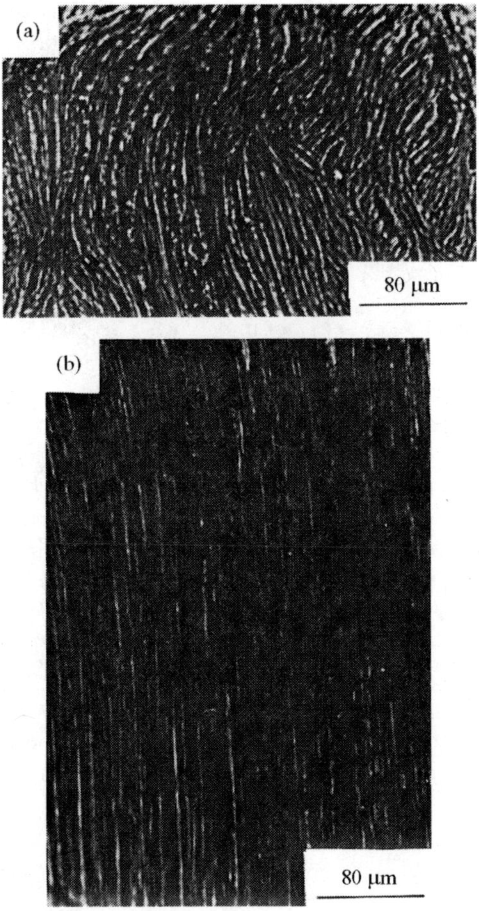

Figure 11.7 Micrographs of the longitudinal section of a frozen extrudate of a blend of polyoxymethylene (POM) and copolyamide (CPA) (a) in the entrance region (region C in Figure 11.6) and (b) in the fully developed region (region D in Figure 11.6). (Reprinted from Tsebrenko et al., *Polymer* 17:831. Copyright © 1976, with permission from Elsevier.)

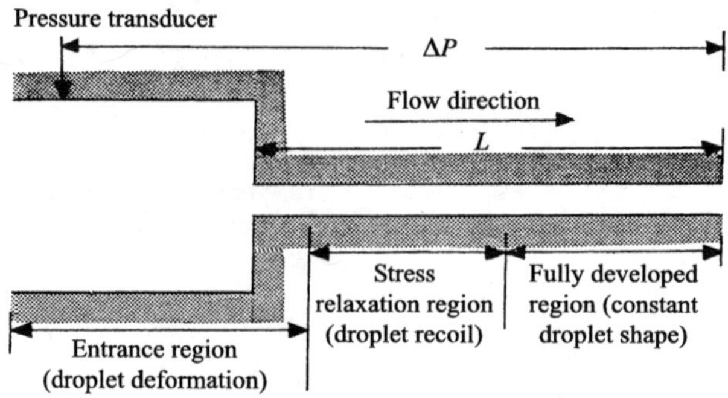

Figure 11.8 Schematic showing the progression of drop deformation, from the entrance region, where drop deformation occurs, through in the stress relaxation region, where the elongated drop recoils, and the fully developed region, where the drop shape remains constant in the cylindrical tube. (Reprinted from Han et al., *Polymer* 36:2451. Copyright © 1995, with permission from Elsevier.)

indicates that the long threadlike liquid cylinders are not perfectly straight, indicative of an occurrence of flow instability, and that indeed some long threadlike liquid drops broke up into smaller sizes.

To elaborate on the explanation we have presented, let us look at the schematic given in Figure 11.8, depicting a long capillary die consisting of three regions: (1) the entrance region, where drop deformation occurs, (2) the stress relaxation region, where drop recoil and drop breakup may occur, and (3) the fully developed region, where the shape of drop no longer changes along the die axis. Suppose that the pressure is measured in the upstream end of the reservoir section of a plunger-type rheometer while a dispersed two-phase polymer blend is forced to flow through the capillary die. Under such circumstances, use of Eq. (11.1) to calculate σ does not make sense because the shape of drops keeps changing in the entrance and stress relaxation regions (in reference to the schematic given in Figure 11.8). This means that even when a very long capillary die is used, under which the Bagley end correction may be neglected, Eq. (11.1) with $n_B = 0$ still is not valid for a dispersed two-phase polymer blend, because not only does the shape of drops keep changing but also, under certain circumstances, drop breakup may occur in the entrance region (Han and Funatsu 1978; Chin and Han 1980). After all, Eq. (11.1) is valid only for homogeneous polymeric fluids. These observations should caution those who wish to use a plunger-type capillary rheometer to determine the rheological properties of immiscible polymer blends.

What is the alternative? The answer clearly lies in the use of a continuous-flow capillary (or slit) rheometer, which makes use of wall normal stress measurements in the fully developed region of a capillary (or slit) die (see Chapter 5). That is, as long as the wall normal stresses along the axis of a die are linear (i.e., in the fully developed region), σ can be calculated from

$$\sigma = \left(-\frac{\partial p}{\partial z}\right)\frac{D}{4} \tag{11.2}$$

in a capillary die, with D being the capillary diameter, or from

$$\sigma = \left(-\frac{\partial p}{\partial z}\right)\frac{h}{2} \tag{11.3}$$

in a slit die, with h being the die opening.

Figure 11.9 gives axial wall normal stress ($T_{rr}(R, z)$) profiles for flow through a capillary die with an L/D ratio of 20 and a diameter of 0.317 cm for a 50/50 PS/high-density polyethylene (HDPE) blend at 200 °C. There should be no argument that PS/HDPE blends form two phases! Notice in Figure 11.9 that the slope of the $T_{rr}(R, z)$ profile is constant over the distance in which wall normal stresses were taken. Under such circumstance, $-\partial T_{rr}/\partial z = -\partial p/\partial z$ holds for fully developed flow (see Chapter 5). We hasten to point out that the constancy of $-\partial p/\partial z$ in capillary flow is only a necessary condition for fully developed flow but does not indicate whether there would be completely stable morphology in immiscible polymer blends (also two-phase liquids in general). This is because wall normal stress measurements are not sensitive enough to detect migrations, if any, of small drops in an immiscible polymer blend in the radial direction of the capillary die. Nevertheless, such an experimental procedure does not involve entrance region flow. Indeed, Han and coworkers (Han 1971, 1974; Han and Kim 1975; Han and Yu 1971a, 1971b, 1972; Han et al. 1973, 1975; Kim and Han 1976) employed such experimental procedures to determine the rheological properties of immiscible polymer blends. Thus, we conclude that the viscosity data of immiscible polymer blends reported in the literature, which were determined from the

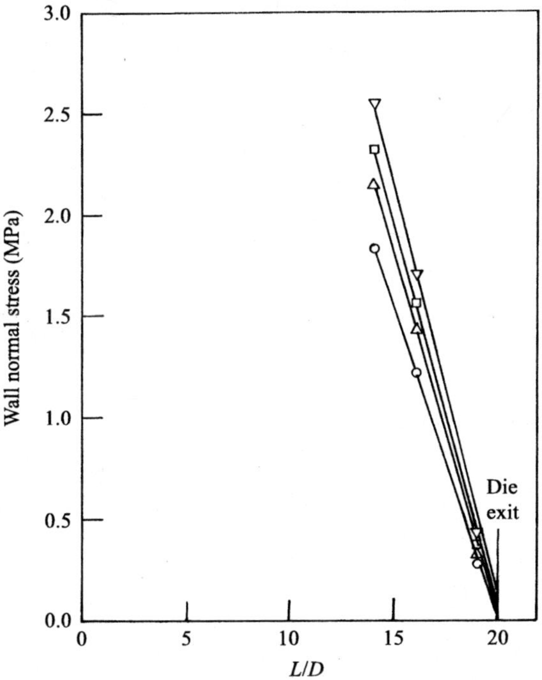

Figure 11.9 Axial wall normal stress distribution during extrusion in a capillary die with a length-to-diameter ratio of 20 and a diameter of 0.317 cm for the 50/50 PS/HDPE blend at 200 °C for various apparent shear rates (s^{-1}): (○) 106, (△) 149, (□) 257, and (▽) 305.

use of a plunger-type capillary rheometer with the aid of Eq. (11.1) with or without end corrections n_B, are subjected to serious criticism.

11.2.2 Effect of Blend Composition on the Steady-State Shear Flow Properties of Immiscible Polymer Blends

Figure 11.10 gives the effect of blend composition on η of PS/polypropylene (PP) blends at 200 °C for different values of σ. The values of η given in Figure 11.10 were obtained from the expression $\eta = \sigma/\dot{\gamma}_{app}$, for which σ was calculated from Eq. (11.2)[1] via wall normal stress measurements in the fully developed region of a capillary die having an *L*/*D* ratio of 20, and $\dot{\gamma}_{app}$ was calculated from $\dot{\gamma}_{app} = 4Q/\pi D^3$, with Q being volumetric flow rate. It is important to emphasize that we cannot calculate "true" shear rate and thus "true" shear viscosity in capillary flow of a two-phase polymer blend because we have no way of determining the velocity profile of a two-phase blend (or any two-phase mixtures) inside the capillary. Figure 11.11 gives plots of exit pressure (P_{Exit}) versus blend composition for the same PS/PP blends at 200 °C for different values of σ. Note that the values of P_{Exit} given in Figure 11.11 were obtained from wall normal stress measurements (see Chapter 5). An independent study on blend morphology has indicated that the 10/90, 20/80, and 50/50 PS/PP blends have the discrete PS phase dispersed in the continuous PP phase, and that the 80/20 PS/PP blend has the discrete PP phase dispersed in the continuous PS phase. It is very interesting to observe that in Figure 11.10, η goes through a minimum, while in Figure 11.11, P_{Exit} goes through a maximum, at a blend composition of approximately 50 wt %. The observation of negative deviation of η from a linear η–blend composition relationship and positive deviation of P_{Exit} from a linear P_{Exit}–blend composition relationship for the PS/PP blends are not mere coincidence, as will be elaborated on next.

Figure 11.12 gives plots of η versus blend composition for PS/PMMA blends, showing that the extent of negative deviation from a linear η–blend composition

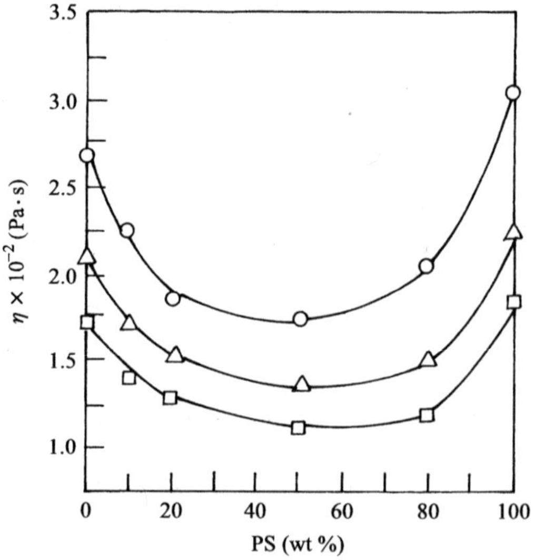

Figure 11.10 Plots of η versus blend composition for PS/PP blends at 200 °C at various wall shear stresses (Pa): (○) 5.5 × 10^4, (△) 6.3 × 10^4, and (□) 7.1 × 10^4.

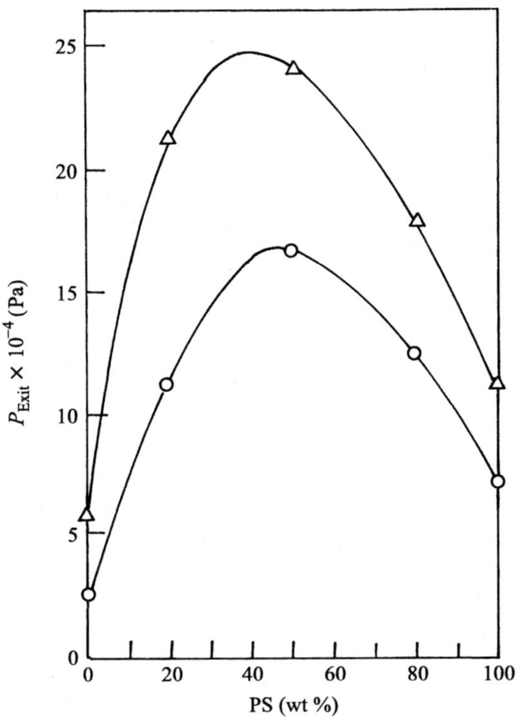

Figure 11.11 Plots of P_{Exit} versus blend composition for PS/PP blends at 200 °C at two different wall shear stresses (Pa): (○) 4.8×10^4 and (△) 6.2×10^4.

relationship (the dashed lines) increases with increasing σ. Figure 11.13 gives plots of steady-state first normal stress difference (N_1) versus blend composition for the same PS/PMMA blends at 200 °C for different values of σ. It should be mentioned that Figures 11.12 and 11.13 were obtained from wall normal stress measurements using a continuous-flow capillary rheometer and that the values of N_1 given Figure 11.13 were calculated from P_{Exit} data using Eq. (5.64). Again, in Figures 11.12 and 11.13 we observe that η goes through a minimum, while N_1 goes through a maximum, at a certain blend composition. This observation is consistent with that in Figures 11.10 and 11.11 for PS/PP blends.

Figure 11.14 shows the effect of blend composition on η for HDPE/PS blends at 200 °C for different values of σ, which was obtained from wall normal stress measurements using a continuous-flow capillary rheometer. In Figure 11.14 we observe that η goes through a minimum at approximately 45 wt % HDPE and a maximum at maximum at approximately 75 wt % HDPE. Figure 11.15 gives the effect of blend composition on N_1 for the HDPE/PS blends at 200 °C at different values of σ. Again, the values of N_1 were determined from exit pressure measurements. It is interesting to observe in Figure 11.15 that N_1 goes through a maximum at approximately 35 wt % HDPE and a minimum at approximately 75 wt % HDPE. Notice further that the extent of the maximum in N_1 increases with increasing σ. What is most interesting, when taking a close look at Figures 11.14 and 11.15, is that a maximum in N_1 occurs at approximately the same blend composition at which a minimum in η occurs, and a minimum in N_1 occurs at approximately the same blend composition at which a maximum

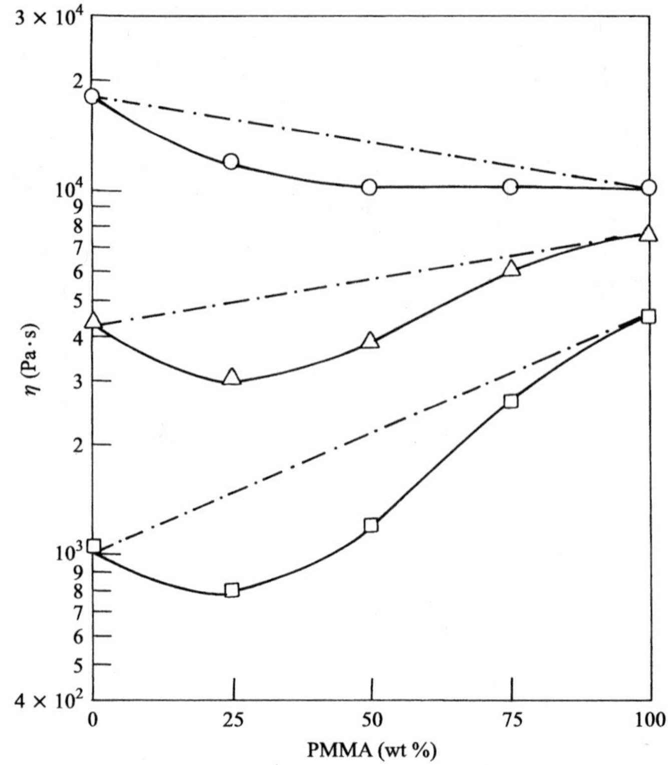

Figure 11.12 Plots of η versus blend composition for PS/PMMA blends at 200 °C at various wall shear stresses (Pa): (○) 10^4, (△) 3×10^4, and (□) 5×10^4. (Reprinted from Han et al., *Applied Polymer Science Symposia* 20:191. Copyright © 1973, with permission from John Wiley & Sons.)

in η occurs. This observation is not a mere coincidence, as will be elaborated on next. Since the seminal studies of Han and coworkers in the 1970s (Han 1971, 1974; Han and Kim 1975; Han and Yu 1971a, 1971b, 1972; Han et al. 1973, 1975; Kim and Han 1976), numerous other research groups have reported similar observations for a variety of other polymer pairs; there are too many studies to cite them here.

Micrographs of the extrudates of three HDPE/PS blends are given in Figure 11.16, in which the dark areas represent the PS phase and the bright areas represent the HDPE phase. Notice in Figure 11.16 that the morphology of the 75/25 HDPE/PS blend, which gives a maximum in η and a minimum in N_1 (see Figures 11.14 and 11.15), is quite different from the morphologies of the other two blends in that the morphology of the 75/25 HDPE/PS blend looks as if the two phases are interlocked. Such a state of dispersion would make the 75/25 HDPE/PS blend very difficult to flow compared with the other two blends, 25/75 and 50/50 HDPE/PS, which have a discrete HDPE phase dispersed in the continuous PS phase. It is clear that the morphologies of the HDPE/PS blends displayed in Figure 11.16 are, qualitatively, reflected in the rheological behavior given in Figures 11.14 and 11.15. The problem of determining quantitative relationships

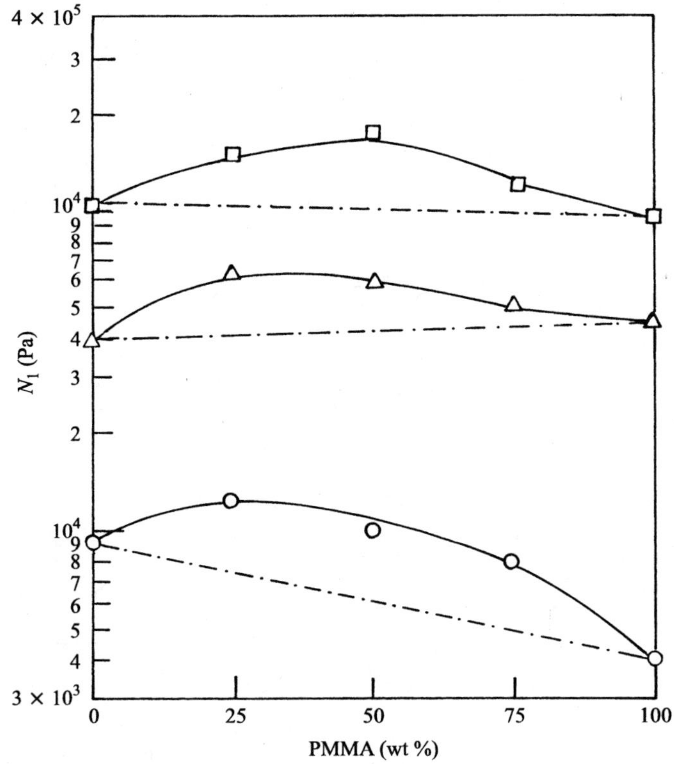

Figure 11.13 Plots of N_1 versus blend composition for PS/PMMA blends at 200 °C at various wall shear stresses (Pa): (○) 10, (△) 3 × 10^4, and (□) 5 × 10. (Reprinted from Han et al., *Applied Polymer Science Symposia* 20:191. Copyright © 1973, with permission from John Wiley & Sons.)

between the rheology and the morphology in immiscible polymer blends as affected by processing conditions (e.g., temperature and shear stress) remains. This subject will be addressed later in this chapter.

The dependence of viscosity on blend composition for polyoxymethylene (POM)/copolyamide (CPA) blends at 190 °C for different values of σ are given in Figure 11.17, showing that the viscosity goes through a maximum at low σ and through a minimum at high σ. While Ablazova et al. (1975) investigated the rheology of POM/CPA blends, Tsebrenko et al. (1976) examined the details of the morphological states of the POM/CPA blends by freezing the extrudates, which were obtained at different σ, in liquid nitrogen. They found that the dispersed phase in POM/CPA blends formed fine fibrils with increasing σ. It is reasonable to speculate that a maximum in viscosity appearing in Figure 11.17 originated from the strong interactions between the dispersed drops at very low σ (i.e., drop–drop aggregation), interfering with the flow of the 80/20 POM/CPA blend.

Figure 11.18 gives plots of η versus composition for binary blends of a fluoroelastomer (Viton) and ethylene-propylene-diene-monomer (EPDM) rubber at

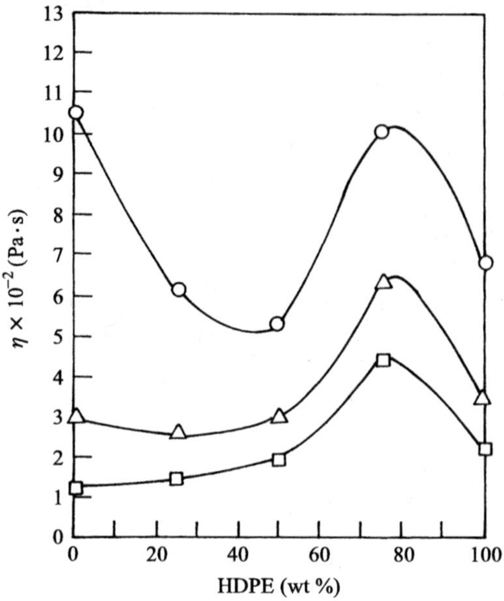

Figure 11.14 Plots of η versus blend composition for HDPE/PS blends at 200 °C at various wall shear stresses (Pa): (□) 4×10^4, (△) 6×10^4, (○) 9×10^4. (Reprinted from Han and Kim, *Transactions of the Society of Rheology* 19:245. Copyright © 1975, with permission from the Society of Rheology.)

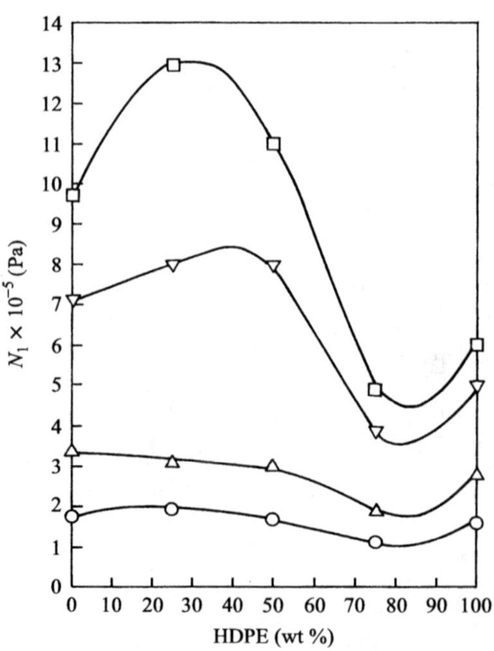

Figure 11.15 Plots of N_1 versus blend composition for HDPE/PS blends at 200 °C at various wall shear stresses (Pa): (○) 4×10^4, (△) 6×10^4, (▽) 9×10^4, and (□) 10^5. (Reprinted from Han and Kim, *Transactions of the Society of Rheology* 19:245. Copyright © 1975, with permission from the Society of Rheology.)

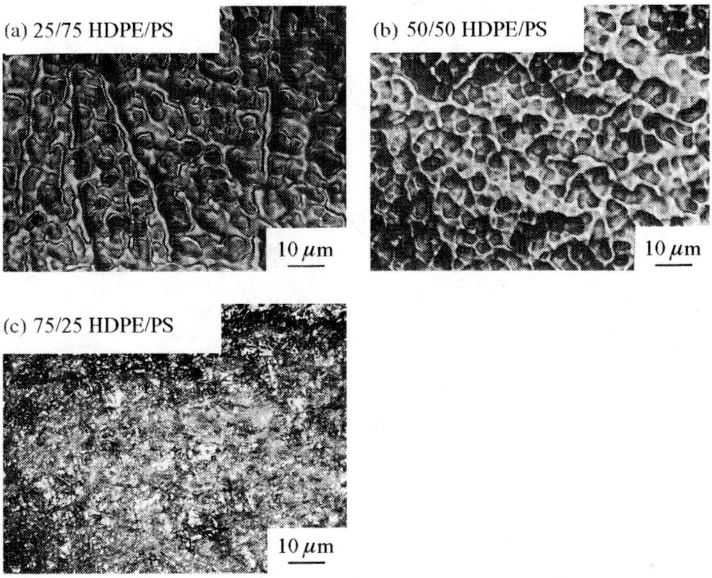

Figure 11.16 Micrographs of the cross section of frozen extrudates, which were collected during extrusion at 200 °C: (a) 25/75 HDPE/PS at $\sigma = 0.63 \times 10^5$ Pa, (b) 50/50 HDPE/PS at $\sigma = 0.61 \times 10^5$ Pa, and (c) 75/25 HDPE/PS at $\sigma = 0.69 \times 10^5$ Pa. (Reprinted from Han and Kim, *Transactions of the Society of Rheology* 19:245. Copyright © 1975, with permission from the Society of Rheology.)

160 °C for $\dot{\gamma} = 14\ \mathrm{s}^{-1}$, showing a broad and flat minimum in η over a very wide range of blend compositions. Notice in Figure 11.18 the magnitude of η drop for the Viton/EPDM blends over a very broad range of blend compositions (0.05–95 wt % EPDM), compared with the magnitude of η drop for the PS/PP blends in Figure 11.10, PS/PMMA blends in Figure 11.12, and HDPE/PS blends in Figure 11.14. Shih (1976) speculated that the slippage at the interface of two polymers and the coated die surface might be responsible for such a very large drop in viscosity for the Viton/EPDM blends. Such a speculation seems reasonable. The mechanism associated with a very large, almost composition-independent viscosity reduction in the Viton/EPDM blends displayed in Figure 11.18 is believed to be quite different from that associated with a moderate composition-dependent viscosity reduction in the PS/PP, PS/PMMA, and HDPE/PS blends observed in Figures 11.10, 11.12, and 11.14. As will be elaborated on later in this chapter, the negative deviation from a linear η–blend composition relationship observed in Figures 11.10, 11.12, and 11.14 and the positive deviation from a linear N_1–blend composition relationship observed in Figures 11.13 and 11.15 are attributable to the deformation of the dispersed phase (i.e., drops) in immiscible polymer blends during flow. Further, the experimental results available suggest that the higher the extent of drop deformation, the greater will be the extent of negative deviation from a linear relationship between η and blend composition in immiscible polymer blends.

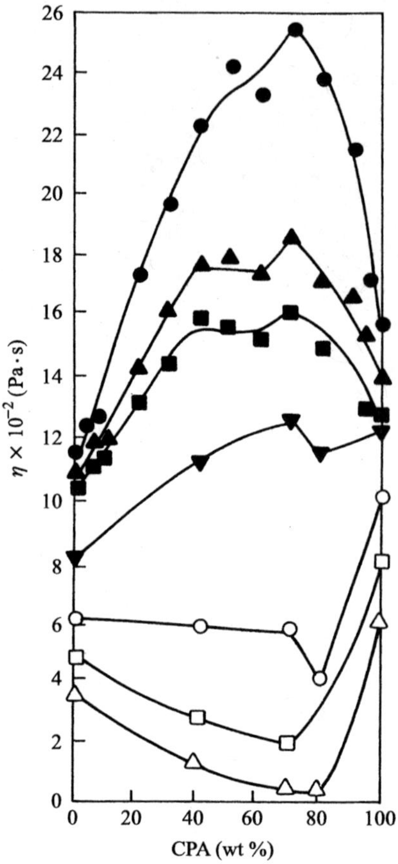

Figure 11.17 Plots of viscosity versus blend composition for blends of polyoxymethylene (POM) and copolyamide (CPA) at 190 °C for various wall shear stresses (Pa): (●) 1.27×10^4, (▲) 3.93×10^4, (■) 5.44×10^4, (▼) 6.30×10^4, (○) 1.26×10^5, (□) 1.93×10^5, and (△) 3.16×10^5. (Reprinted from Ablazova et al., *Journal of Applied Polymer Science* 19:1781. Copyright © 1975, with permission from John Wiley & Sons.)

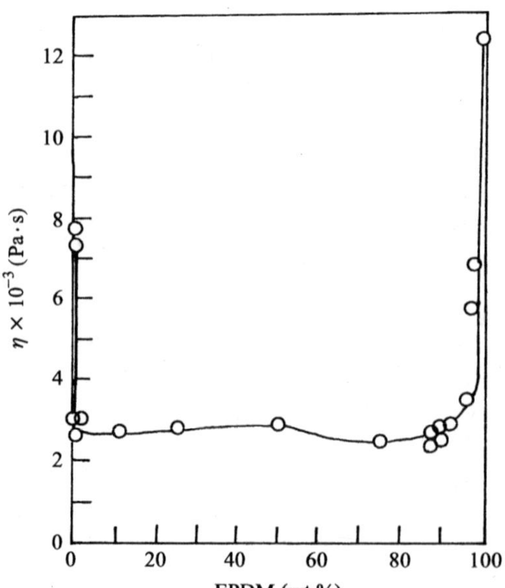

Figure 11.18 Plots of η versus blend composition for Viton/EPDM blends at 160 °C at a shear rate of 14 s^{-1}. (Reprinted from Shih, *Polymer Engineering and Science* 16:742. Copyright © 1976, with permission from the Society of Plastics Engineers.)

It can easily be surmised that the size of drops would play an important role in determining the bulk rheological properties of immiscible polymer blends, because as the drop size becomes smaller, larger forces will be required to deform the drop. It is of interest to note that van Oene (1972) carried out extrusion experiments with HDPE/PS blends by varying the length of the capillary, which then affected the residence time. He found that there was no marked change in extrudate morphology, except that extrusion through a short capillary resulted in a coarser dispersion. Van Oene noted further that a once-extruded blend could be re-extruded without change in extrudate structure, although re-extrusion made the dispersion more uniform. Similar experiments were also carried out by Tsebrenkö et al. (1974), who employed POM/CPA blends. They observed that re-extrusion increased the viscosity of the blends, and re-extruded extrudates had more uniform and finer drops dispersed in the continuous phase. This seems to confirm that the large deviation in viscosity from a linear viscosity–blend composition relationship for the immiscible polymer blends observed in Figures 11.10, 11.12, and 11.14 is attributable to the deformation of the discrete drops dispersed in the continuous phase. What we need is theoretical (or computational) study to establish the relationships between the bulk viscosity of an immiscible polymer blend and its morphology. We shall revisit this subject later in this chapter, after presenting the deformation of drops in capillary flow as affected by flow conditions.

11.2.3 Linear Dynamic Viscoelastic Properties of Immiscible Polymer Blends

Linear dynamic viscoelastic properties of immiscible polymer blends have been reported by numerous investigators; they are too many to cite them all here. Figure 11.19 gives plots of $\log|\eta*|$ versus $\log\omega$ for the homopolymer PMMA and for the 50/50 PMMA/PS blend at 210 and 230 °C. Also given in Figure 11.19, for comparison, are plots of $\log\eta$ versus $\log\dot{\gamma}$ for the same materials. In Figure 11.19 we observe that the Cox–Merz rule (Cox and Merz 1958) holds for the homopolymer PMMA but not for the 50/50 PMMA/PS blend. When considering the fact that the state of the dispersion in the 50/50 PMMA/PS blend in steady-state shear flow would be different from that in oscillatory shear flow, one should not be surprised to observe in Figure 11.19 that the Cox–Merz rule does not hold for the 50/50 PMMA/PS blend. This observation would be true for other blend compositions in the PMMA/PS blend system, and also to multiphase polymer systems in general.

Log G' versus log G'' plots are given in Figure 11.20 for homopolymers PMMA and PS at 210, 220, and 230 °C, having a slope of 2 in the terminal region and exhibiting temperature independence. Such behavior is expected for homopolymers (see Chapter 6). However, the log G' versus log G'' plots in the terminal region at 210, 220, and 230 °C, given in Figure 11.21, for the 70/30 PMMA/PS blend have a slope much less than 2, suggesting that the state of dispersion (the size and shape of the discrete PS drops) of 70/30 PMMA/PS blend might have varied significantly with the imposed angular frequency during the oscillatory shear flow experiments. However, the log G' versus log G'' plot is not a tool that can be used to determine the morphological state of multiphase polymer systems, as emphasized in Chapter 8 (dealing with microphase-separated block copolymers) and in Chapter 9 (dealing with thermotropic liquid-crystalline polymers). Although the log G' versus log G'' plot is very sensitive to

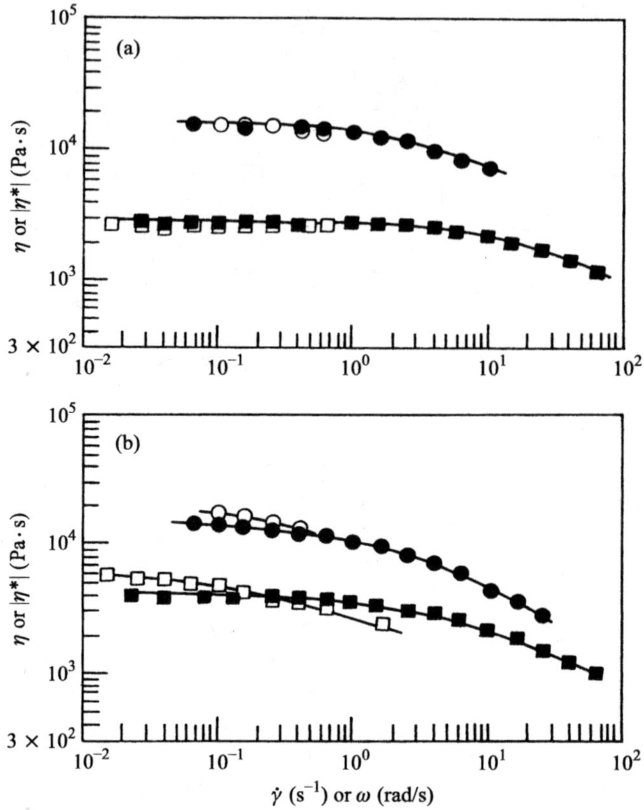

Figure 11.19 (a) Plots of log η versus log $\dot{\gamma}$ and log $|\eta^*|$ versus log ω for (a) PMMA at 210 °C (●, ○) and 230 °C (■, □) and (b) 50/50 PMMA/PS blend at 210 °C (●, ○) and 230 °C (■, □). Open symbols denote η and filled symbols denote $|\eta^*|$. (Reprinted from Han et al., *Polymer* 36:2451. Copyright © 1995, with permission from Elsevier.)

the morphological state, the temperature independence of the log G' versus log G'' plots given in Figure 11.21 suggests that the morphological state of the 70/30 PMMA/PS blend changes little with temperatures between 210 and 230 °C. This can be understood from a thermodynamic point of view. The binodal curve, which separates the homogeneous phase and two-phase region for PMMA/PS blends exhibiting UCST, lies at temperatures much higher than the measurement temperatures employed, 210–230 °C. Practically, it would not be possible for one to raise the measurement temperature sufficiently high to reach a homogeneous phase of PMMA/PS blend without thermal degradation.

11.2.4 Extrudate Swell of Immiscible Polymer Blends

Figure 11.22 shows a photograph of an extrudate of 50/50 PS/HDPE blend that has the PS forming the discrete phase (drops) dispersed in the continuous phase of HDPE.

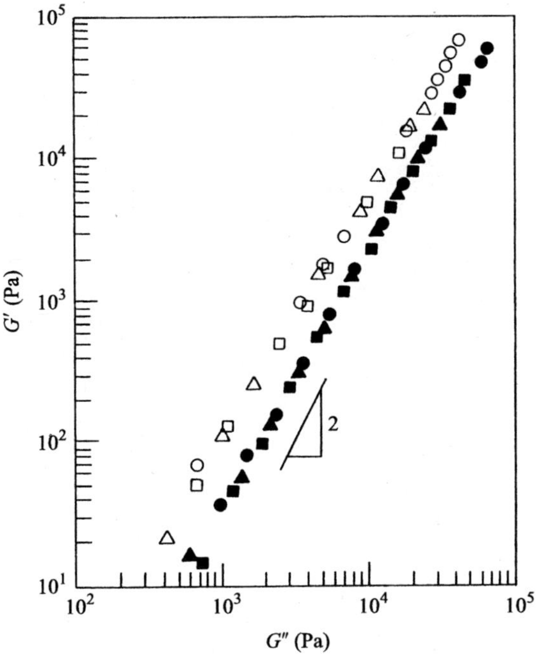

Figure 11.20 Log G' versus log G'' plots for PMMA (filled symbols) and PS (open symbols) at various temperatures (°C): (●, ○) 210, (▲, △) 220, and (■, □) 230. (Reprinted from Han et al., *Polymer* 36:2451. Copyright © 1995, with permission from Elsevier.)

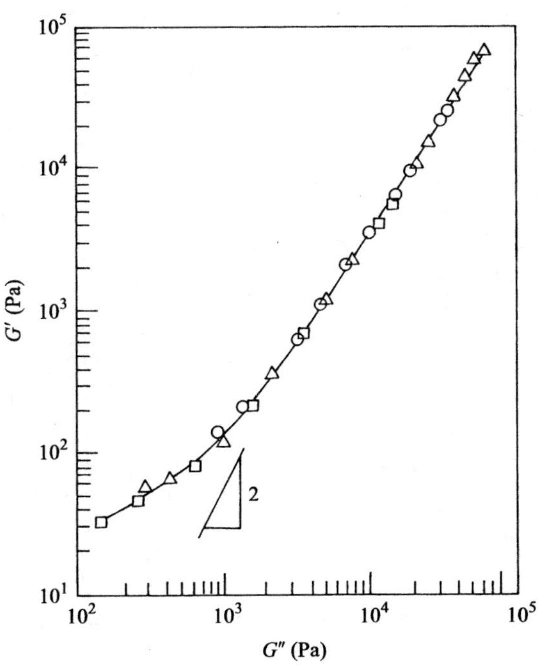

Figure 11.21 Log G' versus log G'' plots for the 70/30 PMMA/PS blend at various temperatures (°C): (○) 210, (△) 220, and (□) 230. (Reprinted from Han et al., *Polymer* 36:2451. Copyright © 1995, with permission from Elsevier.)

Figure 11.22 An extrudate sample of 50/50 HDPE/PS blend, which was extruded at 200 °C from a cylindrical die having an L/D ratio of 4. (Reprinted from Han and Yu, *American Institute of Chemical Engineers Journal* 17:1512. Copyright © 1971, with permission from the American Institute of Chemical Engineers.)

The extrudate was collected after it had been allowed to relax, without cooling, for a sufficiently long time upon simultaneously stopping the extruder and cutting the flowing melt stream. The extrudate hanging at the die exit was about 3 cm long, thus had a minimum influence of gravitational force. It can be seen from Figure 11.22 that the extrudate underwent a tremendous amount of swell while hanging at the die exit. The exceedingly large extrudate swell cannot simply be ascribed to the relaxation of polymer chains upon exiting the die, as is usually interpreted in the extrusion of homopolymers. As a matter of fact, such a large extrudate swell had never been reported for the extrusion of any commercial thermoplastic homopolymers. The unusually large extrudate swell shown in Figure 11.22 is attributable to the recoil of the deformed PS drops upon exiting the die. Figure 11.23 shows the postulated recoil process taking place in an extrudate when a dispersed two-phase blend is extruded through a cylindrical tube. The difference between the part (a) and part (b) of Figure 11.23 is the extent of drop deformation inside the cylindrical tube, resulting in different degrees of extrudate swell. That is, the greater the extent of drop deformation inside the die, the larger will be the extrudate swell. Thus, the physical origin for extrudate swell in a dispersed polymer blend is quite different from that in a homopolymer.

Figure 11.24 shows the effect blend composition on extrudate swell ratio (d_j/D) of HDPE/PS blends at 200 °C using a capillary die having the length-to-diameter (L/D) ratio of 12, where d_j denotes extrudate diameter and D denotes capillary diameter. An independent study on blend morphology has indicated that in the 20/80 and 50/50 HDPE/PS blends, the HDPE formed the drop phase and the PS formed the continuous phase, whereas in the 80/20 HDPE/PS blend the PS formed the drop phase and the HDPE formed the continuous phase. The following observations are worth noting.

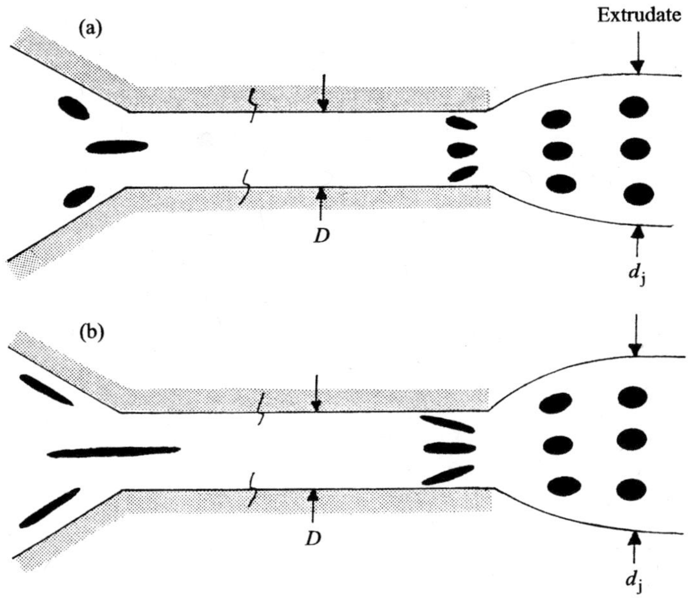

Figure 11.23 Schematic showing the origin of exceedingly large extrudate swell in a blend of two immiscible homopolymers compared with the extrudate swell of a neat homopolymer, where it is assumed that the blend forms a dispersed morphology in which one component forms drops suspended in the other component. The difference between (a) and (b) is the extent of drop deformation, indicating that the greater the drop deformation, the larger the extrudate swell.

First, as the shear stress increases (i.e., as the extrusion rate increases), the d_j/D ratio also increases, which is consistent with the view that the drop recoil, upon exiting the die, is attributable to extrudate swell. The higher the extrusion rate, the greater will be the extent of drop deformation inside the die. Second, the d_j/D ratio goes through a maximum at a blend composition of 20 wt % HDPE, suggesting that the extent of drop deformation is the largest at that particular blend composition.

Figure 11.25 gives plots of η versus σ at 200 °C for PS, PMMA, 25/75 PS/PMMA, 50/50 PS/PMMA, and 75/25 PS/PMMA blends, in which the arrow indicates the crossover shear stress (σ_c) at which the viscosities of PS and PMMA cross each other. The value of σ_c varies with temperature, as follows: $\sigma_c = 1.45 \times 10^4$ Pa at 180 °C, $\sigma_c = 3.26 \times 10^4$ Pa at 200 °C, $\sigma_c = 5.62 \times 10^4$ Pa at 220 °C, and $\sigma_c = 6.55 \times 10^4$ Pa at 240 °C. In Figure 11.25 we observe that the η of PS decreases with increasing σ much faster than the η of PMMA, giving rise to $\eta_{PS}/\eta_{PMMA} > 1$ on the left side of the arrow and $\eta_{PS}/\eta_{PMMA} < 1$ on the right side of the arrow. We also observe that the η of 75/25 PS/PMMA blend is lower than that of the constituent components, as well as that of the 25/75 and 50/50 PS/PMMA blends.

Figure 11.26 gives plots of d_j/D ratio versus log σ for homopolymers PS and PMMA at 180, 200, 220, and 240 °C, showing temperature independence. Similar to plots of N_1 versus σ presented in Chapters 5 and 6, the elastic properties of flexible

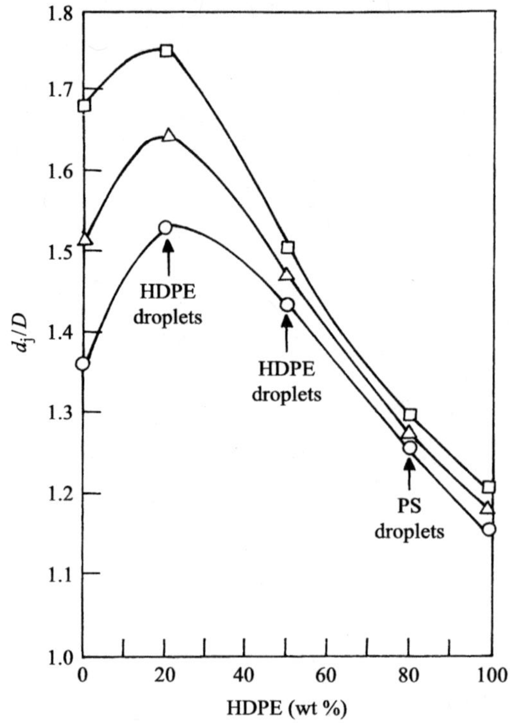

Figure 11.24 Plots of d_j/D ratio versus blend composition for HDPE/PS blends at 200 °C extruded in a capillary die with a length-to-diameter ratio of 12 for various wall shear stresses (Pa): (○) 6.89×10^4, (△) 7.58×10^5, and (□) 8.26×10^5. (Reprinted from Han and Yu, *Polymer Engineering and Science* 12:81. Copyright © 1972, with permission from the Society of Plastics Engineers.)

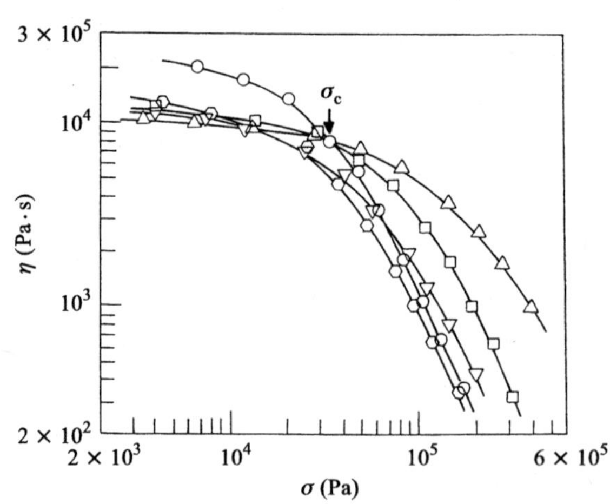

Figure 11.25 Plots of log η versus log σ at 200 °C for (○) PS, (△) PMMA, (□) 25/75 PS/PMMA blend, (▽) 50/50 PS/PMMA blend, and (⬡) 75/25 PS/PMMA blend. (Reprinted from Lyngaae-Jørgensen et al., *Polymer Alloys III*, Klempner D, Frisch KC (eds), p 105. Copyright © 1983, with permission from Springer.)

homopolymers are expected to exhibit temperature independence when plotted against the amount of energy lost per unit volume for the flow.

Plots of d_j/D ratio versus σ at four different temperatures are given in Figure 11.27 for the 25/75 PS/PMMA blend and in Figure 11.28 for the 50/50 PS/PMMA blend, in which the arrows indicate the σ_c at which the viscosities of PS and PMMA cross each

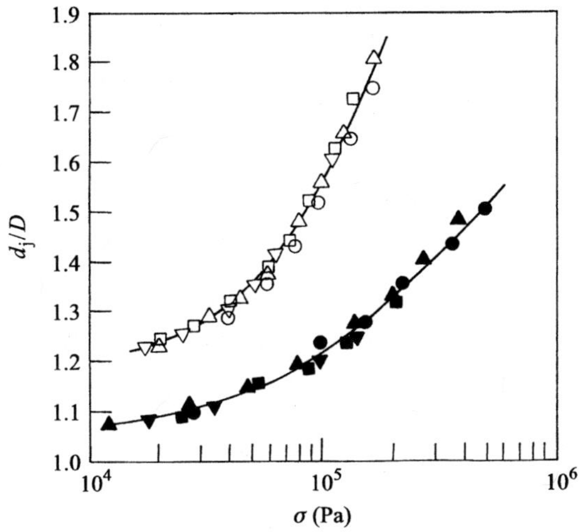

Figure 11.26 Plots of d_j/D ratio versus σ for PS (open symbols) and PMMA (filled symbols) extruded in a capillary die with a length-to-diameter ratio of 81.8 at various temperatures (°C): (○, ●) 180, (△, ▲) 200, (□, ■) 220, and (▽, ▼) 240. (Reprinted from Lyngaae-Jørgensen et al., *Polymer Alloys III*, Klempner D, Frisch KC (eds), p 105. Copyright © 1983, with permission from Springer.)

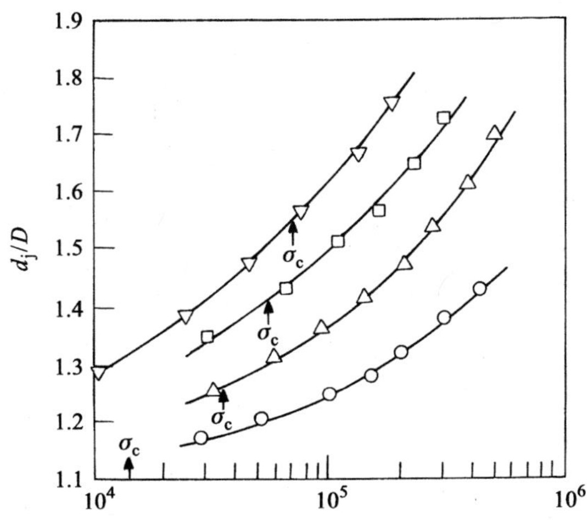

Figure 11.27 Plots of d_j/D ratio versus log σ for 25/75 PS/PMMA blend extruded in a capillary die with a length-to-diameter ratio of 81.8 at various temperatures (°C): (○) 180, (△) 200, (□) 220, and (▽) 240. (Reprinted from Lyngaae-Jørgensen et al., *Polymer Alloys III*, Klempner D, Frisch KC (eds), p 105. Copyright © 1983, with permission from Springer.)

other, namely, $\eta_{PS}/\eta_{PMMA} > 1$ on the left side of an arrow and $\eta_{PS}/\eta_{PMMA} < 1$ on the right side of an arrow. It has been reported that in the 25/75 PS/PMMA blend the minor component PS forms drops with sizes of 0.2–0.6 μm dispersed in the continuous phase of PMMA, and that in the 50/50 PS/PMMA blend the PS forms drops with sizes of 0.4–3.5 μm dispersed in the continuous phase of PMMA (Lyngaae-Jørgensen et al. 1983).

It should be pointed out that the physical origin of the increasing trend of d_j/D ratio with increasing σ observed for the 25/75 and 50/50 PS/PMMA blends

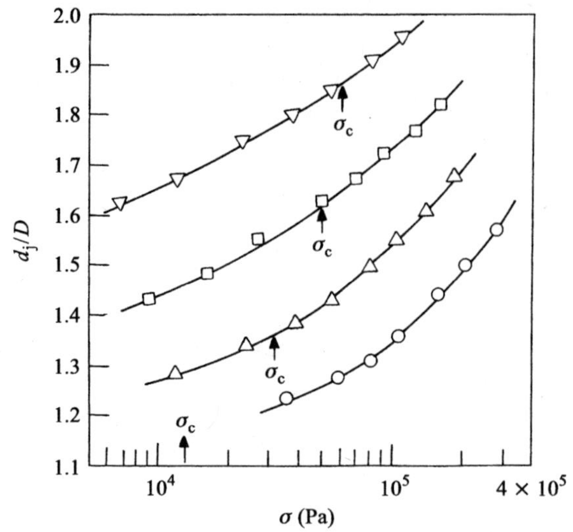

Figure 11.28 Plots of d_j/D ratio versus log σ for 50/50 PS/PMMA blend extruded in a capillary die with a length-to-diameter ratio of 81.8 at various temperatures (°C): (○) 180, (△) 200, (□) 220, and (▽) 240. (Reprinted from Lyngaae-Jørgensen et al., *Polymer Alloys III*, Klempner D, Frisch KC (eds), p 105. Copyright © 1983, with permission from Springer.)

(Figures 11.27 and 11.28), in which the PS forms drops dispersed in the PMMA phase, is quite different from that observed for the homopolymers, PS and PMMA (see Figure 11.26). Namely, the recoil of elongated PS drops upon exiting the die has contributed significantly to the extrudate swell in the 25/75 and 50/50 PS/PMMA blends, whereas the relaxation of residual wall normal stress (i.e., the relaxation of the extended polymer chains) upon exiting the die has contributed to the extrudate swell of the homopolymers. As will be elaborated on later in this chapter, what determines the extent of drop deformation during flow for a given fluid system is the viscosity ratio of the drop phase and the suspending medium (η_d/η_m), the size of drops, and the local shear stress (σ) applied to the suspending medium. That is, for a given fluid system, the drop deformation is expected to increase as the ratio η_d/η_m decreases and σ increases.

Notice in Figure 11.25 that the ratio η_d/η_m decreases with increasing σ at $\sigma > \sigma_c$. This now explains why in Figures 11.27 and 11.28 the d_j/D ratio increases with increasing σ much more rapidly at $\sigma > \sigma_c$ than at $\sigma < \sigma_c$. The extent of drop deformation is expected to increase, barring drop breakup, with increasing temperature when the same σ is applied because viscosity decreases with increasing temperature. As can be seen in Figure 11.29, the temperature dependence of viscosity is not much different between PS and PMMA at 200–230 °C,[2] suggesting that the ratio η_d/η_m would be little affected by increasing the temperature from 180 to 240 °C. Note, however, that the viscosities of both the PS drops and the continuous PMMA phase will decrease with increasing temperature. This observation now explains that the increase in d_j/D ratio given in Figure 11.27 for the 25/75 PS/PMMA blend and in Figure 11.28 for the 50/50 PS/PMMA blend is attributable to the greater extent of deformation of PS drops due to a decrease in viscosity of both PS and PMMA as the temperature is increased from 180 to 240 °C while maintaining the same σ.

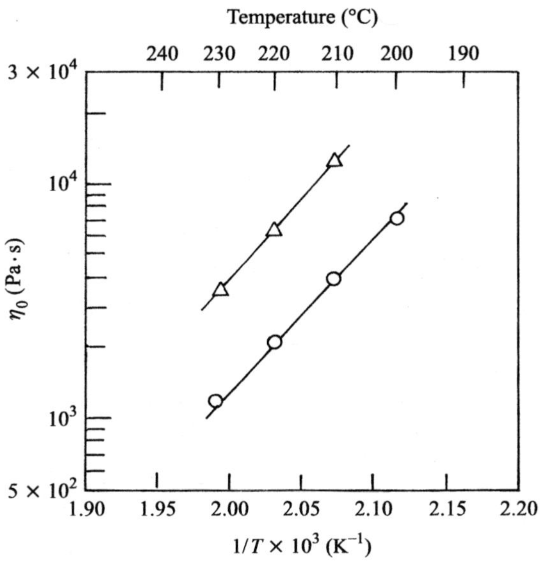

Figure 11.29 Plots of zero-shear viscosity (η_0) versus the reciprocal of absolute temperature for (○) PS and (△) PMMA.

11.3 Consideration of Large Drop Deformation and Bulk Rheological Properties of Immiscible Polymer Blends in Pressure-Driven Flow

Since the seminal experimental study of Taylor (1934), during the past several decades numerous research groups have investigated drop deformation experimentally (Bartok and Mason 1959; Bentley and Leal 1986; Chin and Han 1979; Delaby et al. 1994; Guido and Villone 1998; Han and Funatsu 1978; Levitt and Macosko 1996; Mighir et al. 1997; Olbright and Kung 1992; Olbright and Leal 1983; Phillips et al. 1980; Rumscheidt and Mason 1961; Tavgac 1972; Torza et al. 1972) and theoretically (Barthès-Biesel and Acrivos 1973b; Chaffey et al. 1965; Chaffey and Brenner 1967; Chin and Han 1979; Choi and Schowalter 1975; Cox 1969; Flumerfelt 1980; Hinch and Acrivos 1980; Hyman and Skalak 1972; Turner and Chaffey 1969), and also drop breakup experimentally (Bentley and Leal 1986; Chin and Han 1980; Flumerfelt 1972; Grace 1982; Han and Funatsu 1978; Janssen et al. 1994; Karam and Bellinger 1968; Milliken and Leal 1991; Olbright and Kung 1992; Olbright and Leal 1983; Rumscheidt and Mason 1961, 1962; Stroeve and Varanasi 1984; Tavgac 1972; Taylor 1934; Torza et al. 1972; Varanasi et al. 1994) and theoretically (Acrivos and Lo 1978; Buckmaster 1973; Chin and Han 1980; Goren 1962, 1964; Hinch and Acrivos 1979, 1980; Mikami et al. 1978; Stone and Leal 1989, 1990a, 1990b; Tomotika 1935, 1936). The majority of the experimental studies dealt with steady-state uniform shear flow or steady-state elongational flow, while only a few studies (Chin and Han 1979, 1980; Han and Funatsu 1978) reported on the deformation and breakup of a drop in the entrance region of a cylindrical or slit die. Steady-state uniform shear flow or steady-sate elongation flow are seldom encountered in an internal mixer or in a twin-screw extruder, which are commonly used for compounding two immiscible polymers.

Numerous research groups (Barthès-Biesel and Acrivos 1973a; Choi and Schowalter 1975; Frankel and Acrivos 1970; Fröhlich et al. 1946; Goddard and Miller 1967; Oldroyd 1953; Roscoe 1967; Schowalter et al. 1968) made attempts at predicting theoretically the bulk rheological properties of dilute emulsions. One of the salient features of the studies (Barthès-Biesel and Acrivos 1973a; Choi and Schowalter 1975; Frankel and Acrivos 1970) is that the phenomenological theory predicts shear-thinning behavior and normal stress effects in a dilute emulsion consisting of two Newtonian liquids. Such a prediction has originated from the consideration of the deformability of drops and the presence of a finite value of interfacial tension.

There are experimental studies that support the theoretical predictions of shear-thinning behavior of an emulsion consisting of two Newtonian liquids (Suzuki et al. 1970; Vadas et al. 1976), and also others that support the existence of first normal stress difference in emulsions consisting of two Newtonian liquids (Han and King 1980). It should be pointed out, however, that those experimental studies employed concentrated emulsions, while the theories referred to above are based on dilute emulsions. Figure 11.30 gives plots of log η versus log $\dot{\gamma}$ for nylon 6, poly(ethylene terephthalate) (PET), and the blends 30/70 and 70/30 nylon 6/PET, at 270 °C, which were taken using a cone-and-plate rheometer. It is seen in Figure 11.30 that the 30/70 and 70/30 nylon 6/PET blends exhibit shear-thinning behavior, whereas the constituent components, nylon 6 and PET, exhibit only Newtonian behavior over the entire range of shear rates tested from 0.01 to 10 s^{-1}. It seems appropriate to mention at this juncture that using the stream function method, Hyman and Skalak (1972) calculated the shapes of a string of Newtonian drops, dispersed in another Newtonian liquid, moving along the centerline of a cylindrical tube and observed non-Newtonian effects arising from the deformation of drops. Such a theoretical prediction reinforces our contention, presented in the preceding sections, that the rheological behavior of an immiscible polymer blend and its morphology (the state of dispersion) are inseparable. This means that any attempt to predict the rheological behavior of an immiscible polymer blend must include the prediction of the state of dispersion of the blend.

In the past, several research groups (Carley 1985; Carley and Crossan 1981; Heitmiller et al. 1964; Lin 1979; Lyngaae-Jørgensen et al. 1983; Utracki 1983;

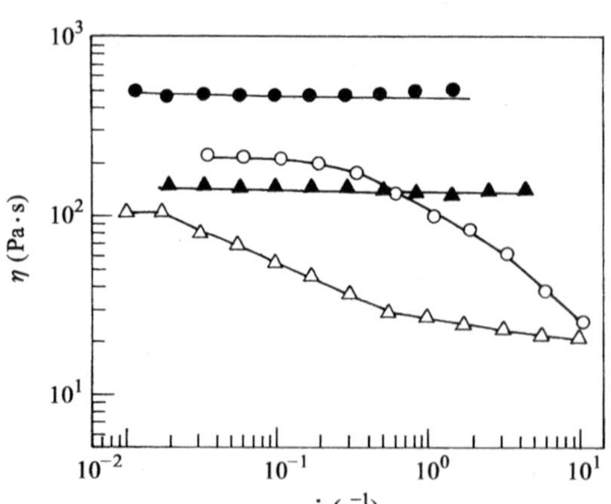

Figure 11.30 Plots of log η versus log $\dot{\gamma}$ for nylon 6 (●), PET (▲), 70/30 nylon 6/PET blend (○), and 30/70 nylon 6/PET blend (△) at 270 °C. The weight-average molecular weight (M_w) and polydispersity index (M_w/M_n) determined by gel permeation chromatography are 3.1×10^4 and 2.3 for nylon 6, and 5.7×10^4 and 4.0 for PET.

Utracki and Kamal 1982) have suggested empirical expressions to describe viscosity–composition relationships in immiscible polymer blends; specifically, some research groups (Heitmiller et al. 1964; Lin 1979) have suggested an inverse additivity relationship, and others (Carley 1985; Carley and Crossan 1981; Lyngaae-Jørgensen et al. 1983; Utracki 1983; Utracki and Kamal 1982) have suggested a polynomial relationship, and still others (Utracki et al. 1982) suggested a logarithmic additivity relationship between blend viscosity and blend composition. The first suggestion of an inverse additivity relationship (Heitmiller et al. 1964) was based on the premise that alternating annular layers of two polymers can represent the state of the dispersion of an immiscible polymer blend flowing through a long cylindrical tube. Such a premise is not tenable because, as can be seen in Figures 11.31 and 11.32, the dispersed phase in an immiscible polymer blend, during flow, consists of many drops with different

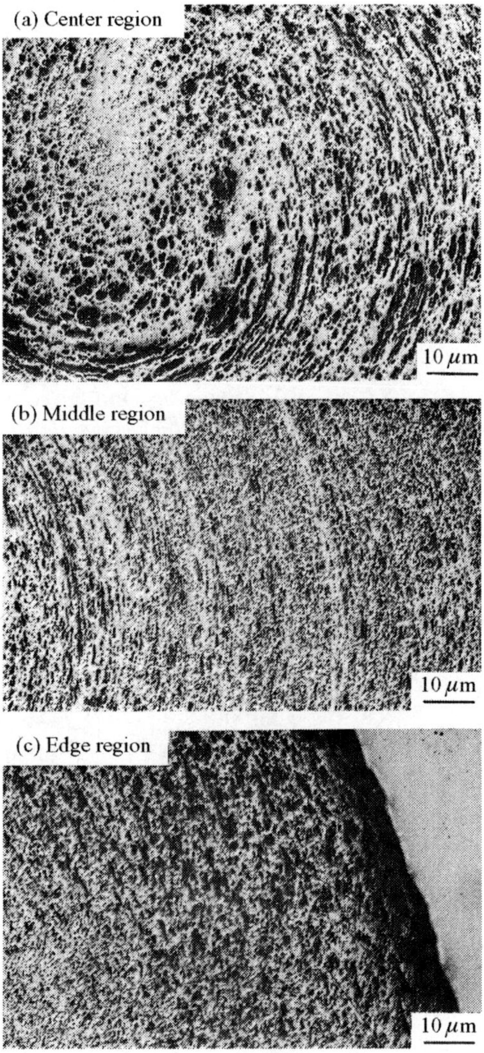

Figure 11.31 Micrographs of the cross section of 20/80 PS/HDPE blend extruded at a shear rate of 664 s^{-1} and 200 °C: (a) center region, (b) middle region, and (c) edge region. The dark areas represent the PS phase and the bright areas represent the HDPE phase. (Reprinted from Han and Yu, *Journal of Applied Polymer Science* 15:1163. Copyright © 1971, with permission from John Wiley & Sons.)

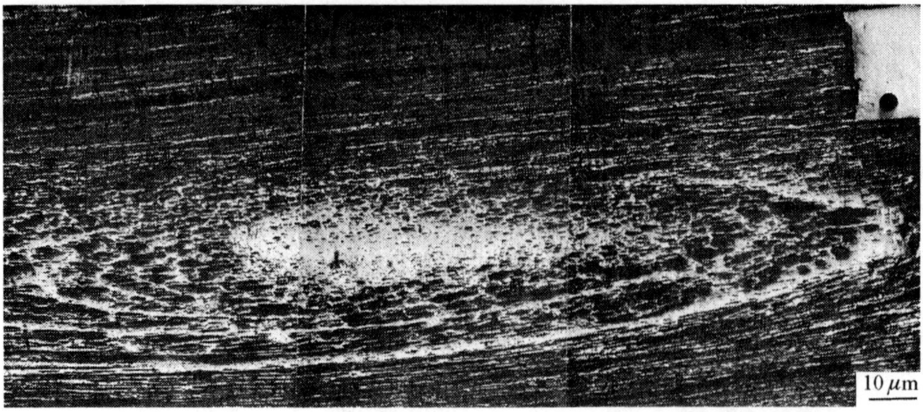

Figure 11.32 Micrograph of the longitudinal section of 20/80 PS/HDPE blend extruded at a shear rate of 664 s^{-1} and 200 °C. The dark areas represent the PS phase and the bright areas represent the HDPE phase. (Reprinted from Han and Yu, *Journal of Applied Polymer Science* 15:1163. Copyright © 1971, with permission from John Wiley & Sons.)

degrees of deformation, depending upon their locations in a cylindrical tube. In Figure 11.31 we observe that the drops are much larger near the center of the tube than near the edge of the tube, and in Figure 11.32 we observe that the drops are deformed more near the wall of the tube than near the center of the tube. These observations are in line with the expectations from the fluid mechanics point of view because the shear stress increases from the center of a tube and reaches a maximum at the tube wall, and thus drops are expected to deform more near the tube wall than near the center of the tube. We thus conclude that the suggestions (Carley 1985; Carley and Crossan 1981; Heitmiller et al. 1964; Lin 1979; Lyngaae-Jørgensen et al. 1983; Utracki 1983; Utracki and Kamal 1982; Utracki et al. 1982) for using such empirical expressions as an inverse additivity relationship, a logarithmic additivity relationship, or a polynomial relationship to describe the bulk viscosities of immiscible polymer blends without including the state of dispersion of an immiscible polymer blend have no physical and rheological significance.

Since the small-deformation theory for a single drop in shear flow was reviewed in the 1980s (Acrivos 1983; Han 1981; Rallison 1984), relatively few new findings have been reported. Thus, we have chosen not to review such theories here. Small-deformation theory for a single drop in uniform shear flow is important in its own right, but such theory is of very limited use for a better understanding of the bulk rheological properties of immiscible polymer blends. This is because in the processing of polymer blends from an industrial point of view, large deformations of the discrete phase (drops) would invariably take place in a very complicated flow geometry. Although it may be regarded as being an oversimplification, we can categorize the various complicated flows encountered in industrial polymer processing equipment into three groups, as depicted schematically in Figure 11.33, namely, (1) uniform shear flow, (2) nonuniform shear flow, and (3) converging flow. In Figure 11.33 we observe different shapes of drop in the three flow geometries: in uniform shear flow the shape of drop is the same across the flow channel, in nonuniform shear flow (i.e., in pressure-driven flow) the shape

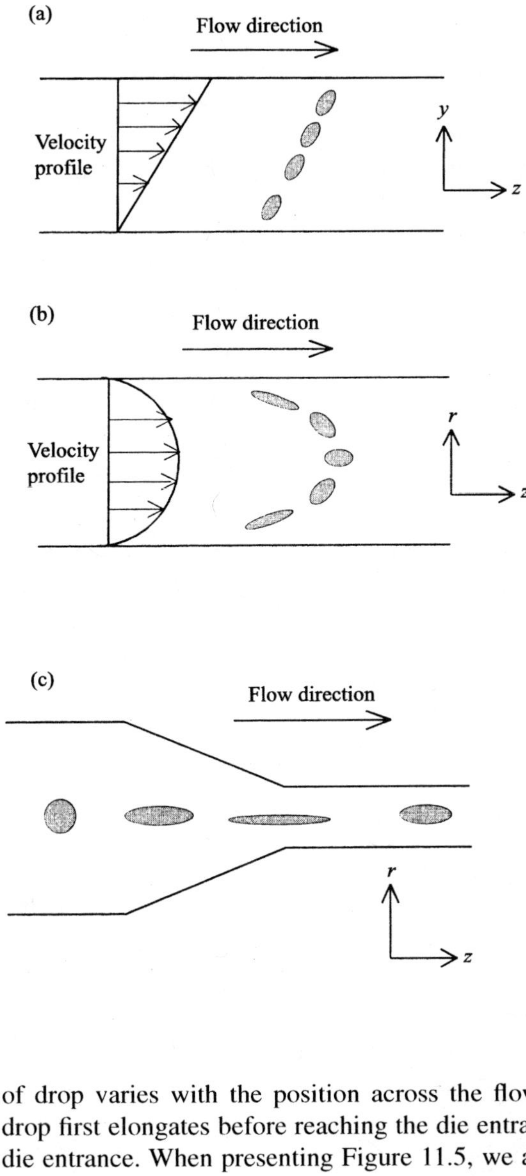

Figure 11.33 Drop shape (a) in uniform shear flow, (b) in nonuniform shear flow, and (c) in converging flow.

of drop varies with the position across the flow channel, and in converging flow a drop first elongates before reaching the die entrance and then recoils upon passing the die entrance. When presenting Figure 11.5, we already speculated that the differences in the shape of drops between uniform and nonuniform shear flows would have large influence on the bulk rheological properties of dispersed two-phase polymer blends. Of particular interest is the shape of a drop moving along the centerline in the entrance region of a cylindrical die (see Figure 11.33c).

Admittedly, the prediction of the large deformation of a drop in shear flow, which requires numerical computations, is not a trivial problem. Nevertheless, a number of research groups (Bozzi et al. 1997; Coulliette and Pozrikidis 1998; Kennedy et al. 1994; Loewenberg and Hinch 1996; Martinez and Udell 1990; Stone and Leal 1990a; Zhou and Pozrikidis 1994; Zinchenko et al. 1997) employed the boundary integral method to investigate the creeping motion of a single Newtonian drop suspended in another Newtonian liquid in a long cylindrical tube or in uniform shear flow. Kim and Han (2001) employed finite element method (FEM) to predict the shape of a non-Newtonian drop that is suspended in another non-Newtonian liquid moving along the centerline of a cylindrical tube. Here, we present their principal findings, namely, the extent of drop deformation in the entrance region of the cylindrical tube as affected by the viscosity ratio of drop phase to suspending medium, apparent shear rate, and the initial size of drop. We then suggest a finite element approach to predict the bulk rheological properties of immiscible polymer blends in pressure-driven flow.

11.3.1 Finite Element Analysis of Large Drop Deformation in the Entrance Region of a Cylindrical Tube

The most general situation describing dispersed liquid–liquid two-phase creeping flow, often encountered in the processing of two immiscible polymers (see Chapter 3 of Volume 2), may be represented by the continuity and momentum balance equations:

$$\nabla \cdot \mathbf{v}_{\mathrm{m}} = 0; \quad \nabla \cdot \mathbf{v}_{\mathrm{d}} = 0 \tag{11.4}$$

$$\rho_{\mathrm{m}}\left(\frac{\partial \mathbf{v}_{\mathrm{m}}}{\partial t}+\mathbf{v}_{\mathrm{m}}\cdot\nabla\mathbf{v}_{\mathrm{m}}\right)=-\nabla\cdot p_{\mathrm{m}}\boldsymbol{\delta}+\nabla\cdot\boldsymbol{\sigma}_{\mathrm{m}};\ \rho_{\mathrm{d}}\left(\frac{\partial \mathbf{v}_{\mathrm{d}}}{\partial t}+\mathbf{v}_{\mathrm{d}}\cdot\nabla\mathbf{v}_{\mathrm{d}}\right)=-\nabla\cdot p_{\mathrm{d}}\boldsymbol{\delta}+\nabla\cdot\boldsymbol{\sigma}_{\mathrm{d}} \tag{11.5}$$

in which ∇ is the gradient operator, $\mathbf{v}$ is the velocity vector, ρ is the density of fluid, p is pressure, $\boldsymbol{\sigma}$ is the stress tensor, $\boldsymbol{\delta}$ is the unit tensor, and subscripts m and d on the variables denote the matrix phase (suspending medium) and dispersed phase (i.e., drop), respectively. In writing Eqs. (11.4) and (11.5) we have assumed that both the suspending medium and the drop are incompressible and that gravity force is negligible.

For an axisymmetric creeping flow of a single non-Newtonian drop suspended in another non-Newtonian fluid in the entrance region of a cylindrical tube, as schematically shown in Figure 11.34, let us assume that a spherical drop is placed in the upstream end of the reservoir section and this drop moves along the centerline of the tube. Using

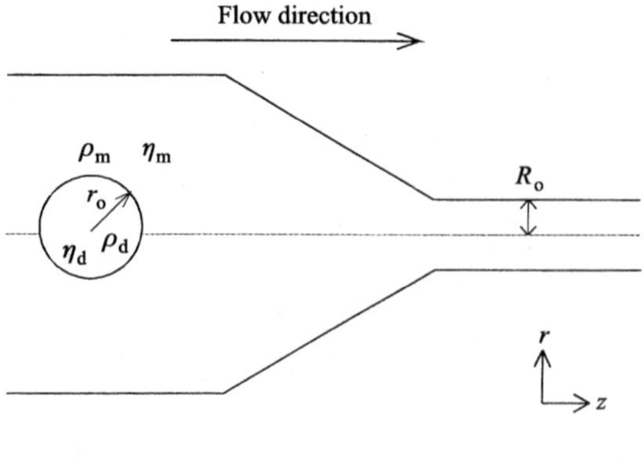

Figure 11.34 Schematic showing a cylindrical tube with a reservoir section, in which a spherical drop suspended in another liquid moves along the central axis from the upstream reservoir section to the downstream tube. (Reprinted from Kim and Han, *Journal of Rheology* 45:1279. Copyright © 2001, with permission from the Society of Rheology.)

dimensionless variables in the cylindrical coordinates, Eqs. (11.4) and (11.5) can be written as

$$\nabla^* \cdot \mathbf{v}_i^* = 0 \tag{11.6}$$

$$Re\left(\frac{\partial \mathbf{v}_i^*}{\partial t^*} + \mathbf{v}_i^* \cdot \nabla^* \mathbf{v}_i^*\right) = -\nabla^* \cdot p_i^* \boldsymbol{\delta} + \nabla^* \cdot \boldsymbol{\sigma}_i^* \tag{11.7}$$

in which $\rho_m = \rho_d$ is assumed and the dimensionless variables (with asterisks) are defined by $\nabla^* = R_o \nabla$, $\mathbf{v}_i^* = \mathbf{v}_i / V_o$, $t^* = t/t_o$, $p_i^* = p_i/(\bar{\eta} V_o / R_o)$, $\boldsymbol{\sigma}_i^* = \boldsymbol{\sigma}_i / (\bar{\eta} V_o / R_o)$ for i = m and d, with R_o (tube radius), V_o, and $t_o = R_o/V_o$ being the characteristic length, velocity, and time, and $\bar{\eta}$ being the reference viscosity. In Eq. (11.7), $Re = \rho_m V_o R_o / \bar{\eta}$ is the Reynolds number.

In order to solve Eqs. (11.6) and (11.7), one must specify a relationship between the stress tensor $\boldsymbol{\sigma}^*$ and the rate-of-deformation tensor $\mathbf{d}^*$. The most general situation would be many viscoelastic drops suspended within another viscoelastic medium, which is extremely complicated to handle, even when using the most sophisticated computational tools available. Therefore, let us consider a simpler model, the truncated power-law model (see Chapter 6), as schematically shown in Figure 11.35, describing the shear-rate dependence of viscosities, $\eta_m(\dot{\gamma})$ and $\eta_d(\dot{\gamma})$, of the suspending medium and the drop phase, respectively. The truncated power-law model for the drop and the suspending medium in terms of dimensionless viscosities $\eta_d^*(\dot{\gamma}^*) = \eta_d(\dot{\gamma})/\bar{\eta}$ and $\eta_m^*(\dot{\gamma}^*) = \eta_m(\dot{\gamma})/\bar{\eta}$ are given by

$$\eta_i^*(\dot{\gamma}^*) = \begin{cases} \eta_{0,i}/\bar{\eta} & \text{for } \dot{\gamma}^* \le (R_o/V_o)\dot{\gamma}_{c,i} \\ (K_i/\bar{\eta})(V_o/R_o)^{n_m - 1}(\dot{\gamma}^*)^{n_m - 1} & \text{for } \dot{\gamma}^* > (R_o/V_o)\dot{\gamma}_{c,i} \end{cases} \tag{11.8}$$

for i = m and d, where $\eta_{0,i}$ is the zero-shear (Newtonian) viscosity, $\eta_i(\dot{\gamma})$ is the shear-rate dependent viscosity, $\dot{\gamma}_{c,i}$ is the critical shear rate at which the viscosities of the drop and the suspending medium start to deviate from Newtonian behavior, K_i is power-law consistency, n_i is power-law index, and $\dot{\gamma}^* = (R_o/V_o)\dot{\gamma}$ is a dimensionless shear rate, where $\dot{\gamma} = (2d_{ij}d_{ij})^{1/2}$ with d_{ij} being the ijth component of the rate-of-deformation tensor $\mathbf{d}$.

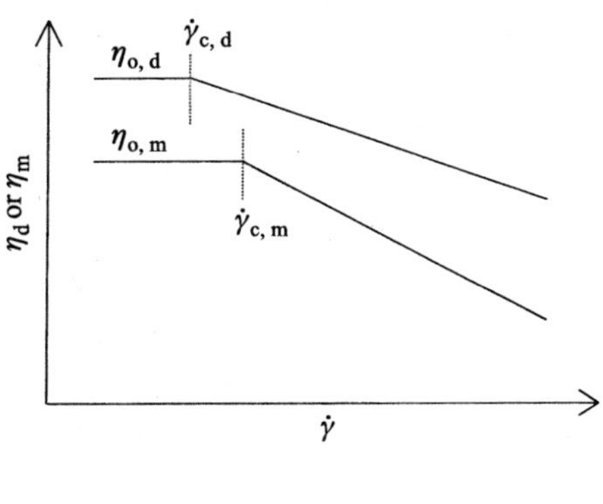

Figure 11.35 Shear-rate dependence of viscosity in the truncated power-law model.

At the interface between the drop and the suspending medium, the following normal stress and shear stress balances must be satisfied:

$$\text{(i)} \quad t_i n_j (\sigma^*_{\mathrm{m},ij} - \sigma^*_{\mathrm{d},ij}) = 0 \tag{11.9}$$

$$\text{(ii)} \quad (-p^*_{\mathrm{m}}\boldsymbol{\delta} + \boldsymbol{\sigma}^*_{\mathrm{m}}) \cdot \mathbf{n} - (-p^*_{\mathrm{d}}\boldsymbol{\delta} + \boldsymbol{\sigma}^*_{\mathrm{d}}) \cdot \mathbf{n} = \frac{H^*}{Ca}\mathbf{n} \tag{11.10}$$

where $\mathbf{t}$ (the components of which appear in Eq. (11.9)) and $\mathbf{n}$ are, respectively, the local unit tangential and outward normal vectors to the interface, H^* is the sum of local mean curvature defined by $H^* = R_o/R$ with R being the sum of local mean radius, and $Ca = \bar{\eta}V_o/\gamma$ is the capillary number with γ being the interfacial tension. We assume that the velocity is continuous (i.e., no slip) at the fluid interface, $\mathbf{v}_{\mathrm{m}} = \mathbf{v}_{\mathrm{d}}$, and at the tube wall, $\mathbf{v}_{\mathrm{m},z} = 0$ and $\mathbf{v}_{\mathrm{m},r} = 0$. Further, fully developed flow is assumed to prevail at the tube inlet and outlet of the circular tube (see Figure 11.34). There are different ways of calculating values of H^* (Ambravaneswaran et al. 2002; Bozzi et al. 1997; Jackson and Tucker 2003).

In this section, we present some representative computational results reported by Kim and Han (2001), who simulated the experimental observed drop shapes summarized in Figure 11.36. Table 11.1 gives a summary of the numerical values of the parameters appearing in the power-law model, Eq. (11.8), for four polymer solutions. Table 11.2 gives a summary of the numerical values of the parameters and flow conditions employed for the numerical computations. The details of the computational procedures employed are given in the original paper.

Figure 11.37 gives the computed shapes of a 2 wt % polyisobutylene (PIB) drop of 0.12 cm in radius, suspended in 2 wt % aqueous solution of polyacrylamide (Separan, Dow Chemical) at $\dot{\gamma} = 75.4\,\mathrm{s}^{-1}$, as it moves along the centerline of the conical reservoir section into the cylindrical tube, simulating the experimental results given in Figure 11.36a. Enlarged drop shapes at four different positions near and just inside the tube inlet are also given. It is seen that the drop elongates considerably at the inlet of the tube and then recoils somewhat after entering the tube. It suffices to state that the computed drop shape in the entrance region shows what is expected intuitively and is also in reasonable agreement with experiment.

Figure 11.38 gives the computed shapes of a 6 wt % PIB drop of 0.09 cm in radius, suspended in 2 wt % Separan solution at $\dot{\gamma} = 102.4\,\mathrm{s}^{-1}$, as it moves along the centerline of the conical reservoir section into the cylindrical tube, simulating the experimental results given in Figure 11.36b. Enlarged drop shapes at four different positions near and just inside the tube inlet are also given. It is seen that the drop elongates at the tube entrance, but that it recoils very little at position 4 compared with the drop at roughly the same position in Figure 11.37 (for a 2 wt % PIB drop). It is worth pointing out that the extent of drop deformation in Figure 11.38 is as large as that in Figure 11.37, in spite of the fact that Figure 11.38 has a smaller drop size (0.09 cm) compared with the drop size (0.12 cm) in Figure 11.37. It is intuitively expected that the smaller the drop size, the less will be the extent of drop deformation under otherwise identical flow conditions. However, the value of $\dot{\gamma}$ ($102.4\,\mathrm{s}^{-1}$) in Figure 11.38 is higher than that ($75.4\ \mathrm{s}^{-1}$) in Figure 11.37. In other words, a smaller size of drop requires a higher shear rate in order to have the same extent of deformation as a larger size of

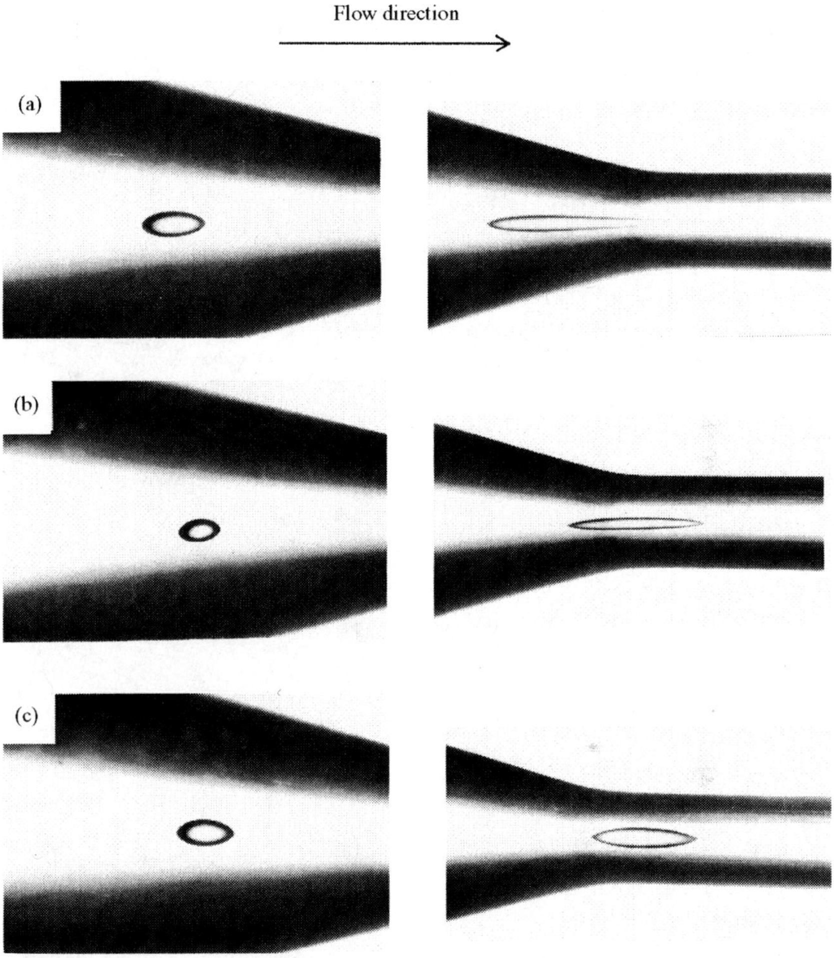

Figure 11.36 Experimentally observed drop shapes in the entrance region of a cylindrical tube having the reservoir radius of 5.35 cm, the tube radius (R_0) of 0.3 cm, and the entrance angle of 30°. (a) A drop of 2 wt % PIB solution suspended in a 2 wt % Separan solution with the undeformed drop radius (r_0) of 0.12 cm and the shear rate ($\dot{\gamma}$) of 75.4 s^{-1}. (b) A drop of 6 wt % PIB solution suspended in a 2 wt % Separan solution with $r_0 = 0.091$ cm and $\dot{\gamma} = 102.4$ s^{-1}. (c) A drop of 10 wt % PIB solution suspended by a 2 wt % Separan solution with $r_0 = 0.12$ cm and $\dot{\gamma} = 102.4$ s^{-1}. (Reprinted from Kim and Han, *Journal of Rheology* 45:1279. Copyright © 2001, with permission from the Society of Rheology.)

Table 11.1 Numerical values of the parameters appearing in the truncated power-law model for an aqueous solution of Separan and PIB solutions in decalin

Fluid		*Rheological Parameters*
2 wt % PIB in decalin	Drop phase	$\eta_{0,\mathrm{d}} = 0.113$ Pa·s for $\dot{\gamma} \le \dot{\gamma}_c$ with $\dot{\gamma}_c = 35\ \mathrm{s}^{-1}$ $n_\mathrm{d} = 0.94$ and $K_\mathrm{d} = 0.14$ Pa·s$^{n_\mathrm{d}}$ for $\dot{\gamma} > \dot{\gamma}_c$
6 wt % PIB in decalin	Drop phase	$\eta_{0,\mathrm{d}} = 5.438$ Pa·s for $\dot{\gamma} \le \dot{\gamma}_c$ with $\dot{\gamma}_c = 4\ \mathrm{s}^{-1}$ $n_\mathrm{d} = 0.59$ and $K_\mathrm{d} = 9.60$ Pa·s$^{n_\mathrm{d}}$ for $\dot{\gamma} > \dot{\gamma}_c$
10 wt % PIB in decalin	Drop phase	$\eta_{0,\mathrm{d}} = 19.30$ Pa·s for $\dot{\gamma} \le \dot{\gamma}_c$ with $\dot{\gamma}_c = 1\ \mathrm{s}^{-1}$ $n_\mathrm{d} = 0.80$ and $K_\mathrm{d} = 19.30$ Pa·s$^{n_\mathrm{d}}$ for $\dot{\gamma} > \dot{\gamma}_c$
2 wt % Separan solution	Suspending medium	$\eta_{0,\mathrm{m}} = 269$ Pa·s for $\dot{\gamma} \le \dot{\gamma}_c$ with $\dot{\gamma}_c = 0.007\ \mathrm{s}^{-1}$ $n_\mathrm{m} = 0.38$ and $K_\mathrm{m} = 12.45$ Pa·s$^{n_\mathrm{m}}$ for $\dot{\gamma} > \dot{\gamma}_c$

Table 11.2 Parameters and flow conditions relevant to the experimental studies of Chin and Han (1979, 1980)

Fluid System	$\dot{\gamma}$ *(s^{-1})*	r_o *(m)*	γ *(N/m)*	η_m *(Pa·s)*	$\eta_\mathrm{d}/\eta_\mathrm{m}$	Ca^a
2 wt % PIB/ 2 wt % Separan	75.4	1.20×10^{-3}	18.2×10^{-3}	0.85	0.13	4.24
6 wt % PIB/ 2 wt % Separan	102.4	0.91×10^{-3}	14.3×10^{-3}	0.71	2.04	4.60
10 wt % PIB/ 2 wt % Separan	102.4	1.20×10^{-3}	13.8×10^{-3}	0.71	10.76	6.29

[a] $Ca = \eta\dot{\gamma}r_o/\gamma$ is the capillary number where η is the viscosity of the suspending medium η_m, $\dot{\gamma}$ is the shear rate, r_o is the drop radius prior to deformation, and γ is the interfacial tension.

drop, as can be expected from the definition of the capillary number $Ca = \eta_\mathrm{m}\dot{\gamma}r_o/\gamma$. As we will show, the viscosity ratio $\eta_\mathrm{d}/\eta_\mathrm{m}$ also plays an important role in determining the extent of drop deformation. Notice that in the computation for Figure 11.38, $\eta_\mathrm{d}/\eta_\mathrm{m}$ is 2.04, while in the computation for Figure 11.37, $\eta_\mathrm{d}/\eta_\mathrm{m}$ is 0.13. This observation also suggests that the extent of deformation of a 6 wt % PIB drop suspended in 2 wt % Separan solution will be less than that of a 2 wt % PIB drop suspended in the same medium under otherwise identical flow conditions and for the same drop size. Thus, a higher shear rate is needed to compensate for the difference in the drop viscosity between 6 wt % PIB drop (Figure 11.38) and 2 wt % PIB drop (Figure 11.37), even if the drop size were the same in both situations. It should be mentioned that the viscosity of 6 wt % PIB in decalin is about an order of magnitude higher than that of 2 wt % PIB in decalin. The apparent difference in the extent of drop recoil between 6 wt % PIB drop (Figure 11.38) and 2 wt % PIB drop (Figure 11.37) right after the die entrance

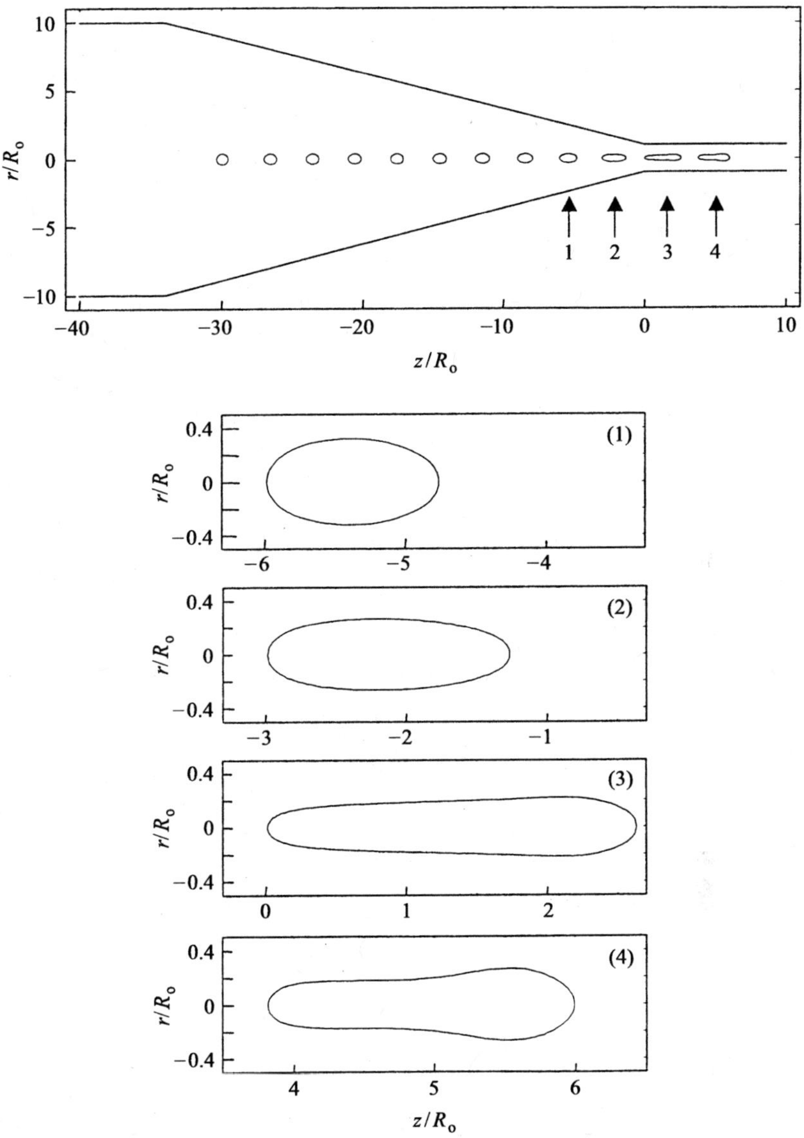

Figure 11.37 Computed drop shapes, simulating via FEM the experimental result given in Figure 11.36a, in the entrance region of a cylindrical tube with $R_o = 0.3$ cm, where the drop phase is a non-Newtonian 2 wt % PIB solution with $r_o = 0.12$ cm and the suspending medium is a non-Newtonian 2 wt % Separan solution flowing at $\dot{\gamma} = 75.4\ \mathrm{s}^{-1}$. For better detail, the enlarged drop shapes at four different positions in the entrance region are given in the lower panels. (Reprinted from Kim and Han, *Journal of Rheology* 45:1279. Copyright © 2001, with permission from the Society of Rheology.)

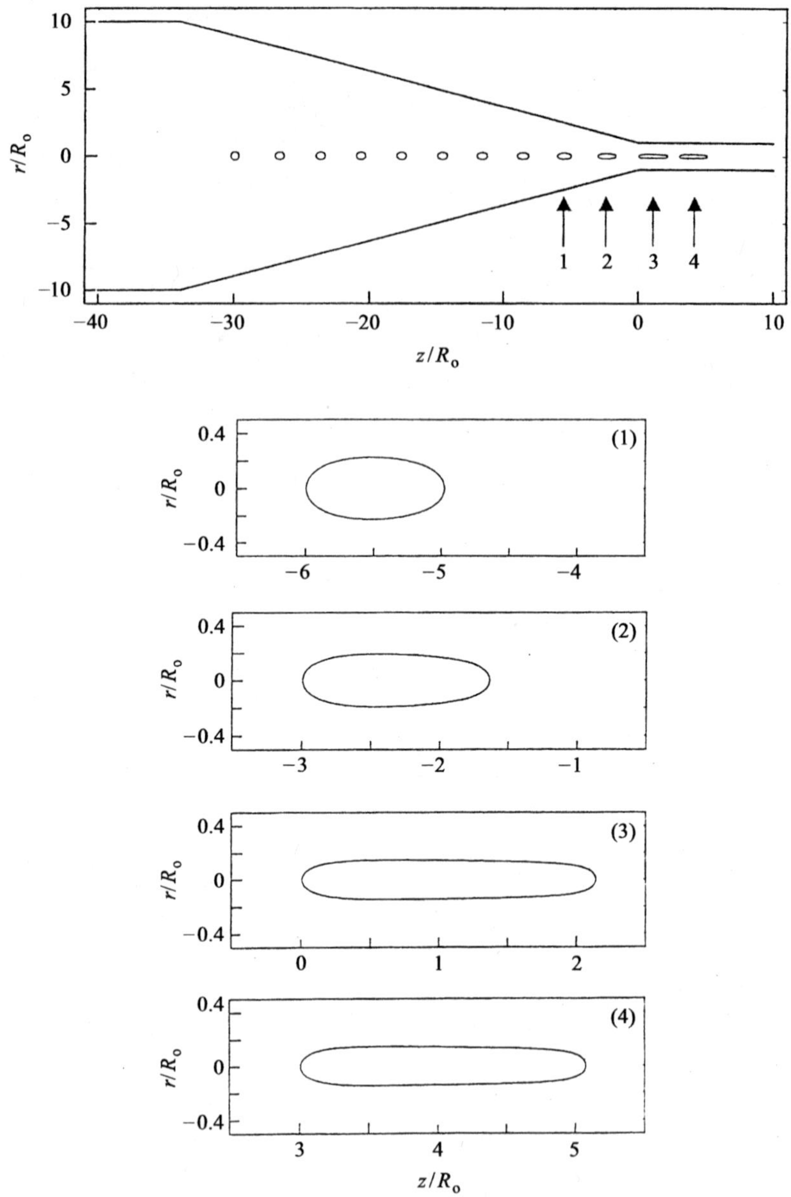

Figure 11.38 Computed drop shapes, simulating via FEM the experimental result given in Figure 11.36b, in the entrance region of a cylindrical tube with $R_o = 0.3$ cm, where the drop phase is a non-Newtonian 6 wt % PIB solution with $r_o = 0.09$ cm and the suspending medium is a non-Newtonian 2 wt % Separan solution flowing at $\dot{\gamma} = 102.4\ \text{s}^{-1}$. For better detail, the enlarged drop shapes at four different positions in the entrance region are given in the lower panels. (Reprinted from Kim and Han, *Journal of Rheology* 45:1279. Copyright © 2001, with permission from the Society of Rheology.)

(at position 4) is attributable to the longer relaxation time of the 6 wt % PIB drop compared with that of the 2 wt % PIB drop. This is because the relaxation time of a polymer solution increases with concentration. Thus, the 6 wt % PIB drop would have a longer relaxation time compared with the 2 wt % PIB drop. The longer the relaxation time of a polymer solution, the slower will be the rate of stress relaxation. This consideration can explain why the 6 wt % PIB drop at position 4 in Figure 11.38 recoiled very little compared with that at position 3, whereas the 2 wt % PIB drop at position 4 in Figure 11.37 recoiled noticeably compared with that at position 3. The extent of drop recoil in the entrance region of a cylindrical tube, as we have discussed, is believed to have strong relevance to the elastic properties of fluids.

Figure 11.39 gives the computed shapes of a 10 wt % PIB drop of 0.12 cm radius, suspended in 2 wt % Separan solution at $\dot{\gamma} = 102.4\,\mathrm{s}^{-1}$, as it moves along the centerline of the conical reservoir section into the cylindrical tube, simulating the experimental results given in Figure 11.36c. Enlarged drop shapes at four different positions near and just inside the tube inlet are also given. Notice that the drop radius (0.12 cm) simulated in Figure 11.39 is the same as that in Figure 11.37 but larger than that (0.09 cm) in Figure 11.38, and yet the extent of drop deformation at the tube entrance in Figure 11.39 is noticeable. This is because the value of $\dot{\gamma}$ ($102.4\,\mathrm{s}^{-1}$) simulated in Figure 11.39 is much higher than that ($\dot{\gamma} = 75.4\,\mathrm{s}^{-1}$) simulated in Figure 11.37. Conversely, the extent of deformation simulated for a 10 wt % PIB drop (Figure 11.39) is less than that for a 6 wt % PIB drop (Figure 11.38) under otherwise identical flow conditions ($\dot{\gamma} = 102.4\,\mathrm{s}^{-1}$). Thus, this difference is attributable to the higher viscosity of 10 wt % PIB solution compared with that of 6 wt % PIB solution. Again, in Figure 11.39 we observe very little recoil occurring in the drop when comparing the computed drop shape at position 4 with that at position 3. This observation is very similar to that made with reference to Figure 11.38.

The value of η_d/η_m employed in the computation for Figure 11.39 is very high (10.76) compared with that (2.04) employed in the computation for Figure 11.38, while the value of $Ca = 6.29$ employed in the computation for Figure 11.39 is slightly higher than that (4.60) employed in the computation for Figure 11.38 (see Table 11.2). Note that the larger the value of Ca, the greater will be the drop deformation. However, comparison of Figure 11.39 with Figure 11.38 reveals that the extent of drop deformation given in Figure 11.39 is less than that given in Figure 11.38, in spite of the fact that the value of $Ca = 6.29$ employed in Figure 11.39 is higher than that ($Ca = 4.60$) employed in Figure 11.38. This is attributed to the fact that the ratio η_d/η_m of 10.76 employed in the computation for Figure 11.39 is much higher than that (2.04) employed in the computation for Figure 11.38.

The effect of shear rate on drop shape along the centerline in the entrance region is given in Figure 11.40 for $\dot{\gamma} = 60\,\mathrm{s}^{-1}$, in Figure 11.41 for $\dot{\gamma} = 120\,\mathrm{s}^{-1}$, and in Figure 11.42 for $\dot{\gamma} = 180\,\mathrm{s}^{-1}$. It is clearly seen that the extent of drop deformation becomes greater with increasing $\dot{\gamma}$. It should be mentioned that for a shear-thinning suspending medium, the value of Ca is no longer proportional to $\dot{\gamma}$, because the value of η_m appearing in the definition of $Ca = \eta_m \dot{\gamma} r_o/\gamma$ decreases with increasing $\dot{\gamma}$. We have $Ca = 3.75$ for $\dot{\gamma} = 60\ \mathrm{s}^{-1}$, $Ca = 4.89$ for $\dot{\gamma} = 120\ \mathrm{s}^{-1}$, and $Ca = 5.70$ for $\dot{\gamma} = 180\ \mathrm{s}^{-1}$. Note that the ratio η_d/η_m increases with increasing $\dot{\gamma}$ for a 6 wt % PIB drop phase suspended in 2 wt % Separan solution because the shear-thinning behavior with increasing $\dot{\gamma}$ is much greater in 2 wt % Separan solution than in 6 wt % PIB

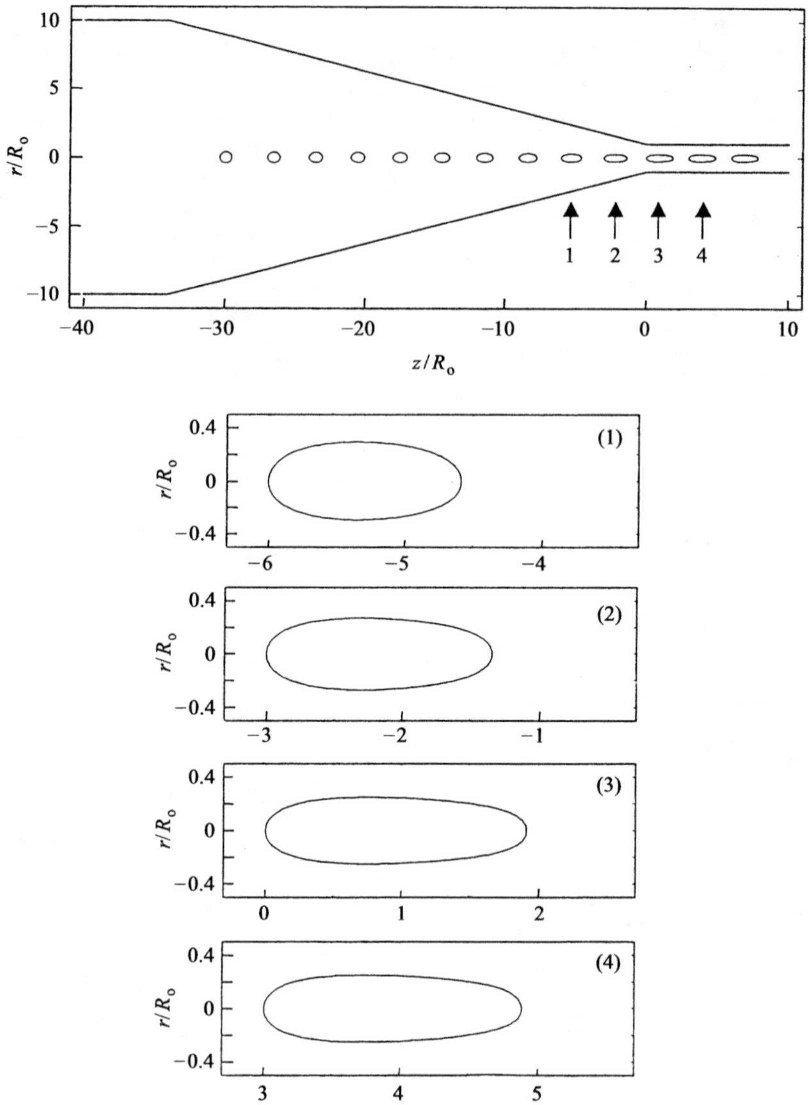

Figure 11.39 Computed drop shapes, simulating via FEM the experimental result given in Figure 11.36c, in the entrance region of a cylindrical tube with $R_o = 0.3$ cm, where the drop phase is a non-Newtonian 10 wt % PIB solution with $r_o = 0.12$ cm and the suspending medium is a non-Newtonian 2 wt % Separan solution flowing at $\dot{\gamma} = 102.4\ \mathrm{s}^{-1}$. For better detail, the enlarged drop shapes at four different positions in the entrance region are given in the lower panels. (Reprinted from Kim and Han, *Journal of Rheology* 45:1279. Copyright © 2001, with permission from the Society of Rheology.)

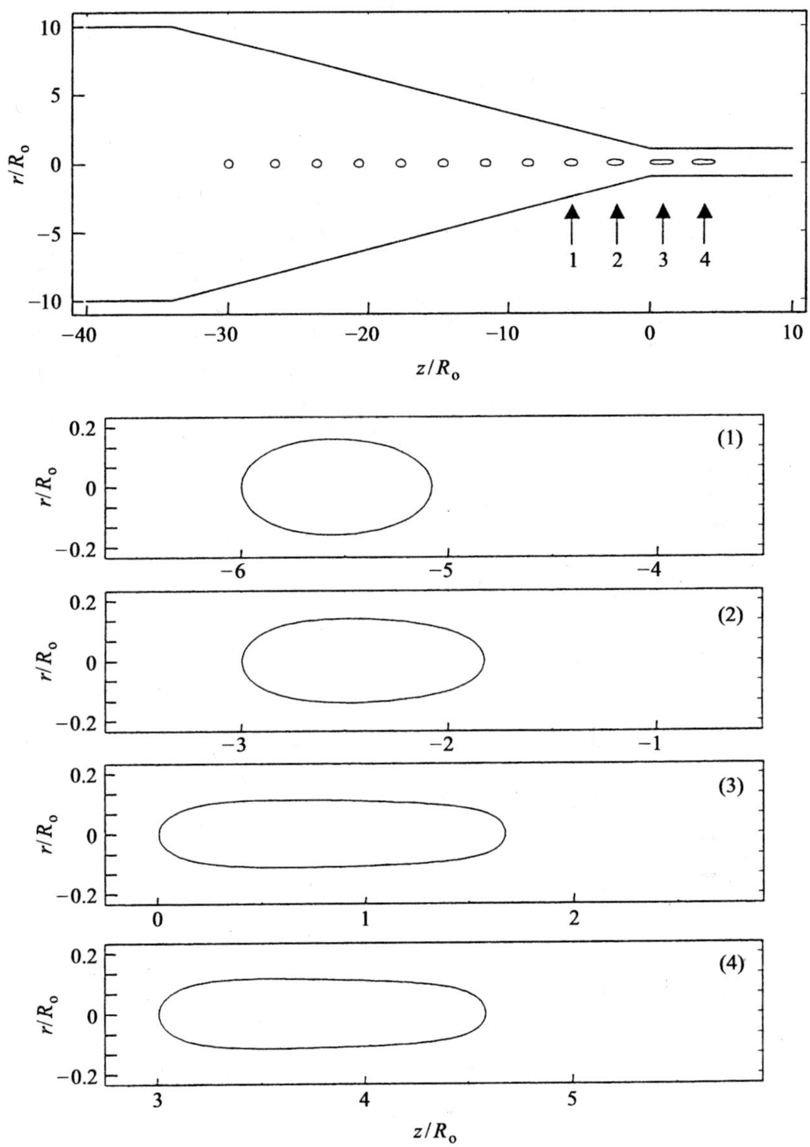

Figure 11.40 Computed drop shapes, via FEM, in the entrance region of a cylindrical tube with $R_o = 0.3$ cm, where the drop phase is a non-Newtonian 6 wt % PIB solution with $r_o = 0.09$ cm and the suspending medium is a non-Newtonian 2 wt % Separan solution flowing at $\dot{\gamma} = 60\ \text{s}^{-1}$. For better detail, the enlarged drop shapes at four different positions in the entrance region are given in the lower panels.

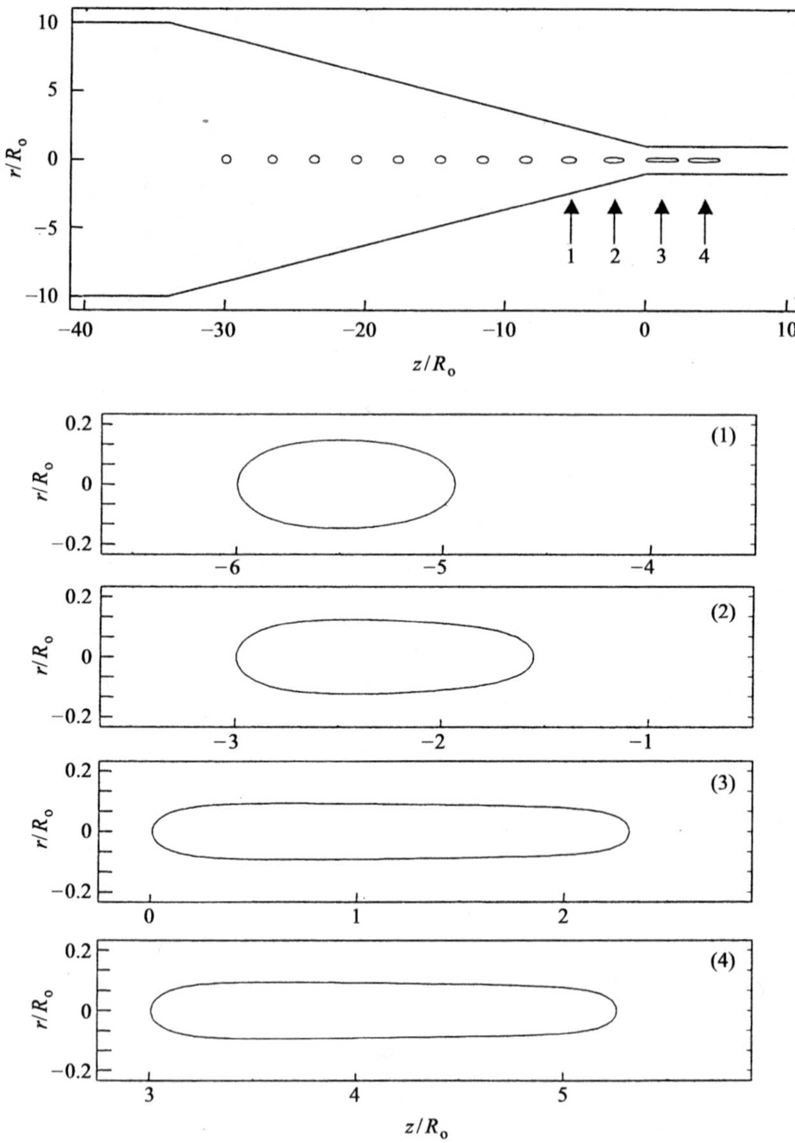

Figure 11.41 Computed drop shapes, via FEM, in the entrance region of a cylindrical tube with $R_o = 0.3$ cm, where the drop phase is a non-Newtonian 6 wt % PIB solution with $r_o = 0.09$ cm and the suspending medium is a non-Newtonian 2 wt % Separan solution flowing at $\dot{\gamma} = 120\ \mathrm{s}^{-1}$. For better detail, the enlarged drop shapes at four different positions in the entrance region are given in the lower panels. (Reprinted from Kim and Han, *Journal of Rheology* 45:1279. Copyright © 2001, with permission from the Society of Rheology.)

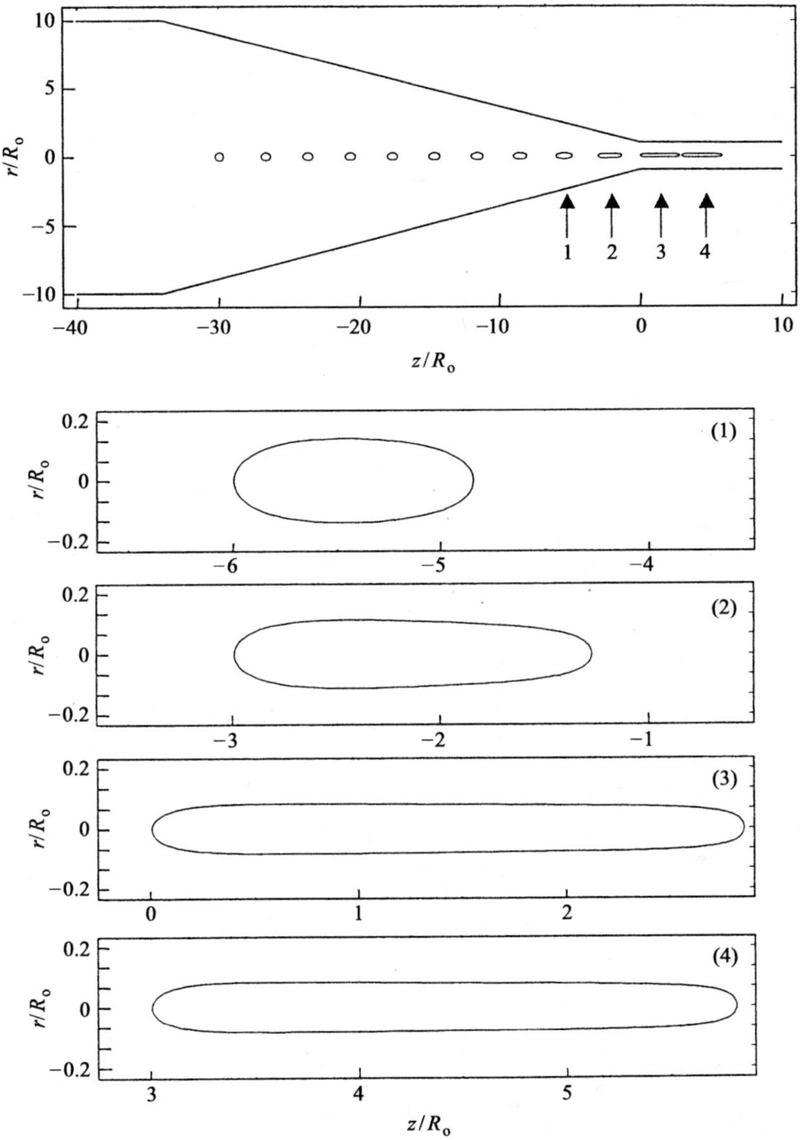

Figure 11.42 Computed drop shapes, via FEM, in the entrance region of a cylindrical tube with $R_o = 0.3$ cm, where the drop phase is a non-Newtonian 6 wt % PIB solution with $r_o = 0.09$ cm and the suspending medium is a non-Newtonian 2 wt % Separan solution flowing at $\dot{\gamma} = 180\ \mathrm{s}^{-1}$. For better detail, the enlarged drop shapes at four different positions in the entrance region are given in the lower panels. (From Kim and Han, *Journal of Rheology* 45:1279. Copyright © 2001, with permission from the Society of Rheology.)

solution. We have $\eta_d/\eta_m = 1.82$ at $\dot{\gamma} = 60\ s^{-1}$, $\eta_d/\eta_m = 2.11$ at $\dot{\gamma} = 120\ s^{-1}$, and $\eta_d/\eta_m = 2.29$ at $\dot{\gamma} = 180\ s^{-1}$. The above observation leads us to conclude that the greater extent of drop deformation observed in Figures 11.40–11.42 for an increase in $\dot{\gamma}$ from 60 to 180 s^{-1} is attributable to a concomitant increase in *Ca,* and a moderate increase in η_d/η_m with increasing $\dot{\gamma}$ from 60 to 180 s^{-1} apparently played little role in determining the extent of drop deformation.

The effect of viscosity ratio on drop shape along the centerline in the entrance region is given in Figure 11.43 for $\eta_d/\eta_m = 1$, in Figure 11.44 for $\eta_d/\eta_m = 0.1$, and in Figure 11.45 for $\eta_d/\eta_m = 0.01$, where a Newtonian drop ($\eta_d = 0.113$ Pa·s) with a radius of 0.12 cm is suspended in a Newtonian medium at $\dot{\gamma} = 75.4\ s^{-1}$. In calculating the shape of drop given in Figures 11.43–11.45 the viscosity of the Newtonian medium (η_m) is increased for the same Newtonian drop with $\eta_d = 0.113$ Pa·s. It is seen that the extent of drop deformation increases dramatically as the ratio η_d/η_m, decreases from 1 to 0.01.

11.3.2 Theoretical Approach to the Prediction of Rheology–Morphology–Processing Relationships in Pressure-Driven Flow of Immiscible Polymer Blends

We have shown in a preceding section that the size and the distribution of the dispersed phase (i.e., drops) and the shape of the drops dictate the bulk rheological properties of an immiscible polymer blend. Therefore, for a given flow field, one must relate the size and the shape of all drops to the bulk rheological properties of immiscible polymer blends. Although such a task seems almost next to impossible, the very complicated problem can be better understood (granted it is an oversimplification) by investigating the deformation and breakup of a single drop in a given flow field. That is, the dynamics of an isolated drop in shear flow can provide some insight into the complex rheological behavior of dispersed two-phase polymer blends.

The inclusion of the state of dispersion of an immiscible polymer blend into the prediction of its rheological behavior can be accomplished only when the shape of drops is predicted in a given flow field. This can be accomplished by developing FEM computer codes. As mentioned in the preceding section, FEM can handle very large deformations of drops. We wish to point out that constitutive equations for dilute emulsions are important in their own right, but they are not useful at all for predicting the bulk viscosity of an immiscible polymer blend in pressure-driven flow as is encountered in many polymer processing operations. In the development of constitutive equations for the bulk rheological properties, it is not possible to include the shapes of many drops undergoing large deformations in a pressure-driven flow. Thus, it is fair to state that such efforts would not bring meaningful results describing the rheology–morphology relationships in immiscible polymer blends. For such purposes, one must solve the momentum equations, as shown in the preceding section, for both the drop phase and the suspending medium, with proper boundary conditions at the interface. The deformation of a single drop has little significance to the real problems of practical interest. In other words, one must solve system equations for many drops. Admittedly, the task here is not trivial by any measure, but it is not insurmountable.

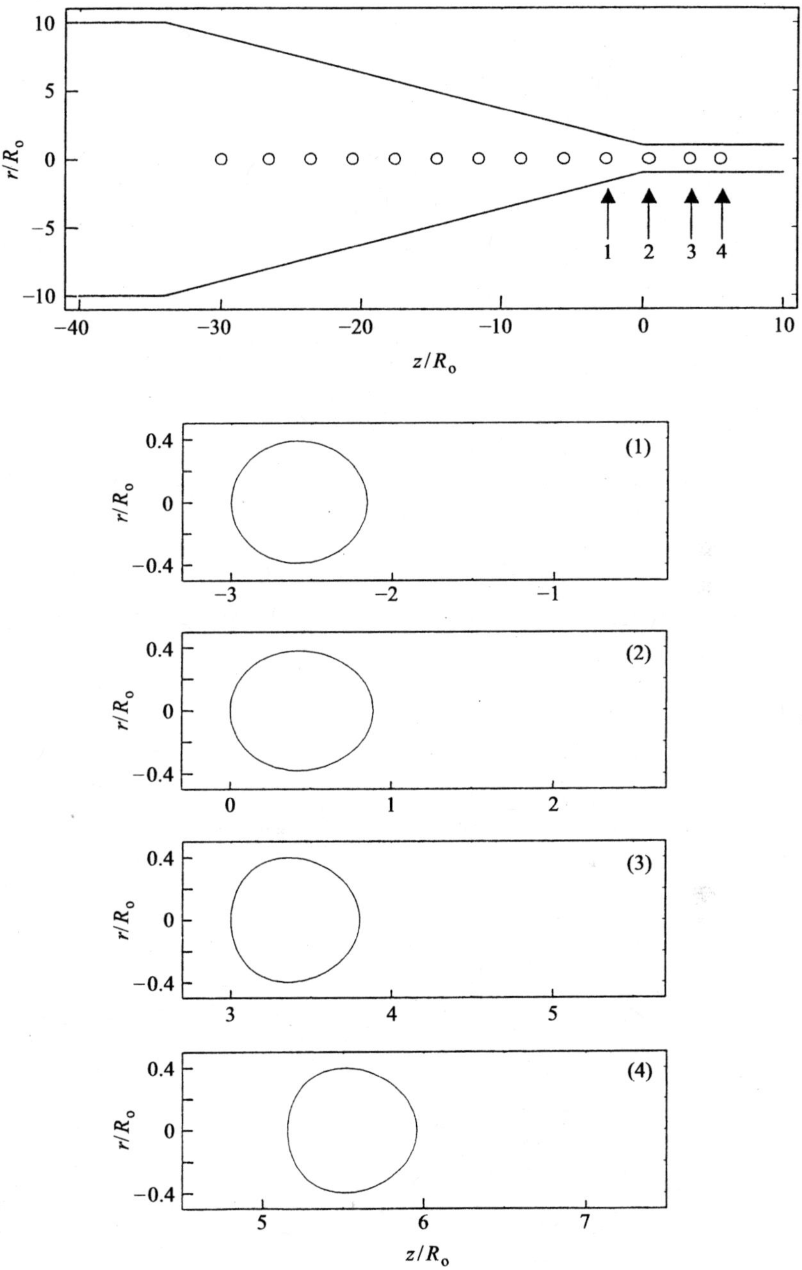

Figure 11.43 Computed drop shapes, via FEM, of a Newtonian drop with $r_o = 0.12$ cm and $\eta_d = 0.113$ Pa·s, suspended in a Newtonian liquid with $\eta_d/\eta_m = 1$, moving along the centerline of the reservoir section and cylindrical tube with $R_o = 0.3$ cm, where $\dot{\gamma} = 75.4\ \text{s}^{-1}$. For better detail, the enlarged drop shapes at four different positions in the entrance region are given in the lower panels.

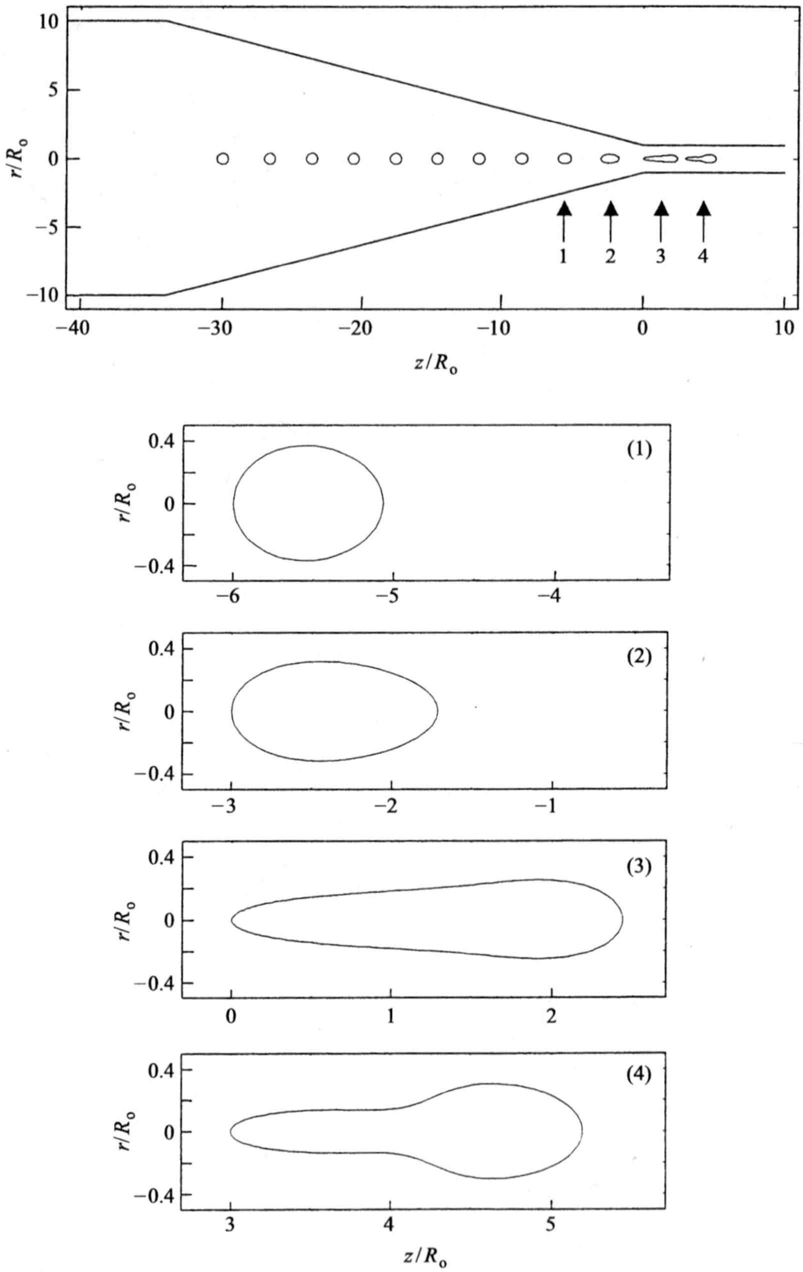

Figure 11.44 Computed drop shapes, via FEM, of a Newtonian drop with $r_o = 0.12$ cm and $\eta_d = 0.113$ Pa·s, suspended in Newtonian liquid with $\eta_d/\eta_m = 0.1$, moving along the centerline of the reservoir section and cylindrical tube with $R_o = 0.3$ cm, where $\dot{\gamma} = 75.4\ s^{-1}$. For better detail, the enlarged drop shapes at four different positions in the entrance region are given in the lower panels. (Reprinted from Kim and Han, *Journal of Rheology* 45:1279. Copyright © 2001, with permission from the Society of Rheology.)

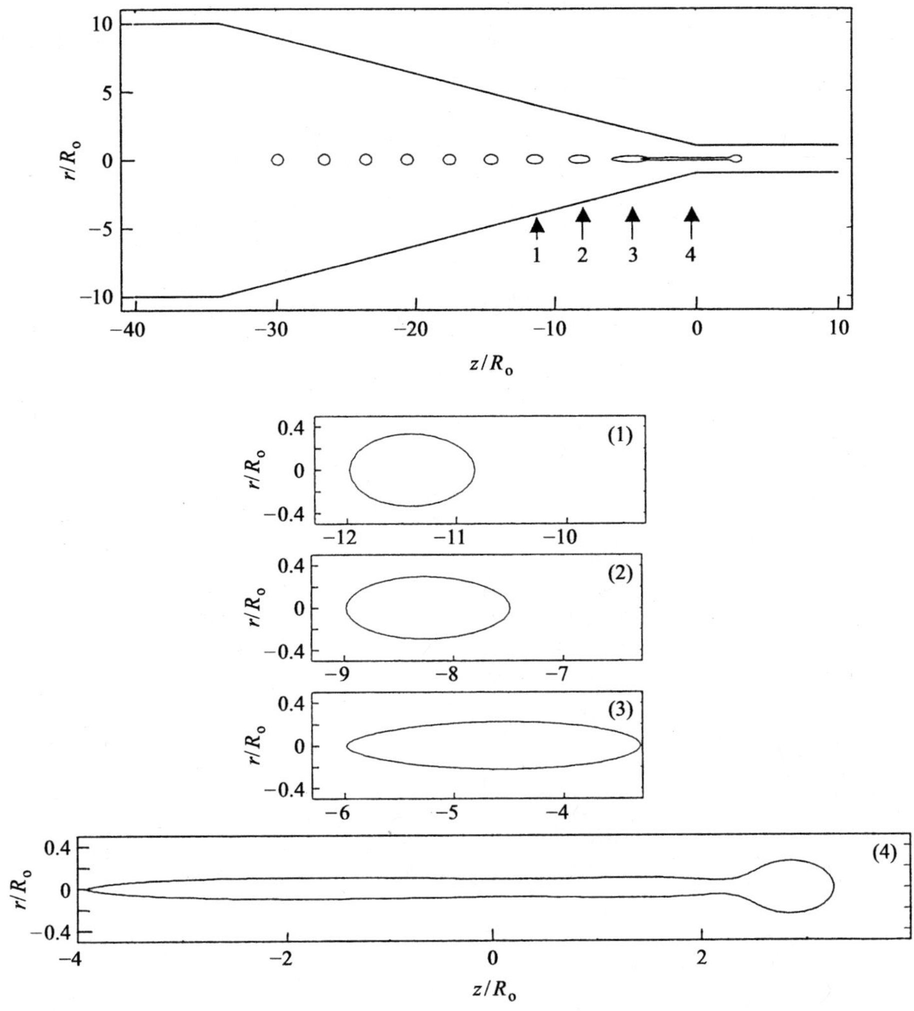

Figure 11.45 Computed drop shapes, via FEM, of a Newtonian drop with $r_o = 0.12$ cm and $\eta_d = 0.113$ Pa·s, suspended in Newtonian liquid with $\eta_d/\eta_m = 0.01$, moving along the centerline of the reservoir section and cylindrical tube with $R_o = 0.3$ cm, where $\dot{\gamma} = 75.4\ s^{-1}$. For better detail, the enlarged drop shapes at four different positions in the entrance region are given in the lower panels.

It is then clear that one can establish rheology–morphology–processing relationships in an immiscible polymer blend by computing the shapes of many drops in a given pressure-driven flow field. Successful computations predicting the shapes of all the drops in a given fluid mixture will automatically lead one to establish rheology–morphology–processing relationships in an immiscible polymer blend. Granted that the problem at hand is a very complicated and formidable task. Nevertheless, we present here a fundamental approach for dealing with the problem.

Let us confine our attention to the pressure-driven flow of a dispersed two-phase mixture through a cylindrical tube. For the purpose of illustration and simplicity, let us consider the situation as shown schematically in Figure 11.46. Let us make the following assumptions. (1) One phase consists of 13 drops in total: one drop at the center, 4 drops in the first row, and 8 drops in the second row, and the other phase forms the continuous phase (suspending medium). (2) Initially, all 13 drops have the same size and are spherical in shape in the upstream end of the reservoir section, where little deformation of drops occurs. (3) No drop breakup occurs in the entrance region. Referring to Figure 11.46, we have three regions: (1) the entrance region, (2) the drop recoil region, and (3) the fully developed region. Let us consider the three regions separately.

(a) *Drop Deformation in the Entrance Region* As the dispersed two-phase mixture flows into the tube entrance, the drop at the center elongates along the tube axis under the influence of elongational flow. At the same time, the drops away from the centerline deform under the influence of combined elongational and shear flows; note that the extent of deformation of these drops will be less as they are away from the centerline in the entrance region, as schematically shown in Figure 11.46. The prediction of the shapes of all 13 drops in the entrance region requires numerical solutions of three-dimensional FEM governing equations with appropriate interfacial boundary conditions. This can be done, in principle, by extending Eqs. (11.4) and (11.5) to three-dimensional flow of multiple drops in the entrance region. Note that the shapes of drops in the entrance region vary along the flow direction. In the preceding section, we have shown the predicted shape of a single drop moving along the centerline of a cylindrical tube, for which two-dimensional governing equations (owing to axisymmetricity) were solved numerically using FEM (Kim and Han, 2001).

(b) *Drop Deformation in the Recoil Region* Upon passing the tube entrance, as schematically shown in Figure 11.46, the drops will recoil because stresses relax and velocity rearrangement occurs in this region. For the Newtonian fluid system, drop

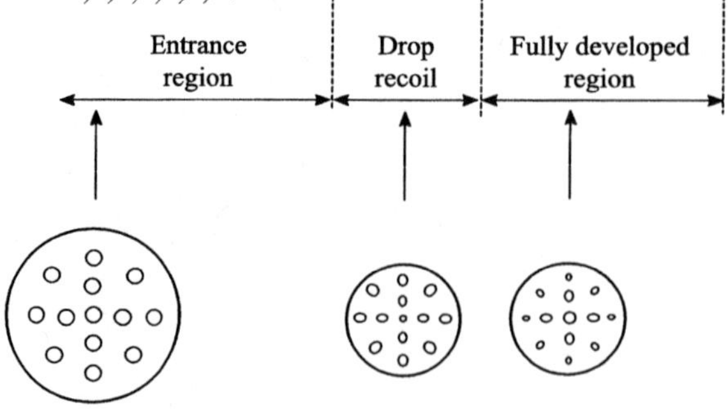

Figure 11.46 Schematic showing the flow of a dispersed two-phase system consisting of 13 drops moving through the entrance region, the drop recoil region, and the fully developed region of a cylindrical tube. The cross-sectional view in each region is also given, in which the cross-sectional areas of the 13 drops are given schematically.

recoil will occur strictly due to the interfacial tension between the drop and suspending medium. However, for the viscoelastic fluid system, the extent of drop recoil (hence the length of recoil region) will depend on, in addition to interfacial tension, fluid elasticities of both the drop phase and the suspending medium. The prediction of the shapes of all 13 drops in the recoil region requires numerical solutions of three-dimensional FEM governing equations with appropriate interfacial boundary conditions. Note that the shapes of drops in the recoil region vary along the flow direction.

(c) *Drop Deformation in the Fully Developed Region* In the fully developed region, the shapes of all 13 drops become independent of the axial position of the cylindrical tube. Thus, the computational effort necessary for predicting the shapes of drops in the fully developed region will be much less than the computational effort necessary for predicting the shapes of drops in the entrance and recoil regions, where the shapes of drops vary along the flow direction. It is important to recognize, as schematically shown in Figure 11.46, that in the fully developed region, the drops near the tube wall elongate much more than those near the center of the tube. This can be understood from the point of view of shear stress distribution inside the tube. Three-dimensional calculation of drop shape in the fully developed region of a cylindrical tube is a formidable task.

Hence, the calculation of bulk pressure gradient $(-\partial p/\partial z)_{\text{bulk}}$ in the fully developed region, via the solution of three-dimensional FEM governing equations with appropriate interfacial boundary conditions, will enable one to calculate the bulk shear stress σ_{bulk} and then bulk (apparent) shear viscosity η_{bulk} of a dispersed two-phase mixture. This can be done, in principle, by extending Eqs. (11.4) and (11.5) to three-dimensional flow of multiple drops in the fully developed region. Note that $(-\partial p/\partial z)_{\text{bulk}}$ depends on the volume fraction (ϕ) of the drop phase (i.e., the number of drops), the initial drop radius (r_{o}), the shapes of all drops (ζ_i: $i = 1, 2, \ldots, N$, with N being the total number of drops), the volumetric flow rate (via apparent shear rate, $\dot{\gamma}_{\text{app}} = 4Q/\pi D^3$ with D being the tube diameter), the viscosity of the suspending medium η_{m}, and the viscosity ratio $\eta_{\text{d}}/\eta_{\text{m}}$, with η_{d} being the viscosity of the drop phase, and interfacial tension γ.

$$(-\partial p/\partial z)_{\text{bulk}} = f(\phi, r_{\text{o}}, \zeta_i, N, \dot{\gamma}_{\text{app}}, \eta_{\text{m}}, \eta_{\text{d}}/\eta_{\text{m}}, \gamma) \tag{11.11}$$

When the calculated value of $(-\partial p/\partial z)_{\text{bulk}}$ agrees with the experimentally measured one, one can then determine blend morphology (via the calculated shapes of drops) of the dispersed two-phase mixture in the fully developed region of a cylindrical tube. At this point, one can calculate (predict) the bulk shear viscosity η_{bulk} of the dispersed two-phase mixture from

$$\eta_{\text{bulk}} = \sigma_{\text{bulk}}/\dot{\gamma}_{\text{app}} \tag{11.12}$$

in which

$$\sigma_{\text{bulk}} = D(-\partial p/\partial z)_{\text{bulk}}/4 \tag{11.13}$$

It is seen that Eq. (11.12) determines bulk viscosity–morphology–processing relationships in dispersed two-phase polymer blends because η_{bulk} depends on both processing variables (temperature and $\dot{\gamma}_{\text{app}}$ via flow rate) and σ_{bulk}, which in turn depends on the morphology (via the size and the shapes of drops) of the dispersed two-phase blend.

We have not considered the effect of fluid elasticity of the constituent components on $(-\partial p/\partial z)_{\text{bulk}}$ or the morphology (via the shape of drops) of dispensed two-phase polymer blends. The suggested analysis can be extended to viscoelastic fluids, however this will make the numerical computation of three-dimensional FEM governing equation very complicated. This subject requires much attention in the future.

11.4 Summary

In this chapter we have presented fundamental concepts associated with the rheology of immiscible polymer blends. We have presented experimental results showing that the melt viscosities of PMMA/PS blends determined with a cone-and-plate rheometer do not overlap those determined with a capillary rheometer. The experimental observation is attributed to the differences in blend morphology because the extent of drop deformation in the entrance region of a cylindrical or slit die and subsequent recoil or breakup, upon passing the entrance region of the die, can give rise to very complicated blend morphology in nonuniform shear flow (i.e., in pressure-driven flow). The interpretation of the rheological data offered on the PMMA/PS blends would equally be applicable to other multiphase polymer systems. Owing to the fact that the morphological state of an immiscible polymer blend depends strongly on the intensity of flow applied, there is no reason why one should expect that the rheological properties determined with a cone-and-plate rheometer should overlap those determined with a capillary or slit rheometer. This leads us to conclude that rheological behavior of immiscible polymer blends (also, multiphase polymer systems in general) is expected to be dependent upon the geometry of the rheometer employed. However, when the dispersed phase is grafted onto or cross-linked with the matrix phase (e.g., graft polymers and reactive polymer blends), where the morphological state is not affected by flow or temperature, the rheological properties determined with a cone-and-plate rheometer may overlap those determined with a capillary or slit rheometer.

We have presented experimental evidence showing clearly that the rheological behavior of an immiscible polymer blend is intimately related to its morphology (i.e., the state of dispersion) and thus the two are inseparable. In order to demonstrate this in a very simplistic framework, we have presented the deformation of a single drop in the entrance region of a cylindrical die. We have emphasized the importance of a better understanding of the rheological behavior of immiscible polymer blends in pressure-driven flow because such a flow field is prevalent in polymer processing operations. Although theoretical prediction of small-deformations of single drops in uniform shear flow is important in its own right, such a flow field is seldom encountered in practical situations dealing with blending two immiscible polymers, for example, when using mixing equipment or processing preblended compounds via extrusion or injection molding, to name only a few.

We have presented the results of numerical computation via FEM, demonstrating that large deformations of a single non-Newtonian drop suspended in another non-Newtonian liquid can be predicted in a complex flow field, namely, the entrance region of a cylindrical tube. Needless to say, such prediction is not possible using asymptotic analysis. We have suggested that such computational effort be extended to many drops in order to simulate situations closer to reality, namely, immiscible polymer blends and concentrated emulsions. Granted that the computational effort required for many drops would be very demanding, but it is perhaps not insurmountable. Numerical computation is the only avenue left to attack such a complex problem.

The evolution of polymer blend morphology in mixing equipment is another very complex problem (see Chapter 3 of Volume 2). Yet, this is one of the most fundamental and important problems in polymer processing operations. The capability of predicting the evolution of polymer blend morphology in mixing equipment is highly desirable. It should be mentioned that in mixing equipment, not only drop deformation but also drop breakup, drop coalescence, and drop migration would occur. Although many studies on the breakup of single drops in uniform shear flow have been reported, there have been very few studies, theoretical or experimental, on the breakup of multiple drops in pressure-driven flow. This is significant because pressure-driven flow in converging and diverging flow geometries is prevalent in many polymer processing operations. Similarly very few studies have been reported on drop coalescence in pressure-driven flow. There is no question that these are very complex and difficult problems, but only difficult problems remain to be solved.

An immiscible polymer blend can be compatibilized with a judicious choice of a compatibilizing agent. The thermodynamic principles of compatibilization of immiscible polymer blends and the rationale behind the choice of effective compatibilizer are presented in Chapter 4 of Volume 2.

Problems

Problem 11.1

In Figure 11.24 we observe that the d_j/D ratio of dispersed two-phase blends goes through a maximum at a certain critical blend ratio, and that measured values of d_j/D ratio for blends are higher than those estimated from the linear additive rule with respect to blend composition. What might be a physical origin(s) of such experimental observations?

Problem 11.2

In Figure 11.30 we observe that the viscosities of the 70/30 nylon 6/PET blend lie between those of neat nylon 6 and PET at $\dot{\gamma} < 0.5\ s^{-1}$, while the viscosities of the same blend are lower than those of neat nylon 6 and PET at $\dot{\gamma} > 0.5\ s^{-1}$. However, at $\dot{\gamma} > 0.5\ s^{-1}$ the viscosities of both 70/30 and 30/70 nylon 6/PET blends lie below those of neat nylon 6 and PET. You may assume that the minor component in each blend forms the discrete phase (drops) and the major component forms the continuous phase. Explain why the viscosities of the 70/30 nylon 6/PET blend lie between those of neat nylon 6 and PET at $\dot{\gamma} < 0.5\ s^{-1}$ and below those of neat nylon 6 and PET at $\dot{\gamma} > 0.5\ s^{-1}$. Explain why the viscosities of the 30/70 nylon 6/PET blend lie below those of neat nylon and PET over the entire range of shear rates tested.

Notes

1. For convenience, we use the notation η interchangeably with the notation η_{app}.
2. Figure 11.29 shows that the slope of log η_0 versus $1/T$ plot is almost the same for both PS and PMMA, meaning that the viscous flow activation energy E in the Arrhenius expression, $\eta_0 \propto \exp(E/RT)$, is almost the same for both PS and PMMA. Thus, the viscosity ratio, $\eta_{0,PS}/\eta_{0,PMMA}$, is expected to be virtually independent of temperature.

References

Ablazova TI, Tsenbrenko MV, Yudin AV, Vinogradov GV, Yarlykov BV (1975). *J. Appl. Polym. Sci.* 19:1781.

Acrivos A (1983). *Ann. New York Acad. Sci.* 404:1.

Acrivos A, Lo TS (1978). *J. Fluid Mech.* 86:641.

Ambravaneswaran B, Wilkes ED, Sasaran OS (2002). *Phys. Fluids* 14:2606.

Barthès-Biesel D, Acrivos A (1973a). *Int. J. Multiphase Flow* 1:1.

Barthès-Biesel D, Acrivos A (1973b). *J. Fluid Mech.* 61:1.

Bartok W, Mason SG (1959). *J. Colloid Sci.* 14:13.

Bentley BJ, Leal LG (1986). *J. Fluid Mech.* 167:241.

Bozzi LA, Feng JQ, Scott TC, Pearlstein AJ (1997). *J. Fluid Mech.* 336:1.

Brizitsky VI, Vinogradov GV, Isayev AI, Podolsky YY (1978). *J. Appl. Polym. Sci.* 22:751.

Buckmaster J (1973). *J. Appl. Mech.* E40:18.

Carley JF (1985). *Polym. Eng. Sci.* 25:1017.

Carley JF, Crossan SC (1981). *Polym. Eng. Sci.* 21:249.

Chaffey CE, Brenner H, Mason SG (1965). *Rheol. Acta* 4:56.

Chaffey CE, Brenner H (1967). *J. Colloid Interface. Sci.* 24:258.

Chin HB, Han CD (1979). *J. Rheol.* 23:557.

Chin HB, Han CD (1980). *J. Rheol.* 24:1.

Choi SJ, Schowalter WR (1975). *Phys. Fluids* 18:420.

Cooper SL, Estes GM (eds) (1979). *Multiphase Polymers*, Adv. Chem. Series, no 176, American Chemical Society, Washington, DC.

Coulliette C, Pozrikidis C (1998). *J. Fluid Mech.* 358:1.

Cox RG (1969). *J. Fluid Mech.* 37:601.

Cox WP, Merz EH (1958). *J. Polym. Sci.* 28:619.

Delaby I, Germain Y, Ernst B, Muller R (1994). *J. Rheol.* 38:1705.

Flumerfelt RW (1972). *Ind. Eng. Fundam.* 11:312.

Flumerfelt RW (1980). *J. Colloid Interface Sci.* 76:330.

Frankel NA, Acrivos A (1970). *J. Fluid Mech.* 44:65.

Fröhlich H, Sack R, Wills HH (1946). *Proc. Roy. Soc.* A185:415.

Goddard JD, Miller C (1967). *J. Fluid Mech.* 28:657.

Goren SL (1962). *J. Fluid Mech.* 12:309.

Goren SL (1964). *J. Colloid Sci.* 19:81.

Grace HP (1982). *Chem. Eng. Commun.* 14:225.

Guido S, Villone M (1998). *J. Rheol.* 42:395.

Han CD (1971). *J. Appl. Polym. Sci.* 15:2579.

Han CD (1974). *J. Appl. Polym. Sci.* 18:481.

Han CD (1981). *Multiphase Flow in Polymer Processing*, Academic Press, New York, Chaps 4 and 5.

Han CD (ed) (1984). *Polymer Blends and Composites in Multiphase Systems*, Adv. Chem. Series, no 206, American Chemical Society, Washington, DC.
Han CD, Funatsu K (1978). *J. Rheol.* 22:113.
Han CD, Kim YW (1975). *Trans. Soc. Rheol.* 19:245.
Han CD, King RG (1980). *J. Rheol.* 24:213.
Han CD, Yu TC (1971a). *AIChE J.* 17:1512.
Han CD, Yu TC (1971b). *J. Appl. Polym. Sci.* 15:1163.
Han CD, Yu TC (1972). *Polym. Eng. Sci.* 12:81.
Han CD, Kim KU, Parker J, Siskovic N, Huang CR (1973). *Appl. Polym. Sci. Symp.* 20:191.
Han CD, Kim YW, Chen SJ (1975). *J. Appl. Polym. Sci.* 19:2831.
Han JH, Choi-Feng C, Li DJ, Han CD (1995). *Polymer* 36:2451.
Heitmiller RF, Naar RZ, Zabusky HH (1964). *J. Appl. Polym. Sci.* 8:873.
Hinch EJ, Acrivos A (1979). *J. Fluid Mech.* 91:401.
Hinch EJ, Acrivos A (1980). *J. Fluid Mech.* 98:305.
Hyman WA, Skalak R (1972). *AIChE J.* 18:149.
Jackson NE, Tucker CL (2003). *J. Rheol.* 47:659.
Janssen JJM, Boon A, Agterof WGM (1994). *AIChE J.* 40:1929.
Karam HJ, Bellinger JC (1968). *Ind. Eng. Chem. Fundam.* 7:576.
Kennedy MR, Pozrikidis C, Skalak R (1994). *Comput. Fluids* 23:251.
Kim YW, Han CD (1976). *J. Appl. Polym. Sci.* 20:2905.
Kim SJ, Han CD (2001). *J. Rheol.* 45:1279.
Lee JK, Han CD (1999). *Polymer* 40:6277.
Lee JK, Han CD (2000). *Polymer* 41:1799.
Levitt L, Macosko CW (1996). *Polym. Eng. Sci.* 36:1647.
Lin CC (1979). *Polym. J.* 11:185.
Loewenberg M, Hinch EJ (1996). *J. Fluid Mech.* 321:395.
Lyngaae-Jørgensen J, Andersen PE, Alle N (1983). In *Polymer Alloys III*, Klempner D, Frisch KC (eds). Plenum Press, New York, p 105.
Martinez MI, Udell KS (1990). *J. Fluid Mech.* 210:565.
Mighir F, Aiji A, Carreau PJ (1997). *J. Rheol.* 41:1183.
Mikami T, Cox RG, Mason SG (1978). *Int. J. Multiphase Flow* 2:113.
Milliken WJ, Leal LG (1991). *J. Non-Newtonian. Fluid Mech.* 40:355.
Olbright WL, Kung DM (1992). *Phys. Fluids* A4:1347.
Olbright WL, Leal LG (1983). *J. Fluid Mech.* 134:329.
Oldroyd JG (1953). *Proc. Roy. Soc.* A218:122.
Paul DR, Newman S (eds) (1978). *Polymer Blends*, Academic Press, New York.
Phillips WJ, Graves RW, Flumerfelt RW (1980). *J. Colloid Interface Sci.* 76:350.
Platzer NAJ (ed) (1971). *Multicomponent Polymer Systems*, Adv. Chem. Series, no 99, American Chemical Society, Washington, DC.
Platzer NAJ (ed) (1975). *Copolymers, Polyblends, and Composites*, Adv. Chem. Series, no 142, American Chemical Society, Washington, DC.
Rallison JM (1984). *Ann. Rev. Fluid Mech.* 16:45.
Roscoe R (1967). *J. Fluid Mech.* 28:273.
Rumscheidt FD, Mason SG (1961). *J. Colloid Sci.* 16:238.
Rumscheidt FD, Mason SG (1962). *J. Colloid Sci.* 17:260.
Schowalter WR, Chaffey CE, Brenner H (1968). *J. Fluid Mech.* 26:152.
Shih CK (1976). *Polym. Eng. Sci.* 16:742.
Sperling LH (ed) (1974). *Recent Advances in Polymer Blends, Grafts, and Blocks*, Plenum Press, New York.
Stone HA, Leal LG (1989). *J. Fluid Mech.* 198:399.
Stone HA, Leal LG (1990a). *J. Fluid Mech.* 211:123.

Stone HA, Leal LG (1990b). *J. Fluid Mech.* 220:161.

Stroeve P, Varanasi PP (1984). *J. Colloid Interface Sci.* 99:360.

Suzuki K, Watanabe T, Ono S (1970). *Proc. Fifth Int. Cong. Rheol.*, Vol 2, University Park Press, Baltimore, Maryland, p 339.

Tavgac T (1972). *Drop Deformation and Breakup in Simple Shear Fields*, Doctoral Dissertation, University of Houston, Houston, Texas.

Taylor GI (1934). *Proc. Roy. Soc.* A146:501.

Tomotika S (1935). *Proc. Roy. Soc. (London)* A150:322.

Tomotika S (1936). *Proc. Roy. Soc. (London)* A153:302.

Torza S, Cox RG, Mason SG (1972). *J. Colloid Interface Sci.* 38:395.

Tsebrenko MV, Jakob M, Kuchinka MY, Yudin AV, Vinogrdov GA (1974). *Intern J. Polymeric Mater.* 3:99.

Tsebrenko MV, Yudin AV, Ablazova TI, Vinogradov GV (1976). *Polymer* 17:831.

Turner BM, Chaffey CE (1969). *Trans. Soc. Rheol.* 13:411.

Utracki LA (1983). *Polym. Eng. Sci.* 23:602.

Utracki LA (1990). *Polymer Alloys and Blends: Thermodynamics and Rheology*, Hanser, New York.

Utracki LA, Kamal MR (1982). *Polym. Eng. Sci.* 22:96.

Utracki LA, Catani AM, Bata GL (1982). *J. Appl. Polym. Sci.* 27:1913.

Vadas EB, Goldsmith HL, Mason SG (1976). *Trans. Soc. Rheol.* 20:373.

van Oene H (1972). *J. Colloid Interface Sci.* 40:448.

van Oene H (1978). In *Polymer Blends*, Vol 1, Paul DR, Newman S (eds), Academic Press, New York, Chap 7.

Varanasi PP, Ryan ME, Stroeve P (1994). *Ind. Eng. Chem. Res.* 33:1858.

Vinogradov GV, Yarlykov BV, Tsenbrenko MV, Yudin A, Ablazova TI (1975). *Polymer* 16:609.

Zhou H, Pozrikidis C (1994). *Phys. Fluids* 6:80.

Zinchenko AZ, Rother MA, Davis RH (1997). *Phys. Fluids* 9:1493.

12

Rheology of Particulate-Filled Polymers, Nanocomposites, and Fiber-Reinforced Thermoplastic Composites

12.1 Introduction

Polymer composites consisting of a thermoplastic polymer forming the matrix phase and a large amount of inorganic particles (commonly referred to as fillers) or glass fibers, which are often referred to as particulate-filled polymers, are very common in the plastics and elastomer's industries (Deanin and Schott 1974; Kraus 1965; Lubin 1969). Polymer composites are developed to achieve a set of properties not possessed by the thermoplastic polymer (i.e., polymeric matrix) alone. Polymeric matrices can be thermoplastics, which soften and behave as viscous liquids when heated to above their glass transition temperatures (in the case of amorphous thermoplastic polymers) or above their melting temperatures (in the case of semicrystalline thermoplastic polymers). Polymeric matrices can also be thermosets, which undergo a transformation from a viscous resinous liquid to a hard or rubbery solid in the presence of heat and/or curing agents. There are numerous industrial products made of particulate-filled polymeric materials; for example, thermoplastic polymers filled with mica or calcium carbonate, carbon-black-filled elastomers, thermoplastic polymers or thermosets reinforced with glass fibers or carbon fibers.

The ultimate goal of adding fillers to a thermoplastic polymer and adding glass fiber or carbon fiber to a thermoset is to improve the mechanical properties of the polymer. However, fillers, glass fibers, or carbon fibers themselves usually supply little or no reinforcement since there is little interfacial interaction between a thermoplastic polymer and fillers, and between a thermoset and glass fiber or carbon fiber. This has led to the development of "coupling agents," chemical additives capable of improving the interfacial bonds between a thermoplastic polymer and fillers, and between a thermoset and glass fibers or carbon fibers (Plueddemann 1982). The use of coupling agents for the surface modification of fillers to reinforce thermoplastics has generally been directed towards improving the mechanical strength and chemical resistance of composites by improving adhesion across the interface.

When inorganic fillers or glass fibers are added to a thermoplastic polymer, the resulting material exhibits a complex rheological behavior, quite different from the rheology of neat homopolymers presented in Chapter 6. From the rheological point of view, particulate-filled thermoplastic polymers may be regarded as concentrated suspensions. However, what distinguishes particulate-filled molten thermoplastic polymers from "conventional" concentrated suspensions lies in that the matrix phase of particulate-filled thermoplastic polymers usually exhibits non-Newtonian, viscoelastic behavior while the matrix phase of conventional concentrated suspensions is usually Newtonian fluids. It is not difficult to surmise that the rheology of particulate-filled thermoplastic polymers would be much more complicated than that of conventional concentrated suspensions. It should be mentioned that as far back as the 1930s through the 1970s, numerous research groups investigated the rheology of conventional concentrated suspensions, but there are too many papers (hundreds) to cite them all here.

Today, it is well established that, in general, the addition of fillers to a molten polymer increases the melt viscosity and decreases the melt elasticity. The volume fraction of the particles as well as other factors, such as the shape of the particles, the particle size and size distribution, and the state of dispersion of the particles (well-dispersed state or agglomerated state), are known to influence the *bulk* rheological properties of particulate-filled polymer melts.

In this chapter, we first present the bulk rheological behavior of particulate-filled molten thermoplastics, carbon-black-filled elastomers, and the effect of chemical treatment of fillers on the bulk rheological properties of particulate-filled molten thermoplastics. In so doing, we point out the limitations of some conventional experimental techniques for measuring the bulk rheological properties of particulate-filled polymer systems. Then, we present theoretical consideration of the yield behavior of particulate-filled polymer systems, followed by the bulk rheological properties of thermoplastic polymer/organoclay nanocomposites and block copolymer/organoclay nanocomposites. We distinguish the nanocomposites from particulate-filled polymers in that the sizes of particles employed to prepare nanocomposites are very small (e.g., 1–10 nm) as compared with those (e.g., 1–10 μm) in particulate-filled polymers, and the volume fractions of particles in a nanocomposite are very low (e.g., 3–7 wt %) as compared with those (up to 50–60 wt %) in particulate-filled polymers. Finally, we present the bulk rheological properties of some fiber-reinforced thermoplastic composites. In this chapter, emphasis is placed on the fundamental concepts associated with the rheology of particulate-filled molten thermoplastics, organoclay nanocomposites, and fiber-reinforced thermoplastic composites.

12.2 Rheology of Particulate-Filled Polymers

From the point of view of polymer processing, the bulk rheological properties of particulate-filled polymers are of great interest. In particulate-filled polymer systems that include molten thermoplastics and elastomers, fine particle size and high surface area generally favor reinforcement of the polymer. Over the past three decades, numerous research groups have investigated experimentally the rheological behavior of particulate-filled polymer systems. Some investigators (Agarwal et al. 1978; Dreval and Borisenkova 1993; Faulkner and Schmidt 1977; Han 1974; Lin and Masuda 1990; Lobe and White 1979; Minagawa and White 1976; Rong and Chaffey 1988a; Suetsugu and White 1983; Tanaka and White 1980; White and Tanaka 1981) reported on the steady-state shear flow properties, and others (Fujima and Kwasaki 1989; Lin and Masuda 1989; Rong and Chaffey 1988b) reported on the oscillatory shear flow properties of particulate-filled thermoplastics. The rheological behavior of carbon-black-filled elastomers was also reported by several research groups (Araki and White 1988; Ertong et al, 1994; Lakdawala and Salovey 1987a, 1987b; Markovic et al. 2000; Montes et al. 1985; Vinogradov et al. 1972; White and Crowder 1974).

In the measurement of bulk rheological properties of particulate-filled polymers, one needs to be very cautious about the choice of experimental techniques. Specifically, when the sizes of fillers are sufficiently large and the concentration of fillers exceeds a certain critical value, the use of a cone-and-plate rheometer, in which the gap opening (the distance between the tip of the cone and the flat plate) is usually set to be about 50 μm, would pose a difficulty for obtaining reproducible results. When a plunger-type capillary rheometer is used, a particulate-filled molten polymer (or a concentrated suspension) can easily jam the die entrance, giving rise to exceedingly large entrance pressure drops (thus exceedingly large entrance corrections). Under such circumstances, use of the Bagley end correction, discussed in Chapter 5 for homopolymers, would not provide meaningful rheological information. Because this subject is so fundamental, in this section we address this issue as we present experimental studies reported in the literature.

Since very few rigorous theories are available in the literature to enable one to predict the bulk rheological properties of particulate-filled viscoelastic polymeric fluids, in this section we present primarily the experimental observations reported in the literature on the rheological behavior of particulate-filled thermoplastic polymers and elastomers.

12.2.1 Rheology of Particulate-Filled Molten Thermoplastics and Elastomers

12.2.1.1 Steady-State Shear Flow Properties

Figure 12.1 gives the axial distribution of wall normal stress $\left(T_{yy}(b, z)\right)$ for polypropylene (PP) filled with 10 wt % calcium carbonate ($CaCO_3$) flowing through a long slit die at 200 °C for two different shear rates. It is seen in Figure 12.1 that at a given shear rate, the $T_{yy}(b, z)$ profile is linear over the distances where $T_{yy}(b, z)$ was measured, satisfying the necessary condition for fully developed flow (see Chapter 5 for the principles of slit rheometry). From the constant slope of the axial distribution of $T_{yy}(b, z)$ in Figure 12.1, one can calculate shear stress (σ) (see Eq. (5.68)) and thus "apparent" shear viscosity[1] of the $CaCO_3$-filled PP composite at 200 °C. Figure 12.2 gives logarithmic plots of steady-state shear viscosity (η) versus shear rate ($\dot{\gamma}$) for $CaCO_3$-filled PP composites with varying amounts of $CaCO_3$ at 200 °C. For comparison, also given in Figure 12.2 is a plot of $\log \eta$ versus $\log \dot{\gamma}$ for neat PP. It can be

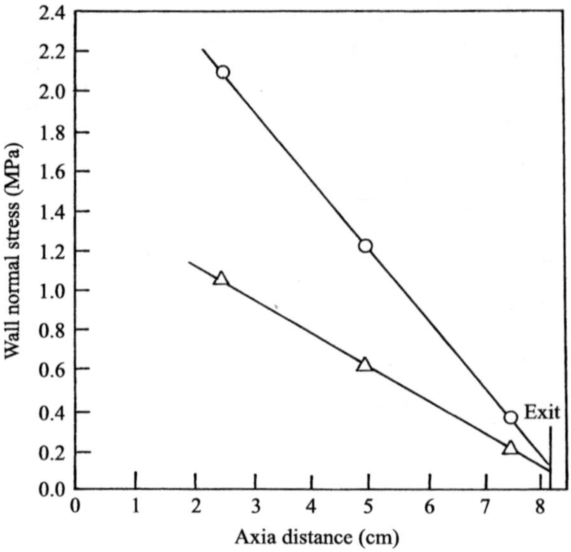

Figure 12.1 Axial distribution of wall normal stress distribution in a slit die for PP composite filled with 10 wt % $CaCO_3$ at 200 °C for two different shear rates (s^{-1}): (○) 49.5 and (△) 9.6. The dimensions of the die employed are: width (w) of 2.54 cm, height (h) of 0.254 cm, length (L) of 8.33 cm. (Reprinted from Han, *Journal of Applied Polymer Science* 18:821. Copyright © 1974, with permission from John Wiley & Sons.)

seen from Figure 12.2 that values of η of $CaCO_3$-filled PP composite increase with increasing concentration of $CaCO_3$.

Figure 12.3 gives $\log \eta$ versus $\log \sigma$ plots for $CaCO_3$-filled PP composites with varying amounts of $CaCO_3$ at 200 °C. Comparison of Figure 12.3 with Figure 12.2 shows that the dependence of η on σ appears much stronger than the dependence of η on $\dot{\gamma}$, and the range of σ variation is much smaller than that of $\dot{\gamma}$ for a given range of

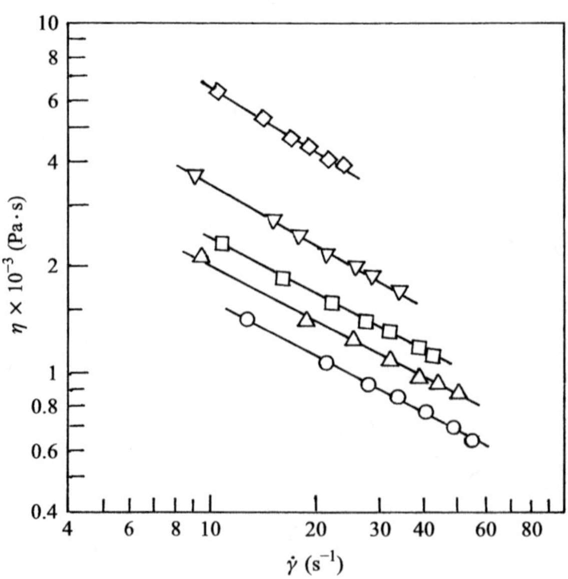

Figure 12.2 Plots of $\log \eta$ versus $\log \dot{\gamma}$ at 200 °C for neat PP (○) and PP composites filled with varying amounts of $CaCO_3$ (wt %): (△) 10, (□) 20, (▽) 40, and (◇) 70. The data were obtained using a slit die with the dimensions given in Figure 12.1. (Reprinted from Han, *Journal of Applied Polymer Science* 18:821. Copyright © 1974, with permission from John Wiley & Sons.)

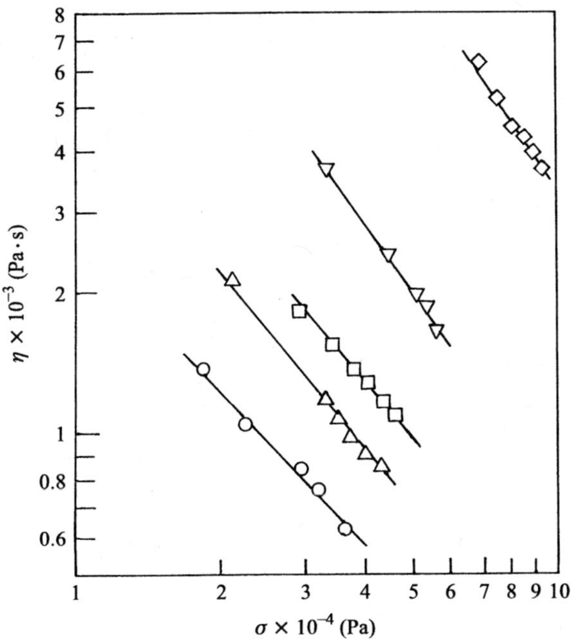

Figure 12.3 Plots of log η versus log σ at 200 °C for neat PP (○) and $CaCO_3$-filled PP composites with varying amounts of $CaCO_3$ (wt %): (△) 10, (□) 20, (▽) 40, and (◇) 70. The data were obtained using a slit die with the dimensions given in Figure 12.1.

volumetric flow rates employed in the experiment. Again, in Figure 12.3 we observe that η increases with increasing concentration of $CaCO_3$.

Figure 12.4 gives log N_1 versus log $\dot{\gamma}$ plots for $CaCO_3$-filled PP composites with varying amounts of $CaCO_3$ at 200 °C, where N_1 denotes steady-state first normal stress difference that was determined from the exit pressure method described in Chapter 5. Note that the experimental data used to calculate values of N_1 given in Figure 12.4

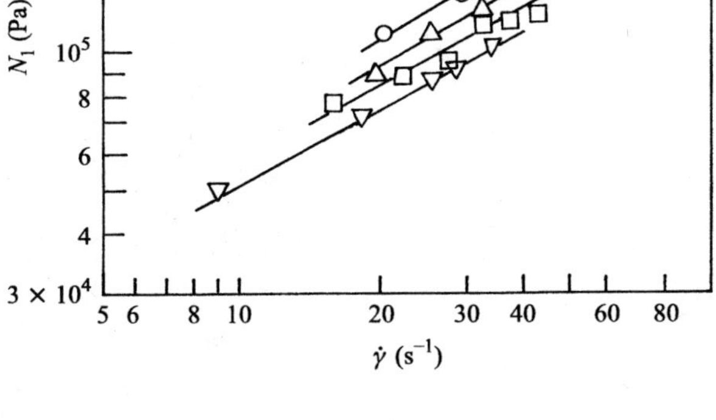

Figure 12.4 Plots of log N_1 versus log $\dot{\gamma}$ at 200 °C for neat PP (○) and $CaCO_3$-filled PP composites with varying amounts of $CaCO_3$ (wt %): (△) 10, (□) 20, and (▽) 40. The exit pressure method described in Chapter 5 was used to determine values of N_1 from a slit die with the dimensions given in Figure 12.1. (Reprinted from Han, *Journal of Applied Polymer Science* 18:821. Copyright © 1974, with permission from John Wiley & Sons.)

are the same as those used to calculate values of η presented in Figures 12.2 and 12.3, because wall normal stress measurements in the fully developed region of a slit die allow one to obtain both pressure gradient and exit pressure, which then enable one to calculate both η and N_1 (see Chapter 5). It is seen in Figure 12.4 that values of N_1 for $CaCO_3$-filled PP composites decrease with increasing concentration of $CaCO_3$. This can be explained by the fact that the addition of $CaCO_3$ to neat PP has diluted the fluid elasticity of the PP.

Figure 12.5 gives log N_1 versus log σ plots for $CaCO_3$-filled PP composites with varying amounts of $CaCO_3$ at 200 °C, showing that N_1 decreases with increasing concentration of $CaCO_3$. The rheological properties of particulate-filled molten polymers depend, among many factors, on the surface characteristics (e.g., the shape) of the particulates, particle size (and thus surface area per unit volume of the particulate), and interactions between particulates and polymer matrix.

Figure 12.6 gives log η versus log $\dot{\gamma}$ plots for carbon-black-filled polystyrene (PS) composites with varying amounts of carbon black at 170 °C from data obtained using a cone-and-plate rheometer and using capillary dies with end correction.[2] It is seen in Figure 12.6 that η of carbon-black-filled PS composites increases rapidly as $\dot{\gamma}$ approaches 0.01 s^{-1}, especially for the PS filled with 20 or 25 wt % carbon black. This observation becomes very dramatic, as can be seen in Figure 12.7, when η is plotted against σ in logarithmic coordinates. Such observation is referred to as "yield behavior," characteristic of particulate-filled polymer systems. The physical origin

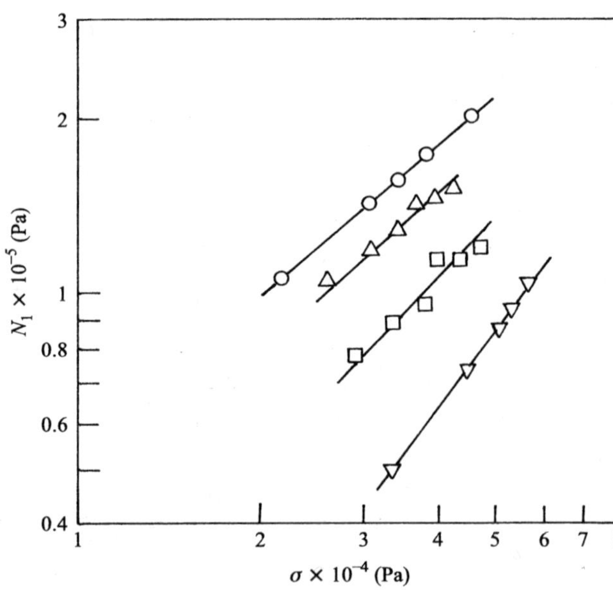

Figure 12.5 Plots of log N_1 versus log σ at 200°C for neat PP (○) and $CaCO_3$-filled PP composites with varying amounts of $CaCO_3$ (wt %): (△) 10, (□) 20, and (▽) 40. The exit pressure method described in Chapter 5 was used to determine values of N_1 from a slit die with the dimensions given in Figure 12.1.

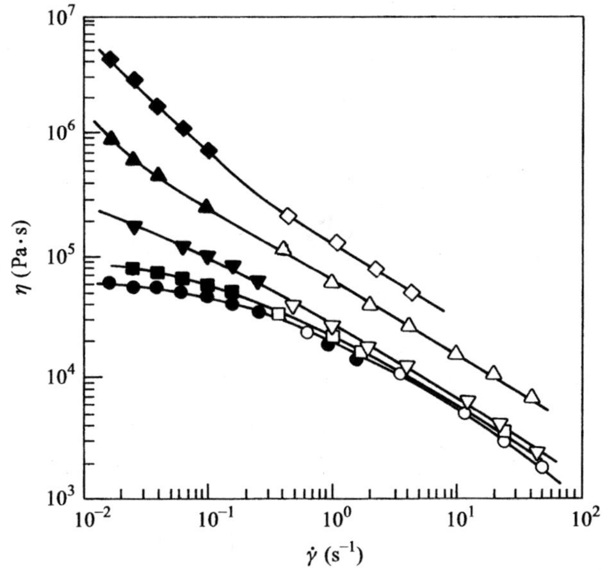

Figure 12.6 Plots of log η versus log $\dot{\gamma}$ at 170 °C for neat PS (○, ●) and carbon-black-filled PS composites with varying amounts of carbon black (wt %): (□, ■) 5, (▽,▼) 10, (△, ▲) 20, and (◇, ◆) 25, where filled symbols represent data obtained using a cone-and-plate rheometer, and open symbols represent data obtained using an Instron capillary rheometer with end correction. (Reprinted from Lobe and White, *Polymer Engineering and Science* 19:617. Copyright © 1979, with permission from the Society of Plastics Engineers.)

of yield behavior in highly-filled polymer systems is believed to lie in the presence of aggregates consisting of many small particles at very low σ, but yield behavior begins to disappear as the applied σ is increased sufficiently to separate small particles from the large aggregates. There are other ways of separating small particles from large aggregates, which will be presented below when we discuss the rheology of nanocomposites.

Figure 12.8 gives log N_1 versus log $\dot{\gamma}$ plots for carbon-black-filled PS composites with varying amounts of carbon black at 180 °C, showing that values of N_1 increase with increasing concentration of carbon black. It has been reported that carbon black has a rather complex morphology which varies with the manufacturing process, and that some carbon blacks interact with certain polymers, especially with elastomers (Medalia 1970). Note that the values of N_1 in Figure 12.8 were obtained using a cone-and-plate rheometer. The dependence of N_1 on the concentration of carbon black given in Figure 12.8 is at variance with Figure 12.4 for $CaCO_3$-filled PP composites. However, interestingly, as can be seen in Figure 12.9, the dependence of N_1 on the concentration of carbon black is reversed when N_1 is plotted against σ in logarithmic coordinates; that is, the N_1 for the carbon-black-filled PS composites decreases with increasing concentration of carbon black, in agreement with the observation made in Figure 12.5 for $CaCO_3$-filled PP composites. We thus conclude that log N_1 versus log σ plots, instead of log N_1 versus log $\dot{\gamma}$ plots, must be used to determine the effect of filler concentration on fluid elasticity in particulate-filled molten thermoplastics.

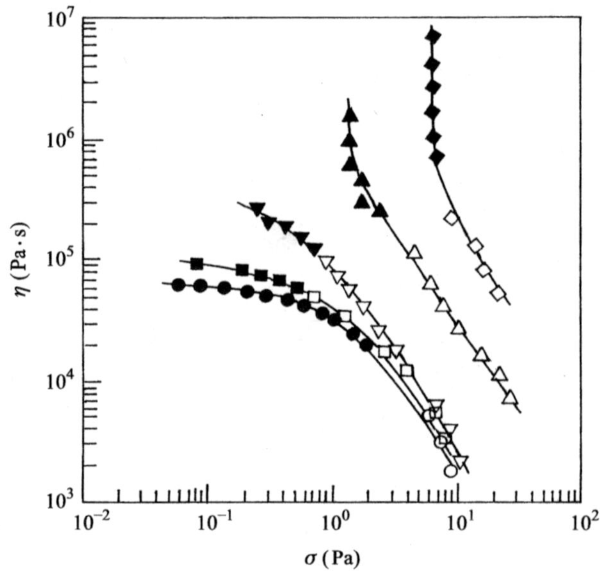

Figure 12.7 Plots of log η versus log σ at 170 °C for neat PS (○, ●) and carbon-black-filled PS composites with varying amounts of carbon black (wt %): (□, ■) 5, (▽,▼) 10, (△, ▲) 20, and (◇, ◆) 25, where filled symbols represent data obtained using a cone-and-plate rheometer, and open symbols represent data obtained using an Instron capillary rheometer with end correction. (Reprinted from Lobe and White, *Polymer Engineering and Science* 19:617. Copyright © 1979, with permission from the Society of Plastics Engineers.)

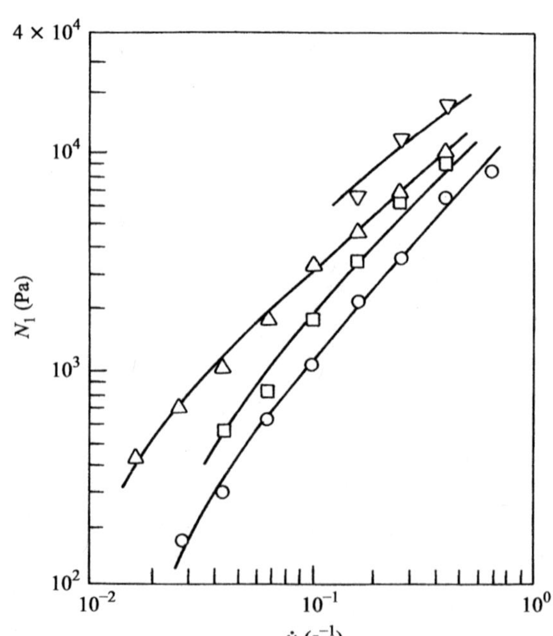

Figure 12.8 Plots of log N_1 versus log $\dot{\gamma}$ at 180 °C for neat PS (○) and carbon-black-filled PS composites with varying amounts of carbon black (wt %): (□) 10, (△) 20, and (▽) 30. The data were obtained using a cone-and-plate rheometer. (Reprinted from Tanaka and White, *Polymer Engineering and Science* 20:949. Copyright © 1980, with permission from the Society of Plastics Engineers.)

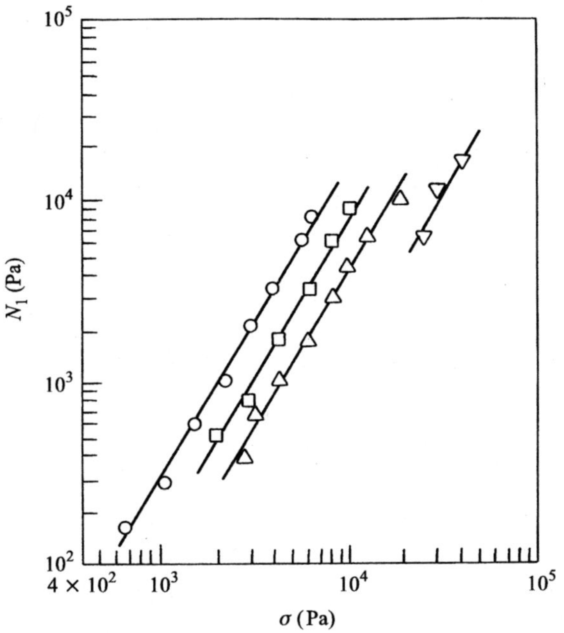

Figure 12.9 Plots of log N_1 versus log σ at 180 °C for neat PS (○) and carbon-black-filled PS composites with varying amounts of carbon black (wt %): (□) 10, (△) 20, and (▽) 30. The data were obtained using a cone-and-plate rheometer. (Reprinted from Tanaka and White, *Polymer Engineering and Science* 20:949. Copyright © 1980, with permission from the Society of Plastics Engineers.)

In Chapter 6 we have discussed the practical significance of log N_1 versus log σ plots to investigate the fluid elasticity of flexible homopolymers in the molten state.

Figure 12.10 gives log η versus log σ plots for carbon-black-filled Guayule rubber[3] with varying amounts of carbon black, in which values of η were obtained using different rheological instruments. It is seen that values of η determined using three different instruments agree well, indicating that wall effects in the use of a cone-and-plate rheometer and any bridging in the entrance region of the capillary die might have been minimal. As we have discussed, this observation made from Figure 12.10 is very similar to that made for the carbon-black-filled molten thermoplastics from Figure 12.7. In the tire and rubber compounding industry, carbon black is an essential element for processing for two reasons. One reason is that rubber is so elastic that it gives rise to melt fracture[4] at relatively low extrusion rates, which is detrimental to the control of the product quality and to obtaining reasonably high extrusion rates. The addition of carbon black to an elastomer decreases the fluid elasticity of the elastomer (see Figure 12.9) during extrusion, enabling one to maintain high extrusion rates. The other reason is that the addition of carbon black to an elastomer dramatically increases the integrity of the elastomer and gives rise to strong resistance to wear and abrasion.

12.2.1.2 Oscillatory Shear Flow Properties

When presenting the rheological behavior of homopolymers in Chapters 3–6, we compared (1) log G' versus log ω plots obtained from oscillatory shear flow experiments with log N_1 versus log $\dot{\gamma}$ plots obtained from steady-state shear flow experiments,

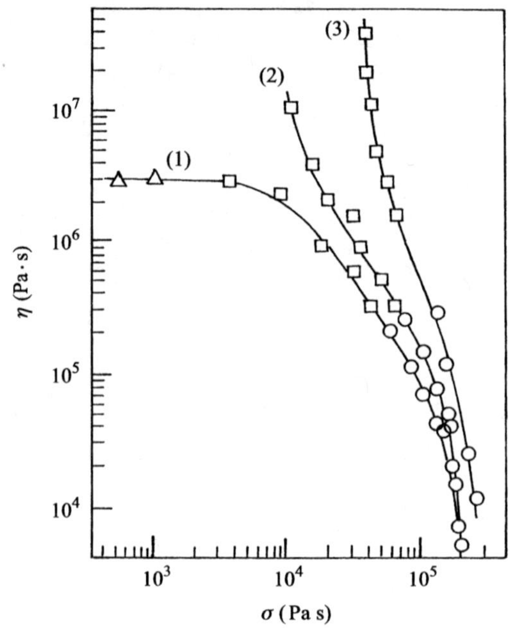

Figure 12.10 Plots of log η versus log σ for (1) Guayule rubber (GR), (2) GR filled with 0.1 volume fraction of carbon black, and (3) GR filled with 0.2 volume fraction of carbon black, where the symbol ○ represents data obtained using an Instron capillary rheometer with end correction, the symbol △ represents data obtained using creep measurement, and the symbol □ represents data obtained using a sandwich rheometer. (Reprinted from Montes et al., *Journal of Polymer Engineering* 5(3):209. Copyright © 1985, with permission from Freund Publishing House.)

(2) log G' versus log G'' plots obtained from oscillatory shear flow experiments with log N_1 versus log σ plots obtained from steady-state shear flow experiments, and (3) log $|\eta^*|$ versus log ω plots obtained from oscillatory shear flow experiments with log η versus log $\dot{\gamma}$ plots obtained from steady-state shear flow experiments. Below, we present oscillatory shear flow properties of $CaCO_3$-filled PP composites and then compare them with the steady-state shear-flow properties we have presented.

Figure 12.11 gives log G' versus log ω and log G'' versus log ω plots and Figure 12.12 gives log $|\eta^*|$ versus log ω plots for $CaCO_3$-filled PP composites at 190 °C.

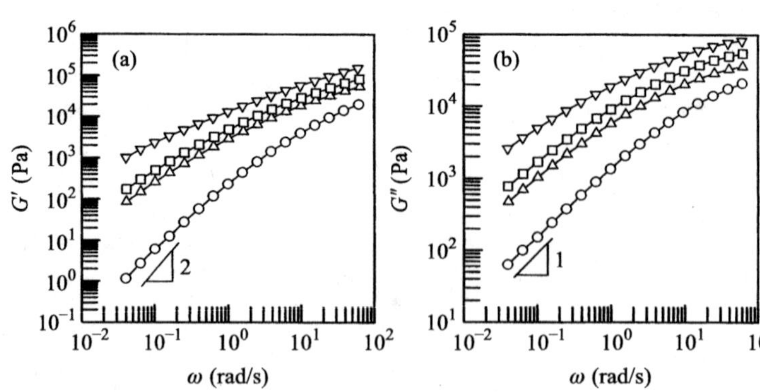

Figure 12.11 (a) Plots of log G' versus log ω and (b) plots of log G'' versus log ω at 190 °C for neat PP (○) and $CaCO_3$-filled PP composites with varying amounts of $CaCO_3$ (wt %): (△) 10, (□) 20, and (▽) 40.

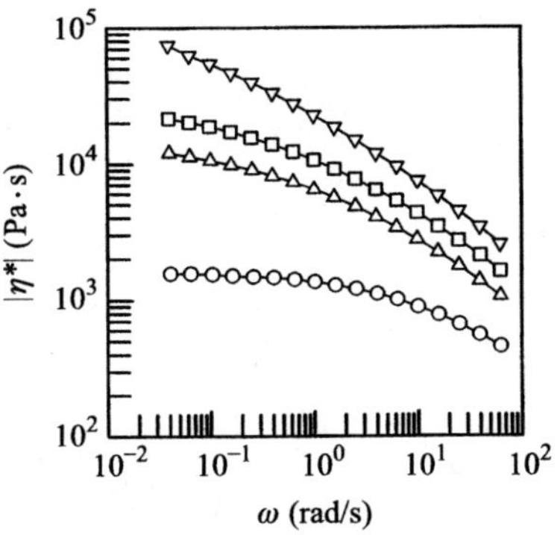

Figure 12.12 Plots of log $|\eta^*|$ versus log ω at 190 °C for neat PP (○) and $CaCO_3$-filled PP composites with varying amounts of $CaCO_3$ (wt %): (△) 10, (□) 20, and (▽) 40.

For comparison, also given in Figures 12.11 and 12.12 are the oscillatory shear flow data for neat PP. The following observations are worth noting in Figures 12.11 and 12.12. Both G' and G'' increase and also $|\eta^*|$ increases with increasing amount of $CaCO_3$. An increase in $|\eta^*|$ with increasing amount of $CaCO_3$ (Figure 12.12) makes sense and is consistent with the steady-state shear flow data, namely, η increases with increasing amount of $CaCO_3$ (Figure 12.2). However, the trend of an increase in G' with increasing amount of $CaCO_3$ (Figure 12.11) is opposite to that observed in steady-state shear data, namely, N_1 decreases with increasing amount of $CaCO_3$ (Figure 12.4). Notice in Figure 12.11 that the slope of log G' versus log ω plot at very low values of ω (i.e., in the terminal region) decreases with increasing amount of $CaCO_3$, approaching solidlike behavior, which makes sense.

Figure 12.13 gives log G' versus log G'' plots for neat PP and $CaCO_3$-filled PP composites at 190 °C, showing that the slope of the plot is less than 2 for the three $CaCO_3$-filled PP composites, while the neat PP has a slope close to 2. Note that when presenting the rheological behavior of immiscible polymer blends in Chapter 11, we showed that disperse two-phase polymer blends also exhibit a slope less than 2 in the terminal region of log G' versus log G'' plots. What is interesting, however, in Figure 12.13 is that the log G' versus log G'' plot is virtually independent of the concentration of $CaCO_3$, while the log N_1 versus log σ plots for the same $CaCO_3$-filled PP composites show that N_1 decreases with increasing amount of $CaCO_3$ (Figure 12.5).

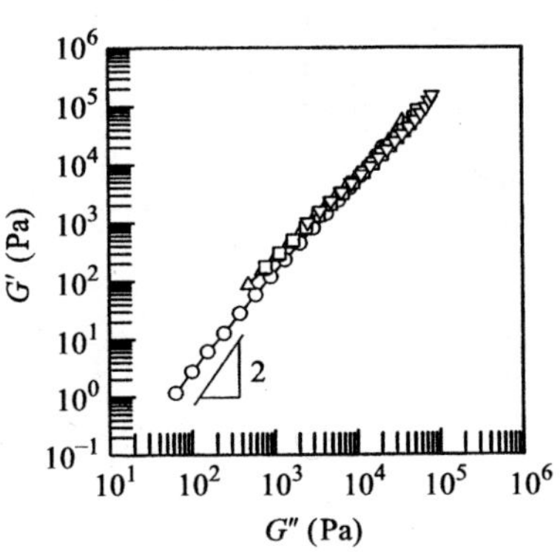

Figure 12.13 Plots of log G' versus log G'' at 190 °C for neat PP (○) and $CaCO_3$-filled PP composites with varying amounts of $CaCO_3$ (wt %): (△) 10, (□) 20, and (▽) 40.

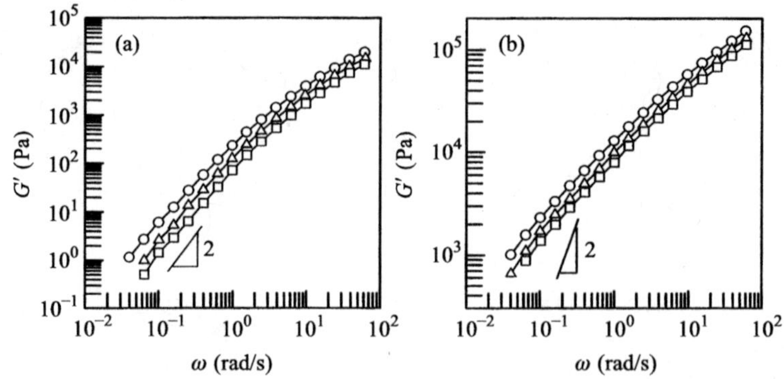

Figure 12.14 Plots of log G' versus log ω for (a) neat PP and (b) PP composite filled with 40 wt % $CaCO_3$ at various temperatures (°C): (○) 190, (△) 210, and (□) 230.

When presenting the rheological behavior of microphase-separated block copolymers in Chapter 8 and the rheological behavior of thermotropic liquid-crystalline polymers in Chapter 9, we showed that the log G' versus log G'' plot is very sensitive to a change in morphological state of a polymer. We can now understand the reason why the log G' versus log G'' plots for the $CaCO_3$-filled PP composites given in Figure 12.13 do not show the dependence of $CaCO_3$ concentration. This is because variations in the amount of $CaCO_3$ would change little the morphological state of $CaCO_3$-filled PP composites.

Figure 12.14 gives the temperature dependence of log G' versus log ω plot for neat PP and a $CaCO_3$-filled PP composite, showing that values of G' decrease with increasing temperature although the temperature dependence of log G' versus log ω plot appears to be weaker in the $CaCO_3$-filled PP composite. Figure 12.15 gives temperature independence of log G' versus log G'' plots for neat PP and a $CaCO_3$-filled PP composite. It is seen that log G' versus log G'' plots for the $CaCO_3$-filled PP composite

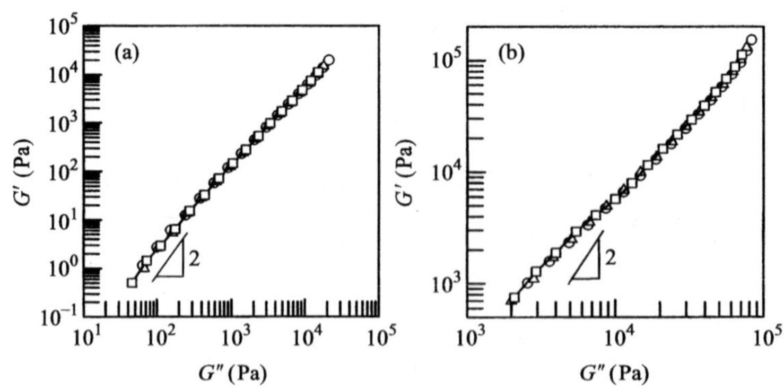

Figure 12.15 Plots of log G' versus log G'' for (a) neat PP and (b) PP composite filled with 40 wt % $CaCO_3$ at various temperatures (°C): (○) 190, (△) 210, and (□) 230.

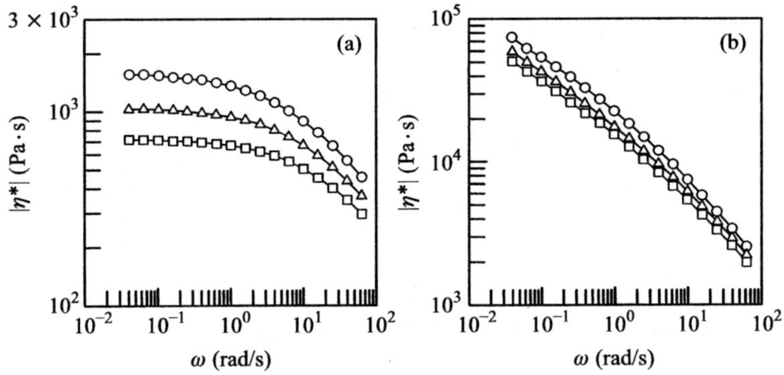

Figure 12.16 Plots of log $|\eta^*|$ versus log ω for (a) neat PP and (b) PP composite filled with 40 wt % $CaCO_3$ at various temperatures (°C): (○) 190, (△) 210, and (□) 230.

are independent of temperature, indicating that the morphological state of the $CaCO_3$-filled PP composite is little affected by the temperature, as for neat PP. Notice in Figure 12.15 that the slope of log G' versus log G'' plot in the terminal region is less than 2 for the $CaCO_3$-filled PP composite, while it is close to 2 for neat PP. This is because the $CaCO_3$-filled PP composite is not a homogeneous polymer.

Figure 12.16 gives temperature dependence of log $|\eta^*|$ versus log ω plot for neat PP and a $CaCO_3$-filled PP composite. It is interesting to observe in Figure 12.16 that the $CaCO_3$-filled PP composite exhibits yield behavior at very low values of ω and no Newtonian behavior over the entire range of ω tested, while neat PP shows Newtonian behavior at low values of ω and then shear thinning behavior as the ω is increased. The frequency dependence of $|\eta^*|$ for the $CaCO_3$-filled PP composite observed in Figure 12.16b is consistent with the shear-rate dependence of η in steady-state shear flow of carbon-black-filled PS composites (Figure 12.6).

12.2.2 Rheology of Molten Thermoplastics with Chemically Treated Fillers

In general, there is little chemical affinity between fillers and polymers, and thus without surface treatment of fillers, particulate-filled polymers have rather weak mechanical properties. But, treatment of fillers with a proper chemical(s), commonly referred to as "coupling agent," can significantly improve the mechanical properties of particulate-filled polymers. Hence, surface treatment of fillers is of great practical importance for enhancing the mechanical properties of particulate-filled polymers. Needless to say, there is a variety of coupling agents. Thus, the judicious choice of a coupling agent is the key to obtaining successful polymer composites. Below, we describe how surface treatment of fillers might influence the rheological properties of particulate-filled molten polymers (Han et al. 1978, 1981). Any discussion of the choice of coupling agent to improve the flow behavior (or processability) of a particulate-filled polymer is beyond the scope of this chapter.

Figure 12.17 gives log η versus log σ and log N_1 versus log σ plots at 200 °C for neat PP, PP composite filled with 50 wt % $CaCO_3$, PP composite filled with 50 wt %

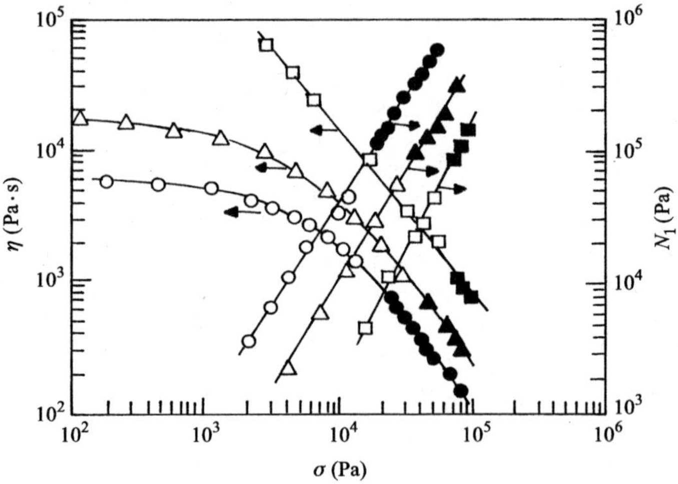

Figure 12.17 Plots of log η versus log σ and log N_1 versus log σ at 200 °C for: (○, ●) neat PP, (□, ■) PP composite filled with 50 wt % $CaCO_3$, (△, ▲) PP composite filled with 50 wt % $CaCO_3$ treated with 1 wt % titanate coupling agent KR-TTS. Open symbols denote data obtained using a cone-and-plate rheometer, and filled symbols denote data obtained with the exit pressure method using a slit die. (Reprinted from Han et al., *Polymer Engineering and Science* 21:196. Copyright © 1981, with permission from the Society of Plastics Engineers.)

$CaCO_3$ treated with 1 wt % titanate coupling agent KR-TTS, which is isopropyl triisostearoyl titanate (Kenrich Petrochemicals) with chemical structure

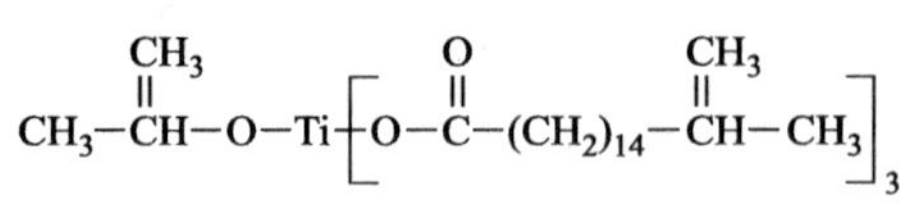

It is interesting to observe in Figure 12.17 that the treatment of $CaCO_3$ with 1 wt % KR-TTS has decreased the η of the PP composites filled with 50 wt % $CaCO_3$ but increased its N_1 over the range of σ investigated.

Figure 12.18 gives log η versus log σ and log N_1 versus log σ plots at 240 °C for neat high-density polyethylene (HDPE) and HDPE composites filled with varying amounts of $CaCO_3$ treated with 1 wt % KR-TTS. It is interesting to observe in Figure 12.18 that the η of the HDPE composite filled with 40 wt % $CaCO_3$ treated with 1 wt % KR-TTS is lower than that of the neat HDPE over the entire range of σ investigated. This observation indicates that the treatment of $CaCO_3$ with 1 wt % KR-TTS is very effective at bringing down the viscosity of the HDPE composite with 40 wt % $CaCO_3$, which is desirable from a processing point of view. Note further that as the concentration of $CaCO_3$ increases, $CaCO_3$-filled HDPE composites exhibit yield behavior and N_1 decreases, which is consistent with the observations made from Figures 12.5 and 12.9.

Figure 12.19 gives the effect of silane coupling agents, which were used to treat $CaCO_3$ or glass beads, on the η of $CaCO_3$-filled PP composites and

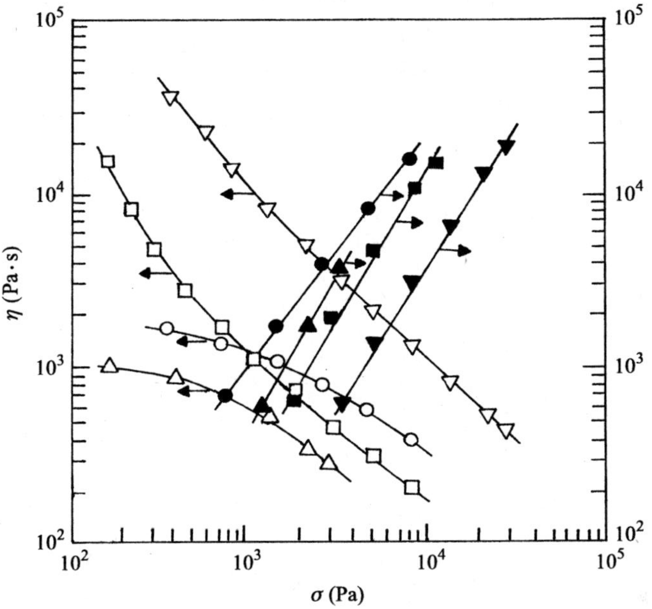

Figure 12.18 Plots of log η versus log σ (open symbols) and log N_1 versus log σ (filled symbols) at 240 °C for HDPE (○, ●) and $CaCO_3$-HDPE composites filled with varying amounts of $CaCO_3$ (wt %): (△, ▲) 40, (□, ■) 55, and (▽, ▼) 70, in which $CaCO_3$ was treated with 1 wt % titanate coupling agent KR-TTS. The data were obtained using a cone-and-plate rheometer. (Reprinted from Han, *Multiphase Flow in Polymer Processing*, Chapter 3. Copyright © 1981, with permission from Elsevier.)

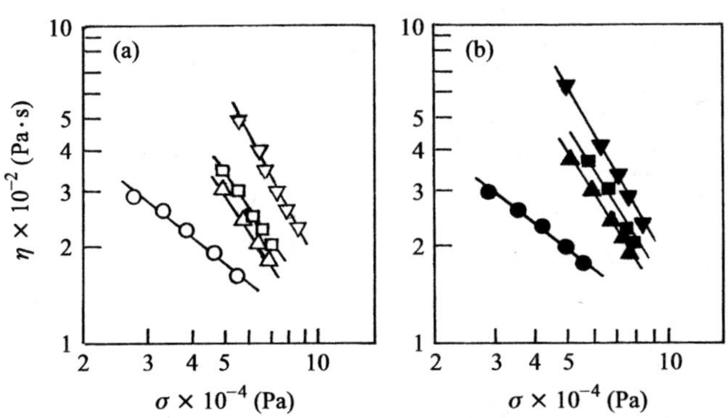

Figure 12.19 Plots of log η versus log σ at 200 °C for PP composites filled with $CaCO_3$ having a nominal size of 2.4 μm or glass beads having diameters of 5–44 μm. (a) $CaCO_3$-filled PP composite: (○) neat PP, (△) PP composite filled with 50 wt % $CaCO_3$ treated with 1 wt % silane coupling agent Y9187, (□) PP composite filled with 50 wt % $CaCO_3$ treated with 1 wt % silane coupling agent A1100, (▽) PP composite filled with 50 wt % $CaCO_3$. (b) Glass-bead-filled PP composite: (●) neat PP, (▲) PP composite filled with 50 wt % glass bead treated with 1 wt % silane coupling agent A1100, (■) PP composite filled with 50 wt % glass bead, (▼) PP composite filled with 50 wt % glass bead treated with 1 wt % silane coupling agent Y9187. The data were obtained by the exit pressure method using a slit die. (Reprinted from Han, *Multiphase Flow in Polymer Processing*, Chapter 3. Copyright © 1981, with permission from Elsevier.)

glass-bead-filled PP composites. The following chemicals were used as coupling agents: *N*-octyltriethoxysilane (Y9187), $CH_3(CH_2)Si(OCH_2CH_3)_3$; and γ-aminopropyltriethoxysilane (A1100), $HN_2(CH_2)Si(OCH_2CH_3)_3$. In Figure 12.19a, we observe that the treatment of $CaCO_3$ with Y9187 or A1100 has decreased the η of $CaCO_3$-filled PP composites. However, in Figure 12.19b we observe that the effect of the two coupling agents, Y9187 and A1100, on the η of glass-bead-filled PP composites is somewhat complex; that is, whereas the treatment of glass beads with A1100 has decreased the η of glass-bead-filled PP composites, the treatment of glass beads with Y9187 has increased the η of glass-bead-filled PP composites.

Figure 12.20 gives the effect of two silane coupling agents, which were used to treat $CaCO_3$ or glass bead, on the N_1 of $CaCO_3$-filled PP composites and glass-bead-filled PP composites. In Figure 12.20a, we observe that the treatment of $CaCO_3$ with Y9187 or A1100 has increased the N_1 of $CaCO_3$-filled PP composites. However, in Figure 12.20b we observe that the effect of the coupling agents Y9187 or A1100 on the N_1 of glass-bead-filled PP composites is somewhat complex; that is, whereas the treatment of glass beads with A1100 has increased the N_1 of the glass-bead-filled PP composites, the treatment of glass beads with Y9187 has decreased the N_1 of the glass-bead-filled PP composites. We can then conclude from Figures 12.19 and 12.20 that the effects of coupling agents on the steady-state shear flow properties of particulate-filled molten polymers depend on both the type of coupling agent and the polymer/filler system under consideration.

Figure 12.21 gives scanning electron micrographs of the tensile fracture surface of injection-molded specimens of glass-bead-filled PP composites, without and with

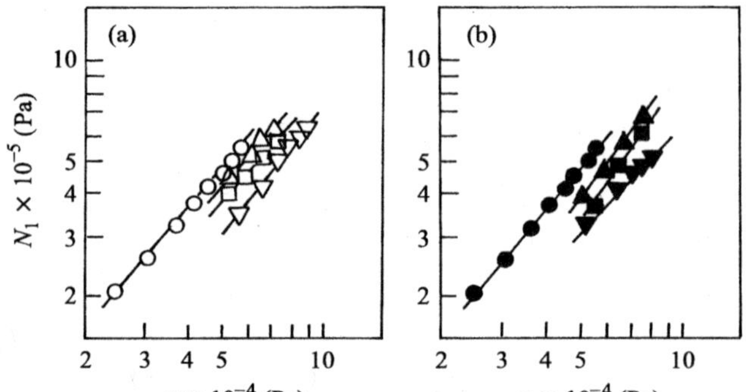

Figure 12.20 Plots of log N_1 versus σ at 200 °C for PP composites filled with $CaCO_3$ having a nominal size of 2.4 μm or glass beads having diameters of 5–44 μm. (a) $CaCO_3$-filled PP composites: (○) neat PP, (△) PP composite filled with 50 wt % $CaCO_3$ treated with 1 wt % silane coupling agent Y9187, (□) PP composite filled with 50 wt % $CaCO_3$ treated with 1 wt % silane coupling agent A1100, (▽) PP composite filled with 50 wt % $CaCO_3$. (b) Glass-bead-filled PP composites: (●) neat PP, (▲) PP composite filled with 50 wt % glass beads treated with 1 wt % silane coupling agent A1100, (■) PP composite filled with 50 wt % glass beads, (▼) PP composite filled with 50 wt % glass beads treated with 1 wt % silane coupling agent Y9187. The data were obtained by the exit pressure method using a slit die. (Reprinted from Han, *Multiphase Flow in Polymer Processing*, Chapter 3. Copyright © 1981, with permission from Elsevier.)

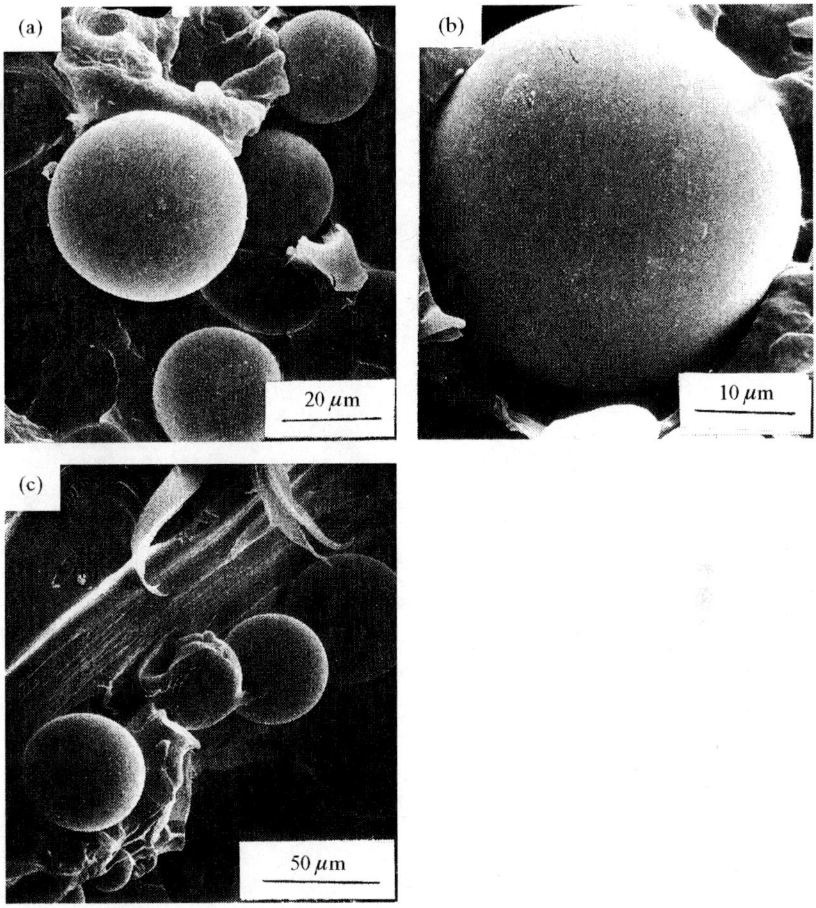

Figure 12.21 Scanning electron micrographs of the tensile fracture surface of PP composites filled with glass beads (a) without being treated with coupling agent, (b) treated with silane coupling agent A1100, and (c) treated with silane coupling agent Y9187. (Reprinted from Han et al., *Polymer Engineering and Science* 21:196. Copyright © 1981, with permission from the Society of Plastics Engineers.)

coupling agent Y9187 or A1100. It is seen that the treatment of glass beads with Y9187 has promoted some interactions between the glass beads and the PP phase, whereas the treatment of glass beads with A1100 has not. Figure 12.22 gives scanning electron micrographs of the tensile fracture surface of injection-molded specimens of $CaCO_3$-filled PP composites, without and with coupling agent KR-TTS, Y9187, or A1100. It is seen that although none of the coupling agents appears to have created adhesion between the $CaCO_3$ particles and the PP phase, the treatment of $CaCO_3$ particles with Y9187 or KR-TTS has certainly changed the morphology of the virgin PP. Specifically, Figures 12.22c and 12.22d clearly show that the PP phase exhibits a morphology of long fibrils, whereas such a morphology is not seen in Figure 12.22a, where

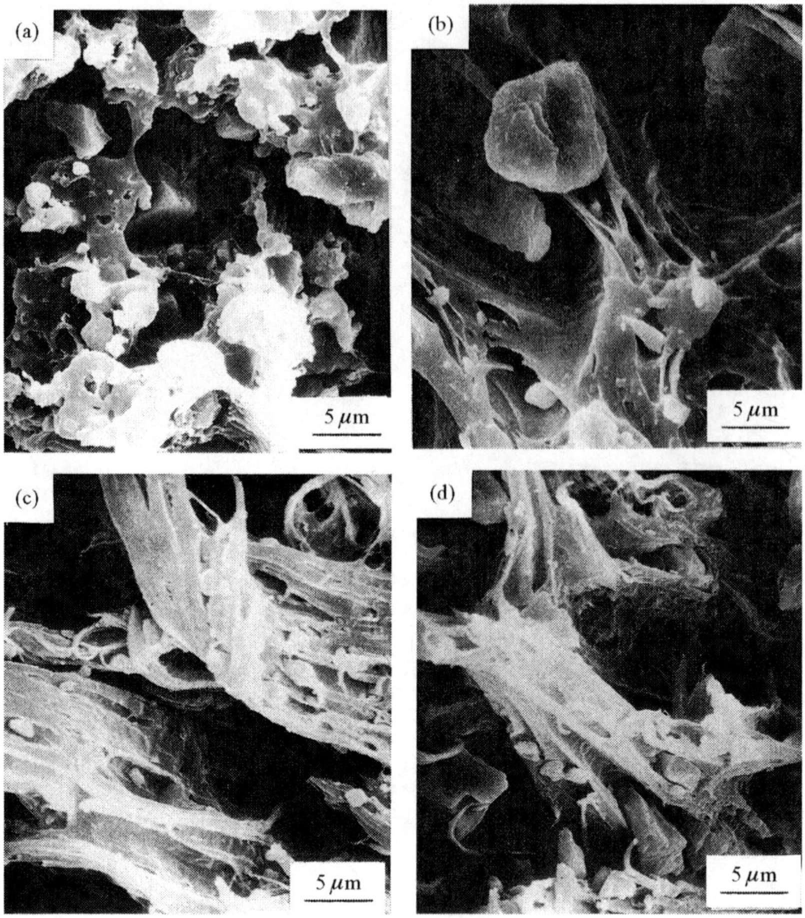

Figure 12.22 Scanning electron micrographs of the tensile fracture surface of PP composites filled with $CaCO_3$ (a) without being treated with coupling agent, (b) treated with silane coupling agent A1100, (c) treated with silane coupling agent Y9187, and (d) treated with titanate coupling agent KR-TTS. (Reprinted from Han et al., *Polymer Engineering and Science* 21:196. Copyright © 1981, with permission from the Society of Plastics Engineers.)

no coupling agent was used to treat $CaCO_3$ particles, or in Figure 12.22b, where A1100 was used to treat $CaCO_3$ particles. Figures 12.22c and 12.22d seem to suggest that the treatment of $CaCO_3$ with Y9187 or KR-TTS influenced the crystallization kinetics of the semicrystalline PP when the specimens were injection molded and subsequently cooled. The modification of the morphology of PP by coupling agent (A1100, Y9187, or KR-TTS) in $CaCO_3$-filled PP composites enhanced the mechanical properties of injected-molded specimens and the spinnnability of the composites during melt spinning (Han et al. 1981). It is clear from the chemical structures given above that A1100, Y9187, and KR-TTS cannot possibly have coupling reactions with PP, which itself has no functional group. In this regard, A1100, Y9187, and KR-TTS cannot be regarded as being "coupling" agents in this case.

A decrease in η, in the presence of a coupling agent, of a particulate-filled molten polymer presented above might have resulted from the coupling agent acting as a lubricant or surfactant, and thus modifying the surfaces of the particulates. If that is the case, under a shear flow polymer molecules will slip between the particulates treated with the coupling agent, encountering a frictional resistance far less than that from untreated particulates. This argument, though speculative, implies that there probably is little true coupling between the particulates and the polymer matrix. It is difficult to imagine how a reduction in η can occur if polymer molecules are interconnected to each other through particulates, when a chemical agent acts as a true coupling agent, instead of as a surface modifier.

As shown in Figures 12.19 and 12.20, there is an instance where the use of a coupling agent increases the η while decreasing the N_1 of a particulate-filled molten polymer. In such a case, it may be conjectured that the particular coupling agent makes the long macromolecules, subjected to a shearing flow, less flexible (or mobile) by connecting (or bridging) them through particulates. If this is so, one would expect to have stiff, long molecules interconnected to each other (similar to the cross-linking of large molecules). It is not difficult to imagine that such a molecular arrangement would have less capability to store elastic energy than flexible macromolecules would have. It is clear from the above observations that under proper combinations of polymer and coupling agent a drastic change in the rheological properties of particulate-filled molten polymers is possible. However, at present, little basic study has been performed to define the mechanism (or mechanisms) that can convincingly explain the changes.

12.2.3 Theoretical Consideration of the Rheology of Particulate-Filled Polymers

We have shown in the preceding section that the rheological properties of particulate-filled molten thermoplastics and elastomers depend on many factors: (1) particle size (d_p), (2) particle shape (α), (3) volume fraction of filler (ϕ), and (4) applied shear rate ($\dot{\gamma}$) or shear stress (σ). The situation becomes more complicated when interactions exist between the particulates and polymer matrix. There is a long history for the development of a theory to predict the rheological properties of dilute suspensions, concentrated suspensions, and particulate-filled viscoelastic polymeric fluids. As early as 1906, before viscoelastic polymeric fluids were known to the scientific community, Einstein (1906, 1911) developed a theory predicting the viscosity of a dilute suspension of rigid spheres and obtained the following expression for the bulk (effective) viscosity η of a suspension:

$$\eta = \eta_0(1 + 2.5\phi) \tag{12.1}$$

where η_0 is the Newtonian viscosity of the suspending medium and ϕ is the volume fraction of the spheres. Equation (12.1) is valid only for extremely dilute suspensions, in which interactions between neighboring particles are negligible (i.e., in the absence of hydrodynamic interactions), and for a Newtonian fluid as the suspending medium.

Following Einstein, Jeffery (1922) investigated the motion of non-spherical particles (rigid ellipsoidal particles) in a shear field of Newtonian liquid, on the basis of the creeping flow equation, and obtained the following expression for the bulk viscosity:

$$\eta = \eta_0(1 + \bar{\nu}\phi) \tag{12.2}$$

where $\bar{\nu}$ is a parameter that depends on the geometry of ellipsoidal particles. Theories for concentrated suspensions based on a Newtonian medium[5] were developed in the 1950s through the 1970s by a number of investigators.

Some theoretical attempts have been reported in the literature describing the yield behavior in particulate-filled molten thermoplastics and elastomers, the experimental observations of which are presented above. It was Bingham (1922) who first reported on yield behavior of concentrated suspension in shear flow of a Newtonian fluid. Experimental observations were fitted to the following expression, referred to as the Bingham plastic model:

$$\left.\begin{array}{ll} \dot{\gamma} = 0, & \text{for } \sigma < Y \\ \sigma = Y + \eta\dot{\gamma}, & \text{for } \sigma \geq Y \end{array}\right\} \tag{12.3}$$

in which σ denotes shear stress, Y denotes yield stress, η denotes the bulk (apparent or effective) viscosity, and $\dot{\gamma}$ denotes apparent shear rate. Later, Herschel and Bulkley (1926) extended Eq. (12.3) to concentrated suspensions in shear flow of a power-law fluid, referred to as the Herschel–Bulkley model:

$$\left.\begin{array}{ll} \dot{\gamma} = 0 & \text{for } \sigma < Y \\ \sigma = Y + K\dot{\gamma}^n & \text{for } \sigma \geq Y \end{array}\right\} \tag{12.4}$$

A tensorial formulation of a Bingham plastic fluid was first introduced by Hohenemser and Prager (1932) and later by Oldroyd (1947). On the other hand, the experimental data presented above show that particulate-filled molten thermoplastics and elastomers exhibit both non-Newtonian viscosity and normal stress effects at large strain rates or large shear stresses, while exhibiting "yield values" at small strain rates or small shear stresses. Therefore, it is desirable to develop a three-dimensional rheological model that can describe such experimental observations.

Extending the approach originated by Hohenemser and Prager (1932) and Oldroyd (1947), White (1979) formulated a rheological model for nonlinear viscoelastic Bingham fluids with an explicit criterion for yield behavior:

$$\boldsymbol{\sigma} = \left\{1 + Y(1/2\,\mathrm{tr}\,\mathbf{H}^2)^{-1/2}\right\}\mathbf{H} \tag{12.5}$$

subject to tr $\boldsymbol{\sigma}^2 > 2Y$ with Y being yield value, where $\boldsymbol{\sigma}$ is the deviatoric stress tensor and $\mathbf{H}$ is a nonlinear memory integral of the form often used for isotropic viscoelastic fluids (see Chapter 3):

$$\mathbf{H} = \int_0^\infty m(s)\left\{\mathbf{C}^{-1} - 1/3(\mathrm{tr}\,\mathbf{C}^{-1})\mathbf{I}\right\}ds \tag{12.6}$$

In which $\mathbf{C}^{-1}$ is the Finger deformation tensor (see Chapter 2) and $m(t)$ is the memory function defined by

$$m(t) = (G/\lambda_1)e^{-t/\lambda_1} \tag{12.7}$$

with G being the elastic modulus and λ_1 being the relaxation time. For steady-state shear flow, White (1979) obtained the following expression for shear stress σ:

$$\sigma = \left\{ 1 + \frac{Y}{G\left[(\lambda_1\dot{\gamma})^2 + \frac{4}{3}(\lambda_1\dot{\gamma})^4\right]^{1/2}} \right\} G\lambda_1\dot{\gamma} \tag{12.8}$$

At very low shear rates (i.e., for $\dot{\gamma} \to 0$), Eq. (12.8) reduces to

$$\lim_{\dot{\gamma}\to 0} \sigma = Y \tag{12.9}$$

Note that Eq. (12.8), hence Eq. (12.9), is based on the explicit criterion for yield behavior, and that they do not offer any explanation as to why yield behavior is observed experimentally.

A very clear manifestation of the existence of yield stress in particulate-filled polymer can be seen in Figure 12.23, where $\log\sigma$ versus $\log\dot{\gamma}$ plots at 60 °C are given for polyisobutylene (PIB) filled with varying amounts of carbon black. It is seen from Figure 12.23 that the experimental data are well represented by Eq. (12.4) and

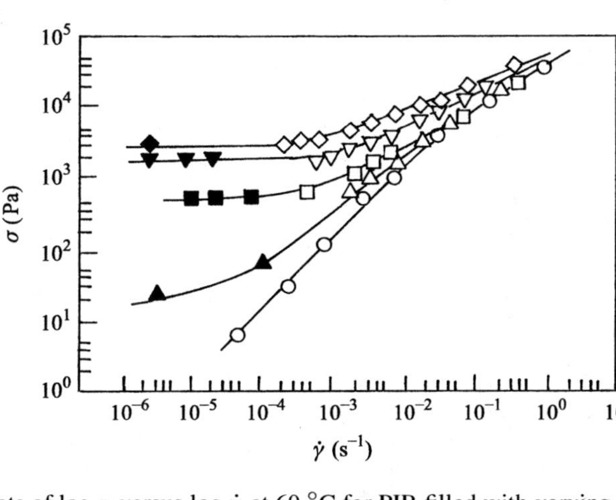

Figure 12.23 Plots of $\log\sigma$ versus $\log\dot{\gamma}$ at 60 °C for PIB filled with varying amounts of carbon black (vol %): (○) 0.0 (neat PIB), (△, ▲) 2.5, (□, ■) 5.0, (▽, ▼) 9.0, and (◇, ◆) 13, where open symbols represent data obtained using a cone-and-plate rheometer and filled symbols represent data obtained using a plane-parallel ribbed plate instrument. (Reprinted from Vinogradov et al., *International Journal of Polymeric Materials* 2:1. Copyright © 1972, with permission from Taylor & Francis Group.)

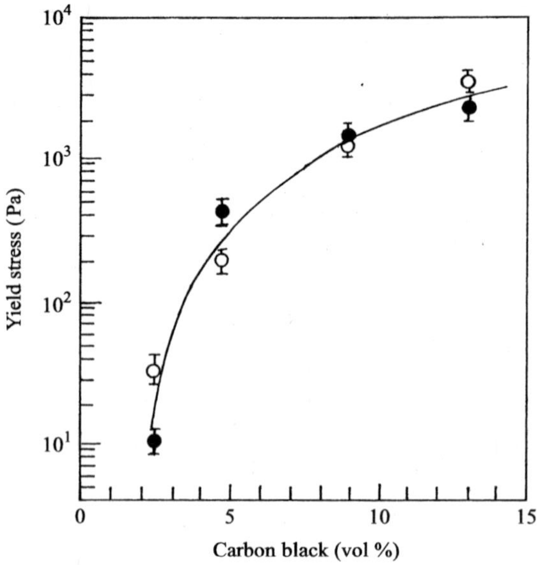

Figure 12.24 Dependence of yield stress on the concentration of carbon black in PIB at 60 °C: (○) low-molecular-weight PIB, and (●) high-molecular-weight PIB. (Reprinted from Vinogradov et al., *International Journal of Polymeric Materials* 2:1. Copyright © 1972, with permission from Taylor & Francis Group.)

that "yield stress" increases with increasing concentration of carbon black. Figure 12.24 shows the effect of carbon black concentration on the yield stress of carbon-black-filled PIBs. It is interesting to observe in Figure 12.24 that yield stress initially increases very rapidly and then tends to level off as the amount of carbon black is increased further above a certain critical value. There is at present no theory predicting the effect of filler concentration on the yield stress of particulate-filled molten thermoplastics or elastomers.

There are two factors that can contribute to the yield behavior observed experimentally in particulate-filled molten thermoplastics and elastomers. One factor is the particle–particle interactions and the other is the particulate–matrix interactions. Recall from the experimental results presented above that yield behavior is observed only when the loading (i.e., the concentration) of the filler is higher than a certain critical value, which seems to suggest that particle–particle interactions play an important role in the observed yield behavior under such a circumstance. Conversely, when particulates are well dispersed in the matrix phase, particulate–matrix interactions may become predominant over particle–particle interactions. Such a situation can occur when particles are adsorbed on the polymer matrix or when some type of physico-chemical interactions exist between the particulates and polymer matrix. Carbon-black-filled elastomers are known to undergo some type of physico-chemical interactions between the carbon black particles and elastomer, the extent of which is believed to depend on the manufacturing process employed for producing the carbon black. In this regard, a phenomenological approach, such as that presented above, is not useful for explaining the origin(s) of the yield behavior observed in particulate-filled molten thermoplastics and elastomers. It is then desirable to develop a rheological model for particulate-filled polymer systems without imposing a priori a mathematical yield criterion. Leonov (1990) took such an approach to formulate a rheological model for particulate-filled polymer systems by including particle–particle interactions and neglecting particulate–matrix interactions.

In the formulation of equations, Leonov included the kinetics of rupture and restoration of the agglomerates of filler particles. However, particle–particle interactions in a particulate-filled polymer system would become less important when the compounding of polymer and particulates is carried efficiently to the extent that the particles are well dispersed in the polymer matrix; that is, particle–particle interactions may become less important when an efficient compounding is accomplished.

Under the assumption that polymer–particle interactions are predominant over particle–particle interactions in a particulate-filled polymer system, Doremus and Piau (1991) formulated a system of equations that include the kinetic expressions for the creation and destruction of the network formed between the particulates and the polymer matrix. A similar study was also reported by Simhambhatla and Leonov (1995). Specifically, Doremus and Piau (1991) postulated that when particulates are mixed with a polymeric liquid, polymer–particle junctions, in addition to polymer–polymer entanglements, are created due to the absorption of the polymer by the particles, and that these newly created junctions are responsible for causing yield stress in a particulate-filled polymer system. In doing so, they formulated two kinetic expressions, one for polymer–particulate interactions and another for polymer–polymer interactions by further postulating that polymer–particulate junctions are strong whereas polymer–polymer interactions are weak. The system of equations thus formulated contain many parameters that must yet be determined by independent experiments. As will be discussed below, when presenting the rheological behavior of nanocomposites, particles and polymer matrix must have attractive interactions, via specific interaction, in order to form polymer–particle junctions. Specific interaction may occur in several different forms, depending on the chemical structure of the polymer matrix and the surface characteristics of the particles. As presented in a previous section, particulates are sometimes treated with a coupling agent, which can be regarded as being a surfactant. This then raises a very fundamental question as to how a specific interaction between the surfactant residing at the surface of particulates and a polymer matrix should be handled in the formulation of a system of equations. This is a challenge awaiting future investigation.

12.3 Rheology of Nanocomposites

Nanocomposites are composed of a thermoplastic polymer or an elastomer and layered silicate platelets of about 1 nm thick and large aspect ratio. In recent years, nanocomposites have attracted much attention from both industry and academia because they may offer unique mechanical and physical properties (e.g., high strength, high modulus, and high heat-distortion temperature) that are not readily available from the conventional particulate-filled thermoplastic polymers. One of the advantages of such nanocomposites lies in that the concentration of nanoparticles required is much lower (e.g., less than 7 wt %) than that (e.g., 40–60 wt %) required for the conventional particulate-filled thermoplastic composites. Thus the weight-to-volume ratio of nanocomposites is much lower than that of the conventional thermoplastic composites, offering a distinct advantage in that less energy would be required when such nanocomposites are processed for automotive or aerospace applications. A large number of publications have reported on the preparation, characterization, and rheological or mechanical properties

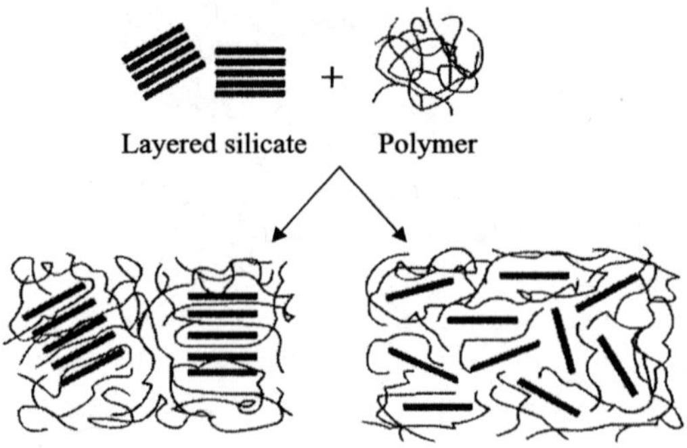

Figure 12.25 Intercalation and exfoliation of layered silicates in a nanocomposite.

of layered silicate-based nanocomposites. There are too many articles to cite them all here, but some selected articles will be cited below.

When a thermoplastic polymer is mixed with layered silicates, it either intercalates or exfoliates the layered silicates, as schematically shown in Figure 12.25. Thermodynamic aspects of intercalation and exfoliation of layered silicate in nanocomposites have been investigated (Balazs et al. 1998; Vaia and Giannelis 1997a, 1997b; Zhulina et al. 1999). The extent of dispersion (intercalation or exfoliation) of the aggregates of layered silicate in nanocomposites is commonly assessed using X-ray diffraction (XRD) or transmission electron microscopy (TEM). From the point of view of obtaining improved physical/mechanical properties of nanocomposites, exfoliation is preferred to intercalation because the greater the degree of exfoliation of layered silicate the larger will be the surface areas of the dispersed layered silicates. In general, intercalation is observed when a polymer matrix and layered silicates do not have sufficient chemical affinity (attractive interactions), while exfoliation is observed when a polymer matrix and layered silicates have strong attractive interactions. Therefore, as in the preparation of compatible polymer blends, attractive interactions between layered silicates and polymer matrix are necessary to achieve a high degree of exfoliation of layered silicate. Thus, a fundamental issue in the preparation of nanocomposites is how to provide strong attractive interactions between layered silicates and the polymer matrix.

When a polymer matrix and layered silicate chosen do not have sufficient chemical affinity, one must modify the surface characteristics of the layered silicate by treatment with a surfactant and/or modify the chemical structure of the polymer matrix by introducing a functional group(s). Much effort has been spent on the development of various surfactants that can be used to treat the surfaces of layered silicates. Table 12.1 gives a summary of some commercial organoclays (Southern Clay Products) together with the chemical structure of the surfactant applied to each organoclay, and the mean interlayer spacing of the (001) plane (d_{001}) for all seven organoclays. Many research groups have employed one or more of the organoclays listed in Table 12.1. Of particular note in Table 12.1 are the chemical structures of the surfactants applied to each of the organoclays. Among the seven organoclays listed in Table 12.1, Cloisite 30B is the only organoclay that is treated with a surfactant (MT2EtOH) that has hydroxyl groups, the other organoclay are treated with a surfactant that has no polar group. In order to help

Table 12.1 Chemical structure of surfactant and the mean interlayer spacing (d_{001}) of organoclays available from Southern Clay Products

Organoclay	*Chemical Structure of Surfactant*	d_{001} (nm)
Cloisite 30B	methyl, tallow, bis-2-hydroxyethyl, quaternary ammonium chloride, MT2EtOH	1.85
Cloisite 6A	dimethyl, dihydrogenated tallow, quaternary ammonium chloride, 2M2HT	3.51
Cloisite 10A	dimethyl, benzyl, hydrogenated tallow, quaternary ammonium chloride, 2MBHT	1.92
Cloisite 15A	dimethyl, dihydrogenated tallow, quaternary ammonium chloride, 2M2HT	3.15
Cloisite 20A	dimethyl, dihydrogenated tallow, quaternary ammonium chloride, 2M2HT	2.42
Cloisite 25A	dimethyl, dihydrogenated tallow, 2ethylhexyl quaternary ammonium, 2MHTL8	1.86
Cloisite 93A	dimethyl, dihydrogenated tallow, ammonium, 2M2HT	2.36

Based on the technical bulletins from Southern Clay Products.

understand the rheological behavior of some organoclay nanocomposites that will be presented below, we give here the chemical structures of two surfactants, MT2EtOH and 2M2HT:

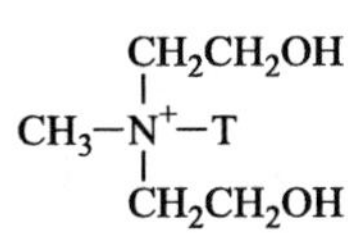

MT2EtOH

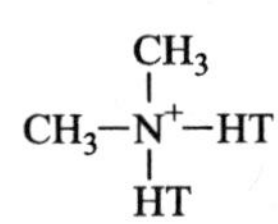

2M2HT

Notice in Table 12.1 that four organoclays, Cloisite 6A, Cloisite 15A, Cloisite 20A, and Cloisite 93A, have the same surfactant 2M2HT in varying amounts. In the chemical structure of MT2EtOH, N^+ denotes quaternary ammonium chloride and T denotes tallow consisting of approximately 65% C18, approximately 30% C16, and approximately 5% C14, and in the chemical structure of 2M2HT, N^+ denotes quaternary ammonium chloride and HT denotes hydrogenated tallow consisting of approximately 65% C18, approximately 30% C16, and approximately 5% C14. The amount of surfactant MT2EtOH residing at the surface of Cloisite 30B is 90 meq/100 g and the amount of surfactant 2M2HT residing at the surface of Cloisite 20A is 95 meq/100 g. In all the organoclays listed in Table 12.1, 100% of Na^+ ions in natural clay (montmorillonite) have been exchanged. The chemical structure of clay is well documented in the literature (Bennett and Hulbert 1986; Grim 1968; van Olphen 1977).

The Fourier transform infrared (FTIR) spectra for MT2EtOH, 2M2HT, Cloisite 30B, and Cloisite 15A are given in Figure 12.26, in which the absorption band at approximately 3360 cm^{-1} represents hydroxyl (–OH) group. It is clearly seen in Figure 12.26 that both MT2EtOH and Cloisite 30B have an –OH group, and that the area under the

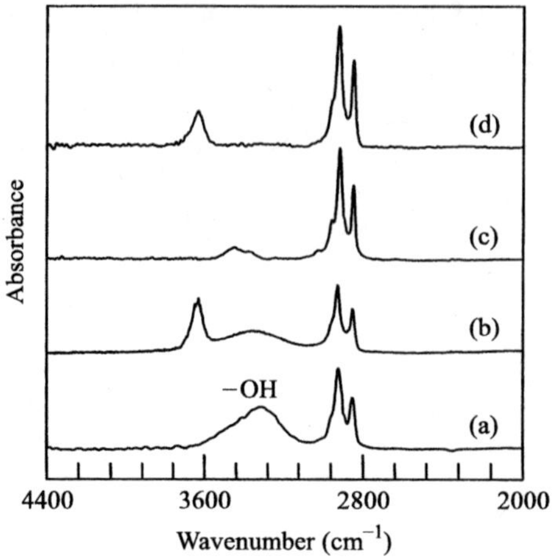

Figure 12.26 FTIR spectra for (a) surfactant MT2EtOH, (b) Cloisite 30B, (c) surfactant 2M2HT, and (d) Cloisite 15A. (Reprinted from Lee and Han, *Macromolecules* 36:7165. Copyright © 2003, with permission from the American Chemical Society.)

absorption band at approximately 3360 cm^{-1} for Cloisite 30B is much smaller. This is reasonable, because the amount of MT2EtOH in Cloisite 30B is approximately 32%. Conversely, in Figure 12.26 we observe no evidence of an –OH group in 2M2HT or Cloisite 15A. The –OH group in the surfactant MT2EtOH residing at the surface of Cloisite 30B can form hydrogen bonds with a polar group(s) in a polymer matrix (Lee and Han 2003a, 2003b, 2003c; Choi et al. 2004).

Besides hydrogen bonding, specific interactions such as ionic interactions, dipole–dipole interactions, and the formation of electron donor–acceptor complexes can also provide attractive interactions between layered silicates and the polymer matrix. Indeed, some research groups (Messersmith and Giannelis 1993, 1995; Hoffmann et al. 2000a; Lepoittevin et al. 2002) have prepared layered silicate nanocomposites, in which the majority of the polymer chains are tethered to the surface of the layered silicates. When chemical modification of a thermoplastic polymer, chosen for mixing with an organoclay, is not possible, one may use a third component as a compatibilizing agent. One good example of such practice is the use of maleated PP in the preparation of nanocomposites composed of PP and organoclay. Some research groups (Hasegawa et al. 1998; Kato et al. 1997; Kawasu et al. 1997; Liu and Wu 2001; Nam et al. 2001) have used such an approach to prepare organoclay nanocomposites based on PP. Nanocomposites have been prepared from a number of other thermoplastic polymers besides PP: polyamides (nylon) (Hoffmann et al. 2000a; Kojima et al. 1993, 1994; Usuki et al. 1993a, 1993b; Yano et al. 1993), polystyrene (Hasegawa et al. 1999; Hoffmann et al. 2000b; Lim and Park 2001; Sikka et al. 1996), poly(ethylene oxide) (Liu et al. 1996; Ogata et al. 1997), poly(ε-caprolacton) (Hoffmann et al. 2000b; Jimenez et al. 1997; Krishnamoorti and Giannelis 1997; Messersmith and Giannelis 1993, 1995; Pantoustier et al. 2001), poly(ethylene-*ran*-vinyl acetate) (Alexandre et al. 2001a, 2001b; Beyer 2001; Riva et al. 2002). The list given above is by no means exhaustive. It should be mentioned, however, that not all nanocomposites reported in the literature had a high degree of exfoliation of organoclay aggregates owing to the lack of attractive interactions between the surfactant treated on the surface of the organoclay and the polymer matrix.

Understandably, the rheological behavior of nanocomposites depends on the degree of dispersion of organoclay aggregates (i.e., morphology), which in turn depends, among many factors, on the degree of attractive interactions between polymer matrix and organoclay. In addition, the gallery distance of organoclay would also play an important role in determining the degree of exfoliation of organoclay aggregates. What determines the gallery distance of an organoclay is the chemistry of the surfactant that is applied to the surface of pristine layered silicates. Note that layered silicates have a large active surface area (700–800 m^2/g in the case of montmorillonite) and a moderately negative surface charge (cation exchange capacity). Upon replacing the hydrated metal cation from the interlayers of the pristine layered silicates with organic cations such as alkyammonium, the layered silicate attains a hydrophobic/organophilic character, which typically results in large interlayer spacing. Because the negative charge originates in the layered silicate, the cationic head groups of the alkylammonium chloride molecule preferentially reside at the surface of the layered silicate (i.e., the quaternary ammonium chloride portion of the surfactants, for example, interacts with the silicate surface), while the oligomeric tallow species, which sometime contain polar groups, extend into the galleries. Therefore, it is very important to match the chemical affinity between a polymer matrix and an organoclay in order to achieve a very high degree of exfoliation of organoclay platelets in the polymer matrix.

There are two ways of preparing thermoplastic polymer/clay nanocomposites: one method is direct intercalation (or exfoliation) of layered silicates during melt compounding, and the other method is through in situ polymerization, wherein the organoclay platelets are first preintercalated with monomer. Nylon 6/clay nanocomposites are preferably prepared by the latter method, while PP/clay nanocomposites are prepared by the former method. Note that nylon 6 has amine end groups that can easily interact with the polar groups grafted onto the surface of the layered silicates of the organoclay. When a thermoplastic polymer does not have a functional group(s), chemical modification of the polymer is necessary.

Enhanced mechanical properties of nylon 6/organoclay nanocomposites were first reported by Toyota researchers (Kojima et al. 1993, 1994; Usuki et al. 1993a, 1993b; Yano et al. 1993). They reported that a doubling of the tensile modulus and strength was achieved for nylon 6/layered silicate nanocomposites containing as little as 2 vol % organoclay. More importantly, the heat distortion temperature of the nanocomposites increased by up to 100 °C, thereby extending the use of the composite to higher temperature environments, such as automotive under-the-hood parts. In their studies, in situ polymerization of ε-caprolactam was carried out in the presence of an organoclay, giving rise to the polymer chains end-tethered on the silicate layers. However, contrary to the impression one might get from the studies of the Toyota researchers, some investigators (Lincoln et al. 2001a, 2001b; Maiti and Okamoto 2003) have shown that the crystalline structure of nylon 6 and the crystallization rate of nylon 6 in the presence of organoclay are quite different from those of pristine nylon 6, indicating that the experimentally observed enhancement in the mechanical properties of nylon 6/organoclay nanocomposites is not entirely due to the attractive interactions between nylon 6 and organoclay. In other words, the mechanism(s) for the experimentally observed enhancement in heat distortion temperature and mechanical properties of nylon 6/organoclay nanocomposites was not obvious.

Maiti and Okamoto (2003) prepared nylon 6/organoclay nanocomposites by in situ polymerization of ε-caprolactam in laurylammonium-intercalated montmorillonite in the presence of a small amount of 6-aminocaproic acid. The amounts of inorganic part and montmorillonite clay, obtained from the burning off of the organic part, were 1.6 and 3.7 wt %, respectively. They compared the crystallization of pristine nylon 6 with that of nylon 6/organoclay nanocomposites using light scattering, and characterized the nanocomposites using wide-angle X-ray scattering (WAXS), dynamic mechanical properties measurement, and transmission electron microscopy. Figure 12.27 describes time variations of reduced invariant $Q_\delta / Q_\delta^\infty$ during isothermal crystallization of pristine nylon 6 and nylon 6 in the presence of an organoclay in the quiescent state at 190 °C, which was obtained from light scattering experiments (Maiti and Okamoto 2003). In Figure 12.27, Q_δ is defined by

$$Q_\delta = \int_0^\infty I(q) q^2 \mathrm{d}q \tag{12.10}$$

with q being the scattering vector and $I(q)$ being the intensity of the scattering light at q, and Q_δ^∞ being Q_δ at an infinitely long time of crystallization (up to full solidification of the melt). Note that Figure 12.27 describes time evolution of crystallization of nylon 6 with or without an organoclay, and the slopes of the curves in Figure 12.27 indicate the overall crystallization rates of the systems.

Maiti and Okamoto (2003) made the following observations from their study. (1) The overall crystallization rate of nylon 6 was enhanced dramatically in the presence of organoclay. (2) Nylon 6 exhibits both an α- and a γ-form, and the γ-form gradually decreases and vanishes as the temperature is increased to 200 °C and higher, whereas the nanocomposites always exhibit predominantly γ-form throughout the whole crystallization temperature range investigated. (3) The improvement in the mechanical properties of the nanocomposites observed is attributable to the epitaxial growth (α′-form) on the clay lamella forming the "shish-kebab" type of structure as determined by WAXS. (4) The heat distortion temperature of the nanocomposite is 80 °C higher than that of pristine nylon 6. (5) The strong interacting nature of nylon 6 with organoclay arises from the exclusive formation of the γ-phase in the presence of organoclay

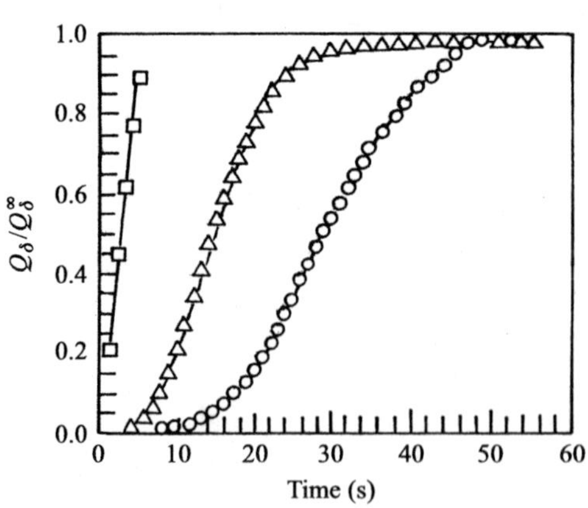

Figure 12.27 Variations of reduced invariant $Q_\delta / Q_\delta^\infty$ with time during isothermal crystallization at quiescent state at 190 °C for: (○) nylon 6, (△) nylon 6 with 1.3 wt % clay, and (□) nylon 6 with 3.7 wt % clay, where Q_δ is defined by Eq. (12.10) and Q_δ^∞ is the value of Q_δ at an infinitely long duration of crystallization. (Reprinted from Maiti and Okamoto, *Macromolecular Materials and Science* 288:440. Copyright © 2003, with permission from Taylor & Francis Group.)

particles, and it also comes from the unique nucleation and growth mechanism of nylon 6 on the silicate layers; once one molecular layer is nucleated on the clay surface, other molecules may form hydrogen bonds to the molecule already hydrogen bonded to the silicate surface.

The conclusions drawn by Maiti and Okamoto (2003) are reminiscent of the conclusions drawn from Figure 12.22 in that use of $CaCO_3$ treated with silane or titanate coupling agents as a filler in PP affected the morphology and crystallization rate of semicrystalline PP during melt blending and consequently enhanced the mechanical properties of the composites (Han et al. 1981).

In this section, we present the dispersion characteristics and rheology of organoclay nanocomposites based on thermoplastic homopolymer, or block copolymer, showing that the presence of attractive interactions between the surfactant-treated surface of the organoclay and the polymer matrix gives rise to a very high degree of exfoliation, and consequently to unusual rheological behavior.

12.3.1 Rheology of Organoclay Nanocomposites Based on Thermoplastic Polymer

There are so many thermoplastic polymers that it is not possible to present the rheological behavior of organoclay nanocomposites based on each of them. Here, we will consider rheological behavior of organoclay nanocomposites based on polycarbonate (PC), which can be regarded as being representative of the majority organoclay nanocomposite systems.

12.3.1.1 XRD Patterns

Figure 12.28 gives XRD patterns of natural clay (montmorillonite, MMT), Cloisite 30B, 95.7/4.3 PC/MMT nanocomposite, and 95.7/4.3 PC/Cloisite 30B nanocomposite, in which 95.7/4.3 refers to the weight percent of PC and MMT or Cloisite 30B.

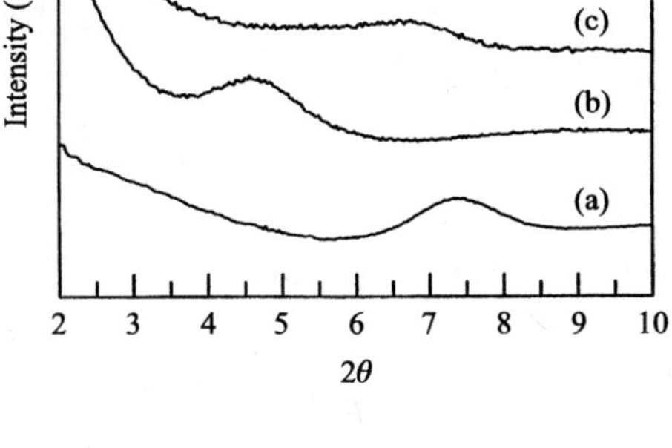

Figure 12.28 XRD diffraction patterns for (a) MMT, (b) Cloisite 30B, (c) 95.7/4.3 PC/MMT nanocomposite, and (d) 95.7/4.3 PC/Cloisite 30B nanocomposite. (Reprinted from Lee and Han, *Polymer* 44:4573. Copyright © 2003, with permission from Elsevier.)

The nanocomposites were prepared by melt blending in a twin-screw extruder. It is seen in Figure 12.28 that (1) MMT has a d_{001} spacing of 1.17 nm, (2) the 95.7/4.3 PC/MMT nanocomposite has a d_{001} spacing of 1.31 nm, which is slightly higher than that of MMT, (3) Cloisite 30B has a d_{001} spacing of 1.85 nm, which is much higher than that of MMT, and (4) the 95.7/4.3 PC/Cloisite 30B nanocomposite has a featureless XRD pattern. The featureless XRD patterns in Figure 12.28 for the 95.7/4.3 PC/Cloisite 30B nanocomposite suggests that a significant degree of dispersion of the Cloisite 30B aggregates might have occurred due to the PC having the carbonyl functional group. The improved dispersion of Cloisite 30B aggregates by the PC is believed to have originated from the presence of strong interactions between the carbonyl groups in PC and the hydroxyl groups in MT2EtOH residing at the surface of the organoclay Cloisite 30B.

12.3.1.2 TEM Images

Figure 12.29 gives TEM images of 95.7/4.3 PC/Cloisite 30B nanocomposite and 95.7/4.3 PC/MMT nanocomposite, where the dark areas represent the clay and the gray/white areas represent the PC matrix. It is clearly seen from Figure 12.29 that PC in the 95.7/4.3 PC/Cloisite 30B nanocomposite has dispersed the organoclay aggregates fairly well, whereas PC in the 95.7/4.3 PC/MMT nanocomposite has not (i.e., large aggregates of natural clay are bundled together), indicating that the carbonyl groups in PC have played the major role in dispersing (or breaking) the organoclay aggregates, leading to a significant degree of dispersion of the silicate layers of Cloisite 30B aggregates. The ability of the PC chains to enter, during melt blending, between the Cloisite 30B platelets is attributed to the presence of attractive interactions between PC and Cloisite 30B, helping disperse the Cloisite 30B aggregates. The better dispersed the organoclay aggregates, the larger will be the surface areas of the layered silicates of Cloisite 30B that become available for attractive interactions with the carbonyl

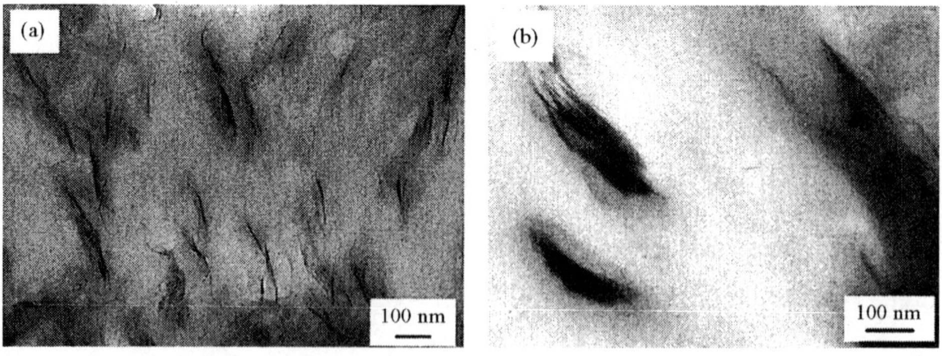

Figure 12.29 TEM images of (a) 95.7/4.3 PC/Cloisite 30B nanocomposite and (b) 95.7/4.3 PC/MMT nanocomposite, where the dark areas represent the clay and the gray/white areas represent the PC matrix. (Reprinted from Lee and Han, *Polymer* 44:4573. Copyright © 2003, with permission from Elsevier.)

groups in PC, which in turn will help increase the extent of dispersion of Cloisite 30B aggregates.

12.3.1.3 FTIR Spectra

Figure 12.30 gives in situ FTIR spectra for neat PC, 95.7/4.3 PC/MMT nanocomposite, and 95.7/4.3 PC/Cloisite 30B nanocomposite at various temperatures ranging from 30 to 280 °C. It is seen in Figure 12.30a that an absorption band appears at approximately 1775 cm^{-1}, representing the free carbonyl stretching vibration peak in neat PC (Abbate et al. 1994; Coleman et al. 1991). Similar observation can be made in Figure 12.30b for 95.7/4.3 PC/MMT nanocomposite, indicating that few positive interactions have taken place between PC and MMT. However, in Figure 12.30c we observe two stretching peaks, one peak at about 1790 cm^{-1} that is attributable to the free carbonyl stretching peak and another peak at about 1750 cm^{-1} that is attributable to the hydrogen bonded carbonyl stretching peak in the 95.7/4.3 PC/Cloisite 30B nanocomposite. The hydrogen-bonded carbonyl stretching peak appearing at about 1750 cm^{-1} is due to the specific interactions between the carbonyl groups in PC and the hydroxyl groups in MT2EtOH of organoclay in the 95.7/4.3 PC/Cloisite 30B nanocomposite. Notice in Figure 12.30c that the hydrogen-bonded band (1750 cm^{-1}) appears as a shoulder and is not well resolved. This is understandable because the amount of hydroxyl groups in the 95.7/4.3 PC/Cloisite 30B nanocomposite is very small. This is because the amount of organoclay Cloisite 30B in the nanocomposite is only 4.3 wt %, and that Cloisite 30B contains 32 wt % MT2EtOH, which in turn has small amounts of hydroxyl groups. It is interesting to observe in Figure 12.30c that the area under the absorption band at about 1750 cm^{-1} for the 95.7/4.3 PC/Cloisite 30B nanocomposite remains more

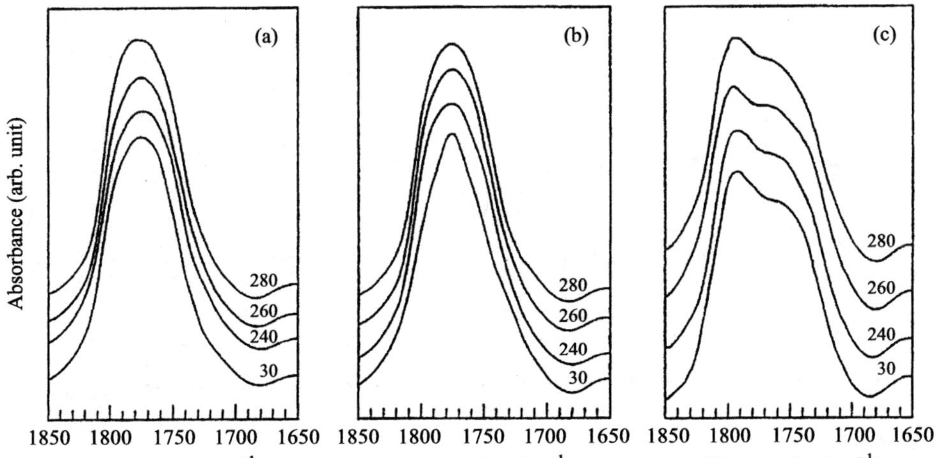

Figure 12.30 In situ FTIR spectra at varying temperatures for (a) PC, (b) 95.7/4.3 PC/MMT nanocomposite, and (c) 95.7/4.3 PC/Cloisite 30B nanocomposite. (Reprinted from Lee and Han, *Polymer* 44:4573. Copyright © 2003, with permission from Elsevier.)

or less constant over temperatures ranging from 30 to 280 °C. The above observation indicates that the strength of the hydrogen bonds between the carbonyl groups in PC and the hydroxyl groups in MT2EtOH of Cloisite 30B persists up to 280 °C. Thus, it can be concluded from Figure 12.30 that specific interactions, via hydrogen bonding, exist between the carbonyl group in PC and the hydroxyl group in MT2EtOH of Cloisite 30B in the 95.7/4.3 PC/Cloisite 30B nanocomposite, whereas no specific interactions exist between PC and MMT in the 95.7/4.3 PC/MMT nanocomposite. The observations made above from Figure 12.30 now explain the origin for the indiscernible XRD peaks for the 95.7/4.3 PC/Cloisite 30B nanocomposite when compared with the 95.7/4.3 PC/MMT nanocomposite (see Figure 12.28) and the high degree of dispersion of Cloisite 30B aggregates in the 95.7/4.3 PC/Cloisite 30B nanocomposite compared with the rather poor dispersion of MMT aggregates in the 95.7/4.3 PC/MMT nanocomposite.

12.3.1.4 Linear Dynamic Viscoelastic Properties

Figure 12.31 shows the effect of temperature on $\log G'$ versus $\log \omega$ plots for 97.7/2.3 PC/Cloisite 30B nanocomposite and 95.7/4.3 PC/Cloisite 30B nanocomposite. The following observations are worth noting in Figure 12.31. For the 97.7/2.3 PC/Cloisite 30B nanocomposite, values of G' in the low-frequency region (for ω lower than about 0.4 rad/s) increase with increasing temperature, while values of G' in the higher frequency region (for ω higher than about 1 rad/s) decrease with increasing temperature. The same trend is observed also for the 95.7/4.3 PC/Cloisite 30B nanocomposite. Interestingly, values of G' in the low-frequency region (for ω lower than about 0.4 rad/s) are increased by about one order of magnitude as the concentration of Cloisite 30B is increased from 2.3 to 4.3 wt %. Such an observation is not made when MMT is added to PC. Notice further in Figure 12.31 that the slope of $\log G'$ versus $\log \omega$ plots for PC/Cloisite 30B nanocomposites is decreased drastically, exhibiting solidlike behavior, as the temperature is increased from 240 to 280 °C.

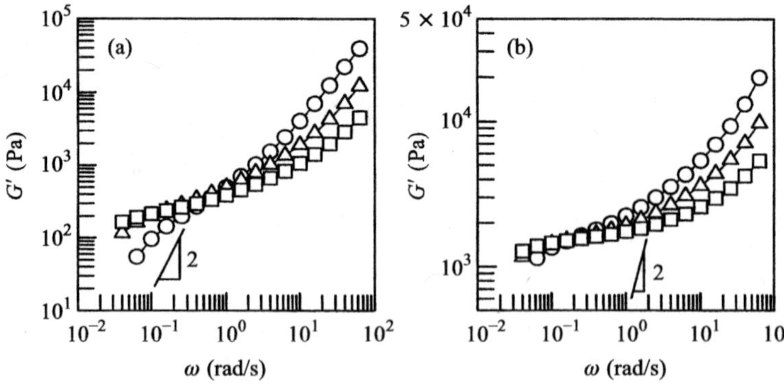

Figure 12.31 Plots of $\log G'$ versus $\log \omega$ for (a) 97.7/2.3 PC/Cloisite 30B nanocomposite and (b) 95.7/4.3 PC/Cloisite 30B nanocomposite at various temperatures (°C): (○) 240, (△) 260, and (□) 280. (Reprinted from Lee and Han, *Polymer* 44:4573. Copyright © 2003, with permission from Elsevier.)

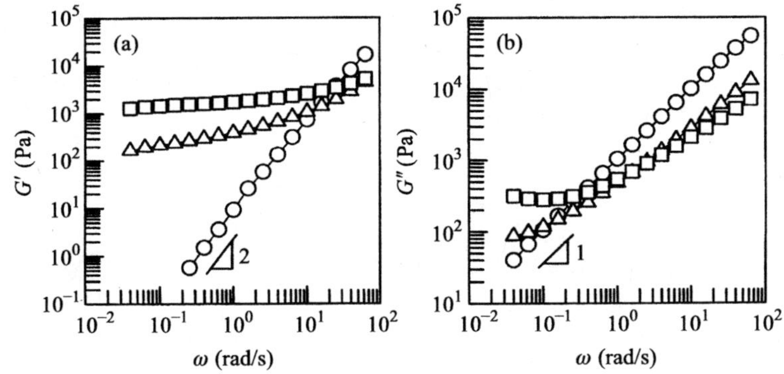

Figure 12.32 Plots of (a) $\log G'$ versus $\log \omega$ and (b) $\log G''$ versus $\log \omega$ at 280 °C for (○) neat PC, (△) 97.7/2.3 PC/Cloisite 30B nanocomposite, and (□) 95.7/4.3 PC/Cloisite 30B nanocomposite. (Reprinted from Lee and Han, *Polymer* 44:4573. Copyright © 2003, with permission from Elsevier.)

Figure 12.32 shows the effect of the concentration of organoclay (Cloisite 30B) on the linear viscoelastic dynamic moduli of PC/Cloisite 30B nanocomposites at 280 °C as functions of ω. For comparison, also given in Figure 12.32 are $\log G'$ versus $\log \omega$, and $\log G''$ versus $\log \omega$ plots for neat PC at 280 °C. In Figure 12.32a we observe a large effect of the concentration of Cloisite 30B on the G' of PC/Cloisite 30B nanocomposites; that is, an increase of Cloisite 30B from 2.3 to 4.3 wt % has increased the values of G' by an order of magnitude for the PC/Cloisite 30B nanocomposite in the low-frequency region (for ω lower than about 0.1 rad/s). Notice that both 97.7/2.3 and 95.7/4.3 PC/Cloisite 30B nanocomposites exhibit solidlike behavior. In Figure 12.32b we also observe a pronounced effect of the concentration of Cloisite 30B on the G'' of PC/Cloisite 30B nanocomposites; that is, in the low-frequency region (for ω about 0.1 rad/s) values of G'' for the 95.7/4.3 PC/Cloisite 30B nanocomposite are about an order of magnitude greater than those for neat PC. This is attributable to the presence of strong attractive interactions, via hydrogen bonding, between the hydroxyl groups in the surfactant MT2EtOH residing at the surface of Cloisite 30B and the carbonyl groups in PC (see Figure 12.30).

Figure 12.33 gives $\log |\eta^*|$versus $\log \omega$ plots for 95.7/4.3 PC/MMT nanocomposite and 95.7/4.3 PC/Cloisite 30B nanocomposite at 240, 260, and 280 °C. The following observations are worth noting in Figure 12.33. The $\log |\eta^*|$ versus $\log \omega$ plot of the 95.7/4.3 PC/MMT nanocomposite exhibits very mild shear-thinning behavior in the low-frequency region (for ω lower than about 1 rad/s) and then Newtonian behavior in the higher frequency region (for ω higher than about 1 rad/s). In the previous section, we showed that $CaCO_3$-filled (say, higher than 40 wt % filler) PP exhibits strong yield behavior at low shear rates in steady-state shear flow or at low angular frequencies in oscillatory shear flow. It is interesting to observe in Figure 12.33 that the 95.7/4.3 PC/MMT nanocomposite containing only 4.3 wt % MMT exhibits yield behavior at ω lower than about 1 rad/s. We attribute this observation to particle–particle interactions of the layered silicates. Whether or not yield behavior may be observed in a mixture of molten polymer and particulates would depend on the surface area of the particulates.

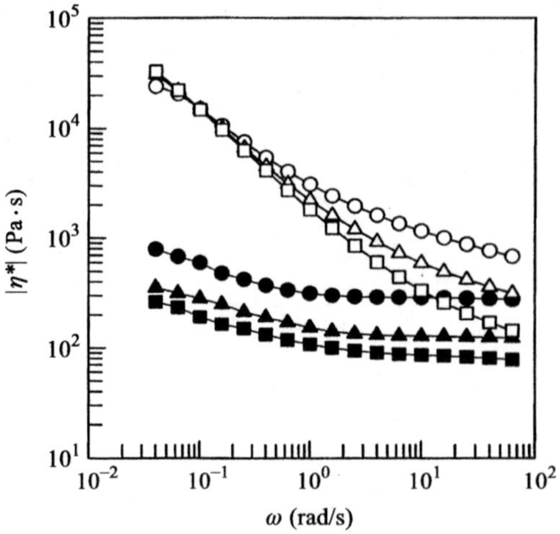

Figure 12.33 Plots of log $|\eta^*|$ versus log ω for 95.7/4.3 PC/MMT nanocomposite at various temperatures (°C): (●) 240, (▲) 260, and (■) 280, and 95.7/4.3 PC/Cloisite 30B nanocomposite at various temperatures (°C): (○) 240, (△) 260, and (□) 280. (Reprinted from Lee and Han, *Polymer* 44:4573. Copyright © 2003, with permission from Elsevier.)

The smaller the size of particulates (thus the larger the surface area of particulates) in a nanocomposite, the lower will be the concentration of the particulates that may give rise to yield behavior. Note that little or no attractive interactions exist between PC and the surface of MMT in the 95.7/4.3 PC/MMT nanocomposite.

The above argument must be modified when attractive interactions exist between the polymer matrix and the surface of particulates, as in the PC/Cloisite 30B nanocomposites, where Cloisite 30B has hydroxyl groups that may interact with the carbonyl groups in PC. In Figure 12.33 we observe very strong "yield behavior" in the 95.7/4.3 PC/Cloisite 30B nanocomposite over the entire range of ω and at all three temperatures, 240, 260, and 280 °C. Notice in Figure 12.33 that the value of $|\eta^*|$ at $\omega = 0.03$ rad/s and at 280 °C for the 95.7/4.3 PC/Cloisite 30B nanocomposite is about two orders of magnitude greater than that for the 95.7/4.3 PC/MMT nanocomposite. We attribute such a dramatic increase in $|\eta^*|$ to the presence of attractive interactions, via hydrogen bonding, between the carbonyl groups in PC and the hydroxyl groups in MT2EtOH residing at the surface of Cloisite 30B (see Figure 12.30).

Figure 12.34 gives log G' versus log G'' plots for 97.7/2.3 PC/MMT and 95.7/4.3 PC/MMT nanocomposites at 240, 260, and 280 °C, and for 97.7/2.3 PC/Cloisite 30B and 95.7/4.3 PC/Cloisite 30B nanocomposite at 240, 260, and 280 °C. We have shown that the log G' versus log G'' plot is independent of temperature for homopolymers (see Chapter 6), and that the log G' versus log G'' plot may be used to determine the order–disorder transition temperature of microphase-separated block copolymers (see Chapter 8) and the clearing temperature of thermotropic liquid-crystalline polymers (see Chapter 9). Thus, we can conclude from Figure 12.34a that the morphological state of both 97.7/2.3 PC/MMT and 95.7/4.3 PC/MMT nanocomposites changes little with temperature because the log G' versus log G'' plots of the nanocomposites have very weak temperature dependence, while we can conclude from Figure 12.34b that the morphological state of both 97.7/2.3 PC/Cloisite 30B and 95.7/4.3 PC/Cloisite 30B nanocomposites varies with temperature because the log G' versus log G'' plots of the nanocomposites exhibit temperature dependence. Notice in Figure 12.34b that

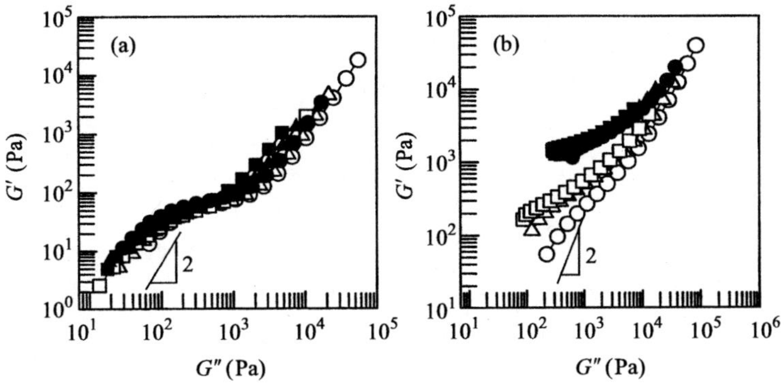

Figure 12.34 (a) Plots of log G' versus log G'' for 97.7/2.3 PC/MMT nanocomposite at various temperatures (°C): (○) 240, (△) 260, and (□) 280, and for 95.7/4.3 PC/MMT nanocomposite at various temperatures (°C): (●) 240, (▲) 260, and (■) 280. (b) Plots of log G' versus log G'' for 97.7/2.3 PC/Cloisite 30B nanocomposite at various temperatures (°C): (○) 240, (△) 260, and (□) 280, and for 95.7/4.3 PC/Cloisite 30B nanocomposite at various temperatures (°C): (●) 240, (▲) 260, and (■) 280. (Reprinted from Lee and Han, *Polymer* 44:4573. Copyright © 2003, with permission from Elsevier.)

the log G' versus log G'' plot is shifted upward with increasing temperature, and in the vicinity of $G'' = 100$ Pa the values of G' at 280 °C for the 95.7/4.3 PC/Cloisite 30B nanocomposite are about one order of magnitude greater than those for the 97.7/2.3 PC/Cloisite 30B nanocomposites. This is not the case for the 97.7/2.3 PC/MMT and 95.7/4.3 PC/MMT nanocomposites (see Figure 12.34a). That is, the surfactant residing at the surface of Cloisite 30B has a profound influence on the log G' versus log G'' plot (thus the morphological state) of the nanocomposites, as compared with the MMT. That is, the differences in the log G' versus log G'' plot between the PC/Cloisite 30B and PC/MMT nanocomposites lies in the presence of attractive interactions, via hydrogen bonding, between the carbonyl groups in PC and the hydroxyl groups in MT2EtOH residing at the surface of Cloisite 30B.

12.3.1.5 Nonlinear Rheological Properties

Figure 12.35 shows variations of shear stress growth $\left(\sigma^+(\dot{\gamma}, t)\right)$ as a function of shear strain ($\dot{\gamma}t$) upon startup of shear flow at 260 °C and $\dot{\gamma} = 1.0\ \mathrm{s}^{-1}$ for 95.7/4.3 PC/MMT nanocomposite and 95.7/4.3 PC/Cloisite 30B nanocomposite, where three separate runs, each using a fresh specimen, were made. Thus, the three curves given in Figure 12.35 represent variability of $\sigma^+(\dot{\gamma}, t)$ on three repeated measurements. The following observations are worth noting in Figure 12.35. Upon startup of shear flow, both 95.7/4.3 PC/MMT and 95.7/4.3 PC/Cloisite 30B nanocomposites exhibit an overshoot followed by monotonic decay reaching steady state, and the magnitude of overshoot peak (σ^+_{peak}) for the 95.7/4.3 PC/Cloisite 30B nanocomposite is several times greater than that for the 95.7/4.3 PC/MMT nanocomposite. We attribute the large σ^+_{peak} observed in the 95.7/4.3 PC/Cloisite 30B nanocomposite to the presence of attractive interactions between PC and Cloisite 30B, and the small σ^+_{peak}

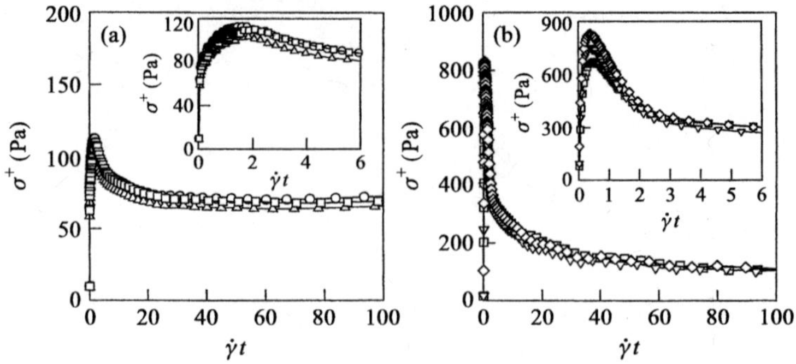

Figure 12.35 Variations of $\sigma^+(\dot{\gamma}, t)$ with $\dot{\gamma}t$ at 260 °C during transient shear flow at $\dot{\gamma} = 1.0\ s^{-1}$ for (a) 95.7/4.3 PC/MMT nanocomposite for three repeated runs: (○) run #1, (△) run #2, and (□) run #3, and for (b) 95.7/4.3 PC/Cloisite 30B nanocomposite for three repeated runs: (○) run #1, (△) run #2, and (□) run #3. (Reprinted from Lee and Han, *Polymer* 44:4573. Copyright © 2003, with permission from Elsevier.)

observed in the 95.7/4.3 PC/MMT nanocomposite to particle–particle interactions in the nanocomposite. Referring to Figure 12.35, variability of σ^+_{peak} on three repeated runs is much greater in 95.7/4.3 PC/Cloisite 30B nanocomposite than in 95.7/4.3 PC/MMT nanocomposite. This observation suggests that a greater variability of the state of dispersion of clay aggregates exists in PC/Cloisite 30B nanocomposite than in PC/MMT nanocomposite. The steady-state value of shear stress (σ) for the 95.7/4.3 PC/Cloisite 30B nanocomposite is about 60% higher than that for the 95.7/4.3 PC/MMT nanocomposite. The above observations clearly indicate that the organoclay Cloisite 30B has a much greater effect on the transient shear response of PC nanocomposite than MMT.

Figure 12.36 gives the rheological behavior of intermittent shear flow, namely, variations of $\sigma^+(\dot{\gamma}, t)$ as a function of $\dot{\gamma}t$ when a shear flow at $\dot{\gamma} = 1.0\ s^{-1}$ was applied to 95.7/4.3 PC/MMT nanocomposite and 95.7/4.3 PC/Cloisite 30B nanocomposite after rest for different periods upon cessation of steady shear flow. It is seen in Figure 12.36a that the evolution of $\sigma^+(\dot{\gamma}, t)$ with $\dot{\gamma}t$ for 95.7/4.3 PC/MMT nanocomposite is virtually independent of the duration of rest after cessation of steady shear flow. Conversely, in Figure 12.36b we observe that the duration of rest after cessation of steady shear flow has a profound influence on the evolution of $\sigma^+(\dot{\gamma}, t)$ with $\dot{\gamma}t$ during intermittent shear flow of 95.7/4.3 PC/Cloisite 30B nanocomposite. Specifically, Figure 12.36b shows that the value of σ^+_{peak} during intermittent shear flow of the 95.7/4.3 PC/Cloisite 30B nanocomposite increases with the duration of rest up to approximately 60 min and then levels off. This transient shear response is quite different from that observed in Figure 12.36a for the 95.7/4.3 PC/MMT nanocomposite. The above observations suggest that during rest up to about 60 min after cessation of steady shear flow, in the 95.7/4.3 PC/Cloisite 30B nanocomposite attractive interactions, via hydrogen bonding, continue under quiescent conditions between the hydroxyl groups in MT2EtOH of Cloisite 30B and the carbonyl groups in PC.

Figure 12.37 describes shear-rate dependence of viscosity, showing that 95.7/4.3 PC/MMT nanocomposite exhibits weak shear-thinning behavior at $\dot{\gamma} < 1\ s^{-1}$ followed by Newtonian behavior at higher shear rates, while 95.7/4.3 PC/Cloisite 30B

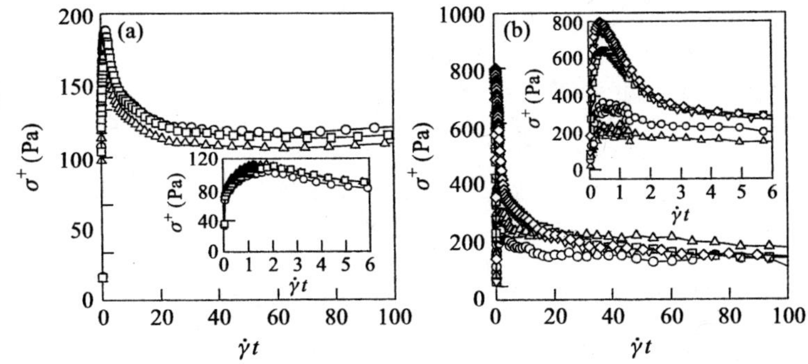

Figure 12.36 Variations of $\sigma^+(\dot{\gamma}, t)$ with $\dot{\gamma}t$ at 260 °C during intermittent shear flow at $\dot{\gamma} = 1.0\ s^{-1}$ for (a) 95.7/4.3 PC/MMT nanocomposite after rest for different periods (min): (○) 5, (△) 30, and (□) 60, and for (b) 95.7/4.3 PC/Cloisite 30B nanocomposite after rest for different periods (min): (○) 5, (△) 15, (□) 30, (▽) 60, and (◇) 90. (Reprinted from Lee and Han, *Polymer* 44:4573. Copyright © 2003, with permission from Elsevier.)

nanocomposite exhibits very strong shear-thinning behavior over the entire range of shear rates investigated. The dependence of η on $\dot{\gamma}$ given in Figure 12.37 is very similar to the dependence of $|\eta^*|$ on ω given in Figure 12.33. Again, the difference in shear-rate dependence of η between the two nanocomposites lies in the presence of attractive interactions, via hydrogen bonding, between the carbonyl groups in PC and the hydroxyl groups in MT2EtOH residing at the surface of Cloisite 30B in the 95.7/4.3 PC/Cloisite 30B nanocomposite.

12.3.2 Rheology of Organoclay Nanocomposites Based on Block Copolymer

Block copolymers have an advantage over flexible homopolymers in introducing functional groups. Thus, block copolymers are very attractive for use in nanocomposites.

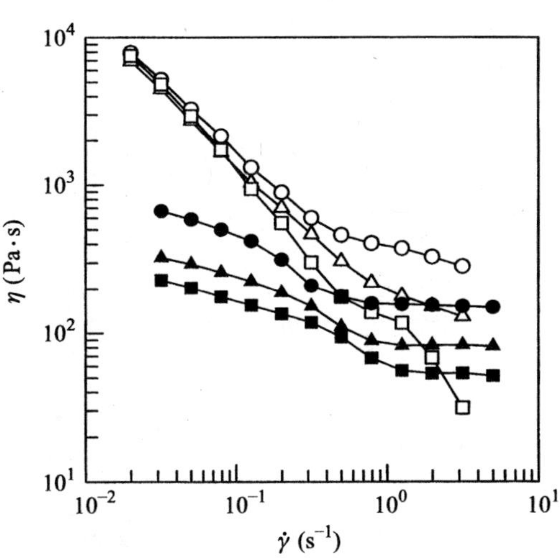

Figure 12.37 Plots of log η versus log $\dot{\gamma}$ for 95.7/4.3 PC/MMT nanocomposite at various temperatures (°C): (●) 240, (▲) 260, and (■) 280, and for 95.7/4.3 PC/Cloisite 30B nanocomposite at various temperatures (°C): (○) 240, (△) 260, and (□) 280. (Reprinted from Lee and Han, *Polymer* 44:4573. Copyright © 2003, with permission from Elsevier.)

In this section, we present the rheological behavior of organoclay nanocomposites based on polystyrene-*block*-polyisoprene (SI diblock) copolymer or polystyrene-*block*-hydroxylated polyisoprene (SIOH diblock) copolymer. This section will first show how important functionalization is for achieving a high degree of exfoliation of organoclay aggregates in nanocomposites, and will then describe the rheological behavior of such nanocomposites. We will show that linear dynamic viscoelastic properties, when properly interpreted, can be used very effectively to assess qualitatively the degree of dispersion of organoclays in nanocomposites.

Lee and Han (2003b) investigated linear dynamic viscoelastic properties of organoclay nanocomposites based on both SI diblock copolymer and SIOH diblock copolymer. An SIOH diblock copolymer (referred to as SI-14/3-OH) was obtained by hydroxylation of the polyisoprene (PI) block of a highly asymmetric SI diblock copolymer (SI-14/3, with $M_n = 1.7 \times 10^4$ and $M_w/M_n = 1.08$). SI-14/3 was a homogeneous (disordered) diblock copolymer at room temperature, while SI-14/3-OH was a microphase-separated diblock copolymer having hexagonally packed, cylindrical microdomains of PS. They employed SI-14/3-OH to prepare nanocomposites with MMT or Cloisite 30B. For comparison, they also employed a nearly symmetric, lamella-forming SI diblock copolymer (SI-10/9, with $M_n = 1.94 \times 10^4$ and $M_w/M_n = 1.02$) to prepare nanocomposites with Cloisite 30B. Below we summarize their findings.

12.3.2.1 XRD Patterns

Figure 12.38 gives XRD patterns of organoclay Cloisite 30B and its nanocomposites with SI-10/9 and SI-14/3-OH. The XRD diffraction in Figure 12.38 shows that Cloisite 30B has a d_{001} spacing of 1.85 nm, and that the d_{001} spacing of the 95/5 (SI-10/9)/Cloisite 30B nanocomposite, in which 95 and 5 refer to the weight percent of SI-10/9 and Cloisite 30B, respectively, is little different from that of Cloisite 30B itself. This suggests that the block copolymer SI-10/9 has not intercalated the Cloisite 30B aggregates. However, in Figure 12.38 we observe *no* sharp reflections in the XRD diffraction pattern of (SI-14/3-OH)/Cloisite 30B nanocomposites although

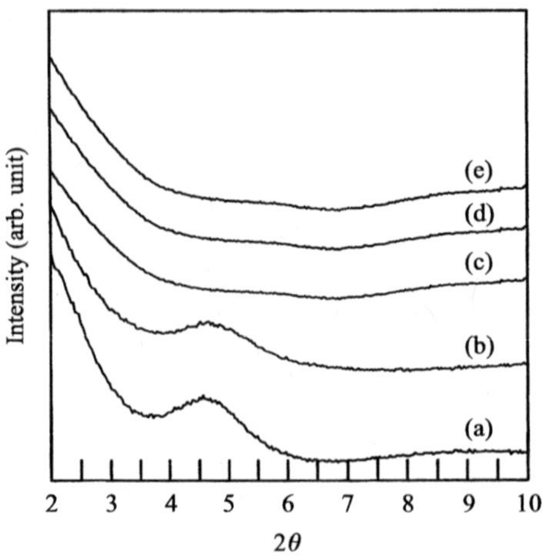

Figure 12.38 XRD patterns for (a) Cloisite 30B, (b) (SI-10/9)/Cloisite 30B nanocomposite, (c) 95/5 (SI-14/3-OH)/Cloisite 30B nanocomposite, (d) 97/3 (SI-14/3-OH)/Cloisite 30B nanocomposite, and (e) 99/1 (SI-14/3-OH)/Cloisite 30B nanocomposite. (Reprinted from Lee and Han, *Macromolecules* 36:804. Copyright © 2003, with permission from the American Chemical Society.)

a subtle difference seems to exist between the three different concentrations of Cloisite 30B. The lack of discernible intensity peak in Figure 12.38 for the three (SI-14/3-OH)/Cloisite 30B nanocomposites with varying amounts of Cloisite 30B indicates that a significant degree of dispersion of the Cloisite 30B aggregates might have occurred within the block copolymer SI-14/3-OH. The observed increase in dispersion of Cloisite 30B aggregates by SI-14/3-OH is believed to have originated from the presence of attractive interactions between the hydroxyl groups in SI-14/3-OH and the hydroxyl groups in MT2EtOH residing on the surface of Cloisite 30B.

12.3.2.2 TEM Images

Figure 12.39 gives TEM images of 95/5 (SI-14/3-OH)/Cloisite 30B nanocomposite and 95/5 (SI-10/9)/Cloisite 30B nanocomposite, where the dark areas represent the Cloisite 30B and the gray/white areas represent the SI-14/3-OH matrix. It is clearly seen in Figure 12.39 that SI-14/3-OH in the 95/5 (SI-14/3-OH)/Cloisite 30B nanocomposite dispersed Cloisite 30B aggregates fairly well, whereas SI-10/9 in the 95/5 (SI-10/9)/Cloisite 30B nanocomposite did not (i.e., large aggregates of Cloisite 30B are bundled together), indicating that the hydroxylation of SI-14/3 has played the major role in dispersing Cloisite 30B aggregates.

12.3.2.3 FTIR Spectra

Figure 12.40 gives in situ FTIR spectra at various temperatures ranging from 30 to 240 °C for neat block copolymer SI-14/3-OH, 95/5 (SI-14/3-OH)/Cloisite 30B nanocomposite, and 95/5 (SI-10/9)/Cloisite 30B nanocomposite, where hydrogen-bonding observations are worth noting. The area under the absorption band at about 3330 cm^{-1} for SI-14/3-OH is largest at 30 °C and then decreases as the temperature is increased. Notice in Figure 12.40a that the absorption band at about 3330 cm^{-1}

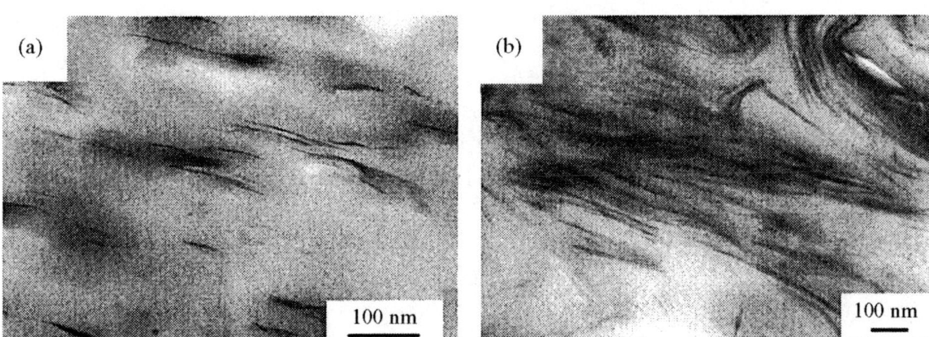

Figure 12.39 TEM images of (a) 95/5 (SI-14/3-OH)/Cloisite 30B nanocomposite and (b) 95/5 (SI-10/9)/Cloisite 30B nanocomposite, where the dark areas represent the organoclay and the gray/white areas represent the block copolymer matrix. (Reprinted from Lee and Han, *Macromolecules* 36:804. Copyright © 2003, with permission from the American Chemical Society.)

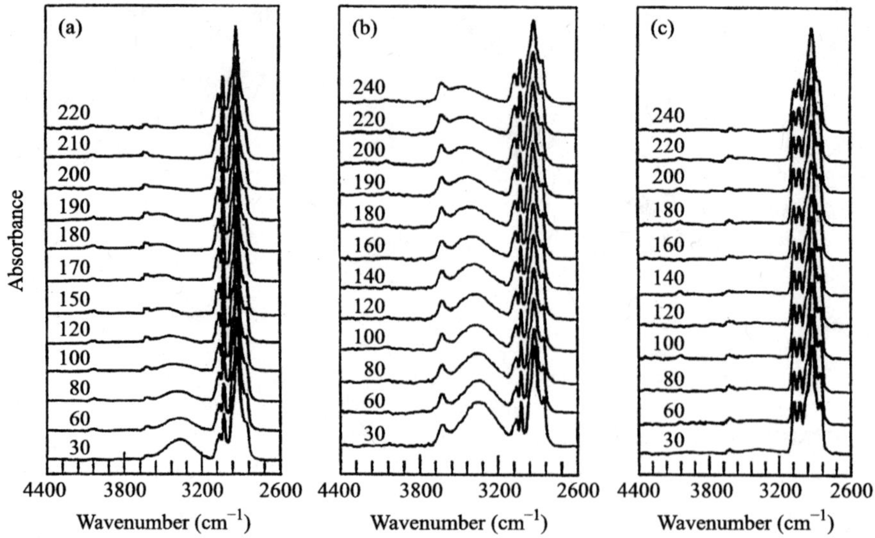

Figure 12.40 In situ FTIR spectra at varying temperatures (°C) for (a) SI-14/3-OH, (b) 95/5 (SI-14/3-OH)/Cloisite 30B nanocomposite, and (c) 95/5 (SI-10/9)/Cloisite 30B nanocomposite. (Reprinted from Lee and Han, *Macromolecules* 36:804. Copyright © 2003, with permission from the American Chemical Society.)

virtually disappears at 200 °C, which is very close to the T_{ODT} of SI-14/3-OH. This observation indicates that the hydroxyl groups in the block copolymer form hydrogen bonds, which are weakened with increasing temperature. In Figure 12.40b, we observe that at 30 °C the area under the absorption band at about 3330 cm^{-1} for the 95/5 (SI-14/3-OH)/Cloisite 30B nanocomposite is indeed very large compared with that for the SI-14/3-OH shown in Figure 12.40a. Very interestingly, the absorption band at about 3330 cm^{-1} for the 95/5 (SI-14/3-OH)/Cloisite 30B nanocomposite persists at temperatures as high as 240 °C, the highest experimental temperature employed. Notice that this is not the case for SI-14/3-OH (compare Figure 12.40b with Figure 12.40a). This observation suggests that there is strong hydrogen bonding between the hydroxyl groups in SI-14/3-OH and the hydroxyl groups in MT2EtOH of Cloisite 30B. In Figure 12.40c, virtually no hydrogen bonding is discernible in the 95/5 (SI-10/9)/Cloisite 30B nanocomposite. This should not surprise us because SI-10/9 does not have any hydroxyl groups. Thus, small amounts of hydroxyl groups present in the 5 wt % Cloisite 30B are diluted by the 95 wt % SI-10/9, making it virtually impossible for us to discern the presence of hydrogen bonds in the 95/5 (SI-10/9)/Cloisite 30B nanocomposite.

12.3.2.4 Linear Dynamic Viscoelastic Properties

Figure 12.41 gives $\log G'$ versus $\log \omega$ plots and $\log G''$ versus $\log \omega$ plots for SI-10/9 at various temperatures. As can be seen from the TEM image given in the inset of Figure 12.41, SI-10/9 has lamellar microdomains. In Figure 12.41 we observe that

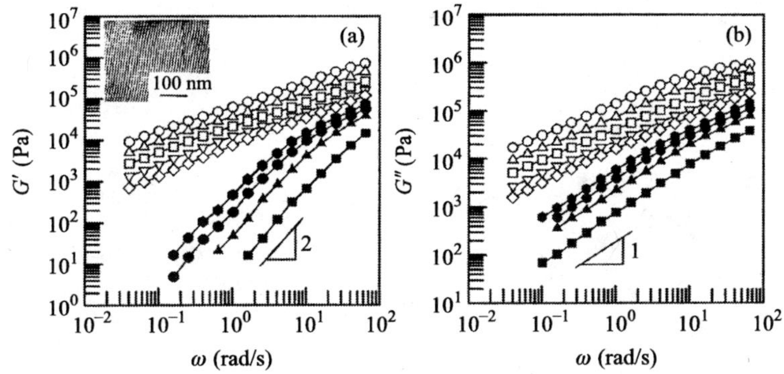

Figure 12.41 (a) Plots of log G' versus log ω and (b) log G'' versus log ω for SI-10/9 diblock copolymer at various temperatures (°C): (○) 95, (△) 100, (□) 105, (▽) 109, (◇) 112, (●) 115, (▲) 118, (■) 121, and (▼) 125. The inset gives a TEM image of SI-10/9 taken at room temperature. (Reprinted from Lee and Han, *Macromolecules* 36:804. Copyright © 2003, with permission from the American Chemical Society.)

values of G' are very large at low temperatures, with a slope much less than 2 in the terminal region, indicating that the block copolymer exhibits linear viscoelasticity somewhere between solidlike and liquidlike states. As the temperature is increased to a certain critical temperature (115 °C), values of G' in the terminal region drop rapidly. The log G' versus log ω plot is then shifted downwards with a slope of 2 in the terminal region, characteristic of liquidlike behavior, while a change in the slope of log G'' versus log ω plot at 115 °C is relatively small. Similar rheological behavior of microphase-separated block copolymers has been presented in Chapter 8.

Figure 12.42 gives log G' versus log ω plots and log G'' versus log ω plots for 95/5 (SI-10/9)/Cloisite 30B nanocomposite at various temperatures. Comparison of Figure 12.42a with Figure 12.41a indicates that the addition of 5 wt % Cloisite 30B to SI-10/9 has changed drastically the temperature dependence of log G' versus log ω plots. Specifically, below 115 °C the slope of log G' versus log ω plot for the 95/5 (SI-10/9)/Cloisite 30B nanocomposite in the terminal region is much greater than that for SI-10/9, indicating that the nanocomposite's behavior is less solidlike as compared with the neat diblock copolymer SI-10/9. This observation suggests that the lamellar microdomains of SI-10/9 became less solidlike in the presence of Cloisite 30B. Also, comparison of Figure 12.42b with Figure 12.41b indicates that the log G'' versus log ω plot is less sensitive to the addition of 5 wt % Cloisite 30B to SI-10/9 than is the log G' versus log ω plot. Of particular interest in Figure 12.42 is that at 115 °C and higher temperatures the log G' versus log ω plots in the terminal region have a slope of 2 and the log G'' versus log ω plots in the terminal region have a slope of 1, indicating liquidlike behavior, although the nanocomposite has 5 wt % Cloisite 30B. That is, at 115 °C and higher temperatures, the linear dynamic viscoelasticity of the 95/5 (SI-10/9)/Cloisite 30B nanocomposite is little different from that of SI-10/9.

Figure 12.43a gives log G' versus log G'' plots for the neat diblock copolymer SI-10/9 at various temperatures, showing that the log G' versus log G'' plot in the

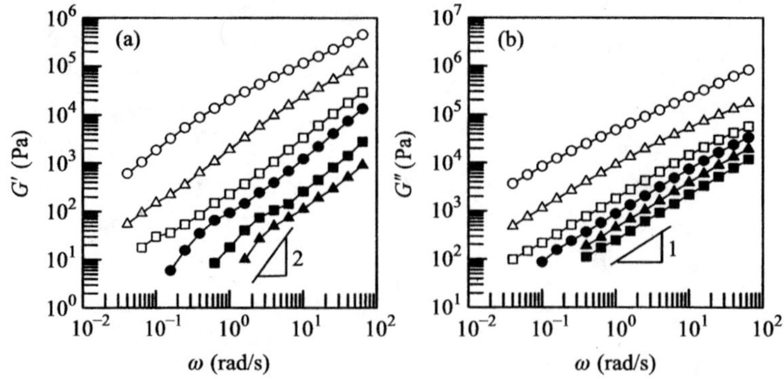

Figure 12.42 (a) Plots of log G' versus log ω and (b) log G'' versus log ω for 95/5 (SI-10/9)/Cloisite 30B nanocomposite at various temperatures (°C): (○) 90, (△) 100, (□) 110, (●) 115, (▲) 120, and (■) 125. (Reprinted from Lee and Han, *Macromolecules* 36:804. Copyright © 2003, with permission from the American Chemical Society.)

terminal region at temperatures below approximately 115 °C has a slope much less than 2 and then shifts suddenly downward with a slope of 2 in the terminal region as the temperature is increased to 115 °C and higher. Following the rheological criterion presented in Chapter 8, from the log G' versus log G'' plots we determine the T_{ODT} of SI-10/9 to be approximately 115 °C. The inset of Figure 12.43a describes the results of isochronal dynamic temperature sweep experiments at $\omega = 0.01$ rad/s. According to the

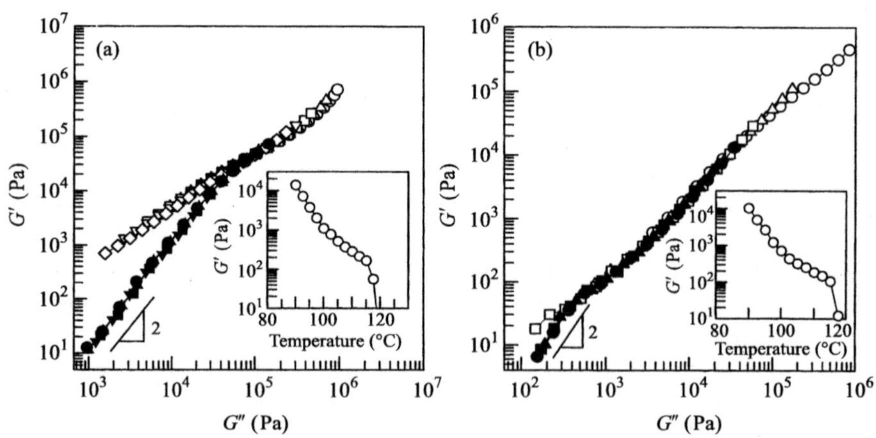

Figure 12.43 (a) Plots of log G' versus log G'' for SI-10/9 diblock copolymer at various temperatures (°C): (○) 95, (△) 100, (□) 105, (▽) 109, (◇) 112, (●) 115, (▲) 118, (■) 121, and (▼) 125. (b) Plots of log G' versus log G'' for 95/5 (SI-10/9)/Cloisite 30B nanocomposite at various temperatures (°C): (○) 90, (△) 100, (□) 110, (●) 115, (▲) 120, and (■) 125. The insets show variations of G' with temperature during isochronal dynamic temperature sweep experiments at $\omega = 0.01$ rad/s. (Reprinted from Lee and Han, *Macromolecules* 36:804. Copyright © 2003, with permission from the American Chemical Society.)

rheological criteria presented in Chapter 8, from the inset of Figure 12.43a we determine the T_{ODT} of SI-10/9 to be approximately 117 °C, which is very close to the T_{ODT} determined from the $\log G'$ versus $\log G''$ plot. Note that two different rheological criteria give rise to a virtually identical value of T_{ODT} for the lamella-forming block copolymer SI-10/9 (see Chapter 8).

Figure 12.43b gives $\log G'$ versus $\log G''$ plots for 95/5 (SI-10/9)/Cloisite 30B nanocomposite, showing that the nanocomposite is virtually independent of temperature over the entire range of temperatures tested. However, the slope of the $\log G'$ versus $\log G''$ plot in the terminal region is less than 2 for values of G'' above approximately 500 Pa, and then becomes very close to 2 for G'' less than about 500 Pa at 115 °C and higher temperatures. According to the rheological criteria presented in Chapter 8, the temperature independence of a $\log G'$ versus $\log G''$ plot is not sufficient for one to conclude that a block copolymer is in the disordered state; that is, in addition to temperature independence, a $\log G'$ versus $\log G''$ plot must have a slope of 2 in the terminal region for one to conclude that a block copolymer is in the disordered state. Following these rheological criteria, from the $\log G'$ versus $\log G''$ plots in Figure 12.43b we determine the T_{ODT} of the 95/5 (SI-10/9)/Cloisite 30B nanocomposite to be approximately 115 °C. Again, the T_{ODT} determined from the $\log G'$ versus $\log G''$ plot agrees very well with that determined from the inset of Figure 12.43b. The above observations indicate that the addition of 5 wt % Cloisite 30B to SI-10/9 did not change its T_{ODT}, suggesting that little interaction, physical or chemical, took place between Cloisite 30B and the lamella-forming SI diblock copolymer SI-10/9. However, the above observation should not surprise us because SI-10/9 does not have any functional groups that can have attractive interactions with the hydroxyl groups in the surfactant MT2EtOH residing at the surface of Cloisite 30B. Note that 5 wt % Cloisite 30B consists of 3.4 wt % clay and 1.6 wt % surfactant MT2EtOH. In the absence of attractive interactions, 3.4 wt % clay in the 95/5 (SI-10/9)/Cloisite 30B nanocomposite is not sufficient to exhibit "filler effect" in the terminal region.

Figure 12.44a gives $\log G'$ versus $\log \omega$ plots and $\log G''$ versus $\log \omega$ plots for SI-14/3–OH, which has hexagonally packed, cylindrical microdomains of PS (see the inset), at various temperatures. In Figure 12.44a we observe that values of G' are very large at temperatures below 198 °C and have a slope much less than 2 in the terminal region. This represents a fluid somewhere between solidlike and liquidlike states, similar to that observed in Figure 12.41a for the lamella-forming block copolymer SI-10/9. As the temperature is increased to 198 °C, values of G' in the terminal region drop dramatically and the $\log G'$ versus $\log \omega$ plot in the terminal region has a slope of 2, again very similar to that observed in Figure 12.41a for SI-10/9. The frequency dependence of G'' for SI-14/3-OH given in Figure 12.44b looks very complicated at temperatures below 198 °C, but $\log G''$ versus $\log \omega$ plots at 198 °C and higher temperatures show liquidlike behavior (a slope of 1 in the terminal region).

Figure 12.45 gives $\log G'$ versus $\log \omega$ plots and $\log G''$ versus $\log \omega$ plots for the 95/5(SI-14/3-OH)/Cloisite 30B nanocomposite at various temperatures, showing that over the entire range of ω tested, both G' and G'' increase as the temperature increases, a trend that is quite opposite to that observed in Figure 12.42 for the 95/5 (SI-10/9)/Cloisite 30B nanocomposite. This observation indicates that 5 wt % Cloisite 30B in the 95/5 (SI-14/3-OH)/Cloisite 30B nanocomposite has induced attractive interactions, via hydrogen bonding, between the hydroxyl groups of SI-14/3-OH and

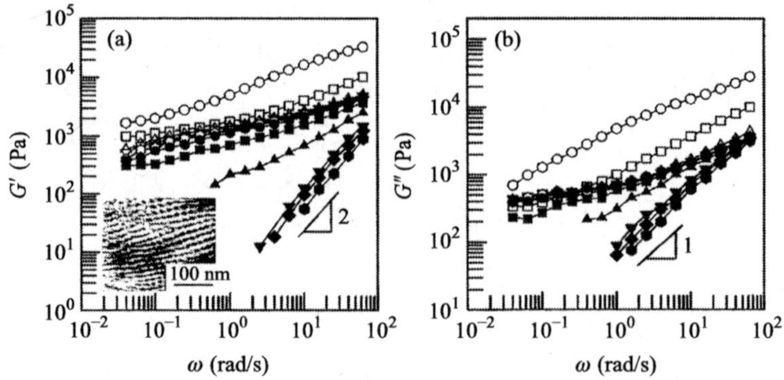

Figure 12.44 (a) Plots of log G' versus log ω and (b) plots of log G'' versus log ω for SI-14/3-OH at various temperatures (°C): (○) 130, (□) 150, (△) 170, (▽) 180, (◇) 183, (⬡) 186, (●) 189, (■) 192, (▲) 195, (▼) 198, (◆) 200, and (⬢) 205. (Reprinted from Lee and Han, *Macromolecules* 36:804. Copyright © 2003, with permission from the American Chemical Society.)

the hydroxyl groups in the surfactant MT2EtOH residing at the surface of Cloisite 30B (see Figure 12.40).

Figure 12.46 gives log G' versus log ω plots for SI-14/3-OH and 95/5 (SI-14/3-OH)/Cloisite 30B nanocomposite at various temperatures, and Figure 12.47 gives log $|\eta^*|$ versus log ω plots for SI-14/3-OH and 95/5 (SI-14/3-OH)/Cloisite 30B nanocomposite at various temperatures. Again, we observe that values of G' and $|\eta^*|$ decrease with an increase in temperature for SI-14/3-OH, whereas values of G' and $|\eta^*|$ increase with an increase in temperature for (SI-14/3-OH)/Cloisite 30B nanocomposite. The difference between SI-14/3-OH and (SI-14/3-OH)/Cloisite 30B nanocomposite is attributable to

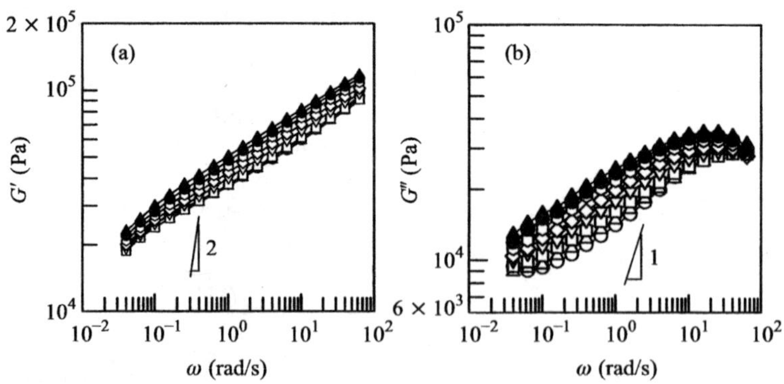

Figure 12.45 (a) Plots of log G' versus log ω and (b) plots of log G'' versus log ω for 95/5 (SI-14/3-OH)/Cloisite 30B nanocomposite at various temperatures (°C): (○) 170, (△) 180, (□) 190, (▽) 200, (◇) 210, (⬡) 220, (●) 230, and (▲) 240. (Reprinted from Lee and Han, *Macromolecules* 36:804. Copyright © 2003, with permission from the American Chemical Society.)

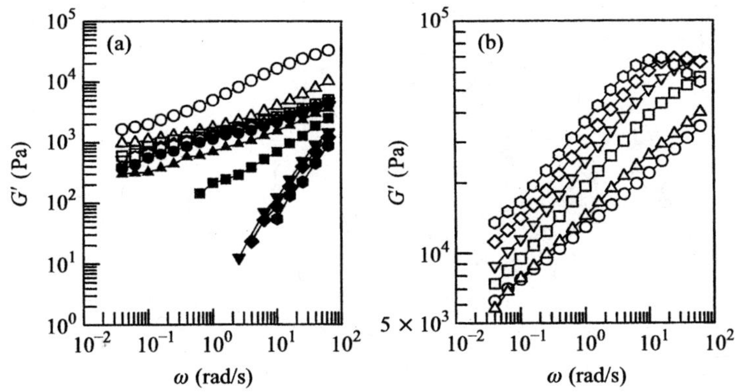

Figure 12.46 (a) Plots of log G' versus log ω for SI-14/3-OH at various temperatures (°C): (○) 130, (□) 150, (△) 170, (▽) 180, (◇) 183, (⬡) 186, (●) 189, (■) 192, (▲) 195, (▼) 198, (◆) 200, and (⬢) 205. (b) Plots of log G' versus log ω for 95/5 (SI-14/3-OH)/Cloisite 30B nanocomposite at various temperatures (°C): (○) 170, (△) 185, (□) 200, (▽) 215, (◇) 225, and (⬡) 240. (Reprinted from Lee and Han, *Macromolecules* 36:804. Copyright © 2003, with permission from the American Chemical Society.)

the presence of attractive interactions, via hydrogen bonding, between the hydroxyl groups of SI-14/3-OH and the hydroxyl groups in the surfactant MT2EtOH residing at the surface of Cloisite 30B (see Figure 12.40).

Figure 12.48 compares log G' versus log G'' plots of SI-14/3-OH with those of 95/5 (SI-14/3-OH)/Cloisite 30B nanocomposite at various temperatures, in which the insets show the results of the isochronal dynamic temperature sweep experiments at

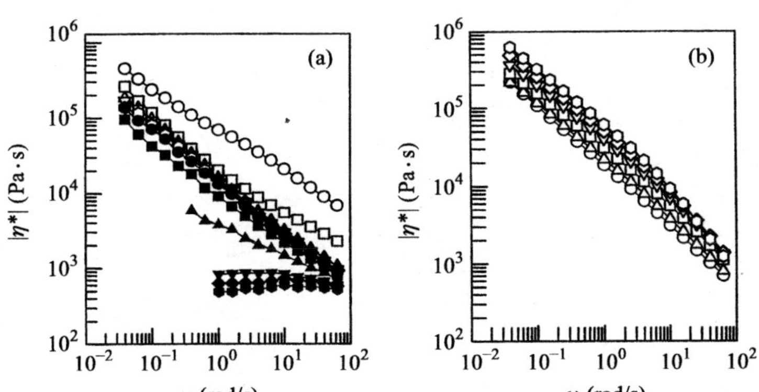

Figure 12.47 (a) Plots of log $|\eta^*|$ versus log ω for SI-14/3-OH at various temperatures (°C): (○) 130, (□) 150, (△) 170, (▽) 180, (◇) 183, (⬡) 186, (●) 189, (■) 192, (▲) 195, (▼) 198, (◆) 200, and (⬢) 205. (b) Plots of log $|\eta^*|$ versus log ω for 95/5 (SI-14/3-OH)/Cloisite 30B nanocomposite at various temperatures (°C): (○) 170, (△) 185, (□) 200, (▽) 215, (◇) 225, and (⬡) 240. (Reprinted from Lee and Han, *Macromolecules* 36:804. Copyright © 2003, with permission from the American Chemical Society.)

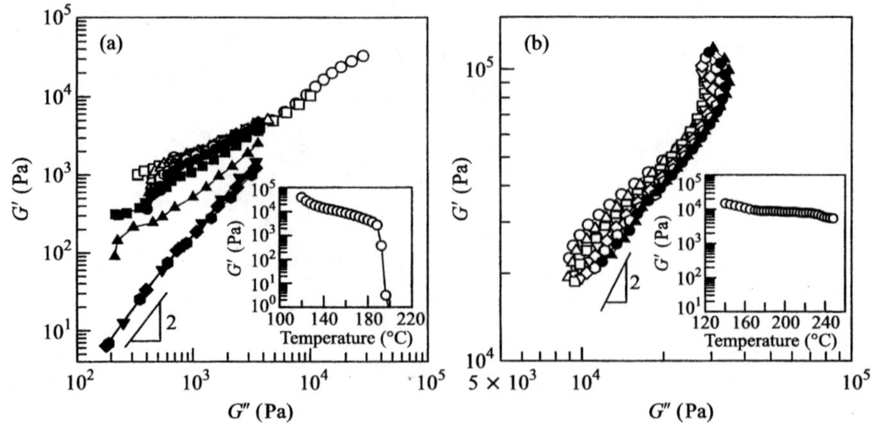

Figure 12.48 (a) Plots of log G' versus log G'' for SI-14/3-OH diblock copolymer at various temperatures (°C): (○) 130, (□) 150, (△) 170, (▽) 180, (◇) 183, (⬡) 186, (●) 189, (■) 192, (▲) 195, (▼) 198, (◆) 200, and (⬢) 205. The inset shows variations of G' with temperature during isochronal dynamic temperature sweep experiments at $\omega = 0.01$ rad/s. (b) Plots of log G' versus log G'' for 95/5 (SI-14/3-OH)/Cloisite 30B nanocomposite at various temperatures (°C): (○) 170, (△) 180, (□) 190, (▽) 200, (◇) 210, (⬡) 220, (●) 230, and (▲) 240. The inset shows variations of G' with temperature during isochronal dynamic temperature sweep experiments at $\omega = 0.01$ rad/s. (Reprinted from Lee and Han, *Macromolecules* 36:804. Copyright © 2003, with permission from the American Chemical Society.)

$\omega = 0.01$ rad/s. From the log G' versus log G'' plot given in Figure 12.48a we determine the T_{ODT} of SI-14/3-OH to be approximately 198 °C, which is very close to the temperature at which G', according to the inset, begins to drop precipitously. Conversely, from the inset of Figure 12.48b we observe that values of G' stay more or less constant as the temperature is increased to approximately 250 °C, the highest experimental temperature employed; that is, in the presence of 5 wt % Cloisite 30B, the T_{ODT} of SI-14/3-OH has become so high that it could not be measured. In other words, in the presence of 5 wt % Cloisite 30B, the hexagonally packed, cylindrical microdomains of PS in SI-14/3-OH have persisted even at such a high temperature as 250 °C. The same observation can be made from the log G' versus log G'' plots in Figure 12.48b because the log G' versus log G'' plot in the terminal region with a slope much less than 2 continues to shift downward as the temperature is increased from 170 to 240 °C, indicating that SI-14/3-OH is still in the ordered state over the entire range of temperatures tested. The above observations indicate that the T_{ODT} of SI-14/3-OH in the 95/5 (SI-14/3-OH)/Cloisite 30B nanocomposite is much higher than 240 °C. Note that the T_{ODT} of SI-10/9 in the presence of the organoclay Cloisite 30B has not changed (see Figure 12.43b). Thus, we can conclude that an increase in T_{ODT} of the 95/5 (SI-14/3-OH)/Cloisite 30B nanocomposite observed in Figure 12.48b is attributable to the presence of attractive interactions, via hydrogen bonding, between the hydroxyl groups of SI-14/3-OH and the hydroxyl groups in MT2EtOH residing at the surface of Cloisite 30B.

It is then clear that the unusual linear dynamic viscoelasticity observed in (SI-14/3-OH)/Cloisite 30B nanocomposites has originated from the presence of hydrogen bonding between the hydroxyl groups in SI-14/3-OH and the hydroxyl groups in MT2EtOH of Cloisite 30B. Specifically, we attribute an increase in G', G'', and $|\eta^*|$ with increasing temperature, observed for the (SI-14/3-OH)/Cloisite 30B nanocomposites, to an increase in the availability of the surface areas of the dispersed silicate layers of the Cloisite 30B platelets as the temperature increases, thereby increasing the extent of hydrogen bonding between the hydroxyl groups in MT2EtOH of Cloisite 30B and the hydroxyl groups in SI-14/3-OH. That is, as the number of sites available for hydrogen bonds between the hydroxyl groups in MT2EtOH of Cloisite 30B and the hydroxyl groups in SI-14/3-OH increases, values of G', G'', and $|\eta^*|$ in the corresponding nanocomposites is expected to increase. Here, we postulate that the number of sites available for hydrogen bonds between the hydroxyl groups in MT2EtOH of Cloisite 30B and the hydroxyl groups in SI-14/3-OH increases as the temperature increases, leading to an increase in attractive interactions. We can use the same argument to explain why the T_{ODT} of SI-14/3-OH is increased in the presence of Cloisite 30B (see Figure 12.48). The segment–segment interactions between PS and PIOH are more repulsive than those between PS and PI (Lee and Han 2002). Thus it is reasonable to expect that repulsive segment–segment interactions between PS and PIOH would increase when the hydroxyl groups in PIOH form hydrogen bonds with the hydroxyl groups in MT2EtOH residing at the surface of Cloisite 30B. As has been discussed in Chapter 8, the T_{ODT} of a block copolymer increases as the segmental interaction parameter increases (i.e., the block components become more repulsive). Thus, we can conclude that the T_{ODT} of SI-14/3-OH in the presence of Cloisite 30B, where hydrogen bonding occurs between the hydroxyl groups in SI-14/3-OH and the hydroxyl groups in MT2EtOH of Cloisite 30B, is expected to be higher than the T_{ODT} of SI-14/3-OH.

12.3.3 Rheology of Organoclay Nanocomposites Based on End-Functionalized Polymer

In this section, we present the dispersion characteristics and rheology of end-functionalized polymers investigated by Zha et al. (2005). In their study, a PS and an SI diblock copolymer were synthesized via anionic polymerization. Then, the chain end of the PS was carboxylated to obtain end-functionalized PS-t-COOH, and the chain end of the PI block in the SI diblock copolymer was carboxylated to obtain end-functionalized SI-t-COOH. Subsequently, both PS-t-COOH and SI-t-COOH were neutralized using sodium hydroxide (NaOH) to obtain PS-t-COONa and SI-t-COONa. Each of the end-functionalized polymers was mixed with Cloisite 30B or Cloisite 20A to prepare nanocomposites. Table 12.2 gives sample codes of the eight organoclay nanocomposites prepared.

12.3.3.1 Rheology of Organoclay Nanocomposites Based on End-Functionalized Polystyrene

Figure 12.49 gives XRD patterns of organoclay nanocomposites based on neat PS and PS terminated by –COOH group or –COONa group. The following observations are worth noting in Figure 12.49. The XRD patterns indicate that the *d*-spacing of PS/Cloisite 30B nanocomposite is 1.85 nm, which is virtually the same as that of

Table 12.2 Sample codes of the nanocomposites based on end-functionalized polymers[a]

Sample Code	*Functionality at Chain End*
(a) End-functionalized PS/organoclay nanocomposites	
(PS-t-COOH)/Cloisite 20A	PS terminated by –COOH group
(PS-t-COOH)/Cloisite 30B	PS terminated by –COOH group
(PS-t-COONa)/Cloisite 20A	PS terminated by –COONa group
(PS-t-COONa)/Cloisite 30B	PS terminated by –COONa group
(b) End-functionalized SI diblock copolymer/organoclay nanocomposites	
(SI-t-COOH)/Cloisite 20A	PI block in SI terminated by –COOH group
(SI-t-COOH)/Cloisite 30B	PI block in SI terminated by –COOH group
(SI-t-COONa)/Cloisite 20A	PI block in SI terminated by –COONa group
(SI-t-COONa)/Cloisite 30B	PI block in SI terminated by –COONa group

[a]All nanocomposites have 5 wt % organoclay. Based on Zha et al. 2005.

Cloisite 30B, and the *d*-spacing of PS/Cloisite 20A nanocomposite is 1.77 nm, which is slightly smaller than that of Cloisite 20A (2.42 nm). This is not surprising because no attractive interaction can be expected between neat PS and the surfactant residing at the surface of Cloisite 30B or Cloisite 20A. In Figure 12.49 we still observe a broad XRD reflection peak having the *d*-spacing of 1.85 nm in both (PS-t-COOH)/Cloisite 30B and (PS-t-COOH)/Cloisite 20A nanocomposites, although the peak height of the XRD patterns for the two nanocomposites is much smaller than those for the PS/Cloisite 30B and PS/Cloisite 20A nanocomposites. This observation suggests that the –COOH group attached to the chain end of PS might not have sufficiently strong attractive interactions with the surfactant residing at the surface of Cloisite 30B and Cloisite 20A, giving rise to a low degree of dispersion of organoclay aggregates in the respective nanocomposites.

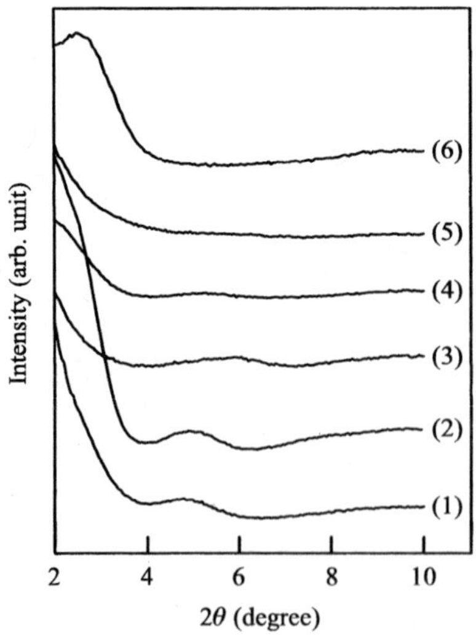

Figure 12.49 XRD patterns for (1) PS/Cloisite 30B nanocomposite, (2) PS/Cloisite 20A nanocomposite, (3) (PS-t-COOH)/Cloisite 30B nanocomposite, (4) (PS-t-COOH)/Cloisite 20A nanocomposite, (5) (PS-t-COONa)/Cloisite 30B nanocomposite, and (6) (PS-t-COONa)/Cloisite 20A nanocomposite. (Reprinted from Zha et al., *Macromolecules* 38:8418. Copyright © 2005, with permission from the American Chemical Society.)

This can be explained by the following observations. The –COOH group present at the chain end of PS-t-COOH is not expected to have any attractive interactions with the surfactant 2M2HT residing at the surface of Cloisite 20A, while it can potentially form hydrogen bonds with the hydroxyl group in the surfactant MT2EtOH residing at the surface of Cloisite 30B. However, only one –COOH group at the end of PS-t-COOH would not be sufficient to form strong attractive interactions with the hydroxyl groups in the surfactant MT2EtOH residing at the surface of Cloisite 30B. The situation would be quite different if many –COOH groups had been present in the entire PS molecule.

Interestingly, in Figure 12.49 we observe featureless XRD patterns in both (PS-t-COONa)/Cloisite 20A and (PS-t-COONa)/Cloisite 30B nanocomposites, suggesting that the organoclay aggregates might be well dispersed in the respective nanocomposites. This observation suggests that sufficiently strong attractive interactions might be present between the –COONa group at the chain end of PS and the surfactant 2M2HT residing at the surface of Cloisite 20A, and between the –COONa group at the chain end of PS and the surfactant MT2EtOH residing at the surface of Cloisite 30B, in the respective nanocomposites.

Figure 12.50 gives TEM images of the nanocomposites based on neat PS and PS terminated by a –COOH group or –COONa group. In Figure 12.50 we observe that the degree of dispersion of organoclays, Cloisite 30B and Cloisite 20A, is rather poor in the matrix of neat PS and PS-t-COOH, while both organoclays, Cloisite 30B and Cloisite 20A, are very well dispersed in the matrix of PS-t-COONa. This observation is consistent with that made above from the XRD patterns given in Figure 12.49.

Figure 12.51 gives FTIR spectra for PS-t-COONa, Cloisite 30B, (PS-t-COOH)/ Cloisite 30B nanocomposite, and PS-t-COONa)/Cloisite 30B nanocomposite. The following observations are worth noting in Figure 12.51. As expected, no absorption band appears at the wavenumber of 3620 cm^{-1} for PS-t-COONa. However, two new absorption bands at the wavenumbers 3620 and 1500 cm^{-1} appear in Cloisite 30B and in the two nanocomposites. This is due to the presence of –OH groups in the surfactant MT2EtOH residing at the surface of Cloisite 30B. To facilitate our discussion here, an inset is given in Figure 12.51 showing the absorption bands at wavenumbers ranging from 1200 to 1800 cm^{-1}. It is clearly seen from the inset that an additional absorption band appears at 1600 cm^{-1} for the (PS-t-COONa)/Cloisite 30B nanocomposite; otherwise, the absorption bands are identical for both the (PS-t-COONa)/Cloisite 30B and (PS-t-COOH)/Cloisite 30B nanocomposites. This observation leads us to conclude that the additional absorption band appearing at 1600 cm^{-1} in the FTIR spectra for the (PS-t-COONa)/Cloisite 30B nanocomposite is attributable to the presence of a newly formed chemical structure, via ionic interactions, between the negatively charged $-COO^{-}$ at the chain end of PS and the positively charged N^{+} in the surfactant MT2EtOH residing at the surface of Cloisite 30B. Interestingly, Yano and coworkers (Kutsumizu et al. 1999; Tachino et al. 1994) conducted infrared spectroscopy studies of the binary blends of sodium and zinc salt ionomers of poly(ethylene-*co*-methacrylate) (PEMMA) and observed the appearance of a new asymmetric carboxylate stretching band at 1569 cm^{-1}. They assigned this stretching band to carboxylate groups bridging sodium and zinc cations. The wavenumber 1569 cm^{-1} at which the carboxylate stretching band was observed by Yano and coworkers is very close to the wavenumber 1600 cm^{-1} at which a new absorption peak was observed in Figure 12.51. Since the sodium carboxylate (–COONa) is also involved in the ionic interaction between

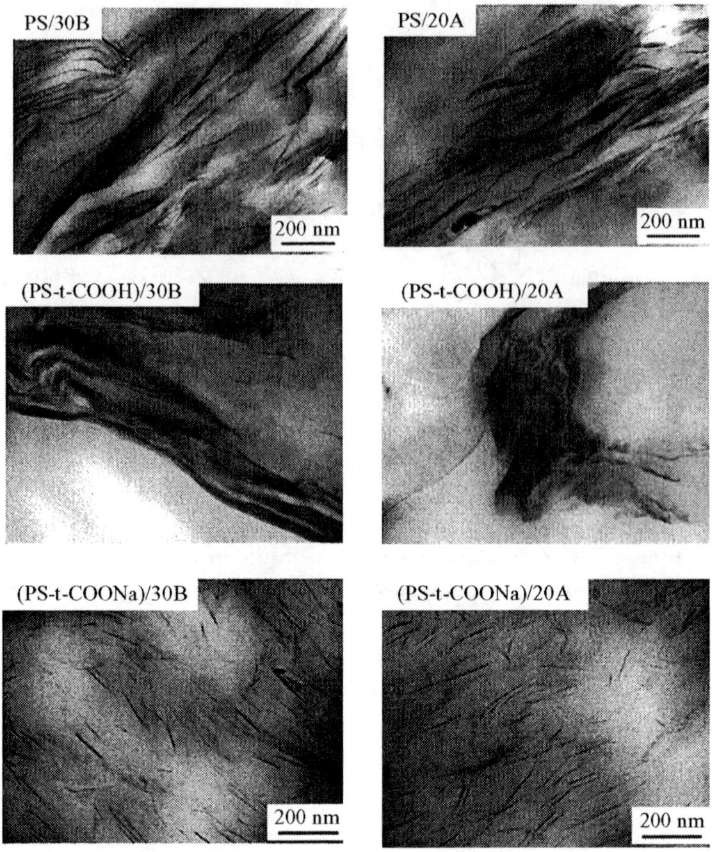

Figure 12.50 TEM images of PS/Cloisite 30B nanocomposite, PS/Cloisite 20A nanocomposite, (PS-t-COOH)/Cloisite 30B nanocomposite, (PS-t-COOH)/Cloisite 20A nanocomposite, (PS-t-COONa)/Cloisite 30B nanocomposite, and (PS-t-COONa)/Cloisite 20A nanocomposite, in which the dark areas represent organoclay, Cloisite 30B or Cloisite 20A, and the gray/white areas represent the polymer matrix. (Reprinted from Zha et al., *Macromolecules* 38:8418. Copyright © 2005, with permission from the American Chemical Society.)

the negatively charged $-COO^-$ at the chain end of PS and the positively charged N^+ in the surfactant (2M2HT or MT2EtOH), we conclude that the formation of ionic clusters is responsible for the appearance of an absorption band at 1600 cm^{-1} in the (PS-t-COONa)/Cloisite 30B nanocomposite.

Figure 12.52 describes the temperature dependence of dynamic storage modulus G' during the isochronal dynamic temperature sweep experiment at an angular frequency (ω) of 0.1 rad/s for PS, PS-t-COONa, (PS-t-COONa)/Cloisite 20A nanocomoposite, and (PS-t-COONa)/Cloisite 30B nanocomposite. The following observations are worth noting in Figure 12.52. Not only is the magnitude of G' for PS-t-COONa much larger than that for neat PS, but also the values of G' for PS-t-COONa decrease slowly, as compared with the values of G' for neat PS, with increasing temperature. We attribute this observation to the formation of ionic clusters in PS-t-COONa. It has been reported

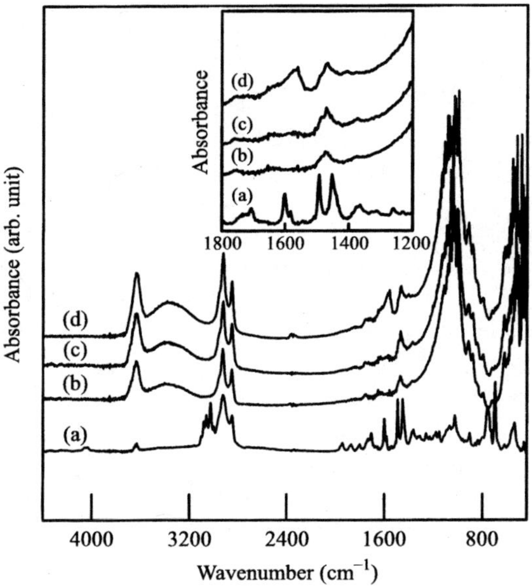

Figure 12.51 FTIR spectra for (a) PS-t-COONa, (b) Cloisite 30B, (c) (PS-t-COOH)/Cloisite 30B nanocomposite, and (d) (PS-t-COONa)/Cloisite 30B nanocomposite. (Reprinted from Zha et al., *Macromolecules* 38:8418. Copyright © 2005, with permission from the American Chemical Society.)

(Eisenberg et al. 1990; Hird and Eisenberg 1992) that ionic groups attached to the chains of an organic polymer form quadruplets, sextuplets, and higher aggregates (multiplets), leading to clusters which then reduce the mobility of the polymer chains and act like cross-links or reinforcing filler particles. Further, it has been reported (Eisenberg and Smith 1982) that the ionic clusters are very stable at elevated temperatures, giving rise to a slow decrease in elastic modulus with increasing temperature.

It is interesting to observe in Figure 12.52 that the values of G' for the (PS-t-COONa)/Cloisite 30B nanocomposite decrease extremely slowly with increasing

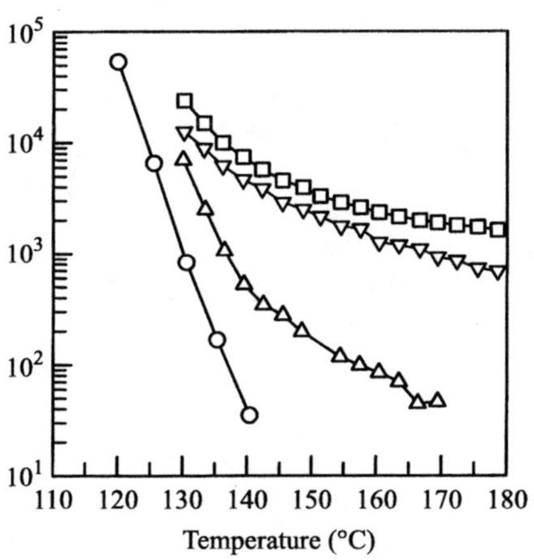

Figure 12.52 Variations of G' with temperature during the isochronal dynamic temperature sweep experiments at $\omega = 0.1$ rad/s for (○) PS, (△) PS-t-COONa, (□) (PS-t-COONa)/Cloisite 30B nanocomposite, and (▽) (PS-t-COONa)/Cloisite 20A nanocomposite. (Reprinted from Zha et al., *Macromolecules* 38:8418. Copyright © 2005, with permission from the American Chemical Society.)

temperature from 130 to 180 °C, and that the value of G' at 180 °C for the nanocomposite is about two orders of magnitude greater than that for the PS-t-COONa. Similar observation can be made for the (PS-t-COONa)/Cloisite 20A nanocomposite from Figure 12.52. We attribute the unusual temperature dependence of G' to the presence of ionic interactions between the $-COO^-$ at the chain end of PS and the N^+ present at the surface of the organoclays, Cloisite 30B or Cloisite 20A. Note that the surfactant MT2EtOH residing at the surface of Cloisite 30B, and also the surfactant 2M2HT residing at the surface of Cloisite 20A, have quaternary ammonium salt with alkyl bonds to a nitrogen atom, and the nitrogen atom bears a full positive charge. It is well established that the ionic interaction is much stronger than hydrogen bonding, although not as strong as covalent bonding. Eisenberg and coworkers (Smith and Eisenberg 1983; Rutkowska and Eisenberg 1984a, 1984b, 1985) utilized ionic interactions to enhance the miscibility of immiscible polymer blends. It therefore seems very reasonable to conclude that ionic interactions between the negatively charged $-COO^-$ at the chain end of PS and the positively charged N^+ in the surfactant (MT2EtOH in Cloisite 30B and 2M2HT in Cloisite 20A) enhanced the compatibility between PS-t-COONa and organoclay (Cloisite 30B or Cloisite 20A), and thus gave rise to a high degree of dispersion of organoclay aggregates in the matrix of PS-t-COONa (see the TEM images given in Figure 12.50). Here, we use the term "a high degree of dispersion" in the qualitative sense because it is very difficult, if not impossible, to describe quantitatively the degree of dispersion (low, medium, or high) from the TEM images given in Figure 12.50.

Figure 12.53 gives $\log G'$ versus $\log G''$ plots for (PS-t-COONa)/Cloisite 30B nanocomposites at various temperatures, ranging from 140 to 175 °C. In Figure 12.53 we observe that the $\log G'$ versus $\log G''$ plots for the (PS-t-COONa)/Cloisite 30B nanocomposite are virtually independent of temperature, suggesting that the morphological state of the nanocomposite remains constant over the entire range of temperatures investigated. Notice in Figure 12.53 that the slope of $\log G'$ versus $\log G''$ plots in the terminal region is much less than 2, which is the signature of the presence of attractive interactions between the N^+ ion in the surfactant MT2EtOH residing at the surface of Cloisite 30B and the $-COO^-$ ion at the chain end of the matrix PS-t-COONa.

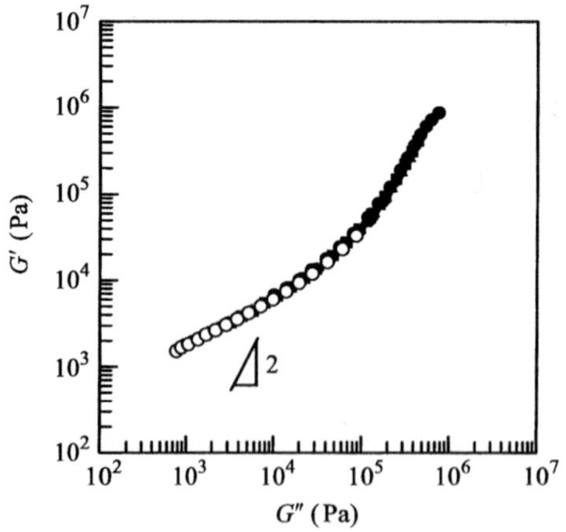

Figure 12.53 Plots of $\log G'$ versus $\log G''$ for (PS-t-COONa)/Cloisite 30B nanocomposite at various temperatures (°C): (●) 140, (■) 150, (▲) 155, (▼) 160, (◆)165, (⬣) 170, and (○) 175. (Reprinted from Zha et al., *Macromolecules* 38:8418. Copyright © 2005, with permission from the American Chemical Society.)

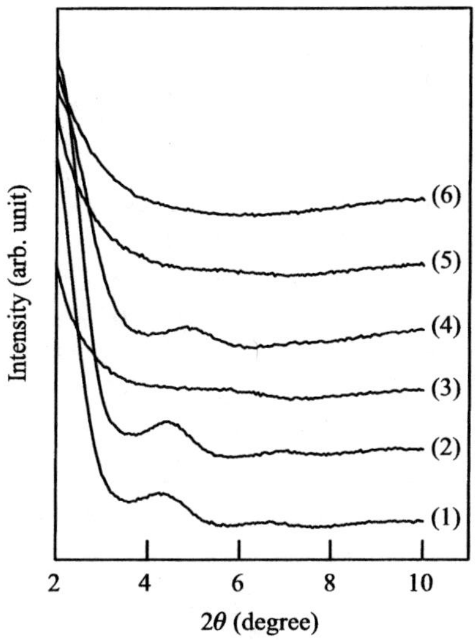

Figure 12.54 XRD patterns for (1) SI/Cloisite 30B nanocomposite, (2) SI/Cloisite 20A nanocomposite, (3) (SI-t-COOH)/Cloisite 30B nanocomposite, (4) (SI-t-COOH)/Cloisite 20A nanocomposite, (5) (SI-t-COONa)/Cloisite 30B nanocomposite, and (6) (SI-t-COONa)/Cloisite 20A nanocomposite. (Reprinted from Zha et al., *Macromolecules* 38:8418. Copyright © 2005, with permission from the American Chemical Society.)

12.3.3.2 Rheology of Organoclay Nanocomposites Based on End-Functionalized SI Diblock Copolymer

Figure 12.54 shows XRD patterns of nanocomposites based on neat SI diblock copolymer SI-1, SI-t-COOH, or SI-t-COONa, in which functional groups, –COOH group or –COONa group, are attached to the chain end of PI block of SI-1. The following observations are worth noting in Figure 12.54. (1) The XRD patterns of SI/Cloisite 30B nanocomposite show a broad reflection peak at $2\theta = 4.6°$ (the *d*-spacing of 1.85 nm) and the XRD patterns for SI/Cloisite 20A nanocomposite show a broad reflection peak at $2\theta = 4.8°$ (the *d*-spacing of 1.77 nm). Notice that the shape of the XRD patterns for SI/Cloisite 30B and SI/Cloisite 20A nanocomposites shown in Figure 12.54 is very similar to that for PS/Cloisite 30B and PS/Cloisite 20A nanocomposites (see Figure 12.49). Thus, we conclude that both SI/Cloisite 30B and SI/Cloisite 20A nanocomposites would have a low degree of dispersion of organoclay aggregates. This is not surprising because no attractive interaction is expected between the SI diblock and the surfactant (MT2EtOH or 2M2HT) residing at the surface of the organoclay employed. (2) The XRD pattern of (SI-t-COOH)/Cloisite 30B nanocomposite shows a very mild reflection peak, while the XRD pattern of (SI-t-COOH)/Cloisite 20A nanocomposite shows a conspicuous reflection peak at $2\theta = 5.0°$ (the *d*-spacing of 1.70 nm). This observation leads us to conclude that (SI-t-COOH)/Cloisite 30B nanocomposite might have a higher degree of dispersion of organoclay aggregates than (SI-t-COOH)/Cloisite 20A nanocomposite. Again, this observation seems reasonable from the point of view that the –COOH group at the chain end of the PI block of SI-1 is expected to have very weak interactions, if any, with the hydroxyl groups in the surfactant MT2EtOH residing at the surface of Cloisite 30B, while no attractive interaction is expected between the –COOH group at the chain end of the PI block of SI-1 and the surfactant 2M2HT residing at the surface of Cloisite 20A. (3) Both (SI-t-COONa)/Cloisite 30B and (SI-t-COONa)/Cloisite 20A nanocomposites show featureless XRD patterns over the entire range of 2θ angles investigated, suggesting that a very high degree of dispersion of organoclay aggregates is expected in the matrix of SI-t-COONa for the respective nanocomposites. This observation is very similar to that made above for the (PS-t-COONa)/Cloisite 30B nanocomposite (see Figure 12.49).

Figure 12.55 gives TEM images of the nanocomposites based on SI, SI-t-COOH, or SI-t-COONa diblock copolymers. In order to help observe clearly the state of dispersion of organoclay aggregates in the matrix phase, the block copolymers in the respective nanocomposites were not stained. Thus, the white/gray areas represent the block copolymer matrix phase and the dark areas represent organoclay. It can be seen in Figure 12.55 that the nanocomposites based on SI and SI-t-COOH diblock copolymers

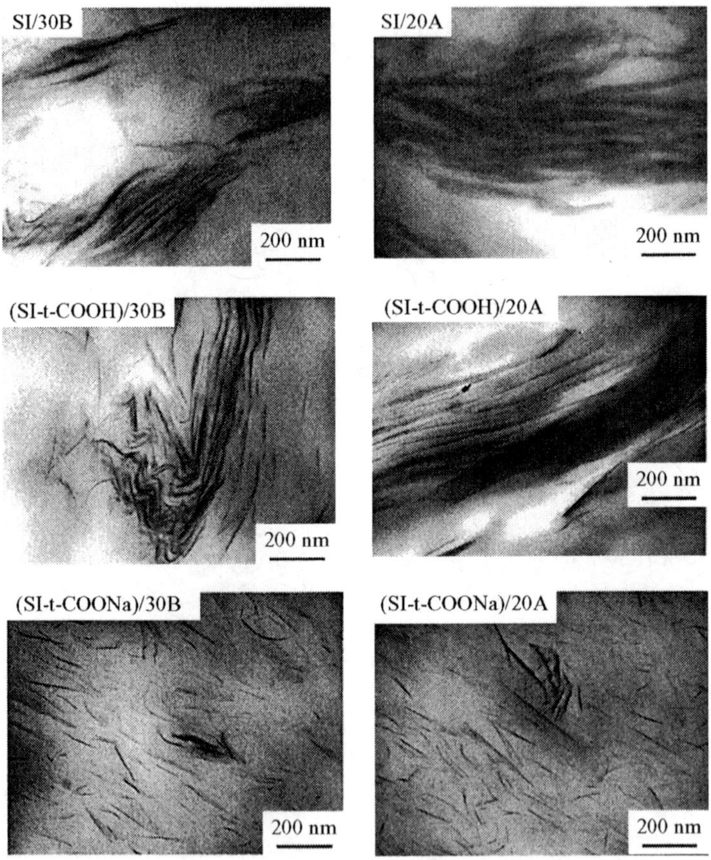

Figure 12.55 TEM images at room temperature of SI/Cloisite 30B nanocomposite, SI/Cloisite 20A nanocomposite, (SI-t-COOH)/Cloisite 30B nanocomposite, (SI-t-COOH)/Cloisite 20A nanocomposite, (SI-t-COONa)/Cloisite 30B nanocomposite, and (SI-t-COONa)/Cloisite 20A nanocomposite, in which the dark areas represent organoclay, Cloisite 20A or Cloisite 30B, and the gray/white areas represent the polymer matrix. The specimens were not stained by osmium tetroxide. (Reprinted from Zha et al., *Macromolecules* 38:8418. Copyright © 2005, with permission from the American Chemical Society.)

have very poor dispersions of organoclay aggregates, whereas the aggregates of Cloisite 30B and Cloisite 20A are very well dispersed in the matrix of SI-t-COONa diblock copolymer. This observation is consistent with that made above from the XRD patterns given in Figure 12.54. Again, we attribute this observation to the presence of ionic interactions between the negatively charged $-COO^-$ present at the chain end of the PI block of an SI-t-COONa and the positively charged N^+ in the surfactant (MT2EtOH or 2M2HT) residing at the surface of organoclay (Cloisite 30B or Cloisite 20A). Conversely, the rather poor dispersion of organoclay aggregates in the matrix of SI and SI-t-COOH is due to the absence (or insufficient) attractive interaction between the polymer matrix and the surfactant, MT2EtOH or 2M2HT, residing at the surface of the respective organoclays, Cloisite 30B and Cloisite 20A.

Figure 12.56 describes the temperature dependence of G' during isochronal dynamic temperature sweep experiments for SI-t-COONa (○), (SI-t-COOH)/Cloisite 30B nanocomposite (▽), (SI-t-COOH)/Cloisite 20A nanocomposite (◇), (SI-t-COONa)/Cloisite 30B nanocomposite (△), and (SI-t-COONa)/Cloisite 20A nanocomposite (□). The following observations are worth noting in Figure 12.56. (1) Values of G' for SI-t-COONa decrease steadily with increasing temperature over the entire range of temperatures investigated from 60 to 90 °C, suggesting that the end-functionalized block copolymer SI-t-COONa has no microdomains. (2) Values of G' for the (SI-t-COOH)/Cloisite 30B and (SI-t-COOH)/Cloisite 20A nanocomposites initially decrease rapidly as the temperature is increased from 60 to approximately 80 °C, and then at a much slower rate as the temperature is increased further to 160 °C. It is interesting to observe in Figure 12.56 that values of G' for the nanocomposites are still substantial even at 150 °C, suggesting that in the presence of organoclay (Cloisite 30B or Cloisite 20A) the block copolymer matrix SI-t-COOH in the respective nanocomposites might still retain microdomains. The temperature dependence of G' for the (SI-t-COONa)/Cloisite 30B and (SI-t-COONa)/Cloisite 20A nanocomposites is similar to that for the (SI-t-COOH)/Cloisite 30B and (SI-t-COOH)/Cloisite 20A nanocomposites, but the magnitude of G' at 100–150 °C is much larger for the nanocomposites based

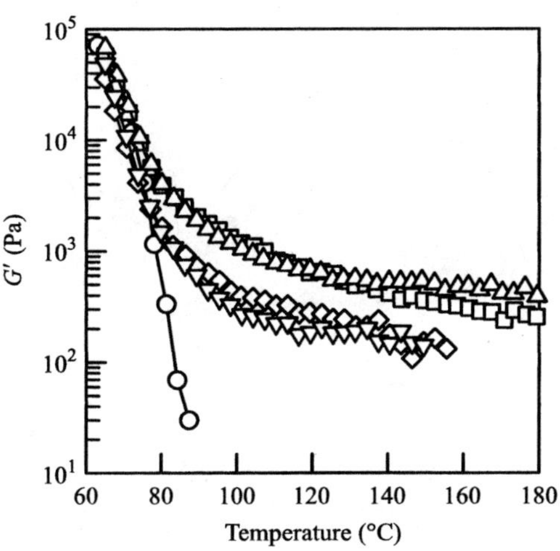

Figure 12.56 Variations of G' with temperature during the isochronal dynamic temperature sweep experiments at $\omega = 0.1$ rad/s for: (○) SI-t-COONa, (▽) (SI-t-COOH)/Cloisite 30B nanocomposite, (◇) (SI-t-COOH)/Cloisite 20A nanocomposite, (△) (SI-t-COONa)/Cloisite 30B nanocomposite, and (□) (SI-t-COONa)/Cloisite 20A nanocomposite. (Reprinted from Zha et al., *Macromolecules* 38:8418. Copyright © 2005, with permission from the American Chemical Society.)

on SI-t-COONa than that for the nanocomposites based on SI-t-COOH. This observation suggests that the ionic interactions between the –COONa group at the chain end of PI block in the SI-t-COONa and the surfactant (MT2EtOH or 2M2HT) residing at the surface of organoclay (Cloisite 30B or Cloisite 20A) are much stronger than the interactions between the –COOH group at the chain end of PI block in the SI-t-COOH and the surfactant. As pointed out above, the interactions between –COOH group in the SI-t-COOH and the surfactant MT2EtOH residing at the surface of organoclay Cloisite 30B would be relatively weak.

Figure 12.57 gives TEM images of (SI-t-COONa)/Cloisite 30B nanocomposite at 70 and 150 °C and (SI-t-COOH)/Cloisite 30B nanocomposite at 70 and 150 °C, for which each specimen was first stained with osmium tetroxide, then annealed at 70 °C for 6 h or at 150 °C for 2 h, followed by rapid quenching in ice water. In Figure 12.57 we observe that at 70 °C the lamellar microdomains of SI-t-COONa and the organoclay Cloisite 30B coexist, indicating that microdomain separation was induced, via ionic interaction between the –COO$^-$ group at the chain end of PI block in SI-t-COONa and the surfactant MT2EtOH residing at the surface of Cloisite 30B, when a disordered (homogenous) diblock copolymer SI-t-COONa was mixed with organoclay Cloisite 30B. Note that according to the results of the isochronal dynamic temperature sweep

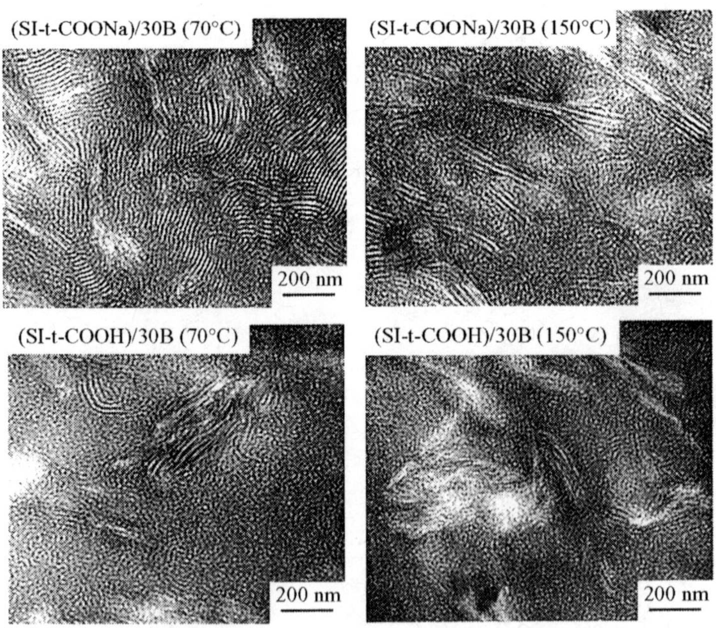

Figure 12.57 TEM images of (SI-t-COONa)/Cloisite 30B nanocomposite at 70 and 150 °C (upper panel), and TEM images of (SI-t-COOH)/Cloisite 30B nanocomposite at 70 and 150 °C (lower panel). Specimens were first annealed at 70 °C for 24 h or at 150 °C for 6 h and then stained with osmium tetroxide, followed by a rapid quench in ice water. (Reprinted from Zha et al., *Macromolecules* 38:8418. Copyright © 2005, with permission from the American Chemical Society.)

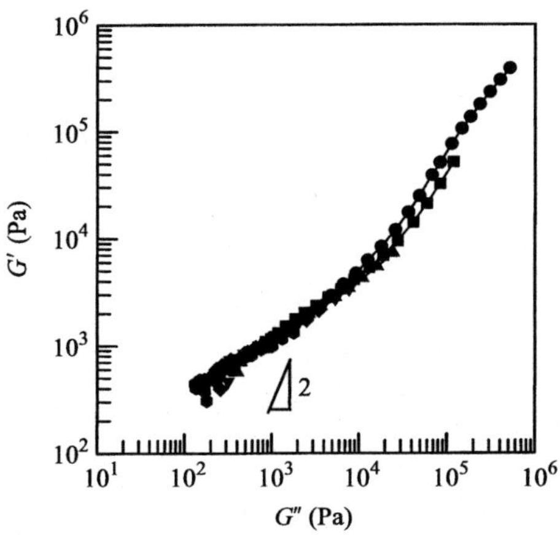

Figure 12.58 Plots of log G' versus log G'' for (SI-t-COONa)/Cloisite 30B nanocomposite at various temperatures (°C): (●) 80, (■) 100, (▲) 120, (▼) 140, (◆)160, and (⬢) 180. (Reprinted from Zha et al., *Macromolecules* 38:8418. Copyright © 2005, with permission from the American Chemical Society.)

experiment given in Figure 12.56, SI-t-COONa is a disordered block copolymer. Interestingly, as the temperature is increased to 150 °C, the (SI-t-COONa)/Cloisite 30B nanocomposite still retains microdomains of the matrix phase SI-t-COONa in the presence of exfoliated organoclay Cloisite 30B. However, the microdomain structure of SI-t-COONa at 150 °C does not look the same as that at 70 °C; that is, at 150 °C little lamellar microdomain structure is seen in the TEM image of (SI-t-COONa)/Cloisite 30B nanocomposite while the aggregates of organoclay Cloisite 30B are well dispersed in the block copolymer matrix.

Figure 12.58 gives log G' versus log G'' plots for (SI-t-COONa)/Cloisite 30B nanocomposite at various temperatures ranging from 80 to 180 °C. From Figure 12.58 we conclude that specific interactions exist in the nanocomposites because the slope of the log G' versus log G'' plots in the terminal region is much smaller than 2.

12.4 Rheology of Fiber-Reinforced Thermoplastic Composites

The use of glass fibers for producing fiber-reinforced thermoplastic composites has long been practiced in industry. The use of glass fibers in thermoplastic polymers requires special attention in that the fibers, when mixed with a molten polymer, orient in certain directions, thus producing some unique mechanical properties not achievable with particulate-filled thermoplastic polymers.

12.4.1 Theoretical Consideration of Fiber Orientation in Flow

The rheological properties of fiber-reinforced thermoplastic composites depend on fiber orientation. Therefore, it is very important to understand flow-induced fiber orientation. For this reason, in this section we first present briefly the approach of Dinh and Armstrong (1984), who developed a constitutive equation for concentrated fiber suspensions in a solvent on the basis of a general formulation proposed earlier by Batchelor (1970). We then present expressions predicting the shear viscosity of fiber-reinforced thermoplastic composite, where the suspending medium is assumed to follow

Newtonian behavior. Finally, we describe a computational procedure to solve a system of equations predicting the extent of fiber orientation in a shear flow field.

Batchelor (1970) formulated a general constitutive equation for a suspension of particles in a solvent and presented the following form for the total extra stress tensor $\boldsymbol{\sigma}$:[6]

$$\boldsymbol{\sigma} = 2\eta_s \mathbf{d} - \frac{1}{V}\sum_{i=1}^{nV}\int_{A_i}[\boldsymbol{\pi}_s \cdot \mathbf{n}]\,\mathbf{r}\,\mathrm{d}A_i \tag{12.11}$$

where η_s is the solvent viscosity, $\mathbf{d}$ is the bulk rate-of-deformation tensor, n is the total number of particles in a volume V, $\boldsymbol{\pi}_s$ is the total stress acting on the surface of the particle, $\mathbf{n}$ is the unit vector normal to the surface area $\mathrm{d}A_i$, and $\mathbf{r}$ is a position vector from a fixed origin to a point on the particle surface. Having recognized the practical difficulty with evaluating the particle surface integrals, Dinh and Armstrong (1984) simplified Eq. (12.11) for a single fiber of length L and diameter d suspended in a solvent, as schematically shown in Figure 12.59, and replaced the second term on the right-hand side of Eq. (12.11) with the expression

$$-\frac{1}{V}\sum_{i=1}^{nV}\int_{A_i}[\boldsymbol{\pi}_s \cdot \mathbf{n}]\,\mathbf{r}\,\mathrm{d}A_i = -\int\int_{A_i}[\boldsymbol{\pi}_{\text{eff}} \cdot \mathbf{n}]\,\mathbf{r}\,f(\mathbf{u},t)\,\mathrm{d}A_i\,\mathrm{d}\mathbf{u} \tag{12.12}$$

where $\boldsymbol{\pi}_{\text{eff}}$ represents the effective stress on the test fiber due to the suspending medium, and $f(\mathbf{u}, t)$ is the orientation distribution function, describing the probability that the test fiber would be oriented somewhere in the range of $\mathbf{u}$ and $\mathbf{u} + \mathrm{d}\mathbf{u}$ at time t (see Figure 12.59) with $\mathbf{u}$ being unit vector.

Dinh and Armstrong (1984) simplified Eq. (12.12) further by replacing the quantity $[\boldsymbol{\pi}_{\text{eff}} \cdot \mathbf{n}]\mathrm{d}A_i$, describing the drag force on the surface of the test fiber, with the product

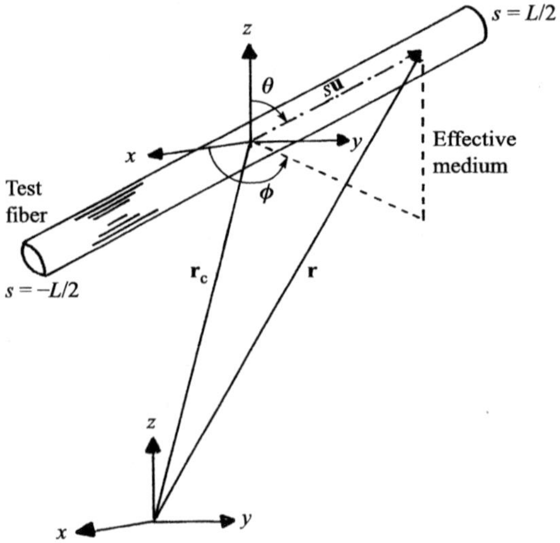

Figure 12.59 Schematic showing a test fiber immersed in an effective medium. (Reprinted from Dinh and Armstrong, *Journal of Rheology* 28:207. Copyright © 1984, with permission from the Society of Rheology.)

of a drag coefficient tensor $\boldsymbol{\zeta}_{\text{eff}}$ and the relative velocity, $(\mathbf{v} - \dot{\mathbf{r}})$, between the effective medium and the test fiber and obtained the expression

$$-\int_{A_i} [\boldsymbol{\pi}_{\text{eff}} \cdot \mathbf{n}]\, dA_i = \frac{1}{L} \int_{-L/2}^{L/2} \zeta_{\text{eff}} \cdot [\mathbf{v} - \dot{\mathbf{r}}]\, ds \tag{12.13}$$

where s is a scalar quantity (see Figure 12.59) and L is the fiber length. Substitution of Eq. (12.13) into (12.12) allows us to rewrite Eq. (12.11) as

$$\boldsymbol{\sigma} = 2\eta_s \mathbf{d} + \frac{n}{L} \iint_{-L/2}^{L/2} \zeta_{\text{eff}} \cdot [\mathbf{v} - \dot{\mathbf{r}}]\mathbf{r} f(\mathbf{u}, t)\, d\mathbf{u}\, ds \tag{12.14}$$

Referring to Figure 12.59, we have $\mathbf{r} = \mathbf{r}_c + s\mathbf{u}$, thus $\dot{\mathbf{r}} = \dot{\mathbf{r}}_c + s\dot{\mathbf{u}}$. Note that $\dot{\mathbf{u}}$ is the average time rate of change of the orientation vector defined by[7]

$$\dot{\mathbf{u}} = \mathbf{L} \cdot \mathbf{u} - \mathbf{L} : \mathbf{uuu} \tag{12.15}$$

with the initial condition, $\mathbf{u} = \mathbf{u}_0$ at $t = 0$ with $\mathbf{u}_0$ being the initial orientation of a fiber, in which $\mathbf{L}$ is the velocity gradient tensor, $(\nabla\mathbf{v})^{\text{T}}$ (see Chapter 2). Equation (12.15) describes that a fiber moves affinely (i.e., the orientation vector $\mathbf{u}$ changes as though it were an element of the fluid), represented by $[\mathbf{L} \cdot \mathbf{u}]$, except that the fiber cannot stretch and thus the stretching part of the motion, $[\mathbf{L} : \mathbf{uuu}]$ is subtracted off. For rigid fibers in homogeneous flows (i.e., velocity gradients independent of position), the bulk stress and the bulk rate of strain are independent of position in the physical space. In such a situation, the velocity of the effective medium[8] can be written as

$$\mathbf{v} = \mathbf{L} \cdot \mathbf{r} \tag{12.16}$$

We then have $\dot{\mathbf{r}}_c = \mathbf{v}_0 + \mathbf{L} \cdot \mathbf{r}_c$, and thus

$$\dot{\mathbf{r}} = \mathbf{v}_0 + [\mathbf{L} \cdot \mathbf{r}_c] + s(\mathbf{L} \cdot \mathbf{u} - \mathbf{L} : \mathbf{uuu}) \tag{12.17}$$

$$\mathbf{v} = \mathbf{v}_0 + \mathbf{L} \cdot \mathbf{r} = \mathbf{v}_0 + \mathbf{L} \cdot (\mathbf{r}_c + s\mathbf{u}) \tag{12.18}$$

Substitution of Eqs. (12.17) and (12.18) into (12.14) yields (see Appendix 12A)

$$\boldsymbol{\sigma} = 2\eta_s \mathbf{d} + \frac{nL^2}{12} \zeta_{\text{eff}} \cdot \left[\mathbf{L} : \langle\mathbf{uuuu}\rangle\right] \tag{12.19}$$

where[9]

$$\langle\mathbf{uuuu}\rangle = \int \mathbf{uuuu}\, f\, d\mathbf{u} \tag{12.20}$$

Note that for semiconcentrated suspensions, the motion of a fiber is limited because of the presence of other fibers. Fibers cannot pass through each other and thus the bulk rheological properties of the suspension would depend strongly on the structural arrangement of the fibers, satisfying $(1/L^3) < n < (1/dL^2)$ or $(d/L)^2 < c < (d/L)$, where d is the fiber diameter and c is the volume fraction of fibers in the suspension. $\boldsymbol{\zeta}_{\text{eff}}$ has two terms defined by

$$\boldsymbol{\zeta}_{\text{eff}} = \zeta_{\text{t}}(\boldsymbol{\delta} - \mathbf{uu}) + \zeta_{\text{p}}\mathbf{uu} \tag{12.21}$$

with ζ_{t} and ζ_{p} being the drag coefficients for the transverse and parallel to the fiber axis. By assuming $\zeta_{\text{p}} \gg \zeta_{\text{t}}$ (i.e., $\boldsymbol{\zeta}_{\text{eff}} \approx \zeta_{\text{p}}\,\mathbf{uu}$) and using the following expression for ζ_{p}:

$$\zeta_{\text{p}} = \frac{2\pi\eta_{\text{s}}L}{\ln(2h/d)} \tag{12.22}$$

where h is the average lateral spacing between fibers, Eq. (12.19) can be rewritten as[10]

$$\boldsymbol{\sigma} = 2\eta_{\text{s}}\left\{\mathbf{d} + \frac{\pi}{12}\,\frac{nL^3}{\ln(2h/d)}\mathbf{d} : \langle\mathbf{uuuu}\rangle\right\} \tag{12.23}$$

The first term represents the contribution from the solvent to the bulk stress, and the second term represents the internal structure and the amount of solids in the suspension.

In concentrated suspensions, the distance between fibers is less than a fiber diameter and thus the entire suspension would deform like a solid. Under such situations, one must include the interparticle interaction potential, the exact form of which is not well defined at the present time. When the suspending medium is other than solvent, η_{s} appearing in Eq. (12.23) may be replaced with another expression (e.g., power-law model) describing the viscosity of the medium.

To obtain the expressions from Eq. (12.23) for bulk shear stress and bulk first normal stress difference in steady-state shear flow, for instance, one must evaluate the quantity $\langle\mathbf{uuuu}\rangle$ defined by Eq. (12.20). For this, information on $f(\mathbf{u}, t)$ must be provided. Once $f(\mathbf{u}, t)$ is known, in principle one can calculate $\langle\mathbf{uuuu}\rangle$ from Eq. (12.20). Several different ways of approximating $\langle\mathbf{uuuu}\rangle$, referred to as "closure (or decoupling) approximation," have been suggested in the literature (Advani and Tucker 1980, 1987; Doi 1981; Hand 1962; Hinch and Leal 1976; Marrucci and Grizzuti 1984) (see Chapter 9). The simplest form of approximation is

$$\tilde{\mathbf{a}}_4 = \langle\mathbf{uuuu}\rangle = \langle\mathbf{uu}\rangle\langle\mathbf{uu}\rangle = \mathbf{a}_2\mathbf{a}_2 \tag{12.24}$$

where

$$\mathbf{a}_2 = \langle\mathbf{uu}\rangle = \int \mathbf{uu}\,f d\mathbf{u} \tag{12.25}$$

Equation (12.24) is often referred as the quadratic approximation of the fourth-order tensor $\langle\mathbf{uuuu}\rangle$. It should be mentioned that Eq. (12.24) has been very successful in describing certain physical phenomena, but it has failed to describe other physical phenomena, as we have discussed in Chapter 9. The following linear closure approximation has been suggested (Hand 1962):

$$\hat{\mathbf{a}}_4 = \langle \mathbf{uuuu} \rangle = \hat{a}_{ijk\ell} = -\tfrac{1}{35}(\delta_{ij}\delta_{k\ell} + \delta_{ik}\delta_{j\ell} + \delta_{i\ell}\delta_{jk}) + \tfrac{1}{7}(a_{ij}\delta_{k\ell} + a_{ik}\delta_{j\ell} + a_{i\ell}\delta_{jk} + a_{k\ell}\delta_{ij} + a_{j\ell}\delta_{ik} + a_{jk}\delta_{i\ell}) \tag{12.26}$$

The linear closure approximations are exact for a completely random distribution of fiber orientation, while the quadratic closure approximations are exact for perfect uniaxial alignment of the fibers (Advani and Tucker 1987).

If we assume that fibers move with the bulk motion of the fluid, then the rate of change of $f(\mathbf{u}, t)$ can be expressed as (Dinh and Armstrong 1984)

$$\frac{\partial f}{\partial t} = -\frac{\partial}{\partial \mathbf{u}} \cdot [\dot{\mathbf{u}} f] \tag{12.27}$$

The solution of Eq. (12.27) requires information on the rate of change of $\mathbf{u}$, $\dot{\mathbf{u}}$. There are different ways of expressing $\dot{\mathbf{u}}$. One simple expression for $\dot{\mathbf{u}}$ is already given by Eq. (12.15). Substitution of Eq. (12.15) into (12.27) gives[11]

$$\frac{\partial f}{\partial t} + (\mathbf{L} \cdot \mathbf{u} - \mathbf{L} : \mathbf{uuu}) \cdot \frac{\partial f}{\partial \mathbf{u}} = 3(\mathbf{L} : \mathbf{uu}) f \tag{12.28}$$

which must be solved for $f(\mathbf{u}, t)$ with an appropriate initial condition. It has been shown that Eq. (12.28) is satisfied by the expression (Dinh and Armstrong 1984)

$$f = \frac{1}{4\pi} \frac{1}{\left(\boldsymbol{\Delta}^{\mathrm{T}} \cdot \boldsymbol{\Delta} : \mathbf{uu}\right)^{2/3}} \tag{12.29}$$

where $\boldsymbol{\Delta} = \mathbf{E}^{-1}$, with $\mathbf{E}^{-1}$ being the reciprocal of a deformation gradient tensor $\mathbf{E}$ (see Chapter 2). Substitution of Eq. (12.29) into (12.23) gives a constitutive equation:

$$\boldsymbol{\sigma} = 2\eta_{\mathrm{s}} \left\{ \mathbf{d} + \frac{1}{48} \frac{nL^3}{\ln(2h/d)} \mathbf{d} : \int \frac{\mathbf{uuuu}\, d\mathbf{u}}{(\boldsymbol{\Delta}^{\mathrm{T}} \cdot \boldsymbol{\Delta} : \mathbf{uu})^{2/3}} \right\} \tag{12.30}$$

It is interesting to observe in Eq. (12.30) that an evaluation of the fourth-order tensor $\langle \mathbf{uuuu} \rangle$ is not necessary to obtain expressions for, for instance, shear stress and first normal stress difference in steady-state shear flow (see Appendix 12B).

Other forms of $\dot{\mathbf{u}}$, which are necessary for solving Eq. (12.27) for the orientation distribution function $f(\mathbf{u}, t)$, have been suggested; for example, the following form of $\dot{\mathbf{u}}$ (Advani and Tucker 1987; Folgar and Tucker 1984):[12]

$$\dot{\mathbf{u}} = -(\boldsymbol{\omega} \cdot \mathbf{u}) + \lambda(\mathbf{d} \cdot \mathbf{u} - \mathbf{d} : \mathbf{uuu}) - \tilde{D}_{\mathrm{r}} \frac{1}{f} \frac{\partial f}{\partial \mathbf{u}} \tag{12.31}$$

where ω_{ij} is the ijth component of the vorticity tensor $\boldsymbol{\omega}$, d_{ij} is the ijth component of the rate-of-deformation tensor $\mathbf{d}$, λ is a parameter related to the shape of the rigid rodlike particle, and $\tilde{D}_{\mathrm{r}}$ is the orientation-dependent rotary diffusivity.[13] Substitution of Eq. (12.31) with λ equal to unity into Eq. (12.27) gives

$$\frac{\mathrm{D}f}{\mathrm{D}t} = \frac{1}{\sin\theta}\frac{\partial}{\partial\theta}\left(\tilde{D}_{\mathrm{r}}\sin\theta\frac{\partial f}{\partial\theta} - f\sin\theta\left(\nabla\mathbf{v}^{\mathrm{T}} : \boldsymbol{\delta}_{\mathrm{r}}\boldsymbol{\delta}_{\theta}\right)\right) + \frac{1}{\sin\theta}\frac{\partial}{\partial\phi}\left(\frac{\tilde{D}_{\mathrm{r}}}{\sin\theta}\frac{\partial f}{\partial\theta} - f\left(\nabla\mathbf{v}^{\mathrm{T}} : \boldsymbol{\delta}_{r}\boldsymbol{\delta}_{\phi}\right)\right) \quad (12.32)$$

where D/Dt is the material derivative, and $\boldsymbol{\delta}_r$, $\boldsymbol{\delta}_\theta$, and $\boldsymbol{\delta}_\phi$ are the unit vectors in spherical harmonics. The solution of Eq. (12.32) allows one to have information on the orientation distribution function $f(\mathbf{u}, t)$ for any flow field, which then can be substituted to evaluate $\mathbf{d} : \langle\mathbf{uuuu}\rangle$ in Eq. (12.23) and hence the components of the stress tensor $\boldsymbol{\sigma}$. However, solution of Eq. (12.32) would require a considerable amount of numerical computation. Even if the numerical solution of Eq. (12.32) were obtained, it would be a formidable task to evaluate $\mathbf{d} : \langle\mathbf{uuuu}\rangle$ in Eq. (12.23).

An alternative approach to determining the degree of fiber orientation without having to solve Eq. (12.32) has been suggested (Advani and Tucker 1987; Folgar and Tucker 1984), namely, a hybrid closure approximation that may cover the entire range of orientation with the form

$$\bar{\mathbf{a}}_4 = (1 - P_2)\hat{\mathbf{a}}_4 + P_2\tilde{\mathbf{a}}_4 \quad (12.33)$$

where $\hat{\mathbf{a}}_4$ and $\tilde{\mathbf{a}}_4$ are defined by Eqs. (12.26) and (12.24), respectively, and P_2 is a scalar measure of fiber orientation, which is expressed in terms of the second-order tensor $\mathbf{a}_2$, defined by Eq. (12.25), as

$$P_2 = A\mathbf{a}_2 : \mathbf{a}_2 - B \quad (12.34)$$

in which $A = 3/2$ and $B = 1/2$ for three-dimensional orientation, and $A = 2$ and $B = 1$ for two-dimensional orientation. Advani and Tucker (1980) have suggested another form of P_2:

$$P_2 = 1 - N \det\mathbf{a}_2 \quad (12.35)$$

where $N = 27$ for three-dimensional orientation and $N = 4$ for two-dimensional orientation, and det $\mathbf{a}_2$ denotes the determinant of the second-order tensor $\mathbf{a}_2$. The advantage of using the Advani–Tucker approach lies in that very cumbersome calculation of the orientation distribution function $f(\mathbf{u}, t)$ can be avoided, provided that the calculation of $\mathbf{a}_2$ can be carried out without having to use $f(\mathbf{u}, t)$. For this, Advani and Tucker (1987) have suggested solving the following rate expression:[14]

$$\frac{\mathrm{D}\mathbf{a}_2}{\mathrm{D}t} = -(\boldsymbol{\omega}\cdot\mathbf{a}_2 - \mathbf{a}_2\cdot\boldsymbol{\omega}) + \lambda(\mathbf{d}\cdot\mathbf{a}_2 + \mathbf{a}_2\cdot\mathbf{d} - 2\mathbf{d} : \mathbf{a}_4) + 2\tilde{D}_{\mathrm{r}}(\boldsymbol{\delta} - \alpha\mathbf{a}_2) \quad (12.36)$$

where $\alpha = 3$ for three-dimensional orientation and $\alpha = 2$ for planar (two-dimensional) orientation. Notice in Eq. (12.36) that the fourth-order tensor $\mathbf{a}_4$ appears. Thus, the computational procedures for calculating the degree of fiber orientation P_2 are as follows. First, one solves, via numerical integration, Eq. (12.36) for $\mathbf{a}_2$ using one of the closure approximations described above for $\mathbf{a}_4$ (preferably Eq. (12.33)). Once $\mathbf{a}_2$ is known, one can then calculate the degree of fiber orientation P_2 using Eq. (12.34) or (12.35) and also any components of the stress tensor $\boldsymbol{\sigma}$ (thus bulk shear viscosity) from Eq. (12.23).

Advani and Tucker (1980) have presented the following general expression for determining the rheological properties of a suspension of fibers in a Newtonian fluid:[15]

$$\boldsymbol{\sigma} = \mathbf{C}:2\mathbf{d} \quad \text{or} \quad \sigma_{ij} = C_{ijk\ell}d_{k\ell} \tag{12.37}$$

where $\mathbf{C}$ is an anisotropic viscosity tensor that has the form

$$\begin{aligned} C_{ijk\ell} = {} & B_1 a_{ijk\ell} + B_2(a_{ij}\delta_{k\ell} + a_{k\ell}\delta_{ij}) + B_3(a_{ik}\delta_{j\ell} + a_{i\ell}\delta_{jk} + a_{j\ell}\delta_{ik} + a_{jk}\delta_{i\ell}) \\ & + B_4\delta_{ij}\delta_{k\ell} + B_5(\delta_{ik}\delta_{j\ell} + \delta_{i\ell}\delta_{jk}) \end{aligned} \tag{12.38}$$

where B_i are material constants. Equation (12.37) reduces to (12.23) for the following choices of B_i appearing in Eq. (12.38):

$$B_1 = \frac{\pi}{12}\frac{\eta_s n L^3}{\ln(2h/d)}; \quad B_2 = \eta_s/2; \quad B_3 = B_4 = B_5 = 0 \tag{12.39}$$

It is clear from the materials presented above that coupling between flow and orientation of fibers is necessary to accurately describe or predict the rheological properties of concentrated suspensions of fibers. Several other research groups (Altan et al. 1989, 1990, 1992; Ranganathan and Advani 1991; Shanker et al. 1991; Shaqfeh and Fredrickson 1990; Tucker and Advani 1974) reported on flow-induced fiber orientation in semiconcentrated or concentrated suspensions closely related to the processing of thermoplastic composite materials. In Chapter 13 of Volume 2 we discuss the importance of fiber orientation in the processing of thermoset/fiber composites.

12.4.2 Experimental Observations

In this section, we present some experimental results of fiber-reinforced thermoplastic composite reported in the literature, which were obtained using a cone-and-plate rheometer or a plunger-type capillary rheometer. In so doing, we point out practical difficulties with obtaining reproducible rheological data when using such experimental methods. We will then present bulk viscosity in the flow of a glass-fiber-reinforced thermoplastic composite.

Figure 12.60 gives a photograph describing the orientation of fibers in the flow of a molding compound reinforced with short glass fibers in the entrance region of

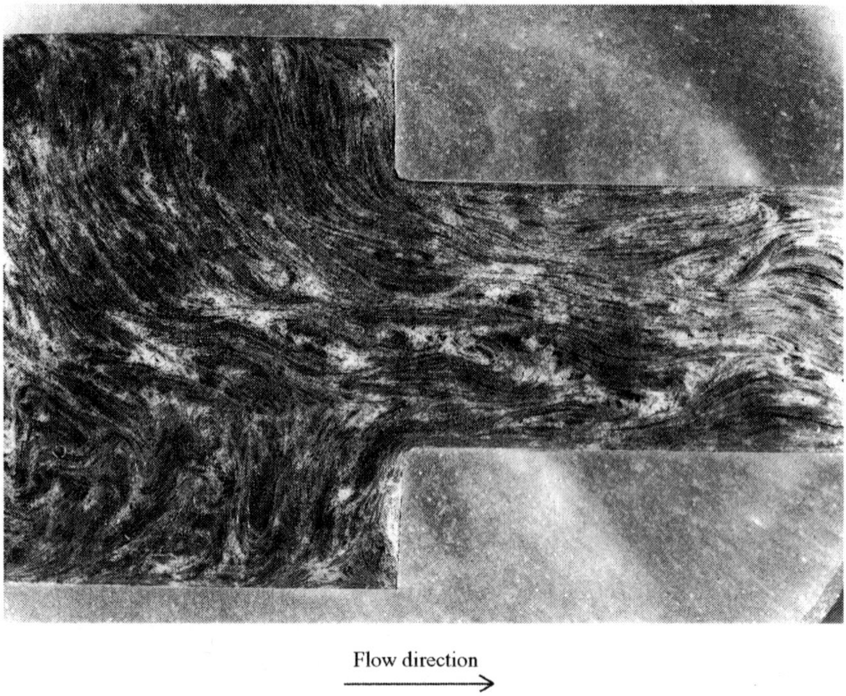

Figure 12.60 Photograph showing the orientation of fibers of an epoxy molding compound reinforced with short glass fibers flowing in the entrance region of a rectangular channel. (Reprinted from Goettler, *Modern Plastics* 47(4):140. Copyright © 1970, with permission from Modern Plastics.)

a rectangular die. It is seen that the glass fibers begin to orient as they approach the die entrance, the flow pattern of which is very similar to that observed in the entrance region flow of a slit die for a molten HDPE (see Figure 1.21 in Volume 2). However, there is a very important difference between the two situations, namely, in the entrance region a fiber-reinforced epoxy molding compound (Figure 12.60) is a highly anisotropic fluid, while the HDPE is a homogeneous, isotropic fluid. As will be elaborated on below, this difference raises a serious question about the accuracy and reproducibility of some of the experimental data reported in the literature on the bulk rheological properties of fiber-reinforced thermoplastic composites when the data were obtained using a plunger-type capillary rheometer. Below, we present some experimental observations (Chan et al. 1978a, 1978b; Czarnecki and White 1980; Han 1991; Kim and Song 1997; Kim and Park 2000) of the rheological behavior of glass-fiber-reinforced thermoplastic polymers.

Figure 12.61 gives log η versus log $\dot{\gamma}$ plots, and Figure 12.62 gives log N_1 versus log σ plots, for glass-fiber-reinforced PS composites, which were obtained using a cone-and-plate rheometer. Figure 12.63 shows photographs of two glass-fiber-reinforced PS composites prior to rheological measurements, the results of which are given in Figures 12.61 and 12.62. The glass fibers shown in Figure 12.63 have a nominal diameter

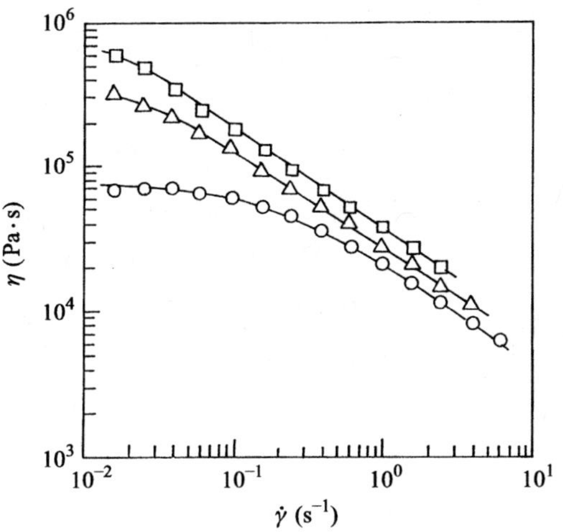

Figure 12.61 Plots of log η versus log $\dot{\gamma}$ at 180 °C for glass-fiber-reinforced PS composites with varying amounts of glass fibers (vol %): (○) 0 (neat PS), (△) 9.5, and (□) 22. The experimental data were obtained using a cone-and-plate rheometer. (Reprinted from Czarnecki and White, *Journal of Applied Polymer Science* 25:1217. Copyright © 1980, with permission from John Wiley & Sons.)

of 12.7 μm and lengths ranging from 1 to 2 mm. A 25 mm cone diameter, with a cone angle of 0.1 radians, was used (Czarnecki and White 1980). Since a typical distance between the apex of the cone and the plate (h_c) is about 50 μm, the value of h_c employed in obtaining the data given in Figures 12.61 and 12.62 is approximately only four times the fiber diameter, and the length of glass fiber is 200–400 times the value of h_c. The following observations are worth noting in Figures 12.61 and 12.62. At very low shear rates (say, at $\dot{\gamma} = 0.01 s^{-1}$), η for neat PS tends to exhibit Newtonian behavior, and values of N_1 for glass-fiber-reinforced PS composites are higher than that for neat PS and they increase with increasing fiber content. On the basis of the experimental results for particulate-filled PP composites and carbon-black-filled PS

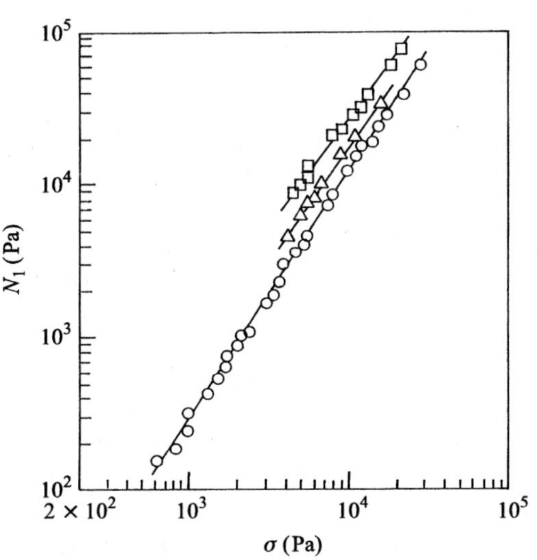

Figure 12.62 Plots of log N_1 versus log σ at 180 °C for glass-fiber-reinforced PS composites with different amounts of glass fibers (vol %): (○) 0 (neat PS), (△) 9.5, and (□) 22. The experimental data were obtained using a cone-and-plate rheometer. (Reprinted from Czarnecki and White, *Journal of Applied Polymer Science* 25:1217. Copyright © 1980, with permission from John Wiley & Sons.)

Figure 12.63 Photographs of glass-fiber-filled PS composite prior to rheological measurement: (a) 9.5 vol % glass fiber, and (b) 22 vol % glass fiber. (Reprinted from Czarnecki and White, *Journal of Applied Polymer Science* 25:1217. Copyright © 1980, with permission from John Wiley & Sons.

presented in a preceding section, we would expect to observe "yield behavior" at such a low shear rate (say, at $\dot{\gamma} = 0.01\ s^{-1}$). Further, there is no obvious reason why values of N_1 (i.e., melt elasticity) should increase when glass fiber is added to PS. As a matter of fact, an opposite trend was observed when $CaCO_3$ was added to a PP (see Figures 12.5) or when carbon black was added to a PS (see Figure 12.9). In view of the fact that the bulk rheological properties given in Figures 12.61 and 12.62 were obtained using the value of h_c which is only approximately four times the fiber diameter, it seems reasonable to suspect that there might have been very significant wall effects during the rheological measurements using a cone-and-plate rheometer. The above observations should caution those who wish to make rheological measurements of fiber-filled molten polymers using a cone-and-plate rheometer.

Figure 12.64 compares η with elongational viscosity (η_E) for a sheet molding compound (SMC), which was measured at temperatures ranging from 35 to 80 °C, over which little curing reaction is expected to take place. η was determined using a very long capillary die ($L/D = 30$) and thus entrance corrections were neglected, and η_E was determined using a constant-force squeeze-plate rheometer (Han 1991). It should be mentioned that typical formulation of an SMC has 25–30 wt % glass fiber, 15–20 wt % unsaturated polyester, approximately 10 wt % styrene monomer, 25–30 wt % $CaCO_3$, approximately 20 wt % low-profile additive, and 2–5 wt % viscosity thickener (e.g., MgO). It is interesting to observe in Figure 12.64 that η_E is about two orders of magnitude larger than η. One reason for the large difference between η_E and η is the differences in fiber orientation distributions. Namely, in the η measurement using a long capillary die, the fibers were forced to align in the direction of die axis, minimizing the orientation effects on the increase in η of the suspending medium. Conversely, in the η_E measurement using a parallel squeeze-type rheometer, the orientation of fibers was essentially random in the plane parallel to the disks, maximizing

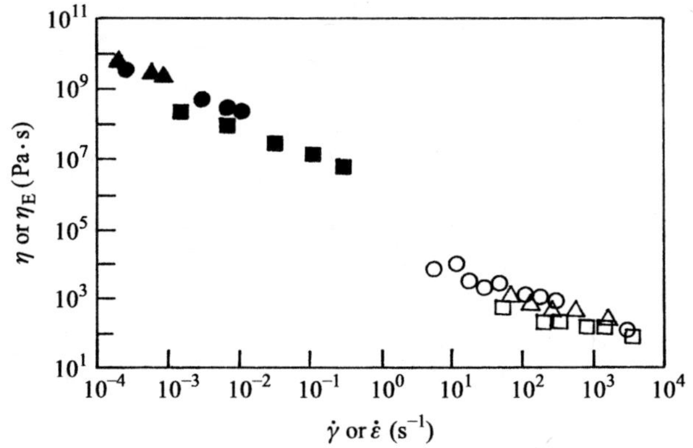

Figure 12.64 Comparison of shear viscosity (η) and elongational viscosity (η_E) of a sheet molding compound containing 20 wt % glass fibers in the matrix of an unsaturated polyester. Plots of log η versus log $\dot{\gamma}$ at various temperatures (°C): (○) 35, (△) 45, and (□) 55, and plots of log η_E versus log $\dot{\varepsilon}$ at various temperatures (°C): (●) 45, (▲) 60, and (■) 80. (Reprinted from Han, Masters Thesis at the University of Akron. Copyright © 1991, with permission from Peter K. Han.)

the orientation effects on the increase in η_E of the suspending medium. Figure 12.64 indicates that the extent of fiber orientation has a profound influence on the rheological behavior of glass-fiber-reinforced composites.

Figure 12.65 gives log $|\eta^*|$ versus log ω plots at 240 °C for 20/80 glass-fiber-reinforced PS composite, which was obtained by applying multiple frequency sweeps to the same specimen. A parallel-plate fixture with the gap opening (h_{gap}) of approximately 2 mm was used to obtain the results given in Figure 12.65, and the glass fibers

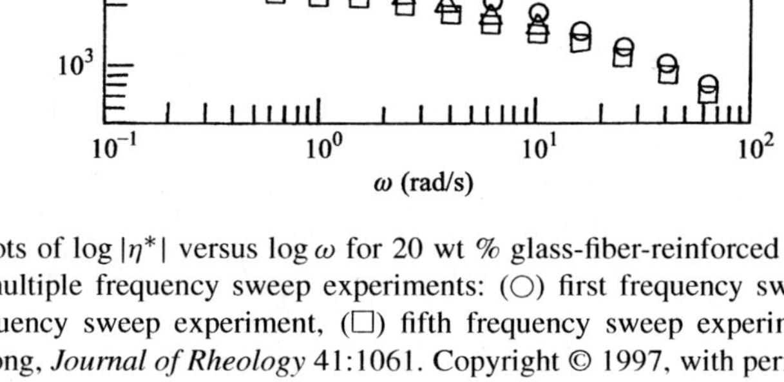

Figure 12.65 Plots of log $|\eta^*|$ versus log ω for 20 wt % glass-fiber-reinforced PS composite at 240 °C during multiple frequency sweep experiments: (○) first frequency sweep experiment, (△) second frequency sweep experiment, (□) fifth frequency sweep experiment. (Reprinted from Kim and Song, *Journal of Rheology* 41:1061. Copyright © 1997, with permission from the Society of Rheology.)

had a nominal diameter (d) of 13 μm and a nominal length of about 3 mm prior to compounding in a twin-screw extruder (Kim and Song 1997). It was found that the average length of the glass fibers decreased to 0.7 mm after compounding. In view of the fact that the value of h_{gap} is very large compared with the fiber diameter (i.e., $h_{gap}/d = 154$), most likely there might have been negligible wall effect on the rheological measurement. It is interesting to observe in Figure 12.65 that at low values of ω, $|\eta^*|$ decreases as the number of frequency sweep experiments is increased and then tends to level off after the fifth frequency sweep. Since it is reasonable to expect that the degree of fiber orientation would increase with an increase in the number of frequency sweep experiments, the observed decrease in $|\eta^*|$ with increasing number of frequency sweep experiments is attributable to an enhanced fiber orientation in the specimen.

12.5 Summary

In this chapter, we have presented the rheological behavior of particulate-filled polymers, organoclay nanocomposites based on thermoplastic polymer or block copolymer, and fiber-reinforced thermoplastic composites. We have pointed out that different flow geometries (i.e., cone-and-plate rheometer versus capillary rheometer) can give rise to different rheological responses in particulate-filled polymers, nanocomposites, and fiber-reinforced thermoplastic composites. Specifically, when using a pressure-driven rheometer (slit or capillary rheometer), measurements of wall normal stress in the fully developed region give the most reliable and reproducible rheological information for particulate-filled polymers. Note that such an experimental method does not require end corrections.

In this chapter, we have shown that the rheological behavior of organoclay nanocomposites is very sensitive to the state of dispersion (exfoliation versus intercalation) of organoclay aggregates by presenting experimental results in the literature. Specifically, we have presented linear dynamic viscoelastic properties of (1) PC/organoclay nanocomposites, (2) SI diblock copolymer/organoclay nanocomposites, and (3) SIOH diblock copolymer/organoclay nanocomposites. We have pointed out the importance of having attractive interactions between an organoclay and a polymer in order to obtain a high degree of exfoliation of organoclay aggregates in nanocomposites. This means that when preparing nanocomposites via melt blending or solution blending, a judicious choice of a pair of organoclay and polymer is necessary to achieve a high degree of exfoliation; otherwise, the nanocomposites prepared would have a poor degree of dispersion (intercalation at best) of organoclay aggregates. We have shown that when a nanocomposite has a high degree of exfoliation of organoclay aggregates, the G' and $|\eta^*|$ of the nanocomposite may increase with increasing temperature, an opposite trend to that usually observed in ordinary flexible polymers. The seemingly unusual temperature dependence of G' and $|\eta^*|$ for the nanocomposite is ascribed to an enhanced dispersion (thus increased surface area) of the organoclay nanoparticles with increasing temperature, which in turn is ascribed to the presence of attractive interactions (e.g., hydrogen bonding) between the surfactant residing at the surface of the organoclay and the polymer matrix. An independent study of in situ FTIR spectroscopy has confirmed the formation of hydrogen bonds in such organoclay nanocomposites, and the results of XRD and TEM support the conclusion drawn.

We have shown that the organoclay nanocomposites based on polystyrene or SI diblock copolymer that were end-functionalized with –COONa group have a very high degree of exfoliation of organoclay aggregates, as determined by XRD and TEM, when mixed with Cloisite 20A having methyl, hydrogenated tallow, quaternary ammonium chloride (2M2HT) as surfactant, and Cloisite 30B having methyl, tallow, bis-2-hydroxyethyl, quaternary ammonium chloride (MT2EtOH) as surfactant. We have shown that there is little or no difference between Cloisite 20A and Cloisite 30B in achieving a very high degree of dispersion (exfoliation) of organoclay aggregates. We have attributed the exfoliation of organoclay (Cloisite 20A or Cloisite 30B) to the presence of ionic interactions between the negatively charged $-COO^-$ at the chain end of polymer matrix and the positively charged N^+ in the surfactant (MT2EtOH or 2M2HT) residing at the surface of the organoclay. We have also shown that the organoclay nanocomposites based on PS or SI diblock copolymer, each being end-functionalized with –COOH group, yielded only intercalation of organoclay aggregates. One might expect that the –COOH group at the chain end of PS or SI diblock copolymer can form hydrogen bonds with the –OH groups in the surfactant MT2EtOH residing at the surface of organoclay Cloisite 30B. Thus, we conclude that no or extremely weak hydrogen bonds might have been formed in the respective nanocomposites, suggesting that the presence of only one –COOH group at the chain end of homopolymer PS and SI diblock copolymer is not sufficient to give rise to strong hydrogen bonds with the –OH groups in the surfactant MT2EtOH residing at the surface of organoclay Cloisite 30B. This demonstrates that ionic association is much more effective in achieving an exfoliated structure of organoclay aggregates than hydrogen bonding.

Finally, we have presented the rheological behavior of fiber-reinforced thermoplastic composites. We have concluded that the degree of fiber orientation plays a very important role in the rheological behavior of glass-fiber-reinforced thermoplastic composites.

Appendix 12A: Derivation of Equation (12.19)

Substituting Eqs. (12.17) and (12.18) into (12.13) gives

$$-\int_{A_i}[\boldsymbol{\pi}_{\text{eff}}\cdot\mathbf{n}]\mathrm{d}A_i = \frac{1}{L}\int_{-L/2}^{L/2}\boldsymbol{\zeta}_{\text{eff}}\cdot\Big[\mathbf{v}_0+\mathbf{L}\cdot(\mathbf{r}_c+s\mathbf{u}) - \big\{\mathbf{v}_0+[\mathbf{L}\cdot\mathbf{r}_c]+s\mathbf{L}\cdot\mathbf{u}-s\mathbf{L}:\mathbf{uuu}\big\}\Big]\mathrm{d}s = \frac{1}{L}\int_{-L/2}^{L/2}(\boldsymbol{\zeta}_{\text{eff}}\cdot s\mathbf{L}:\mathbf{uuu})\,\mathrm{d}s \tag{12A.1}$$

Then

$$-\int_{A_i}[\boldsymbol{\pi}_{\text{eff}}\cdot\mathbf{n}]\mathbf{r}\,\mathrm{d}A_i = \frac{1}{L}\int_{-L/2}^{L/2}(\boldsymbol{\zeta}_{\text{eff}}\cdot s\mathbf{L}:\mathbf{uuu})\mathbf{r}\,\mathrm{d}s$$

$$= \frac{1}{L}\int_{-L/2}^{L/2}\big[(\boldsymbol{\zeta}_{\text{eff}}\cdot s\mathbf{L}:\mathbf{uuu})\mathbf{r}_c\,\mathrm{d}s + (\boldsymbol{\zeta}_{\text{eff}}\cdot s^2\mathbf{L}:\mathbf{uuuu})\big]\,\mathrm{d}s$$

$$= \frac{1}{L}\left[(\boldsymbol{\zeta}_{\text{eff}} \cdot \frac{s^2}{2}\mathbf{L} : \mathbf{uuu})\mathbf{r}_{\text{c}} + (\boldsymbol{\zeta}_{\text{eff}} \cdot \frac{s^3}{3}\mathbf{L} : \mathbf{uuuu}) \right]\Bigg|_{-L/2}^{L/2}$$

$$= \frac{L^2}{12}\zeta_{\text{eff}} \cdot [\mathbf{L} : \langle \mathbf{uuuu} \rangle] \tag{12A.2}$$

in which $\mathbf{r} = \mathbf{r}_{\text{c}} + s\mathbf{u}$ was used. Substitution of Eq. (12A.2) into Eq. (12.14) yields Eq. (12.19).

Appendix 12B: Derivation of Three Material Functions for Steady-State Shear Flow from Equation (12.30)

For shear flow, $\mathbf{E}$, $\boldsymbol{\Delta}$, and $\boldsymbol{\Delta}^{\text{T}}$, respectively, are given by

$$\mathbf{E} = \begin{pmatrix} 1 & \gamma & 0 \\ 0 & 1 & 0 \\ 0 & 0 & 1 \end{pmatrix}; \quad \boldsymbol{\Delta} = \begin{pmatrix} 1 & -\gamma & 0 \\ 0 & 1 & 0 \\ 0 & 0 & 1 \end{pmatrix}; \quad \boldsymbol{\Delta}^{\text{T}} = \begin{pmatrix} 1 & 0 & 0 \\ -\gamma & 1 & 0 \\ 0 & 0 & 1 \end{pmatrix} \tag{12B.1}$$

thus

$$\boldsymbol{\Delta}^{\text{T}} \cdot \boldsymbol{\Delta} = \begin{pmatrix} 1 & -\gamma & 0 \\ -\gamma & 1+\gamma^2 & 0 \\ 0 & 0 & 1 \end{pmatrix} \tag{12B.2}$$

Using the polar coordinates $\mathbf{u} = (\sin\theta \cos\phi, \sin\theta \sin\phi, \cos\theta)$, we have

$$\mathbf{uu} = \begin{Vmatrix} \sin^2\theta\cos^2\phi & \sin^2\theta\sin\phi\cos\phi & \sin\theta\cos\phi\cos\theta \\ \sin^2\theta\sin\phi\cos\phi & \sin^2\theta\sin^2\phi & \sin\theta\sin\phi\cos\theta \\ \sin\theta\cos\phi\cos\theta & \sin\theta\sin\phi\cos\theta & \cos^2\theta \end{Vmatrix} \tag{12B.3}$$

thus we have

$$\boldsymbol{\Delta}^{\text{T}} \cdot \boldsymbol{\Delta} : \mathbf{uu} = \gamma^2 \sin^2\theta \sin^2\phi - \gamma \sin^2\theta \sin^2\phi + 1 \tag{12B.4}$$

Since $\mathbf{d}$ is given by Eq. (2.60)

$$2\mathbf{d} = \begin{Vmatrix} 0 & \dot{\gamma} & 0 \\ \dot{\gamma} & 0 & 0 \\ 0 & 0 & 0 \end{Vmatrix} \tag{12B.5}$$

we have

$2\mathbf{d} : \mathbf{uuuu}$

$$= 2\dot{\gamma}\sin^2\theta\cos\phi\sin\phi \begin{Vmatrix} \sin^2\theta\cos^2\phi & \sin^2\theta\sin\phi\cos\phi & \sin\theta\cos\phi\cos\theta \\ \sin^2\theta\sin\phi\cos\phi & \sin^2\theta\sin^2\phi & \sin\theta\sin\phi\cos\theta \\ \sin\theta\cos\phi\cos\theta & \sin\theta\sin\phi\cos\theta & \cos^2\theta \end{Vmatrix} \tag{12B.6}$$

$$(2\mathbf{d}:\mathbf{uuuu})_{xy} = 2\dot{\gamma}\sin^4\theta\cos^2\phi\sin^2\phi = (1/2)\dot{\gamma}\sin^4\theta\sin^2 2\phi \tag{12B.7}$$

$$\begin{aligned}(2\mathbf{d}:\mathbf{uuuu})_{xx} - (2\mathbf{d}:\mathbf{uuuu})_{yy} &= 2\dot{\gamma}\sin^2\theta\cos\phi\sin\phi\sin^2\theta(\cos^2\phi - \sin^2\phi)\\ &= (1/2)\dot{\gamma}\sin^4\theta\sin 4\phi\end{aligned} \tag{12B.8}$$

$$\begin{aligned}(2\mathbf{d}:\mathbf{uuuu})_{yy} - (2\mathbf{d}:\mathbf{uuuu})_{zz} &= 2\dot{\gamma}\sin^2\theta\cos\phi\sin\phi(\sin^2\theta\sin^2\phi - \cos^2\theta)\\ &= \dot{\gamma}\sin^4\theta\sin 2\phi(\sin^2\phi - \cot^2\theta)\end{aligned} \tag{12B.9}$$

Using the definitions $\sigma^+(t,\dot{\gamma}) = \eta^+(t,\dot{\gamma})\dot{\gamma}$, $N_1{}^+(t,\dot{\gamma}) = \Psi_1{}^+(t,\dot{\gamma})\dot{\gamma}^2$, and $N_2{}^+(t,\dot{\gamma}) = \Psi_2{}^+(t,\dot{\gamma})\dot{\gamma}^2$, from Eq. (12.30) we obtain (Dinh and Armstrong 1984)

$$\left(\frac{\eta^+(t,\dot{\gamma})}{\eta_s} - 1\right)\frac{\ln(2h/d)}{nL^3} = \int_0^{2\pi}\int_0^{\pi}\frac{\sin^5\theta\sin^2 2\phi\,d\theta\,d\phi}{96(\gamma^2\sin^2\theta\sin^2\phi - \gamma\sin^2\theta\sin 2\phi + 1)^{3/2}} \tag{12B.10}$$

$$\Psi_1^+(t,\dot{\gamma})\dot{\gamma}\frac{\ln(2h/d)}{nL^3} = \int_0^{2\pi}\int_0^{\pi}\frac{\sin^5\theta\sin 4\phi\,d\theta\,d\phi}{96(\gamma^2\sin^2\theta\sin^2\phi - \gamma\sin^2\theta\sin 2\phi + 1)^{3/2}} \tag{12B.11}$$

$$\Psi_2^+(t,\dot{\gamma})\dot{\gamma}\frac{\ln(2h/d)}{nL^3} = \int_0^{2\pi}\int_0^{\pi}\frac{\sin^5\theta\sin 2\phi(\sin^2\phi - \cot^2\theta)\,d\theta\,d\phi}{48(\gamma^2\sin^2\theta\sin^2\phi - \gamma\sin^2\theta\sin 2\phi + 1)^{3/2}} \tag{12B.12}$$

Problems

Problem 12.1

A $CaCO_3$-filled PP melt is found to obey the relationship given by Eq. (12.4), in which Y is 2.5×10^4 Pa, K is 4.5×10^4 Pa · s$^{0.5}$, and $n = 0.5$ at 200 °C. (a) Sketch the viscosity versus shear rate curve. (b) Sketch velocity and shear stress distributions in a long cylindrical tube through which this material is extruded. (c) What will the pressure drop be when this material is extruded in a long flat film die (60 cm wide, 0.2 cm thick, and 10 cm long) at a volumetric flow rate of 100 cm^3/s at 200 °C?

Problem 12.2

Consider that a viscous liquid containing small, spherical glass beads of uniform size (10 vol %) is injected into a position somewhere between the tube wall and the centerline of a long cylindrical tube. Assume that the diameter d_p of the glass beads is very small compared with the diameter D of the tube (i.e., $d_p \ll D$). Will there be a radial migration of the glass beads as they move along the tube axis? If so, in which direction will the glass beads migrate (i.e., toward the tube wall or toward the centerline of the tube)? In answering the question, consider the following situations: (1) the suspending medium is a Newtonian liquid, (2) the suspending medium may be represented by the ZFD model (see Chapter 3), (3) the suspending medium may be represented by the Coleman–Noll second-order fluid (see Chapter 3). Assume that the viscosity of the suspension is so large that the inertia effect is negligible compared with the viscous force.

Notes

1. Since a particulate-filled molten polymer consists of particulates and polymer matrix (i.e., it is a two-phase system), like an immiscible polymer blend presented in Chapter 11, the velocity gradient is not continuous at the interface between the particle and polymer matrix in a given particulate-filled polymer. Therefore, it is not possible to define "true" shear rate for such a fluid system inside a slit die, whereas we can define shear rate when dealing with homogeneous polymeric liquids. Thus, actually, we have calculated "apparent" shear viscosity, defined by $\eta_{\text{app}} = \sigma/\dot{\gamma}_{\text{app}}$ with σ being the shear stress and "apparent" shear rate defined by $\dot{\gamma}_{\text{app}} = 6Q/wh^2$ with w being the width and h being the height of the slit die (see Chapter 5). However, for convenience, throughout this chapter we use the notation $\dot{\gamma}_{\text{app}}$ and η_{app} for particulate-filled polymers interchangeably with the notation $\dot{\gamma}$ and η for homopolymers.
2. The authors of Figure 12.6 employed a plunger-type capillary rheometer and then used Eq. (5.48) to calculate shear stress. However, as described in Chapter 11 when we discussed the rheology of dispersed two-phase polymer blends, use of a plunger-type capillary rheometer is not warranted to calculate the shear stress, via Eq. (5.48), for carbon-filled PS composites, because they are multiphase systems. Note that Eq. (5.48) is valid only for homopolymers.
3. Guayule rubber is a natural rubber composed of *cis*-1,4 polyisoprene.
4. Melt fracture is a physical phenomenon observed in extrusion when the extrusion rate exceeds a certain critical value, giving rise to distorted extrudate or rough extrudate surface. The type of extrudate distortion or roughness of extrudate surface varies with the chemical structure of polymer, and the severity of melt fracture increases with decreasing temperature and increasing extrusion rate. This subject has been discussed very extensively in the literature during the past five decades. Interested readers are referred to standard textbooks (for example, Han 1976) dealing with polymer processing.
5. Theories for concentrated suspensions having a Newtonian liquid as the suspending medium are not applicable to particulate-filled molten polymers when the suspending medium is a viscoelastic fluid.
6. Note that the first term on the right-hand side of Eq. (12.11) has $2\eta_s\mathbf{d}$, whereas the first term on the right-hand side of Eq. (2) in the paper by Dinh and Armstrong (1984)

has $\eta_s\dot{\boldsymbol{\gamma}}$. The difference between the two comes from the use of slightly different definitions of the rate-of-deformation tensor, $\mathbf{d}$ or $\dot{\boldsymbol{\gamma}}$. Namely, in this section we use the definition $\mathbf{d} = (1/2)(\mathbf{L} + \mathbf{L}^{\mathrm{T}})$, given by Eq. (2.60), with $\mathbf{L}$ being the velocity gradient tensor, whereas Dinh and Armstrong used the definition $\dot{\boldsymbol{\gamma}} = \boldsymbol{\kappa} + \boldsymbol{\kappa}^{\mathrm{T}}$ with $\boldsymbol{\kappa}$ being the velocity gradient tensor. Thus, $2\mathbf{d} = \dot{\boldsymbol{\gamma}}$. To maintain consistency throughout this book, in Eq. (12.11) we use the definition of $\mathbf{d}$ given by Eq. (2.60).

7. Equation (12.15) is identical to Eq. (9.3) and its derivation is given in Appendix 9A in Chapter 9.
8. The effective medium is meant to represent the continuous medium (solvent in the present case) surrounding the test fiber. When dealing with fiber-reinforced thermoplastic polymers, the effective medium represents the polymer melt suspending the fibers.
9. The fourth-order tensor $\langle\mathbf{uuuu}\rangle$ also appeared many times in Chapter 9.
10. In obtaining Eq. (12.23) from (12.19), use was made of $\mathbf{L} = \mathbf{d} + \mathbf{d}^{\mathrm{T}}$.
11. Recalling that $\mathbf{A} \cdot \alpha\mathbf{B} = \mathbf{A}\alpha \cdot \mathbf{B} + \alpha\mathbf{A} \cdot \mathbf{B}$, where α is a scalar, we have

$$\frac{\partial}{\partial \mathbf{u}} \cdot [\mathbf{L} \cdot \mathbf{u} - \mathbf{L} : \mathbf{uuu}] f = \frac{\partial f}{\partial \mathbf{u}} \cdot [\mathbf{L} \cdot \mathbf{u} - \mathbf{L} : \mathbf{uuu}] + f \frac{\partial}{\partial \mathbf{u}} \cdot [\mathbf{L} \cdot \mathbf{u} - \mathbf{L} : \mathbf{uuu}]$$

Note that

$$\frac{\partial}{\partial \mathbf{u}} \cdot [\mathbf{L} \cdot \mathbf{u} - \mathbf{L} : \mathbf{uuu}] = \frac{\partial}{\partial \mathbf{u}} \cdot [\mathbf{L} \cdot \mathbf{u}] - \frac{\partial}{\partial \mathbf{u}} \cdot [\mathbf{L} : \mathbf{uuu}] = -3(\mathbf{L} : \mathbf{uu})$$

in which use was made of the following relationships:

$$\frac{\partial}{\partial \mathbf{u}} \cdot [\mathbf{L} \cdot \mathbf{u}] = \mathbf{L} : \frac{\partial}{\partial \mathbf{u}} \mathbf{u} = \mathbf{L} : (\boldsymbol{\delta} - \mathbf{uu})$$

$$\frac{\partial}{\partial \mathbf{u}} \cdot [\mathbf{L} : \mathbf{uuu}] = 2(\mathbf{L} : \mathbf{uu})$$

12. Note that Eq. (12.31) is slightly different from Eq. (30) in the paper by Advani and Tucker (1987) who used the following definitions:

$$\dot{\gamma}_{ij} = \partial v_i/\partial x_j + \partial v_j/\partial x_i \quad \text{and} \quad \omega_{ij} = \partial v_i/\partial x_j - \partial v_j/\partial x_i \ (\omega_{ij} = -\omega_{ij})$$

which are different by a factor 2 from the definitions of the rate-of-deformation tensor $\mathbf{d}$ (Eq. (2.60)) and the vorticity tensor $\boldsymbol{\omega}$ (Eq. (2.61)) used in this chapter.

13. See Eq. (9.4) for the definition of the rotary diffusion coefficient $\tilde{D}_{\mathrm{r}}$.
14. Equation (12.36) is slightly different from Eq. (33) in the paper by Advani and Tucker (1987) because we use the definitions of $\mathbf{d}$ and $\boldsymbol{\omega}$ given by Eqs. (2.60) and (2.61), respectively.
15. In Eq. (12.37) we use $\mathbf{d}$ defined by Eq. (2.60), whereas Eq. (5) in the paper by Advani and Tucker (1980) used the definition of the rate-of-deformation tensor $\dot{\boldsymbol{\gamma}} = \boldsymbol{\kappa} + \boldsymbol{\kappa}^{\mathrm{T}}$ (or $\dot{\gamma}_{ij} = \mathrm{d}v_i/\mathrm{d}x_j + \mathrm{d}v_j/\mathrm{d}x_j$). Thus, $2\mathbf{d} = \dot{\boldsymbol{\gamma}}$.

References

Abbate M, Martuschelli E, Musto P, Ragosta G, Scarinzi G (1994). *J. Polym. Sci., Polym. Phys. Ed.* 32:395.

Advani SG, Tucker CL (1980). *J. Rheol.* 41:367.

Advani SG, Tucker CL (1987). *J. Rheol.* 31:751.

Advani SG, Tucker CL (1990). *Polym. Compos.* 11:164.

Agarwal PK, Bagley EG, Hill CR (1978). *Polym. Eng. Sci.* 18:282.

Alexandre M, Beyer G, Henrist C, Cloots R, Rulmont A, Jérôme R, Dubois P (2001a). *Macromol. Rapid Commun.* 22:643.

Alexandre M, Beyer G, Henrist C, Cloots R, Rulmont A, Jérôme R, Dubois P (2001b). *Chem. Mater.* 13:3830.

Altan MC, Advani SG, Güceri SI, Pipes RB (1989). *J. Rheol.* 33:1129.

Altan MC, Subbiah S, Güceri SI, Pipes RB (1990). *Polym. Eng. Sci.* 30:848.

Altan MC, Güceri SI, Pipes RB (1992). *J. Non-Newtonian Fluid. Mech.* 42:65.

Araki T, White JL (1988). *Polym. Eng. Sci. 38*:616.

Balazs AC, Singh C, Zhulina E (1998). *Macromolecules* 31:8370.

Batchelor GK (1970). *J. Fluid Mech.* 41:545.

Bennett RH, Hulbert MH (1986). *Clay Microstructure*, International Human Resources Development, Boston, Massachusetts.

Beyer G (2001). *Fire Mater.* 25:193.

Bingham JC (1922). *Fluidicity and Plasticity*, McGraw-Hill, New York.

Chan Y, White JL, Oyanagi Y (1978a). *Polym. Eng. Sci.* 18:268.

Chan Y, White JL, Oyanagi Y (1978b). *J. Rheol.* 22:507.

Choi S, Lee KM, Han CD (2004). *Macromolecules* 37:7649.

Coleman MM, Graf JF, Painter PC (1991). *Specific Interactions and the Miscibility of Polymer Blends*, Technomic, Lancaster, Pennsylvania.

Czarnecki L, White JL (1980). *J. Appl. Polym. Sci.* 25:1217.

Deanin RD, Schott NR (eds) (1974). *Fillers and Reinforcements for Plastics*, Adv. Chem. Series. no 134, American Chemical Society, Washington, DC.

Dinh SM, Armstrong RC (1984). *J. Rheol.* 28:207.

Doi M (1981). *J. Polym. Sci. Polym. Phys. Ed.* 19:229.

Doremus P, Piau JM (1991). *J. Non-Newtonian Fluid Mech.* 39:335.

Dreval VE, Borisenkova EK (1993). *Rheol. Acta* 32:337.

Einstein A (1906). *Ann. Physik* 19:289.

Einstein A (1911). *Ann. Physik* 34:591.

Eisenberg A, Smith P (1982). *Polym. Eng. Sci.* 22:1117.

Eisenberg A, Hird B, Moore RB (1990). *Macromolecules* 23:4098.

Ertong S, Eggers H, Schümmer P (1994). *Rubber. Chem. Technol.* 67:207.

Faulkner DL, Schmidt LR (1977). *Polym. Eng. Sci.* 17:657.

Folgar F, Tucker CL (1984). *J. Reinf. Plast. Compos.* 3:98.

Fujima M, Kwasaki Y (1989). *Rheol. J. (Japan)* 17:60.

Goettler LA (1970). *Modern Plast.* 47(4):140.

Grim RE (1968). *Clay Minerology*, 2nd ed, McGraw Hill, New York.

Han CD (1974). *J. Appl. Polym. Sci.* 18:821.

Han CD (1976). *Rheology in Polymer Processing*, Academic Press, New York, Chap 12.

Han CD (1981). *Multiphase Flow in Polymer Processing*, Academic Press, New York, Chap 3.

Han CD, Sandford C, Yoo HJ (1978). *Polym. Eng. Sci.* 18:849.

Han CD, Van den Weghe T, Shete P, Haw JR (1981). *Polym. Eng. Sci.* 21:196.

Han PK (1991). *A Fundamental Study on the Rheology and Mold Filling of Sheet Molding Compounds*, Masters Thesis, The University of Akron, Akron, Ohio.

Hand GL (1962). *J. Fluid Mech.* 13:33.
Hasegawa N, Kawasumi M, Kato M, Usuki A, Okada A (1998). *J. Appl. Polym. Sci.* 67:87.
Hasegawa N, Okamoto H, Kawasumi M, Usuki A (1999). *J. Appl. Polym. Sci.* 74:3359.
Herschel WH, Bulkley R (1926). *Kolloid Z.* 39:291.
Hinch EJ, Leal LG (1976). *J. Fluid Mech.* 76:187.
Hird B, Eisenberg A (1992). *Macromolecules* 25:6466.
Hoffmann B, Kressler J, Stöppelmann G, Fredrich C, Kim GM (2000a). *Colloid. Polym. Sci.* 278:629.
Hoffmann B, Dietrich C, Thomann R, Fredrich C, Mülhaupt R (2000b). *Macromol. Rapid Commun.* 21:57.
Hohenemser VK, Prager W (1932). *ZAMM* 12:216.
Jackson WC, Advani SG, Tucker CL (1986). *J. Compos. Mater.* 20:539
Jeffery GB (1922). *Proc. Roy. Soc.* A102:161.
Jimenez G, Okata N, Kawai H, Ogihara T (1997). *J. Appl. Polym. Sci.* 64:2211.
Kato M, Usuki A, Okada A (1997). *J. Appl. Polym. Sci.* 66:1781.
Kawasu M, Hasegawa N, Kato M, Usuki A, Okada A (1997). *Macromolecules* 30:6333.
Kim JK, Song JH (1997). *J. Rheol.* 41:1061.
Kim JK, Park SH (2000). *J. Mat. Sci.* 35:1069.
Kojima Y, Usuki A, Kawasumi M, Okada A, Kurauchi T, Kamigaito O (1993). *J. Polym. Sci., Polym. Chem. Ed.* 31:983.
Kojima Y, Usuki A, Kawasumi M, Okada A, Kurauchi T, Kamigaito O, Kaji K (1994). *J. Polym. Sci., Polym. Phys. Ed.* 32:62.
Kraus G (ed). (1965). *Reinforcement of Elastomers*, Interscience, New York.
Krishnamoorti R, Giannelis EP (1997). *Macromolecules* 30:4097.
Kutsumizu S, Hara H, Tachino H, Shimabayashi K, Yano S (1999). *Macromolecules* 32:6340.
Lakdawala K, Salovey R (1987a). *Polym. Eng. Sci.* 27:1035.
Lakdawala K, Salovey R (1987b). *Polym. Eng. Sci.* 27:1043.
Lee KM, Han CD (2002). *Macromolecules* 35:760.
Lee KM, Han CD (2003a). *Polymer* 44:4573.
Lee KM, Han CD (2003b). *Macromolecules* 36:804.
Lee KM, Han CD (2003c). *Macromolecules* 36:7165.
Leonov AI (1990). *J. Rheol.* 34:1039.
Lepoittevin B, Pantoustier N, Devalckenaere M, Alexandre M, Kubies D, Calberg D, Jérôme R, Dubois P (2002). *Macromolecules* 35:8385.
Lim YT, Park OO (2001). *Rheol. Acta* 40:220.
Lin L, Masuda T (1989). *Rheol. J. (Japan)* 17:145.
Lin L, Masuda T (1990). *Rheol. J. (Japan)* 18:190.
Lincoln DM, Vaia RA, Wang ZG, Hsiao BS (2001a). *Polymer* 42:1621.
Lincoln DM, Vaia RA, Wang ZG, Hsiao BS, Krishnamoorti Z (2001b). *Polymer* 42:9975.
Liu X, Wu Q (2001). *Polymer* 42:10013.
Liu YJ, Schindler JL, DeGroot DC, Kannewurf CR, Hirpo W, Kanatzidis MG (1996). *Chem. Mater.* 8:525.
Lobe VM, White JL (1979). *Polym. Eng. Sci.* 19:617.
Lubin G (ed) (1969). *Handbook of Fiberglass and Advanced Plastics Composites*, Van Nostrand Reinhold, New York.
Maiti P, Okamoto M (2003). *Macromol. Mater. Sci.* 288:440.
Markovic MC, Choudhury NR, Dimopoulos M, Matisons JG, Dutta NK, Bhattacharya AK (2000). *Polym. Eng. Sci.* 40:1065.
Marrucci G, Grizzuti N (1984). *J. Non-Newtonian Fluid Mech.* 14:103.
Medalia AI (1970). *J. Colloid Interf. Sci.* 32:115.
Messersmith PB, Giannelis EP (1993). *Chem. Mater.* 5:1064.

Messersmith PB, Giannelis EP (1995). *J. Polym. Sci., Polym. Phys. Ed.* 33:1047.
Minagawa N, White JL (1976). *J. Appl. Polym. Sci.* 20:501.
Montes SA, Ramos de Valle LF, White JL (1985). *J. Polym. Eng.* 5(3):209.
Nam PH, Maiti P, Okamoto M, Kotaka T, Hasegawa N, Usuki A (2001). *Polymer* 42:9633.
Ogata N, Kawakage S, Ohihara T (1997). *Polymer* 38:5115.
Oldroyd JG (1947). *Proc. Camb. Phil Soc.* 43:100.
Pantoustier N, Alexandre M, Degée P, Calberg C, Jérôme R, Henrist C, Cloots R, Rulmont A, Dubois P (2001). *e-Polymer* 9:1.
Physical Properties Bulletin from Southern Clay Products.
Plueddemann EP (1982). *Silane Coupling Agents,* Plenum Press, New York.
Ranganathan S, Advani SG (1991). *J. Rheol.* 35:1499.
Riva A, Zanetti M, Braglia M, Camino G, Flaqui L (2002). *Polym. Degrad. Stab.* 77:299.
Rong SD, Chaffey CE (1988a). *Rheol. Acta.* 27:179.
Rong SD, Chaffey CE (1988b). *Rheol. Acta.* 27:186.
Rutkowska M, Eisenberg A (1984a). *Macromolecules* 17:821.
Rutkowska M, Eisenberg A (1984b). *J. Appl. Polym. Sci.* 29:755.
Rutkowska M, Eisenberg A (1985). *J. Appl. Polym. Sci.* 30:3317.
Shanker R, Gillespie JW, Güceri SI (1991). *Polym. Eng. Sci.* 31:161.
Shaqfeh SG, Fredrickson GH (1990). *Phys. Fluids* 2:7.
Sikka M, Cerini LN, Ghosh SS, Winey KI (1996). *J. Polym. Sci., Polym. Phys. Ed.* 34:1443.
Simhambhatla MV, Leonov AI (1995). *Rheol. Acta* 34:329.
Smith P, Eisenberg A (1983). *J. Polym. Sci., Polym. Lett. Ed.* 21:223.
Suetsugu Y, White JL (1983). *J. Appl. Polym. Sci.* 28:1481.
Tachino H, Hara H, Hirasawa E, Kutsumizu S, Yano S (1994). *Macromolecules* 27:372.
Tanaka H, White JL (1980). *Polym. Eng. Sci.* 20:949.
Tucker CL, Advani SG (1974). In *Flow and Rheology in Polymer Composite Manufacturing*, Advani SG (ed), Elsevier, New York, p 147.
Usuki A, Kawasumi M, Kojima Y, Fukushima Y, Okada T, Kurauchi T, Kamigaito O (1993a). *J. Mat. Res.* 8:1174.
Usuki A, Kojima Y, Kawasumi M, Okada A, Fukushima Y, Kurauchi T, Kamigaito O (1993b). *J. Mater. Res.* 8:1179.
Vaia RA, Giannelis EP (1997a). *Macromolecules* 30:7990.
Vaia RA, Giannelis EP (1997b). *Macromolecules* 30:8000.
van Olphen H (1977). *Clay Colloid Chemistry*, 2nd ed, John Wiley & Sons, New York.
Vinogradov GA, Malkin AYa, Plotnikova EP, Sabsai OYu, Nikolayeva NE (1972). *Intern. J. Polym. Mater.* 2:1.
White JL (1979). *J. Non-Newtonian Fluid Mech.* 5:177.
White JL, Crowder JW (1974). *J. Appl. Polym. Sci.* 18:1013.
White JL, Tanaka H (1981). *J. Non-Newtonian Fluid Mech.* 8:1.
Yano K, Usuki A, Okada A, Kurauchi T, Kurauchi T, Kamigaito O (1993). *J. Polym. Sci., Polym. Chem. Ed.* 31:2493.
Zha W, Choi S, Lee KM, Han CD (2005). *Macromolecules* 38:8418.
Zhulina E, Singh C, Balazs AC (1999). *Langmuir* 15:3935.

13

Rheology of Molten Polymers with Solubilized Gaseous Component

13.1 Introduction

Polymer melts (or polymer solutions) with a solubilized gaseous component (which occur under sufficiently high pressures, thus forming homogeneous mixtures), and polymer melts (or polymer solutions) with dispersed gas bubbles (thus forming heterogeneous mixtures of polymeric fluid and gas bubbles) are encountered in thermoplastic foam processing and polymer devolatilization. Thus, a good understanding of the rheological behavior of such mixtures is very important to the design of processing equipment and successful optimization of such polymer processing operations. From the 1950s through the 1970s, the dynamics of a single, spherical gas bubble dispersed in a stationary Newtonian or viscoelastic medium was extensively reported in the literature (Barlow and Langlois 1962; Duda and Vrentas 1969; Epstein and Plesset 1950; Folger and Goddard 1970; Marique and Houghton 1962; Plessst and Zwick 1952; Rosner and Epstein 1972; Ruckenstein and Davis 1970; Scriven 1959; Street 1968; Street et al. 1971; Tanasawa and Yang 1970; Ting 1975; Yang and Yeh 1966; Yoo and Han 1982; Zana and Leal 1975). While such investigations are of fundamental importance in their own right, they are not much help to describe bubble dynamics in thermoplastic foam extrusion or structural foam injection molding, for instance. There is no question that an investigation of bubble dynamics in a flowing molten polymer with dispersed gas bubbles is a very difficult subject by any measure. Thus, understandably, a relatively small number of research publications on bubble dynamics in a flowing molten polymer have been reported (Han and Villamizar 1978; Han et al. 1976; Yoo and Han 1981). The complexity of the problem arises from other related issues, such as the solubility and diffusivity of gaseous component(s) in a flowing molten polymer, which in turn depend on temperature and pressure of the system. Further, a gaseous component solubilized in molten polymer in the upstream side of a die, for instance, may nucleate as the pressure of the fluid stream decreases along the die axis, after which they could grow continuously as the molten polymer with dispersed gas bubbles flows through the rest of the die. Under such circumstances, any study of the rheological behavior of a molten polymer with dispersed gas bubbles must deal with both the dynamics of gas bubbles and the mass transfer between the gas phase and the liquid phase of molten polymer. The subject of solubility and diffusivity of gaseous components in a molten polymer is discussed in Chapter 10 of Volume 2, in conjunction with thermoplastic foam processing. The subject of bubble nucleation in polymer solutions or polymer melts is also discussed in Chapter 10 of Volume 2.

One of the most challenging issues, from the point of view of thermoplastic foam processing, when dealing with molten polymers with solubilized gaseous component or with dispersed gas bubbles has been the measurement of the rheological properties of such mixtures as functions of shear rate, temperature, and concentration of the gaseous component. Several research groups (Elkovitch et al. 1999; Foster and Lindt 1987; Gerhardt et al. 1997, 1998; Han and Ma 1983a, 1983b; Kwag et al. 1999; Lee et al. 1999; Mendelson 1979, 1980; Royer et al. 2000, 2001) have reported on measurements of the viscosities of polymer melts or polymer solutions with a solubilized gaseous components(s). In this chapter, we describe experimental methods for determination of the viscosity of polymer melts or polymer solutions with a solubilized gaseous component(s) without bubble nucleation under a sufficiently high pressure (i.e., homogeneous mixtures of polymeric fluid and gaseous component) using a continuous-flow capillary (or slit) rheometry. Then, we discuss some useful correlations, which will enable one to estimate, prior to measurement, the viscosity of molten polymers with solubilized gaseous components. The material presented in this chapter provides a foundation upon which thermoplastic foam extrusion will be presented in Chapter 10 of Volume 2.

13.2 Rheological Behavior of Molten Polymers with Solubilized Gaseous Component

13.2.1 Experimental Methods for Rheological Measurements of Molten Polymers with Solubilized Gaseous Component

One would expect that conventional rheological instruments without modifications may not be suitable for rheological measurements of polymer melts or solutions with a solubilized gaseous component or volatile solvent. Specifically, when using a cone-and-plate rheometer, as discussed in Chapter 5, the rheometer must be placed in an enclosed chamber under a sufficiently high pressure, such that the solubilized gaseous component or volatile solvent cannot escape from the polymer melt or solution. This means that the pressure of the enclosed chamber must be sufficiently high, so that the formation of gas bubbles (bubble nucleation) can be suppressed at a specified measurement temperature. Such an experimental technique will become increasingly difficult as the measurement temperature is increased, for example, to 200 °C. This is because the vapor pressure of solubilized gaseous component or volatile solvent in a polymer solution will becomes exceedingly high.

Similarly, when using a plunger-type capillary rheometer, as discussed in Chapter 5, the rheometer must be attached with a high-pressure chamber, so that at the die exit the polymer melt with solubilized gaseous component can exceed the critical pressure for bubble nucleation (P_{nc}). If gas bubbles were to be formed along the axis of a capillary die, such an experimental method would be of no rheological significance for calculating shear stress (hence viscosity). Some investigators (Blyler and Kwei, 1971; Oyanagi and White 1979) determined the viscosities of mixtures of a molten polymer and a chemical blowing agent[1], using a plunger-type capillary rheometer that was not attached with a high-pressure chamber at the die exit. Such an experimental approach is subject to serious criticism because there is no way of assuring that foaming did not occur inside the capillary die as the mixture of molten polymer and blowing agent approached the die exit. Some investigators (Foster and Lindt, 1987; Mendelson 1979, 1980) used a plunger-type capillary rheometer attached with a pressurized isothermal chamber to measure the viscosities of polymer solutions with volatile solvent, and others (Gehardt et al. 1997, 1998; Kwag et al. 1999) employed the same method to measure the viscosities of polymer melts with solubilized gaseous components.

Another method is to use continuous-flow capillary (or slit) rheometry, as discussed in Chapter 5. Han and coworkers (Han and Ma 1983a, 1983b; Han and Villamizar 1978; Han et al. 1976) were the first to use a continuous-flow capillary (or slit) rheometer to determine the viscosity of polymer melts with solubilized gaseous components. Later, other investigators (Elkovitch et al. 1999; Lee et al. 1999; Royer et al. 2000, 2001) employed the same method.

Figure 13.1 gives a schematic of a process describing an apparatus that supplies a continuous stream of molten polymer with solubilized gaseous component. The apparatus consists of a plasticating screw extruder, a high-pressure diaphragm-type metering pump for injection of a physical blowing agent, two static mixers connected in series, a capillary die, and pressure measuring devices. Here, a continuous stream of molten polymer with solubilized gaseous component (i.e., a homogeneous mixture) at sufficiently high pressure is introduced to the capillary die, such that bubble nucleation in the melt stream would not occur at the die entrance.

Using the cylindrical coordinate system (r, θ, z) we can describe the r-component of stress $T_{rr}(r, z)$ in the capillary and reservoir sections by

$$T_{rr}(r, z) = p(r, z) + \sigma_{rr}(r, z) \tag{13.1}$$

Note in Eq. (13.1) that the total stress $T_{rr}(r, z)$ consists of two terms, pressure $p(r, z)$ and deviatoric stress $\sigma_{rr}(r, z)$, and all three quantities vary with the radial (r) and axial (z) directions. When pressure transducers are mounted on the die wall along the die axis, as schematically shown in Figure 5.10, the pressure transducers measure total normal stress $T_{rr}(R, z)$ at the die wall ($r = R$) and at position z (hereafter referred to as wall normal stress):

$$T_{rr}(R, z) = -p(R, z) + \sigma_{rr}(R, z) \tag{13.2}$$

As discussed in Chapter 5, $\sigma_{rr}(R, z)$ becomes independent of z when flow is fully developed, reducing Eq. (13.2) to

$$T_{rr}(R, z) = -p(R, z) + \sigma_{rr}(R) \tag{13.3}$$

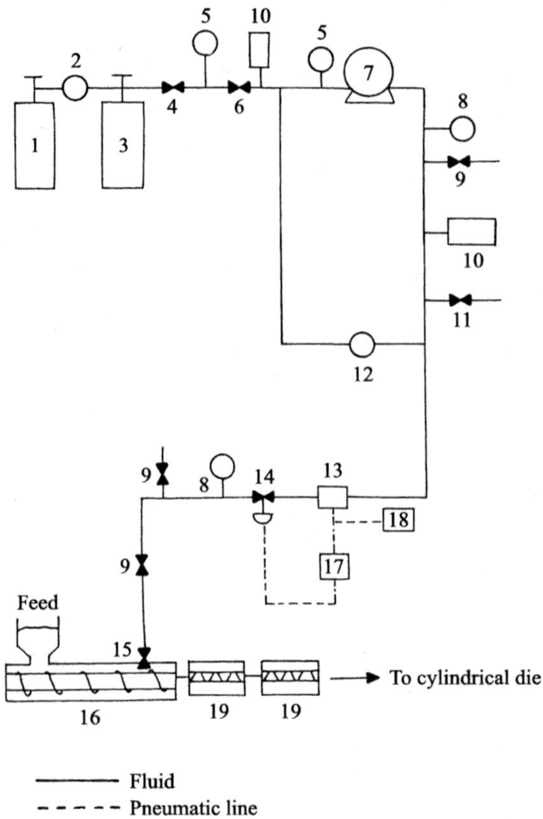

Figure 13.1 Schematic of a process supplying a continuous stream of molten polymer with dissolved gaseous component: (1) nitrogen tank, (2) pressure regulator, (3) volatile liquid tank, (4) ball valve, (5) pressure gauge, (6) filter, (7) diaphragm pump, (8) pressure gauge, (9) bleed valve, (10) accumulator, (11) adjustable relief valve, (12) back-pressure regulator, (13) flow meter, (14) control valve, (15) check valve, (16) screw extruder, (17) pneumatic controller, (18) pneumatic recorder, (19) static mixer with hot oil temperature control unit. (Reprinted from Han and Ma, *Journal of Applied Polymer Science* 28:831. Copyright © 1983, with permission from John Wiley & Sons.)

It is very important to understand that we have no way of separating $p(R, z)$ and $\sigma_{rr}(R, z)$ or $\sigma_{rr}(R)$ from the measured $T_{rr}(R, z)$ at any position z, and thus even an estimate of pressure $p(R, z)$ in the capillary section from the measurement of $T_{rr}(R, z)$ is not possible. However, in the upstream end of the reservoir section, where the deformation of fluid can be regarded to be negligibly small (thus $\sigma_{rr}(R, z) \approx 0$), Eq. (13.2) is simplified to

$$T_{rr} = -p \tag{13.4}$$

that is, measured wall normal stress in the upstream end of the reservoir section provides us with information on pressure p of the fluid there. For fully developed flow, from Eq. (13.3) we obtain

$$\frac{\partial T_{rr}(R, z)}{\partial z} = -\frac{\partial p(R, z)}{\partial z} \tag{13.5}$$

indicating that the wall normal stress gradient is the same as the pressure gradient.

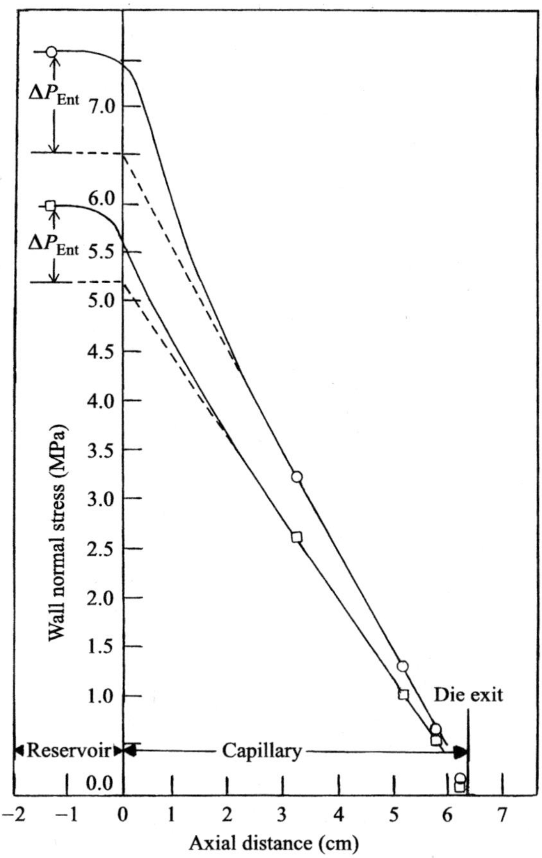

Figure 13.2 Wall normal stress profiles along the axis of a capillary die (diameter of 0.3175 cm) for an 80/20 Rexene 143/(FC-114) mixture at 110 °C for two different shear rates: (□) 156.8 s^{-1} and (○) 260.3 s^{-1}. (Reprinted from Han and Ma, *Journal of Applied Polymer Science* 28:831. Copyright © 1983, with permission from John Wiley & Sons.)

Figure 13.2 gives wall normal stress profiles along a capillary die for an 80/20 (by weight) mixture of a low-density polyethylene (LDPE) (Rexene 143) and dichlorotetrafluoroethane (FC-114)[2] at 110 °C, for two different shear rates, 156.8 and 260.3 s^{-1}. In Figure 13.2, we observe that the wall normal stress near the die exit plane is very low, deviating considerably from the extrapolated linear relationship based on the three data points measured within the capillary. Under such circumstances, the extrapolation of wall normal stresses to the exit plane is of no rheological significance when the wall normal stress profile begins to deviate from linearity somewhere within the die. The deviation of wall normal stress from the linear relationship observed in Figure 13.2 is attributed to significant foaming near the die exit.

The exact location of gas bubble formation inside the die depends on the shear rate and temperature (thus, the solubility of the gaseous component in the molten polymer), the concentration of gaseous component, and the type (i.e., molecular structure) of gaseous component. Therefore, for a given combination of polymer and gaseous component, there must exist a critical pressure for bubble nucleation. To illustrate the point, Figure 13.3 gives wall normal stress profiles along the axis of a capillary die for a high-density polyethylene (HDPE) at 200 °C without and with the chemical blowing

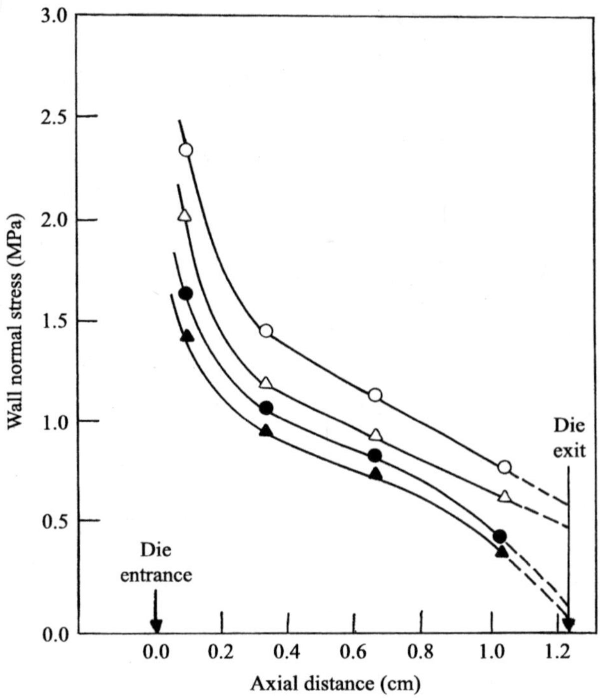

Figure 13.3 Wall normal stress profiles along the axis of a capillary die (diameter of 0.3175 cm) for an HDPE at 200 °C without blowing agent (open symbols) and with blowing agent (Celogen CB, generating N_2) (filled symbols), flowing through a cylindrical die with a length-to-diameter ratio of 4 at two different shear rates (s^{-1}): (○, ●) 377.8 and (△, ▲) 149.2. (Reprinted from Han et al., *Journal of Applied Polymer Science* 20:1583. Copyright © 1976, with permission from John Wiley & Sons.)

agent Celogen CB, which generates mostly nitrogen (N_2) gas at elevated temperature. It is seen in Figure 13.3 that the wall normal stress profile for the HDPE/N_2 mixture starts to deviate from linearity inside the capillary die, whereas the wall normal stress profile for neat HDPE does not. Notice in Figure 13.3 that the HDPE/N_2 mixture gives rise to lower wall normal stresses than the neat HDPE.

Figure 13.4 gives a schematic of the axial wall normal stress profile, $T_{rr}(R, z)$, in a capillary or slit die, through which a molten polymer with solubilized gaseous component is forced to flow. A constant pressure gradient $-\partial p/\partial z$ exists after passing the entrance region, but $T_{rr}(R, z)$, shows curvature before reaching the die exit. Under such circumstances, one can determine shear stress (hence viscosity) only in the region where constant $-\partial p/\partial z$ holds. Here, we assume that no measurable bubble formation has occurred in the region where constant $-\partial p/\partial z$ is observed. Therefore, in the use of a continuous-flow capillary (or slit) rheometer, it is essential to first check the linearity of $T_{rr}(R, z)$, profiles before proceeding to determine the viscosities of the molten polymer with solubilized gaseous component. It is then fair to say that a measurement of wall normal stress along the axis of a capillary or slit die is a very powerful experimental technique for determining the viscosity of molten polymer with solubilized gaseous component. Such an experimental method can be applied not only to molten polymer with solubilized gaseous component but also to polymer solution with volatile solvent, without having to install a pressurized chamber at the die exit, which would be required when using a plunger-type capillary rheometer.

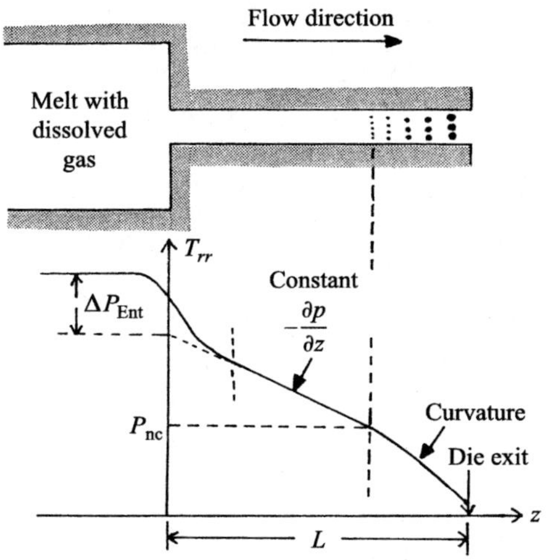

Figure 13.4 Schematic showing the wall normal stress profile along the axis of a capillary (or slit) die for a molten polymer with dissolved gaseous component that undergoes bubble nucleation inside the die. It is assumed in the schematic that a deviation of wall normal stress profile from linearity begins at a position where gas bubbles nucleate from the mixture.

13.2.2 Experimental Observations of Reduction in Melt Viscosity by Solubilized Gaseous Component

Han and Ma (1983a, 1983b) conducted extensive viscosity measurements for molten polystyrene (PS) or LDPE solubilized with fluorocarbon (FC) blowing agent using a continuous-flow capillary rheometer with four pressure transducers mounted along the die axis. In so doing, they used only the data points showing linearity of $T_{rr}(R, z)$ profile (i.e., constant pressure gradient $-\partial p/\partial z$) along the die axis to calculate viscosity. Below, we use the notation $\overline{\eta}$ to denote the viscosity of homogeneous mixtures consisting of a molten polymer and a physical blowing agent (e.g., FCs or other gaseous components) that under sufficiently high pressure is completely solubilized in the molten polymer.

Figure 13.5 gives plots of $\overline{\eta}$ versus shear rate ($\dot{\gamma}$) for mixtures of LDPE (Rexene 143) and FC-114 for varying concentrations of FC-114 at three different temperatures. It is seen in Figure 13.5 that for a given concentration of FC-114, the shear-rate dependence of $\overline{\eta}$ for Rexene 143/(FC-114) mixtures follows a power-law behavior, and that $\overline{\eta}$ decreases with increasing temperature. Such a consistent trend of $\overline{\eta}$ with respect to $\dot{\gamma}$ and temperature would not have been observed if the wall normal stress data points deviated from linearity inside the capillary die; that is, if the pressure gradient $-\partial p/\partial z$ were not constant (see Figure 13.4).

Figure 13.6 compares the $\overline{\eta}$ of Rexene 143/(FC-114) mixtures with the viscosity (η) of neat Rexene 143 at three different temperatures, showing that the values of $\overline{\eta}$ for Rexene 143/(FC-114) mixture are lower than the values of η for neat Rexene 143, and that $\overline{\eta}$ decreases with increasing concentration of FC-114. Figure 13.7 describes the effect of the chemical structure of blowing agent on the $\overline{\eta}$ of mixtures of Rexene 143 and FC blowing agent. It is seen in Figure 13.7 that the $\overline{\eta}$ of Rexene 143/(FC-12) mixtures is lower than that of Rexene 143/(FC-114) mixtures, indicating that the chemical structure of the FC blowing agent plays an important role in the extent of viscosity reduction of neat Rexene 143.

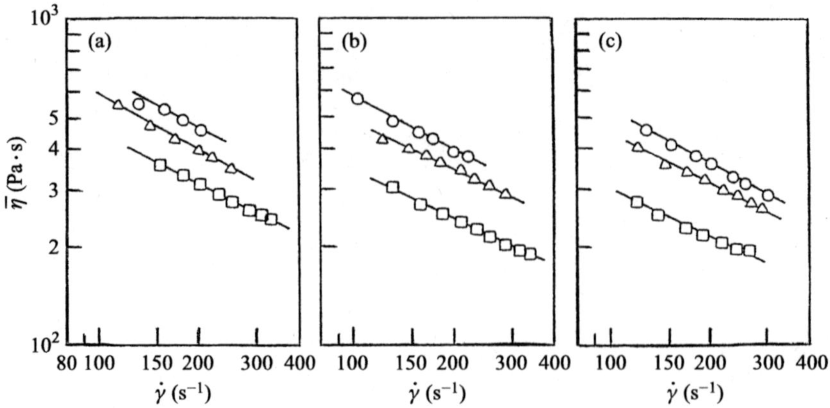

Figure 13.5 Plots of log $\bar{\eta}$ versus log $\dot{\gamma}$ at various temperatures (°C): (○) 110, (△) 120, and (□) 140, for (a) 90/10 Rexene 143/(FC-114) mixture, (b) 85/15 Rexene 143/(FC-114) mixture, and (c) 80/20 Rexene 143/(FC-114) mixture. (Reprinted from Han and Ma, *Journal of Applied Polymer Science* 28:831. Copyright © 1983, with permission from John Wiley & Sons.)

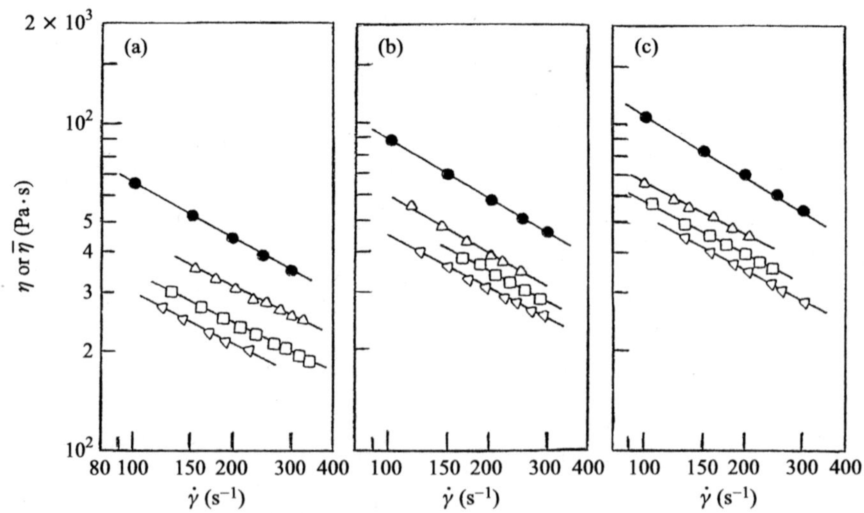

Figure 13.6 Plots of log $\bar{\eta}$ or log η versus log $\dot{\gamma}$ at various temperatures (°C): (a) 140, (b) 120, (c) 110, for (●) neat Rexene 143, (△) 90/10 Rexene 143/(FC-114) mixture, (□) 85/15 Rexene 143/(FC-114) mixture, and (▽) 80/20 Rexene 143/(FC-114) mixture. (Reprinted from Han and Ma, *Journal of Applied Polymer Science* 28:831. Copyright © 1983, with permission from John Wiley & Sons.)

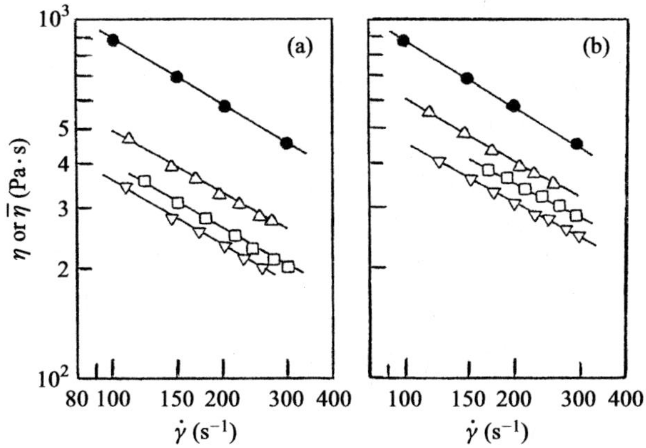

Figure 13.7 (a) Plots of log $\bar{\eta}$ or log η versus log $\dot{\gamma}$ at 120 °C for (●) neat Rexene 143, (△) 90/10 Rexene 143/(FC-12) mixture, (□) 85/15 Rexene 143/(FC-12) mixture, and (▽) 80/20 Rexene 143/(FC-12) mixture. (b) Plots of log $\bar{\eta}$ or log η versus log $\dot{\gamma}$ at 120 °C for (●) neat Rexene 143, (△) 90/10 Rexene 143/(FC-114) mixture, (□) 85/15 Rexene 143/(FC-114) mixture, and (▽) 80/20 Rexene 143/(FC-114) mixture. (Reprinted from Han and Ma, *Journal of Applied Polymer Science* 28:831. Copyright © 1983, with permission from John Wiley & Sons.)

Figure 13.8 gives log $\bar{\eta}$ versus log $\dot{\gamma}$ plots for mixtures of PS (Styron 678) and FC blowing agent (dichlorodifluoromethane (FC-12)[3] or trichlorofluoromethane(FC-11)[4]) at four different temperatures. It is seen in Figure 13.8 that the $\bar{\eta}$ of Styron 678/(FC-11) mixtures is considerably lower than that of Styron 678/(FC-12) mixtures at the identical temperatures tested. Note that the solubility of an FC blowing agent (or any gaseous components) in a molten polymer depends on temperature and pressure. As the temperature is increased, the solubility of FC blowing agents will decrease and hence the FC blowing agent, once solubilized, tends to form a separate phase (i.e., forms gas bubbles) unless a higher pressure (or a higher shear rate) is exerted on the mixture. Since all the viscosity data presented above were obtained in the range of wall normal stress measurements that gave constant $-\partial p/\partial z$, it is reasonable to assume that, over

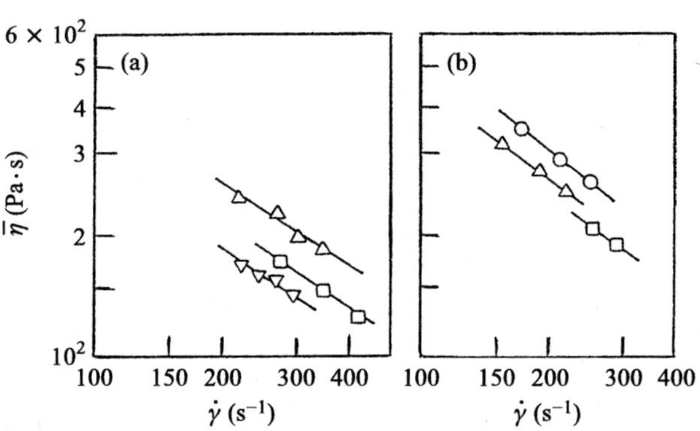

Figure 13.8 Plots of log $\bar{\eta}$ versus log $\dot{\gamma}$ at various temperatures (°C): (○) 140, (△) 150, (□) 160, and (▽) 170 for (a) 90/10 Styron 678/(FC-11) mixture and (b) 90/10 Styron 678/(FC-12) mixture. (Reprinted from Han and Ma, *Journal of Applied Polymer Science* 28:851. Copyright © 1983, with permission from John Wiley & Sons.)

the range of temperatures tested, the mixtures of molten polymer and FC blowing agent had a homogeneous phase.

The reduction in melt viscosity of Rexene 143 or Styron 678 with increasing FC blowing agent concentration observed in Figures 13.5–13.8 might have been caused by the plasticization of the molten polymers with the solubilized FC blowing agent (i.e., by the dilution of the entangled polymer with the FC blowing agent) under the experimental conditions employed. This interpretation, of course, assumes that the FC blowing agent is completely solubilized and distributed uniformly in the molten polymer. The extent to which a blowing agent (or any gaseous components) reduces the viscosity of a molten polymer depends on the molecular structure of the blowing agent, which in turn determines its solubility in the polymer.

What is common in Figures 13.5–13.8 is that the viscosity of a neat polymer, LDPE or PS, is decreased considerably by the solubilization of an FC blowing agent. It would be of practical interest to determine the extent of viscosity reduction of a molten polymer by a solubilized gaseous component. For this, let us define a viscosity reduction factor a_{RF} by (Han and Ma 1983a, 1983b)

$$a_{RF} = \overline{\eta}(\dot{\gamma}, T, c)/\eta(\dot{\gamma}, T) \tag{13.6}$$

in which T denotes temperature and c denotes the concentration of solubilized gaseous component; that is, a_{RF} is the ratio of the viscosity $\overline{\eta}(\dot{\gamma}, T, c)$ of mixture of molten polymer and solubilized gaseous component and the viscosity $\eta(\dot{\gamma}, T)$ of neat molten polymer. In general, for a given pair of polymer and solubilized gaseous component or volatile solvent, a_{RF} would depend on shear rate, temperature, and the concentration of solubilized gaseous component. Table 13.1 gives a summary of the a_{RF} for mixtures of LDPE (Rexene 143) and FC blowing agent, and Table 13.2 gives a summary of the a_{RF} for mixtures of PS (Styron 678) and FC blowing agent. It is interesting to observe in Tables 13.1 and 13.2 that a_{RF} is very weakly dependent upon shear rate and temperature, indicating that over the limited range of shear rates and temperatures tested, a_{RF} can be regarded as being virtually independent of shear rate and temperature. Notice in Tables 13.1 and 13.2 that the value of a_{RF} decreases as the FC blowing agent concentration increases, and that it varies with the chemical structure of FC blowing agent. For the limited amount of experimental data given in Tables 13.1 and 13.2, a_{RF} appears to vary only with the concentration of FC blowing agent. We hasten to point out that such a limited experimental observation may not hold for other mixtures of molten polymer and solubilized gaseous component.

Figure 13.9 gives plots of a_{RF} versus weight percent of FC blowing agent for Rexene 143/(FC-12) and Rexene 143/(FC-114) mixtures, and Figure 13.10 gives similar plots for Styron 678/(FC-11) and Styron 678/(FC-12) mixtures. Figures 13.9 and 13.10 indicate that the size of the FC molecules plays an important role in the extent of plasticization of large polymer molecules in the molten state under high pressure; for example, values of a_{RF} for Rexene 143/(FC-12) mixtures are smaller than those for Rexene 143/(FC-114) mixtures. The significance of Figure 13.9 lies in that one can estimate, with the aid of Eq. (13.6), the $\overline{\eta}(\dot{\gamma}, T, c)$ of Rexene 143/(FC-114) or Rexene 143/(FC-12) mixtures at any shear rate in the shear-thinning region and temperature once information on the $\eta(\dot{\gamma}, T)$ of neat polymer is available. Similarly, one can estimate from Figure 13.10, with the aid of Eq. (13.6), the $\overline{\eta}(\dot{\gamma}, T, c)$ of Styron 678/(FC-11)

Table 13.1 Summary of the viscosity reduction factor a_{RF} for mixtures of Rexene 143 and FC blowing agent

	Concentration of Blowing Agent (wt %)		
Shear Rate (s^{-1})	10	15	20
(a) Rexene 143/(FC-114) mixture at 110 °C			
100	0.682	0.552	0.500
200	0.693	0.550	0.507
300	0.688	0.566	0.528
(b) Rexene 143/(FC-114) mixture at 120 °C			
100	0.701	0.607	0.501
200	0.689	0.620	0.521
300	0.703	0.613	0.520
(c) Rexene 143/(FC-114) mixture at 140 °C			
100	0.723	0.553	0.461
200	0.716	0.568	0.466
300	0.728	0.571	0.471
(d) Rexene 143/(FC-12) mixture at 120 °C			
100	0.568	0.455	0.400
200	0.569	0.448	0.396
300	0.566	0.455	0.404
(e) Rexene 143/(FC-12) mixture at 110 °C			
100	0.568		
200	0.557		
300	0.559		
(f) Rexene 143/(FC-12) mixture at 140 °C			
100	0.582		
200	0.577		
300	0.569		

and Styron 678/(FC-12) mixtures at any shear rate in the shear-thinning region. We hasten to point out that such a prospect is based on the limited experimental observations that a_{RF} is virtually independent of shear rate and temperature for the particular set of experimental data. Even if a_{RF} is a function of shear rate and temperature as well as the concentration of solubilized gaseous component, the $\bar{\eta}(\dot{\gamma}, T, c)$ of molten polymers with solubilized gaseous component can be estimated once values of a_{RF} and information on the $\eta(\dot{\gamma}, T)$ of the neat polymer is available.

On the basis of the observations made above, one can calculate the average values of a_{RF} at various blowing agent concentrations and temperatures for the different chemical structures of FC blowing agent, the results of which are summarized in Table 13.3. Note that the values of a_{RF} given in Table 13.3 are independent of shear rate and temperature.

Table 13.2 Summary of the viscosity reduction factor a_{RF} for mixtures of Styron 678 and FC blowing agent

	Concentration of Blowing Agent (wt %)		
Shear Rate (s^{-1})	5	10	15
(a) Styron 678/(FC-12) mixture at 160 °C			
200	0.660	0.597	0.524
300	0.661	0.596	0.523
400	0.653	0.596	0.500
(b) Styron 678/(FC-12) mixture at 150 °C			
200	0.571		
300	0.567		
400	0.568		
(c) Styron 678/(FC-12) mixture at 140 °C			
100	0.555		
200	0.544		
300	0.541		

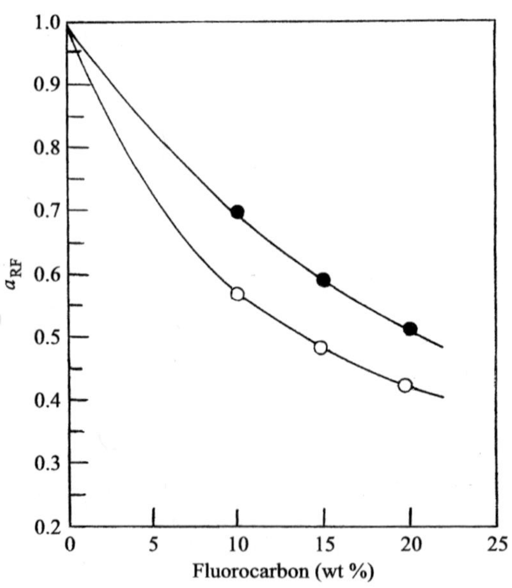

Figure 13.9 Plots of a_{RF} versus weight percent of fluorocarbon blowing agent in (○) Rexene 143/(FC-12) mixtures and (●) Rexene 143/(FC-114) mixtures. (Reprinted from Han and Ma, *Journal of Applied Polymer Science* 28:831. Copyright © 1983, with permission from John Wiley & Sons.)

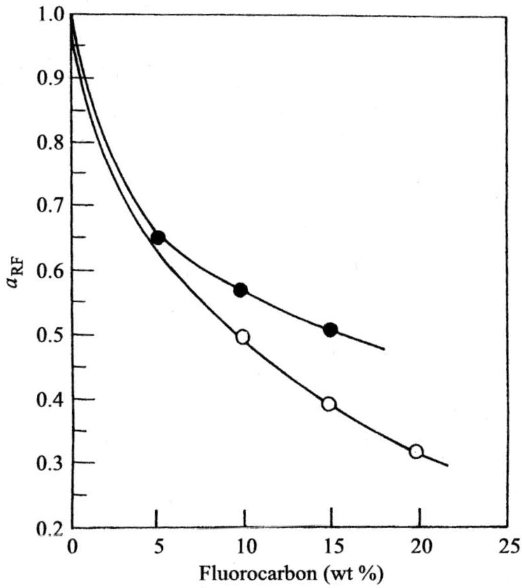

Figure 13.10 Plots of a_{RF} versus weight percent of fluorocarbon blowing agent in (○) Styron 678/(FC-11) mixtures and (●) Styron 678/(FC-12) mixtures. (Reprinted from Han and Ma, *Journal of Applied Polymer Science* 28:851. Copyright © 1983, with permission from John Wiley & Sons.)

When a_{RF} is plotted against moles of FC blowing agent/kg of polymer, we obtain an interesting correlation (given in Figure 13.11), which does not seem to depend on the chemical structure of the FC blowing agent but only on the chemical structure of the polymer. Such a correlation would be extremely helpful for one to estimate the $\bar{\eta}(\dot{\gamma}, T, c)$ of homogeneous mixtures consisting of a molten polymer and a solubilized

Table 13.3 Average value of viscosity reduction factor a_{RF} for mixtures of Rexene 143 or Styron 678 with FC blowing agent.

Concentration of FC Blowing Agent (wt %)		
wt %	*Moles of FC/kg of Polymer*	a_{RF}
(a) Rexene 143/(FC-12) mixture		
10	0.917	0.567
15	1.457	0.487
20	2.066	0.423
(b) Rexene 143/(FC-114) mixture		
10	0.649	0.694
15	1.031	0.587
20	1.462	0.510
(c) Styron 678/(FC-12) mixture		
5	0.435	0.658
10	0.917	0.581
15	1.457	0.516

Based on Han and Ma 1983a and 1983b.

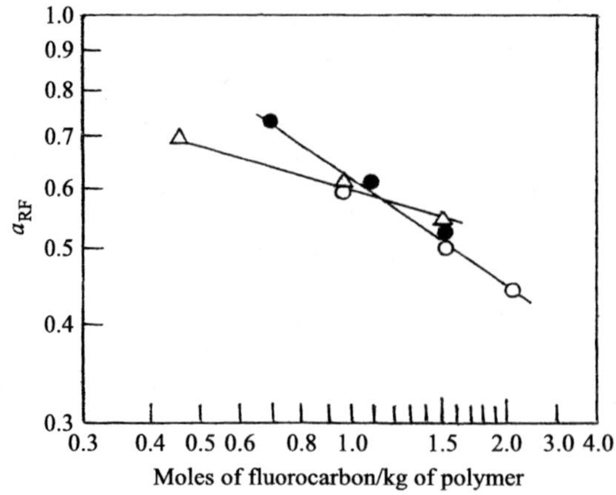

Figure 13.11 Plots of a_{RF} versus moles of FC blowing agent/kg of polymer in (○) Rexene 143/(FC-12) mixtures, (●) Rexene 143/(FC-114) mixtures, and (△) Styron 678/(FC-12) mixtures. (Based on Han and Ma 1983a and 1983b.)

gaseous component if the correlation holds for all types of solubilized gaseous components. Specifically, one can use Figure 13.11 to obtain the value of a_{RF} for a specified concentration of an FC blowing agent and then estimate, with the aid of Eq. (13.6), the $\overline{\eta}(\dot{\gamma}, T, c)$ of mixtures of LDPE and FC blowing agent with information only on the $\eta(\dot{\gamma}, T)$ of an LDPE as functions of shear rate and temperature. In view of the practical difficulties of measuring viscosities of homogeneous mixtures of molten polymer with a solubilized gaseous component at elevated temperature, such a procedure would be very useful. Needless to say, the correlation given in Figure 13.11 applies only to the homogeneous mixtures consisting of an LDPE or PS and an FC blowing agent over a limited range of shear rates and temperatures. It is highly desirable to obtain such a correlation for other pairs of molten polymer and solubilized gaseous component.

Not all experimental data in the literature reporting the $\overline{\eta}(\dot{\gamma}, T, c)$ of mixtures of molten polymer and a solubilized gaseous component were obtained with proper care using a continuous-flow capillary rheometer. Lee et al. (1999) used the average value of wall normal stress readings from two pressure transducers, one at the position of $L/D = 4$ from the die entrance and the other at the position of $L/D = 2$ from the die exit, to calculate the $\overline{\eta}(\dot{\gamma}, T, c)$ of PS/carbon dioxide (CO_2) mixtures at 220 °C with varying concentrations (2–4 wt %) of CO_2. Taking an average value of the wall normal stress readings from the two pressure transducers is tantamount to the assumption that the wall normal stress profile along the axis of a capillary die, where the two pressure transducers were mounted, was linear. Such an assumption is not warranted, as can be seen very clearly from Figures 13.2 and 13.3. When using a continuous-flow capillary rheometer to determine the $\overline{\eta}$ of a polymer melt with solubilized gaseous component, a minimum of three pressure transducers must be employed, with each transducer mounted along the die axis far away from the die entrance, and also sufficiently far away from the die exit plane, in order to ensure that the molten polymer with solubilized gaseous component is free from gas bubbles (i.e., no bubble nucleation) in the region where wall normal stresses are measured.

Mixtures of molten polymer and CO_2 have also been used in studies measuring melt viscosities (Elkovitch et al. 1999; Gehardt et al. 1997, 1998; Kwag et al. 1999; Royer et al. 2000, 2001). Needless to say, the solubility of CO_2 in a molten polymer is quite different from that of the FC blowing agents considered in Figures 13.5–13.8. The subject of solubility of a gaseous component in polymer melt or polymer solution is presented in Chapter 10 of Volume 2.

It is very difficult to determine, using a continuous-flow capillary/slit rheometer, the $\overline{\eta}$ of mixtures of molten polymer and gaseous component (or volatile liquid) at very low shear rates (e.g., $\dot{\gamma} < 1.0\,\mathrm{s}^{-1}$) without encountering the possibility of bubble nucleation. This is because as shear rate decreases, the location of bubble nucleation in a capillary or slit die moves toward the die entrance, decreasing the possibility of having a constant pressure gradient in the capillary or slit die over which wall normal stresses can be measured. This is particularly so when the solubility of a gaseous component in a molten polymer is rather low. Note that the lower the solubility of a gaseous component in a molten polymer, the higher will be the critical pressure for bubble inflation. The only way to overcome such a difficulty is to extrude a mixture of molten polymer and solubilized gaseous component into a pressurized, isothermal chamber, the pressure of which is kept above the critical pressure for bubble nucleation. Such an approach was used by Mendelson (1979, 1980), and also by Foster and Lindt (1987) who determined the $\overline{\eta}$ of solutions of PS and methylbenzene at elevated temperatures (60–225 °C) and various concentrations (50–90 wt %) of PS. Later, Gerhardt et al. (1997) also employed the same method to determine the $\overline{\eta}$ of poly(dimethyl siloxane) (PDMS) approach was used by Mendelson (1979, 1980), and also by Foster 2300 s^{-1}.

Figure 13.12 gives log $\overline{\eta}$ versus log $\dot{\gamma}$ plots for PDMS with solubilized CO_2 and log η versus log $\dot{\gamma}$ plot for neat PDMS at 50 °C. Note that PDMS is a liquid at room temperature. The data used in Figure 13.12 were obtained using a plunger-type capillary rheometer (Instron capillary rheometer) fitted with a pressurized isothermal chamber, with the pressure of the isothermal chamber maintained above the critical pressure for bubble nucleation, and using two different capillary dies having very large length-to-diameter (*L*/*D*) ratios (79.69 and 159.86). It can be seen in Figure 13.12 that at high shear rates ($\dot{\gamma} \geq 400\ \mathrm{s}^{-1}$), the log $\overline{\eta}$ versus log $\dot{\gamma}$ plots for PDMS/CO_2 mixtures and

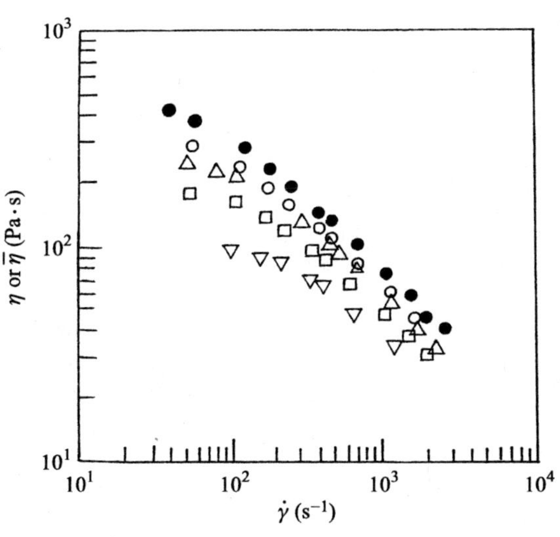

Figure 13.12 Log η versus log $\dot{\gamma}$ plot for neat PDMS (●) and log $\overline{\eta}$ versus log $\dot{\gamma}$ plots for PDMS/CO_2 mixtures at 50 °C for various concentrations of CO_2 (wt %): (○) 4.84, (△) 9.03, (□) 14.4, and (▽) 20.7. All data points for the PDMS/CO_2 mixtures were obtained using a plunger-type capillary rheometer fitted with a pressurized isothermal chamber at the end of the capillary die. (Reprinted from Gehardt et al., *Journal of Polymer Science, Polymer Physics Edition* 35:523. Copyright © 1997, with permission from John Wiley & Sons.)

the log η versus log $\dot{\gamma}$ plot for neat PDMS have virtually identical slopes, which is very similar to the observation made from Figures 13.6 and 13.7 for Rexene 143/(FC-114) and Rexene 143/(FC-12) mixtures. However, for $\dot{\gamma} < 400\,\text{s}^{-1}$, the difference between $\overline{\eta}$ and η increases as $\dot{\gamma}$ decreases. Notice in Figure 13.12 that the extent of viscosity reduction in the power-law region at high shear rates is not the same as that in the Newtonian region.

To obtain concentration-independent viscosity curves for PDMS/CO_2 mixtures from Figure 13.12, Gerhardt et al. (1997) introduced a concentration-dependent shift factor a_c, defined by

$$a_c = \overline{\eta}_0(T, c)/\eta_0(T) \tag{13.7}$$

in which $\overline{\eta}_0(T, c)$ is the zero-shear viscosity of a PDMS/CO_2 mixture at temperature T and concentration c of CO_2, and $\eta_0(T)$ is the zero-shear viscosity of neat PDMS at T. The shift factor a_c defined by Eq. (13.7) may be considered to be a special case of the viscosity reduction factor a_{RF} defined by Eq. (13.6) in that a_{RF} at $\dot{\gamma} = 0$ reduces to a_c. It is very difficult, if not impossible, to practically measure $\overline{\eta}_0(T, c)$ at elevated temperatures without encountering bubble nucleation in a capillary die. Under such circumstances, one usually approximates $\overline{\eta}_0(T, c)$ to be the viscosity of a mixture of molten polymer/solubilized gaseous component measured at the lowest shear rate employed without encountering bubble nucleation. With the aid of a_c, Gerhardt et al. (1997) obtained a concentration-independent correlation, log $\overline{\eta}/a_c$ versus log $a_c\dot{\gamma}$ plots (displayed in Figure 13.13) for PDMS/CO_2 mixtures at 50 °C

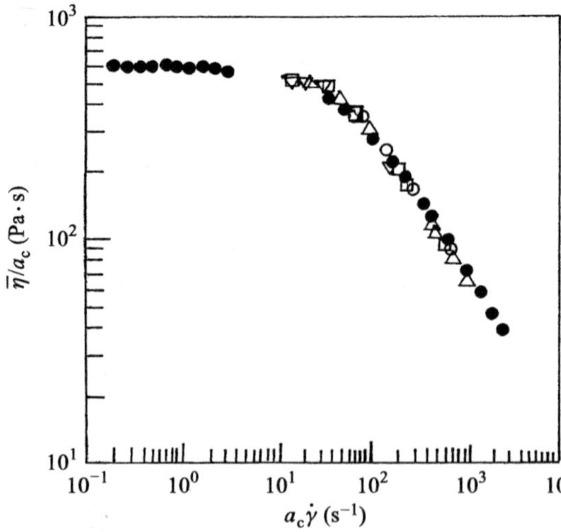

Figure 13.13 Log $\overline{\eta}/a_c$ versus log $a_c\dot{\gamma}$ plots at 50 °C for neat PDMS (●) and PDMS/CO_2 mixtures with various concentrations of CO_2 (wt %): (○) 4.84, (△) 9.0, (□) 14.4, and (▽) 20.7. The data points with the symbol ● (neat PDMS) for $a_c\dot{\gamma} < 3\ s^{-1}$ were obtained using a cone-and-plate rheometer, and all other data points for $a_c\dot{\gamma} > 10\ s^{-1}$ were obtained using a plunger-type capillary rheometer with a pressurized isothermal chamber attached at the end of the capillary die. (Reprinted from Gehardt et al., *Journal of Polymer Science, Polymer Physics Edition* 35:523. Copyright © 1997, with permission from John Wiley & Sons.)

with varying concentrations of CO_2. Notice in Figure 13.13 that both the viscosity ($\overline{\eta}$) and shear rate ($\dot{\gamma}$) are scaled with a concentration-dependent shift factor a_c, and $\overline{\eta}/a_c$ reduces to η_0 when $\dot{\gamma}$ approaches zero.

13.3 Theoretical Consideration of Reduction in Melt Viscosity by Solubilized Gaseous Component

Since the publication of the seminal paper by Doolittle (1951) it has been well accepted that the viscosity of a polymeric fluid is related to its free volume, which in turn is related to the glass transition temperature (T_g) of the polymer. Today, it is well documented (Ferry 1980) that the viscosity of a molten polymer is related to its T_g. It is then reasonable to expect that the free volume of a polymeric fluid will be increased when low-molecular-weight diluents (e.g., organic compounds, gaseous components under high pressure) are solubilized in the polymer. The extent of an increase in free volume of polymer/diluent mixtures will depend on the solubility of diluents, which in turn depends on temperature and pressure, especially for soluble gaseous components (e.g., FC blowing agent, carbon dioxide).

13.3.1 Depression of Glass Transition Temperature of Amorphous Polymer by the Addition of Low-Molecular-Weight Soluble Diluent

Gibbs and DiMarzio (1958) appear to have been the first to offer a molecular interpretation, via statistical mechanics, of the depression of glass transition temperature (T_g) of an amorphous polymer by the addition of low-molecular-weight soluble diluents, in terms of the size and stiffness of diluents and the diluent concentration. Subsequently, using both classical and statistical thermodynamics, Chow (1980) derived the following expression, which predicts the T_g depression of polymer/diluent mixtures:

$$\ln\left(T_g/T_{go}\right) = \beta\left[(1-\theta)\ln(1-\theta) + \theta\ln\theta\right] \tag{13.8}$$

in which T_g is the glass transition temperature of polymer/diluent mixture, T_{go} is the glass transition temperature of neat polymer, and θ and β, respectively, are defined by

$$\theta = \frac{M_p}{zM_d}\frac{w}{1-w} \quad \text{and} \quad \beta = \frac{zR}{M_p \Delta C_p} \tag{13.9}$$

where M_p and M_d are the molecular weights of polymer and diluent, respectively, z is a lattice coordinate number, w is the weight fraction of diluent, R is the universal gas constant, and ΔC_p is the change in specific heat of the polymer at its glass transition temperature. Using Eq. (13.8), Chow (1980) prepared plots of T_g/T_{go} versus θ for PS solubilized by 13 different diluents and observed that values of T_g/T_{go} decrease monotonically with increasing θ; that is, the T_g of PS decreases with increasing concentration of diluents.

Note that Eq. (13.8) was derived on the basis of the Gibbs–DiMarzio criterion, which states that the entropy is zero at T_g, and that it was developed for liquid plasticizers and thus it does not contain pressure as an independent variable. However, pressure is a very important processing variable in foam extrusion, in which a blowing agent in the liquid state is injected at very high pressure (i.e., well above the critical pressure for bubble nucleation) into molten polymer so that the formation of gas bubbles can be prevented (see Chapter 10 of Volume 2). To overcome the practical limitations of the Chow theory, Condo et al. (1992) employed a lattice fluid model (Panayiotou, 1986, 1987; Sanchez and Lacombe 1976, 1978) to develop a theory that predicts the T_g depression of polymer/diluent mixtures caused by a compressible fluid. The advantage of using lattice fluid models lies in that they treat temperature and pressure as the independent variables. Thus, the theory of Condo et al. (1992) can predict the T_g depression of polymer/diluent mixtures in which a gaseous component is solubilized under a sufficiently high pressure in a molten polymer.

Some researcher groups (Chiou et al. 1985; Condo and Jonston 1992; Condo et al. 1994; Wissinger and Paulaitis 1991) conducted experimental studies on the measurements of T_g of PS and poly(methyl methacrylate) (PMMA) as affected by the solubilization of CO_2 at elevated pressures and temperatures, and they interpreted the experimental results with theory. Some investigators (Chiou et al. 1985; Condo and Jonston 1992; Wissinger and Paulaitis 1991) observed that the Chow theory (1980) predicts experimental results reasonably well for low concentrations (e.g., up to approximately 12 wt %) of CO_2, while others (Condo et al. 1994) observed that the theory of Condo et al. (1992) predicts experimental results reasonably well for concentrations of CO_2 as high as 20 wt %. Figure 13.14 compares theory with experiment on the extent of

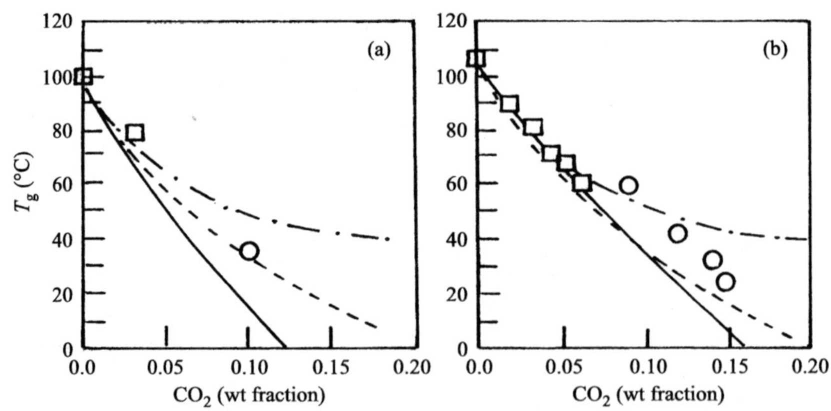

Figure 13.14 Plots of glass transition temperature (T_g) versus weight fraction of carbon dioxide (CO_2) in (a) PS/CO_2 mixtures and (b) PMMA/CO_2 mixtures. The symbol ○ represents the experimental data from Condo et al. (1994), the symbol □ represents experimental data from Chiou et al. (1985), the solid line represents the theoretical prediction from the model of Condo et al. (1992), the dashed line represents the theoretical prediction from the Chow model (1980) with the lattice coordinate number (z) of 1, and the broken line represents the theoretical prediction from the Chow model (1980) with $z = 2$. (Reprinted from Condo et al., *Macromolecules* 27:365. Copyright © 1994, with permission from the American Chemical Society.)

reduction in T_g of PS in the PS/CO_2 mixtures and of PMMA in the PMMA/CO_2 mixtures. It is seen in Figure 13.14 that the reduction in T_g for the respective mixtures is indeed considerable as the concentration of CO_2 increases.

13.3.2 Depression of Melting Point of Semicrystalline Polymer by the Addition of Low-Molecular-Weight Soluble Diluent

The following expression has been suggested for estimating the melting point depression of semicrystalline polymer by the addition of low-molecular-weight soluble components (Olabisi et al. 1979):

$$\frac{1}{T_m} - \frac{1}{T_m^o} = \frac{RV_2}{\Delta H_2 V_1}\left[(1-\phi_2) - \chi(1-\phi_2)^2\right] \tag{13.10}$$

where T_m is the melting temperature of the polymer/diluent mixture, T_m° is the melting temperature of the semicrystalline polymer, R is the universal gas constant, V_2 is the molar volume of polymer repeat unit, V_1 is the molar volume of diluent, ΔH_2 is the heat of fusion of 100% crystalline polymer, ϕ_2 is the volume fraction of semicrystalline polymer, and χ is the Flory–Huggins interaction parameter. At present, information on χ for the LDPE/(FC-12) and LDPE/(FC-114) pairs as a function of temperature is not available. Note that χ depends on temperature and composition (see Chapter 7). Nevertheless, it is clear from Eq. (13.10) that T_m is smaller than T_m° (thus melting point depression occurs) as long as $\chi \leq 0$. When an FC blowing agent (FC-114 or FC-12) is solubilized in an LDPE under a sufficiently high pressure, it is reasonable to state that $\chi < 0$ for LDPE/(FC-12) and LDPE/(FC-114) pairs under such high pressures. Precise determination of the melting point depression of LDPE in the presence of solubilized FC-114 or FC-12 can be determined if we have information on χ as functions of temperature and concentration of the blowing agent.

13.3.3 Theoretical Interpretation of Reduction in Melt Viscosity by Solubilized Gaseous Component

We have shown above experimental results of reduction in melt viscosity of LDPE and PS by the solubilization of an FC blowing agent (FC-114, FC-11, and FC-12) or CO_2, and also the T_g depression of PS and PMMA by the solubilization of CO_2. Both phenomena, reduction in melt viscosity and T_g depression, are due to an increase in free volume by the addition of low-molecular-weight diluent. Specifically, an increase in the concentration of solubilized gaseous component (diluent) decreases both the T_g and the viscosity of polymer melts containing diluent. We can then conclude that the T_g depression of PS and the T_m depression of LDPE by the addition of diluent (e.g., solubilized FC blowing agent or CO_2) are reflected in the reduction in melt viscosity. In this regard, it is reasonable to interpret the experimentally observed viscosity reduction using the expressions based on free-volume theory. After all, it has long been recognized that the viscosity of an amorphous polymer can be related to free-volume parameters.

Specifically, on the basis of free-volume theory, Doolittle (1951) has related the zero-shear viscosity (η_0) of an amorphous polymer to relative (fractional) free volume f by

$$\ln \eta_0 = \ln A + B/f \tag{13.11}$$

in which A and B are constants. Assuming the following relationship between f and temperature T:

$$f = f_g + \alpha(T - T_g) \tag{13.12}$$

in which T_g is the glass transition temperature, f_g is the fractional free volume at T_g, and α is the relative free volume thermal expansion coefficient (or difference in volume expansion between the temperature above T_g and the temperature below T_g), Williams et al. (1955) obtained the following expression, referred to as the Williams–Landel–Ferry (WLF) equation:

$$\log a_T = \log\left(\frac{\eta_T}{\eta_{T_g}}\right) = \frac{-C_1(T - T_g)}{C_2 + T - T_g} \tag{13.13}$$

where a_T is referred to as the "temperature shift factor," η_T is the viscosity of a polymer at temperature T, η_{T_g} is the viscosity at T_g, and C_1 and C_2 are constants. It is then clear that the WLF equation has an origin in free-volume theory. In Chapter 6, we used the WLF equation to obtain temperature-independent reduced viscosity plots for flexible homopolymers.

Several research groups used free-volume theory, including the WLF equation and its analogues for concentration and pressure, to explain the experimentally observed reduction in viscosity of molten polymers with solubilized gaseous component. However, there is a great deal of confusion in the literature that has dealt with this problem. Because of its importance, we address this issue below.

In analyzing the viscosity data at various temperatures and shear rates, which were obtained via slit rheometry[5] for molten PS, PMMA, polypropylene (PP), and LDPE with solubilized CO_2 at varying concentrations, Royer et al. (2000, 2001) used the following expressions, analogues of Eq. (13.13), for the concentration-dependent shift factor a_c at temperatures below $T_g + 100\,°C$:

$$\log a_c = \log\left(\frac{\eta_{T,p_o,c_o}}{\eta_{T,p_o,c}}\right) = \frac{C_1(T - T_{g,p_o})}{C_2 + T - T_{g,p_o}} - \frac{C_1(T - T_{g,\text{mix},p_o})}{C_2 + T - T_{g,\text{mix},p_o}} \tag{13.14}$$

in which η_{T,p_o,c_o} refers to the viscosity data at the measurement temperature T, at a pressure corrected to atmospheric pressure p_o, and at a concentration corrected to zero concentration c_o of diluent, and $\eta_{T,p_o,c}$ refers to the viscosity data at the measurement temperature T, at a pressure corrected to atmospheric pressure p_o, and at the concentration c of diluent employed, and for the pressure-dependent shift factor a_{p} at temperatures below $T_{\mathrm{g}} + 100\,°\mathrm{C}$:

$$\log a_{\mathrm{p}} = \log\left(\frac{\eta_{T,p,c}}{\eta_{T,p_{\mathrm{o}},c}}\right) = \frac{C_1(T - T_{\mathrm{g,mix},p_{\mathrm{o}}})}{C_2 + T - T_{\mathrm{g,mix},p_{\mathrm{o}}}} - \frac{C_1(T - T_{\mathrm{g,mix},p})}{C_2 + T - T_{\mathrm{g,mix},p}} \tag{13.15}$$

in which $\eta_{T,p,c}$ refers to the viscosity data at the measurement temperature T, pressure p, and concentration c of diluent, and $\eta_{T,p_{\mathrm{o}},c}$ refers to the viscosity data at the measurement temperature T, at a pressure corrected to atmospheric pressure p_{o}, and at the concentration c of diluent employed.

Applying the shift factors defined above to the experimentally observed viscosities of PS, PMMA, PP, and LDPE with solubilized CO_2, Royer et al. (2000, 2001) prepared reduced plots, log $\overline{\eta}/a_{\mathrm{T}}a_{\mathrm{c}}a_{\mathrm{p}}$ versus log $a_{\mathrm{T}}a_{\mathrm{c}}a_{\mathrm{p}}\dot{\gamma}$ plots, showing temperature-, concentration-, and pressure-independent correlations. However, it is not at all clear how they obtained information on pressure p, which was needed to calculate values of the shift factors a_{c} and a_{p}, from the measurements of wall normal stress $T_{yy}(b, z)$ via pressure transducers, along the axis of a slit die. This is because the $T_{yy}(b, z)$ measured in the fully developed region of a slit die consists of two terms, pressure $p(b, z)$ and deviatoric stress $\sigma_{yy}(b)$ (see Chapter 5):

$$T_{yy}(b, z) = -p(b, z) + \sigma_{yy}(b) \tag{13.16}$$

where the subscript y refers to the direction perpendicular to the flow direction z, and b is half of the slit height. Royer et al. might, mistakenly, have thought that the pressure transducer mounted on the wall of a slit die actually measured p. There is no way of separating $p(b, z)$ from the measured $T_{yy}(b, z)$, because pressure transducers do not measure the $p(b, z)$ in the fully developed region of a slit die! Notice that both $T_{yy}(b, z)$ and $p(b, z)$ vary with the flow direction z. As described in Chapter 5, as long as the $T_{yy}(b, z)$ profile along the z direction is linear, there is no reason why one has to consider the effect of pressure on the calculated viscosity. However, when the $T_{yy}(b, z)$ profile shows curvature for whatever the reasons (e.g., viscous shear heating, pressure effect), the viscosity of a fluid cannot be calculated from $T_{yy}(b, z)$ measurements because pressure gradient $(\partial p/\partial z)$ would not be constant (simply because Eq. (13.16) is not valid). It seems contradictory for Royer et al. to have evaluated the effect of pressure on melt viscosity inside the slit die without being able to calculate $p(b, z)$ from the measured $T_{yy}(b, z)$ whilst they calculated the viscosities of PS melts with solubilized CO_2 from the slope of linear $T_{yy}(b, z)$ profile in the slit die. In other words, a linear $T_{yy}(b, z)$ profile would not have existed if the effect of pressure were significant enough to affect the melt viscosity inside the slit die. Note from Eq. (13.16) that $\partial T_{yy}(b, z)/\partial z = -\partial p(y, z)/\partial z =$ constant in the fully developed region; that is, only when a linear $T_{yy}(b, z)$ profile existed.

Using two pressure transducers, Lee et al. (1999) measured wall normal stress $T_{rr}(R, z)$ of molten PS with solubilized CO_2 along the axis of a capillary die and then interpreted their experimental results by including the effect of pressure p on melt viscosity with the expression

$$\eta_0 = A \exp\left(\frac{\alpha}{T - T_r} + \beta p\right) \tag{13.17}$$

which is based on a modified version of Eq. (13.11) with the aid of Eq. (13.12), where η_0 is the zero-shear viscosity of molten PS with solubilized CO_2, A is a preexponential factor, T is the measurement temperature, T_r is a reference temperature, and α and β are constants. In calculating η_0 from Eq. (13.17), Lee et al. took the average value of the pressure transducer readings between two positions in the capillary die. Apparently, they mistakenly thought that the pressure transducers mounted on the wall of a capillary die measured pressure p. As shown by Eqs. (13.2) and (13.3), the pressure transducers mounted on the wall of a capillary die measure $T_{rr}(R, z)$ but not $p(R, z)$. There is no way of determining $p(R, z)$ from the measurement of $T_{rr}(R, z)$ in the capillary die! Therefore, we conclude that the pressure corrected viscosities reported by Lee et al., which were based on Eq. (13.17), of molten PS with solubilized CO_2 are of little rheological significance.

Fujita and Kishimoto (1958) have shown that the effect of diluting a polymer with small amounts of solute can be described by

$$\ln a_c = \frac{1}{f_m} - \frac{1}{f_p} \tag{13.18}$$

where f_m is the relative free volume of a diluted mixture with concentration c and f_p is the relative free volume of pure polymer. They obtained an expression that is analogous to Eq. (13.12) by using the following relationship for small proportions of diluent:

$$f_m = f_p + \beta'(1 - c/\rho_m) \tag{13.19}$$

where ρ_m is the density of the pure polymer, c is the concentration of diluent, and β' is approximately equal to the fractional volume of diluent f.

Gerhardt et al. (1998) analyzed the viscosity data (see Figures 13.12 and 13.13) for PDMS melts solubilized with CO_2 at 50 and 80 °C for concentrations of CO_2 ranging from 4.84 to 20.7 wt %. Note that the viscosities were determined using a plunger-type capillary rheometer with a pressurized isothermal chamber attached at the end of a capillary die. For the analysis, they used the following form of a free-volume expression for the concentration shift factor a_c:

$$a_c = \frac{\bar{\eta}_0(T, c)}{\eta_0(T)} = (1 - w_c)^n \left(\frac{V_p}{V_m}\right)^n \exp\left[\frac{1}{f_m} - \frac{1}{f_p}\right] \tag{13.20}$$

in which $\bar{\eta}_0(T, c)$ is the zero-shear viscosity of polymer melt solubilized with diluent with T being the measurement temperature and c being the concentration of diluent,

$\eta_0(T)$ is the zero-shear viscosity of neat polymer melt, w_c is the weight fraction of diluent, the exponent n is 3.4–3.5 for entangled polymer systems, V_p and V_m are the specific volumes of the neat polymer melt and polymer/diluent mixture, respectively, and f_p and f_m are the fractional free volumes of neat polymer melt and polymer/diluent mixture, respectively. Eq. (13.20) can be regarded as being a modified version of Eq. (13.18). Gerhardt et al. determined values of f_p and f_m using the lattice fluid model of Sanchez and Lacombe (1976, 1978). In their experiments, Gerhardt et al. (1998) measured the force applied to the plunger of an Instron capillary rheometer, which then allowed them to calculate the pressure p in the upstream end of the reservoir section of the rheometer, as defined by Eq. (13.4). Since Eq. (13.4) is based on the premise that the deformation of a fluid is negligibly small (i.e., $\sigma_{rr} \approx 0$) in the upstream end of the reservoir section, use of the measured pressure p to calculate, via the Sanchez–Lacombe equation-of-state, values of f_p and f_m appearing in Eq. (13.20) is warranted. The empirically defined a_c given by Eq. (13.7) then has a theoretical interpretation on the basis of free-volume theory.

At this juncture, it is very important to emphasize that Eq. (13.20), or any modified version of Eq. (13.20), is not appropriate for calculating the shift factor a_c for shear-rate dependent viscosities of polymer/diluent mixtures in the capillary section where the deformation of fluid is significant (i.e., $\sigma_{rr} \neq 0$) and information on pressure p in the capillary section is not available from experiment, as discussed in a previous section of this chapter (see Eqs. (13.2) and (13.3)) and also in Chapter 5. For this reason, use of the state-of-equation approach to interpret the shear-rate dependent viscosity data for polymer/diluent mixtures from $T_{rr}(R, z)$ measurements in a capillary die is not appropriate because information on pressure p, which is needed to calculate f_p and f_m, for instance, from Eq. (13.20), is not available.

Penwell et al. (1971) proposed the following form of pressure-dependent viscosity $\eta(p, T, \dot{\gamma})$ in the shear-thinning region of a flow curve in a capillary die:

$$\eta(p, T, \dot{\gamma}) = K'\dot{\gamma}^{n-1}\,\mathrm{e}^{bp} \tag{13.21}$$

where K' and n are power-law constants, and p is a pressure somewhere inside a capillary die. Equation (13.21) is based on the expression

$$\eta(p, T) = \eta_1 \mathrm{e}^{bp} \tag{13.22}$$

where $\eta_1 = B\eta_{0,T_g}$ with B being a constant and η_{0,T_g} being the zero-shear viscosity at T_g, and b is a constant. Note that Eq. (13.22) was obtained, with some approximations, by substituting the pressure-dependent glass transition temperature from the experimental study of Gee (1966):

$$T_g = T_{go} + A_1 p \tag{13.23}$$

into Eq. (13.13), where T_{go} is the glass transition temperature at atmospheric pressure and A_1 is a constant. Since Eq. (13.23) was obtained under static conditions, η defined by Eq. (13.22) must be construed as zero-shear viscosity. It is not at all clear how

Eq. (13.22) can be generalized to Eq. (13.21) for determining shear-rate dependent viscosity in a capillary die, where information on p is not available, experimentally or theoretically, as pointed out in a previous section in conjunction with Eqs. (13.2) and (13.3), and also in Chapter 5.

In analyzing the viscosity data, which were obtained using a plunger-type capillary rheometer (Instron capillary rheometer), for PS melts solubilized with CO_2 or FCs (1,1-difluoroethane or 1,1,1,2-tetrafluoroethane) at elevated temperatures, Kwag et al. (1999) defined the pressure shift factor a_p by

$$a_p = \overline{\eta}(p, c)/\overline{\eta}(p_o, c) \tag{13.24}$$

in which $\overline{\eta}(p, c)$ is the viscosity of a polymer/diluent mixture at pressure p and diluent concentration c, and $\overline{\eta}\,(p_o, c)$ is the viscosity of polymer/diluent mixture at an arbitrary reference pressure p_o (e.g., the atmospheric pressure) and diluent concentration c. They prepared pressure-independent logarithmic plots of $\overline{\eta}(T, c)/a_c a_p$ versus $a_c a_p \dot{\gamma}$ using a combined shift factor $a_c a_p$, and also logarithmic plots of $\overline{\eta}(T, p, c)/a_T a_c a_p$ versus $a_T a_c a_p \dot{\gamma}$ using a combined shift factor $a_T a_c a_p$ for the molten PS/diluent mixtures. In doing so, they calculated values of a_p from the expression

$$a_p = \frac{\eta\,(p_o, T,\ \dot{\gamma})}{\eta\,(p^*, T,\ \dot{\gamma})} = \exp\left[b(p_o - p^*)\right] \tag{13.25}$$

in which b is a constant, $\eta(p_o, T,\ \dot{\gamma})$ is the viscosity at a reference pressure p_o, measurement temperature T, and shear rate $\dot{\gamma}$ in the capillary section, $\eta(p^*, T, \dot{\gamma})$ is the viscosity at $T, \dot{\gamma}$, and pressure p^* defined by

$$p^* = \frac{1}{b}\ \ln\left[\frac{b(p_1 - p_2}{e^{-bp_2} - e^{-bp_1}}\right] \tag{13.26}$$

where p_1 is the pressure at the entrance of a capillary die having the length L and p_2 is the pressure at the exit of the capillary die attached with a pressurized isothermal chamber. Note that Eq. (13.25) is a modified form of Eq. (13.21) in that according to Kwag et al. (1999), p^* appearing in Eq. (13.25) is associated with the pressure somewhere in the capillary die. However, at the entrance (at $z = 0$) of a capillary die, one cannot determine p_1 because a molten polymer/diluent mixture undergoes considerable deformation at $z = 0$ and thus there is no way one can determine, experimentally or theoretically, values of p_1 at $z = 0$. In other words, the pressure p in the upstream end of the reservoir section, which is measured in a plunger-type rheometer (e.g., Instron capillary rheometer), is not the same as pressure $p(r, 0)$ at the entrance ($z = 0$) of a capillary die. Further, there is no way of knowing (or even estimating) the pressure $p(r, z)$ inside a capillary die, as described by Eqs. (13.2) and (13.3). Thus, one may question the validity of Eq. (13.26) and thus Eq. (13.25).

However, Eq. (13.25) can be modified as

$$a_p = \frac{\eta_0(p_o, T)}{\eta_0(p, T)} = \exp\left[b(p_o - p)\right] \tag{13.27}$$

where $\eta_0(p_o, T)$ is the zero-shear viscosity at measurement temperature T and a reference pressure p_o, and $\eta_0(p, T)$ is the zero-shear viscosity at measurement temperature T and pressure p in the upstream end of the reservoir section, where the use of $T_{rr} = T_{zz} = -p$ is warranted to describe the pressure shift factor a_p, which is not, however, associated with $\dot{\gamma}$.

It is worth mentioning that Ferry and Stratton (1960) proposed the following form of pressure shift factor a_p:

$$\log a_p = \log\left(\frac{\eta_{0,p}}{\eta_{0,p_o}}\right) = \frac{(1/2.303 f_o)\,(p, -p_o)}{f_o/\beta_f + p - p_o} \tag{13.28}$$

which is the analogue of the temperature shift factor a_T given by Eq. (13.13), where $\eta_{0,p}$ is the zero-shear viscosity at measurement pressure p, η_{0,p_o} is the zero-shear viscosity at a reference pressure p_o, f_o is the fractional free volume at p_o, and $\beta_f = -(1/v)(\partial v_f/\partial p)$ with v being the specific volume and v_f being the free volume per gram (excess over the specific volume extrapolated to absolute zero).

13.4 Summary

This chapter has described the rheological behavior of molten polymers with a solubilized gaseous component. Emphasis was placed on the importance of suppressing bubble nucleation during rheological measurements of molten polymers with a solubilized gaseous component; otherwise, the analysis of experimental data would be very complicated in that bubble dynamics would need to be considered. In an effort to prevent the vaporization of a solubilized gaseous component in a polymer melt or the solvent in a polymer solution at elevated temperatures, some investigators have determined the viscosity of such mixtures by fitting a pressurized isothermal chamber to the exit of a plunger-type capillary viscometer and measuring pressure difference between the capillary entrance and exit. In such an approach, polymer solutions must be prepared by first dissolving a polymer in a volatile solvent at room temperature, then charging the polymer solution into the barrel of the viscometer, and finally extruding it at elevated temperature into a pressurized isothermal chamber. From the point of view of sample preparation, such an experimental technique is not easy to apply to molten polymers containing a volatile, solubilized gaseous component.

In this chapter, we have presented experimental results showing that the viscosity of a molten polymer with solubilized gaseous component is lower than that of neat molten polymer, and that the viscosity reduction increases with increasing concentration of solubilized gaseous component. Such experimental results are interpreted in terms of the reduction in glass transition temperature for amorphous polymers or the reduction in melting temperature for semicrystalline polymers; that is, in terms of plasticization of the molten polymer by the solubilized gaseous component. We have shown that the plasticizing efficiency (i.e., the extent of reduction in T_g and viscosity) depends on the molecular weight, size, and concentration of solubilized gaseous component (or diluent), and also on the chemical structures of diluent and polymer. A fundamental understanding of this subject requires serious theoretical investigation in the future.

We have also presented some useful correlations between the extent of viscosity reduction and the concentration of solubilized gaseous component. We have described how such correlations may be used to estimate the viscosities of an untested molten polymer containing a solubilized gaseous component. The experimentally observed reduction in melt viscosity in the presence of solubilized gaseous component is interpreted using the concept of free-volume theory. We have pointed out that extreme caution is necessary when trying to include the effect of pressure on the viscosities of polymer melts with and without a solubilized gaseous component when a plunger-type capillary rheometer or a continuous-flow rheometer is used to determine the viscosity in the flow regime, where significant deformation of fluid occurs, giving rise to shear-thinning behavior.

Problems

Problem 13.1

Figure 13.3 gives wall normal stress profiles of an HDPE, with and without a chemical blowing agent (generating nitrogen), flowing through a cylindrical tube. It is seen that the wall normal stress profile of the HDPE melt with solubilized gaseous component (filled symbols) begins to show curvature (concave down) at about the middle of the cylindrical tube, whereas the wall normal stress profile of the neat HDPE (open symbols) shows a constant slope.

(a) Explain what might have occurred in the cylindrical tube while the HDPE melt with solubilized gaseous component approached the die exit.

(b) Why does the shape of the pressure gradient of the HDPE melt with dispersed gas bubbles concave downward?

Problem 13.2

The shear-rate dependent viscosity of an LDPE (Rexene 143) may be described by Eq. (6.10) with $\eta_0 = 1.06 \times 10^4$ Pa · s, $\tau_1 = 0.62$ s, and $n = 0.224$ at 160 °C. Note that the shear-rate dependent viscosity ($\overline{\eta}$) of LDPE/(FC-114) mixtures at any temperature for a specified blowing agent (FC-114) concentration can be estimated once the viscosity (η) of Rexene 143 is known because plots of $\overline{\eta}/\eta$ given in Figure 13.11 are independent of shear rate and temperature. Calculate, with the aid of Eq. (13.6), the known because plots of $\overline{\eta}/\eta$ given in from 10^{-2} to 10^3 s^{-1} and prepare plots of $\overline{\eta}(\dot{\gamma})/\eta_0$ versus $\tau_1\dot{\gamma}$ for the mixture.

Problem 13.3

The shear-dependent viscosity of a PS (Styron 678) may be described by Eq. (6.10) with $\eta_0 = 7.36 \times 10^3$ Pa·s, $\tau_1 = 0.28$ s, and $n = 0.137$ at 160 °C. Note that the shear-rate dependent viscosity ($\overline{\eta}$) of PS/(FC-12) mixtures at any temperature for a specified blowing agent (FC-12) concentration can be estimated once the viscosity (η) of PS is known because the plot of $\overline{\eta}/\eta$ given in Figure 13.11 is independent of shear rate and temperature. Calculate, with the aid of Eq. (13.6), the viscosities of 75/25 PS/(FC-12) shear rate and temperature. Calculate, with the aid of Eq. (13.6), prepare plots of $\overline{\eta}(\dot{\gamma})/\eta_0$ versus $\tau_1\dot{\gamma}$ for the mixture.

Notes

1. Most of the chemical blowing agents are organic azo compounds which generate, upon thermal decomposition, mainly nitrogen, carbon monoxide, and carbon dioxide.
2. Dichlorotetrafluoroethane (FC-114) is a physical blowing agent having a critical pressure of 32.2 atm (473 psig or 3.26 MPa) and a critical temperature of 146 °C. Although, owing to environmental problems, the use of fluorocarbon blowing agents in thermoplastic foam processing will be banned in the future, the experimental results presented in this chapter will still be valid for other types of blowing agent.
3. Dichlorodifluoromethane (FC-12) is a physical blowing agent having a critical pressure of 40.7 atm (598 psig or 4.12 MPa) and a critical temperature of of 112 °C.
4. Trichlorofluoromethane (FC-11) is a physical blowing agent having a critical pressure of 43.5 atm (640 psig or 4.40 MPa) and a critical temperature of of 198 °C.
5. See Chapter 5 for the principles of slit rheometry.

References

Barlow EJ, Langlois WE (1962). *IBM J.* 6:329.
Blyler LL, Kwei TK (1971). *J. Polym. Sci. Part C* 35:165.
Chiou J S, Barlow JW, Paul DR (1985). *J. Appl. Polym. Sci.* 30:2633.
Chow TS (1980). *Macromolecules* 13:362.
Condo PD, Jonston KP (1992). *Macromolecules* 25:6730.
Condo PD, Sanchez IC, Panayiotou CG, Johnston KP (1992). *Macromolecules* 25:6119.
Condo PD, Paul DR, Jonston KP (1994). *Macromolecules* 27:365.
Doolittle AK (1951). *J. Appl. Phys.* 22:1471.
Duda JL Vrentas JS (1969). *AIChE J.* 15:351.
Elkovitch MD, Tomasko DL, Lee LJ (1999). *Polym. Eng. Sci.* 39:2075.
Epstein PS, Plesset MS (1950). *J. Chem. Phys.* 18:1505.
Ferry JD (1980). *Viscoelastic Properties of Polymers*, 3rd ed, Wiley, New York.
Ferry JD, Stratton RA (1960). *Kollod-Z.* 171:107.
Fogler HS, Goddard JD (1970). *Phys. Fluids* 13:1135.
Foster RW, Lindt JT (1987). *Polym. Eng. Sci.* 27:1292.
Fujita H, Kishimoto A (1958). *J. Polym. Sci.* 28:547.
Gee G (1966). *Polymer* 7:177.
Gerhardt LJ, Manke CW, Gulari E (1997). *J. Polym. Sci., Polym. Phys. Ed.* 35:523.
Gerhardt LJ, Garg A, Manke CW, Gulari E (1998). *J. Polym. Sci. Polym. Phys. Ed.* 36:1911.
Gibbs JH, DiMarzio EA (1958). *J. Chem. Phys. 28*:373.
Han CD, Kim YW, Malhotra KD (1976). *J. Appl. Polym. Sci.* 20:1583.
Han CD, Villamizar CA (1978). *Polym. Eng. Sci.* 18:687.
Han CD, Ma CY (1983a). *J. Appl. Polym. Sci.* 28:831.
Han CD, Ma CY (1983b). *J. Appl. Polym. Sci.* 28:851.
Kwag C, Manke CW, Gulari E (1999). *J. Polym. Sci., Polym. Phys. Ed.* 37:2771.
Lee M, Park CB, Tzoganakis C (1999). *Polym. Eng. Sci.* 39:99.
Marique LA, Houghton G (1962). *Can. J. Chem. Eng.* 40:122.
Mendelson RA (1979). *J. Rheol. 23*:545.
Mendelson RA (1980). *J. Rheol.* 24:765.

Olabisi O, Robeson LM, Shaw MT (1979). *Polymer-Polymer Miscibility*, Academic Press, New York, Chap 3.

Oyanagi Y, White JL (1979). *J. Appl. Polym. Sci.* 23:1013.

Panayiotou CG (1986). *Macromol. Chem.* 187:2867.

Panayiotou CG (1987). *Macromolecules* 20:861.

Penwell RC, Porter RS, Middleman, S (1971). *J. Polym. Sci,. A-2* 9:731.

Plesset MS, Zwick SA (1952). *J. Appl. Phys.* 23:95.

Rosner DE, Epstein M (1972). *Chem. Eng. Sci.* 27:69.

Royer JR, Gay YJ, DeSimone JM, Khan SA (2000). *J. Polym. Sci., Polym. Phys.Ed.* 38:3168.

Royer JR, DeSimone JM, Khan SA (2001). *J. Polym. Sci., Polym. Phys. Ed.* 39:3055.

Ruckenstein E, Davis EJ (1970). *J. Colloid Interface Sci.* 31:142.

Sanchez IC, Lacombe RH (1976). *J. Phys. Chem.* 80:2352.

Sanchez IC, Lacombe RH (1978). *Macromolecules* 11:1145.

Scriven LE (1959). *Chem. Eng. Sci.* 10:1.

Street JR (1968). *Trans. Soc. Rheol.* 12:103.

Street JR, Fricke AL, Reiss LP (1971). *Ind. Eng. Chem. Fundam.* 10:54.

Tanasawa L, Yang WJ (1970). *J. Appl. Phys.* 41:4526.

Ting RY (1975). *AIChE J.* 21:810.

Williams ML, Landell RF, Ferry JD (1955). *J. Amer. Chem. Soc.* 77:3701.

Wissinger RG, Paulaitis ME (1991). *J. Polym. Sci., Polym. Phys. Ed.* 29:631.

Yang WJ, Yeh HC (1966). *AIChE J.* 12:927.

Yoo HJ, Han CD (1981). *Polym. Eng. Sci.* 21:69.

Yoo HJ, Han CD (1982). *AIChE J.* 28:1002.

Zana E, Leal LG (1975). *Ind. Eng. Chem. Fundam.* 14:175.

14

Chemorheology of Thermosets

14.1 Introduction

Thermosets (e.g., unsaturated polyester, epoxy, urethane) are small molecules containing functional groups, which undergo chemical reactions (commonly referred to as "cure") in the presence of an initiator(s) or a catalyst(s). In a broader sense, thermosets can be regarded as being parts of reactive polymer systems, which include pairs of polymers (e.g., blends of maleated polyolefin and nylon 6, as presented in Chapter 11) that undergo chemical reactions during compounding, and mixtures of an elastomer and a vulcanizing agent that undergo cross-link reactions (commonly referred to as vulcanization) at an elevated temperature. The subject of investigating the rheological behavior of reactive polymer systems is referred to as "chemorheology." Since chemorheology is such a very broad field of investigation, one must specify the polymer system under consideration, classifying as chemorheology of thermosets, chemorheology of reactive polymer blends, chemorheology of elastomer vulcanization, and so on. In this chapter, for a number of reasons we restrict our presentation to the chemorheology of thermosets only. These reasons include (1) the limited space available here, meaning that it is not possible to present the chemorheology of every reactive polymer system, (2) thermosets play a very important role in polymer processing from an industrial point of view, and (3) the presentation of the chemorheology of thermosets in this chapter lays the foundation for the presentation of processing of thermosets in Chapters 11–13 of Volume 2.

In the 1970s and 1980s, considerable amounts of effort were spent on investigating the chemorheology of thermosets. There are many experimental techniques that have been used to investigate the cure kinetics of thermosets: differential scanning calorimetry (DSC), Fourier transform infrared (FTIR) spectroscopy, dielectric measurements, and rheokinetic measurements. There are monographs (Kock 1977; May 1983; Turi 1981) and a comprehensive review article (Halley and Mackay 1996) on the subject.

A better understanding of the chemorheology of thermosets requires an understanding of the kinetics of chemical reactions during cure. It can then easily be surmised that an understanding of the chemorheology of thermosets is much more complex than the rheology of thermoplastics presented in Chapter 6 through Chapter 12. What complicates the matter is that a generalization of the rheological behavior learned from one type of thermoset to other types is rather difficult because it depends on the type of chemical reactions taking place during cure. Nevertheless, there are some features that can be regarded as being common to many thermosets, if not all, regardless of their chemical structures. Cure reaction of a thermoset generates heat by exothermic chemical reactions, making the rheological measurement under isothermal conditions rather difficult. However, rheological measurements under isothermal conditions can be carried out by ensuring that the sample has a large surface-to-volume ratio, and/or by reducing the rate of cure reaction. It is then clear that the cure reaction and rheological behavior of thermosets are inseparable. As illustrated schematically in Figure 14.1, the rheological properties (η, N_1, G', and G'') of thermosets must include the extent of cure reaction α (or degree of conversion). This means that a chemorheological expression for thermosets requires information on an expression for cure kinetics.

Basically, there are two forms of kinetic expressions (or models) describing the cure reactions of thermosets: empirical and mechanistic models. An empirical model assumes an overall reaction order and is fit to experimental data to determine numerical values of the parameters appearing in the model. Such empirical models cannot provide information on the mechanism(s) of reaction kinetics. Different research groups,

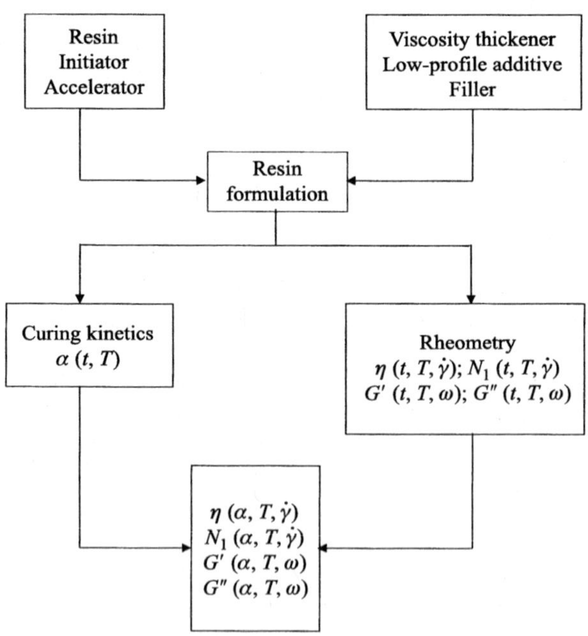

Figure 14.1 Schematic showing the relationships between cure kinetics and rheological properties of thermoset resins.

as reviewed in a paper of Halley and Mackay (1996), have suggested a number of empirical kinetic models for cure reactions of thermosets. Conversely, a mechanistic model is derived from an analysis of the individual reactions involved during cure, requiring detailed measurements of the concentration of reactants, intermediates, and the final products. Thus, a mechanistic model for one type of thermoset would not be applicable to other types of thermosets, suggesting that a mechanistic model must be developed for each type of thermoset. Below, we will show the advantages of using a mechanistic model over empirical model in describing the viscosity variations of a thermoset during cure.

Before presenting the chemorheology of specific thermosets, let us first consider the factors that affect the cure kinetics of thermosets in general. For this, we consider the following empirical model of cure kinetics (Kamal and Sourour 1973):

$$\frac{\mathrm{d}\alpha}{\mathrm{d}t} = (k_1 + k_2\alpha^m)(1-\alpha)^n \tag{14.1}$$

with

$$k_1 = k_{10}\mathrm{e}^{-E_1/RT}; \quad k_2 = k_{20}\mathrm{e}^{-E_2/RT} \tag{14.2}$$

in which α denotes the degree of cure, k_{10} and k_{20} are preexponential factors, E_1 and E_2 are the activation energies, m and n are empirical constants, R is the universal gas constant, and T denotes absolute temperature.

The viscosity of a thermoset increases as the material is transformed into a cross-linked network. Once the gelation of a resin has occurred, the resin loses its fluidity and cannot flow thereafter. Gelation and vitrification usually characterize the isothermal cure of thermosets. "Gelation" refers to the incipient formation of an infinite molecular network of cross-linked molecules, and "vitrification" refers to a transformation from a liquid or rubbery state to a glassy state as a result of an increase in molecular weight. Near vitrification, the kinetics is affected by the local viscosity, which in turn is affected by the extent of chemical reaction and temperature.

It should be mentioned that the molecular weight (M_w) increases with increasing α until encountering vitrification, at which the cure reaction virtually stops. Thus, it is important to recognize that α is also a function of M_w and the glass transition temperature (T_g); that is, $\alpha = f_1(T, M_\mathrm{w}, T_\mathrm{g})$. Further, when the M_w increases as the cure reaction progresses, the concomitant increase in viscosity (η) would slow down the rate of cure reaction because the diffusion (or mobility) of the reacting species becomes increasingly difficult. Thus, in the more general picture we have $\alpha = f_2(T, M_\mathrm{w}, T_\mathrm{g}, \eta)$ or $\eta = f_3(T, M_\mathrm{w}, T_\mathrm{g}, \alpha)$, which includes the late stage diffusion-controlled cure reaction.

In the early stage of cure reaction, when diffusion control can be neglected and where the cure reaction is assumed to be first order (i.e., $k_2 \ll k_1$), Eq. (14.1) reduces to

$$\frac{\mathrm{d}\alpha}{\mathrm{d}t} = k_{10}\mathrm{e}^{-E_1/RT}(1-\alpha) \tag{14.3}$$

from which the time for vitrification, t^*_{vit}, can be determined from

$$t^*_{vit} = e^{A_r/T^*} \int_0^{\alpha_g(T^*_g)} \frac{d\alpha}{(1-\alpha)} \tag{14.4}$$

where $t^*_{vit} = k_{10}t_{vit}$ is the dimensionless time to reach vitrification (i.e., T_g), $T^* = T/T_{g0}$ is a dimensionless temperature with T_{g0} being the glass transition temperature of the freshly mixed uncured resin, $A_r = E_1/RT_{g0}$, and $\alpha_g(T^*_g)$ is the degree of conversion at the dimensionless glass transition temperature, $T^*_g = T_g/T_{g0}$. Note that at $T < T_{g0}$, no reaction occurs because the reactive species are immobilized in the glassy state. In order to estimate the value of t^*_{vit} from Eq. (14.4), one must have information on α_g, the degree of conversion at T_g.

According to Adabbo and Williams (1982), we have the following relationship (Enns and Gillham 1983):

$$\frac{T_g - T_{g0}}{T_{g0}} = \frac{(E_x/E_m - F_x/F_m)\alpha_g}{1-(1-F_x/F_m)\alpha_g} \tag{14.5}$$

where E_x/E_m is the ratio of lattice energies for cross-linked and uncross-linked polymers and F_x/F_m is the corresponding ratio of segmental mobilities. The ratios E_x/E_m and F_x/F_m in Eq. (14.5) can be determined by fitting Eq. (14.5) to an experimental plot of α_g versus T_g. When the glass transition temperature is equal to the cure temperature, rearrangement of Eq. (14.5) gives

$$\alpha_g = \frac{T^*_g - 1}{E_x/E_m - 1 + (1 - F_x/F_m)T^*_g} \tag{14.6}$$

Figure 14.2 gives a plot of α_g versus T_g/T_{g0} for an epoxy with $T_{g0} = -19\,°C$, $k_{10} = 4.51 \times 10^6$ min^{-1} and $E_1 = 5.29 \times 10^4$ J/mol, in which the solid line is the best fit of data to Eq. (14.6) with the values $E_x/E_m = 0.337$ and $F_x/F_m = 0.194$. Using these values of E_x/E_m and F_x/F_m, $\alpha_g(T_g)$, is calculated from Eq. (14.6), and then the time to reach vitrification, t_{vit}, is estimated from Eq. (14.4). Table 14.1 gives a summary of estimated t_{vit} for the first-order ($n = 1$) kinetics at various cure temperatures together with experimental results, where the glass transition temperature is assumed to be equal to the cure temperature ($T_{cure} = T_g$) It is seen in Table 14.1 that the higher the cure temperature, the sooner vitrification is reached. The differences between the estimated t_{vit} and experimentally determined t_{vit} may be attributable to the assumption of first-order kinetics in the calculation of t_{vit} (see Eq. (14.4)).

Figure 14.3 gives a generalized time–temperature–transformation cure diagram (Enns and Gillham 1983), showing four distinct states: liquid state, gelled rubbery state, gelled glassy state, and ungelled glassy state. Notice in Figure 14.3 that t_{vit} becomes shorter as the cure temperature is increased. The material presented above demonstrates clearly that glass transition temperature plays a very important role in the cure, and hence the chemorheology, of thermosets. Below, we will show how the

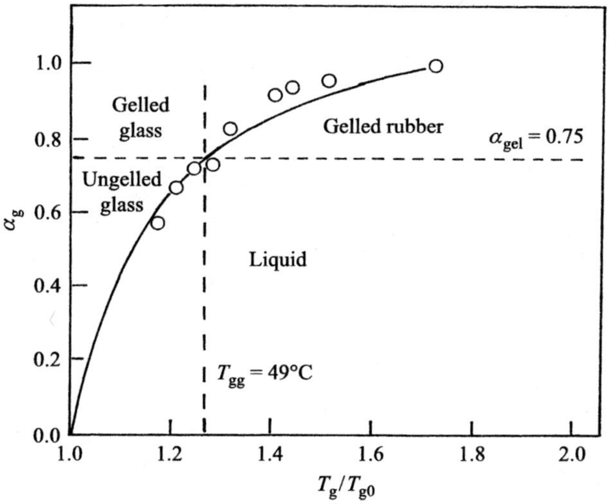

Figure 14.2 Schematic showing the degree of cure of an epoxy at glass transition α_g as a function of dimensionless temperature T_g/T_{g0}, where T_g denotes the glass transition temperature, T_{g0} denotes the glass transition temperature of uncured resin, T_{gg} denotes the temperature at which gelation and vitrification cross each other, and the solid line is the best fit to Eq. (14.6). (Reprinted from Enns and Gillham, *Journal of Applied Polymer Science* 28:2567. Copyright © 1983, with permission from John Wiley & Sons.)

Table 14.1 Summary of the calculated and experimentally determined vitrification times of an epoxy using Eq. (14.4)

T_{cure} (°C)	T^{*a}	α_g	t_{vit} *(cal)* *(min)*	t_{vit}^{b} *(exp)* *(min)*
151	1.669	0.980	2.8	7
133	1.598	0.957	4.3	12
113	1.519	0.925	8.1	11
92	1.437	0.882	17.0	18
71	1.350	0.823	42.3	35
51	1.279	0.758	96.9	76
42	1.244	0.715	148.6	149

[a]Dimensionless temperature defined by $T^* = T/T_{g0}$. [b]From Enns and Gillham (1983).

glass transition temperature may be incorporated into the expression, describing the variations in the viscosity of thermoset during cure.

In this chapter, we present the chemorheology of three very important thermosets, namely, unsaturated polyester, epoxy, and urethane, placing emphasis on the necessity of connecting the rheological behavior of thermosets with the chemical reactions taking

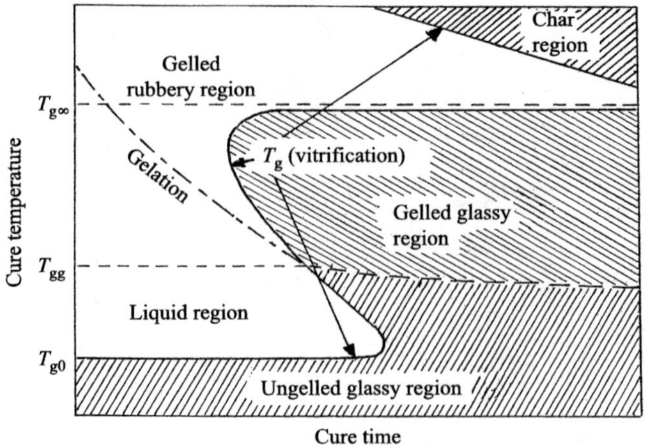

Figure 14.3 Generalized time–temperature–transformation cure diagram showing four distinct states: liquid state, gelled rubbery state, gelled glassy state, and ungelled glassy state in the cure of thermoset. (Reprinted from Enns and Gillham, *Journal of Applied Polymer Science* 28:2567. Copyright © 1983, with permission from John Wiley & Sons.)

place during cure. For the reason of limited space available here, we do not discuss the chemorheology of other thermosets. The material presented in this chapter is utilized when we present the reaction injection molding of thermosets in Chapter 11 of Volume 2, the pultrusion of thermoset/fiber composites in Chapter 12 of Volume 2, and the compression molding of thermoset/fiber composites in Chapter 13 of Volume 2.

14.2 Chemorheology of Unsaturated Polyester

Unsaturated polyesters are prepared commercially by the reaction of a saturated diol with an unsaturated dibasic acid (or corresponding anhydrides) (Bruins 1976; Parkyn et al. 1967). The unsaturated acid provides sites for subsequent cross-linking. Often, a modifying acid or anhydride (e.g., saturated dibasic acid, phthalic anhydride, adipic acid, sebacic acid) is used to reduce the number of reactive unsaturated sites along the polymer, and hence to reduce the cross-link density and brittleness of the final product. Propylene glycol is the diol most widely used for the manufacture of unsaturated polyesters. Diols other than propylene glycol, such as diethylene glycol and 2,2-dimethylpropane-1,3-diol, are also utilized, though to a lesser extent, to impart certain properties (e.g., greater flexibility, improved resistance to thermal stability). Maleic anhydride is the most important unsaturated component used in the manufacture of unsaturated polyester. Maleic anhydride is preferred to maleic acid because it is more reactive and gives rise to less water on esterification. Acids or anhydrides other than maleic anhydride, such as fumaric acid and phthalic anhydride, are also utilized. Phthalic anhydride is used as a modifying component to decrease the number of reactive sites. It is possible to cross-link unsaturated polyester chains directly one to another, but reaction is very slow and a low degree of cross-linking is achieved.

These limitations are overcome by the introduction of styrene; that is, styrene is used as cross-linking monomer for unsaturated polyester.

The reaction scheme between maleic anhydride and propylene glycol is

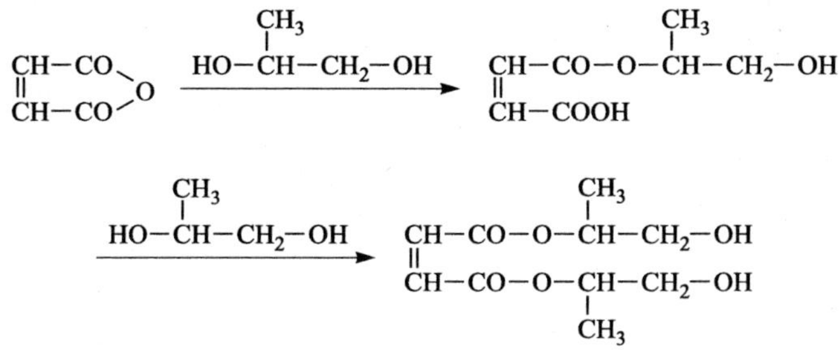

which proceeds in two distinct steps. In the first step, esterification of maleic anhydride occurs to form a free acid group, which is then esterified in the second step. The first step proceeds more rapidly than the second step, because the anhydride group is more reactive than the free acid group. A diol and an anhydride may interact through a sequence of reactions of the foregoing type to yield an unsaturated polyester. Thus, a segment of the linear polyester obtained from propylene glycol, maleic anhydride, and phthalic anhydride might have the structure

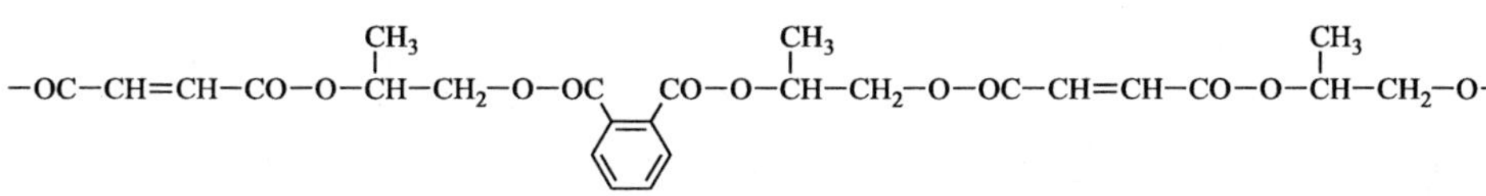

As mentioned above, the cross-linking reaction of unsaturated polyesters involves the reaction of the unsaturated sites in the polymer chain with styrene. This reaction is analogous to conventional vinyl copolymerization and proceeds by an essentially similar mechanism; thus, cross-linking of unsaturated polyester is a free radical polymerization, for which an initiator is needed. Two types of initiators are used: low-temperature initiator and high-temperature initiator. The most important initiators used at elevated temperatures are peroxides, such as benzoyl peroxide (BPO) and *tert*-butyl perbenzoate (TBPB), which generate free radicals as a result of thermal decomposition. Initiators that are effective at room temperature normally consist of mixtures of a peroxide and an activator ("accelerator"). In the presence of an accelerator, the peroxide rapidly decomposes, without the application of heat, into free radicals. The most common accelerators are cobalt octoate and *N,N*-dimethylaniline. Variation in the type and amount of individual components allow for great flexibility in the design of resin formulations. For example, it is generally known that isophthalic acid gives higher heat distortion temperatures.

Unsaturated polyester is one of the most versatile thermosets in that it is used in the fabrication of a broad range of products, including structural parts of automobiles, building materials, coating materials, electrical parts, appliances, and boat hulls. Consequently, a variety of resin formulations have been developed, depending on specific applications as well as fabrication processes. Unsaturated polyester and its fiber-reinforced composites are processed using, for instance, spray-up, decorative coating, compression molding, transfer molding, injection molding, pultrusion, and reaction injection molding. For the use in fiber-reinforced composites, unsaturated polyester is used for preparation of various molding compounds, such as bulk molding compound (BMC), sheet molding compound (SMC), and thick molding compound (TMC). If the molding compounds are not prepared correctly, one will encounter many difficulties when using the materials to achieve the desired mechanical properties of the finished products.

Depending upon the application and manufacturing process selected, a number of other additives are employed to provide specific products or end-use properties. These include inert fillers, flame retardants, compounds to enhance surface finish and reduce shrinkage in the mold (low-profile additive), release agents, and viscosity control materials (viscosity thickener). No-shrink unsaturated polyester systems for use in BMC, SMC, and TMC are commercially available. The goal of development work in this area has been molded parts with smooth surfaces, no warpage and no sink marks. Thus, successful processing operations of unsaturated polyester, such as in compression molding, transfer molding, pultrusion, and reaction injection molding, require a better understanding of the chemorheology of the resin system. In this section, we present the chemorheology of unsaturated polyester, without or with additives.

14.2.1 Viscosity Rise during Cure of Neat Unsaturated Polyester

Figure 14.4 gives plots of viscosity ($\log \eta$) versus cure time for different shear rates ($\dot{\gamma}$) when a general-purpose unsaturated polyester (Aropol 7030, Ashland Chemical Company) was subjected to steady-state shear flow in a cone-and-plate rheometer under an isothermal condition at 60 °C. The resin had been prepared by the reaction of propylene glycol with a mixture of maleic anhydride and isophthalic anhydride. The resin system used in Figure 14.4 was cured with BPO as initiator and a solution of 5 wt % *N,N*-dimethylaniline, diluted in styrene, as accelerator. The following observations are worth noting in Figure 14.4. At an early stage of curing, the η increased slowly, but then

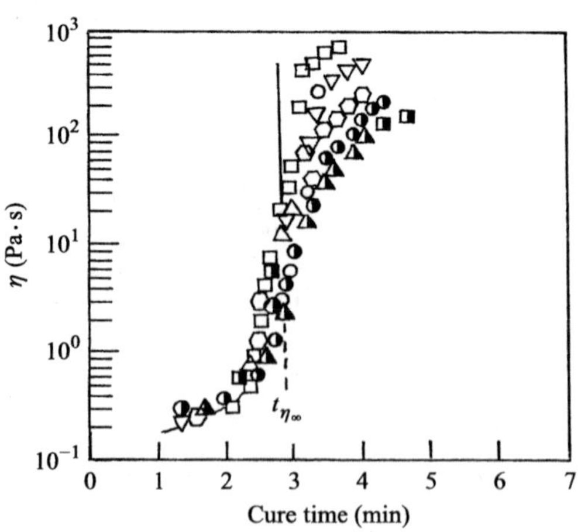

Figure 14.4 Plots of $\log \eta$ versus cure time with 3.16 wt % BPO as initiator and 1.9 wt % *N,N*-dimethylaniline as accelerator, during isothermal cure in a cone-and-plate rheometer at 60 °C, for Aropol 7030 subjected to different shear rates (s^{-1}): (○) 0.269, (△) 1.07, (□) 4.27, (▽) 6.77, (◑) 10.7, (▲) 17.0, and (◨) 26.9. (Reprinted from Han and Lem, *Journal of Applied Polymer Science* 28:3155. Copyright © 1983, with permission from John Wiley & Sons.)

increased very rapidly, approaching a very large value as cure progressed. After the η started to increase very rapidly, the rate of increase in η began to diminish at a critical value of cure time. When this happened, the torque output signal became irregular and the material exuded from the gap between the cone and plate, indicating that the flow had become unstable (Han and Lem 1983). It is interesting to observe in Figure 14.4 that the time t_{η_∞}, at which the η approaches a very large value, is virtually independent of $\dot{\gamma}$, and that the value of η that starts to deviate from the vertical line decreases as $\dot{\gamma}$ increases. The observed shear-thinning behavior at cure time beyond t_{η_∞} is largely attributable to the macromolecules formed during cure. It has been reported that values of t_{η_∞} increased from 2.9 to 4.7 min when cure temperature was decreased from 60 to 50 °C. This is understandable because the rate of cure reaction is expected to decrease as the temperature is decreased. In other words, variation of η during cure depends on the cure kinetics of the unsaturated polyester, suggesting that the chemorheology of unsaturated polyester must be related to the degree of cure α.

Figure 14.5 gives plots of log η versus α for Aropol 7030 for various isothermal cure temperatures, showing that larger values of α are achieved as the cure temperature is increased, before η approaches a very large value. According to Flory (1953), one would expect to observe the same value of α at the incipient gel point, regardless of the temperature at which cure reaction takes place. Conversely, vitrification can occur before gelation when the cure is carried out at a temperature below T_{gg}, the temperature at which gelation and vitrification can occur simultaneously (Gillham 1982). Since the glass transition temperature (T_{g_∞}) for the fully cured unsaturated polyester is approximately 110 °C, which is much higher than the isothermal cure temperatures (30–60 °C) employed in Figure 14.5, one may be tempted to explain the observed temperature dependence of the η–α relationship using an argument of vitrification and gelation. It has been argued that vitrification could not have occurred at t_{η_∞}, before gelation, in the unsaturated polyester. The observed increase in α with increasing cure temperature is characteristic of styrene-unsaturated polyester systems, which give rise to a heterogeneous network structure containing gel particles (Han and Lem 1983). From the experimental results presented in Figures 14.4 and 14.5, it is clear that, during cure,

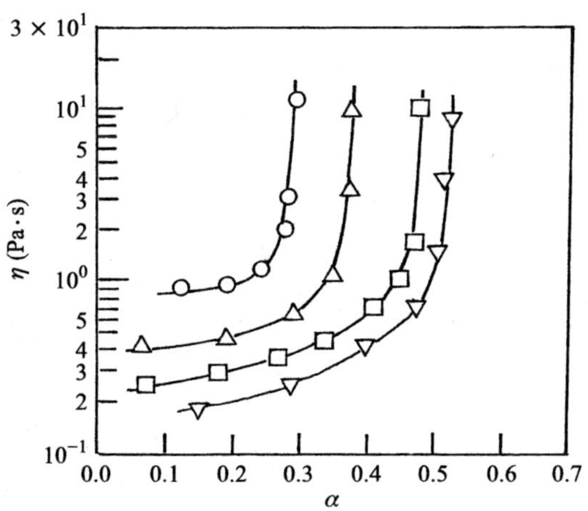

Figure 14.5 Plots of log η versus α for Aropol 7030 with 3.16 wt % BPO as initiator and 1.9 wt % *N,N*-dimethylaniline as accelerator at various isothermal cure temperatures (°C): (○) 30, (△) 40, (□) 50, and (▽) 60. (Reprinted from Han and Lem, *Journal of Applied Polymer Science* 28:3155. Copyright © 1983, with permission from John Wiley & Sons.)

the η of unsaturated polyester depends on the degree of cure and cure temperature. What is then needed is to obtain an expression that describes the variations of η with t, α, and T.

14.2.2 Chemorheological Model for Neat Unsaturated Polyester

The following form of an empirical expression describing the variation of η during the cure of thermoset has been suggested (Roller 1975):

$$\ln \eta(t, T) = \ln \eta_{\infty} + E_{\eta}/RT + \int_0^t k_{\infty} \exp(E_k/RT)dt \tag{14.7}$$

where $\eta(t, T)$ is the viscosity at time t and at temperature T, η_{∞} is the viscosity at very high temperature (T_{∞}), E_{η} is the flow activation energy, R is the universal gas constant, k_{∞} is the rate constant for cure reactions at T_{∞}, and E_k is the activation energy for the cure reaction. Note that the last term on the right-hand side of Eq. (14.7) describes the effect of the cure reaction on $\eta(t, T)$. The use of the Arrhenius expression, Eq. (14.7), in describing the variation of $\eta(t, T)$ during the cure of thermoset implies that cure reactions are kinetically controlled. A serious deficiency of Eq. (14.7) lies in that E_{η} is assumed to be constant during cure reactions. Such an assumption is not tenable because the molecular weight increases during cure and thus the value of E_{η} is expected to increase as the cure reaction progresses. This then suggests that an alternative approach is needed to describe the variations of η with cure time t, α, and T during the cure of unsaturated polyester. In this regard, the free-volume approach seems to be appropriate because it is consistent with the diffusion-controlled reaction mechanism. Indeed, some investigators (Feger and MacKnight 1985; Tajima and Crozier 1983) have suggested that the Williams–Landel–Ferry (WLF) equation be used to relate the T_g of partially cured thermosets, formed during isothermal cure, to α. As will be shown below, the cure of unsaturated polyester may best be described by the diffusion-controlled polymerization mechanism (Biesenberger and Sebastian 1983; Malkin et al. 1984; Soh and Sundberg 1984), even at an early stage of cure reaction.

Using a general-purpose unsaturated polyester (OC-E701, Owens-Corning Fiberglas) cured in the presence of TBPB as initiator, Lee and Han (1987a) measured (1) the quantity of ethylenic double bonds, via Fourier transform infrared (FTIR) spectroscopy, to determine the degree of cure before and after the cure reactions, (2) the glass transition temperature (T_g) using DSC, and (3) the viscosity during isothermal cure as a function of shear rate ($\dot{\gamma}$), at several temperatures, using a cone-and-plate rheometer. In so doing, they collected, at predetermined time intervals, samples with varying degrees of cure by stopping further cure reaction with an addition of p-benzoquinone as inhibitor and quenching it in a mixture of dry ice and acetone. Table 14.2 gives a summary of the time intervals for sample collection, the degree of cure, and the T_g of seven partially cured samples. Figure 14.6 gives plots of α versus cure time and Figure 14.7 gives plots of log η versus log $\dot{\gamma}$ for varying cure times or α (see Table 14.2) for several partially cured unsaturated polyesters using TBPB as initiator. It is seen in Figure 14.7 that all samples exhibit Newtonian behavior. This is not surprising because the partially cured unsaturated polyesters have relatively low values of α (see Table 14.2). Figure 14.8 gives plots of η versus cure time for several partially cured

Table 14.2 Sample code and the degree of cure and glass transition temperature of partially cured OC-E701 using TBPB as initiator

Sample Code	*Cure Time (min)*	α	T_g (°C)
Sample 1	0.00	0.000	−78.0
Sample 2	8.00	0.050	−76.5
Sample 3	8.25	0.078	−73.0
Sample 4	8.50	0.114	−71.0
Sample 5	8.67	0.140	−68.0
Sample 6	8.83	0.165	−65.0
Sample 7	9.00	0.195	−61.0

Reprinted from Lee and Han, *Polymer Engineering and Science* 27:955. Copyright © 1987, with permission from the Society of Plastics Engineers.

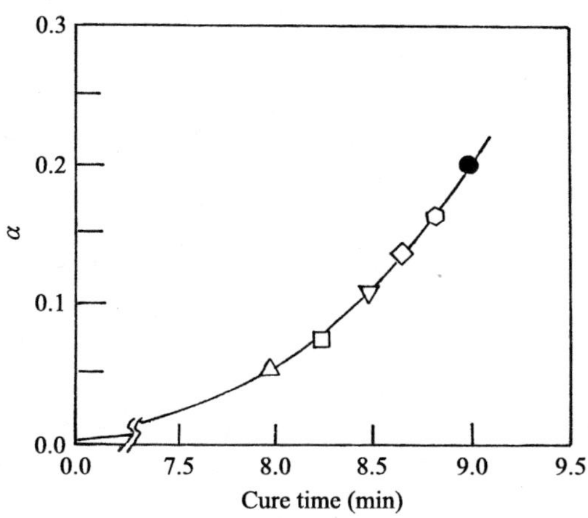

Figure 14.6 Plot of α versus cure time for OC-E701 with 0.4 wt % TBPB as initiator: (△) sample 2, (□) sample 3, (▽) sample 4, (◇) sample 5, (⬡) sample 6, and (●) sample 7. Values of α were determined from the measurements of ethylenic bonds before and after the cure using FTIR spectroscopy. (Reprinted from Lee and Han, *Polymer Engineering and Science* 27:955. Copyright © 1987, with permission from the Society of Plastics Engineers.)

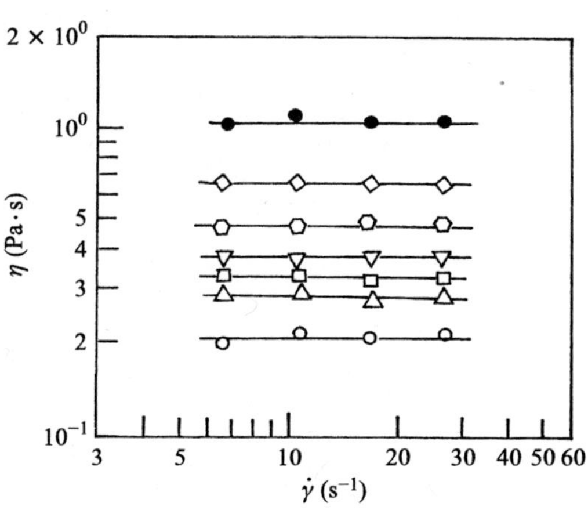

Figure 14.7 Plots of log η versus log $\dot{\gamma}$ at 40 °C for partially cured OC-E701 with 0.4 wt % TBPB as initiator: (○) sample 1, (△) sample 2, (□) sample 3, (▽) sample 4, (⬡) sample 5, (◇) sample 6, and (●) sample 7. Viscosity measurements were conducted in a cone-and-plate rheometer for each sample. Refer to Table 14.2 for the cure times at which samples were taken during cure. (Reprinted from Lee and Han, *Polymer Engineering and Science* 27:955. Copyright © 1987, with permission from the Society of Plastics Engineers.)

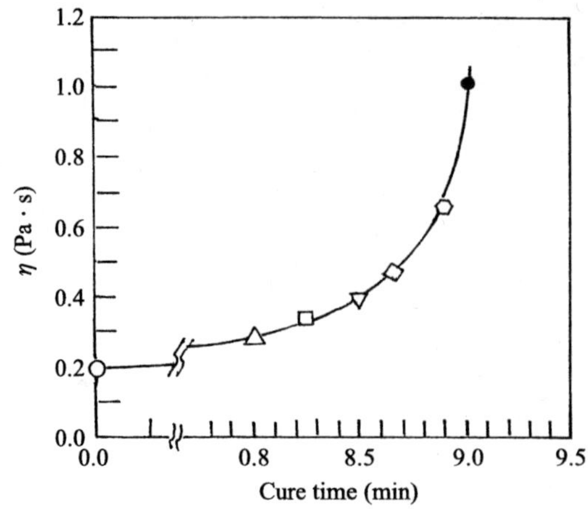

Figure 14.8 Plots of log η versus cure time for OC-E701 with 0.4 wt % TBPB as initiator: (○) sample 1, (△) sample 2, (□) sample 3, (▽) sample 4, (◇) sample 5, (⬠) sample 6, and (●) sample 7. Values of η were determined at 40 °C using a cone-and-plate rheometer. (Based on unpublished data by Lee and Han 1987.)

unsaturated polyesters using TBPB as initiator, showing that η increases very rapidly with increasing cure time. A plot of log η versus $1/T$ (an Arrhenius plot) for the seven samples having different value of α (see Table 14.2) enables one to determine the flow activation energy E_η from $\eta = A\exp(E_\eta/RT)$. Figure 14.9 gives the dependence of E_η on α, showing a linear relationship, $E_\eta = a + b\alpha$. This is reasonable because the molecules grow during cure. This observation demonstrates clearly that the assumption, made in Eq. (14.7), that the value of E_η is constant was not justified.

To analyze experimental data, Lee and Han (1987a) employed a WLF expression to describe the viscosity of unsaturated polyester:

$$\log \eta(T) = \log \eta_{T_g} - \frac{C_1(T - T_g)}{51.6 + T - T_g} \tag{14.8}$$

where C_1 is a free-volume parameter and η_{T_g} is the viscosity at T_g. In using Eq. (14.8), they measured the T_gs of partially cured unsaturated polyester samples. Figure 14.10 gives plots of T_g versus α for the samples, from which the following expression is

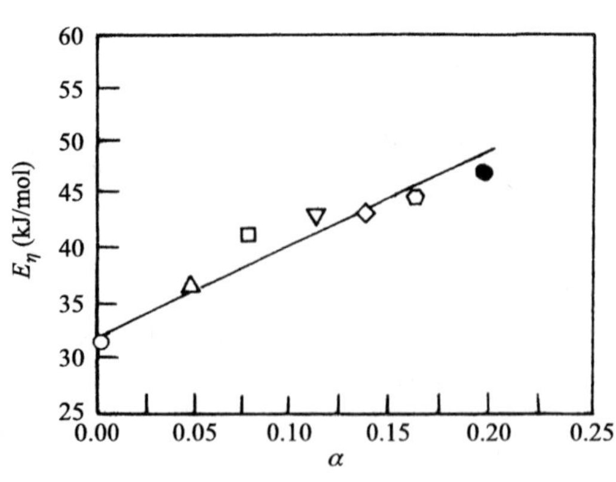

Figure 14.9 Plots of flow activation energy E_η versus α for partially cured OC-E701 with 0.4 wt % TBPB as initiator: (○) sample 1, (△) sample 2, (□) sample 3, (▽) sample 4, (◇) sample 5, (⬠) sample 6, and (●) sample 7. Refer to Table 14.2 for the cure times at which samples were taken during cure. (Reprinted from Lee and Han, *Polymer Engineering and Science* 27:955. Copyright © 1987, with permission from the Society of Plastics Engineers.)

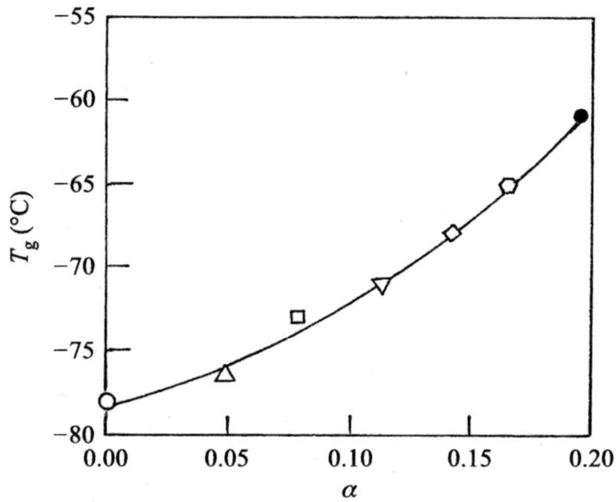

Figure 14.10 Plots of T_g versus α for partially cured OC-E701 with 0.4 wt % TBPB as initiator: (○) sample 1, (△) sample 2, (□) sample 3, (▽) sample 4, (◇) sample 5, (⬡) sample 6, and (●) sample 7. Refer to Table 14.2 for the cure time at which samples were taken during cure. (Reprinted from Lee and Han, *Polymer Engineering and Science* 27:955. Copyright © 1987, with permission from the Society of Plastics Engineers.)

obtained:

$$T_g = 194.0 + 33.06\alpha + 281.4\alpha^2 \tag{14.9}$$

To determine relationships between the η_{T_g} and α, and between C_1 and the α, Eq. (14.8) may be rewritten in the form

$$\log \eta(T) = \log \eta_{T_g} - C_1 + \frac{51.6C_1}{51.6 + T - T_g} \tag{14.10}$$

From the slope and the intercept of the plots of log $\eta(T)$ versus $1/(51.6 + T - T_g)$, the dependence of C_1 on α is given in Figure 14.11, and the dependence of log η_{T_g} on

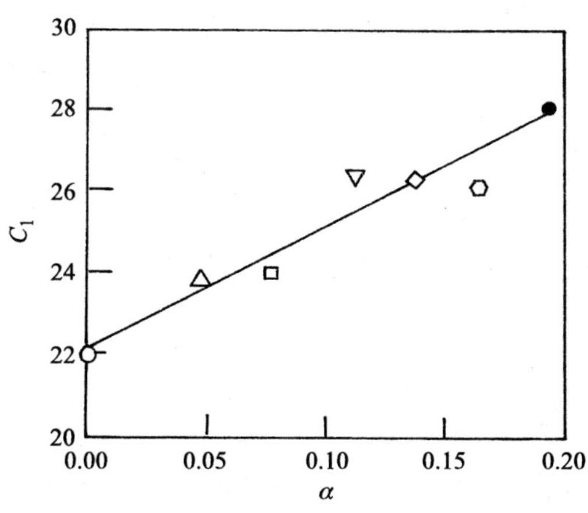

Figure 14.11 Plots of free-volume parameter C_1 versus α for partially cured OC-E701 with 0.4 wt % TBPB as initiator: (○) sample 1, (△) sample 2, (□) sample 3, (▽) sample 4, (◇) sample 5, (⬡) sample 6, and (●) sample 7. Refer to Table 14.2 for the cure times at which samples were taken during cure. (Reprinted from Lee and Han, *Polymer Engineering and Science* 27:955. Copyright © 1987, with permission from the Society of Plastics Engineers.)

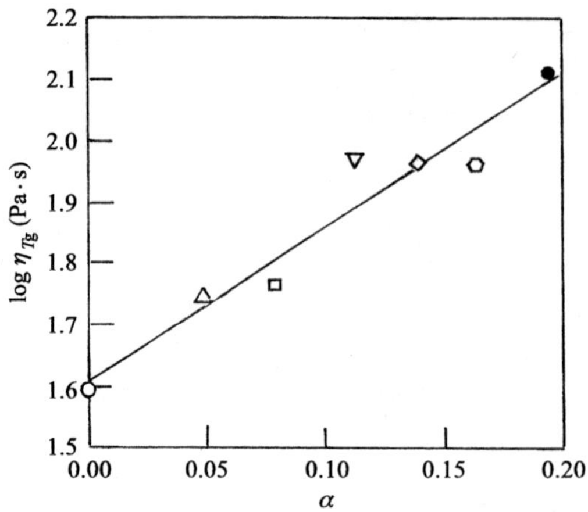

Figure 14.12 Plots of $\log \eta_{T_g}$ versus α for partially cured OC-E701 with 0.4 wt % TBPB as initiator: (○) sample 1, (△) sample 2, (□) sample 3, (▽) sample 4, (◇) sample 5, (⬡) sample 6, and (●) sample 7. Refer to Table 14.2 for the cure times at which samples were taken during cure. (Reprinted from Lee and Han, *Polymer Engineering and Science* 27:955. Copyright © 1987, with permission from the Society of Plastics Engineers.)

α is given in Figure 14.12. We now have the following relationships:

$$C_1 = 22.11 + 29.92\alpha \tag{14.11}$$

and

$$\log \eta_{T_g} = 16.05 + 25.14\alpha \tag{14.12}$$

using least-squares analysis. Substitution of Eqs. (14.9), (14.11), and (14.12) into (14.10) yields

$$\log \eta(T, \alpha) = -6.60 - 4.78\alpha + \frac{51.6(22.11 + 29.92\alpha)}{T - 142.14 - 33.06\alpha - 281.4\alpha^2} \tag{14.13}$$

Using Eq. (14.13), we can now predict the viscosity of unsaturated polyester during cure once information on α during cure is available to us. In order to use Eq. (14.13) to describe the variation of η during cure of unsaturated polyester, one must have an expression describing the kinetics of the cure reactions. Below, we present experimental methods that can be used to determine an expression for cure kinetics.

14.2.3 Cure Kinetics of Neat Unsaturated Polyester

In the 1960s through the 1980s, a number of research groups (Han and Lem 1983; Han and Lee 1987a, 1987b; Horie et al. 1969, 1970a; Kamal and Sourour 1973; Pusatcioglu et al, 1979; Stevenson 1986) investigated the cure reactions of unsaturated polyester. Most of the studies employed isothermal DSC to measure the rate of heat generation $(dQ/dt)_T$ at a predetermined temperature T and then assumed that $(dQ/dt)_T$ was directly proportional to the rate of cure $d\alpha/dt$:

$$\frac{d\alpha}{dt} = \frac{1}{Q_{UT}} \left(\frac{dQ}{dt} \right)_T \tag{14.14}$$

in which Q_{UT} is the ultimate heat that can be generated when the cure reaction attains completion (100%) during isothermal cure. Note that Q_{UT} consists of two parts, one part from the amount of heat (Q_T) that is generated during the isothermal DSC run at a predetermined temperature T, and the other part from the residual heat (Q_R) that is released when the sample is heated further, upon completion of an isothermal cure reaction, to a much higher temperature (e.g., 200 °C), at a given scan rate; that is, $Q_{UT} = Q_T + Q_R$. It should be mentioned that the completion (100%) of cure is not possible during an isothermal DSC run at a predetermined temperature T, and thus $Q_T < Q_{UT}$ Integrating Eq. (14.14) yields

$$\alpha = \frac{1}{Q_{UT}} \int_{t_i}^{t} \left(\frac{dQ}{dt} \right)_T dt \tag{14.15}$$

or $\alpha = Q_t(t, T)/Q_{UT}$, with $Q_t(t, T)$ being the heat generated at a particular time t and at an isothermal cure temperature T.

Figure 14.13 gives typical plots of dQ/dt versus cure time for a general-purpose unsaturated polyester (OC-P340, Owens-Corning Fiberglas) with TBPB as initiator, at three different isothermal cure temperatures. It is seen that the peak value of dQ/dt appears earlier as the cure temperature is increased from 110 to 130 °C. By integrating the dQ/dt versus t curve in Figure 14.13, one can obtain Q_T during the isothermal cure. Figure 14.14 gives variations of Q_T and Q_R with cure temperature during isothermal DSC runs, showing that $Q_{UT} = Q_T + Q_R$ is constant. Figure 14.15 gives plots of $d\alpha/dt$ versus cure time for OC-P340 with TBPB as initiator, which is obtained from Figure 14.13 with the aid of Eq. (14.14), and Figure 14.16 gives plots of α versus

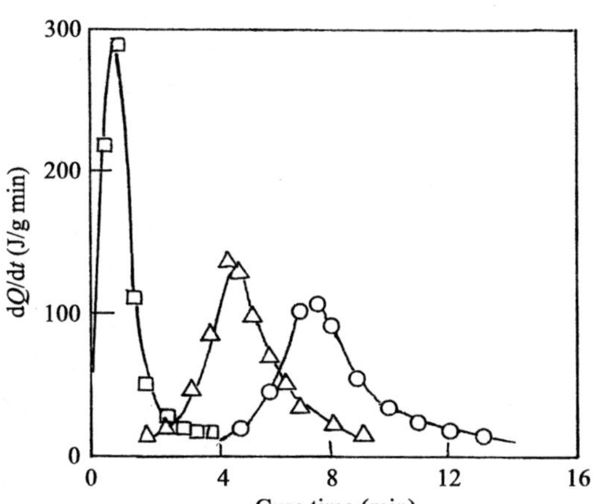

Figure 14.13 Plots of dQ/dt versus cure time for OC-P340 with 1.0 wt % TBPB as initiator at different isothermal cure temperatures (°C): (○) 110, (△) 120, and (□) 130. (Reprinted from Han and Lee, *Journal of Applied Polymer Science* 33:2859. Copyright © 1987, with permission from John Wiley & Sons.)

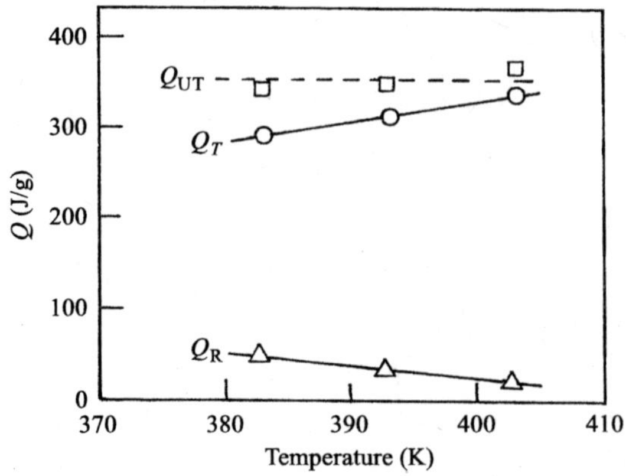

Figure 14.14 Heat generated Q by cure reaction of OC-P340 with 1.0 wt % TBPB as initiator at various isothermal temperatures, where Q_T denotes the total heat generated during isothermal cure, Q_R denotes the heat generated after raising the temperature upon completion of isothermal cure, and Q_{UT} denotes the ultimate heat generated, which is the sum of Q_T and Q_R. (Reprinted from Han and Lee, *Journal of Applied Polymer Science* 33:2859. Copyright © 1987, with permission from John Wiley & Sons.)

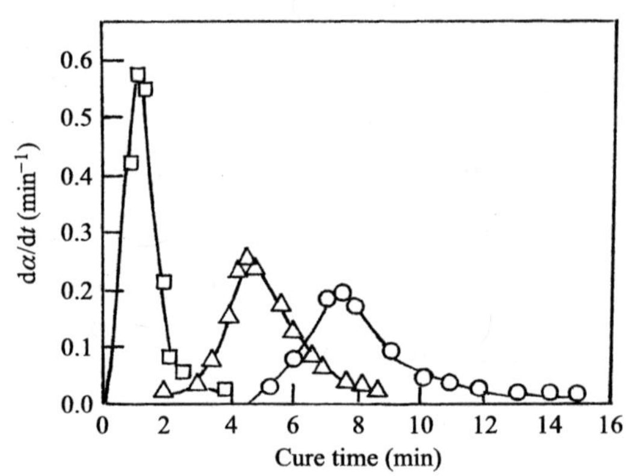

Figure 14.15 Plots of dα/dt versus cure time for OC-P340 with 1.0 wt % TBPB as initiator at different isothermal cure temperatures (°C): (○) 110, (△) 120, and (□) 130. (Based on unpublished data by Han and Lee 1987.)

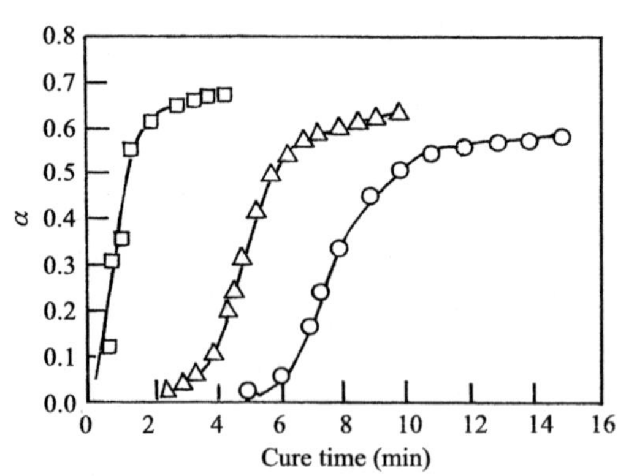

Figure 14.16 Plots of α versus cure time for OC-P340 with 1.0 wt % TBPB as initiator at different isothermal cure temperatures (°C): (○) 110, (△) 120, and (□) 130. (Based on unpublished data by Han and Lee 1987.)

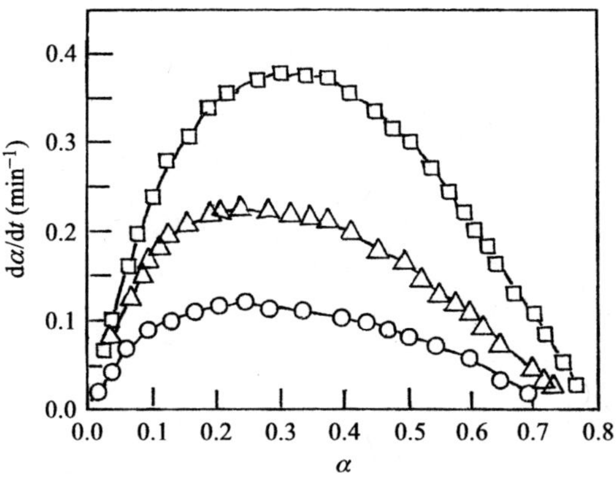

Figure 14.17 Plots of $d\alpha/dt$ versus α for OC-E701 with 1.0 wt % TBPB as initiator at various isothermal cure temperatures (°C): (○) 110, (△) 120, and (□) 130. (Based on unpublished data by Han and Lee 1987.)

cure time for OC-P340 with TBPB as initiator, which is obtained from Figure 14.13 with the aid of Eq. (14.15).

Figure 14.17 gives plots of $d\alpha/dt$ versus α for OC-E701 with 1.0 wt % TBPB as initiator at three isothermal cure temperatures. Notice in Figure 14.17 that the peak value of $d\alpha/dt$ increases with increasing cure temperature, the area under the curve increases with increasing cure temperature, and the initial value of $d\alpha/dt$ is zero, that is, $(d\alpha/dt)_{t=0} = 0$, although this is not apparent from the figure. Following the procedures suggested by Ryan and Dutta (1979), numerical values of the parameters (k_1, k_2, m, and n) appearing in Eq. (14.1) are determined by assuming a second-order reaction ($m + n = 2$) and they are summarized in Table 14.3 for three general-purpose unsaturated polyesters, OC-E701, OC-P340, and OC-E980, in which we observe that over the range of isothermal cure temperatures tested, values of m and n are relatively insensitive to temperature. The value of k_1 is determined from $k_1 = (d\alpha/dt)_{t=0}$ (Ryan and Dutta 1979). Figure 14.18 gives the temperature dependence of the rate constant k_2 for three general-purpose unsaturated polyesters, OC-E701, OC-P340, and OC-E980, showing that the Arrhenius relation holds over the range of temperatures tested (see Eq. (14.2)).

Figure 14.19 gives plots of $d\alpha/dt$ versus α for Aropol 7030 with 3.16 wt % BPO as initiator and 1.9 wt % *N,N*-dimethylaniline as accelerator at five isothermal cure temperatures. Figure 14.20 gives the temperature dependence of the rate constants, k_1 and k_2, for Aropol 7030 with 3.16 wt % BPO as initiator and 1.9 wt % *N,N*-dimethylaniline as accelerator, showing that the Arrhenius relation holds over the range of temperatures tested (see Eq. (14.2)). Applying least-squares analysis to Figure 14.20, we obtain $k_{10} = 2.139 \times 10^{19}$ min^{-1}, $E_1 = 1.29 \times 10^5$ J/mol, $k_{20} = 3.428 \times 10^{14}$ min^{-1}, and $E_2 = 4.94 \times 10^4$ J/mol (see Eq. (14.2)). It should be mentioned that one encounters $k_1 = 0$ when the rate of cure reaction of an unsaturated polyester is relatively slow and $k_1 \neq 0$ when an accelerator (e.g., *N,N*-dimethylaniline) is added to speed up the cure reaction of an unsaturated polyester, even at relatively low cure temperatures.

Horie et al. (1969) pointed out that the cure reaction of unsaturated polyester can be analyzed using the approach of free-radical polymerization. It should be pointed out that the cure reaction of unsaturated polyester involves the formation of three-dimensional

Table 14.3 Summary of the kinetic parameters appearing in Eq. (14.1) for different unsaturated polyesters

Temperature (°C)	α_p[a]	$\dot{\alpha}_p$[b] (min^{-1})	k_1[c] (min^{-1})	k_2 (min^{-1})	m	n
OC-E701 with 1.0 wt % TBPB as initiator						
110	0.250	0.116	0.000	0.357	0.50	1.50
120	0.244	0.230	0.000	0.700	0.49	1.51
130	0.299	0.381	0.000	1.296	0.60	1.40
OC-P340 with 1.0 wt % TBPB as initiator						
100	0.179	0.052	0.000	0.124	0.36	1.64
105	0.207	0.081	0.000	0.161	0.41	1.59
110	0.193	0.123	0.000	0.241	0.39	1.61
115	0.219	0.169	0.000	0.342	0.44	1.56
120	0.251	0.207	0.000	0.432	0.50	1.50
OC-E980 with 1.0 wt % TBPB as initiator						
110	0.266	0.064	0.000	0.185	0.45	1.55
120	0.284	0.105	0.000	0.345	0.57	1.43
130	0.185	0.239	0.000	0.621	0.37	1.63

[a] α_p is the value of α at which the plot of $d\alpha/dt$ versus cure time goes through a maximum.
[b] $\dot{\alpha}_p$ is the peak value of $d\alpha/dt$ in the plot of $d\alpha/dt$ versus cure time. [c] k_1 is determined from the relationship, $k_1 = (d\alpha/dt)_{t=0}$.

Based on unpublished data by Han and Lee 1987.

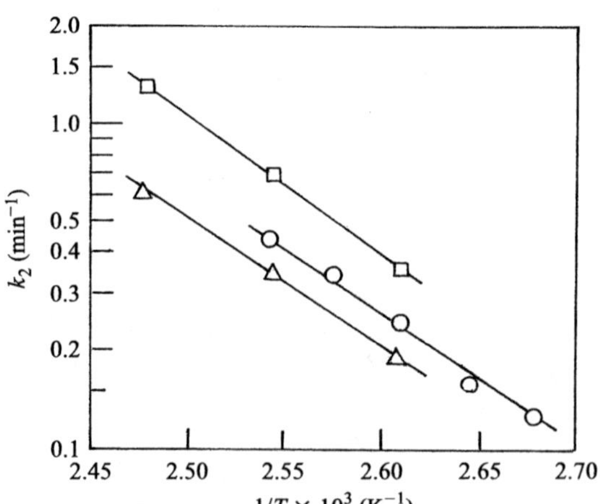

Figure 14.18 Plots of rate constant k_2 versus $1/T$ for three general-purpose unsaturated polyesters, (○) OC-P340, (△) OC-E980, and (□) OC-E701, each with 1.0 wt % TBPB as initiator. (Based on unpublished data by Han and Lee 1987.)

networks, whereas the polymerization of methyl methacrylate or styrene, for instance, gives rise to linear (uncross-linked) macromolecules. Having realized the complexity of the curing reactions of unsaturated polyester, which involves copolymerization, Stevenson (1986) made some simplifying assumptions to develop a mechanistic kinetic model, via the concepts of free-radical polymerization, for unsaturated polyester.

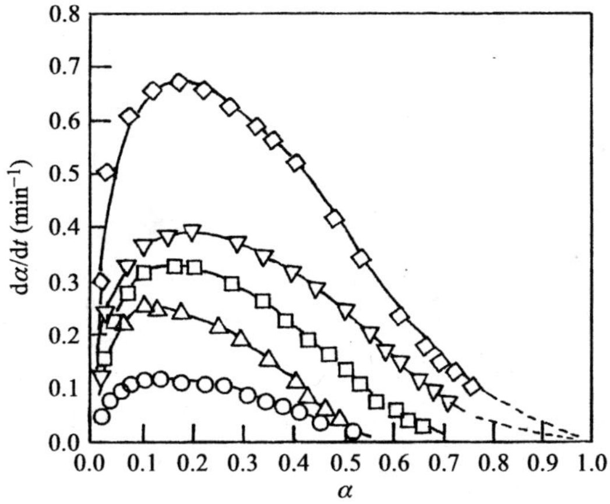

Figure 14.19 Plots of $d\alpha/dt$ versus α for Aropol 7030 with 3.16 wt % BPO as initiator and 1.9 wt % *N,N*-dimethylaniline as accelerator at various isothermal cure temperatures (°C): (○) 40, (△) 45, (□) 50, (▽) 55, and (◇) 60. (Reprinted from Han and Lem, *Journal of Applied Polymer Science* 28:3155. Copyright © 1983, with permission from John Wiley & Sons.)

Later, other investigators (Han and Lee 1987a, 1987b; Huang et al. 1990b) employed the same approach to obtain slightly different rate expressions.

The curing reaction of unsaturated polyester resin involves the copolymerization of unsaturated polyester and styrene in the presence of an organic initiator. Like any other free-radical polymerization (e.g., bulk polymerization of methyl methacrylate), the organic initiator first decomposes to form an initiator radical $[I\cdot]$, which will then react with unsaturated polyester to form a polyester radical $[E\cdot]$ and with styrene monomer to form a styrene radical $[S\cdot]$. Therefore, the curing reaction can be treated by a free-radical polymerization approach. Below, we describe briefly the mechanistic kinetic model for unsaturated polyester (Han and Lee 1987a, 1987b; Stevenson 1986).

Using the assumptions that (1) any carbon double bond in the unsaturated polyester and styrene monomer has the same rate constant for reaction with an initiator radical $[I\cdot]$, (2) the rate of initiation can be lumped with the rate of conversion to give a rate expression of the total radical concentration $[R\cdot]$, (3) the propagation reactions can be expressed by a single average rate constant, and (4) all termination reactions can be combined in a single effective termination constant, the following rate

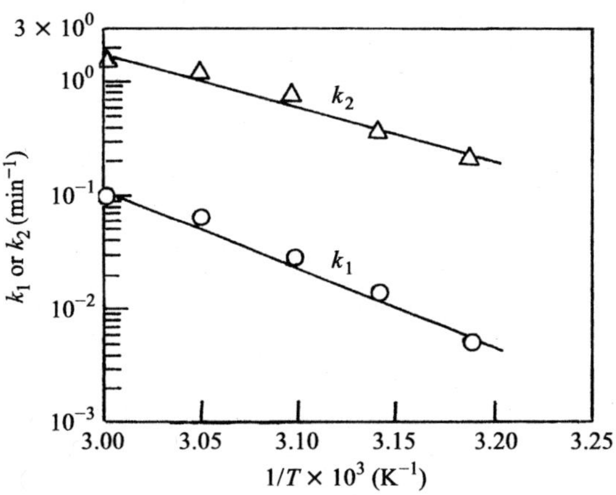

Figure 14.20 Plots of rate constants k_1 and k_2 versus $1/T$ for Aropol 7030 being cured with 3.16 wt % BPO as initiator and 1.9 wt % *N,N*-dimethylaniline as accelerator. (Based on Table III of Han and Lem 1983.)

expressions may be written:

$$d[Z]/dt = -k_z[Z][R\cdot] \quad (14.16)$$

$$d[I_j]/dt = -k_{dj}[I_j];\ j = 1, 2, \ldots, N \quad (14.17)$$

$$d[M]/dt = -k_p[M][R\cdot] \quad (14.18)$$

$$d[R\cdot]/dt = 2\sum_{j=1}^{N} f_j - k_{dj}[I_j] - k_z[Z][R\cdot] - k_t[R\cdot]^2 \quad (14.19)$$

in which [Z] is the inhibitor concentration, $[I_j]$ is the concentration of the *j*th initiator, [*M*] is the total monomer concentration that is assumed to be the sum of the concentration of polyester [*E*] and styrene monomer [*S*], $[R\cdot]$ is the radical concentration that is assumed to be the sum of the concentrations of polyester radical $[E\cdot]$, styrene radical $[S\cdot]$, and the initiator radical $[I_j\cdot]$ (i.e., $[R\cdot] = [E\cdot]+[S\cdot]+[I_j\cdot]$), k_z is the rate constant of the inhibition reaction, k_{dj} is the rate constant of the decomposition reaction of the *j*th initiator, k_p is the rate constant of the propagation reaction, k_t is the rate constant of the termination reaction, and f_j is the efficiency of the *j*th initiator.

With reference to Eqs. (14.16)–(14.19), the following observations are worth noting: (1) more than one initiator may participate in the free-radical polymerization, and they are represented by $[I_j]$, $j = 1, 2, \ldots, N$, and (2) no steady-state assumption is made for the generation of free radicals. Equations (4.16)–(4.19) may be solved numerically for [Z], $[I_j]$, [*M*], and $[R\cdot]$ with the initial conditions; that is, $[Z(0)] = [Z]_0$, $[M(0)] = [M]_0$, $[I_j(0)] = [I_j]_0$ and $[R\cdot(0)] = 0$. However, before Eqs. (14.16)–(14.19) are solved, one must specify the rate constants k_z, k_p, k_t and the initiator efficiency $f_j (j = 1, 2, \ldots, N)$.

The cross-linking reaction of unsaturated polyester with styrene causes gelling to begin at a very early stage, that is, at very low levels of conversion, say 3–5% (Horie et al. 1969). This implies that this reaction is diffusion-controlled over almost the entire range of conversion. This is not the case, for instance, with the bulk polymerization of methyl methacrylate. This observation suggests that the rate constants, k_p and k_t must be represented in such a way that the reaction is identified as diffusion-controlled. In view of the practical difficulty with introducing the free-volume parameters into the representation of the rate constants k_p and k_t, the following empirical expressions may be used:

$$k_p = k_{p0}\exp(-E_p/RT)\left[1 - (\alpha/\alpha_f)\right]^m \quad (14.20)$$

$$k_t = k_{t0}\exp(-E_t/RT)\left[1 - (\alpha/\alpha_f)\right]^n \quad (14.21)$$

where α is the degree of conversion defined by $\alpha = ([M]_0 - [M])/[M]_0$ with $[M]_0$ being the initial concentration of monomer, and α_f is the final degree of conversion defined by $\alpha_f = ([M]_0 - [M]_\infty)/[M]_0$ with $[M]_\infty$ being the monomer concentration after the cure, and k_{p0} and k_{t0} are preexponential factors, E_p is the activation energy of the propagation reaction, E_t is the activation energy of the termination reaction, *R* is the universal gas constant, and *T* is the absolute temperature.

In solving Eqs. (14.16)–(14.19), one must also specify a value for the initiator efficiency $f_j (j = 1, 2, \ldots, N)$. It seems reasonable to regard f_j as being dependent upon the concentration of the initiator radicals and the degree of cure because some of the initiators may be trapped as the curing reaction progresses; this is often referred to as the "caging effect." Since a rigorous description of the caging effect is very difficult, empirical expressions have been suggested for free-radical bulk polymerization (Biesenberger and Sebastian 1983). In modeling the cure reaction of unsaturated polyester, the following expression for f_j has been suggested (Han and Lee 1987a):

$$f_j = \frac{f_{j0}^2[1-(\alpha/\alpha_f)]^2}{2(1-f_{j0})([I_j]/[I_j]_0)} \left\{ \left[1 + \frac{4(1-f_{j0})([I_j]/[I_j]_0)}{f_{j0}^2[1-(\alpha/\alpha_f)]^2} \right]^{1/2} - 1 \right\} \tag{14.22}$$

where $f_{j0} (j = 1, 2, \ldots, N)$ is the initial value of the jth initiator efficiency and $[I_j]_0$ is the initial concentration of jth initiator. Note that Eq. (14.22) indicates that the initiator efficiency f_j decreases with increasing α (i.e., as the curing reaction progresses), which is reasonable from the point of view of the diffusion-controlled curing reaction of unsaturated polyester. Han and Lee (1987a) solved numerically Eqs. (14.16)–(14.19), together with Eqs. (14.20)–(14.22). In doing so, they replaced Eq. (4.18) with the expression

$$\frac{d\alpha}{dt} = k_p(1-\alpha)[R\cdot] \tag{14.23}$$

with the initial condition, $\alpha = 0$ at $t = 0$, and employed the expression

$$k_d = k_{d0} \exp(-E_d/RT) \tag{14.24}$$

for the decomposition rate constant of TBPB as initiator with $k_{d0} = 8.524 \times 10^{15}$ min^{-1} L/M and $E_d = 8.55 \times 10^4$ J/mol and the expression

$$k_z = k_{z0} \exp(-E_z/RT) \tag{14.25}$$

for the inhibition rate constant $k_{z0} = 3.024 \times 10^{17}$ min^{-1} L/M and $E_z = 8.69 \times 10^4$ J/mol. They determined values of k_{p0}, k_{t0}, m, and n until the sum of the squares of the differences between the computed values of both $d\alpha/dt$ and α and the experimentally measured values reached a minimum throughout the entire period of cure. After considerable computations, they found that the termination reaction plays a minor role in the simulation, and thus in subsequent computations dropped the last term on the right-hand side of Eq. (14.19) by setting $k_{t0} = 0$. They concluded that the use of nonzero values of the parameter m appearing in Eq. (14.20) gave rise to predictions of $d\alpha/dt$ much closer to the experimental results than did the use of $m = 0$. This observation reinforces the view that the propagation reaction in the curing of unsaturated polyester is diffusion-controlled.

Mechanistic kinetic models have a number of significant advantages over empirical ones for simulating various processing operations of an unsaturated polyester or

Table 14.4 Numerical values of the parameters appearing in the mechanistic model for OC-E701 with TBPB as initiator

Parameter	*Value*
k_{d10}	8.524×10^{15} min^{-1} L/M for TBPB as initiator
E_{d1}	8.562×10^{4} J/mol for TBPB as initiator
k_{z0}	3.024×10^{17} min^{-1} L/M for inhibitor
E_z	8.734×10^{4} J/mol for inhibitor
k_{p0}	3.441×10^{14} min^{-1} L/M for the propagation reaction of OC-E701
E_p	8.323×10^{4} J/mol for the propagation reaction of OC-E701
f_{10}	0.17 for initial efficiency of initiator TBPB
$[M]_0$	0.352 kg/L as the initial concentration of unsaturated polyester
$[Z]_0$	5.09×10^{-4} M/L for the initial concentration of inhibitor
$[I]_0$	5.716×10^{-2} M/L for 1.0 wt % TBPB as initiator
m	0.8, parameter appearing in Eq. (14.20)

Based on Han and Lee 1987a and 1987b.

its fiber-reinforced composites. When using empirical models, for instance Eq. (14.1), one must determine the values of the rate constants k_1 and k_2 each time the type, concentration or number of initiators is varied. However, when using mechanistic models, one does not have to conduct curing experiments each time the type, concentration or number of initiators is changed, once the values of the rate constants are determined. Table 14.4 gives a summary of the numerical values of the kinetic parameters appearing in the mechanistic model for OC-E701 with TBPB as initiator.

Figure 14.21 gives predicted plots of α versus cure time, Figure 14.22 gives predicted plots of η versus cure time, and Figure 14.23 gives predicted plots of α versus η at six different cure temperatures for OC-E701 with 1 wt % TBPB as initiator. In obtaining Figures 14.21–14.23, Eqs. (14.16)–(14.19), together with Eqs. (14.20), (14.22), and (14.23), were numerically solved with the numerical values of the kinetic parameters given in Table 14.4.

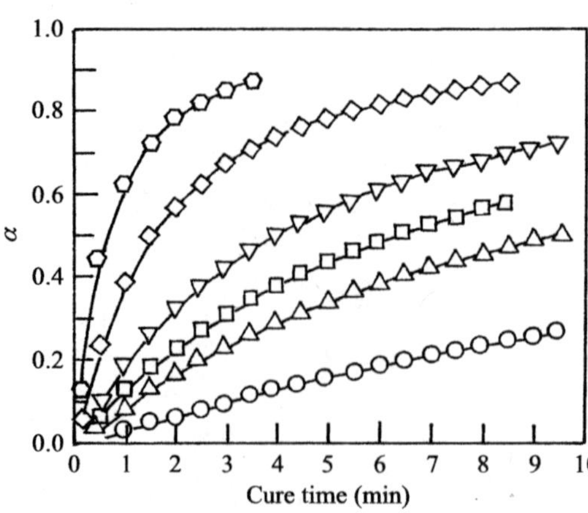

Figure 14.21 Prediction of α during cure of OC-E701 with 1.0 wt % TBPB as initiator at various temperatures (°C): (○) 30, (△) 40, (□) 45, (▽) 50, (◇) 55, and (⬡) 60. The prediction is made from the numerical solution of Eqs. (14.16)–(14.19) together with Eqs. (14.20), (14.22), and (14.23), using the values of the parameters given in Table 14.4.

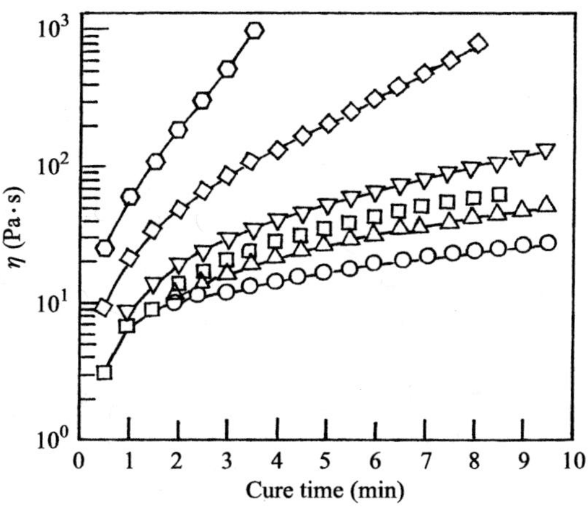

Figure 14.22 Prediction of viscosity rise during cure of OC-E701 with 1.0 wt % TBPB as initiator at various temperatures (°C): (○) 30, (△) 40, (□) 45, (▽) 50, (◇) 55, and (⬠) 60. The prediction is made from the numerical solution of Eqs. (14.16)–(14.19) together with Eqs. (14.20), (14.22), and (14.23), using the values of the parameters given in Table 14.4, and then from Eq. (14.13).

14.2.4 Effects of Particulates on the Chemorheology of Unsaturated Polyester

Inorganic particulates are often used in thermosets as a heat sink to achieve better temperature control across a molded part during cure. They are also used to reduce the amount of resin to be used, the overall shrinkage, and the cost. Poor temperature control during cure often gives rise to one, or several, of the following defects: warpage, sink, shrinkage, motley surface resulting from overcure, and blisters resulting from undercure. In particulate-filled thermoset composites, the resin matrix is the major load-bearing constituent (although the particulates also bear some load). It is generally accepted that interfacial adhesion between the resin matrix and particulates is the basic requirement for the improvement of the end-use properties of the material at service conditions. In order to improve interfacial adhesion between the resin matrix and the particulates, it is a common practice to treat the particulates with an organic additive,

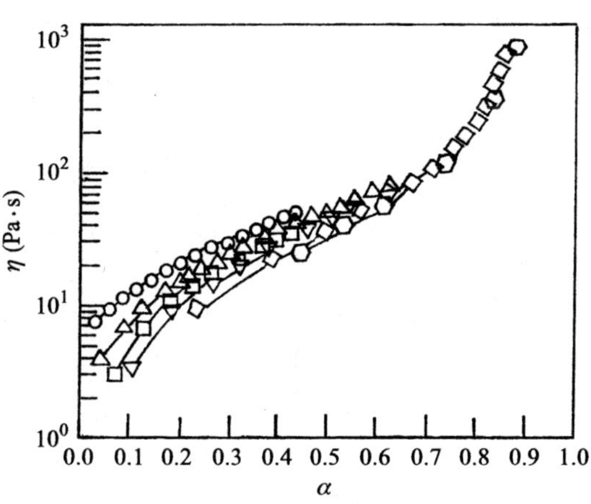

Figure 14.23 Prediction of η–α relationship for OC-E701 with 1.0 wt % TBPB as initiator at various temperatures (°C): (○) 30, (△) 40, (□) 45, (▽) 50, (◇) 55, and (⬠) 60. The results here are cross plots of Figures 14.25 and 14.26.

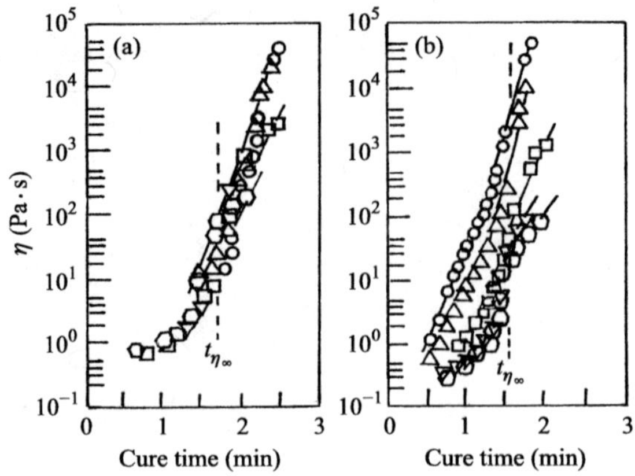

Figure 14.24 (a) Plots of log η versus cure time at 60 °C for Aropol 7030/25 wt % $CaCO_3$/BPO/accelerator mixture at various shear rates (s^{-1}): (○) 0.03, (△) 0.11, (□) 1.0, (▽) 6.77, and (⬡) 17.0. (b) Plots of log η versus cure time at 60 °C for Aropol 7030/50 wt % $CaCO_3$/BPO/accelerator mixture at various shear rates (s^{-1}): (○) 0.03, (△) 0.11, (□) 1.07, (▽) 6.77, and (⬡) 17.0. The accelerator employed was *N,N*-dimethylaniline diluted in styrene. (Reprinted from Lem and Han, *Journal of Applied Polymer Science* 28:3185. Copyright © 1983, with permission from John Wiley & Sons.)

often referred to as a "coupling agent," which also helps improve the dispersion of the particulates in the resin matrix. Kubota (1975) reported on the effect of particulates on the curing behavior of unsaturated polyester molding compounds, while Lem and Han (1983a) reported on the effects of particulates on both the rheological behavior during cure and the curing kinetics of unsaturated polyester. Below, we present the effect of particulates on the chemorheology of unsaturated polyester.

Figure 14.24 shows the rise in η during the cure of Aropol 7030 filled with 25 wt % and 50 wt % $CaCO_3$ (Camel-Wite, Flintkote Company) at various $\dot{\gamma}$. The $CaCO_3$ particles, which were treated with 1.0 wt % of a silane coupling agent, γ-methacryloxyl propyl-trimethoxysilane, has an average diameter of 3.0 μm. Note that the rise in η during cure of neat resin, Aropol 7030, is already given in Figure 14.4. Comparison of Figure 14.24 with Figure 14.4 indicates that $t_{\eta\infty}$ is decreased from 2.9 to 1.8 min when 25 wt % $CaCO_3$ is added, and from 2.9 to 1.6 min when 50 wt % $CaCO_3$ is added to the neat resin. Another interesting observation worth noting in Figure 14.24 is that at the loading of 25 wt % $CaCO_3$ the rise in η during cure was virtually independent of $\dot{\gamma}$, while at the loading of 50 wt % $CaCO_3$, the value of η decreased, during cure, with increasing $\dot{\gamma}$.

Figure 14.25 gives plots of the rate of heat generated, dQ/dt, versus cure time for Aropol 7030 loaded with 25 wt % $CaCO_3$ particles at various isothermal cure temperatures. It is seen that the peak value of dQ/dt appears sooner as the cure temperature is increased from 30 to 60 °C.

Figure 14.26 gives plots of dα/dt versus α for Aropol 7030 filled with 25 wt % $CaCO_3$ particles at various isothermal cure temperatures. On the basis of Figure 14.26, the kinetic parameters (k_1, k_2, m, and n) were determined, and they are summarized in Table 14.5. It is seen that the magnitude of k_2 is much greater than that of k_1,

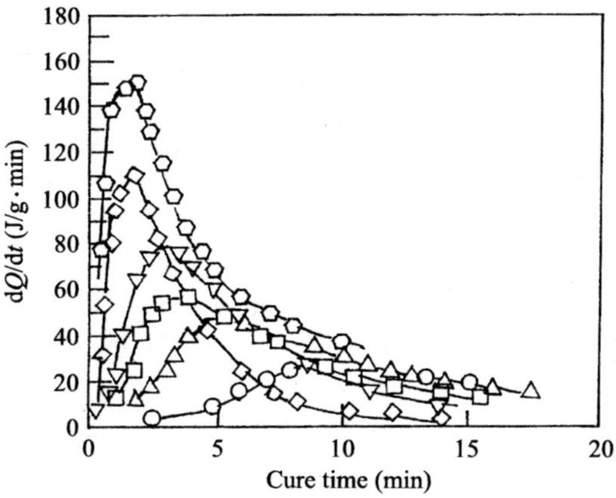

Figure 14.25 Plots of dQ/dt versus cure time for Aropol 7030/25 wt % $CaCO_3$/BPO/accelerator mixture at various isothermal cure temperatures (°C): (○) 30, (△) 40, (□) 45, (▽) 50, (◇) 55, and (⬠) 60. The accelerator employed was *N,N*-dimethylaniline diluted in styrene. (Reprinted from Lem and Han, *Journal of Applied Polymer Science* 28:3185. Copyright © 1983, with permission from John Wiley & Sons.)

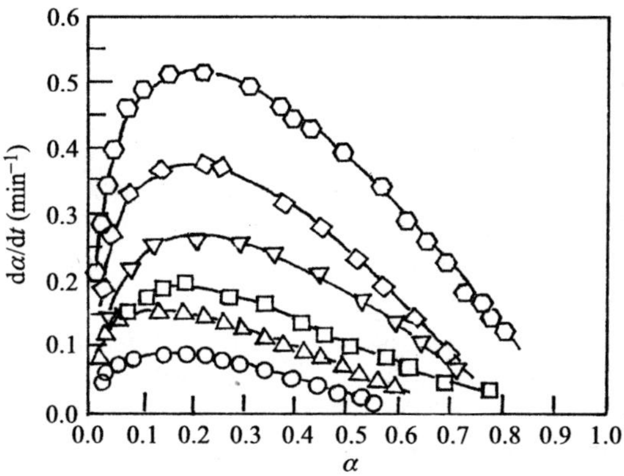

Figure 14.26 Plots of dα/dt versus α for Aropol 7030/25 wt % $CaCO_3$/BPO/accelerator mixture at various isothermal cure temperatures (°C): (○) 35, (△) 40, (□) 45, (▽) 50, (◇) 55, and (⬠) 60. The accelerator employed was *N,N*-dimethylaniline diluted in styrene. (Reprinted from Lem and Han, *Journal of Applied Polymer Science* 28:3185. Copyright © 1983, with permission from John Wiley & Sons.)

Table 14.5 Summary of the kinetic parameters for Aropol 3070 filled with $CaCO_3$ particles

Temp (°C)	k_1 *(min^{-1})*	k_2 *(min^{-1})*	m	n
Resin/25 wt % $CaCO_3$				
45	0.014	0.439	0.39	1.61
50	0.028	0.796	0.43	1.57
60	0.074	1.490	0.55	1.45
Resin/50 wt % $CaCO_3$				
55	0.053	1.470	0.41	1.59
60	0.088	1.540	0.41	1.59
65	0.104	2.050	0.49	1.51
75	0.257	2.700	0.48	1.52

Reprinted from Lem and Han, *Journal of Applied Polymer Science* 28:3185. Copyright ©1983, with permission from John Wiley & Sons.

Table 14.6 Summary of the activation energy of aropol 7030 and Aropol filled with $CaCO_3$ particles

Material	E_1 *(J/mol)*	E_2 *(J/mol)*
Aropol 7030	1.282×10^5	9.773×10^4
Aropol 7030 with 25 wt % $CaCO_3$	9.773×10^4	6.986×10^4
Aropol 7030 with 50 wt % $CaCO_3$	7.171×10^4	3.081×10^4

Based on Lem and Han 1983a.

suggesting that the contribution of k_2 to the cure kinetics is much greater than that of k_1. The activation energies E_1 and E_2 of cure reactions corresponding to the rate constants k_1 and k_2, respectively, are summarized in Table 14.6. It is seen that E_2 decreased considerably from 6.99×10^4 to 3.08×10^4 J/mol as the loading of $CaCO_3$ particles was increased from 25 to 50 wt %. The decrease in the activation energy of the cure reaction implies an increase in the rate of cure, $d\alpha/dt$.

Having obtained information on the variations in both η and α with cure time, plots of log η versus α at $\dot{\gamma} = 1.07\ s^{-1}$ were constructed for Aropol 7030 filled with 25 wt % $CaCO_3$ particles, which are given in Figure 14.27. It is seen that the addition of 25 wt % $CaCO_3$ particles to neat resin increases the η and decreases the ultimate value of α at which η approaches infinity. Figure 14.28 gives plots of η versus α for Aropol 7030 filled with 50 wt % $CaCO_3$ particles at various $\dot{\gamma}$. Owing to the strong shear-thinning behavior exhibited by the Aropol 7030 filled with 50 wt % $CaCO_3$ particles (see Figure 14.24b), shear-rate dependence is observed in Figure 14.28. Nevertheless, the ultimate value of α at which η approaches infinity is about 0.32, independent of $\dot{\gamma}$, and this value is decreased as the loading of $CaCO_3$ particles is increased from 25 to 50 wt % (compare Figure 14.28 with Figure 14.27). Notice further that at a comparable $\dot{\gamma}$, the η of $CaCO_3$-filled Aropol 7030 is increased as the loading of $CaCO_3$ particles is increased from 25 to 50 wt %. It is thus concluded that the addition of particulates greatly influences the chemorheology of unsaturated polyester.

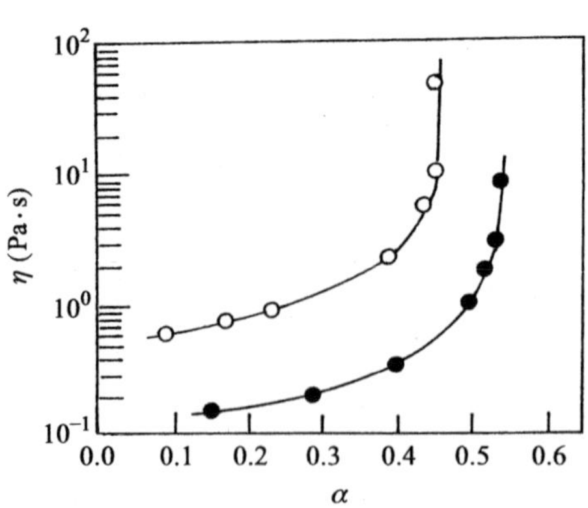

Figure 14.27 Plots of log η versus α at 60 °C and $\dot{\gamma} = 1.07\ s^{-1}$ for (○) Aropol 7030/BPO/accelerator and (●) Aropol 7030/25 wt % $CaCO_3$/BPO/accelerator mixture, in which the accelerator employed was *N,N*-dimethylaniline diluted in styrene. (Reprinted from Lem and Han, *Journal of Applied Polymer Science* 28:3185. Copyright © 1983, with permission from John Wiley & Sons.)

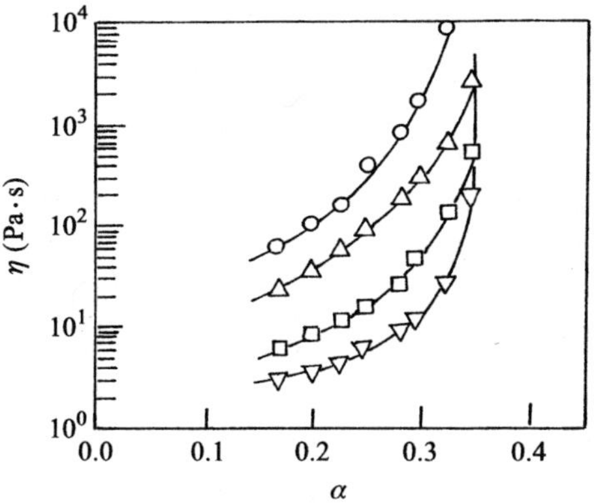

Figure 14.28 Plots of log η versus α at 60 °C for Aropol 7030/50 wt % $CaCO_3$/BPO/accelerator mixture at various shear rates (s^{-1}): (○) 0.0, (△) 0.11, (□) 1.07, and (▽) 6.77. The accelerator employed was *N,N*-dimethylaniline diluted in styrene. (Reprinted from Lem and Han, *Journal of Applied Polymer Science* 28:3185. Copyright © 1983, with permission from John Wiley & Sons.)

14.2.5 Effects of Low-Profile Additive on the Chemorheology of Unsaturated Polyester

Without a certain thermoplastic additive, commonly referred to as a "low-profile additive," molded parts made of fiber-reinforced unsaturated polyester composites suffer from wavy surfaces, warpage, internal cracks, voids, and deep sink marks (Atkins 1978; Bartkus and Kroekel 1970; Pattison et al. 1974, 1975; Siegmann et al. 1978). With the development of low-profile additives, many of these problems have been alleviated, and it is now possible to obtain the desired products with relatively little shrinkage. Three mechanisms have been proposed for the controlled shrinkage that low-profile additives provide (Bartkus and Kroekel 1970), namely, (1) optical heterogeneity and boiling monomer, (2) strain relief through stress cracking, and (3) thermal expansion. Pattison et al. (1974, 1975) have proposed the mechanism of "strain relief through stress cracking," in which they stipulate that the unreacted styrene monomer thermally expands to compensate for the loss of volume due to polymerization shrinkage. At the same time, as the cure progresses, shrinkage causes strain to develop in the system, which increases to such an extent that cracks are formed to relieve it. Such stress cracking propagates through the weakest part of the material, which is either in the thermoplastic additive phase or at the interface between the dispersed phase and the continuous phase, and consequently voids are formed to compensate for the loss of volume due to polymerization. Atkins (1978) noted that in order for a thermoplastic polymer to function effectively as a low-profile additive, it must be incompatible, during cure, with the styrene cross-linked polyester matrix. The effect of two low-profile additives, poly(methyl methacrylate) (PMMA) in styrene and poly(vinyl acetate) (PVAc) in styrene, on the chemorheology of unsaturated polyester were investigated by Lem and Han (1983b) and Lee and Han (1987b).

Figure 14.29 shows the rise of η during cure for Aropol 7030 containing (a) 10 wt % PVAc and (b) 20 wt % PVAc in styrene at 60 °C for various values $\dot{\gamma}$. According to Han and Lem (1983), the resin formulation has resin/initiator/accelator = 60/2.0/1.2 on a weight basis, where BPO is the initiator and *N,N*-dimethylaniline is the accelator. It is seen in Figure 14.29 that as the amount of PVAc solution was increased from 10 to 20 wt %, t_{η_∞} increased from 3 to 4.3 min, indicating that the presence of low-profile

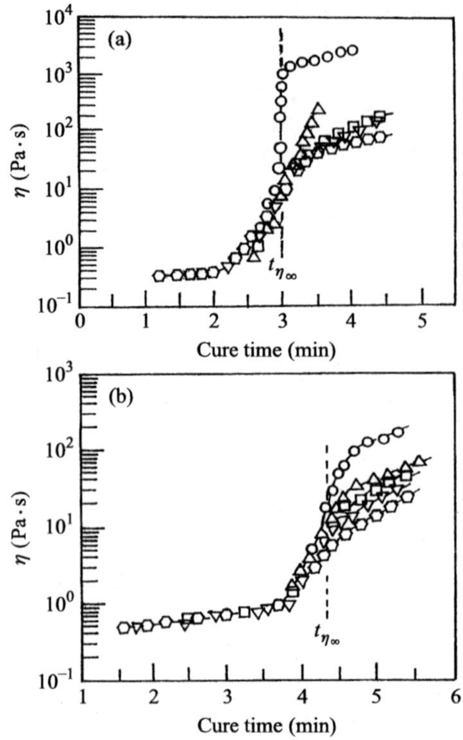

Figure 14.29 Plots of log η versus cure time at 60 °C for Aropol 7030/BPO/accelerator mixture with (a) 10 wt % PVAc and (b) 20 wt % PVAc at various shear rates (s^{-1}): (○) 0.27, (△) 2.69, (□) 6.77, (▽) 10.7, and (⬡) 17.0. The accelerator employed was *N,N*-dimethylaniline diluted in styrene. (Reprinted from Lem and Han, *Journal of Applied Polymer Science* 28:3207. Copyright © 1983, with permission from John Wiley & Sons.)

additive reduces the rate of cure. Note that t_{η_∞} for neat resin, Aropol 7030, at 60 °C is 2.9 min (see Figure 14.4). This observation indicates that the rise of η is slowed down as the amount of the PVAc solution is increased. This is attributed to the fact that since the PVAc solution contains 60 wt % styrene, an increase in the amount of PVAc solution makes more styrene available in the mixture, thus diluting the initiator and accelerator concentrations, and consequently slows down the cure reaction.

Figure 14.30 shows the rise of η during cure for Aropol 7030 containing (a) 10 wt % PMMA and (b) 20 wt % PMMA in styrene at 60 °C for various values of $\dot{\gamma}$. It is seen that as the amount of PMMA solution was increased from 10 to 20 wt %, t_{η_∞} increased from 4.2 to 5.6 min. Note that the value of t_{η_∞} for the resin/PMMA mixture is greater than that for the resin/PVAc mixture for approximately the same amount of styrene available (compare Figure 14.30 with Figure 4.29). Therefore, the observed difference in t_{η_∞} between Aropol/PMMA and Aropol/PVAc mixtures must have its origin in the nature of the compatibility of the respective low-profile additive with the unsaturated polyester. The resin/PMMA mixtures exhibit shear-thinning behavior even before cure has progressed much, suggesting that the PMMA might have formed a separate phase in the form of small drops. This speculation is based on the experimental evidence that a two-phase emulsion can exhibit shear-thinning behavior when either the aggregates of drops are broken up or the individual drops deform when the emulsion is subjected to intensive shearing motion (see Chapter 11). Therefore, a slower rate of η rise for the resin/PMMA mixtures, compared with that for the resin/PVAc mixtures, may be

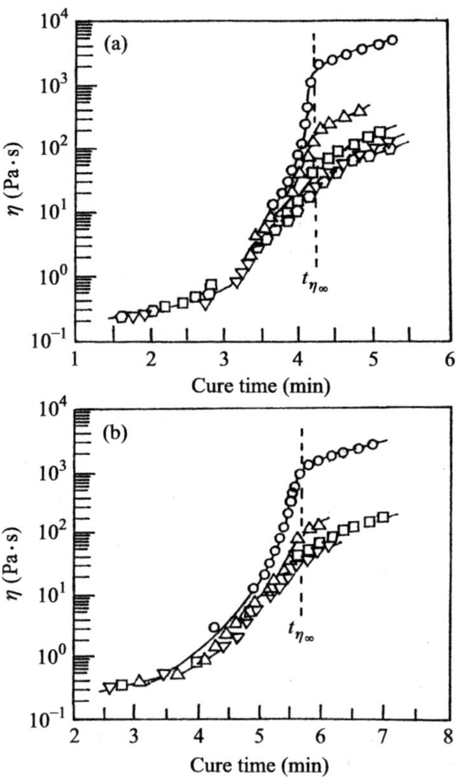

Figure 14.30 Plots of log η versus cure time at 60 °C for Aropol 7030/BPO/accelerator mixture with (a) 10 wt % PMMA and (b) 20 wt % PMMA at various shear rates (s^{-1}): (○) 0.27, (△) 2.69, (□) 6.77, (▽) 10.7, and (⬡) 17.0. The accelerator employed was *N,N*-dimethylaniline diluted in styrene. (Reprinted from Lem and Han, *Journal of Applied Polymer Science* 28:3207. Copyright © 1983, with permission from John Wiley & Sons.)

attributable to the formation of a PMMA/styrene drop phase in the resin/PMMA mixture. Under such a circumstance, part of the accelarator (*N,N*-dimethylaniline) and the initiator (BPO) might have been occluded in the PMMA/styrene drops, consequently decreasing the rate of cure.

The kinetic parameters (k_1, k_2, m, and n) for the resin/PVAc and resin/PMMA mixtures are summarized in Table 14.7. It is seen that the magnitude of k_2 is much greater than that of k_1, suggesting that the contribution of k_2 to the cure kinetics is much greater than that of k_1. As the amount of low-profile additive increased from 10 to 20 wt %, the activation energy for the rate constant k_2 decreased for the resin/PVAc mixtures, whereas it increased for the resin/PMMA mixtures (Lem and Han 1983b). Such an observation seems to indicate that there is a substantial difference in the roles that the two low-profile additives play during cure. Thus, in the cure of a mixture consisting of unsaturated polyester and low-profile additive, a decrease in the activation energy of the cure reaction does not necessarily imply an increase in the rate of cure, because the degree of cure (also, the rate of cure) is decreased as the amount of low-profile additive is increased. The decrease in the activation energy for cure reaction may be attributable to the differences in the morphology of styrene cross-linked polyester network between the unsaturated polyester/PVAc mixtures and the unsaturated polyester/PMMA mixtures.

Table 14.7 Summary of the kinetic parameters for Aropol 7030 with low-profile additive

Temp (°*C*)	k_1 (*min*$^{-1}$)	k_2 (*min*$^{-1}$)	*m*	*n*
Mixture of Aropol 7030 and 10 wt % PVAc				
45	0.005	0.228	0.27	1.73
50	0.006	0.458	0.29	1.71
55	0.014	0.345	0.27	1.73
60	0.050	0.780	0.35	1.65
65	0.060	0.751	0.34	1.66
75	0.122	1.140	0.33	1.67
Mixture of Aropol 7030 with 20 wt % PVAc				
45	0.008	0.144	0.27	1.73
55	0.012	0.270	0.29	1.71
60	0.026	0.282	0.27	1.73
65	0.034	0.416	0.24	1.76
85	0.100	0.577	0.24	1.76
Mixture of Aropol 7030 with 10 wt % PMMA				
40	0.019	0.170	0.33	1.67
50	0.027	0.342	0.34	1.66
55	0.026	0.423	0.30	1.70
60	0.037	0.736	0.36	1.64
65	0.078	0.566	0.23	1.77
70	0.144	0.914	0.28	1.72
Mixture of Aropol 7030 and 20 wt % PMMA				
60	0.053	0.354	0.28	1.72
65	0.080	0.576	0.26	1.74
70	0.105	0.433	0.14	1.86
75	0.140	0.942	0.32	1.68
80	0.175	0.836	0.23	1.77

Reprinted from Lem and Han, *Journal of Applied Polymer Science* 28:3207. Copyright © 1983, with permission from John Wiley & Sons.

Figure 14.31 gives plots η versus α at 60 °C for Aropol 7030/PVAc mixtures at $\dot{\gamma} = 17.0\ \text{s}^{-1}$, and Figure 14.32 gives similar plots at 60 °C for Aropol 7030/PMMA mixtures at $\dot{\gamma} = 17.0\ \text{s}^{-1}$. It is seen that the addition of low-profile additive to neat resin increases η and decreases the ultimate α at which η approaches infinity. The effect of different types of low-profile additive on the rate of cure is not that simple to interpret. The effect of isothermal cure temperature on the rate of cure of a mixture consisting of unsaturated polyester and low-profile additive can be very complicated, depending upon the morphological state of the mixture. As presented in an earlier section, the cure reaction of an unsaturated polyester can be described by the free-radical polymerization between an unsaturated acid in the polyester resin and a reactive diluent (e.g., styrene). Therefore, the mole ratio of styrene to the unsaturated acid influences the rate of cure. When a low-profile additive is added, the additional amount of styrene that becomes available will increase the mole ratio of styrene to the unsaturated acid, decreasing

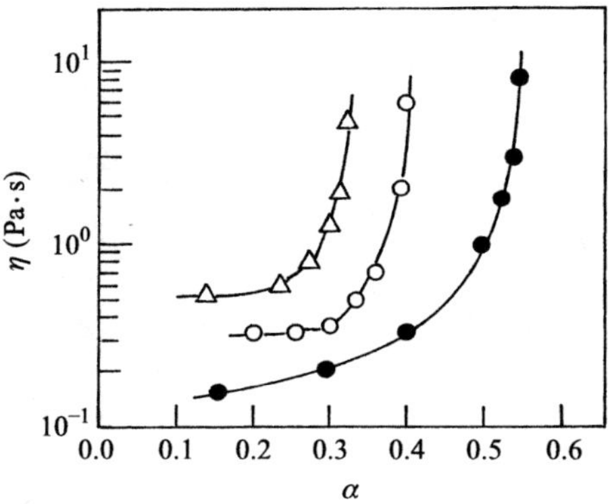

Figure 14.31 Plots of log η versus α at 60 °C and = $17 s^{-1}$ for (●) Aropol 7030/BPO/accelerator, (○) Aropol 7030/10 wt % PVAc/BPO/accelerator mixture, and (△) Aropol 7030/20 wt % PVAc/BPO/accelerator mixture, in which the accelerator employed was *N,N*-dimethylaniline diluted in styrene. (Reprinted from Lem and Han, *Journal of Applied Polymer Science* 28:3207. Copyright © 1983, with permission from John Wiley & Sons.)

the rate of cure and, also, the final degree of cure by changing the ratio of resin to initiator/accelerator. However, a thermoplastic additive is needed to control shrinkage of the unsaturated polyester during cure. Therefore, one must determine an optimum amount of low-profile additive to be added that will minimize the sacrifice of the rate of cure and yet maximize the shrinkage control.

The viscosity measurements and cure study presented above cannot describe whether the addition of PVAc or PMMA to an unsaturated polyester has actually decreased shrinkage during cure. Lee and coworkers (Hsu et al. 1991; Kinkelaar et al. 1994; Li and Lee 1998, 2000a, 2000b) investigated the shrinkage of unsaturated polyester resins or sheet molding compounds in the presence of low-profile additives. Figure 14.33 shows the effect of the concentration of PVAc on the variation in volume change, via dilatometry, during the cure of unsaturated polyester systems having methyl ethyl ketone peroxide (MEKP) as an initiator and cobalt octoate as an accelerator. It can be seen in Figure 14.33 that the volume of the resin system having 3.5 wt % PVAc starts to expand at approximately 310 min after cure began, while the resin system having 6 wt % PVAc starts to expand at 460 min after cure began, and no expansion of volume occurs in the resin system having 10 wt % PVAc. It has been observed that the molecular

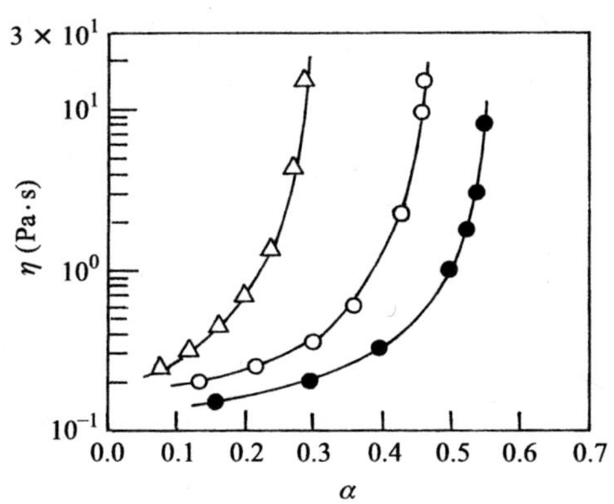

Figure 14.32 Plots of log η versus α at 60 °C and $\dot{\gamma} = 17 s^{-1}$ for (●) Aropol 7030/BPO/accelerator, (○) Aropol 7030/10 wt % PMMA/BPO/accelerator mixture, and (△) Aropol 7030/20 wt % PMMA/BPO/accelerator mixture, in which the accelerator employed was *N,N*-dimethylaniline diluted in styrene. (Reprinted from Lem and Han, *Journal of Applied Polymer Science* 28:3207. Copyright © 1983, with permission from John Wiley & Sons.)

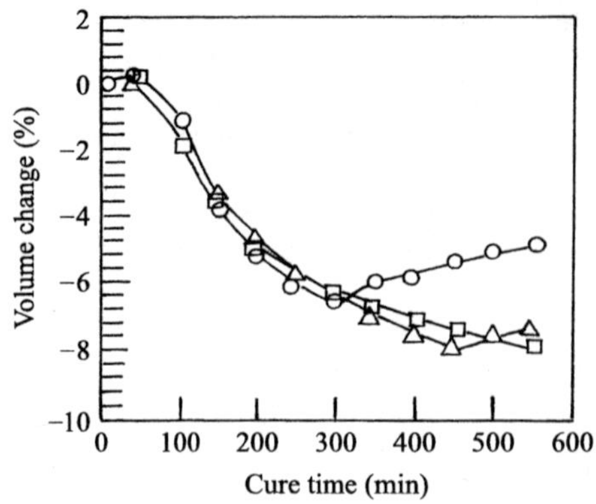

Figure 14.33 Variation of volume change during cure at 35 °C of unsaturated polyester resin systems having MEKP as an initiator and cobalt octoate as an accelerator in the presence of PVAc at different concentrations (wt %): (○) 3.5, (△) 6.0, and (□) 10. (Reprinted from Li and Lee, *Polymer* 39:5677. Copyright © 1998, with permission from Elsevier.)

weight of PVAc, the concentration of PVAc, and cure temperature affect greatly the effectiveness of shrinkage control of unsaturated polyester resins during cure.

14.2.6 Oscillatory Shear Flow during Cure of Unsaturated Polyester

While steady-state shear flow measurements are regarded as being effective to characterize the rheological properties of a material in the liquid state, oscillatory shear flow measurements are regarded as being effective to characterize the rheological properties of a material in both the rubbery and glassy states, implying that it can be used after a thermoset enters the gelation stage (beyond the gel point). Thus, it has been suggested that gel time of a thermoset during cure may be determined from the variations of dynamic storage and loss moduli (G' and G'') with time by identifying the time at which G' and G'' cross each other, $t_{G'=G''}$ or $t_{\tan\delta=1}$ or by identifying the time at which G' has the greatest (or maximum) value, that is, $t_{G''_{\max}}$ (Tung and Dynes 1982; Yap and Williams 1982). It has been reported that $t_{\tan\delta=1}$ during isothermal cure of epoxy resins coincided with the gel time independently measured by the standard gel time test (Tung and Dynes 1982).

Figure 14.34 shows the time evolution of G' and G'' during cure of Aropol 7030 at 30 °C and at an angular frequency (ω) of 1.89 rad/s, showing that the G' increases

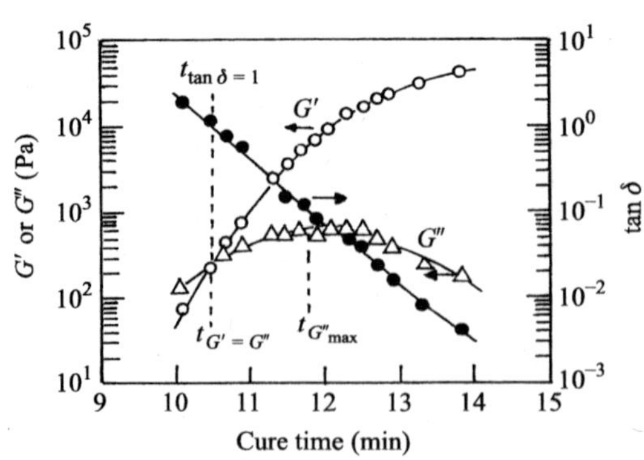

Figure 14.34 Plots log G' (○), log G'' (△) and tan δ (●) versus cure time during isothermal cure of Aropol 7030/BPO/accelerator at $\omega = 1.89$ rad/s and 30 °C, in which the accelerator employed was *N,N*-dimethylaniline diluted in styrene. (Reprinted from Han and Lem, *Journal of Applied Polymer Science* 28:3155. Copyright © 1983, with permission from John Wiley & Sons.)

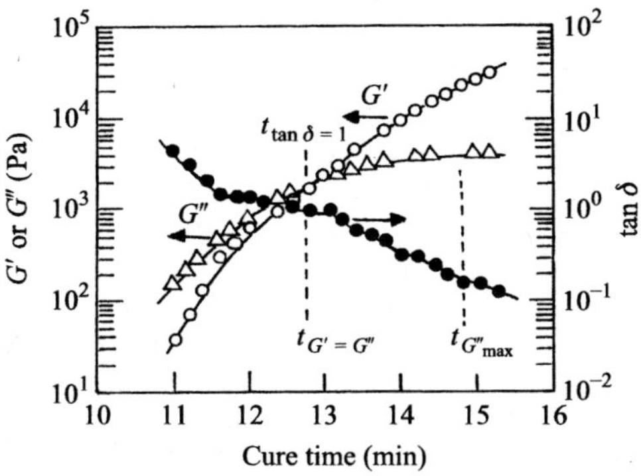

Figure 14.35 Plots of log G' (○), log G'' (△), and tan δ (●) versus cure time during isothermal cure of Aropol 7030/BPO/accelerator at $\omega = 29.94$ rad/s and 30 °C, in which the accelerator employed was *N,N*-dimethylaniline diluted in styrene. (Reprinted from Han and Lem, *Journal of Applied Polymer Science* 28:3155. Copyright © 1983, with permission from John Wiley & Sons.)

monotonically with cure time while the G'' goes through a maximum. Figure 14.35 describes the time evolution of G′ and G'' during cure of Aropol 7030 at 30 °C and at $\omega = 29.94$ rad/s, showing that G' increases monotonically with cure while G'' first increases and then approaches a constant value as cure continues. On each figure, the time at which the G' and G'' cross each other ($t_{G'=G''} = t_{\tan\delta=1}$) and the time at which G″ exhibits a maximum ($t_{G''_{max}}$) are indicated. Table 14.8 gives a summary of t_{η_∞} at four different isothermal cure temperatures and Table 14.9 gives a summary of the comparison between $t_{\tan\delta=1}$ and $t_{G''_{max}}$ at two different isothermal cure temperatures for Aropol 7030. It is seen in Tables 14.8 and 14.9 that at 30 °C the values of t_{η_∞} and $t_{\tan\delta=1}$ come reasonably close to each other, while at 50 °C the values of t_{η_∞} and $t_{G''_{max}}$ come reasonably close to each other, and values of $t_{\tan\delta=1}$ are smaller than those of $t_{G''_{max}}$.

14.3 Chemorheology of Epoxy Resin

Epoxies have a variety of commercial applications, such as surface coatings, encapsulation of electronic components, adhesives, coatings, and laminates (Ranney 1977).

Table 14.8 Summary of t_{η_∞} for Aropol 7030 at various isothermal cure temperatures

Temperature (°C)	t_{η_∞} *(min)*
30	11.0–12.3
40	7.7–8.1
50	4.7
60	2.9

Reprinted from Han and Lem, *Journal of Applied Polymer Science* 28:3155. Copyright © 1983, with permission from John Wiley & Sons.

Table 14.9 Summary of $t_{\tan\delta=1}$ and $t_{G''_{max}}$ for Aropol 7030 at various angular frequencies

ω (rad/s)	$t_{\tan\delta=1}$ (min)	$t_{G''_{max}}$ (min)
at 30 °C		
1.19	12.9	13.8
1.89	10.5	11.8
2.99	11.4	13.2
4.74	11.4	12.8
7.52	13.1	14.3
11.92	12.4	14.0
29.94	12.8	14.8
at 50 °C		
1.89	3.3	4.1
2.99	3.6	4.4
4.74	3.6	4.8
7.52	3.5	4.2
11.92	3.7	5.1
18.89	3.8	4.8
29.94	3.4	4.6

Reprinted from Han and Lem, *Journal of Applied Polymer Science* 28:3155. Copyright ©1983, with permission from John Wiley & Sons.

Today, most commercial epoxies (also referred to as epoxy resins) are prepared by the reaction of 2,2-bis(4′-hydroxyphenyl) propane (bisphenol A) and epichlorohydrin, but other types of products are also available. Many of the commercial liquid epoxy resins consist essentially of the low-molecular-weight diglycidyl ether of bisphenol A, with the chemical structure

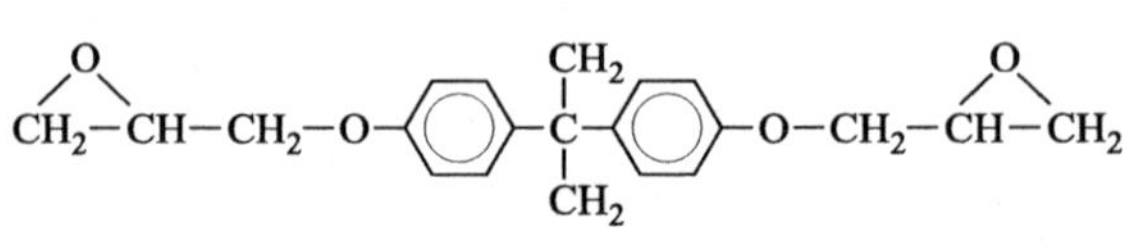

The cure of epoxy resins may be carried out either through the epoxy groups or hydroxyl groups. Two types of curing agent may be used, a catalytic system and polyfunctional cross-linking agents that link the epoxide resin molecules together. Sometimes, a curing agent contains both the catalytic and cross-linking systems. The epoxy ring can be readily attacked by active hydrogen in the system. The epoxy–hydroxyl reaction may be expressed as

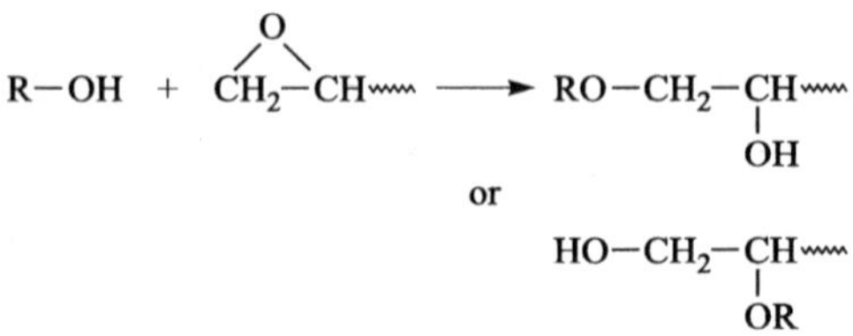

The product will contain new hydroxyl groups that can react with other epoxy rings, generating further active hydroxyl groups, such as,

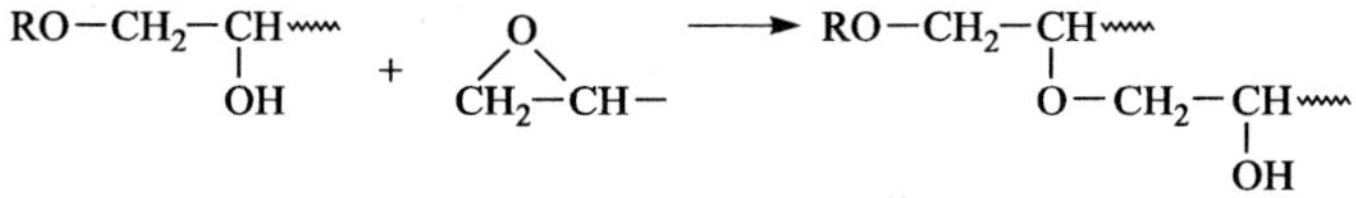

The predominance of one reaction over the other is greatly influenced by the catalyst system employed. Amines and acid anhydrides are extensively used as curing agents. Tertiary amine systems are often used in practice. In addition to the catalytic reactions, the epoxy resins may be cross-linked by agents that link across the epoxy molecules. These reactions may be via the epoxy ring or through the hydroxyl groups, as illustrated below.

(1) With amines

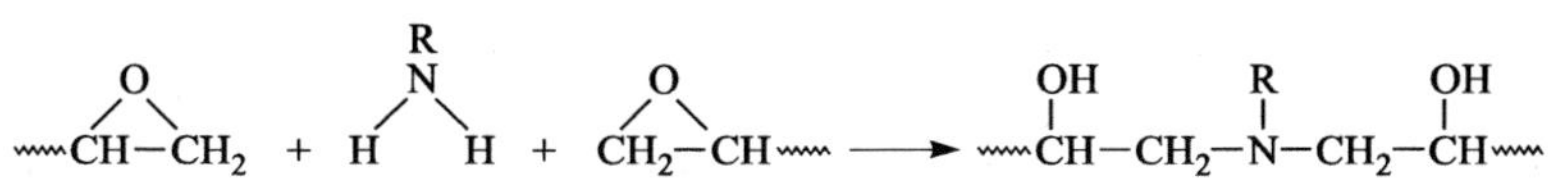

(2) With acids

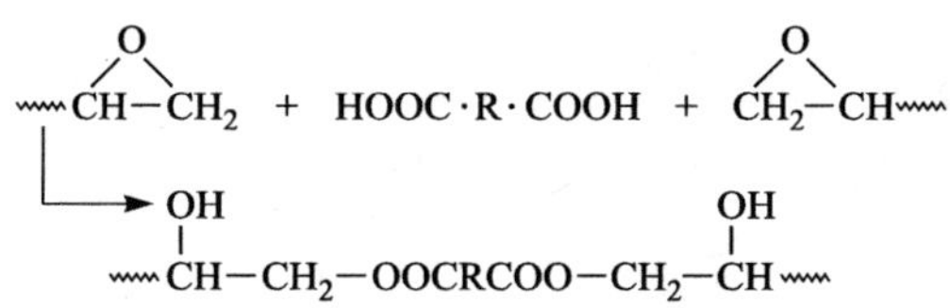

The reactions indicated above lead only to chain extension. In practice, polyamines are used so that the number of active hydrogen atoms exceeds two and so cross linkage occurs.

Understandably, the cure mechanisms are quite complex. The cure kinetics of epoxy were investigated extensively in the 1970s and 1980s (Acitelli et al. 1971; Dusek et al. 1975; Dutta and Ryan 1979; Enns and Gillham 1983; Horie et al. 1970b; Mijovic et al. 1984; Osinski 1983; Peyser and Bascom 1977; Prime 1973; Sacher 1973; Senich et al. 1979; Sourour and Kamal 1976; Sung et al. 1986). Most of the studies employed isothermal DSC to monitor the rate of cure, $d\alpha/dt$, from the measurements of the total amount of heat released with increasing cure time, dQ/dt, and the degree of cure, α, was calculated by integrating dQ/dt, using the procedure described in the preceding section. Due to the rather complicated cure mechanisms involved with epoxy, for simplicity an empirical kinetic expression, such as Eq. (14.1), has widely been used to describe the cure kinetics of epoxy. It has been reported that the order of cure kinetics of epoxy depends very much on the type of curing agent (e.g., primary amine or secondary amine) employed. Dusek et al. (1975) developed mechanistic models for cure reactions of epoxy resin and Sung et al. (1986) investigated, via UV-visible and fluorescence spectroscopy, cure reactions of model compounds consisting of *p,p'*-diaminoazobenzene and 1,2-epoxy-3-phenoxypropane (glycidyl phenyl ether).

To investigate the chemorheology of epoxy resins, one needs an expression describing the variation of α during cure. A number of research groups (Apicella et al. 1984;

Kim and Kim 1987a; Mussatti and Macosko 1973; Roller 1975; Ryan and Dutta 1979; Tajima and Crozier 1983, 1988; Tung and Dynes 1982; White 1974) have investigated the chemorheology of epoxy during cure. Initially, an empirical expression, such as Eq. (14.7), was suggested to describe viscosity variations in epoxy during cure (Roller 1975). Since Eq. (14.7) does not contain an explicit term describing the cure kinetics (i.e., α), Eq. (14.7) was soon abandoned. Subsequently, a WLF expression, similar to Eq. (14.13), was developed to describe viscosity variations in epoxy during cure (Tajima and Crozier 1983, 1988). One such expression developed by Tajima and Crozier (1988) is written as

$$\log \eta(T) = \log \eta(T_s) - \frac{15.46(T - T_s)}{54.40 + (T - T_s)} \tag{14.26}$$

where

$$T_s = 307.53\alpha^2 - 53.613 \tag{14.27}$$

$$\log \eta(T_s) = 2.0874\alpha + 10.372 \tag{14.28}$$

It is seen in Eq. (14.26) that η is related to α through Eqs. (14.27) and (14.28).

Applying dynamic DSC at different scan rates to an epoxy, Kim and Kim (1987a) obtained the temperature dependence of exotherm dQ/dt shown in Figure 14.36.[1] Using the procedures described in a preceding section (see Eq. (14.14)), they obtained an nth-order empirical kinetic expression

$$\frac{d\alpha}{dt} = k_{10} \exp(-E_1/RT)(1 - \alpha)^n \tag{14.29}$$

with the numerical values of the parameters summarized in Table 14.10, in which k_{10} is a preexponential factor, E_1 is the activation energy, R is the universal gas constant, and T is the absolute temperature. Note that Eq. (14.29) is a special case of Eq. (14.1). Using a parallel plate rheometer, they measured the viscosity variations and temperature simultaneously during cure of an epoxy at heating rates of 3, 5, 7, and 10 °C/min, and the experimental results are displayed in Figure 14.37. It is interesting to observe in Figure 14.37 that the viscosity initially decreases, going through a minimum, and then

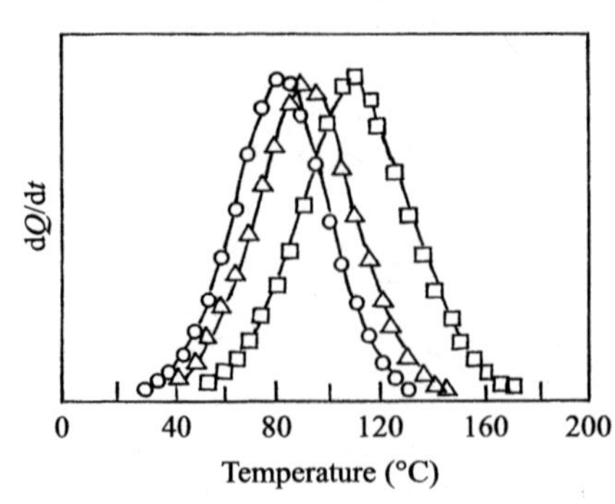

Figure 14.36 Plots of dQ/dt versus temperature for an epoxy at different scan rates (°C/min): (○) 5, (△) 10, and (□) 20. (Reprinted from Kim and Kim, *Polymer Composites* 8:208. Copyright © 1987, with permission from the Society of Plastics Engineers.)

Table 14.10 Summary of the kinetic parameters for the cure reaction of epoxy and the chemorheological model parameters

(a) Kinetic parameters
$k_{10} = 1.0 \times 10^{11}\ \text{min}^{-1}$; $E_1 = 7.795 \times 10^4\ \text{J/mol}$;
$n = 1.64$; $\Delta H_R = 217\ \text{J/g}$ (heat of reaction)

(b) Chemorheological model parameter
$\eta_1 = 1.0 \times 10^{-8}\ \text{Pa·s}$; $E_\eta = 5.080 \times 10^4\ \text{J/mol}$;
$\alpha_g = 0.894$; $a = -0.155$; $b = 5.25$.

Based on Kim and Kim 1987a.

increases very sharply as the temperature increases. They used the following form of an empirical expression for η:

$$\eta(T, \alpha) = \eta_1 \exp(E_\eta/RT) \left(\frac{\alpha_g}{\alpha_g - \alpha} \right)^{a+b\alpha} \tag{14.30}$$

where α_g denotes the degree of cure at T_g, η_1, a, and b are constants, and E_η is the flow activation energy, R is the universal gas constant, and T is the absolute temperature, with the numerical values of the parameters summarized in Table 14.10. The solid lines in Figure 14.37 were calculated using Eqs. (14.29) and (14.30). In doing so, Eq. (14.29) was rewritten as

$$\frac{d\alpha}{dT} = \frac{k_{10}}{S_r} \exp(-E_1/RT)(1-\alpha)^n \tag{14.31}$$

where $S_r = dT/dt$ is the experimental scanning rate (3, 5, 7, and 10 °C/min). Tajima and Crozier (1988) also reported the temperature dependence of viscosity of an epoxy, very similar to Figure 14.37, at various heating rates during cure. Using DSC experiments, they obtained an empirical second-order cure kinetics, $k_2 = 0$ and $n = 2$ in Eq. (14.1), and calculated viscosity variation during cure using Eq. (14.26).

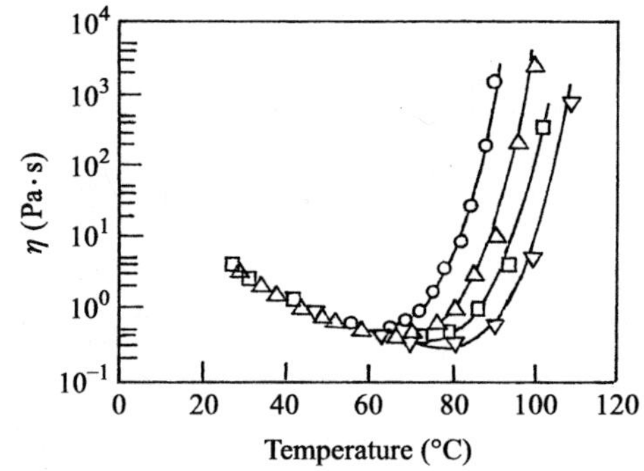

Figure 14.37 Plots of log η versus temperature for an epoxy at different scan rates (°C/min): (○) 3, (△) 5, (□) 7, and (▽) 10. (Reprinted from Kim and Kim, *Polymer Composites* 8:208. Copyright © 1987, with permission from the Society of Plastics Engineers.)

14.4 Chemorheology of Thermosetting Polyurethane

Several kinds of polyurethanes are commercially produced and they have a wide range of applications, such as rigid foams, flexible foams, elastomers, fibers, surface coatings, and adhesives (Oertel, 1985). In Chapter 10, we discussed the rheology of thermoplastic polyurethanes (linear polymers) that are produced by the chemical reactions between a diisocyanate and a diol. Here, we present the chemorheology of thermosetting liquid polyurethanes that are produced by the cross-linking reactions between a diisocyanate and a polyol. The two types of polyurethanes must be distinguished on the basis of reaction kinetics. Here, for brevity we use the word "polyurethane" (PU).

Thermosetting polyurethanes, in general, are prepared by two routes: (1) by the step-growth polymerization of a glycol and a diisocyanate in the presence of a small amount of polyfunctional alcohol or (2) by the reaction of low-molecular-weight polymers containing hydroxyl endgroups with an excess of a diisocyanate and subsequent cross-linking of the resulting higher molecular weight polymer by secondary reactions of urethane groups (Lenz 1967). The reaction between a diol and a diisocyanate gives rise to linear polyurethane

$$n(\mathrm{HO{-}R{-}OH}) + n(\mathrm{OCN{-}R'{-}NCO}) \longrightarrow \left[\mathrm{O{-}R{-}O}\overset{\mathrm{O}}{\overset{\|}{\mathrm{C}}}\mathrm{{-}NH{-}R'{-}NH{-}}\overset{\mathrm{O}}{\overset{\|}{\mathrm{C}}}\mathrm{O}\right]_n$$

where the diol can be 1,2-propanediol, 1,4-butanediol, poly(tetramethylene adipate)diol, poly(ethylene adipate)diol, and so forth, and the diisocynate can be 4,4′ diphenylmethane diisocyanate (MDI), which is widely used in the formulations for reaction injection molding (RIM) that are presented in Chapter 11 of Volume 2, 1,6-hexamethylene diisocynate, 2,6-tolylene diisocyanate (TDI), 2,4-tolylene diisocyanate, and so forth. When a small amount of triol (e.g., trimethylol propane) is added to the mixture of a diol and a diisocynate shown above, cross-linking reactions will take place forming three-dimensional networks, and will eventually lead to a system having an infinite molecular weight. It is not difficult to imagine that the viscosity will rise rapidly as the molecular weight increases rapidly. The reaction mechanisms of thermosetting polyurethanes are indeed very complex (Lenz 1967; Saunders and Frisch 1964).

In the 1970s and 1980s, a number of research groups investigated cure kinetics (Lipshitz and Macosko 1977; Richter and Macosko 1978; Hartley and Williams 1981; Huang et al. 1990a) and chemorheology (Castro 1980; Castro and Macosko 1982; Castro et al. 1984; Kim 1987; Kim and Kim 1987b) of thermosetting PUs. To investigate the cure kinetics of thermosetting PUs, some research groups (Castro 1980; Lipshitz and Macosko 1977; Richter and Macosko 1978) measured adiabatic temperature rise and/or employed infrared (IR) spectroscopy, others (Huang et al. 1990a) employed ultraviolet (UV) absorption and fluorescence spectroscopy, and still others (Kim 1987; Kim and Kim 1987b) employed DSC. Owing to the very complex nature of the reaction mechanisms involved, the development of a mechanistic model for the cure of thermosetting urethane has not been reported, while an empirical model, such as Eq. (14.29), has been suggested (Lipshitz and Macosko 1977; Richter and Macosko 1978). Such an empirical kinetic expression is based on the simplified rate

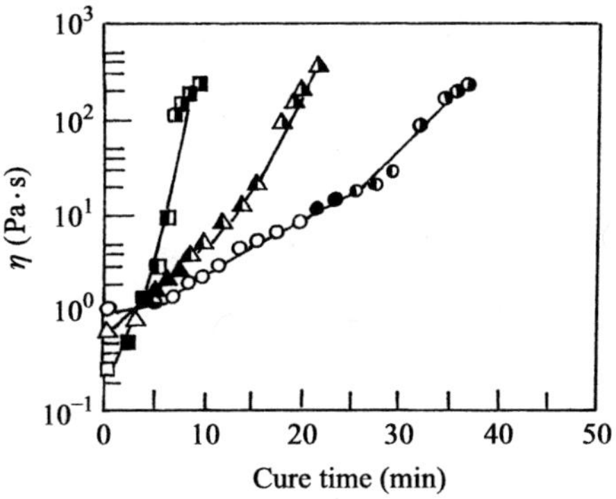

Figure 14.38 Plots of log η versus cure time for a urethane (RIM 2200) at three different isothermal cure temperatures. At 30 °C for different shear rates (s^{-1}): (◒) 250, (○) 62.5, (●) 40, (◐) 25.5, and (◑) 2.5. At 46 °C for different shear rates (s^{-1}): (△) 25, (▲) 100, (▲) 25, and (▲) 2.5. At 65 °C for different shear rates (s^{-1}): (■) 250, (◧) 25, and (◨) 2.5. (Based on the doctoral dissertation of Jose Castro 1980.)

expression:

$$\frac{d[NCO]}{dt} = -k_1[NCO]^n \tag{14.32}$$

where $[NCO]$ is the concentration of diisocyanate and k_1 is a temperature-dependent rate constant lumped with the catalyst concentration.

Figure 14.38 describes the rise in η during isothermal cure at three different temperatures for a thermosetting urethane system consisting of MDI in liquid form (RIM 2200, Union Carbide) and polyol (85 wt% polyether prepolymer having $M_n \approx 5000$ and 15 wt% 1,4 butane diol).[2] Figure 14.39 gives plots of α versus cure time for different cure temperatures, and Figure 14.40 gives plots of η versus α for the same urethane system. It is interesting to observe that the shear rate applied to the specimen in a cone-and-plate rheometer has little effect on the viscosity, similar to that shown in Figure 14.4 for an unsaturated polyester, and the temperature dependence of α during cure is similar to that shown in Figure 14.16 for unsaturated polyester.

Castro (1980) has shown that use of reduced viscosity, η/η_r, suppress the temperature dependence, where η_r is the viscosity before cure begins. Figure 14.41 gives plots of log η/η_r versus α for the same urethane system, showing indeed almost temperature independence. Figure 14.41 suggests that the viscosity rise during isothermal cure of the

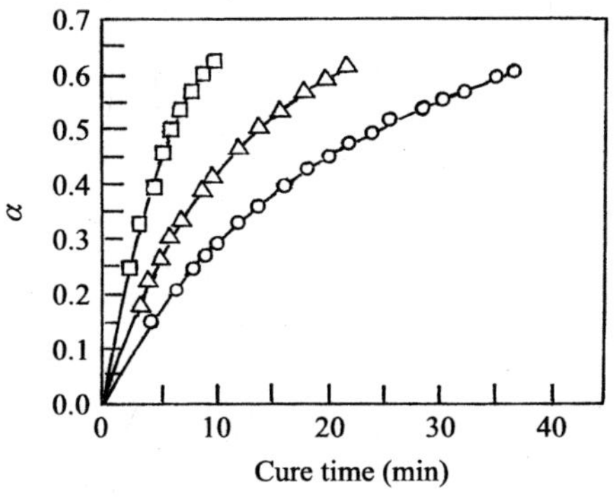

Figure 14.39 Plots of α versus cure time for a urethane (RIM 2200) at various isothermal cure temperatures (°C): (○) 30, (△) 46, and (□) 65. (Based on the doctoral dissertation of Jose Castro 1980.)

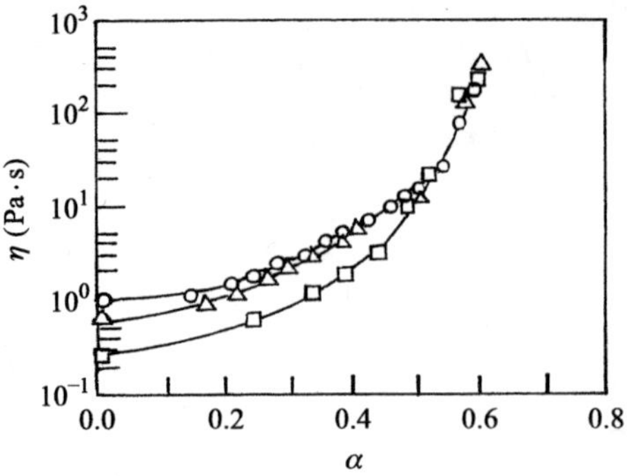

Figure 14.40 Plots of log η versus α for a urethane (RIM 2200) at various isothermal cure temperatures (°C): (○) 30, (△) 46, and (□) 65. (Based on the doctoral dissertation of Jose Castro 1980.)

system can be expressed as $\eta(T, \alpha) = \eta_r(T) f(\alpha)$, which is essentially Eq. (14.30) with $\eta_r(T) = \eta_1 \exp(E_\eta/RT)$ and $f(\alpha) = [\alpha_g/(\alpha_g - \alpha)]^{a+b\alpha}$. For the cure of RIM 2200, Castro and Macosko (1982) reported: $\eta_1 = 1.03 \times 10^{-7}$ Pa·s, $E_\eta = 4.13 \times 10^4$ J/mol, $\alpha_g = 0.65, a = 1.5$, and $b = 1$.

Kim (1987) used cobalt naphthenate to accelerate the cure reaction of a urethane system, consisting of poly(propylene glycol) as the long diol, with trimethylol propane having three hydroxyl groups as the chain extender/cross-linking agent, and MDI as diisocyanate. Figure 14.42 gives plots of d*Q*/d*t* versus temperature obtained from DSC for four different concentrations of cobalt naphthenate. In Figure 14.42, we observe that the peak value of d*Q*/d*t* occurs at lower temperatures as the concentration of cobalt naphthenate is increased, indicating that the rate of cure is accelerated with increasing concentration of cobalt naphthenate. Using the procedures described in a preceding section (see Eq. (14.14)), the DSC thermogram was curve-fitted to the following form of empirical kinetic expression:

$$\frac{d\alpha}{dt} = \left(k_{10} e^{-E_1/RT} + k_{20} e^{-E_2/RT} [CoNa]^m\right)(1 - \alpha)^n \tag{14.33}$$

where $[CoNa]$ denotes the concentration (parts per hundred resin, phr) of cobalt naphthenate, and the numerical values of the parameters are summarized in Table 14.11. Note that when $[CoNa] = 0$, Eq. (14.33) reduces to (14.29). Figure 14.43 describes the

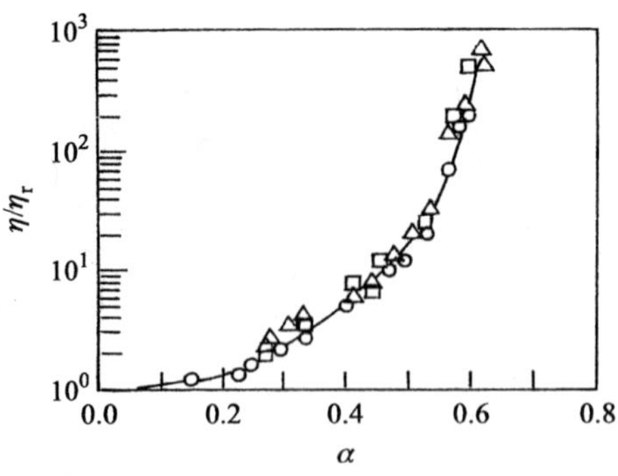

Figure 14.41 Plots of η/η_r versus α for a urethane (RIM 2200) at various isothermal cure temperatures (°C): (○) 30, (△) 46, and (□) 65. (Based on the doctoral dissertation of Jose Castro 1980.)

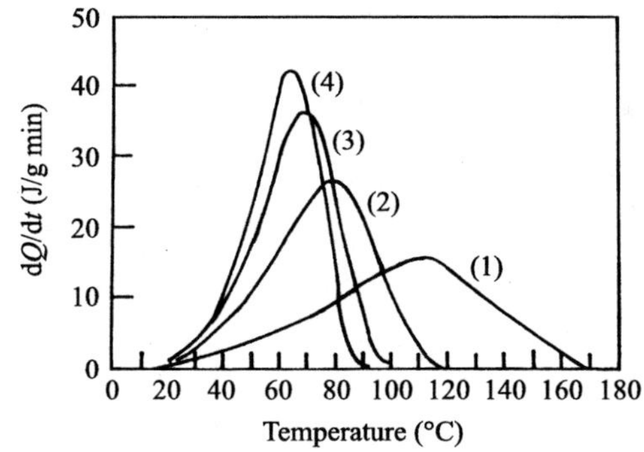

Figure 14.42 Plots of dQ/dt versus temperature during isothermal cure of a urethane (RIM 2200) at various concentrations of catalyst [*CoNa*] (phr): (1) without catalyst, (2) 0.1, (3) 0.3, and (4) 0.5. (From the doctoral dissertation of Jin Hak Kim. Copyright © 1987, with permission from Korea Advanced Institute of Science and Technology.)

temperature dependence of α during the cure of a catalyzed thermosetting urethane for different concentrations of cobalt naphthenate, in which the solid lines were calculated using Eq. (14.33). In Figure 14.43, we observe that the completion of cure is achieved at lower temperatures as the concentration of cobalt naphthenate is increased.

14.5 Summary

In this chapter, we have presented some fundamental aspects of chemorheology of thermosets. A better understanding of the chemorheology of thermosets requires information on the kinetic expression for the chemical reactions taking place during cure. Ideally, one needs a mechanistic model describing cure reactions in terms of the rate constants for each chemical reaction taking place during cure. The advantages of a mechanistic model over an empirical one were demonstrated using the curing reactions of unsaturated polyesters, which happen to follow the well-established free-radical polymerization mechanism. Unfortunately, cure reactions of epoxy resins and thermosetting polyurethanes are so complicated, that, to date, no mechanistic model has been developed. What is needed is a continuing effort to develop mechanistic models for the cure kinetics of epoxy resins and thermosetting polyurethanes.

Table 14.11 Summary of the kinetic parameters for the cure reaction of urethane and the chemorheological model parameters

(a) Kinetic parameters

$k_{10} = 2.90 \times 10^5\ \text{min}^{-1}$; $E_1 = 4.484 \times 10^4\ \text{J/mol}$;
$k_{20} = 1.04 \times 10^{12}\ \text{min}^{-1}$; $E_2 = 7.837 \times 10^4\ \text{J/mol}$;
$m = 0.86$; $n = 1.3$; $\Delta H_R = 259\ \text{J/g}$ (heat of reaction)

(b) Chemorheological model parameters

$\eta_1 = 1.9 \times 10^{-10}\ \text{Pa·s}$; $E_\eta = 5.406 \times 10^4\ \text{J/mol}$;
$\alpha_g = 0.694$; $a = -1.99$; $b = 6.67$.

Based on Kim 1987.

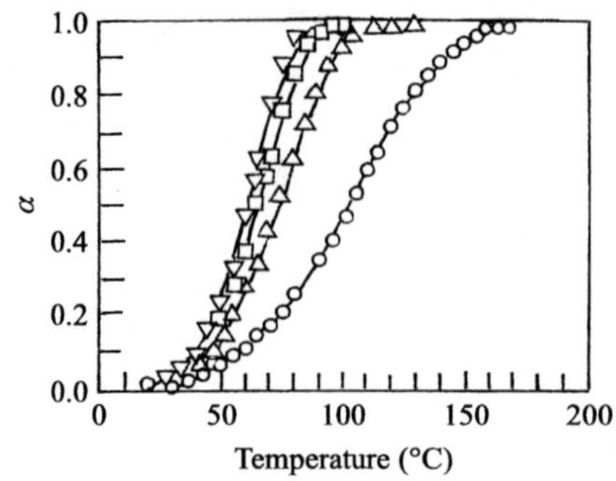

Figure 14.43 Plots of α versus temperature for during isothermal cure of a urethane at different concentrations of catalyst [*Co Na*] (phr): (○) without catalyst, (△) 0.1, (□) 0.3, and (▽) 0.5. (From the doctoral dissertation of Jin Hak Kim. Copyright © 1987, with permission from Korea Advanced Institute of Science and Technology.)

Particulates, low-profile additives, or flame retardant are often added to thermosets, and glass fibers or carbon fibers are mixed with a thermoset to produce fiber-reinforced composites. Processing of such thermoset systems requires information on chemorheology. In this chapter, we have shown how the addition of particulates or low-profile additives to an unsaturated polyester might affect the chemorheology of such mixtures. A generalization of chemorheology of a particular class of thermosets or its composite materials is virtually impossible because different chemical reactions during cure give rise to different chemorheological responses. In this regard, the materials presented in this chapter should not be regarded as universal characteristics of all types of thermosets. This makes predictions of the chemorheology of thermosets and their composites very difficult on the basis of limited experimental observations. Nevertheless, the materials presented in this chapter will provide some general ideas about the unique nature of the chemorheology of thermosets, which is quite different from and far more complex than the rheology of thermoplastic polymers that do not undergo chemical reactions during rheological measurements.

The materials presented in this chapter will be used in the discussions on the processing of thermosets in Chapters 11–13 of Volume 2, which cover: reaction injection molding, pultrusion of thermoset/fiber composites and compression molding of thermoset/fiber composites, respectively.

Problems

Problem 14.1

Calculate viscosity rise for the first 10 minutes after cure begins at 60 °C for an unsaturated polyester using Eq. (14.13), for which you are asked to calculate α using an empirical kinetic expression given by Eq. (14.1). For the calculation of α, use the numerical values of the rate constants and other parameters for Aropol 7030 given in Table 14.3.

Problem 14.2

Calculate viscosity rise for the first 10 minutes after cure begins at 60 °C for an unsaturated polyester using Eq. (14.13), for which you are asked to calculate α using the mechanistic model described in this chapter. For the calculation of α, use the numerical values of rate constants and other parameters given in Table 14.4.

Problem 14.3

Verify the solid lines in Figure 14.37 by first integrating Eq. (14.31) to calculate α as a function of temperature and then substituting the values of α into Eq. (14.30).

Notes

1. The authors did not provide numerical values of dQ/dt in the ordinate of Figure 14.36.
2. Figures 14.38–14.40 were prepared using the tabulated data in Appendix B of the doctoral dissertation of Castro (1980).

References

Acitelli MA, Prime RB, Sacher E (1971). *Polymer* 12:335.

Adabbo HE, Williams RJJ (1982). *J. Appl. Polym. Sci.* 27:1327.

Apicella A, Iannone NM, Passerini P (1984). *J. Appl. Polym. Sci.* 29:2083.

Atkins KE (1978). In *Polymer Blends*, Vol 2, Paul DR, Newman S (eds), Academic Press, New York. p 391.

Bartkus EJ, Kroekel CH (1970). In *Polyblends and Composites*, Bruins PF (ed), John Wiley & Sons, New York, p 113.

Biesenberger JS, Sebastian DH (1983). *Principles of Polymerization Engineering*, Wiley, New York, p 193.

Bruins PF (ed) (1976). *Unsaturated Polyester Technology*, Gordon Breach Science, New York.

Castro, J (1980). *Mold Filling and Curing Studies for the Polyurethane Reaction Injection Molding Process*, Doctoral Dissertation, University of Minnesota, Minneapolis, Minnesota.

Castro JM, Macosko CW (1982). *AIChE J.* 28:250.

Castro JM, Macosko CW, Perry SJ (1984). *Polymer* 25:82.

Dusek K, Ilavsky M, Lunak S (1975). *J. Polym. Sci. Symp.* 53:29, 45.

Dutta A, Ryan ME (1979). *J. Appl. Polym. Sci.* 24:635.

Enns JB, Gillham JK (1983). *J. Appl. Polym. Sci.* 28:2567.

Feger C, MacKnight WJ (1985). *Macromolecules* 18:280.

Flory PJ (1953). *Principles of Polymer Chemistry*, Cornell University Press, Ithaca, New York.

Gillham JK (1982). In *Developments in Polymer Characterization*, Dawkins JV (ed), Applied Science, London, Chap 5.

Halley PJ, Mackay ME (1996). *Polym. Eng. Sci.* 36:593.

Han CD, Lem KW (1983). *J. Appl. Polym. Sci.* 28:3155.

Han CD, Lee DS (1987a). *J. Appl. Polym. Sci.* 33:2859.

Han CD, Lee DS (1987b). *J. Appl. Polym. Sci.* 34:793.

Hartley MD, Williams HL (1981). *Polym. Eng. Sci.* 21:135

Horie K, Mita I, Kambe H (1969). *J. Polym. Sci., A-1* 7:2561.

Horie K, Mita I, Kambe H (1970a). *J. Polym. Sci., A-1* 8:2839.

Horie K, Hiura H, Sawada I, Mita I, Kambe H (1970b). *J. Polym. Sci., A-1* 8:1357.

Hsu CP, Kinkelaar M, Hu P, Lee LJ (1991). *Polym. Eng. Sci.* 31:1450.

Huang XY, Yu W, Sung CSP (1990a). *Macromolecules* 23:390

Huang YJ, Fan JD, Lee LJ (1990b). *Polym. Eng. Sci.* 30:684.

Kamal MR, Sourour S (1973). *Polym. Eng. Sci.* 13:59.

Kim JH (1987). *Reaction Injection Molding Process of Polyurethane-Unsaturated Polyester Blends*, Doctoral Dissertation, Korea Advanced Institute of Science and Technology, Taejeon, Korea.

Kim DH, Kim SC (1987a). *Polym. Compos.* 8:208.
Kim JH, Kim SC (1987b). *Polym. Eng. Sci.* 27:1243.
Kinkelaar M, Wang B, Lee LJ (1994). *Polymer* 35:3011.
Kock E (1977). *Non-isothermal Reaction Analysis*, Academic Pres, New York.
Kubota H (1975). *J. Appl. Polym. Sci.* 19:2279.
Lee DS, Han CD (1987a). *Polym. Eng Sci.* 27:955.
Lee DS, Han CD (1987b). *Polym. Eng. Sci.* 27:964.
Lem KW, Han CD (1983a). *J. Appl. Polym. Sci.* 28:3185.
Lem KW, Han CD (1983b). *J. Appl. Polym. Sci.* 28:3207.
Lenz RW (1967). *Organic Chemistry of Synthetic High Polymers*, Interscience, New York.
Li W, Lee LJ (1998). *Polymer* 39:5677.
Li W, Lee LJ (2000a). *Polymer* 41:685.
Li W, Lee LJ (2000b). *Polymer* 41:697.
Lipshitz SD, Macosko CW (1977). *J. Appl. Polym. Sci.* 21:2029.
Malkin AYa, Kulichikhin SG, Emel'yanov DN, Smetanina IE, Ryabokon NV (1984). *Polymer* 25:778.
May CA (ed) (1983). *Chemorheology of Thermosetting Polymers*, ACS Symposium Series no 227, American Chemical Society, Washington, DC.
Mijovic J, Kim J, Slaby J (1984). *J. Appl. Polym. Sci.* 29:1449.
Mussatti FG, Macosko CW (1973). *Polym. Eng. Sci.* 13:236.
Oertel, G (ed) (1985). *Polyurethane Handbook*, Hanser, Munich.
Osinski JS (1983). *Polym. Eng. Sci.* 23:756.
Parkyn B, Lamb F, Clifton BV (1967). *Polyesters:Unsaturated Polyesters and Polyester Plasticisers*, Vol 2, American Elsevier, New York.
Pattison VA, Hindersinn RR, Schwartz WT (1974). *J. Appl. Polym. Sci.* 18:2763
Pattison VA, Hindersinn RR, Schwartz WT (1975). *J. Appl. Polym. Sci.* 19:3045.
Peyser P, Bascom WD (1977). *J. Appl. Polym. Sci.* 21:2359.
Prime RB (1973). *Polym. Eng. Sci.* 13:365.
Pusatcioglu SY, Fricke AL, Hassler JC (1979). *J. Appl. Polym. Sci.* 24:937.
Ranney MW (1977). *Epoxy Resins and Products*, Noyes Data, Park Ridge, New Jersey.
Richter EB, Macosko CW (1978). *Polym. Eng. Sci.* 18:1012.
Roller MR (1975). *Polym. Eng. Sci.* 15:406.
Ryan ME, Dutta A (1979). *Polymer* 20:203.
Sacher E (1973). *Polymer* 14:91.
Saunders JH, Frisch KC (1964). *Polyurethanes, Chemistry and Technology*, Interscience, New York.
Senich GA, MacKnight WJ, Schneider NS (1979). *Polym. Eng. Sci.* 19:313.
Siegmann A, Narkis M, Kost J, DiBenedetto AT (1978). *Intern. J. Polymeric Mater.* 6:217.
Soh SK, Sundberg DC (1984). *J. Polym. Sci., Polym. Chem. Ed.* 20:1299, 1315, 1331, 1345.
Sourour S, Kamal MR (1976). *Thermochimica* 14:41.
Stevenson JF (1986). *Polym. Eng. Sci.* 26:746.
Sung CSP, Pyun E, Sun HL (1986). *Macromolecules* 19:2922.
Tajima YA, Crozier D (1983). *Polym. Eng. Sci.* 23:186.
Tajima YA, Crozier D (1988). *Polym. Eng. Sci.* 28:491.
Tung CM, Dynes PJ (1982). *J. Appl. Polym. Sci.* 27:569.
Turi EA (ed) (1981). *Thermal Characterization of Polymeric Materials*, Academic Press, New York.
White RP (1974). *Polym. Eng. Sci.* 14:50.
Yap CY, Williams HL (1982). *Polym. Eng. Sci.* 22:254.

Author Index

Subject Index